Current Topics in

ELASTOMERS RESEARCH

Current Topics in
ELASTOMERS RESEARCH

Edited by
Anil K. Bhowmick

CRC Press
Taylor & Francis Group
Boca Raton London New York

CRC Press is an imprint of the
Taylor & Francis Group, an **informa** business

CRC Press
Taylor & Francis Group
6000 Broken Sound Parkway NW, Suite 300
Boca Raton, FL 33487-2742

First issued in paperback 2019

© 2008 by Taylor & Francis Group, LLC
CRC Press is an imprint of Taylor & Francis Group, an Informa business

No claim to original U.S. Government works

ISBN-13: 978-0-8493-7317-6 (hbk)
ISBN-13: 978-0-367-40351-5 (pbk)

Library of Congress Cataloging-in-Publication Data

Bhowmick, Anil K., 1954-
   Current topics in elastomers research / Anil K. Bhowmick.
      p. cm.
   Includes bibliographical references and index.
   ISBN-13: 978-0-8493-7317-6 (alk. paper)
   ISBN-10: 0-8493-7317-4 (alk. paper)
   1. Elastomers. I. Title.

TS1925.B46 2008
678--dc22   2007040724

Visit the Taylor & Francis Web site at
http://www.taylorandfrancis.com

and the CRC Press Web site at
http://www.crcpress.com

Dedication

Dedicated to my only son BUNA

Contents

SECTION I Introductory Chapter

SECTION II New Elastomers and Composites

SECTION III Rubber Ingredients

SECTION IV New Characterization Techniques

SECTION V   *Physics and Engineering*

SECTION VI Tires

SECTION VII Eco-Friendly Technology and Recycling

Preface

Rubber is an important engineering material and is used in many critical applications. An example of its importance is the space shuttle *Challenger* disaster because of a failed rubber O-ring. Rubber has implications ranging from birth control to the growth of modern civilization. Can we imagine fast transportation without rubber tires? Artificial hearts can be made of rubber. Hence, research activities are always intense in the biomedical field in spite of new and continuous challenges. Research has also burgeoned because of the development of innovative materials and stringent application requirements. However, many of these activities are concentrated in the industry and are not published in the open literature. Any researcher or beginner in rubber science and technology will always have a question, "What are the topics of rubber research in the world?" It is with this perspective that the book has been compiled to give its readership a broad view of rubber research activities in the world.

The book is divided into 7 sections with 38 chapters. An introductory chapter, the first section, sets the tone of the book. The second section, "New Elastomers and Composites," deals with a variety of newly developed rubbery materials and composites, and also their chemistry and structure–property relationship. The third section, "Rubber Ingredients," covers the important ingredients, current thought in this field, and the mechanisms of action of these ingredients. The fourth section, "New Characterization Techniques," predominantly describes two techniques, 3D-TEM and AFM, currently used by leading researchers. The fifth section, "Physics and Engineering," discusses reinforcement mechanisms; effects of time, temperature, and fluids; viscoelastic properties; fatigue life; abrasion; adhesion; rheology; mixing and processing; and the state of the art in these areas. The sixth section discusses, "Tires," one of the major applications of rubber, whose critical requirements have guided research for more than 50 years, and which are also described. Only tires are discussed as an example of a typical rubber product as it was not possible to include recent developments of other products. The last section, "Eco-Friendly Technology and Recycling," discusses important topics in any engineering field in the globalization era.

Although this book covers a large number of topics by experts from various countries, critics might say that a few more topics need to be included. This was not possible because information in certain areas is still classified. Also, some ideas have been repeated in the book to stress the importance of the subject and also to emphasize the views of the contributors engaged in growing or nascent fields of research.

All topics have been contributed by leading experts and stalwarts in their fields. I am grateful to the contributors for their time and efforts to make the venture successful. The book took shape because of continued support from my former as well as present students: Anirban Ganguly, Dr. Susmita Sadhu, Madhuchhanda Maiti, Dr. Abhijit Bandyopadhyay, Jinu J. George, Dr. Sritama Kar, Dr. Francis R. Costa, Dr. Naba K. Dutta, Dr. Namita R. Roychowdhury, Dr. Sudip Ray, Dr. Rajatendu Sengupta, Dr. Indranil Banik, Dr. Papiya Sen Majumar, Dr. V. Vijayabaskar, Dr. Arup K. Chandra, Saikat Dasgupta, Dr. Anand Srinivasan, Dr. A.M. Shanmugharaj, Mithun Bhattacharya, Dinesh Kumar Kotnees, Suman Mitra, Pradip K. Maji, Ganesh Basak, Anusuya Choudhury, Anjan Biswas, Haimanti Datta, and Samik Gupta. I also thank my colleagues at IIT Kharagpur for their support, and the outstanding cooperation of my research collaborators and funding agencies in India and abroad. I would like to thank various companies, authors, editors, and journals for giving me permission to reproduce material and also for providing other necessary assistance.

Finally, I would like to thank my late parents Jatindra M. Bhowmick and Hemprova Bhowmick, and also Dr. S.K. Biswas (late) and K. Biswas for inspiring me during the early stages of preparation of the book. I also express my gratitude to my son Asmit (BUNA) and my wife Dr. Kundakali, for sparing me to work in the office, when actually I should have been at home in their company.

Anil K. Bhowmick
Indian Institute of Technology
Kharagpur

Editor

Anil K. Bhowmick is a professor and former head of the Rubber Technology Center, and was former dean of postgraduate studies and dean of sponsored research and industrial consultancy at IIT Kharagpur. He was previously associated with the University of Akron, Ohio, the London School of Polymer Technology, and the Tokyo Institute of Technology. His main research interests are in the fields of nanocomposites, thermoplastic elastomers and polymer blends, polymer modification, rubber technology, failure and degradation of polymers, and adhesion and adhesives. He has more than 400 publications in these fields, written 35 book chapters, and has coedited 5 books. He was also coeditor of the *Journal of Macromolecular Science* special issue on polymer and composite characterization. He is the 2002 winner of the Chemistry of Thermoplastic Elastomers Award, the 1997 winner of the George Stafford Whitby Award of the rubber division, American Chemical Society, for distinguished teaching and innovative research, and the 2001 K.M. Philip Award of the All India Rubber Industries Association for outstanding contributions to the growth and development of rubber industries in India. He has also received the 1991 NOCIL Award, the 1990 JSPS Award, the 1990 Commonwealth Award, the 1989 MRF Award, and the 1989 Stanton–Redcroft ITAS Award. He is on the editorial boards of the *Journal of Adhesion Science and Technology*, the *Journal of Applied Polymer Science*, the *Journal of Materials Science*, *Polymers and Polymer Composites*, the *Journal of Materials Science*, and *Rubber Chemistry and Technology*.

Contributors

John M. Baldwin
Exponent Failure Analysis Associates
Farmington Hills, Michigan
and
Department of Chemistry
Oakland University
Rochester Hills, Michigan

Abhijit Bandyopadhyay
Rubber Technology Center
Indian Institute of Technology
Kharagpur, India

Indranil Banik
Centre for Biocomposites and
    Biomaterials Processing
University of Toronto
Toronto, Ontario, Canada

David R. Bauer
Exponent Failure Analysis Associates
Farmington Hills, Michigan

Satinath Bhattacharya
Rheology and Materials Processing
    Centre
RMIT University
Melbourne, Australia

Anil K. Bhowmick
Rubber Technology Centre
Indian Institute of Technology
Kharagpur, India

Robert P. Campion
Materials Engineering Research
    Laboratory Ltd.
Hitchin, Hertfordshire, United Kingdom

Arup K. Chandra
Apollo Tyres Ltd., R&D Centre
Limda, Gujarat, India

Namita Roy Choudhury
Ian Wark Research Institute
University of South Australia
Mawson Lakes, Australia

Francis R. Costa
Leibniz-Institut für Polymerforschung
Dresden, Germany

Rabin N. Datta
Department of Rubber Technology
University of Twente
Enschede, the Netherlands

Sudhin Datta
R&D Polymers
Exxon Mobil Chemical Co
Baytown, Texas

Des (C.J.) Derham
Materials Engineering Research
    Laboratory Ltd.
Hitchin, Hertfordshire, United Kingdom

Wilma Dierkes
Department of Elastomer Technology
    and Engineering
University of Twente
Enschede, the Netherlands

Herman Dikland
DSM Elastomers
Geleen, the Netherlands

Martin van Duin
DSM Elastomers
Geleen, the Netherlands

Naba K. Dutta
Ian Wark Research Institute
University of South Australia
Mawson Lakes, Australia

Christopher M. Elvin
CSIRO Livestock Industries
Queensland Bioscience Precinct
St Lucia, Australia

Ali Fatemi
Mechanical, Industrial and Manufacturing
    Engineering Department
The University of Toledo
Toledo, Ohio

Ephraim Feinblum
Dimona Silica Industries Ltd.
Be′er-Sheva, Israel

Yoshihide Fukahori
Materials Department
Queen Mary, University of London
London, United Kingdom

Anirban Ganguly
Rubber Technology Centre
Indian Institute of Technology
Kharagpur, India

Alan N. Gent
Department of Polymer Science
The University of Akron
Akron, Ohio

Jinu Jacob George
Rubber Technology Centre
Indian Institute of Technology
Kharagpur, India

Karl A. Grosch
VMI Holland BV
Geiriaweg, the Netherlands

Saikat Das Gupta
Hari Shankar Singhania Elastomer
    and Tyre Research Institute
Kankroli, Rajsamand
Rajasthan, India

Ryan J. Harbour
Mechanical, Industrial and Manufacturing
    Engineering Department
The University of Toledo
Toledo, Ohio

Gert Heinrich
Leibniz-Institut für Polymerforschung
Dresden, Germany

Nico M. Huntink
Teijin Aramid BV
Arnhem, the Netherlands

Mickey G. Huson
CSIRO Textile and Fibre Technology
Geelong, Australia

Frederick Ignatz-Hoover
Flexsys America L.P.
Akron, Ohio

Yuko Ikeda
Graduate School of Science
    and Technology
Kyoto Institute of Technology
Matsugasaki, Kyoto, Japan

Sritama Kar
Rubber Technology Centre
Indian Institute of Technology
Kharagpur, India

Atsushi Kato
NISSAN ARC, Ltd.
Yokosuka, Kanagawa, Japan

Manfred Klüppel
Deutsches Institut für
    Kautschuktechnologie
Hannover, Germany

Shinzo Kohjiya
Mapua Institute of Technology
Intramuros, Manila, Philippines

Doug J. Kohls
University of Cincinnati
Cincinnati, Ohio

Raissa Kosso
Dimona Silica Industries Ltd.
Be′er-Sheva, Israel

Jean L. Leblanc
Polymer Rheology and Processing
Universite Pierre et Mariè Curie
Vitry-sur-Seine, France

Michael V. Lewan
Materials Engineering Research
    Laboratory Ltd.
Hitchin, Hertfordshire, United Kingdom

Sergei N. Magonov
Nanotechnology Measurements Division
Agilent Technologies
Chandler, Arizona

Madhuchhanda Maiti
Reliance Industries Limited
Vadodara Manufacturing Division
Vadodara, Gujarat, India

Papiya Sen Majumder
Leibniz-Institut für Polymerforchung
Dresden, Germany

Duryodhan Mangaraj
Innovative Polymer Solutions
Delaware, Ohio

Will V. Mars
Cooper Tire and Rubber Company
Findlay, Ohio

José Miguel Martín-Martínez
Adhesion and Adhesives Laboratory
University of Alicante
Alicante, Spain

Hans-Georg Meyer
Berstorff Rubber Processing Machinery
Hannover, Germany

Glyn J. Morgan
Materials Engineering Research
    Laboratory Ltd.
Hitchin, Hertfordshire, United Kingdom

Michael D. Morris
Business and Technology Center
Cabot Corporation
Billerica, Massachusetts

Rabindra Mukhopadhyay
Hari Shankar Singhania Elastomer
    and Tyre Research Institute
Kankroli, Rajsamand
Rajasthan, India

Ken Nakajima
Department of Organic and Polymeric Materials
Graduate School of Science and Engineering
Tokyo Institute of Technology
Meguro, Tokyo, Japan

Gerard Nijman
Department of R&D
Vredestein Banden BV
Enschede, the Netherlands

Toshio Nishi
Department of Organic and Polymeric Materials
Graduate School of Science and Engineering
Tokyo Institute of Technology
Meguro, Tokyo, Japan

Jacques Noordermeer
Department of Elastomer Technology
    and Engineering
University of Twente
Enschede, the Netherlands

Muthukumaraswamy Pannirselvam
Rheology and Materials Processing
    Centre
RMIT University
Melbourne, Australia

Alice Bope Parsons
Battelle Memorial Institute
Columbus, Ohio

Judit E. Puskas
Department of Polymer Science
The University of Akron
Akron, Ohio

R.S. Rajeev
School of Mechanical and Aerospace
    Engineering
Queen's University
Belfast, Northern Ireland, United Kingdom

C.M. Roland
Chemistry Division
Naval Research Laboratory
Washington, DC

Susmita Dey Sadhu
Department of Polymer Science
Bhaskaracharya College of Applied Sciences
University of Delhi
Delhi, India

Dale W. Schaefer
University of Cincinnati
Cincinnati, Ohio

Rajatendu Sengupta
Hari Shankar Singhania Elastomer
    and Tyre Research Institute
Kankroli, Rajsamand
Rajasthan, India
and
Rubber Technology Center
Indian Institute of Technology
Kharagpur, India

Dipak K. Setua
Defence Research & Development
    Organization
Ministry of Defence
Kanpur, India

A.M. Shanmugharaj
Institute for Materials Chemistry
    and Engineering
Kyushu University
Fukuoka, Japan

Rui Shi
Department of Polymer Engineering
School of Material Science and Engineering
Beijing University of Chemical Technology
Beijing, China

Anandhan Srinivasan
Asian Institute of Medicine
Science and Technology
Sungal Petani, Kedah, Malaysia

Byron To
Flexsys America L.P.
Akron, Ohio

V. Vijayabaskar
Product Development Centre
Balmer Lawrie & Co. Ltd.
Manali, Chennai, India

Thomas A. Vilgis
Max-Planck-Institut für Polymerforschung
Mainz, Germany

Meng-Jiao Wang
Business and Technology Center
Cabot Corporation
Billerica, Massachusetts

Natalya A. Yerina
Veeco Instruments
Santa Barbara, California

Li-Qun Zhang
Department of Polymer Engineering
School of Material Science and Engineering
Beijing University of Chemical Technology
Beijing, China

Section I

Introductory Chapter

1 Some Outstanding Problems in the Mechanics of Rubbery Solids

Alan N. Gent

CONTENTS

1.1 INTRODUCTION

Several aspects of the deformation and fracture of rubbery materials are reviewed in this chapter. They reveal some important gaps in our present understanding of the behavior of simple rubbery solids.

1.2 END EFFECTS IN RUBBER SPRINGS

When a rubber block of rectangular cross-section, bonded between two rigid parallel plates, is deformed by a displacement of one of the bonded plates in the length direction, the rubber is placed in a state of simple shear (Figure 1.1). To maintain such a deformation throughout the block, compressive and shear stresses would be needed on the end surfaces, as well as on the bonded plates [1,2]. However, the end surfaces are generally stress-free, and therefore the stress system necessary

3

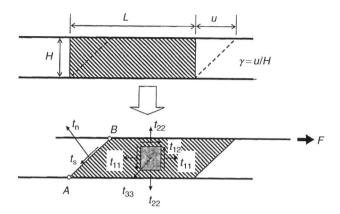

FIGURE 1.1 Shear of a bonded block. The points A and B denote regions in tension and compression, respectively.

to maintain a simple shear throughout the block is incomplete. Surprisingly large consequences follow from this failure to apply the requisite stresses on the end surfaces, even when the block is long and thin and the end conditions might be thought unimportant. The effects have been studied by finite element analysis (FEA) for a long thin block, sheared in the length direction [3], and for rubber tube springs sheared azimuthally and axially [4].

For a rectangular rubber block, plane strain conditions were imposed in the width direction and the rubber was assumed to be an incompressible elastic solid obeying the simplest nonlinear constitutive relation (neo-Hookean). Hence, the elastic properties could be described by only one elastic constant, the shear modulus μ. The shear stress t_{12} is then linearly related to the amount of shear γ [1,2]:

$$t_{12} = \mu\gamma \tag{1.1}$$

However, normal stresses were also predicted by Rivlin [1]:

$$t_{11} = \mu\gamma^2 + p \tag{1.2}$$

$$\text{and } t_{22} = t_{33} = p \tag{1.3}$$

where p denotes a reference pressure. If it is assumed that the end faces in the length direction are free from stress, then $t_{11} = 0$ and

$$t_{22} = t_{33} = -\mu\gamma^2 \tag{1.4}$$

Note that the normal (vertical) stress t_{22} is a second-order compressive stress in this case. However, as pointed out by Rivlin [1] and Ogden [2], stresses need to be applied to the block end surfaces to maintain the shear deformation, consisting of a stress t_{n} normal to the end surface in the deformed state, and a shear stress t_{s} (Figure 1.1):

$$t_{\mathrm{n}} = p - \mu\gamma^2/(1 + \gamma^2) \tag{1.5}$$

$$t_{\mathrm{s}} = \mu\gamma/(1 + \gamma^2) \tag{1.6}$$

But it is generally impractical to apply such stresses. When they are absent we must reconsider the entire system of stresses. For example, an estimate of the horizontal stress t_{11} in the bulk

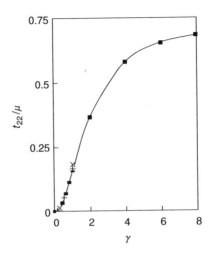

FIGURE 1.2 Normal stress t_{22} in the central region of the bond line versus imposed shear strain γ. (From Gent, A.N., Suh, J.B., and Kelly, III, S.G., *Int. J. Non-Linear Mech.*, 42, 241, 2007. With permission.)

of the block can be obtained by putting $t_n = 0$ in Equation 1.5, yielding a new value for the reference pressure p:

$$p = \mu\gamma^2/(1 + \gamma^2) \tag{1.7}$$

The stresses t_{11} and t_{22} then become:

$$t_{11} = \mu\gamma^2(2 + \gamma^2)/(1 + \gamma^2) \tag{1.8}$$

$$t_{22} = t_{33}(=p) = \mu\gamma^2/(1 + \gamma^2) \tag{1.9}$$

These are quite different from Equations 1.2 and 1.3, but the general form of the predicted stresses have now been corroborated by FEA calculations [3]. The normal stress t_{22} was found to be tensile, in agreement with Equation 1.9 (Figure 1.2), while the longitudinal stress t_{11} increased rapidly with increasing shear γ, approximately in proportion to γ^2 (Figure 1.3). Such profound consequences of the boundary conditions do not appear to have been noted previously.

The normal stresses t_{11} and t_{22} were also found to be markedly affected by the shape of the free surfaces [3]. If the end surfaces of the block in the undeformed state were slanted at an angle of 45°, so that when a shear strain of unity was imposed the block became substantially rectangular in cross-section, then the stresses t_{11} and t_{22} were in good agreement with Rivlin's solution for a volume element in simple shear (Equation 1.4). The normal stress t_{22} was now negative, instead of positive. Also, when the material of the block was allowed to undergo slight changes in volume on shearing, by making Poisson's ratio slightly lower than 0.5, the normal stresses were reduced; t_{22} eventually became negative in the center of the block at a value of Poisson's ratio of 0.495 (Figure 1.4).

We conclude that high internal stresses are generated by simple shear of a long incompressible rectangular rubber block, if the end surfaces are stress-free. These internal stresses are due to restraints at the bonded plates. One consequence is that a high hydrostatic tension may be set up in the interior of the sheared block. For example, at an imposed shear strain of 3, the negative pressure in the interior is predicted to be about three times the shear modulus μ. This is sufficiently high to cause internal fracture in a soft rubbery solid [5].

Similar internal stresses can develop in the interior of a rubber tube spring, consisting of a cylindrical rubber tube bonded on its inner and outer curved surfaces between rigid cylindrical

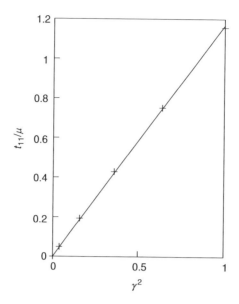

FIGURE 1.3 Normal stress t_{11} in the central region versus γ^2. (From Gent, A.N., Suh, J.B., and Kelly, III, S.G., *Int. J. Non-Linear Mech.*, 42, 241, 2007. With permission.)

surfaces [4,6]. When the rubber is sheared by rotating the inner tubular surface about its axis, with the outer surface held fixed (azimuthal shear), then the shape of the free surfaces is not greatly altered when the tube is relatively long. The longitudinal (axial) stress is then small and the normal (radial) stress in the center of the tube is in accord with Rivlin's solution [1]. However, when the tube is sheared with respect to the other tubular surface by displacing one of the bounding surfaces parallel to its axis (axial shear), failure to apply the requisite stresses on the end surfaces leads to high longitudinal stresses, as in sheared rectangular blocks. Also, radial stresses become tensile, instead of compressive [4,6].

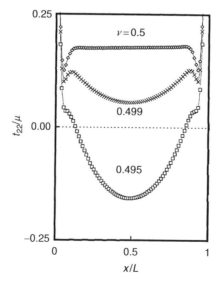

FIGURE 1.4 Effect of Poisson's ratio on the distribution of normal stress t_{22} along the centerline; shear strain $\gamma = 1$. (From Gent, A.N., Suh, J.B., and Kelly, III, S.G., *Int. J. Non-Linear Mech.*, 42, 241, 2007. With permission.)

1.3 BOND FAILURES IN SHEARED BLOCKS

Bond failure would be expected to start at one of the two corners, A and B in Figure 1.1, where there are high stress concentrations. Because the rubber is mainly in tension at point A, whereas it is under compression at B, we would expect bond failure to begin at A. But finite element analysis suggests otherwise, as shown by Muhr et al. [7–9], following previous work by Lindley and Teo [10]. We consider here energy release rates G for a crack growing at the bond line of a long rectangular neo-Hookean block (length $L = 10\,H$, where H is the block thickness) subjected to a moderate shear strain of 0.5 [3]. Crack growth was simulated by sequential release of nodes at the bond line, starting at one edge, by steps Δc of one element in length. Corresponding changes ΔW in the stored strain energy W were calculated by FEA. Values of the strain energy G available to propagate a crack through unit area were then obtained from the ratio $(\Delta W/\Delta c)$, and assigned to a crack of effective length $[c + (\Delta c/2)]$. Results are plotted in Figure 1.5 in nondimensional form, as G/UH, where U is the stored elastic energy per unit volume, for a crack propagating from the "tension" edge, point A in Figure 1.1, and from the compression edge, point B in Figure 1.1.

When the crack length is much greater than the block thickness, it might be assumed that all the strain energy in a volume $H\Delta c$ is released as the crack grows by a further distance Δc, and thus G would be expected to reach a limiting value of UH [3,7–10]. It is clear in Figure 1.5 that this value is attained for a crack at the compression side when the length c is about 25% of the length of the block (about three times the thickness). However, G rises to only about one-half of UH for a crack on the tension side, even when the crack length is one-half of the block width and five times the thickness. This striking difference was pointed out by Muhr and his colleagues [7–9], who concluded that cracks would grow preferentially at the compression side. They also carried out some fatigue experiments on blocks with an initial crack, a few millimeters long, placed at both edges, and observed that the crack at the compression side grew more rapidly, eventually causing fracture. While this observation is consistent with the results shown in Figure 1.5, we note that, for a small crack, the energy released initially by crack growth at the tension side is considerably greater than that at the compression side. Thus, a small initial crack or defect seems more likely to grow at the tension side, but it will presumably slow down or stop when the energy release rate becomes small. Thus, the conclusions of Muhr and his colleagues are probably correct: a crack at the compression edge is likely to prevail. Nonetheless, the growth of small cracks needs further study, since crack initiation at bonded interfaces and at the sides, edges, and corners of free surfaces is still not well understood.

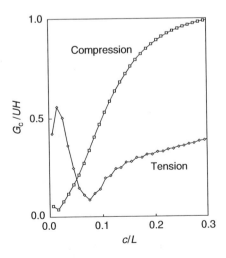

FIGURE 1.5 Energy release rate G_c versus length c of a bond line crack. U is the strain energy density. Block thickness $H = 2$ mm, length $L = 20$ mm. (From Gent, A.N., Suh, J.B., and Kelly, III, S.G., *Int. J. Non-Linear Mech.*, 42, 241, 2007. With permission.)

1.4 RELATION BETWEEN THE STRENGTH OF RUBBER AND ITS MOLECULAR STRUCTURE

1.4.1 VISCOELASTICITY

Consider a deformation consisting of repeated sinusoidal oscillations of shear strain. The relation between stress and strain is an ellipse, provided that the strain amplitude is small, and the slope of the line joining points where tangents to the ellipse are vertical represents an effective elastic modulus, termed the storage modulus μ'. The area of the ellipse represents energy U_d dissipated in unit volume per cycle of deformation, expressed by the equation

$$U_d = \pi \mu^2 \gamma_m^2 \tag{1.10}$$

where
   γ_m is the amplitude of strain
   μ'' is termed the loss modulus

The ratio μ''/μ' is the tangent of a phase angle δ by which the strain lags behind the stress. When the ellipse degenerates into a straight line ($\tan \delta = 0$), the material is a perfectly elastic solid. Values of $\tan \delta$ for typical rubber compounds at room temperature range from about 0.03 for a highly resilient, "springy" material with low internal viscosity to about 0.2 for a compound with a relatively large internal viscosity and consequently high energy dissipation.

1.4.2 EFFECTS OF TEMPERATURE AND FREQUENCY OF OSCILLATION

A rubbery solid consists of a three-dimensional network of molecular strands that are in rapid motion (Brownian motion) from thermal agitation. The frequency ϕ of molecular jumps of a single physical unit (segment) of a strand to new positions depends strongly on temperature, falling to a low figure, of only about 0.1 movements per second, at the glass transition temperature T_g. At this point the segments, and consequently the molecular strands, appear to be stationary and the material is glass-hard.

At a temperature T above T_g, the response of an elastomer depends markedly on the frequency of deformation, but only on the frequency *relative* to the frequency of molecular Brownian motion at T. The dependence of μ' and $\tan \delta$ on frequency is shown schematically in Figure 1.6. At low frequencies the dynamic modulus μ' is low and nearly constant but, as the frequency is raised, the compound becomes stiffer, until at high frequencies it is hard and glasslike (Figure 1.6). This striking transition, by about three orders of magnitude, reflects a change from facile response of molecular segments under low-frequency stresses to complete immobility under high-frequency stresses. The transition is centered at an oscillation frequency of the same order of magnitude as the natural frequency of thermal motion of entire molecular strands.

Because the loss modulus μ'' is a direct measure of viscous resistance to segmental motion, it increases with frequency to an even more marked degree than μ', usually becoming larger than μ' in the transition range of frequencies. At sufficiently high frequencies, however, the segments become unable to respond to oscillatory stresses, and internal motion ceases. Energy dissipation that is associated with the longer-range motion of molecular segments in a viscous environment also ceases, and μ'' falls to the relatively low value characteristic of polymeric glasses. In consequence, the ratio $\tan \delta (= \mu''/\mu')$ passes through a pronounced maximum as the oscillation frequency is raised toward the glassy state, as shown schematically in Figure 1.6.

When the test temperature is raised, the rate of Brownian motion increases by a certain factor, denoted a_T, and it would therefore be necessary to raise the frequency of oscillation by the same factor a_T to obtain the same physical response, as shown in Figure 1.6. The dependence of a_T upon the temperature difference $T - T_g$ follows a characteristic equation, given by Williams, Landel, and Ferry (WLF) [11]:

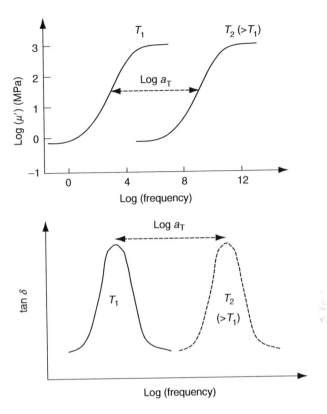

FIGURE 1.6 Sketch of dynamic shear modulus μ' and tan δ versus log (frequency) at two temperatures T_1 and $T_2(T_2 > T_1)$.

$$\log[(a_T/a_{T_g})] = A(T - T_g)/(B + T - T_g) \qquad (1.11)$$

where A and B are constants, 17.5°C and 52°C, respectively, for nearly all rubbery substances. T_g is the glass temperature at which molecular segments move only about once in 10 s, so that in most experiments they do not move at all and the material appears to be a rigid glass. Thus, the reference factor a_{T_g} is assigned the value 0.1 s^{-1}.

For motion of entire molecular strands, consisting of n segments, to take place in 0.1 s, the frequency of segmental motion must be much faster than 0.1 s^{-1}, by a factor of n^2 or more. This rate is achieved only at a temperature well above T_g for typical values of n, of the order of 100. Thus, fully rubber-like response will not be achieved until the test temperature is $T_g + 30$°C, or even higher. (On the other hand, for sufficiently slow movements that take place over several hours or days, an elastomer would still be able to respond at temperatures below the conventionally defined glass transition temperature.)

Equation 1.11 can be used to relate the dynamic behavior at one temperature T_1 to that at another, T_2, as illustrated in Figure 1.6. When the temperature is raised to T_2, the curves are displaced laterally by the distance, log a_T, on the logarithmic frequency axis, where log a_T now reflects the change in characteristic response frequency of molecular segments when the temperature is changed from T_1 to T_2. Thus, log a_T is given by

$$\log a_T = 17.5 \times 52(T_2 - T_1)/(52 + T_2 - T_g)(52 + T_1 - T_g) \qquad (1.12)$$

from Equation 1.11.

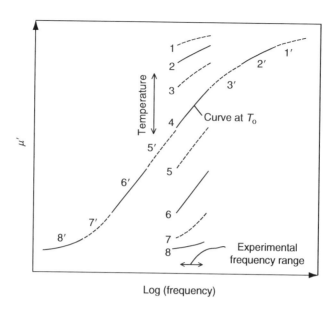

FIGURE 1.7 Construction of a "master curve" of dynamic modulus μ' versus log (frequency) by lateral shifting of experimental results made over a small frequency range but at several different temperatures.

Measurements taken over a limited frequency range at many different temperatures can be superposed by lateral shifts of log a_T along the logarithmic frequency axis to construct a "master curve," as shown schematically in Figure 1.7. This curve shows the expected response over an extremely wide frequency range at the chosen temperature. Moreover, it can be made to apply at any other temperature by an appropriate lateral shift along the log frequency axis, calculated from Equation 1.12. This is a powerful method of predicting the viscoelastic response over a wide range of frequency from measurements over a limited range of frequency, but at many temperatures. Results confirm the general principle that a change in temperature is equivalent to a certain change in frequency.

Dynamic mechanical measurements for elastomers that cover wide ranges of frequency and temperature are rather scarce. Payne and Scott [12] carried out extensive measurements of μ' and μ'' for unvulcanized natural rubber as a function of test frequency (Figure 1.8). He showed that the experimental relations at different temperatures could be superposed to yield master curves, as shown in Figure 1.9, using the WLF frequency–temperature equivalence, Equation 1.11. The same "shift factors," log a_T, were used for both experimental quantities, μ' and μ''. Successful superposition in both cases confirms that the dependence of the viscoelastic properties of rubber on frequency and temperature arises from changes in the rate of Brownian motion of molecular segments with temperature.

We now turn to the tear strength of rubbery solids, and its dependence on rate of tearing and temperature.

1.4.3 TEAR STRENGTH OF SIMPLE RUBBERY SOLIDS

The elastic strain energy G available to propagate an interfacial crack in a sheared block was evaluated in Section 1.2. The basic assumption of fracture mechanics is that G must reach a critical value, denoted G_c, for a crack to grow. This critical value characterizes the strength of an interface, or of the material itself, when a crack propagates through the interior. In the latter case it is termed fracture energy or tear energy. We now consider actual values of tear energy for some simple rubbery solids.

Tearing through a flexible sheet (Figure 1.10) is a common method of measuring fracture energy G_c [13]. Work ΔW is expended in tearing through an incremental distance Δc, i.e., through an area $T\Delta c$, where T is the sheet thickness:

$$\Delta W = G_c T \Delta c \qquad (1.13)$$

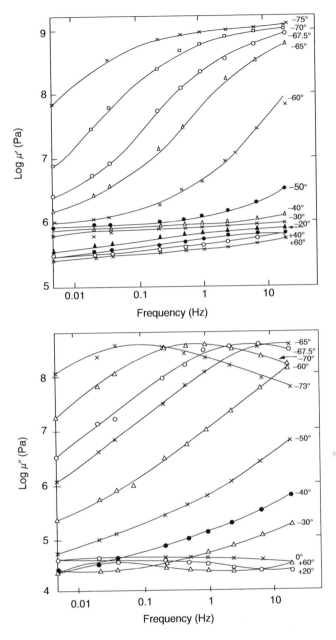

FIGURE 1.8 In-phase (μ') and out-of-phase (μ'') components of the complex shear modulus of uncross-linked natural rubber versus log (frequency) at various temperatures. (From Payne, A.R. and Scott, J.R., *Engineering Design with Rubber*, Interscience Publishers, New York, 1960.)

Assuming that the arms of the test specimen are not stretched significantly, the energy supplied is approximately $2F\Delta c$. Hence, equating the energy expended with that supplied, the tear energy is given by

$$G_c = 2F/T \tag{1.14}$$

Values of tear energy obtained in this way for some simple rubbery solids are discussed in Section 1.4.4, and compared with theoretical estimates for a network of long flexible molecules with C–C backbones.

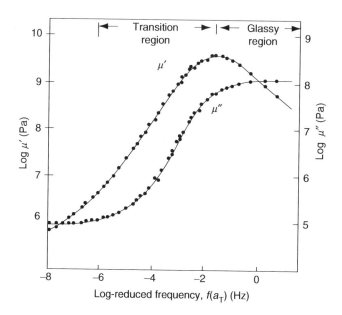

FIGURE 1.9 Master curves of in-phase (μ') and out-of-phase (μ'') components of the complex shear modulus of uncross-linked natural rubber versus log (frequency) at T_g. (From Payne, A.R. and Scott, J.R., *Engineering Design with Rubber*, Interscience Publishers, New York, 1960.)

1.4.4 THRESHOLD TEAR STRENGTH

Berry [14] deduced the fracture energy for C–C polymers by summing the C–C bond dissociation energies for all the bonds that cross unit area, and obtained a value of about 1 J/m^2. This is about 20 times larger than the energy of van der Waals bonds in a simple liquid, reflecting the higher dissociation energies of covalent bonds, but it is still quite small. The measured tear energy for rubber ranges from about 300 to 100,000 J/m^2 or more, depending on the material, the crack speed, and the test temperature [13].

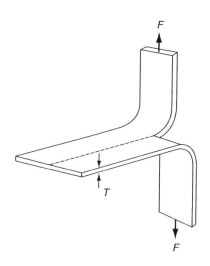

FIGURE 1.10 Tear of a flexible strip, thickness T.

By taking into account the high extensibility of rubber molecules, Lake and Thomas [15] showed that Berry's theoretical value should be raised by a factor of approximately $n^{1/2}$, where n is the number of segments in a molecular strand between cross-links. Using a representative value for n of 100, this increases the theoretical value for G_c to about 10 J/m², but this is still far below normal values of tear strength. Thus, a fundamental question in the mechanics of rubber fracture is: Why is the observed strength so much higher than expected? The answer is found by considering the effects of temperature and crack speed on the tear resistance of simple rubbery solids.

1.4.5 EFFECT OF TEAR RATE AND TEST TEMPERATURE ON TEAR ENERGY G_c

Measurements of tear energy G_c as a function of tear rate at various temperatures are plotted in Figure 1.11 [16]. The material examined was a simple unfilled rubbery solid, a cross-linked sample of a copolymer of styrene and butadiene containing about 40% styrene, with a glass temperature of about $-30°C$. Values of G_c are seen to change by a large factor, about three orders of magnitude, as the test conditions change from low tear speeds at high temperatures to high tear speeds at low temperatures. Plotted using a logarithmic scale for tear speed, they appear to be superposable by horizontal shifts in the same way that μ', μ'', and tan δ are superposable (Figures 1.8 and 1.9). Indeed, using WLF shift factors calculated from Equation 1.11, as shown schematically in Figure 1.7, the results from Figure 1.11 are accurately superposable (see Figure 1.12). The success of this superposition shows that internal dissipation from segmental motion governs the strength of elastomers [17–19], and that high strength is associated with reduced molecular mobility, not with a change in the intrinsic strength of the rubber molecule. For example, it is well known that the molecular mobility of butyl rubber is anomalous—it increases more slowly as the temperature is raised above T_g than for other elastomers. We would therefore expect the tear strength of butyl rubber compounds to decrease more slowly as the temperature is raised, and this turns out to be the case.

At low speeds and high temperatures, viscous energy dissipation is minimized, and the tear strength approaches a lower limit G_o of about 20 J/m² (Figure 1.11). This is of the same order of

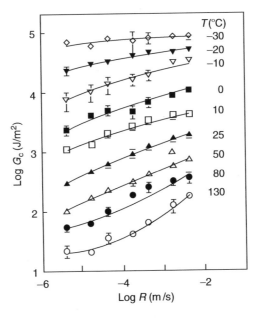

FIGURE 1.11 Tear energy G_c versus rate R of tear propagation for a cross-linked sheet of a high-styrene copolymer of butadiene and styrene (48% styrene; $T_g = -30°C$). (From Gent, A.N. and Lai, S.-M., *J. Polymer Sci., Part B: Polymer Phys.*, 32, 1543, 1994. With permission.)

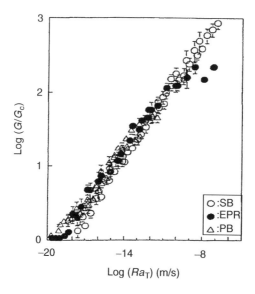

FIGURE 1.12 Master curve of tear energy G_c versus rate R of tear propagation at T_g for three cross-linked elastomers: polybutadiene (BR, $T_g = -96°C$); ethylene–propylene copolymer (EPR, $T_g = -60°C$); a high-styrene–styrene–butadiene rubber copolymer (HS-SBR, $T_g = -30°C$). (From Gent, A.N. and Lai, S.-M., *J. Polymer Sci.*, *Part B: Polymer Phys.*, 32, 1543, 1994. With permission.)

magnitude as the theoretical value for the intrinsic strength of a network of C–C molecular chains. Moreover, G_o increases with the length M_c of the molecular strands in the network, roughly in proportion to $M_c^{1/2}$, in good agreement with the Lake–Thomas theory of threshold strength [15]. (Note that the threshold strength is *inversely* related to the degree of cross-linking; it decreases as the degree of cross-linking is increased.)

The tear strength of the molecular network under nonequilibrium conditions appears to be a product of two terms, the inherent strength G_o and a factor reflecting dissipative processes. Indeed, when the ratio G_c/G_o is plotted against the equivalent tear speed Ra_T at a temperature T_g, obtained by applying shift factors calculated from Equation 1.11 to experimental measurements of G_c at various temperatures, a single curve is obtained for three elastomers having quite different chemical structures and different values of T_g (Figure 1.12 [16]). They are indistinguishable in Figure 1.12. Thus, when the molecular mobility of network strands is similar, so is the tear strength, regardless of the detailed chemical structure. Moreover, using the WLF rate–temperature equivalence, the tear strength of a simple rubber compound can be predicted at any tear speed and at any temperature from two basic parameters, the glass temperature T_g and the degree of cross-linking, represented (inversely) by the length of network strands. This is a remarkable success of the application of molecular mechanics to rubbery solids.

However, although we conclude that, in the simplest cases, the tear strength is governed by energy dissipation from viscous processes, the form of the "universal" curve relating tear strength G_c to the reduced rate of tearing has not yet been well understood. The strength curve, shown in Figure 1.12, somewhat resembles the dependence of elastic modulus μ' on reduced frequency of deformation, shown in Figure 1.9 [12], but the linear dimension required to bring the two scales for tear rate and frequency into accord is unreasonably small, only about 1 nm. In fact, the wide range of log (tear rate) needed to cover the complete rubber-to-glass transition (Figure 1.12) is about twice as large as the corresponding range of log (oscillation frequency) needed to cover the change in elastic and dissipative response functions between the rubbery and glassy states (Figure 1.9). Molecular rearrangements are apparently much slower in fracture than under an imposed deformation.

1.5   CRACK SPLITTING AND TEAR DEVIATION

1.5.1   ANISOTROPY OF STRENGTH IN STRETCHED RUBBER

When a strip of rubber is broken in tension, a crack is initiated by a stress-raising flaw or defect, usually in one edge, and grows across the specimen, breaking it into two. But the crack often takes a surprising path, especially in the initial stages, and especially for compounds reinforced by particulate fillers, such as carbon black. A typical sequence of crack-growth steps is shown in Figure 1.13, where two pieces of a strip that was broken in tension have been placed back together [20]. The test strip had a sharp horizontal cut made in one edge to initiate fracture. As the strip was stretched to break in the vertical direction, a crack grew from the tip of the cut, but instead of continuing to go forward, it turned upwards and grew for some distance in this direction, and even curved backwards to some degree before coming to a halt. This left a relatively smooth surface parallel to the applied tensile stress in which a new crack tip had to develop before the sample could be broken. But, again, the new crack quickly turned sideways, grew for some distance in this direction, and then stopped. Several micro-cracks developed near the tip of the initial cut, and each grew predominantly sideways and stopped, before the final crack appeared, and caused complete fracture of the strip. Even then it set off sideways (downwards) before turning to run across the strip, breaking it into two.

This surprising feature of crack growth appears to be a major aspect of reinforcement by fillers [21]. Busse [22] showed that "knotty" tearing (crack splitting or turning) is common in natural rubber compounds containing large amounts of carbon black, and occurs both in simple tearing and in crack growth under repeated stretching (mechanical fatigue). He attributed it to the development of marked anisotropy of strength in stretched rubber, demonstrated by easy tearing in the direction parallel to an applied tensile strain. Measurements on a number of rubber compounds of the energy needed to propagate a tear in the direction of stretching showed that it is reduced markedly by a quite moderate extension, of the order of 100% [23].

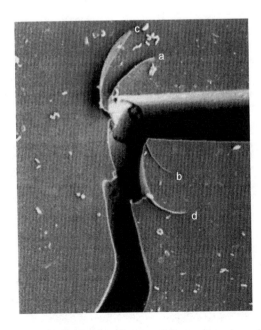

FIGURE 1.13 "Knotty" tearing at the tip of an edge cut, 2.8 mm long, in a sheet of a carbon-black-filled natural rubber vulcanizate, broken in tension. (Reproduced from Hamed, G.R. and Park, B.H., *Rubber Chem. Technol.*, 72, 946 (Figure 6), 1999. With permission.)

Crack splitting or turning from a straight path will obviously increase the energy required for fracture. First, the transverse crack formed at the crack tip will effectively blunt it. Also, the larger area torn through will require a greater amount of fracture energy. We thus reach the paradoxical conclusion that the presence of a weak plane in the transverse direction increases the resistance to fracture.

It has been calculated that the strength of a weak plane in the transverse (vertical) direction must be less than one-fifth of the tensile breaking stress in order to induce a crack to turn sideways [24]. Lake et al. [25] considered the energy release rate G_S available for further growth of a sideways crack that is already long, and concluded that G_S is a rather small fraction of the value G_F for forward growth:

$$G_S/G_F = \lambda^{1/2}/2\pi \tag{1.15}$$

where λ is the extension ratio imposed on the sample. For low applied strains (λ close to unity), G_S is predicted by Equation 1.15 to be only about one-sixth of G_F. Moreover, Equation 1.15 is based on a value for G_F for an edge crack that is too large by a factor of 1.26 [26], so that the actual value of G_S at small strains is only about $G_F/8$. Even at a relatively high strain, when λ is given a value of 3, G_S is only increased to about $G_F/5$. Thus, crack splitting or turning appears to require a material at the crack tip that is highly anisotropic in strength—much weaker for a crack growing sideways than for one growing forwards.

But these calculations are valid only for a sideways crack that is relatively long. Recent FEA calculations of the strain energy release rate G for cracks growing sideways in a sheet of a neo-Hookean elastic solid, stretched to various extents, show that the initial value of G_S is much larger when the length of the sideways crack is small, approaching zero, i.e., for *initiation* of a sideways crack [27]. Only a modest degree of strength anisotropy would therefore be sufficient to bring about crack deviation.

1.5.2 Calculation of G_S and G_F by Finite Element Analysis

Calculations of strain energy W for a thin rubber sheet with a horizontal crack in one edge were made as before, using Abaqus software. The sheet was assumed to be made of an incompressible neo-Hookean solid, and stretched under plane stress conditions. A uniform grid of elements was employed, consisting of 200 equally spaced elements in the vertical direction and 100 in the horizontal direction, as shown in Figure 1.14. The length of the initial crack was represented by at least five elements. Nodes ahead of the initial crack were released sequentially, allowing the crack to grow by steps Δc of one element in length. Corresponding changes ΔW in the stored strain energy were calculated, and a value of G_F was obtained for a crack of mean length $[c + (\Delta c/2)]$ from the ratio $\Delta W/t\,\Delta c$, where t is the sheet thickness. To determine the strain energy release rate G_S for a sideways crack, nodes at the tip of the initial horizontal crack were released in a vertical plane, as shown schematically in Figure 1.14. Thus, the length c_S of the sideways crack started at zero and increased in steps of one element length, until it became comparable to the height of the rubber strip.

1.5.3 Energy Release Rates for a Crack Growing Sideways

Values of the energy G_S available for propagating a crack sideways from the tip of an edge crack were calculated for two values of the imposed tensile strain, 2% and 200%, corresponding to brittle materials that break at low strains, and to highly extensible materials, like rubber. The results are plotted in Figure 1.15 against the length c_S of the sideways crack, relative to the length c_F of the initial edge crack. They are given as a ratio of G_S to the energy G_F available for further forward growth of the initial crack. The ratio G_S/G_F is seen to be quite high initially, about 0.4 at the low extension (2%) and about 0.6 at the

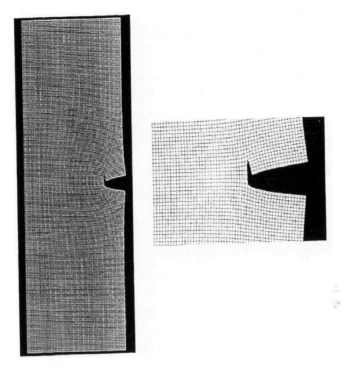

FIGURE 1.14 Sketch of the finite element grid with a plane of symmetry at the left edge and a short crack in the center of the right edge. Close-up showing a sideways crack obtained by release of nodes in the vertical direction. (Reproduced from Gent, A.N., Razzaghi-Kashani, M., and Hamed, G.R., *Rubber Chem. Technol.*, 76, 122 (Figure 3), 2003. With permission.)

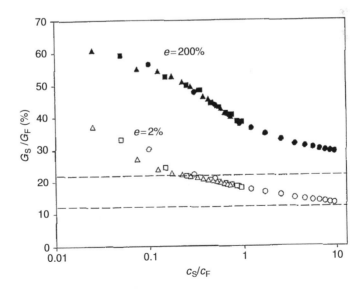

FIGURE 1.15 Energy G_S available for growth of a sideways crack relative to energy G_F available for forward growth, for sideways cracks of various lengths c_S at the tip of an edge crack of length c_F. Lower curve: results for an imposed tensile strain of 2%. Upper curve: results for an imposed tensile strain of 200%. $c_F = 0.5$ mm, O; 1 mm, □; 2 mm, Δ. (Reproduced from Gent, A.N., Razzaghi-Kashani, M., and Hamed, G.R., *Rubber Chem. Technol.*, 76, 122 (Figure 7), 2003. With permission.)

high extension (200%), but in both cases it decreased rapidly as the sideways crack grew. For comparison, the values predicted by Equation 1.15 for a long sideways crack, after incorporating the missing factor of 1.26 in the denominator, are shown as the two horizontal broken lines in Figure 1.15. The FEA values are seen to decrease toward the predicted values as the length c_S of the sideways crack increased. Initially, the ratio G_S/G_F is about three times as large as that predicted.

Thus, even for a brittle solid with a small breaking strain, it appears that a crack will turn sideways if the strength in that direction is about 40% or less of that in the forward direction. For a highly extensible neo-Hookean solid, the permitted strength in the transverse direction can be even higher, approaching 60% of that in the forward direction, and crack splitting would still be expected to occur. In both cases, as the crack grows sideways the energy available for further growth falls markedly, so that the sideways crack will come to a halt unless the transverse plane is much weaker. Moreover, because higher extensions are necessary to induce fracture at the crack tip when the initial crack is short, we also infer that crack deviation or splitting will occur more readily from a short initial edge crack than from a long one, as is commonly observed.

Note that results for longer forward cracks, and correspondingly longer sideways cracks, consisting of more elements, superposed rather well with the results for shorter cracks when plotted against the crack length ratio (Figure 1.15). The success of this superposition implies that errors due to using few elements must be relatively small, and that the trend of the results is correct even though the actual values of G_S for *initiating* a sideways crack are rather uncertain.

We conclude that the energy available to initiate crack deviation at the tip of an edge crack in a stretched rubber sheet is larger than hitherto realized, about 40% of that available for forward growth when the imposed strain is small, and about 60% at higher strains. Thus, when the fracture energy for a crack growing in the direction of the applied tension is less than about 40% of that for forward growth, a crack will inevitably turn sideways. But as it grows in this direction, the available energy falls significantly, and the sideways crack will tend to stop after growing a distance comparable with the length of the precursor edge crack, as is observed (Figure 1.13).

These conclusions are based on the rather simple model shown in Figure 1.14, where the sideways crack is accurately perpendicular to the original edge crack. In practice, as seen in Figure 1.13, the sideways crack generally follows a curved path, presumably because the available energy is then even greater than the values calculated here. The degree of strength anisotropy required for sideways cracking is probably thus even smaller than the relatively modest degree indicated by the present calculations.

Up to this point we have been discussing crack *splitting*, where a new crack forms at the crack tip and propagates in a new direction. This appears to be a common occurrence in reinforced rubber compounds (see Figure 1.14 and [27,28]). But cracks have also been observed to follow gradually curving paths which eventually result in marked deviations from the straight-ahead direction [29]. Crack curving has been attributed to the development of a ''frozen'' stress pattern at the crack tip, as a result of hysteresis or softening [29]. It is not clear which mechanism—crack splitting due to strength anisotropy or crack deviation due to a stationary stress distribution at the crack tip—is more important in rubber reinforcement. A detailed study of crack growth is needed to elucidate these various modes of crack growth, the range of conditions under which they occur, and the special physical characteristics of rubbery solids that induce them. Moreover, it seems a necessary step in developing a better understanding of the strength of rubbery solids.

1.6 SUMMARY AND CONCLUSIONS

1.6.1 END EFFECTS IN RUBBER SPRINGS

The stresses set up in a long rubber block or tube under simple shear deformations are found to depend on the *shapes* of the end surfaces, even when the block or tube is quite long.

This phenomenon is attributed to the absence on the end surfaces of the shear and compressive stresses that are needed in order to maintain a state of simple shear. As a result, the stresses throughout the block are affected (in contrast to a conventional "end effect" that would have a negligible effect far from the ends). One consequence is that high internal stresses can develop, sufficient in principle to cause failure. It is clear that the effect of the special conditions obtaining at the ends should be taken into consideration in the rational design of rubber springs.

1.6.2 Bond Failures in Sheared Blocks

Fracture energies have been determined for bond failure at either end of the bond line for a sheared rubber block, but the results are inconclusive—it is not clear where bond failure will occur, or at what load, even when the fracture properties of the rubber are known. Thus, the initiation of cracks, especially at interfaces and corners, needs further study.

1.6.3 Relation between the Strength of Rubber and Its Molecular Structure

A direct relation is known to hold between tear strength and internal energy dissipation for all rubbery materials. In the simplest case, energy dissipation is due to a viscous process, and therefore the tear energy obeys the well-known WLF rate–temperature equivalence—a change in rate (in this case, of tear propagation) is equivalent to a corresponding change in test temperature. This equivalence is known to hold for all physical properties of elastomers that are governed by internal viscosity, including measurements of strength. However, the molecular mechanics whereby high internal viscosity results in high strength is not at all clear. No theoretical treatment relates the two properties successfully. This gap in understanding hinders the development of stronger materials, which are currently produced only as a result of rather empirical research strategies.

1.6.4 Crack Splitting and Tear Deviation

An important strengthening feature of rubbery solids, especially evident in particle-filled (reinforced) compounds, is "knotty" tearing, where cracks may split, turn abruptly at 90°, or follow a continuously curving path. Although the conditions under which "knotty" tearing occurs are well known, the cause is obscure. Even the detailed process is unclear. As a result, this strengthening feature cannot be predicted.

ACKNOWLEDGMENTS

Helpful discussions with J.B. Suh, Dr. O.H. Yeoh, Dr. A.H. Muhr, and professor G.R. Hamed are gratefully acknowledged.

REFERENCES

1. R.S. Rivlin, Large elastic deformations of isotropic materials. IV. Further developments of the general theory, *Philos. Trans. R. Soc. (London) Ser. A*, **241**, 379–397, 1948.
2. R.W. Ogden, *Non-Linear Elastic Deformations*, Halsted Press/Wiley, New York, 1984; Dover Publications, Mineola, NY, 1997, p. 227.
3. A.N. Gent, J.B. Suh, and S.G. Kelly, III, Mechanics of rubber shear springs, *Int. J. Non-Linear Mech.*, 42, 241–249, 2007.
4. J.B. Suh, A.N. Gent, and S.G. Kelly, III, Shear of rubber tube springs, *Int. J. Non-Linear Mech.*, 42, 1116–1126, 2007.
5. A.N. Gent and P.B. Lindley, Internal rupture of bonded rubber cylinders in tension, *Proc. R. Soc. (London)*, **A249**, 195–205, 1958.

6. A. Wineman, Some results for generalized neo-Hookean elastic materials, *Int. J. Non-Linear Mech.*, **40**, 271–279, 2005.

7. A.H. Muhr, A.G. Thomas, and J.K. Varkey, A fracture mechanics study of natural rubber-to-metal bond failure, *J. Adhesion Sci. Technol.*, **10**, 593–616, 1996.

8. I.H. Gregory and A.H. Muhr, Stiffness and fracture analysis of bonded rubber blocks in simple shear, in *Finite Element Analysis of Elastomers*, ed. by D. Boast and V.A. Coveny, Professional Engineering Publications, Bury St. Edmunds, United Kingdom, 1999, pp. 265–274.

9. J. Gough and A.H. Muhr, Initiation of failure of rubber close to bondlines, in *Proceedings of International Rubber Conference*, Maastricht, The Netherlands, June 2005, IOM Communications Ltd., London, 2005, pp. 165–174.

10. P.B. Lindley and S.C. Teo, Energy for crack growth at the bonds of rubber springs, *Plast. Rubber. Mat. Appl.*, **4**, 29–37, 1979.

11. M.L. Williams, R.F. Landel, and J.D. Ferry, The temperature dependence of relaxation mechanisms in amorphous polymers and other glass-forming liquids, *J. Am. Chem. Soc.*, **77**, 3701–3707, 1955.

12. A.R. Payne and J.R. Scott, Dynamic and related time-dependent properties of rubber, Chap. 2 in *Engineering Design with Rubber*, Interscience Publishers, New York, 1960.

13. G.J. Lake and A.G. Thomas, Strength, Chap. 5 in *Engineering with Rubber*, 2nd. edn., ed. by A.N. Gent, Hanser Publishers, Munich, 2001.

14. J.P. Berry, Fracture processes in polymeric materials. I. The surface energy of poly(methyl methacrylate), *J. Polymer Sci.*, **50**, 107–115, 1961.

15. G.J. Lake and A.G. Thomas, The strength of highly elastic materials, *Proc. R. Soc. (London)*, **A300**, 108–119, 1967.

16. A.N. Gent and S.-M. Lai, Interfacial bonding, energy dissipation and adhesion, *J. Polymer Sci., Part B: Polymer Phys.*, **32**, 1543–1555, 1994.

17. T.L. Smith, Dependence of the ultimate properties of a GR-S rubber on strain rate and temperature, *J. Polymer Sci.*, **32**, 99–113, 1958.

18. L. Mullins, Rupture of rubber. Part 9. Role of hysteresis in the tearing of rubber, *Trans. Inst. Rubber Ind.*, **35**, 213–222, 1959.

19. H.W. Greensmith, L. Mullins, and A.G. Thomas, Strength of rubbers, Chap.10 in *The Chemistry and Physics of Rubber-Like Substances*, ed. by L. Bateman, Wiley, New York, 1963.

20. G.R. Hamed and B.H. Park, The mechanism of carbon black reinforcement of SBR and NR vulcanizates, *Rubber Chem. Technol.*, **72**, 946–959, 1999.

21. L. Mullins, Effect of fillers in rubber, Chap. 11 in *The Chemistry and Physics of Rubber-Like Substances*, ed. by L. Bateman, Wiley, New York, 1963.

22. W.F. Busse, Tear resistance and structure of rubber, *Ind. Eng. Chem.*, **26**, 1194–1199, 1934.

23. A.N. Gent and H.J. Kim, Tear strength of stretched rubber, *Rubber Chem. Technol.*, **51**, 35–44, 1978.

24. J. Cook, J.E. Gordon, C.C. Evans, and D.M. Marsh, A mechanism for the control of crack propagation in all-brittle systems, *Proc. R. Soc. (London)*, **A282**, 508–520, 1964.

25. G.J. Lake, A. Samsuri, S.C. Teo, and J. Vaja, Time-dependent fracture in vulcanized elastomers, *Polymer*, **32**, 2963–2975, 1991.

26. J.P. Bentham and W.T. Koiter, *Mechanics of Fracture*, ed. by G.C. Sih, Noordhoff International Publishing, Leiden, 1972, pp. 131–178; referred to in *Stress Intensity Factors Handbook* by Y. Murakami, Pergamon Press, New York, 1987, pp. 6–9.

27. A.N. Gent, M. Razzaghi-Kashani, and G.R. Hamed, Why do cracks turn sideways?, *Rubber Chem. Technol.*, **76**, 122–131, 2003.

28. A.N. Gent and C.T.R. Pulford, Micromechanics of fracture in elastomers, *J. Materials Sci.*, **19**, 3612–3619, 1984.

29. E.H. Andrews, Crack propagation in a strain-crystallizing elastomer, *J. Appl. Phys.*, **32**, 542–548, 1961.

Section II

New Elastomers and Composites

2 Elastomer–Clay Nanocomposites

Susmita Dey Sadhu, Madhuchhanda Maiti, and Anil K. Bhowmick

CONTENTS

2.1 INTRODUCTION

Fillers are solid substances that can be incorporated in a polymer to improve mechanical properties, hardness, tear resistance, and processing; to control density, optical properties, thermal conductivity, thermal expansion, electrical and magnetic properties, flame retardance, etc.; and to reduce cost. They can be distinguished as nonfunctional or nonreinforcing fillers and functional or reinforcing fillers. Nonreinforcing fillers are mostly added to polymers to reduce cost. They do not have any functional groups that can interact with polymers. Reinforcing fillers, on the other hand, might interact chemically or physically with polymer chains to improve the tensile properties like tensile strength, elongation at break and modulus. Incorporation of fillers in a polymeric material increases density and modulus of composites. Other properties like flow behavior, and thermal and barrier properties might also change.

Structurally, filler particles can be globular, rodlike, or platelet type. For example, carbon black particles are globular in nature, while fiber is an ideal example of unidirectional rodlike filler. Globular fillers like carbon black or silica particles remain agglomerated and form a chain-like structure. The particle size of carbon black varies from 20 to 90 nm depending on the manufacturing conditions [1]. Silica fillers can be of mainly two types, fumed silica and precipitated silica. They differ from each other mainly in particle dimensions. Fumed silica has a particle size of 10 nm, while the other is nearly 20 nm [2]. Aluminosilicates are known to have platelet-like structures, where two of the dimensions (length and breadth) are very large compared to that of the third one. Depending on the clay type, the nature of layered structure and chemical compositions also vary [3]. The globular fillers by virtue of their isomorphic structure give similar properties in all directions. But the unidirectional structure of the rodlike or platelet-like fillers imparts unidirectional property to composites. In brief, it can be concluded that many crucial parameters are determined by the filler geometry (size, shape, and aspect ratio) and filler–matrix interactions. In the case of reinforcing fillers, rubbery polymer mostly adheres to the surface of the filler forming an interphase, which is known as the bound rubber. The mechanism of interaction of reinforcing fillers with rubber is dealt with in later chapters in detail.

2.2 NANOCOMPOSITES

2.2.1 DEFINITION

A nanocomposite is defined as the composite of two materials, one having the dimension of nanometric level at least in one dimension. In polymer nanocomposites (PNC), the fillers are dispersed on a nanolevel.

2.2.2 CLASSIFICATION

Depending on the nature of filler, type of dispersion, and method of preparation, the nanocomposites can be divided into subclasses.

With the variation in nanofillers, mainly the following types of nanocomposites can be obtained:

1. Clay-based nanocomposites
2. Silica-based nanocomposites
3. Polyhedral oligomeric silsesquioxane (POSS)-based nanocomposites
4. Carbon nanotube-filled nanocomposites and
5. Nanocomposites based on other nanofillers like metal oxides, hydroxides, and carbonates

Different clays having different structures and compositions give different types of nanocomposites. In this chapter, clay-based nanocomposites will be discussed in detail.

2.3 NANOFILLERS

2.3.1 DEFINITION AND GENERAL IDEA ABOUT NANOFILLERS

Nanofiller is a class of new-generation fillers, which have at least one characteristic length scale in the order of nanometer with varying shapes ranging from isotropic to highly anisotropic needle-like or sheet-like elements. Uniform dispersion of these nanosized particles can lead to ultra-large interfacial area between a polymer and the fillers. This large interfacial area between the filler and a polymer and the nanoscopic dimension differentiate PNC from traditional composites. The major characteristics, which control the performance of nanocomposites, are nanoscale-confined matrix polymer, nanoscale inorganic and organic fillers, and nanoscale arrangements of these constituents [4].

2.3.2 ADVANTAGES

Besides their improved properties, these nanocomposite materials are also easily extrudable or moldable to near-final shape. Since high degrees of stiffness and strength are realized with little amount of high-density inorganic materials, they are much lighter compared to conventional polymer composites This weight advantage could have significant impact on environmental concerns among many other potential benefits. In addition to all these, their outstanding combination of barrier and mechanical properties may eliminate the need for a multipolymer layer design in packaging materials.

Thus, it can be concluded that nanocomposites have four major advantages over conventional ones: (1) lighter weight due to low filler loading, (2) low cost due to lesser amount of filler used, (3) improved properties compared to these conventional filler-based composites at very low loading of filler, and (4) combination of specific properties.

2.3.3 TYPES OF NANOFILLERS

Among the conventional fillers that have been used widely in polymers are carbon black of different sizes and shapes, silica, and clay. Among all these, carbon black and silica are known as reinforcing fillers. Carbon black particles, which are commonly used in the rubber industry, have a particle size of nanometric level, but mostly remain agglomerated. Particle dimension of silica also ranges between 10 and 20 nm depending on its nature. Clay is mostly used as nonreinforcing filler in rubber industries. These clays are mostly wollastonite or kaolinite. The size of these unmodified clay particles is in micrometer range. Since these are very cheap and widely available, they are used to reduce the cost of a product. But in 1980, the Toyota Research Group of Japan invented nylon 6-based nanocomposite with *smectite*-group clay particles dispersed as fillers in the matrix by *in situ* polymerization method [5]. This invention revealed an entirely new area of research in polymer composite field. These clays having a characteristic structure can form nanocomposites by the breakdown of the clay layers in nanometer range. Afterwards, nanocomposites have been prepared based on various polymers and nanoclays having different structures. Layered carbon and graphite are other nanofillers having platelet structure. Carbon nanotubes, carbon fibers, and nanosilica have been introduced subsequently.

Nano-dispersion of these fillers enhances the degree of improvement in properties further. Carbon nanotubes and nanofibers owing to their very high aspect ratio show a directional property and they are also conductive in nature. Silica, on the other hand, is globular. When their loading is higher, they form clusters. Nanoclays and graphite have platelet-like structure. The layers of these nanofillers break down in a polymer matrix as the polymer gets into the layers more and more. The typical thickness of a single layer is ~1 nm. As the aluminosilicates have two-dimensional giant structures, the aspect ratio in this case is very high. Graphite having layered structure can also be intercalated by polymers and can form nanocomposites by exfoliation of the layers.

2.4 PHYSICAL, CHEMICAL, AND STRUCTURAL CHARACTERISTICS OF CLAYS

1. Clay minerals tend to form microscopic to submicroscopic crystals.
2. They can absorb water or lose water from simple humidity changes.
3. When mixed with limited amounts of water, clays become plastic and can be molded.
4. When water is absorbed, clays will often expand as the water fills the spaces between the stacked silicate layers.
5. Due to the absorption of water, the specific gravity of clays is highly variable and is lowered with increased water content.
6. The hardness of clays is difficult to determine due to the microscopic nature of the crystals, but hardness is usually 2–3 in Mohs' scale and many clays give a hardness of 1 in field tests.

2.5 MAJOR GROUPS OF CLAY MINERALS

Clay minerals are divided into three major groups. These are the important clay mineral groups.

2.5.1 THE KAOLINITE GROUP

This group has three members (kaolinite, dickite, and nacrite) and a formula of $Al_2Si_2O_5(OH)_4$ [6]. The different minerals are polymorphs, meaning that they have the same chemistry but different structures. The general structure of the kaolinite group is composed of silicate sheets (Si_2O_5) bonded to aluminum oxide/hydroxide layers ($Al_2(OH)_4$). The silicate (tetrahedral layer, T) and aluminum oxide (octahedral layer, O) layers are tightly bonded together with bonding existing between the T and O paired layers (Figure 2.1) [7]. These are mostly used in ceramics, as filler for paint, rubber, and plastics. The largest use is in the paper industry that uses kaolinite to produce a glossy paper such as that used in most magazines.

Due to the 'TO' structure of the clay, the interlayer H-bonding is very strong. This hinders the intercalation of any molecule or chain into the gallery. Hence, the kaolinite clay cannot be used for nanocomposite preparation.

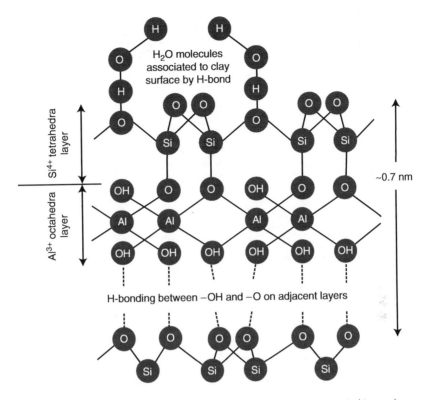

FIGURE 2.1 The layered structure of kaolinite clay. (From http://jan.ucc.nau.edu/doetqp/courses/env440/env440_2/lectures/lec19/Fig19_3.gif, access date 4.11.06.)

2.5.2 The Montmorillonite/Smectite Group

This group is composed of several minerals including pyrophyllite, talc, vermiculite, sauconite, saponite, nontronite, and montmorillonite (MMT). They differ mostly in chemical content. The general formula is $(Ca,Na,H)(Al,Mg,Fe,Zn)_2(Si,Al)_4O_{10}(OH)_2 - xH_2O$, where x represents the variable amount of water that members of this group could contain [3]. Talc's formula, for example, is $Mg_3Si_4O_{10}(OH)_2$. The octahedral (gibbsite) layers of the kaolinite group can be replaced in this group by a similar layer that is analogous to $(Mg_2(OH)_4)$. The structure of this group is composed of silicate layers sandwiching a gibbsite layer in between, in a T–O–T stacking sequence (Figure 2.2) [8]. The variable amounts of water molecules would lie between the T–O–T sandwiches. The usage includes facial powder (talc), filler for paints and rubbers, and an electrical, heat- and acid-resistant porcelain in drilling mud and as a plasticizer in molding sands and other materials.

2.5.3 The Illite (or the Clay–Mica) Group

This group is basically a hydrated microscopic muscovite. The mineral illite is the only common mineral represented; however, it is a significant rock-forming mineral being a main component of shales and other argillaceous rocks. The general formula is $(K,H)Al_2(Si,Al)_4O_{10}(OH)_2 - xH_2O$, where x represents the variable amount of water that this group could contain [3]. The structure of this group is similar to the MMT group with silicate layers sandwiching a gibbsite-like layer in between, in a T–O–T stacking sequence (Figure 2.3) [9]. The variable amounts of water molecules as well as potassium ions would lie between the T–O–T sandwiches. This is a common constituent in shale and is used as filler and in some drilling mud.

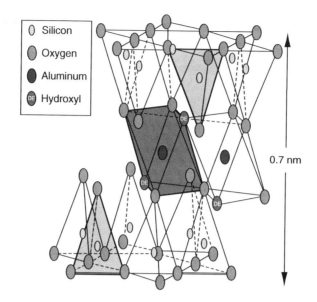

FIGURE 2.2 The smectite clay structure. (From http://www.pslc.ws/macrog/mpm/composit/nano/struct3_1. htm, access date 4.11.06.)

2.6 MONTMORILLONITE AND HECTORITE

Many varieties of clay are aluminosilicates with a layered structure which consists of silica (SiO_4^{4-}) tetrahedral sheets bonded to alumina (AlO_6^{9-}) octahedral ones. These sheets can be arranged in a variety of ways; in smectite clays, a 2:1 ratio of the tetrahedral to the octahedral is observed. MMT and *hectorite* are the most common of smectite clays.

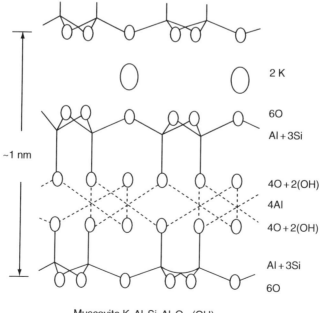

Muscovite $K_2Al_2Si_2Al_4O_{20}(OH)_4$

FIGURE 2.3 Structure of mica. (From http://grunwald.ifas.ufl.edu/Nat_resources/mineral_components/mica. gif, access date 4.11.06.)

The structure of MMT $((Na,Ca)_{0.33}(Al,Mg)_2(Si_4O_{10})(OH)_2 - nH_2O)$ is derived from the original pyrophyllite structure by partial substitution of the trivalent Al-cation in the octahedral layer by the divalent Mg-cation. Similarly, the structure of hectorite $(Na_{0.3}(Mg,Li)_3Si_4O_{10}(OH)_2)$ is derived by partial substitution of bivalent Mg-cation by univalent Li-cation. Because of the difference in charge between the Al and Mg ions or Mg and Li ions, the center layer of these 2:1 silicates is negatively charged and the negative charge is balanced by group I or II metal ions present between the 2:1 sheets. These ions do not fit in the tetrahedral layers such as in mica and the negative charge is located in the octahedral layer, thus making the attractive forces between the layers weaker. Therefore, the layers are not collapsed upon each other such as in mica. The MMT and hectorite can absorb water between the charged layers because of this weak bonding and the large spacing and are therefore members of a group of water-expandable clay minerals known as smectites or smectite clays. As clearly shown in Figure 2.4, in MMT, oxygen atoms from each alumina octahedral sheet also belong to the silica tetrahedral ones, the three of them consisting of ~1 nm thin layer [10]. These layers are in turn linked together by van der Waals bonds and organized in stacks with a regular gap between them called *interlayer* or *gallery*. Within the layers, isomorphic substitution of Al^{3+} with Mg^{2+} or Fe^{2+} generates an excess of negative charge, the amount of which characterizes each clay type and is defined through the *cation exchange capacity* (CEC). The CEC value for smectites depends on its mineral origin and is typically 65 meq/100 g–150 meq/100 g. In natural clays, ions such as Na^+, Li^+, or Ca^{2+} in their hydrated form balance this excess negative charge; this means natural MMT or hectorite is only compatible with hydrophilic polymers. General characteristics of these materials include 20 layer thicknesses of ~1 nm and lateral dimensions ranging from ~25 nm to ~5 μm.

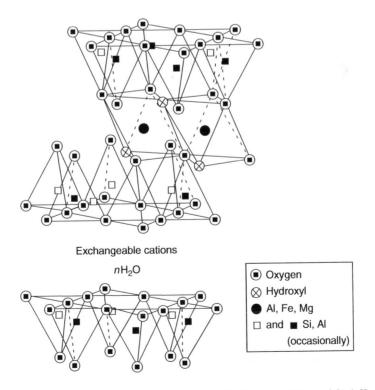

FIGURE 2.4 T–O–T structure of montmorillonite clay. (From Sadhu, S. and Bhowmick, A.K., *J. Polym. Sci., Part B: Polym. Phys.*, 42, 1573, 2004. Courtesy of Wiley InterScience.)

2.6.1 PHYSICAL CHARACTERISTICS OF MONTMORILLONITE

1. Color is usually white, gray or pink with tints of yellow or green.
2. Luster is dull.
3. Crystals are translucent and masses are opaque.
4. Crystal system is monoclinic.
5. It is never found in large individual crystals, usually found in compact or lamellar masses and also seen as inclusions in quartz as fibers and powder-like masses.
6. Cleavage is perfect in one direction, basal and not seen in massive specimens.
7. Hardness is 1–2 in Mohs' scale (can sometimes leave marks on paper).
8. Specific gravity is variable from 2 to 3 (average).
9. Streak is white.
10. Crystals expand to many times their original volume when added to water.
11. Associated minerals include other clays, garnets, biotite and quartz.
12. These are widely available in France, Italy, the United States and many other localities worldwide.
13. Best field indicators are softness, color, soapy feel, luster and expandability when added to water.

2.6.2 CATION EXCHANGE CAPACITY

For a given clay, the maximum amount of cations that can be taken up is constant and is known as the cation exchange capacity (CEC). It is measured in milliequivalents per gram or per 100 g (meq/100 g). The CEC of MMT generally varies from 80 to 150 meq/100 g. Determination of the CEC is generally done by saturating the clay with NH_4^+ or Ba^{+2} and determining the amount held at pH 7 by conductometric titration [11,12]. Another method consists of saturating the clay with alkylammonium ions and evaluating the quantity of ions intercalated by ignition of the sample [13].

2.7 NANOCLAY AND CLAY MODIFICATIONS: STRUCTURAL ASPECT

The MMT clays are naturally occurring minerals and are subject to natural variability in their constitution. The chemical composition of the clay can affect the final nanocomposite properties. These naturally occurring aluminosilicates have a sheet-like (layered) structure and consist of layers of silica, SiO_4, tetrahedra bonded to alumina AlO_6 octahedra in a variety of ways. A 2:1 ratio of the tetrahedra to the octahedra results in smectite clays, the most common of which is MMT. Other metals such as magnesium may replace the aluminum in the crystal structure. Depending on the precise chemical composition of the clay, the sheets bear a surplus charge on the surface and edges, this charge being balanced by counter-ions, which reside in part in the interlayer spacing of the clay. The thickness of the layers (platelets) is of the order of 1 nm and aspect ratios are high, typically 100–1500. It is important to note that the molecular weight of the platelets (ca. 1.3×10^8) is considerably greater than that of polymers. In addition, platelets also have a degree of flexibility. The clays often have very high surface areas, up to hundreds of square meters per gram. The clays are also characterized by their ion (e.g., cation) exchange capacities, which can vary widely. Due to the charged nature of the clays, they are generally highly hydrophilic species and therefore naturally incompatible with a wide range of polymer types. An important prerequisite for successful formation of polymer–clay nanocomposites is therefore alteration of the clay polarity to make the clay "organophilic." Organophilic clay can be produced from normally hydrophilic clay by ion exchange with an organic cation such as alkylammonium ion.

The way in which this is done has a major effect on the formation of a particular nanocomposite and this is discussed further below. A wide range of ω-amino acids have been intercalated between the layers of MMT [14]. Amino acids have been successfully used in the synthesis of polyamide

6-clay hybrids [15] because the acid function has the ability to polymerize the ε-caprolactum intercalated between the layers. The most widely used alkylammonium ions are based on primary alkyl amines put in an acidic medium to protonate the amine function. Their basic formula is $CH_3-(CH_2)_n-NH_3^+$ where n is between 1 and 18. It is important to note that the length of the ammonium ions may have a strong impact on the resulting structure of nanocomposites. Lan et al. [16] showed, for instance, that those alkyl ammonium ions with chain length larger than eight carbon atoms were favoring the synthesis of delaminated nanocomposites, whereas alkylammonium ions with shorter chain lengths lead to the formation of intercalated nanocomposites. Alkyl ammonium ions based on secondary and tertiary amine have also been used [17]. Although the organic pretreatment adds to the cost of the clays, the clays are still relatively cheap feedstocks with minimal limitation on supply. The MMT is the most common type of clay used for nanocomposite formation; however, other types of clays can also be used depending on the precise properties required from the product. These clays include hectorites (magnesiosilicates), which contain very small platelets and synthetic clays (e.g., hydrotalcite), which can be produced in a very pure form and can carry a positive charge on the platelets.

2.8 FACTORS AFFECTING THE TYPE OF ORGANOCLAY HYBRID FORMED

Since clay nanocomposites can produce dramatic improvements in a variety of properties, it is important to understand the factors which affect delamination of the clay. The properties are controlled to a great extent by the structure of the hybrid, for example, the synthesized hybrid has an intercalated or exfoliated structure. These factors can be affected by the exchange capacity of the clay, the polarity of the reaction medium, and the chemical nature of the interlayer cations (e.g., onium ions). By modifying the surface polarity of the clay, onium ions allow thermodynamically favorable penetration of polymer precursors into the interlayer. The ability of the onium ions to assist in delamination of the clay depends on its chemical nature such as its polarity. The loading of the onium ions on the clay is also crucial for success. For positively charged clays, such as hydrotalcite, the onium salt modification is replaced by use of a cheaper anionic surfactant. Other types of clay modifications can be used depending on the choice of polymer, including ion–dipole interactions, use of silane coupling agents, and use of block copolymers.

An example of ion–dipole interactions is the intercalation of a small molecule such as dodecylpyrrolidone into the clay. Entropically driven displacement of the small molecules then provides a route to introducing polymer molecules. Unfavorable interactions of clay edges with polymers can be overcome by use of silane coupling agents to modify the edges. These can be used in conjunction with the onium ion-treated organoclay. An alternative approach to compatibilizing clays with polymers has been introduced [18], on the basis of use of block or graft copolymers where one component of the copolymer is compatible with the clay and the other with the polymer matrix. This is similar in concept to compatibilization of polymer blends. A typical block copolymer would consist of a clay-compatible hydrophilic block and a polymer-compatible hydrophobic block (Figure 2.5).

The block length must be controlled and must not be too long. High degrees of exfoliation are claimed using this approach.

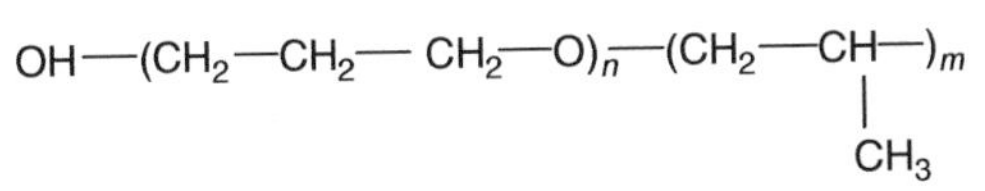

$$OH-(CH_2-CH_2-CH_2-O)_n-(CH_2-CH-)_m$$
$$| $$
$$CH_3$$

FIGURE 2.5 Structure of a typical polymer-compatible hydrophobic block.

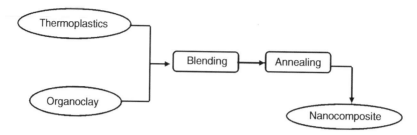

SCHEME 2.1 The melt-intercalation method for nanocomposite preparation.

2.9 PREPARATION OF NANOCOMPOSITES

Nanocomposites can be prepared in various ways:

1. Melt intercalation: Melt intercalation was first reported by Vaia et al. in 1993 [19]. A thermoplastic is melted with organoclay in order to optimize polymer–clay interaction. The mixture is then annealed at a temperature above the glass transition temperature of the polymer to form a nanocomposite. The polymer can undergo center of mass transport in between the clay layers, although the unperturbed radius of gyration of the polymer is of magnitude greater than the interlamellar spacing [20]. This method has become increasingly popular because of its great potential for application in industry [21–23]. The method is represented in Scheme 2.1.

2. Solution blending: Polar as well as nonpolar solvents can be used in this method. The polymer is solubilized in a proper solvent and then mixed with the filler dispersion. In solution, the chains are well separated and easily enter the galleries or the layers of the fillers. After the clay gets dispersed and exfoliated, the solvent is evaporated usually under vacuum. High-density polyethylene [24], polyimide (PI) [25], and nematic liquid crystal [26] polymers have been synthesized by this method. The schematic presentation is given in Scheme 2.2.

3. *In situ* polymerization: This is the first method used to synthesize polymer–clay nanocomposites based on polyamide 6 [5]. Monomers are added to the fillers and allowed to enter the clay galleries or mix with the fillers in a proper solvent medium. Then they are polymerized using some initiator to get the nanocomposite *in situ*. This method requires a certain time period, which depends on the polarity of monomer, surface treatment of monomer, and swelling temperature. Then the reaction is initiated. During the swelling phase, the high surface energy of the clay attracts polar monomer molecules so that they diffuse between the clay layers. When certain equilibrium is reached, the diffusion stops and the clay is swollen in the monomer to a certain extent corresponding to a particular orientation of the alkyl-ammonium ions. The process is schematically given in Scheme 2.3.

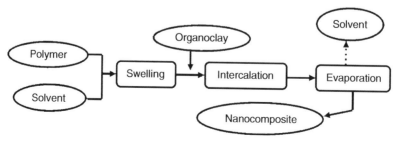

SCHEME 2.2 The solution intercalation method for nanocomposite preparation.

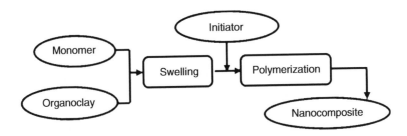

SCHEME 2.3 The *in situ* polymerization method for nanocomposite preparation.

2.10 MICROSTRUCTURE OF CLAY NANOCOMPOSITES

Nanocomposites based on clay can be of three different types depending on the extent of intercalation and dispersion, which are different from conventional composites.

- Conventional composites: In conventional composites the clay layers remain stacked. The polymer chains cannot intercalate into the gallery and instead remain attached to the surface of the clay layers.
- Intercalated nanocomposites: When the polymer chains intercalate into the clay gallery gaps but are unable to break down the layered structure, they are called intercalated nanocomposites (Figure 2.6a).
- Exfoliated nanocomposites: Exfoliated nanocomposites are those where the degree of intercalation is so high that distance between the layers increases to a great extent and they no longer remain as stacked structure (Figure 2.6b).

2.11 ADVANTAGES OF NANOCLAYS

Thorough study of nanocomposites has revealed clearly that nanoclays can provide certain advantages in properties in comparison to their conventional filler counterparts. Properties which have been shown to undergo substantial improvements include:

- Mechanical properties, for example, strength, modulus, and dimensional stability
- Decreased permeability to gases, water, and hydrocarbons
- Thermal stability and heat distortion temperature
- Flame retardancy and reduced smoke emissions

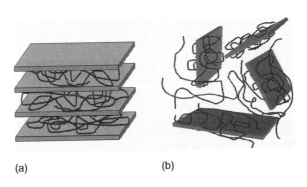

(a)   (b)

FIGURE 2.6 (a) Intercalated structure. (b) Exfoliated structure.

- Chemical resistance
- Surface appearance
- Optical clarity in comparison to conventionally filled polymers

It is important to recognize that nanoparticle/nanofiber loading confers significant property improvements with very low loading. Traditional microparticle additives require much higher loading levels to achieve similar performance. This in turn can result in significant weight reductions (of obvious importance to various military and aerospace applications) for similar performance, greater strength for similar structural dimensions, and increased barrier performance for similar material thickness.

The data provided by Toyota Research Group of Japan on polyamide–MMT nanocomposites indicate tensile strength improvements of approximately 40%–50% at 23°C and modulus improvement of about 70% at the same temperature. Heat distortion temperature has been shown to increase from 65°C for the unmodified polyamide to 152°C for the nanoclay-modified material, all the above having been achieved with just a 5% loading of MMT clay. Similar mechanical property improvements were presented for polymethyl methacrylate–clay hybrids [27].

Further data provided by Honeywell relating the polyamide-6 polymers confirm these trends in property [28]. In addition, further benefits of short/long glass fiber incorporation, together with nanoclay incorporation, are clearly revealed.

One of the few disadvantages associated with nanoparticle incorporation concerns the loss of some properties. Some of the data presented have suggested that nanoclay modification of polymers such as polyamide could reduce impact performance [28]. Nanofillers are sometimes very matrix-specific. High cost of nanofillers prohibits their use.

2.12 POLYMERS USED FOR NANOCOMPOSITE FORMATION AND PROPERTIES OF THE NANOCOMPOSITES

Many different polymers have already been used to synthesize polymer–clay nanocomposites. In this section, an overview of the advances that have been made during the last 10 years in the intercalation and the delamination of organoclay in different polymeric media is given. The discussion mainly covers the work involving thermoset nanocomposites along with a brief discussion about thermoplastic-based nanocomposites.

2.12.1 THERMOSETS

This is a less investigated area of nanocomposites. Only recently elastomer nanocomposites have been prepared. Elastomers have very low glass transition temperature and their viscosity is high. Hence, it was a big challenge to achieve nanosized dispersion of fillers, especially clay, in rubbers. Various methods, namely (1) solution blending, (2) *in situ* polymerization, and recently (3) melt intercalation, have been used for the preparation. Mostly MMT and other metal oxides have been used as nanofillers. Many research groups from all over the world have investigated physical, mechanical, barrier, rheological, and morphological properties and reported also simulations, kinetic thermodynamic studies. Some of the important results are mentioned below.

2.12.1.1 Natural Rubber

This is the most widely used naturally occurring rubber. The literature search shows that many research groups have prepared nanocomposites based on this rubber [29–32]. Varghese and Karger-Kocsis have prepared natural rubber (NR)-based nanocomposites by melt-intercalation method, which is very useful for practical application. In their study, they have found increase in stiffness, elongation, mechanical strength, and storage modulus. Various minerals like MMT, bentonite, and hectorite have been used.

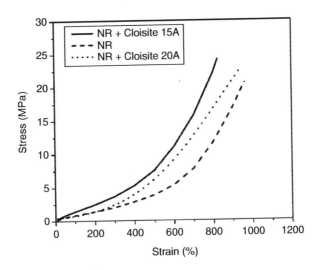

FIGURE 2.7 Effect of various clays on natural rubber (NR)-based nanocomposites (at 4 phr loading). (From Bhattacharya and Bhowmick, Unpublished data.)

They have studied the properties of NR-epoxidized natural rubber (ENR) blend nanocomposites also. Vulcanization kinetics of natural rubber-based nanocomposite was also studied. The effect of different nanoclays on the properties of NR-based nanocomposite was studied. The tensile properties of different nanocomposites are shown in Figure 2.7 [33].

2.12.1.2 Epoxidized Rubber and Epoxy Resin

Literature search shows that epoxy-based nanocomposites have been prepared by many researchers [34–38]. Becker et al. have prepared nanocomposites based on various high-functionality epoxies. The mechanical, thermal, and morphological properties were also investigated thoroughly [39–43]. The cure characteristics, effects of various compatibilizers, thermodynamic properties, and preparation methods [16,17,44–49] have also been reported. ENR contains a reactive epoxy group. ENR–organoclay nanocomposites were investigated by Teh et al. [50–52].

2.12.1.3 Acrylic Rubber

This rubber is very tacky in nature and contains acrylic group, which makes it polar in nature. Nanocomposites have been prepared based on this elastomer with a wide range of nanofillers. Layered silicates [53–55] have been used for this preparation. Sol–gel method [56,57], *in situ* polymerization [58], and nanocomposites based on different clays like bentonite [59] and mica [60] have been described. The mechanical, rheological, and morphological behaviors have been investigated thoroughly.

2.12.1.4 Ethylene Propylene Diene Rubber

This is a nonpolar rubber with very little unsaturation. Nanoclays as well as nanotubes have been used to prepare nanocomposites of ethylene–propylene–diene monomer (EPDM) rubber. The work mostly covers the preparation and characterization of these nanocomposites. Different processing conditions, morphology, and mechanical properties have been studied [61–64]. Acharya et al. [61] have prepared and characterized the EPDM-based organo-nanoclay composites by X-ray diffractogram (XRD), Fourier transform infrared spectroscopy (FTIR), scanning electron microscopy

TABLE 2.1

Thermal Degradation Results of EPDM and Its Nanocomposites

Sample	Initial Thermal Decomposition T_i (°C)	Final Decomposition Temperature T_f (°C)	Weight Loss (%)
Pure EPDM	325	460	96.4
EPDM + 2 wt% 16 Me-MMT[a]	341	459	95.2
EPDM + 3 wt% 16 Me-MMT	355	460	92.2
EPDM + 4 wt% 16 Me-MMT	372	461	92.1
EPDM + 5 wt% 16 Me-MMT	372	460	92.0
EPDM + 6 wt% 16 Me-MMT	374	462	91.8

Source: Acharya, H., Pramanik, M., Srivastava, S.K., and Bhowmick, A.K., *J. Appl. Polym. Sci.*, 93, 2429, 2004. Courtesy of Wiley InterScience.

[a] 16 Me-MMT = hexadecyl amine modified MMT.

(SEM), and transmission electron microscopy (TEM). Study of mechanical and thermal properties shows significant improvement over the gum. The thermal results are given in Table 2.1.

The results suggest that the thermal stability improves with higher loading till 6 phr of nanoclay and this improvement is attributed to the barrier effect of the exfoliated and the intercalated nanoclay particles.

2.12.1.5   Ethylene Vinyl Acetate

This is another important and widely used polymer. Nanocomposites have been prepared based on this rubber mostly for flame-retardancy behavior. Blends with acrylic functional polymer and maleic anhydride-grafted ethylene vinyl acetate (EVA) have also been used both with nanoclays and carbon nanotubes to prepare nanocomposites [65–69].

2.12.1.6   ENGAGE

ENGAGE is an ethylene–octene copolymer. Ray and Bhowmick [70] have prepared nanocomposites based on this copolymer. In this study, the nanoclay was modified *in situ* by polymerization of acrylate monomer inside the gallery gap of nanoclay. ENGAGE was then intercalated inside the increased gallery gap of the modified nanoclay. The nanocomposites prepared by this method have improved mechanical properties compared to that of the conventional counterparts. Preparation and properties of organically modified nanoclay and its nanocomposites with ethylene–octene copolymer were reported by Maiti et al. [71]. Excellent improvement in mechanical properties and storage modulus was noticed by the workers. The results were explained with the help of morphology, dispersion of the nanofiller, and its interaction with the rubber.

2.12.1.7   Polyurethane

Nanocomposite based on polyurethane (PU) is prepared using silica, clay, and Polyhedral Oligomeric Silsesquioxane (POSS). Preparation, characterization, mechanical and barrier properties, morphology, and effect of processing conditions have been reported on polyurethane-based nanocomposites [72,73].

2.12.1.8   Acrylonitrile–Butadiene Rubber

Nanocomposites have been prepared with this polymer and mechanical and barrier properties and fracture behavior have been studied [74–76]. The latex of this rubber has also been used for the same [77]. Sadhu and Bhowmick [78–81] have studied the preparation, structure, and various

TABLE 2.2
Mechanical Properties with Varying Degrees of Polarity of Rubber

Sample Designation	Tensile Strength (MPa)	Elongation at Break (%)	Work Done at Break (J/m²)	Modulus at 50% Elongation (MPa)	Young's Modulus × 10³ (kPa)	Volume Fraction (Vᵣ)
19NBR	3.00	362	2531	0.66	14.6	0.332
19NBRN4[a]	2.73	147	915	1.00	20.5	0.359
19NBROC4[a]	3.60	154	1301	1.23	25.0	0.436
34NBR	2.10	399	2138	0.53	11.1	0.215
34NBRN4	2.00	279	1778	0.68	14.6	0.219
34NBROC4	4.85	923	8268	0.52	12.5	0.150
50NBR	2.75	553	3337	0.55	12.4	0.178
50NBRN4	2.10	414	2161	0.57	12.7	0.224
50NBROC4	5.60	679	7567	0.65	15.1	0.264

Source: Sadhu, S. and Bhowmick, A.K., *J. Polym. Sci., Part B: Polym. Phys.*, 42, 1573, 2004. Courtesy of Wiley InterScience.

[a] 19NBROC4 = nitrile rubber with 19% acrylonitrile + 4 phr octadecyl amine modified MMT, 19NBRN4 = nitrile rubber with 19% acrylonitrile + 4 phr MMT.

properties like mechanical, dynamic mechanical, barrier, and thermal degradation of nitrile rubber-based nanocomposites and compared these properties with variation in structural parameters of the base rubbers. Studies have been carried out using different grades of nitrile rubber with varying polarity to understand the effect of comonomer content and polarity on the structure and properties of these nanocomposites. The investigation reveals that the degree of intercalation inside the clay galleries are combined effects of the (1) polarity, (2) bulkiness of the polymer, and (3) H-bond-forming ability. By balancing all these factors the optimum intercalation can be achieved. Nitrile rubbers with lower polarity and acrylonitrile content show minimum increment in mechanical strength while acrylonitrile–butadiene rubber (NBR) with 50% acrylonitrile content shows the maximum. The mechanical properties with varying acrylonitrile contents are tabulated in Table 2.2. The TEM photomicrographs depict mostly exfoliated structure in the case of nitrile rubber with 19% acrylonitrile content while at higher acrylonitrile content the clays remain intercalated. The TEM photos are given in Figure 2.8a and b for these two samples.

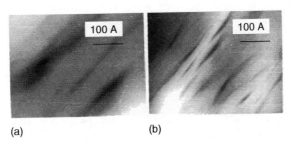

(a)   (b)

FIGURE 2.8 Transmission electron microscopy (TEM) photographs of clay nanocomposites with acrylonitrile–butadiene rubber (NBR) having (a) 50% and (b) 19% acrylonitrile content, respectively (magnification = 13,500). (From Sadhu, S. and Bhowmick, A.K., *J. Polym. Sci., Part B: Polym. Phys.*, 42, 1573, 2004. Courtesy of Wiley InterScience.)

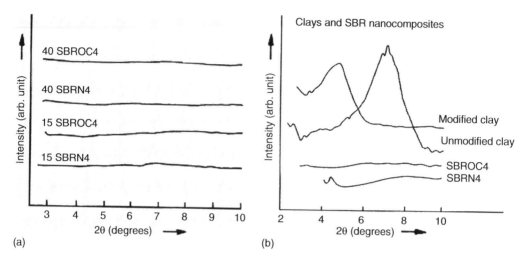

FIGURE 2.9 X-ray diffractogram (XRD) spectra of unmodified and modified nanoclays and styrene–butadiene rubber (SBR)-based nanocomposites with styrene content of (a) 15% and 40% and (b) 23%. (From Sadhu, S. and Bhowmick, A.K., *J. Polym. Sci., Part B: Polym. Phys.*, 42, 1573, 2004. Courtesy of Wiley InterScience.)

2.12.1.9   Styrene–Butadiene Rubber

This is another commonly used synthetic rubber. This rubber shows very low strength. Nanocomposites based on this can offer very large improvement in properties. The morphology and mechanical properties have been studied on styrene–butadiene rubber (SBR) nanocomposites [78,82–87]. The shear-controlled morphology has been studied by Schön et al. [84]. Sadhu and Bhowmick [78,86,88] have studied the effect of styrene content on nanocomposite properties. The effects of bulkiness on the mechanical properties and structure have been studied. Although the extent of intercalation in the unmodified Na-MMT varies with the styrene content of SBR, the modified organoclays in all cases show complete exfoliation. The XRD spectra of the SBR-based nanocomposites with unmodified and modified clay are given in Figure 2.9a and b. The improvement in tensile strength varies depending on the nature of rubber. With higher styrene content, the tensile strength increases in nanocomposites. A plot of variation in magnitude of properties with percent styrene content is given in Figure 2.10.

2.12.1.10   Brominated Poly(Isobutylene-Co-Para-Methylstyrene)

Brominated poly(isobutylene-co-para-methylstyrene) (BIMS)-based nanocomposites have been prepared by Maiti et al. [89] by solution intercalation method using various organoclays and their mechanical, dynamic mechanical and rheological properties have been measured and explained with reference to the XRD and TEM results. The increment in strength in the modified-clay-filled BIMS is tremendous when compared with that of the gum. Maiti et al. reported excellent improvement in oxygen permeability just by incorporation of four parts of organoclay (Figure 2.11). The rubber gives exfoliated nanocomposites with organoclays. They also studied the effect of nanoclays in conventional carbon black and silica-filled systems. The nanoclays exhibited a synergistic effect with conventional fillers [90].

Tsou and Measmer examined the dispersion of organosilicates on two different butyl rubbers, namely BIMS and brominated poly(isobutylene-co-isoprene) (BIIR) with the help of small angle X-ray scattering (SAXS), wide angle X-ray scattering (WAXS), atomic force microscopy (AFM), and TEM [91]. There is also a patent on BIMS nanocomposites for low permeability and their uses in tire inner tubes [92].

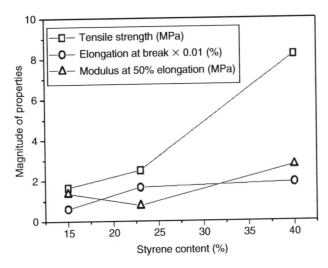

FIGURE 2.10 Variation in mechanical properties with styrene content in styrene–butadiene rubber (SBR)-based nanocomposites.

2.12.1.11 Fluoroelastomers

Fluoroelastomer-based nanocomposites were prepared using various nanoclays and their different properties were studied [93–98].

Maiti and Bhowmick reported exciting results that a polar matrix like fluoroelastomer (Viton B-50) was able to exfoliate unmodified clay (Cloisite NA^+) as well as the modified one (Cloisite 20A) [93]. They studied morphology, mechanical, dynamic mechanical and swelling properties of fluoroelastomer nanocomposites. The unmodified-clay-filled systems showed better properties than the modified ones (Table 2.3).

For example, the increment in maximum stress over the neat polymer is 100% and 53% in the case of the unmodified- and the modified-clay-filled samples, respectively. The extraordinary results obtained with the unmodified clays were explained with the help of thermodynamics and surface energetics. They explained it as follows.

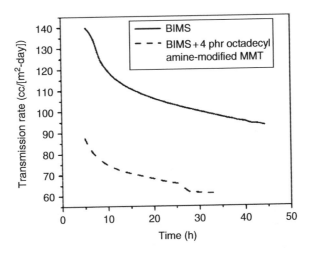

FIGURE 2.11 Plot of oxygen transmission rate versus time for brominated poly(isobutylene-co-isoprene) (BIMS) nanocomposite.

TABLE 2.3

Mechanical Properties of Different Fluoroelastomer Nanocomposites

Sample Name	Modulus at 100% Elongation (MPa)	Elongation at Break (%)	Maximum Stress (MPa)
Fluoroelastomer (F)	0.35	107	0.46
Fluoroelastomer + 4 phr Cloisite NA + (FNA4)	0.88	519	0.90
Fluoroelastomer + 4 phr Cloisite 20A (F20A4)	0.64	444	0.70

Source: Maiti, M. and Bhowmick, A.K., *J. Polym. Sci., Part B: Polym. Phys.*, 44, 162, 2006. Courtesy of Wiley InterScience.

The free energy change of the system after mixing the clay in fluoroelastomers may be given as

$$\Delta G_E = \Delta H_E - T\Delta S_E \text{ for elastomers} \tag{2.1}$$

$$\Delta G_C = \Delta H_C - T\Delta S_C \text{ for clays} \tag{2.2}$$

Therefore, total free energy change of the system is

$$\Delta G_S = \Delta H_S - T\Delta S_S = \Delta H_S - T(\Delta S_E + \Delta S_C) \tag{2.3}$$

From the expression, ΔG_S value will be negative and hence the most favorable interaction between the clay and the rubber will take place when ΔH_S is negative and ΔS_S is positive.

When polymer chains enter into the gallery of the clay, they reside in a restrained form, i.e., ΔS_E is negative. In contrast, the expansion of the gallery by elastomer chains causes the entropy change in the clay, ΔS_C, to be positive. If the clays are exfoliated, this may probably compensate the entropy loss associated with the confinement of elastomer chains. Hence, in this condition, negative ΔH_S value makes ΔG_S negative. They calculated ΔH_S for different clay systems from the infrared (IR) spectra using Fowkes's equation, $\Delta H_S = 0.236 \times \Delta \nu$. ΔH_S is negative (-2.60 kcal/mol) for unmodified-clay-based system. But for the modified-clay-filled samples, it is zero or has small positive value. As a result, the mixing of the unmodified clay with the fluoroelastomer will be more favorable than that of the modified one.

The better interaction observed with the unmodified clay was also explained in terms of surface energy. The values of surface energy of the fluoroelastomer and the clays, along with work of adhesion, spreading coefficient and interfacial tension are reported in Table 2.4.

TABLE 2.4

Different Surface Properties of Fluoroelastomer and Clays

Sample	Work of Adhesion (mJ/m²)	Spreading Coefficient (mJ/m²)	Interfacial Tension (mJ/m²)
Fluoroelastomer and Cloisite NA+	67.63	5.47	1.10
Fluoroelastomer and Cloisite 20A	51.42	−9.91	2.47

Source: Maiti, M. and Bhowmick, A.K., *J. Polym. Sci., Part B: Polym. Phys.*, 44, 162, 2006. Courtesy of Wiley InterScience.

The surface energy of unmodified clay (37.22 mJ/m^2) is higher than that of fluorocarbon rubber (31.51 mJ/m^2). Assuming fluoroelastomer as wetting polymer, its lower surface energy helps in wetting the solid unmodified clay, following the Zisman approach. This is, however, not the case with the fluoroelastomer-modified clay system, where the surface energy of the modified clay (22.38 mJ/m^2) is much lower. These results are more apparent from the $\Delta\gamma$ values. The $\Delta\gamma$ value is much lower (5.71 mJ/m^2) in the case of Viton B-50 and the unmodified clay than that (9.13 mJ/m^2) of Viton B-50 and the modified clay. A lower surface energy mismatch should give better wetting, better interfacial adhesion, and increased diffusion of the polymers across the interface. The interfacial tension between fluoroelastomer and the unmodified clay is also much less. The positive spreading coefficient for this system definitely indicates better diffusion of the polymers into the clay interface. Hence, interaction will be more in the case of the unmodified clay. Besides, the work of adhesion is also higher in the case of the rubber-unmodified clay system. Hence, the polymer chains can spread more easily on the surface of the unmodified clay than that of the modified clay.

Maiti and Bhowmick also investigated the diffusion and sorption of methyl ethyl ketone (MEK) and tetrahydrofuran (THF) through fluoroelastomer–clay nanocomposites in the range of 30°C–60°C by swelling experiments [98]. A representative sorption-plot (i.e., mass uptake versus square root of time, $t^{1/2}$) at 45°C for all the nanocomposite systems is given in Figure 2.12.

The overall sorption value tends to decrease with the addition of the nanoclays. The decrease is maximum for the unmodified-clay-filled sample. As the temperature of swelling increases, the penetrant uptake increases in all the systems (Table 2.5). The rate of increase of solvent uptake is slower for the unmodified-clay-filled sample compared to the modified one. From Table 2.5 it can be seen that the M_∞ values are higher for THF compared to MEK in every composite system. The higher sorption can be explained from the difference in solubility parameter of solvent and rubber $(\partial_s - \partial_r)$ and polarity. The solubility parameter value of MEK, THF, and the rubber is 19.8, 18.6, and 14.8 MPa$^{1/2}$, respectively. This difference is lower (3.8 MPa$^{1/2}$) in the case of THF than that of MEK (5.0 MPa$^{1/2}$).

They also established a model to predict the aspect ratio of nanoclays from the swelling studies. They proposed the model in the following way [98].

Addition of layered nanoclays to a neat polymer restricts the permeability due to the following two phenomena:

1. The available area for diffusion will decrease as a result of impermeable nanoclays replacing the permeable polymer.

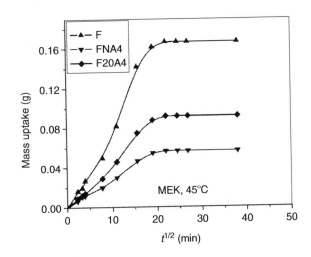

FIGURE 2.12 Sorption curves for different nanocomposites at 45°C (solvent methyl ethyl ketone [MEK]). (From Maiti, M. and Bhowmick, A.K., *J. Appl. Polym. Sci.*, 105, 435, 2007. Courtesy of Wiley InterScience.)

TABLE 2.5

Equilibrium Sorption Values of Different Nanocomposite-Solvent Systems

Sample	M_∞ (g) MEK			M_∞ (g) THF		
	30°C	45°C	60°C	30°C	45°C	60°C
F	0.144	0.167	0.181	0.168	0.188	0.206
FNA4	0.051	0.057	0.065	0.080	0.086	0.090
F20A4	0.082	0.092	0.117	0.124	0.137	0.148

Source: Maiti, M. and Bhowmick, A.K., *J. Appl. Polym. Sci.*, 105, 435, 2007. Courtesy of Wiley
InterScience.

2. When nanoclays are added to the system, we may assume that the clay layers are randomly
    placed in the matrix. The diffusion of the solvent will detour around the impermeable clay
    layers. Diffusion will be diverted to pass a clay platelet in every layer and, hence, the
    solvent must have to travel a longer path (d_f) in the filled system compared to that (d_0) for
    the neat polymer.

Using scaling concept, permeability P can be written as

$$P \sim A/d \tag{2.4}$$

where
  A is the cross-sectional area available for diffusion
  d is the path length the solvent must travel to cross the sample

As a result, the permeability of nanocomposites (P_f) is reduced from that of the neat polymer
(P_0) by the product of the decreased area and the increased path length as follows:

$$\frac{P_0}{P_f} = \left(\frac{A_0}{A_f}\right)\left(\frac{d_f}{d_0}\right) \tag{2.5}$$

where A_0 is the cross-sectional area available for diffusion in a neat polymer sample, A_f is the cross-
sectional area available for diffusion in a nanocomposite, d_0 is the sample thickness (i.e., the
distance a solvent molecule must travel to cross the neat polymer sample), and d_f is the distance a
solvent molecule must travel to cross the nanocomposite sample.

  Now,

$$\frac{P_0}{P_f} = \frac{V_0/d_0}{(V_0 - V_f)/d_f}\left(\frac{d_f}{d_0}\right) \tag{2.6}$$

where
  V_0 is the total volume of the neat polymer sample
  V_f is the volume of nanoclays in the nanocomposite sample

$$\frac{P_0}{P_f} = \frac{V_0}{V_0 - V_f}\left(\frac{d_f}{d_0}\right)^2 \tag{2.7}$$

$$= \frac{1}{1-\phi}\left(\frac{d_f}{d_0}\right)^2 \tag{2.8}$$

where ϕ is the volume fraction of filler.

When a solvent diffuses across a neat polymer, it must travel the thickness of the sample (d_0). When the same solvent diffuses through a nanocomposite film with nanoclays, its path length is increased by the distance it must travel around each clay layer it strikes. According to Lan et al. [99] the path length of a gas molecule diffusing through an exfoliated nanocomposite is

$$d_f = d_0 + d_0 L\phi/2d_c \tag{2.9}$$

where L and d_c are the length and thickness of a clay layer, respectively.

Substituting this value in Equation 2.8,

$$\frac{P_0}{P_f} = \frac{1}{1-\phi}\left(1 + \frac{L\phi}{2d_c}\right)^2 \tag{2.10}$$

$$= \frac{1}{1-\phi}\left(1 + \frac{\alpha\phi}{2}\right)^2 \tag{2.11}$$

where aspect ratio $\dfrac{L}{d_c} = \alpha$.

The aspect ratio of the nanoclays in different samples has been calculated using Equation 2.11 and is reported in Table 2.6.

It may be mentioned that under a variety of experimental conditions (i.e., solvent, temperature, swelling, or reswelling) the aspect ratio values are very close for a particular system. In the table, the average of aspect ratios calculated from the permeability data in two different solvents at three different temperatures for swelling and reswelling experiments has been reported. It is observed that the aspect ratio is higher (146 ± 14) in the case of FNA4. Similarly, the aspect ratio is 63 ± 5 for F20A4. These values are in good accord with those measured from the transmission electron micrographs (shown in Figure 2.13a and b).

2.12.1.12 Silicone Rubber

Silicone rubber nanocomposites have also been reported [100–107]. Silicone rubber–clay nanocomposites were synthesized by a melt-intercalation process using synthetic Fe-MMT and organically modified natural Na-MMT by Kong et al. [103]. This study was designed to determine if the presence of structural iron in the matrix could result in radical trapping and then enhance thermal stability, and affect the cross-linking degree and elongation. It was found that the iron acted as an antioxidant and radicals trap not only in thermal degradation, but also in the vulcanization process.

TABLE 2.6

Average Aspect Ratio of Clay Layers Present in Different Nanocomposites

| | Aspect Ratio | |
Sample	Swelling	Morphology
FNA4	146 ± 14	145 ± 6
F20A4	63 ± 5	53 ± 6

Source: Maiti, M. and Bhowmick, A.K., *J. Appl. Polym. Sci.*, 105, 435, 2007. Courtesy of Wiley InterScience.

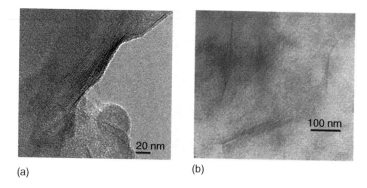

FIGURE 2.13 Transmission electron microscopy (TEM) photographs of (a) FNA4 and (b) F20A4. (From Maiti, M. and Bhowmick, A.K., *J. Appl. Polym. Sci.*, 105, 435, 2007. Courtesy of Wiley InterScience.)

Novel room-temperature-vulcanized silicone rubber–organo-MMT nanocomposites were prepared by a solution intercalation process by Wang et al. [104]. A new strategy was developed by Ma et al. [105] to prepare disorderly exfoliated nanocomposites, in which a soft siloxane surfactant with a weight-average molecular weight of 1900 was adopted to modify the clay.

2.12.2 THERMOPLASTIC ELASTOMERS

2.12.2.1 Poly[Styrene–(Ethylene-Co-Butylene)–Styrene]

Thermoplastic elastomer (TPE)–clay nanocomposites based on poly[styrene–(ethylene-co-butylene)–styrene] triblock copolymer (SEBS) were prepared by various workers using natural Na-MMT and different organically modified nanoclays [108–110]. Mechanical, dynamic mechanical properties, and morphology of these nanocomposites were also studied. Modified clay showed better properties than the unmodified one. Ganguly et al. also compared two processing techniques, namely, solution and melt blending. With AFM, they showed distinctly different morphologies in nanocomposites prepared through solution and melt processing. Extensive morphological investigations were also done by Ganguly et al. [111] using AFM. The lamellar thickness of the soft phases of SEBS was

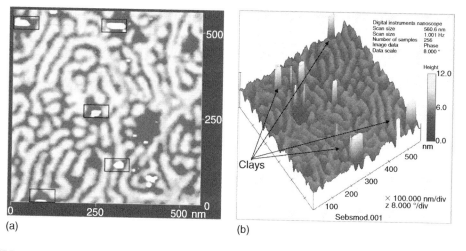

FIGURE 2.14 Tapping mode phase morphology of the nanocomposites (a) poly[styrene-(ethylene-co-butylene)-styrene] (SEBS)–Cloisite 20A and (b) its 3D image. (From Ganguly, A., Sarkar, M.D., and Bhowmick, A.K., *J. Polym. Sci., Part B: Polym. Phys.*, 45, 52, 2006. Courtesy of Wiley InterScience.)

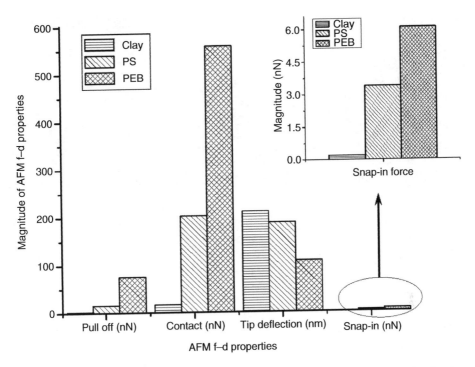

FIGURE 2.15 Forces of interactions on blocks and nanoclay of poly[styrene-(ethylene-co-butylene)-styrene] (SEBS)–clay nanocomposite taken from force–volume experiments. (From Ganguly, A., Sarkar, M.D., and Bhowmick, A.K., *J. Polym. Sci., Part B: Polym. Phys.*, 45, 52, 2006. Courtesy of Wiley InterScience.)

widened in nanocomposites, where the layered clay silicates were embedded in the soft rubbery phases in the block copolymeric matrix of the nanocomposite (shown in Figure 2.14a and b).

Qualitative and quantitative investigation of surface forces of interaction for the neat SEBS and its nanocomposite measured at constituting blocks and clay regions by force distance plots was also done on single points and on entire force volume by the same workers. Maximum adhesive force of 25 nN was found in rubbery poly(ethylene-co-butylene) (PEB) segments and cantilever deflection was found to be maximum (210 nm) for clay regions both in single-point force mapping and entire-volume force mapping of SEBS–clay nanocomposite under investigation (Figure 2.15) [111].

The same research group functionalized SEBS at the mid-block by means of chemical grafting by two polar moieties—acrylic acid and maleic anhydride, and subsequently synthesized nanocomposites based on hydrophilic MMT clay at very low loadings [112]. The mid-block was grafted with 3 and 6 wt% acrylic acid through solution grafting and 2 and 4 wt% maleic anhydride though melt grafting reactions which were confirmed by spectroscopic techniques. The nanocomposites derived from the grafted SEBS and MMT clay conferred dramatically better mechanical, dynamic mechanical, and thermal properties as compared to those of SEBS and its clay-based nanocomposites. The morphology also changed with modification. SEBS with 4 phr MMT nanocomposites showed agglomerated microstructure while acrylated and maleated SEBS with same amount of MMT exhibited exfoliation and exfoliation–intercalation, respectively (Figure 2.16a through c).

2.12.3 Thermoplastics

The first nanocomposite prepared by Toyota Group of Japan was based on nylon 6. *In situ* polymerization of caprolactum inside the gallery of 5% MMT resulted in the first nylon 6–clay nanocomposite. Besides nylon, polypropylene (PP) is probably the most thoroughly investigated system. Excepting the study of the various properties, theoretical aspects and simulations have also

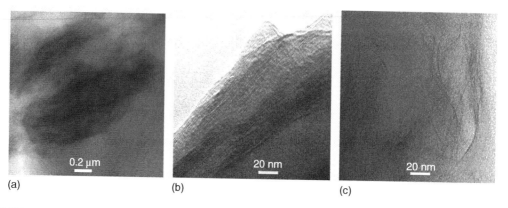

FIGURE 2.16 Transmission electron microscopy (TEM) micrographs of (a) agglomerated poly[styrene-(ethylene-co-butylene)-styrene] (SEBS)–montmorillonite (MMT4), (b) exfoliated–acrylated SEBS–MMT4, and (c) intercalated–exfoliated mixed maleated SEBS–MMT4. (From Ganguly, A. and Bhowmick, A.K., 170th Meeting Rubber Division, ACS at Cincinnati, OH, October 10–12, 2006. Courtesy of ACS, Rubber Division.)

been done with various plastic nanocomposites. Among the different base polymers nylon, PP, polystyrene (PS), polyacrylonitrile (PAN), polyimide (PI), polyesters (PET/PBT), etc. are common. Some of the important works done on these systems are cited below for further development of similar elastomeric systems.

2.12.3.1 Polyamide

This is a highly polar polymer and crystalline due to the presence of amide linkages. To achieve effective intercalation and exfoliation, the nanoclay has to be modified with some functional polar group. Most commonly, amino acid treatment is done for the nanoclays. Nanocomposites have been prepared using *in situ* polymerization [85] and melt-intercalation methods [113–117]. Crystallization behavior [118–122], mechanical [123,124], thermal, and barrier properties, and kinetic study [125,126] have been carried out. Nylon-based nanocomposites are now being produced commercially.

2.12.3.2 Polypropylene

PP is probably the most thoroughly investigated system in the nanocomposite field next to nylon [127–132]. In most of the cases isotactic/syndiotactic-PP-based nanocomposites have been prepared with various clays using maleic anhydride as the compatibilizer. Sometimes maleic anhydride-grafted PP has also been used [127]. Nanocomposites have shown dramatic improvement over the pristine polymer in mechanical, rheological, thermal, and barrier properties [132–138]. Crystallization [139,140], thermodynamic behavior, and kinetic study [141] have also been done.

2.12.3.3 Polystyrene

PS is another general-purpose plastic which is widely used in daily life. This is a nonpolar plastic having bulky styrene groups. Generally alkyl amine-modified nanoclays are used to prepare the nanocomposites. As usual, these nanocomposites can be prepared by various methods, namely, radiation polymerization, suspension polymerization, *in situ* method, and melt process [142–145]. Mechanical, rheological, morphological, thermal, and magnetic properties [146–152] have also been measured by researchers. The effects of premelting temperature, thickness, etc. [153] on morphology have also been studied.

2.12.3.4 Polyimide

PI nanocomposites have been prepared by various methods with different fillers. The nanocomposites might have many applications starting from barrier and thermal resistance to a compound with low coefficient of thermal expansion (CTE) [154–167]. These hybrid materials show very high thermal and flame retardation as well as barrier resistance and adhesion. Tyan et al. [158] have shown that depending on the structure of the polyimide the properties vary. Chang et al. [159] have also investigated the dependency of the properties on the clay modifiers.

2.12.3.5 Ethylene Vinyl Acetate (EVA)

Owing to the good barrier property, the EVA-based clay nanocomposites are used for packaging, bottle-making, etc. [168].

2.12.3.6 Miscellaneous Plastics

Apart from the above discussion, some other work has been done on different plastics. Giannelis et al. [169,170] have done extensive study on clay-based nanocomposites. Vaia et al. [171–173] have prepared several nanocomposites based on various thermoplastic and elastomers like polyethylene oxide (PEO), high-density polyethylene, polyvinyl chloride, and PP. The work includes extensive study of the morphology of nanocomposite by x-ray diffraction technique and study of their kinetics, thermal degradation, and influence of structural parameters on nanocomposite properties. Kornmann et al. [174], Fröhlich et al. [175], Cser and Bhattacharya [176], Shah and Paul [177], Maity and Biswas [178], and Tanaka and Goettler [179] also have made significant contribution to the understanding of the behavior of nanocomposites. The work of Nah et al. on preparation and properties of polyimide [180]-based nanocomposites has resulted in good understanding of the behavior of this class of materials. Besides, Kim et al. [181], Maiti et al. [182], and Wang et al. [183] have done some pioneering work in this field. They have synthesized various nanoclays and nanocomposites. Various structural and application-based studies have been done by them too.

2.12.4 Latex Nanocomposites

Recently a lot of attention is being given to the field of latex-based nanocomposites. Various organoclays as well as pristine clays have been intercalated in aqueous medium with NR latex, SBR latex, NBR latex, as well as carboxylated nitrile rubber (XNBR) latex [184–187], to achieve a good degree of dispersion.

2.12.5 Polymer Blends and Other TPEs

Polymeric blends are popularly known to show desired properties based on the components. But their nanocomposites have been prepared to further modify the properties. Various blends like NR–ENR and PP–EPDM [188,189] with nanofillers have also been studied.

TPE–clay nanocomposites based on rubber–plastic blends such as Engage–PP, nylon 6–BIMS, EPDM–PP, EVA–styrene acrylonitrile (EVA–SAN), polyamide 6–maleated SEBS, PP–SEBS, polyamide 6–silicone rubber were reported by different researchers [190–195]. Based on the results of X-ray diffraction measurements, dynamic mechanical testing, and thermal characterization, Naderi et al. [191] found that the microstructure of the prepared nanocomposites was sensitive to the viscosity ratio of PP/EPDM and the nanoclay content. Dong et al. [195] prepared a novel flame-retardant nanocomposite of nylon 6–unmodified clay–silicone rubber with high toughness, heat resistance, stiffness, and good flowability and proposed a mechanism of synergistic flame retardancy.

There are few reports on block-copolymeric TPE (namely, polyurethane, EVA, SBS, poly (styrene-*b*-butyl acrylate) (PSBA))–clay nanocomposites also [196–199]. Choi et al. [196] studied the effect of the silicate layers in the nanocomposites on the order–disorder transition temperature of the thermoplastic polyurethane (TPU) from the intensity change of the hydrogen-bonded and free carbonyl stretching peaks and from the peak position change of the N–H bending peak. Hasegawa and Usuki [198] observed under TEM that the triblock copolymer microdomain structures were arranged along the dispersed silicate layers. They suggested that the formation of the arranged microdomain structures was induced and controlled by the interaction of the silicate layers, which acted as templates. It was thought that these controlled nanostructures were formed through the selective absorption of the polystyrene segments on dispersed silicate surfaces followed by the segregation of each segment.

2.13 MECHANISMS FOR EXFOLIATION

There are different mechanisms for explaining exfoliation of organically modified clays in the polymer matrices. One such mechanism is the Lattice model proposed by Vaia and Giannelis [200]. The mechanism is explained as follows.

The polymer chains diffuse from the bulk polymer melt into the galleries between the silicate layers. Depending on the degree of penetration of the polymer into the silicate structure, hybrids are obtained with structures ranging from intercalated to exfoliated (Figure 2.17). Polymer penetration resulting in finite expansion of the silicate layers produces intercalated hybrids consisting of well-ordered multilayers composed of alternating polymer and silicate layers. Extensive polymer penetration resulting in disorder and eventual delamination of the silicate layers produces exfoliated hybrids consisting of individual nanometer-thick silicate layers suspended in a polymer matrix. In general, an interplay of entropic and energetic factors determines the outcome of polymer intercalation. The model indicates that the entropic penalty of polymer confinement may be compensated for gallery heights up to the length of the fully extended tethered chains by the increased conformational freedom of the tethered chains as the layers separate. When the total entropy change is small, small changes in the system's internal energy will determine if intercalation is thermodynamically possible. Complete layer separation, though, depends on the establishment of very favorable polymer–organo-layered silicate interactions to overcome the penalty of polymer confinement. For alkylammonium-modified layered silicates, a favorable energy change is accentuated by maximizing the magnitude and number of favorable polymer–surface interactions while minimizing the magnitude and number of unfavorable polar interactions between the polymer and the tethered surfactant chains.

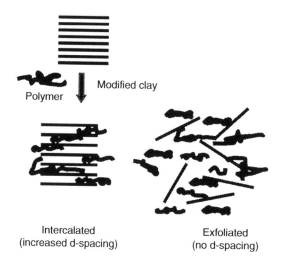

Polymer

Modified clay

Intercalated
(increased d-spacing)

Exfoliated
(no d-spacing)

FIGURE 2.17 Mechanism of exfoliation.

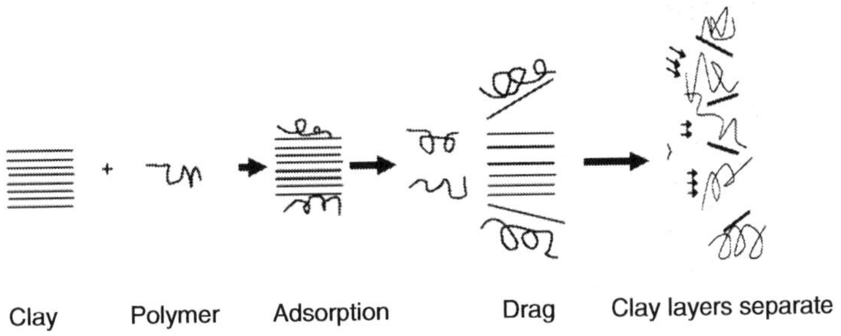

FIGURE 2.18 Adsorption, followed by drag and separation mechanism for exfoliation of silicate layers.

The authors have proposed another mechanism. They have shown that it is not necessary to modify the clays organically to get exfoliated morphology in the resultant nanocomposites. Their model can explain exfoliation of both the unmodified and modified clays. They have proposed that if the work of adhesion between clay surface and polymer is strong enough, the polymer will be adsorbed on the surface. When the work of adhesion between the polymer and the clay is sufficiently higher than the cleavage energy of clay layers, the outer clay layers separate. This is called "drag." Once the surface layers are separated, the polymer chains can adsorb on the next layers and operate by the same mechanism, thus exfoliating the clay layers (shown in Figure 2.18). Simultaneously, the polymer chains will intercalate inside the clay galleries and help in exfoliation as proposed by many workers.

2.14 AREAS OF APPLICATION

Improvements in mechanical properties have resulted in major interest in nanocomposite materials in numerous automotive and general/industrial applications. These include potential for utilization as mirror housings on various vehicle types, door handles, engine covers, and belt covers. More general applications currently being considered include usage as impellers and blades for vacuum cleaners, power tool housings, mower hoods, and covers for portable electronic equipment such as mobile phones and pagers.

2.14.1 Gas Barriers

The improvement in gas barrier property that can result from incorporation of relatively small quantities of nanoclay materials is shown to be substantial. The data provided by various sources indicate oxygen transmission rates for polyamide–organoclay composites which are usually less than half of that of the unmodified polymer. Further data reveal the extent to which both the amount of clay incorporated in the polymer and the aspect ratio of the filler contribute to overall barrier performance. In particular, aspect ratio is shown to have a major effect, with high ratio quite dramatically enhancing gas barrier properties. Such excellent barrier characteristics have resulted in considerable interest in nanoclay composites in food-packaging applications, both flexible and rigid. Specific examples include packaging for processed meats, cheese, confectionery, cereals, and boil-in-the-bag foods, also extrusion-coating applications in association with paperboard for fruit juice and dairy products, together with co-extrusion processes for the manufacture of beer and carbonated drink bottles. The use of nanocomposite formulations would be expected to enhance considerably the shelf life of many types of foods.

Honeywell has also been active in developing a combined active–passive oxygen barrier system for polyamide-6 materials [201]. Passive barrier characteristics are provided by nanoclay particles incorporated via melt processing techniques, while active contribution comes from an oxygen-scavenging ingredient (undisclosed). Oxygen transmission results reveal substantial benefits provided by nanoclay incorporation in comparison to the base polymer (rates approximately 15%–20% of the bulk polymer value, with further benefits provided by the combined active–passive system).

2.14.2 FOOD PACKAGING

Triton Systems and the U.S. Army [202] have conducted further work on barrier performance in a joint investigation. The requirement here is for a nonrefrigerated packaging system capable of maintaining food freshness for 3 years. Nanoclay polymer composites are currently showing considerable promise for this application.

It is likely that excellent gas barrier properties exhibited by nanocomposite polymer systems will result in their substantial use as packaging materials in future years.

A somewhat more esoteric possibility arises from enhanced barrier performance of blown films recently suggested for artificial intestines.

2.14.3 FUEL TANKS

The ability of nanoclay incorporation to reduce solvent transmission through polymers such as polyamides has been demonstrated. Data provided by de Bièvre and Nakamura [203] of UBE Industries reveal significant reductions in fuel transmission through polyamide–6/66 polymers by incorporation of a nanoclay filler. As a result, considerable interest is now being shown in these materials as both fuel tank and fuel line components for cars. Of further interest, the reduced fuel transmission characteristics are accompanied by significant material cost reductions.

2.14.4 FILMS

The presence of filler incorporation at nanolevels has also been shown to have significant effects on the transparency and haze characteristics of films. In comparison to conventionally filled polymers, nanoclay incorporation has been shown to significantly enhance transparency and reduce haze. With polyamide-based composites, this effect has been shown to be due to modifications in the crystallization behavior brought about by the nanoclay particles—spherulitic domain dimensions being considerably smaller. Similarly, nano-modified polymers have been shown, when employed to coat polymeric transparency materials, to enhance both toughness and hardness of these materials without interfering with light transmission characteristics. An ability to resist high velocity impact combined with substantially improved abrasion resistance was demonstrated by Haghighat of Triton Systems [204].

2.14.5 ENVIRONMENTAL PROTECTION

Water-laden atmospheres have long been regarded as one of the most damaging environments which polymeric materials can encounter. Thus, the ability to minimize the extent to which water is absorbed can be a major advantage. Data provided by Beall from Missouri Baptist College [205] indicate the extent to which nanoclay incorporation can reduce the extent of water absorption in a polymer. Similar effects have been observed by van Es [201] of DSM with polyamide-based nanocomposites. In addition, van Es noted a significant effect of nanoclay aspect ratio on water diffusion characteristics in a polyimide nanocomposite. Specifically, increasing aspect ratio was found to diminish substantially the amount of water absorbed, thus indicating the beneficial effects likely from nanoparticle incorporation in comparison to conventional microparticle loading. Hydrophobic enhancement would clearly promote both improved nanocomposite properties and diminish the extent to which water would be transmitted through to an underlying substrate. Thus, there are applications in which contact with water or moist environments could clearly benefit.

2.14.6 FLAMMABILITY REDUCTION

The ability of nanoclay incorporation to reduce the flammability of polymeric materials was a major theme of the paper presented by Gilman et al. [206] of the National Institute of Standards and Technology in the United States. In his work, Gilman demonstrated the extent to which flammability behavior could be restricted in polymers such as PP with as little as 2% nanoclay loading. In particular, heat release rates, as obtained from cone calorimetry experiments, were found to diminish substantially by nanoclay incorporation. Although conventional microparticle filler incorporation, together with the use of flame-retardant and intumescent agents would also minimize flammability behavior, this is usually accompanied by reduction in various other important properties. With the nanoclay approach, this is usually achieved while maintaining or enhancing other properties and characteristics.

There are many other applications coming up, which are not covered here for space limitation.

2.15 CONCLUSIONS

The above discussion on the work done so far in the field of nanocomposites reveals that addition of nanofillers increases the tensile, thermal, and barrier properties, and flame retardancy in general. The literature search has also shown that the thermodynamic parameters are dependent on the nature and amount of nanofillers and their interaction with the matrix. Till now, the field of thermoplastics has been explored to its maximum and some of the products are now being commercially produced and used. Nanocomposites by virtue of their properties can become very useful in aeronautics, packaging, and many other industries. The elastomer field is now being explored, although some important work has already been reported. Research on various preparation methods and understanding of various parameters affecting the final product morphology and correlation between the structure and property of a particular composite help us achieve a final product with desired properties.

REFERENCES

1. J.T. Bayers, *Rubber Technology*, Chapter 3, Part I, Third Edition, Edited by Maurice Morton, Van Nostrand Reinhold, New York, 1985.
2. M.P. Wagner, *Rubber Technology*, Chapter 3, Part II, Third Edition, Edited by Maurice Morton, Van Nostrand Reinhold, New York, 1985.
3. http://mineral.galleries.com/minerals/silicate/clays.htm, access date 9.8.2004.
4. E. Bunesberg and C.V. Clemency, *Clays Clay Miner.*, **21**, 213, 1973.
5. Y. Kojima, A. Usuki, M. Kawasumi, A. Okada, T. Kurauchi, and O. Kamigaito, *J Polym. Sci., Part A: Polym. Chem.*, **31**, 1755, 1993.
6. R. Grim, *Clay Mineralogy*, McGraw-Hill, New York, 1968.
7. http://jan.ucc.nau.edu/doetqp/courses/env440/env440_2/lectures/lec19/Fig19_3.gif, access date 4.11.06.
8. http://www.pslc.ws/macrog/mpm/composit/nano/struct3_1.htm, access date 4.11.06.
9. http://grunwald.ifas.ufl.edu/Nat_resources/mineral_components/mica.gif, access date 4.11.06.
10. U. Hoffman, K. Endell, and D. Wilm, *Z. Kristallogr.*, **86**, 340, 1933.
11. E. Bunesberg and C.V. Clemency, *Clays Clay Miner.*, **21**, 213, 1973.
12. M.M. Mortland and J.L. Mellor, *Soil Sci. Soc. Proc.*, **18**, 363, 1954.
13. J.L. McAtee, *Am. Miner.*, **44**, 1230, 1959.
14. A. Usuki, M. Kawasumi, Y. Kojima, A. Okada, T. Kurauchi, and O. Kamigaito, *J. Mater. Res.*, **8**, 1174, 1993.
15. A. Usuki, M. Kawasumi, Y. Kojima, A. Okada, T. Kurauchi, and O. Kamigaito, *J. Mater. Res.*, **8**, 1179, 1993.
16. T. Lan, P.D. Kaviratna, and T.J. Pinnavaia, *Chem. Mater.*, **7**, 2144, 1995.
17. Z. Wang and T.J. Pinnavaia, *Chem. Mater.*, **10**, 1820, 1998.

18. mtc.tpd.tue.nl/nanocomposites/nano_home.html, access date 11.05.2004.
19. R.A. Vaia, H. Ishii, and E.P. Giannelis, *Chem. Mater.*, **5**, 1694, 1993.
20. R.A. Vaia, K.D. Jandt, E.J. Kramer, and E.P. Giannelis, *Macromolecules*, **28**, 8080, 1995.
21. L. Liu, Z. Qi, and X. Zhu, *J. Appl. Polym. Sci.*, **71**, 1133, 1999.
22. R.A. Vaia, K.D. Jandt, E.J. Kramer, and E.P. Giannelis, *Chem. Mater.*, **8**, 2628, 1996.
23. M. Kawasumi, N. Hasegawa, M. Kato, A. Usuki, and A. Okada, *Macromolecules*, **30**, 6333, 1997.
24. H.G. Jeon, H.T. Jung, and S.D. Hudson, *Polym. Bull.*, **41**, 107, 1998.
25. K. Yano, A. Usuki, A. Okada, T. Karauchi, and O. Kamigaito, *J. Polym. Sci., Part A*, **31**, 2493, 1993.
26. M. Kawasumi, N. Hasegawa, A. Usuki, and O. Aken, *Mater. Eng. Sci. C*, **6**, 135, 1998.
27. P.B. Messersmith and E.P. Giannelis, *Chem. Mater.*, **6**, 1719, 1994.
28. Y. Ou, F. Yang, and Z.-Z. Yu, *J. Polym. Sci., Part B: Polym. Phys.*, **36**, 789, 1998.
29. M. López-Manchado, B. Herrero, and M. Arroyo, *Polym. Int.*, **52**, 1070, 2003.
30. S. Varghese and J. Karger-Kocsis, *J. Appl. Polym. Sci.*, **91**, 813, 2004.
31. M.A. López-Manchado, M. Arroyo, B. Herrero, and J. Biagiotti, *J. App. Polym. Sci.*, **89**, 1, 2003.
32. S. Varghese, K.G. Gatos, A.A. Apostolov, and J. Karger-Kocsis, *J. Appl. Polym. Sci.*, **92**, 543, 2004.
33. M. Bhattacharya, and A.K. Bhowmick, unpublished data.
34. K.H. Chen and S.M. Yang, *J. Appl. Polym. Sci.*, **86**, 414, 2002.
35. C. Chen and D. Curliss, *J. Appl. Polym. Sci.*, **90**, 2276, 2003.
36. K. Zhang, L. Wang, F. Wang, G. Wang, and Z. Li, *J. Appl. Polym. Sci.*, **91**, 2649, 2004.
37. D. Ratna, N.R. Manoj, R. Varley, R.K.S. Raman, and G.P. Simon, *Polym. Int.*, **52**, 1403, 2003.
38. O. Becker, G.P. Simon, R.J. Varley, and P.J. Halley, *Polym. Eng. Sci.*, **43**, 850, 2003.
39. S. Balakrishnan and D. Raghavan, *Macromol. Rapid Commun.*, **25**, 481, 2004.
40. A. Lee and J.D. Lichtenhan, *J. Appl. Polym. Sci.*, **73**, 1993, 1999.
41. C. Wan, X. Qiao, Y. Zhang, and Y. Zhang, *J. Appl. Polym. Sci.*, **89**, 2184, 2003.
42. Y. Deng, A. Gu, and Z. Fang, *Polym. Int.*, **53**, 85, 2004.
43. T.B. Tolle and D.P. Anderson, *J. Appl. Polym. Sci.*, **91**, 89, 2004.
44. B. Guo, X. Ouyang, C. Cai, and D. Jia, *J. Polym. Sci., Part B: Polym. Phys.*, **42**, 1192, 2004.
45. W.B. Xu, S.-P. Bao, S.-J. Shen, G.-P. Hang, and P.-S. He, *J. Appl. Polym. Sci.*, **88**, 2932, 2003.
46. H. Lu and S. Nutt, *Macromol. Chem. Phys.*, **204**, 1832, 2003.
47. L. Torre, E. Frulloni, J.M. Kenny, C. Manferti, and G. Camino, *J. Appl. Polym. Sci.*, **90**, 2532, 2003.
48. S.C. Jana and J.H. Park, *Macromolecules*, **36**, 2758, 2003.
49. W. Feng, A. Ait-kadi, and B. Riedl, *Polym. Eng. Sci.*, **42**, 1827, 2002.
50. P.L. Teh, Z.A. Mohd Ishak, A.S. Hashim, J. Karger-Kocsis, and U.S. Ishiaku, *Eur. Polym. J.*, **40**, 2513, 2004.
51. P.L. Teh, Z.A. Mohd Ishak, A.S. Hashim, J. Karger-Kocsis, and U.S. Ishiaku, *J. Appl. Polym. Sci.*, **100**, 1083, 2006.
52. P.L. Teh, Z.A. Mohd Ishak, A.S. Hashim, J. Karger-Kocsis, and U.S. Ishiaku, *J. Appl. Polym. Sci.*, **94**, 2438, 2004.
53. F. Dietsche, Y. Thomann, R. Thomann, and R. Mülhaupt, *J. Appl. Polym. Sci.*, **75**, 396, 2000.
54. L. Bokobza, A. Burr, G. Garnaud, M.Y. Perrin, and S. Pagnotta, *Polym. Int.*, **53**, 1060, 2004.
55. H.R. Jung, M.S. Cho, J.H. Ahn, J.D. Nam, and Y. Lee, *J. Appl. Polym. Sci.*, **91**, 894, 2004.
56. P. Wojciechowski, A. Joachimiak, and T. Halamus, *Polym. Adv. Technol.*, **14**, 826, 2003.
57. E. Müh, H. Frey, J.E. Klee, and R. Mülhaupt, *Adv. Funct. Mater.*, **11**, 425, 2001.
58. C.-S. Wu and H.-T. Liao, *J. Polym. Sci., Part B: Polym. Phys.*, **41**, 351, 2003.
59. W.-F. Lee and Y.-C. Chen, *J. Appl. Polym. Sci.*, **91**, 2934, 2004.
60. J. Lin, J. Wu, Z. Yang, and M. Pu, *Macromol. Rapid Commun.*, **22**, 422, 2001.
61. H. Acharya, M. Pramanik, S.K. Srivastava, and A.K. Bhowmick, *J. Appl. Polym. Sci.*, **93**, 2429, 2004.
62. L. Valentini, J. Biagiotti, J.M. Kenny, and M.A. López Manchado, *J. Appl. Polym. Sci.*, **89**, 2657, 2003.
63. W. Li, Y.D. Huang, and S.J. Ahmadi, *J. Appl. Polym. Sci.*, **94**, 440, 2004.
64. K.G. Gatos, R. Thomann, and J. Karger-Kocsis, *Polym. Int.*, **53**, 1191, 2004.
65. M. Okoshi and H. Nishizawa, *Polym. Eng. Sci.*, **42**, 2156, 2004.
66. M. Pramanik, H. Acharya, and S.K. Srivastava, *Macromol. Mater. Eng.*, **289**, 562, 2004.
67. W. Gianelli, G. Camino, N.T. Dintcheva, S.L. Verso, and F.P. La Mantia, *Macromol. Mater. Eng.*, **289**, 238, 2004.

68. Y. Tang, Y. Hu, J. Wang, R. Zong, Z. Gui, Z. Chen, Y. Zhuang, and W. Fan, *J. Appl. Polym. Sci.*, **91**, 2416, 2004.
69. B. Han, A. Cheng, G. Ji, S. Wu, and J. Shen, *J. Appl. Polym. Sci.*, **91**, 2536, 2004.
70. S. Ray and A.K. Bhowmick, *Rubber Chem. Technol.*, **74**, 0835, 2001.
71. M. Maiti, S. Sadhu, and A.K. Bhowmick, *J. Appl. Polym. Sci.*, **101**, 603, 2006.
72. X. Ma, H. Lu, G. Liang, and H. Yan, *J. Appl. Polym. Sci.*, **93**, 608, 2004.
73. V.Y. Kramarenko, T.A. Shantalil, I.L. Karpova, K.S. Dragan, E.G. Privalko, V.P. Privalko, D. Fragiadakis, and P. Pissis, *Polym. Adv. Technol.*, **15**, 144, 2004.
74. C. Nah, H.J. Ryu, W.D. Kim, and Y.-W. Chang, *Polym. Int.*, **52**, 1359, 2003.
75. C. Nah, H.J. Ryu, W.D. Kim, and S.-S. Choi, *Polym. Adv. Technol.*, **13**, 649, 2002.
76. C. Nah, H.J. Ryu, S.H. Han, J.M. Rhee, and M.-H. Lee, *Polym. Int.*, **50**, 1265, 2001.
77. L. Zhang, Y. Wang, Y. Wang, Y. Sui, and D. Yu, *J. Appl. Polym. Sci.*, **78**, 1873, 2000.
78. S. Sadhu and A.K. Bhowmick, *J. Polym. Sci., Part B: Polym. Phys.*, **42**, 1573, 2004.
79. S. Sadhu and A.K. Bhowmick, *J. Polym. Sci., Part B: Polym. Phys.*, **43**, 1854, 2005.
80. S. Sadhu and A.K. Bhowmick, *Rubber Chem. Technol.*, **78**, 321, 2005.
81. S. Sadhu and A.K. Bhowmick, *J. Mater. Sci.*, **40**, 1633, 2005.
82. M.A. de Luca, T.E. Machado, R.B. Notti, and M.M. Jacobi, *J. Appl. Polym. Sci.*, **92**, 798, 2004.
83. V.P. Privalko, S.M. Ponomarenko, E.G. Privalko, F. Schön, W. Gronski, R. Staneva, and B. Stühn, *Macromol. Chem. Phys.*, **204**, 2148, 2003.
84. F. Schön, R. Thomann, and W. Gronski, *Macromol. Symp.*, **189**, 105, 2002.
85. A. Mousa and J. Karger-Kocsis, *Macromol. Mater. Eng.*, **286**, 260, 2001.
86. S. Sadhu and A.K. Bhowmick, *J. Appl. Polym. Sci.*, **92**, 698, 2004.
87. G. Heideman, J.W.M. Noordermeer, R.N. Datta, and B. van Baarle, *Kautsch. Gumni. Kunstst.*, **56**, 650, 2003.
88. S. Sadhu and A.K. Bhowmick, *Rubber Chem. Technol.*, **76**, 860, 2003.
89. M. Maiti, S. Sadhu, and A.K. Bhowmick, *J. Polym. Sci., Part B: Polym. Phys.*, **43**, 4489, 2004.
90. M. Maiti, S. Sadhu, and A.K. Bhowmick, *J. Appl. Polym. Sci.*, **96**, 443, 2005.
91. A.H. Tsou and M.B. Measmer, *Rubber Chem. Technol.*, **79**, 281, 2006.
92. A.J. Dias, C. Gong, W. Weng, D.Y. Chung, and A.H. Tsou, Exxon Mobil Chemical Patents Inc., 2001, US 2001-296873 20010608.
93. M. Maiti and A.K. Bhowmick, *J. Polym. Sci., Part B: Polym. Phys.*, **44**, 162, 2006.
94. A.M. Kader and C. Nah, *Polymer*, **45**, 2237, 2004.
95. M. Maiti and A.K. Bhowmick, *J. Appl. Polym. Sci.*, **101**, 2407, 2006.
96. M. Maiti and A.K. Bhowmick, *Polymer*, **47**, 6156, 2006.
97. M.A. Kader, M. Lyu, and C. Nah, *Compos. Sci. Technol.*, **66**, 1431, 2006.
98. M. Maiti and A.K. Bhowmick, *J. Appl. Polym. Sci.*, **105**, 435, 2007.
99. T. Lan, P.D. Kaviratna, and T.J. Pinnavaia, *Chem. Mater.*, **6**, 573, 1994.
100. S. Wang, C. Long, X. Wang, Q. Li, and Z. Qi, *J. Appl. Polym. Sci.*, **69**, 1557, 1998.
101. P.C. LeBaron and T.J. Pinnavaia, Abstracts of Papers, 222nd ACS National Meeting, Chicago, IL, August 26–30, 2001.
102. N. Zhou, X. Xia, Y. Wang, J. Ma, Z. Yu, H. Kuan, A. Dasari, and Y. Mai, *Macromol. Rapid Commun.*, **26**, 830, 2005.
103. Q. Kong, Y. Hu, L. Song, Y. Wang, Z. Chen, and W. Fan, *Polym. Adv. Technol.*, **17**, 463, 2006.
104. J. Wang, Y. Chen, and Q. Jin, *High Perform. Polym.*, **18**, 325, 2006.
105. J. Ma, Z. Yu, H. Kuan, A. Dasari, and Y. Mai, *Macromol. Rapid Commun.*, **26**, 830, 2005.
106. D. Shia, C.Y. Hui, S.D. Burnside, and E.P. Giannelis, *Polym. Compos.*, **19**, 608, 1998.
107. S.D. Burnside and E.P. Giannelis, *J. Polym. Sci., Part B: Polym. Phys.*, **38**, 1595, 2000.
108. A. Ganguly, M. De Sarkar, and A.K. Bhowmick, *J. Appl. Polym. Sci.*, **100**, 2040, 2006.
109. S.T. Lim, C.H. Lee, Y.K. Kwon, and H.J. Choi, *J. Macromol. Sci. Phys.*, **B43**, 577, 2004.
110. C. Lee, S. Lim, Y.C. Kwon, and J. Hyoung, *Polym. Prepr.*, **44**, 825, 2003.
111. A. Ganguly, M. De Sarkar, and A.K. Bhowmick, *J. Polym. Sci., Part B: Polym. Phys.*, **45**, 52, 2006.
112. A. Ganguly and A.K. Bhowmick, 170th Meeting Rubber Division, ACS at Cincinnati, OH, October 10–12, 2006.
113. F. Yang, Y. Ou, and Z. Yu, *J. Appl. Polym. Sci.*, **69**, 355, 1998.
114. X. Liu, Q. Wu, L.A. Berglund, H. Lindberg, J. Fan, and Z. Qi, *J. Appl. Polym. Sci.*, **88**, 953, 2003.

115. S.C. Tjong, Y.Z. Meng, and Y. Xu, *J. Polym. Sci., Part B: Polym. Phys.*, **40**, 2860, 2002.
116. P. Reichert, J. Kressler, R. Thomann, R. Müllhaupt, and G. Stöppelmann, *Acta Polym.*, **49**, 116, 1998.
117. S. Idemura and J. Preston, *J. Polym. Sci., Part A: Polym. Chem.*, **41**, 1014, 2003.
118. X. Liu and Q. Wu, *Macromol. Mater. Eng.*, **287**, 180, 2002.
119. Y. Lu, G. Zhang, M. Feng, Y. Zhang, M. Yang, and D. Shen, *J. Polym. Sci., Part B: Polym. Phys.*, **41**, 2313, 2003.
120. C.-C.M. Ma, C.-T. Kuo, H.-C. Kuan, and C.-L. Chiang, *J. Appl. Polym. Sci.*, **88**, 1686, 2003.
121. X. Liu, Q. Wu, L.A. Berglund, and Z. Qi, *Macromol. Mater. Eng.*, **287**, 515, 2002.
122. Q. Wu, X. Liu, and L.A. Berglund, *Macromol. Rapid Commun.*, **22**, 1438, 2001.
123. H. Kharbas, P. Nelson, M. Yuan, S. Gong, L.-S. Turng, and R. Spindler, *Polym. Compos.*, **24**, 655, 2003.
124. K. Masenelli-Varlot, E. Reynaud, G. Vigier, and J. Varlet, *J. Polym. Sci., Part B: Polym. Phys.*, **40**, 272, 2002.
125. F. Dabrowski, M.L. Bras, R. Delobel, J.W. Gilman, and T. Kashiwagi, *Macromol. Symp.*, **194**, 201, 2003.
126. M.R. Kamal, N.K. Borse, and A. Garcia-Rejon, *Polym. Eng. Sci.*, **42**, 1883, 2002.
127. N. Hasegawa and A. Usuki, *J. Appl. Polym. Sci.*, **93**, 464, 2004.
128. P. Svoboda, C. Zeng, H. Wang, L.J. Lee, and D.L. Tomasko, *J. Appl. Polym. Sci.*, **85**, 1562, 2002.
129. Q. Zhang, Q. Fu, L. Jiang, and Y. Lei, *Polym. Int.*, **49**, 1561, 2000.
130. S. Hambir, N. Bulakh, and J.P. Jog, *Polym. Eng. Sci.*, **42**, 1800, 2002.
131. A. Sasaki and J.L. White, *J. Appl. Polym. Sci.*, **91**, 1951, 2004.
132. S.C. Tjong and Y.Z. Meng, *J. Polym. Sci., Part B: Polym. Phys.*, **41**, 2332, 2003.
133. T.S. Ellis and J.S. D'Angelo, *J. Appl. Polym. Sci.*, **90**, 1639, 2003.
134. G. Marosi, P. Anna, A. Márton, G. Bertalan, A. Bóta, A. Tóth, M. Mohai, and I. Rácz, *Polym. Adv. Technol.*, **13**, 1103, 2002.
135. W.S. Chow, Z.A. Mohd Ishak, U.S. Ishiaku, J. Karger-Kocsis, and A.A. Apostolov, *J. Appl. Polym. Sci.*, **91**, 175, 2004.
136. C.M. Koo, M.J. Kim, M.H. Choi, S.O. Kim, and I.J. Chung, *J. Appl. Polym. Sci.*, **88**, 1526, 2003.
137. Y. Tang, Y. Hu, S. Wang, Z. Gui, Z. Chen, and W. Fan, *Polym. Int.*, **52**, 1396, 2003.
138. J. Li, C. Zhou, G. Wang, and D. Zhao, *J. Appl. Polym. Sci.*, **89**, 13, 3609, 2003.
139. J. Ma, S. Zhang, Z. Qi, G. Li, and Y. Hu, *J. Appl. Polym. Sci.*, **83**, 1978, 2002.
140. P. Maiti, P.H. Nam, M. Okamoto, T. Kotaka, N. Hasegawa, and A. Usuki, *Polym. Eng. Sci.*, **42**, 1864, 2002.
141. W. Xu, M. Ge, and P. He, *J. Polym. Sci., Part B: Polym. Phys.*, **40**, 408, 2002.
142. C.-R. Tseng, J.-Y. Wu, H.-Y. Lee, and F.-C. Chang, *J. Appl. Polym. Sci.*, **85**, 1370, 2002.
143. J.M. Hwu, T.H. Ko, W.-T. Yang, J.C. Lin, G.J. Jiang, W. Xie, and W.P. Pan, *J. Appl. Polym. Sci.*, **91**, 101, 2004.
144. W.-A. Zhang, X.-F. Shen, M.-F. Liu, and Y.-E. Fang, *J. Appl. Polym. Sci.*, **90**, 1692, 2003.
145. Z.M. Wang, T.C. Chung, J.W. Gilman, and E. Manias, *J. Polym. Sci., Part B: Polym. Phys.*, **41**, 3173, 2003.
146. D. López, I. Cendoya, F. Torres, J. Tejada, and C. Mijangos, *Polym. Eng. Sci.*, **41**, 1845, 2001.
147. J. Fan, S. Liu, G. Chen, and Z. Qi, *J. Appl. Polym. Sci.*, **83**, 66, 2002.
148. T.-M. Wu, S.-F. Hsu, and J.-Y. Wu, *J. Polym. Sci., Part B: Polym. Phys.*, **40**, 736, 2002.
149. H.-W. Wang, K.-C. Chang, J.-M. Yeh, and S.-J. Liou, *J. Appl. Polym. Sci.*, **91**, 1368, 2004.
150. Y.T. Lim and O.O. Park, *Macromol. Rapid Commun.*, **21**, 231, 2000.
151. T.H. Kim, S.T. Lim, C.H. Lee, H.J. Choi, and M.S. Jhon, *J. Appl. Polym. Sci.*, **87**, 2106, 2003.
152. C.I. Park, W.M. Choi, M.H. Kim, and O. Park, *J. Polym. Sci., Part B: Polym. Phys.*, **42**, 1685, 2004.
153. T.-M. Wu, S.-F. Hsu, and J.-Y. Wu, *J. Polym. Sci., Part B: Polym. Phys.*, **41**, 1730, 2003.
154. Y.-H. Yu, J.-M. Yeh, S.-J. Liou, C.-L. Chen, D.-J. Liaw, and H.-Y. Lu, *J. Appl. Polym. Sci.*, **92**, 3573, 2004.
155. H.-L. Tyan, K.-H. Wei, and T.-E. Hsieh, *J. Polym. Sci., Part B: Polym. Phys.*, **38**, 2873, 2000.
156. J.-M. Yeh, C.-L. Chen, T.-H. Kuo, W.-F. Su, H.-Y. Huang, D.-J. Liaw, H.-Y. Lu, C.-F. Liu, and Y.-H. Yu, *J. Appl. Polym. Sci.*, **92**, 1072, 2004.
157. O. Khayankarn, R. Magaraphan, and J.W. Schwank, *J. Appl. Polym. Sci.*, **89**, 2875, 2003.
158. H.-L. Tyan, C.-Y. Wu, and K.-H. Wei, *J. Appl. Polym. Sci.*, **81**, 1742, 2001.
159. J.-H. Chang, K.M. Park, D. Cho, H.S. Yang, and K.J. Ihn, *Polym. Eng. Sci.*, **41**, 1514, 2001.

160. J.-H. Chang and K.M. Park, *Polym. Eng. Sci.*, **41**, 2226, 2001.
161. G.-H. Hsiue, J.-K. Chen, and Y.-L. Liu, *J. Appl. Polym. Sci.*, **76**, 1609, 2000.
162. J.-H. Chang, D.-K. Park, and K.J. Ihn, *J. Appl. Polym. Sci.*, **84**, 2294, 2002.
163. J. Liu, Y. Gao, F. Wang, and M. Wu, *J. Appl. Polym. Sci.*, **75**, 384, 2000.
164. V.A. Bershtein, L.M. Egorova, P.N. Yakushev, P. Pissis, P. Sysel, and L. Brozova, *J. Polym. Sci., Part B: Polym. Phys.*, **40**, 1056, 2002.
165. L.-Y. Jiang, C.-M. Leu, and K.-H. Wei, *Adv. Mater.*, **14**, 426, 2002.
166. A. Gu, S.-W. Kuo, and F.-C. Chang, *J. Appl. Polym. Sci.*, **79**, 1902, 2001.
167. A. Gu and F.-C. Chang, *J. Appl. Polym. Sci.*, **79**, 289, 2001.
168. http://www.marketresearch.com/map/prod/276277.html, access date 21.8.2004.
169. D. Shah, P. Maiti, E. Gunn, D.F. Schmidt, D.D. Jiang, C.A. Batt, and E.P. Giannelis, *Adv. Mater.*, **16**, 1173, 2004.
170. E.P. Giannelis, *App. Organomet. Chem.*, **12**, 675, 1998.
171. R.A. Vaia and W. Liu, *J. Polym. Sci., Part B: Polym. Phys.*, **40**, 1590, 2002.
172. R.A. Vaia, W. Liu, and H. Koerner, *J. Polym. Sci., Part B: Polym. Phys.*, **41**, 3214, 2003.
173. A. Sinsawat, K.L. Anderson, R.A. Vaia, and B.L. Farmer, *J. Polym. Sci., Part B: Polym. Phys.*, **41**, 3272, 2003.
174. X. Kornmann, L.A. Berglund, J. Sterte, and E.P. Giannelis, *Polym. Eng. Sci.*, **38**, 1351, 1998.
175. J. Fröhlich, R. Thomann, O. Gryshchuk, J. Karger-Kocsis, and R. Mülhaupt, *J. Appl. Polym. Sci.*, **92**, 3088, 2004.
176. F. Cser and S.N. Bhattacharya, *J. Appl. Polym. Sci.*, **90**, 3026, 2003.
177. R.K. Shah and D.R. Paul, *Polymer*, **45**, 2991, 2004.
178. A. Maity and M. Biswas, *J. Appl. Polym. Sci.*, **94**, 803, 2004.
179. G. Tanaka and L.A. Goettler, *Polymer*, **43**, 541, 2002.
180. Y.-W. Chang, Y.L Yang, S. Ryu, and C. Nah, *Polym. Int.*, **51**, 319, 2002.
181. J.W. Kim, S.G. Kim, H.J. Choi, and M.S. Jhon, *Macromol. Rapid Commun.*, **20**, 450, 1999.
182. P. Maiti, P.H. Nam, M. Okamoto, T. Kotaka, N. Hasegawa, and A. Usuki, *Polym. Eng. Sci.*, **42**, 1864, 2002.
183. X. Wang, J. Lin, X. Wang, and Z. Wang, *J. Appl. Polym. Sci.*, **93**, 2230, 2004.
184. Y. Wang, H. Zhang, Y. Wu, J. Yang, and L. Zhang, *J. Appl. Polym. Sci.*, **96**, 318, 2005.
185. I. González, J.I. Eguiazábal, and J. Nazábal, *J. Polym. Sci., Part B: Polym. Phys.*, **43**, 3611, 2005.
186. L. Zhang, Y. Wang, Y. Wang, Y. Sui, and D. Yu, *J. Appl. Polym. Sci.*, **78**, 1873, 2000.
187. Y.-P. Wu, Y.-Q. Wang, H.-F. Zhang, Y.-Z. Wang, D.-S. Yu, L.-Q. Zhang, and J. Yang, *Compos. Sci. Technol.*, **65**, 1195, 2005.
188. M. Frounchi, S. Dadbin, Z. Salehpour, and M. Noferesti, *J. Membr. Sci.*, **282**, 142, 2006.
189. M. Arroyo, M.A. López-Manchado, J.L. Valentín, and J. Carretero, *Compos. Sci. Technol.*, **67**, 1330, 2007.
190. M. Maiti, A. Bandyopadhyay, and A.K. Bhowmick, *J. Appl. Polym. Sci.*, **99**, 1645, 2006.
191. G. Naderi, P.G. Lafleur, and C. Dubois, Annual Technical Conference—Society of Plastics Engineers, 63rd, 3587, 2005.
192. J. Patel, M. Maiti, K. Naskar, and A.K. Bhowmick, *Polym. Polym. Compos.*, **14**, 515, 2006.
193. I. Gonzalez, J.I. Eguiazabal, and J. Nazabal, *Eur. Polym. J.*, **42**, 2905, 2006.
194. C.M. Small, G.M. McNally, P. McShane, and I. Kenny, Annual Technical Conference—Society of Plastics Engineers, 64th, 1740, 2006.
195. W. Dong, X. Zhang, Y. Liu, Q. Wang, H. Gui, J. Gao, Z. Song, J. Lai, F. Huang, and J. Qiao, *Polymer*, **47**, 6874, 2006.
196. M. Choi, S. Anandhan, J. Youk, D. Baik, S. Seo, L. Won, and S. Han, *J. Appl. Polym. Sci.*, **102**, 3048, 2006.
197. M. Pramanik, S.K. Srivastava, B.K Samantaray, and A.K. Bhowmick, *J. Polym. Sci., Part B: Polym. Phys.*, **40**, 2065, 2002.
198. N. Hasegawa and A. Usuki, *Polym. Bull.*, **51**, 77, 2003.
199. M. Laus, O. Francescangeli, and F. Sandrolini, *J. Mater. Res.*, **12**, 3134, 1997.
200. R.A. Vaia and E.P. Giannelis, *Polymer*, **47**, 5196, 2006.
201. http://www.azom.com/Details.asp?ArticleID = 921, access date 15.4.2007.
202. http://www.natick.army.mil/soldier/jocotas/ColPro_Papers/Donahue.pdf, access date 21.8.2004.

203. B. de Bièvre and K. Nakamura, UBE Europe. Polyamide Nanoclay Hybrids at UBE. Nanocomposites 2002. Conference Paper. January 2002.
204. http://www.navysbir.brtrc.com/SuccessStories/TritonSystems.pdf, access date 21.8.2004.
205. A.Y. Goldman, J.A. Montes, A. Barajas, G. Beall, and D.D. Eisenhour, Annual Technical Conference—Society of Plastics Engineers, 56th(Vol. 2), 2415, 1998.
206. J.W. Gilman, C.L. Jackson, A.B. Morgan, R. Harris, E. Manias, E.P. Giannelis, M. Wuthenow, D. Hilton, and S.H. Philips, *Chem. Mater.*, **12**, 1866, 2000.

3 Rubber–Silica Hybrid Nanocomposites

Abhijit Bandyopadhyay and Anil K. Bhowmick

CONTENTS

3.1 INTRODUCTION

Materials and their development are fundamental to society. Major historical periods of society are ascribed to materials (e.g., Stone Age, Bronze Age, Iron Age, Steel Age, Polymer Age, Silicon Age, and Silica Age). However, scientists will open the next societal frontiers not by understanding a particular material, but by optimizing the relative contributions afforded by a combination of different materials.

The invasion of nanoscale materials and associated nanoscience and technology in recent years affords unique opportunities to create revolutionary material combinations. The beginning of nanotechnology and nanoscience research can be traced back over 40 years. However, it is during the last decade that the world has seen bigger strides of this technology in various disciplines. From chemistry to biology, from materials science to electrical engineering, several tools have been created and expertise has been developed to bring nanotechnology out of the research laboratories into the commercial market.

By definition, a polymer nanocomposite is either an interacting or noninteracting combination of two phases resulting in a hybrid material, of which at least one of the phases is in the nanometer size range in any one of its dimension [1]. Hybrid nanocomposites can either have organic polymer matrices, which capture inorganic reinforcing filler, or an inorganic matrix with organic nano-filler particles. In nature, organic–inorganic hybrid materials play important roles as structural composites and represent some of the finest examples of optimized interfacial interaction between a matrix and filler particles via small-scale design [2].

Hybrid formation is an important and evolutionary route for the growth of a strong polymer–filler interface. Increasing interest in hybrid formation of organic–inorganic materials stems from the ability to control the nano-architecture of materials at a very early stage of preparation. These can often be directed at a molecular level, almost a billionth of a meter in length. Several routes such as intercalation [3], electro-crystallization [4], and sol–gel processing [5,6] can be employed to synthesize hybrid materials.

Depending on the level of interaction between these organic–inorganic phases, hybrid materials can either possess weak interaction between these phases such as van der Waals, hydrogen bonding, or electrostatic interaction [7,8], or be of strong, chemically bonded (covalent or coordinate) types [9].

3.1.1 CLASSIFICATION OF POLYMER–INORGANIC HYBRIDS AND NANOCOMPOSITES

Figure 3.1 gives a hand-drawn pictorial representation of the general classes of polymer–inorganic hybrids.

The nanocomposites described in Type I are prepared by dispersing nanofillers into a polymer matrix. Polymer–metal oxide combinations are included in this class of composites. The nanofillers (metal oxide) are generally formed via sol–gel reaction. Type II nano-reinforced systems are synthesized by dispersing nanometric clay fillers into a polymer matrix through intercalation and/or delamination, which were discussed in an earlier chapter. Layered graphite and carbon nanotube-reinforced polymer hybrids are also included in this series. In these two types (Types I and II) weak noncovalent interactions (e.g., hydrogen bonding or van der Waals attractions) are generally established between a polymer matrix and the nanofiller. Type III systems are tailor-made materials and are synthesized from nanostructured chemical building blocks to copolymerize or graft onto a polymer backbone. These blocks are attached to the polymer matrix through covalent bonds. Polyhedral oligomeric silsesquioxane (POSS) is an example of this class of hybrid.

3.1.2 POLYMER–INORGANIC HYBRID NANOCOMPOSITES THROUGH SOL–GEL REACTION

Sol–gel chemistry offers a unique advantage in the creation of novel organic–inorganic hybrids. The sol–gel process begins with a solution of metal alkoxide precursors $[M(OR)_n]$ and water, where M is a network-forming element, and R is typically an alkyl group. Hydrolysis and

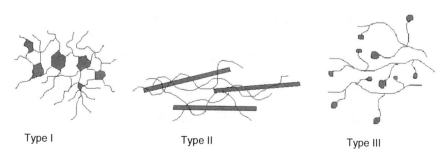

Type I   Type II   Type III

FIGURE 3.1 Classification of nanocomposite materials.

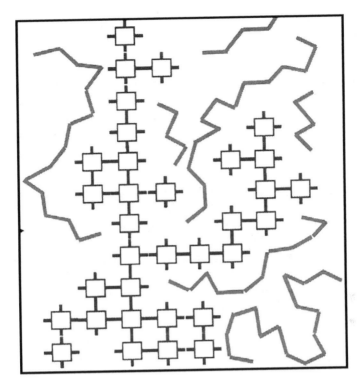

FIGURE 3.2 A model showing the presence of inorganic phase dispersed within a polymer matrix.

condensation of the alkoxide are the two fundamental steps to produce a network (Figure 3.2) in the presence of an acidic or a basic catalyst.

The entire reaction is illustrated through Scheme 3.1.

Where M is a silicon or transition metal and R is an alkyl group. From the reactions shown in Scheme 3.1, it is obvious that metal alkoxide is first hydrolyzed in the presence of water and then the product in the hydrolysis reaction condenses to produce metal oxide. Water and alcohol are the two by-products of the sol–gel reaction. Silicon alkoxide (e.g., tetraethoxysilane (TEOS)) [10] is the most commonly used metal alkoxide due to its mild reaction condition. Various routes are used to prepare polymer–inorganic hybrids using sol–gel process. These are solution method, soaking method, *in situ* polymerization, and simultaneous sol–gel reaction method, and the multi-step method of formation of functionalized cluster of metal oxide and its subsequent polymerization with monomers. The degree of molecular association of the metal alkoxide (oligomerization) has a significant influence on the sol–gel reaction and hence on the resultant morphology and properties. The nature and size of the ligand and the central metal atom can also dictate the reaction rate. For example, with the same tetraethoxy ligand $(OC_2H_5)_4$, the degree of association follows the order thorium > hafnium > zirconium > titanium > silicon [11].

Wen and Wilkes [12] enlisted the names and types of most of the polymers employed for the preparation of hybrid composites using sol–gel technique. In the field of rubber, wide-scale research

$$M(OR)_n + nH_2O \longrightarrow M(OH)_n + nROH \text{ (hydrolysis)}$$

$$MOH + HOM \longrightarrow M - O - M + H_2O \text{ (condensation)}$$

SCHEME 3.1 Hydrolysis and condensation reactions of metal alkoxide forming metal oxide.

has not been carried out so far to develop new hybrid materials after Mark et al. [13], who had reported poly(dimethyl siloxane) (PDMS)–silica hybrid composites in 1984. In the last five years, more effort has been offered toward the development of *in situ*-generated hybrids, where the organic and the inorganic components are allowed to grow simultaneously. The monomers like acrylates and methacrylates or their derivatives [14,15] are mostly used as the organic components. The inorganic components are alkoxysilane and their derivatives. Even though the resultant hybrids are targeted for optical and optoelectronic applications for their high transparency, generation of high-strength composites seems to be difficult using this method as there are a number of intervening parameters acting in tandem. Apart from the polymer–silica hybrids, other hybrids based on Ti, Zr, Al, etc. have also been reported [16,17]. However, the nature of interaction in those cases is only little explored. Svergun et al. [18] and Carotenuto and Nicolais [19] have reported the formation of gold- and platinum-based hybrid materials. Reports are available on the use of silica obtained from silicic acid for modification of polymers. This is generally obtained as silicic acid oligomer (SAO) from water glass. Abey and Misono [20] and Gunji et al. [21] first studied the formation of silica network from SAO in tetrahydrofuran (THF) [20,21]. Wang et al. [22] have used silica (silicic acid) and titania in polyimide matrix and generated Si–O–Ti inorganic composite phase via nonhydrolytic sol–gel route. Earlier, they synthesized polyimide–nanosilica hybrids starting from SAO via sol–gel route [23]. Zhao et al. [24] synthesized nanoporous silica with average particle diameter of 50 nm within polyacrylamide interpenetrating network gel. Nanosilica synthesis in cubic templates of thermoreversible block copolymer has been reported by Pozzo and Walker [25]. Use of functional silicic acid for simultaneous synthesis of the organic and the inorganic phases has been reported [26]. Walldal et al. [27] studied the nature of organic–inorganic interfaces in these hybrids. Takeuchi [28] showed that polymer-coated silica particles, synthesized via silicic acid route, are more stable and are suitable as packing materials in high-performance liquid chromatography. However, the present authors are not aware of any reports made on rubbers and rubbery materials in this area.

Although very popular, the sol–gel technique has not been used extensively in studying the composites based on rubbers after pioneering research in this field by Mark [13]. Later on, Kohjiya et al. [29–31] engaged in forming some hybrids using several nonpolar rubbers like natural rubber (NR), styrene–butadiene rubber (SBR), and polybutadiene rubber (BR) through sol–gel soaking method. This is explained later. Recently, number of reports have been published by the present authors on rubber–*in situ* silica hybrid nanocomposites synthesized from sol–gel solution process [32–38]. A research group in South Australia has reported ionomers–silica hybrid nanocomposites using similar technology [11]. In almost all the cases, silica has been preferred as the inorganic nano-phase mainly because of its control of self-aggregation tendency as already mentioned. Moreover, it is a commercially important filler for several rubbers and improves viscoelastic responses and beneficiates thermal, aging, and wear characteristics of the composites. This chapter is principally dedicated to provide a deep insight into the rubber–silica hybrid nanocomposites starting from their synthesis to end-use properties.

3.1.3 NANOCOMPOSITES FORMED BY PREFORMED SILICA

Published reports are available on the use of nanosilica in polymers/elastomers, where the silica has not been prepared through *in situ* sol–gel route. Garcia et al. [39] have reported the effect of different nanosilica concentrations (from very high to very low) on the thermal degradation behavior of poly(methyl methacrylate). Zhang et al. [40] have studied the effect of nanosilica concentrations on electrical conductivity of epoxy–carbon black composite. These nanosilica particles have been reported to offer superiority in several physico-mechanical properties including processing [41] and toughening behaviors [42–44]. A reference on structure–property correlation in surface modified nanosilica and polymer is also available in the literature [45].

3.2 METHODS OF RUBBER–SILICA HYBRID PREPARATION THROUGH SOL–GEL CHEMISTRY

There are various routes to synthesize rubber-based organic–inorganic hybrid composites through sol–gel process. These are mentioned in brief below.

3.2.1 SOLUTION METHOD

This method is suitable for rubbers, which are readily soluble in common solvents. The catalytic sol–gel reaction has been carried out in homogeneous solution phase in order to disperse the inorganic oxide within the rubber matrices. Although this technique possesses the biggest advantage of intimate interphase mixing between the two energetically incompatible components, the hybrid formation is largely affected by several other factors like the nature of solvents, the hydrophobic–hydrophilic balance of the rubber, alkoxysilane/H_2O mole ratio, reaction temperature, and pH. Mark et al. [13] and other workers [46] have reported the formation of low-molecular weight PDMS–silica hybrid composites in solution. Recently, Bandyopadhyay et al. [32,33] have published synthesis of acrylic rubber (ACM)–in situ silica and epoxidized natural rubber (ENR)–in situ silica hybrid nanocomposites using solution process. A suitable technique for synthesizing rubber–silica hybrid nanocomposites in solution is summarized in Scheme 3.2.

In general, the experiments have been largely accomplished under ambient conditions; experiments at higher temperature have also been reported so far (e.g., at 40°C [47]).

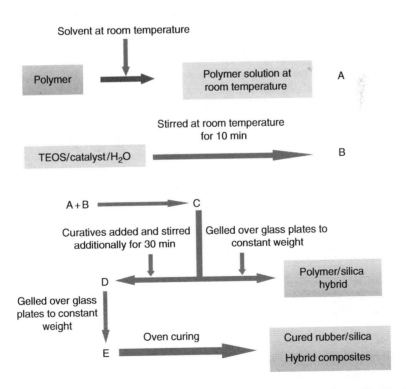

SCHEME 3.2 Preparation chart of rubber–*in situ* silica hybrid nanocomposites reported by Bandyopadhyay et al. (From Bandyopadhyay, A., Bhowmick, A.K., and De Sarkar, M., *J. Appl. Polym. Sci.*, 93, 2579, 2004. Courtesy of Wiley InterScience.)

3.2.2 SOAKING METHOD

This process is highly suitable for rubbers with poor solubility. In this process, the rubber sheet is soaked in TEOS or quite often in TEOS–solvent mixture and the *in situ* silica generation is conducted by either acid or base catalysis. The sol–gel reaction is normally carried out at room temperature. Kohjiya et al. [29–31] have reported various nonpolar rubber–silica hybrid nanocomposites based on this technique. The network density of the rubber influences the swelling behavior and hence controls the silica formation. It is very likely that there has been a graded silica concentration from surface to the bulk due to limited swelling of the rubber. This process has been predominantly used to prepare ionomer–inorganic hybrids by Siuzdak et al. [48–50].

3.2.3 *IN SITU* POLYMERIZATION AND THE SOL–GEL REACTION OF ORGANIC–INORGANIC COMPONENTS

The simultaneous polymerization and sol–gel reaction often brings complexity to the overall reaction. Moreover, it is difficult to control the molecular weight of the sample. Recently, Patel et al. [51] have synthesized the rubber grade acrylic copolymers and terpolymers–*in situ* silica hybrid nanocomposites using this technique.

3.3 EFFECT OF ALKOXYSILANE ON STRUCTURE AND PROPERTIES OF RUBBER–SILICA HYBRID NANOCOMPOSITES

Sol–gel solution technique provides an outstanding advantage of intimate interphase mixing and also generates optically clear composites. Various types of alkoxysilanes are available depending upon the number of silicon valences free to condense, but TEOS has been the universal choice because of its mild reaction conditions. Also, upon hydrolysis, polycondensation can take place utilizing all the four valences of silanol (SiOH) moiety. The silica particles have been found to generate from TEOS within rubber solution as spherical dispersants with an average diameter less than 100 nm, although the concentration of TEOS used to generate silica plays a key role in it. Figure 3.3a shows the transmission electron microscopic (TEM) image of an ACM–*in situ* silica hybrid nanocomposite, synthesized by dissolving the rubber in THF with 10 wt% of TEOS (with respect to the rubber) [32].

The average diameter of the silica particles is close to 10 nm. Similar silica morphology has also been reported by Kohjiya et al. [30]. When the TEOS concentration is increased to 30 wt% with respect to ACM, the same silica particles grow much bigger in size, displaying an average diameter of 70 nm, as illustrated in Figure 3.3b [32].

This trend is still continued at 50 wt% of TEOS concentration where the average diameter of the *in situ*-generated silica particles reaches almost 100 nm (Figure 3.3c [32]). Similar observations on silica morphology have been reported by Patel et al. [52] in rubber grade acrylic copolymer matrices. Figure 3.4a and b shows the high resolution scanning electron microscopic (SEM) images of these acrylic copolymer–*in situ* silica hybrid nanocomposites synthesized by simultaneous polymerization and silica formation technique using 10 and 50 wt% TEOS, respectively. The figure shows that the average size of the silica particles increases from 60 to 120 nm on increasing the TEOS concentration.

Concentration of TEOS in all these cases has been restricted up to 50 wt% with respect to the rubber. Beyond 50 wt%, all the hybrids show phase separation which may be due to higher amount of water condensate that is continuously generated and acts as nonsolvent for the rubbers. This is easily understood from the visual appearance of the samples; phase-separated composites slowly turn opaque in the course of gelation.

Greater size of the silica particles indicates higher extent of polycondensation that has taken place in the inorganic phase. The generation of silica particles is directly proportional to the alkoxysilane concentration and higher concentration of TEOS used for these investigations allows higher extent of

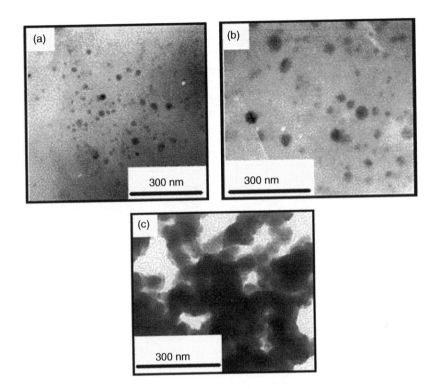

FIGURE 3.3 (a) Transmission electron microscopic (TEM) image of acrylic rubber (ACM)–silica hybrid nanocomposite synthesized from 10 wt% of tetraethoxysilane (TEOS). (From Bandyopadhyay, A., Bhowmick, A.K., and De Sarkar, M., *J. Appl. Polym. Sci.*, 93, 2579, 2004. Courtesy of Wiley Interscience.) Transmission electron microscopic (TEM) photographs of acrylic rubber (ACM)–silica hybrid nanocomposites prepared from (b) 30 wt% and (c) 50 wt% tetraethoxysilane (TEOS) concentrations. (From Bandyopadhyay, A., Bhowmick, A.K., and De Sarkar, M., *J. Appl. Polym. Sci.*, 93, 2579, 2004. Courtesy of Wiley InterScience.)

polycondensation of the SiOH moieties to occur, producing bigger silica particles. Higher proportions of the inorganic component in the plastic grade hybrids turn it into a brittle one. Zoppi et al. [53] have increased the TEOS proportion up to 80 wt% by using 1,1,3,3 tetramethyl-1,3-diethoxydisiloxane

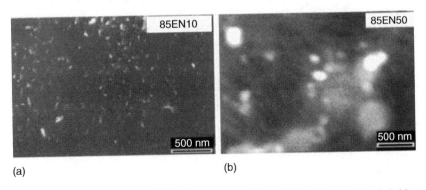

FIGURE 3.4 Scanning electron microscopic (SEM) images of acrylic copolymer–silica hybrid nanocomposites synthesized from (a) 10 wt% and (b) 50 wt% tetraethoxysilane (TEOS) concentrations. The first number in the legend indicates the wt% of ethyl acrylate (EA) (85) in the ethyl acrylate–butyl acrylate (EA–BA) copolymer, N stands for nanocomposite, and the last number (10, 50) is indicative of the tetraethoxysilane (TEOS) concentration. (From Patel, S., Bandyopadhyay, A., Vijayabaskar, V., and Bhowmick, A.K., *J. Mater. Sci.*, 41, 926, 2006. Courtesy of Springer.)

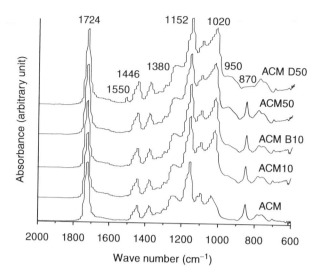

FIGURE 3.5 Fourier Transform infrared (FTIR) spectra of acrylic rubber (ACM)–silica hybrid nanocomposites. The numbers after ACM (10 and 50) indicate the wt% tetraethoxysilane (TEOS) concentration. The letters preceding the numbers indicate the ACM–silica samples cross-linked from benzoyl peroxide (B) and a mixed cross-linker hexamethylene diamine carbamate and ammonium benzoate (D). The numbers over the absorption peaks are the wave numbers corresponding to absorbance of those peaks. (From Bandyopadhyay, A., Bhowmick, A.K., and De Sarkar, M., *J. Appl. Polym. Sci.*, 93, 2579, 2004. Courtesy of Wiley InterScience.)

(TMDES) in conjunction with TEOS to resist brittleness in Nafion–silica hybrids. Contrastingly, rubber grade organic–inorganic hybrids can comfortably take fairly high TEOS loading without any coprecursor, possibly due to the inherent flexibility of the rubbers.

The Fourier Transform infrared (FTIR) spectra of the ACM–silica hybrids synthesized from 10 and 50 wt% initial TEOS concentrations, in Figure 3.5, demonstrate higher extent of silica generation with higher TEOS concentrations.

The numbers over the absorption peaks are the wave numbers corresponding to the absorptions of different functional groups and bonds in the hybrid nanocomposites. The absorption spectra of both cross-linked and uncross-linked hybrid composites are represented in this figure, and at particular TEOS concentrations they hardly differ in nature. The noticeable aspect in the figure is the increase in the absorbance peak area at 1020 and 950 cm^{-1} (which indicates the Si–O–Si asymmetric and Si–OH stretching vibrations, respectively) in the hybrid nanocomposites on increasing TEOS concentrations. Higher TEOS generates higher amount of silica and hence the absorbance peak area increases. Moreover, the change in the nature of the peak at 1020 cm^{-1} (a combination of several small peaks apart from a sharp peak at 1020 cm^{-1}) from that of the virgin ACM is due to the various siloxane structures formed in the inorganic phase within the hybrid nanocomposites. Table 3.1 correlates this TEOS–silica interconversion from ash content studies in the gelled rubber–silica films.

It is interesting to note that there is only slight difference in silica contents within chemically interactive (ENR) and noninteractive (ACM) rubber matrices at similar TEOS concentrations.

Higher extent of silica generation with high TEOS concentration improves the mechanical properties severalfolds as illustrated by the tensile stress–strain plots on ACM–silica hybrid nanocomposites on increasing TEOS concentrations in Figure 3.6.

Unfortunately, precipitated silica used in the rubber industry under these conditions does not give similar reinforcement. This is compared in Figure 3.7.

The nanosilica-filled rubber composites show a steep rise in the tensile strength. This is principally due to better dispersion of silica in comparison to precipitated silica within the rubber matrix. This is clearly demonstrated through the SEM pictures presented in Figure 3.8.

TABLE 3.1

TEOS–Silica Interconversion in the Rubber–Silica Hybrid Nanocomposites

TEOS (wt%)	Silica in ACM (wt%)	Silica in ENR (wt%)
10	2.3	2.5
20	5.2	5.5
30	8.1	8.3
40	11.2	11.2
50	14.1	14.2

Source: Bandyopadhyay, A., De Sarkar, M., and Bhowmick, A.K., *J. Appl. Polym. Sci.*, 95, 1418, 2005 and Bandyopadhyay, A., De Sarkar, M., and Bhowmick, A.K., *J. Mater. Sci.*, 40, 53, 2005. Courtesy of Springer.

Fine silica network is visible in the *in situ* silica-filled rubber composite synthesized from 50 wt% of TEOS, producing almost 15 wt% of silica. On the other hand, addition of only 10 wt% of precipitated silica externally gives distinct aggregations.

Dispersion of nanosilica within the rubber matrices usually generates optically transparent materials. All the ACM–silica and ENR–silica hybrid composites are completely transparent up to 50 wt% of TEOS concentrations. Following are the figures (Figure 3.9) which show the visual appearance of the representative hybrid nanocomposites. The logos over which the films (average film thickness 0.25 mm) are placed are clearly visible.

Formation of optically transparent silicone rubber–silica hybrid composites has also been reported by Rajan and Sur [54].

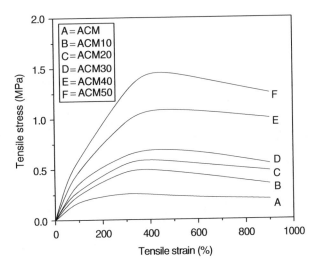

FIGURE 3.6 Tensile stress–strain plots of acrylic rubber (ACM)–silica hybrid nanocomposites using different tetraethoxysilane (TEOS) concentrations. The number in the legends indicates wt% TEOS concentrations. (From Bandyopadhyay, A., Bhowmick, A.K., and De Sarkar, M., *J. Appl. Polym. Sci.*, 93, 2579, 2004. Courtesy of Wiley InterScience.)

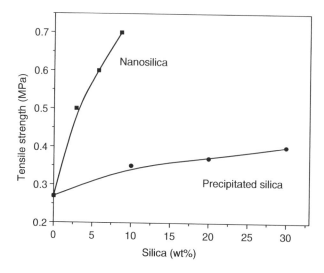

FIGURE 3.7 Comparative plots of tensile strength values of nanosilica and precipitated silica-filled acrylic rubber (ACM) composites. (From Bandyopadhyay, A., Bhowmick, A.K., and De Sarkar, M., *J. Appl. Polym. Sci.*, 93, 2579, 2004. Courtesy of Wiley InterScience.)

3.4 EFFECT OF POLARITY OF THE RUBBER ON STRUCTURE AND PROPERTIES

Polarity of the rubber matrices plays an important role in assigning the final structure and properties of the resultant hybrid nanocomposites. Silica is a polar filler and needs to be compatibilized with the rubber to aid dispersion and also to achieve substantial reinforcement. In fact, the silane-coupling agents are used in silica-filled rubber compounds for similar reasons [55]. Patel et al. [52] have nicely demonstrated the effect of polarity during hybrid formation from acrylic copolymers and terpolymers, both via *in situ* and solution processes. They found that, on increasing the polarity by introducing acrylic acid (AA) in a copolymer of ethyl acrylate (EA) and butyl acrylate (BA), the average diameter of the *in situ* silica particles significantly decrease (Figure 3.10a and b). Another example is the ACM–silica and ENR–silica systems reported by Bandyopadhyay et al. [32,33]

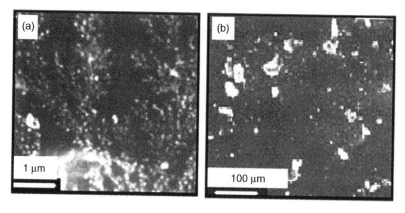

FIGURE 3.8 Scanning electron microscopic (SEM) pictures of acrylic rubber (ACM)–silica hybrid composites (a) synthesized from 50 wt% tetraethoxysilane (TEOS) and (b) synthesized from 10 parts of precipitated silica. (From Bandyopadhyay, A., Bhowmick, A.K., and De Sarkar, M., *J. Appl. Polym. Sci.*, 93, 2579, 2004. Courtesy of Wiley InterScience.)

(a) (b)

FIGURE 3.9 Visual appearance of (a) ACM30 and (b) ENR30 hybrid nanocomposite films. The number indicates the wt% tetraethoxysilane (TEOS) concentration. (From Bandyopadhyay, A., Thesis submitted for PhD degree to Indian Institute of Technology, Kharagpur, India, August 2005.)

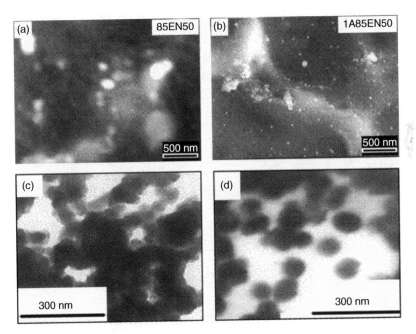

FIGURE 3.10 Microscopic images of rubber–silica hybrid nanocomposites: (a) ethyl acrylate–butyl acrylate (EA–BA) copolymer (85 wt% EA) synthesized from 50 wt% tetraethoxysilane (TEOS), (b) ethyl acrylate–butyl acrylate–acrylic acid (EA–BA–AA) terpolymer (85 wt% EA and 1 wt% AA) synthesized from 50 wt% TEOS, (c) acrylic rubber (ACM)–silica, and (d) epoxidized natural rubber (ENR)–silica hybrid nanocomposites synthesized from 50 wt% TEOS. 85EN50 indicates the copolymer composition of 85 wt% EA and 15 wt% BA and the nanocomposite has been prepared starting from 50 wt% TEOS. "1A" before 85EN50 indicates terpolymer using 1 wt% AA. (From Bandyopadhyay, A., Bhowmick, A.K., and De Sarkar, M., *J. Appl. Polym. Sci.*, 93, 2579, 2004; Bandyopadhyay, A., De Sarkar, M., and Bhowmick, A.K., *Rubber Chem. Technol.*, 77, 830, 2004; and Patel, S., Bandyopadhyay, A., Vijayabaskar, V., and Bhowmick, A.K., *J. Mater. Sci.*, 41, 926, 2006. Courtesy of Wiley Interscience, Springer, and ACS Rubber Division.)

where, because of higher polarity of ENR, the silica particles are more finely distributed. All these are illustrated in Figure 3.10c and d.

Polar rubbers can restrict the silica particles from self-aggregation in their very early stage of synthesis, forming more rubber-coated silica domains by virtue of greater rubber–silica interactions. This is mediated through the attraction between SiOH groups on oligomeric silica particles and the polar, interactive groups on the rubbers. Otherwise, self-aggregation predominates due to higher interparticle interactions and results in macro-phase separation. ACM does not contain interactive hydroxyl groups (OH) like ENR and hence cannot restrict the interparticle interactions as displayed in Figure 3.10c, while, due to greater interfacial interaction in ENR–silica system of the same composition, the silica particles are more discrete (Figure 3.10d). Similar reasons could also be cited for explaining the morphology of the nanocomposites depicted in Figure 3.10a and b. It may be mentioned here that the self-aggregation tendency of silica is predominant at higher TEOS concentrations. At low concentrations, this is not quite evident and could easily be understood by comparing the images of ACM10 and ACM50, displayed in Figure 3.3b and c, respectively. Polarity of the rubbers is also essential to maintain a substantial hydrophilic–hydrophobic balance so that it does not cause macro-phase separation in the course of hydrolysis and polycondensation of the alkoxysilane.

Finer dispersion of silica improves the mechanical and dynamic mechanical properties of the resultant composites. Figure 3.11a and b compares the tensile properties of the acrylic copolymer and terpolymers in the uncross-linked and cross-linked states, respectively.

Introducing AA into the copolymer backbone improves interaction between the SiOH groups of the silica and the carboxyl groups (COOH) in the terpolymer. Also, the finer distribution of the silica provides higher surface area to interact with the terpolymer and results in almost 500 times improvement in tensile strength even in the uncured state. On cross-linking, this strength is further increased in the terpolymer–silica system. The elongation leading to failure is also significantly high in the terpolymer than in the copolymer in the uncross-linked state, which may be due to relatively big silica particles in the copolymer–silica nanocomposites, acting as stress concentration sites and causing early failure.

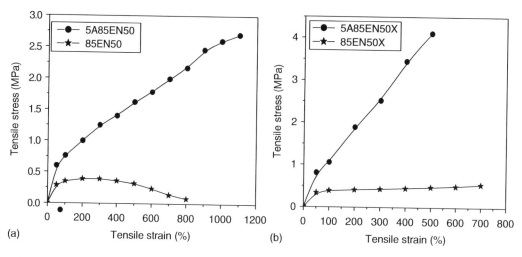

FIGURE 3.11 Tensile stress–strain plots for *in situ* synthesized acrylic copolymer and terpolymer–silica hybrid nanocomposites in (a) uncross-linked and (b) cross-linked states. 85EN50 indicates the copolymer composition of 85 wt% ethyl acrylate (EA) and 15 wt% BA and the nanocomposite has been prepared starting from 50 wt% tetraethoxysilane (TEOS). "5A" before 85EN50 indicates terpolymer using 5 wt% acrylic acid (AA). The letter "X" in the legends indicates cross-linked samples. (From Patel, S., Bandyopadhyay, A., Vijayabaskar, V., and Bhowmick, A.K., *Polymer*, 46, 8079, 2005. Courtesy of Elsevier.)

3.5   EFFECT OF SOLVENTS ON STRUCTURE AND PROPERTIES

Different characteristics of solvents seriously affect the sol–gel reaction in solution. This in turn influences the physico-mechanical properties of the resultant rubber–silica hybrid composites. Bandyopadhyay et al. [34,35] have carried out extensive research on structure–property correlation in sol–gel-derived rubber–silica hybrid nanocomposites in different solvents with both chemically interactive (ENR) and noninteractive (ACM) rubber matrices. Figure 3.12 demonstrates the morphology of representative ACM–silica and ENR–silica hybrid composites prepared from various solvents. In all the instances, the concentration of TEOS (45 wt%), TEOS/H_2O mole ratio (1:2), pH (1.5), and the gelling temperature (ambient condition) were kept unchanged.

ACM–silica system has been synthesized from four different solvents whose characteristics are described in Table 3.2. Three different solvents have been employed in studying ENR–silica system and the solvent characteristics are also denoted in the same table (Table 3.2).

As Figure 3.12 shows, both the rubbers exhibit distributed silica morphology (ENR has shown slightly better distribution than ACM due to the reason described in the previous section) when the composites are prepared from THF rather than from other solvents with fairly high TEOS concentration (45 wt%). Otherwise, the silica particles have been found to be aggregated in rest (except *N,N*-dimethyl-formamide (DMF) for ACM–silica) of the solvents studied [34] (morphology of the hybrids in other solvents has not been provided as they do not contain additional information). The dielectric constant and the solubility of different solvents in water are considered as the critical factors while explaining the results. Greater the solubility, more homogeneous is the precursor rubber–TEOS solution and better is the distribution of silica, keeping aside the rubber–silica interaction factor. An exception is the ACM–silica hybrid formed in DMF, which, in spite of

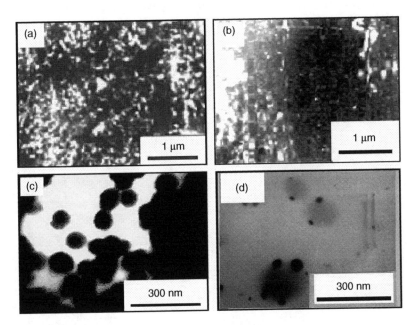

FIGURE 3.12 Morphology of rubber–silica hybrid composites synthesized from solution process using different solvents: (a) and (b) are the scanning electron microscopic (SEM) pictures of acrylic rubber (ACM)–silica hybrid composites prepared from THF (T) and ethyl acetate (EAc) (E) and (c) and (d) are the transmission electron microscopic (TEM) pictures of epoxidized natural rubber (ENR)–silica hybrid composites synthesized from THF and chloroform (CH). (From Bandyopadhyay, A., De Sarkar, M., and Bhowmick, A.K., *J. Appl. Polym. Sci.*, 95, 1418, 2005 and Bandyopadhyay, A., De Sarkar, M., and Bhowmick, A.K., *J. Mater. Sci.*, 40, 53, 2005. Courtesy of Wiley InterScience and Springer, respectively.)

TABLE 3.2

Characteristics of Different Solvents Used in Sol–Gel Reaction for ACM–Silica and ENR–Silica Systems

Solvents	Boiling Point (°C)	Density (kg/m³)	Solubility in Water (kg/100 kg)	Dielectric Constant
N,N-Dimethyl-formamide (DMF)	153.0	940	Miscible	36.70
Methyl ethyl ketone (MEK)	79.6	800	25.60	18.40
Ethyl acetate (EAc)	77.0	810	8.70	6.00
Tetrahydrofuran (THF)	66.00	880	30.00	7.60
Chloroform (CHCl$_3$)	61.20	1500	0.80	4.81
Carbon tetrachloride (CCl$_4$)	76.70	1600	0.05	2.24

Source: Bandyopadhyay, A., De Sarkar, M., and Bhowmick, A.K., *J. Appl. Polym. Sci.*, 95, 1418, 2005 and Bandyopadhyay, A., De Sarkar, M., and Bhowmick, A.K., *J. Mater. Sci.*, 40, 53, 2005. Courtesy of Wiley InterScience and Springer, respectively.

having complete miscibility in water, gives silica particles with average diameter close to 10 nm (almost nine times less than that obtained with THF). The plausible reason may be the basic nature of DMF, which hinders the polycondensation of SiOH by strongly interacting with the acidic hydrogen through acid–base reaction [34]. This has also reduced the average amount of silica (12%) generated in the system [34], measured from ash content studies (silica in THF-based hybrid is 13.8%). Solvents with poor water miscibility, displayed in the table, form heterogeneous precursor sol to begin with and ultimately end up with poor organic–inorganic phase miscibility due to aggregated silica structures [34] (alcohol solubility of the solvents in the latter stage of the reaction matters little because it volatilizes at room temperature). These solvents produce composites which do not contain nanosilica particles (exception DMF) and are strictly not recommended as nanocomposites. They are not transparent in appearance due to scattering of visible light from the silica aggregates unlike the composites prepared from THF.

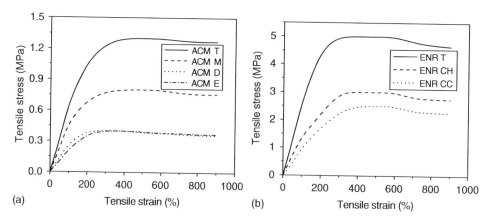

(a)   (b)

FIGURE 3.13 Tensile stress–strain plots for acrylic rubber (ACM)–silica and epoxidized natural rubber (ENR)–silica hybrid composites synthesized from various solvents: (a) ACM–silica and (b) ENR–silica. The letters after the rubbers in the legend indicate solvents used: T = THF, M = methyl ethyl ketone (MEK), D = DMF, E = EAc, CH = CHCl$_3$, CC = CCl$_4$. (From Bandyopadhyay, A., De Sarkar, M., and Bhowmick, A.K., *J. Appl. Polym. Sci.*, 95, 1418, 2005 and Bandyopadhyay, A., De Sarkar, M., and Bhowmick, A.K., *J. Mater. Sci.*, 40, 53, 2005. Courtesy of Wiley InterScience and Springer, respectively.)

Finer dispersion of silica in THF provides higher surface area to interact with the rubber molecules and the resultant nanocomposites show better mechanical properties than their macro-counterparts. Figure 3.13 illustrates these results.

All the data provided in the figure are on uncross-linked rubbers. This clearly shows the tremendous impact of the finer distribution of the silica particles on the performance of the composites from the tensile strength point of view.

3.6 EFFECT OF DIFFERENT TEOS/H₂O MOLE RATIOS, pH, AND GELATION TEMPERATURE ON STRUCTURE AND PROPERTIES OF THE HYBRID COMPOSITES

The first step in sol–gel processing is the catalytic hydrolysis of TEOS and the second step is the polycondensation of SiOH moieties framing into silica (Scheme 3.1). In the first step of the reaction, water is present as a reactant while it is the by-product in the second step. It is likely that the molar ratio of TEOS/H_2O would influence the sol–gel chemistry and hence the end properties of the resultant hybrids. The most interesting part of the sol–gel chemistry is that the catalytic hydrolysis of TEOS is an ion-controlled reaction, while polymerization of silica is not. Usually, the ionic reactions are much faster than the condensation reactions. The stoichiometric equation showing the silica formation from TEOS is presented in Scheme 3.3.

Tian et al. [56] have studied poly($\in$-caprolactone)–silica and Sengupta et al. [57] have investigated nylon 66–silica hybrid systems and have observed that the phase separation started when Si/H_2O mole ratio is increased above 2 and the resultant hybrid films become opaque. Gao [11] has reported similar observations on sol–gel-derived ionomeric polyethylene–silica system. A wide range of literatures is not available on this topic of rubber–silica hybrid nanocomposites, though Bandyopadhyay et al. [34,35] have reported the hybrid formation with different TEOS/H_2O mole ratios from ACM and ENR and also demonstrated detailed structure–property correlation in these systems. The hybrids have been prepared with 1:1, 1:2, 1:4, 1:6, 1:8, and 1:10 TEOS/H_2O mole ratios. Figure 3.14 shows the morphology of the ACM–silica hybrid composites prepared from different TEOS/H_2O mole ratios.

It is very clear from these pictures that with lower proportions of water than the stoichiometric amount (1:2), the size of the *in situ*-generated silica particles is very small, in spite of having very high TEOS concentration initially. On the other hand, aggregation of silica particles results quite comprehensively on increasing the proportion of water beyond the stoichiometric requirement. The size of the silica particles is around 90 nm with some signs of interparticle aggregation observed on using the stoichiometric amount (1:2) of water for forming the composite.

Silica particles of very small size as well as of low concentrations (the amount of silica has been measured by ash content studies) are obtained due to insufficient amount of water available in the system to substantiate sol–gel processing. The rate of hydrolysis becomes slower and so does the condensation, principally due to lower concentration of SiOH groups. Higher amount of water makes the hydrolysis of TEOS much faster and generates a huge amount of SiOH in a very quick time. These SiOH groups do not allow sufficient time for the interdiffusion of polymer molecules between themselves and therefore starts self-aggregation (Figure 3.14b). This has also slightly increased the amount of silica generated within the system. Using stoichiometric amount of water, the self-aggregation has been avoided to a greater extent. Following are the FTIR spectra of these hybrid composites (Figure 3.15) which also show abrupt increase in the concentration of silica (peak at 1060 cm^{-1}) using higher amount of water due to aggregation.

$$Si(OC_2H_5)_4 + 2H_2O \longrightarrow SiO_2 + 4C_2H_5OH$$

SCHEME 3.3 Overall reaction for silica synthesis from tetraethoxysilane (TEOS).

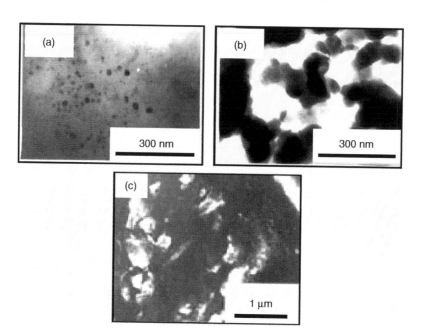

FIGURE 3.14 Transmission electron microscopic (TEM) pictures of (a) acrylic rubber (ACM)–silica hybrid prepared from 1:1 tetraethoxysilane (TEOS)/water (H_2O) and (b) 1:2 TEOS/H_2O mole ratios and (c) scanning electron microscopic (SEM) picture of ACM–silica hybrid composite synthesized from 1:6 TEOS/H_2O mole ratio. The concentration of TEOS has been kept constant at 45 wt% and the samples have been gelled at room temperature. (From Bandyopadhyay, A., De Sarkar, M., and Bhowmick, A.K., *J. Appl. Polym. Sci.*, 95, 1418, 2005. Courtesy of Wiley InterScience.)

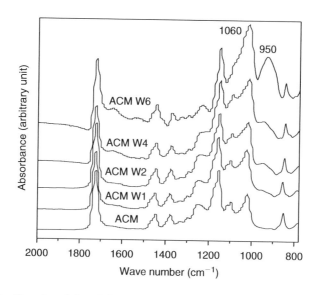

FIGURE 3.15 Fourier Transform infrared (FTIR) spectra of acrylic rubber (ACM)–silica hybrid composites synthesized from different tetraethoxysilane (TEOS)/water (H_2O) mole ratios. "W" in the legends indicates water and the numbers following "W" are the moles of water used per mole of TEOS. (From Bandyopadhyay, A., De Sarkar, M., and Bhowmick, A.K., *J. Appl. Polym. Sci.*, 95, 1418, 2005. Courtesy of Wiley InterScience.)

The figure demonstrates substantial increase in concentration of SiOH groups concurrently on using higher proportion of water. This clearly indicates that many of the SiOH groups are yet to polymerize under these conditions but could not do so as the samples as a whole complete gelation. Visually these composites are not transparent as earlier and also show comparatively less mechanical reinforcement when tested.

ENR–silica system has shown quite similar results on varying the molar proportions of water in identical fashion in forming the composites. All the hybrids formed with 1:1 and 1:2 TEOS/H_2O mole ratios contain nanosized silica particles, whereas on increasing the water beyond this, the silica particles aggregate due to rapid hydrolysis of the silica precursor. The composites synthesized from 1:1 and above 1:2 TEOS/H_2O mole ratios in both the rubbers do not exhibit much improvement in mechanical properties, partly due to low silica generation (TEOS/H_2O 1:1) and the rest due to silica aggregation (TEOS/H_2O > 1:2). Maximum improvement has been obtained from the composite synthesized by using stoichiometric amount of water (TEOS/H_2O 1:2).

pH of the reaction medium plays a key role in determining the nature of the sol–gel hybrids. Iler [58] has demonstrated the effects of pH in the colloidal silica–water system. He found that there are three regions: a metastable region at pH < 4.0, a particle growth region at pH > 7.0, and a rapid aggregation region at 4.0 < pH < 7.0. Landry et al. [59] have studied the effects of pH to form polymethylmethacrylate (PMMA)–silica hybrids. The hybrids formed in both acidic and basic environments show that silica uniformly disperses in the polymer matrix in particles smaller than 100 nm when an acid catalyst is used, while spherical silica domains of 0.1–0.2 µm aggregate to form clusters of various sizes (1–5 µm in diameter) and shapes in the presence of a basic catalyst. Huang et al. [46] have prepared PDMS–silica hybrids using various hydrogen chloride (HCl)/TEOS ratios. They observed that the structure and morphology of the hybrids vary on using different amounts of HCl. Kohjiya et al. [30,31] synthesized rubber–silica hybrid composites from amine catalysts. Extensive studies on structure–property correlation have been carried out by Bandyopadhyay et al. [36]. The series of hybrids have been formed at high and low pH ranges using HCl and NaOH catalysts on ACM and ENR matrices at fixed proportions of TEOS and water (TEOS: 45 wt% and TEOS/H_2O 1:2). All the hybrids synthesized within the pH 1.0–2.0 are transparent and contain nanoscale silica dispersion. On the other hand, the hybrids prepared in the pH > 2.0 are not transparent and show aggregated silica domains. The series of pictures on ACM–silica and ENR–silica hybrid composites in Figure 3.16 provides a nice illustration of this phenomenon.

At very high pH, the dispersion of the silica particles becomes very poor as observed from these figures. Even the aggregation of silica at pH > 2.0 could not be restricted by using strongly interacting 98% hydrolyzed grade poly(vinyl alcohol) as the sol–gel matrix. This shows that Iler's observation made on colloidal silica–water system could well be extended to both strongly and moderately interactive high polymer matrices. At low pH, the rubber molecules are adsorbed over the metastable silica particles and do not allow aggregation to a large extent. This is how the rubber-coated silica domains are formed. At moderate pH, due to rapid growth of the silica, the rubbers could not adsorb on the faster-growing silica particles, which ultimately leads to aggregated silica domains. At still higher pH, growth of ring and long-chain silica particles occurs quite comfortably irrespective of the nature of the rubber–polymer matrices and the resultant composites become completely opaque. In fact, most of the sol–gel nano-hybrids have been reported in high acidic pH, where the growth of the silica particles could be controlled. Some hybrid formation with dispersion of nanoscale silica has also been reported in solutions of mild basicity, e.g., in polyethylene ionomers by Gao [11].

Figure 3.17 shows the mechanical properties of the ACM–silica and ENR–silica hybrid composites synthesized from various pH, reproduced from the data reported by Bandyopadhyay et al. [36]. As morphology indicates, all the samples prepared within the pH range 1.0–2.0 are transparent, contain nanosilica particles, and are superior in tensile strength and modulus

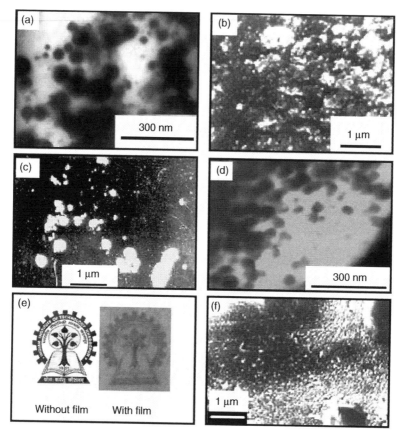

FIGURE 3.16 Morphology and visual appearance of acrylic rubber (ACM)–silica and epoxidized natural rubber (ENR)–silica hybrid composites prepared from different pH ranges: (a) transmission electron microscopic (TEM) picture of ACM–silica in pH 1.0–2.0, (b) scanning electron microscopic (SEM) picture of ACM–silica in pH 5.0–6.0, (c) SEM image of ACM–silica in pH 9.0–10.0, (d) TEM picture of ENR–silica in pH 1.0–2.0, (e) Visual appearance of ENR–silica film (average thickness 0.25 mm) in pH 5.0–6.0, and (f) SEM picture of ENR–silica in pH 9.0–10.0. (From Bandyopadhyay, A., De Sarkar, M., and Bhowmick, A.K., *J. Mater. Sci.*, 41, 5981, 2006. Courtesy of Springer.)

compared to other composites prepared at higher pH. Aggregation of silica at higher pH decreases filler surface area as well as the concentration of free-interacting SiOH groups over the surface. Both these factors become instrumental in deteriorating the mechanical properties for both these systems.

The structure and properties of the hybrid composites are found to vary quite significantly with the gelation temperature of the hybrids. Polymerization of SiOH occurs during gelation and on increasing the temperature, vigorous condensation takes place to give aggregated silica domains. The rate of condensation increases following Arrhenius-type rate equation and the composites gelled at higher temperature than the ambient become completely opaque. Following are the representative pictures of ACM–silica and ENR–silica hybrid composites gelled at 100°C (Figure 3.18). The TEOS concentration has been fixed at 45 wt%.

Both these composites were optically transparent when gelled at room temperature. Accelerated gelling of both the phases (organic and inorganic) primarily affects the inorganic phase causing self-aggregation. Improper and poor dispersion of silica at higher gelling temperature provides poorer physico-mechanical properties in both the rubber–silica hybrid systems than from the composites where the gelation occurs naturally.

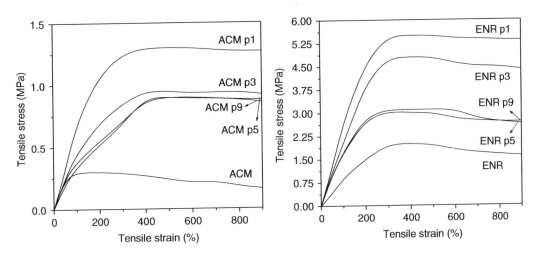

FIGURE 3.17 Tensile stress–strain plots for uncross-linked acrylic rubber (ACM)–silica and epoxidized natural rubber (ENR)–silica hybrid composites synthesized from various pH. Letter "p" in the legends indicates pH and the numbers following the letter are indicative of the pH ranges; e.g., 1 means pH range of 1.0–2.0, similarly 3 means 3.0–4.0. (From Bandyopadhyay, A., De Sarkar, M., and Bhowmick, A.K., *J. Mater. Sci.*, 41, 5981, 2006. Courtesy of Springer.)

3.7 ANALYSIS OF RUBBER–SILICA INTERACTION IN THE NANOCOMPOSITES

Understanding the physics of rubbers and their interaction with solid, elastic filler particles has generated a lot of interest. Reinforcement of rubbers using various grades of carbon black and silica has been extensively studied in the recent years [60]. In general, the rubber–silica interactions within a composite depends on the size and shape of the filler, their volume fraction, and also on their interparticle interaction [61]. Rubber–silica interaction featuring in rubber nanocomposites generates tremendous inquisitiveness simply because of distinctly different filler configuration. Studies on solvent swelling and viscoelastic properties of these hybrid nanocomposites provide an insight into this topic of interest.

Kraus equation and Kraus plots based on swelling data are largely used to explore the rubber–filler interaction in conventional composites [62]. Bandyopadhyay et al. [38] have employed the same equation for understanding the reinforcement behavior in ACM–silica and ENR–silica hybrid

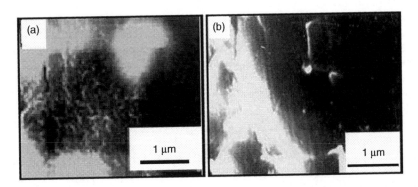

FIGURE 3.18 Scanning electron microscopic (SEM) pictures of (a) acrylic rubber (ACM)–silica and (b) epoxidized natural rubber (ENR)–silica hybrid composites gelled at 100°C. (From Bandyopadhyay, A., De Sarkar, M., and Bhowmick, A.K., *J. Appl. Polym. Sci.*, 95, 1418, 2005 and Bandyopadhyay, A., De Sarkar, M., and Bhowmick, A.K., *J. Mater. Sci.*, 40, 53, 2005. Courtesy of Wiley InterScience and Springer, respectively.)

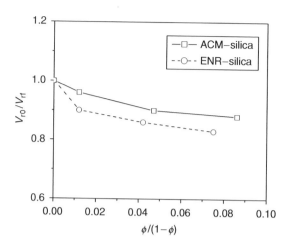

FIGURE 3.19 Comparative Kraus plots of acrylic rubber (ACM)–silica and epoxidized natural rubber (ENR)–silica hybrid nanocomposites at different *in situ* silica concentrations. (From Bandyopadhyay, A., De Sarkar, M., and Bhowmick, A.K., *J. Polym. Sci., Part B: Polym. Phys.*, 43, 2399, 2005. Courtesy of Wiley InterScience.)

nanocomposites. Figure 3.19 shows the comparative Kraus plots [V_{ro}/V_{rf} versus $\phi/(1-\phi)$] of cross-linked ENR–silica and ACM–silica hybrid nanocomposites with various silica concentrations. Here, V_{ro} is the volume fraction of the unfilled rubber in the swollen gel and V_{rf} is the same nominated to filled rubbers. ϕ represents the volume fraction of silica. Both these systems were swollen in THF under ambient conditions.

The swelling of the filled rubber composites are suppressed by the presence of solid fillers. The rubber molecules get adsorbed over the filler surface and alter the molecular weight distribution. The extent of this adsorption is a strong function of rubber–filler interaction. The more negative slope in ENR–silica system than in ACM–silica implies that the ENR molecules undergo stronger immobilization in the swollen gel than ACM molecules. Mukhopadhyay and De [63] have studied numerous rubber–silica composites and have reported that commercial silica in rubber initially shows a positive slope due to vacuole generation and after certain higher-volume fractions turns to negative slope reflecting better rubber–silica interaction. Surprisingly, with nanosilica, the slopes are negative throughout, demonstrating much better interaction in this case. At the molecular level, greater rubber–silica interaction alters the molecular mobility of the rubbers. The relative proportion of the rubber at the vicinity of the interface is expected to be inversely proportional to the size of the inclusion [64].

Swelling is related to the rubber–silica adhesion but the filler structure is of course a significant factor. Spherical nanosilica particles exist both at high and low TEOS concentrations in ENR, while the same do not remain strictly spherical and show strong interparticle interaction (silica aggregation) specially at higher TEOS concentrations for slightly less interactive rubber, ACM. Discrete particles are indicative of rubber-coated silica domains which clearly predominate in ENR–silica. On the other hand, the silica particles do not remain strictly spherical at higher concentration in ACM and hence create lot of free volumes between silica and rubber molecules, where the rubber chains could not penetrate but the smaller solvent molecules could easily do so. This is nicely demonstrated through the model described in Figure 3.20.

The model in Figure 3.20a applies to ACM–silica, while Figure 3.20b suits the ENR–silica system. Kraus constant, C, determined from the slope of the plots in Figure 3.19, quantifies the rubber–silica interaction in these systems. $C_{ACM/silica}$ is 1.85 and $C_{ENR/silica}$ is 2.30 and these values are significantly higher than the reinforcing black-filled rubber composites [65]. Hydrogen-bonded interaction between the SiOH and the vicinal diols in ENR is responsible for this, whereas dipolar interaction between ester and SiOH in ACM–silica only results in weaker adsorption of the rubber over the filler surfaces.

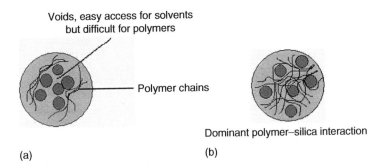

FIGURE 3.20 Occurrence of silica in (a) less interactive and (b) more interactive rubber matrices. (From Bandyopadhyay, A., De Sarkar, M., and Bhowmick, A.K., *J. Polym. Sci., Part B: Polym. Phys.*, 43, 2399, 2005. Courtesy of Wiley InterScience.)

An attempt has also been made for qualitative estimation of the interaction at the rubber–silica interfaces. This has been accomplished by recording the solution viscosity of the precursor sols of these hybrid composites continuously for five days, with an interval of 24 h in the course of *in situ* silica generation. Figure 3.21 shows the result.

The viscosity increases with time due to the generation and dispersion of the nanosilica particles in the rubber matrices. Interestingly, ACM10 and ENR10 show almost similar trends; the others are different. Morphologically, these two compositions are quite similar but they change significantly with silica concentrations: ACM–silica shows more silica aggregation than ENR–silica due to weaker interaction in the former. Moreover, the rise in viscosity is more uniform in ENR–silica than in ACM–silica; ACM50 equilibrates only after 48 h but ENR50 does the same after 96 h of experiment. The latter system contains more rubber-coated silica domains which display uniform rise in viscosity in the course of silica generation, while more interparticle interaction among the silica particles ceases the viscosity enhancement in the ACM–silica system quite early. Also, the net change in viscosity after five days is greater in ENR–silica than in ACM–silica due to the same reasons.

Adsorption of rubber over the nanosilica particles alters the viscoelastic responses. Analysis of dynamic mechanical properties therefore provides a direct clue of the rubber–silica interaction. Figure 3.22 shows the variation in storage modulus (log scale) and tan δ against temperature for ACM–silica, ENR–silica, and *in situ* acrylic copolymer and terpolymer–silica hybrid nanocomposites.

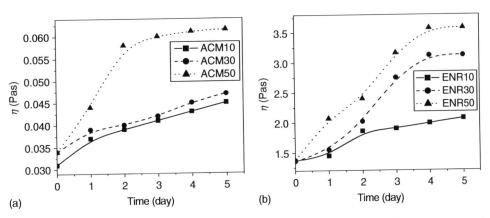

FIGURE 3.21 Plots of shear viscosity versus time for (a) acrylic rubber (ACM)–silica and (b) epoxidized natural rubber (ENR)–silica hybrid nanocomposites in solution at different tetraethoxysilane (TEOS) concentrations, continuously for five days. The numbers in the legends indicate wt% TEOS concentration. (From Bandyopadhyay, A., De Sarkar, M., and Bhowmick, A.K., *J. Polym. Sci., Part B: Polym. Phys.*, 43, 2399, 2005. Courtesy of Wiley InterScience.)

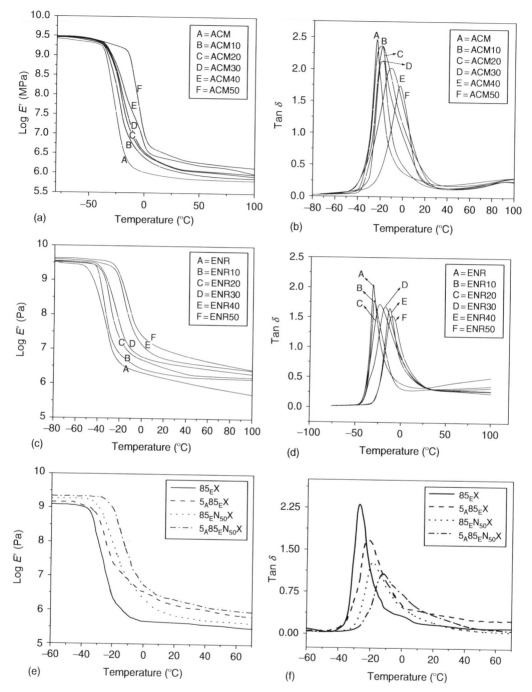

FIGURE 3.22 Plots of (a) storage modulus and (b) tan δ against temperature for acrylic rubber (ACM)–silica, (c) storage modulus and (d) tan δ against temperature for epoxidized natural rubber (ENR)–silica, and (e) storage modulus and (f) tan δ against temperature for *in situ* acrylic copolymer–terpolymer–silica hybrid crosslinked (X) nanocomposites. (From Bandyopadhyay, A., Bhowmick, A.K., and De Sarkar, M., *J. Appl. Polym. Sci.*, 93, 2579, 2004; Bandyopadhyay, A., De Sarkar, M., and Bhowmick, A.K., *Rubber Chem. Technol.*, 77, 830, 2004; and Patel, S., Bandyopadhyay, A., Vijayabaskar, V., and Bhowmick, A.K., *Polymer*, 46, 8079, 2005. Courtesy of Wiley InterScience, ACS Rubber Division, and Elsevier, respectively.)

Mark et al. [66] and Kohjiya et al. [31] have also reported the temperature sweep experiments on rubber–silica hybrids. Interestingly, the modulus values are distinctly higher with more interactive rubber matrices than in the less interactive systems. The moduli in glassy and rubbery regions of the dynamic storage modulus versus temperature plots are distinctively high in those systems. Moreover, the modulus drops in much steeper fashion during glass–rubber transition in the latter than in the former. The modulus is a direct function of rubber–silica interaction and, due to stronger interfaces in ENR–silica and in the terpolymer–silica than in ACM–silica, the former shows fairly high modulus throughout the experiment at a fixed TEOS concentration or on increasing silica percentage. The molecular vibration of the ENR and the terpolymer shows more damping behavior compared to ACM molecules by virtue of stronger rubber–silica interactions. As a result, the loss tangent values are less and the plots become much wider. Greater adsorption of interactive rubber molecules over polar nanosilica particles preceded by greater rubber–silica interaction causes more vibration damping and restricts interchain mobility. Broadening of the loss tangent curves indicates wider molecular weight distribution due to adsorption of rubber molecules over the filler surfaces. The glass–rubber transition temperatures are also shifted to higher values for similar reasons. Figure 3.23 shows the strain sweep experiment results on ACM–silica and ENR–silica hybrid nanocomposites.

The experiments have been carried out at a higher temperature (50°C) so that the rubber molecules remain in a state of high fluidity. The pure rubbers in both the cases do not show any change in modulus on straining, whereas the nanosilica-filled composites record a drop in modulus at higher amplitude of strain and its extent is in direct proportion with the increasing nanosilica content in the rubbers and in reverse proportion with the polarity of the rubbers. Payne had observed this phenomenon way back in 1962 [67] in carbon black-filled rubber composites and it is known by his name as Payne effect. The reason that Payne had suggested was the destructuring of the secondary network of carbon black on straining. But the beauty of these results on nanocomposites is that neither is the overall magnitude of drop so high nor are the silica particles extremely aggregated. Bandyopadhyay et al. [38] have referred to this as the slippage of the rubber molecules from the silica surfaces at higher strain mediated by the stick-slip molecular motions of the rubber on and from the filler surface. The nonlinear viscoelastic response of nanosilica-filled rubbers is therefore quite different from what Payne had suggested. Reducing polarity of the rubbers,

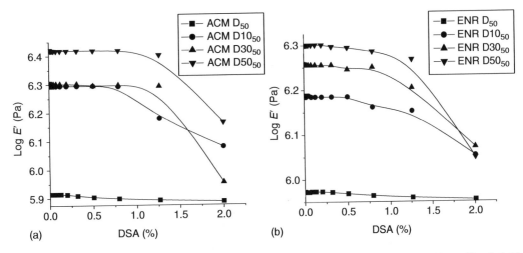

FIGURE 3.23 Plots of storage modulus against variable strain for the cross-linked rubber–silica hybrid nanocomposites: (a) (ACM)–silica and (b) epoxidized natural rubber (ENR)–silica at different tetraethoxysilane (TEOS) concentrations. (From Bandyopadhyay, A., De Sarkar, M., and Bhowmick, A.K., *J. Polym. Sci., Part B: Polym. Phys.*, 43, 2399, 2005. Courtesy of Wiley InterScience.)

the extent of slippage increases simply because of relatively poor rubber–silica adhesion. The present authors [38] have also correlated the drop in dynamic modulus to the volume fraction of the silica by a power law equation and have assigned a new parameter called *rubber–filler attachment parameter*. This has greater values for highly interactive polymer matrices like polyvinyl alcohol (PVA) or ENR and lower values for less interactive matrices like ACM. This parameter has also been compared with that of Kluppel's exponent [68] and has been found to be much smaller than that. This further suggests the existence of unusual filler–rubber network in these nanosilica-filled rubbers.

3.8 STUDIES ON DEGRADATION OF RUBBER NANOCOMPOSITES

Kohjiya et al. [30] have reported the thermal degradation studies on rubber–silica hybrid nanocomposites synthesized from soaking technique. These nanocomposites always record some early weight losses due to retainment of moisture and alcohol (if any). They have also calculated the amount of silica generated in the sol–gel reaction from the TGA residue and found that the experimental weight of silica is always less than the theoretical weight. This is probably due to the incomplete condensation of the SiOH moieties. Bandyopadhyay et al. [69] have calculated the silica from ash content measurements and inferred the same. They have also found that the practical weight of silica is slightly different under variable reaction conditions.

Presence of nanosilica and its interaction with the rubber matrices strongly affect the low and high temperature degradation behaviour of the hybrid nanocomposites. Figure 3.24 shows the post-aging swelling analysis of the cross-linked ACM–silica and ENR–silica hybrid nanocomposites. The data points are collected after aging of the samples at 50°C, 70°C, and 90°C for 72 h.

ACM–silica hybrid nanocomposite does not show any change against the variable aging conditions while a tremendous increase in the gel content is noticed in ENR–silica system on aging at fairly high temperature mainly because of the greater interaction in this case. The pre- and post-aging views of these hybrid composites are presented in Figure 3.25.

The ACM–silica film looks darker after aging than the ENR–silica film. Better dispersion and interaction with the silica provides better aging resistance in the latter than in the former. The observations on the high temperature degradation of these composites are quite similar [69].

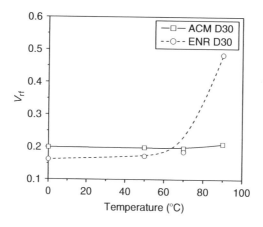

FIGURE 3.24 Plots of volume fraction of the filled rubber in the swollen gel (V_{rf}) against aging temperature for acrylic rubber (ACM)–silica and epoxidized natural rubber (ENR)–silica hybrid nanocomposites. (From Bandyopadhyay, A. and Bhowmick, A.K., *Plastic Rubber Comp. Macromol. Eng.*, 35, 210, 2006. Courtesy of Maney Publishers.)

FIGURE 3.25 Visual appearance of the pre- and post-aged cross-linked rubber–silica hybrid nanocomposites synthesized from 30 wt% tetraethoxysilane (TEOS): (a) acrylic rubber (ACM)–silica and (b) epoxidized natural rubber (ENR)–silica films of average thickness 0.25 mm. (From Bandyopadhyay, A. and Bhowmick, A.K., *Plastic Rubber Comp. Macromol. Eng.*, 35, 210, 2006. Courtesy of Maney Publishers.)

Figure 3.26 demonstrates the TGA and DTG plots of the cross-linked ACM–silica and ENR–silica hybrid nanocomposites at different silica concentrations. The initial weight loss in the vicinity of 100°C due to removal of moisture occurs in both the systems but is slightly greater in ENR–silica than in ACM–silica, which may be due to greater polarity of ENR. The rate of degradation has been improved in both the systems but in the latter the maximum degradation temperature has been shifted by a magnitude of 23°C to the higher side because of the presence of nanosilica. This does not take place in ACM–silica due to relatively poor rubber–silica interphase mixing. Similar observations are also obtained with the rubber grade acrylic copolymers and terpolymer–silica systems.

3.9 RHEOLOGICAL BEHAVIOR OF THE RUBBER–SILICA NANOCOMPOSITES

Understanding the melt rheology of rubber nanocomposites is crucial from the processing perspective. Bandyopadhyay et al. [37] have studied the melt flow behavior of rubber–silica hybrid nanocomposites in a capillary rheometer.

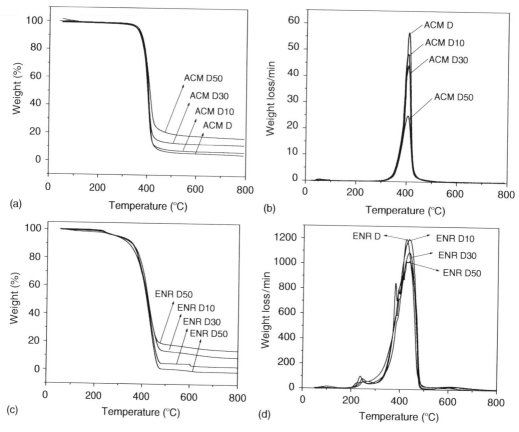

FIGURE 3.26 TGA (under nitrogen) and DTG plots of cross-linked rubber–silica hybrid nanocomposites: (a) and (b) TGA and DTG plots of acrylic rubber (ACM)–silica and (c) and (d) TGA and DTG plots of epoxidized natural rubber (ENR)–silica. (From Bandyopadhyay, A. and Bhowmick, A.K., *Plastic Rubber Comp. Macromol. Eng.*, 35, 210, 2006. Courtesy of Maney Publishers.)

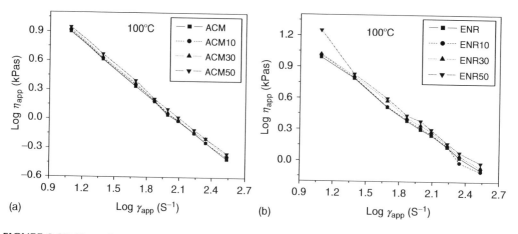

FIGURE 3.27 Plots of shear viscosity against shear rate at 100°C for (a) acrylic rubber (ACM)–silica and (b) epoxidized natural rubber (ENR)–silica hybrid nanocomposites at various silica concentrations. (From Bandyopadhyay, A., De Sarkar, M., and Bhowmick, A.K., *Rubber Chem. Technol.*, 78, 806, 2005. Courtesy of ACS Rubber Division.)

Figure 3.27 shows the log–log plots of melt shear viscosity against shear rate for these nanocomposites. Both these systems are pseudoplastic in nature and show quite a steady decrease in viscosity on increasing shear rate. Interestingly, the rise in viscosity due to increased concentration of nanosilica on both the occasions is not great. This is assigned to better dispersion of the filler particles within the rubber matrix. The nanosilica particles act as *dead rocks* over which the rubber molecules slip during melt flow. The extrudate roughness has been found to depend on the state of rubber–silica adhesion; ENR–silica is more interactive than ACM–silica and gives smoother extrudates. Greater rubber–filler interaction enhances viscoelastic response and decreases extrusion swelling in these hybrid nanocomposites. The rise in activation energy on increasing nanosilica concentrations has been found to be marginal.

3.10   RUBBER–SILICA NANOCOMPOSITE ADHESIVES

Patel et al. [70] in a recent publication have explored the adhesive action of the rubber–silica hybrid nanocomposites on different substrates. The rubber–silica hybrid nanocomposites are synthesized through *in situ* silica formation from TEOS in strong acidic pH within acrylic copolymer (EA–BA) and terpolymer (EA–BA–AA) matrices. The transparent nanocomposites have been applied in between the aluminum (Al), wood (W), and biaxially oriented polypropylene (PP) sheets separately and have been tested for peel strength, lap shear strength, and static holding power of the adhesive joints.

The nanocomposites show superior peel and lap shear strengths than their virgin counterparts. Due to greater polarity, the improvement in the strength of the joints is significantly high with high-energy substrates like Al, but little low with nonpolar, low-energy substrates like PP. Figure 3.28 shows the static holding power test results on the terpolymer–silica hybrid nanocomposites applied in between Al, W, and PP substrates. The cohesive failure behavior of the adhesives is almost similar in polar substrates like Al and W, but much lower in PP (nonpolar).

The copolymer–silica adhesives also follow a similar trend but fail much earlier than the terpolymer-based adhesives. This is because of two factors: (1) the increase in the inherent strength of the adhesive due to more favorable terpolymer rubber–silica interaction and (2) chemisorption in much higher magnitude between the polar substrates and the nanocomposites.

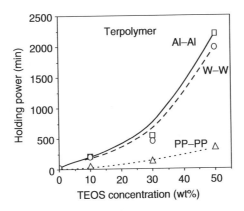

FIGURE 3.28 Static holding power test results of acrylic terpolymer–silica nanocomposite adhesives carried out on Al, W, and PP substrates. (From Patel, S., Bandyopadhyay, A., Ganguly, A., and Bhowmick, A.K., *J. Adhes. Sci. Technol.*, 20, 371, 2006. Courtesy of VSP–Brill Academic Publishers.)

3.10.1 RUBBER–SILICA SOL–GEL HYBRIDS AND HEALTH ISSUE

Exposure to fume and the precipitated silica (amorphous silica) causes fatal bronchial response as these are respirable. Repeated exposure dries the skin and causes mechanical irritation to eyes. Sol–gel silica, as an alternative to this, ceases the floating problem and avoids unnecessary contamination.

3.11 POLYHEDRAL OLIGOMERIC SILSESQUIOXANE (POSS)

POSS are cage-like molecules of silicon and oxygen with similarities to both silica and silicone. POSS molecules can be thought of as the smallest particles of silica possible (size range 0.7–30 nm). However, unlike silica or modified clays, each POSS molecule contains covalently bonded reactive functionalities suitable for polymerization or grafting. The nonreactive organic functionalities in POSS are aided for solubility and compatibility of the POSS segments with the various polymer systems. The chemical diversity of POSS molecules is very broad and a large number of POSS monomers and polymers are currently available or undergoing development. Figure 3.29 shows the structural composition of a POSS molecule.

POSS can be added nearly to all types of polymers. These are physically large with respect to polymer dimensions and are equivalent in size to most polymer segments and coils. This has been the latest commercially available nanoadditive for the polymers. Hybrid Plastics in the United States are the global supplier of POSS.

Literatures are available on POSS–polymer composites synthesized from different thermoplastics [71–74]. These composites are lightweight and show good fire retardancy, thermal stability, and mechanical reinforcement. Literatures on POSS–rubber composites are yet to come in a big way.

Recently Sahoo and Bhowmick [75] synthesized hydroxyl-terminated POSS in their laboratory starting from (3-aminopropyl) triethoxysilane (APS) and phenylglycidylether (PGE) and used it as a curative in carboxylated nitrile rubber (XNBR). This has been a newer class of material where the nanofiller simultaneously cures the rubber and promotes solvent resistance, as well as mechanical and dynamic mechanical properties. Table 3.3 illustrates some of these findings.

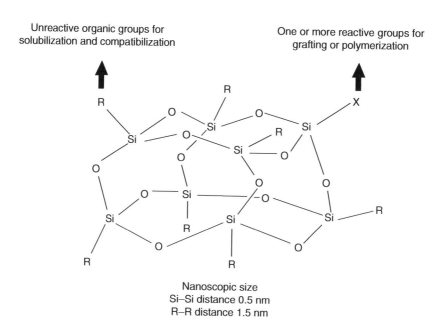

FIGURE 3.29 Structure of a POSS.

TABLE 3.3
Comparative Mechanical Properties between Neat XNBR and POSS-Treated XNBR

Mechanical Properties	Neat XNBR	POSS-Treated XNBR
Tensile strength (MPa)	0.1	0.6
Elongation at break (%)	>1500	2100
Modulus at 100% elongation	0.09	0.16
Modulus at 200% elongation	0.07	0.16
Modulus at 300% elongation	0.04	0.15

3.12 CONCLUSIONS

This chapter represents a comprehensive and in-depth analysis on synthesis and characterization of various rubber–silica hybrid nanocomposites. Sol–gel synthesis, although popular for ceramers, could be judiciously used for generating rubber–*in situ* silica hybrid nanocomposites. This technique, in comparison to others, has been so far the best method and is extensively dealt with in this chapter. The rubber–silica nanocomposites could be framed either using a rubber solution (sol–gel solution method) or applying *in situ* technique (sol–gel *in situ* polymerization method) with silica precursor mix or even soaking the cured rubber sheet in the precursor solution. Several factors like nature of polymer (polarity), nature of solvent (polarity and water miscibility), nature of catalyst and medium pH, alkoxysilane/H_2O mole ratio, and gelling temperature have been found to be the deciding factors in framing the rubber–silica hybrid composites. Silica particles are found to exist in nanodimensions (1–100 nm) within the rubber matrices and their shape and size (secondary structure formation via aggregation) depend on the reaction parameters just mentioned; slight deviation causes aggregation of the filler. Finer dispersion of silica in the rubber produces superior mechanical, dynamic mechanical, solvent resistance and adhesion properties, while when aggregated these properties are inferior. The *in situ*-generated silica, although little costly, beneficiates the mechanical properties without the use of any coupling agent and also resists health problems which are usually caused by bronchial infection of floating silica particles. The commercially available nanosilica (POSS) has been used to form polymer composites with superior mechanical, thermal, and cross-linkability. Recently, this material has been successfully synthesized in the laboratory and applied in rubber for the first time with multipurpose activity.

REFERENCES

1. Roy, R. *Mater. Res. Soc. Symp. Proc.* 286, 241, 1993.
2. Calvert, P.D. *Mater. Res. Soc. Bull.* 17, 37, 1992.
3. Pinnavaia, T.J. and Beall, G.W. *Polymer-Clay Nanocomposites*, Wiley: United Kingdom, 2000.
4. Naka, K. and Chujo, Y. *Chem. Mater.* 13, 3245, 2001.
5. Brinker, C.J. and Scherer, G.W. *Sol–Gel Science: The Physics and Chemistry of Sol–Gel Processing*, Academic Press: California, 1990.
6. Wilkes, G.L., Huang, H.-H., and Glaser, R.H. *Polym. Prepr.* 431, 2001.
7. Wen, J. and Wilkes, G.L. *Chem. Mater.* 8, 1667, 1996.
8. Nell, J.L., Wilkes, G.L., and Mohanty, D.K., *J. Appl. Polym. Sci.* 40, 1177, 1990.
9. Lee, R.-H., Hsiue, G.-H., and Jeng, R.-J. *J. Appl. Polym. Sci.* 79, 1852, 2001.
10. Grady, B.P., Start, P.R., and Mauritz, K.A. *J. Polym. Sci., Part B: Polym. Phys.* 39, 197, 2001.
11. Gao, Y. *A study of a polyethylene ionomer, its nanocomposites and hybrids*, PhD Thesis, The University of South Australia, April 2003.
12. Wen, J. and Wilkes, G.L. *Chem. Mater.* 8, 1667, 1996.

13. Mark, J.E., Jiang, C.Y., and Tang, M.Y. *Macromolecules* 17, 2613, 1984.
14. Xu, J., Pang, W., and Shi, W. *Thin Solid Films* 514, 69, 2006.
15. Blanc, D., Zhang, W., Massard, C., and Mugnier, J. *Opt. Mater.* 28, 331, 2006.
16. Shao, P.L., Mauritz, K.A., and Moore, R.B. *Chem. Mater.* 7, 192, 1995.
17. Shao, P.L., Mauritz, K.A., and Moore, R.B. *J. Polym. Sci., Part B: Polym. Phys.* 34, 873, 1996.
18. Svergun, D.I., Kozin, M.B., Konarev, P.V., Shtykova, E.V., Volkov, V.V., Chernyshov, D.M., Valetsky, P.M., and Bronstein, L.M. *Chem. Mater.* 12, 3552, 2000.
19. Carotenuto, G. and Nicolais, L. *Composites: Part B* 35, 385, 2004.
20. Abe, Y. and Misono, T. *J. Polym. Sci.* 21, 41, 1983.
21. Gunji, T., Nagao, Y., Misono, T., and Abe, Y. *J. Polm. Sci., Part A: Polym. Chem.* 30, 1779, 1992.
22. Wang, H., Zhong, W., Xu, P., Du, Q. *Composites Part A* 36, 909, 2005.
23. Wang, H., Zhong, W., Du, Q., Yang, Y., Okamoto, H., and Inoue, S. *Polym. Bull.* 51, 63, 2003.
24. Zhao, Q., Chen, W., and Zhu, Q. *Mater. Lett.* 57, 3606, 2003.
25. Pozzo, D.C. and Walker, L.M. *Colloids Surf. A* 294, 117, 2007.
26. Hoebbel, D., Endres, K., Reinert, T., and Pitsch, I. *J. Non-Cryst. Solids* 176, 179, 1994.
27. Walldal, C., Wall, S., and Biddle, D. *Colloids Surf. A* 131, 203, 1998.
28. Takeuchi, T., Hu, W., Haraguchi, H., and Ishii, D. *J. Chromatogr.* 51, 257, 1990.
29. Hashim, A.S., Kawabata, N., and Kohjiya, S. *J. Sol–Gel Sci. Tech.* 5, 211, 1995.
30. Ikeda, Y., Tanaka, A., and Kohjiya, S. *J. Mater. Chem.* 7, 1497, 1997.
31. Hashim, A.S., Ikeda, Y., and Kohjiya, S. *Polym. Int.* 38, 111, 1995.
32. Bandyopadhyay, A., Bhowmick, A.K., and De Sarkar, M. *J. Appl. Polym. Sci.* 93, 2579, 2004.
33. Bandyopadhyay, A., De Sarkar, M., and Bhowmick, A.K. *Rubber Chem. Technol.* 77, 830, 2004.
34. Bandyopadhyay, A., De Sarkar, M., and Bhowmick, A.K. *J. Appl. Polym. Sci.* 95, 1418, 2005.
35. Bandyopadhyay, A., De Sarkar, M., and Bhowmick, A.K. *J. Mater. Sci.* 40, 53, 2005.
36. Bandyopadhyay, A., De Sarkar, M., and Bhowmick, A.K. *J. Mater. Sci.* 41, 5981, 2006.
37. Bandyopadhyay, A., De Sarkar, M., and Bhowmick, A.K. *Rubber Chem. Technol.* 78, 806, 2005.
38. Bandyopadhyay, A., De Sarkar, M., and Bhowmick, A.K. *J. Polym. Sci. Part B: Polym. Phys.* 43, 2399, 2005.
39. Garcia, N., Corrales, T., Guzman, J., and Teimblo, P. *Polym. Degrad. Stab.* 92, 635, 2007.
40. Zhang, W., Blackburn, R.S., and Dehghani-Sanij, A.A. *Scripta Mater.* 56, 581, 2007.
41. Huang, H., You, B., Zhou, S., and Wu, L. *J. Colloid Interface Sci.* (available online), 2007, accessed on 20.4.07.
42. Johnsen, B.B., Kinloch, A.J., Mohammed, R.D., Taylor, A.C., and Sprenger, S. *Polymer* 48, 530, 2007.
43. Liu, Y. and Kontopoulou, M. *Polymer* 47, 7731, 2006.
44. Obberdisse, J., Harrak, A.E., Carrot, G., Jestin, J., and Boue, F. *Polymer* 46, 6695, 2005.
45. Vega-Bandrit, J., Sibaja-Ballestero, M., Vazquez, P., Torregrosa-Macia, R., and Martin-Martinez, J.M. *Int J. Adhes. Adhes.* 27, 469, 2007.
46. Huang, H.-H., Orler, B., and Wilkes, G.L. *Macromolecules* 20, 1322, 1987.
47. Liu, M., Zeng, Z., and Fang, H. *J. Chromatogr.* 1076, 16, 2005.
48. Siuzdak, D.A. and Mauritz, K.A. *J. Polym. Sci., Part B: Polym. Phys.* 37, 143, 1999.
49. Siuzdak, D.A. and Mauritz, K.A. *Polym. Prepr.* 245, 1997.
50. Siuzdak, D.A., Start, P.R., and Mauritz, K.A. *J. Appl. Polym. Sci.* 77, 2832, 2000.
51. Patel, S., Bandyopadhyay, A., Vijayabaskar, V., and Bhowmick, A.K. *Polymer* 46, 8079, 2005.
52. Patel, S., Bandyopadhyay, A., Vijayabaskar, V., and Bhowmick, A.K. *J. Mater. Sci.* 41, 926, 2006.
53. Zoppi, R.A., Yoshida, I.V.P., and Nunes, S.P. *Polymer* 39, 1309, 1997.
54. Rajan, G.S. and Sur, G.S. *J. Polym. Sci., Part B: Polym. Phys.* 41, 1897, 2003.
55. Brinke, J.W., Litvinov, V.M., Wijnhoven, J.E.G.J., and Noordermeer, J.W.M. *Macromolecules* 35, 10026, 2002.
56. Tian, D., Blacher, S., and Jerome, R. *Polymer* 40, 951, 1999.
57. Sengupta, R., Bandyopadhyay, A., Sabharwal, S., Chaki, T., and Bhowmick, A.K. *Polymer* 46, 3343, 2005.
58. Iler, R.K. *The Chemistry of Silica*, Wiley: New York, 1979.
59. Landry, C.J.T., Coltrain, B.K., and Brady, B.K. *Polymer* 33, 1486, 1992.
60. Heinrich, G., Kluppel, M., and Vilgis, T.A. *Curr. Opin. Solid State Mater.* 6, 195, 2002.
61. Sarvestani, A.S. and Picu, C.R. *Polymer* 45, 7779, 2004.

62. Shanmugharaj, A.M. and Bhowmick, A.K. *Rubber Chem. Technol.* 76, 299, 2003.
63. Mukhopadhyay, R. and De, S.K. *Rubber Chem. Technol.* 52, 263, 1979.
64. Berriot J., Martin, F., Montes, H., Monneric, L., and Sotta, P. *Polymer (Guildford)* 44, 1437, 2003.
65. Kraus, G. *J. Appl. Polym. Sci.* 7, 861, 1963.
66. Wen, J. and Mark, J.E. *J. Appl. Polym. Sci.* 58, 1135, 1995.
67. Payne, A.R. *J. Polym. Sci.* 6, 57, 1962.
68. Kluppel, M. and Heinrich, G. *Rubber Chem. Technol.* 68, 623, 1995.
69. Bandyopadhyay, A. and Bhowmick, A.K. *Plastic Rubber Comp. Macromol. Eng.* 35, 210, 2006.
70. Patel, S., Bandyopadhyay, A., Ganguly, A., and Bhowmick, A.K. *J. Adhes. Sci. Technol.* 20, 371, 2006.
71. Kannan, R.Y, Salacinski, H.J., Odlyha, M., Butler, P.E., and Seifalian, A.M. *Biomaterials*, 27, 1971, 2006.
72. Chen, R., Feng, W., Zhu, S., Botton, G., Ong, B., and Wu, Y. *Polymer*, 47, 1119, 2006.
73. Ni, Y., Zheng, S., and Nie, K. *Polymer*, 45, 5557, 2004.
74. Joshi, M. and Butola, B.S. *Polymer*, 45, 4953, 2004.
75. Sahoo, S. and Bhowmick, A.K. *Rubber Chem. Technol.* 80, 826, 2007.

4 Rubber Nanocomposites Based on Miscellaneous Nanofillers

Anirban Ganguly, Jinu Jacob George, Sritama Kar,
Abhijit Bandyopadhyay, and Anil K. Bhowmick

CONTENTS

4.1   INTRODUCTION

Nanoscale technology is a suite of techniques used to manipulate matter at the scale of atoms and molecules. "Nano" is a measurement—not an object. A "nanometer" (nm) equals one-billionth of a meter. The key to understanding the unique power and potential of nanotech is that at the nanoscale (below about 100 nm), a material's property can change dramatically; these unexpected changes are often called "quantum effects." With only a reduction in size and no change in the substance itself, materials can exhibit new properties such as electrical conductivity, insulating behavior, elasticity, greater strength, different color, and greater reactivity—characteristics that the very same substances do not exhibit at the micro- or macroscale. For example,

1. Carbon in the form of graphite is soft and malleable; at the nanoscale, carbon can be stronger than steel and is six times lighter.
2. Zinc oxide is usually white and opaque; at the nanoscale, it becomes transparent.
3. Aluminum—the material for soft drink cans—can spontaneously combust at the nanoscale and could be used as rocket fuel.

Effects of nanoclay and silica in rubber matrices have been discussed in earlier chapters. Recently, several other nanofillers have been investigated and have shown a lot of promise. All these fillers have not been investigated on rubbers extensively, although they have great potential to do so in the days to come. In this chapter, we have compiled the current research on rubber nanocomposites having nanofillers other than nanoclay and nanosilica. Further, this chapter provides a snapshot of the current experimental and theoretical tools being used to advance our understanding of rubber nanocomposites.

4.2 DIFFERENT NANOFILLERS

4.2.1 CARBON-BASED NANOFILLERS

The carbon-based nanofillers are mainly layered graphite, nanotube, and nanofibers. Graphite is an allotrope of carbon, the structure of which consists of graphene layers stacked along the c-axis in a staggered array [1]. Figure 4.1 shows the layered structure of graphite flakes.

High crystallinity of graphite is disadvantageous in forming the nanocomposites with polymers, as the giant polymer molecules do not find spaces within the graphene sheets. This has been overcome by modifying the graphite flakes with several oxidizing agents [2]. The effective method of preparing the polymer–expanded graphite composite is by rapidly heating the pretreated (oxidized) graphite to a high temperature. The exfoliation of graphite is a process in which graphite expands by up to hundreds of times along the c-axis, resulting in a puffed-up material with a low density and a high temperature resistance [3]. The graphite thus resulted contains nanodimensional flakes (expanded graphite) providing greater surface/volume ratio for interaction with suitable rubber matrices. Recently, George et al. [4] have synthesized ethylene–vinyl acetate (EVA) (having 60 wt% vinyl acetate content)–graphite nanocomposites using both tailor-made and commercially available expanded graphites. Figure 4.2 demonstrates the TEM image of EVA-expanded graphite nanocomposite containing 4 wt% expanded graphite [5].

In both the occasions, well-dispersed, fine graphite flakes are visible within the EVA matrices and this results in superior mechanical, dynamic mechanical, and processibility characteristics over the systems having natural graphite. An example is given in Figure 4.3 [5].

A carbon nanotube (CNT) is a hexagonal network of carbon atoms rolled up into a seamless, hollow cylinder, with each end capped with half a fullerene molecule [6]. Although similar in chemical composition to graphite, CNTs are highly isotropic, and it is this topology that distinguishes them from other carbon structures and gives them their unique properties. There are many possibilities for rolling a slice of graphene into a seamless cylinder (Figure 4.4), because when rolled into a nanotube, the hexagons may spiral around the cylinder, giving rise to "chirality," a twist that determines whether the CNT behaves like a metal or a semiconductor. The diameter, chirality, and form of the nanotube determine its properties [6].

CNTs can be single-walled or multiwalled depending on the number of graphitic cylinders with which they are formed. The single-walled carbon nanotubes (SWNT) and multiwalled carbon nanotubes (MWNT) generally exist with diameter of 1–2 and 10–40 nm, respectively, and length of few micrometers. Functionalization of CNT has been found as an important aspect in developing rubber–CNT composites for better physico-mechanical properties. Many references are available on this topic [7–12]. CNTs exhibit unique mechanical, electronic, and magnetic properties, which have caused them to be widely studied [13–15]. CNTs are probably the strongest substances that

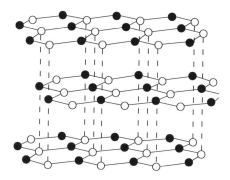

FIGURE 4.1 3D structure of graphene layers in graphite. (From www.benbest.com/cryonics/lessons.html, accessed May 21, 2007.)

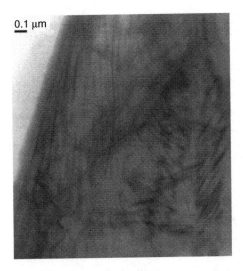

FIGURE 4.2 Transmission electron microscopic (TEM) image of ethylene–vinyl acetate (EVA)-expanded graphite (EG) (4 wt%) nanocomposites. (From George, J.J. and Bhowmick, A.K., *J. Mater. Sci.*, 43, 702, 2008. Courtesy of Springer.)

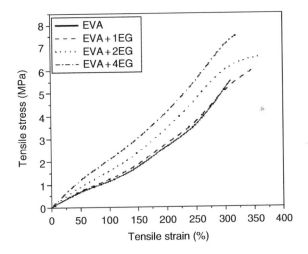

FIGURE 4.3 Stress–strain characteristics of ethylene–vinyl acetate (EVA)-expanded graphite (EG) nanocomposites at different loadings. (From George, J.J. and Bhowmick, A.K., *J. Mater. Sci.*, 43, 702, 2008. Courtesy of Springer.)

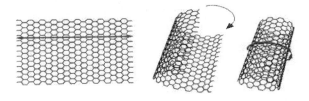

FIGURE 4.4 Graphene sheet and single-walled carbon nanotube.

will ever exist, with a tensile strength greater than steel, but only one-sixth the weight of steel [16]. Iijima first discovered CNTs using arc discharge method [17,18]. Following this discovery, a number of scientific research projects have been initiated and a variety of methods have been used to synthesize CNTs, namely, arc discharge, laser vaporization [19], and catalytic chemical vapor deposition of hydrocarbons [20–22]. Nanotubes are strong and resilient structures that can be buckled and stretched into shapes without major structural failure [23,24]. The Young's modulus and tensile strength rival that of diamond (1 TPa and ~200 GPa respectively) [25].

This fantastic property of mechanical strength allows these structures to be used as possible reinforcing materials. Just like current carbon fiber technology, this nanotube reinforcement would allow very strong and light materials to be produced. These properties of CNTs attracted the attention of scientists all over the world because of their high capability for absorbing the load which is applied to nanocomposites [23–25].

Chemical pretreatments with amines, silanes, or addition of dispersants improve physical disaggregation of CNTs and help in better dispersion of the same in rubber matrices. Natural rubber (NR), ethylene–propylene–diene–methylene rubber, butyl rubber, EVA, etc. have been used as the rubber matrices so far. The resultant nanocomposites exhibit superiority in mechanical, thermal, flame retardancy, and processibility. George et al. [26] studied the effect of functionalized and unfunctionalized MWNT on various properties of high vinyl acetate (50 wt%) containing EVA–MWNT composites. Figure 4.5 displays the TEM image of functionalized nanotube-reinforced EVA nanocomposite.

Dynamic mechanical properties of the nanocomposites are shown in Figure 4.6. There is 10% improvement of the storage modulus at 20°C by incorporating only 4 wt% of the nanotube.

Better results achieved with functionalized CNTs are probably due to better interaction between the polar rubber matrix and the polar groups chemically bonded to the filler surface. Atieh et al. [27] designed and fabricated a floating catalyst chemical vapor deposition (FC-CVD) method to produce high-quality CNTs. MWNTs were used to prepare NR nanocomposites. They achieved nanostructures in NR-MWNT nanocomposites by incorporating CNTs in a polymer solution and subsequently evaporating off the solvent. Using this technique, nanotubes can be dispersed homogeneously in the NR matrix in an attempt to increase the mechanical properties of

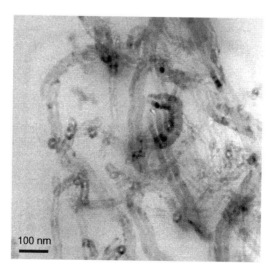

FIGURE 4.5 Transmission electron microscopic (TEM) image of amine-modified nanotube (ANT)-reinforced ethylene–vinyl acetate (EVA) nanocomposite. (From George, J.J., Sengupta, R., and Bhowmick, A.K., *J. Nanosci. Nanotechnol.*, 8, 1, 2007. Courtesy of American Scientific Publishers.)

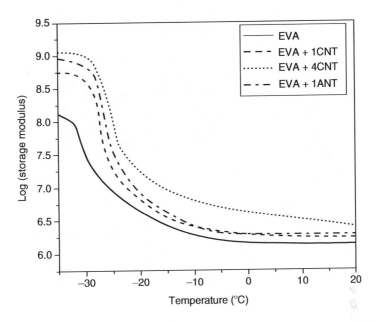

FIGURE 4.6 Variation of storage modulus against temperature for ethylene–vinyl acetate (EVA) nanocomposites having different loadings of carbon nanotube (CNT) and ANT. (From George, J.J., Sengupta, R., and Bhowmick, A.K., *J. Nanosci. Nanotechnol.*, 8, 1, 2007. Courtesy of American Scientific Publishers.)

these nanocomposites. Mechanical test results show an increase in the initial modulus for up to 12-fold in relation to neat NR. The composites containing 10–20 phr uniformly dispersed CNT exhibited significant drop of volume resistivity [28]. The effect of CNTs on the mechanical and electrical properties of ethylene–propylene–diene monomer (EPDM) rubber in comparison to that of high-abrasion furnace (HAF) carbon black was studied by Ying et al. [29].

Vapor-grown carbon nanofibers (VGCNFs) have a graphitic structure similar to that of CNTs, with typical diameters of 50–200 nm. The inner diameter is 30–90 nm and the length is in the range of 50–100 μm, so that the aspect ratios are in the 100–500 range. This high aspect ratio leads to higher mechanical properties of the VGCFs.

4.2.2 METAL OXIDES, HYDROXIDES, AND CARBONATES

Recently, several metal oxides apart from silica have been investigated and reported for rubber-based nanocomposites. Some important and commercially meaningful oxides used in rubber are zinc oxide (ZnO), magnesium hydroxide (MH), calcium carbonate, zirconate, iron oxide, etc.

ZnO has been an important semiconductor material because of its wide applications in preparing solar cells [30], gas sensors [31,32], catalysts [33], and electrical and optical devices [34]. ZnO has been an important material for the rubber industry since long. The first published reference to this application was patented in 1849. It is used both as curing agent and cure activator for natural and synthetic rubbers. ZnO nanoparticles can be synthesized by numerous methods like high temperature solid vapor deposition [35], solid phase methods [36], colloidal chemistry techniques [37], sol–gel method [38], synthesis in reversed micelles [39], and precipitation method [40]. Sahoo and Bhowmick [41] have reported nano-ZnO synthesis from precipitative double decomposition reaction of zinc nitrate and ammonium carbonate in water followed by calcination of zinc carbonate. Figure 4.7 shows the TEM image of this synthesized nano-ZnO with average particle diameter of 50 nm.

FIGURE 4.7 Transmission electron microscopic (TEM) image of nano-zinc oxide (ZnO). (From Sahoo, S. and Bhowmick, A.K. *J. Appl. Polym. Sci.*, 106, 3077, 2007. Courtesy of Wiley InterScience.)

ZnO nanoparticles possess greater surface/volume ratio. When used in carboxylated nitrile rubber as curative, ZnO nanoparticles show excellent mechanical and dynamic mechanical properties [41]. The ultimate tensile strength increases from 6.8 MPa in ordinary rubber grade ZnO-carboxylated nitrile rubber system to 14.9 MPa in nanosized ZnO-carboxylated nitrile rubber without sacrificing the elongation at failure values. Table 4.1 compares these mechanical properties of ordinary and nano-ZnO-carboxylated nitrile rubbers, where the latter system is superior due to more rubber–ZnO interaction at the nanolevel.

Better cross-linking with the latter also improves post T_g viscoelastic responses of the rubber vulcanizates. Similar effect has also been observed with polychloroprene as investigated by Sahoo and Bhowmick [41]. Figure 4.8 represents the comparative tensile stress–strain behavior of polychloroprene rubber (CR) vulcanizates, highlighting superiority of the nanosized ZnO over conventional rubber grade ZnO [41].

Different polymer–nano-ZnO hybrid systems based on epoxy12 [42], poly(styrene-co-acrylic acid) [43], polyurethane [44], etc. have been reported by several other researchers.

TABLE 4.1

Comparative Mechanical Properties Data on Ordinary ZnO and Nano-ZnO-Filled Carboxylated Nitrile Rubber Systems

Physical Properties	With Ordinary ZnO	With Nano-ZnO
Tensile strength (MPa)	6.8	14.9
Elongation at break (%)	680	800
Tensile modulus (100%) (MPa)	1.3	1.7

Source: Sahoo, S. and Bhowmick, A.K., *J. Appl. Polym. Sci.*, 106, 3077, 2007. Courtesy of Wiley InterScience.

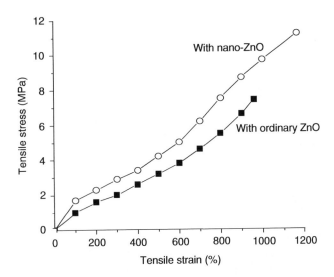

FIGURE 4.8 Comparative tensile stress–strain plot of polychloroprene–ordinary zinc oxide (ZnO) and polychloroprene–nano-ZnO system. (From Sahoo, S., Kar, S., Ganguly, A., Maiti, M., and Bhowmick, A.K., *Polym. Polym. Compos.*, 2007 (in press), Courtesy of Smithers Rapra Technology Ltd.)

One of the most important phenomena in material science is the reinforcement of rubber by rigid entities, such as carbon black, clays, silicates, calcium carbonate, zinc oxide, MH, and metal oxide [45–47]. Thus, these fillers or reinforcement aids are added to rubber formulations to optimize properties that meet a given service application or sets of performance parameters [48–53]. Although the original purpose is to lower the cost of the molding compounds, prime importance is now attached to the selective active fillers and their quantity that produce specific improvements in rubber physical properties.

The synthesis of MH nanoparticles with special morphologies has recently attracted much attention because of the expectation of novel properties and applications in electronics, catalysis, ceramics, nanostructured composites, or halogen-free flame retardants [54]. For example, MH nanorods could be used as precursors for the synthesis of MgO nanorods [55], which are expected to have novel mechanical, catalytic, and electronic properties due to their extremely small size, large anisotropy, and perfect crystallinity [56]. MH nanoneedles and nanolamellas could be good candidates for functional polymeric composites and fiber hybrid materials as reinforcing agents or halogen-free flame retardants [57,58]. However, the synthesis of MH nanoparticles with special morphological structure is difficult, because nanoparticles may adopt various shapes [58]. Some research effort has been devoted to the synthesis of MH nanoparticles with special morphology. Recently, Li et al. [59] proposed a route to synthesize MH nanorods based on the hydrothermal reaction of pure magnesium powder in an autoclave at higher temperature using long reaction time and low reactant concentration. Similarly, Yu et al. [60] produced MH nanoplates by the hydrothermal reaction of MgO crystals. Henrist and coworkers [61] produced nanoplates by a wet precipitation method using dilute aqueous solution. All these chemical syntheses were, however, characterized by problems such as being limited to one or two morphological preparations, relatively low yields, and being economically unacceptable. A simple fabrication method of MH nanoparticles with several morphological structures is still a challenge. Lv et al. [62] reported a relatively simple, effective, and size-controllable solution precipitation method to prepare well-dispersed MH nanocrystals with different morphologies by using gelatin and lauryl sodium sulfate as complex water-soluble polymer dispersants. Three morphological structures with needle-, lamellar-, and rodlike nanocrystals were obtained by controlling the reaction temperature, alkaline-injection rate, and reactant concentration, respectively.

TABLE 4.2
Effect of Nano-MgO on Thermal Properties of CR

Sample	T_i (°C)	T_{max} (°C)	Residue (%)	Maximum Rate of Degradation (%/°C)
CR + ZnO + MgO (conventional)	352	353	28	6.3
CR + ZnO + MgO (nanoparticles)	367	381	27	1.7

The results obtained by Kuila et al. and Acharya et al. [63,64] from the EVA elastomer blended with lamellar-like Mg-Al layered double hydroxide (LDH) nanoparticles demonstrate that MH nanocrystals possess higher flame-retardant efficiency and mechanical reinforcing effect by comparison with common micrometer grade MH particles. Kar and Bhowmick [65] have developed MgO nanoparticles and have investigated their effect as cure activator for halogenated rubber. The results as shown in Table 4.2 are promising.

CaCO$_3$ particles are generally supplied as agglomerates and during processing they are broken and dispersed into primary particles. Large particle–particle interactions result in inhomogeneous distribution of the filler, processing problems, poor appearance, and inferior properties. This fact may emphasize the importance of homogeneity where the increasing amount of aggregates leads to a decrease of tensile properties of the rubber composites [62,66,67]. It is well known that the overall performance of composites is achieved by the addition of a low-molecular weight organic compound, such as stearic acid [68], resulting in a better dispersion of the particles in the rubber matrix.

A special category of precipitated calcium carbonates (PCCs) is the ultrafine PCC or, as it is now more commonly known, the nano PCC. Specialty Minerals Inc. (SMI) has manufactured nano PCCs for more than 25 years. SMI's ultrafine PCCs range in particle size from 0.06 μm or 60 nm to 0.15 μm or 150 nm in median particle size. This is an order of magnitude smaller than the so-called ultrafine ground calcium carbonates (GCCs), which are typically 0.7 μm. SMI's PCC with tightly controlled precipitation process results in ultrafines that are uniform in shape, size, and particle size distribution. With these extremely small particles, true thixotropic structure can be built in a sealant or other moderately to highly filled product in which control of viscosity, sag, slump, and other rheological properties are needed. These ultrafine PCC particles also act as semireinforcing filler, for strong physical performance. PVC plastisols, urethanes, silicones, polysulfides, and silylated polyethers are some of the types of high-performance, long-lived automotive and construction sealants that use nano PCCs.

When formulating with a nano PCC, replacing a larger-sized GCC, the amount of expensive acrylic or chlorinated polyethylene (CPE) impact modifier used can be substantially reduced, saving money. Nano PCCs also give the highest gloss and best surface finish to PVC window profile extrusions. Nano PCCs thicken PVC plastisol silk screen inks. Gravure inks need very low-abrasion fillers. Small-sized PCC particles are excellent here. Rheology and filling are also the reasons to use ultrafine PCCs in epoxy and other adhesives as well as in unsaturated polyester gel coats. Other uses of PCC include high-strength reinforcement of rubber, detackification, and mold release for thin-gauge surgical and other medical gloves, nucleating agent in emulsion polymerization, carrier for catalysts, peroxides, fragrances, and powder flow control additive for acrylic modifiers and other sticky products.

Recent development of rubber nanocomposites by other nanofillers includes piezo-rubber application by incorporating lead–zirconate by Tandon et al. [69], Fe-containing silicone rubber by Yurkov et al. [70], and crab shell whisker-reinforced natural rubber nanocomposites by Nair and Durfreshe [71].

4.3 MARKET TREND AND FUTURE

Reinforcement of polymer matrices using various types of nanofillers is being extensively studied nowadays. The reinforcement mechanisms as well as enhancement of properties are different with different types of fillers. This field is quite green and many more developments are yet to come to enrich our science and technology in the near future.

According to a new and updated technical market research report from BCC Research [72], the global market for CNT was $50.9 million in 2006 and will grow to $79.1 million by the end of 2007. At a compound annual growth rate (CAGR) of 73.8%, the market will reach $807.3 million by 2011. Composite CNTs by far hold the highest share of the market through the forecast period. In 2006, they were worth more than 84.5% of the total market. Though they retain the highest share throughout the forecast period, the percentage will drop dramatically by the end of 2011 to just below 56% [73]. On this basis and the current level of their technical maturity, cost, and limited availability, CNTs are initially projected to generate only modest revenue streams. However, a more optimistic upswing in revenue is anticipated starting in 2008 and beyond. At the same time, metal oxide and other nanofillers are coming strongly to stay as very competitive members in this world.

Many leading tire manufacturers are now developing engineered nanoparticles to further extend tire life. Currently around 50% of a car tire is made from natural rubber. Small particles of carbon black (including nanoparticles) have long been mixed with rubber to improve the wear and strength of tires. At present, 16.5 million tires are retreaded every year in the United States alone. Cabot, one of the world's leading tire rubber producers, successfully tested "PureNano" silica carbide nanoparticles designed by Nanoproducts Co. of Longmont, Colorado [74]. Added to tires, the "PureNano" particles reduce abrasion by almost 50%—a simple improvement that if widely adopted should help tires last up to twice as long and thereby significantly reduce the need for new tire rubber.

Aeromet Technologies, Inc. of Sandy, Utah [73], will craft a cleaner process for bonding steel tire cords to rubber during tire manufacturing, the largest commercial application requiring bonding of metal to rubber. The new bonding material based on nanoparticles will eliminate the use of hazardous chemicals such as cobalt and cyanide. Intelligent Optical Systems, Inc. of Torrance, California [75], is at the threshold of adapting immunoassays, highly sensitive biological laboratory tests, to develop field monitors that test more precisely for organic pollutants in water supplies and food processing. Strong, lightweight rubber–carbon nanotube composites [76] are coming strongly in market applications.

REFERENCES

1. Dresselhaus, M.S. *Graphite Fibers and Filaments*, Springer-Verlag: London, 1988.
2. Inagaki, M., Tashiro, R., Washino, Y., and Toyoda, M. *J. Phys. Chem. Solids* 65, 133, 2004.
3. Chung, D.D.L. *J. Mater. Sci.* 22, 4190, 1987.
4. George, J.J., Bandyopadhyay, A., and Bhowmick, A.K. *J. Appl. Polym. Sci.* 2007, DOI 10.1002/APP.25067.
5. George, J.J. and Bhowmick, A.K. *J. Mater. Sci.* 43, 702, 2008.
6. Dresselhaus, M.S., Lin, Y.M., Rabin, O., Jorio, A., Souza Filho, A.G., Pimenta, M.A., Saito, R., Samsonidze, G.G., and Dresselhaus, G. *Mater. Sci. Eng.* C23, 129, 2003.
7. Zhu, Y., Liang, J., Wang, J., and Zhou, X. (Tsinghua University, Peoples Republic of China). *Faming Zhuanli Shenqing Gongkai Shuomingshu* 8, 2005.
8. Shanmugharaj, A.M. and Ryu, S.H. Fall Technical Meeting—American Chemical Society, Rubber Division, 168th Meeting, Pittsburgh, PA, Nov. 1–3, 2005.
9. Sui, G., Liang, J., Zhu, Y., and Zhou, X. *Gaojishu Tongxun* 14, 41, 2004.
10. Yu, Y., Zeng, Y., Zhang, L., and Jia, Z. *Tanxingti* 12, 1, 2002.
11. Beyer, G. and Kabelwerk, Eupen, A.G., and Eupen, Belg. *Gummi, Fasern, Kunststoffe* 55, 596, 2002.
12. Yoshikai, K., Hasuo, H., Ohsaki, T., and Hokoku, K. *Fukuoka-ken Kogyo Gijutsu Senta* 15, 35, 2005.

13. Dresselhaus, M.S., Dresselhaus, G., and Eklund, P.C. *Science of Fullerenes and Carbon Nanotubes*, Academic Press: San Diego, CA, 1996.
14. Wong, E.W., Sheehan, P.E., and Lieber, C.M. *Science* 277, 1971, 1997.
15. Treacy, M.M.J., Ebbesen, T.W., and Gibson, J.M. *Nature* 381, 678, 1996.
16. Iijima, S. *Nature* 354, 56, 1991.
17. Saito, R., Dresselhaus, G., and Dresselhaus, M.S. *Physical Properties of Carbon Nanotubes*, Imperial College Press: London, 1–4, 1999.
18. Scott, C.D., Arepalli, S., Nikolaev, P., and Smalley, R.E. *Appl. Phys. A* 72, 573, 2001.
19. Ebbesen, T.W. *Carbon Nanotubes: Preparation and Properties*, CRC Press: Boca Raton, FL, 139–162, 1997.
20. Andrews, R., Jacques, D., Rao, A.M., Derbyshire, F., Qian, D., Fan, X., Dickey, E.C., and Chen, J. *Chem. Phys. Lett.* 303, 467, 1999.
21. Falvo, M.R., Clary, G.J., Taylor II, R.M., Chi, V., Brooks Jr., F.P., Washburn, S., and Superfine, R. *Nature* 389, 582, 1997.
22. Dalton, A.B., Collins, S., Munoz, E., Razal, J.M., Ebron, V.H., Ferraris, J.P., Coleman, J.N., Kim, B.G., and Baughman, R.H. *Nature* 423, 703, 2003.
23. Dujardin, E., Ebbesen, T.W., Krishnan, A., Yianilos, P.N., and Treacy, M.J. *Phys. Rev. B* 58, 14, 013, 1998.
24. Baughman, R.H., Zakhidov, A.A., and Heer, W.A. *Science* 279, 787, 2002.
25. Qian, D., Dickey, E.C., Andrews, R., and Rantell, T. *Appl. Phys. Lett.* 76, 2868, 2000.
26. George, J.J., Sengupta, R., and Bhowmick, A.K. *J. Nanosci. Nanotechnol.* 8, 1, 2007.
27. Atieh, M.A., Girun, N., Ahmadun, F.R., Guan, C.T., Mahdi, El. S., and Baik, D.R. *J. Nanotech.* 1, 1, 2005.
28. Kazumasa, Y., Haruumi, H., Tetsuro, O., and Hokoku, K. *Fukuoka-ken Kogyo Gijutsu Senta* 15, 35, 2005.
29. Ying, Y., Yan, Z., Li-sha, Z., and Zhi-jie, J. *Tanxingti* 12, 1, 2002.
30. Wang, Z.S., Huang, C.H., Huang, Y.Y., Hou, Y.J., Xie, P.H., Zhang, B.W., and Cheng, H.M., *Chem. Mater.* 13, 678, 2001.
31. Lin, H.M., Tzeng, S.J., Hsiau, P.J., and Tsai, L. *Nanostruct. Mater.* 10, 465, 1998.
32. Xu, J.Q., Pan, Q.Y., Shun, Y.A., and Tian, Z.Z., *Sens. Actuators, B, Chem.* 66, 277, 2000.
33. Curridal, M.L., Comparelli, R., Cozzli, P.D., Mascolo, G., and Agostiano, A. *Mater. Sci. Eng. C* 23, 285, 2003.
34. Wu, R. and Xie, C.S., *Mater. Res. Bull.* 39, 637, 2004.
35. Gao, P.X., Lao, C.S., Hughes, W.L., and Wang, Z.L., *J. Chem. Phys. Lett.* 408, 174, 2005.
36. Hu, Z., Oskam, G., Lee, R., Penn, I., Pesika, N., and Searson, P.C. *Phys. Chem. B* 107, 3124, 2003.
37. Wong, E.M., Bonevich, J.E., and Searson, P.C. *J. Phys. Chem. B* 102, 7770, 1998.
38. Bahnemann, D.W., Kormann, C., and Hoffmann, M.R. *J. Phys. Chem.* 91, 3789, 1987.
39. Kaneko, D., Shouji, H., Kawai, T., and Kon-No, K. *Langmuir* 16, 4086, 2000.
40. Rodríguez-Paéz, J.E., Caballero, A.C., Villegas, M., Moure, C., Durán, P., and Fernández, J.F. *J. Eur. Ceram. Soc.* 21, 925, 2001.
41. Sahoo, S. and Bhowmick, A.K. *J. Appl. Polym. Sci.* 106, 3077, 2007; Sahoo, S., Kar, S., Ganguly, A., Maiti, M., and Bhowmick, A.K. *Polym. Polym. Compos.*, 2007 (in press).
42. Liufu, S., Xiao, H., and Li, Y.P. *Polym. Degrad. Stab.* 87, 103, 2005.
43. Ali, H.A. and Iliadis, A.A. *Thin Solid Films* 471, 154, 2005.
44. Zheng, J., Ozisik, R., and Siegel, R.W., *Polymer* 46, 10873, 2005.
45. Kraus, G. (Ed.) *Reinforcement of Elastomers*, Interscience: New York, 1965.
46. Blow, C.M. *Rubber Technology and Manufacture*, Butterworth: London, 1971.
47. Wagner, M.P. *Rubber Technology*, M. Morton (Ed.), Reinhold: New York, 1987.
48. Ishida, H., *Interfacial Phenomena in Polymer, Ceramic and Metal Matrix Composites*, Elsevier: New York, 1988.
49. Medalia, A.I. and Kraus, G. *Science and Technology of Rubber*, 2nd ed., Mark, J.E., Erman, B., and Eirich, F.R. (Eds.), Academic Press: San Diego, CA, 1994.
50. Wolff, S. and Wang, M.-J. *Rubber Chem. Technol.* 65, 329, 1992.
51. Wolff, S. *Rubber Chem. Technol.* 69, 325, 1996.
52. Hamed, G.R. *Rubber Chem. Technol.* 73, 524, 2000.
53. Kohjiya, S. and Ikeda, Y. *Rubber Chem. Technol.* 73, 534, 2000.
54. Xie, Y., Huang, J., Li, B., Liu, B., and Qian, Y.T. *Adv. Mater.* 12, 1523, 2000.

55. Mckelvy, M., Sharma, R., Chizmeshya, A.G., Carpenter, R.W., and Streib, K. *Chem. Mater.* 13, 921, 2002.
56. Lieber, C.M., Morales, A.M., Sheehan, P.E., Wong, E.W., and Yang, P., Chemistry on the Nano-meter Scale Proceedings of Robert A. Welch Foundation 40th Conference on Chemical Research (Houston, TX), 1996.
57. Xie, R.C. and Qu, B.J. *Preparation of Needle-Shaped Mg(OH)₂ Nanoparticles.* Patent Application Number 00135436 of P.R. China, 2001.
58. Alexandre, M., Beye, G., Henrist, C., Cloots, R., Rulmont, A., Jérôme, R., and Dubois, P. *Macromol. Rapid Commun.* 22, 643, 2001.
59. Li, Y.D., Sui, M., Ding, Y., Zhang, G., Zhuang, J., and Wang, C., *Adv. Mater.* 12, 818, 2000.
60. Yu, J.C., Xu, A.W., Zhang, L.Z., Song, R.Q., and Wu, L., *J. Phys. Chem. B* 108, 64, 2004.
61. Henrist, C., Mathieu, J.P., Vogels, C., Rulmont, A., and Cloots, R. *J. Cryst. Growth* 249, 321, 2003.
62. Lv, J., Qiu, L., and Qu, B. *Nanotechnology* 15, 1576, 2004.
63. Kuila, T., Acharya, H., Srivastava, S.K., and Bhowmick, A.K. *J. Appl. Polym. Sci.* 104, 1845, 2007.
64. Acharya, H., Srivastava, S.K., and Bhowmick, A.K. *Nanosci. Res. Lett.* 2, 1, 2007.
65. Kar, S. and Bhowmick, A.K. IIT Kharagpur, unpublished data, 2007.
66. Poh, B.T., Kwok, C.P., and Lim, G.H. *Eur. Polym. J.* 31, 223, 1995.
67. Poh, B.T., Ismail H., and Tan, K.S. *Polym. Test.* 21, 801, 2002.
68. Silva, A.L.N., da., Rocha, M.C.G., Moraes, M.A.R., Valente, C.A.R., and Coutinho, F.M.B. *Polym. Test.* 21, 57, 2002.
69. Tandon, R.P., Choubey, D.R., Singh, R., and Soni, N.C. *J. Mat. Sci. Lett.* 12, 1182, 1993.
70. Yurkov, G.Y., Astaf'ev, D.A., Nikitin, L.N., Koksharov, Y.A., Kataeva, N.A., Shtykova, E.V., Dembo, K.A., Volkov, V.V., Khokhlov, A.R., and Gubin, S.P. *Inorg. Mat.* 42, 496, 2006.
71. Nair, K.G. and Durfreshe, A. *Biomacromolecules* 4, 666, 2003.
72. http://www.bccresearch.com/nan/NAN024C.asp access date April 14, 2007.
73. http://www.voyle.net/Nano%20Biz/NanoBiz-2004 access date April 6, 2007.
74. www.azonano.com/Details.asp?ArticleID=1351 access date April 25, 2007.
75. http://www.nanovip.com/node/2945 access date April 25, 2007.
76. Strano, M.S. *Nat. Mat.* 5, 433, 2006.

5 Thermoplastic Elastomers

*Francis R. Costa, Naba K. Dutta, Namita Roy Choudhury,
and Anil K. Bhowmick*

CONTENTS

5.1 INTRODUCTION

The emergence of thermoplastic elastomers (TPEs) provided a new horizon to the field of polymer technology. Their development and growth have reached a high level of commercial significance, and they have become an important segment of polymer science and technology in the last four decades. In the simplest way, TPEs can be defined as a class of polymers, which combine the service properties of elastomers with the processing properties of thermoplastics. TPEs consist of two structural units: (1) amorphous segment that is soft in nature and has a low glass-transition temperature (T_g) and (2) a crystalline and/or hard segment with high T_g that acts as physical cross-link for the soft segments. These two segments, i.e., soft and hard segments, must be thermodynamically incompatible to each other at service temperature to prevent the interpenetration of the segments. Therefore, a TPE possesses biphasic morphological structure. The hard segments that act as physical cross-links are thermally labile, and this allows TPEs to soften and flow under shear force at elevated temperature as in the case of true thermoplastics. TPEs bridge the gap between conventional rubbers and thermoplastics. The key factors responsible for their continuous acceptance and sustained growth are listed in Table 5.1.

TABLE 5.1

Advantages of Thermoplastic Elastomer over Thermoset Rubber Processing

Variables	Thermoplastic Elastomer	Thermoset Rubber
1. Fabrication	Rapid (s)	Slow (min)
2. Scrap	Reusable	High percentage waste
3. Curing agent	None	Required
4. Machinery	Conventional thermoplastic equipments	Special vulcanizing equipments
5. Additives	Minimal or none	Numerous processing aids
6. Remold parts	Yes	Impossible
7. Heating, sealing	Yes	No

Source: Walker, B.M. and Rader, C.P. (eds.), *Handbook of Thermoplastic Elastomers*, Van Nostrand Reinhold, New York, 1988.

5.2 HISTORY AND GROWTH

TPEs appeared in the market as commercial products in the late 1950s with the introduction of thermoplastic polyurethane (TPU, B.F. Goodrich Co.). This was preceded by activity in 1930s and 1940s that led to the introduction of TPU in Germany and polyvinyl chloride (PVC) in the United States. By the 1960s, TPEs were in their infancy. During this period enormous research efforts were devoted to the development of novel block copolymer TPE and a variety of blends of thermoplastic polyolefins (TPO) with elastomers. In 1965, the company Shell developed and introduced commercial styrene diene block copolymer, Kraton. The 1970s saw the emergence of commercial copolyester, Hytrel from DuPont and blends of polypropylene (PP) and ethylene–propylene–diene monomer (EPDM) rubber from Uniroyal. During this period, it was realized that TPE would have a bright and promising future in both the rubber and the plastic industries. Thus extensive research efforts were devoted toward their development and commercialization. As a consequence, the 1980s witnessed the growth of TPE to maturity. Table 5.2 illustrates the milestones for the commercialization of TPEs.

TABLE 5.2
Milestones for Commercialization of Thermoplastic Elastomer

Roots

1933	Semon: Flexible PVC, Goodrich patent
1940	Henderson: PVC–NBR blends, Goodrich patent
1947	Commercial PVC–NBR blends, Goodrich
1952	Synder: Elastic thread from linear copolyester, DuPont patent
1954	PU Spandex fiber, DuPont patent
1955–1957	Schollenberger: PU TPE, patent and paper, Goodrich
1957	Bateman, Merrett: NR–PMMA grafts, BRPRA

TPE first decade of progress

1958–1959	Tobolosky suggestion of crystalline and amorphous polyolefin copolymers
1959	Commercial PU elastic fiber, DuPont
1960	Commercial PU TPEs
1961	Ionomeric TPE, DuPont (Surlyn)
1962	Kontons α-olefin TPE research, Uniroyal
1962	Gessler patent on PP-CIIR blends, dynamically cured
1965	Commercial triblock TPEs, Shell (Kratons)
1967	California Institute of Technology and ACS, Polymer Group, Symposium on TPE theory

Second decade of progress

1968	Radial styrenic block polymers, Phollips (Solprene)
1972	Polyolefin blends, Uniroyal (TPR)
1972	Copolyester TPEs, DuPont (Hytrel)
1972	SEBS TPEs, Shell (Kraton G)

Third decade of progress

1978	A.Y. Coran's research on the melt–mixed blends of elastomer and thermoplastics with dynamic vulcanization
1981	Commercial melt–mixed blends of EPDM and PP, dynamically vulcanized, Monsanto (Santroprene)
1982	Polyamide TPE, Atochem (Pebax)
1985	Commercial blends of NBR and PP, dynamically vulcanized, Monsanto (Geolast)
1985	Single phase melt-processable rubber, DuPont (Alcryn)
1988	Functionalization hydrogenated styrenics TPEs, Shell (Shell FG)

(continued)

TABLE 5.2 (continued)
Milestones for Commercialization of Thermoplastic Elastomer

Fourth Decade of Progress

1988–2006	Blends of TPEs with existing polymers for enhancement of properties. Academic research in various fields of TPEs worldwide
	New family of TPV having heat and oil resistance based on ACM and polyamide
	Development of crystalline–amorphous block copolymers (Engage), mettalocene catalyzed TPEs, Polyolefin elastomer (POEs), application research on TPEs
	Protein-based block-copolymer

Source: Modified from Legge, N.R., *Rubber Chem. Technol.*, 62, 529, 1989.

5.3 GENERIC CLASSIFICATION OF THERMOPLASTIC ELASTOMERS

TPEs may be rationally classified into the following classes according to their chemistry and morphology:

1. Block Copolymers
   a. Styrenic block copolymers
   b. Thermoplastic copolyesters (COPE)
   c. Thermoplastic polyurethanes (TPU)
   d. Thermoplastic polyamides (COPA)
2. Blends and elastomeric alloys
   a. Elastomeric rubber–plastic blends
   b. Thermoplastic vulcanizates (TPVs)
   c. Melt processable rubber
3. Crystalline–amorphous block copolymers
4. Ionomers
   a. Sulfonated-EPDM rubber (S-EPDM)
   b. Zn or Na salt of ethylene acrylic acids
5. Miscellaneous

A number of other polymers have the characteristics of TPE and some are available commercially, such as (1) 1,2-polybutadiene, (2) *trans*-polyisoprene (PI), (3) modified polyethylene (PE) (e.g., ethylene vinyl acetate [EVA] and ethylene ethyl acrylate [EEA]), (4) nonhydrocarbon elastomer-based TPEs, (5) metallocene elastomers/TPEs (MEs/TPEs), and (6) graft copolymeric TPEs.

Recently, a new method of nomenclature of TPEs was produced jointly by the International Organization for Standardization (ISO), the Society of Automotive Engineers (SAE), the Association of the Automotive Industries (VDA), Germany, and the producers of raw materials. Nevertheless, for this chapter the above-mentioned abbreviations for the TPEs have been utilized. Table 5.3 provides a list of significant suppliers of different generic classes of TPEs with respective trade names.

5.4 POLYMER SYNTHESIS

5.4.1 BLOCK COPOLYMERS

5.4.1.1 Styrenic Block Copolymer

Triblock copolymers of ABA type, where B is the central elastomeric block and A is the rigid end-block, are well-known commercially available polymers [7,8]. The chemical structures of some common TPEs based on styrenic block copolymers are given in Figure 5.1. Synthesis of such ABA-type polymers can be achieved by three routes [9]:

TABLE 5.3
Significant Thermoplastic Elastomer Suppliers and Their Products

TPE Type	Supplier	Trade Name
Styrenic	Kraton Polymers	Kraton
	Alpha Gary	Evoprene
	Concept Polymers	C-Flex
	Dexco Polymers LP (Dow/Exxon)	Vector
	Fina Oil	Finaprene
	J-Von	J-Plast
	Nippon Zeon	Quantic
	Firestone (Bridgestone)	Stereon
	Kraiburg	Thermoplast K
	PolyOne	Synprene
	Enichem	Europrene
	BASF	Styroflex
	Technor Apex	Elexar, Monoprene, Tekron, Tekbond
	Philips	K-Resin, Solprene
	Asahi Kasei	Tufprene, Asaprene, Asaflex
Polyurethane	BASF	Elastollan, Elastocell
	Akzo	Urafil
	Borealis	Daplen
	Noveon	Tecoflex, Tecothane, Carbothane, Tecophilic, Tecoplast, EstaGrip, Estane
	B.F. Goodrich	Estane, Estaloc
	Bayer	Texin, Baytec, Vulkollan, Desmopan
	A. Schulam	Polypur
	Morton Int.	Morthane
	DuPont	Hyelene
	GLS Corp.	Versollan
	J-Von	J-Plast
	Discas Inc.	Prolastic
	K.J. Quinn	Q-Thane
	Dow	Pellethane, Isoplast
Copolyesters	DSM	Arnitel
	DuPont	Hytrel
	General Electric	Lomod
	Eastman Chemical	Ecdel
	GE Plastics	Lomod
	Hoechst Celanese	Riteflex
	Toyobo	Pelprene
	Celanese	Gaflex
	Wilden	Saniflex
Polyamides	Atochem (now AtoFina)	Pebax
	ESM-American Garilon	Grilamid
	Huls	Vestamid
	DuPont	Zytel FN
	Dow Chemical	Pebax
TPOs or TEOs EPDM rubber/polyolefin	DSM	Sarlink
	Ferro Corp.	Feroflex
	Himont	Dytral/TP
	Montell	Hifax

(continued)

TABLE 5.3 (continued)
Significant Thermoplastic Elastomer Suppliers and Their Products

TPE Type	Supplier	Trade Name
	So.F.Ter. S.p.A.	Forprene
	Mitsui Petrochemicals	Milastomer
	Research Polymers Int.	Renflex
	Nippon Petrochemicals	Softlex
	A. Schulman	Polytrope
	Teknor Apex	Telcar
	Union Carbide	Flexomar
	Colonial Rubber	Solmel
NBR/PVC	A. Schulman/Mitsubishi	Sunprene, Sunfrost
	Polyone	Proflex
Thermoplastic vulcanizates	Advanced Elastomer Systems, LP	Santoprene, Geolast, Vyram, Dytron, Trefsin
	Teknor Apex	Uniprene
	DuPont	ETPV
	Zeon Chemicals L. P.	Zeotherm
Ionomer	DuPont	Alcryn
	Exxon	Iotek
	DuPont	Surlin

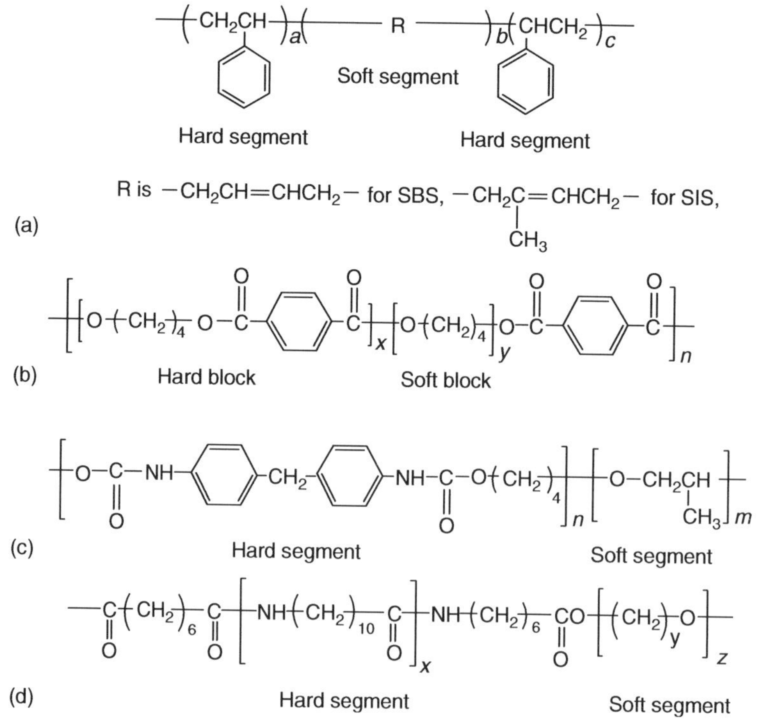

FIGURE 5.1 Chemical structure of block copolymeric thermoplastic elastomers (TPEs) (a) styrenic, (b) COPE, (c) thermoplastic polyurethane, and (d) thermoplastic polyamide.

1. Sequential monomer addition
2. Synthesis of AB diblock followed by linking or coupling reaction
3. Successive polymerization of monomers starting from a difunctional initiator

Linear triblock copolymers of the type styrene–butadiene–styrene (SBS) and styrene–isoprene–styrene (SIS) are produced commercially by anionic polymerization through sequential addition of monomers in the reaction chamber [10] as shown below:

$$Styrene + Initiator \rightarrow SSS^-$$

$$SSS^- + Butadiene \rightarrow SSSSBBBB^-$$

$$SSSBBBB^- + Styrene \rightarrow SSSBBBBSSSSS^-$$

Saturated styrene–ethylene–butylene–styrene (SEBS) and styrene–ethyl–propyl–styrene (SEPS) are produced by hydrogenating SBS and SIS block copolymers, respectively, before they are recovered from the solution [3].

Though living anionic polymerization is the most widely used technique for synthesizing many commercially available TPEs based on styrenic block copolymers, living carbocationic polymerization has also been developed in recent years for such purposes [10,11]. Polyisobutylene (PIB)-based TPEs, one of the most recently developed classes, are synthesized by living carbocationic polymerization with sequential monomer addition and consists of two basic steps [10] as follows:

1. Living polymerization of isobutylene (IB) by di- and trifunctional initiators to make the nearly uniform rubber mid-block
2. Sequential addition of a second monomer (e.g., styrene, p-terbutyl styrene, p-methyl styrene, indene, p-chloro styrene, isoprene, and α-methyl styrene) to make glassy outer-block

Poly(styrene-b-isobutylene-b-styrene) (PS–PIB–PS), triblock copolymers can be prepared via coupling of living PS–PIB diblock copolymers in a one-pot procedure [12].

Most common styrene-based block copolymers behaving as TPE show limited range of service temperature, below glass-transition temperature (T_g) of hard segment (around 100°C). For example, mechanical properties of these block copolymers are highly affected at elevated temperature and at T_g of polystyrene (PS) domains. This is often related to the lack of stereoregularity in polystyrene and the rubbery domains resulted from the anionic polymerization technique [13]. To solve this problem, several modifications have been proposed by researchers to enhance the T_g of the glassy domain; e.g., use of poly(α-methyl styrene) [14,15], poly(ethylene sulfide) [16] and syndiotactic poly(methyl methacrylate) (PMMA) (sPMMA) [17] as hard segments is well known. There have been some recent developments in this regard; like a triblock copolymer based on isobornyl methacrylate (IBMA) and butadiene has been synthesized, which shows T_g of the hard segment in the range of 170°C–206°C depending on the chain tacticity [18]. TPEs based on polyacrylates such as poly(methacrylate-b-dienes-b-methacrylate) [9] also show high T_g of the hard segment. Ban et al. have very recently reported a new approach to introduce crystalline PS domains in SBS-type triblock copolymer through a stereospecific sequential triblock copolymerization of S with B using a specialized catalyst. The introduction of syndiotactic or isotactic PS and elastic cis-polybutadine block (instead of low cis-PB obtained in anionic SBS) causes improvement of several properties like satisfactory mechanical and chemical resistance at elevated temperatures [13].

Branched copolymers or star-block copolymers are new entrants in TPE family. There is a growing interest in the synthesis of star-block TPEs because of their unique mechanical and rheological properties. They can be represented by a general structural formula $(A - B)_n - X$, where X represents the core that holds the block copolymeric arms and n is an integer greater that 2. The arms comprise of inner soft rubbery segment (B) connected to hard glassy outer segments (A). The core (X) may be of different types, like calex[8]arene [19], cross-linked polydivinyl benzene [20], siloxane [21,22],

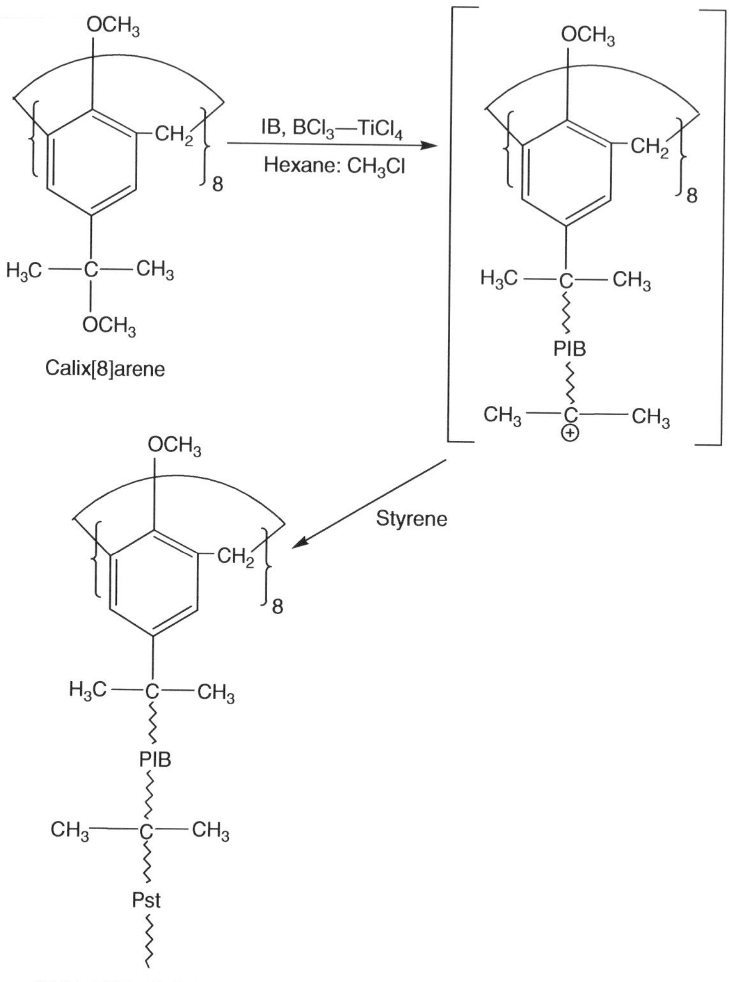

FIGURE 5.2 Synthesis of polyisobutylene (PIB)-based star-block thermoplastic elastomer (TPE). (From Jacob, S. and Kennedy, J.P., *Adv. Polym. Sci.*, 146, 1, 1999.)

etc. The synthetic procedure of star-block copolymers is well known for many years [20–27]. Different approaches are usually used to synthesize star-block copolymers such as anionic, cationic, free radical, condensation polymerization, and often combinations of them. However, the synthesis by anionic polymerization using living di- or triblock copolymer is the most popular method. Kennedy et al. have extensively reported the synthetic strategy (Figure 5.2) and characterization of PIB-based star-block copolymers [20,21,28]. They exhibit high tensile strength (18–26 MPa) and elongation at break (500%– 600%). Star copolymer with 32% PS content shows tensile strength of about 26 MPa.

5.4.1.2 Copolyester

The first commercial polyester TPE (called thermoplastic copolyester, COPE) was introduced by DuPont with the trade name Hytrel, and at present it is the single-largest-volume TPE material which comes under the category of "engineering thermoplastic elastomer." Polyester TPEs possess hard segments generally based on oligourethanes or aromatic polyesters and soft segments derived from aliphatic polyethers. By varying the concentration and the molecular weight of the soft segment, a wide range of products with different properties could be obtained. The segmented polyester–polyether block copolymers are prepared readily and economically by melt transesterification. A mixture of phthalate ester, a low-molecular-weight diol, and a poly(alkylene ether) diol is heated to 160°C in the presence of transesterification catalyst [29]. This has been covered elsewhere in this book.

A disadvantage of the commercial poly(ether–ester) is a partial compatibility of the hard and soft segments. Even a poor compatibility reduces the crystallinity and mechanical stability of the hard segments and raises the T_g of the soft segments. Hard segments containing imide-building blocks should favor the microphase separation because of their high polarity and planarity of the aromatic imide group [30]. New TPEs prepared from polytetramethylene oxide (PTMO), 1,4-dihydroxybutane, and a bisimide dicarboxylic acid based on 1,4-diaminobutane and trimellitic anhydride show very good microphase separation. When PTMO of high polydispersity is used, good mechanical properties are obtained, which are better than those of classical TPEs based on PBT hard segments [30]. However, all poly(ether–ester–imides) crystallize slowly, which could be a hindrance for injection molding but allow their application as films.

Polyether-based thermoplastic copolyesters show a tendency toward oxidative degradation and hydrolysis at elevated temperature, which makes the use of stabilizer necessary. The problem could be overcome by incorporation of polyolefinic soft segments in PBT-based copolyesters [31,32]. Schmalz et al. [33] have proposed recently a more useful technique to incorporate nonpolar segments in PBT-based copolyesters. This involves a conventional two-step melt polycondensation of hydroxyl-terminated PEO–PEB–PEO (synthesized by chain extension of hydroxyl-terminated hydrogenated polybutadienes with ethylene oxide) and PBT-based copolyesters.

5.4.1.3 Polyurethanes

Polyurethane chemistry is a very broad field and encompasses a large number of chemical reactions of isocyanate with various active hydrogen-containing compounds [34]. TPUs are linear-segmented copolymers with alternating hard and soft segments. The soft segment is usually a high-molecular-weight macroglycol (polyether or polyester type) and the hard segment is formed by the reaction between a diisocyanate with a low-molecular-weight diol, such as 1,4-butanediol. Primarily, TPU linear polymers are synthesized by the condensation of diisocyanate with short-chain diol (chain extender) and polyester or polyether diol (polyol). The preparation of TPU can be done either by the one-shot [35] or the two-shot (prepolymer) [36] process. The latter involves preparation of a low-molecular-weight linear isocyanate-terminated prepolymer followed by its chain extension to produce the final TPU. There have been a number of patents and literatures available on synthesis of thermoplastic polyurethanes [37–41].

5.4.1.4 Copolyamides (Thermoplastic Polyamides)

Thermoplastic polyamide elastomers consist of a regular linear chain of rigid polyamide segments interspaced with flexible polyether segments. They are basically segmented block copolymers having general structure $(AB)_n$. The hard segments may be based on partially aromatic polyamide or aliphatic polyamide. Polyester amide (PESA) and polyether ester amide (PEEA) are prepared from the aromatic polyamide, while in the polyether block amide (PEBAX or VESTAMID) the hard segments are derived from aliphatic polyamide. In both PEEA and PEBA, the soft segment of aliphatic polyesters is linked to the hard segment by an ester group. The three principal methods for their preparation of thermoplastic polyamide elastomers [2] are as follows:

1. The Dow process consists of the formation of an acid-terminated soft segment, first by the esterification of polyoxyalkylene or other glycol, followed by the reaction of a diisocyanate and additional diacid to form PESA.
2. The ATO Chemie process involves the formation of an adipic acid-capped hard segment block of poly(11-aminoundecanoic) of molecular weight 800–1500, joined with a soft segment of polyol in a polyesterification process.

3. The Emser Industries process has no ester linkage, the bonds between the two segments being amides. An amine-terminated soft bis(3-aminopropyl)polyoxytetramethylene glycol is reacted with a dimer acid (Empol 1010) and caprolactum to form polyetheramide (PETA).

5.4.2 BLENDS AND ELASTOMERIC ALLOYS

5.4.2.1 Elastomeric Rubber–Plastic Blends

TPEs prepared from rubber–plastic blends have gained considerable interest in last two decades [3,42,43]. They provide the simplest way of achieving outstanding properties at low cost and with scope to tailor-make the properties by simple variation of the blend composition, viscosity of the components, and compounding ingredients. The first useful rubber–plastic blends were those obtained from nitrile rubber (NBR) and polyvinylchloride (PVC). There have been constant researches going on in this area to develop thermoplastic elastomeric blends with improved properties [44–46]. Elastomer–plastic blends containing large amount of elastomers have generally been prepared by melt mixing, which is easy and economical. The basic factors to be considered for the preparation of TPE of this kind are (1) high shear stress during mixing [47], (2) complete homogenization of the dispersed component particles, and (3) avoidance of local overheating and polymer degradation. Thus, mixing could be carried out in (i) a high-speed, two-rotor continuous mixer, (ii) a twin-screw extruder, or (iii) a single-screw extruder (length/diameter ratio, L:D > 30) [48].

TPEs prepared from rubber–plastic blends usually show poor high-temperature properties. This problem could be solved by using high-melting plastics like polyamides and polyesters. But, often they impart processing problems to the blends. Jha and Bhowmick [49] and Jha et al. [50] have reported the development and properties of novel heat and oil-resistant TPEs from reactive blends of nylon-6 and acrylate rubber (ACM). The properties of various thermoplastic compositions are shown in Table 5.4. In this kind of blend, the plastic phase forms the continuous phase, whereas

TABLE 5.4
Typical Properties of Thermoplastic Elastomers Developed from Nylon-6–Acrylate Rubber Blends

TPE Composition	A	AB_{30}	AC_{30}	AS_{30}
Nylon-6	40	40	40	40
ACM	60	60	60	60
HMDC	0.5	0.5	0.5	0.5
N770	—	30	—	—
China clay	—	—	30	—
Ultrasil VN_3	—	—	—	30
Tensile strength (MPa)	12	11	11	15
Elongation at break (%)	114	94	115	183
Young's modulus	52	40	52	41
Tension set	35	25	28	23
Oil aging at 150°C in ASTM oil 3				
Volume swell (%)	3	5	7	2
Tensile strength (MPa)	12	13	11	16
Elongation at break (%)	70	62	60	113
Aging at 175°C for 72 h				
Tensile strength (MPa)	10	12	13	18
Elongation at break (%)	100	30	46	60

Source: Jha, A., Dutta, B., and Bhowmick, A.K., *J. Appl. Polym. Sci.,* 74, 1490, 1999.

the rubber phase remains as dispersed particles. The modulus, tensile strength, elongation at break, and hardness of these blends increase with increasing interaction between nylon-6 and ACM. The blends show excellent resistance to oil swelling at elevated temperature (e.g., 150°C), and their service temperature range can be extended up to 170°C–200°C without much deterioration of the mechanical properties. The mechanical properties of these blends are improved a lot with the addition of lower amounts of carbon black (i.e., 10–20 phr) and higher percentage of silica filler (30 phr). However, swelling properties are affected slightly on the addition of filler in such blends [51]. On dynamic vulcanization of such blends, hardness and modulus fall slightly, but tensile strength and elongation at break improve.

De Sarkar et al. [52] have reported a series of new TPEs from the blends of hydrogenated SBR and PE. These binary blends are prepared by melt mixing of the components in an internal mixer, such as Brabender Plasticorder. The tensile strength, elongation at break, modulus, set, and hysteresis loss of such TPEs are comparable to conventional rubbers and are excellent. At intermediate blend ratio, the set values show similarity to those typical of TPEs (Table 5.5).

Jha and Bhowmick [51] have reported the development and properties of thermoplastic elastomeric blends from poly(ethylene terephthalate) and ACM by solution-blending technique. For the preparation of the blend the two components, i.e., poly(ethylene terephthalate) and ACM, were dried first in vacuum oven. The ACM was dissolved in nitrobenzene solvent at room temperature with occasional stirring for about three days to obtain homogeneous solution. PET was dissolved in nitrobenzene at 160°C for 30 min and the rubber solution was then added to it with constant stirring. The mixture was stirred continuously at 160°C for about 30 min. The blend was then drip precipitated from cold petroleum ether with stirring. The ratio of the petroleum ether/nitrobenzene was kept at 7:1. The precipitated polymer was then filtered, washed with petroleum ether to remove nitrobenzene, and then dried at 100°C in vacuum.

Interesting TPEs can be derived from binary and ternary blends of polyfunctional acrylates, ACM, and fluorocarbon rubber (FKM) [53]. During the blend preparation, the liquid multifunctional acrylate monomer used is polymerized and forms the continuous matrix encapsulating the

TABLE 5.5
Properties of Thermoplastic Elastomeric Composition Based on Hydrogenated Styrene–Butadiene Rubber and Low-Density Polyethelene

Blend Composition	HL$_0$	HL$_{20}$	HL$_{30}$	HL$_{40}$	HL$_{50}$	HL$_{60}$	HL$_{70}$
HSBR (pbw)	100	80	70	60	50	40	30
LDPE (pbw)	0	20	30	40	50	60	70
Transition temperature (°C)							
(i) α	−106	−107	−106	−104	−108	−105	−103
(ii) β	−18	−18	−21	−19	−20	−20	−21
(iii) γ	—	—	72	66	64	64	64
T_g of rubber phase (°C)	−28	−26	−31	−32	−32	−32	−32
Tensile strength (MPa)	5.7	7.0	8.4	8.6	9.0	10.5	11.4
Elongation at break (%)	957	743	735	600	580	560	555
Work at break (kJ · m^{-2})	12.1	13.9	15.5	15.8	15.8	16.8	19.3
Modulus (MPa)							
100%	1.5	2.7	3.3	3.8	4.7	5.7	6.2
200%	2.3	3.7	4.4	5.0	5.8	7.0	7.5
Set at 100% elongation (%)	2.0	6.3	7.5	10.0	12.5	15.0	17.5
Hysteresis (J · cm^{-3})	3.04	6.14	7.43	9.73	11.55	13.55	14.13

Source: De Sarkar, M., De, P.P., and Bhowmick, A.K., *Polymer*, 39, 1201, 1998.

TABLE 5.6

Mechanical Properties of Thermoplastic Elastomers Based on Acrylate Rubber–Fluorocarbon Rubber–Polyacrylate Monomer

Blend Composition	AT30	AFT30	AFH30	AFD30
ACM	100	70	50	50
FKM	—	30	50	50
HDDA[a]	—	—	30	—
TMPTA[b]	30	30	—	—
DPHA[c]	—	—	—	30
Modulus (100%) (MPa)	4.3	7.8	4.7	—
Tensile strength (MPa)	8.5	9.2	6.5	12.8
Elongation at break (%)	450	138	270	52

Source: Kader, M.A., Bhowmick A.K., Inoue, T., and Chiba, T., *J. Mat. Sci.*, 37, 6789, 2002.

[a] HDDA = Hexanediol diacrylate.
[b] TMPTA = Trimethylolpropane triacrylate.
[c] DPHA = Dipentaerythritol hexacarylate.

rubber matrix. This procedure generates a very fine morphology of the dispersed rubber phase. The TPEs obtained from such blends show very good mechanical properties (Table 5.6).

5.4.2.2 Thermoplastic Vulcanizates

This is a special class of TPEs resulting from the synergistic interaction of two or more polymers to give properties apparently better than those of simple blends [54]. This interaction can be brought about in an elastomer–thermoplastic system by cross-linking the elastomer and generating a fine dispersion of cross-linked elastomer particles in a continuous matrix of thermoplastics. In most cases, fully cross-linked droplets of 0.1–2.0 μm diameter are dispersed in thermoplastic matrix [55]. TPVs offer properties of cross-linked rubber, coupled with the processability of thermoplastics. There may be several combinations of conventional rubbers, like EPDM, natural rubber (NR), butyl rubber, and NBR, with conventional plastics, like PP, PE, polyesters, and polyamides, to form TPV compositions—the vulcanization of the elastomer phase being carried out under dynamic shear during mixing (dynamic vulcanization). This results in enhancement in the impact and failure properties of the blend. The conventionally used vulcanization systems, e.g., sulfur and accelerator, peroxide, and resin, are employed for carrying out dynamic vulcanization. Among nonconventional vulcanizing systems, hydrosilylation cross-linking system is commercially used, where a silicon hydride having at least two SiH groups is reacted with carbon–carbon multiple bond present in the elastomer component in the presence of a platinum-containing catalyst [56,57].

For the compatible elastomer–thermoplastic blends, melting of the two polymers is the first step followed by subsequent vulcanization of the elastomeric phase. A typical mixing cycle for dynamically vulcanized NR–PE blend (DVNR) in a Brabender mixer is as follows [58]:

- At 0 min add PE
- After 1–2 min add masticated NR
- In 2–3 min add ZnO and stearic acid
- In 3–4 min add accelerator
- In 4–5 min add sulfur
- After 5–6 min dump

For the elastomer–thermoplastic combinations, which are not technologically compatible, a uniform blend could be obtained by incorporating a suitable compatibilizing agent [59–61]. The effects of the type and amount of curing system for NBR-based or polyamide-based TPVs have been studied in details by Mehrabzadeh et al. [62]. Recently, Han et al. [63] developed a new method of EPDM-based or PP-based TPV using supercritical propane. In this method, PP and EPDM are first dissolved in supercritical fluid in certain concentrations to form homogeneous solution followed by the addition of peroxide curatives. The vulacanization of the EPDM phase was carried out by stirring the solution at 175°C and 650 bar pressure for 1 h. The polymer was then precipitated by reducing the pressure to below the cloud-point and subsequently reducing the temperature at constant pressure to the ambient temperature. The advantage of this method is that it provides much smaller size of the dispersed EPDM domains as compared to the melt blending technique.

However, when softening point of thermoplastic phase is very high, e.g., PET, nylon, polycarbonates, the conventional, that is, simple melt mixing of the components followed by dynamic curing with peroxide may not be very effective. In these cases, an indirect two-step masterbatch process can be used. Here, indirect means that the peroxide is premixed and dispersed in rubber phase; and a two-step masterbatch means that the dynamic vulcanizate is prepared first, followed by blending with the thermoplastic [55]. The flow diagram of preparation of TPV based on PET–rubber blend compatibilizer with EPR-*g*-glycidylmethacrylate is shown in Figure 5.3 [55].

A new TPV based on PP matrix in which a metallocene-type ethylene–octene (EO) copolymer is in dispersed phase has been synthesized and characterized by Fritz et al. [64]. The EO copolymers

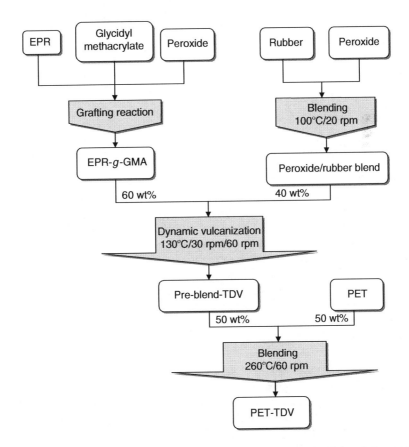

FIGURE 5.3 Flow diagram for the production of polyethylene terepthalate (PET)-based thermoplastic elastomer (TPE). (From Papke, N. and Kargar-Kocsis, J., *Polymer*, 42, 1109, 2001. Courtesy of Elsevier.)

are grafted with vinyl organosilanes, then blended with PP, hydrolyzed, and water cross-linked in a corotating, intermeshing twin-screw extruder. The resulting TPV grades are easy to process into articles showing an attractive range of properties. They can easily be colored to any desired shades. The reduced cost of ingredients is another advantage of the newly developed TPVs. High ultimate tensile strength and elongation at break values as well as a broad spectrum of different Shore A grades open up many fields of application. This new TPV differs from closely related members in TPV family (Santroprene) in the following aspects:

1. Replacement of EPM copolymers and EPDM ter-polymers by saturated metallocene-catalyzed EO copolymer
2. Modification of rubber phase cross-linking by the use of organosilanes

These two-phase polymers are generated in an optimized single-stage process using a corotating, close-intermeshing, twin-screw extruder.

5.4.2.3 Melt Processable Rubber

The melt processable rubber (MPR) blends are single-phase system. The single commercially available MPR is Alcryn, currently sold by Advanced Polymer Alloys. It comprises alloys of ethylene interpolymer and chlorinated polyolefins, in which the ethylene polymer component is partially cross-linked *in situ* [65]. Halogenated polyolefin is miscible with a large number of ethylene interpolymers that are used as intermediate in MPRs. These are designed specially to contain strongly proton-accepting functional groups, in order to promote hydrogen bonding with the alpha hydrogen atom of halogenated polyolefins.

The Alcryn product line is grouped into four series—1000, 2000, 3000, and 4000—differing in hardness and mechanical properties. For example, Alcryn series 1000 has general plastic-processing characteristics and excellent weather resistance and is not suitable for injection molding, for which series 2000 has been designed. Series 2000 has high flow and is excellent for complex extrusion and is not suitable for calendering. The 1000 series has the highest viscosity and the 2000 series has the lowest. The 3000 and 4000 series have intermediate viscosity.

5.4.3 Crystalline–Amorphous Block Copolymers

The polyolefin elastomers based on olefins with distinct crystalline and amorphous blocks belong to a new class of TPEs. These kinds of polymers are basically copolymers synthesized from one major olefin, like ethylene or propylene and a second minor comonomer, like butene or octene, that is incorporated to create amorphous block in the polymer backbone. ENGAGE polyolefin elastomer based on ethylene and octene is commercially available, which is synthesized by using INSITE catalyst and process technology that allows extraordinary control of polymer structure, properties, and rheology [66]. The ethylene/octene ratio could be varied to produce a range of polymeric materials from rubber to TPE. For example, the polymers with 25%, 19%, and 14% octene content have DSC melting temperatures of 55°C, 76°C, and 95°C, respectively. These copolymers present a broad range of solid-state structures from highly crystalline lamellar morphologies to fringed micellar morphologies of low crystallinity.

Recently, Bhowmick [3] reported the properties and thermoplastic elastomeric behavior of a series of ENGAGE copolymers and their blends (Table 5.7). The copolymer with low octene content (Engage 8440) displays highest tensile strength, elongation at break, modulus and permanent set, while Engage 8150 exhibits excellent permanent set, tensile strength, and modulus. Blending of Engage 8150 with low-density PEs (LDPE) in a ratio of 70:30 improves the modulus at the compromise of set and elongation properties. All the above properties are, however, further improved with the addition of 0.6 phr dicumyl peroxide as a cross-linker in the dynamic vulcanization process.

TABLE 5.7
Properties of Different Blends with ENGAGE

Blend	T.S. (MPa)	E.B. (%)	100% Modulus (MPa)	200% Modulus (MPa)	300% Modulus (MPa)	Permanent Set (%)
Engage 8150	16.5	991	2.31	3.10	3.79	6
Engage 8440	30.7	1012	6.00	6.58	8.02	24
Engage 8150/LDPE 70:30	16.8	918	3.56	4.11	4.86	11
Engage 8150/LDPE 70:30, 0.6 phr DCP	19.4	943	3.67	4.28	5.15	9
Engage 8150/EVA12 70:30	20.3	999	3.14	3.92	4.17	10
Engage 8440/LDPE 70:30	28.7	933	6.09	6.66	8.07	24
Engage 8150/EVA12 70:30, 0.6 phr DCP	18.9	783	4.07	5.45	6.82	9

Copolymerization of ethylene and styrene by the INSITE technology from Dow generates a new family of ethylene–styrene interpolymers. Polymers with up to 50-wt% styrene are semicrystalline. The stress–strain behavior of the low-crystallinity polymers at ambient temperature exhibits elastomeric characteristics with low initial modulus, a gradual increase in the slope of the stress–strain curve at the higher strain and the fast instantaneous recovery [67]. Similarly, ethylene–butylene copolymers may also be prepared.

5.4.4 IONOMERS

Ionic polymers are a special class of polymeric materials having a hydrocarbon backbone containing pendant acid groups. These are then neutralized partially or fully to form salts. Ionomeric TPEs are a class of ionic polymers in which properties of vulcanized rubber are combined with the ease of processing of thermoplastics. These polymers contain up to 10 mol% of ionic group. These ionomeric TPEs are typically prepared by copolymerization of a functionalized monomer with an olefinic unsaturated monomer or direct functionalization of a preformed polymer [68–71]. The methods of preparation of various ionomeric TPEs are discussed below.

5.4.4.1 Metal-Sulfonated EPDM

The preparation of metal-sulfonated EPDM consists of two steps: sulfonation of EPDM and neutralization of the free acid groups [69,72]. Typical diene monomers that can be sulfonated include 5-ethylidene-2-norbornene (ENB), end-dicyclopentadiene (DCPD) and 1,4-hexadiene. The free form of S-EPDM is prepared from EPDM by using a sulfonating agent, which is added to the hydrocarbon (aliphatic) solution of the former at room temperature. After about half an hour a chain-terminating agent (i.e., isopropanomethyl-6-*tert*-butylphenol) is added to terminate sulfonation reaction. The resultant polymeric sulfonic acid is separated through steam stripping, washed with water, and dried. Neutralization of S-EPDM can be carried out in one of the two ways. In the first case, the polymer is redissolved in a mixture of toluene–methanol or hexane–alcohol to which metal acetate solution in water or water–methanol is added to neutralize the polymer. In second procedure (direct neutralization), the resultant solution after sulfonation is treated with an alcohol or water–alcohol mixture or an aqueous solution of metal hydroxide or acetate [73].

5.4.4.2 Zinc-Maleated EPDM

Blends of ionomeric Zn-maleated EPDM rubber (Zn-mEPDM) and Zn-maleated PP (Zn-mPP) show behaviors typical of TPEs [74]. The composition ranges of 90–10 (Zn-mEPDM–Zn-mPP, wt%)

to 60–40 is found to be more suitable for this purpose. Zn-mEPDM is prepared by the addition of ZnO into maleated EPDM. The mixing processes depend on the ethylene content of maleated EPDM. The mixing of ZnO into maleated EPDM of low ethylene content is done on an open two-roll mill, while that in high ethylene-containing EPDM is done in a Brabender Plasticorder at a temperature of 70°C and at a rotor speed of 60 rpm.

5.4.4.3 Sulfonated-Styrene(Ethylene-Co-Butylene)-Styrene Triblock Ionomer

Weiss et al. [75] have synthesized Na and Zn salt of sulfonated styrene(ethylene-co-butylene)-styrene triblock ionomer. The starting material is a hydrogenated triblock copolymer of styrene and butadiene with a rubber mid-block and PS end-blocks. After hydrogenation, the mid-block is converted to a random copolymer of ethylene and butylene. Ethyl sulfonate is used to sulfonate the block copolymer in 1,2-dichloroethane solution at 50°C using the procedure developed by Makowski et al. [76]. The sulfonic acid form of the functionalized polymer is recovered by steam stripping. The neutralization reaction is carried out in toluene–methanol solution using the appropriate metal hydroxide or acetate.

5.4.5 Thermoplastic Elastomers from Waste Rubbers

Thermoplastic elastomeric compositions can also be obtained from the blends of recycled rubbers and plastics. Nevatia et al. [77] have developed recently thermoplastic elastomeric compositions from reclaimed rubber and scrap plastics. The physical and dynamic mechanical properties, rheological behavior, and phase morphology of such blends have also been reported extensively. The blends are prepared from reclaimed NR and scrap LDPE using dicumyl peroxide and sulfur accelerators as curing agent in an open two-roll mill at high temperature. A 50–50 rubber–plastic blend is found to be the best for processability, ultimate elongation, and set properties. A sulfur-accelerator system is found to be better than a peroxide system for dynamic cross-linking of such blends. The waste rubber can be used to replace virgin rubber substantially in many existing thermoplastic elastomeric blends without much loss in properties. Recently, Jacob et al. [78] have reported the replacement of virgin rubber by ground rubber EPDM vulcanizates in EPDM–PP thermoplastic elastomeric composition. The blends of virgin EPDM-waste EPDM–PP at a constant rubber/plastic ratio of 70:30 show typical TPE-like morphology with a finely dispersed rubber phase in a continuous PP phase. The virgin EPDM and the rubber component present in waste EPDM together constitute the rubber phase in the EPDM–PP blends. Although the addition of waste-EPDM to virgin-EPDM–PP blend causes an initial drop in the mechanical properties, TPEs with enhanced properties are obtained at higher loading of waste-EPDM. Table 5.8 shows the mechanical properties of various blend compositions.

Naskar et al. have reported replacement of 50% of EPDM by ground rubber tire (GRT) in a thermoplastic–elastomeric blend of EPDM–poly(ethylene-co-acrylic acid) blend [79]. It is believed that surface modification of GRT imparts better physical properties of the GRT–plastic composites through improved adhesion between the composites. Adam et al. [80,81] and Furhrmann and Karger-Kocsis [82] have reviewed different surface modification techniques of polymers. Grafting of maleic anhydride (MA) is well-established and commercially exploited technique for modification of polymers [83–85]. Recently, Ray et al. [86] have developed thermoplastic elastomeric composition based on MA-grafted GRT. GRT powder is maleated in an internal mixture using MA and dicumyl peroxide at 160°C. The blend is prepared by mixing maleated GRT and EPDM first in Brabender Plasticorder at room temperature using cam type rotor and 60 rpm rotor speed followed by the addition of this mix to acrylated high-density PE (HDPE) at 160°C in a Brabender Plasticorder at a screw speed of 30 rpm. In the final stage of mixing, DCP is added and mixed thoroughly. Physical properties of 60:40 (rubber/plastic) thermoplastic elastomeric composition based on EPDM and acrylated-HDPE, where 50% EPDM is replaced by rubber hydrocarbon of maleated GRT, are shown in Table 5.9.

TABLE 5.8

Mechanical Properties of Thermoplastic Elastomer Composition with Varying Waste-Rubber Loading at Constant Rubber/Plastic Ratio of 70:30 (w/w)

Blend Compositions	F_0	F_{10}	F_{20}	F_{30}	F_{40}
R-EPDM[a]	70	63.0	56.0	49.0	42.0
W-EPDM[b,c]	0.0	22.7	45.4	68.0	90.7
	(0)	(7.0)	(14.0)	(21.0)	(28.0)
PP	30	30	30	30	30
DCP	1.2	1.2	1.2	1.2	1.2
Tensile strength (MPa)	4.83	4.30	4.20	4.20	5.50
Modulus (100%) (MPa)	3.51	3.77	3.32	2.99	3.79
Elongation at break (%)	220	183	215	216	239
Tear strength (kN/m)	30.2	25.8	26.4	24.5	29.0
Tension set at 100% elongation (%)	14	26	18	14	14
Toughness (J/m^2)	3323	2794	3079	3136	4039

Source: Jacob, C., De, P.P., Bhowmick, A.K., and De, S.K., *J. Appl. Polym. Sci.*, 82, 3304, 2001.

[a] Raw EPDM.

[b,c] Values in parentheses are the rubber content in W-EPDM. Ratios of W-EPDM to R-EPDM in different mixes were as follows (total rubber content in each case being constant at 70 parts): F_0,0:100, F_{10},10:90, F_{20},20:80; F_{30},30:70; F_{40},40:60.

5.4.6 MISCELLANEOUS NEW THERMOPLASTIC ELASTOMERS

In addition to the two-phase TPEs, two new technologies have emerged. They are the metallocene-catalyzed polyolefin plastomers (POPs, the name given to Exxon's EXACT product line) and polyolefin elastomers (POEs, DuPont Dow Elastomer's ENGAGE), and reactor-made thermoplastic polyolefin elastomers (R-TPOs). These new types of TPEs are often called metallocene elastomers–TPEs (MEs–TPEs) [87]. The new POPs and POEs are essentially very low-molecular-weight–linear low-density PEs (VLMW–LLDPE). These new-generation TPEs exhibit rubber-like properties and can be processed on

TABLE 5.9

Physical Properties of the 60–40 Rubber–Plastic Blends Containing Ground Rubber Tire and Maleated-GRT

GRT Type in the Blend Composition[a]	Tensile Strength (MPa)	Ultimate Elongation (%)	Young's Modulus (MPa)	100% Modulus (MPa)	Work of Rupture (kJ/m^2)	Tear Strength (kN/m)	Tension Set at 100% Elongation (%)	Hardness (Shore A)
GRT (acetone extracted)	8.2	152	36.6	7.8	4.5	56.6	24	82
GRT (unextracted)	7.1	203	22.3	6.2	5.3	60.8	26	80
m-GRT (acetone extracted)	10.3	158	41.2	9.9	5.9	62.5	24	85
m-GRT (unextracted)	9.0	190	37.2	8.5	6.2	62.0	26	82

Source: Naskar, A.K., Bhowmick, A.K., and De, S.K., *Polym. Eng. Sci.*, 41, 1087, 2001.

[a] EPDM/GRT/acrylated-HDPE 30/68/40 (w/w).

conventional thermoplastic equipments. They are synthesized using metallocene-based catalyst and could be tailored to desired specific properties not achievable with the conventional TPOs available in the market from 1980s. Some of their characteristics are (1) low temperature toughness, (2) chemical resistance, (3) ozone and UV resistance, (4) wide service temperature range from −55°C to 120°C, (5) low density, (6) no plasticizer migration, (7) high tear strength, (8) good printability, etc.

TPEs can be obtained from many conventional homopolymerizing olefins, like propylene, ethylene, and 1-octene, using suitable metallocene catalyst or cocatalyst composition. A typical catalyst composition is metallocene of Ti, Zr, or Hf and methylaluminoxane (MAO) as cocatalyst. In this process, polymerization occurs along two propagation paths, wherein one path causes homopolymerization of the monomer stereoselectively into crystallizable blocks in the resulting polymer chain, and the other path creates irregular noncrystallizable amorphous blocks [88]. Tullock et al. [89] synthesized PP exhibiting elastomeric properties using bis(arene)Ti, Zr, and Hf catalyst with materials of heterogeneous composition. A complete review on propene-based elastomers by transition metal catalyst is available [90].

Graft copolymers are not as common as TPE family. It has already been demonstrated by Jacob et al. [91] that graft copolymer molecular architecture can enhance TPE properties. To achieve acceptable mechanical properties with graft copolymers, it is necessary to use a rubbery backbone with glassy branches at two or more branch points and a fairly high molecular weight of rubbery backbone and branches [92]. Kennedy et al. reported that graft copolymers with trifunctional branch points randomly distributed along the backbone can lead to TPEs with improved properties [91,92]. Weidish et al. [93] reported PS–PI multigraft copolymers with tetrafunctional branch architecture. The materials comprise of PI branches with two PS branches grafted at each of several tetrafunctional branch point per molecule. Multigraft copolymers, with 8%–67% PS and 7–9 branch points per molecules, show unoriented morphology and, thus, their mechanical properties do not depend on the direction of stressing. Materials between 8% and 37% PS exhibit the behavior of TPEs as indicated by large increase of stress at higher strain. While the materials containing 8% PS and eight branch points show a low tensile strength, multigraft copolymers with 22% PS and seven branch points reveal both high tensile strength and high elongation at break. With increasing PS content, strain at break decreases dramatically without corresponding increase in strength. Modern synthetic techniques [94,95] even allow to control backbone molecular weight, and number and placement of branch points along the backbone. The presence of two glassy PS branches at each branch point provides enhanced stress transfer between the rubbery matrix and PS domains as compared to triblock copolymers and trifunctional multigraft copolymers. Figure 5.4 shows PI backbone has

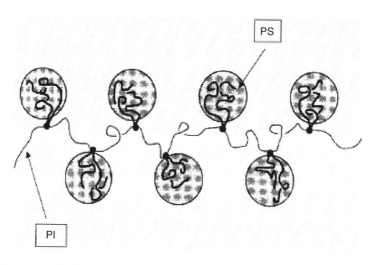

FIGURE 5.4 Polystyrene (PS) branches grafted on to the multiple branch points along the polyisoprene backbone.

multiple branch points connected with PS chains. In microphase-separated morphology, the junction points reside at the interface between PS cylindrical domains and the PI matrix [93]. Linear chain-branched PEs having crystallizable branches grafted on an amorphous and plastomeric copolymer backbone show TPE character. Recently, Markel et al. [96] have reported the synthesis and characterization of such TPEs. These are synthesized using mixed metallocene catalyst, for instance Cp_2ZrCl_2 and $(C_5Me_4SiMe_2NC_{12}H_{23})TiCl_2$ activated with MAO. Conceptually, the synthesis can be divided into two steps: (1) generation of vinyl terminated, crystallizable macromonomers and (2) incorporation of these macromonomers onto the copolymer backbone.

5.4.6.1 Protein-Based Thermoplastic Elastomers

TPEs traditionally have been synthesized with well-defined synthetic organic blocks of compositional dissimilarity that display phase-separated morphology and offer unique technical properties. However, the synthetic repertoire of the block copolymers has been so far limited to the tapered blocks of uniform sequence, which restricts preparation of controlled architecture of organizational and functional complexity of protein-based structural materials that exist in nature [97]. There is an increasing demand to develop biocompatible smart building blocks and functional molecular devices through self-assembly and natural functional materials, particularly peptides and proteins remain an inspiration for the quest. All the natural proteins are biopolymers based on unique linear arrangement of just 20 different unique amino acid monomers (Table 5.10) [98]. However, the self-organized folded proteins achieve a tremendous breadth of physical and chemical activities, ranging from exquisitely specific room temperature catalysis to the formation of unusually strong and tough biomaterials, such as collagen and spider silk. The complexity of the chemical structure of biopolymers and the general inability of amino acids to yield to standard synthetic protocol and physical processing methods restrict their production synthetically. However, the emergence of genetic engineering of synthetic polypeptides has recently enabled to synthesize protein-based analogues of conventional TPEs, artificially, by incorporation of two or more incompatible protein-based tandem sequences of different chemical, physical, and biological characteristics [99–102]. The resulting block copolypeptides would confer highly sophisticated chemical structure, unique self-assembly, functionality, and physiochemical properties to the resultant materials. This novel technique has opened up unlimited possibilities to synthesize new functional materials. Protein polymers can be designed to display a wide variety of properties including stimuli-responsive block copolymer assemblies; gelation under physiological conditions; film and fiber formation; adhesion to synthetic surface; controlled restorability; micelles for drug delivery; biological recognition; and chemical, enzymatic, and therapeutic activity.

A significant research has been dedicated in elastin–mimetic protein polymer and a versatile class of TPE based on block copolypeptide has been developed. It has been identified in early 1980s that the key structural feature that introduces elasticity to the protein of short repeat sequence contains proline and glycine residue. For example, the elastic proteins elastin and spider flagelliform silk are dominated by the pentapeptide repeats valine–proline–glycine–valine–glycine (VPGVG) [103,104] and glycine–proline–glycine–glycine–x (GPGGX) [105–107], respectively. Elastin is the major structural protein of those tissues, which required rapid extension and complete recovery (e.g., aorta, nuchal ligament). Thus significant focus has been placed on repeat sequence VPXYG [97,108]. It has been observed that the lower critical solution temperature (LCST) of this polypeptide can be manipulated by adjusting the identity of the fourth residue Y. In addition substitution of an ala residue for the consensus gly residue in the third position (X) of the repeat results in the change in the mechanical response of the biopolymer from elastomeric to plastic. The pentapeptide repeat residue has been the basis for the development of many ABA-type peptide-based triblock copolymer; some of them behave like hydrocarbon-based TPEs. Spontak and Patel [109] synthesized elastin–mimetic triblock polypeptides, ABA-type block copolymers using biological methods with a sequence based on the key elastin repeat sequence VPGVG that undergoes phase separation

TABLE 5.10

Standard Amino Acid Abbreviations

Amino Acid	Structure	3-Letter Code	1-Letter Code
Alanine		ala	A
Arginine		arg	R
Asparagine		asn	N
Aspartic acid		asp	D
Cysteine		cys	C
Glutamic acid		glu	E
Glutamine		gln	Q
Glycine		gly	G
Histidine		his	H
Isoleucine		ile	I

TABLE 5.10 (continued)
Standard Amino Acid Abbreviations

Amino Acid	Structure	3-Letter Code	1-Letter Code
Leucine		leu	L
Lysine		lys	K
Methionine		Met	M
Phenylanine		Phe	F
Proline		Pro	P
Serine		ser	S
Theonine		Thr	T
Tryptophan		trp	W
Tyrosine		tyr	Y
Valine		val	V

above a critical temperature. The gel reliquifies upon cooling and thus behaves like TPE. The A blocks chosen in the polymer are more hydrophobic than VPGVG. These hydrogels are also ionically and thermally responsive (i.e., they undergo large volume changes as a result of these responses). By combining polypeptide blocks with different inverse temperature transition values due to hydrophobicity differences of the blocks, it is possible to produce amphiphilic polypeptides that self-assemble into hydrogels and exhibit TPE-like behavior.

Capello et al. developed a technology for successfully combining sequential block copolymers consisting of tandem repetition based on the natural repeating structure of silk (hard block) and elastin-like amino acid sequence blocks in a high-molecular-weight protein polymer using genetic engineering and biological production method [110]. This new class of TPEs was described as "ProLastins." They consist of silklike and elastin-like blocks and may be defined using the general formula $\{[S]_m[E]_n\}_o$, where "S" is the silklike block having the general amino acid sequence "GAGAGS," "E" is an elastic-like block having the general amino acid sequence "GVGVP," m is from 2 to 8 (but may be as high as 16), n is from 1 to 16 (but may have no limit as long as there is at least one silklike domain on both ends of each elastic-like domain). O was adjusted from 2 to 100 with respect to n and m to yield protein polymer of molecular weight ranging from 60,000 to 85,000 [111,112]. The compositions of ProLastins used in the study are described in Table 5.11 and their expected primary sequences of amino acids are shown in Table 5.12.

Thermal analysis and rheological investigation confirmed that "ProLastin" solution undergoes self-organization resulting in gelation due to nonreversible crystallization event and the process can be controlled by adjusting copolymer structure and conditions that affect crystallization of the protein polymer. Figure 5.5a shows the increase in viscosity as a function of time for three representative block structures. It is clearly demonstrated that the gelation time is a function of the presence of number of silklike blocks. ProLastic 816 solution (20% w/w) which contains eight silklike blocks starts gelling within 30 min. It has also been observed that higher temperature promotes gelation for the polymer (Figure 5.5b). It has also been demonstrated that hydrogen-bond disrupters such as urea affect the rate of crystallization, whereas, nucleating agents accelerate the gelation process. The suitability of the polymer gel for localized drug release has been examined in detail and predicted that ProLastin can be used as a framework on which to custom design and build functional protein-based biomaterial system for drug delivery.

Elizabeth et al. [108] reported an elastin–mimic polypeptide sequence that mimics triblock copolymers (ABA) with the sequence of the respective blocks as

$$\{VPAVG[(IPAVG)_4(VPAVG)]_{16}IPAVG\} - \{X\} - \{VPAVG[(IPAVG)_4(VPAVG)]_{16}IPAVG\} \text{ where}$$
$$[X] = VPGVG[VPGVG)_2VPGEG(VPGVG)_2]_{30}VPGVG$$

TABLE 5.11
ProLastin Polymer Compositions

Polymer	Monomer Structure	Number of Repeats	Molecular Weight
ProLastin 27K (SELPOK)	$[S]_2[E]_4[EK][E]_3$	18	80,734
ProLastin 37K (SELP9K)	$[S]_3[E]_4[EK][E]_3$	12	60,103
ProLastin 47K (SELPSK)	$[S]_4[E]_4[EK][E]_3$	13	69,814
ProLastin 48 (SELP8)	$[S]_4[E]_8$	13	69,977
ProLastin 816 (SELP5)	$[S]_8[E]_{16}$	7	84,609

Source: Capello, J.W. Crissman, M. Crissaman, F.A. Ferrari, G. Textor, O. Wallis, J.R. Whitledge, X. Zhou, D. Burman, L. Aukerman, E.R., and Stedronsky, J., *Controlled Release*, 53, 105, 1998, Table 1.

TABLE 5.12

Amino Acid Sequence[a] of ProLastin Polymers

ProLastin 27K	Amino acids: 1000	MW: 80,734

MDPVVLQRRDWENPGVTQLNRLAAHPPFASDPM
GAGAGS GAGAGS[(GVGVP)$_4$GKGVP(GVGVP)$_3$(GAGAGS)$_2$]$_{12}$GA
MDPGRYQDLRSHHHHHH

ProLastin 37K	Amino acids: 749	MW: 60,103

MDPVVLQRRDWENPGVTQLNRLAAHPPFASDPM
[GAGAGS(GVGVP)$_4$GKGVP(GVGVP)$_3$(GAGAGS)$_2$]$_{12}$GAGA
MDPGRYQDLRSHHHHHH

ProLastin 47K	Amino acids: 884	MW: 69,814

MDPVVLQRRDWENPGVTQLNRLAAHPPFASDPM
GAGSGAGAGS[(GVGVP)$_4$GKGVP(GVGVP)$_3$(GAGAGS)$_4$]$_{12}$(GVGVP)$_4$
GKGVP(GVGVP)$_3$(GAGAGS)$_2$GAGA
MDPGRYQDLRSHHHHHH

ProLastin 48	Amino acids: 889	MW: 69,977

MDPVVLQRRDWENPGVTQLNRLAAHPPFASDPM
GAGSGAGAGS[(GVGVP)$_8$(GAGAGS)$_4$]$_{12}$(GVGVP)$_8$(GAGAGS)$_2$GAGA
MDPGRYQLSAGRYHYQLVWCQK

ProLastin 816	Amino acids: 953	MW: 84,609

MDPVVLQRRDWENPGVTQLNRLAAHPPFASDPM
GAGS (GAGAGS)$_3$[(GVGVP)$_{16}$(GAGAGS)$_8$]$_6$(GVGVP)$_{16}$(GAGAGS)$_3$GAGA
MDPGRYQLSAGRYHYQLVWCQK

Source: Capello, J.W. Crissman, M. Crissaman, F.A. Ferrari, G. Textor, O. Wallis, J.R. Whitledge, X. Zhou, D. Burman, L. Aukerman, E.R., and Stedronsky J., *Controlled Release*, 53, 105, 1998, Table 2.

[a] Single amino acid abbreviations used.

This polypeptide is structurally identical to ABA-type triblock copolymer with a central hydrophilic elastomeric end-block capped with two hydrophobic plastic end-blocks and exhibits amphiphilic characteristics. The end-blocks of the polymer were chosen in such a way that their LCST would reside at or near room temperature. Thus the polymer exhibits phase separation, which is analogue to conventional TPEs, and offers TPE gels under physiological relevant conditions [104]. Glutamic acid residue is placed periodically in the elastomeric mid-block to increase its affinity towards the aqueous

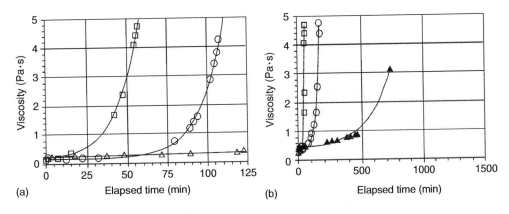

FIGURE 5.5 (a) Viscosity versus time plot of 20% solution of ProLastin 816 (□), ProLastin 47 (○), and ProLastin 27 (△). (b) Viscosity versus time plot of 20% solution of ProLastin 816. 37 (□), 23 (○), 4 (▲).

solvent resulting in enhanced segregation between the two blocks above an LCST. Differential scanning calorimetry (DSC) and variable temperature (VT) ^{1}H–^{13}C heteronuclear correlation through multiple quantum coherence (HMQC) nuclear magnetic resonance spectroscopy (NMR) confirm selective phase separation. The rheological behavior of the protein indicated that it behaved as gel at 25°C in that the G' exceeded that of G'' by nearly two orders of magnitude under these conditions. The morphology of the gel state was evaluated using cryo-high-resolution scanning electron microscopy (cryo-HRSEM), which exhibits microphase-separated morphologies with spherical micellar segregation of 20–30 nm in diameter. The underlying morphology of the protein aggregates within the gels can be more clearly observed from specimen in which vitreous ice has been etched from the surface via controlled sublimation.

Nagapudi et al. reported synthesis of high-molecular-weight recombinant protein ABA-type triblock copolymers that exhibit tunable properties that may find applications in novel scaffolds for tissue engineering and new biomaterials for controlled drug delivery and cell encapsulation [113]. They employed a modular convergent biosynthetic strategy which may greatly enhance the synthesis of triblock copolymer with significant flexibility in the selection and assembly of blocks of diverse size and structure. Development of large molecular weight ABA-type block-type protein-based biopolymers offers a unique opportunity to systematically modify microstructure on both nanoscale and mesoscale. They also demonstrated that through the rational choice of processing conditions, including solvent type, temperature, and pH, an array of TPE materials could be produced systematically from the ABA block copolymer that displays wide range of physical properties. Figure 5.6a shows the stress–strain relation for the films casted at different temperatures and pH from water or 2,2,2-trifluoroethanol (TFE) and significant difference in the films' stress–strain characteristics is clearly observed. TFE solvates both the blocks and renders the material plastic, whereas water preferentially dissolves the hydrophilic mid-block and generates elastomeric materials. Changing the pH of the solvent by addition of NaOH further enhances the elastic properties. Hysteresis behavior of the film cast from TFE and water is shown in Figure 5.6b. Petka et al. [114], Minich et al. [115], Breedveld et al. [116], Pochan et al. [117], and Nowak et al. [118] have recently reported the development of many different protein-based block copolymers that exhibit TPE-like behavior. The elastomeric behavior of the film cast from water is clearly observed from the figure. These functional materials expect to find applications in novel scaffold for tissue engineering and as new biomaterials for controlled drug delivery and cell encapsulation.

5.4.6.2 Hybrid Molecular Structure with Peptide and Nonpeptide Sequences

In protein-based biopolymers, particularly structural proteins, the conformational stability and precise chain sequences work together to hierarchically assemble three-dimensional complex fine structure from the linear macromolecules, which impart these materials with specific biological function. However, it is challenging to produce them in commodity volume, synthetically. Attempts to synthesize polypeptides with well-defined amino acid sequences, using synthetic protocol, have been plagued by unwanted side reactions, polydispersity, and poor organization. On the contrary, synthetic polymers can be manufactured in bulk at low cost. However, due to the lack of precise control over composition, architecture, and dimension, well-defined complex folding organization cannot be designed [119]. The access to a variety of synthetic building blocks in hydrocarbon polymers holds tremendous potential to create an improved class of nonnatural bioinspired polymers of hybrid molecular structure with peptide and nonpeptide sequences. These polymers can not only mimic the protein, structure, activity, and properties with novel backbone and side-chain chemistry, but also can be obtained at low cost, chemical diversity, and biological activity. The spectrum of polymeric materials and the wide range of possibilities of nonnatural sequence-specific block copolymers are unlimited [120–123].

Recently, synthesis of many compositionally and topologically different block copolypeptides has been reported [119,124]. In most cases, the copolypeptide block is composed of γ-benzyl

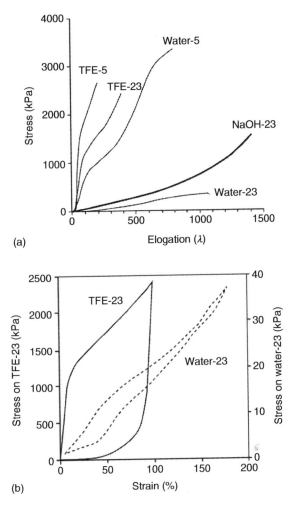

FIGURE 5.6 (a) Stress–strain relation for the films cast at different temperatures and pH from water or 2,2,2-trifluoroeyhanol (TFE). Significant difference in the films' stress–strain characteristics is clearly observed. TFE solvates both the blocks and renders the material plastic whereas water preferentially dissolves the hydrophilic mid-block and generates an elastomeric material. Changing the pH of the solvent by addition of NaOH further enhances the elastic properties. (b) Hysteresis behavior of the film cast from 2,2,2-trifluoroeyhanol (TFE) and water. Significant difference in the films' stress–strain characteristics is clearly observed. TFE solvates both the blocks and renders the material plastic whereas water preferentially dissolves the hydrophilic mid-block and generates an elastomeric material.

L-glutamate, β-benzyl L-aspartate, or Nε-benzyloxycarbonyl-L-lysine as the polymerization of these NCAs is the best controlled of all. The chemical nature of the synthetic block segment has been varied to a much greater extent using both hydrophobic [such as PS, polybutadiene, PI, poly(methyl methacrylate), poly(propylene oxide), poly(dimethyl siloxane)], and hydrophilic segment [poly (ethylene oxide), poly(vinyl alcohol), poly(2-methyl oxazoline), etc]. Highly-effective zero-valent organo nickel, such as 2,2′-bipyridyl Ni(1,5-cyclooctadiene) and cobalt complex initiators, which are able to eliminate significant competing termination and transfer steps from NCA polymerization to allow preparation of well-defined peptides, has recently been demonstrated by Deming et al. [125]. Recently, Seidel et al. [126] also reported the use of chiral ruthenium and iridium amido-sulfona-midate complex for controlled enantioselective polypeptide synthesis. These new methods can now be employed successfully to realize stereocontrol in NCA polymerization that has been difficult to achieve before.

Recently, unique vesicle-forming (spherical bilayers that offer a hydrophilic reservoir, suitable for incorporation of water-soluble molecules, as well as hydrophobic wall that protects the loaded molecules from the external solution) self-assembling peptide-based amphiphilic block copolymers that mimic biological membranes have attracted great interest as polymersomes or functional polymersomes due to their new and promising applications in drug delivery and artificial cells [122]. However, in all the cases the block copolymers formed are chemically dispersed and are often contaminated with homopolymer.

5.5   STRUCTURAL AND PHASE CHARACTERISTICS

TPEs demonstrate a number of unique properties as a result of their morphological features and have one feature in common—they generally exhibit a phase-separated system in bulk. One phase is hard at room temperature while the other phase is soft and elastomeric. The microphase or mesophase-separated morphology is made possible by the type of macromolecular architecture. TPEs should have rubber-like plateau, which is as wide and as flat as possible and that should be adjustable according to certain applications. The properties of TPEs are strongly sensitive to which phases are present and how they are spatially arranged (for instance, disperse versus co-continuous). Important parameters in this respect are crystallinity and crystal imperfection of the hard segment, composition of the amorphous phase, and continuity of different phases. There exist many important structural differences among the different block copolymers. Styrenics are represented by SES (S—PS, E—elastomer) or a branched structure of general formula $(S - E)_n X$ ($n =$ multifunctional junction points). The remaining block copolymers TPU, TPA, and COPE have formulae (H − E), H = hard block. An important structural parameter that affects the ultimate properties of such polymer is the block frequency distribution. The average block length depends on the composition as well as molecular weight of the starting materials. The hard segment length is very important in most of these materials because the extent of phase separation depends significantly on block length. Monte Carlo [127–129] simulation has also been used to derive hard segment molecular weight distribution, MWD under ideal and nonideal conditions. In general, longer average sequence length for hard segment block offers increased crystallinity and modulus. Other parameters of chain microstructure such as tacticity, cis–trans-isomerization, and copolymerization content also have very important influence on the performance of the TPE. These parameters are usually characterized by using NMR [130] and, in some polymers, by infrared spectroscopy (IR) technique.

In block copolymers, incompatibility does exist, but phase segregation is restricted to microscopic or mesoscopic dimensions and the equilibrium morphological state of the microphase is a highly organized domain structure. There is a critical molecular weight of each block below which phase separation is insufficient [131]. Phase separation also becomes more difficult as the number of blocks increases [132]. In linear triblock copolymer of the type ABA (where A is the hard and B is the soft segment), the free chain ends are associated together to form discrete microdomains of the hard phase (as are large number of particles of rigid glass polymer that are dispersed in a soft rubbery matrix), which eventually behaves as thermo reversible physical cross-links. Continuous three-dimensional network structure is obtained, as many flexible rubber chains emanate from each domain and terminate in the neighboring domain as shown in Figure 5.7. At temperature above the softening point of the glassy domains the material can be made to flow (Figure 5.8a). The lower bound of the service temperature of these materials is governed by the glass-transition temperature, T_g value of the continuous rubbery phase.

The morphology of the ABA-type linear block copolymers is strongly influenced by the volume fraction of the two components. For example, in PS–EB–PS-type block copolymer as the volume fraction of PS is increased, the shape of the dispersed PS phase changes from spherical (comprising body-centered cubic spheres of PS dispersed in continuous soft phase) to cylindrical form (hexagonal packed cylinders of PS) [10,133,134]. When the volume fraction of the two phases

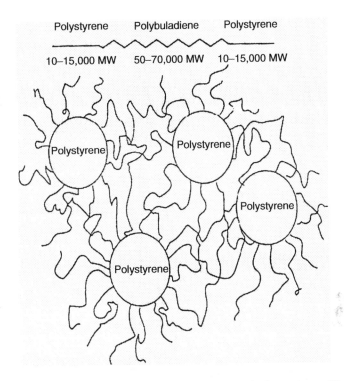

Polystyrene   Polybuladiene   Polystyrene

10–15,000 MW   50–70,000 MW   10–15,000 MW

FIGURE 5.7 Phase separation in styrene–butadiene–styrene (SBS) triblock copolymer. The isolated spherical styrene domains form the hard phase, which act both as intermolecular tie points and filler. The continuous butadiene imparts the elastomeric characteristics to this polymer. MW = molecular weight. (From Grady, B.P. and Cooper, S.L., *Science and Technology of Rubber*, Mark, J.E., Erman, B., and Eirich, F.R. (eds.), Academic Press, San Diego, CA, 1994. With permission.)

becomes comparable a lamellar morphology is developed containing alternating layers of the components and the materials lose their TPE characteristics [134]. With the change in temperature, these various forms of morphology may undergo reversible transition from one to another [135,136]. For example, SIS triblock copolymer has cylindrical morphology of the dispersed phase at 150°C, which changes to spherical microdomains at 200°C [135]. In addition to dispersed glassy domain, multifunctional physical cross-links can be formed by regions of crystallinity, ionic attraction, or strong hydrogen bonding in case of functionalized materials.

There are several models for describing the microphase separation and domain morphologies, e.g., confined chain model [137,138] and mean field model [139–141]. However, these models are restricted to simple diblock or triblock copolymers. Suitable models for microphase separation in multiblock segmented copolymers, in particular, to those containing crystallizable components such as TPU or COPE are still lacking in wide scale. It is generally observed that in the melt state these materials have a homogenous mixed phase of the two segments. Upon cooling to below the melting temperature, the microphase separation is facilitated by the crystallization process of the hard segments and no phase separation precedes the crystallization process [142]. Recently, a new model has been proposed by Gabrieelse et al. to describe the microstructure and phase behavior of block co-poly(ether–ester) TPEs [143]. Unlike previous concepts, which conclude into a distinct biphasic morphology of block co-poly(ether–ester) (a crystalline polyester phase and a homogeneous amorphous phase of polyether mixed with noncrystalline polyesters [142,144–147]), this new model proposes the amorphous phase as inhomogeneous and segmented into microphases. This effect is pronounced in co-poly(ether–ester) with a relatively large concentration of soft segment and relatively long soft-segment block lengths.

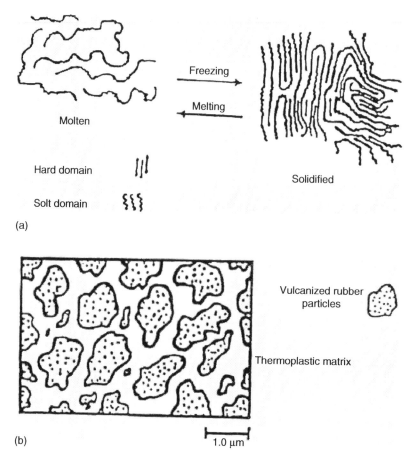

(a)

(b) |———| 1.0 μm

FIGURE 5.8 (a, b) Morphology of block copolymer thermoplastic elastomers (TPE) in molten and solidified rubber state. The transformation is reversible as the material passes through the melting point. Morphology of ethyl acrylate (EA) thermoplastic elastomers (TPE). The fine highly cross-linked rubber particles are distributed homogenously throughout the thermoplastic matrix. (From Arnold, R.L. and Rader, C.P., *Handbook of Plastics, Elastomer and Composites*, C.A. Harper (ed.), McGraw-Hill, New York, 1992. With permission.)

Unlike the block copolymers, the blends and alloys consist of two or more polymer systems. The phase morphology in rubber–plastic blend changes as a function of blend composition, intrinsic viscosity of the components, and the mixing conditions. Blends with a greater disparity in the component melt viscosity show larger domains. In some cases, the lower viscosity of the plastic at mixing temperature accompanying high proportion of rubber in the mixes favors continuity in both the phases [148]. The mixing time is a determinant of particle size and shape. Sometimes morphological changes after mixing occur by simply coarsening of rubber–plastic emulsion due to melt agglomeration [149]. In elastomeric alloys, dynamic vulcanization offers a route to control the particle size by resisting agglomeration, as a result of which the size of the rubber particles is reduced. Figure 5.8b represents the morphology of dynamically cross-linked blends. It consists of finely divided particles of about 1 μm diameter of highly cross-linked EPDM in PP. Bhowmick and Inoue studied the structure development during dynamic vulcanization of hydrogenated nitrile rubber (HNBR)–nylon blends through light scattering technique [150]. They found that the size of the dispersed HNBR particle in a 50:50 blend passes through a minimum with increasing mixing time at constant temperature. The increase in gel fraction of the HNBR and delaying the addition of vulcanizing agent also increased the dispersed particle when mixing was carried out at same temperature and for same duration. Kresge [151] reviewed the morphological changes that can result from processing blend compositions into finished product. Optimum dispersion of the rubber particles in plastic is achieved by matching the melt viscosities of the two phases and a closeness of their surface energies and significant interaction between them. For optimum properties, cross-linked rubber particles should be approximately 1–2 μm in diameter [140]. However, the MPR polymer blends are single-phase systems. Strong hydrogen bonding exists in MPR; thus miscibility is exhibited over the entire composition range and no phase separation occurs with the miscible blends of MPR polymer constituents [65,152,153].

Kader et al. [53] have described the morphology development of thermoplastic composition based on ternary blends of ACM, FKM, and polyfunctional acrylates (p-TMPTA). The changes in the blend morphology with varying compositions of ACM and FKM at a constant level of p-TMPTA are shown in Figure 5.9. The blends containing 70–30 ACM–FKM and p-TMPTA show phase separation. The dark phase in Figure 5.9a represents FKM, which forms a dispersed phase with elongated ellipsoid particle. On the other hand, ACM combines with acrylate to form the continuous phase (bright region). The sizes of the dispersed phase show polynodal distribution with varying particle dimension. It has been observed that at higher magnification, there is no sharp boundary between the dispersed phase and the matrix. Some of the ACM or polyacrylate are also included in the form of islands in the dispersed phase. This could be due to the miscibility of ACM–FKM blend [53]. With the increasing amount of FKM in the ternary blend of ACM–FKM–p-TMPTA to the level of 50–50–30 (w/w), the population of dispersed phase particles per unit area is increased with retention of phase-separated morphology (Figure 5.9b). When FKM concentration in such a ternary blend is increased further, more and more dispersed phase particles combine together and form macrophase leading to formation of co-continuous phase formation (Figure 5.9c). Figure 5.9d shows a transmission electron

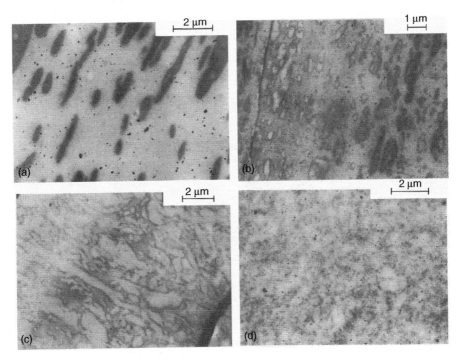

FIGURE 5.9 Transmission electron micrograph of (a) 70–30–30 (w/w) ACM–FKM–p-TMPTA blend; (b) 50–50–30 (w/w) ACM–FKM–p-TMPTA blend; (c) 30–70–30 (w/w) ACM–FKM–p-TMPTA blend; and (d) 100–30 (w/w) FKM–p-TMPTA blend. (From Kader, M.A., Bhowmick, A.K., Inoue, T., and Chiba, T. *J. Mat. Sci.*, 37, 6789, 2002. Courtesy of Springer.)

microscopy (TEM) photomicrograph of FKM–p-TMPTA blend. As the concentration of rubber (bright portion) is higher than that of the plastic, the blend behaves like a filled raw rubber.

Theoretical and experimental evidences on morphological features of ionomers show that salt groups in them exist in two different environments' multiplets and clusters [154,155]. The multiplets are considered to consist of small numbers of ion dipoles associated together to form higher multipoles—quadrupoles, hexa or octapoles, etc. These multiplets are dispersed in the hydrocarbon matrix and are not phase separated from it. In addition to acting as cross-links, they affect the properties of the matrix, e.g., T_g and water sensitivity. The clusters are considered to be small (<5 nm) microphase-separated zones rich in ion pairs in conjunction with ordered crystalline region, both of which are embedded in an amorphous matrix. There are few models that have been suggested for cluster morphologies. In one model, crystallites are represented as regularly folded section with disordered amorphous region acting as the third interconnecting phase (Figure 5.10a). They possess some properties of phase-separated systems. The proportion of salt group residing in the two environments in a particular ionomer is determined by the nature of the backbone, the total amount of salt group, and their chemical nature. Figure 5.10b represents a typical shell-core model for clusters [156,157]. It postulates that in the dry state a cluster of ~0.1 nm radius is shielded from surrounding matrix ion not incorporated in the cluster by hydrocarbon chain. The surrounding matrix ions that are separated from the cluster by hydrocarbon chains will be attracted to it by electrostatic forces.

The morphology is not very well defined in industrially produced TPEs, which are far from equilibrium. Prediction of precise morphology requires knowledge of sample history. Deviation from equilibrium morphology can also arise due to polydispersity in block length. There are various analytical methods to characterize the morphology of TPEs. TEM [158–161] IR [162–165], wide angle X-ray scattering (WAXS) and small angle X-ray scattering (SAXS) [166–170], small angle neutron scattering (SANS) [171], and NMR [172–175] techniques have been used efficiently and extensively by several authors for morphological characterization of TPEs. Further evidence for the two-phase structure can be deduced from thermal analysis of these materials [45]. They exhibit a glass-transition temperature (T_g) for the elastomeric phase and a melting endotherm corresponding to melting temperature (T_m) for the crystalline phase. Table 5.13 shows the T_g and T_m values for various TPEs.

(a)

FIGURE 5.10 (a) Scientific representation of the three-phase structure of a dry ionomer, consisting of cation clusters, lamellae, and disordered regions. (From Cowie, J.M.G., *Polymer: Chemistry and Physics of Modern Materials*, Intertext, London, 1973.)

(*continued*)

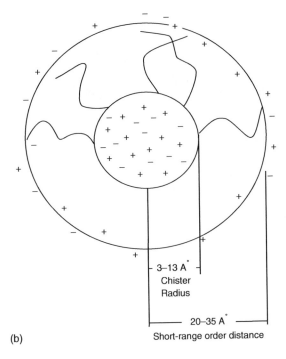

(b)

FIGURE 5.10 (continued) (b) Shell-core model for clusters. (From Macknight, W.J., Taggart, W.P., and Stein, R.S., *J. Polym. Symp.*, 45, 127, 1974. With permission of Wiley, New York.)

TABLE 5.13
Glass-Transition and Crystal-Melting Temperature[a]

TPE Type	Soft Rubber Phase T_g (°C)	Hard Phase T_g or T_m (°C)
PS–elastomer block copolymer		
SBS	−90	95 (T_g)
SIS	−60	95 (T_g)
SEBS	−60	95 (T_g)
Polyurethane–elastomer block copolymer	−40 (polyether)	190 (T_m)
	−60 (polyester)	
Polyester–elastomer block copolymer	−65 to −40	185 to 220 (T_m)
Polyamide–elastomer block copolymer	−65 to −40	120 to 275 (T_m)
Hard thermoplastic–elastomer blends[b]	−60	165 (T_m)
Alcryn	−40	—

Source: Holden, G., *Rubber Technology*, Morton, M., ed., Van Nostrand Reinhold, New York, 1987.

[a] Measured by differential scanning calorimetry.
[b] Values quoted are for polypropylene–EPDM blend.

5.6 THERMODYNAMICS OF PHASE SEPARATION

Phase separation between block components occurs since the dissimilar block components, such as homopolymers, are typically incompatible with one another as a result of positive heat of mixing. There is a fundamental difference between phase separation in a system of incompatible homopolymers and that in a corresponding block copolymer system. In the latter, the two incompatible components are chemically bonded to one another and hence the segregated phases are restricted from growing indefinitely in size. As a result, the domain sizes are much smaller in block copolymer in comparison to that observed in a physical mixture of the two corresponding incompatible homopolymers. Most theories take into account four factors that influence the phase separation of block copolymers: the Flory–Huggins interaction parameter λ, the overall degree of polymerization N, architectural constrains, and weight fraction of one component. By controlling appropriately the segment nature and length of each constituent of the block in block copolymers, a wide variety of microdomain structures of high degree of richness and complexity in bulk as well in solution phase are possible. Their complex structure has significant effect on the static, dynamic, and other functional properties of the polymer. In the ordered state, the block copolymers form "microdomains" with long-range order. They can also form a homogeneous structure in which the two segments are molecularly mixed in the "disordered state."

There are several theoretical models for describing microphase separation. The confined chain model presented by Meier [137] describes domain structure by random flight chains being constrained by barrier. The barriers represent the effect of the incompatibility between the components and the resulting restriction of the components to region of space. The mean field model developed by Helfand and Wasserman [140] discusses theoretical approaches of phase separation containing all the necessary parameters for a complete description in strong segregation limit (SSL). Three energetic contributions are reported in the theory. Thus during microphase separation a loss of system entropy occurs in various ways: (1) confinement entropy loss as a consequence of confinement of joint degree of freedom, (2) conformational entropy loss resulting from various numbers of polymer conformations that are not allowed as they produce a lower density near the domain than at the boundaries, and (3) enthalpy caused by mixing of A and B segments. With the growth of microdomains, the entropy loss increases until it balances the decrease in surface energy and a thermodynamically stable size is attained, e.g., spherical, cylindrical, or lamellar microdomains (actually, meso-domains with size ca. several tens of nanometers). Self-consistent mean field theory (SCFT) has been the most successful, in describing the phase behavior of high-molecular-weight polymers [176–179]. Using SCFT approximation, where thermal fluctuations are ignored, the phase diagram for conformationally symmetric diblock may be parameterized using only two parameters: χN—the segregation parameter, and f—the block length ratio, and is shown in Figure 5.11a. The

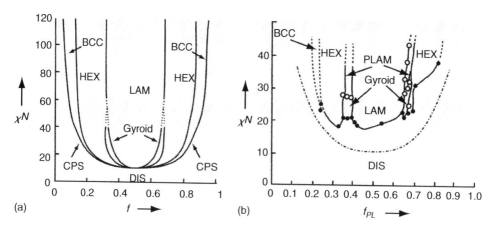

FIGURE 5.11 Theoretical (a) and experimental phase diagram (b) for block copolymers. Various self-organized morphologies can be obtained through variation of the block length, f and the segregation parameter, χN. (Adopted from (a) Foster, S., Khandpur, A.K., Zhou, J., Bates, F.S., Hamley, I.W., Ryan, A.J., and Bras, W., *Macromolecules*, 27, 6922, 1994. With permission. (b) Khandpur, A.K., Foster, S., Bates, F.S., Hamley, I.W., Ryan, A.J., Bras, W., Almdal, K., and Mortensen, K., *Macromolecules*, 28, 8796, 1995. With permission.)

experimentally observed phase diagram of a typical diblock copolymer, PS–PI, is also shown in the same figure (Figure 5.11b) [180,181]. The phase diagram exhibits several regimes of the phase segregation. For any fixed value of f, with increase in χN, one passes through a disordered regime, where the melt exhibits a disordered state, to a weak segregation regime (WSR), to an intermediate segregation regime, to finally the strong segregation regime (SSR). The sizes of the A- and B-rich domains in WSR are of the same order as the interfacial regions around the bonding points, and in SSR the domain size is much larger than the interfacial length. Thus, from the thermodynamic concept there are two limits: the strong segregation limit (SSL) and weak segregation limit (WSL). In SSL, the equilibrium state of the material consists of relatively pure phase of A separated from relatively pure phase of B (ordered state). In the WSL, the two phases are intimately mixed. The transition from the disordered state (WSL) to ordered state (SSL) is known as microphase separation transition (MST) or order–disorder transition (ODT) and the reverse is called microphase dissolution [179,182,183]. The rich polymorphism near the ODT in WSL, where $\chi N \sim 10$, is remarkable (Figure 5.11b). In SSL, at large values of χN, limited morphologies are stable. In diblock copolymers the microdomain structures formed can be controlled in a systematic fashion between 10 and 200 nm depending on the composition, molecular dimension, and architecture.

Microdomain structure is a consequence of microphase separation. It is associated with processability and performance of block copolymer as TPE, pressure sensitive adhesive, etc. The size of the domain decreases as temperature increases [184,185]. At processing temperature they are in a disordered state, melt viscosity becomes low with great advantage in processability. At service temperature, they are in ordered state and the dispersed domain of plastic blocks acts as reinforcing filler for the matrix polymer [186]. This transition is a thermodynamic transition and is controlled by counterbalanced physical factors, e.g., energetics and entropy.

Figure 5.12 schematically depicts the transition in "AB" diblocks. As the percentage of "A" hard block increases, the equilibrium morphology changes from body-centered cube (BCC) to hexagonally packed cylinder (HEX) to lamellar morphology (LAM) according to the sequence shown. Besides these simple morphology various complex minimal surfaces (gyroid, F-surface, P-surface), simple modulated lamellae (MLAM), and perforated lamellae (PLAM) are also observed in many specific diblock copolymers (Figure 5.11). However, this is also to note that the self-organization in block copolymer is a spontaneous process based on the information intrinsic to the assembling unit; and the complexity and multiplicity of the morphology increases with the number of blocks present in the copolymer. The complexity of the morphological states will increase even further if one of the synthetic blocks in the multiblock copolymer is replaced by protein-peptide segments (Figure 5.13) [123,187]. To understand the availability of the numerous morphological organizations available in multiblock complex materials, significant research efforts need to be dedicated in the area.

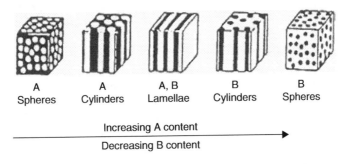

| A Spheres | A Cylinders | A, B Lamellae | B Cylinders | B Spheres |

Increasing A content

Decreasing B content

FIGURE 5.12 Change in the morphology of an A–B–A block copolymer as a function of composition. (From Walker, M. and Rader, C.P. (eds.), *Handbook of Thermoplastic Elastomers*, Van Nostrand Reinhold, New York, 1988.)

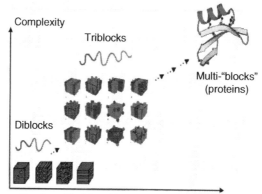

Number of building blocks along the chain

FIGURE 5.13 Complexity diagram for block copolymers. (Adapted from Morton, M., *Thermoplastic Elastomer: A Comprehensive Review*, Legge, N.R., Holden, G., and Schroeder, H.E. (eds.), Hanser, Munich, 1987 and Simon, P.F.W., Ulrich, R., Spiess, H.W., and Wiesner, U., *Chem. Mater.*, 13, 3464, 2001. With permission.)

5.7 PROPERTIES OF THERMOPLASTIC ELASTOMERS

5.7.1 Factors Influencing the Properties

Since TPEs typically consist of two incompatible phases at service temperature, the final properties of TPEs are influenced by the properties of the individual phases and the interface. It is therefore necessary to consider several factors, like the constituents' nature, their molecular weight and relative amounts, processing history, etc., to understand the properties of TPE. The influences of various factors on TPEs' properties are discussed below.

5.7.1.1 Constituents of Thermoplastic Elastomers

The constituents of the TPEs play a vital role in determining the properties. In triblock ABA-type copolymer, the change in the center-block or the end-block results in difference in properties. In SIS, where isoprene is the center-block, the tensile strength does not vary with the styrene content, whereas in SBS, it does so with the styrene content. Despite the fact that T_gs for butadiene and isoprene segments are different, no marked effect due to this difference is observed in room temperature properties of the triblock TPEs [188]. However, substitution of α-methyl styrene gives rise to a tougher polymer that is partly due to higher T_g values of this block. This emphasizes the major role of the hard phase on the mechanical properties of TPEs. For segmented TPEs, e.g., COPE, TPU, and for blends, changing hard segment type will affect the crystallinity of the hard phase. In TPUs, symmetric diisocyanates produce strong TPEs, whereas substituents on the aromatic ring tend to reduce the mechanical properties. The properties of polyether–polyester rubbers depend on the structure and relative amounts of polyether soft segments and polyester soft segments used [189]. The soft segment type also influences the driving force for phase separation and hence the physical properties. It has been observed that soft segments, which are capable of strain crystallizing, produce tough material with higher tensile strength and tear resistance. Improved phase separation does not necessarily lead to improved properties. Though polyester-based materials do not show complete phase separation as both polyether [190] and polybutadiene soft segments [191] do, yet the polyester-based materials show better mechanical properties. The reason is the poor interfacial adhesion [192].

5.7.1.2 Molecular Weight

The properties of TPUs, COPEs, and TPAs are also influenced by the molecular weight and molecular weight distribution, e.g., molecular weight influences the kinetics of phase separation, which affect the mechanical properties. Increasing the molecular weight of the soft segment promotes phase separation in block copolymers during polymerization, which will reduce the fractional conversion. Soft segment with average molecular weight between 1000 and 5000 Dalton gives materials with optimum properties, below which adequate phase separation is not possible. Moreover, low-molecular-weight material does not allow strain crystallization of soft segment that imparts the strengthening process. The effect of hard segment block length is also important. The longer blocks lead to phase separation and better properties. Apostolov and Fakirov [193] studied the effect of block length on the deformation behavior of polyether esters by SAXS. The overall molecular weight of TPE has little effect on mechanical properties, provided it is greater than the threshold value. However, soft segment polydispersity does not remarkably alter the physical properties. Monodispersed hard segments give higher modulus and tensile strength presumably because of better packing of hard segments and improved phase separation [194].

Recently, Tong et al. [195] have shown that in methyl methacrylate-b-alkyl acrylate-b-methyl methacrylate (MAM) and SIS-type triblock copolymers, the ultimate tensile strength is inversely proportional to the molecular weight between the chain entanglements in the middle soft block at comparable proportion of the outer block.

5.7.1.3 Relative Amount of Hard and Soft Segments

The relative proportion of hard and soft segments is an important factor that decides whether the elastic properties of a TPE would be more close to those of cross-linked elastomer or of hard plastics. The material changes from a flexible elastic rubber to semirigid plastic with increasing hard segment content and at very high level of the hard segments the materials behave more like toughened plastic. Therefore, TPE features are obtained within certain composition range of the components. In case of multiblock-segmented block copolymers (like COPE, TPU, COPA, etc.), the performance characteristics of TPE depend on the weight fraction of crystallinity of hard phase and its T_m. Crystallinity affords a mean for rather large deformation in hard phase. The useful temperature range for a TPE is between T_m and T_g. Within this range it is elastomeric, below which it is brittle and above which the hard phase melts. Passage from T_g to T_m is reversible. The elastic properties of a TPE blend, e.g., Youngs and shear moduli are functions of elastic properties of the components. Strength of the hard phase represents limit for strength of such blend.

5.7.1.4 Effect of Processing

The conditions applied during processing and fabrication of TPEs strongly influence their morphological features and hence their final properties. ABA-type materials often show marked difference in the shape and orientation of the hard domains between the samples prepared by extrusion and film casting from solution. For example, SBS block copolymers when cast from solution usually show spherical shape of the dispersed PS domains with no orientation, whereas the same materials when extruded develop cylindrical nature of the PS domain with preferred orientation in the direction of flow. This results in highly anisotropic mechanical properties in the latter [133]. The rapid cooling of the melt after processing cycles freezes the structures induced in the melt by shear stresses. Arcozzi et al. [196] investigated the correlation between physical properties and processing condition on a styrene–butadiene block copolymer under different molding conditions. If unidirectional shear strains are frozen into the polymer as in injection and extrusion molding, the material will be anisotropic due to statistically fibrilar structure. If the shear strains applied are high enough, the domain structure is fine textured, an isotropic article will be produced, e.g., injection molding. If nondirectional shear strains are frozen into the material, the material will be isotropic.

5.7.2 General Properties

The majority of TPEs function as rubber at temperature as low as $-40°C$ or even lower as measured by their brittle point. The upper temperature limit is determined by the maximum temperature at which it can give satisfactory retention of tensile stress–strain and hardness properties. Table 5.14a

TABLE 5.14a
Key Properties of Generic Classes of Thermoplastic Elastomers

Property	Styrenic	Copolyester COPE	Polyurethane TPU	Polymide TPA	TPO	EA Two Phase	EA Single Phase	Ionomer
Specific gravity	0.9–1.1	1.1–1.3	1.1–1.3	1.0–1.2	0.89–1.0	0.94–1.0	1.2–1.3	0.95–1.95
Hardness, Shore	30A–75D	35A–72D	60A–55D	75A–65D	60A–75D	55A–50D	55A–80A	49A–90A
Low-temperature (continuous) limit (°C)	−70	−65	−50	−40	−60	−60	−40	−57
High-temperature limit (°C)	100	125	120	170	120	135	125	180
Compression set resistance at 100°C	P	F	F/G	F/G	P	G/E	F	G
Resistance to aqueous fluid	G/E	P/G	F/G	F/G	G/E	G/E	F/G	F/G
Resistance to hydrocarbon fluid	P	G/E	F/E	G/E	P	F/E	G/E	G

Source: Arnold, R.L. and Rader, C.P., *Handbook of Plastics, Elastomers and Composites*, Harper, C.A., ed., McGraw-Hill, New York, 1992. With permission.

Note: P, poor; F, fair; G, good; E, excellent.

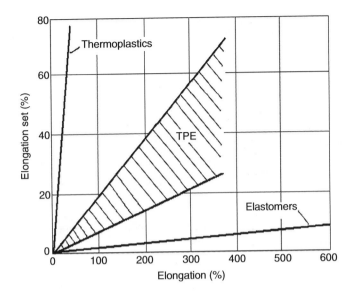

FIGURE 5.14 Elongation set properties of elastomers, thermoplastic elastomers (TPEs), and thermoplastics. (From Hofmann, W., *Kunststoffe*, 77, 767, 1987.)

reports various properties of different TPEs. The upper service temperature increases with cost. Styrenics with saturated soft block have higher heat resistance than those with diene soft blocks. Elastomeric alloys from saturated elastomer give better high-temperature performance character-istics than those from an unsaturated elastomer backbone. TPEs generally extend to high elongation and often in some cases with residual elongation or permanent set. Their set properties are in between elastomers and thermoplastics as shown in Figure 5.14. TPEs, toward the bottom of the curve scattered ranges, are close to elastomers and those toward the top are close to thermoplastics. Mullins effect or stress softening [197] occurs in most TPEs. Choudhury and Bhowmick [198] have observed a stress softening and strain hardening in NR–PE elastomeric alloys. This phenomenon occurs presumably because of two irreversible effects: (1) breakdown of weaker points of continu-ous hard or glassy phase and (2) orientation of the hard phase in the stretching direction. Pedemonte et al. [199] have reported similar stress softening and strain hardening behavior of thermoplastic block copolymers. A model proposed by the same authors explains that strain hardening is a result of interaction of rigid plastic domain.

In TPE, the hard domains can act both as filler and intermolecular tie points; thus, the toughness results from the inhibition of catastrophic failure from slow crack growth. Hard domains are effective fillers above a volume fraction of 0.2 and a size <100 nm [200]. The fracture energy of TPE is characteristic of the materials and independent of the test methods as observed for rubbers. It is, however, not a single-valued property and depends on the rate of tearing and test temperature [201]. The stress-strain properties of most TPEs have been described by the empirical Mooney–Rivlin equation

$$\sigma = (\rho RT/M_c + 2C_2/\lambda)(\lambda - 1/\lambda^2) \tag{5.1}$$

where
$\sigma =$ stress
$R =$ gas constant
$T =$ temperature
$C_2 =$ experimental constant
$M_c =$ molecular weight between cross-links
$\lambda =$ extension ratio

Swelling data and stress–strain results in SBS have shown that M_c is approximately the molecular weight between entanglements. Choudhury and Bhowmick [201] have shown in the case of different NR-based TPEs the following relation:

$$\sigma' \propto E \quad \text{and} \quad (1 + e_{bmax})^2 \propto 1/E \tag{5.2}$$

$$(1 + e_{bmax}) \propto M_c^{1/2} \tag{5.3}$$

where

M_c = the cross-link density of melt
σ' = stress
E = modulus
e_{bmax} = maximum elongation

When plastics act as a physical cross-link and strength properties are indirectly related to the modulus of hard phase and morphology of the blend, the filler effect is analyzed by the following equation:

$$E_F/E = (1 + 2.5\phi + 14.1\phi^2) \tag{5.4}$$

where

E_F/E = modulus of filler/modulus of unfilled elastomer
ϕ = volume fraction of filler

The above equations gave reasonably reliable M_c value of SBS. Another approach to modeling the elastic behavior of SBS triblock copolymer has been developed [202]. The first one, the simple model, is obtained by a modification of classical rubber elasticity theory to account for the filler effect of the domain. The major objection was the simple application of rubber elasticity theory to block copolymers without considering the effect of the domain on the distribution function of the rubber matrix chain. In the derivation of classical equation of rubber elasticity, it is assumed that the chain has Gaussian distribution function. The use of this distribution function considers that all spaces are accessible to a given chain. However, that is not the case of TPEs because the domain also takes up space in block copolymers.

Another comprehensive model [202] accounts for a possible contribution from trapped entanglement to the equilibrium elastic behavior, in addition to the factors considered by the other models. Inclusion of trapped entanglement causes the modulus to increase an order of magnitude than considering only the domains as cross-links. Hergenrother and Doshak [203] have reported the determination of the entanglement molecular weight of TPE (triblock copolymer) by tensile retraction measurement. Origin of rubber elasticity in EAs has been modeled by Kikuchi et al. [204] in the two-dimensional model with many rubber inclusions in ductile matrix. Finite element method (FEM) of the elastic plastic analysis on the deformation mechanism revealed that even at the highly deformed state at which almost whole matrix has been yielded by stress concentration, the ligament matrix between rubber inclusion in stretching direction is locally preserved within an elastic limit and it acts as an *in situ* formed adhesive for inter-connecting the two phase systems and also provides a key mechanism of strain recovery in the two phase systems.

The typical tensile behavior of TPE, with change in temperature, could be best represented by single parabolic curve, "Failure envelop." Figure 5.15 shows such a curve for NR–PE system. There are two extremes—one corresponds to an increase in strength with increase in breaking extension at higher temperature and low strain rates and the other showing high breaking strength and low extensibility. As the temperature rises in TPE, modulus and strength decrease due to softening of hard domain, in the vicinity of softening point, the properties decrease dramatically and

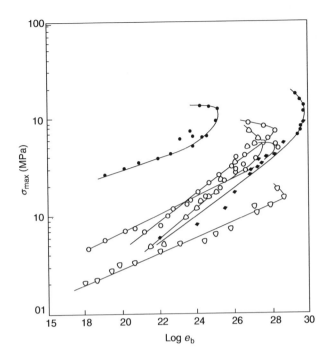

FIGURE 5.15 "Failure envelope" of various mixes: Δ, natural rubber–polyethylene (NR–PE) vulcanizate (peroxide cured); •, NR–PE vulcanizate (sulfur cured); □, NR–PE vulcanizate with CPE as compatibilizer; ▽, EPDM–PE vulcanizate; ○, EPDM–PP vulcanizate (sulfur cured); ◆, NR–ENR–PE$_m$–PE. (From Roy Choudhury, N. and Bhowmick, A.K., *J. Mat. Sci.*, 25, 161, 1990. With permission from Chapman & Hall.)

the material cannot be used as TPE [205]. Additives could be added to alter this temperature, but they may influence the domain structure. Most of the TPEs are in the high rubber-hardness range (above 80 Shore A). Styrenics are available in the soft hardness range (below 55 Shore A). EA with 40 Shore A hardness has been developed. The tear strengths of COPE and TPU are excellent. COPE also offers high fatigue resistance. TPU offers very good abrasion resistance. In general, TPE will be less rubbery and more like a thermoplastic as hardness is increased. Most TPEs have fair to good compression set resistance at ambient temperature. The set properties of the styrenics and TPOs become poorer at elevated temperature. The set resistance will decrease with increase in hardness. Marasch has reported a new, relatively soft TPU (polyether-based), which is claimed to have rubber-like behavior in terms of recovery after deformation [206].

The resistance of a TPE to different chemicals is influenced greatly by its chemical similarity to the fluid. The nonpolar styrenics, TPOs, and EAs have high resistance to polar chemical. Their resistance to nonpolar hydrocarbon ranges from poor to fair. The polar COPE, TPA, and TPU have better resistance to hydrocarbon fluid, but poorer resistance to polar chemicals than that of conventional rubbers. They are also very susceptible to oxidation at elevated temperature. Antioxidant and other additives could improve the chemical resistance of these materials. The resistance to many oils and greases is high for more polar TPEs. Hydrolytic stability is poor for the polyester-based TPU and copolyamides because the ester and amide linkages can be attacked by water. The addition of small amount of 1%–2% of hindered aromatic diamide can increase the lifetime significantly in hot water or steam. Ionomers swell when exposed to water. The electrical properties of the generic classes of TPEs are listed in Table 5.14b.

The electrical properties of styrenics, TPVs, and TPOs are very good. Their nonpolar nature allows their use as a primary electrical insulation where temperature and fluid resistance are

TABLE 5.14b
Electrical Properties of Generic Classes of Thermoplastic Elastomers

Properties	Test Method	Styrenic Elexar 8421	COPE Hytrel 4056	TPU Texin 480AR	TPA Estamid 90A	TPO TPR 1600	EA, Two Phase, Santoprene 101–73/201–73	EA, Single Phase, Alcryn R12028–80A
Volume resistivity (ohm · cm)	ASTM D257	3.5×10^{16}	8.2×10^{12}	2.1×10^{12}	8.1×10^{10}	3×10^{16}	2×10^{16}	1.4×10^{10}
Surface resistivity (ohm)	ASTM D257	6.2×10^{16}	—	3.3×10^{13}	3.1×10^{12}		-2×10^{16} (D275)	—
Dielectric constant								
60 Hz		2.4	—	6.34	10.3	2.21		—
1 kHz		2.4	5.1	6.08	9.3	2.22	2.95	11.1
1 MHz	ASTM D150	2.4	4.7	5.15	5.7	2.20	3.15	—
Dissipation factor								
60 Hz		0.0008	—	0.024	0.092	0.0005		—
1 kHz	ASTM D150	0.0008	0.008	0.019	0.066	0.0008	0.0039	0.40
1 MHz		0.0008	0.060	0.063	0.100	0.0001	0.0079	—
Dielectric strength, V/mil	ASTM D149	660	410	330	—	600	590	140

Source: Walker, B.M. and Rader, C.P. (eds.), *Handbook of Thermoplastic Elastomers*, Van Nostrand Reinhold, New York, 1988.

not very critical. These nonpolar TPEs are excellent starting materials because of their compatibility with wide variety of polymers, which provide low-cost and effective compounds. Polar TPEs, e.g., COPE, PU, and TPA, are also compatible with a wide variety of fillers, however, the modified compounds do not provide an outstanding balance of overall properties. Therefore, they have not made major penetration in stringent electrical applications compared to styrenics and TPVs.

5.8 RHEOLOGY

The melt rheological behaviors of TPEs are strongly dependent on shear rate, temperature, and the composition. TPEs exhibit wide variation in viscosity and elasticity. A wide variation in morphology (i.e., copolymers, physical blends, side-chain branching, etc.) results in semi-Newtonian to highly shear-thinning behaviors. There are fundamental differences in rheological responses between a conventional rubber and TPE. In the vicinity of T_g of a conventional rubber or T_g of the soft segment in TPEs, both are very resistant to stress and below this temperature they behave as a single-phase brittle glassy polymer. On increasing temperature, the hard domain starts weakening and flows. The melt rheology of TPE is thus related to that of hard segment and/or plastic materials. Vinogradov and coworkers [207] have reported the viscoelastic properties of SBS triblock copolymers at wide range of shear rate (hence shear stress) and their comparison with statistical copolymer and polybutadiene of comparable molecular weight. It was observed that SBS block copolymers show three distinct regions of melt flow behavior in the temperature range of 110°C–170°C. In the intermediate shear rate the block copolymers show pseudo-Newtonian flow behavior closely resembling the polybutadiene. At low and very high shear rates they show pronounced non-Newtonian flow behavior usually not observed in homopolymer or statistical copolymers. The block copolymers also show much higher melt viscosity and its dependency on temperature. The non-Newtonian behavior at low shear rate in SBS copolymer indicates its resemblance with filled polymeric system, where the dispersed filler particle shows structure formation [207]. Such unusual rheological behaviors are mostly attributed to the complex morphological features observed in triblock copolymers. The dispersed PS domains may undergo reversible breakdown or deformation under shearing at temperature close to the T_g of PS [133,207]. Such breakdown could further intensify with increase in temperature and shear rate causing significant lowering of viscosity of the material. The apparent viscosity depends strongly on the shear rate (and corresponding shear stress), which decreases sharply with shear rate (Figure 5.16). This implies that molding should be carried out at higher shear rate so that the low viscosity allows quick filling of the mold. Also parts could be removed rapidly from the mold. TPE viscosity is less sensitive to temperature than shear rate indicating the fact that input of mechanical energy is more effective during processing than the input of thermal energy. In an injection molding operation, higher injection pressure will have a significant impact on flow rate and fill time than a change in temperature. Wallace [208] has reported on the unique rheological characteristics of processing of MPR and polyolefin alloys in injection and blow moldings. Recently, Borgaonkar and Ramani [209] reported extrusion behavior of TPEs. Energy input to extrude and viscous dissipation are related to process parameters and polymer rheological characteristics. Metering zone temperature and revolutions per minute (rpm) are the most significant parameters to influencing melt temperature, extruder pressure, torque, and output.

The processing characteristics of TPE are intimately related to the ODT temperature or the MST temperature as the viscosity becomes lower above ODT. The styrene–diene triblock systems have ODTs above a reasonable processing temperature. Melt viscosities are usually higher in block copolymer than in homopolymer of the same molecular weight. Rheological measurements can give information of the ODT. Dynamic mechanical experiments are used as a macroscopic characterization of phase separation because highly phase-separated systems will have larger temperature difference between hard and soft segments and a feature plateau modulus. In one method, the

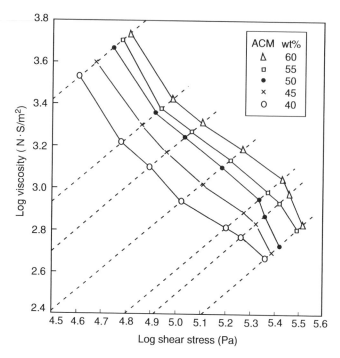

FIGURE 5.16 Viscosity–shear stress relationships for various compositions of nylon-6–acrylate rubber (ACM) blends at 240°C.

storage and loss moduli are measured as a function of frequency at different temperatures above and below ODT. The higher frequency data are put into one master plot by applying WLF equation [210]. Though WLF is not generally applicable to TPE as two phases make the material thermorheologically complex, entanglement dominates the response at higher frequency and microphase-separated nature has little influence on the rheological behavior [210]. However, below ODT, these properties are highly strain dependent. The critical frequency (ω_c) is first found out where the curves start diverging. Then, the dynamic measurements are carried out below ω_c. The temperature at which discontinuity occurs in the modulus marks the ODT as represented in the Figure 5.17. This method is based on different behaviors of storage and loss moduli at low frequencies for ordered ($G' \sim G'' \sim \omega^5$) and disorder ($G' \sim \omega^2$, $G'' \sim \omega$) systems. The most attractive property of TPEs is their ease of processing. They can be inexpensively processed into simpler or complex goods by a variety of methods. The extrusion and injection molding techniques are two of the most important processes for their fabrication. Other methods include blow molding, calendering, and solution fabrication. The saving in part fabrication cost by thermoplastic processing and from low specific gravity can often outweigh the difference in material cost of a TPE compared to a thermoset.

5.9 PROCESSING

5.9.1 DRYING

Moisture may cause both processing and aesthetic problems. So many TPE resins need to be dried before they are processed. A moisture content that is too high in molding or extruding can cause appearance of defects, such as splay marks and bubbles on the surface of the finished part. Besides, water may react with the polymer during processing, changing the physical performance of the final part. Hygroscopic polymers like polyamides, polycarbonates, polyesters, polyurethanes, etc., require careful drying prior to processing. Because of the chemical differences between SBCs,

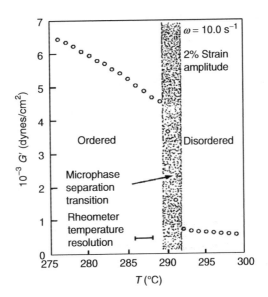

FIGURE 5.17 Temperature versus G′—the shear storage modulus at a frequency of 1.6 Hz for diblock copolymer poly(ethylene propylene)–poly(ethylethylene) (PEP–PEE). The order–disorder transition (ODT) calculated to be 291°C ± 1°C. (From Rosedale, J.H. and Bates, F.S., *Macromolecules*, 23, 2329, 1990. With permission of American Chemical Society.)

TPVs, TPUs, etc., different drying techniques are required. The drying of TPE is directly dependent upon the hygroscopic characteristics of the TPEs. TPVs, TPUs, thermoplastic polyesters, and TPAs are more hygroscopic than SBCs and recommended to be dried before processing using a desiccant drier. SBCs can be molded without drying. However, where aesthetic requirement is very stringent drying is a must [211].

5.9.2 EXTRUSION: A CONTINUOUS PROCESS

Extrusion can be used to manufacture a variety of TPE shapes, e.g., tubing, hoses, sheets, and complex profiles. A thermoplastic extruder with a screw length to diameter ratio L/D of 20:1 and preferably 24:1 to 32:1 is recommended for most TPEs since longer screw leads to better melting and mixing of polymers. Coextrusion of a TPE with a thermoplastic or another TPE is possible provided they are compatible and have melting points in the same range [212]. It is a cost-effective process by which soft TPE could be used as a sealing surface and a rigid hard plastic can act as a support. Crosshead extrusion is a popular technique of applying TPE insulating or jacketing layer on an electrical wire or cable or a cover over a reinforced hose.

5.9.2.1 Screw Design

The screw is the heart of the extruder, and screw geometry greatly affects the efficiency of extrusion. Extrusion of TPEs can be accomplished using general-purpose screws. Other aspects to be considered during extrusion are given below [213]:

1. Single-stage, three-zone, polyolefin-type metering screw
2. L/D ratio 24:1 minimum (for TPVs and SBCs)
3. L/D ratio 24:1 minimum for heavy wall extrusion (for TPUs)
4. L/D ratio 30:1 minimum for thin wall extrusion (for TPUs)
5. 2.5:1 to 3.1 compression ratio

6. Cooling at feed throat recommended
7. Screw cooling should be avoided
8. Dead or stagnation zones with that flat plate dies should be avoided

Screws having equal length of feed, compression, and metering zones are very popular in extruding TPEs. However, it is always advisable to use the suitable design for different goods. A variety of screw designs—polyolefin type, flighted barrier, pin mixing, and Maddock mixing—have been used. For even mold flow and clean melt stream, screen packs of 20–40–60 mesh are used.

5.9.2.2 Die Design

The main design criteria of most TPE dies are to ensure that changes in flow channel diameter from the extruder barrel bore to the die exit are equal. Most of the viscoelastic materials exhibit a die swell on exit from a die. TPEs tend to show die swell significantly lower than that of typical thermoplastics. This swell must be taken into consideration in designing dies and adjusting extrusion condition to achieve a perfect profile. The die swell normally increases with increasing hardness and shear rate and decreasing temperature.

5.9.2.3 Temperature Profile

The temperature has a strong influence on the viscosity of the melt, so its fluctuation should be kept in a narrow range to obtain good extrudate. The polymer melt temperature should be kept at about 30°C–70°C above the melting point of TPE. In general, the temperature profile for extrusion gradually increases from the feeding zone to compression zone and then gradually drops off in the metering zone. The die is kept at polymer melt temperature.

5.9.3 Injection Molding

Injection molding is the most widely used process for fabricating TPE parts [214]. It fully exploits the processing advantages of thermoplastics. Elimination of scrap and short cycles of this molding outweighs the higher material cost of TPE compared to a thermoset. TPEs can be molded in the same type of machine as used for conventional thermoplastics. Although ram injection molding has been successfully used for some TPEs, conventional reciprocating screw injection machines for these types of materials are recommended. The operating temperature for processing normally lies in the range of 20°C–50°C above the melting point of the TPE to allow adequate mold packing and minimum shrinkage. Good mold packing will give shrinkage as low as 1%–2.5% for most TPEs.

5.9.3.1 Injection Pressure and Rate

Too high or too low injection pressure will cause overfilling or incomplete filling of the mold, respectively. The typical injection pressure is kept in the medium range (34.5–103.4 MPa) for rapid mold filling. However, the hold pressure is kept at nearly half of the first stage injection pressure, for a period of 1–10 s to allow gate freeze off. The rate of injection depends mainly on the configuration of the part. Thicker parts require a slower speed whereas thinner sections require faster injection.

5.9.3.2 Screw Speed and Back Pressure

TPEs are quite fluid at shear rate above $500 \ s^{-1}$. A moderate speed is always preferred. A screw-back pressure below 1.38 MPa is sufficient for homogenization.

5.9.3.3 Mold Temperature and Cycle Time

Mold temperature and cycle time vary with part thickness, part configuration, and hardness and the temperature is kept normally in the range of 40°C–60°C. Proper mold temperature will ensure proper release of parts. The harder the TPE, shorter is the cycle time; thicker and complicated the part, longer is the cycle time. The cycle time contributes to the cost calculation of the molding application.

5.9.3.4 Part Design and Mold Design Considerations

In addition to the process parameters, there are large numbers of factors that decide the success of an injection molding operation. Mold design, product design, position, and size of gates, type and positioning of runners, etc., are extremely important factors to be taken care of. A detail consideration regarding these factors has been discussed below [213].

Mold Type
Two- and three-plate mold designs are commonly used for TPEs. Hot runner tools and stack molds may also be used. Molds made of metal with high thermal conductivities enable TPEs to achieve shorter cycle times when compared to lower thermal conductivity molds. For prototyping and limited-use tools, aluminum molds are suitable. Beryllium copper molds are preferred over steel for blow molds and some injection molds because of their higher heat conductivity. Prehardened steel (P20) injection molds are very durable and have low part cost for long run. Posthardened and ground molds have the highest durability for extended usage. TPEs can have a high coefficient of friction to most tooling materials and will tend to stick in nondrafted cavities and shrink onto long cores. TPEs release better from sand-blasted or EDM tool surfaces. Good results have been achieved with steel shot blasted surfaces. High draft (over 3° per side) may be helpful, or a release coating on the mold may be necessary. Teflon or molybdenum disulfide-type impregnated mold coatings offer some aid with softer TPEs.

Part Design
To meet the products' functional needs the wall thickness of the part can be varied. However, to obtain optimum processing cycles and reduce the chance of sink- and flow-mark formation, wall thickness should be kept as uniform as possible. Where variation in wall thickness is unavoidable, gradual transitions in the wall thickness should be provided to ease the polymer melt flow transitions. Part design, mold design, and gate location should work together to allow the material from heavier section into thinner walls for best appearance and process latitude.

Gate Design and Location
SBCs and TPVs are shear-thinning materials. Therefore, as such they are benefited from the flow restrictions at the entry of molds and this allows the provision for small diameter, short landed gates. For TPVs and SBCs, it is suggested that gates should be 20%–30% of the wall thickness of the part to be filled.

TPUs are not shear thinning and so gates with generous cross-sectional area allow the material to flow freely with minimum pressure loss. Usually, gate size should be 50% of the thickest section of the part. Fan gates and tab gates are the most, and pin gates are the least, suitable for TPU processing. The land length for gates should be kept as short as possible.

Common problems like insufficient filling–packing and poor dimensional control are often related to the gate size and design. Similarly, gate location is another important factor. They should be located in areas having heaviest cross-section of the part to assure fill-out and elimination of sink marks. Also their position should not facilitate the residual molded stress formation in the part, knit line formation.

Runners and Sprues
Full round, trapezoidal, or modified-trapezoidal runners are preferred for TPE processing, while half round runners give poor results. The length of the runners should be kept as short as possible. Hot runner systems using valve gates are typically used to reduce both gate freezing and cycle time. Cold slug catchers should be placed at all runner transition. The sprue should be as short as possible and polished in the direction of ejection. Hot sprue bushings are often used to reduce cycle time and

eliminate sprue sticking with softer grades of TPEs. For harder grades of TPEs tapered cone, "Z" pullers and "Christmas tree" sprue pullers are used depending upon grades and hardness [214].

5.9.4 BLOW MOLDING

By using injection and extrusion blow moldings, hollow and thin parts could be uniquely manufactured from TPEs [215,216]. Blow molding cannot be used for conventional thermoset rubber. Injection blow molding is a capital-intensive process whereas extrusion is the simpler process. The blow ratio (cavity diameter/parison diameter) of TPEs may be limited and parison diameter should be as large as possible related to mold cavity. Because of high viscosity at low shear rates, TPE melt retains their shapes in parison prior to blowing. The processing temperature, in the same range of extrusion or injection, is normally used. Cycle time is 10–240 s and could be reduced by rapid water circulation in the mold. Cakmak and Wang [217] reported that the tubular blown film of PP–EPDM elastomeric alloys exhibits complex superstructural behavior. The inner surface appears relatively featureless, while the outer surface exhibits bundle-like fibrillar structure probably due to rapid cooling of this surface during processing. The various technological advances in the blow molding of TPE have been discussed by Rader et al. [215].

5.9.5 OTHER FABRICATION METHODS

5.9.5.1 Thermoforming

TPEs in comparison to thermoset are also suited for thermoforming, which is a fast, low-cost production. The maximum draw ratio decreases with decrease in hardness. Draw ratio up to 0.1 can be attained with TPEs in the hardness range of 40–50 Shore D. The major limitations are to avoid undercut sharp corners, and adequate taper angle for rapid ejection of parts.

5.9.5.2 Foam Preparation

TPEs have been applied with considerable success in foaming extruded profiles, tubings, and sheeting with a thin solid skin over closed cell interior foam. TPVs could be foamed by water injection, compounded chemical blowing agent, and dry blended chemical blowing agents. SBC foams are made by dry blending chemical blowing agents and gas injection. Chemical foaming agent (azodicarbonamides) can reduce the bulk specific gravity of extrudates from 1.0 to 0.7 and physical blowing agents, e.g., fluorocarbons can lower down to even further about 0.3. Die swell in TPE foam extrusion is 10–50 times than that in dense TPE.

5.9.5.3 Combining TPEs with Other Materials: Bonding and Overmolding

TPEs are often employed to produce a resilient and soft ergonomic layer over a hard structural part. This could be done in a number of ways [213]. TPEs can also be overmolded onto an engineering material through insert molding or multiple-shot injection molding.

Bonding
TPEs can be bonded to other materials by adhesive, heat bonding, electromagnetic filling, radio frequency, heat-sealing lamination, friction and spin welding, and ultrasonic welding. For TPUs, the most widely used techniques are radio frequencies, and ultrasonic and hot stamping. A few typical applications include football bladders, valves, and conveyer belts.

Overmolding
Overmolding is the process by which two different materials are joined into one assembly without using secondary operations like gluing or welding. In case the materials are chemically compatible, chemical bonds may form between them and so mechanical interlocks are not required. There are two common techniques of overmolding—insert molding and multiple-shot injection molding.

Insert molding is accomplished in two stages. In the first step, the more rigid part is made in one machine and in the second step the elastomer is injection molded over the rigid part in a second machine. The tool and part design in this kind of process has few limitations. The rigid part should be stiff enough to resist the high hydraulic forces experienced when elastomer enters the tool and impinge on the parts. Shuttle press and rotary press are commonly used. Insert molding is used when the rigid part is metal or thermoset or the substrate requires a primer. This technique is also used for thermoplastic substrate when multiple-shot equipment is not available.

Multiple-shot overmolding process uses a specialized machine with multiple plastic processing barrels and a set of pivoting molds. In this process, the substrate on which TPEs are injected are thermoplastic in nature.

5.9.6 REGRINDING OF TPES

Because of their heat fusible nature, TPEs provide option of having sprues, runners, and scrap parts reground. This provides an exceptional economic advantage as compared to traditional elastomers. Since TPEs are often supplied in a variety of soft grades with high elongation and tear strength, high-quality grinders are required to get efficient regrinding. Only grinders with high-quality support bearings and a rigid frame can maintain the tolerance necessary to achieve the necessary rotor knife to bed knife clearance [213]. During grinding process, small feed into the grinder is to be maintained at a time to avoid heat buildup into the materials inside the grinder, which makes grinding more difficult. Excessive heat buildup may cause reagglomeration of the ground materials and can cause blocking of the grinder knife.

The level of regrind that could be blended with virgin polymer to obtain optimum properties depends on the nature of both regrind and virgin polymer. Like virgin polymers, regrinds of TPVs, TPUs, and other polar TPEs should be dried properly before processing. It has been found that a good starting point in determining optimum levels of regrind that could be blended with the same virgin polymer is 25% for styrenic block copolymers, 20% for TPVs, and 15% for TPUs [213]. This level can be increased or decreased based on the properties achieved and performance expected from the blends.

5.10 SURFACE AND INTERFACE OF TPES

The presence of two phases in TPE influences its applications sensitive to surface or interfacial properties. The interphase between the hard and soft segments is often considered as a separate phase and it can be substantially different from the bulk of the material. The presence of interface introduces thermodynamic factors that can alter the morphology near the interphase. The understanding of the influence of the interface on the properties of TPE still needs in-depth investigation, although it is well known that many stress-relieving processes, e.g., deflection and bifurcation of crack and sharing of loads or stress transfer, occur at the interface. At the interface between the glassy and rubbery phases, some degrees of interpenetration of the constitutive blocks occur in the case of diblock systems [218]. Thus, lower the difference in solubility parameter $(\sigma_A - \sigma_B)$, the higher will be the thickness of the interface where covalent bonds between two types of blocks are generally found. The adhesion between the interface and the polymer matrix is very important in the case of rubber–plastic blends as often the failure occurs in such blends because of poor interfacial adhesion. Systems that have narrow interface usually show complete phase separation. The important parameter that characterizes the interfacial behavior of a material is the surface energy γ_s. It can be estimated from contact angle measurement of various liquids on given polymer surface [219]. The difference between the surface tension for wetting between elastomer and plastics, called surface energy mismatch gives an estimation of interfacial tension between them during melt mixing. Interfacial tension determines the size of one phase dispersed in the matrix; lower mismatch gives finer dispersion. Jha et al. [220] has determined interphase thickness of several rubber–plastic combinations used in preparation of TPEs. They observed a thickness of about 50 nm between nylon and ACM. It has also been found that peel adhesion strength depends sharply on the rate of separation. Several authors [221] have reported surface energy mismatch analysis for thermoplastic elastomeric rubber–plastic blends. The relative adhesion at the interface between polymer pairs can provide information about the interface. Various adhesion tests have been performed and reported in literature [222–225]. However, there are several problems associated with the lap shear adhesion information because it does not reflect the exact stress at which joint failure occurs. Heikens et al. [226–228] have done pioneering work on the measurement of volume dilation during mechanical stressing of the blend using a relatively simple stress dilatometer, which gives indication of *in situ* interfacial debonding during mechanical testing of the blends.

5.11   BLENDS OF TPEs WITH OTHER POLYMERS

Polymer blending has become the more attractive and versatile means to generate new materials from the existing polymeric materials as well as to improve some deficiencies of common thermoplastics. TPEs have been blended, particularly with a variety of polymer either as a minor component to serve as impact modifier or to improve some physico-mechanical or rheological characteristics of TPE. All these features have been reviewed by Kresge [229,230] recently. The styrenic TPEs can be blended with many polymers, e.g., PS, PP, and LDPE and HDPE, to modify their properties [229,231–233] but more polar polymers, e.g., PVC, do not give satisfactory products. Blends with PS restore the impact resistance of PS, which is lost when flame retardants are mixed into high-impact PSs (HIPS) and upgrading HIPS to super high-impact product.

Impact properties of various thermoplastics (polycarbonate, polyphenylene ether, polyesters) can be modified by blending them with styrenic block copolymers and poly[styrene(ethylene-co-butylene)styrene] (SEBS) block copolymers. They may be used to produce tough blends with nylon [234]. Blends with PE are mostly used to make blown film of improved impact resistance and tear strength for seal application. Blends with PP have improved impact resistance at low temperature. The styrenics are also used as a compatibilizer to improve adhesion between two incompatible polymers, e.g., blends of PO with PS, and polyethylene terepthalate. Addition of low-molecular-weight SEBS to these blends changes them to more ductile material because of improved interfacial adhesion [235,236]. A high-molecular-weight SEBS has been used in a blend with PC and PP with improved solvent resistance, lower density, and cost. Blend of Kraton with a thermoplastic liquid crystal polymer (LCP), Vectra A900 has been reported by Verhooght et al. in terms of their formation and stability of LCP fiber [237,238].

The good melt stability and low melt viscosity of polyester elastomers facilitate their use in blends to improve impact resistance at low temperature and also to provide elastomeric character. Lower-melting polyester with mixed short-chain ester can be blended with temperature-sensitive polymer, e.g., PVC to improve the low-temperature flexibility, brittle point, and heat distortion temperature. Dynamic mechanical properties, NMR, linear thermal expansion of blends of soft polyester elastomers with rigid PVC, in the absence of conventional plasticizers, suggest a high degree of miscibility directly after mixing at 150°C; however, following annealing at 130°C the same blend shows evidence of phase separation with several folds increases in impact strength [239]. This observation suggests the existence of an upper critical solution temperature (UCST) for the system. Blends of NBR provide softer composition for blow molding [240]. Blends of SBS and polyester elastomer have been reported to give enhanced properties [241].

High modulus blends can be developed by mixing TPU with acetal copolymer (trioxane ethylene oxide copolymers) [242–244]. The highly crystalline acetal forms a second continuous phase. Kumar et al. studied behavior of such blends [245]. TPU retains none of its physical properties after immersion in water at 70°C for three weeks. The hydrolysis resistance of TPU can be improved by blending with polycarbodiimides [246]. Two parts of carbodiimide with TPU offer 87% retention of its strength, 93% of elongation, and 75% of modulus under the same condition. Blends of plasticized PVC and low-softening TPU offer materials with wide range of hardness and modulus; the softer compositions are particularly suitable for shoe sole applications. Blends of TPU and ABS are reported by Farrissey and Shah [247] to combine TPUs' excellent abrasion resistance, good heat resistance, low temperature properties, heat-sag resistance, toughness, impact resistance, chemical resistance with the higher rigidity and excellent processability of ABS. Blends of TPU with SAN, PS, and acrylonitrile styrene acrylic ester (ASA) have been reported by Demma et al. [243]. Ahn et al. have studied the miscibility of TPU with chlorine-containing polymers [248]. Reinforcement of TPU with glass fiber provides a material with increased stiffness, while maintaining high-impact strength [230,249].

PESA can be blended with various thermoplastics to alter or enhance their basic characteristics. Depending on the nature of thermoplastic, whether it is compatible with the polyamide block or with the soft ether or ester segments, the product is hard, nontacky or sticky, soft, and flexible. A small amount of PESA can be blended to engineering thermoplastics, e.g., polyethylene terepthalate (PET), polybutylene terepthalate (PBT), polypropylene oxide (PPO), polyphenylene sulfide (PPS), or poly-ether amide (PEI) for impact modification of the thermoplastic, whereas small amount of thermoplastic, e.g., nylon or PBT, can increase the hardness and flex modulus of PESA or PEEA [247].

A route to compatibility involving ionomers has been described recently by Eisenberg and coworkers [250–252]. The use of ionic interactions between different polymer chains to produce new materials has gained tremendous importance. Choudhury et al. [60] reported compatibilization of NR–polyolefin blends with the use of ionomers (S-EPDM). Blending with thermoplastics and elastomers could enhance the properties of MPR. The compatibility of copolyester TPE, TPU, flexible PVC, with MPR in all proportions, enables one to blend any combination of these plastics with MPR to cost performance balance. Myrick has reported on the effect of blending MPR with various combinations and proportions of these plastics and provided a general guideline for property enhancement [253].

5.12 NOVEL FUNCTIONAL NANOMATERIALS BASED ON THERMOPLASTIC ELASTOMERS

Nanotechnology is the control and manipulation of matter in nanometer length scale and is an emerging technology. The essence of nanotechnology is the ability to work at the nanolevel to generate large structure with fundamentally new molecular organisation and functionality. It is predicted that in the future nanodevices will have an enormous impact on enhanced energy conversion, controlled pollution, production of food, and improvement of human health and longevity [254]. TPEs particularly block copolymer-based TPEs are characterized by fluid disorder on the molecular scale and a very high degree of order at longer-length scales. In such block copolymers, phase-segregated polymer with formation of meso-domains with size ca. several tens of nanometers is observed [255]. This complex phase-separated structure and the supramolecular self-assemblies of mesoscale or nanoscale dimension are of great recent interest in the design of novel functional materials and template for nano-fabrication, using "bottom-up" approach [256,257]. By controlling appropriately the segment nature and length of each constituent of the block in block copolymers, a wide variety of microdomain structures of high degree of richness and complexity in bulk as well in solution phase are possible. Significant theoretical developments in predictive statistical theories and recent breakthroughs in polymer synthesis have created an unparallel opportunity to combine multiple block in novel molecular architectures in order to precisely manipulate the compositional (hydrophilic–hydrophobic, polar–nonpolar, charged–uncharged, linear–branched), conformational (rigid–flexible, oriented–unoriented, long–short), morphological (amorphous–crystalline, glassy–rubbery, swollen–contracted) and functional (reactive–unreactive, photoresponsive–passive, electroactive–nonconducting, neutral–biocompatible) contrast among the blocks. The so-produced controlled structure materials will provide tailored mechanical, optical, electrical, ionic, barrier, and other physical properties.

Specific block copolymers may be optimum nanomaterials, either for their intrinsic properties as self-organized assemblies or for their ability to template other organic, inorganic, semiconducting, metallic, or biologically relevant materials. Highly ordered self-assembling block copolymer film may be used for the creation of regular pore size and spacing and fabrication of nanoelectrode arrays (NEAs), which have applications ranging from *in vivo* sensors to smart electronic nose [258]. Selective etching of bicontinuous microdomains of block copolymers and successive nonelectrolytic plating have been used to produce nano-channels with metal coating to prepare highly effective membrane reactors [259]. Nano-thin chemically tethered layers from functionalized (SEBS) block-copolymers have been fabricated to produce robust and superior microtribological properties for boundary lubrication [260]. Nano-patterned surface, which can provide novel platforms for the assembly of complicated functional nano-structures and nanodevices, has also been reported [261]. A method of ordered deposition of inorganic clusters for generating quasi-regular arrays of nanometer-sized noble metal and metal oxide cluster using block copolymer as template has been described by Spatz et al. [262]. Catalytic properties of palladium nano-cluster synthesized with block copolymer have been reported by Ciebien et al. [263]. Stable gold nanoparticles have been fabricated using star-block copolymers by Youk et al. [264]. Fink et al. reported the potential future possibilities of the use of block copolymers as the polymeric photonic bandgap materials that are unattainable from rigid monolithic photonic materials [265]. Metallodielectric photonic crystals based on block copolymers have been reported by Bockstaller et al. [266]. Block copolymer-mediated synthesis of gold quantum dots has recently been reported by Tamil et al. [267] and Selvan et al. [268]. Sohn et al. demonstrated magnetic properties of iron oxide within microdomain of block copolymers [269].

Tailoring block copolymers with three or more distinct type of blocks creates more exciting possibilities of exquisite self-assembly. The possible combination of block sequence, composition, and block molecular weight provides an enormous space for the creation of new morphologies. In multiblock copolymer with selective solvents, the dramatic expansion of parameter space poses both experimental and theoretical challenges. However, there has been very limited systematic research on the phase behavior of triblock copolymers and triblock copolymer-containing selective solvents. In the future an important aspect in the fabrication of nanomaterials by bottom-up approach would be to understand, control, and manipulate the self-assembly of phase-segregated system and to know how the selective solvent present affects the phase behavior and structure offered by amphiphilic block copolymers.

5.13 APPLICATIONS

Over the last 10 years, TPEs have gained widespread recognition as an ideal material for a broad range of applications. The principal successes of TPEs have derived from their replacement of thermoset rubbers, either directly in the existing application or in new applications that otherwise would have specified thermoset compounds. The fact that TPEs do not require vulcanization is of course the key point of its growth. TPEs also offer several environmental benefits in the production of items and in the disposal of worn-out products. Scraps generated during production may be usually reground and recycled. Since 1970, TPEs have enjoyed a compounded annual growth rate of 8%–9% in the market in contrast to the low growth rate of rubbers (0%–2%) and plastics (3%–6%) industries [270,271]. Table 5.15 lists the past, present, and projected future consumption of TPEs in the European market area [272]. Growth of TPE usage is due to three main factors: (1) replacement for other materials, (2) new processing technologies, and (3) new applications and markets. TPEs have proven themselves in meeting a wide range of demanding engineering requirements and automotive applications. These applications will continue to grow because of the cost savings provided and the performance delivered. TPEs will continue to replace thermoset rubbers for applications in which they offer cost advantages and design flexibility to the automotive engineer [272]. World demand for thermoplastic (TPEs) has grown 6.4% per year in 2006. In addition to

TABLE 5.15
Growth in European Demand for Thermoplastic Elastomers

	2000 (000 t)	2005 (000 t)	Increase (%/Annum)
SBCs	195	226	3
TPOs	135	172	5
TPVs	36	59	10.5
TPUs	62	79	5
COPEs	20	30	8
COPAs	7.5	10	5
Total	455.5	576	4.8

Source: Dufton, P.W., *Thermoplastic Elastomers*, Rapra Industry Analysis Report, 2001

displacing competitive materials, TPEs are also being overmolded onto rigid consumer goods to enhance ergonomic or "soft-touch" features. The TPE industry will remain concentrated in the United States, Western Europe, and Japan, while Asian markets such as China grows faster [272]. Figure 5.18 compares the performance and cost of the different classes of TPEs to that of the different thermoset rubbers. Each class of TPEs is a rational competitor for those thermoset rubbers at its position on the two-dimensional chart. Though the TPEs are more costly than the thermoset on a cost-per-kilogram

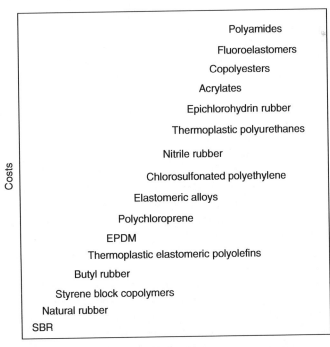

FIGURE 5.18 Comparison of different thermoplastic elastomer (TPE) classes with corresponding cross-linked elastomers. (From Rader, C.P., *Kunstst. Ger. Plast.*, 83, 777, 1993.)

basis it is competitive in cost per finished product (up to 20%) due to immense cost-saving in the processing cost of a thermoplastic material. The principal market segments where TPEs made significant inroads are automotive, footwear soling, wire and cable, general mechanical goods, electrical and electronic products, hose industries, medical applications, adhesives and sealants, etc.

TPEs have been finding an ever-expanding home in the worldwide automotive market by combining high performance, better processability, physical properties over a broad temperature range, part aesthetics with extended design possibilities, and cost-reduced manufacturing technology. The application of TPEs in automotive industries may be broken down into three categories: under the bonnet, interior, body, and trim. Typical applications of the TPEs under the bonnet are clamps, electrical wires, control cables, electrical motor housing, connectors, fuel lines, vibration dampers, seals, buffer blocks for bonnets, etc. Boots and bellows (rack and pinion bellows, constant velocity boots) form another very important application. In this application, TPEs (elastomeric alloys and copolyesters) replaced vulcanized rubber because its superior tensile and tear strength and resilience permitted redesign of the boots to reduce polymer volume without sacrifice of essential properties. Fast economical processing in plastic injection molding and extrusion blow molding makes MPRs and TPVs the materials of choice for such complex molding parts. Soft touch and the possibility of making them colorful make TPEs eminently suitable for the interior application such as side panels, back support, car phone cables, car lock parts, heating channels, and soft touch buttons. Due to their ultraviolet (UV), abrasion and impact resistance, TPEs can be used for exterior applications, such as body panels, spoilers, bumper blocks, body parts, sunroof parts, door seals and door cover panels, styling trims, and body plugs. Weather strips are another major automotive market for TPEs. The use of a low set TPE gives a major weight reduction from a PVC or a thermoset rubber. The ability of TPE to form heat-welded corner and coextrusions will simplify assembly job and reduce cost. They can also be flocked with textile fibers.

In building construction, TPEs have made highly significant inroads in uses such as tube and cable glands, seal and tube plugs, building-site hoses, door and window seals and glazing, expansion joints, membranes, installation seals, and plumbing applications. Adhesive and sealants are another large TPE market, with the styrene TPEs controlling most of the applications. The market for pressure sensitive adhesives, construction adhesives and sealants, products assembling adhesives, and hot melt processable adhesives have been developed from TPEs, because they have the ability to achieve good adhesion while enjoying the inherent cost, safety, and environmental advances of hot melt processing.

Opportunity for TPEs will continue to be realized in electrical applications. SEBS compounds provide the superior electrical properties combined with low-temperature flexibility—down to −40°C—and straightforward coloring. Transparent formulations are also possible, e.g., for switch covers. The coextrusion process permits product of soft cable sheathing and hard plugs in a simple operation. Other applications are switches, covers, cable glands, hand lamps, etc. Copolyester TPEs are widely used in fiber optics as the primary coating, secondary coating, buffer tube, and jacket. They are also widely used in spring telephone cables. Mechanical strength combined with flexibility and the ability to perform in various environments are the key advantages of copolyester TPEs for such applications. The inertness and durability of EPDM–TPVs have enabled their uses in several demanding electrical applications including computer software and hardware, telephones, electrical appliances, and office equipments.

A wide variety of hoses and tube products are produced from copolyester elastomers ranging from compact garden hose to high-pressure hydraulic hose designed for the most stringent strength and chemical-resistance requirements. A new and growing market of TPEs is power drive belts. Unreinforced copolyester elastomers have sufficient strength to replace fiber-reinforced rubber. Positive engagement "toothed" belts produced by simple injection molding are finding utilities in office equipment and computers.

The resistance to ozone and weathering and the excellent grip even in wet condition make SEBS compounds the ideal material for sports equipment, especially for outdoor use such as golf club grips. Further applications are diving goggles and flippers, bike handlebar grips, shock-absorbing elements for tennis, rackets, rifle butts, skiing goggles, grip for ski sticks, ice pick, etc. The top professional soccer ball today makes use of copolyester elastomer that is equivalent of leather, coated with a very durable and hard-wearing TPU. The finished ball remained consistent irrespective of the playing conditions. Footwear market is another bulk user of TPEs. TPE-coated fabrics or resin-impregnated felt systems find application in such areas as inflatable life rafts and escape slides, life jackets, floating roof seals, food quality conveyor belts, sewer pipe repair, and submarine escape suits.

There are many medical devices where use of TPEs in place of thermoset rubbers could offer many advantages to the users. TPEs have distinct toxicological advantage over thermoset rubber in the uses involving direct contact with food, beverages, pharmaceutical preparations, and living tissues. They have found application in syringe plunger tips, aerosol valve seals, drug vile stoppers, disposable kits and trays, etc. The ability to fabricate thin-wall parts by TPEs will encourage material saving and replacement of older heavy-walled rubber parts like blood pressure cuffs, bed pads, and reusable tubings. Foaming technology will be important for wheelchair pads, orthopedic appliances, instrument handles, grips, and other thicker pieces. Low extractable alloys to replace flexible highly plasticized polymers are gaining more attention in blood contact devices [273]. New transparent and low-tensile set biocompatible TPEs based on polysiloxane-modified polyolefin block copolymers and TPUs are finding applications in many medical applications such as blood pump tubing, catheters, urethral stints, internal feeding sets, and myringotomy tubes [274]. The high degree of surface tack of some TPEs, notably styrenics, offers self-sealing properties that could replace natural rubbers for the protection of injection sites. Long-term implants made of TPE have long history of clinical use, including the development of thermoplastic elastomeric lumber disc spacers [275].

The sign of TPE vitality and dynamism is demonstrated by the growth in the intra-material competition. Some low-priced TPEs like TPOs are challenging higher-priced engineering TPEs in many applications. Extensive research is going on and incremental developments are also expanding TPE performance. Producers are also broadening the properties envelope of various grades of TPEs, thus improving the balance of cost and performance.

5.14 FUTURE

Traditionally, the major issues considered in the design of product have been related only to the function, appearance, and production cost. However, the recent changes in environmental regulations and consumer pressure are forcing manufacturers to become more responsible for the safe disposal and recycling of products after finishing their useful life. The goal of the new production trend, "design for disassembly" (DFD) [276,277], is to close the production loop, to conceive, develop, and build a product with a long-term view of how its components can be recycled and reused. It can be predicted with confidence that this newfound environmentalism and the take-back laws (new laws across Europe and the United States to compel manufacturers to take back the used products) will open significant new market for TPEs. It is also expected that their penetration in electrical wire and electrical product market will grow rapidly in the near future. In most of the general-purpose thermoset materials the flame-retardant properties are provided by halogen. The styrenic TPEs, the TPOs, most of the elastomeric alloys, and copolyesters can be flame-retarded and halogen-free, though they get restricted acceptance due to their high cost, compared to the traditional systems based on halogens. However, society will not continue to accept excessive smoke and toxicity of combustion products at the expense of personal safety.

In terms of applications, TPE industries will remain closely linked with the motor vehicle industry. The solid gains forecast for TPEs in this market are due to the development of new products for exterior (e.g., body seal) and interior (e.g., instrument and door panel skins) applications at the expense of EPDM and thermoplastics such as PVC [278]. The new environmental consensus will make everything color-coded. It is likely that an industrial standard would be adopted where brake hose, air condition hose, fuel line hose, and power steering hose will all be of different colors and their cover materials in each and every case will be a TPE with life-of-the-car performance. This will facilitate maintenance and improve reliability. The appearance of new and varied chemicals in treated and waste water and the toxicological issues outlined in recent potable water standard will also lead to an increase in the use of TPEs in plumbing application. An increasing trend away from solvent toward hot melt and water-based systems due to environmental impact concerns and regulations will enhance the use of TPEs in such applications.

TPEs are gaining and will capture more ground in the multimillion-dollar medical supply and artificial organs market as replacement materials for thermosets with all the performance advantages and low processing costs. However, TPEs have to be specially made for such applications, particularly to withstand the physiological environment *in vivo*.

The recent developments in biosynthesis and polymer synthesis have also revealed very attractive potential to synthesize block copolypeptides (synthetic polypeptides of unique composition), of controlled structures, dimension (narrow molecular weight distribution, control monomer sequence, well-defined molecular architectures and conformations), and self-organization. A potential way of combining all the advantages of synthetic block copolymer and peptide structures has also been recently developed to create "molecular chimeras" or "hybrids," that are block copolymers of synthetic segments and amino acid sequences. These new materials hold the promise for development of biomimetic polymers that can not only closely match the functionality of those of biological molecules in one aspect, but which are enhanced in other novel functionalities. These new biocompatible block copolymers and their hydrogels that behave like TPE materials have significant potential application in tissue engineering, drug delivery, and biomimetic composite formation. However, significant research using multidisciplinary approach is crucial to develop this new class of bioinspired materials that capture the advantages of both protein and polymer systems. Therefore, it seems that the future of TPEs is very bright—as a replacement for thermoset rubber in numerous applications, and as new soft and flexible material for new design and new and novel applications in industrial and health care products.

REFERENCES

1. Holden G., Legge N.R., Quirk R., and Schroeder H.E. (eds.), *Thermoplastic Elastomers*, 2nd edn., Hanser Publishers, Munich, 1996.
2. Walker M. and Rader C.P. (eds.), *Handbook of Thermoplastic Elastomers*, Van Nostrand Reinhold, New York, 1988.
3. Bhowmick A.K. and Stephens H.L. (eds.), *Handbook of Elastomers, New Developments and Technology*, Marcel Dekker, New York, 2001.
4. De S.K. and Bhowmick A.K. (eds.), *Thermoplastic Elastomers from Rubber Plastic Blends*, Ellis Horwood, London, 1990.
5. Legge N.R., Thermoplastic elastomers: Three decades of progress, *Rubber Chem. Technol.*, 62, 529, 1989.
6. Gorski E., The nomenclature of thermoplastic elastomers, *Kunstst. Ger. Plast.*, 83, 29, 1993.
7. Shell Chemical Co., Kraton, Cariflex TR, Bel Pat. 671460, 1966.
8. Haw J.R., Solprene rubbers, *Rubber World*, 27, 1973.
9. Varshney S.K. et al., Synthesis of ABA type thermoplastic elastomers based on polyacrylates, *Macromolecules*, 32, 235, 1999.
10. Puskas J.E. and Kaszas G., Polyisobutylene-based thermoplastic elastomers: A review, *Rubber Chem. Technol.*, 69, 462, 1996.
11. Keneddy J.P. et al., Thermoplastic elastomers of isobutylene and process of process of preparation, *US Patent 4946899*, Aug, 1990.
12. Cao X. and Faust R., Polyisobutylene based thermoplastic elastomer 5. Poly(styrene-b-isobutylene-b-styrene) tri-block copolymers by coupling of living poly(styrene-b-isobutylene) di-block copolymers, *Macromolecules*, 32, 5487, 1999.

13. Ban H.T., Kase T., Kawabe M., Miyazawa A., Ishihara T., Hagihara H., Tsunogae Y., Murata M. and Shiono T.A. New approach to styrenic thermoplastic elastomers: Synthesis and characterization of crystalline styrene–butadiene–styrene triblock copolymers, *Macromolecules*, 39, 171, 2006.

14. Fetters L.J. and Morton M., Synthesis and properties of block polymers. I. Poly(a-methylstyrene)-polyisoprene-poly(a-methylstyrene), *Macromolecules*, 2, 453, 1969.

15. Tsunogae Y. and Kennedy J.P., Thermoplastic elastomers by sequential monomer addition. VI. Poly(p-methylstyrene-b-isobutylene-b-*p*-methylstyrene), *Polym. Bull.*, 31, 1436, 1993.

16. Morton M. and Miksell S.L., ABA block copolymer of dienes and cyclic sulfides, *J. Macromol. Sci. Chem.*, A7(7), 1391, 1993.

17. Kennedy J.P. and Price J.L., Synthesis, characterization and physical properties of poly(methyl methacrylate-b-isobutylene-b-methyl methacrylate) triblock copolymers, *Polym. Mater. Sci. Eng.*, 64, 40, 1991.

18. Yu J.M., Dubios P., and Jerome R., Synthesis and properties of poly[isobornylmethacrylate (IBMA)–b–butadiene (BD)–b–IBMA] copolymers: New thermoplastic elastomers of a large service temperature range, *Macromolecules*, 29, 7316, 1996.

19. Jacob S. and Kennedy J.P., Synthesis, characterization and properties of octa-arm polyisobytylene-based star polymers, *Adv. Polym. Sci.*, 146, 1, 1999.

20. Asathana S., Majoros I., and Kennedy J.P., TPEs: Star-block comprising multiple polystyrene-b-PIB arms radiating from a crosslinked polydivinylbengene core, *Rubber Chem. Technol.*, 71, 949, 1998.

21. Shim J.S. and Kennedy J.P., Novel thermoplastic elastomers. II. Properties of star-block copolymers of PST-b-PIB arms emanating from cyclosiloxane cores, *J. Polym. Sci., Part A, Polym. Chem.*, 37, 815, 1999.

22. Kennedy J.P. and Shim J.S., Star-block polymers having multiple polyisobutylene containing diblock copolymers arm radiating from a siloxane core and a method of synthesis thereof, *Disclosure 318, US Patent*, Notice of Allowance, 2001.

23. Kabayashi N. and Yoshida M., Star-shaped block copolymers and production process thereof, *US Patent* 6310175, 2001.

24. Wunsch J. and Geprags M., Star polymers and their preparation, *US Patent* 6303806, 2001.

25. Yijin X. and Caiyaun P., Block and star-block copolymers by mechanism transformation. 3. S-(PTHF-PSt)4 and S-(PTHF-PSt-PMMA)4 from living CROP to ATRP, *Macromolecules*, 33, 4750, 2000.

26. Feldthusen J., Ivan B., and Mueller A.H.E., Synthesis of linear and star-shaped block copolymers of isobutylene and methacrylates by combination of living cationic and anionic polymerizations, *Macromolecules*, 31, 578, 1998.

27. Pitsikalis M., Pispas S., Mays J.W., and Hadjichristidis N. Nonlinear block copolymer architectures, *Adv. Polym. Sci.*, 135, 1, 1998.

28. Kennedy J.P., Fenyvesi G., and Keszler B., Three arm star composition of matter having diblock arms based on polyisobutylene and method of preparation, *US Patent* 6228945, May 8, 2001.

29. Wolfe J.R., Jr, *Block-Copolymers: Science and Technology* (Meier D.J., ed.), MMI Press Symposium Series, Ellis Horwood, New York, 1983, 145.

30. Kricheldorf H.R., Wollheim T., Koning C.E., Werumeus B.H.G., and Altstadt V. Thermoplastic elastomers 1. Poly(ether-ester-imide)s based on 1,4-diaminobutane, trimellitic anhydride, 1,4-dihydroxybutane and poly(tetramethylene oxide) diols, *Polymer*, 42, 6699, 2001.

31. Walch E. and Gaymans R.J., Synthesis and properties of poly(butylenes terephthalate)-b-polyisobutylene segmented block copolymers, *Polymer*, 35, 636, 1994.

32. Deak G. and Kennedy J.P., Copolyesters containing polyisobutylene soft segments and polybutylene terephthalate as crystalline segments, *Macrolmol. Rep.*, A33(Suppls 7 & 8), 439, 1996.

33. Schmalz H., Abetz V., Lange R., and Soliman, M. New thermoplastic elastomers by incorporation of nonpolar soft segments in pbt-based copolyesters, *Macromolecules*, 34, 775, 2001.

34. Saunders J.H. and Frisch K.C., *Polyurethanes: Chemistry and Technology*, Part I, *Chemistry High Polymer Series*, Vol. 16, Wiley, New York, 1962.

35. Shollenberger C.S., Simulated vulcanizates of polyurethane elastomers, *US Patent* 2871218, 1955.

36. Carvey R.M. and Witenhafer D.E., *British Patent* 1087743, 1965

37. Singer S.M. and Allott M.T., Thermoplastic polyurethane elastomer based on a saturated hydroxyl terminated polyol, difunctional aromatic chain extender and 1,5-naphthalene diisocyanate, *US Patent* 5 599 874, 1997.

38. Ishimaru F. et al., Thermoplastic polyurethane, *US Patent* 5695884, 1997.

39. Handlin D.L. et al., Methods for producing mixed polyol thermoplastic polyurethane compositions, *US Patent* 6323299, 2001.

40. Martin D.J., Warren L.A.P., Gunatillake P.A., McCarthy S.J., Meijs G.F., and Schindhelm, K. Polydimethylsiloxane/polyether-mixed macrodiol-based polyurethane elastomer, *Biomaterials*, 21, 1021, 2000.

41. Kim H.D., Lee T.J., Huh J.H., and Lee D.J. Preparation and properties of segmented thermoplastic polyurethane elastomers with two different soft segments, *J. Appl. Polym. Sci.*, 73, 345, 1999.

42. Rader C.P. and Abdou-Sabet S., *Thermoplastic Elastomer from Rubber–Plastic Blends* (De S.K. and Bhowmick A.K., eds.), Ellis Horwood, London, 1990.

43. Wolfe J.R., *Thermoplastic Elastomers* (Holden G., Legge N.R., Quirk R.P., and Schroeder H.E., eds.), Hanser Publishers, Munich, 1996.

44. Chattopadhyay S., Chaki T.K., and Bhowmick A.K., New thermoplastic elastomers from poly(ethyleneoctene) (engage), poly(ethylene-vinyl acetate) and low-density polyethylene by electron beam technology structural characterization and mechanical properties, *Rubber Chem. Technol.*, 74, 815, 2001.

45. Roy Choudhury N. and Dutta N.K., Thermoplastic elastomeric natural rubber–polypropylene blends with reference to interaction between the components, *Advances in Polymer Blends and Alloys Technology*, Vol. 5 (K. Finlayson, ed.), Technomic Publishers, Pensylvania, 1994, 161.

46. Olabisi O. and Farman A.G., *Multiphase Polymers, Advances in Chemistry Series*, Vol. 176 (Cooper S.L. and Estes G.M., eds.), American Chemical Society, Washington DC, 1993, 559.

47. Roy Choudhury N., Bhowmick A.K., and De S.K., Thermoplastic natural rubber, *Natural Rubber: Biology, Cultivation and Technology* (Sethuraj M.R. and Mathew N.M., eds.), Elsevier, New York, 1992.

48. Elliot D.J., Natural rubber systems, *Development of Rubber Technology 3* (Whelan A. and Lee K.S., eds.), Applied Science, New York, 1982.

49. Jha A. and Bhowmick A.K., Thermoplastic elastomeric blends of nylon 6/acrylate rubber: Influence of interaction of mechanical and dynamic mechanical thermal properties, *Rubber Chem. Technol.*, 70, 798, 1997.

50. Jha A., Dutta B., and Bhowmick A.K., Effect of fillers and plasticizers on the performance of novel heat and oil-resistant thermoplastic elastomers from nylon-6 and acrylate rubber blends, *J. Appl. Polym. Sci.*, 74, 1490, 1999.

51. Jha A. and Bhowmick A.K., Thermoplastic elastomeric blends of poly(ethyleneterephthalate) and carylate rubber: 1. Influence of interaction on thermal, dynamic mechanical and tensile properties, *Polymer*, 38, 4337, 1997.

52. De Sarkar M., De P.P., and Bhowmick A.K., New polymeric blends from hydrogenated styrene–butadiene rubber and polyethylene, *Polymer*, 39, 1201, 1998.

53. Kader M.A., Bhowmick A.K., Inoue T., and Chiba T. Morphology, mechanical and thermal behavior of acrylate rubber/fluorocarbon elastomer/polyacrylate blends, *J. Mat. Sci.*, 37, 6789, 2002.

54. Rader C.P. and Abdou Sabet S., Two phase elastomeric alloys, *Thermoplastic Elastomers from Rubber Plastic Blends* (De S.K. and Bhowmick, A.K., eds.), Ellis Horwood, London, 1990, 159.

55. Papke N. and Kargar-Kocsis J., Thermoplastic elastomer based on compatibilised poly(ethyleneterphthalate) blend: Effect of rubber type and dynamic curing, *Polymer*, 42, 1109, 2001.

56. Medsker R.E. and Patel R., Hydrosilylation crosslinking, *US Patent* 5672660, 1997.

57. Umpleby J.D., Process of preparation of thermoplastic elastomers, *US Patent* 4803224, 1989.

58. Roy Choudhury N., De P.P., and Bhowmick A.K., Thermoplastic elastomeric natural rubber–polyolefin blends, *Thermoplastic Elastomers from Rubber Plastic Blend* (De S.K. and Bhowmick A.K., eds.), Ellis Horwood, London, 1990, 11.

59. Coran A.Y., Patel R.P., and Williams-Head D., Rubber–thermoplastic compositions. IX. Blends of dissimilar rubbers and plastics with technological compatibilization, *Rubber Chem. Technol.*, 58, 1014, 1985.

60. Roy Choudhury N. and Bhowmick A.K., Compatibilization of natural rubber–polyolefin thermoplastic elastomeric blends by phase modification, *J. Appl. Polym. Sci.*, 30, 1091, 1989.

61. Bhowmick A.K., Chiba T., and Inoue T., Reactive processing of rubber–plastic blend: Role of a chemical compatibilizer, *J. Appl. Polym. Sci.*, 50, 2055, 1993.

62. Mehrabzadeh M. and Delfan N., Thermoplastic elastomers of butadiene–acrylonitrile copolymer and polyamide. VI. Dynamic crosslinking by different systems, *J. Appl. Polym. Sci.*, 77, 2057, 2000.

63. Han S.J., Lohse D.J., Radosz M., and Sperling L.H. Thermoplastic vulcanizates from isotactic olypropylene and ethylene–propylene–diene terpolymer in supercritical propane: Synthesis and morphology, *Macromolecules*, 31, 5407, 1998.

64. Fritz H.G., Boel U., and Cai Q., Innovative TPV two-phase polymers: Formulation, morphology formation, property profiles and processing characteristics, *Polym. Eng. Sci.*, 39, 1087, 1999.

65. Wallace J.G., Single phase melt processible rubber, *Handbook of Thermoplastic Elastomer* (Walker B.M. and Rader C.P., eds.), Van Nostrand Reinhold, New York, 1988, 141.

66. DuPont Dow Elastomers, *Product Information Literature*, Freeport, TX, 1999.

67. Chen H., Guest M.I., Chum S., Hiltner A., and Baer E. Classification of ethylene–styrene interpolymers based on comonomer content, *J. Appl. Polym. Sci.*, 70, 109, 1998.

68. Macknight W.J. and Lundberg R.D., Research on ionomeric systems, *Thermoplastic Elastomers* (Holden, G., Legge N.R., Quirk R.P., and Schroeder H.E., eds.), Hanser Publishers, Munich, 1996.

69. Xie H. and Ma B., Properties of sulfonated EPDM ionomers obtained by sulfonation in the presence of phase transfer catalyst, *J. Macromol. Sci. Phys.*, 51, 1328, 1989.

70. Upaeglis A. and O'Shea F.X., Thermoplastic elastomer compounds from sulfonated EPDM ionomers, *Rubber Chem. Technol.*, 61, 223, 1988.

71. Fitzgerald J.J. and Weiss R.A., Synthesis, properties and structure of sulfonate ionomers, *J. Macromol. Sci. Rev. Macromol. Chem. Phys. C*, 28, 99, 1988.

72. Macknight W.J. and Lundberg R.D., Elastomeric ionomers, *Rubber Chem. Technol.*, 57, 652, 1984.

73. Kar K.K. and Bhowmick A.K., *Handbook of Elastomers*, 2nd edn. (Bhowmick A.K. and Stephens H.L., eds.), Marcel Dekker, New York, 2001.

74. Antony P., Bandyopadhyay S., and De S.K., Thermoplastic elastomers based on ionomeric polyblends of zinc salts of maleated polypropylene and maleated EPDM rubber, *Polym. Eng. Sci.*, 39, 963, 1999.

75. Weiss R.A., Sen A., Pottick L.A., and Willis C.L. Block copolymer ionomers. Thermoplastic elastomers possessing two distinct physical networks, *Polym. Commun.*, 31, 220, 1990.

76. Makowski H.S., Lundberg R.D., and Singhal G.S., *US Patent 3870841*, 1975.

77. Nevatia P., Banerjee T.S., Dutta B., Jha A., Naskar A.K., and Bhowmick A.K. Thermoplastic elastomer from reclaimed rubber and waste particles, *J. Appl. Polym. Sci.*, 83, 2035, 2002.

78. Jacob C, De P.P., Bhowmick A.K., and De S.K. Recycling of EPDM waste. II. Replacement of virgin rubber by ground EPDM vulcanizates in EPDM/PP thermoplastic elastomeric composition, *J. Appl. Polym. Sci.*, 82, 3304, 2001.

79. Naskar A.K., Bhowmick A.K., and De S.K., Thermoplastic elastomeric composition based on ground rubber tire, *Polym. Eng. Sci.*, 41, 1087, 2001.

80. Adam G., Sebenik A., Osredkar U., Veksli Z., and Ranogajec F. Grafting of waste rubber, *Rubber Chem. Technol.*, 63, 660, 1990.

81. Adam G., Sebenik A., Osredkar U., Ranogajec F., and Veksli Z. The possibility of using grafted waste rubber—photochemically induced grafting, *Rubber Chem. Technol.*, 64, 133, 1991.

82. Fuhrmann I. and Karger-Kocsis J., Promising approach to functionalization of ground tire rubber, *Plast. Rubber Compos.*, 28, 500, 1999.

83. Waddell W.H., Evans L.R., Gillick J.G., and Shuttleworth D. Polymer surface modification, *Rubber Chem. Technol.*, 65, 687, 1992.

84. Choudhury N.R. and Bhowmick A.K., Compatibilization of natural rubber–polyolefin thermoplastic elastomeric blends by phase modification, *J. Appl. Polym. Sci.*, 38, 1091, 1989.

85. Sen A.K., Mukherjee B., Bhattacharyya A.S., De P.P., and Bhowmick A.K. Functionalization of polyethylene and ethylenepropylene diene terpolymer in the bulk through dibutyl maleate grafting, *Angew. Makromol. Chem.*, 191, 5, 1991.

86. Ray A.K., Jha A., and Bhowmick A.K., Chemical modification of EPDM rubber and its properties, *J. Elast. Plast.*, 29, 201, 1997.

87. Schlechter M., *Metallocene Elastomers/TPEs*, Business Communication, Wellesley, MA, 1996, 159.

88. Figuly G.D. and Goldfinger M.B., Thermoplastic elastomers, *US Patent 6300463*, 2001.

89. Tullock C.W. et al., Polyethylene and elastomeric polypropylene using alumina-supported bis(arene) titanium, zirconium, and hafnium catalysts, *J. Polym. Sci., Part A, Polym. Chem.*, 27, 3063, 1989.

90. Mueller G. and Rieger R., Propene based thermoplastic elastomers by early and late transition metal catalysis, *Prog. Polym. Sci.*, 27, 815, 2002.

91. Jacob S., Majoros I., and Kennedy J.P., Novel thermoplastic elastomers: Star-blocks consisting of eight poly(styrene-b-isobutylene) arms radiating from a calex[8]arenecore, *Rubber Chem. Technol.*, 71, 708, 1998.

92. Kennedy J.P., *Thermoplastic Elastomers* (Legge N.R., Holden G., Quirk R., and Schroeder H.E., eds.), Hanser Publishers, Munich, 1996, 365.

93. Weidisch R., Gido S.P., Uhrig D., Iatrou H., Mays J., and Hadjichristidis N. Tetrafunctional multigraft copolymer as novel thermoplastic elastomer, *Macromolecules*, 34, 6333, 2001.

94. Xenidou M. and Hadjichristidis N., Synthesis of model nultigraft copolymers of butadiene with randomly placed single and double polystyrene branches, *Macromolecules*, 31, 5690, 1998.

95. Itarou H., Mays I.W., and Hadjichri-Stidis N., Regular comb polystyrene and graft polyisoprene/ polystyrene copolymers with double branches ("Centipedes"). Quality of (1,3-phenylene)bis(3-methyl-1-phenylpentylidene)dilithium initiator in the presence of polar additives, *Macromolecules*, 31, 6697, 1998.

96. Markel E.J., Weng W., Peacock A.J., and Dekmezian A.H. Metallocene-based branched-block thermoplastic elastomers, *Macromolecules*, 33, 8541, 2000.

97. Wright E.R., McMillan R., Andrew C.A., Apkarian, R.P., and Conticello, V.P. Thermoplastic elastomer hydrogels via self-assembly of an elastin–mimetic triblock polypeptide, *Adv. Func. Mater.*, 12, 149, 2002.

98. Voet D. and Voet J.G., *Biochemistry*, 2nd edn., Wiley, New York, 1995, 59.

99. Jones J., *The Chemical Synthesis of Peptides—International Series of Monographs on Chemistry*, Oxford University Press, Oxford, 1994.

100. Nediljko B., *Engineering the Genetic Code: Expanding the Amino Acid Repertoire for the Design of Novel Proteins*, Wiley, New York, 2006.

101. Jane K. (ed.), *Genetic Engineering: Principles and Methods*, Vol. 25, Kluwer, Norwell, MA, 2003.

102. Pühler A. (ed.), *A Multi-Volume Comprehensive Treatise*, Vol. 2, *Genetic Fundamentals and Genetic Engineering*, 2nd edn., Wiley, New York, 1996.

103. Urry D.W., Whai is elastin; what is not, *Ultrastructural Pathology*, 4, 227, 1983.

104. Urry D.W. et al., *Protein-Based Materials* (McGrath K. and Kaplan D., eds.), Birkhauser, Boston, MA, 1997.

105. Hayashi C. and Lewis R., Molecular architecture and evolution of a modular spider silk protein gene, *Science*, 287, 1477, 2000.

106. Hayash C.I., Shipley N., and Lewis R., Hypotheses that correlate the sequence, structure, and mechanical properties of spider silk proteins, *Int. J. Biol. Macromol.*, 24, 271, 1999.

107. Hayashi C.Y. and Lewis R.V., Spider flagelliform silk: Lessons in protein design, gene structure, and molecular evolution, *BioEssays*, 23, 750, 2001.

108. Lee T.A.T., Cooper A., Apkarian R.P., and Conticello V.P. Thermo-reversible self-assembly of nanoparticles derived from elastin–mimetic polypeptides. *Advanced Materials*, 12, 1105, 2000.

109. Spontak R.J. and Patel N.P., Thermoplastic elastomers: Fundamentals and applications, *Curr. Opin. Colloid Interface Sci.*, 5, 334, 2000.

110. Cappello J. and Ferrari F., Microbial production of structural protein polymers, (Mobler D.P., ed.), *Plastics from Microbes*, Hanser Publishers, Munich, 1994, 35.

111. Ferrari F.A., Richardson C., Chambers J., Causey S.C., and Pollock T.J. Construction of synthetic DNA and its use in large polypeptide synthesis, *US Patent* 5243038, 1993.

112. Cappello J., Crissman J.W., Crissman M., Ferrari F.A., Textor G., Wallis O., Whitledge J.R., Zhou X., Burman D., Aukerman L., and Stedronsky E.R. *In-situ* self-assembling protein polymer gel systems for administration, delivery, and release of drugs, *J. Contr. Rel.*, 53, 105, 1998.

113. Nagapudi K., Brinkman W.T., Leisen, J., Thomas B.S., Wright E.R., Haller C., Wu X., Apkarian R.P., Conticello V.P., and Chaikof E.L. Protein-based thermoplastic elastomers, *Macromolecules*, 38, 345, 2005.

114. Petka W.A., Hardin J.L., McGrath K.P., Wirtz D., and Tirrell, D.A. Reversible hydrogels from self-assembling artificial proteins, *Science*, 281, 389, 1998.

115. Minich E.A., Nowak A.P., Deming T.J., and Pochan, D.J. Rod–rod and rod–coil self-assembly and phase behavior of polypeptide diblock copolymers, *Polymer*, 45, 1951, 2004.

116. Breedveld V., Nowak A.P., Sato J., Deming T.J., and Pine D.J. Rheology of block copolypeptide solutions: Hydrogels with tunable properties, *Macromolecules*, 37, 3943, 2004.

117. Pochan D.J., Pakstis L., Ozbas B., Nowak A.P., and Deming T.J. SANS and cryo-TEM study of self-assembled diblock copolypeptide hydrogels with rich nano through microscale morphology, *Macromolecules*, 35, 5358, 2002.

118. Nowak A.P., Breedveld V., Pakstis L., Ozbas B., Pine D.J., Pochan D., and Deming T.J. Rapidly recovering hydrogel scaffolds from self-assembling diblock copolypeptide amphiphiles, *Nature*, 417, 424, 2002.

119. Schlaad H. and Antonietti M., Block copolymers with amino acid sequences: Molecular chimeras of polypeptides and synthetic polymers, *Eur. Phys. J. E*, 10, 17, 2003.
120. Barron A.E. and Zuckermann R.N., Bioinspired polymeric materials: In-between proteins and plastics, *Curr. Opin. Chem. Biol.*, 3, 681, 1999.
121. Deming T.J., Methodologies for preparation of synthetic block copolypeptides: Materials with future promise in drug delivery, *Adv. Drug Deliv. Rev.*, 54, 1145, 2002.
122. Taubert A., Napoli A., and Meier W., Self-assembly of reactive amphiphilic block copolymers as mimetics for biological membranes, *Curr. Opin. Chem. Biol.*, 8, 598, 2004.
123. Lazzari M., Liu G., and Lecommandoux S. (eds.), *Block Copolymers in Nanoscience*, Wiley–VCH, Weinheim, Germany, 2006.
124. Gallot B., Comb-like and block liquid crystalline polymers for biological applications, *Prog. Polym. Sci.*, 21, 1035, 1996.
125. Deming T.J., Facile synthesis of block copolypeptides of defined architecture, *Nature*, 390, 386, 1997.
126. Seidel S.W. and Deming T.J., Use of chiral ruthenium and iridium amido-sulfonamidate complexes for controlled, enantioselective polypeptide synthesis, *Macromolecules*, 36, 969, 2003.
127. Sorta E. and Melis L.A., Block length distribution in finite polycondensation copolymers, *Polymer*, 19, 1153, 1978.
128. Miller J.A., Speckhard T.A., Homan J.G., and Cooper S.L. Monte Carlo simulation study of the polymerization of polyurethane block co-polymers. 4. Modeling of experimental data, *Polymer*, 28, 758, 1987.
129. Speckhard T.A., Miller J.A., and Cooper S.L., Monte Carlo simulation study of polymerization of polyurethane block co-polymers. 1. Natural compositional heterogeneity under ideal polymerization condition, *Macromolecules*, 19, 1550, 1986.
130. Sperling L.H., *Introduction to Physical Polymer Science*, Wiley, New York, 1987.
131. Mistrali F. and Proni A., Styrenic block copolymers, *Development in Rubber Technology* 3 (Whelan A. and Lee K.S., eds.), Applied Science, New York, 1982.
132. Krause S., Microphage separation in block copolymers: Zeroth approximation including surface free energies, *Macromolecules*, 3, 84, 1970.
133. Aggarwal S.L., Structure and properties of block polymers and multiphase polymer systems: An overview of present status and future potential, *Polymer*, 17, 938, 1976.
134. Puskas J.E., Antony P., ElFray M., and Altstadt V. The effect of hard and soft segment composition and molecular architecture on the morphology and mechanical properties of polystyrene–polyisobutylene thermoplastic elastomeric block copolymers, *Eur. Polym. J.*, 39, 2041, 2003.
135. Sakurai S., Kawada H., Hashimoto T., and Fetters L.J. Thermoreversible morphology transition between spherical and cylindrical microdomains of block copolymers, *Macromolecules*, 26, 5796, 1993.
136. Hajduk D.A., Gruner S.M., Rangarajan P., Register R.A., Fetters L.J., Honeker C., Albalak, R.J., and Thomas E.L. Observation of a reversible thermotropic order–order transition in a diblock copolymer, *Macromolecules*, 27, 490, 1994.
137. Meier D.J., Theory of block copolymers: Domain formation in A–B block copolymers, *J. Polym. Sci., Part C*, 26, 81, 1969.
138. Hashimoto T., Shibayama M., Kawai H., and Meier D.J. Confined chain statistics of block polymers and estimation of optical anisotropy and domain size, *Macromolecules*, 18, 1855, 1985.
139. Helfand E., Block copolymer theory. I. Statistical thermodynamics of the microphases, *ACS Poly. Prep.*, 14, 970, 1973.
140. Helfand E. and Wasserman Z.R., Block copolymers theory. 6. Cylindrical domain, *Macromolecules*, 13, 994, 1980.
141. Hong K.M. and Noolandi J., Theory of inhomogeneous multicomponent polymer systems, *Macromolecules*, 14, 727, 1981.
142. Veenstra H., Hoogvliet R.M., Norder B., De B., and Abe P. Microphase separation and rheology of a semicrystalline poly(ether–ester) multiblock copolymer, *J. Polym. Sci. B. Polym Phys.*, 36, 1795, 1998.
143. Garbrieelse W., Soliman M., and Dijkstra K., Microstructure and phase behaviour of block copoly(ether ester) thermoplastic elastomers, *Macromolecules*, 34, 1685, 2001.
144. Adams R.K., Hoeschele G.K., and Witsiepe W.K., *Thermoplastic Elastomers*, 2nd edn. (Holden G., Legge N.R., Quirk R., and Schroeder H.E., eds.), Hanser Publishers, Munich, 1996, 191.
145. Zhu L.L. and Wegner G. The morphology of semicrystalline segmented poly(ether ester) thermoplastic elastomers, *Macromol. Chem.*, 182, 3625, 1981.

146. Cella R.J., Morphology of segmented polyester thermoplastic elastomers, *J. Polym. Sci. Symp.*, 42, 727, 1973.

147. Philips R.A., McKenna J.M., and Cooper S.L., Glass transition and melting behavior of poly(ether–ester) multiblock copolymers with poly(ethyleneterephthalate) hard segments, *J. Polym. Sci. Part B*, 32, 791, 1994.

148. Elliott D.J., Natural rubber–polypropylene blends, *Thermoplastic Elastomers from Rubber–Plastic Blends* (Bhowmick A.K. and De S.K., eds.), Ellis Horwood, London, 1990.

149. Stehling F.C., Huff T., Speed C.S., and Wissler G. Structure and properties of rubber modified polypropylene impact blends, *J. Appl. Polym. Sci.*, 26, 2693, 1981.

150. Bhowmick A.K. and Inoue T., Structure development during dynamic vulcanization of hydrogenated nitrile rubber/nylon blends, *J. Appl. Polym. Sci.*, 49, 1893, 1993.

151. Kresge E.N., Elastomeric blends, *J. Appl. Polym. Sci., Appl. Polym. Symp.*, 39, 37, 1984.

152. Brydson J.A., Thermoplastic rubbers, *Rubbery Materials and Their Compounds*, Elsevier, London, 1988, 312.

153. Holden G., *Thermoplastic Elastomers*, 2nd edn. (Holden G., Legge N.R., Quirk R., and Schroeder H.E., eds.), Hanser Publishers, Munich, 1996, 481.

154. Eisenberg A., Clustering of ions in organic polymers. A theoretical approach, *Macromolecules*, 3, 147, 1970.

155. Forsman W.C., Effect of segment–segment association on chain dimension, *Macromolecules*, 15, 1032, 1982.

156. Macknight W.J., Taggart W.P., and Stein R.S., A model for the structure of Ionomers, *J. Polym. Sci., Polym. Symp.*, 45, 113, 1974.

157. Fujimura M., Hashimoto T., and Kawai H., Small-angle x-ray scattering study of perfluorinated ionomer membranes. 2. Models for ionic scattering maximum, *Macromolecules*, 15, 136, 1982.

158. Spontak R.J., Williams M.C., and Agard D.A., Interphase composition profile in SB/SBS block copolymers, measured with electron microscopy, and microstructural implications, *Macromolecules*, 21, 1377, 1988.

159. Chen S., Cao T., and Jin Y., Ruthenium tetraoxide staining technique for transmission electron microscopy of segmented block copoly(ether–ester), *Polym. Commun.*, 28, 314, 1987.

160. Li C. and Cooper S.L., Direct observation of micromorphology of polyether polyurethane using high voltage electron microscopy, *Polymer*, 31, 3, 1990.

161. Li C., Register R.A., and Cooper S.L., Direct observation of ionic aggregates in sulfonated polystyrene ionomers, *Polymer*, 30, 1227, 1989.

162. Harthcock M.A., Probing the complex hydrogen bonding structure of urethane block co-polymers and various acid containing copolymers using infra red spectroscopy, *Polymer*, 30, 1234, 1989.

163. Sung C.S.P. and Schneider N.S., Infrared studies of hydrogen bonding in toluene diisocyanate based polymethylene, *Macromolecules*, 8, 68, 1975.

164. Koberstein J.T., Gancarz I., and Clarke T.C., The effect of morphological transition on hydrogen bonding in PU: Preliminary results of simultaneous DSC-FTIR experiments, *J. Polym. Sci. B*, 24, 2487, 1986.

165. Skrovanek D.J., Painter P.C., and Coleman M.M., Hydrogen bonding in polymers 2. Infra red temperature studies on nylon 11, *Macromolecules*, 19, 699, 1986.

166. Briber R.M. and Thomas E.L., The structure of MDI/BDO based polyurethanes: Diffraction studies on model compounds and oriented thin film, *J. Polym. Sci., Polym. Phys.*, 23, 1915, 1985.

167. StarkWeather H.W., Cocrystallization and polymer miscibility, *J. Appl. Polym. Sci.*, 25, 139, 1980.

168. Seguela R. and Prud'homme J., Affinity of grain deformation in mesomorphic block polymers submitted to simple elongation, *Macromolecules*, 21, 635, 1988.

169. Pakula T., Saijo K., Kawai H., and Hashimoto T. Deformation behaviour of styrene butadiene styrene triblock copolymer with cylindrical morphology, *Macromolecules*, 18, 1294, 1985.

170. Tyagi D., McGrath J.E., and Wilkes G., Small angle x-ray studies of siloxane urea segmented copolymers, *Polym. Eng. Sci.*, 26, 1371, 1986.

171. Hasegawa H., Hashimoto T., Kawai H., Lodge T.P., Amis E.J., Glinka C.J., and Han C. SANS and SAXS studies on molecular conformation of a block copolymer in microdomain space. 2. Contrast matching technique, *Macromolecules*, 18, 67, 1985.

172. Tanaka H. and Nishi T., Study of block copolymer interface by pulsed NMR, *J. Chem. Phys.*, 82, 4326, 1985.

173. Kornfield J.A., Spiess H.W., Nefzger H., Hayen H., and Eisenbach C.D. Deuteron NMR measurement of order and mobility in the hard segments of a model polyurethane, *Macromolecules*, 24, 4787, 1991.

174. Meltzer A.D., Spiess H.W., Eisenbach C.D., and Hayen H. Motional behaviour within the hard domain of segmented polyurethane: A ^{2}H NMR study of a triblock model system, *Macromolecules*, 25, 993, 1992.

175. Okamoto D.T., Cooper S.L., and Root T.W., ^{1}H and ^{13}C NMR investigation of a polyurethane urea system, *Macromolecules*, 25, 1068, 1992.

176. Matson M.W. and Bates F.S., Block copolymer microstructures in the intermediate-segregation regime, *J. Chem. Phys.*, 106, 2436, 1997.

177. Bates F.S. and Fredricson F.S., Block copolymer thermodynamics: Theory and Experiment, *Ann. Rev. Phys. Chem.*, 41, 525, 1990.

178. Matson M.W. and Bates F.S., Unifying weak- and strong-segregation block copolymer theories, *Macromolecules*, 29, 1091, 1996.

179. Leibler L., Theory of microphase separation in block copolymers, *Macromolecules*, 13, 1602, 1980.

180. Foerster S., Khandpur A.K., Zhao J., Bates F.S., Hamley I.W., Ryan A.J., and Bras W. Complex phase behavior of polyisoprene–polystyrene diblock copolymers near the order–disorder transition, *Macromolecules*, 27, 6922, 1994.

181. Khandpur A.K., Foerster S., Bates F.S., Hamley I.W., Ryan A.J., Bras W., Almdal K., and Mortensen K. Polyisoprene–polystyrene diblock copolymer phase diagram near the order–disorder transition, *Macromolecules*, 28, 8796, 1995.

182. Hashimoto T., Shibayama M., and Kawai H., Ordered structure in block copolymer solution. 4. Scaling rules on size of fluctuations with block molecular weight, concentration temperature in segregation and homogeneous regimes, *Macromolecules*, 16, 1093, 1983.

183. Hashimoto T., Order disorder transition in block copolymers, *Thermoplastic Elastomers: A Comprehensive Review* (Legge N.R., Holden G., and Schroeder H.E., eds.), Hanser Publishers, Munich, 1987.

184. Bianchi U. and Pedemonte E., Morphology of styrene butadiene styrene copolymer, *Polymer*, 11, 268, 1970.

185. Grady B.P. and Cooper S.L., *Science and Technology of Rubber* (Mark J.E., Erman B., and Eirich F.R., eds.), Academic Press, London, 1994.

186. Kraus G., *Block Copolymers: Science and Technology* (Neier D.J., eds.), MMI Press, Gordon and Breach, London, 1983.

187. Simon P.F.W., Ulrich R., Spiess H.W., and Wiesner U. Block copolymer-ceramic hybrid materials from organically modified ceramic precursors, *Chem. Mater.*, 13, 3464, 2001.

188. Morton M., Research on anionic triblock copolymers, *Thermoplastic Elastomer: A Comprehensive Review* (Legge N.R., Holden G., and Schroeder H.E., eds.), Hanser Publishers, Munich, 1987.

189. Witsiepe, W.K., Segmented polyester thermoplastic elastomers, *Adv. Chem. Ser.*, 129, 39, 1973.

190. Srichatrapimuk V.W. and Cooper S.L., Infrared thermal analysis of polyurethane block polymers, *J. Macromol. Sci. Phys. B*, 15, 267, 1978.

191. Speckhard T.A. and Cooper S.L., Ultimate tensile properties of segmented polyurethane elastomers: Factors leading to reduced properties for polyurethane based on nonpolar soft segments, *Rubber Chem. Technol.*, 59, 405, 1986.

192. Schneider N.S. and Matton W., Thermal transition behaviour of polybutadiene containing polyurethane, *Polym. Eng. Sci.*, 19, 1122, 1979.

193. Apostolov A.A. and Fakirov S., Effect of the length on the deformation behaviour of polyetheresters as revealed by small angle x-ray scattering, *J. Macromol. Sci., Phys. B*, 31, 329, 1992.

194. Miller J.A., Lin S.B., Hwang K.K.S., Wu K.S., Gibson P.E., and Cooper S.L. Properties of polyether–polyurethane block copolymers: Effect of hard segment length distribution, *Macromolecules*, 18, 32, 1985.

195. Tong J.D. and Jerome R., Dependence of the ultimate tensile strength of thermoplastic elastomers of the triblock type on the molecular weight between chain entanglements of the central block, *Macromolecules*, 33, 1479, 2000.

196. Arcozzi A., Diani E., and Vitali R., *Proceedings of International Rubber Conference. The Society of Rubber Industry, Japan*, Tokyo, 1975, 135.

197. Mullins L., Thixotropic behavior of carbon black in rubber, *J. Phys. Chem.*, 54, 239, 1950.

198. Roy Choudhury N. and Bhowmick A.K., Hysteresis of thermoplastic rubber vulcanizates, *Plast. Rubber Process. Appl.*, 11, 1989, 185.

199. Pedemonte E., Alfonso G.C., Dondero G., De C.F., and Araimo L. Correlation between morphology and stress strain properties of three block copolymers. 2. The hardening effect of the second deformation, *Polymer*, 18, 191, 1977.

200. Smith T.L., *Encyclopedia of Materials Science and Engineering* (Bever, M.B. ed.), Pergamon Press, New York, 1986, 1341.

201. Roy Choudhury N. and Bhowmick A.K., Strength of thermoplastic elastomers from rubber–polyolefin blend, *J. Mat. Sci.*, 25, 161, 1990.

202. Pearson D.S. and Graessley W.W., Elastic properties of well characterized ethylene propylene copolymer network, *Macromolecules*, 13, 1001, 1980.

203. Hergenrother W.L. and Doshak J.M., Determination of entanglement molecular weight of thermoplastic elastomers by tensile retraction measurement, *J. Appl. Polym. Sci.*, 40, 989, 1990.

204. Kikuchi Y., Fukui T., Okada T., and Inoue T. Origin of rubber elasticity in thermoplastic elastomers consisting of crosslinked rubber particles and ductile matrix, *J. Appl. Polym. Sci., Appl. Polym. Symp.*, 50, 261, 1992.

205. Choudhury N.R., Chaki T.K., Dutta A., and Bhowmick A.K. Thermal, x-ray and dynamic mechanical properties of thermoplastic elastomeric natural rubber–polyethylene blends, *Polymer*, 30, 2047, 1989.

206. Marasch M.J., TPU's: Growth from versatility, 53rd Annual Tech. Conference, Antech'95 4088, Boston, May 7–11, 1995.

207. Vinogradov G.V., Dreval V.E., Malkin A.Y., Yanovskii Yu G., Barancheeva V.V., Borisenkova E.K., Zabugina M.P., Plotnikova E.P., and Sabsai O.Y. Viscoelastic properties of butadiene–styrene block copolymers, *Rheol. Acta*, 17, 258, 1978.

208. Wallace J.G., A fabrication guide to processing MPRs and polyolefin alloys, *Elastomerics*, 120, 25, 1988.

209. Borgaonkar H. and Ramani K., Stability analysis in single screw extrusion of thermoplastic elastomers using simple design of experiments, *Adv. Polym. Technol.*, 17, 115, 1998.

210. Gehlsen M.D., Almdal K., and Bates F.S., Order disorder transition: Diblock vs triblock copolymers, *Macromolecules*, 25, 939, 1992.

211. ISOPLAS, *Processing Guidelines for ISOPLAS Resins*, Dow Plastics.

212. Arnold R.L. and Rader C.P., Thermoplastic elastomers, *Handbook of Plastics, Elastomers and Composites* (Harper C.A., ed.), McGraw-Hill, New York, 1992.

213. *TPE Processing Guide*, Thermoplastic Elastomer Division, GLS Corporation, McHenry, IL.

214. Heineck D.W. and Rader C.P., Thermoplastic elastomers: Economical rubber products for the plastic processors, *Plastics Eng.*, 45, 87, 1989.

215. Auteuil J.G.D', Peterson D.E., and Rader C.P., Blow molding of thermoplastic elastomers: A major opportunity for the plastic processors, Antec'89, 47th Annual Tech. Conference of SPE, New York, May 1–4, 1989, 1740.

216. Van I.E., Peterson D.E., and Weider B.K., Antec'86, 44th Annual Tech. Conference of SPE, Boston, MA, 1986.

217. Cakmak M. and Wang M.D., Structure development in the tubular blown film of PP/EPDM thermoplastic elastomer, Antec'89, 47th Annual Tech. Conference of SPE, New York, May 1–4, 1989, 1756.

218. Hashimoto T., Todo A., Itoi H., and Kawai H. Domain boundary structure of styrene–isoprene block copolymer films cast from solution. 2. Quantitative estimation of the interfacial thickness of lamellar microphase systems, *Macromolecules*, 10, 377, 1977.

219. Roy Choudhury N. and Bhowmick A.K., Adhesion between individual components and mechanical properties of natural rubber–polypropylene thermoplastic elastomeric blends, *J. Adhes. Sci. Technol.*, 2(3), 167, 1988.

220. Jha A., Bhowmick A.K., Fujitsuka R., and Inoue T. Interfacial interaction and peel adhesion between polyamide and acrylate rubber in thermoplastic elastomeric blends, *J. Adhes. Sci. Technol.*, 13(6), 649, 1999.

221. Coran A.Y., Patel R., and Williams D., Rubber–thermoplastic composition. VI. The swelling of vulcanized rubber–plastic composition in fluids, *Rubber Chem. Technol.*, 55, 1063, 1982.

222. Roy Choudhury N. and Bhowmick A.K., Studies on adhesion between natural rubber and polyethylene and the role of adhesion promoters, *J. Adhes. Sci. Technol.*, 1, 27, 1990.

223. Wu S., Interfacial energy, structure and adhesion between polymers, *Polymer Blends*, Vol. 1 (Paul D.R. and Newman S., eds.), Academic Press, New York, 1978.

224. Wu S., *Polymer Interface and Adhesion*, Marcel Dekker Inc., New York, 1982.
225. Frenkel R., Duchacek V., Kirillova T., and Kuz'min E. Thermodynamic and structural properties of acrylonitrile butadiene rubber/polyethylene blends, *J. Appl. Polym. Sci.*, 34, 1301, 1987.
226. Couman W.J., Heikens D., and Sjoerdsma S.D., Dilatometric investigation of deformation mechanism in polystyrene–polyethylene block copolymer blend: Correlation between Poisson ratio and adhesion, *Polymer*, 21, 103, 1980.
227. Coumans W.J. and Heikens D., Dilatometer for use in tensile tests, *Polymer*, 21, 957, 1980.
228. Sjoerdsma S.D., Bleijenberg A.C.A.M., and Heikens D., The Poisson ratio of polymer blend, effects of adhesion and correlation with the Kerner packed grain model, *Polymer*, 22, 619, 1981.
229. Kresge E.N., Rubbery thermoplastic blends, *Polymer Blends*, Vol. 2 (Paul D.R. and Newman S., eds.), Academic Press, New York, 1978.
230. Kresge E.N., Polyolefin thermoplastic elastomer blends, *Rubber Chem. Technol.*, 64, 469, 1991.
231. Piazza S., Arcozzi A., and Verga G., Modified bitumen containing thermoplastic polymers, *Rubber Chem. Technol.*, 53, 994, 1980.
232. Bull A.L. and Holden G., The use of thermoplastic rubbers in blends with other plastics. Paper presented in Rubber Division, American Chemical Society, April 1976 and *J. Elast. Plast.*, 9, 281 1977.
233. Inoue T., Soen T., Hashimoto T., and Kawai, H. Studies on domain formation of the A-B type block copolymer with polystyrene and polyisoprene, *Macromolecules*, 13, 87, 1970.
234. Kirkpatrick J.P. and Preston D.T., Polymer modification with styrenic block copolymers, *Elastomerics*, 120, 30, 1988.
235. Lindsey C.R., Paul D.R., and Barlow J.W., Mechanical properties of HDPE–PS–SEBS blends, *J. Appl. Polym. Sci.*, 26, 1, 1981.
236. Bartlett D.W., Paul D.R., and Barlow, J.W. Additive improves properties of scrap PP/PS blends, *Mod. Plastics*, 58, 60, 1981.
237. Verhoogt H., Willems C.R.J., Dam V.J., and De Boer P.A., Blends of a thermotropic LCP and a thermoplastic elastomer. II: Formation and stability of LCP fibers, *Polym. Eng. Sci.*, 34, 453, 1994.
238. Verhoogt H., Langelaan H.C., Van Dam J., and De Boer P.A., Blends of a thermotropic liquid crystalline polymer and a thermoplastic elastomer. I: Mechanical properties and morphology, *Polym. Eng. Sci.*, 33, 754, 1993.
239. *Hytrel Bulletin 1–25*, Blends of Hytrel polyester elastomer with PVC, E.I. DuPont de Nemours and Co., Delaware, 1976.
240. Whitlock K.H., Elastomer blend, *US Patent* 4124653, 1978.
241. Masanori N. and Noritaka O., Thermoplastic elastomer composition of excellent heat fusibility, *Japanese Patent* 03-100045, 1991.
242. Schoullenberger C.S., *Polyurethane Technology* (Bruim P.F., ed.), Wiley, New York, 1969, 197.
243. Demma G., Martuscelli E., Zanetti A., and Zorzetto M. Morphology and properties of polyurethane-based blends, *J. Mater. Sci.*, 18, 89, 1983.
244. Chang J.C.C., Tsai F.J., and Chang M.C., SPE Antec'87, 1348. Proceedings of the 45th Annual Technical Conference and Exhibit, Los Angeles, 4–7 May, 1987.
245. Kumar G., Neelakantan N.R., and Subramanian N., Mechanical behaviour of polyacetal and thermoplastic polyurethane elastomer toughened polyacetal, *Polym. Plastics Technol. Eng.*, 32, 33, 1993.
246. Newmann W. et al., *Preprints, 4th Rubber Technology Conference*, London, May 22–25, 1962.
247. Farrissey W.J. and Shah T.M., *Handbook of Thermoplastic Elastomers* (Walker B.M. and Rader C.P., eds.), Van Nostrand Reinhold, New York, 1988.
248. Ann T.O., Han K.T., Jeong H.M., and Lee S.W. Miscibility of thermoplastic polyurethane elastomers with chlorine containing polymers, *Polym. Int.*, 29, 115, 1992.
249. Lausberg D.R., Arenz S., and Teinberger R.S, *Polyurethane World Congress*, Nice, France, September 24–26, 1991, 680.
250. Eisenberg A., Smith P., and Zhou L.L., Ionomeric blends. I. Compatibilization of polystyrene–polyethyl acrylate styrene via ion interaction, *Polym. Eng. Sci.*, 22, 1117, 1982.
251. Smith P. and Eisenberg A., Ionomeric blends. I. Compatibilization of the polystyrene–poly(ethyl acrylate) system via ionic interactions, *J. Polym. Sci., Polym Lett.*, 21, 223, 1983.
252. Zhou L.L. and Eisenberg A., Ionomeric blends. II. Compatibility and dynamic mechanical properties of sulfonated cis-1,4-polyisoprenes and styrene/4-vinylpyridine copolymer blends, *J. Polym. Sci., Polym. Phy.*, 21, 595, 1983.

253. Myrick R.E., Property enhancement of melt processible rubber by blending with thermoplastics and elastomers, *Proceeding of the Regional Tech. Conference of SPE*, Brookfield, CT, 1993, 55.
254. Nanotechnology Research Directions, IWGN Workshop Report: Vision of Nanotechnology in Next Decade, National Science and Technology Council, Washington DC, September 1999.
255. Bates F.S. and Fredrickson G.H., Block copolymers-designer soft materials, *Phys. Today*, 52, 32, 1999.
256. Alexandridis P. and Lindman B. (eds.), *Amphiphilic Block Copolymers: Self-Assembly and Applications*, Elsevier, Amsterdam, 2000.
257. Pralle M.U., Whitaker C.M., Braun P.V., and Stupp S.I. Molecular variables in the self-assembly of supramolecular nanostructure, *Macromolecules,* 33, 3550, 2000.
258. Jeoung E., Galow T.H., Schotter J., Bal M., Ursache A., Tuominen M.T., Stafford C.M., Russell T.P., and Rotello V.M. Fabrication and characterization of naoelectrode arrays formes via block copolymers self-assembly, *Langmuir*, 17, 6396, 2001.
259. Hashimoto T., Tsutsumi K., and Funaki Y., Nanoprocessing based on bicontinuous microdomains of block copolymers: Nanochannel coated with metals, *Langmuir*, 13, 6869, 1997.
260. Luzinov I. et al., Microtribological behaviour of tethered block copolymer monolayers, *Polym. Prepr.*, 41, 1499, 2000.
261. Böker A., Müller A.H.E., and Krausch G., Functional ABC triblock copolymers for controlled surface patterns of nanometer scale, *Polym. Mater. Sci. Eng.*, 84, 312, 2001.
262. Spatz J.P., Moessmer S., Hartmann C., Moeller M., Herzog T., Krieger M., Boyen H-G., Ziemann P., and Kabius B. Ordered deposition of inorganic clusters from micellar block copolymer films, *Langmuir*, 16, 407, 2000.
263. Ciebien J.F., Cohen R.E., and Duran A., Catalytic properties of palladium nanoclusters synthesized within diblock copolymer films: Hydrogenation of ethylene and propylene, *Supramol. Sci.*, 5, 31, 1998.
264. Youk J.H., Park M.K., Locklin J., Advincula R., Yang J., and Mays J. Preparation of aggregation stable gold nanoparticles using star-block copolymers, *Langmuir*, 18, 2455, 2002.
265. Fink Y., Urbas A.M., Bawendi M.G., Joannopoulos J.D., and Thomas E.L. Block copolymers as photonic bandgap materials, *J. Lightwave Tech.*, 17, 1963, 1999.
266. Blockstaller M., Kolb R., and Thomas E.L., Metallodielectric photonic crystals based on diblock copolymers, *Adv. Mater.*, 13, 1783, 2001.
267. Tamil S.S., Tomokatsu H., Masayuki N., and Martin M. Block copolymer mediated synthesis of gold quantum dots and novel gold-polypyrole nanocomposites, *J. Phys. Chem.*, 103, 7441, 1999.
268. Tamil S.S., Spatz, J.P., Klok H.A., and Martin M. Gold-polypyrrole core-shell particles in diblock copolymer micells, *Adv. Mater.*, 10, 132, 1998.
269. Sohn B.H., Cohen R.E., and Papaefthymiou G.C., Magnetic properties of iron oxide nanoclusters within microdomains of block copolymers, *J. Magn. Magn. Mater.*, 182, 216, 1998.
270. School R.J., Markets for thermoplastic elastomers, *Elastomer Technology Handbook* (Chremisinoff N.P., ed.), CRC Press, Boca Raton, FL, 1993.
271. Radar C.P., Thermoplastic elastomers, *Kunstst. Ger. Plast.*, 83, 777, 1993.
272. Dufton P.W., *Thermoplastic Elastomers*, Rapra Industry Analysis Report, 2001.
273. School R.J., Markets for thermoplastic elastomers, *Handbook of Thermoplastic Elastomers* (Walker B.M. and Rader C.P., eds.), Van Nostrand Reinhold, New York, 1988, 285.
274. Isqacs P.A., Elastomer alloys in medical products reduce costs and toxicity concern, *Elastomerics*, 121, 26, 1989.
275. Anonymous, New silicone modified TPE combines best of both worlds, *Elastomerics*, 120, 28, 1988.
276. Hill B., Industry's integration of environmental product design, IEEE International Symposium on Electronics and the Environment, Virginia, IEEE, 1993.
277. Design for disassembly challenge of the future, DFMA 1992, Rhode Island, BDI.
278. World Thermoplastic Elastomers, Freedona Industry Study, 2002, 1553.

6 Plastomers

Sudhin Datta

CONTENTS

6.1 CONCEPTS OF E-PLASTOMERS

Plastomer, a nomenclature constructed from the synthesis of the words plastic and elastomer, illustrates a family of polymers, which are softer (lower flexural modulus) than the common engineering thermoplastics such as polyamides (PA), polypropylenes (PP), or polystyrenes (PS). The common, current usage of this term is restricted by two limitations. First, plastomers are polyolefins where the inherent crystallinity of a homopolymer of the predominant incorporated monomer (polyethylene or isotactic polypropylene [iPP]) is reduced by the incorporation of a minority of another monomer (e.g., octene in the case of polyethylene, ethylene for iPP), which leads to amorphous segments along the polymer chain. The minor commoner is selected to distort and, in the limit, not incorporate into the crystalline lattice of the predominant monomer. Second, the most common designators for plastomers are copolymers, which are intermolecularly and intramolecularly uniform in composition. This requires their synthesis to be with a single-site polymerization in a uniformly back-mixed polymerization process. These functional limitations on the definition of plastomers eliminates copolymers such as styrene–butadiene copolymers since they are not polyolefins or compositionally broad polyolefins such as those made with Ziegler–Natta process by incorporation of a minority of ethylene into iPP (random copolymers). Within the classes of polyolefins, which meet these composition, crystallinity, and synthesis limitations, the sequence of the monomers within the linear chains is dictated by the reactivity ratios of monomer incorporation in the chain-propagation mechanisms. These limitations in the definition of plastomers are not arbitrary; they arise from an industrial and academic recognition that molecularly uniform plastomers of reduced crystallinity, compared to the corresponding crystalline homopolymer, have unique properties of miscibility, toughening, and elasticity, which the molecularly dispersed copolymers of similar average composition do not replicate. Further, the introduction of plastomers is closely aligned with the development of discrete, single-sited, organometallic polymerization catalysts (e.g., metallocene), which enable the formation of plastomers to be a commercial reality.

Structurally, plastomers straddle the property range between elastomers and plastics. Plastomers inherently contain some level of crystallinity due to the predominant monomer in a crystalline sequence within the polymer chains. The most common type of this residual crystallinity is ethylene (for ethylene-predominant plastomers or E-plastomers) or isotactic propylene in meso (or m) sequences (for propylene-predominant plastomers or P-plastomers). Uninterrupted sequences of these monomers crystallize into periodic structures, which form crystalline lamellae. Plastomers contain in addition at least one monomer, which interrupts this sequencing of crystalline mers. This may be a monomer too large to fit into the crystal lattice. An example is the incorporation of 1-octene into a polyethylene chain. The residual hexyl side chain provides a site for the dislocation of the periodic structure required for crystals to be formed. Another example would be the incorporation of a stereo error in the insertion of propylene. Thus, a propylene insertion with an *r* dyad leads similarly to a dislocation in the periodic structure required for the formation of an iPP crystal. In uniformly back-mixed polymerization processes, with a single discrete polymerization catalyst, the incorporation of these interruptions is statistical and controlled by the kinetics of the polymerization process. These statistics are known as reactivity ratios.

A significant feature of the properties of plastomers is their inherent crystallinity, which arises from the intramolecular and intermolecular crystallization of the interrupted monomer sequences discussed above. Crystallization of the sequences occurs predominantly for longer sequences of monomers: shorter sequences of monomers cannot participate in crystal formation at ambient temperatures. It is believed that uninterrupted sequences of 12–20 monomers of ethylene or propylene are required for the formation of the most rudimentary crystals. Longer sequences lead to the formation of more robust structures. The statistics of copolymerization for most known discrete catalysts which have a reactivity ratio between 0.5 and 10 indicate a kinetic model where the addition of the interrupting monomer leads to rapid loss in the population of the extended sequences required for the formation of crystals. As an illustration, about 20 mol% of octene is adequate for the elimination of all crystallinity at room temperature in copolymers of ethylene and octene. A consequence of this artifact of crystallization of polyolefins and statistics of monomer incorporation is that most plastomers are close in average composition to the homopolymer of the predominant monomer. An alternate view would be that the incorporation of modest amounts of the interrupting monomer in either polyethylene or PP leads to the large changes in the crystallinity and, thus, the properties of the plastomers. An illustration of this effect is shown in Figure 6.1, where the stress–strain properties of a related set of propylene–ethylene polymers, with increasing amounts of ethylene leading to less-crystalline tensile characteristics, are revealed.

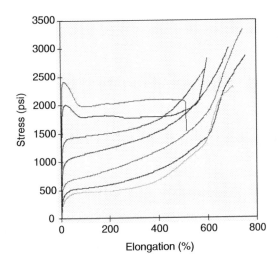

FIGURE 6.1 A set of consistent stress–strain data for P-plastomers with ethylene content between 8.2 wt% (highest) and 16.0 wt% (lowest). The data for P-plastomers with intermediate composition is intermediate within these extremes.

The loss of bulk crystallinity in the plastomers with increasing amounts of the comonomer is also accompanied by microscopic changes in the crystal structure. These changes affect the melting point of the crystals. While the unit cells of these crystalline structures are similar to the known crystal structure of the predominant monomer, the plastomers crystals are smaller, both due to thinner lamellae and due to lack of radial extension of the lamellae. The lamellar dimensions decrease with increasing comonomer content, leading to increasing imperfection in the crystal structure. A consequence of this effect is that the melting point of the crystalline point of plastomers rapidly decreases to ambient temperature with increasing levels of comonomer. Figure 6.2 illustrates this effect. In summary, plastomers are partly crystalline due to introduction of small amounts of an interrupting comonomer into a crystalline sequence of olefin mers. However, both the extent of crystallinity and the melting point of the crystals rapidly diminish by addition of the comonomer. Figure 6.3 illustrates the location of the E-plastomers and the P-plastomers within this continuum of the polyolefin compositions. Within this relatively narrow range of composition, plastomers can display a wide range of mechanical behaviors—from typical plastics to traditional elastomers, as well as a host of intermediate behavior.

Most of the above-mentioned discussion has dealt with the softness of the plastomers compared to the thermoplastics. An important technical advancement was a study of the deformation of E-plastomers' uniaxial tension as a function of comonomer content and molecular weight [1]. Within the melting range of E-plastomers, temperature was used as an experimental variable to reveal the relationship between crystallinity and stress response. The concept of a network of flexible chains with fringed micelle crystals serving as the multifunctional junctions provided the structural basis for analysis of the elastic behavior. The rubber modulus scaled with crystallinity. Furthermore, the dimension of the fringed micelle junction obtained from the modulus correlated well with the average crystallizable sequence length of the copolymer. Because classical rubber theory could not account for the large strain dependence of the modulus, a theory, which incorporates the contribution of entanglements to the network response, was considered. Slip-link theory described the entire stress–strain curve. The slip-link correlated with crystallinity; the cross-link density did not depend on crystallinity and appeared to represent a permanent network. The latter was further revealed by the effect of molecular weight on the stress–strain behavior. The study

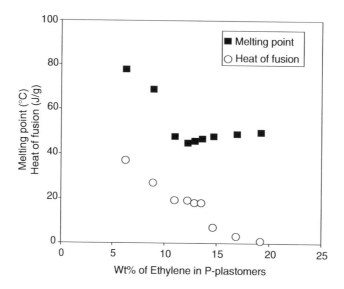

FIGURE 6.2 Heat of fusion and melting point for a set of P-plastomers made under similar process conditions but with difference in the composition.

proposed that lateral attachment and detachment of crystallizable chain segments at the crystal edges provide the sliding topological constraint attributed to slip-link, and entanglements that tighten into rigid knots upon stretching function as permanent network junctions.

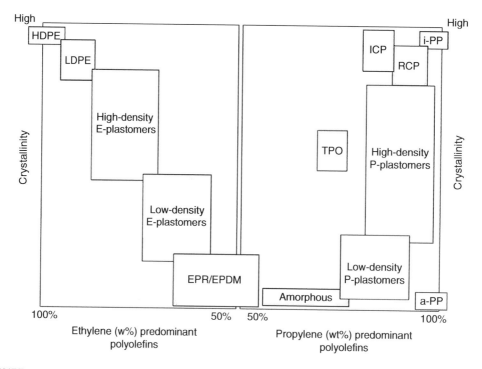

FIGURE 6.3 The crystallinity and composition continuum for ethylene- and propylene-dominated polyolefins. Note the dispersion for the propylene-dominated polyolefins due to much-greater prevalence of blends and the presence of tacticity derived changes in crystallinity.

The commercial growth of plastomers has been unexpectedly robust for an entirely new polymer. Commercial production by a single manufacturer, Dow Chemical, has grown from commercial introduction in 1992 to about 2×10^9 lb of capacity by 2003 [1]. This growth is almost entirely due to E-plastomers, which appeared in early 1990s; the growth of the P-plastomers is very much in the future since they were introduced in 2003. Other manufacturers are believed to have similar growth [1]. A cursory analysis of the utility of these application areas shows these developments are concentrated in two collective groups. In the first group, higher crystallinity plastomers are often used as soft thermoplastics in traditional plastic applications such as films and nonwoven fabrics. These plastomers have enough crystallinity that they are amenable to plastics processing, which typically requires rapid crystallization at elevated temperatures. However, the reduced levels of crystallinity compared to the thermoplastics provide the desirable attributes of softness and elongation. The less-crystalline plastomers have been used as blend components to impart a soft and impact-resistance feel to thermoplastics (mostly iPP).

Industrially, the first E-plastomers were Tafmers from Mitsui [2], which were introduced, in a limited volume, in the late 1980s. Tafmer is an E-plastomer with minor amounts of propylene and other higher alpha olefins as comonomer. They were developed in a modified Ziegler–Natta process, which produced approximately a single-polymerization mode. The Tafmer range of products exemplified some of the common themes that presaged the advent of plastomers. Most of the products were in a range of viscosities that were intermediate between the highly viscous rubbery elastomers and the comparatively lower viscosities of the thermoplastics. The products were concentrated in a range of intermediate crystallinity and densities—less than polyethylene yet more than ethylene–propylene rubber (EPR). Most importantly, they were offered as free-flowing pellets which enabled the user to compound these materials and impart rubbery properties to the other blend components without using the expensive and slow rubber processing. This seemingly unimportant yet commercially significant milestone of obtaining soft rubbery properties from a pelletized polyolefin of viscosity low enough (melt index at 190°C of >0.3), in contrast to rubbers which are sold as bales and have a viscosity much greater than the above-mentioned limit, has been a theme in the development of plastomers from these early developments. This allows the compounders to use plastics feeders which are designed to handle free-flowing pellets and twin-screw extruder-mixing equipment for plastomers instead of the more-traditional bale dispersators and choppers in combination with intensive batch mixers, such as Banbury mixers, which have been used for rubbers.

6.2 BLENDS OF E-PLASTOMERS

The melt rheology and the corresponding solid-state morphological studies indicate that plastomers of similar composition are essentially miscible with each other as well as the homopolymer of the predominant monomer in the melt [3]. Phase separation due to differential crystallization occurs on cooling the melt to the solid state. Melt morphology was inferred from melt rheology using E-plastomers in combination with (1) high-density polyethylene (HDPE), (2) iPP, and (3) copolymers of propylene with small amounts of ethylene or ethylene and butane (EB) combined (RCP) in separate binary experiments. The rheological analysis was based on three methods. The first analyzed the shape and diameter of the semicircular curves in the plots of the storage melt viscosity (η) versus the loss melt viscosity (η'') representing the Cole–Cole plot. The second was based on the slopes in logarithmic plots of the dynamic storage modulus (G') versus the dynamic loss modulus (G') between the blend components and the pure components. The third analyzed the deviation of the plot of the log complex viscosity (log η^*) versus the blend compositions from the linearity. The results indicated that the E-plastomers with HDPE blend showed semicircular curves with the same diameter in the Cole–Cole plot, linearity in the slope of the log G' versus log G'' between the pure components and the blend compositions, and obeyed the rule of mixtures between the melt viscosity and the blend compositions. In contrast, the other binary blends did not follow any of these rheological criteria. This indicates that only the E-plastomer–HDPE binary blends are compatible. This result is in keeping with the similarity of composition of E-plastomers and HDPE. Additionally, the morphology of the blends was determined microscopically at room temperature, after the melt had been frozen. The phase morphology of E-plastomers–HDPE binary blend appeared homogeneous, whereas the other three blends were very heterogeneous with phase domains in the range of several micrometers (μm).

A more detailed study of the compatibility of similar E-plastomers at different compositions of ethylene contained within the E-plastomer compared blends and single polymer of similar average crystallinity. The blends contained a higher and lower ethylene-content E-plastomer to achieve the target level of crystallinity. Thermal analysis (differential scanning calorimetry [DSC]) indicated that the components crystallized separately in all the blends. The stress–strain behavior of the copolymers and their blends was evaluated as a function of temperature. At ambient temperature the total amount of crystallinity was the primary indicator of the stress–strain relationship regardless of whether the material was a single copolymer or a copolymer blend. Any effects of phase morphology were subtle at ambient temperature. This indicates that any phase separation of the different crystallinity (as indicated by DSC) in essentially incomplete. At higher temperatures, where the network junctions started to melt, miscibility of the noncrystalline regions produced a synergistic effect on the tensile strength. This effect was limited since the copolymers were immiscible in the molten state if their ethylene content was very different and the copolymer blend had a lower tensile strength than the target at higher temperatures.

Blends of PS and polymethyl methacrylate (PMMA) with E-plastomers have been investigated. The experiments were performed in a counterrotating twin-screw extruder. The evaluation of mechanical properties and morphology were completed to determine stress–strain behavior, impact, domain size, and interfacial adhesion for the blends. The results of tensile strength, modulus of elasticity, and impact properties for the blends show that incompatible and synergistic behaviors cover a wide range. Morphology studies using scanning electron microscope (SEM) indicate that the domain sizes of the rubber phase in the blend range from 1 to 15 μm and a variety of interfacial adhesion behaviors. All the blends displayed complete phase separation. The compatibilized blends of PS–E-plastomer, which are compatibilized with styrene–ethylene–butylene–styrene block copolymer (SEBS), illustrate that not only the domain size of the elastomer phase can be reduced but also the interfacial adhesion can be enhanced. This improvement in interfacial adhesion is correlated to improved impact strength and tensile strength.

6.2.1 CROSS-LINKED E-PLASTOMERS

E-plastomers, which traditionally lack an abundance of pendant unsaturated reactive sites, due to lack of incorporation of nonconjugated diolefins, such as 5-ethylidene-2-norbornene (ENB) and 5-vinyl-2-norbornene (VNB), are amenable to chain extension and eventually cross-linking by free radical sources such as peroxides which can abstract secondary and tertiary hydrogen (attached to carbon atoms), leading to the formation of free radicals [4]. These can participate in chain-extension reaction by reacting with other polymer molecules with elimination of a second hydrogen atom. The predominance of the ethylene within E-plastomers structurally leads to a greater probability that the initial abstraction of hydrogen is from a secondary carbon atom rather than a tertiary one. This difference leads to a net chain extension and molecular weight growth rather than chain scission—which is the more commonly observed effect for P-plastomers which are ethylene-deficient in composition and, thus, contain a relative abundance of tertiary hydrogen compared to the secondary ones. E-plastomers can thus be readily cross-linked with peroxide. Other free-radical-generating agents such as heat, oxygen, and radiation are also quite effective in inducing chain extension in E-plastomers. The peroxide reactivity can also be diverted to induce condensation reactions of E-plastomers with easily condensable products substrates. Vinyl alkoxy silanes are a common example. The condensation of E-plastomers with these compounds leads to polymeric products which contain substantial amounts of pendant alkoxy silanes on the chain that they can be cross-linked with moisture.

In an initial study, peroxide cross-linking of E-plastomers (octene comonomers; Engage 8150 and Engage 8180) showed that they can be compounded and cross-linked very efficiently using traditional rubber equipment, and the cross-linked materials can compete with high-performance EPR. Compounds based on polyolefin elastomers display excellent electrical and mechanical properties and outstanding heat-aging performance. These advantages make ENGAGE polyolefin elastomers well suited for a variety of rubber applications where ethylene–propylene–diene mono-mer (EPDM) is usually used. The initial applications considered are low-voltage electrical flexible cable and automotive hose.

The initial study was soon followed by cross-linking of E-plastomers with gamma radiation. E-plastomers (octene copolymers) of differing molecular weights and comonomer content were exposed to 25 kGy of gamma radiation from a ^{60}Co source in an inert atmosphere. Tests were performed to evaluate the changes in the chemical, physical, and mechanical properties. The effect of gamma irradiation is most noticeable in the mechanical properties such as the tensile modulus and elongation to break. The combination of the low octene-1 content and high-molecular weight leads to higher postirradiation tensile modulus. Spectroscopy studies indicate that formation of unsaturated chemical moieties after irradiation of samples with same molecular weight occurs more readily in the lower octene-1-content E-plastomers. The low-molecular-weight and low-octane-content copolymers showed some evidence of polymer-chain scission as the dominant effect. The study has evaluated elongation, mechanical properties, melting point, tensile strength, and thermal properties as affected by the extent of gamma radiation.

A more detailed study has looked at the structural model of cross-linked E-plastomers. The purpose of this study was to develop correlations between dynamic properties such as the hysteretic loss and compression set of E-plastomers with key microstructural variables such as density and molecular weight.

Moisture cross-linkable, silane-grafted E-plastomers are made by grafting the blend with a silane cross-linker (typically vinyl triethoxy silane in the presence of peroxide), fabricating the silane-grafted blend and curing the shaped, silane-grafted blend with water, preferably in the presence of a cure catalyst. In a specific example, a blend of 93% E-plastomer (contains 41 wt% octene) and 7% HDPE having melt index 1.4 g (190°C) was grafted with vinyltrimethoxysilane, and cured to give test samples having tensile strength 481 psi and compression set 44%. A cross-linkable composition of an E-plastomer formed by moisture hydrolysis of silane-grafted polymers has been developed to be useful as footwear soles and uppers, gaskets, membranes, bushings, hinges, and drawer slides.

In an alternate synthesis, the formation of thermoplastic vulcanizate (TPV) is accomplished with two polymers, wherein one polymer is grafted with a carboxylic acid anhydride, which then is reacted with an aminosilane, which reacts with the acid anhydride and then cross-links with moisture. The vulcanizates exhibit good mechanical properties and lower melt-flow index than the starting polymers.

In a very innovative result, the ability of the E-plastomers to undergo chain extension has been used to introduce limited amounts of high-molecular-weight fractions (at a concentration substantially below the gel point) to induce a longer relaxation time and, thus, a better melt strength for better processing in some fabrication processes. This is important for E-plastomers, which are typically narrow in molecular weight distribution and lack the high-molecular-weight fraction, which is required for these processing conditions. These branched, but not gelled, polymers have improved rheological performances and/or melt-strength attributes relative to the unmodified polymer, while having the same or better physical properties. These properties are achieved in the E-plastomers which are treated with a peroxide in an amount less than that which would cause minimal gel formation in melt-processing conditions.

6.2.2 Cross-Linked E-Plastomer Blends

The ability of the E-plastomers to participate in the peroxide-mediated chain-extension processes can be augmented in blends with EPDM, where the mixture is homogeneously cross-linked with free radicals [5]. The use of these blends of EPDM and E-plastomers leads to improved processing and physical properties of the combination, compared to the EPDM alone, though the resulting vulcanizates are somewhat harder than the EPDM vulcanizates alone due to the presence of the semicrystalline plastomers in the vulcanized mixture.

A prominent example of this technology is shown in the work from ExxonMobil Chemical where an EPDM with incorporated VNB (which is a particularly potent cross-linking agent with peroxides) was blended with 20 wt% of E-plastomer and the resulting combination showed better extrusion processing and better physical properties than the EPDM alone. It is noteworthy that these blends had an improved cure rate compared to an EPDM alone, which indicates a very significant participation by the plastomer within the vulcanized network. This technology was applied to form an electrical cable insulator in a combination of EPDM and plastomers in compounds containing clay filler. Comparison of the EPDM compound in the cable-insulator compositions which had a tensile strength 8.5 MPa, and elongation 250% to an 85–15 EPDM–E-plastomer blend which had a tensile strength 9.9 MPa, and elongation 319% showed the advantage of these blends.

Elastomeric composition for dynamic application of cross-linked E-plastomers has been made with filer-reinforced systems which contain a metal salt (typically zinc) of an alpha, beta unsaturated acid. These additives improve the tensile and tear strength of the elastomer and are cured with a peroxide cure system. These cross-linked articles are suitable for dynamic loading applications such as belting, including power transmission and flat belting.

6.2.3 Blend with Isotactic Polypropylene and Thermoplastic Olefin

An important development in the use of E-plastomers was an unexpected breakthrough—that the blends of E-plastomers with iPP were surprisingly compatible [6]. This compatibility, which has been the subject of theoretical calculations, leads to iPP–E-plastomer-based compositions for automotive exterior and interior applications. E-plastomers based on single-site catalysts exhibit a number of physical properties that make them extremely useful for automotive interior applications. Due to the low level of unsaturation these polymers exhibit outstanding heat and ultraviolet (UV)-aging resistance. Their molecular structures enable these polymers to exhibit low glass transition temperature (T_g). Thus, compositions containing these polymers exhibit very good low-temperature impact properties. Furthermore, these products impart inherent flexibility and soft touch to blends and eliminate the need for plasticizers. E-plastomers can be compounded to produce flexible TPO compositions with elevated temperature performance and desirable softness suitable for automotive interior applications.

A study on the effectiveness of the E-plastomers as impact modifiers for iPP was carried out in relation to the traditional modifier EPDM. In this study, the flow properties of the E-plastomer–iPP and EPDM–PP blends were also evaluated. The blends were analyzed by solid-state 13C-nuclear magnetic resonance (NMR) spectroscopy, microscopy (SEM), and DSC. The results showed that E-plastomer–PP and EPDM–PP blends present a similar crystallization behavior, which resulted in a similar mechanical performance of the blends. However, the E-plastomer–PP blend presents lower torque values than the EPDM–PP blend, which indicates a better processability when E-plastomer is used as an impact modifier for iPP.

The thermal, mechanical, and morphological behaviors of two binary blends, HDPE–E-plastomer (Engage 8200) and iPP–E-plastomer (Engage 8200) have been investigated to compare the compatibility and molecular mechanistic properties of the blends. Both systems are thermodynamic-ally immiscible but mechanically compatible. Thermal studies indicate that both blends exhibit two distinct melting peaks and there is depression of the HDPE melting peak in the blend with high

E-plastomer content. Two discernible beta transitions have been observed in the iPP–E-plastomer blends with lower content of octene (up to 50%). However, the beta relaxation has been shifted to a lower temperature with the E-plastomer content in HDPE–E-plastomer system. The impact strength of HDPE–E-plastomer blends is higher than that of the PP–E-plastomer blend. The morphology of tensile fractured surfaces indicates that the HDPE–E-plastomer system is more homogeneous than the PP–E-plastomer system. The degree of compatibility in HDPE–E-plastomer blends is larger than that of the iPP–E-plastomer system, which may be due to the similar chemical structures of HDPE and E-plastomer. In a companion but enlarged study, these properties were also investigated for four binary blends—HDPE–E-plastomer, iPP–E-plastomer, RCP–E-plastomer, and poly(propylene-co-ethylene-co-1-butene) Ter RCP–E-plastomer. The thermal and mechanical properties showed that all the blends become immiscible but are mechanically compatible. A lower content of E-plastomer (up to 50%) in PP–E-plastomer, RCP–E-plastomer, and Ter RCP–E-plastomer blends showed two beta transitions, whereas beta relaxation was shifted to a lower temperature with the E-plastomer content in the HDPE–E-plastomer system.

In hard, automotive exterior thermoplastic olefins (TPOs) where the proportion of the iPP in the blend is the predominant fraction of a representative formulation, with 100 parts of a RCP, E-plastomer 24 parts, and talc 37 parts, showed Izod impact strength 20.1 kJ/m^2. In another representative formulation for an impact-modified iPP a 30:50:20 blend of three components (1) an EPDM (69% ethylene, 2% ENB) copolymer rubber with an intrinsic viscosity at 135°C of 2.3 dL/g, (2) an iPP (mass flow rate [MFR] 60 g/10 min at 135°C, sol fraction 100%, isotactic pentad fraction 98.6%, M_w/M_n 11.0), and (3) talc (particle size 2.4 μm) gave on compounding a blend with MFR 12 g/10 min, flexural modulus 1850 MPa, Izod impact strength 650 J/m, Rockwell R-scale hardness 47, brittle temperature −40°C. In a representative formulation, an ICP (containing 25 wt% EPR) having a molecular weight of 380,000 and an MFR at 230°C was blended with 5 wt% of an E-plastomer-containing butene as a comonomer; MFR 4.0; density 0.862 g/cm^3) 4, pelletized and injection molded into test parts having MFR 28, bending modulus (JIS K 7203) 1080 MPa, notched Izod impact strength 32 kJ/m^2, Rockwell hardness 72. In another representative formulation an RCP containing 4.3 wt% ethylene was blended with 30 wt% of an amorphous EPR. This blend was recompounded with 10 wt% of Engage 8200, an E-plastomer, and a small amount of fillers to produce a soft, smooth sheet.

The use of E-plastomers has been extended to soft iPP blends, in contrast to TPOs where the relative composition is weighed to have a majority of the soft plastomer rather than the much more rigid iPP. In a representative composition, a blend of ICP containing 12% xylene-extractable ethylene–propylene copolymer, 50 wt% iPP, 15 wt% Engage 8200 (an E-plastomer containing octene as comonomer) and 14 wt% talc (average particle diameter 2.5 μm) showed MFR 23, flexural modulus 26,500 kg/cm^2, notched Izod value (JIS K 7110, 23°C) 38, and ultimate elongation 500%. In a separate example, injection molding a blend of 54 wt% Polypro J-785H (MFR 11), 4 wt% Engage 8189 (E-plastomer-containing octene as comonomer with MI 0.6, density 0.863), 4 wt% HDPE, 23 wt% talc parts and other additives gave test pieces with Izod impact strength 53 kJ/m, flexural modulus 2720 MPa, and resistance to scratch whitening.

Morphological (microscopic) and spectroscopic analysis of the blends (infrared [IR] and electron spectroscopy) indicates that in spite of the deficiency of the E-plastomer in the blends containing iPP, these surfaces contain an enriched amount of the E-plastomers. It is surmised that that during the process of fabrication by high shear equipments, such as injection molding, the incompatible E-plastomer aggregates near the surface of the part which is in the area of the least shear—leading to the artificial concentration of the E-plastomer near the surface of the molded part. Industrially, this effect is important in the painting of these blends where the presence of the amorphous E-plastomer near the surface leads to strong adhesion of the paint to the polyolefin blend. This is in contrast to the attempts at painting iPP alone where the lack of adhesion of the paint is apparent in the poor resistance to mechanical or chemical removal. In a specific example, 65 wt% iPP, 14 wt% of EPDM, and 21 wt% of a mixture of two E-plastomers (containing 51% of M_w

12,700 and 49% of M_w 227,000 polymer) were blended and then injection molded to give a shaped article, which was painted with a two-component polyurethane coating. The paint showed gasoline resistance for more than 120 min and no peel at 30 min.

The compatibly of the E-plastomers (particularly the ethylene–octene copolymers) and iPP is demonstrated in reports of compounds and blends which can be bent without stress whitening. Stress whitening of the blends is a process of delaminating of the blend components due to applied mechanical strain. Under these conditions, the weakest part of the blend, typically the iPP–E-plastomer interface fails. In these reports, the absence of this failure indicates that the interfaces are extremely adept at accepting and responding to mechanical strain without failure. Further advances in this area have been reported by using a mixture of two plastomers differing in their composition and, thus, their miscibility with iPP. Theory suggests that the E-plastomers with a greater proportion of comonomer are more miscible with iPP. Thus, in these ternary systems, this E-plastomer acts as a compatibilizer or a bridging agent between the iPP and the other E-plastomers which have a greater proportion of ethylene. The advantage of these ternary blend systems is that they combine the economy of the E-plastomers which have a greater proportion of ethylene and the miscibility of the E-plastomers which have a greater proportion of the comonomer. These ternary blend systems have been used in automotive interiors and other injection-molded parts where appearance and resistance to strain-induced failure are of critical importance.

Alternate processes for improving compatibility of the iPP–E-plastomer blends have been investigated. The use of block polymers of ethylene and butene where the segments differ in the composition along the chain has been suggested. These block polymers are obtained by the hydrogenation of an anionically polymerized butadiene polymer where the segments differ in the amount of 1,2 versus 1,4 insertion of the butadiene. Hydrogenation of this polymer leads to a surrogate ethylene–butene copolymer which differs in the amount of pendant ethyl group arising from the 1,2 insertion of butadiene. The use of the block polymers where the end rich in butene is more miscible with the iPP and the end deficient in the butene is more miscible with the E-plastomers has been shown to have a minor improvement in the impact strength. A more detailed investigation, where a composition contained a higher-density E-plastomer, is indicated in these studies but has not been described in detail. Typically, the use of the compatibilizing block polymer is restricted to about 2–5 wt% of the blend. Using these formulation processes, blends containing about 20–30 wt% of the E-plastomer show a modest improvement in impact strength even in the presence of about 20 wt% of the inert filler.

6.2.4 BLEND WITH ISOTACTIC POLYPROPYLENE AND THERMOPLASTIC OLEFIN—THEORY AND CHARACTERIZATION

Industrially, the poor-impact but the favorable-stiffness properties of IPP over a wide temperature range had prompted a large amount of technological studies aimed at incorporating, in blends and compounds, an elastomeric phase within the matrix of the iPP to improve this key deficiency [7]. Historically, these were blends of iPP with EPR or EPDM. With the availability of E-plastomers in wide variety of molecular weights and densities, as well as analogous E-plastomers with different comonomers, much of the research and development focus shifted to these materials. The physics of the compatibility and the process of phase separation in blends of iPP and E-plastomers have been studied by a number of authors. These studies were expected to replicate the processes that occur during the cooling of the injection-molded or extruded compounds of iPP and E-plastomers. The ultimate aim in these blends is prediction of the bulk and surface properties from these correlations.

This development was technologically successful and E-plastomers are widely recognized to be very effective impact modifiers for iPP. In initial experiments, binary blends of iPP were compounded with EPDM, E-plastomers, or SEBS as the elastomeric phase. In comparison to the known modifiers, such as EPDM or SEBS, the binary blends with E-plastomers have properties which strongly depend on the amount and the identity of the E-plastomer. Thus, the addition of the about 20–30 wt% of the E-plastomer leads to improvement in impact performance adequate for most industrial uses. The more-effective E-plastomers were those that contain a larger amount of the comonomer or those that have lower density. The effect of molecular weight and molecular weight distribution are believed to of a much lower magnitude.

Glass transition temperature (T_g), measured by means of dynamic mechanical analysis (DMA) of E-plastomers has been measured in binary blends of iPP and E-plastomer. These studies indicate some depression in the T_g in the binary, but incompatible, blends compared to the T_g of the corresponding neat E-plastomer. This is attributed to thermally induced internal stress resulting from differential volume contraction of the two phases during cooling from the melt. The temperature dependence of the specific volume of the blend components was determined by PVT measurement of temperatures between 30°C and 270°C and extrapolated to the elastomer T_g at −50°C.

In one study, phase separation and crystallization behavior in extruded-iPP blend containing EPR and analogous E-plastomers were analyzed in binary and ternary blends. The blends were compounded and fabricated into parts by melt extrusion. The study looked at liquid–liquid phase separation which was investigated by time-resolved light scattering (TR-LS) and optical microscopy. Liquid–liquid phase separation through spinodal decomposition was confirmed by TR-LS data. After liquid–liquid phase separation at 250°C for various intervals, the blend sample was cooled to 130°C for isothermal crystallization of the iPP phase, and the effects of liquid–liquid phase separation on crystallization was investigated. Memory of liquid–liquid phase separation through spinodal decomposition remained during crystallization. The crystallization rate decreased with increasing liquid–liquid phase separation time at 250°C. Slow crystallization for the long liquid–liquid phase separation time can be ascribed to decrease in chain mobility of iPP by decrease in rubbery components in the iPP-rich region. In particular, the ethylene–octene E-plastomer exhibited affinity with iPP, leading to a slow growth of a concentration fluctuation during annealing.

The mechanism of formation of morphology structures in iPP–E-plastomers blends via shear-dependent mixing and demixing was investigated by optical microscopy and electron microscopy. A single-phase structure is formed under high shear condition in injection machine; after injection, namely under zero-shear environments, spinodal decomposition proceeds and leads to the formation of a bicontinuous phase structure. The velocity of spinodal decomposition and the phase separation depend on the molecular structure of iPP and E-plastomer components.

In a complementary rheological experiment, studies have been performed on binary blends based on octene-containing E-plastomers and iPP. The flow behavior of E-plastomers, iPP, and their blends were analyzed using an Instron capillary rheometer and a Rheometrics dynamic stress rheometer. A non-Newtonian flow behavior was observed in all the samples in the shear rate range from 27 to 2700 s^{-1}, whereas at shear rates in the range from 0.01 to 0.04 s^{-1}, a Newtonian flow behavior was verified. The SEM showed that the size of the elastomer domains increases as the E-plastomer content increases. The SEM analysis also showed that phase continuity of the elastomer phase may occur between 50 and 60 wt% of E-plastomer. This result is consistent with the Sperling's model. The mechanical analysis that showed the blend with 5 wt% of E-plastomer, showed an increase in mechnical properties. However, as the E-plastomer increased, a negative deviation in properties was observed. Thermal analysis showed that there were no changes in the crystallization behavior of the matrix when different elastomer contents were added. The dynamic mechanical thermal analysis (DMTA) data illustrated that samples with low E-plastomer showed a single transition, indicating a certain degree of miscibility between the components of these blends.

In addition to the properties noted above, the formulation parameter in iPP–E-plastomer blends have a profound influence on the dynamic loading (e.g., vibration) performance. The load limits of the blend for applications in which dynamic stresses are predominant were studied by using the hysteresis measurement method. However, their technical application requires knowledge of critical load values.

A large amount of technical investigation has been undertaken in the area of understanding the mechanistic and molecular basis of the essential work of fracture by simulating the toughness response of deeply double-edge notched specimens. A variety of blends of polyolefin-based matrixes with different rubber types, and also kaolin particles, were used to validate the proposed function. The empirical function used for the simulation gave satisfactory results over a range of crack sizes. In a development of the earlier work, fracture mechanical terms related to the crack initiation and propagation, respectively, were assessed by the essential work of fracture (EWF) method for compression-molded polypropylene (iPP) blends. The resistance to crack initiation (RCI) was characterized by the EWF. The use of the PP blends with 20 vol% E-plastomer (butene comonomer) of various butene contents (58 and 90 wt%) and morphologies (EB58 is amorphous, whereas EB90 is semicrystalline) increased with increasing butene and, thus, amorphous content. At the same time, in the slope of the EWF resistance curves, which inform about the resistance to crack propagation (RCP) of the PP–EB blends, an adverse change was observed. The above-mentioned findings were interpreted by considering the morphology and its rearrangement due to loading. It was suggested that the RCI is improved by the order of the crystal structure and by the density of tie molecules that transfer the stress from the amorphous toward the crystalline phase. On the other hand, the RCP is controlled by a stress-redistribution process that is influenced by possible changes in the local morphology.

The influence of comonomer incorporation on morphology and thermal and mechanical properties of blends based upon iPP and E-plastomers (butene comonomer) have been investigated. Blends of iPP with E-plastomers (butene comonomer) containing 10, 24, 48, 58, 62, 82, and 90 wt% 1-butene were prepared in order to examine the influence of the EB molecular architecture on the morphological development as well as on the thermal and mechanical properties. Compatibility between iPP and E-plastomers increased with increasing 1-butene content in E-plastomers to afford single-phase blends at a 1-butene content exceeding 82 wt%. The morphology was investigated using atomic force microscopy (AFM) and transmission electron microscopy (TEM). Improved compatibility accounted for enhanced E-plastomers dispersion and interfacial adhesion.

6.2.5 Blend with Isotactic Polypropylene and Thermoplastic Elastomer

The formation of essentially compatible blends of iPP and E-plastomers soon led to the invention of the blends where the dispersed E-plastomer phase was substantially or completely cross-linked by the addition of a dynamic vulcanization step in the presence of chemical cross-linking agent [8]. These cross-linked blends are the analogs of the TPV which are blends of iPP and EPDM, where the latter is dispersed and cross-linked. The intent was to replace some of the TPVs with these novel blends of iPP and E-plastomer. The anticipated improvement of these blends over the existing TPVs is that, in general, because of the better processing character of the E-plastomer–iPP blend, it was not necessary to add the full complement of the process oil, which is a part of the formulation of the TPV. These process oils, which lead to much-improved processing, do exude subsequently during long-term storage and use. The exudation of the process oil leads to eventual shrinkage of the part as well as surface contamination. It was believed that the new E-plastomer–iPP blend would not suffer from these disadvantages. In addition, because of the better compatibility of the E-plastomer with the iPP in comparison to the inherently poor compatibility of the EPDM to iPP, it was also expected that the particle size of the dispersed E-plastomer would be smaller leading to better optical transmittance in the blend.

The new blends are composed of an iPP matrix in which an E-plastomer (octene comonomer) is dispersed and cross-linked. The E-plastomer is grafted with vinyl organosilanes, and then blended with iPP, hydrolyzed and water-cross-linked in a corotating, intermeshing twin-screw extruder. Grafting, hydrolysis, and condensation reactions with the organosilanes are carried out in a twin-screw extruder in the course of a single-stage process. The resulting blends are easy to process into articles showing an attractive range of properties. The advantages arise for the cross-linking of the

E-plastomer phase. The new blends can be processed without predrying. Moldings are characterized by excellent surface quality and absence of color and odor due to the absence of chromophores in these formulations. They can easily be colored to any desirable shade. Compared to the uncross-linked blends of similar composition, these blends have high tensile strength and elongation. A broad spectrum of different hardness grades whose main focus has been on the automotive applications is possible. The advantage of these blends is the absence of process oil which leads to significant improvement in the amount of automotive interior glass-fogging.

Similar blends have been made by cross-linking the E-plastomer with peroxides. This process suffers from an inherent degradation of the iPP by peroxide. In a representative formulation, a mixture of 60 parts of E-plastomer (octene commoner), 15 parts maleated (0.6%) iPP, 25 parts of EPDM, 10 parts of paraffinic plasticizer, 5 parts of dicumyl peroxide, and 1 part of stabilizer was treated at 170°C for 5 min to give a cross-linked blend with Shore A hardness 66, tensile strength 5.5 MPa, and elongation 190%. Similar blends have been made with the incorporation of a limited amount of a SEBS polymer to act as a compatibilizer between the E-plastomer and the iPP.

The dynamics of vulcanization-induced phase separation in PP–E-plastomer blends have been studied using TR-LS and optical microscopy. Using syndiotactic PP (sPP) as the matrix, solvent cast films show no phase separation in the melt. Upon cooling, phase separation takes place in the vicinity of crystallization temperature of sPP. The phase diagram of the blend is essentially an overlap of an upper critical solution temperature (UT) and the melting temperature of sPP in the blends. A lower critical solution temperature (LT) was observed in the melt, that is, about 30 K above the UT. Vulcanization was undertaken in the single phase using dicumyl peroxide (DCP). Temporal evolution of structure factors and the emergence of phase-separated domains in these blends have been investigated by TR-LS and optical microscopy. The temporal evolution of structure factors has been analyzed in the context of nonlinear dynamical scaling laws to elucidate mechanism(s) of vulcanization-induced phase separation of sPP–E-plastomer blends.

6.2.6 BLEND WITH ISOTACTIC POLYPROPYLENE AND STYRENE–ETHYLENE–BUTYLENE–STYRENE BLOCK COPOLYMER

Hydrogenated styrene–butadiene/isoprene block polymers have been used as surface agents or compatibilizers in the blends of E-plastomers with iPP [9]. The technology evolves around the known benefits of the blending of SEBS with iPP, which leads to improvement in the impact strength as well as the scratch and mar resistance of the iPP. In these ternary blend systems it is presumed that the hydrogenated E–B block of the SEBS also acts as compatibilizer between the iPP and the E-plastomer. In one study, the influence of E-plastomer (containing 10 wt% butene) on the compounding of iPP in the presence of SEBS and talc was studied. With increasing E-plastomer and SEBS volume fraction, the stiffness and yield stress of blend decreased, due to the low stiffness and tensile strength of the dispersed components. However, the low impact strength iPP–E-plastomers blends, which are due to weak interfacial interactions between these components, can be improved significantly by simultaneously blending with SEBS. This is due to the compatibilizer function of SEBS, as demonstrated by TEM. The stiffness of the ternary blends can be enhanced remarkably by addition of talc, which also acts as nucleating agent for the PP crystallization.

6.2.7 BLEND WITH ISOTACTIC POLYPROPYLENE AND SOFT THERMOPLASTIC OLEFIN

Extremely soft and fluid blends of E-plastomers with iPP have been made by incorporating a large amount of a low-density E-plastomer with added process oil for fluidity and softness with a minor amount of iPP [10]. In a representative formulation, 56 parts of an impact copolymer (ICP) which had 62 wt% of iPP and the balance an EPR, 22 parts of an E-plastomer (hexene comonomer), 11 parts of an E-plastomer (octene comonomer), and 5 parts of talc showed a Rockwell R hardness 77, flexural modulus 197 Kpsi, ductile failure D, ductility index 0.38, and peel strength 1050 g cm^{-1}. These soft formulations have also been developed for roofing membranes.

6.2.8 BLENDS WITH ISOTACTIC POLYPROPYLENE AND SURFACE

A noteworthy aspect of the blends of the iPP with E-plastomers is the apparent concentration or migration of the E-plastomer to the surface of the molded or fabricated part [11]. This migration process and the effect of this segregation have been studied, in particular with respect to adhesion and other surface properties. In one study, the influence of compositional variations in compounded E-plastomer–iPP blends on the physical and mechanical attributes of injection-molded plaques has been examined. Both the choice of compounding ingredients and the injection-molding conditions through which the desired plastic part is fabricated are known to influence the properties of the molded parts. The influences of the iPP homopolymer, and E-plastomer utilized in the compounded blend, in conjunction with the molding conditions used to produce plaques, are studied as they relate to the physical and mechanical properties of the plaques. Paint adhesion and the friction-induced paint-damage resistance of coated plaques are shown to be directly related to PP molecular weight and elastomer crystallinity. The physical and mechanical properties of the injection-molded TPO as a function of shear stress produced by molding at slow- and fast-injection velocities are discussed. In a further study, the melt temperature and injection speed were the most significant molding parameters that affected adhesion, as measured by fluorescent dye penetration, birefringence, and TPO-surface plastic modulus. In order to attain a low shear stress in molded parts, the ability of the semicrystalline-incompatible TPO blend to crystallize and phase separate must be controlled. The injection-molding process was also studied by varying melt temperature, injection speed, gate dimensions, and part thickness in relation to the physical attributes of plastic parts. Attributes examined included optical birefringence, surface microhardness, dye penetration, and adhesion of paint on topcoated parts. Correlations to mold-filling analysis indicate that shear, as effected by melt temperature, injection speed, and gate dimensions, play a significant role in the ability to achieve paint adhesion to the molded article.

In one study, Fluorescence and Raman chemical imaging of TPO has been used to study surface adhesion. The surface properties of TPO tend to prohibit adhesion of paints used to increase the longevity and enhance the cosmetic appearance of exterior TPO components. To increase adhesion, the TPO surface is coated with a primer. The primer increases the TPO-surface free energy and promotes paint adhesion. A generally applicable strategy was developed for quantitative monitoring of primer film uniformity, thickness, and adhesion with the use of an environmentally sensitive fluorescent dye. The dye-tagged primer is analyzed in real time with a macroscopic imaging system employing liquid crystal tunable filter (LCTF) imaging spectrometers. The fluorescent-chemical-imaging approach is applicable to online monitoring of the TPO-surface-modification process. Controlled modification of the TPO surface is contingent on detailed understanding of the underlying TPO substrate. Raman chemical imaging is employed as a non-invasive approach to characterize TPO-surface architecture. Specifically, a method was developed for visualizing the surface-to-bulk distribution of primer, iPP and E-plastomer phases in TPO without the use of dyes or stains. Blend components are visualized on the basis of their intrinsic vibrational spectrum. In general, Raman microscopy represents a powerful method for polymer characterization that is competitive with time-of-flight secondary ion mass spectrometry (TOF-SIMS) and TEM.

The need to paint the blends has also resulted in new painting systems. Recently, a nonpolar color coat based on a hydrogenated polybutadiene diol and melamine resin for TPO bumper fascia was invented. The breakthrough technology allows the elimination of the TPO pretreatment step such as adhesion promoter, flame, or plasma during manufacturing. The paintability of two different types of E-plastomers was evaluated. The olefinic white paint was found to provide excellent paint adhesion for both types of metallocene plastomers. Paint peeling was not observed in any of the test conditions. For the conventional white paint, only the ethylene–butene E-plastomer was found to pass the additional friction test.

6.2.9 Blend with Isotactic Polypropylene and Filler

Recent advances in the application of ultrafine talc for enhanced mechanical and thermal properties have been studied [12]. A particularly important use is of finely divided filler in TPO as a flame-retardant additive. In a representative formulation, 37 parts of E-plastomer, MI 2.0, density 0.92, 60 parts of amorphous EPR, and 4 parts of fine carbon black were dry blended, kneaded at 180°C, pelletized, and press molded into test pieces, which showed oxygen index 32 versus 31 in the absence of a filler. The oxygen index is a measure of flame retardancy.

6.3 BLEND—OTHER POLYOLEFINS

The interaction of the E-plastomer with a number of other hydrocarbon and polyolefin resins has been studied [13]. In one study, the toughness mechanism in semicrystalline polymer blends of HDPE blended with E-plastomer. The results are compared to similar blends of HDPE with EPR. The E-plastomer was blended into HDPE at volume fractions of up to 0.22. These rubbers were in the form of finely dispersed spherical inclusions with sizes less than 1 μm. The incorporation of rubber into HDPE does not substantially change its crystallinity, but produces special forms of preferential crystallization around the rubber particles. The notch toughness of the rubber-modified HDPE increases by more than 16-fold as a result. The single parameter controlling the notch toughness of these blends was found to be the matrix-ligament thickness between rubber inclusions. When this thickness is above a certain critical value, the notch toughness of the material remains as low as that of the unmodified HDPE. When the average ligament thickness is less than the critical value, there is a substantial increase in the toughness. The critical ligament thickness for the HDPE-rubber systems was found to be around 0.6 μm, independent of the type of the rubber used. The sharp toughness threshold in the rubber-modified HDPE results from a specific micro-morphology of the crystalline component of HDPE surrounding the rubber particles. The poly-ethylene (PE) crystallites of approximately 0.3 μm length perpendicular to the interface are primarily oriented with their (100) planes parallel to the particle interfaces. Material of this constitution has an anisotropic plastic resistance of only about half that of randomly oriented crystallites. Thus, when the interparticle ligaments of PE are less than 0.6 μm in thickness, the specially-oriented crystalline layers overlap, and percolate through the blend, resulting in overall plastic resistance levels well under that which result in notch brittle behavior, once rubbery particles cogitate in response to the deformation-induced internal negative pressure. This renders ineffective the usual strength-limiting microstructural flaws and results in superior toughness at impact strain rates.

The effect of E-plastomers on recycled HDPE was compared to the data for the effect of EPR on HDPE. The rheological, mechanical, and thermal properties of their blends with a recycled HDPE were studied. The results show effective improvements in engineering properties of the recycled polyethylene (HDPE). Low percentages of these modifiers gave rise to higher elastic properties of the blends. Addition of all modifiers resulted in decreases in crystallinity and in melting point of the blends. Depending on modifier type, the intensity of the changes was different. It is concluded from IR spectra of the blends that no chemical interaction occurs during melt mixing of the ingredients. These improvements are attributed to morphological changes and physical interactions between modifiers and HDPE matrix. In melt state, as the modification percentage increases and depending on the nature of modifier, the resulting blends show different rheological behaviors. The results presented in terms of weighed relaxation time spectra show different behaviors depending on the additive. For all modifiers, the 7% modification was always found to have the most-pronounced effect on almost all properties of the blends.

Another study looked at the miscibility of E-plastomer–polyisobutylene blends. Blends were prepared from linear E-plastomers and a polyisobutylenes in the entire composition range. Flory–Huggins interaction parameters were determined from DMTA and DSC measurements. The usual technique had to be modified in the case of DSC data, since the T_g of E-plastomers cannot be detected by this technique. The two methods yielded identical results and indicated good interaction of the components, which was supported also by a SEM study and the mechanical properties of the blends.

In another unexpected study, the morphology and mechanical properties of natural rubber (NR)–E-plastomer blends have been investigated. NR–E-plastomer blends were compounded in a intensive internal mixer with different amounts of filler and cure system for dynamic vulcanization. Morphology and mechanical properties were examined. Processing parameters including processing speed, cycle time, and temperature strongly affected the properties of the blend. The ozone resistance of the blends was very high. In a further study, liquid natural rubber (LNR) was used as a compatibilizer in NR–linear low-density polyethylene (LLDPE) blends. Rheological studies and mechanical properties of the blends showed that the molecular weight of LNR played a very important role in defining blend performance. The blends that contained 15%–25% LNR with M_w 4.8×10^5 and M_n 1.66×10^5 showed the best mechanical properties. Outside this range of molecular weight, the compatibilizing property of LNR is no longer effective due to inhomogeneous blends. Improvements in the mechanical properties were in accord with the increase in cross-linking of the blends. The addition of LNR in the blends was able to reduce the interfacial tension, thus improving the interaction between the phases of the blends.

6.4   BLEND—ENGINEERING PLASTICS

There have a number of efforts which would extend the beneficial effects the E-plastomers have in blends with iPP to other, more-polar engineering thermoplastics [14]. In one study, morphology and impact property of poly(phenylene oxide) (PPO)–polyamide (PA6)–E-plastomer blends has been studied. Compatibility between different binary blends in PPO–PA6–E-plastomer (Engage 8150) and the effects of maleic anhydride (MA)-grafting degree on the E-plastomer and the PPO were experimentally determined. The impact property of PPO–PA6 blends was studied through microscopy (TEM and SEM) and impact testing. The parent binary blends are incompatible. However, the addition of PPO-g-MA and E-plastomer-g-MA has *in situ* compatibilization effect on PPO–PA6 blends. PPO and E-plastomer are dispersed in PA6 continuous phase within the composition limits. The brittle–ductile transition of the blends is controlled by interfacial adhesion strength between PA6 and E-plastomer. In a companion study, mechanical properties and morphology of poly (butylene terephthalate) (PBT)–E-plastomer blends were prepared by *in situ* reactive extrusion. E-plastomer was reactively grafted with MA through extrusion before blending. The impact and tensile strength, thermal properties, and morphologic structures of the blends were investigated. The grafting rate as well as extrusion conditions had significant effects on the mechanical properties of the blends.

Carbon black and carbon nanotubes have also become an important area for the development of blend components for E-plastomers. In one study, carbon-black-filled E-plastomers were evaluated. Carbon black contents from 8 to 10 wt% in E-plastomers were shown to be suitable for antistatic applications. Furthermore, carbon-black-filled E-plastomers in the range from 12 to 20 wt% of carbon black contents were suitable for electrostatic-discharge (ESD) applications. The results showed a significant change in the dependence on carbon loading at a critical concentration around 12 wt%, which is near the percolation threshold. The relative resistivity of the 15 and 20 wt% filled compositions showed a strong dependence on elongation. In a companion experiment, composites that incorporate conductive filler into an E-plastomer matrix were evaluated for DC electrical and mechanical properties. Comparing three types of fillers (carbon fiber, low- and high-structure carbon black), it was found that the composite with high-structure carbon black exhibited a combination of properties not generally achievable with this type of filler in an elastomeric matrix. A decrease in resistivity at low strains is unusual and has only been reported previously in a few instances. Reversibility in the resistivity upon cyclic deformation is a particularly unusual feature of E-plastomers with high-structure carbon black. The mechanical and electrical performances of the high-structure carbon black composites at high strains were also impressive. Mechanical reinforcement in accordance with the Guth model attested to good particle–matrix adhesion. The E-plastomers matrix also produced composites that retained the inherent high elongation of the unfilled elastomer even with the maximum amount of filler (30% by volume). The E-plastomers matrix with other conducting fillers did not exhibit the exceptional properties of E-plastomers with high-structure carbon black. Composites with carbon fiber and low-structure carbon black did not maintain good mechanical properties, generally exhibited an increase in resistivity with strain, and irreversible changes in both mechanical and electrical properties after extension to even low strains. An explanation of the unusual properties of E-plastomers with high-structure carbon black required unique features of both filler and the matrix. The proposed model incorporates the multifunctional physical cross-links of the E-plastomers matrix and dynamic filler-matrix bonds.

6.4.1 FOAMS

Optimized curative packages (peroxide/coagents) for E-plastomers (octene comonomer) foams have been elucidated in a study [15]. This study also showed that the polymer molecular structure (density, molecular weight) has a significant effect of the foam characteristics. Cross-linked poly-olefin foams are used in a variety of applications for high-compression or dynamic-cushioning requirements. In an interesting study, polyolefin elastomer foams having enhanced physical properties were prepared with dual cross-linking systems. An optimum level of elevated-temperature elastic properties via cross-linking, allows the decomposition gas to expand controllably to produce foams with desirable cell sizes. Cross-linked systems also exhibit enhanced physical properties, such as compression-set resistance. However, an upper cross-link level exists for foam manufacture since excessively cross-linked compounds will exhibit minimal expansion. A technique was developed to impart additional curing to foams after expansion. Foamed articles having enhanced tensile, elastic recovery, and creep and fatigue resistance properties are prepared using a dual cure system of heat or radiation-activated first-stage cure (which follows or coincides with foaming) followed by a moisture-activated second-stage cure. Vinyl-functional silane was grafted onto E-plastomers (octene comonomer) during the foaming process. The grafted-E-plastomer-based foam was then further cured via moisture. These moisture-cured foams exhibited enhanced physical properties, notably compression set, over foams cured with peroxide alone. In a representative formulation, 100 parts of Engage 8200, 4.5 parts of azodicarbonamide, 1.35 parts of ZnO, 2 parts of dicumyl peroxide, 2.5 parts of trimethylolpropane trimethacrylate, 0.1 part of antioxidant, 5 parts of filler, 2 parts of vinyltrimethoxysilane, and 3.01 parts of pigments were conditioned at 100°C at 1 t pressure for 20 min and foamed at 175°C at 25 t pressure for 8 min and moisture cured in water bath at 40°C for three–four days to give a cross-linked foam.

The effect of oxidative irradiation on mechanical properties on the foams of E-plastomers has been investigated. In this study, stress relaxation and dynamic rheological experiments are used to probe the effects of oxidative irradiation on the structure and final properties of these polymeric foams. Experiments conducted on irradiated E-plastomer (octene comonomer) foams of two different densities reveal significantly different behavior. Gamma irradiation of the lighter foam causes structural degradation due to chain scission reactions. This is manifested in faster stress-relaxation rates and lower values of elastic modulus and gel fraction in the irradiated samples. The incorporation of O_2 into the polymer backbone, verified by IR analysis, confirms the hypothesis of chain scission occurring at the labile peroxide linkages. In contrast, the denser foam shows a small amount of cross-linking and a concomitant improvement in mechanical properties after oxidative irradiation.

6.4.2 FIBERS

Melt spinning of the E-plastomers has been the source of a commercial development directed to woven cloth of cross-linked E-plastomers [16]. Recent work on the rheological and theoretical estimation of the spinnability of polyolefins is a part of this development.

Melt spinning is a polymer-processing technique that makes great demands on the extensibility of the polymer melt in the distance between die exit and solidification point. The polymer material is exposed to a rapidly growing deformation rate over a large range of deformation within a short time of approximately 100 μs. Simultaneously, an extreme cooling occurs with cooling rates of approximately 1000°K s^{-1}. For this reason only a few polymer materials are usable for this kind of polymer processing with sufficient take-up speeds. Most polymers show a fiber break in the molten state either by brittle cohesive rupture or ductile failure when approaching critical conditions of deformation. The rheological behavior of a polymer melt at the critical conditions of deformation in the fiber-forming process cannot be predicted by means of usual rheological material functions. This study reports the attempt to find out material functions, which describe the critical deformation states of the melt-spinning process. In addition, the thermal and mechanical properties of melt-spun fibers of E-plastomers (octene comonomer) were influenced by chemical cross-linking using graft-copolymerized trimethoxyvinylsilane. The combined twin-screw extruder and spinning head used to carry out grafting, cross-linking, and melt spinning in one step is described, and the changes in the mechanical and thermal behavior of the copolymer fibers have been investigated.

Cross-linking the fibers with radiation lead to durable-stretch fabrics. The fabrics can be made by any process, such as weaving and knitting, and from any combination of cross-linked, heat-resistant olefin elastic and inelastic (hard) fibers, e.g., cotton and wool. These fabrics exhibit excellent chemical resistance (e.g., chlorine or caustic resistance) and durability, that is, they retain their shape and feel (hand) over repeated exposure to processing conditions, such as stone-washing, dye-stripping, and PET-dyeing.

6.4.3 FILMS

E-plastomers, particularly the high- and medium-density materials, have found extensive use in films [17]. They are valued for their excellent seal character which allows the formation of mechanically strong seals at relatively low temperatures compared to traditional low-density polyethylene (LDPE). In addition, these E-plastomers can be obtained in a range of crystallinities and softness. These higher-density materials are typically made in the blown-film process and are used for protective film covers and disposable bags.

Lower-density E-plastomers have found alternate use in cast film processes to make elastic film laminates with good breathability which contain laminates of liquid impermeable extensible polymeric films with extensible-thermoplastic-polymer-fiber nonwovens and nonwoven webs of polyethylene-elastomer fibers as the intermediate layers. The development relates to a breathable film including an E-plastomer and filler that contributes to pore formation after fabrication and distension of the film. The method and extent of distension is designed to produce a breathable film by stretching the film to form micropores by separation of the film of the E-plastomer from the particulate solids. This film is useful for manufacture of absorbent personal-care articles, such as disposable diapers and sanitary napkins and medical garments. In detail, these constructions comprise a liquid impermeable extensible film comprising polyolefins. The outer layer contains extensible-thermoplastic-polymer-fiber nonwovens, and an elastic intermediate layer contains nonwoven webs of fiber E-plastomers. The intermediate layer is bonded to the film layer and the outer nonwoven layer, and exhibits water vapor transmission >300 g m^{-2} h^{-1}. These elastic laminates are useful as diaper outer covers, training pants, incontinence garments, and feminine hygiene products. In a representative example, a compound containing 47% CaCO$_3$ and 53% Affinity PL-1280 was extruded, stretched 300% at roll temperature 160°F, and heat-treated at 180°F to give an oriented elastic film. The film was pressed together with spun-bonded PP-fiber nonwoven fabric having E-plastomer (Engage 5280) fiber melt-blown nonwoven web as an overlaid layer at pattern roll temperature of 180°F to give a laminate having LDPE fiber nonwoven as the intermediate layer showing elastic properties on extending the laminate to 150% in the machine direction. Anistropic elasticity in these films (machine versus transverse direction) can be induced by differential extension in the directions.

6.4.4 ADHESIVES

E-plastomers, particularly those that contain large amounts of a long-chain alpha olefin as a comonomer, have entanglement molecular weights large enough to have plateau modulus low enough for use as an adhesive at elevated temperatures above the melting point of the E-plastomers [18]. Development of these materials toward these applications has occurred and is mostly concentrated in the use of hot melt adhesives used for box and package closures. In addition, hot melt adhesives are also used in construction and elastic attachment applications on nonwoven disposable articles. A representative formulation is a hot melt pressure-sensitive adhesive for use with a medical absorbent article. The adhesive contains 5%–15% a blend of a SEBS, about 60 wt% of a low-molecular-weight E-plastomer, as well as a tackifying resin and a plasticizer.

6.5 DERIVATIVES OF E-PLASTOMERS

Sulfonation and chlorination of E-plastomers has been demonstrated by treating the E-plastomers with the appropriate reagent [19]. The sulfonated E-plastomers are expected to have excellent mechanical toughness and be resistant to scratch. The chlorinated elastomers are expected to be vulcanizable with metal salts. In a representative synthesis, 30 g of an E-plastomer (16.7 wt% butene) and M_w 7400 g mol^{-1} was dissolved in heptane, heated to 89°C–95°C and sulfonated with 0.9 mL H$_2$SO$_4$ for 30 min, to have an S content of 0.43%. A blend comprising 3.0% Na-neutralized sulfonated material mixed with E-plastomer with melt index 4.5 and density 0.873 was mixed in a Brabender mixer at 193°C, cooled to room temperature, and compression molded to give specimens exhibiting modulus 708.0 psi, energy at break 12.7 in. lb^{-1}, elongation at break 1013.0%, and tensile strength 1723.0 psi.

E-plastomer containing 10 wt% butene (intrinsic viscosity 135°C [decalin 1.5]) was chlorinated, compounded with the additives and ZnO curative, and vulcanized at 160°C for 20 min to give a sheet having JIS A hardness 72, elongation 390%, 100% modulus, 50 kg cm^{-2}, and tensile strength 190 kg cm^{-2}. In an alternate synthesis of chlorinated or chlorosulfonated E-plastomer, 1-butene-ethylene copolymer (I) having crystallinity 24% and pore volume 0.36 cm^3 g^{-1} was chlorosulfonated by reaction with a gaseous mixture of SO$_2$ and C$_{12}$ at 60°C–100°C for 3–10 h to produce a chlorosulfonated E-plastomer. This had a Cl content 15%, S content 1.2%, crystallinity 4.7%, tensile modulus 920 psi, and tensile strength 820 psi, versus 27–35% Cl content, 0.9–1.4% S content, <2% crystallinity, <500 psi tensile modulus, and 400–500 psi tensile strength, respectively for commercial chlorosulfonated polyethylene.

6.6 CONCEPTS FOR P-PLASTOMERS

E-plastomers, based on ethylene and its copolymers with alpha olefins such as propylene or octene, combine the desirable attributes of processibility, economy, and resistance to environmental degradation [20]. These properties, in conjunction with excellent elastomeric attributes such as large tensile elongations and quick recovery from large strain deformation, have led to widespread introduction in a host of fabricated forms and products. However, for these copolymers the elastomeric properties are only accessible for high-molecular-weight polymers after vulcanization. In the absence of both these conditions, the properties rapidly degrade. This requirement imposes both chemical and engineering limitations on the elastomers. First, they require a convenient vulcanization site on the polymer chain, typically introduced by the incorporation of a cross-linkable monomer during polymerization. Second, the high-molecular weight requires the use of specialized procedures and machines for the compounding and fabrication of articles containing these elastomers. The structural and handling requirements of these conventional polyolefin elastomers have been instrumental in retarding the use of simple, high-volume fabrication processes such as those used in the formation of thermoplastic films and nonwoven fabrics for these polyolefin elastomers.

The advent of P-plastomers, announced independently in 2003 by ExxonMobil and Dow was a significant solution to this feature of E-plastomers. In contrast to most known E-plastomers and EPRs which contain at least 50 mol% of ethylene, these P-plastomers contain at least 70 mol% of propylene. A significant difference between this polymer and the established polyolefin elastomers (e.g., EPR), which are either amorphous or contain small amounts of ethylene crystallinity, is that the new elastomers necessarily contain isotactic propylene (mm) crystallinity. This crystallinity arises from the aggregation of long sequences of propylene residues within chains where the pendant methyl group is meso to each other. The successful synthesis of these elastomers, which requires both intramolecular control of the tacticity of the insertion of the propylene units, as well as intermolecular control of the composition of the polymer, is only possible through the use of advanced polymerization catalysts. These catalysts, particularly in combination with a solution polymerization process, lead to a detailed control of the polymer characteristics, which define the properties of the polymer. The presence of a limited amount of the isotactic propylene crystallinity implies that the new polymers are thermoplastic elastomers (TPEs). Structurally, at a molecular level, they display a phase separation of crystalline propylene-rich sequences within the continuum of a majority of the amorphous copolymer. These crystalline segments, which act in concert between the different chains, are virtual vulcanization sites at room temperatures leading to very high viscosities and a comparative lack of creep or deformation. However, at elevated temperatures beyond the melting point of the crystalline sites, this virtual cross-linking is removed and the material is easy to process.

The new polymers are intermediate in composition and crystallinity between the essentially amorphous EPR and the semicrystalline iPP. The presence of the complementary blocks of elastomers for both ethylene and propylene crystallinity should not indicate a similarity, beyond the levels of the crystallinity in the properties of the E-plastomers and the P-plastomers. The E-plastomers and the P-plastomers differ in their structural, rheological, as well as their thermal, mechanical, and elastic properties. In a comparison of the tensile strength and tensile recovery (tension set) from a 100% elongation for a range of P-plastomers and E-plastomers, the former have lower tension set than EPR and iPP. However, for comparative E-plastomers and P-plastomers at equivalent tensile strength, the latter have significantly better tension set. In summary, P-plastomers are tough polyolefins which are uniquely soft and elastic.

6.6.1 P-Plastomers: Composition and Tacticity

P-plastomers are semicrystalline, elastomeric copolymers composed predominantly of propylene with limited amounts of ethylene [21]. The concentration of ethylene is typically less than 20 wt%. The placement of the propylene residues is predominantly in a stereoregular isotactic manner. This leads to the crystallinity (which is critical) in the copolymer. The extent of the crystallinity is attenuated by errors in the placement of the propylene and by the incorporation of ethylene. These two structural features contribute to lower the crystallinity, as measured by the heat of fusion, to less than $40 \, \mathrm{J \, g^{-1}}$. Copolymers of propylene and ethylene, which have higher levels of crystallinity are thermoplastics and are not elastomers. During a uniaxial tensile elongation, these thermoplastic polymers undergo an irreversible deformation (commonly referred to as the yield point) and, thus, do not retract to their original dimensions on removal of the force of distention. Conversely, totally amorphous polymers of propylene and ethylene will creep during elongation and will not retract to their original dimensions. Thus, the elasticity and the elastic recovery of P-plastomers are limited by an upper and a lower, greater than zero bounds on the crystallinity. The polymerization catalysts and the exact condition of the polymerization determine the errors in the propylene placements which are responsible for a part of this diminution in the crystallinity. The balance of the reduction in the crystallinity arises from the insertion of ethylene. Thus, within the family of P-plastomers polymers that is made under similar synthesis conditions, the crystallinity is strongly and inversely correlated to the ethylene content of the polymer.

The melting point of the P-plastomers decreases monotonically with the introduction of ethylene, except at ethylene levels of >12 wt% it shows no further decrease though the heat of fusion does decrease throughout this range. It is believed that these discordant behaviors could be due to either, or both, of the factors given below:

- Crystallization of the same sequence length—such that, independent of comonomer content, a minimal sequence length always exists that is capable of crystallization at room temperature but this fraction decreases as comonomer content increases.
- Change in crystal structure above a certain comonomer content—such that the new crystal structure could accommodate comonomer units in the crystal. It is possible that on adding increased amounts of comonmer, the crystal structure changes in the commonly observed monoclinic (PP) to a hexagonal structure.

The T_g of P-plastomers changes as a function of ethylene content. The T_g decreases with increasing ethylene content, primarily due to an increase in chain flexibility and loss of pendant methyl residues due to incorporation of ethylene units in the backbone. It is well known that PP has a T_g of 0°C, and polyethylene a $T_g < -65$°C. The addition of ethylene to a propylene polymer would therefore be expected to decrease the T_g, as is observed here. A secondary effect would be the reduction in the level of crystallinity associated with increasing ethylene content, which is expected to reduce the constraints placed upon the amorphous regions in proximity to the crystallites. Thus, an increase in ethylene content will result in a lower T as well as an increase in magnitude and a decrease in breadth of the glass transition.

While the previous description of the structural attributes of the P-plastomer polymers indicates some of the limits of the composition and the crystallinity, the key distinguishing feature of these polymers, compared to the earlier attempts, is the essential uniformity of the composition and the crystallinity of the polymer along each chain (intramolecular homogeneity) and between different chains (intermolecular homogeneity). Intramolecularly, the P-plastomers are random copolymers of propylene and ethylene. They are not, and are not intended to be, block copolymers. This uniform synthesis is possible through the development of novel metallocene polymerization catalysts and the use of uniform solution polymerization procedures which have a single-polymerization environment for all the polymer chains. The intramolecular random distribution of the monomers in the P-plastomer polymers can be calculated from an evaluation of the r_1/r_2 reactivity ratio, by 13C NMR. The ratio is in the vicinity of 1, indicating a uniform distribution of propylene and ethylene residues. Uniform intermolecular distribution of both crystallinity and composition requires a more extensive analysis. It is theoretically possible that in multisited catalyst systems there could be independent differences in the composition and crystallinity among different polymer chains. These would arise from a difference in the stereoregularity of insertion of the propylene on different chains, which emanate from different catalyst sites. Thus, two polymer chains which have the same composition of propylene and ethylene could have different crystallinity, because of the different levels of tacticity in the insertion of propylene in these polymer chains. The unequivocal analysis of these effects is twofold. First, is a detailed temperature-rising thermal fractionation in a solvent to resolve differences in the crystallinity, followed by IR analysis of the fractions to isolate the differences in composition. P-plastomers, over the entire range of crystallinity and compositions, are narrow in the distribution of crystallinity, and narrow within the limits of analysis, in distribution of composition. We have sometimes observed finite breadth in the solvent fractionation, where a single polymer fractionates in two adjacent fractions with the same analytical crystallinity and composition. We believe that this small dispersion arises from the finite breadth of the molecular weight distribution (M_w/M_n near 2.0) for the P-plastomer polymer. It is expected that the results of thermal solvent fractionation are sometimes contaminated by the difficulty in solubilizing a higher-molecular-weight polymer compared to a similar polymer having a lower molecular weight. We note that this structural characteristic of uniform intramolecular and intermolecular composition is extremely difficult to achieve in practice. Ziegler–Natta catalysts and even some of the earlier work on propylene polymerization using single-sited metallocene catalysts lead to broad intermolecular differences in composition and crystallinity. These products do not display the characteristic elongation, and more importantly the elastic recovery, of P-plastomers.

6.6.2 ERRORS IN PROPYLENE INSERTION

The properties of the P-plastomers containing propylene crystallinity are, secondarily after composition, dependent on the extent of the creation of stereo and regio insertion errors in the insertion of propylene [22]. Stereo errors are those that are created by the 1,2 insertion of propylene, but with a racemic orientation of the methyl group with relation to the adjacent methyl groups. This arises almost entirely from an insertion of the propylene in the "wrong face of the planar sp^2 olefin structure." The regio errors are those where the 1,2 insertion process is temporarily violated by the formation of a 2,1 insertion error or a 1,3 insertion error. In the former, the olefin is inverted as it enters the chain and leads to a structure where vicinal carbon atoms each contain a methyl group. These are often referred to as the head-to-head structure. In the other principal type of regio errors, the insertion of the propylene occurs by a 1,3 insertion such that the section of the chain contains only incorporated methylene units. These errors disrupt the crystallization of the isotactic propylene units and lead to polymers with lower levels of crystallinity. These errors, which are created by the multiple stereochemical pathways for the insertion of propylene, are primarily a reflection of the catalyst structure and the polymerization process. While several lines of characterization have analyzed and resolved the differences in the type and the frequency of errors in the polymerization of propylene, for the P-plastomer polymers, mechanical and physical properties are most closely aligned with the absolute residual stereoregularity of the polymer.

The polymerization of P-plastomers at high polymerization temperatures makes these mistakes in the insertion of propylene more frequent. The stereoregularity in the insertion of propylene is measured by a C13-NMR procedure, which analyzes the sets of three adjacent propylene residues (triads) for the relative orientation of the methyl groups. Isotactic propylene crystallinity corresponds to meso orientation, and thus the prevalence of the mm triads indicates an essentially iPP polymer. Conversely, the diminution in the amount of the triads leads to a large amount of atactic insertion of the propylene monomer. The correlation of the triad meso stereoregularity to the crystallinity of the polymer is true if the only contribution to the error is the stereo errors. This is generally, though not absolutely, true, since other errors such as the regio errors of 2,1 insertion and 1,3 insertions, which contribute to the loss of crystallinity, are not accounted for in this calculation of the prevalence of mm triads. In addition, this procedure does not account for the loss of crystallinity of the propylene polymer that is due the addition of other monomers, such as ethylene. However within the narrow boundaries of this experimental determination, where only a single P-plastomer composition is attained, the crystallinity of the copolymer is closely and inversely aligned with the increasing amounts of errors in the propylene incorporation.

6.6.3 ELASTICITY AND ELASTIC RECOVERY

Within the bounds of composition and crystallinity that we described, P-plastomers are uniquely elastic and processable [22]. Their performance is further enhanced by their ease of compatibility with iPP. The incorporation of iPP, in small amounts as a separate blend component, leads to enhanced tensile strength, and enhanced abrasion and temperature resistance for the copolymer. In this it acts much like a reinforcing filler such as carbon black in compounds of general-purpose rubbers. Traditional elastomers, such as NR and EPDM which are both essentially amorphous and have high-molecular weights, owe their properties to strong entanglements. These provide the desired elasticity, but also provide a tremendous barrier to processing and fabrication. Rubber processing, compared to the processing of thermoplastics, is thus extremely time-consuming and involved. By comparison, thermoplastics which owe their properties at room temperature to extensive crystallinity, which provides a virtual network between the chains, are easy to process at temperatures above the melting point of the polymer but fail to provide an elastomeric tensile response at ambient temperatures.

P-plastomers provide a unique combination of ease of processing, such that conventional thermoplastic-processing routines' and arid equipment can be adapted to this polymer as well as for a final fabricated product that is elastic. This combination of properties leads to the easy fabrication of elastic materials such as fibers and films, which traditionally have only been made inelastic by the use of thermoplastics. This advance opens the pathway to the introduction of desirable elastic properties to a host of fabrication processes very different from either the conventional rubber-processing equipment or the conventional rubber products, such as tires. P-plastomers and their fabricated products are not only soft, but also elastic.

Elasticity is a combination of the ability of piece of a polymer to be uniaxially deformed to several times its original length and the ability to quickly retract back to its original dimension when the distending force is removed. The tensile elongations for P-plastomer polymers are typically greater than 1000%. This extensibility is a significant difference and much larger than noted for compositionally heterogeneous blends of propylene–ethylene copolymers which have the same average composition as the corresponding P-plastomers. This fundamental property change between materials of ostensibly the same average composition is due to the lack of dispersity in composition and tacticity for the P-plastomers. In these polymers, the entire elastomeric phase is semicrystalline and in a single phase; while in the blends, there is segregation into amorphous and semicrystalline phases. It is easy to accept that, in the absence of strong interphase interactions in these blend polymers, tensile deformation occurs exclusively in the less-crystalline phase and leads to an easy rupture of the analytical specimen. For the majority of the P-plastomers, polymers with crystallinities less than 20 J g^{-1}, corresponding to approximately 12% ethylene, the tensile-elongation curves do not show any evidence of irreversible deformations during the elongation cycle. We would thus expect that on removal of the stress, these specimens should retract essentially to their original dimensions. It is also noteworthy that all the P-plastomers except the ones which have the least amount of crystallinity, show strong evidence for increasing tensile modulus at elongations above about 300%. We believe that this increase in the tensile modulus for polymers with intermediate levels of crystallinity implies a process of strain-induced crystallization that appears for these polymers and contributes both to the extreme elongations, as well as the large value of the tensile modulus.

The tensile and permanent set properties of P-plastomers exhibit remarkable mechanical-annealing characteristics. This phenomenon is similar to the Mullins effect for elastomer compounds, but occurs even in pure P-plastomers polymers in addition to their compounds. This effect has two distinct components. First, the repeated cyclic elongation (to the same distention) results in an observed strain softening in the second and subsequent cycles. This effect has been observed for cyclic elongations in the range of 100%–400%, but there is very little evidence to suggest that this does not persist throughout the entire elongation spectrum. The most-marked strain softening is observed in the first one or two cycles. Further cycling does not lead to any observable changes in the tensile-elongation characteristics. The second feature of the mechanical-annealing process is that the tension set, which is the extent of the transient deformation in the sample during an extension and contraction cycle, rapidly diminishes between the first and second cycles to the same elongation. This transient deformation is the deformation of the sample in the direction of the elongation when the force in the retractive portion of the cycle reaches zero. These two data, in combination, indicate that for the P-plastomers, the initial mechanical distension leads to a small amount of change in the polymer morphology or crystal structure. This change persists when the extensional forces are removed and leads to a highly elastic polymer that has low extensional tensile modulus and almost complete recovery from elongation. Thus, the mechanical properties of P-plastomer polymers are improved significantly by mechanical work, and the measure of repeatable properties should be conducted on samples that have undergone at least one extensional cycle. We speculate that the irreversible changes that lead to this improvement may be reorientation or agglomeration of crystallites.

The P-plastomers polymers combine an unusual and unexpected degree of elongation and elastic recovery without the need for cross-linking. These properties mimic those of cross-linked elastomers while the processing and fabrication should be similar to that of the conventional polyolefins such as PP. Based on the structural information obtained from various techniques, a coherent picture of the morphology of P-plastomers polymers with increasing ethylene content has been generated. The details of the structure and morphology have been used to explain the observed physical–mechanical properties.

P-plastomer polymer catalysts introduce substantial error in the insertion of propylene, leading to much diminished crystallinity at equivalent ethylene content than more stereoregular PP polymerization catalysts. The stereoregular polymers are shown to be different from the P-plastomer polymers at equivalent ethylene content due to differences in defect content in the PP sequences. This appears to be an important parameter in the determination of the elastic properties of the polymer. The most-striking changes in the morphology correlate to the appearance of elastic properties, since the morphology of the PP crystallites changes from spherulitic to individual lamellar to fringed micellar with increasing C_2 content, as observed by TEM, optical microscopy, and SAXS. Wide-angle x-ray scattering (WAXS) showed that increasing C_2, content led to a decrease in alpha crystallinity and an increase in gamma crystallinity. Although the degree of crystallinity obtained by DSC and WAXS decreased monotonically with C_2 content, mechanical properties like modulus and permanent set decreased rapidly up to the morphological transition, and thereafter are weakly dependent (or independent) of C_2 content. The mechanical properties of interest have been explained on the basis of the morphology of the polymer. Differences in properties above and below a specific C_2 content have been attributed to the morphological transition. For P-plastomer and stereoregular samples, the transitions occur at 12 wt% and room temperature. As a result, the noncrystalline phase is soft and mobile, thereby making available an entropic mechanism for recovery. Upon application of a stress, low levels of deformation can occur simultaneously throughout the length of the sample, resulting in affine deformation. The crystallites act as physical cross-link, endowing the sample with good elastic recovery. In addition to crystallites, entanglements can also act as cross-links on the timescale of the experiment and are principally active at higher strains, contributing to strain-hardening behavior.

6.6.4 P-PLASTOMERS: BLENDS

P-plastomers, even more than the E-plastomers, have been blended with a number of substrates [23]. The most-important one is blend with iPP which forms compatible blends with P-plastomer for a wide range of relative weights fractions of P-plastomer and iPP as well as a wide range of molecular weights for both of the components. The formation of the blends with iPP leads to changes in the elastic and tensile response with elongation modulus, monotonically increasing with the amount of iPP.

The range of elastic modulus covers the spectrum of the properties from elastomeric to thermoplastics. This wide latitude in blending and the resulting wide space of accessible properties including variation in tensile strength and elastic recovery are invaluable in compounding applications This basic binary blend character can be extended to other alloys and blends with iPP: thus, iPP blends with elastomers can also be successfully compounded to form softer products with P-plastomers. Random propylene thermoplastic copolymers can be used to increase the elongation to break and toughness of TPVs. In an important development, a small amount of P-plastomers in iPP–E-plastomer blends has been shown to have a very advantageous effect on the impact properties.

Blends of P-plastomer with EPR, SEBS, and E-plastomers have been separately detailed. A most interesting development was the realization that P-plastomers were compatible with isobutylene-based polymer. These blends had improved green strength compared to the butyl polymer alone. The blends have improved green strength, green elongation, and green relaxation properties of isobutylene-based elastomers at elevated temperatures along with improved aging and barrier properties that are achieved by blending semicompatible, semicrystalline polymers with the isobutylene elastomers.

6.6.5 P-PLASTOMER FABRICATIONS

One of the important aspects of the development of P-plastomers was the expectation that these materials were amenable to plastics processing such as fiber and film formation and yet would yield soft elastic fabrication. This combination was hitherto unknown [24]. The formation of nonwoven fabrics including spun-bond and melt-blown nonwoven fabrics as well as their laminated forms has been documented. Similarly, cast film operation to form elastic monolithic films or composite structures which are not only amenable to these processes, but also to a variety of postfabrication processes have been described.

6.7 CONCLUSION

Plastomers represent a major advancement for polyolefins. Their success allows polyolefins to have a continuum of products from amorphous EPR to thermoplastic PE and iPP. This development coincides with the advent of single-site catalysts: these are necessary for copolymers of components of widely different reactivity such as ethylene and octene. Their rapid introduction into the mainstream polymer use indicates that this spectrum of properties and the inherent economy, stability and processibility of polyolefins are finding new applications to enter.

REFERENCES

1. (a) Bensason, S. Ph.D. dissertation. Case Western Reserve University, Cleveland, OH (1996), 156 pp. Avail.: UMI, Order No. DA9723501. From: *Diss. Abstr. Int.*, B (1997), 58(2), 728DT. (b) Penfold, *J. Caoutch. Plast.* (1999), 76(775), 53–56. (c) Pillow, J.G. Kautsch. *Gummi Kunstst.* (1998), 51(12), 855–859. (d) Sylvest, R.T., Lancester, G., Betso, S.R. *Kautsch. Gummi Kunstst.* (1997), 50(3), 186–191. (e) Bensason, S., Stepanov, E.V., Chum, S., Hiltner, A., Baer, E. *Macromolecules* (1997), 30(8), 2436–2444.

2. Reference for tafmer and other types of plastomer compositions. (a) Utracki, L.A., (ed), *Polymer Blends Handbook*, Vol 1, (2002), p. 6, Kluwer Academic Publishers, (b) see also www.mfpforum.com for the most current listing.

3. (a) Tang, H., Beatty, C.L. *Ann. Tech. Conf.—Soc. Plast. Eng.* (2000), 58(2), pp. 2474–2478. (b) Kwak, H., Rana, D., Cho, K., Rhee, J., Woo, Taewoo, L., Byung, H., Choe, S. *Polym. Eng. Sci.* (2000), 40(7), 1672–1681. (c) Kwak, H., Rana, D., Choe, S. *J. Ind. Eng. Chem.* (2000), 6(2), 107–114. (d) Bensason, S., Nazarenko, S., Chum, S., Hiltner, A., Baer, E. *Polymer* (1997), 38(15), 3913–3919.

4. (a) Walton, K.L., Karande, S.V. *Annu. Tech. Conf.—Soc. Plast. Eng.* (1997), 55(3), 3250–3254. (b) Fanichet, L., Clayfield, T. *Elastomery* (1997), 1(4), 16–21. (c) Benson, R.S., Moore, E.A., Martinez-Pardo, M.E., Luna Z.D. *USA Nucl. Instrum. Methods Phys. Res., Sect. B* (1999), 151(1–4), 174. (d) Minick, J., Sehanobish, K. *Annu. Tech. Conf.—Soc. Plast. Eng.* (1996), 54(2), 1883–1886.

(e) Brann, J.E., Hughes, M.M., Cree, S.H., Penfold, J., Dow Chemical Co., US Patent 6048935 (04/11/2000). (f) Schombourg, J.F., Kraxner, P., Furrer, W., Abderrazig, A., PA Witco Corporation, WO Patent 9967330 (12/29/1999). (g) Rowland, M.E., Turley, R.R., Hill, J.J., Kale, L.T., Kummer, K.G., Lai, S., Chum, P.-W.S. Dow Chemical Co., PCT Patent WO 9732922 A2 (09/12/1997).

5. (a) Chee, H.J., Hwang, M.J., Lim, B.Y., Choi, C.H. Honam Petrochemical Corporation, US Patent 6143828 (11/07/2000). (b) Dharmarajan, N.R., Ravishankar, P.S., Burrage, C.D. Exxon Chemical Co., European Patent 9856012 A1 (12/10/1998). (c) Weaver, L., Hughes, M.M. Dupont Dow Elastomers, European Patent 9826001 A1 (06/18/1998). (d) Dharmarajan, N.R., Ravishankar, P.S., Burrage, C.D., Exxon Chemical, US Patents 5763533 A (06/09/1998), US Patents 5952427A (09/14/1997), US Patents 6150467A (11/21/2000). (e) Yarnell, L., South, B.E. The Gates Corporation, US Patent 5610217 A (03/11/1997).

6. (a) Walton, K.L., Clayfield, T. *Annu. Tech. Conf.—Soc. Plast. Eng.* (2000), 58(3), 2623–2627. (b) Rana, D., Lee, C.H., Choe, S., Cho, K., Lee, B.H. *Polym. J.* (1998), 6(2), 158–166. (c) Rana, D., Lee, C.H., Cho, K., Lee, B.H., Choe, S., *S.J. Appl. Polym. Sci.* (1998), 69(12), 2441–2450. (d) Da Silva, A.L.N., Tavares, M.I.B., Politano, D.P., Coutinho, F.M.B., Rocha, M.C. *J. Appl. Polym. Sci.* (1997), 66(10), 2005–2014. (e) Shimojo, M., Ohkawa, K., Kanzaki, S., Nagai, T., Nomura, T., Matsuda, M., Iwai, H. Sumitomo Chemical Co., European Patent WO2000064972 (11/02/2000). (f) Sadatoshi, H., Osakawa, K., Kanzaki, S., Iwai, H., Nomura, T., Ichikawa, S. Sumitomo Chemical, European Patent WO200002049 (04/13/2000). (g) Kuramochi, H.K., Osamu, S., Yoshitaka, N., Takanori, A.S. Chisso Corporation, European Patent WO2000011081 (03/02/2000). (h) Sobashima, Y.I., Masaaki, A.A., Hamaura, M., Niimi, T. Nippon Polychemicals Co., Japanese Patent JP2000026697 (01/25/2000). (i) Ratzsch, M., Reichelt, N., Hesse, A., Borealis, A.-G., European Patent EP972802 (01/19/2000). (j) Srinivasan, S., Szczepaniak, E., Her, J., Laughner, M.K., Karjala, T. Solvay Engineered Polymers, PCT application WO 9820069 A1 (05/14/1998). (k) Tomomatsu, R., Ishii, T., Ohnishi, S., Nara, T. Idemitsu Petrochemical Co., PCT application WO 9749765 (12/31/1997). (l) Shimojo, M., Kondo, S., Sumitomo Chemical Co., European Patent DE-19821937 (11/19/1998). (m) Sakai, I., Takaoka, T., Moriya, S., Hashimoto, M., Nakagawa, H., Yoshikawa, H. Grand Polymer Co., PCT Patent WO 9847959 (10/29/1998). (n) Pellegati, G., Braga, V., Bonari, R., Montell Technology PCT Patent WO 9837144 A1 (08/27/1998). (o) Sobajima, Y., Hayakawa, Y., Banno, Y. Japan Polychem, European Patent EP 953602 A1 (11/03/1999). (p) Sanpei, A., Shimomura, Y. Chisso Corporation, PCT Patent WO 9854257 A1 (12/03/1998). (q) Sobajima, Y., Fujii, H., Amano, A., Akashige, E. Japan Polychem Corporation, European Patent EP 896028 A1 (02/10/1999). (r) Nagai, T., Jagawa, Y., Nishio, T., Zanka, Y., Tsutsumi, I., Ishii, I., Satou, H., Sano, H. Japan Polychem Corporation, European Patent EP 774489 A1 (05/21/1997). (s) Sugimoto, H., Nakatsuji, Y., Hozumi, H. Sumitomo Chemical Co., PCT Patent WO 9733940A1 (09/18/1997). (t) Brennan, J.V., Ginkel, S.T., Skogland, T. Minnesota Mining and Manufacturing, PCT Patent WO 9732314 A2 (09/04/1997).

7. (a) Lee, J.H., Lee, J.K., Lee, K.H., Lee, C.H. *Polym. J.* (Tokyo) (2000), 32(4), 321–325. (b) Da Silva, A.L.N., Rocha, M.C.G., Coutinho, F.M.B., Bretas, R., Scuracchio, C. *Polym. Test.* (2000), 19(4), 363–371. (c) Raue, F., Ehrenstein, G.W., 148 (7th International Conference on Polymer Characterization [POLY-CHAR-7]) (1999), 229–240. (d) Da Silva, A.L.N., Rocha, M.C.G., Coutinho, F.M.B., Bretas, R., Scuracchio, C. *J. Appl. Polym. Sci.* (2000), 75(5), 692–704. (e) Raue, F., Ehrenstein, G.W., *Annu. Tech. Conf.—Soc. Plast. Eng.* (1999), 57(2), 2752–2756. (f) Sano, H., Matsuda, M., Satou, H., Nomura, T. *Japan Kobunshi Ronbunshu* (1999), 56(10), 693–701. (g) Raue, F., Ehrenstein, G.W. *J. Elastomers Plast.* (1999), 31(3), 194–204. (h) Maeder, D., Bruch, M., Maier, R.-D., Stricker, F., Muelhaupt, R. *Macromolecules* (1999), 32(4), 1252–1259. (i) Ou, Y.-C., Guo, T.-T., Fang, X.-P., Yu, Z.-Z. *J. Appl. Polym. Sci.* (1999), 74(10), 2397–2403. (j) Mouzakis, D.E., Karger-Kocsis, J. *Polym. Bull.* (Berlin) (1999), 43(4–5), 449–456. (k) Karger-Kocsis, J. *J. Macromol. Sci., Phys.* (1999), B38(5 & 6), 635–646. (l) Ganter, M., Brandsch, R., Thomann, Y., Malner, T., Bar, G., *Kautsch. Gummi Kunstst.* (1999), 52(11), 717–718, 720–723. (m) Mader, D., Thomann, Y., Suhm, J., Mulhaupt, R. *J. Appl. Polym. Sci.* (1999), 74(4), 838–848. (n) Nitta, K.-H., Takayanagi, M.J. *Polym. Sci., Part B: Polym. Phys.* (1999), 37(4), 357–368. (o) Yamaguchi, M., Miyata, H., Nitta, K.-H. *J. Appl. Polym. Sci.* (1996), 62(1), 87–97.

8. (a) Kyu, T., Kaewwattana, W., Ramanujam, A., Chiu, H.-W. *Annu. Tech. Conf.—Soc. Plast. Eng.* (1999), 57(2), 1699–1702. (b) Fritz, H.G., Bolz, U., Cai, Q. *Polym. Eng. Sci.* (1999), 39(6), 1087–1099. (c) Fritz, H.G., Cai, Q., Bolz, U. *Kautsch. Gummi Kunstst.* (1999), 52(4), 272–274, 276–278, 280–281. (d) Fritz, H.G. *Polym. Mater. Sci. Eng.* (1998), 79, 92–93. (e) Garois, N. Hutchinson, PCT Patent WO 9744390 A1

(11/27/1997). (f) Tasaka, M., Tamura, A. Riken Vinyl Industry, European Patent EP 845498 A1 (06/03/1998).

9. (a) Stricker, F., Maier, R-D., Muelhaupt, R. *Angew. Makromol. Chem.* (1998), 256, 95–99. (b) Sanpei, A., Shimomura, Y. Chisso Corporation, PCT Patent WO 9854257 A1 (12/03/1998).

10. (a) Enami, H., Ono, K., Hioki, K. Mitsuboshi Belting, European Patent EP 811657A2 (12/10/1997). (b) Srinivasan, S., Szczepaniak, E., Shah, S.S., Edge, D.K.D & S Plastics, PCT Patent WO 9606132 A1 (02/29/1996). (c) Paeglis, A.U., Boysen, R.L., Lynn, T.R., Collins, J. Union Carbide, European Patent EP 729986 A2 (09/04/1996). (d) Nishio, T., Nomura, T., Okishio, Y., Nishi, I., Yamamoto, T., Mayumi, J., Usami, T., Yazaki, T. Mitsubishi Chemical, Eurpean Patent EP 682074 A1 (11/15/1995).

11. (a) Morris, H.R., Munroe, B., Ryntz, R.A., Treado, P.J. *Langmuir* (1998), 14(9), 2426–2434. (b) Ryntz, R.A. Vinyl Addit. Technol. (1997), 3(4), 295–300. (c) Ryntz, R.A., McNeight, A. *Plast. Eng. (Brookfield, CT)* (1996), 52(9), 35–38. (d) Srinivasan, S., Her, J. Solvay Engineered Polymers, US Patent 5783629 A (07/21/1998). (e) Yu, T.C. *Annu. Tech. Conf.—Soc. Plast. Eng.* (1998), 56(3), 2636–2641. (f) Ryntz, R.A., McNeight, A., Ford, A. *Annu. Tech. Conf.—Soc. Plast. Eng.* (1996), 54(3), 2707–2711.

12. (a) Radosta, J.A.,TPOs Automot. '98, Int. Conf., 5th (1998). (b) Nomura, M. . (Idemitsu Petrochemical), Japan Patent JP 63008445 A2 (01/14/1988).

13. (a) Dahlan, H.M., Zaman, M.D., Khairul I. *J. Appl. Polym. Sci.* (2000), 78(10), 1776–1782. (b) Szabo, P., Pukanszky, B. *Macromol. Symp.* (1998), 129 (Modified Polyolefins for Advanced Polymeric Materials), 29–42. (c) Bartczak, Z., Argon, A.S., Cohen, R.E., Weinberg, M. *Polymer* (1999), 40(9), 2331–2346. (d) Michaeli, W., *Kautsch. Gummi Kunstst.* (1991), 44(9), 827–832.

14. (a) Flandin, L., Chang, A., Nazarenko, S., Hiltner, A., Baer, E. *J. Appl. Polym. Sci.* (2000), 76(6), 894–905. (b) Feng, W., Li, H., Zhang, Q., Wu, D., Jin, R. *Huagong Daxue Xuebao* (1999), 26(3), 30–33. (c) Tai, H., Jin, R., Zhang, L., *Hebei Gongye Daxue Xuebao* (1999), 28(2), 37–42. (d) Huang, J.-C., Huang, H. *J. Polym. Eng.* (1997), 17(3), 213–229. (e) Tung, J.F., Moh, S.L., Liew, K.F. *Annu. Tech. Conf.—Soc. Plast. Eng.* (1995), 53(1), 962–965. (f) Steinbrecher, B. TECHNOMER '97, Fachtag. Verarb. Anwend. Polym., Vortr., 15th 1997, P33/1–P33/6. Editor(s): Mennig, G., Meyer, F. (g) Yousefi, A.A., Ait-Kadi, A., Roy, C. *Adv. Polym. Technol.* (1998), 17(2), 127–143.

15. (a) Walton, K.L., Karande, S.V. *Tech. Conf.—Soc. Plast. Eng.* (1996), 54(2), 1926–1930. (b) Karande, S.V., Walton, K.L. *Polym. Prepr. (Am. Chem. Soc., Div. Polym. Chem.)* (1996), 37(1), 271–272. (c) Bhatt, C.U., Hwang, C.R., Khan, S.A. *Radiat. Phys. Chem.* (1998), 53(5), 539–547. (d) Walton, K.L., Karande, S.V. Dow Chemical Company, PCT Patent WO 9711985 A1 (04/03/1997). (e) Enami, H., Ono, K., Kubomoto, K., Okazawa, T. Mitsuboshi Belting, European Patent EP-950490 A1 (10/20/1999).

16. (a) Vogel, R., Bruenig, H., Beyreuther, R., Taendler, B., Voigt, D. *Polym. Process.* (1999), 14(1), 69–74. (b) Hoffmann, M., Beyreuther, R., Vogel, R., Tandler, B. *Chem. Fibers Int.* (1999), 49(5), 410–412. (c) Reid, R.L., Ho, T.H., Bensason, S., Patel, R.M., Batistini, A. Dow Chemical, PCT Patent WO 03/078723A1.

17. (a) Jaeger, J.T., Sipinen, A.J. Minnesota Mining and Manufacturing, PCT Patent WO 9951666 A1 (10/14/1999). (b) Topolkaraev, V., Soerens, D.A., Thomas, O. Kimberly Clark, PCT Patent WO 9933651 A1 (07/08/1999). (c) Gwaltney, S.W., Milicevic, C.J., Shawver, S., Elaine, E., Paul, W., Haffner, W.B., McCormack, A.L., Hetzler, K.G., Jacobs, R.L., Kimberly-Clark, WO 829246 A1 (07/09/1998). (d) Gwaltney, S.W., Milicevic, C.J., Shawver, S.E., Estey, P.W., Haffner, W.B., McCormack, A.L., Hetzler, K.G., Jacobs, R.L., Kimberly-Clark, PCT Patent WO 9829479 A1 (07/09/1998). (e) Abuto, F.P., Haffner, W.B., Jordan, J.F., McCormack, A.L., Uitenbroek, D.G. Kimberly-Clark, US Patent US-6096668 A (08/01/2000).

18. (a) De, K.N.R.M., Stoner, C.A. Shell, PCT Patent WO2001000257 A1 (01/04/2001). (b) Wang, B.L., Malcolm, R. Ato Findley, US Patent US-6143818 A (11/07/2000).

19. (a) Peiffer, D.G., Chludzinski, J.J., Lue, C.-T., Erderly, T. Exxon, PCT Patent WO 9703125 A1 (01/30/1997). (b) Peiffer, D.G., Chludzinski, J.J., Lue, C.-T., Erderly, T. Exxon, WO 9703123 A1 (01/30/1997). (c) Tojo, T., Matsumoto, M., Kikuchi, Y. Mitsui, European Patent EP 638594 A1 (02/15/1995). (d) Ito, N., Okayama, K., Karasuda, T., Miyagawa, Y. Tosoh Corp, Japanese Patent JP 06157642 A2 (06/07/1994). (e) Rifi, M.R. Union Carbide, European Patent EP 131938 A1 (01/23/1985).

20. Datta, S., Srinivas, S., Cheng, C.Y., Hu, W., Tsou, A., Lohse, D.J. Rubber World 229(1), 332–341.

21. Cozewith, C., Datta, S., Hu, W. US Patent US6525157B2 (11/20/2003).

22. (a) Datta, S., Cozewith, C., Ravishankar, P., Stachowski, E.J. US Patent US6867260B2 (04/22/2004). (b) Datta, S., Gadkari, A.C., Cozewith, C. US Patent US6635715B1 (08/12/1997).

23. (a) Finerman, T., Ellul, M.D., Abdou Sabet, S., Datta, S., Gadkari, A. US Patent US 6288171 B2 (09/11/2001). (b) Wang, H.C., Duvdevani, I., Datta, S., Qian, C.R. US Patent US6326433B1 (05/11/2000). (c) Datta, S., Cozewith, C., Periagaram, P., Stachowski, E.J., US Patent US6642316B1 (06/29/1999). (d) Dharmarajan, N.R., Datta, S., Bulawa, M.C., Srinivas, S., Reynolds, T.J. US Patent US6852424B2. (e) Schauder, J.R., Datta, S. US Patent US6884850B2 (10/27/2003).

24. (a) Cheng, C.Y., Datta, S., Agarwal, P.K. US6342565B1 (05/12/2000). (b) Datta, S., Middlesworth, J. US Patent US6500563B1 (05/11/2000). (c) Dharmarajan, N.R., Bulawa, M.C., Datta, S., Tsou, A.H., US Patent US6750284B1 (05/15/2000). (d) Dharmarajan, N.R., Datta, S., Meka, P., Williams, M.G., Srinivas, S. US Patent US7105603B2 (07/21/2004).

7 New Elastomers: Biomacromolecular Engineering via Carbocationic Polymerization

Judit E. Puskas

CONTENTS

7.1 INTRODUCTION

We recently defined "biomacromolecular engineering" as the precision design, synthesis, and characterization/testing of macromolecules (polymers), combined with biomedical engineering.[1] Polymer science and engineering have made tremendous progress in the last few decades. With the development of various controlled/living polymerization techniques we are in the position to synthesize a wide variety of architectures, e.g., stars, brush-like chains, centipedes, dendritic (hyperbranched and arborescent) structures, self-assembling block copolymers, and supramolecular structures.[1–13] This makes us capable of delivering material properties to the biomedical engineer for almost any application; however, we need precise definition of those material properties. Elastomers and rubbers are desirable biomaterials; after all, with the exception of bone and cartilage, we are made of soft tissues.[14–18] Carbocationic polymerization has gained new significance in biomacromolecular engineering; one of the most-promising thermoplastic elastomeric (TPE) biomaterial that emerged in the last few years, polystyrene-*b*-polyisobutylene-*b*-polystyrene (SIBS), can only be synthesized by living carbocationic polymerization.[1,6,7,17–32] The recent Food and Drug Administration (FDA) approval of SIBS, used as the polymeric coating on the Taxus coronary stent,[33–35] opened new avenues for polyisobutylene-based TPEs in biomedical engineering.

This chapter will highlight biomacromolecular engineering via carbocationic polymerization using two case studies. The first will discuss an application where the desired properties were precisely defined by the biomedical engineer. The second will highlight a reverse case where the desired properties are not defined in engineering terms, so the macromolecular engineer will have to provide a variety of materials for clinical trials. The third part of this chapter will discuss our latest results in the design, synthesis, and characterization of novel SIBS TPEs with dendritic (arborescent or hyperbranched) polyisobutylene core (DIB stands for dendritic polyisobutylene, and SDIBS stands for DIB capped with polystyrene blocks).

7.2   BIOMACROMOLECULAR ENGINEERING VIA CARBOCATIONIC POLYMERIZATION

7.2.1   CASE STUDY 1: LESS-LETHAL AMMUNITION

The use of lethal force by law-enforcement agencies is becoming politically and socially unacceptable. It has always been desirable by the corrections community to achieve control without causing serious injuries by using various less-lethal weapons.[36] The demand for less-lethal weapons has led manufacturers to develop new devices for law-enforcement agencies. Unfortunately, less-lethal technologies to date have proven to be either impractical or ineffective, or a combination of both. The most important legal issue that faces law enforcement, correctional administration, and public managers is the liability issue for misusing less-lethal weapons. The importance of developing improved less-lethal ammunitions for peacekeeping applications (e.g., a handgun that can target a hijacker in a cockpit without penetrating the airplane's fuselage, or can be used in crowd control) was clearly recognized after 9/11. Unfortunately, it appears that less-lethal ammunition such as plastic and rubber bullets has been developed on a trial-and-error basis, with the error resulting in tragic circumstances. Sometimes, these are referred to as "nonlethal"—we prefer the term "less lethal," as improper use can always be lethal. There are barely any scientific articles in the field of less-lethal ammunition from the material science point of view, with most of the material-related information available in patents.

Baton rounds made of polyurethane-based hard thermoplastics were first used by the police in Northern Ireland in 1973 and their application has steadily evolved over the years. Similar thermoplastic resin-based projectiles have been disclosed in the patent literature. Projectiles based on filled thermoset rubbers such as ethylene–propylene–diene monomer (EPDM) rubber, natural rubber (NR) and synthetic polyisoprene (IR), and polybutadiene (PBD) are also mentioned. Projectiles based on block-type TPEs (styrene-b-isoprene-b-styrene [SIS] or styrene-b-butadiene-b-styrene [SBS]) and their blends with thermoplastic resins (polystyrene, polypropylene, or polyvinylchloride) were also disclosed. Projectiles consisting of an elastomer- or thermoplastic-based casing (styrene–butadiene rubber [SBR], silicone rubber, or thermoplastic polypropylene resin), filled with lead shots or powder, are also known. In North America, beanbags and sock rounds are the most widely used less-lethal projectiles. Both of these consist of a fabric pouch filled with lead shots; the sock round has a tail to enhance flight stability. Shortcomings of these projectiles include inaccuracy, and failure to open up when ejected from the shell and hitting the target on edge (i.e., "Frisbee effect"). In addition, the fabric may tear, causing serious injuries and environmentally toxic lead pellets to spill over the target (perhaps this is the last legal application of lead). All these "less-lethal ammunition" have been linked to fatal accidents or serious injuries (Figure 7.1).

It was emphasized that any thoracic penetration may result in acute life-threatening injuries such as tension pneumothorax, hemothorax, massive cardiac injury with tamponade, great vessel injury, hemoptysis, and lung collapse.[37] Thermoset rubber bullets can also cause serious injuries, as shown in Figure 7.2.

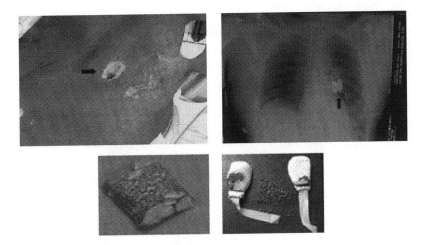

FIGURE 7.1 Thoracic penetration of bean bags and sock rounds. (From Charles, A. et al., *J. Trauma*, 53, 997, 2002. With permission.)

"Inaccuracy of rubber bullets and improper aiming and range of use resulted in severe injury and death in a substantial number of people. This type of ammunition therefore should not be considered a safe method of crowd control."[38]

One of the major concerns for law-enforcement agencies in using less-lethal ammunition projectiles is the delicate balance between their effectiveness and their lethality. Impact biomechanics studies of a human body, including the use of less-lethal projectiles, are critically important when considering the protection of military personnel and civilians from accidental injury.[37–40] Bir et al.[41–45] have analyzed the effect of blunt ballistic impact of a baton-type, less-lethal projectile on the thoracic region using human cadavers. She determined human-response corridors and developed biomechanical surrogates, which can be used for testing different new projectiles for their blunt ballistic impact. The human-response force corridor for blunt thoracic impact by a thermoplastic polyurethane-based baton round with 37 mm diameter, weighing 30 g, and traveling

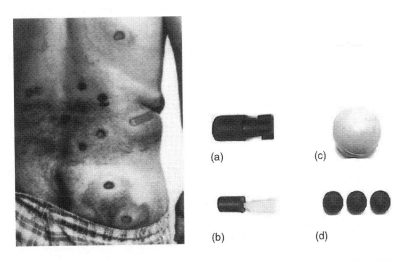

FIGURE 7.2 Penetration of rubber bullets. (From Mahajna, A. et al., *Lancet*, 359, 1795, 2002. With permission.) (a) Rocket, (b) tube, (c) mono-ball, and (d) multiple-ball. (From Puskas, J.E., *Kautsch. Gummi Kunstst.*, 58, 288, 2005. With permission.)

with a velocity of 60 m s^{-1} (~200 ft s^{-1}) was found to be between 1800 and 3700 N (400–830 lb$_f$) for 0.275 m s^{-1} impact duration. The corresponding thorax deflection corridor was between 1.43 and 6.93 mm. The biomedical surrogates (3-Rib Ballistic Impact Device (3-RBID) and a modeling-clay impactor) yielded similar response corridors. Bir's data clearly outlined the requirements that a rubber bullet should meet in order to be able to temporarily incapacitate a person, without serious injury.

When designing new projectiles, several considerations should be taken into account. There are optimum material-density requirements for any bullet, lethal or less-lethal. In addition, the propellant powders are intended to work with a projectile of a certain mass that provides a typical pressure versus time curve. Using a lighter projectile could lead to too low energy transfer. Lighter projectiles would have to be much faster to provide sufficient energy transfer, which requires a propellant powder that would burn faster. The combination of fast powders and lightweight projectiles can dangerously exceed recognized gun chamber pressure standards. In addition, if higher-than-desired linear velocity of a projectile is inherently coupled with higher-than-desired spin velocity, it may result in poorer accuracy of the bullet. For projectiles, lower density almost invariably translates into poorer performance in terms of accuracy. With less-dense materials, the bullet cannot be brought up to its desired weight by increasing its size due to the dimensional limitations imposed by the standard gun chambers into which the bullet must fit. In terms of impact, the most important design criterion is the maximum tolerable energy or force per unit area that an average human could withstand without serious injury. This is highly individual as it generally depends on the individual's body fat and muscle mass. Bir considered the average male (about 5 ft 10 in, 160 lb). Based on her results, we calculated the tolerable energy transfer at muzzle as 44 J (32 ft lb^{-1}) for the 20-gauge projectile and 73 J (54 ft lb^{-1}) for the 12-gauge projectile. As discussed above, the maximum allowable deflection of 6.93 mm and the maximum tolerable energy transfer on impact without serious injury are limited to 3700 N. With the desirable 76 m s^{-1} (250 ft s^{-1}) muzzle velocity, this translates into about 15 and 25 g weight for the 20- and 12-gauge projectiles respectively. Thus, the minimum density of the projectiles with the specified dimensions was calculated to be 2.4 g cm^{-3}.

The next criterion was the selection of materials that would yield the desired mechanical properties. Butyl elastomer (IIR), a copolymer of isobutylene with a small amount of isoprene, has outstanding low-temperature properties and very high damping, but has very high creep without cross-linking.[46] It was theorized that a blend of SIBS TPE and butyl elastomer, filled to achieve the required minimum density of 2.4 g cm^{-3}, would be a promising composite for less-lethal ammunition. Based on extensive studies,[36,47–50] the SIBS50 blend (IIR/SIBS/Iron 50/50/233), where IIR is regular butyl rubber,[46] was shown to have optimum hardness, compression creep, and dynamic mechanical properties for less-lethal ammunition projectile application.

The accuracy and the impact energy of less-lethal projectile made of SIBS50 were compared with sock rounds for 12- and 20-gauge projectiles. The sock round was selected for this comparative test since it is the most widely used and has proven to be the best of all existing less-lethal projectiles, despite being associated with some serious injuries. The calculated impact energies were compared with the impact-energy corridors proposed by Bir.[41,42] The clay impactor used by Bir is routinely employed by the military as a "surrogate" for testing ammunition, to mimic the "viscoelasticity" of human soft tissues. Clay requires significant conditioning at a very narrow temperature and humidity range, thus its use is rather cumbersome. We tested the projectiles using a novel Styrofoam impactor. Styrofoam (polystyrene foam) used as an insulator in buildings maintains constant mechanical properties over a broader range of temperatures and humidity. Table 7.1 summarizes the test results for the new 20- and 12-gauge less-lethal projectiles, in comparison with sock rounds.

One of the main factors in judging the performance of less-lethal projectiles is velocity consistency. A velocity deviation less than 10% is regarded excellent when using less-lethal projectiles. The velocity deviation for the new 20-gauge projectiles was 3%, in comparison with 7.8% for the sock rounds. The second main factor is the impact area of the projectile. Both the new 20-gauge projectiles and the sock rounds impacted in the 10.16 cm (4 in.) diameter circle from about

TABLE 7.1

Impact Results from 20- and 12-Gauge Less-Lethal Projectiles

Type/Gauge (SD%)[a]	Distance from Target (cm)	Velocity (m s^{-1})	Penetration (cm)	Muzzle Energy (J)	Impact Energy (J)
SIBS50/20	9.3	84.2	19.4	47.7	8.8
	(21.6)	(3.0)	(6.3)	(12.5)	(19.2)
Sock/20	7.2	92.6	17.4	141.0	21.9
	(45.3)	(7.8)	(19.2)	(17.8)	(7.5)
SIBS50/12	7.4	86.2	9.8	89.2	12.3
	(43.6)	(4.0)	(25.6)	(8.1)	(25.0)
Sock/12	8.4	72.6	12.8	120.0	22.8
	(45.8)	(7.0)	(15.6)	(13.6)	(8.5)

Source: Adapted from Puskas, J.E., *Kautsch. Gummi Kunstst.*, 58, 288, 2005.

[a] SD% = standard deviation from 10 measurements (%).

5 m (15 ft), but the sock rounds showed much higher standard deviation. In addition, it was observed during this test that some of the sock rounds failed to unfold upon launching and ended up hitting the foam impactor while still folded. Under optimum conditions, a sock round must open within 2 m of launching in order to allow for a higher air resistance and therefore slow down in order to transfer less energy before hitting the target. In the case of the 12-gauge projectiles, tests were carried out from a common police and military firing distance of 10.7 m (~35 ft) using a standard 12-gauge shotgun. The velocity deviation for the 12-gauge new projectiles and sock rounds was 4% and 7% respectively, and all projectiles impacted in the 10.16 cm (4 in) diameter circle. These results are excellent in terms of velocity consistency. The new Styrofoam Impactor employed for measuring the impact energy of the less-lethal projectiles yielded favorable results in comparison to the comprehensive study conducted by Bir et al.[41,42] who employed the Abbreviated Injury Score (AIS), a scale developed in the 1960s to standardize the severity of injuries such as blunt thoracic impact. An AIS value of 0 is equivalent to a minor injury and 6 is equivalent to untreatable injury. It was noted, however, that injuries with an AIS of 3 or larger result in a high probability of death. For example, an injury of lung contusion with internal hemorrhage in the thorax region is considered an AIS 4 injury. This scale was used to evaluate the injuries caused by the 37 mm ammunitions. The tolerable force corridor of 1800–3700 N for a 37 mm projectile weighing 30 g and traveling at 60 m s^{-1} with an impact time of 0.275 m s^{-1} was translated into an impact energy range of 2.6–25.6 J. Under these conditions, the average muzzle energy, Bir observed, was approximately 54 J. These data correspond to an AIS injury of less than 3. The average impact energy for the newly developed 20-gauge less-lethal projectile was 8.8 ± 1.7 J at muzzle energy of 47.7 ± 6.0 J. This impact energy value is closer to the lower boundary set by Bir's actual cadaver testing. For the sock rounds, the average impact energy was 21.9 ± 1.7 J at muzzle energy of 141.0 ± 25.2 J, closer to the upper boundary. This implies that sock rounds may cause serious injuries if projectiles are fired from shorter distances.

The 12-gauge rounds were fired from a shotgun with more muzzle energy to compensate for the longer distances they travel before impacting the target. They also weigh more than the 20-gauge rounds. The 23 g, 12-gauge, new less-lethal projectile left the muzzle with 89.1 ± 7.2 J energy to impact with 12.3 ± 3.1 J energy from approximately 10 m distance. The impact energy here was still near the lower boundary limits published by Bir. The 12-gauge sock round with 119.1 ± 16.1 J muzzle energy impacted with 22.8 ± 1.9 J. This was slightly lower than, but close to, the upper boundary limit of 25.6 J, set by Bir. With this type of energy transfer, there is greater probability of causing serious injuries to human targets. Subsequently, the new projectiles were tested in Bir's laboratory, and performed very well.

In summary, the newly developed less-lethal projectiles performed well in comparison with the most frequently used sock rounds in terms of accuracy and velocity consistency, while delivering impact energies close to the lower boundary of the tolerable energy corridors developed by Bir. The impact energy delivered by the sock rounds was closer to the upper limit, with the 12-gauge round surpassing the limit.

7.2.2 CASE STUDY 2: BREAST IMPLANTS

In this case, the biomedical engineer practically has no material property data that the macromolecular engineer can match. Biocompatibility in terms of chemistry and engineering remains undefined. The European Society for Biomaterials "defined" biocompatibility as "The ability of a material to perform with an appropriate host response in a specific application."[17] The definition of biostability is also vague, stating that "A biostable material does not degrade in the human body." Unfortunately, this gives very little direction to biomaterial research. According to the current "definition" in the medical literature, a breast implant needs to be soft and deformable while maintaining the appropriate shape.[17] This reflects the history of soft materials used in medicine. Implantable and indwelling soft materials (e.g., catheters, tubes, membranes) currently in clinical practice derive from products originally developed for industrial uses, e.g., nylon sutures from lady's hose, polyurethane catheters from furniture upholstery, silicones and silicone tubing from automotive applications, polyester (Dacron) sheets and mesh from noniron apparel, and polyethylene implants from electrical insulation. Polymeric biomaterials that have clinically been demonstrated to exhibit biocompatibility during short- or long-term use, saved millions of lives, but all show some shortcomings. The "ideal" implant would be the reproduced biological tissue itself. While genetic and tissue engineering have made incredible progress during the last decade, it is acknowledged that synthetic biomaterials will play a very important role in the next few decades. The most critical problem with all synthetic biomaterials is adverse tissue reaction.[15–18] For example, over 100 million urinary catheters and stents are used annually in the United States alone, and 50% of the patients catheterized for more than 30 days experience encrustation and/or blockage, requiring removal of the device. In 2005, more than 350,000 breast reconstructions and augmentations were performed in the United States, which is a sixfold increase since 1990. According to the statistics of the American Cancer Society,[51] 34% of these patients reported complications, the most critical being adverse tissue reaction leading to calcification and capsular contracture (~10%; see Figure 7.3).

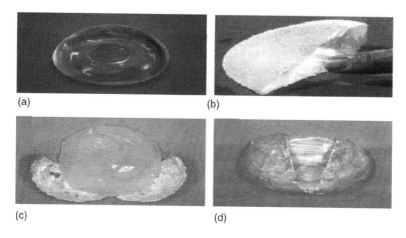

(a)   (b)

(c)   (d)

FIGURE 7.3 Implants before and after implantation. Before: (a) gel-filled round implant and (b) cohesive gel implant. After: (c and d) distorted implant in a calcified capsule. (From Pittet, B. et al., *Lancet Infect. Dis.*, 5, 94, 2005. With permission.)

Currently, breast cancer is the most frequently diagnosed cancer in both white and African-American women. According to latest reports,[51] of the 662,870 cancer cases reported for women in 2005, 32% were breast cancer. The incidence of breast cancer increased from 1 in 20 in 1960 to 1 in 8 today. More than 100,000 women per year require mastectomy for treatment, and every year 75% decide to have reconstruction. About half the number of these women select prostheses made of silica-reinforced silicone–rubber shell filled with silicone gel, while the other half have the same shell filled with physiological saline. Reportedly, gel-filled prostheses feel more natural, but are associated with true or perceived health problems and remain highly controversial.

Silicone breast implants were introduced in 1963;[53] their history has been reviewed.[17,18,52] A breast implant consists of a rubbery shell, and a filler material—silicone gel or saline solution. The shell can be prefilled and sealed before it is inserted into the body, or can be filled after insertion via a valve. The shell is expected to provide strength and barrier properties, whereas the filling supplies bulk and consistency. The first-generation silicone gel implants (1960s–1970s) had the thickest shells and the most viscous gel. They felt too firm compared with natural breast tissue, so the layer thickness and gel viscosity were reduced in the second generation (early 1980s). This led to gel bleed and rupture of the shell. The third-generation implants consist of shells of intermediate thickness filled with gels of medium viscosity.[17,18] Cohesive gel implants introduced recently are considerably firmer, and perceived to be harder than the natural breast tissue (Figure 7.3). The shape of the implant can be round, or "anatomically correct," and the surface can be smooth or textured.

Breast implants have been the subject of intense controversy over the last decade. The debate is driven by legal advocacy as well as by "objective" science. The scientific literature is also divided on the effects of silicone on the body. Thirty-four percent of women with breast prostheses report complications, with 32% being material-related such as capsular contracture (deformation of the implant by the surrounding fibrous/calcified envelope), gel bleed, shell rupture, wrinkling, sagging, asymmetry, etc. Capsular contracture, the leading long-term complication, is frequently not reported because the patient is able to tolerate the discomfort (see Figure 7.3). The mechanism of reported systemic immunological reactions to silicone (e.g., acute renal insufficiency, respiratory compromise, fibromyalgia, and other connective-tissue diseases) is not understood.[17,52] Recent findings indicated a statistically significant link between Magnetic Resonance Imaging (MRI)-diagnosed extracapsular silicone gel and fibromyalgia and other connective-tissue diseases, but the majority of scientific studies state that no such link can be supported. Figure 7.4 shows a plot of the average

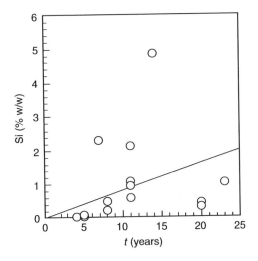

FIGURE 7.4 Silicon in the surrounding tissues versus years of implantation. (From Garrido, L. and Young, V.L., *Magn. Res. Med.*, 42, 436, 1999. With permission.)

amount of silicon found in the capsular tissue of intact implants measured by[29] silicon nuclear magnetic resonance spectroscopy (Si NMR).[54] The authors apparently saw no correlation between silicon content in the tissue and time.

In 1992, the FDA issued a moratorium on silicone-gel-filled implants, and restricted their use to reconstruction and clinical studies. In 2000, they approved saline-filled implants. In 2003, the General and Plastic Surgery Devices (GPSD) Advisory Panel recommended reapproval of gel-filled implants, but the FDA decided to wait for more clinical evidence of safety. In late 2005, the Panel recommended conditional approval of Mentor's and Inamed's gel implants. In October 2006, Health Canada approved the use of silicone-gel-filled implants, with a warning that "no medical device is 100% safe."*

To understand the complications related to breast implants, understanding the chemistry and material properties of silicones is important.[55,56] Silicone gel is produced from low molecular weight (MW) chains by cross-linking. The silicone rubber shell is produced by cross-linking high-MW long chains. Cross-linking is induced by peroxides and heat or catalyzed by metals such as platinum. Similarly to other thermoset rubbers, cross-linked silicone rubber is very weak and must be reinforced.[46] Medical-grade silicone is reinforced with about 30 wt% SiO_2 particles to reach a tensile strength of about 10 MPa with about 400%–600% elongation, less than half of the strength of other filled rubbers.[46] The bond angle in Si–O–Si is much higher than in C–C–C; as a consequence, silicones are known to be highly permeable. This is the reason for "gel bleed": when the non-cross-linked low-MW polymer and silicone oil plasticizer seep through the shell. In addition, the shell can rupture, and then the released gel can migrate.[17,52,54] The extent of gel bleed is dependent on the MW of the components and the degree of cross-linking of both the gel and shell, and the surface area of the prosthesis. It was shown that diffusion of the dimethylsiloxane small molecules out of the envelope can be reduced by using an additional layer of another elastomer, such as poly(methylphenylsiloxane) or fluorosilicone. However, the shell itself has been found to release low-MW moieties even without contacting the gel. Gel bleed and capsular contracture are believed to be correlated by some, although the trigger for capsule formation and calcification remains unproven. Two hypotheses have been suggested: hypertropic scar theory and infectious stimulus. To support this latter theory, substantial reduction in capsule thickness was demonstrated by the use of sodium 2-mercaptoethane sulfonate in a New Zealand White Rabbit Model.[57] For the material scientist, there is indication that capsule formation is somehow material-specific. For example, a remarkable decrease of capsular contracture was reported with polyurethane-foam-covered implants.[17] Texturing the surface of silicone is suggested to have a positive influence, but experimental data (micron-scale texturing) do not seem to support this hypothesis.[58,59] At sites of implants where silicone was absent, no mineral deposition was found.[60] Hydrophilization of the implant surface is believed to improve tissue interaction, but contradictions exist—for instance, poly(acrylic-acid)-grafted surface showed poorer cell-binding capability than silicone itself, while both covalent and adsorptive binding of fibronectin and Gly-Arg-Gly–Asp-Ser (GRDGS) improved cell–material interactions.[59] A recent study claimed poly(2-hydroxyethyl methacrylate) hydrogel as a capsule-resistant material *in vivo* in rats.[61] Unfortunately, no clinical solution exists today to prevent capsule formation.

The GPSD Advisory Panel expressed special concern about "silent" or asymptotic rupture and the consequences of extracapsular gel. The testing methodology currently used for fatigue rupture testing of silicone breast implants showed that these devices did not rupture under loads well in excess of those expected *in vivo*, thus the validity of the current compression method is questioned by the FDA.[62] Even though reports about the possible adverse effects of silicones used in implantation have surfaced shortly after their introduction, silicone rubber remains the only material used in clinical practice today; all other materials (poly[vinyl-alcohol], polyethylene, poly[tetrafluoroethylene], and polyurethane) have been recalled from the market. Implants with alternative fillers (soybean oil

---

* In November of 2006, the FDA lifted its 14-year ban of silicone gel filled implants.

and poly[vinyl pyrrolidone] hydrogel) were also recalled.[63,64] Surface modification of biomedical devices based on polymeric materials has been perhaps the most important area of research in recent years. The mechanisms of cell proliferation/capsule formation and calcification are not understood, and data in the medical literature are often controversial. It has been proposed that hydrophilic and amphiphilic surfaces are preferred, but poorer cell growth was reported on silicone with polyacrylic-acid-grafted surface than on nonmodified silicone. Surface structuring of breast implants has not produced unambiguous clinical evidence for improved tissue response. Heparin and phosphorylcholine have shown promise, but did not reduce encrustation significantly as expected. In general, the ultimate challenge is to design and prepare biomaterials that combine desired bulk mechanical properties with biocompatible surfaces, and we have very little guidance from biology and medicine what properties are needed for a specific application.

Section 7.3 will describe tools we developed to synthesize and characterize soft dendritic nanostructured TPE biomaterials via living carbocationic polymerization, and decorate their surfaces with "tissue-friendly" groups.

7.3 SYNTHESIS AND CHARACTERIZATION OF NOVEL BIOELASTOMERS

7.3.1 SYNTHESIS

The synthetic strategy for producing SDIBS is shown in Figure 7.5.[1,7] The first examples of this new class of TPEs were produced by initiating the living carbocationic polymerization of styrene from dendritic (hyperbranched or arborescent) polyisobutylene (PIB), or DIB for short, produced by 4-(2-methoxy-isopropyl) styrene *inimer* (= *ini*tiator-mono*mer*).[7] Key to the synthesis of high-MW DIB is that the *tertiary* cumyl-type and the para-substituted *secondary* benzyl carbocation, generated from the *tertiary* ether and the vinyl group of the *inimer*, respectively, are expected to have similar reactivities under polymerization conditions (Scheme 7.1).

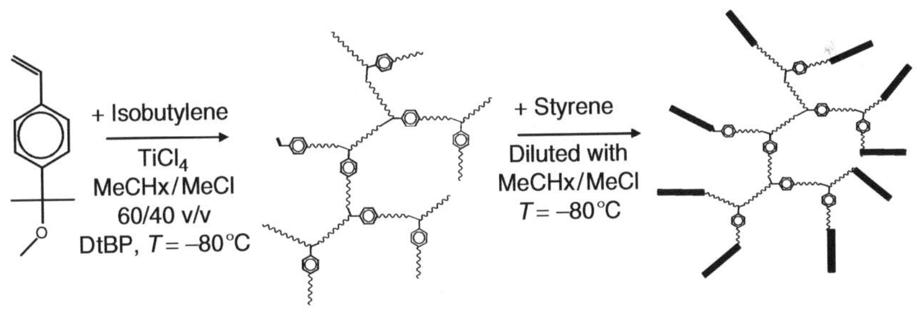

FIGURE 7.5 Synthetic strategy. (From Puskas, J.E., Kwon, Y., Antony, P., and Bhowmick, A.K., *J. Polym. Sci. Chem.*, 43, 1811, 2005. With permission.)

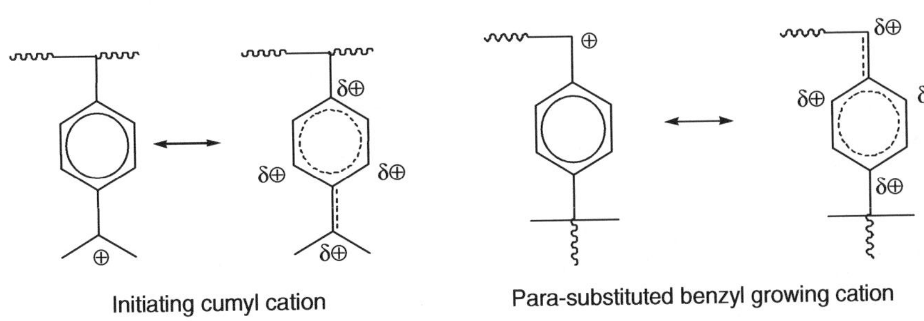

Initiating cumyl cation   Para-substituted benzyl growing cation

SCHEME 7.1 Carbocations generated from the 4-(2-methoxyisopropyl) styrene *inimer*/TiCl$_4$/isobutylene (IB) system. (Adapted from Puskas, J.E., Kwon, Y., Antony, P., and Bhowmick, A.K., *J. Polym. Sci. Chem.*, 43, 1811, cover page, 2005. With permission.)

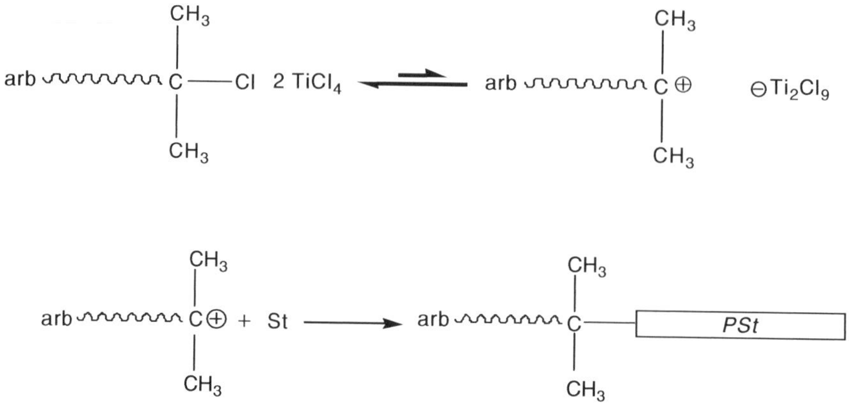

SCHEME 7.2 Reactivation of dendritic polyisobutylene (DIB) chain ends. (From Puskas, J.E., Kwon, Y., Antony, P., and Bhowmick, A.K., *J. Polym. Sci. A.*, 43, 1811, 2005, cover page. With permission.)

Indeed, cumyl carbocations are known to be effective initiators of IB polymerization, while the *p*-substituted benzyl cation is expected to react effectively with IB (*p*-methylstyrene and IB form a nearly ideal copolymerization system[65]). Severe disparity between the reactivities of the vinyl and cumyl ether groups of the *inimer* would result in either linear polymers or branched polymers with much lower MW than predicted for an *inimer*-mediated living polymerization. Styrene was subsequently blocked from the *tert*-chloride chain ends of high-MW DIB, activated by excess TiCl$_4$ (Scheme 7.2).

In this two-step synthesis, the DIB can be fully characterized prior to blocking.[66–74] In the one-pot or sequential monomer-addition method, first the PIB core is synthesized, and after the consumption of the IB monomer styrene is added.[1,75] The synthetic strategy shown in Figure 7.5 has numerous advantages over the synthetic strategy used to produce linear triblocks (SIBS). The PIB core in SDIBS can reach high MW; DIB with M_n close to 10^6 g mol^{-1} has been synthesized with the *inimer* technique.[67] In contrast, the highest M_n values reported for linear PIBs synthesized by living carbocationic polymerization using a difunctional initiator have never exceeded 200,000 g mol^{-1}. Unsuccessful blocking of styrene from both ends of a linear living chain may lead to undesirable diblocks, while blocking from several living chain ends of a DIB will always yield A–B–A type blocks, even if some ends do not have polystyrene (PSt) blocks. The process is reproducible, as demonstrated in Figure 7.6.

7.3.2 CHARACTERIZATION

SDIBS polymers have been characterized by a variety of methods. The average number of branches (*BR* in our earlier publications, *B* here to conform with other publications) in the SDIBS core was determined by selective link destruction (see Figure 7.7) using the following equation:[66,67]

$$B = M_{n,\,total}/M_{n,\,arms} \tag{7.1}$$

The results agreed well with the "kinetic" *B*, assuming random copolymerization and similar reactivities of the initiating vinyl sites of the *inimer*, calculated from the following equation:[67]

$$B = \left(\frac{M_n}{M_{n\,theo}}\right) - 1 \tag{7.2}$$

where

$$M_{n\,theo} = \frac{IB\ (g)}{IM\ (moles)} \tag{7.3}$$

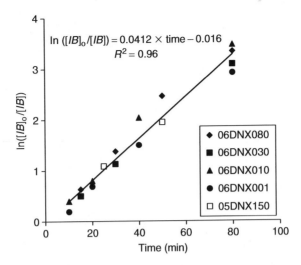

FIGURE 7.6 Reproducibility of the synthesis of dendritic polyisobutylene capped with polystyrene blocks (SDIBS) on a pound scale. The various symbols represent different experiments (details will be published elsewhere).

B was also obtained from equilibrium-swelling data of selected block copolymers; the results showed excellent agreement with the other methods (Equations 7.1 and 7.2).[7]

Size exclusion chromatograph (SEC) analysis of a series of SDIBS showed that the correlation between the radii of gyration measured from the angle dependence of light scattering and M_w of linear PIBs and SDIBS was nearly identical.[71,72] Their hydrodynamic radii (R_h) from SEC/viscometry (VIS), on the other hand, were smaller than the R_h of linear PIBs. The T_g of DIBs having branch MW higher than the critical entanglement MW was higher than that of linear PIBs, and was found to correlate with both the total M_n and B up to $B \sim 15$, and described with a modified Fox–Flory equation shown in Figure 7.8 and Equation 7.4.

$$T_g = T_g(\infty) - \frac{1}{B} \cdot \left(\frac{A}{M_n} \right) \tag{7.4}$$

Investigation of the linear viscoelastic properties of SDIBS with branch MWs exceeding the critical entanglement MW of PIB (about ~7000 g/mol[69]) revealed that both the viscosity and the length of the entanglement plateau scaled with B rather than with the length of the branches, a distinctively different behavior than that of star-branched PIBs.[76] However, the magnitude of the plateau modulus and the temperature dependence of the terminal zone shift factors were found to

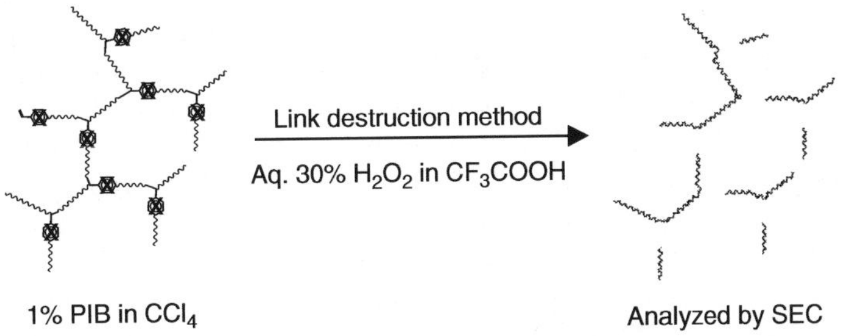

FIGURE 7.7 Selective link destruction.

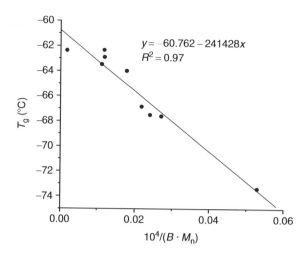

FIGURE 7.8 Correlation between M_n and branching. (From Puskas, J.E. et al., *J. Polym. Sci. A*, 44, 1770, 2006. With permission.)

be the same for linear, star-branched, and SDIBS. The nonlinear shear rheology of SDIBSs and linear PIBs was similar.[68,73] At lower shear rates, SDIBSs were less shear thinning, while at high shear rates they displayed a less-elastic response leading to lower die swell. The nonlinear viscoelastic properties of SDIBS in solution were more sensitive to branch length than to B,[70] in contrast to the linear regime. SDIBS had higher Young moduli than comparable MW linear PIBs, and the moduli varied with both MW and MW distribution (MWD).[68,73] In the case of SDIBS with narrow MWD ($M_w/M_n = 1.2$–1.3), moduli and tensile strength scaled with the MW. The SDIBS displayed interesting DMTA responses,[1,7,19] similar to those reported for near-critical gels and other arborescent structures.[77–80] SDIBSs synthesized by both the two-step or the one-pot methods[1,7,19,81] were characterized by ^{1}H NMR, SEC, tensile test, atomic force microscopy (AFM), photoelectron spectroscopy (XPS), and dynamic mechanical thermal analysis (DMTA). M_n values obtained assuming 100% recovery on the SEC columns and those computed using dn/dc values from NMR composition data showed good agreement. The block copolymers with ~10–30 wt% PSt and a wide variety of MW ($M_n \sim 10^4$–10^6 g mol^{-1}) showed surface morphologies ranging from spherical to cylindrical/lamellar, nanometer-sized discreet PSt phases dispersed in a continuous PIB matrix with a 10 nm PIB layer on the surface, and good mechanical properties (4–10 MPa and exceptionally high—1000%–1800%—elongation). DMTA revealed distinctive differences between linear triblock SIBS and arborescent SDIBS. For this latter, E' was independent of frequency in a broad range (rubbery plateau zone, 10^{-1}–10^1 Hz) and E'' was either constant or decreased at intermediate frequencies, typical behavior of vulcanized rubbers and microgels or cross-linking systems at the critical gel point. Similar decrease in loss modulus with frequency at intermediate level was reported by Roland et al.[69,70,76] for star-branched and DIB, by Hempenius et al.[79] for arborescent polystyrenes, by Pryke et al.[82] for star 1,2-PBDs, and by McLeish et al.[83] for H-shape IR. These polymers showed different relaxation mechanisms due to their branched structures. Relaxation at high frequencies is associated with the motions of outer branches, while relaxation at low frequencies is associated with the constrained segments of the polymer. The "dip" in loss modulus in the intermediate frequency range was attributed to restrictions in the motion of polymer chains by entanglements. In contrast, both E' and E'' increased with frequency for a linear triblock SIBS. The solvent swelling behavior of SDIBS was also unusual—selected blocks absorbed almost 300% hexane relative to their own weight,[7] which may be significant in terms of drug loading.

Our work with SDIBS demonstrated that TPEs based on amorphous plastic–rubber–plastic blocks do not necessarily require narrow MWD for good phase separation and mechanical properties; some of these SDIBS blocks exhibited $MWD > 2$, and irregular phase morphology. This disagrees with earlier "conventional wisdom"[19] and opens new avenues in TPE research.

Polyisobutylene-based TPEs are among the most-promising new biomaterials. Linear SIBS was confirmed to be biocompatible and infection-free in ultra-long-term endoluminal or vascular device applications.[84,85] The oxidation and acid-hydrolysis resistance of a PIB-PS was compared with other well-known implant materials. The physical properties of these TPEs are positioned between polyurethanes and silicone rubber, with tensile strength about equivalent to the former when it is well hydrated in the body. The physical properties of SIBS did not change after boiling it in concentrated nitric acid for 30 min. In contrast, polyurethanes were destroyed, and silicone rubber became brittle and retained only 10% of its tensile strength under the same test conditions. The excellent biostability of SIBS was further confirmed by six-month and two-year implantation/explantation and scanning electron microscope (SEM) analysis of an SIBS porous membrane attached to an Elgiloy braided-wire stent. After explantation, no cracked microfibers were found in the membranes. Also, no encrustation and inflammation were found in conjunction with the implantation of the SIBS sample.[84,85]

Using the hysteresis method adopted for soft biomaterials, the dynamic fatigue properties of SIBS30 were found to be between polyurethane and silicone rubber, with twice as long fatigue life as silicone. Under Single Load Testing (SLT, 1.25 MPa), SIBS30 displayed less than half the dynamic creep compared to silicone, both in air and *in vitro* (37°C, simulated body fluid).[86–88] Preliminary results[89] demonstrated that SDIBS has superior fatigue properties, on account of the branched structure of the PIB core. Hemolysis and 30- and 180-day implantation studies using a breast implant silicone shell as reference material revealed excellent biocompatibility of SIBS.[86,90–92] The biocompatibility of both SIBS and SDIBS most likely is related to the fact that their surface is covered with a thin (10 nm) pure polyisobutylene layer, on account of the lower surface energy of the rubber phase. This was suggested first by Krausch et al. for block copolymers with PBD rubber segments.[93] Subsequently, XPS was used to verify this for SIBS30.7. Thus, only polyisobutylene is in direct contact with biological tissues, while the polystyrene phases are buried within the material, providing reinforcement of the rubber phase. Drug release profiles of paclitaxel from selected arborescent blocks were found to be similar to that measured from Translute.[1]

7.4 LATEST DEVELOPMENTS

The unique properties of SDIBS are due to the branched structure of the DIB core, and consequently the "double-network" structure in which a covalent network is embedded into a self-assembling thermolabile network, as shown in Figure 7.9.

Our latest efforts have been to concentrate on investigating the architecture of the DIB core. As discussed before, the average number of branches per molecule (B) determined by selective link destruction and equilibrium swelling showed good agreement with the kinetic B (Equation 7.2). However, branching analysis by SEC proved to be a challenge.

7.4.1 SEC ANALYSIS OF THE ARCHITECTURE OF THE DIB CORE

In a seminal and seemingly forgotten paper, Burchard et al.[94] discussed the analysis of various polymer architectures based on integrated light scattering (LS) and quasielastic light scattering (QELS). They considered mono- and polydisperse linear and star-branched polymers with f number of arms ("rays"), and "random polycondensates" of A_f or ABC type (identical or different

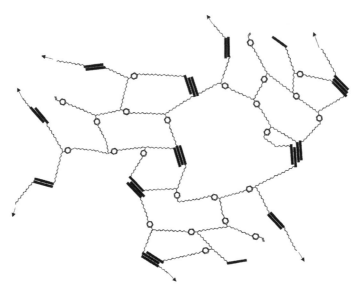

FIGURE 7.9 Architecture of dendritic polyisobutylene capped with polystyrene blocks (SDIBS).

functional groups respectively). Dendrimers and other dendritic (hyperbranched, arborescent, etc.) structures have not been considered at that time. The comparison of these structures is shown in Figure 7.10.[95–97]

The major difference between these structures is the regularity, the number of branches B, and the branch length. The authors considered data from LS and QELS, respectively, that can be obtained without reference to any architecture: (a) LS: the weight-average molecular weight M_w, the mean square radius of gyration $\langle R_g^2 \rangle$, and the particle-scattering function $P_z(q)$ where $q = (4\pi/\lambda)\sin(\theta/2)$ and (b) QELS: the z-average of the translational diffusion constant D_z, $\lim_{q \to 0} d(\Gamma/q^2)/dq^2$ and $\Gamma/q^2 = f(q)$, where $\Gamma = -[d \ln S(q,t)/dt]_{t=0}$ is the first cumulant of the dynamic structure factor $S(q,t)$. From this they identified four parameters, g, h, ρ, and C, which can give us insight into polymer architectures.

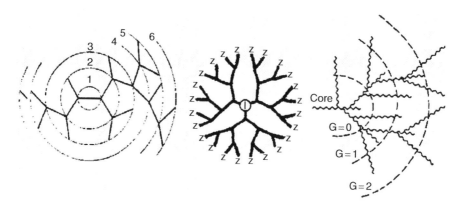

FIGURE 7.10 Comparison of dendritic architectures. (From Flory, P.J., *J. Am. Chem. Soc.*, 63, 3083, 1941; Tomalia, D.A. et al., *Macromolecules*, 191, 2466, 1986; Gauthier, M. and Möller, M., *Macromolecules*, 24, 4548, 1991. With permission.)

The branching ratio for "geometric" dimensions (g_z) was defined by Zimm and Stockmayer[98] as

$$g = \left(\frac{\left\langle R_g^2 \right\rangle_{br}}{\left\langle R_g^2 \right\rangle_{lin}} \right)_{M_Z} \tag{7.5}$$

where $\langle R^2 \rangle_{br}$ and $\langle R^2 \rangle_{lin}$ are the mean square radii of branched and linear polymers of the same z-average MW. Bruchard et al.[94] computed the ratios at the same weight-average MW (M_w), because this is directly measured by light scattering. It was pointed out that the difference between g_z (g at the same z average MW) and g_w can be substantial for highly polydisperse samples. The hydrodynamic branching ratio was subsequently defined by Stockmayer and Fickmann[99] as

$$h = \left(\frac{D_{lin}}{D_{br}} \right)_{M_W} = \left(\frac{R_{hbr}}{R_{hlin}} \right)_{M_W} \tag{7.6}$$

at the same weight-average MW. Burchard et al.[94] wrote

It is commonly presumed that g and h are continuously decreasing functions of the branching density and always smaller than unity. This statement is certainly correct for monodisperse samples and also holds for regular stars. In the other cases, however, g and h are larger than unity for low branching densities, and in the case of randomly branched polycondensates or randomly cross-linked molecules they even increase with the number of functional groups, in contrast to the decrease for star molecules. The reason for this behavior is that polydispersity causes a larger increase of the z-average mean-square radius of gyration than the corresponding increase of the weight-average molecular weight. Thus g and h embody two effects with converse behavior: polydispersity, which causes an increase, and branching, which causes the familiar decrease. (See Figure 7.11, where f is the number of branches or "rays.") (From Burchard, W., Schmidt, M., and Stockmayer, W.H. Macromolecules, 13, 1265, 1980. With permission.)

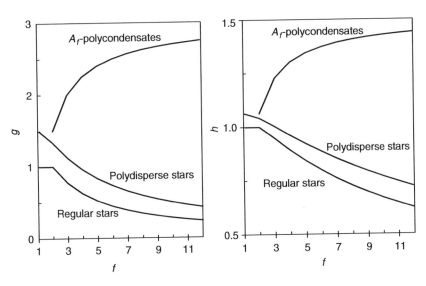

FIGURE 7.11 Dependence of g and h on f. (From Burchard, W., Schmidt, M., and Stockmayer, W.H., *Macromolecules*, 13, 1265, 1980. With permission.)

The equations lead to identical solutions for stars and ABC "polycondensates" for $B = (f-1)/2$.

ρ (Equation 7.7) was derived from the effective z-average Stokes–Einstein hydrodynamic radius (Equation 7.8).

$$\rho = \left(\frac{\left\langle R_g^2 \right\rangle_z^{1/2}}{\left\langle R_h \right\rangle_z} \right) \tag{7.7}$$

$$D_z = \frac{k_B \cdot T}{6 \cdot \pi \cdot \eta_0} \left\langle \frac{1}{R_h} \right\rangle_z \tag{7.8}$$

where ρ is a function of branching, polydispersity, and branch flexibility, but independent of bond angles and the degree of polymerization. This parameter was marked as the most reliable and accurate to give information on polymer architecture. The relationship between ρ and branching for various architectures is shown in Figure 7.12.[94]

C, the fourth parameter, represents the relationship between the first cumulant and the particle-scattering factor. For values of $1/P(q) < 10^2$, the double logarithmic plot of the first cumulant against the reciprocal particle-scattering factor yields a straight line, and the exponent v is related to the initial slope C of $\Gamma/q^2 D$, against u^2 by the equation

$$\Gamma/q^2 D_z = [P_z(q)]^{-v} \tag{7.9a}$$

with

$$v = 3C \tag{7.9b}$$

where C can give additional information on the architecture, provided the polymers are large enough.

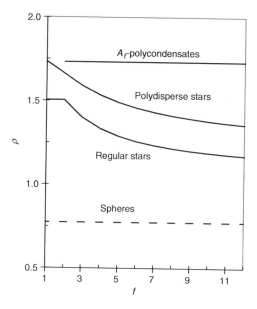

FIGURE 7.12 Dependence of ρ on f. (From Burchard, W., Schmidt, M., and Stockmayer, W.H., *Macromolecules*, 13, 1265, 1980. With permission.)

TABLE 7.2
Examples of Polymer Architecture for Various Polymers

Polymer	$M_w \times 10^3$ (g/mol)	g	h	ρ	C	Architecture
—	—	0.77	0.94	1.4	0.158	Monodisperse three-arm star
PVAc	<1000	—	—	1.84	0.2	Randomly branched
	~1000	—	—	1.7	—	—
	>1500	—	—	0.55	—	Sphere
Amylopectin	—	—	—	0.89	0.4	Not ABC

Source: Adapted from Burchard, W., Schmidt, M., Stockmayer, W.H., *Macromolecules*, 13, 1265, 1980. With permission.

Thus, g, h, ρ, and C are parameters that can give us insight into polymer architectures (Table 7.2).[94] The authors found striking similarities between the scattering behavior of linear chains with the most-probable distribution ($M_w/M_n \sim 2$) and A_f-type "polycondensates." They explained this with the most-probable distribution of the branch lengths in these latter architectures. This does not hold for monodisperse linear and star molecules and ABC polycondensates. The formulas for polydisperse stars and "ABC polycondensates" give identical results with $B = (f-1)/2$ where B is the average number of branching points in the randomly branched structures, except that f is constant across the distribution in stars while B is increasing with MW in the randomly branched polymers.

For example, the parameters $g = 0.77$, $h = 0.94$, $\rho = 1.4$, and $C = 0.158$ measured for a polymer sample and compared with the plots in Figures 7.11 through 7.13 were most consistent with a three-arm star monodisperse polymer; a polydisperse three-arm star would have $g = 1.12$, $h = 1.05$, $\rho = 1.6$, and C close to 0.2.[100] The second example was poly(vinyl acetate) (PVAc) prepared by emulsion polymerization. Since no data for linear equivalent were available, g and h were not calculated. At lower conversion/MW $\rho = 1.84$ was found, only slightly higher than the theoretically expected $\rho = 1.73$ for a randomly branched architecture. ρ slightly decreased with increasing M_w, indicating

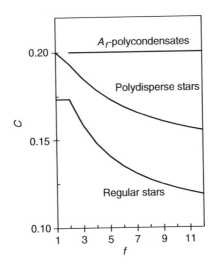

FIGURE 7.13 Coefficient C in the initial slope of $\Gamma = q^2 D_z(1 + Cu^2 - \cdots)$ for three branching models. (From Burchard, W., Schmidt, M., and Stockmayer, W.H., *Macromolecules*, 13, 1265, 1980. With permission.)

that branching was not completely random. Above $M_w \sim 15 \times 10^6$ g/mol, ρ dropped sharply to 0.55, consistent with a more compact spherical shape. The authors also pointed out that $g > 2$ and $h > 1$ would clearly indicate random branching. The third example was amylopectin, which was believed to be an ABC polycondensate. Again, in the absence of data for equivalent linear analogs, g and h could not be calculated. The measured $C = 0.4$ (Equation 9.9b) was consistent with this model, but the measured $\rho = 0.89$ was not, as ρ can never be below 1.22 for this architecture (see Figure 7.12). Enzymatic degradation studies confirmed this conclusion.[101] However, the authors did not propose a specific architecture for amylopectin. In general, they stated that g and h can decrease with M_w in ABC polycondensates. The authors pointed out that for structure analysis by light scattering at least two of the four parameters discussed above should be measured, preferably ρ and h. They also emphasized that the only direct way to prove the architecture of a polymer is to chemically "disassemble" its building blocks.

Various PIB architectures with aromatic "links" are ideal model polymers for branching analysis, since they can be "disassembled" by selective link destruction (see Figure 7.7). For example, a monodisperse star would yield linear PIB arms of nearly equal MW, while polydisperse stars will yield linear arms with a polydispersity similar to the original star. Both a monodisperse and polydisperse randomly branched structure would yield linear PIB with the most-probable distribution of $M_w/M_n = 2$, provided the branches have the most-probable distribution. Indeed, this is what we found after selective link destruction of various DIBs with narrow and broad distribution.[7,66] Recently we synthesized various PIB architectures for branching analysis.

7.4.2 ANALYSIS OF BRANCHING BY SEC COMBINED WITH LS, QELS, AND VIS

We assembled a new SEC system equipped with multiangle LS, QELS, and VIS to analyze branching in our dendritic (arborescent or hyperbranched) DIB and SDIBS samples. With this system, we can conveniently obtain z-average g, h, and ρ data, corresponding to the batch data of Stockmayer et al.[94] R_h obtained for linear PSt and PIB standards by VIS and QELS agreed, with the former being more precise and reproducible (details will be published elsewhere). In addition, we can compute ρ across the whole distribution using the ASTRA software. The radius method computes g from Equation 7.5, where the radii of gyration are obtained from the angle dependence of light scattering. From the measured g values, f for star polymers or B for randomly branched polymers with "three-functional" branch points (T-junctions) is then computed by ASTRA using Equation 7.10 or Equation 7.11 for monodisperse or polydisperse slices, respectively.

$$g = \left[\left(1 + \frac{B}{7}\right)^{1/2} + \frac{4B}{9\pi} \right]^{-1/2} \tag{7.10}$$

$$g = \frac{6}{B} \left\{ \frac{1}{2} \left(\frac{2+B}{B}\right)^{1/2} \ln \left[\frac{(2+B)^{1/2} + B^{1/2}}{(2+B)^{1/2} - B^{1/2}}\right] - 1 \right\} \tag{7.11}$$

where $R_{g,z}$ of the branched DIB samples was measured, and g was then calculated using R_g data generated with linear PIB standards, taking the actual R_g measured for the nearly monodisperse slice at the same MW as that of the M_w of the branched polymer. The z-average value of R_h was obtained the same way, using viscometry (Viscostar) online, and ρ was calculated as $R_{g,z}/R_{h,z}$. Table 7.3 lists data for representative samples (detailed information on the syntheses and branching analysis will be published elsewhere).

TABLE 7.3

Size Exclusion Chromatography (SEC) Analysis of Various Polyisobutylene PIB Architectures

Sample	$M_n \times 10^3$ (g/mol)	M_w/M_n	η_w (mL/g)	R_{hz} (nm)	R_{gz} (nm)	g	h	ρ	B Kin.[a]	B LD[b]
SH061307-4	63.2	1.27	38.2	8.7	8.8	0.74	1.04	1.01	—	—
EF141205–6	97	3.40	58.9	21.9	28.7	0.93	1	1.30	0	5.3
06-DNX040-1	57	1.53	34.3	9.9	16.3	5.40	1.16	1.65	2.8	—
05-DNX190-3	260	2.12	96.3	27.9	34.5	1.32	1.21	1.24	2.6	2.8

[a] Kinetic B (Equation 7.2).

[b] Link destruction (Equation 7.1).

The g, h, and ρ values of the first sample are consistent with a polydisperse star in Figures 7.11 through 7.13. This sample was prepared with a six-functional initiator.

The second and third samples represent *inimer*-type polymerizations. Figure 7.14 shows a typical SEC trace, with populations of distinctively different hydrodynamic volumes.

In the second sample, 06-DNX040-1, $g > 2$ and $h > 1$ clearly indicate random branching. The $\rho = 1.65$, close to the theoretical $\rho = 1.73$ predicted for this architecture, supports this conclusion. Figure 7.15 shows that B steadily increases with MW, providing further evidence.

In the third sample, 05-DNX190-3, $g > 1$ and $h > 1$, and $\rho = 1.25$. These values indicate that in this case branching was not completely random. B computed with the radius method increased with MW to around $B = 15$. The ultimate chemical test, link destruction yielded an average $B = 2.8$, in good agreement with the kinetic B calculated for this sample. The linear chains after link destruction

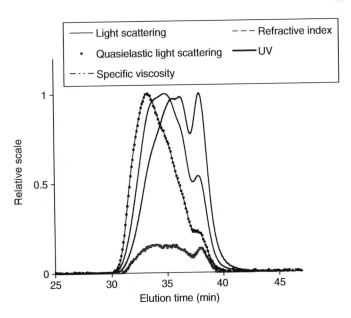

FIGURE 7.14 Size exclusion chromatography (SEC) trace of a dendritic polyisobutylene (DIB) (05-DNX-190-3 in Table 7.3).

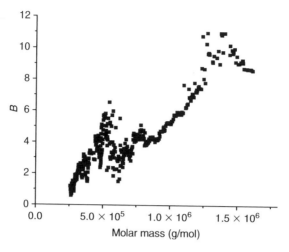

FIGURE 7.15 B versus molecular weight (MW) of 05-DNX190-3.

had $M_w/M_n = 1.6$, somewhat narrower than the most-probable distribution. This also indicates more living character and not completely random branching, similarly to the g, h, and ρ values.

In summary, our new SEC system provides very useful information about branching. Together with selective link destruction, it is a very powerful tool in the analysis of polymer architectures. Further analysis of various model architectures is in progress in our laboratories.

7.4.3 The Chameleon TPE

Among the many unusual properties that the arborescent architecture leads to, most notable is the discovery that block copolymers with a high MW dendritic (arborescent) polyisobutylene core and poly(*para*-methylstyrene) end blocks can manifest themselves either as a rubber, or as a plastic, depending on their environment (Figures 7.16 and 7.17).[75] The behavior is thermally irreversible.

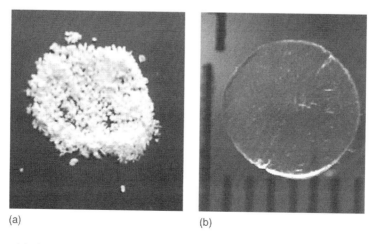

(a)   (b)

FIGURE 7.16 polyisobutylene (PIB)-P*p*MeSt block copolymers precipitated into acetone (a) and methanol (b). (Adapted from Puskas, J.E., Dos Santos, L., and Kaszas, G., *J. Polym. Sci. Chem. A.*, 44, 6494, 2006. With permission.)

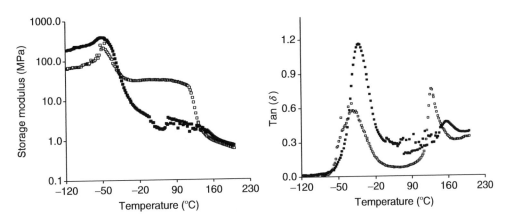

FIGURE 7.17 Storage modulus and loss factor—temperature plots of the "chameleon" *arb*PIB-*b*-P(*p*-MeSt) block copolymer. ■ = precipitated into methanol, □ = precipitated into acetone. (From Puskas, J.E., Dos Santos, L., and Kaszas, G., *J. Polym. Sci. Chem. A.*, 44, 6494, 2006. With permission.)

We called this material the "chameleon TPE." This material represents a new concept in material science: entropy-driven TPE (ETPE).

Biocompatibility is a key factor in the optimum performance of synthetic biomaterials.[14–18] Surface-induced adverse reactions such as thrombosis, calcification/encrustation, and capsule formation remain the most important factors that hinder the successful use of polymers in biomedicine. The surface characteristics of polymeric materials, such as wettability, hydrophilicity/hydrophobicity ratio, bulk chemistry, surface charge and charge distribution, surface roughness and rigidity, have a profound influence on the adsorption and desorption of blood proteins or adhesion and proliferation of different types of mammalian cells. Therefore, it is very important to investigate the surface properties of new biomaterials.[102–106] Surface coatings such as phospholipids or heparin may offer thromboresistance. Phospholipid such as phosphorylcholine, a normal constituent of cell membranes, may orient polar head groups toward the aqueous phase and locally organize water molecules. These surfaces may minimize protein adsorption and improve biocompatibility. Glycosaminoglycan (heparin), a naturally occurring anticoagulant, can suppress thrombus development, and inhibit arterial smooth-muscle cell proliferation. Many researchers have studied immobilization of heparin and its analogs onto polymeric surfaces, but these were rather unstable. Immobilization of fibronectin, laminin, collagen, and peptides containing arginine–glycine–asparagine acid (RGD) sequences on poly(ethylene terephthalate), poly(tetrafluoroethylene), and poly(carbonate urethane) surfaces has been shown to encourage endothelial cell adhesion and promote tissue growth.[17,18]

Surface functional groups that influence cell adhesion can be grouped into oxygen-containing groups such as hydroxyl–CH_2OH, as well as nitrogen-containing groups such as amide–$CONH_2$, and amine–CH_2NH_2. In particular, nitrogen-containing groups, specifically amide groups, were optimal for cell adhesion and adsorption to specific adhesion proteins. Amide and hydroxyl plasma coatings have also been proposed to improve endothelial cell adhesion onto PTFE vascular grafts.

We are actively working on surface modification of SIBS and SDIBS biomaterials to further improve its biocompatibility. We have seen that *E. coli* adhesion in saline was greatly reduced on first-generation SIBS as compared to medical-grade silicone.[107–110] In comparative investigations, both SIBS and medical-grade silicone were coated with a recombinant form of a 29 kDa protein (p29) that can prevent bacterial infection. The coating significantly reduced the attachment of two common uropathogenic bacteria. Unfortunately, in a urine environment the coating

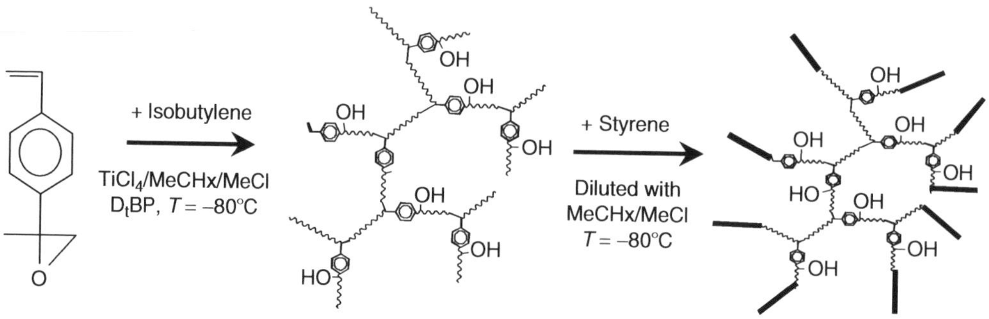

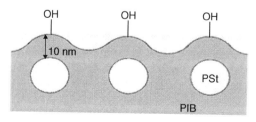

FIGURE 7.18 Synthetic strategy of direct surface functionalization.

was desorbed. This highlighted the necessity of covalently bonding the bioactive groups to the surface.

Our discovery that epoxides can initiate carbocationic polymerization[111–116] led to the effective direct functionalization of PIBs with hydroxyl groups. Figure 7.18 shows our novel method of direct surface functionalization of SDIBSs using 4-(1,2-oxirane-isopropyl)-styrene, a new *inimer*.

SDIBS-OH is expected to self-assemble into a nanostructured TPE whose PIB surface is decorated with –OH groups, while the reinforcing hard PSt phases are embedded in the continuous dendritic PIB matrix. Figure 7.19 helps to visualize the phase morphology.

We have successfully synthesized an inventory of polymers with this method.[118] Table 7.4 summarizes viscosity and size data for select samples. The hydrodynamic radii (R_h) measured by VIS and QELS agree within experimental error.

The radii of gyration (R_g) are larger than R_h for the first and third samples. The M_n of the first sample is lower than that of the third, while their viscosities are nearly identical but R_g is much larger for the more-branched third sample. The R_g values are higher than the corresponding linear PIBs, in agreement with our recent results and Stockmayer's prediction that $R_g > R_h$ and should increase with B for dendritic samples with broad MWD.[94]

Preliminary data are consistent with the presence of –OH groups on the surface of SDIBS.[118] These polar surface groups can also be used to reversibly hydrogen bond drugs onto the surface, gaining control over subsequent drug release profiles.

In summary, SIBS and SDIBS materials have a bright future as biomaterials. Due to their unique combination of properties, they also hold promise as specialty TPEs.

FIGURE 7.19 Self-assembly of *arb*IBS-OH. (From Puskas, J.E., *Poly. Adv. Technol.*, 7, 1, 2006. With permission.)

TABLE 7.4

Characteristic Data of Selected Dendritic Polyisobutylene Capped with Polystyrene Blocks (SDIBS)–Hydroxide (OH) Samples

M_n (g/mol)	η (mL/g)	R_h (nm)		R_g (nm) MALLS
		VIS	QELS	
94,600	56.4	16.2	16.8	19.1
62,800	34.7	10.2	12.8	11.6
102,500	56.8	35.2	32.1	45.0

ACKNOWLEDGMENT

The contribution of L. Dos Santos and E. Foreman to this chapter is acknowledged.

REFERENCES

1. Puskas, J.E. Biomacromolecular engineering: Design, synthesis and characterization. One-pot synthesis of block copolymers of arborescent polyisobutylene and polystyrene, *Polym. Adv. Technol.*, 7, 1, 2006.
2. Ivan, B. and Kennedy, J.P. *Designed Polymers by Carbocationic Marcromolecular Engineering: Theory and Practice*, Hanser Publishers, Munich, 1991.
3. Matyjaszewski, K. and Sawamoto, M. Controlled/living carbocationic polymerization, *Plast. Eng.(NY)*, 35, 265, 1996.
4. Malmstrom, E.E. and Hawker, C.J. Macromolecular engineering via "living" free-radical polymerizations, *Macromol. Chem. Phys.*, 199, 923, 1998.
5. Weidisch, R. et al. Tetrafunctional multigraft copolymers as novel thermoplastic elastomers, *Macromolecules*, 34, 6333, 2001.
6. Puskas, J.E., Antony, P., Paulo, C., Kwon, J., Kovar, M., Norton, P., and Altstädt, V. Macromolecular engineering via carbocationic polymerization: Branched and hyperbranched structures, block copolymers and nanostructures, *Macromol. Mater. Eng.*, 286, 565–582, 2001.
7. Puskas, J.E. et al. Synthesis and characterization of novel dendritic (arborescent) polyisobutylene–polystyrene thermoplastic elastomers, *J. Polym. Sci. A*, 43, 1811, 2005.
8. Papadopoulos, P. et al. Nanodomain-induced chain folding poly(gamma-benzyl-L-glutamate)-b-polyglycine diblock copolymers, *Biomacromolecules*, 6, 2352, 2005.
9. Matyjaszewski, K. Macromolecular engineering: From rational design through precise macromolecular synthesis and processing to targeted macroscopic material properties, *Prog. Polym. Sci.*, 30, 858, 2005.
10. Percec, V. et al. Synthesis and retrostructural analysis of libraries of AB3 and constitutional isomeric AB2 phenyl propyl ether-based supramolecular dendrimers, *J. Am. Chem. Soc.*, 128, 3324, 2005.
11. Percec, V. et al. Principles of self-assembly of helical pores from dendritic dipeptides, *Proc. Natl. Acad. Sci. USA.*, 103, 2518, 2006.
12. Tian, L. and Hammond, P.T. Comb-dendritic block copolymers as tree-shaped macromolecular amphiphiles for nanoparticle self-assembly, *Chem. Mater.*, 18, 3976, 2006.
13. Newkome, G.R., Kotta, K.K., and Moorefield, C.N. Design, synthesis, and characterization of conifer-shaped dendritic architectures, *Chem.—Eur. J.*, 12, 3726, 2006.
14. Ratner, B.D., Hoffmann, A.S., Schoen, F.J., and Lemons, J.E., Eds., *Biomaterials Science: An Introduction to Materials in Medicine*, Academic Press, New York, 1996.
15. Wintermantel, E. and Ha, S.-W. *Medizintechnik mit biocompatiblen Werkstoffen und Verfahren*, 3. Auflage, Springer-Verlag, Berlin, 2002.
16. Elices, M. *Structural Biological Materials: Design and Structure–Property Relationships*, Elsevier, New York, 2000.

17. Puskas, J.E. and Chen, Y. Biomedical application of commercial polymers and novel polyisobutylene-based thermoplastic elastomers for soft tissue replacement, *Biomacromolecules*, 5, 1141, 2004; *GAK-GV*, 7, 455; 8, 526, 2004 (German).

18. Puskas, J.E. et al. Polyisobutylene-based biomaterials, *J. Polym. Sci. Chem.*, 42, 3091, 2004.

19. Kennedy, J.P. and Puskas, J.E. Thermoplastic elastomers by carbocationic polymerization, in *Thermoplastic Elastomers*, 3rd ed., Holden, G., Kricheldorf, H.R., and Quirk, R., Eds., Hanser Publishers, Munich, 285–321, 2004.

20. Puskas, J.E. and Kaszas, G. Polyisobutylene-based thermoplastic elastomers: A review, *Rubber Chem. Technol.*, 69, 462, 1996.

21. Puskas, J.E., Antony, P., Fray, M., and Altstädt, V. Effect of hard and soft segment composition on the mechanical properties of polyisobutylene–polystyrene thermoplastic elastomeric block copolymers, *Eur. Polym. J.*, 39, 2041–2049, 2003.

22. St. Lawrence, S., Shinozaki, D.M., Puskas, J.E., Gerchcovich, M., and Myler, U. Micro-mechanical testing of polyisobutylene–polystyrene block-type thermoplastic elastomers, *Rubber Chem. Technol.*, 74, 601–613, 2001.

23. Antony, P., Puskas, J.E., and Kontopoulou, M. The Rheological and Mechanical Properties of Blends Based on Polystyrene–Polyisobutylene–Polystyrene Triblock Copolymer and Polystyrene. *Proceedings of MODEST, International Symposium on Polymer Modification, Degradation and Stabilization*, Budapest, Hungary, 2002.

24. Antony, P., Puskas, J.E., and Kontopoulou, M. The Rheological and Mechanical Properties of Blends Based on Polystyrene–Polyisobutylene–Polystyrene Triblock Copolymer and Polystyrene. *Proceedings of the Polymer Processing Society Meeting*, May 21–24, Montreal, Canada, 2001.

25. Neagu, C., Puskas, J.E., Singh, M.A., and Natansohn, A. Domain sizes and interface thickness determination for styrene–isobutylene block copolymer systems using solid-state NMR spectroscopy, *Macromolecules*, 33, 5976–5981, 2000.

26. Puskas, J.E., Pattern, W.E., Wetmore, P.M., and Krukonis, A. Synthesis and characterization of novel six-arm star polyisobutylene–polystyrene block copolymers, *Rubber Chem. Technol.*, 72, 559–568, 1999.

27. Puskas, J.E., Wetmore, P.M., and Krukonis, A. Supercritical fluid fractionation of polyisobutylene–polystyrene block copolymers, *Polym. Prepr.*, 40, 1037–1038, 1999.

28. Puskas, J.E., Pattern, W.E., Wetmore, P.M., and Krukonis, A. Multiarm-star polyisobutylene–polystyrene thermoplastic elastomers from a novel multifunctional initiator, *Polym. Mater. Sci. Eng.*, 82, 42–43, 1999.

29. Brister, L.B., Puskas, J.E., and Tzaras, E. Star-branched PIB/poly(p-t-bu-Styrene) block copolymers from a novel epoxide initiator, *Polym. Prepr.*, 40, 141–142, 1999.

30. Chattopadhyay, S., Kwon, Y., Naskar, A.K., Bhowmick, A.K., and Puskas, J.E. Novel Dendritic (Arborescent) Polyisobutylene–Polystyrene Thermoplastic Elastomers. Paper #27, *ACS Rubber Division, 162th Technical Meeting*, October 8–11, Pittsburgh, PA, 2002.

31. Puskas, J.E. and Chattopadhyay, S. Novel Polyisobutylene-Based Thermoplastic Elastomers. *Proceedings of Rubcon* 2002, January 9–20, Chaki, T.K. and Khastgir, D., Eds. India, 2002.

32. Kwon, Y., Antony, P., Paulo, C., and Puskas, J.E. Arborescent polyisobutylene–polystyrene block copolymers—a new class of thermoplastic elastomers, *Polym. Prepr.*, 43, 266–267, 2002.

33. US FDA. Taxus Express 2 Paclitaxel-Eluting Coronary Stent System (Monorail and Over the Wire), PO30025, 2004.

34. Ranade, S.V., Richard, R.E., and Helmus, M.N. Styrenic block copolymers for biomaterial and drug delivery applications, *Acta Biomater.*, 1, 137, 2005.

35. Ranade, S.V. et al. Physical characterization of controlled release of paclitaxel from the TAXUS Express 2 drug-eluting stent, *J. Biomed. Mater. Res.*, 71A, 625, 2004.

36. Puskas, J.E. Novel butyl composite for less-lethal ammunition, *Kautsch. Gummi Kunstst.*, 58, 288, 2005.

37. Charles, A. et al. Penetrating bean bag injury: Intrathoracic complication of a nonlethal weapon, *J. Trauma*, 53, 997, 2002.

38. Mahajna, A. et al. Blunt and penetrating injuries caused by rubber bullets during the Israeli–Arab conflict in October 2000: A retrospective study, *Lancet*, 359, 1795–1800, 2002.

39. Atkinson, P.J., Ewers, B.J., and Haut, R.C. Blunt injuries to the patellofemoral joint resulting from transarticular loading are influenced by impactor energy and mass, *J. Biomech. Eng.*, 123, 293, 2001.

40. Cooper, G.J. et al. The biomechanical response of the thorax to nonpenetrating impact with particular reference to cardiac injuries, *J. Trauma*, 22, 994, 1982.

41. Bir, C.A., Vino, D.C., and King, A.I. *Proceedings of the Non-lethal Conference IV*, Tysons Corner, VA, 2000.
42. Bir, C.A. PhD Thesis, Wayne State University, Detroit, MI, 2000.
43. Bir, C.A. and Viano, D.C. Design and injury assessment criteria for blunt ballistic impacts, *J. Trauma*, 57, 1218, 2004.
44. Viano, D.C. et al. Ballistic impact to the forehead, zygoma, and mandible: Comparison of human and frangible dummy face biomechanics, *J. Trauma*, 56, 1305, 2004.
45. Bir, C., Viano, D., and King, A. Development of biomechanical response corridors of the thorax to blunt ballistic impacts, *J. Biomech.*, 37, 73, 2004.
46. Morton, M., Ed., *Rubber Technology*, Van Nostrand Reinhold, New York, 1987.
47. Puskas, J.E., Kumar, B., Ebied, A., and Lamperd, B. Novel Butyl Composites for Less-Lethal Ammunition. Paper #126, *ACS Rubber Division, 164th Technical Meeting*, October 14–17, Cleveland, OH, 2003.
48. Puskas, J.E., Kumar, B., Ebied, A., and Lamperd, B. Comparison of the Performance of Vulcanized Rubbers and Elastomer/TPE Composites for Specialty Applications. *Proceedings of the Polymer Processing Society*, 2004.
49. Puskas, J.E., Kumar, B., Ebied, A., Lamperd, B., Kaszas, G., Sandler, J., and Altstädt, V. Comparison of the performance of vulcanized rubbers and elastomer/TPE composites for specialty applications, *Polym. Eng. Sci.*, 45, 966–974, 2005.
50. Puskas, J.E., Ebied, A., and Lamperd, B. Less Lethal Ammunition for Peace Keeping. US Patent filed (provisional: 2003, final: 2004).
51. American Cancer Society, 2005, Available at http://www.breastcancer.org and www.fda.gov/cdrh/breastimplants
52. Pittet, B. et al. Infection in breast implants, *Lancet Infect. Dis.*, 5, 94, 2005.
53. Cronin, T. et al. Augmentation mammaplasty: A new "natural feel" prosthesis, *Transactions 3rd International Congress Plastic. Reconstruction Surgery*, Broadband, T.R., Ed., Excerpta Medica Foundation, Amsterdam, 1963, 41.
54. Garrido, L. and Young, V.L. Analysis of periprosthetic capsular tissue from women with silicone breast implants by magic-angle spinning NMR, *Magn. Res. Med.*, 42, 436, 1999.
55. Calrson, S. et al. *Siloxane Polymers*, Prentice-Hall, Englewood Cliffs, NJ, 1993.
56. Dumitriu, S. et al. *Polymeric Biomaterials*, Marcel Dekker, New York, 2002.
57. Ajmal, N. et al. The effectiveness of sodium 2-mercaptoethane sulfonate (Mesna) in reducing capsular formation around implants in a rabbit model, *Plast. Reconstr. Surg.*, 112, 1455, 2003.
58. Fagrell, D. et al. Capsular contracture around saline-filled fine textured and smooth mammary implants: A prospective 7.5 year follow-up, *Plast. Reconstr. Surg.*, 108, 2108, 2001.
59. Siggelkow, W. et al. *In vitro* analysis of modified surfaces of silicone breast implants, *Intl. J. Artif. Org.*, 27, 1100, 2004.
60. Peters, W. et al. Calcification properties of saline-filled breast implants, *Plast. Reconstr. Surg.*, 107, 356, 2001.
61. Vacanti, F.X. PHEMA as a fibrous capsule-resistant breast prosthesis, *Plast. Reconstr. Surg.*, 113, 949, 2004
62. Draft Guidance for Industry and FDA Staff: *Saline, Silicone Gel and Alternative Breast Implants*. Washington DC, January 13, 2004.
63. Gherardini, G. et al. Trilucent breast implants, voluntary removal following the medical device agency recommendation: Report on 115 consecutive patients, *Plast. Reconstr. Surg.*, 113, 1024, 2004.
64. Piza-Katzer, H. et al. Long-term results of MISTI gold breast implants: A retrospective study, *Plast. Reconstr. Surg.*, 110, 1455, 2002.
65. Nagy, A., Orszagh, I., and Kennedy, J.P. Living carbocationic copolymerizations. II. Reactivity ratios and microstructures of isobutylene/p-methylstyrene copolymers, *J. Phys. Org. Chem.*, 8, 273, 1995.
66. Puskas, J.E. and Paulo, C. Synthesis and Characterization of Hyperbranched Polyisobutylenes. *Proceedings of the World Polymer Congress* (IUPAC Macro 2000), 384, 2000.
67. Paulo, C. and Puskas, J.E. Synthesis of hyperbranched polyisobutylenes by inimer-type living polymerization. I. Investigation of the effect of reaction conditions, *Macromolecules*, 34, 734, 2001.
68. Puskas, J.E., Paulo, C., and Altstädt, V. Mechanical and Viscoelastic Characterization of Hyperbranched Polyisobutylenes. Paper #76, *ACS Rubber Division,160th Technical Meeting*, October 16–19, Cleveland, OH, 2001.

69. Roland, C.M. et al. Linear viscoelastic properties of hyperbranched polyisobutylene, *J. Rheol.*, 45, 759, 2001

70. Robertson, C.G., Roland, C.M., and Puskas, J.E. Nonlinear rheology of hyperbranched polyisobutylene, *J. Rheol.*, 46, 307, 2002.

71. Puskas, J.E. et al. Effect of the molecular weight and architecture on the size and glass transition of arborescent polyisobutylenes, *J. Polym. Sci. Chem.*, 44, 1770, 2006.

72. Puskas, J.E. et al. Comparison of molecular weight and size measurement of polyisobutylenes by SEC–MALLS and viscometry, *J. Polym. Sci. Chem.*, 44, 1777, 2006.

73. Puskas, J.E., Paulo, C., and Altstädt, V. Mechanical and viscoelastic characterization of hyperbranched polyisobutylenes, *Rubber Chem. Technol.*, 75, 853–864, 2002.

74. Sui, C., McKenna, G.B., and Puskas, J.E. Nonlinear viscoelastic response of dendritic (arborescent) polyisobutylenes in single- and reversing double-step shearing flows, *J. Rheol*, 51, 1143, 2007.

75. Puskas, J.E., Dos Santos, L., and Kaszas, G. Innovation in material science: The chameleon block copolymer, *J. Polym. Sci. Chem. A*, 44, 6494, 2006.

76. Santangelo, P.G., Roland, C.M., and Puskas, J.E. Rheology of star-branched polyisobutylene, *Macromolecules*, 32, 1972, 1999.

77. Chambon, F. and Winter, H.H. Linear viscoelasticity at the gel point of crosslinking PDMS with imbalanced stoichiometry, *J. Rheol.*, 31, 683, 1987.

78. Winter, H.H. Evolution of rheology during chemical gelation, *Prog. Colloid Polym. Sci.*, 75, 104–110, 1987.

79. Hempenius, M.A. et al. Melt rheology of arborescent graft polystyrenes, *Macromolecules*, 31, 2299, 1998.

80. Simon, P.F.W., Müller, A.H.E., and Pakula, T. Characterization of highly branched poly(methyl methacrylate) by solution viscosity and viscoelastic spectroscopy, *Macromolecules*, 34, 1677, 2001.

81. Puskas, J.E., Antony, P., and Paulo, C. US Patent 6,747,098, 2004.

82. Pryke, A. et al. Synthesis, hydrogenation, and rheology of 1,2-polybutadiene star polymers, *Macromolecules*, 35, 467, 2002.

83. McLeish, T.C.B. et al. Dynamics of entangled H-polymers: Theory, rheology, and neutron scattering, *Macromolecules*, 32, 6734, 1999.

84. Pinchuk, L. US Patent 5,741,331, 1998; US Patent 6,102,939, 2000; *Corvita Corp., now Boston Scientific Inc.* US Patent, 6,197,240, 2001.

85. Pinchuk, L. et al. Polyisobutylene-Based Thermoplastic Elastomers for Ultra Long-Term Implant Applications, *6th World Biomaterials Congress Transactions*, 2001, 1452.

86. El Fray, M., Prowans, P., Puskas, J.E., and Altstädt, V. Biocompatibility and fatigue properties of polystyrene–polyisobutylene–polystyrene, an emerging thermoplastic elastomeric biomaterial, *Biomacromolecules*, 7, 844–850, 2006.

87. El Fray, M., Puskas, J.E., Tomkins, M., and Altstädt, V. Evaluation of the Fatigue Properties of a Novel Polyisobutylene–Polystyrene Thermoplastic Elastomer in Comparison with other Rubbery Biomaterials. Paper #76, *ACS Rubber Division, 166th Technical Meeting*, October 5–8, Columbus, OH, 2004.

88. Puskas, J.E. and Chen, Y. Novel Thermoplastic Elastomers for Biomedical Applications. Paper #40, *ACS Rubber Division, 163nd Technical Meeting*, April 28–30, San Francisco, CA, 2003.

89. Tomkins, M.M.E. Thesis, Queen's University, Kingston, Ontario, Canada, 2006.

90. Puskas, J.E. Dendritic (arborescent) polyisobutylene–polystyrene block copolymers: DMTA analysis and swelling studies, *Polym. Mater. Sci. Eng.*, 91, 875–876, 2004.

91. Puskas, J.E. Biocompatibility studies of novel dendritic polyisobutylene-based block copolymers, *Polym. Prepr.*, 45, 412–413, 2004.

92. Chen, Y. and Puskas, J.E. Polyisobutylene-Based Polymers as Prospective Implants. *Proceedings of the Meeting of the Canadian Biomaterials Society*, Toronto, Canada, 2002, 138.

93. Knoll, A., Magerle, R., and Krausch, G. Tapping made atomic force microscopy on polymers: Where is the true simple surface? *Macromolecules*, 34, 4159, 2001.

94. Burchard, W., Schmidt, M., and Stockmayer, W.H. Information on polydispersity and branching from combined quasielastic and integrated scattering, *Macromolecules*, 13, 1265, 1980.

95. Flory, P.J. Molecular size distribution in three-dimensional polymers I. Gelation, *J. Am. Chem. Soc.*, 63, 3083, 1941.

96. Tomalia, D.A. et al. Dendritic macromolecules—Synthesis of starburst dendrimers, *Macromolecules*, 19, 2466–2468, 1986.

97. Gauthier, M. and Möller, M. Uniform highly branched polymers by anionic grafting: Arborescent graft polymers, *Macromolecules*, 24, 4548, 1991.

98. Zimm, B.H. and Stockmayer, W.H.J. The dimensions of chain molecules containing branches and rings, *Chem. Phys.*, 17, 1301, 1949.

99. Stockmayer, W.H. and Fickmann, M. Dilute solutions of branched polymers, *Ann. N.Y. Acad. Sci.*, 57, 334, 1953.

100. Whitney, R.S. and Burchard, W. Molecular size and gel formation in branched poly(methyl methacrylate) copolymers, *Makromol. Chem.*, 181, 869, 1980.

101. Yu, L.P. and Rollings, J.E.J. Low-angle laser light scattering–aqueous size exclusion chromatography of polysaccharides: Molecular weight distribution and polymer branching determination, *Appl. Polym. Sci.*, 33, 1909, 1987.

102. Antony, P., Kwon, Y., Puskas, J.E., Kovar, M., and Norton, P.R. Atomic force microscopic studies of polystyrene–polyisobutylene block copolymers, *Eur. Polym, J.*, 40, 149–157, 2003.

103. Puskas, J.E., Antony, P., Kwon, Y., Kovar, M., and Norton, P.R. Study of the surface morphology of polyisobutylene-based block copolymers by atomic force microscopy, *J. Macromol. Sci., Macromol. Symp.*, 183, 191–197, 2002.

104. Puskas, J.E. Study of the Surface Morphology of Polyisobutylene-Based Block Copolymers by Atomic Force Microscopy. *Proceedings of the International Conference on Progress in Disperse System*, India, 2002.

105. Puskas, J.E., Antony, P., Kovar, M., Norton, P.R., Watterson, J.D., Cadieux, P., Harbottle, R., Reid, G., and Denstedt, J.D. Study of the Surface Morphology of Polyisobutylene-Based Block Copolymers by Atomic Force Microscopy. *4th International Conference on Advanced Polymers via Macromolecular Engineering (APME 2001)*, August 18–23, Gatlinburg, TN, 2001.

106. Antony, P., Puskas, J.E., Ott, H., Altstädt, V., Kovar, M., and Norton, P.R. Effect of Hard and Soft Segment Composition on the Morphology and Mechanical Properties of Polystyrene–Polyisobutylene Thermoplastic Elastomeric Block Copolymers. *Proceedings of the Polymer Processing Society Meeting*, May 21–24, Montreal, Canada, 2001.

107. Puskas, J.E. et al. Atomic force microscopic and encrustation studies of novel prospective polyisobutylene-based thermoplastic elastomeric biomaterials, *Polym. Adv. Technol.*, 14, 763, 2003.

108. Cadieux, P., Watterson, J.D., Harbottle, R., Howard, J., Puskas, J.E., Denstedt, J., Gan, B.S., James, V., and Reid, G. A new biomaterial for application to the urinary tract, and potential for lactobacillus proteins to reduce the risk of device-associated infections, *J. Colloids Surf. B: Biointerfaces*, 28, 95, 2002.

109. Cadieux, P., Harbottle, R., Reid, G., and Denstedt, J.D. Study of the Surface Morphology of Polyisobutylene-Based Block Copolymer Biomaterials by Atomic Force Microscopy. *Proceedings of MODEST, International Symposium on Polymer Modification, Degradation and Stabilization*. Budapest, Hungary, 2002.

110. Watterson, J.D., Cadieux, P., Puskas, J.E., Harbottle, R., Reid, G., and Denstedt, J.D. Coating of Silicone and a Novel Polymer with Lactobacillus Fermentum Rc-14 Collagen Binding Protein Reduces Bacterial Adhesion. *J. Urol.*, 165 (Suppl.), 2001.

111. Song, J., Bodis, J., and Puskas, J.E. Direct functionalization of polyisobutylene by living initiation with alpha-methylstyrene epoxide, *J. Polym. Sci., Polym. Chem.*, 40, 1005, 2002.

112. Puskas, J.E. US Patent 6,495,647, 2002; US Patent 2,268,446, 2001.

113. Chen, Y., Puskas, J.E., and Tomkins, M. Investigation of the effect of epoxide structure on the initiation efficiency in isobutylene polymerizations initiated by epoxide/$TiCl_4$ systems, *Eur. Polym. J.*, 39, 2147–2153, 2003.

114. Puskas, J.E., Brister, L.B., Michel, A.J., Lanzendörfer, M.G., Jamieson, D., and Pattern, W.G. Novel substituted epoxide initiators for the carbocationic polymerization of isobutylene, *J. Polym. Sci.*, 38, 444–451, 2000.

115. Puskas, J.E. and Michel, A.J. New epoxy initiators for the controlled synthesis of functionalized polyisobutylenes, *Makromol. Chem., Macromol. Symp.*, 161, 141–148, 2000.

116. Michel, A., Brister, L.B., and Puskas, J.E. Novel epoxide initiators for carbocationic polymerizations, *NATO ASI Ionic Polymerizations and Related Processes*, NATO Science Series 359, Kluwer, Dordrecht, the Netherlands, 1999.

117. Foreman, E.A. and Puskas, J.E. Direct surface functionalization of novel biomaterials, *Polym. Prepr.*, 47, 45, 2006.

118. Foreman, E.A., Puskas, J.E., and Kaszas, G., Synthesis and characterization of Arborescent (hyperbranched) polyisobutylenes from the 4-(1,2-oxirane-isopropyl) styrene inimer. *J. Polym. Sci. Polym. Chem.*, 45, 5847, 2007.

8 Novel Elastomers for Biomedical Applications

Li-Qun Zhang and Rui Shi

CONTENTS

8.1   INTRODUCTION

As many tissues in the body have elastomeric properties, successful replacement or tissue engineering of these tissues will require the development of compliant biodegradable elastomeric
scaffolds that can sustain and recover from multiple deformations without irritation to the surrounding tissue. The development of elastomeric scaffolds is desirable because mechanical conditioning
regimens have been shown to promote improved tissue formation and would allow gradual stress
transfer from the degrading synthetic matrix to the newly formed tissue.[1,2] Within the last five years,
investigators have recognized the need for elastomeric biodegradable scaffolds, particularly in the
area of soft tissue (such as cardiovascular) engineering. This review will focus on some novel
biodegradable polyester elastomers, biodegradable polyurethanes (PUs), silicone rubbers with some
surface and bulk modification, and polyphosphazenes (PNs), which are promising in tissue engineering and other biomedical applications. The designation, synthesis and preparation, structure and
properties, characterization, and present or potential applications of bioelastomers are involved.

8.2   NETWORK POLYESTERS SYNTHESIZED WITH POLYATOMIC ACID
AND POLYATOMIC ALCOHOL

8.2.1   POLY(GLYCEROL SEBACATE)

Glycerol is the basic building block for lipids, while sebacic acid is the natural metabolic intermediate in ω-oxidation of medium- to long-chain fatty acids.[3–5] In addition, glycerol and copolymers containing sebacic acid have been approved for their medical applications by Food and Drug
Administration (FDA).[6] Although the poly(glycerol sebacate) (PGS) has not been used in tissue
engineering currently, their ease of synthesis, adjustable mechanical properties, good biocompatibility, and biodegradability make them excellent candidate materials for soft tissue engineering.

The copolymers which were formed by glycerol and sebacic acid were first studied by Natta and
coworkers to obtain environment-friendly plastics.[7,8] In their work, the molar ratio of glycerol to
sebacic acid was 2:3. However, the products obtained by them were highly cross-linked and
nonelastomeric.

Wang and coworkers[9] first reported the use of these monomers as a novel elastomeric material
for potential application in soft tissue engineering in 2002. The molar ratio of glycerol to sebacic
acid they used was 1:1. The equimolar amounts of the two monomers were synthesized by
polycondensation at 120°C for three days. The reaction scheme is shown in Scheme 8.1. To obtain
the elastomers, they first synthesized a prepolymer and then poured an anhydrous 1,3-dioxolane
solution of the prepolymer into a mold for curing and shaping under a high vacuum.

Recent work in the authors' laboratory investigated the effects of the molar ratio and the reactive
conditions on the elasticity, strength, biodegradation, and thermal processing abilities. The different

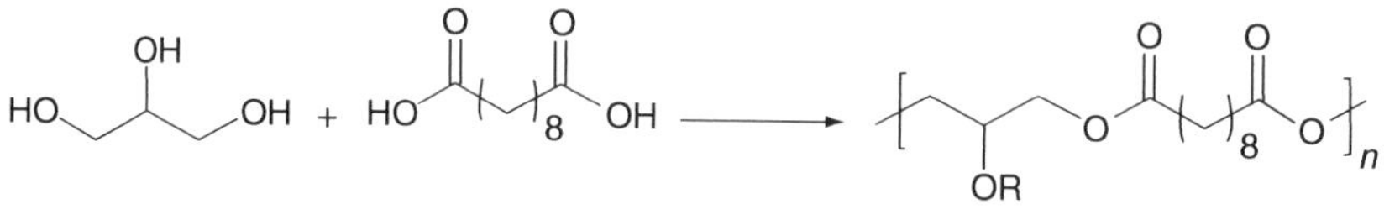

SCHEME 8.1 Synthesis of the poly(glycerol sebacate) (PGS).

molar ratios (glycerol/sebacic acid) in our work were 2:2, 2:2.5, 2:3, 2:3.5, and 2:4.[10] We also prepared the thermoplastic PGS elastomer by two-step method.[11] Firstly, non-cross-linked PGS prepolymers at the 1:1 molar ratio of glycerol to sebacic acid were synthesized through condensation reaction; secondly, the prepolymer keep on reacting after adding some sebacic acid. Results demonstrated that the mechanical properties and the degradation rates of products could be flexibly adjusted by the alteration of the molar ratio.

The PGS obtained by Wang and coworkers was a kind of thermoset elastomer with the Young's modulus of 0.282 ± 0.025 MPa, a tensile strain of at least $267 \pm 59.4\%$, and a tensile strength was at least 0.5 MPa. The mechanical properties of PGS were well consisted with that of some common soft tissues. Although PGS is a thermoset polymer, its prepolymer can be processed into various shapes by solving it in common organic solvents such as 1,3-dioxolane, tetrahydrofuran, isopropanol, ethanol, and *N,N*-dimethylformamide. Porous scaffolds can be fabricated by salt leaching.

The biocompatibility and biodegradation ability of PGS was assessed in comparison with poly (lactide-co-glycolide) (PLGA) (50:50, carboxyl ended, molecular mass = 15,000 from Boehringer Ingelheim, Ingelheim, Germany). The *in vitro* tests showed that NIH 3T3 fibroblasts, human aortic smooth muscle cells (SMCs), and human aortic endothelial cells (ECs) can all attach to PGS. The cells showed normal morphology, and have growth rates that are at least as good as those on PLGA and tissue culture polystyrene.

In vivo biocompatibility was assessed through subcutaneous implantation in Sprague–Dawley rats. PLGA was used as a control polymer. PGS and PLGA implants with the same surface area/volume ratio were implanted in dorsal subcutaneous pockets. A fibrous capsule around PGS (45 μm thick after 35 days implantation) appeared later than that around PLGA (140 μm thick after 14 days implantation). After 60 days of implantation, the implant was completely absorbed with no signs of granulation or scar formation.[9]

The *in vitro* degradation was carried out in PBS at 37°C under agitation. PGS lost $17 \pm 6\%$ of its mass, but during the same time *in vivo*, the copolymer was completely absorbed. The degradation rate *in vivo* is always quicker than that *in vitro* because of the enzymes originating from inflammatory cells or macrophages may be attacking the polymer bonds and catalyzes the degradation process. The *in vivo* experiments showed that the mechanical properties (measured as compression modulus) decreased linearly and in parallel with the degradation of PGS, suggesting surface erosion as the mechanism of degradation. Another important finding was that the shape of PGS was maintained during the degradation unlike that of PLGA. It was also proposed that PGS may perform well as a conduit material in peripheral nerve regeneration applications.[12]

Generally speaking, PGS demonstrated a favorable tissue response profile compared with PLGA, with significantly less inflammation and fibrosis and without detectable swelling during degradation. It is considered to be an excellent candidate material for neural reconstruction applications given its lack of *in vitro* Schwann cell toxicity and minimal *in vivo* tissue response.[13]

8.2.2 POLY(DIOL CITRATES)

Citric acid is a nontoxic metabolic product of the body (Krebs or citric acid cycle), and it has been approved by FDA for its use in humans. It was found that the citric acid can be reacted with a variety of hydroxyl-containing monomers at relatively mild conditions.[14] Citric acid can also participate in hydrogen bonding interactions within a polyester network. Citric acid was chosen as a multifunctional monomer to enable network formation.

Yang and coworkers did the most efforts on the development of network polyester based on citric acid.[14] They investigated the reaction of citric acid with a series of aliphatic diols (from 3–16 carbon chains) and polyether diols such as polyethylene oxide (PEO), in which 1,8-octanediol (POC) and 1,10 decanediol (PDC) have been studied the most.

The significant advantage of poly(diol citrates) when compared to existing biodegradable elastomers is that the synthesis of POC is simple and can be conducted under very mild conditions.

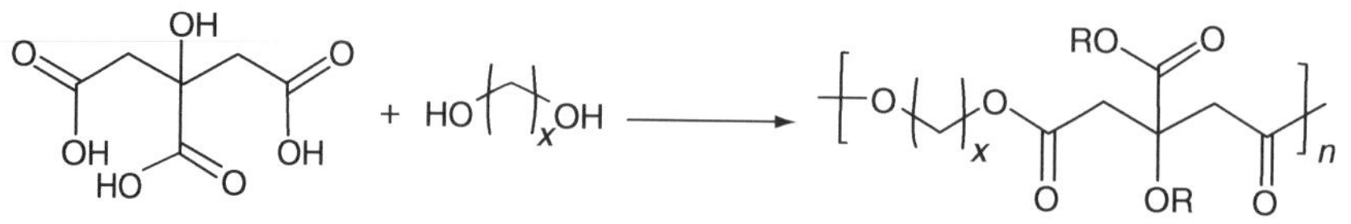

SCHEME 8.2 Synthesis of poly(diol citrates).

Poly(diol citrates) were synthesized by reacting citric acid with various diols to form a covalent cross-linked network via a polycondensation reaction without using exogenous catalysts. The reaction scheme is shown in Scheme 8.2. Firstly, a prepolymer solution is prepared by carrying out a controlled condensation reaction between 1,8-octanediol and citric acid with equimolar amounts. At first, the reaction temperature was controlled at 160°C–165°C with a continuous stir under a flow of nitrogen gas, and then it was lowered to 140°C for another hour's stir. This prepolymer solution can be dissolved in several solvents, including ethanol, acetone, dioxane, and 1,3-dioxolane, facilitating subsequent processing into various scaffold shape. The prepolymer was purified via precipitation in purified water followed by freeze-drying, and than used as for further postpolymerization or cross-linking. The prepolymer has been cross-linked at 120°C, 80°C, 60°C, and 37°C, with and without vacuum to create copolymers with various degrees of cross-linking.

The mechanical properties, degradation, and surface characteristics of poly(diol citrates) could be controlled by choosing different diols and by controlling synthesis conditions such as cross-linking temperature and time, vacuum, and initial monomer molar ratio.

The ultimate tensile strength of poly(diol citrates) was as high as 11.15 ± 2.62 MPa. Young's modulus of poly(diol citrates) ranged from 1.60 ± 0.05 to 13.98 ± 3.05 MPa under the synthesis conditions. Elongation was as high as $502 \pm 16\%$. The equilibrium water-in-air contact angles of measured poly(diol citrates) films ranged from 15°C to 53°C.[15] A glass transition temperature (T_g) was observed below 0°C ($-10°C < 0°C$), confirming that POC is totally amorphous at 37°C. Increasing cross-linking temperatures and reaction time can increase the tensile strength and Young's modulus while decreasing the elongation at break.

POC was chosen as a representative poly(diol citrates) for evaluation of the biocompatibility. *In vitro* cell culture tests showed that human aortic smooth muscle and ECs cultured in low serum media can attach, proliferate and achieve confluence without surface modification. Cell proliferation on POC was similar to that on poly-L-lactic acid films which is a material normally used in clinical practice.[16] In order to learn the *in vivo* biocompatibility of POC, subcutaneous implantation was carried out in Sprague–Dawley rats to assess inflammation and foreign body response. Based on histology evaluation, there were no detectable signs of a chronic inflammatory response. POC samples that were implanted for one month had a thin fibrous capsule surrounding the copolymer. The thickness of the fibrous capsule was approximately 50 μm and did not increase significantly with time after one month of implantation. This thickness is similar to that reported for PGS and smaller than that reported for PLGA (140 μm)[9]. Blood vessels were present throughout the fibrous capsule.

The *in vitro* degradation was tested in PBS. Enzyme-catalyzed hydrolysis was also tested in Rhizopus delemar lipase in PBS. It was reported that the degradation rates of poly(diol citrates) could also be modulated by changing the molar ratio, postpolymerization conditions, and the kinds of diols. Compared to POC with an equal molar ratio between citric acid and glycerol, excess octanediol decreased the degradation rate in PBS while producing a strong elastomeric material (tensile stress $= 2.90 \pm 0.40$ MPa, Young's modulus $= 3.98 \pm 0.09$ MPa for 1.2:1.0 octanediol/citric acid). The degradation rate of the POC can also be decreased by increasing the postpolymerization time. Varying the number of methylene units in the diol monomer can also adjust the degradation rates. Diols with decreasing number of methylene units resulted in faster degradation rates. Degradation in lipase solution was little quicker than that in PBS alone.[14,15] Recently, Yang and coworkers developed an implantable elastomeric and biodegradable biphasic tubular scaffold which was fabricated by sequential postpolymerization of concentric nonporous and porous layers of poly (diol citrates).[17] The biomimic Scaffolds were characterized for tensile and compressive properties, burst pressure, compliance, foreign body reaction, and cell distribution and differentiation. Biphasic tubular scaffold showed the potential usage to implement co-culture of SMCs and ECs *in vitro*, thereby shortening culture time and reducing the risk of contamination. The co-culture method could also potentially reduce the negative side effects associated with compliance mismatch between a graft and the host vessel for long-term implantation *in vivo*.

8.3 BIODEGRADABLE POLY(ETHER ESTER) ELASTOMER

Poly(ether ester) (PEE) copolymers were consisted of soft segments of polyethers and hard crystalline segments of polyesters. Depending on the polyether/polyester ratio, PEE copolymers exhibit a wide range of mechanical behavior combined with solvent resistance, thermal stability, and ease of melt process ability.

In all kinds of PEEs, poly(ethylene glycol)/poly(butylenes terephthalate) (PEG/PBT) have been widely researched in biomedical field. Segmented PEG/PBT multiblock copolymers are thermoplastic elastomers, which were both abbreviated as PEG/PBT or PEO/PBT in references. We use PEG/PBT in this chapter. The composition of the block copolymers is abbreviated as aPEGbPBTc, in which a is the molecular weight of the PEG used, b the weight percentage of PEG soft segments and c the weight percentage of PBT hard segments.

8.3.1 SYNTHESIS

Poly(ester ether)s are chiefly prepared by ring-opening copolymerization of cyclic acid anhydride with cyclic ether. Ring-opening polymerization of cyclic ester-ethers such as 1,4-dioane-2-one, 1,5-dioxepan-2-one, and 4-methyl-1,5-dioxepan-2-one gives alternating poly(ester ether)s. Block copolymers of polyether and polyether are prepared by two-step ring-opening living polymerization of cyclic ether and lactone, or by ring-opening polymerization of lactone initiated with polyether, or by polycondensation of telechelic polyester with telechelic polyether.[18]

8.3.2 PEEs FOR TISSUE-ENGINEERING APPLICATION

In 1972, the potential medical application of PEG/PBT was accidentally discovered by Annis et al.[19] When the PEG/PBT was implanted in animal body, they found that collagenic encystations was connected more closely with the tissue because of the PEG content, which indicated the potential medical application of the copolymer. After the concept of tissue engineering formally proposed in 1988, the degradable PEEs elastomers attracted many researchers' eyes. In the early 1990s, Fakirov and coworkers[20–30] prepared a series of PEG/PBT copolymers with different soft and hard segment content by using ester exchange method. Poly(ethylene glycol) (PEG 1000), 1,4-(butanediol), and dimethyl terephthalate were chosen as the raw material, and titanate as the catalyst in their research. The effects of the initial ratio of the raw material on the physical and chemical properties were studied. At the same time, Scholars in Holland such as Blitterswijk, Feijen, Bezemer, and others widely studied the application of PEG/PBT copolymers in bone, cartilage, and skin tissue engineering, as well as in drug delivery system. Few years later, the commercial product of PEG/PBT copolymers named PolyActive was firstly developed by HC Co. in Holland. After that the Isotis biotechnology Co. and some research organizations (like Leiden University) investigated together on developing tissue-engineering scaffold by PolyActive. The primer study of them showed that the PolyActive could be used in bone replacement, artificial periosteum, wound dressing, artificial skin, and drug control release carrier. Some items (like artificial skin) have already passed the animal experiment, and being tested in the clinical experiments. Another kind of commercial biomedical material named Politerefate has also been developed by inducing the phosphatide groups into the PBT segments. The Politerefate has been widely investigated in nerve tissue engineering and control-releasing drug delivery carrier.[31,32] The structures of PolyActive (Scheme 8.3) and Politerefate (Scheme 8.4) were as follows:

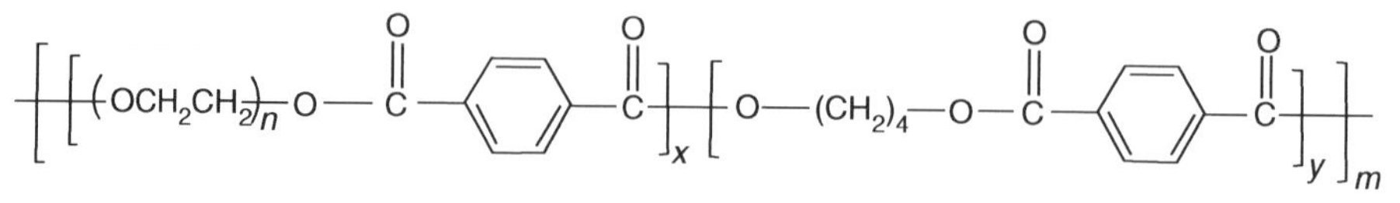

SCHEME 8.3 Structure of the PolyActive.

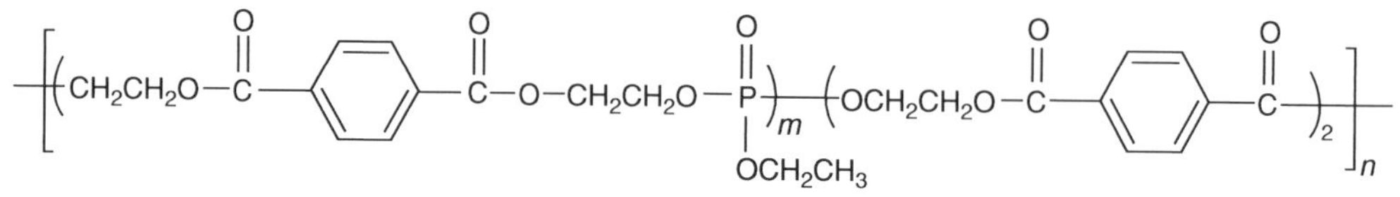

SCHEME 8.4 Structure of the Polyterefate.

8.3.3 PHYSICAL PROPERTIES

Polyactive are a series of segmented block copolymers that are composed of alternating hydrophilic PEG and hydrophobic PBT segments. A major advantage of this system is the ability to vary the amount and length of each of the two building blocks, so as to create a diverse family of customized polymers. Polymer matrix characteristics such as rate of controlled release, degradation, swelling, and strength can be precisely controlled by the appropriate combination of the two copolymer segments. The segmented copolymers are phase separated, which is more obvious for polymers with high hard segment contents and polymers containing high-molecular weight PEG. Tensile strengths were reported, which varies from 8 to 23 MPa and elongations at break changes from 500% to 1300%. Water uptake also has a wide range from 4% to as high as 210%.[33]

When keeping the PEG at a constant molecular weight of 1000, an increase in soft segment content causes a decrease in E-modulus and maximum stress. The E-modulus was reduced from over 300 MPa for 1000 $PEG_{30}PBT_{70}$ to only 40 MPa for 1000 $PEG_{72}PBT_{28}$ and the maximum stress from 23.2 to 13.9 MPa. At the same time, the elongation at break increases from 600% to 1300%. In the case of constant soft/hard segment ratios of 30–70, the E-modulus decreases only slightly when PEG 1000 is used instead of PEG 300, but the elongation at break and the maximum stress increase significantly. In case of a constant soft/hard segment ratio, the mechanical properties are considerably enhanced by a more pronounced phase separation between the soft and hard segments for the polymer prepared with the higher PEG molecular weight.[34]

The water absorbing ability of PEG/PBT copolymers is determined by the hydrophilic PEG segments. Both the PEG molecular weight and the PEG segment content influence the water uptakes greatly. The swelling properties endow the copolymers with particular properties in the body implantation. When swelling in fluid is prohibited by mechanical confinement from the surrounding tissues, the copolymers exert a swelling pressure on surrounding structures. The swelling pressure exerted by dry press-fit-implanted PEG/PBT implants is an important factor in creating a strong interface bond between PEG/PBT and bone.

8.3.4 DEGRADATION

The *in vitro* degradation of PEG/PBT copolymers occurs both by hydrolysis and oxidation, in which hydrolysis is the main degradation mechanism. In both cases, degradation is more rapid for copolymers with high PEG content.[35] Reed et al. investigated the degradation mechanics of PEG/PET copolymers whose structure is similar to that of the PEG/PBT. Although there are ether bonds, ester bonds among the PET, and among the PEG/PET segments, only the last one was easy to hydrolyze.[36] Van Blitterswijk et al.[37] deduced that the degradation mechanics of PEG/PBT was similar to that of the PEG/PET. They pointed out that the ester bonds among the PEG and PBT segments hydrolyzed at first. The degradation products were PEG segments and low-molecular PBT segments. The degradation rates were influenced by the composition, temperature, pH, enzyme content, etc. The degradation rates increased as the PEG content, temperature, and pH value increased, in which the component's content is the most important factor on the degradation rates. Copolymers with high PEG contents degraded more rapidly than those with lower PEG contents. The initial degradation is restricted to the amorphous soft PEG segment. The fact that PEG was detected alone as major degradation product proves this assumption.[38]

In vivo two degradation pathways are also expected to take place. The first route involves hydrolysis of ester bonds in the PBT part or of ester bonds connecting PEG segments and terephthalate units. Besides hydrolysis, it has been suggested that in vivo degradation of the polymers with aliphatic ether groups involves phagocyte-derived oxidative degradation of PEG by a radical mechanism initiated at random along the chain.[39] Implantation of the copolymer devices provokes a foreign body response, during which specific activated cells such as macrophages release enzymes and superoxide anion radicals, which can combine with protons to form hydroperoxide radicals. So the oxidation degradation happens.

From the results of the accelerated hydrolysis experiments of 1000 PEG71PBT29 in PBS at 100°C, the degradation products that only residue with high PEG contents. The monoester of butanediol and terephthalate were soluble at room temperature, which indicate that part of the PBT fraction might remain in the body at later stages of degradation leading to incomplete degradation of PEG/PBT copolymers in vivo on the long term. However, the crystalline PBT fragments seem well tolerated by the body.[40]

8.3.5 BIOCOMPATIBILITY

The PEG/PBT copolymers have been extensively studied for in vitro and in vivo biocompatibility. From in vitro experiments, the PEO/PBT copolymers have shown satisfactory biocompatibility.[41] This kind of copolymer is capable of inducing hydroxylcarbonate apatite (HCAp) formation from physiological solution. The calcified surface may have acted to promote HCAp growth from the solution, bringing about the formation of an HCAp layer on top of the calcified layer. It is proposed that the COOH groups produced during hydrolysis of PEG/PBT play an important role in nucleating hydroxyapatite. A remarkable affinity of the PEG segment for calcium ions may facilitate moving calcium and phosphate from the solution into the polymer for the HCAp growth.[42]

Various in vivo studies have been performed to assess the biocompatibility of PEG/PBT copolymers upon implantation in soft and hard tissues.[43–47]

Results can be concluded as that PEG/PBT copolymers did not induce any adverse affect on the surrounding tissue and thus showed a satisfactory biocompatibility. In experimental studies on rats, where deep dermal wounds were treated with cell-seeded skin substitutes consisting of PEG/PBT copolymers, no marked foreign body reaction was seen either locally or systematically. PEO/PBT copolymers as bone-filling material have shown good bone-bonding ability due to the swelling pressure and a promising calcification effect when implanted in vivo.[48–51]

However, more recent work about using the PEG/PBT as a bone substitute in critical size defects in the iliac bone of goats and humans did not show the expected good bone-bonding and calcification behaviors.[52,53] Reasons for the discrepancy with the earlier results in small animals may be caused by the differences in regenerative capacity between the species, the size of the defect, and the type of bone into which the substitute was implanted.

PEG/PBT copolymers are also very good matrix materials for the release of growth factors in tissue engineering. Proteins have been delivered from PEG/PBT microspheres with preservation of protein delivery of complete activity. In the case of protein delivery from PLGA and poly(ortho ester) microspheres, the protein activity was significantly reduced.[54]

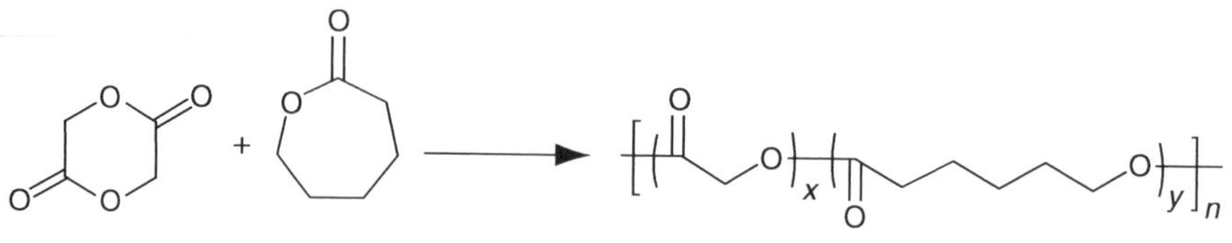

SCHEME 8.5 Synthesis of poly(glycolide-co-caprolactone) (PGCL).

Other clinical studies have been focused on the artificial tympanic membrane application,[55,56] ventilation tubes, an adhesion Barrier,[57] elastic bioactive coatings on load-bearing dental and hip implants,[58] and also for wound healing purposes.[59]

8.4 POLY(ε-CAPROLACTONE) COPOLYMERS WITH GLYCOLIDE OR LACTIDE

Poly(α-hydroxy acids), mainly polylactide (PLA) and polyglycolide (PGA) homo and copolymer are being the first choice compared with other biodegradable polymers in preparation of biomaterials for tissue repair. The osteogenic potential of the poly(α-hydroxy acids) has been widely discussed.[60] However, the stiffness and plastic deformation characteristics limited their application in soft tissue engineering. Compared to the poly(α-hydroxy acids), PCL is a relatively flexible and biodegradable biomaterial. The safety of the PCL as several medical and drug delivery devices has been approved by FDA. However, it seems that the degradation rate of PCL is too slow for most tissue-engineering applications. The total degradation of PCL may take 2–3 years. The copolymerizations of PCL with PLA or PGL can produce a novel material which will overcome the drawbacks of every single constituent, such as to provide better control over the degradation and mechanical properties without sacrificing biocompatibility. Specially, the copolymers are elastomeric which made themselves good candidate for soft tissue engineering.

The poly(glycolide-co-caprolactone) (PGCL) copolymer was mainly synthesized by the ring-opening polymerization. A copolymer with 1:1 mole ratio was synthesized by the ring-opening polymerization in the presence of the catalyst Sn(Oct)$_2$ by Lee and coworkers.[61] The polymerization was under vacuum, and heated in an oil bath at 170°C for 20 h. The copolymer was then dried under vacuum at room temperature for 72 h. The schematic reaction equations are shown in Schemes 8.5 and 8.6.

The porous PGCL scaffolds were fabricated by solvent casting and salt leaching.[61] The PGCL scaffolds produced in this fashion had an open pore structures with an average pore size of 250 μm. The mechanical properties of PGCL scaffold (1:1 mole ratio) was detected in a model system which was relevant to cardiovascular tissue engineering. The elongation at break of this scaffold was up to 250%. Recovery from extension was 98% at applied strains of 120%. Under cyclic loading test, the PGCL scaffolds remained elastic, with a permanent deformation of less than 4% of the applied strain magnitude (applied strains of 5–20%, frequencies of 0.1 and 1 Hz, six days). The low permanent deformation was also displayed for PGCL scaffolds incubated in PBS.

Kwon and coworkers[62] prepared a series of nano- to microstructured biodegradable PCLA porous fabrics by electrospinning. The nanoscale-fiber porous fabrics were electrospun with PCLA (1:1 mole ratio, approximately 0.3–1.2 mm in diameter) using 1,1,1,3,3,3-hexafluoro-2-propanol as a solvent.

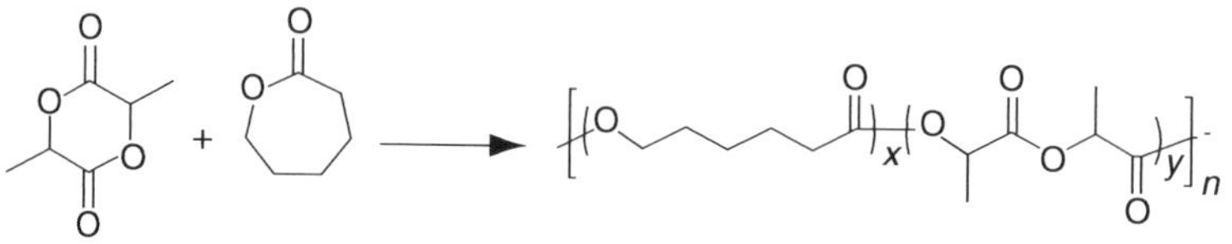

SCHEME 8.6 Synthesis of PCLA.

The decrease in the fiber diameter of fabric resulted in a decrease in porosity and pore size, but an increase in fiber density and mechanical strength. The microfiber fabric made of PCLA (1:1 mole ratio) was elastomeric with a low Young's modulus and an almost linear stress–strain relationship under the maximal stain (500%) in this measurement.

Recently, Cohn and Salomon[63] synthesized and characterized a series of PLCL thermoplastic bioelastomers by two-step synthesis procedure. First, ring-opening polymerization of L-lactide initiated by the hydroxyl terminal groups of the PCL chain. Second, chain extension polymerization of these PLA–PCL–PLA triblocks initiated by the hexamethylene diisocyanate (HDI).

It was observed that the morphology of the copolymers gradually changed as the length of the PLA blocks increased. These PLCL copolymers generated a microseparated morphology. The poly (caprolactone) amorphous chains performed as a molecular spring, while the crystalline PLA blocks functioned as strong noncovalent cross-linking domains. The mechanical properties can be adjusted by the length of the PLA and PCL segments. These thermoplastic elastomers combined remarkable strength (UTS values around 32 MPa), high flexibility (tensile moduli as low as 30 MPa), and enhanced extendibility (above 600%).

A star copolymer (SCP) of PCLA was synthesized by Younes and coworkers.[64] This kind of SCP PCLA elastomer was also synthesized in two steps. First, the small molecular SCP was produced by ring-opening polymerization of ε-caprolactone (ε-CL) with glycerol as initiator and stannous 2-ethylhexanoate as catalyst. Second, the living SCP was further reacted with different ratios of a cross-linking monomer, such as 2,2-bis(ε-CL-4-yl)-propane (BCP) and ε-CL. The SCP elastomers had very low glass transition temperature (−32°C). It was reported that the SCPs were soft and weak with physical properties similar to those of natural bioelastomers such as elastin. A logarithmic decrease in each tensile property with time was observed in this SCP PCLA.

The degradation rate of the PGCL or PCLA can vary over a wide time range by tightly controlling the ratio between each kind of the monomers. Although the degradation products were not characterized, as degradation occurs through the hydrolysis of the ester bonds, the degradation products should contain short-chain oligomers and the corresponding monomer hydroxyl acids (glycolic acid and 6-hydroxyhexanoic acid). All the copolymers degraded faster than each of the homopolymers. In Lee's work, PGCL scaffold (1:1 mole ratio) lost 3% of its initial mass after incubated in PBS for two weeks and lost 50% after six weeks.[61] The result of D. Cohn's work showed that the tensile strength of PCLA copolymers gradually dropped from their high initial levels (around 30–34 MPa) to 2–3 MPa after incubation in PBS for three months. Most of these copolymers remained as flexible elastomers for a significant period of time in PBS.[63]

The compatibility has been detected mostly by Lee and coworkers. They confirmed the ability of rat aortic SMC to attach and proliferate on the PGCL surface,[61] and then the seeded scaffolds were implanted subcutaneously in nude mice and evaluated histologically and immunohistochemically. SMCs were proliferated and differentiated immunohistochemically.

The adherence and proliferation of human umbilical vein endothelial cells (HUVECs) were evaluated on the fabricated PCLA scaffold.[62] Results showed that the HUVECs were adhered and proliferated well on the small-diameter-fiber fabrics (0.3 and 1.2 mm in diameter), whereas markedly reduced cell adhesion, restricted cell spreading, and no signs of proliferation were observed on the large-diameter-fiber fabric (7 mm in diameter). That may be due to the high-surface-density fibers provide an extremely high surface/volume ratio, which favors cell attachment and proliferation.

The PGCL and PCLA copolymers have already been researched in tissue engineering.[65] PCLA was recently used as scaffolds for tissue engineering of vascular autografts.[66,67] Autologous vascular cells or bone marrow were seeded onto PCLA (1:1 mole ratio) scaffold reinforced by PLLA or PGA mesh. In particular, bone marrow-seeded scaffolds were implanted into the inferior vena cava of adult beagles.[68] The implanted grafts functioned without occlusion or stenosis and cells displayed both endothelial and smooth muscle specific makers. The use of PCLA (1:1 mole ratio) grafts was also documented for cardiovascular tissue engineering in humans. The autologous cell-seeded scaffold was implanted into the pulmonary artery of a four-year-old girl. Postoperative examinations seven months after transplantation revealed no dilation or rupture of the grafts or complications related to the tissue-engineered autografts.

8.5 POLY(1,3-TRIMETHYLENE CARBONATE) AND THEIR COPOLYMERS WITH D,L-LACTIC ACID ANDE-CAPROLACTONE

Poly(1,3-trimethylene carbonate) (PTMC) is a rubbery and amorphous polymer.[69] It has already been investigated for the preparation of biodegradable elastomeric porous structures for soft tissue-engineering scaffolds, specifically nerve and heart tissues.[70,71] Although PTMC is rubbery by itself, it is not an ideal candidate for implanted biomedical applications, because its degradation rates are difficult to predict. For example, although high-molecular weight PTMC (M_w: 500,000–600,000 g/mol) degrades very slowly (more than two years) when tested *in vitro*,[72] it degrades within three to four weeks when implanted *in vivo* due to enzymatic activity.[73] In addition, the material also endures significant creep and not recover well from deformation unless cross-linked. In order to obtain materials with suitable mechanical properties and degradation rates, TMC was copolymerized with either D,L-lactide (DLLA) or ε-CL.[65,70,71]

8.5.1 SYNTHESIS

Ring-opening polymerization was used to synthesize PTMC. The polymerization was carried out in evacuated and sealed glass ampoules with stannous octoate as a catalyst. The schematic reaction equations are shown in Schemes 8.7 and 8.8. The reaction time for all homo- and copolymerizations were three days and the reaction temperature at $130 \pm 2°C$. The obtained polymers were purified by dissolution in chloroform and precipitation in isopropanol. The precipitated polymers were collected, and washed with fresh isopropanol, and dried under reduced pressure at room temperature until constant weight.

The cylindrical porous scaffolds based on TMC-DLLA copolymers were prepared by salt-leaching method. Two-ply porous nerve guides were prepared by a combination of dip-coating (inner layer) and fiber-winding (outer layer) techniques.

8.5.2 MECHANICAL PROPERTIES

The mechanical properties of the PTMC copolymers can be varied to a large extent by adjusting the comonomer composition. TMC-DLLA copolymers ranged from hard and brittle to rubbery and soft as the percentage of DLLA in the polymer decreases, while the TCM-CL copolymers are all flexible with higher CL content improving strength.[71] The mechanical properties of the copolymers were listed in Table 8.1. It was reported that the copolymers with 20–50 mol% TMC are suitable to be used in soft tissue engineering. The TMC-DLLA copolymers with 50 mol% are highly flexible and strong with an elongation at break of 570% and an ultimate tensile strength of 10 MPa.

In contrast to TMC-DLLA, the mechanical properties at equilibrium water uptake of TMC-CL copolymers are not significantly different from those in the dry state as shown in Table 8.1.

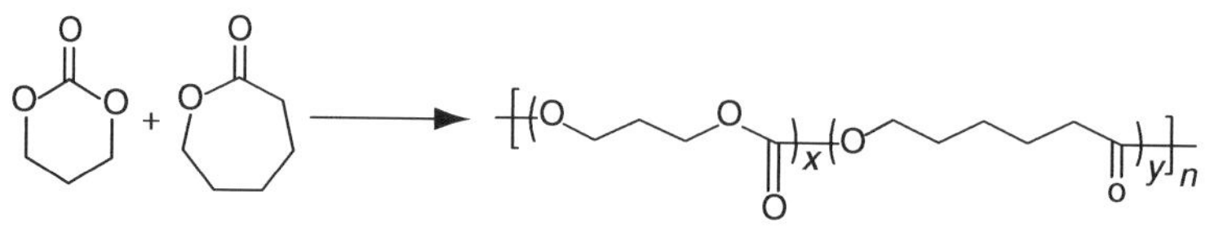

SCHEME 8.7 Synthesis of statistical poly(trimethylene carbonate-co-caprolactone).

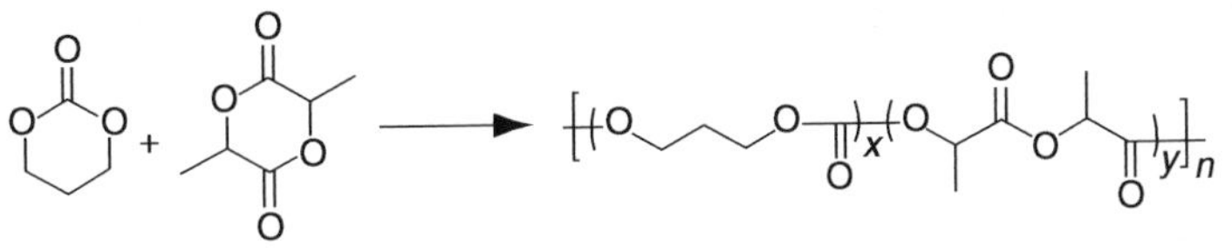

SCHEME 8.8 Synthesis of statistical poly(trimethylene carbonate-co-D,L-lactide) copolymers.

8.5.3 BIODEGRADATION AND BIOCOMPATIBILITY

In vitro degradation of TMC-DLLA was examined by incubating samples in PBS for up to two years.[74] The copolymers degraded much faster than the parent homopolymers, because the ester bonds are more labile to hydrolysis than the carbonate bonds. For D,L-PLA, a continuously decrease of the mechanical properties was observed during the period of study, which could be related to the decrease in molecular weight (8% of its initial value after 60 weeks). In the case of PTMC, it did not show any significant decrease in molecular weight, and only a relatively small deterioration of the tensile strength (53%) was observed during this 60-week study.

It was observed that the degradation speed of poly(TMC-DLLA) (50:50) was the fastest. It degraded into the small pieces only after three months. Poly(TMC-DLLA) (20:80) became brittle, but still integrates after four months, while poly(TMC-DLLA) (79:21) lost its tensile strength after five months of degradation. Although the degradation products were not examined, one would expect that hydrolysis of the ester bonds will yield short-chain oligomers and the original monomers.

The *in vitro* degradation rate of the TMC-CL-based copolymers was much slower than that of the TMC-DLLA-based copolymers. In accordance with the small decrease in molecular weight during the time of the study, no significant changes of water uptake, molecular weight distribution, or mass loss were observed for the TMC-CL copolymers. The mechanical properties of TMC-CL copolymers as a function of degradation time decreased as the molecular weight decreased. All polymers maintained good mechanical performance even after one year of degradation.[74] The *in vivo* degradation and tissue response of TMC-DLLA or TMC-CL copolymers were investigated by implanting the copolymers with 52 mol% DLLA or 89 mol% ε-CL subcutaneously in rats for up to one year.[73] The TMC-DLLA copolymer became smaller and lost 96% of its total mass over the time course of this study. Histological results showed a normal inflammatory response and foreign body reaction. However, in the later stages of the study, the TMC-DLLA copolymers showed a second inflammatory reaction triggered by the cellular removal of residual polymer.

Like the *in vitro* studies, the *in vivo* degradation of the TMC-CL copolymers containing 89% CL showed no change in shape or size after one year. As the degradation of the TMC-CL copolymers was so slow, a mature fibrous capsule remained throughout the year. Histological results of the TMC-CL copolymers were similar to those obtained from TMC-DLLA or other commonly used biodegradable polymers.

TABLE 8.1
Mechanical Properties of P(TMC-DLLA) and P(TMC-CL) Copolymers

Polymer	Tensile Strength (MPa)	Modulus (MPa)	Elongation (%)
P(TMC-DLLA) dry (50:50)	10	16	570
P(TMC-DLLA) dry (20:80)	51	1900	7
P(TMC-DLLA) wet (50:50)	11	13	900
P(TMC-DLLA) wet (20:80)	38	1100	7
P(TMC-CL) (10:90)	23	140	—

8.5.4 POTENTIAL USAGE OF TMC-BASED COPOLYMERS IN TISSUE ENGINEERING

8.5.4.1   TMC-DLLA Copolymers in Heart Tissue Engineering

The use of three-dimensional scaffolds based on copolymers of TMC and DLLA for heart tissue engineering is currently under investigation. In an animal model of the rat, the presence and the mechanical support of the scaffold need to be maintained for at least three months because the process of regeneration was shown to take up to three months. Pego and coworkers showed that the elastomeric TMC-DLLA copolymers with 20 and 50 mol% TMC were amorphous and relatively strong in physiological conditions,[73] which showed suitable mechanical properties under hydrolytic degradation *in vitro* for up to three months and totally resorb *in vivo* within 11 months.[74] Although these materials show signs of chronic inflammation in the course of implantation,[73] their mechanical and degradation properties still make them good candidates for their evaluations as scaffolds for cardiac patches.

8.5.4.2   TMC-CL Copolymers for Nerve Regeneration

The required scaffold for nerve regeneration should perform at least 12 months *in situ* without scar formation or chronic inflammation, maintain its shape and integrity, and possess adequate permeability to surrounding extracellular fluids.[75] The mechanical and degradation properties of the TMC-CL copolymers were very suitable for approaching the requirements of the nerve guide. The Schwann cells were seeded on TMC-CL copolymers of various compositions, including 100:0, 10:90, 82:18, and 0:100 of TMC/CL molar ratio by Pêgo and coworkers, while gelatin was used as a control surface in these studies.[73] These cells (human cell lines NCN61 and NCN68) were attached and proliferated on all the surfaces to various degrees. The best cell affinity and growth rate was on the TMC-rich surfaces. The TMC-CL copolymers were processed into porous tubes using two techniques to obtain a two-ply microarchitechture. The two-ply nerve guide tubes was consisted of an inner layer (pore size between 1 and 10 μm) based on poly(TMC) and an outer layer (pore size between 20 and 60 μm) based on 10 mol% TMC-CL copolymer. The *in vivo* studies are still undergoing to confirm the ability to support nerve regeneration.

8.6 POLYHYDROXYALKANOATES ELASTOMERS

The most extensively studied biodegradable elastomers for tissue engineering are polyhydroxyalkanoates (PHAs). PHAs are polyesters produced by microorganisms under unbalanced growth conditions.[76] The majority of PHAs are aliphatic polyesters composed of carbon, oxygen, and hydrogen. Their general chemical formula is shown in Schemes 8.9. The composition of the side chains or atom R and the value of x will determine together the identity of a monomer unit (poly (3-hydroxybutyrate) (PHB): R = methyl, $X = 1$; poly-4-hydroxybutyrate (P4HB) (see Scheme 8.10): R = hydrogen, $X = 2$; PHO: R = pentyl, $X = 1$). Despite the method of preparation, the structure of P4HB (Scheme 8.10) strongly resembles that of chemically derived polyesters.

Because the PHAs are biodegradable, biocompatible, and synthesized from biorenewable resources without catalyzer and other poisonous chemical substances, they obtained initial interest as a green alternative to chemical synthetic polymers. Over 150 possible monomer units offer a broad range of polymer properties that can compete with petrochemical-derived plastics such as polypropylene and polyethylene.[77] For these reasons, there is continuing effort to use metabolic engineering to produce PHAs directly in plant crops and to link the low-cost/high-volume sustainable production capability of agriculture with the vast and growing polymers and chemicals markets.

In all types of PHAs, P4HB is of the most interest because it was used in the degradable scaffold that resulted in the first successful demonstration of a tissue-engineered tri-leaflet heart valve in a sheep animal model.[78] Its copolymers with PHB and polyhydroxyoctanoate (PHO) are also promising in tissue engineering because of their nontoxic degradation products, stability in tissue culture media, and the potential to tailor the mechanical and degradation properties to match soft tissue.

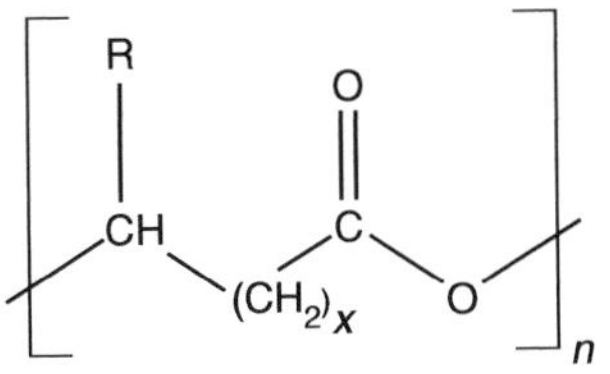

SCHEME 8.9 Chemical structure of polyhydroxyalkanoates (PHAs).

P4HB is strong yet flexible, and elongation at break for P4HB can exceed 1000%. The tensile strength (MPa) and modulus (MPa) are as high as 50 and 70,[79] which was close to the ultrahigh-molecular weight polyethylene. Elastomeric compositions of P4HB can be achieved by copolymerization of 4HB with other hydroxyacids, such as 3-hydroxybutyrate (3HB). As stated above, copolymerization of 4HB with other hydroxyl acids can further extend the range of properties. For example, higher contents of 4HB result in more elastomeric polyester. Random copolymers of 4HB and 3HB are elastomeric at approximately 20%–35% 4HB content providing materials with good strength and flexibility.

The growing interests in finding tissue-engineering solutions to the devastating worldwide problem of cardiovascular disease has prompted the attractions of PHAs in the heart valves and vessel patches tissue engineering.[80–83]

8.6.1 Synthesis and Physical Properties

PHAs are synthesized by various microorganisms as a carbon and energy storage compound during unbalanced growth. These polyesters are produced as storage granules inside cells to regulate energy metabolism. Because it is biologically produced, it does not contain residual metal catalysts that are used in the chemical synthesis of other polyesters. Over 100 different types of PHAs can be produced from a variety of monomers, which results in a wide range of physical properties from crystalline polymers to rubberlike elastomers. Their properties are modulated by the length of the side chain and by the distance between ester linkages in the polymer backbone. PHAs with short side chains tend to be hard and crystalline, whereas PHAs with long side chains are elastomeric.

As opposed to biological synthesis by fermentation, chemical synthesis allows one to more easily and predictably accommodate variations in molecular weight, comonomer type, and comonomer content. The approach is especially attractive for the obtaining of new classes of PHAs that have not yet been discovered by biosynthesizing with microbes. Ring-opening polymerization of β-lactones is the typical method as shown in Scheme 8.11. The difunctional zinc was selected as the initiator,[84] such as aluminoxanes,[85] distannoxanes,[86] and alkylzinc alkoxide.[87] However, the chemical synthesis was generally considered impossible to produce the polyester with sufficiently high-molecular weight which is necessary for most applications. Typically, ring-opening polymerization of γ-butyrolactone (GBL) produced only low-molecular weight oligomers that were viscous fluids rather than the strong and flexible plastic that is derived from the fermentation process.

Tepha, Inc. (Cambridge, MA) is currently the largest producer of P4HB for medical applications. Their product is referred to as PHA4400. Researchers at Tepha, Inc. used an *Escherichia coli* K12 microorganism by new biosynthetic pathways to produce P4HB. A proprietary transgenic fermentation process was specifically engineered to produce the homopolymer. During fermentation, P4HB accumulates inside the cells as inclusion bodies or granules.

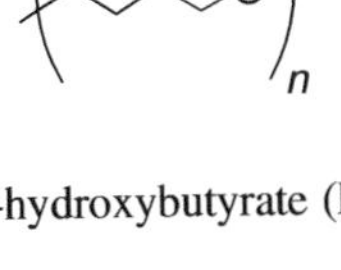

SCHEME 8.10 Chemical structure of poly-4-hydroxybutyrate (P4HB).

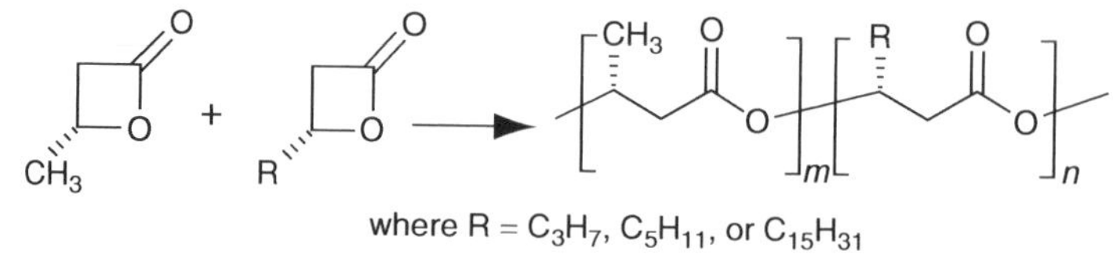

SCHEME 8.11 Ring-opening polymerization of β-lactones to produce polyhydroxyalkanoates (PHAs).

PHAs after fermentation may contain residual protein, surfactants, and high levels of endotoxin, which is a strong pyrogen. Foreign contaminating proteins may induce immune reactions, while the presence of the bacterial endotoxin lipopolysaccharide may induce fever. Endotoxins may also cause complement activation resulting in an acute inflammatory response. For medical devices, the FDA requires that endotoxin levels should not exceed 20 U.S. Pharmacopoeia (USA) endotoxin (USP) units per device.[88] Devices in contact with cerebrospinal fluid should meet more stringent specifications (less than 2.15 USP endotoxin per device). Tepha, Inc. has tackled this problem to produce PHAs with very low levels of endotoxin. They have developed methods to depyrogenate PHAs and provide materials of high purity. The procedure involves depyrogenation of the material using oxidizing agents. Choice of oxidizing agents include inorganic and organic peroxides such as hydrogen peroxide, sodium hypochlorite, and benzoyl peroxide.[88] Purification steps were also improved by using hexane or acetone as extraction solvents. Other promising method for the extraction of PHAs for medical application is the use of pure supercritical CO_2.

8.6.2 BIOCOMPATIBILITY

P4HB has been evaluated by Tepha, Inc. in a series of biocompatibility tests including cytotoxicity, sensitization, irritation and intracutaneous reactivity, hemocompatibility, endotoxin, and implantation. Besides the ubiquitous presence of endotoxin during the fermentation of P4HB, Tepha, Inc. could meet the standards set by the FDA for endotoxin (less than 20 USP EU per device). Now a large number of data indicate that P4HB has good biocompatibility *in vivo*. That is because the hydrolysis product of P4HB is mainly 4HB, which is a natural human metabolite present in the brain, heart, lung, liver, kidney, and muscle tissue.[89] This metabolite can rapidly eliminate from the body within 35 min (by the Krebs cycle). Furthermore, 4HB is less acidic than the α-hydroxy acids such as glycolic and lactic acids that are released from PGA and PLLA implants.

The Tepha, Inc. has also tested the biocompatibility of polyhydroxyoctanoates (PHOH) which was used in the first-generation PHA-based scaffolds for heart valve tissue engineering.[90] Biocompatibility was assessed by determining the degree of fibrosis around the samples and the presence of inflammatory cells. No macrophages and loose connective tissue containing three to six fibroblast cell layers rounded by collagen was observed. The sensitivity of skin to PHOH was evaluated using ASTM F270. There was no formation of erythema in the skin sensitivity test.

During degradation, no abrupt mechanical changes happens. This is potentially advantageous in applications where a sudden loss of a mechanical property is not desirable, and a gradual loss of implant mass with concomitant growth of new tissue is beneficial. Degradation in this way can also avoid large amounts of acidic degradation products from P4HB so as to avoid bad effects on cells. When the P4HB has been used as a coating on PGA mesh, this composite was utilized as a tissue-engineering heart valve scaffold in a sheep model, and complete absorption of P4HB was noted six to eight weeks after implantation.[91]

8.6.3 BIODEGRADATION

The degradation rate of P4HB is slower than that of PGA, but faster than PLLA, PCL, and some other PHAs, such as P3HB. P4HB are likely to undergo gradual changes in mechanical properties rather than the more.[92,95] A remarkable improvement to develop the heart valve was achieved by using PHO scaffold. The experimental sheep were all alive during the study, the valves showed minimal regurgitation. The tissue can be seen covered on the tissue-engineered constructs. No thrombus was observed on any of the specimens. It was concluded that the tissue-engineered heart valve scaffolds fabricated with PHO can be used for implantation in the pulmonary position with an appropriate function for 120 days in lambs.[92]

They also studied the porous P4HB scaffold as a faster degrading alternative scaffold material.[93] They found that a porous P4HB coated with PGA nonwoven mesh scaffold was more stable *in vitro* than PGA alone. After seeding with vascular cells and cell culture under dynamic flow conditions, a tissue-engineered heart valve construct was formed. The scaffold was planted in place of the native pulmonary valve. No stenosis, thrombus, or aneurysms were found during the implantation. After 8 weeks, they found that the scaffold composite in juvenile sheep had completely degraded, and 20 weeks later, a new tissue-engineered heart valve that closely resembled the native valve was formed. The size of the valve had increased from 19 mm at implant to 23 mm after 20 weeks as the lamb had grown. This discovery was a benefit to develop a valve that can grow for children, and does not need replacing.

8.6.4 BIOMEDICAL APPLICATIONS

As a typically flexible material, P4HB has been widely researched in cardiovascular, wound healing, orthopedic, drug delivery, and tissue-engineering fields.[94]

8.6.4.1 Heart Valves

In order to develop a tissue-engineered heart valve, a group at Children's Hospital in Boston evaluated several synthetic absorbable polyesters as potential scaffolding materials for heart valves. Unfortunately, the most synthetic polyesters proved to be too stiff to be function as flexible leaflets inside a tri-leaflet valve.[94] In the late 1990s, a much more flexible PHAs called poly-3-hydroxyoctanoate-co-3-hydroxyhexanoate (PHO) was used as the scaffold material for the valve leaflet, and then the entire heart valve.[95]

8.6.4.2 Blood Vessel Augmentation

It is estimated that approximately 40,000 babies are born with congenital defects which need to be repaired by using surgical patching material in the United States.[96] However, Results showed that using living tissue is better than nonviable synthetic materials such as polytetrafluoroethylene (PTFE).[97] Further more, though large-diameter blood vessels can be replaced by synthetic materials, such as DacronTM (polyethylene terephthalate) or expanded PTFE (ePTFE), these materials do not perform well when a small-diameter graft is required as the grafts rapidly occlude.[98]

Stock et al.[97] used P4HB scaffolds and tissue engineered the patch with a porosity of 95% and pore sizes in the range of 180–240 μm by salt-leaching and solvent evaporation. The sheep autologous cells (endothelial, smooth muscle, and fibroblast cells) were seeded on the scaffold before implantation. Results confirmed that the cell-seeded implants induced progressive tissue regeneration with no thrombus formation, stenosis, or dilatation.

Another example was done by Opitz et al.[99] They utilized P4HB scaffolds to produce viable ovine blood vessels, and then implanted the blood vessels in the systemic circulation of sheep. Enzymatically derived vascular smooth muscle cells (vSMC) were seeded on the scaffolds both under pulsatile flow and static conditions. Mechanical properties of bioreactor-cultured blood vessels which were obtained from tissue engineering approached those of native aorta.

They also seeded autologous vSMC and ECs obtained from ovine carotid arteries to study autologous tissue-engineering blood vessels in the descending aorta of juvenile sheep. They found that after three months' implantation, grafts were fully patent, without dilatation, occlusion, or intimal thickening. A continuous luminal EC layer was formed. However, after six months' implantation, the graft displayed remained functional, but there was significant dilatation and partial thrombus formation. Histology testing displayed layered tissue formation which was similar to native aorta.[100]

8.7   BIODEGRADABLE POLYURETHANES

After almost half a century of use in the health field, PU remains one of the most popular biomaterials for medical applications. Their segmented block copolymeric character endows them with a wide range of versatility in tailoring their physical properties, biodegradation character, and blood compatibility. The physical properties of urethanes can be varied from soft thermoplastic elastomers to hard, brittle, and highly cross-linked thermoset material.

The typical biomedical applications include blood-contact materials,[101] heart valves,[102] insulators for pacemaker electrical leads,[103] cardiovascular catheters,[104] and implants for the knee joint meniscus.[105] While traditionally, investigators were using PUs as long-term implant materials and were attempting to prevent them from the biodegradation processes, more recent work by investigators has utilized the elastomeric and diverse mechanical properties of PU materials to design degradable polymers for tissue replacements.[106,107]

Commercial medical-grade segmented PUs, such as Biomer, Elasthane, and ChronoFlex AR,[108] are typically synthesized from 4,4′-methylenebis(phenylisocyanate) (MDI). Carcinogenic and mutagenic aromatic diamines have been reported as degradation products from PUs incorporating aromatic diisocyanates; however, the question of whether the concentrations of these harmful degradation products attain physiologically relevant levels is currently unresolved and strongly debated.[109] To avoid the potential release of toxic degradation products to the extracellular matrix, it is desirable to synthesize new medical-grade PUs from less toxic intermediates.

8.7.1   SOFT SEGMENTS USED IN BIODEGRADABLE POLYURETHANES

For tissue-engineering applications, polyester macrodiols whose degradation and toxicity have already been well known are always used. Hydroxyl acid oligomers have also been used to form urethane-ureas. Because the degradation in poly(ester-urethane)s typically begins in the soft segment, the degradation rate of the soft segment determine the whole degradation rate. The hydrolyse is the most important degradation mechanism of poly(ester-urethane)s. The degradation rates highly depend on the nature of the chemical links.

There are a limited number of oligomers that have been used in biodegradable PU soft segments. They are PCL,[110–114] PLA,[115] PGA,[116] poly(hydroxybutyrate), poly(ethylene glycerol) (PEG), and PEO.[117]

Different soft segments endow different physical and mechanical properties to the PUs that contain them. For example, PEO is used to enhance degradation and PCL to provide greater hydrolytic stability and elastomeric mechanical properties. PCL usually imparts enhanced crystallinity to the PU while PEO increases hydrophilicity and water uptake. The PEO-based PUs were weak, tacky, amorphous materials, while the PCL-based PUs were relatively strong and elastomeric.[112,118] Results from *in vitro* degradation test showed that the PEO-based PUs exhibited substantial mass loss within 56 days in buffer with the formation of a porous material with little strength; conversely the PCL-containing polymers had modest mass loss and surface alteration and retained their strength.[119]

In later work, PEO and PCL PUs were blended by simple mixing to form three-dimensional sponges using solvent cast/particulate leaching methods.[120]

All blends were semicrystalline in nature and the mechanical properties were dominated by the PCL. The initial mass loss was very quickly in the degradation test which was thought to be associated with the rapid degradation of PEO.

A series of poly(ester-urethane) urea triblock copolymers have been synthesized and characterized by Wagner et al.[121] using PCL, polyethylene glycol, and 1,4 diisocyanatobutane with either lysine ethyl ester or putrescine, as the chain extender. These materials have shown the elongation at break from 325% to 560% and tensile strengths from 8 to 20 MPa. Degradation products of this kind of materials did not show any toxicity on cells.

Saad et al.[122,123] synthesized a series of biodegradable PUs named DegraPol. In this material the soft segments were composed with a block tripolymer of PCL/PEO/PCL and a crystallizable soft segment, α, ω dihyroxy-oligo [((R)-3-hydroxybutyrate-co-(R)-3-hydrocyalerate)-ethylene glycol. These materials have been extensively characterized for their use as bone substitutes. The DegraPol is elastic and forms highly porous foams from 100 to 400 mm. Osteoblasts cultured on these foams proliferate, phagocytose degradation products, retain their phenotype, and induce bone formation in a rat model.

Gogolewski et al.[124] combined Pluronic F-68(poly(ethylene-propylene-ethylene oxide) with PCL within the same soft segment. BD or 2-amino-l-butanol was used as the chain extender and HDI as the hard segment. They observed that all the materials calcified and calcification increased with material hydrophilicity. The structure and composition of the calcium crystals formed on the materials depended on the PU chemistry.

Agarwal et al.[125,126] used glycerol in combination with PEO, lysine methyl ester diisocyanate (LDI), and water to produce a group of biodegradable PU foams. The interconnected pores varied in size from 10 to 2 mm in diameter. Rabbit bone-marrow stromal cells cultured on the materials for up to 30 days formed multilayers of confluent cells and were phenotypically similar to those grown on tissue culture PS. It supported the adherence and proliferation of both bone-marrow stromal cells and chondrocytes *in vitro*. In subdermal implants the investigators found that the material showed infiltration of both vascular cells and connective tissue.

8.7.2 Diisocyanates Used in Biodegradable Polyurethanes

Diisocyanates can be either used as a major component of the hard segment or as a minor component to link two macrodiols in the synthesis of PUs. A major limitation on the type of diisocyanate that can be used in biodegradable polymers for tissue-engineering applications is the toxicity of the degradation products. As alternatives to MDI, LDI, and 1,4-diisocyanatobutane (BDI) have been used to synthesize biomedical PUs. Potential degradation products from these aliphatic diisocyanates are the amino acid lysine and the biological diamines putrescine, respectively. Some of the diisocyanates that have been in the synthesis of biodegradable PUs are listed in Figure 8.1.

Woodhouse et al.[127] designed a unique family of PUs by introducing the enzyme-sensitive linkages into the hard segment. In their study, the PUs were composed of a phenylalanine diester chain extender, LDI, and either PEO or PCL. *In vitro* degradation studies showed that upon exposure to chymotrypsin and trypsin, the polymer was more easier to depredate in enzyme-mediated when compared to the control PU. Chymotrypsin cleaved the ester bonds adjacent to the phenylalanine in the novel chain extender. Woodhouse and collaborators have electrospun the enzyme-degradable PUs to develop a cardiac patch. They coated the degradable PUs with collagen IV and incubated them with stem-cell-derived cardiomyocytes. Data showed that the cells adhere to the matrix and the stem-cell-derived cardiomyocytes extended along the same lines as the fibers of the PU mesh scaffolding.

Santerre and coworkers have designed a group of biodegradable PUs as drug polymers (Epidel) with linear aliphaticdiis ocyanates, PCL, and fluoroquinolone antimicrobial drugs as hard segment monomers.[128–130] This polymer is designed to be degraded by enzymes that are generated by inflammatory white blood cells. Once healing occurs, the level of enzyme production will decrease and thus antibiotic release would decrease. Changing diisocyanates and the distribution of drug monomers would modulate drug delivery doses.

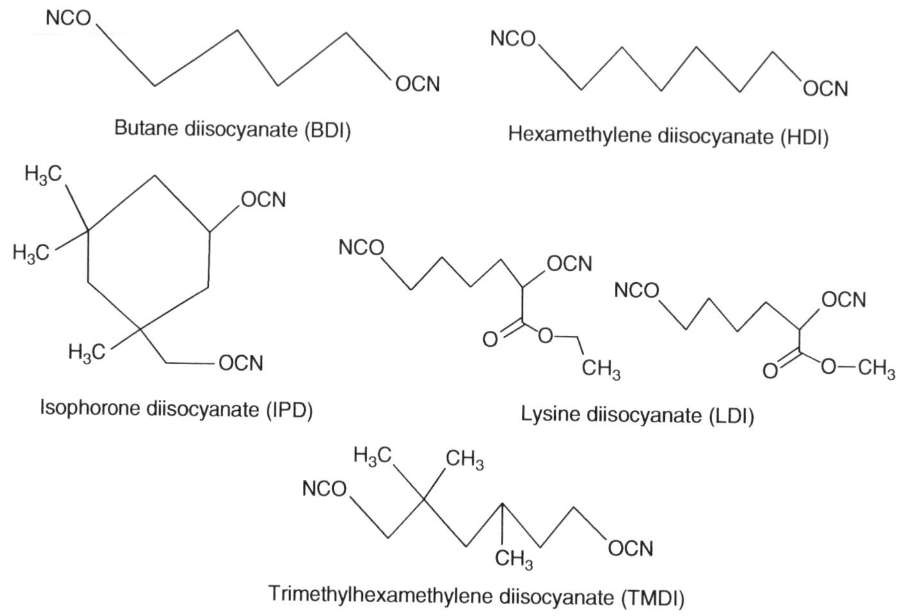

FIGURE 8.1 Diisocyanates in biodegradable polyurethanes (PUs).

8.7.3 CHAIN EXTENDERS USED IN BIODEGRADABLE POLYURETHANES

Chain extenders in thermoplastic PUs are typically low-molecular weight diols. There are several degradable chain extenders that have been previously developed which incorporate amino acids. These degradable chain extenders are diamines (see, e.g., Figure 8.2).

Besides the bulk changes of the PUs, there have been a number of studies that have introduced the use of amphiphilic copolymer additives to modify the surface chemistry of polymers while yielding no change to the chemistry of the bulk polymer. For example, poly(NVP) is known to be a highly biocompatible water-soluble polymer; it finds use in numerous cosmetic, pharmaceutical, and medical products, and has also been used as a coating for catheters. Covering a poly(urethane) surface with a coating of poly(NVP) would render the surface fully hydrated, while the physical/mechanical properties of the materials remain (largely) unaffected. 2-Methacryloyloxyethyl phosphorylcholine (MPC) polymers have been synthesized with attention to the biomembrane structure and are known as excellent blood-compatible polymers.[131,132] SPUs have been modified with various MPC polymers by coating, grafting, or blending.[133,134] They effectively reduced protein adsorption and platelet adhesion compared with the original SPU.

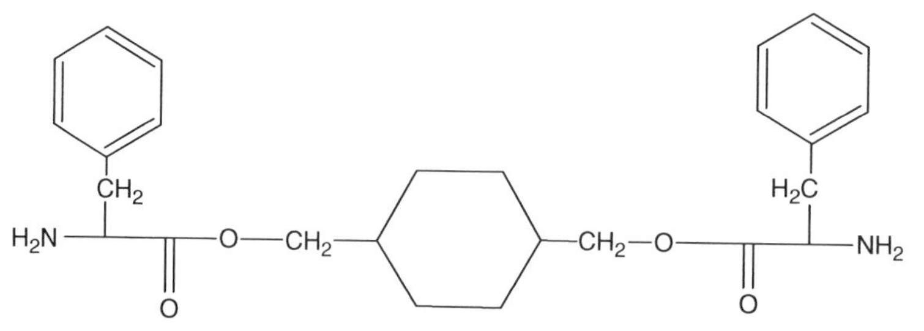

FIGURE 8.2 The structure of degradable chain extender used by Skarja and Woodhouse. (From Skarja, G.A. and Woodhouse, K.A., *J. Biomater. Sci. Polym. Ed.*, 9, 271, 1998.)

8.8 POLYPHOSPHAZENES

Polyorganophosphazenes are a class of polymers of great potential for biomedical applications and biomaterials. These polymers consist of an inorganic backbone of alternating phosphorus and nitrogen atoms bearing side-group substituents at the phosphorus atom. Interest in these polymers relies on the fact that their properties are mainly dictated by the nature of the phosphorus substituents so that polyorganophosphazenes for a wide range of applications can be obtained by the appropriate choice of the backbone derivatives.

8.8.1 SYNTHESIS

Preparation of the P–N backbone can be achieved by the ring-opening polymerization method and the condensation approaches.

In general, PNs are prepared in three stages: the synthesis of the precursor, the polymerization of this precursor leading to the formation of polydichlorophosphazene, and the substitution of the chlorine atoms in this entirely inorganic polymer by organic groups, giving a polyorganophosphazene (Figures 8.3 and 8.4).[135]

As early as 1895, the synthesis of polydichlorophosphazene was attempted by H.N. Stokes by thermal ring-opening polymerization of "hexachloro-triphosphazene" $[(NPCl_2)_3]$.[136] The product obtained by H.N. Stokes was a high-molecular weight cross-linked rubbery material called "inorganic rubber" which is insoluble in all solvents and hydrolytically decomposes into phosphates, ammonia, and hydrochloric acid in the presence of moisture. Because of its insolubility and hydrolytic instability, the polymer found no technological application and remained as a laboratory curiosity.

In 1964, H.R. Allcock filed a patent claiming the preparation of a soluble polydichlorophosphazene.[137] The polymer obtained can be soluble in organic solvents such as benzene and toluene. This thermal ring-opening polymerization of hexachloro-triphosphazene in the melt at 250°C is the most fully developed and commonly used method for polydichlorophosphazene synthesis to date.

By this method, different classes of PNs were prepared by replacing chlorine atoms of the polydichlorophosphazene intermediate by primary or secondary amines,[138] alkoxide or aryl-oxide,[139] or organ metallic reagents.[140]

A procedure which allowed polydichlorophosphazene to be obtained directly from basic chemicals: PCl_3, Cl_2, and NH_4Cl without requiring any particular precursor to be isolated was suggested by Hornbaker and Li.[141] R. De Jaeger, M. Heloui, and E. Puskaric[142] filed a patent in 1979 about a new route to polydichlorophosphazene based upon the polycondensation of N-dichlorophosphoryl-P-tri-chloromonophosphazene.

While the polymerization of hexachloro-triphosphazene results in high-molecular weight polydichlorophosphazene, with little or no molecular weight control and with broad polydispersities. Several attempts were made to synthesize polydichlorophosphazene with narrow polydispersities and controlled molecular weight,[143,144] and a new method was reported by Allcock et al. that allowed the synthesis of the macromolecular intermediate with molecular weight control and narrow polydispersities by living cationic polymerization of phosphoranimines at ambient temperature.[145] This living cationic polymerization method has also been used for the direct synthesis of PNs by the PCl_5-induced polymerization of mono- and di-substituted organophosphoranimines, such as $PhCl_2P = NSiMe_3$, at ambient temperature.[146] The presence of living sites at the polymer chain ends in living cationic polymerization offers the possibility for the formation of block copolymers either with other phosphazene monomers or with organic monomers.[147]

8.8.2 STRUCTURES AND PROPERTIES

The physical properties of polyphosphazene depend on the nature and the number of substitutes. However, the flexibility of the P–N backbone is the property in common. Because of the weakness of the rotation energy around the N–P bond (3.38 and 21.8 kJ/mol, respectively for

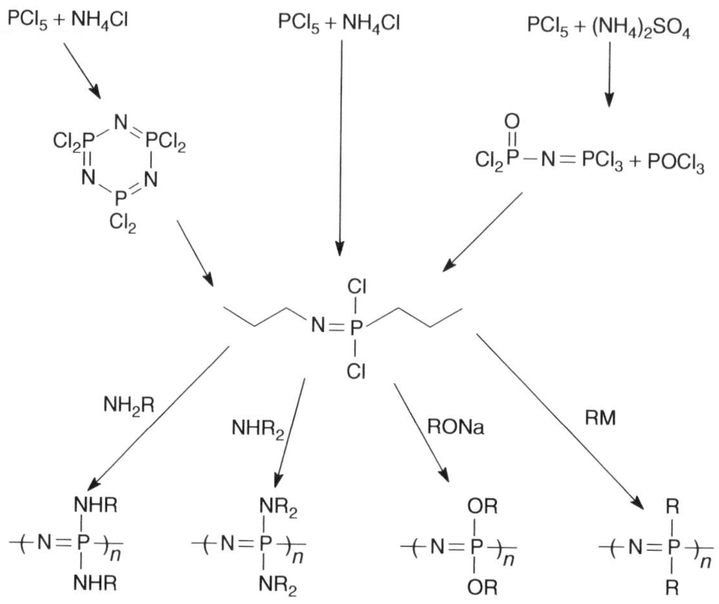

FIGURE 8.3 Access routes to polydichlorophosphazene.

polyditrifluoroethoxy and polydiphenoxyphosphazene),[148] the structure of polyphosphazene has a high degree of freedom and a low glass transition temperature. Small and unhindered substituents such as alkoxys give very low T_g, below $-60°C$ ($-105°C$ for n-butyl). Aromatic rings give more stiffness, leading to a transition temperature of between $-34°C$ and $0°C$ with one ring and between $0°C$ and $100°C$ with two rings.[149] The freedom of movement of the chains can also be limited by interatomic interactions. This is particularly true of primary amine substituted polymers which exhibit a T_g about $100°C$ higher than for polymers substituted with the corresponding alcohols. This effect is explained by segment immobilization due to hydrogen bonding. The P–N backbone flexibility can also make these polymers undergo structure changes in the solid state.[150]

The structural features necessary for applications requiring hydrophilic PNs are summarized by the following diagram:

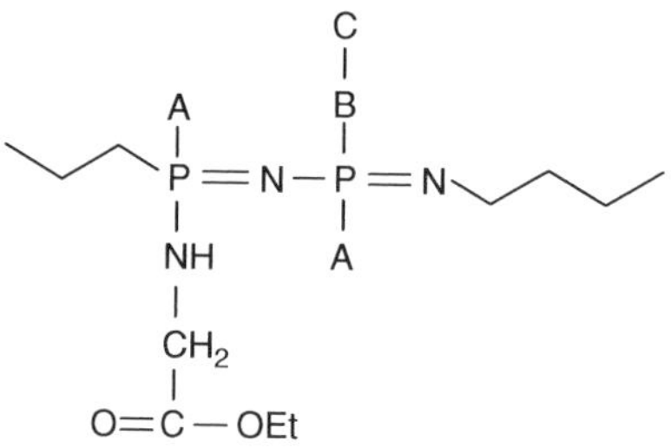

where

   A is hydrophilic

   B is bonding group

   C is active ingredient[150]

These polymers generally comprise a hydrophilic substituent in a ratio corresponding to the required hydrophilicity, and an active principle either grafted or merely trapped in the polymeric

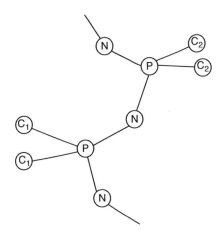

FIGURE 8.4 Structure of polyphosphazene.

network. Optionally, a substituent capable of accelerating the hydrolysis of the chain by an assistance mechanism can also be added. The most frequently mentioned hydrophilic substituent is monomethylamine. Polyethers,[151] glucosy,[152] and imidazole.[153] The presence of amino acid esters among the substituents, in particular ethyl glycinate, helps the hydrolysis of the polyphosphazene backbone.

8.8.3 BIOMEDICAL APPLICATIONS

Because a large number of different side groups can be introduced in these reactions, a wide range of properties may be generated with this polymer system. Specific side groups such as amino acid esters, glucosyl, glyceryl, glycolate, lactate, and imidazole sensitize the polymer backbone to hydrolysis.[8–13] On the other hand, hydrophobic side groups such as aryloxy, fluoroalkoxy, and C4 and higher alkoxy units protect the polymer backbone against hydrolysis. Therefore, a cosubstituted polymer, with both hydrolysis-sensitizing and hydrolysis-retarding groups offers considerable opportunities for controlling the rate of degradation through changes in the ratio of the two side groups.[154]

Among the various classes of degradable PNs, poly[(amino acid ester) phosphazenes] have met with the most success in terms of potential biomedical applications including drug delivery vehicles and tissue-engineering scaffolds.[155] Allcock and coworkers, who extensively explored the field of PNs,[156–160] reported that amino-substituted PNs are susceptible to hydrolytic degradation. They suggested that polymers substituted with amino acid esters hold promise as biodegradable materials for biomedical applications. Poly[(amino acid ester)phosphazenes] have been used for the preparation of macromolecular prodrugs as well as for the production of drug-loaded particles. The amino acidic side groups have been demonstrated to confer them biodegradability and biocompatibility.[161] Moreover, the presence of imidazole as cosubstituent could increase the hydrolytic degradation that occurs through relatively stable intermediates,[162] finally yielding low-molecular weight physiological molecules.

Poly[(di-ethylglycinate)phosphazene] as well as imidazolyl-substituted PNs[153,160] have been tested as controlled-release agents for a number of drugs, including proteins. Macromolecule release was studied using degradable p-methylphenoxy-ethyl glycinate based PNs[163] while soluble polyphosphazene substituted with carboxyphenoxy derivatives, that can be insolubilized by di- or trivalent cations, were also used for vaccine entrapment.[164] Polyfluoroalkoxy and polyaryloxyphosphazenes are well suited as prosthesis materials since they are inert in biological media, and are well tolerated as subcutaneous implants.[165] Heparin and dopamine can be fixed onto these polymers.

A great deal of well-supported evidence exists that one of the initial steps of the cellular adhesion mechanism is the deposition of a biomacromolecule film. This film may consist of polysaccharides and proteins and is deposited by a cell before adhesion. Extracellular matrix proteins such as laminin, fibronectin, tenascin, and collagen are commonly used as cell-friendly materials. Also proteoglycans, glycosaminoglycans, and cell adhesion molecules (e.g., N-CAM, cardherins, and F-spondin) have been utilized.[166] PNs form a broad class of inorganic and organic polymers with a P–N backbone and two organic side groups covalently attached to each phosphorus. The final material properties are determined, to a large degree, by the physical and chemical properties of the side-group structures. By changing the side-group composition and ratio, the opportunity exists to create a range of PNs with characteristics that vary from poor surface cell adhesion to good cell adhesion. Various micropatterning techniques can then be used to pattern regions of interest with such materials. In the work of E.W. Barrett[167] various PNs were developed to promote or prevent cellular adhesion, and both photolithography and microcontact printing were studied as candidate methods to pattern the materials. In their work, the polyphosphazene materials which showed negative cellular adhesive properties (–CAPs) were poly [bis (trifluoroethoxy) phosphazene] (TFE) and poly [bis (methoxyethoxyethoxy) phosphazene] (MEEP). The PNs which showed positive cellular adhesive properties (+CAPs) were poly [(methoxyethoxyethoxy)$_{1.0}$ (carboxylatophenoxy)$_{1.0}$ phosphazene] (PMCPP), poly [(methoxyethoxyethoxy)$_{1.0}$ (cinnamyloxy)$_{1.0}$ phosphazene] (PMCP), and poly [(methoxyethoxyethoxy)$_{1.0}$ (p-methylphenoxy)$_{1.0}$ phosphazene] (PMMP).

8.9 ADVANTAGES IN MODIFICATION OF NONDEGRADABLE SILICONE RUBBER FOR BIOMEDICAL APPLICATIONS

As a preeminent biomaterial, silicones have been the most thoroughly studied polymer over the last half century. From lubrication for syringes to replacements for soft tissue, silicones have set the standard for excellent blood compatibility, low toxicity durability, and bioinertness. Many medical applications would not have been possible without this unique polymer.

One of the most interesting attributes of silicone polymers is their bioinertness, the body's cellular and biomolecular systems cannot readily break down a material. The bioinertness of silicones can be attributed to both its hydrophobicity and chemical inertness. There are typically two ways of degradation of implanted materials: hydrolysis and enzymatic degradation. For the former, water absorptivity is a key character, and for the latter, specific chemical interactions are important. Silicones seem to possess neither. So silicone materials can be used as highly tolerated implanted materials. Other especially desirable properties such as[168] high-degree physical and chemical purity, inactivity to immunity mechanism of body, antiadhesive properties, minimum negative tissue response, inability to initiate formation of blood clots, and not to disturb normal mechanism of blood coagulation. Because of these advantages, up to now, there is no material available that completely fulfils the above requirements.

8.9.1 PROBLEMS OF SILICONE RUBBER FOR MEDICAL APPLICATIONS

Although silicone rubber has been widely used as the replacement of the soft tissue, some problems have occurred when the silicone device were implanted for a long time.

Silicone elastomers have a relatively high coefficient of friction and the resulting lack of lubricity can cause pain and tissue damage on device insertion and removal. Therefore, the tissue side of the prosthesis must be smoothed, polished, and even also lubricated to eliminate friction.

In the early time, the excellent gas permeability seemed to make them promising to be contact lenses. However, the poor hydrophilicity of silicone elastomers diminished their apparent utility.

Modification of silicones with hydrophilic polymers has proved to be a more commercially viable solution.

Approaches that utilize simple mixing of hydrophilic drugs with silicones have been used to create matrix delivery devices for antimicrobials and vaccines. Although the matrix methods have got a certain success for the drug delivery usage, it is commonly held that more delicate therapeutics do not maintain a high level of activity during processing and/or storage in the hydrophobic environments afforded by silicones.[169]

Moreover, it has been noticed that in clinical applications, devices such as catheters, which have to pass through the skin from the outside to the interior of the body. It always exposes the patient to the risk of infection. Because there are dead space between the catheter surface and the surrounding tissue, the bacteria can enter the body along the device surface in direct contact with the skin. Such dead space would not be formed if the material surface were able to bond to the skin tissue at the microscopic level.[170]

In addition, silicone polymers in both gel and elastomer forms have been shown to induce a typical subcutaneous foreign body reaction and fibro sarcoma development in rats.[171] The lower-molecular weight silicone polymers may migrate into the adjacent tissues or stimulate other cellular and immunological response.

8.9.2 Modification of Silicone Rubbers for Medical Applications

8.9.2.1 Surface Modification

The resultant tailored interface is often vastly superior for biomedical applications over the native silicone interface.[172] Furthermore, surface modification maintains the low materials cost and favorable bulk properties of the original silicone elastomer. The modification methods can be divided into physical and chemical techniques.

8.9.2.1.1 Physical Techniques

Chemical alternation of the surface layer and deposition of a new layer on top of the silicone rubber can be achieved by physical techniques. For the inert surface of silicone rubber, the former requires the generation of high-energy species, such as radicals, ions, or molecules in excited electronic states. In the latter case, coatings of atoms or atomic clusters are deposited on polymer surfaces using technique such as plasma (sputtering and plasma polymerization) or energy-induced sublimation, like thermal or electron beam-induced evaporation.

Some physical techniques can be classified into flame treatments, corona treatments, cold plasma treatments, ultraviolet (UV) treatment, laser treatments, x-ray treatments, electron-beam treatments, ion-beam treatments, and metallization and sputtering, in which corona, plasma, and laser treatments are the most commonly used methods to modify silicone polymers. In the presence of oxygen, high-energy-photon treatment induces the formation of radical sites at surfaces; these sites then react with atmospheric oxygen forming oxygenated functions.

It can be reported that low-pressure plasmas can be used for surface activation by the introduction of oxygen-containing functions. It has been claimed that plasma treatment with inert gases greatly improves fluorosiloxane-acrylate rigid gas permeable (RGP) lenses. The results showed that after several minutes of plasma treatment, the contact angle was reduced from more than 100° to about 30°. The effect of oxygen plasma treatment on the wettability and the surface composition of siloxane-acrylate lenses have been extensively investigated by Fakes and coworkers.[173,174] Plasma treatment of silicone rubbers produces highly wettable surfaces, but the effect is only temporary. A more durable hydrophilicity can be obtained by plasma-deposited thin polymeric films. Deposition of a fluorocarbon polymer on the surface of silicone rubber has been shown to be a useful method to improve the blood compatibility.

Williams and coworkers[175] utilized low-powered plasma treatment in the presence of four different gases (O_2, Ar, N_2, and NH_3) to treat with the surface of the medical-grade silicone rubber.

Besides the changes in wettability, they found that the hemocompatibility was significantly affected by plasma treatment. Treatment of PDMS with both Ar and O_2 induced a decrease in hemocompatibility, leading to shorter clotting times. The N_2 and NH_3 treatments had a significantly beneficial effect on the activation of the coagulation cascade.

Infection is the most common cause of biomaterial implant failure in modern medicine. When a covalently coupled 3-(trimethoxysilyl)-propyldimethyloctadecylammonium chloride (QAS) was coated on the surface of the silicone elastomer, a highly adhesive surface with antimicrobial effects towards adhering Gram-positive and Gram-negative bacteria was yielded.[176] Silicone rubber was oxidized by argon plasma treatment firstly, and subsequently reacted with QAS in water. This QAS-coated silicone rubber showed antimicrobial properties against adhering bacteria, both *in vitro* and *in vivo*.

Photolithography and microfabrication can be combined to manufacture well-defined, micrometer-sized, three-dimensional structures in the surface of silicone and other polymers.[177,178] Samuel Boateng and coworkers have produced a variety of micropegged and microgrooved silicone membranes with dimensions in the micrometer-size range using this method. Focused ion beams can also be used to create nanometer-sized structures in polymer surfaces. Cardiac myocytes cultured on similarly microfabricated silicone membranes displayed greater attachment and cell height than those grown on flat polymer surfaces.[179]

Laser treatment is also an energy-efficient process. After treated by the CO_2-pulsed laser beam without photo initiator, the samples showed significant differences in hydrophilicity depending on the number of the laser pulses. The friction coefficient of the PDMS decreased drastically when the surface was treated with the laser, and low platelet spreading and aggregation was observed on the laser-treated PDMS surface.[180]

8.9.2.1.2 Chemical Techniques

The chemical modification techniques refer to the treatments used to modify the chemical compositions of polymer surfaces. Those can also be divided into two categories: modification by direct chemical reaction with a given solution (wet treatment) and modification by covalent bonding of suitable macromolecular chains to the polymer surface (grafting). Among these techniques, surface grafting has been widely used to modify the surface of PDMS.

Radiation-induced graft polymerization is a highly promising method of surface modification in practically all known polymers and allows one to change the surface morphology and structure of the modifying polymers. The thickness of the grafted layer can be controlled by radiation-induced graft polymerization. The common radiation sources are cobalt 60, corona, plasma, or glow-discharge and UV or laser-energy sources.[181] The radiation-induced grafting can be divided into two catalogs: simultaneous irradiation and pre-irradiation. The former is that the radiation is carried out in the presence of monomer, while the later is that the radiation before exposure to the monomer.

Hoffman and his coworkers have done a lot of work on the application of radiation-induced graft polymerization for medical application.[182,183] The hydrophilic polymers that have been used for radiation-induced grafting are *N*-vinyl pyrolidone (NVP), 2-hydroxyethyl methacrylate (HEMA), acrylamide (AAm), acrylic acid (AAc), glycidyl methacrylate (GMA), ethyleneglycol dimethacrylate (EGDMA), and ethyl methacrylate (EMA) onto silicone rubber were widely studied.

Chitosan is the main structural component of crab and shrimp shells. Chitosan contains both reactive amino and hydroxyl groups, which can be used to chemically alter its properties under mild reaction conditions. *N*-acyl chitosans were already reported as blood-compatible materials. UV irradiation grafting technique was utilized to introduce obutyrylchitosan (OBCS) onto the grafted SR film in the presence of the photosensitive heterobifunctional cross-linking agent.[184] The platelet adhesion test revealed that films grafted on OBCS show excellent antiplatelet adhesion.

Covalent binding of peptides to polymer surfaces is now a standard method to improve their biocompatibility. The primary amino acid sequence of a peptide can be chosen to mimic the putative binding site of a protein known to participate in cellular adhesion or some other cell-signaling event. Samuel Boateng and coworkers[185] recently utilized the covalent binding of peptides to modified silicone membranes as improved flexible cell culture substrates in studies of cardiac mechanobiology. The surface of flat silicone membranes was first treated with a water plasma treatment, followed by reaction with 3-aminopropyltriethoxysilane. Their treatment leaves the silicone surface with terminal amine groups, subsequently for the attachment of cysteine-terminated peptides by a maleimide cross-linker. Cardiac fibroblasts showed enhanced adhesion to the peptide-bound silicone surface over unmodified silicone or tissue culture polystyrene (the standard polymeric cell culture material).

Ozonization-induced grafting polymerization is another efficient method to modify the surface of the silicone rubber.[186] It was reported that the ozonization was utilized to introduce active peroxide groups onto the silicone film surface at first, and then a kind of zwitterionic sulfobetaine was graft-polymerized onto the ozone-activated silicone surface. The grafted film possessed relatively hydrophilic surface revealed by contact angle measurement. In the blood compatibility text, no platelet adhesion was observed for the grafted films. Protein adsorption was greatly reduced after incubation in bovine fibrinogen for 120 min.

Chen et al.[187] utilized a direct chemical reaction with a given solution (wet treatment) to modify the surface of the silicone rubber. The presence of a layer of PEO on a biomaterial surface is accompanied by reductions in protein adsorption, and cell and bacterial adhesion. In order to obtain a PEO layer on top of the silicone rubber surface, the surface was firstly modified by incorporating an Si–H bond using (MeHSiO)n, and followed by PEO grafting to the surface using a platinum-catalyzed hydrosilylation reaction. These PEO-modified surfaces were demonstrated by fibrinogen adsorption both from buffer and plasma, as well as albumin adsorption from buffer. Reductions in protein adsorption of as much as 90% were noted on these surfaces.

8.9.2.2 Bulk Modification

The ultimate goal of bulk modification endows with the polymer-specific surface composition or a specific property for a given application.[172] The bulk modification can be classified into blending, copolymerization, interpenetrating polymer networks (IPNs), etc.

8.9.2.2.1 *Blending*

Blending of polymers provides a convenient way of combining the different properties of individual polymers. Hydrophilization of the silicone rubber can be obtained by blending silicone rubber with hydrogels. These kinds of composites combine the good mechanical properties with the hydrophilicity.

Silicone rubber–hydrogel composite is a two-phase system that is capable of swelling in water. The hydrogels prepared have different chemical compositions, size and shape of particles, and correspondingly different specific surfaces. It was found that the mechanical properties of silicone rubber–hydrogel composites depend mostly on the magnitude of the contact surface of both phases.[188]

Poly(HEMA) is one of the commonly used hydrogels based on poly(HEMA) and slightly cross-linked by multifunctional monomers. They have excellent biocompatibility and poor mechanical properties. Composites of silicone rubber and fine particles of hydrated poly(HEMA) were found to combine advantages of both excellent biocompatibility of poly(HEMA) and good mechanical properties of silicone rubber. The permeability of the composites can be varied over a broad range depending on the composition. The results of compatibility, chronic inflammatory reaction, level of calcification, and histological findings of the surrounding tissues of the silicone–hydrogel composites are as good as those of poly(HEMA).[189]

Polymer composites of rubber–hydrogel are permeable for water-soluble low-molecular weight compounds in a swollen state. Therefore, it is suitable for drug delivery carrier. Many papers have researched the basic role of composite morphology in low-molecular weight component transport through two-phase composites where only one phase played the main role.

Lednicky et al.[190] studied the structure of silicone rubber–hydrogel based on poly(HEMA) by SEM. They confirmed the suggestion that the probability of mutual contact of particles, and thus the permeability, grows with growing hydrogel concentration. The existence of aggregates decreased the permeability.

8.9.2.2.2 Copolymerization

The main drawbacks of silicone rubber include poor interfacial adhesion between the two phases and leaching out of the hydrogel particles into the biological media. For this reason, it is required to synthesize properly engineered copolymers.

PDMS-co-PS has been proposed to have the antithrombogenicity. PDMS-PEO-heparin has been synthesized to achieve better blood compatibility.[191] Silicone-PC copolymers are always used as blood oxygenation, dialysis, and microelectrode membranes.

The polyether-PDMS soft segments in segmented polyurea-urethanes have been synthesized to combine the good mechanical properties of PUs and the excellent blood-contacting properties of PDMS. The most widely recognized material in this category is Avcothane [Arkles BC, Med Device Diagn Ind 3:30(1981)], which is characterized as a block copolymer of aromatic polyether urethane and PDMS.

8.9.2.2.3 Interpenetrating Polymer Networks

IPN technique leads to the formation of two interpenetrating three-dimensional networks that are not covalently bonded. The cross-linking of at least one of the polymer systems distinguishes an IPN from a chemical blend.

Biomedical materials prepared from polysiloxane/PU IPNs have been studied, and it has been reported that these materials can be useful as steam-sterilizing medical tubing.[192] The mechanical properties showed a very wide range. Silon-TSR temporary skin, which was composed with polysiloxane/PTFE IPNs, has been proposed for assisting burn healing.[193]

The hydrophilic/hydrophobic SIN composition of PDMS with poly(HEMA) and poly(AAC) were proposed as a potential application for high-permeability soft contact lenses.[194] Other silicone-containing IPNs for contact lenses include polymerization of MMA in the presence of polymerized methacryloxypropyl trimethoxysilane, the cross-linking of a polymeric hydrogel of a copolymer of NVP during the final compression or injection-moulding process.[172]

8.9.3 New Applications in Biomedical Field

The first documented use of silicone as biomaterial was silicone rubber tubing as shunts for treatment of hydrocephalus in about 1955.[195] From 1960 to 1990, the biomedical application of silicone rubber was steadily grown. One of the most well-known silicone product is the Norplant contraceptive implant. The first clinical experiment with this device was reported in 1966. FDA approved its use in 1990.

Today, silicone medical devices are ranging from the therapeutic products (catheters shunts, adhesives, wound dressings, tubing, plugs, cuffs, contact lenses, and cardiac pacemaker leads) to those engineered to replace functional tissues (breast, joints, artificial skin, replacement esophogus, etc.). Recent research involving silicones as biomaterials has included historical applications such as wound healing, ophthalmic devices, drug delivery, and coatings. Furthermore, new application areas such as tissue engineering are the hot subjects of current study. Although this area of research is dominated by biodegradable polymers, silicone rubber's role was also under consideration. In recent study, the silicone tubing is finding use in one of the most challenging area of tissue engineering, nerve regeneration. The nerves regenerate spontaneously into highly differentiated structures leading to the practice of inserting the nerve ends into tubular structures. Tubes filled with a porous support serve as scaffolding for the growing nerve. Due to the requirement that the tubing be biocompatible and no adherent to the nerve and surrounding tissue, silicones were a very good choice. Two recent studies have shown that this type of porous tube supports axonal regrowth in a rat model.[196,197]

REFERENCES

1. Niklason Le, Gao J, and Abbott WM, et al. Function arteries grown *in vitro. Science*, 1999, 284, 489–493.
2. Yang J, Webb AR, Pickerill SJ, Hageman G, and Ameer GA. Synthesis and evaluation of poly(diol citrates) biodegradable elastomer. *Biomaterials*, 2006, 27, 1889–1898.
3. Liu G, Hinch B. and Beavis AD. *J Biol Chem*, 1996, 271, 25–38.
4. Grego, AV and Mingrone, G. Dicarboxylic acids, an alternate fuel substrate in parenteral nutrition: An update. *Clin Nutr*, 1995, 14, 143–148.
5. Mortensen PB. C6–C10-dicarboxylic aciduria in starved, fat-fed and diabetic rats receiving decanoic acid or medium-chain triacylglycerol. An *in vivo* measure of the rate of beta-oxidation of fatty acids. *Biochim Biophys Acta*, 1981, 664(2), 349–355.
6. Fu J, Fiegel J, Krauland E, and Hanes J. New polymeric carriers for controlled drug delivery following inhalation or injection [J]. *Biomaterials*, 2002, 23, 4425–4433.
7. Nagata M, Kiyotsukuri T, Ibuki H, Tsutsumi N, and Sakai W. Synthesis and enzymatic degradation of regular network aliphatic polyesters. *React Funct Polym*, 1996, 30, 165–171.
8. Nagata M, Machida T, Sakai W, and Tsutsumi N. Synthesis characterization and enzymatic degradation of network aliphatic copolyesters. *J Polym Sci A Polym Chem*, 1999, 37, 2005–2011.
9. Wang Y, Ameer GA, Sheppard BJ, and Langer R. A tough biodegradable elastomer. *Nat Biotechnol*, 2002, 20, 602–606.
10. Liu QY, Tian M, Ding T, Shi R, and Zhang LQ. Preparation and characterization of a biodegradable polyester elastomer with thermal processing abilities. *J Appl Polym Sci*, 2005, 98, 2033–2041.
11. Liu QY, Tian M, Ding T, Shi R, Feng YX, Zhang LQ, Chen DF, and Tian W. Preparation and characterization of a thermoplastic poly(glycerol-sebacate) elastomer by two-step method, 2007, 103, 1412–1419.
12. Wang Y, Kim YM, and Langer R. *In vivo* degradation characteristics of poly(glycerol sebacate). *J Biomed Mater Res*, 2003, 66A, 192–197.
13. Sundback CA, Shyu JY, Wang YD, Faquin WC, Langer RS, Vacanti JP, and Hadlock TA. Biocompatibility analysis of poly(glycerol sebacate) as a nerve guide material. *Biomaterials*, 2005, 26, 5454–5464.
14. Yang J, Webb A, and Ameer G. Novel citric acid-based biodegradable elastomers for tissue engineering. *Adv Mater*, 2004, 16, 511–516.
15. Yang J, Webb AR, Pickerill SJ, Hageman G, and Ameer GA. Synthesis and evaluation of poly(diol citrates) biodegradable elastomer. *Biomaterials*, 2006, 27, 1889–1898.
16. Wang Y, Ameer GA, Sheppard BJ, and Langer R. A tough biodegradable elastomer. *Nat Biotechnol*, 2002, 20, 602–606.
17. Yang J, Motlagh D, Webb AR, and Ameer GA. Novel biphasic elastomeric scaffold for small-diameter blood vessel. *Tissue Eng*, 2005, 11, 1876–1886.
18. Okada M. Chemical syntheses of biodegradable polymers. *Prog polym Sci*, 2002, 27, 87–133.
19. Langer R. *Biomaterials* in drug delivery and tissue engineering: One laboratory's experience. *Acc Chem Res*, 2000, 30, 94–101.
20. Fakirov S and Gogeva T. Poly(ether ester)s based on poly(butylene terephthalate) and poly(ethylene oxide)glycols, 1. *Makromol Chem*, 1990, 191, 603–614.
21. Fakirov S and Gogeva T. Poly(ether ester)s based on poly(butylene terephthalate) and poly(ethylene oxide)glycols, 2. *Makromol Chem*, 1990, 191, 615–624.
22. Fakirov S and Gogeva T. Poly(ether ester)s based on poly(butylene terephthalate) and poly(ethylene oxide)glycols, 3. *Makromol Chem*, 1990, 191, 2341–2354.
23. Fakirov S and Gogeva T. Poly(ether ester)s based on poly(butylene terephthalate) and poly(ethylene oxide)glycols, 4. *Makromol Chem*, 1990, 191, 2355–2365.
24. Fakirov S, Apostolov AA, Boeseke P, and Zachmann HG. Structure of segmented poly(ether ester)s as revealed by synchrotron radiation. *J Macromol Sd Phys*, 1990, B29(4), 379–395.
25. Fakirov S, Fakirov C, Fischer EW, and Stamm M. Deformation behaviour of poly(ether ester) thermoplastic elastomers as revealed by SAXS. *Polymer*, 1991, 32, 1173–1180.
26. Apostolov AA and Fakirov SJ. *Macromol Sci Phys*, 1992, B31(3), 329–355.
27. Fakirov S, Fakirov C, Fisher EW, et al. *Polymer*, 1992, 33(18), 3818–3827.
28. Fakirov S, Fakirov C, Fischer EW, et al. *Colloid Polym Sci*, 1993, 271, 811–823.

29. Fakirov S, Denchev Z, Apostolov AA, et al. *Colloid Polym Sci*, 1994, 272, 1363–1372.
30. Stribeck N, Sapoundjieva D, Denchev Z, et al. *Macromolecules*, 1997, 30, 1329–1339.
31. Wang S, Andrew CA, et al. A new nerve guide conduit material composed of a biodegradable poly (phosphoester). *Biomaterials*, 2001, 22, 1157–1169.
32. Zhao Z and Eong LKW. Polyphosphoestes in drug and gene delivery. *Adv Drug Deliv Rev*, 2003, 55, 483–499.
33. Deschamps AA, Claase MB, Sleijster WJ, Bruijn JD, Grijpma DW, and Feijen J. Design of segmented poly(ether ester) materials and structures for the tissue engineering of bone. *J Control Rel*, 2002, 78, 175–186.
34. Deschamps AA, Grijpma DW, and Feijen J. Poly(ethylene oxide)/poly(butylenes terephthalate) segmented block copolymers: The effect of copolymer composition on physical properties and degradation behavior. *Polymer*, 2001, 42, 9335–9345.
35. Sakkers RJB, Dalmeyer RAJ, Wijn JR, and Blitterswijk CA. Use of bone-bonding hydrogel copolymers in bone: An *in vitro* and *in vivo* study of expanding PEO-PBT copolymers in goat femora. *J Biomed Mater Res*, 2000, 49, 312–318.
36. Reed AM and Gilding DK. Biodegradable polymers for use in surgery-poly(ethylene oxide) poly (ethylene terephthalate) (PEO/PET)copolymers: 2. *in vitro* degradation [J]. *Polymer*, 1981, 22, 499–504.
37. Van Blitterswijk CA, Brink JVD, Leenders H, et al. The effect of PEO ratio on degradation, calcification and bone bonding of PEO/PBT copolymer (Polyactive). *Cell Mater*, 1993, 3(1), 23–36.
38. Hayen H, Deschamps AA, Grijpma DW, Feijen J, and Karst U. Liquid chromatographic–mass spectrometric studies on the *in vitro* degradation of a poly(ether ester) block copolymer. *J Chromatogr A*, 2004, 1029, 29–36.
39. Stokes K, Urbanski P, and Upton J. The *in vivo* auto-oxidation of polyether polyurethane by metal ions. *J Biomater Sci Polym Ed*, 1990, 1, 207–230.
40. Deschamps AA, van Apeldoorn AA, Hayen H, et al. *In vivo* and *in vitro* degradation of poly(ether ester) block copolymers based on poly(ethylene glycol) and poly(butylene terephthalate)[J]. *Biomaterials*, 2004, 25, 247–258.
41. Bakker D, Bakker D, van Blitterswijk CT, Daems WTH, and Grote JJ. Biocompatibility of six elastomers *in vitro*. *J Biomed Mater Res*, 1988, 22, 423–439.
42. Li P, Bakker D, and van Blitterswijk CA. The bone-bonding polymer Polyactive® 80/20 induces hydroxycarbonate apatite formation *in vitro*. *J Biomed Mater Res*, 2002, 34(1), 79–86.
43. van Blitterswijk CA, Bakker D, Leenders H, van den Brink J, Hesseling SC, Bovell Y, Radder AM, Sakkers RJ, Gaillard ML, Heinze PH, and Beumer GJ. Interfacial reactions leading to bonebonding with PEO/PBT copolymers (Polyactive). In: Ducheyne P, Kokubo T, and van Blitterswijk CA (eds). *Bone-bonding biomaterials*. Leiderdorp, The Netherlands: Reed Healthcare Communications, 1992, pp. 13–30.
44. van Blitterswijk CA, van den Brink J, Leenders H, and Bakker D. The effect of PEO ratio on degradation, calcification and bonebonding of PEO/PBT copolymer (Polyactive). *Cell Mater*, 1993, 3, 23–36.
45. Beumer GJ, van Blitterswijk CA, Bakker D, and Ponec M. Cell seeding and *in vitro* biocompatibility evaluation of polymeric matrices of PEO/PBT copolymers and PLLA. *Biomaterials*, 1993, 14, 598–604.
46. Beumer GJ, van Blitterswijk CA, and Ponec M. Biocompatibility of a biodegradable matrix used as a skin substitute: An *in vitro* evaluation. *J Biomed Mater Res*, 1994, 28, 545–552.
47. van Loon JA, Leenders H, Goedemoed JH, and van Blitterswijk CA. Tissue reactions during long-term implantation in relation to degradation. A study of a range of PEO/PBT copolymers. In Proceedings of the 20th Annual Meeting of the Society for Biomaterials, 1994, Boston, MA, p. 370.
48. Radder AM, Leenders H, and van Blitterswijk CA. Interface reactions to PEO/PBT copolymers (Polyactive) after implantation in cortical bone. *J Biomed Mater Res*, 1994, 28, 141–151.
49. Kuijer R, Bouwmeester SJM, Drees MMWE, Surtel DAM, Terwindt-Rouwenhorst EAW, van der Linden JA, van Blitterswijk CA, and Bulstra SK. The polymer Polyactive as a bone-filling substance: An experimental study in rabbits. *J Mater Sci Mater Med*, 1998, 9, 449–455.
50. Sakkers RJ, Dalmeyer RA, de Wijn JR, and van Blitterswijk CA. Use of bone-bonding hydrogel copolymers in bone: an *in vitro* and *in vivo* study of expanding PEO/PBT copolymers in goat femora. *J Biomed Mater Res*, 2000, 49, 312–318.
51. Du C, Meijer GJ, van de Valk C, Haan Re, Bezemer JM, Hesseling SC, Cui FZ, de Groot K, and Layrolle P. Bone growth in biomimetic apatite coated porous Polyactive 1000PEGT70PBT30 implants. *Biomaterials*, 2002, 23, 4649–4656.

52. Anderson MLC, Dhert WJA, de Bruijn JD, Dalmeijer RAJ, Leenders H, van Blitterswijk CA, and Verbout AJ, Critical size defect in goat's os ilium. *Clin Orthop*, 1999, 364, 231–239.

53. Roessler, Wilke A, Griss PM, and Kienapfel H. Missing osteoconductive effect of a resorbable PEO/PBT copolymer in human bone defects: A clinically relevant pilot study with contrary results to previous animal studies. *Biomed Mater Res (Appl Biomater)*, 2000, 53, 167–173.

54. Bezemer JM, Radersma R, Grijpma DW, Dijkstra PJ, Feijen J, and van Blitterswijk CA. Zero-order release of lysozyme from poly(ethylene glycol)/poly(butylenes terephthalate) matrices. *Control Rel*, 2000, 64, 179–192.

55. Grote JJ. Reconstruction of the middle ear with hydroxylapatite implants: Long-term results. *Ann Otol Rhinol Laryngol*, 1990, 99(144), 12–16.

56. Grote JJ, Bakker D, Hesseling SC, and van Blitterswijk CA. New alloplastic tympanic membrane material. *Am. J Otol*, 1991, 12, 329–335.

57. Bakkum EA, Trimbos JB, Dalmeyer RAJ, and van Blitterswijk CA. Preventing postoperative intraperitoneal adhesion formation with Polyactive: A degradable copolymer acting as a barrier. *J Mater Sci Mater Med*, 1995, 6, 41–45.

58. Radder AM, Leenders H, and van Blitterswijk CA. Bone-bonding behaviour of PEO/PBT copolymer coatings and bulk implants: A comparative study. *Biomaterials*, 1994, 21, 532–537.

59. Beumer GJ, van Blitterswijk CA, Bakker D, and Ponec M. A new biodegradable matrix as a part of a cell seeded skin substitute for the treatment of deep skin defects: A physico-chemical characterisation. *Clin Mater*, 1993, 14, 21–27.

60. Hollinger JO. Preliminary report on the osteogenic potential of a biodegradable copolymer of polyactide (PLA) and polyglycolide (PGA). *J Biomed Mater Res*, 1983, 17, 71–82.

61. Lee SH, Kim BS, Kim SH, Choi SW, Jeong SI, Kwon IK. Kang SW, Nikolovski J, Mooney DJ, Han YK, and Kim YH. *J Biomed Mater Res*, 2003, 66A, 29–37.

62. Kwon K, Kidoaki S, and Matsuda T. Electrospun nano- to microfiber fabrics made of biodegradable copolyesters: Structural characteristics, mechanical properties and cell adhesion potential. *Biomaterials*, 2005, 26, 3929–3939.

63. Cohn D and Salomon AH. Designing biodegradable multiblock PCL/PLA thermoplastic elastomers. *Biomaterials*, 2005, 26, 2297–2305.

64. Younes HM, Bravo-Grimaldob E, and Amsden BG. Synthesis, characterization and *in vitro* degradation of a biodegradable elastomer. *Biomaterials*, 2004, 25, 5261–5269.

65. Webb A, Yang J, and Ameer G. Biodegradable polyester elastomers for tissue engineering. *Exp Opin Biol Ther*, 2004, 4(6), 801–812.

66. Matsumura G, Miyagawa-Tomita S, Shin'Oka T, Ikada Y, and Kurosawa H. First evidence that bone marrow cells contribute to the construction of tissue-engineered vascular autografts *in vivo*. *Circulation*, 2003, 108, 1729–1734.

67. Matsumura G, Hibino N, Ikada Y, Kurosawa H, and Shin'oka T. Successful application of tissue engineered vascular autografts: Clinical experience. *Biomaterials*, 2003, 24, 2303–2308.

68. Watanabe M, Shin'oka T, Tohyama S, et al. Tissue-engineered vascular autograft: Inferior vena cava replacement in a dog model. *Tissue Eng*, 2001, 7, 429–439.

69. Engelberg I and Kohn J. Physio-mechanical properties of degradable polymers used in medical applications: A comparative study. *Biomaterials*, 1991, 12, 292–304.

70. Pêgo AP, Poot AA, Grijpma DW, and Feijen J. Biodegradable elastomeric scaffolds for soft tissue engineering. *J Control Rel*, 2003, 87, 69–79.

71. Pêgo AP, Poot AA, Grijpma DW, and Feijen J. Copolymers of trimethylene carbonate and epsilon-caprolactone for porous nerve guides: Synthesis and properties. *J Biomater Sci Polym Ed*, 2001, 12(1), 35–53.

72. Pêgo AP, Siebum SB, Luyn MJAV, et al. Preparation of degradable porous structures based on 1,3-trimethylene carbonate and D,L-lactide(co)polymers for heart tissue engineering. *Tissue Eng*, 2003, 9, 981–994.

73. Pêgo AP, Luyn MJAV, Brouwer LA, et al. *In vivo* behaviour of poly(1,3-trimethylene carbonate) and copolymers of 1,3-trimethylene carbonate with D,L-lactide or ε-caprolactone: Degradation and tissue response. *J Biomed Mater Res*, 2003, 67A, 1044–1054.

74. Pêgo AP, Poot AA, Grijpma DW, and Feijen J. *In vitro* degradation of trimethylene carbonate based (co) polymers. *Macromol Biosci*, 2002, 2, 411–419.

75. den Dunnen WF, Meek MF, Grijpma DW, Robinson PH, and Schakenraad JM. *In vivo* and *in vitro* degradation of poly[(50)/(50) ((85)/(15)(L)/(D))LA/epsilon-CL], and the implications for the use in nerve reconstruction. *J Biomed Mater Res*, 2000, 51, 575–585.

76. Doi Y and Steinbüchel A. 2002. *Biopolymers—Polyesters III (Applications and Commercial Products)*, vol. 4. Weinheim, Germany: Wiley-VCH.

77. Steinbuchel A and Valentin HE., Diversity of bacterial polyhydroxyalkanoic acids. *Fems Microbiol Lett*, 1995, 128(3), 219–228.

78. Sodian R, Hoerstrup SP, Sperling JS, Daebritz S, Martin DP, Moran AM, Kim BS, Schoen FJ, Vacanti JP, and Mayer JE. Early In. vivo experience with tissue-engineered trileaflet heart valves. *Circulation*, 2000, 102(III), 22–29.

79. Martin DP and Williams SF. Medical applications of poly-4-hydroxybutyrate: A strong flexible absorbable biomaterial. *Biochem Eng*, 2003, 16, 97–105.

80. Hoerstrup SP, Sodian R, Sperling JS, Vacanti JP, and Mayer JE Jr. New pulsatile bioreactor for *in vitro* formation of tissue engineered heart valves[J]. *Tissue Eng*, 2000, 6(1), 75–79.

81. Kim BS, Nikolovski J, Bonadio J, et al. Cyclic mechanical strain regulates the development of engineered smooth muscle tissue. *Nature Biotech*, 1999, 17, 979–983.

82. Kim BS and Mooney DJ. Scaffolds for engineering smooth muscle tissues under cyclicmechanical strain conditions. *J Biomech Eng*, 2000, 122, 210–215.

83. Nikolovski J, Kim BS, and Mooney DJ. Cyclic strain inhibits switching of smooth muscle cells to an osteoblast-like phenotype. *FASEB J*, 2003, 17, 455–457.

84. Noda I, Green PR, Satkowski MM, and Schechtman LA. Preparation and properties of a novel class of polyhydroxyalkanoate copolymers. *Biomacromolecules*, 2005, 6, 580–586.

85. Benvenuti M and Lenz RW. Polymerization and copolymerization of beta-batyrolactone and benzyl-beta-malolactonate by aluminoxane catalyst. *J Polym Sci, Part A: Polym Chem*, 1991, 29, 793.

86. Hori Y, Suzuki M, Yamaguchi A, and Nishishita T. Ring-opening polymerization of optically active 62-butyrolactoone using distannoxane catalysts: Synthesis of high molecular weight poly(3-hydroxybutyrate). *Macromolecules*, 1993, 26, 5533.

87. Reith, LR., Moore, DR., Lobkovsky, EB., and Coates, GW. *J Am Chem Soc*, 2002, 124, 15239.

88. Williams SF, Martin DP, Horowitz DM, et al. PHA applications: Addressing the price performance issue. *Int J Biol Macromol*, 1999, 25, 111–121.

89. Relkin N and Nelson T. Regulation and. properties of an NADP+ oxidoreductase which functions as a. y-hydroxybutyrate dehydrogenase. *J Neurochem*, 1983, 40, 1639.

90. Shum-Tim D, Stock U, Hrkach J, Shinoka T, Lien J, Moses MA, Stamp A, Taylor G, Moran A, Landis W, Langer R, Vacanti JP, and Mayer JE Jr. Tissue engineering of autologous aorta using a new biodegradable polymer. *Ann Thorac Surg*, 1999, 68(6), 2298–304.

91. Shum-Tim D, Stock U, Hrkach J, Shinoka T, Lien J, Moses MA, Stamp A, Taylor G, Moran A, Landis W, Langer R, Vacanti JP, and Mayer JE Jr. Tissue engineering of autologous aorta using a new biodegradable polymer. *Ann Thorac Surg*, 1999, 68(6), 2298–2304.

92. Sodian R, Hoerstrup SP, Sperling JS, Daebritz S, Martin DP, Moran AM, Kim BS, Schoen FJ, Vacanti JP, and Mayer JE Jr. Early *in vivo* experience with tissue engineered trileaflet heart valves. *Circulation*, 2000, 102(Suppl), 22–29.

93. Hoerstrup SP, Sodian R, Daebritz S, Wang J, Bacha EA, Martin DP, Moran AM, Guleserian KJ, Sperling JS, Kaushal S, Vacanti JP, Schoen FJ, and Mayer JE Jr. Functional living trileaflet heart valves grown *in vitro*. *Circulation*, 2000, 102(III), 44–49.

94. Martin DP and Williams SF. Medical applications of poly 4-hydroxybutyrate: A strong flexible absorbable biomaterial. *Biochem Eng J*, 2003, 16, 97–105.

95. Sodian R, Loebe M, Hein A, Martin DP, Hoerstrup SP, Potapov EV, Hausmann H, Lueth T, and Hetzer R. Application of stereolithography for scaffold fabrication for tissue engineered heart valves. *ASAIO*, 2002, 48, 12–16.

96. Heart and Stroke Statistical Update, American Heart Association, Dallas, TX, 2002.

97. Stock UA, Sakamoto T, Hatsuoka S, Martin DP, Nagashima M, Moran AM, Moses MA, Khalil PN, Schoen FJ, Vacanti JP, and Mayer JE Jr. Patch augmentation of the pulmonary artery with bioabsorbable polymers and autologous cell seeding. *J Thorac Cardiovasc Surg*, 2000, 20, 1158–1168.

98. David PM and Simon FW. Medical applications of poly-4-hydroxybutyrate: A strong flexible absorbable biomaterial. *Biochem Eng J*, 2003, 16, 97–105.

99. Opitz F, Schenke-Layland K, Cohnert TU, Starcher B, Halbhuber KJ, Martin DP, and StockUA. Tissue engineering of aortic tissue: Dire consequence of suboptimal elastic fiber synthesis *in vivo*. *Cardiovasc Res*, 2004, 63, 719–730.

100. Opitz F, Schenke-Layland K, Richter W, Martin DP, Degenkolbe I, Wahlers T, and Stock UA. Tissue engineering of ovine aortic blood vessel substitutes using applied shear stress and enzymatically derived vascular smooth muscle cells. *Ann Biomed Eng*, 2004, 32, 212–222.

101. Thomas V, Kumari TV, and Jayabalan M. *In vitro* studies on the effect of physical cross-linking on the biological performance of aliphatic poly(urethane urea) for blood contact applications. *Biomacromolecules*, 2001, 2, 588–596.

102. Hoffman D, Gong G, Pinchuk L, and Sisto D. Safety and intracardiac function of a silicone–polyurethane elastomer designed for vascular use. *Clin Mater*, 1993, 13, 95–110.

103. Capone CD. Biostability of a non-ether polyurethane. *J Biomater Appl*, 1992, 7, 108–129.

104. Szycher M. *Szycher's Handbook of Polyurethanes*. Boca Raton, FL: CRC Press, 1999.

105. De Groot JH, de Vrijer R, Pennings AJ, Klompmaker J, Veth RPH, and Jansen HWB. Use of porous polyurethanes for meniscal reconstruction and meniscal prosthesis. *Biomaterials*, 1996, 17, 163–174.

106. Borkerhagen M, Stoll RC, Neuenschwander P, Suter UW, and Aebischer P. *In vivo* performance of a new biodegradable polyester urethane system used as a nerve guidance channel. *Biomaterials*, 1998, 19, 2155–2165.

107. Saad B, Hirt TD, Welti M, Uhlschmid GK, Neuenschwander P, and Suter UW. Development of degradable polyester urethanes for medical applications. *J Biomed Mater Res*, 1997, 36, 65–74.

108. Reed AM, Potter J, and Szycher M. A solution grade biostable polyurethane elastomer: ChronoFlex AR. *J Biomater Appl*, 1994, 8, 210–236.

109. Coury A. Chemical and biochemical degradation of polymers. In: Ratner B, Hoffman A, Schoen F, and Lemons J (eds). *Biomaterials Science: An Introduction to Materials in Medicine*. Boston, MA: Elsevier Academic Press, 2004, pp. 411–430.

110. Borkerhagen M, Stoll RC, Neuenschwander P, Suter UW, and Aebischer P. *In vivo* performance of a new biodegradable polyester urethane system used as a nerve guidance channel. *Biomaterials*, 1998, 19, 2155–2165.

111. Saad B, Hirt TD, Welti M, Uhlschmid GK, Neuenschwander P, and Suter UW. Development of degradable polyester urethanes for medical applications. *J Biomed Mater Res*, 1997, 36, 65–74.

112. Skarja GA and Woodhouse KA. Synthesis and characteization of degradable polyurethane elastomers containing an amino acid based chain extender. *J Biomater Sci Polym Ed*, 1998, 9, 271–295.

113. Cohn D, Stern T, Gonzales MF, and Epstein J. Biodegradable poly(ethylene oxide)/poly(ε-caprolactone) multiblock copolymers. *J Biomed Mater Res*, 2002, 59, 273–281.

114. Guan JJ, Sacks MS, Beckman EJ, and Wagner WR. Synthesis, characterization, and cytocompatibility of elastomeric, biodegradable poly(ester-urethane) ureas based on poly(caprolactone) and putrescine. *J Biomed Mater Res*, 2002, 61, 493–503.

115. Bruin P, Veenstra GJ, Nijenhuis AJ, and Pennings AJ. Design and synthesis of biodegradable poly(ester-urethane) elastomer networks composed of non toxic building blocks. *Makromol Chem Rapid Commun*, 1988, 9, 589.

116. Storey RF and Hickey TP. Degradable polyurethane networks based on D,L-lactide, glycolide, e-caprolactone, and trimethylene carbonate homopolyester and copolyester triols. *Polymer*, 1994, 35, 830–838.

117. Gorna K and Gogolewski S. Biodegradable polyurethane implants II *in vitro* degradation and calcification of materials from poly (e-caprolactone)-polyethylene) diols and various chain extenders. *J Biomater Res*, 2002, 60, 592–606.

118. Skarja GA and Woodhouse KA. Structure–property relationships of degradable polyurethane elastomers containing an amino acid based chain extender. *J Appl Polym Sci*, 2000, 75, 1522–1534.

119. Skarja GA and Woodhouse KA. *In vitro* degradation and erosion of degradable, segmented polyurethanes containing amino acidbased chain extender. *J Biomater Sci Polym Ed*, 2001, 12, 851–873.

120. Fromstein JD and Woodhouse KA. Elastomeric biodegradable polyurethane blends for soft tissue application. *J Biomater Sci Polym Ed*, 2002, 13, 391–406.

121. Guan JJ, Sacks MS, Beckman EJ, and Wagner WR. Biodegradable poly(ether ester urethane urea) triblock copolymers and purtrescine: Synthesis, characterisation and cytocompatibility. *Biomaterials*, 2004, 25, 85–96.

122. Saad B, Ciardelli G, Matter S, Welti M, Uhlschmid GK, Neuenschwander P, and Suter UW. Degradable and highly porous polyestherurethane foam as biomaterial: Effects and phagocytosis of degradation products in osteoblasts. *J Biomed Mater Res*, 1998, 39, 594–602.

123. Saad B, Kuboki M, Matter S, Welti M, Uhlschmid GK, Neuenschwander P, and Suter UW. DegraPolfoam: A degradable and highly porous polyesterurethane as a new substrate for bone formation. *Artif Organs*, 2000, 24, 939–945.

124. Grad S, Kupcsik L, Gorna K, Gogolewski S, and Alini M. The use of biodegradable polyurethane scaffolds for cartilage tissue engineering: Potential and limitations. *Biomaterials*, 2003, 24, 5163–5171.

125. Ganta SR, Piesco NP, Long P, Gassner R, Motta LF, Papworth GD, Stolz DB, Watkins SC, and Agarwal S. Vascularisation and tissue infiltration of a biodegradable polyurethane matrix. *J Biomed Mater Res*, 2003, 64A, 238–242.

126. Zhang JY, Beckman EJ, Hu J, Yang GG, Agarwal S, and Hollinger JO. Synthesis, biodegradability, and biocompatibility of lysine discarnate-glucose polymers. *Tissue Eng*, 2002, 8, 771–785.

127. Elliott SL, Fromstein JD, Santerre JP, and Woodhouse KA. Identification of biodegradation products formed by L-phenylalanine based segmented polyurethaneureas. *J Biomater Sci Polym Ed*, 2001, 13(6), 691–711.

128. Ernsting MJ, Labow RS, and Santerre JP. Surface modification of a polycarbonate-urethane using a vitamin-E derivatized fluoroalkyl surface modifier. *J Biomater Sci Polym Ed*, 2003, 14, 1411–1426.

129. Woo GLY, Mittelman MW, and Santerre JP. Synthesis and characterization of a novel biodegradable antimicrobial polymer. *Biomaterials*, 2000, 21, 1235–1246.

130. Yang M and Santerre JP. Utilization of quinolone drugs as monomers: Characterization of the synthesis reaction products for poly(norfloxacin diisocyanatododecane polycaprolactone). *Biomacromolecules*, 2001, 2, 134–141.

131. Ishihara K, Hanyuda H, and Nakabayashi N. Synthesis of phospholipids polymers having a urethane bond in the side chain as coating material on segmented polyurethane and their platelet adhesion resistant properties. *Biomaterials*, 1995, 16(11), 873–879.

132. Sugiyama K, Fukuchi M, Kishida A, Akashi M, and Kadoma Y. Preparation and characterization of poly (2-methacryloyloxyethyl phosphorylcholine-co-methyl methacrylate) graft copolyetherurethanes. *Kobunshi Ronbunshu*, 1996, 53(1), 48–56.

133. Ishihara K, Tanaka S, Furukawa N, Kurita K, and Nakabayashi N. Improved blood compatibility of segmented polyurethanes by polymeric additives having phospholipid polar groups. I. Molecular design of polymeric additives and their functions. *J Biomed Mater Res*, 1996, 32(3), 391–399.

134. Ishihara K, Shibata N, Tanaka S, Iwasaki Y, Kurosaki T, and Nakabayashi N. Improved blood compatibility of segmented polyurethane by polymeric additives having phospholipids polar group. II. Dispersion state of polymeric additive and protein adsorption on the surface. *J Biomed Mater Res*, 1996, 32(3), 401–418.

135. Potin P and Jaeger RD. Polyphosphazenes: Synthesis, structures, properties, applications. *Eur Polym J*, 1991, 415, 341–348.

136. Stokes HN. On the chloronitrides of phosphorous. *Am Chem J*, 1895, 17, 275–290.

137. Allcock HR and Kugel RL. Synthesis of high polymericalkoxy and aryloxy phosphonitriles. *J Am Chem Soc*, 1965, 87, 4216–4217.

138. Allcock HR, Kugel RL, and Valan KJ. Phosphonitrilic compounds. VII. High molecular weight poly (diaminophosphazenes). *Inorg Chem*, 1966, 5, 1716–1718.

139. Allcock HR, Kugel RL, and Valan KJ. Phosphonitrilic compounds. VI. High molecular weight polyalkoxy- and aryloxyphosphazenes. *Inorg Chem*, 1966, 5, 1709–1715.

140. Allcock HR and Chu CTW, Reaction of phenyl lithium withpolydichlorophosphazene. *Macromolecules*, 1979, 12, 551–555.

141. Hornbaker, ED and Li HM. U.S. Patent 4 198 381, 1978.

142. De Jaeger R, Heloui M, and Puskaric E. Fr. Patent 466453, 1981.

143. Potin P and Jaeger RD. Polyphosphazenes: Synthesis, structures, properties, applications. *Eur Polym J*, 1991, 415, 341–348.

144. Neilson RH and Wisian-Neilson P. Poly(alkyl/aryl phos-phazenes) and their precursors. *Chem Rev*, 1988, 88, 541–562.

145. Allcock HR, Crane CA, Morrissey CT, Nelson JM, Reeves SD, Honeyman CH, and Manners I. Living cationic polymerization of phosphoranimines as an ambient temperature route to polyphosphazenes with controlled molecular weights. *Macromolecules*, 1996, 29, 7740–7747.

146. Allcock HR, Nelson JM, Reeves SD, Honeyman CH, and Manners I. Ambient-temperature direct synthesis of poly(organophosphazenes) via the living cationic polymerization of organo-substituted phosphoranimines. *Macromolecules*, 1997, 30, 50–53.

147. Allcock HR, Reeves SD, Nelson JM, Crane CA, and Manners I. Polyphosphazene block copolymers via the controlled cationic, ambient temperature polymerization of phosphoranimines. *Macromolecules*, 1997, 30, 2213–2215.

148. Allen RW and Allcock HR. Conformational analysis of poly(alkoxy- and aryloxyphosphazene) [J]. *Macromolecules*, 1976, 9(6), 956–961.

149. Allcock HR, Mang MN, Dembek AA, et al. Poly[(aryloxy)phosphaz enes] with phenylphenoxy and related bulky side groups, synthesis, thermal transition behavior, and optical properties [J]. *Macromolecules*, 1989, 22, 4179–4190.

150. Potin P and Jaeger RD. Polyphosphazenes: Synthesis, structures, properties, applications. *Eur Polym J*, 1991, 415, 341–348.

151. Allcock HR, Kwon S, Riding GH, Fitzpatrick RJ, and Bennett JL. Hydrophilic polyphosphazenes as hydrogels: Ration cross-linking and hydrogel characteristics of poly [bis(methoxyethoxyethoxy)phosphazene. *Biomaterials*, 1988, 9, 509.

152. Allcock HR, Austin PE, Neenan TX, Langex R, and Shriver DF. Polyphosphazenes with etheric side groups: Prospective biomedical and solid electrolyte. *Macromolecules*, 1986, 19, 1508.

153. Laurencin CT, Koh HJ, Neenan TX, Allcock HR, and Langer R. Controlled release using a new bioerodible polyphosphazene matrix system. *J Biomed Mater Res*, 1987, 21, 1231.

154. Anurima S, Nicholas RK, Swaminathan S, Lakshmi SN, Jacqueline LS, Paul WB, Cato TL, and Harry RA. Effect of side group chemistry on the properties of biodegradable L-alanine cosubstituted polyphosphazenes. *Biomacromolecules*, 2006, 7, 914–918.

155. Allcock HR. *Chemistry and Applications of Polyphosphazenes*. Hoboken, NJ: Wiley-Interscience, 2003, p. 504.

156. Allcock HR. Small-molecule phosphazene rings as models for high polymeric chains. *Acc Chem Res*, 1979, 12, 351–358.

157. Allcock HR. Controlled synthesis of organic–inorganic polymers that possess a backbone of phosphorus and nitrogen atoms. *Makromol Chem*, 1981, (Suppl 4), 3–19.

158. Allcock HR. Organometallic and bioactive phosphazenes. *J Polym Sci Polym Symp*, 1983, 70, 71–77.

159. Allcock HR. Poly(organophosphazenes): Synthesis, unique properties and applications. *Makromol Chem Makromol Symp*, 1986, 6, 101–108.

160. Allcock HR and Fuller TJ. Synthesis and hydrolysis of hexakisumidazoly1)-cyclotriphosphazene. *J Am Chem Soc*, 1981, 103, 225C2256.

161. Allcock HR, Shawn R, and Scopelianos AG. Poly [(amino acid ester) phosphazenes] as substrates for the controlled release of small molecules. *Biomaterials*, 1994, 1, 5563–5569.

162. Caliceti P, Veronese FM, Marsilio F, Lora S, Seraglia R, and Traldi P. Fast atom bombardment in the structural identification of intermediates in the hydrolytic degradation of polyphosphazenes. *Org Mass Spectrom*, 1992, 27, 1199–1202.

163. Payne LG, Jenkino SA, Adrianov A, Langer R, and Robert BE. Xenobiotic polymers as vaccine vehicles. In: Mestecky J (ed), *Advances in Mucosal Immunology*. New York: Plenum Press, pp. 1475–1480.

164. Ibim SM, Ambrosio AA, Larrier D, Allcock HR, and Laurencin CT. Controlled macromolecule release from poly(phosphazene) matrices. *J Control Release*, 1996, 40, 31–39.

165. Wade CWR, Gourlay S, Rice R, Heggli A, Singler RE, and White J. *Organo Metallic Polymers*. New York: Academic Press, 1978, p. 283.

166. Stenger DA. and McKenna TM. *Enabling Technologies for Cultured Neural Networks*. New York: Academic Press, 1994.

167. Barrett EW, Phelps MVB, Silva RJ, Gaumond RP, and Allcock HR. Patterning poly(organophosphazenes) for selective cell adhesion applications. *Biomacromolecules*, 2005, 6, 1689–1697.

168. Hron P. Hydrophilisation of silicone rubber for medical applications. *Polym Int*, 2003, 52, 1531–1539.

169. Tcholakian RK and Raad. Durability of anti-infective effect of long term silicone sheath catheters impregnated with antimicrobial agents. *Antimicrob Agents Chemother*, 2001, 45(7), 1990–1993.

170. Ikada Y. Surface modification of polymers for medical application. *Biomaterials*, 1994, 15, 725–736.

171. James SJ, Pogribna M, Miller BJ, Bolon B, and Muskhelishvili L. Characterization of cellular response to silicone implants in rats: Implications for foreign-body carcinogenesis. *Biomaterials*, 1997, 18, 667–675.

172. Abbasi F, Mirzadeh H, and Katbab AA. Modification of polysiloxane polymers for biomedical applications: A review [J]. *Polym Int*, 2001, 50(12), 1279–1287.

173. Fakes DW, Newton JM, Watts JF, and Edgel MJ. Surface modification of contact lens co-polymer by plasma-discharge treatments. *Surf Interf Anal*, 1987, 10, 416.

174. Fakes DW, Watts JF, and Newton JM. In: Ratner BD (ed), *Surface Characterization of Biomaterials*. Amsterdam: Elsevier, 1998, p. 193.

175. Williams RL, Wilson DJ, and Rhodes, NP. Stability of plasma-treated silicone rubber and its influence on the interfacial aspects of blood compatibility. *Biomaterials*, 2004, 25, 4659–4673.

176. Gottenbosa B, Henny C, van der Meia, Klatterb F, Nieuwenhuisb P, and Busscher HJ. *In vitro* and *in vivo* antimicrobial activity of covalently coupled quaternary ammonium silane coatings on silicone rubber. *Biomaterials*, 2002, 23, 1417–1423.

177. Folch A and Tomer M. Microengineering of cellular interactions. *Annu Rev Biomed Eng*, 2000, 2, 227–256.

178. Desai TA. Micro- and nanoscale structures for tissue engineering constructs [J]. *Med Eng Phys*, 2000, 22, 595–606.

179. Deutsch J, Motlagh D, Russell B, and Desai TA. Fabrication of micro textured membranes for cardiac myocyte attachment and. orientation. *J Biomed Mater Res*, 2000, 53(3), 267–275.

180. Mirzadeh H, Khorasani MT, and Sammez P. Laser surface modification of polymers: A novel technique for the preparation of blood compatible materials-II *In vitro* assay. *Iranian Polym*, 1998, 7, 5.

181. Kabanov VY, Aliev RE, and Kudryavtsev VN. Present status and development trends of radiation-induced graft polymerization. *Radiat Phys Chem*, 1991, 37, 175.

182. Hoffman AS. A review of the use of radiation plus chemical and biochemical processing treatments to prepare novel biomaterials. *Radiat Phys Chem*, 1981, 18, 323–342.

183. Uenoyama, S and Hoffman AS. Synthesis and characterization of acrylamide-N-isopropyl acrylamide copolymer grafts on silicone rubber substrates. *Radiat. Phys. Chem.*, 1988, 32, 605–608.

184. Mao C, Zhu AP, Qiu YZ, Shen J, and Lin SC. Introduction of *O*-butyrylchitosan with a photosensitive hetero-bifunctional crosslinking reagent to silicone rubber film by radiation grafting and its blood compatibility. *Colloids Surf B: Biointerf*, 2003, 30, 299–306.

185. Boateng S, Lateef SS, Crot C, Motlagh D, Desai T, Samarel AM, Russell B, and Hanley L. Peptides bound to silicone membranes and 3D microfabrication for cardiac cell culture. *Adv Mater*, 2002, 14, 461–463.

186. Yuan YL, Zang XP, Ai F, Zhou J, Shen J, and Lin SC. Grafting sulfobetaine monomer onto silicone surface to improve haemocompatibility. *Polym Int*, 2004, 53, 121–126.

187. Chen H, Zhang Z, Chen Y, Brook MA, and Sheardown H. Protein repellant silicone surfaces by covalent immobilization of poly(ethylene oxide). *Biomaterials*, 2005, 26, 2391–2399.

188. Lopour P, Plichta Z, Volfová Z, Hron P, and Vondráček P. Silicone rubber–hydrogel composites as polymeric biomaterials. *Biomaterials*, 1993, 14(14), 1051–1055.

189. Cifkova I, Lopour P, Vondracek P, and Jelínek F. Silicone rubber–hydrogel composites as polymeric biomaterials. *Biomaterials*, 1990, 11, 393–396.

190. Lednicky F, Janatova V, and Lopour P. Silicone rubber–hydrogel composite as polymeric biomaterials [J]. *Biomaterials*, 1991, 12, 848.

191. Grainger DW, Kim SW, and Feijen J. Poly(dimethyl siloxane)–poly(ethylene oxide)–heparin block copolymers. I: Synthesis and characterization. *J Biomed Mater Res*, 1988, 22, 231–242.

192. Arkles B, US patent 4500 688, 1985.

193. Dillon ME, US patent 4 832 009, 1989.

194. Robert C, Bunel C, and Vairon JP, Eur Patent Appl, 643 083, 1995.

195. Mark E. and Van Dyke. Silicone biomaterials: A review. *Polym Prepr*, 2004, 45, 600–601.

196. Price C, Waters MGJ, Williams DW, Lewis MAO, and Stickler D. Surface modification of an experimental silicone rubber aimed at reducing initial candidal adhesion. *J Biomed Met Res*, 2002, 63, 122–128.

197. Chamberlain LJ, Yannas IV, Arrizabalaga A, Hsu HP, Norregaard TV, and Spector M. Early peripheral nerve healing in collagen and silicone tube imolants: Myofibroblasts and the cellular response. *Biomaterials*, 1998, 19, 1393–1403.

9 Recombinant Resilin—A Protein-Based Elastomer

Mickey G. Huson and Christopher M. Elvin

CONTENTS

9.1 ELASTOMERIC PROTEINS

Elastomeric proteins are a diverse group of structural proteins present in widely divergent organisms from insects and mollusks to higher mammals. They are structural proteins displaying reversible deformation and serve an important functional role in providing long-range elasticity, storing kinetic energy, acting as shock absorbers and as antagonists to muscles. The family of elastic proteins includes the vertebrate muscle and connective tissue proteins, elastin, titin, and fibrillin, as well as the seed storage proteins gluten and gliadin, the bivalve ligament protein abductin, mussel shell byssal threads, spider silks, and the rubbery protein resilin from arthropods.

Despite diverse protein sequences amongst the members of this protein family, all elastomeric proteins share a number of common features. Firstly, they all contain repeat peptide motifs comprising both elastic domains, which are rich in the amino acid glycine and other hydrophobic amino acids, as well as nonelastic or "cross-linking domains," involved in the formation of obligatory chemical cross-links in the mature polymer. Secondly, they always occur in nature in a hydrated state, the water acting as both lubricant and plasticizer, and resulting in structures that are highly mobile and conformationally free. Finally, all of the proteins are covalently cross-linked, although the chemical nature of these cross-links varies.

In most cases, biological activity of proteins is dependent on specific three-dimensional (3D) tertiary structures. Not surprisingly, as a consequence of this interest in protein structure and function, attempts were made to explain the rubber-like properties of this group of proteins on the basis of their structure. Gosline [1] invoked hydrophobic interactions in which stretching increases exposure of hydrophobic side chains to the aqueous environment. This has the effect of decreasing the entropy of surrounding water molecules. The restoring force is due to the hydrophobic forces that stabilize the folded protein structures. Urry [2] first proposed the β-spiral structure as providing the basis of elastin elasticity. Regularly spaced beta turns in the protein act as spacers between turns of the spiral, thus maintaining the protein chains in a kinetically free state. The restoring force in this model is also entropic. A more commonly accepted model these days, first proposed by Weis-Fogh [3,4] and backed up by recent studies on elastin [5,6,7], is based on classical rubber theory [8] in which elasticity is attributed to a decrease in conformational entropy on deforming an amorphous kinetically free random polymer network. Stress orders the chains and decreases entropy by limiting their conformational freedom. This is the source of the restoring force, which allows a return to the relaxed state [8,9]. An obvious consequence of this is that there is no secondary or tertiary structure, and indeed Tompe [10,11] has recently published a series of papers describing Intrinsically Unstructured Proteins (IUPs). It has now been shown that for many proteins and protein domains the functional state is intrinsically unstructured. For almost 200 proteins and protein domains the lack of a unique 3D structure has been convincingly demonstrated by using mostly three techniques: x-ray crystallography, multidimensional nuclear magnetic resonance (NMR), and circular dichroism (CD) spectroscopy [12].

There are several excellent reviews and a comprehensive book dealing with the structure and properties of elastomeric proteins, and readers are urged to consult these references for a more detailed background to the field [9,13,14].

Natural resilin is crosslinked and exhibits two outstanding materials properties: high rubber efficiency (resilience) and a very high fatigue lifetime. It is found in specialized regions of the cuticle of most insects, providing low stiffness, high strain, and efficient energy storage [15,16]; it is best known for its roles in insect flight [17,18] (Figure 9.1) and the remarkable jumping ability of fleas [19,20] and spittle bugs [21]. Apart from its role in flight and locomotion, it is also used for other insect functions where efficient energy storage and repetitive movement are required, e.g., in the sound-producing organs of cicadas [22] and moths [23]. While resilin and elastin are highly

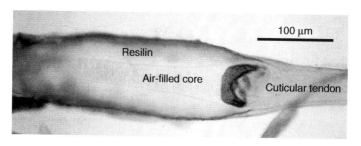

FIGURE 9.1 Adult dragonfly (*Zyxomma* sp.) showing the location of the wing tendon (top) and an optical micrograph showing the location and detailed structure of the resilin in the tendon (bottom). (From Elvin, C.M., Carr, A.G., Huson, M.G., Maxwell, J.M., Pearson, R.D., Vuocolo 1, T., Liyon, N.E., Wong, D.C.C., Merritt, D.J., and Dixon, N.E., *Nature*, 437, 999, 2005.)

resilient, other elastomers are exquisitely designed to have low resilience. These materials are designed to dissipate mechanical energy through molecular friction and include byssus and spider capture silk. Like other hydrogel materials, the properties of protein elastomers are heavily affected by the degree of hydration. Most are soft, flexible, and resilient in the fully hydrated state and become leathery and finally hard and glassy as they dry out.

As a consequence of the remarkable properties of elastomeric and structural proteins, and the rapid advances in molecular biology over the last decade, there has been a surge of interest in reproducing these proteins in a recombinant form. This interest in biomimetics has resulted in a number of biomaterials being produced as recombinant proteins, including spider silks [24,25], collagen [26], wheat gliadin [27], elastin [28], and resilin [29]. In each case, the genes encoding the constituent biopolymer proteins are cloned, or synthesized *in vitro*, and introduced into cells (mammalian or bacterial) for biosynthesis. The goal, of course, is to use a synthetic route to produce a new material with properties that are equal to, or even superior to, its natural precursor.

Resilin has a remarkably high fatigue lifetime (probably >500 million cycles) and our aim is to reproduce this desirable mechanical property in synthetic materials derived from our studies of resilin structure and function. We believe that recombinant resilin-like materials may be used, in the future, in the medical device field as components of prosthetic implants, including spinal disks and synthetic arteries. Spinal disks, for example, must survive for at least 100 million cycles of contraction and relaxation [30].

9.2 PRODUCTION OF CROSS-LINKED RECOMBINANT RESILIN

9.2.1 SYNTHESIS

All elastic proteins contain distinct domains, of which at least one is made up of elastic repeat sequences, and they all contain cross-links between residues in either the nonelastic or elastic domains [9]. Previously, the *Drosophila CG15920* gene was tentatively identified as one encoding a resilin-like protein [31]. To prepare recombinant resilin, we chose to express the first exon of the *Drosophila CG15920* gene [29], which encodes an N-terminal domain in the native protein comprising 17 copies of the putative elastic repeat motif GGRPSDSYGAPGGGN [31].

To express the *CG15920* gene fragment, genomic DNA from adult *D. melanogaster* was used as a template for polymerase chain reaction (PCR) with primers designed to amplify a 946-bp fragment (including codons for an N-terminal His$_6$ affinity tag) that was inserted into a T7-promoter expression vector.

Recombinant resilin production was induced, with the nonmetabolizable lactose analogue IPTG, in the *Escherichia coli* bacterial strain BL21(DE3)/pLysS. Cells were collected by centrifugation (10,000 *g*, 20 min at 4°C) and the cell pellet frozen at −80°C.

9.2.2 PURIFICATION

Soluble proteins were isolated following lysis and sonication of the resuspended pellet in a buffer solution containing protease inhibitor cocktail. The soluble protein fraction was collected after centrifugation at 100,000 *g* for 1 h at 4°C. The cell-free extract contains a large number of molecular mass species (Figure 9.2a, lane 1) and was fractionated by a combination of anion exchange and metal affinity (Ni-nitrilotriacetic acid, Ni-NTA) chromatography. This process gave very pure recombinant resilin as judged by acrylamide gel electrophoresis (sodium dodecyl sulfate polyacrylamide gel electrophoresis, SDS-PAGE) analysis (see Figure 9.2a, lane 3); yields were about 15 mg/L of culture. More recently it has been shown [32] that facile purification can also be effected by a combination of heat treatment and high concentrations of salt.

While the apparent molecular weight was about 47,000 g/mol or daltons (Da) by mobility on SDS-PAGE, separate analysis by sedimentation equilibrium measurements and capillary high-performance liquid chromatography (HPLC) in SDS buffer gave values near 23,000 Da.

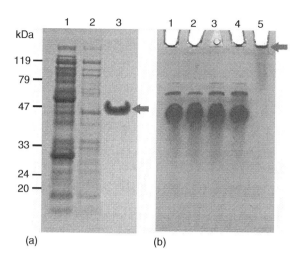

(a) (b)

FIGURE 9.2 10% SDS-PAGE showing changes during purification and cross-linking of soluble recombinant resilin. (a) Steps in the purification. Lane 1, cleared supernatant from lysed cells; lane 2, after anion exchange; lane 3, after Ni-nitrilotriacetic acid (Ni-NTA) affinity chromatography (arrow). (b) Photo-cross-linking. Lane 1, with Ru(II) and ammonium persulfate (APS) (no light); lane 2, recombinant resilin only; lane 3, with Ru(II) only; lane 4, with APS only; lane 5, with Ru(II) and APS after 20 s irradiation. (From Elvin, C.M., Carr, A.G., Huson, M.G., Maxwell, J.M., Pearson, R.D., Vuocolo 1, T., Liyon, N.E., Wong, D.C.C., Merritt, D.J., and Dixon, N.E., *Nature*, 437, 999, 2005.)

From the DNA sequence of the resilin recombinant clone, the amino acid sequence of recombinant resilin can be deduced as

MHHHHHHPEPPVNSYLPPSDSYGAPGQSGPGGRPSDSYGAPGGGNGGRPSDSYGAPGQG-
QGQGQGQGGYAGKPSDSYGAPGGGDGNGGRPSSSYGAPGGGNGGRPSDTYGAPGGGN-
GGRPSDTYGAPGGGGNGNGGRPSSSYGAPGQGQGNGNGGRPSSSYGAPGGGNGGRPSD
TYGAPGGGNGGRPSDTYGAPGGGNNGGRPSSSYGAPGGGNGGRPSDTYGAPGGGNGNG
SGGRPSSSYGAPGQGQGGFGGRPSDSYGAPGQNQKPSDSYGAPGSGNGNGGRPSSSYGAP
GSGPGGRPSDSYGPPASG

The calculated value for the His$_6$-tagged protein is 28,492 Da. N-terminal amino acid sequence analysis (Procise) was carried out on purified recombinant resilin. The following sequence (at 120 pmol yield) was obtained for the first 12 amino acid residues: MHHHHHHPEPPV, as expected from DNA sequence analysis.

9.2.3 CROSS-LINKING

The chemical cross-links in natural resilin from insect joints and tendons occur between tyrosine residues, generating di- and trityrosine [15,33] (Figure 9.3a). A number of methods are available to form dityrosine cross-links in the recombinant material. Quantitative cross-linking of soluble recombinant resilin was achieved by use of *Arthromyces ramosus* peroxidase, an enzyme known to catalyse dityrosine formation [34]. A solid biomaterial was formed; however, the rate of the enzyme-catalyzed reaction was uncontrollably fast and the catalase activity of the peroxidase formed oxygen bubbles in the sample making it of limited value for physical testing. On the other hand, Ru(II)-mediated photo-cross-linking [35] resulted in rapid, quantitative, and controllable conversion of the soluble, low-molecular-weight polypeptide to a high-molecular-weight cross-linked species which barely enters a 10% SDS-PAGE gel (see Figure 9.2b, lane 5, arrow). The method used was a modification of that described by Fancy and Kodadek [35]. Solutions of recombinant resilin (200 mg/mL), 2 mM [Ru(bpy)$_3$]$^{2+}$ (Ru(II)) and 10 mM ammonium persulfate (APS) in phosphate-buffered saline (PBS) were irradiated in glass moulds for 10–20 s at 15 cm with a 600 W white light

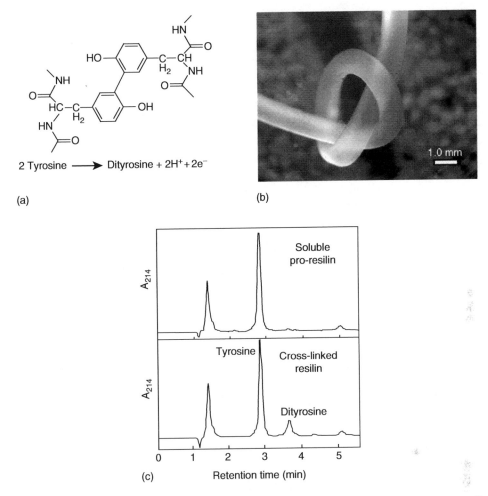

FIGURE 9.3 Cross-linked recombinant resilin. (a) Structure of the dityrosine adduct. (b) A cross-linked moulded rod. (c) HPLC analysis of acid-hydrolyzed uncross-linked recombinant resilin and cross-linked recombinant resilin. (From Elvin, C.M., Carr, A.G., Huson, M.G., Maxwell, J.M., Pearson, R.D., Vuocolo 1, T., Liyon, N.E., Wong, D.C.C., Merritt, D.J., and Dixon, N.E., *Nature*, 437, 999, 2005.)

source, resulting in solid material of sheets, disks, and rods (Figure 9.3b). This method is quick and easy compared to those traditionally used to cross-link recombinant elastin polymers: gamma irradiation [36], lengthy curing using a copper redox system, or enzymatic procedures [37].

Amino acid analysis, by reverse-phase HPLC, of acid-hydrolyzed uncross-linked recombinant resilin and cross-linked recombinant resilin clearly shows the presence of dityrosine in the cross-linked sample (Figure 9.3c). Further evidence of the presence of dityrosine was obtained by UV irradiation (λmax,ex 315 nm; λmax,em 409 nm). Dityrosine endows natural resilin with pH-dependent blue fluorescence [38] on UV irradiation. The cross-linked recombinant resilin material was similarly fluorescent, strongly suggesting dityrosine cross-links.

9.3 CHARACTERIZATION AND PROPERTIES

9.3.1 STRUCTURE

For recombinant resilin, we found that data from the CD spectra (Figure 9.4), x-ray diffraction pattern (Figure 9.5), and NMR were consistent with resilin being an amorphous random network

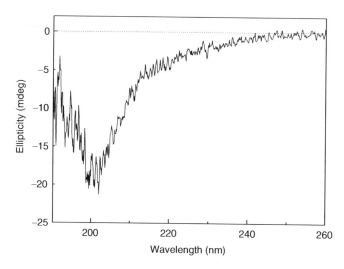

FIGURE 9.4 Circular dichroism (CD) spectrum of dilute solution (0.24 mg/mL) of recombinant resilin in phosphate-buffered saline (PBS). (From Whelan, A.J. and Robinson, A.J., unpublished data).

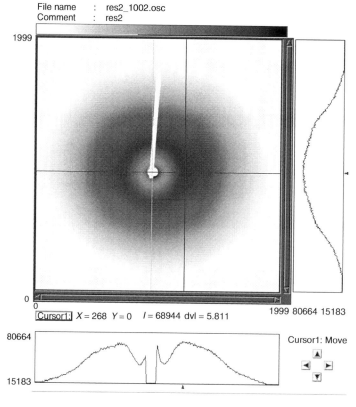

FIGURE 9.5 X-ray diffraction pattern derived from a solid piece of cross-linked recombinant resilin. The lack of any sharp rings or features indicates that there is no distinct ordering in the sample. Some water scatter at about 3.5 Å and the highest scattering around 6 Å is observed. (From Pilling, P. and Varghese, J., unpublished data).

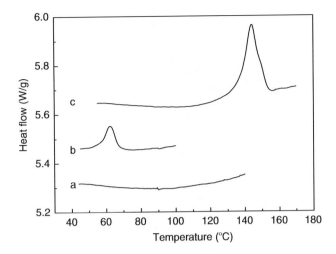

FIGURE 9.6 DSC of: (a) recombinant resilin in water showing no enthalpic events, (b) bovine serum albumin in phosphate-buffered saline (PBS) showing denaturing occurring at 62°C, and (c) wool fiber in water showing denaturing of the α-helix at 145°C (Endotherm up).

polymer (the CD spectrum is also consistent with a polyproline II structure [39]). Analysis of the primary sequence of recombinant resilin using the IUP prediction software (http://iupred.enzim.hu/) shows it to be completely unstructured offering further evidence of its amorphous state.

Structured proteins have also been investigated by thermal analysis [40,41], denaturing resulting in an endotherm which is readily detected by differential scanning calorimetry (DSC). DSC of recombinant resilin in the swollen state showed no transitions over a wide temperature range (25°C–140°C), further evidence of the absence of any structure. This is in contrast to the structured proteins wool and bovine serum albumin, which show denaturation endotherms at 145°C and 62°C, respectively (Figure 9.6).

In spite of the overwhelming evidence suggesting that recombinant resilin is amorphous, there are some results that suggest that a level of defined structure cannot be completely ruled out. In particular, the fact that the protein solution coacervates when cooled (Figure 9.7) suggests that there is a degree of self-association between protein molecules.

Coacervation occurs in tropoelastin solutions and is a precursor event in the assembly of elastin nanofibrils [42]. This phenomenon is thought to be mainly due to the interaction between hydrophobic domains of tropoelastin. In scanning electron microscopy (SEM) pictures, nanofibril structures are visible in coacervate solutions of elastin-based peptides [37,43]. Indeed, Wright et al. [44] describe the self-association characteristics of multidomain proteins containing near-identical peptide repeat motifs. They suggest that this form of self-assembly occurs via specific intermolecular association, based on the repetition of identical or near-identical amino acid sequences. This specificity is consistent with the principle that ordered molecular assemblies are usually more stable than disordered ones, and with the idea that native-like interactions may be generally more favorable than nonnative ones in protein aggregates.

Other clues to the self-association of recombinant resilin in solution, and thus a degree of defined structure, include the propensity of the monomer proteins to covalently cross-link very rapidly through dityrosine side chains using a ruthenium-based photochemical method [29]. Proteins which do not naturally self-associate do not form biomaterials when exposed to the Ru(II)-based photochemical procedure (Elvin, C.E. and Brownlee, A.G., personal communication). Furthermore, Kodadek and colleagues showed that only intimately associated proteins are cross-linked via this "zero-Å" photochemistry procedure [45].

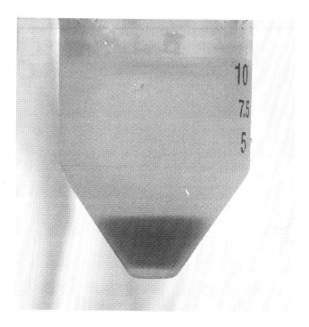

FIGURE 9.7 Coacervated purified recombinant resilin. The lower phase of the aqueous solution contains about 300 mg/mL of purified recombinant resilin protein. The upper aqueous phase contains about 20 mg/mL protein. The protein solution separates at 4°C into a protein-rich lower phase.

9.3.2 Tensile Properties

Tensile tests were carried out on recombinant resilin in PBS buffer on an Instron Tensile Tester (model 4500) at a rate of 5 mm/min and a temperature of 21°C. The swollen strip samples (7 × 1 mm) had a gauge length of 5 mm and were cycled initially up to a strain of about 200%. The

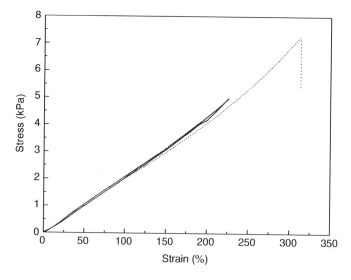

FIGURE 9.8 Typical stress–strain plots for a strip of recombinant resilin tested in phosphate-buffered saline (PBS). Sample cycled to 225%, showing resilience of 97% (solid curve) and later tested to failure showing extension at break of 313% (dotted curve). (From Elvin, C.M., Carr, A.G., Huson, M.G., Maxwell, J.M., Pearson, R.D., Vuocolo 1, T., Liyon, N.E., Wong, D.C.C., Merritt, D.J., and Dixon, N.E., *Nature*, 437, 999, 2005.)

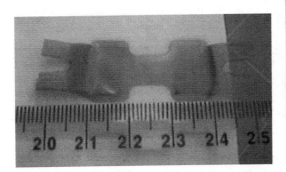

FIGURE 9.9 Small dumbbell-shaped sample of recombinant resilin prepared with embedded fine stainless steel mesh tabs (left) and being tested in phosphate-buffered saline (PBS) (right).

maximum strain was increased successively in steps of 25%–50% until failure occurred after stretching to over 300% of its original length (Figure 9.8). It should be noted that recombinant resilin is highly swollen (80% water at equilibrium) and consequently is quite soft. Figure 9.8 shows the recombinant resilin to have a secant modulus at 100% of about 2.5 kPa, considerably below that of typical unfilled synthetic elastomers (700 kPa) [46] and native elastin (1200 kPa) [47]. The resilience properties of the biomaterial in PBS were shown to be 97%, consistent with the scanning probe microscope (SPM) studies (see later). These values are similar to those reported (90%) for elastin [16].

Conducting tensile tests on soft hydrogel materials poses a number of challenges, not least of which is how to clamp the sample. Initial samples were prepared by gluing plastic tabs to strips of recombinant resilin that had been allowed to dry out. Unfortunately, the constraint imposed on the sample by the tabs during rehydration results in less than uniform samples. Recently small dumbbell-shaped samples have been prepared whereby fine stainless steel mesh tabs are incorporated prior to the photo-cross-linking. The protein solution readily penetrates the mesh and after cross-linking the tab is firmly embedded in the test piece (Figure 9.9). This results in much more uniform samples that can be tested without any dehydration/rehydration step.

9.3.3 BIOCOMPATIBILITY

The biocompatibility, biodegradability, and toxicity of recombinant resilin, including leachables, need to be evaluated prior to use in humans in a similar fashion to any other biopolymer or synthetic material. As discussed in Section 9.2, the solid cured resilin biomaterial is derived from purified soluble recombinant resilin which is cross-linked using tris(bipyridyl) ruthenium II chloride and ammonium persulfate. At the concentrations needed for effective curing of the soluble recombinant resilin, all three components showed no cytotoxicity against primary human chondrocytes (Werkmeister, J.A., Tebb, T., and Ramshaw, J.A.M., unpublished results). Figure 9.10 shows a live/dead viability assay where Calcein AM and Ethidium homodimer-1 (EthD-1) are used to distinguish live and dead cells, respectively [48]. Live cells can be detected by the presence of intracellular esterases that can convert nonfluorescent cell-permeant Calcein AM to a strong green fluorescence. EthD-1 enters cells with damaged cell membranes and upon binding DNA emits a strong red fluorescence. The light gray areas in Figure 9.10 represent green fluorescence, evidence

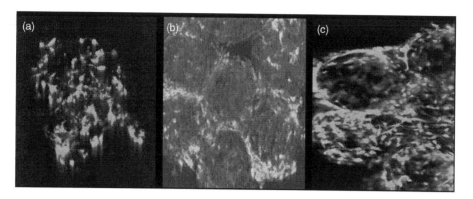

FIGURE 9.10 *In vitro* cytotoxicity testing of individual components of recombinant resilin curing polymer system. The light gray areas represent green fluorescence, evidence of live cells. (a) Ammonium persulphate 10 mM—noncytotoxic. (b) Ruthenium red complex $[Ru(Bpy)_3]^{2+}$, 2 mM—noncytotoxic. (c) Recombinant resilin, 200 mM—noncytotoxic. (From Werkmeister, J.A., Tebb, T., and Ramshaw, J.A.M., unpublished data).

of live cells in all three samples. An insignificant amount of red fluorescence was also visible for the resilin (sample C, not visible in the grayscale image).

In summary, none of the individual components necessary for recombinant resilin curing were cytotoxic, and there were no leachables from the cured resilin that caused cell death. The cured resilin polymer was not a good surface for cell adhesion, but cells can survive and proliferate in the resilin on a gelatine bead. The curing of recombinant resilin in the presence of cells on beads has no effect on the cells' ability to migrate and proliferate with new tissue formation. The resilin is seen to degrade with time, but it is believed that this could be controlled by the type and extent of cross-linking.

9.3.4 MOISTURE UPTAKE

The recombinant resilin is cross-linked from a 20% solution of protein in PBS buffer (see Section 9.2) and this appears to determine the equilibrium swelling of the sample. The solution gels upon irradiation with no apparent change of volume. When the cross-linked sample of recombinant resilin was dried on filter paper to constant weight and then rehydrated in saline solution it took up approximately 80% water/saline solution (Figure 9.11) [49]. The dehydration/rehydration step is quite reversible (Figure 9.11), and mechanical properties too have been shown to be unaffected by hydration cycling. Like all hydrogel materials the properties are heavily dependent on the plasticizing effect of water. Testing at 80% RH has shown recombinant resilin to have less than 10% resilience and be more than 1000 times stiffer than the fully hydrated material. Under ambient conditions it is tough and leathery.

9.3.5 FATIGUE LIFETIME

No results have been acquired to date on the fatigue lifetime of recombinant resilin; however, it is informative to consider the performance of natural resilin.

Firstly, investigation of the expression of the gene encoding the CG15920 protein shows that it is only actively transcribed during the pupal stage of development. Total RNA was prepared from cells at each life stage and first-strand CG15920 cDNA was quantified by real-time quantitative PCR; data were standardized against expression of the ribosomal protein RPLPO gene. The results (Figure 9.12) clearly show that CG15920 is expressed only during the pupal stage, when its expression is several thousandfold greater than in both larval and adult stages. This result is consistent with the developmental profile of holometabolous insects such as *D. melanogaster* which

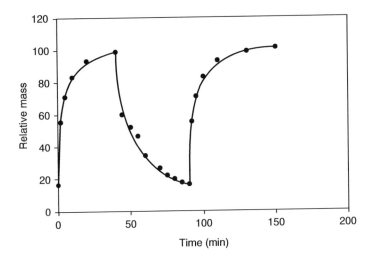

FIGURE 9.11 Relative mass change as a result of water absorption and loss in a solid piece of recombinant resilin prepared from a 20% protein solution in phosphate-buffered saline (PBS), cross-linked using peroxidase (see [29] supplementary material). The fully swollen sample is designated "100." (Data courtesy of Shekibi, Y., Nairn, K., Bastow, T.J., and Hill, A.J., The states of water in a protein based hydrogel, Internal CSIRO Report.)

lay down their adult cuticle in the pupal stage. Thus, the resilin synthesized at the pupal stage must survive for the lifetime of the adult insect. The consequence of this is that when the adult insects emerge from their cocoons, they have their lifetime's supply of resilin, since they have no way of renewing the supply of native protein.

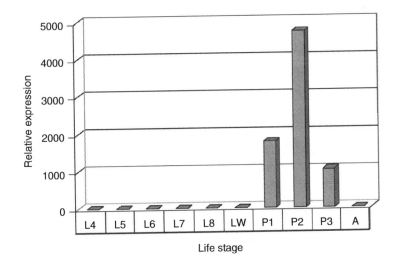

FIGURE 9.12 Real-time quantitative polymerase chain reaction (RT–PCR) analysis of expression of the *Drosophila CG15920* gene. Total RNA was prepared from various life stages (L4–L8, larvae day 4 to day 8; P1, newly pupated; P2, contracted light-colored pupae; P3, dark pupae; A, adult flies) of *Drosophila melanogaster* (Oregon R) and quantitative real-time RT–PCR carried out on first-strand cDNA. CG15920 mRNA levels were normalized to that of the ribosomal protein gene RPLPO, and are shown relative to that in the adult. (From Elvin, C.M., Carr, A.G., Huson, M.G., Maxwell, J.M., Pearson, R.D., Vuocolo 1, T., Liyon, N.E., Wong, D.C.C., Merritt, D.J., and Dixon, N.E., *Nature*, 437, 999, 2005.)

Secondly, if one considers the number of contraction/extension cycles resilin must accomplish during the course of an insect's life, then whether it is in the tymbal organ of a cicada producing sound vibrations at 4 kHz [22] or in a fly flapping its wings at 720,000 cycles/h [50], these calculations sum up to rather big numbers. In the case of a honeybee, for example, the adults, which live for about eight weeks and fly for about 8 h per day, are likely to flap their wings about 500 million times. In this respect, resilin resembles cross-linked elastin in human arteries, which must also survive for the entire lifetime of the organism [16].

9.3.6 NANO-MECHANICAL PROPERTIES BY SPM

It has become routine these days to use an SPM to measure the modulus or stiffness of materials with spatial resolution of nanometers [51]. Adhesion between the tip and the sample is another physical property also commonly measured. These physical properties can be measured because of the ability of the instrument to generate force curves, where the tip is driven toward the sample and then retracted and the cantilever deflection is measured as a function of distance moved by the base of the cantilever. The SPM force curve is similar to a conventional compression test, although the rate of deformation is not constant; rather it decreases as penetration by the cone-shaped indenter increases, and then increases again as the indenter is withdrawn. Analysis often involves fitting the unloading curve to the Sneddon model [52] to calculate an elastic modulus, but given the difficulty of accurately determining the cantilever stiffness and probe radius, it is sometimes just as useful to report maximum penetration as an inverse measure of sample stiffness [53]. A few workers have used the area under the curve to define sample properties. A-Hassan et al. [54] showed a good correlation between areas and Shore A hardness for a range of silicone polymers, whilst Briscoe et al. [55], using a nano-indentor, used the area under the loading and unloading curves to define a plasticity index to characterize the relative elastic/plastic behavior of thermoplastic materials. Recently we reported on the new technique of measuring resilience by means of an SPM [56].

SPM force curves are acquired by moving the tip toward the sample and recording the cantilever deflection as a function of the so-called Z position. Cantilever deflection is directly proportional to the force exerted on the sample by the tip. If the spring constant (k) of the cantilever is known, the force can be calculated. The Z position defines the distance from the sample to the piezo, to which the base of the cantilever is attached (Figure 9.13). By convention the closest point of approach by the piezo is designated as zero on the x-axis. Note that for some instruments the piezo is attached to the sample stage and thus moves the sample up toward the tip; however, this does not change the analysis.

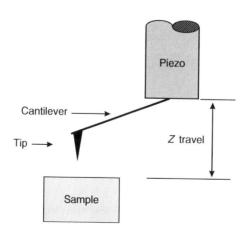

FIGURE 9.13 Schematic diagram of a force–distance experiment, illustrating the principle components and their movement. (From Huson, M.G. and Maxwell, J.M., *Polym. Test.*, 25, 2, 2006.)

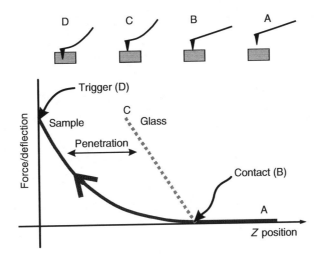

FIGURE 9.14 Typical approach force curve (solid line) for a sample which is penetrated by the scanning probe microscope (SPM) tip. Also shown is the force curve (dashed line) when the tip encounters a hard surface (glass) and schematic drawings of the relative positions of the SPM tip and the sample surface as related to the force curves. (From Huson, M.G. and Maxwell, J.M., *Polym. Test.*, 25, 2, 2006.)

Figure 9.14 shows a typical approach force curve along with schematic drawings of the relative positions of the SPM tip and the sample surface, as related to the force curve. At the start of the experiment, i.e., position A on the right-hand side of the figure, the tip is above the surface of the sample. As it approaches the surface the Z value decreases until at position B the tip contacts the surface. With further downward movement of the piezo the cantilever starts to be deflected by the force imposed on it by the surface. If the surface is much stiffer than the cantilever, we get a straight line with a slope of −1, i.e., for every 1 nm of Z travel we get 1 nm of deflection (line BC in Figure 9.14). If the surface has stiffness similar to that of the cantilever, the tip will penetrate the surface and we get a nonlinear curve with a decreased slope (line BD in Figure 9.14). The horizontal distance between the curve BD and the line BC is equal to the penetration at any given cantilever deflection or force. The piezo continues downward until a preset cantilever deflection is reached, the so-called trigger. The piezo is then retracted a predetermined distance, beyond the point at which the tip separates from the sample.

In the current work a Digital Instruments Dimension 3000 SPM was operated in force–volume mode using a probe with stiffness selected to match the stiffness of the sample. Standard silicon nitride probes with a nominal spring constant of 0.12 or 0.58 N/m were used for recombinant and native resilin samples. These samples were characterized in a PBS bath at a strain rate of 1 Hz. For synthetic rubbers, silicon probes with a nominal spring constant of 50 N/m were used and the material was characterized in air. Typically, at least three force–volume plots (16×16 arrays of force–displacement curves taken over a 10×10 μm area) were recorded for each of the samples.

Figure 9.15 shows typical force curves for a chlorobutyl rubber (CIIR) and a natural rubber (NR) sample. It is immediately obvious that the CIIR sample is softer and, as expected, shows much greater hysteresis and hence poorer resilience than the NR sample.

It has been shown [56] that if we measure the areas under the approach and retract curves of the force–distance plot we can get quantitative values of the resilience. Resilience is closely related to the ability of the polymer chain to rotate freely, and thus will be affected by rate and extent of deformation, as well as temperature. Different materials will respond differently to changes in these variables [46]; hence, changing the conditions of testing will result in a change in absolute values of resilience and may even result in a change in ranking of the materials. Compared to more traditional methods of resilience measurement such as the rebound resiliometer or a tensile/compression tester,

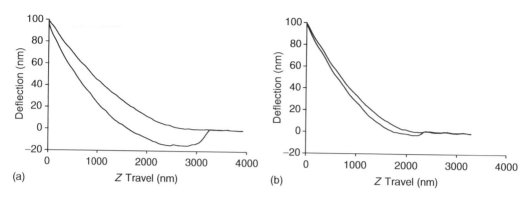

FIGURE 9.15 Typical force curves on (a) chlorobutyl rubber (CIIR) and (b) natural rubber (NR) samples showing much greater hysteresis and hence poorer resilience for the CIIR sample. (From Huson, M.G. and Maxwell, J.M., *Polym. Test.*, 25, 2, 2006.)

the SPM operates at a much slower rate and imposes only a very small force on the sample but because of the probe size the stress imposed is relatively large. In spite of these differences, an SPM investigation of unfilled NR, CIIR, butadiene (BR), and polyurethane (PU) rubber, as well as a filled NR bung (SRB) ranked them in the same order as the rebound test and the compression test although the absolute values were quite different (Figure 9.16, [56]).

The resilience of the PU elastomer as measured by the SPM test, in particular, was surprisingly low; however, examination of the sample after several force–volume tests in the same area revealed a grid of small indentation marks roughly associated with the 16×16 grid of force curves applied during the force–volume tests (Figure 9.17). Thus, on indentation, the relatively high maximum stress imposed by the SPM tip exceeded the elastic limit of the PU, causing plastic deformation. The plastic flow adds to the energy losses and explains the lower measured resilience when compared to the results from the Instron compression test. Denny [57] reports a similar lowering

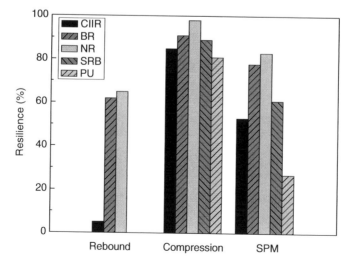

FIGURE 9.16 Resilience values for chlorobutyl rubber (CIIR), butadiene rubber (BR), unfilled natural rubber (NR), filled natural rubber (SRB), and polyurethane (PU) samples tested using a Shore rebound resiliometer, an Instron compression tester and a scanning probe microscope (SPM). (From Huson, M.G. and Maxwell, J.M., *Polym. Test.*, 25, 2, 2006.)

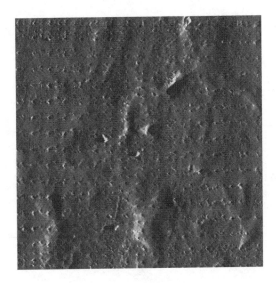

FIGURE 9.17 A 20×20 μm amplitude image of the PU sample after several 16×16 force–volume tests in the same area. (From Huson, M.G. and Maxwell, J.M., *Polym. Test.*, 25, 2, 2006.)

of the resilience due to plastic deformation during the first loading cycle of dragline silk. Obviously, in this case the maximum load could have been reduced until no plastic deformation occurred; however, the purpose of including this result is to highlight the potential for exceeding the elastic limit when using the SPM. Fortunately one of the benefits of using an SPM is that any permanent deformation is readily seen.

A tendon from the wing region of the dragonfly was dissected and a cross-section with exposed resilin was mounted in PBS such that SPM force measurements could be made. The sample was shown to be 92% resilient [29]. When the technique was applied to recombinant resilin, the approach and retract curves were almost superimposed (Figure 9.18a). Analysis yielded a value

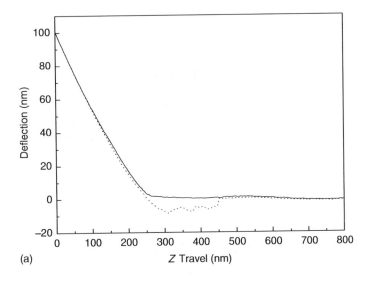

(a)

FIGURE 9.18 Elastic properties of cross-linked recombinant resilin. (a) A single force–extension curve recorded for a sample of cross-linked recombinant resilin (approach curve solid, retract curve dotted).

(continued)

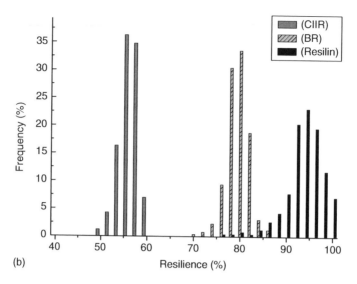

(b)

FIGURE 9.18 (continued) (b) Resilience measurements of elastomers. Samples of chlorobutyl rubber (CIIR), polybutadiene rubber (BR), and cross-linked recombinant resilin. (From Elvin, C.M., Carr, A.G., Huson, M.G., Maxwell, J.M., Pearson, R.D., Vuocolo1, T., Liyou, N.E., Wong, D.C.C., Merritt, D.J., and Dixon, N.E., *Nature*, 437, 999, 2005.)

of 92% resilience, consistent with natural resilin and considerably better than CIIR (56%) and BR (80%) (Figure 9.18b).

9.3.7 CROSS-LINK DENSITY

9.3.7.1 Chemical

The extent of dityrosine formation increased with time of irradiation to a maximum near 20% (relative to total tyrosine) after 30 s (Figure 9.19a). Samples with >17% dityrosine were solid, while those with lesser degrees of cross-linking were not. By comparison, Andersen [33] showed that about 25% of tyrosine occurs as the dityrosine dimer in natural locust wing hinge resilin. The deduced amino acid sequence of recombinant resilin (molecular weight 28,492 Da) contains 17 tyrosine residues; thus, from these data, the molecular mass between cross-links (Mc) can be estimated as 8500 Da. This assumes that all dityrosine molecules form effective cross-links; however, many may be intramolecular leading to a higher actual Mc. The effect of increased cross-linking on physical properties is shown in Figure 9.19b and c. The penetration of an SPM tip into the sample is inversely proportional to the modulus and shows the material getting stiffer with increased exposure time. The resilience increases to a plateau initially but decreases at long exposure times, suggesting some change to the network, even though this is not reflected in the dityrosine level or modulus.

9.3.7.2 Physical

It is also possible to estimate the cross-link density from the stress–strain data, using the statistical theory of rubber-like elasticity [47,58]. For a swollen rubber the relationship is

$$\sigma = NRTv^{1/3}*\left(1 - \frac{2M_c}{M}\right)*\left(\alpha - \frac{1}{\alpha^2}\right) \tag{9.1}$$

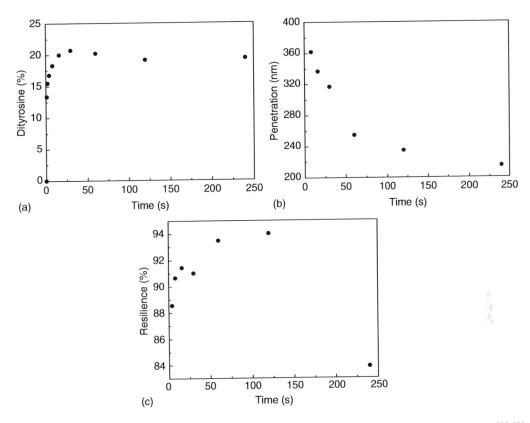

FIGURE 9.19 Time course studies of recombinant resilin showing changes occurring on exposure to a 600 W tungsten-halide source. (a) Level of dityrosine. (b) Penetration of a scanning probe microscope (SPM) tip ($k = 0.12$ N/m) into the samples. (c) Resilience. (From Elvin, C.M., Carr, A.G., Huson, M.G., Maxwell, J.M., Pearson, R.D., Vuocolo1, T., Liyou, N.E., Wong, D.C.C., Merritt, D.J., and Dixon, N.E., *Nature*, 437, 999, 2005.)

where

σ is the stress (force per swollen, unstressed cross-sectional area)

N the cross-link density (equal to ρ/M_c)

R the gas constant

T the absolute temperature

v the volume fraction of rubber in the swollen sample (0.2)

ρ the unswollen density (1.33 g/cc)

M_c the molecular mass between cross-links

M the primary molecular mass (28,492 Da)

α the extension ratio

For an ideal elastomer, a plot of σ versus $(\alpha - 1/\alpha^2)$ should yield a straight line with a slope that can be used to determine the mass between cross-links (M_c). Figure 9.20 shows the stress–strain data for the extension (up to 100% strain) part of the first cycle, plotted according to Equation 9.1. Whilst the first section is reasonably linear with a slope of 733 Pa, the data deviate from a linear relationship after $(\alpha - 1/\alpha^2)$ reaches about 0.6 (25% strain). This is slightly sooner than that found for natural resilin by Weis-Fogh [3], who showed the stress–strain data deviating from the theory at about 50% strain. On average the slope of the stress versus $(\alpha - 1/\alpha^2)$ plots was 814 Pa, which results in an M_c of 14,161 Da, in reasonable agreement with the estimate from chemical analysis data. Note that the

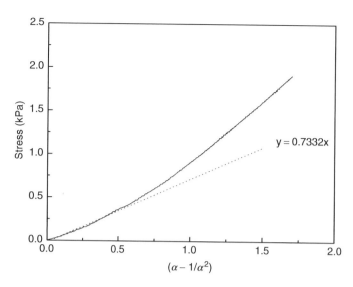

FIGURE 9.20 Stress–strain data from the extension part (up to 100% strain) of the first cycle of a test, plotted according to Equation 9.1. The straight line (dotted) is the least squares fit of the data up to 20% strain $(\alpha - 1/\alpha^2 = 0.5)$. (From Elvin, C.M., Carr, A.G., Huson, M.G., Maxwell, J.M., Pearson, R.D., Vuocolo1, T., Liyou, N.E., Wong, D.C.C., Merritt, D.J., and Dixon, N.E., *Nature*, 437, 999, 2005.)

value of M_c is not very sensitive to the slope of the curve in Figure 9.20. Even if we double the value of the slope, the value of M_c only decreases to 14,077 Da. Note too that the term $(1 - 2M_c/M)$ accounts for chain ends which do not contribute to the load-bearing capacity of the material. For most rubbers, both natural and synthetic, the primary molecular mass (M) is of the order of 10^5 or 10^6 Da and $(1 - 2M_c/M)$ is close to 1. In the case of the recombinant resilin the relatively low primary molecular mass results in a lot of chain ends and may be a major reason for the lower than expected modulus.

9.4   FUTURE WORK

We are planning to investigate the production of various forms of recombinant resilin, including the biosynthesis of resilin proteins from insects such as flea, mosquito, and cicada. We are also carrying out complete synthesis of *in vitro* genes encoding synthetic versions of resilin genes (such as *Drosophila*, *Anopheles*, and *Ctenocephaledes*). Ultimately, we need to develop mimetic forms of resilin that retain the remarkable mechanical properties (high fatigue lifetime, high resilience) whilst surviving degradation in the body. We are currently exploring means to modify the stiffness of the cross-linked hydrogels, by increasing cross-link density and/or increasing the primary molecular mass of the uncross-linked polymer. We believe these stiffer, biocompatible materials would be ideal candidates for spinal disk implants and vascular prostheses. Other forms of biomimetic resilin might find application in industry as microelectromechanical systems (MEMS) actuator devices or as components in accelerometers.

REFERENCES

1. Gosline, J.M., Hydrophobic interaction and a model for the elasticity of elastin, *Biopolymers*, 17, 677–695, 1978.
2. Urry, D.W., Entropic elastic processes in protein mechanisms. 1. Elastic structures due to an inverse temperature transition and elasticity due to internal chain dynamics, *J. Prot. Chem.*, 7, 1–34, 1988.

3. Weis-Fogh, T., Molecular interpretation of the elasticity of resilin, a rubber-like protein, *J. Mol. Biol.*, 3, 520–531, 1961.

4. Weis-Fogh, T., Power in flapping flight, in *The Cell and the Organism,* Ramsay, J.A. and Wigglesworth, V.B. (Eds.), Cambridge University Press, Cambridge, 1961, 283–300.

5. Li, B. and Daggett, V., Molecular basis for the extensibility of elastin, *J. Musc. Res. Cell Motil.*, 23, 561–573, 2002.

6. Rousseau, R., Schreiner, E., Kohlmeyer, A., and Marx, D., Temperature-dependent conformational transitions and hydrogen-bond dynamics of the elastin-like octapeptide GVG(VPGVG): A molecular-dynamics study, *Biophys. J.*, 86, 1393–1407, 2004.

7. Pometun, M.S., Chekmenev, E.Y., and Wittebort, R.J., Quantitative observation of backbone disorder in native elastin, *J. Biol. Chem.*, 279, 7982–7987, 2004.

8. Hoeve, C.A.J. and Flory, P.J., The elastic properties of elastin, *Biopolymers*, 13, 677–686, 1974.

9. Tatham, A.S. and Shewry, P.R., Comparative structures and properties of elastic proteins, *Phil. Trans. R. Soc. Lond. B. Biol. Sci.*, 357, 229–234, 2002.

10. Tompa, P., The interplay between structure and function in intrinsically unstructured proteins, *FEBS Lett.*, 579, 3346–3354, 2005.

11. Tompa, P., Szasz1, C., and Buday, L., Structural disorder throws new light on moonlighting, *TIBS*, 30, 484–489, 2005.

12. Vucetic, S., Obradovic, Z., Vacic, V., Radivojac, P., Peng, K., Iakoucheva, L.M., Cortese, M.S., Lawson, J.D., Brown, C.J., Sikes, J.G., Newton, C.D., and Dunker, A.K., DisProt: A database of protein disorder, *Bioinformatics*, 21, 137–140, 2005.

13. Shewry, P.R., Tatham, A.S., and Bailey, A.J. (Eds.), *Elastomeric Proteins: Structures, Biomechanical Properties and Biological Roles*, Cambridge University Press, Cambridge, 2002.

14. Tatham, A.S. and Shewry, P.R., Elastomeric proteins: Biological roles, structures and mechanisms, *Trends Biochem. Sci.*, 25(11), 567–571, 2000.

15. Andersen, S.O., The crosslinks in resilin identified as dityrosine and trityrosine, *Biochim. Biophys. Acta*, 93, 213–215, 1964.

16. Gosline, J.M., Lillie, M., Carrington, E., Gerette, P., Ortleppa, C., and Savage, K., Elastic proteins: Biological roles and mechanical properties, *Philos. Trans. R. Soc. Lond. B Biol. Sci.*, 357, 121–132, 2002.

17. Weis-Fogh, T., A rubber like protein in insect cuticle, *J. Exp. Biol.*, 37, 887–907, 1960.

18. Gorb, S.N., Serial elastic elements in the damselfly wing: Mobile vein joints contain resilin, *Naturwissenschaften*, 86, 552–555, 1999.

19. Neville, A.C. and Rothschild, M., Fleas—insects which fly with their legs, *Proc. R. Ent. Soc. Lond. (C)*, 32, 9–10, 1967.

20. Rothschild, M. and Schlein, J., The jumping mechanism of *Xenopsylla cheopis*. I. Exoskeletal structures and musculature, *Philos. Trans. R. Soc. Lond. B Biol. Sci.*, 271, 457–490, 1975.

21. Burrows, M., Biomechanics: Froghopper insects leap to new heights, *Nature*, 424, 509, 2003.

22. Young, D. and Bennet-Clark, H.C., The role of the tymbal in cicada sound production, *J. Exp. Biol.*, 198, 1001–1019, 1995.

23. Skals, N. and Surlykke, A., Sound production by abdominal tymbal organs in two moth species: The green silver-line and the scarce silver-line (Noctuoidea: Nolidae: Chloephorinae), *J. Exp. Biol.*, 202, 2937–2949, 1999.

24. Scheller, J., Guhrs, K.H., Grosse, F., and Conrad, U., Production of spider silk proteins in tobacco and potato, *Nat. Biotechnol.*, 19(6), 573–577, 2001.

25. Lazaris, A., Arcidiacono, S., Huang, Y., Zhou, J.F., Duguay, F., Chretien, N., Welsh, E.A., Soares, J.W., and Karatzas, C.N., Spider silk fibers spun from soluble recombinant silk produced in mammalian cells, *Science*, 295(5554), 472–476, 2002.

26. Ito, H., Steplewski, A., Alabyeva, T., and Fertala, A., Testing the utility of rationally engineered recombinant collagen-like proteins for applications in tissue engineering, *J. Biomed. Mater. Res. A.*, 76 (3), 551–560, 2006.

27. Elmorjani, K., Thievin, M., Michon, T., Popineau, Y., Hallet, J.N., and Gueguen, J., Synthetic genes specifying periodic polymers modelled on the repetitive domain of wheat gliadins: Conception and expression, *Biochem. Biophys. Res. Commun.*, 239(1), 240–246, 1997.

28. Miao, M., Cirulis, J.T., Lee, S., and Keeley, F.W., Structural determinants of cross-linking and hydrophobic domains for self-assembly of elastin-like polypeptides, *Biochemistry*, 44(43), 14367–14375, 2005.

29. Elvin, C.M., Carr, A.G., Huson, M.G., Maxwell, J.M., Pearson, R.D., Vuocolo1, T., Liyou, N.E., Wong, D.C.C., Merritt, D.J., and Dixon, N.E., Synthesis and properties of crosslinked recombinant pro-resilin, *Nature*, 437(7061), 999–1002, 13 October 2005.

30. Taksali, S., Grauer, J.N., and Vaccaro, A.R., Material considerations for intervertebral disc replacement implants, *Spine J.*, 4, 231S–238S, 2004.

31. Ardell, D.H. and Andersen, S.O., Tentative identification of a resilin gene in *Drosophila melanogaster*, *Insect Biochem. Mol. Biol.*, 31, 965–970, 2001.

32. Lyons, R.E., Lesieur, E., Kim, M., Wong, D., Brownlee, A., Pearson, R., and Elvin, C., The development of synthetic genes encoding repetitive resilin-like polypeptides: Construct design, bacterial expression and rapid purification, *Protein Eng. Design Select.*, 20(1), 25–32, 2007.

33. Andersen, S.O., Covalent cross-links in a structural protein, resilin, *Acta Physiol. Scand. Suppl.*, 263, 1–81, 1966.

34. Malencik, D.A. and Anderson, S.R., Dityrosine formation in calmodulin: Cross-linking and polymerization catalyzed by *Arthromyces* peroxidase, *Biochemistry*, 35, 4375–4386, 1996.

35. Fancy, D.A. and Kodadek, T., Chemistry for the analysis of protein–protein interactions: Rapid and efficient cross-linking triggered by long wavelength light, *Proc. Natl. Acad. Sci. USA*, 96, 6020–6024, 1999.

36. Urry, D.W., Hugel, T., Seitz, M., Gaub, H.E., Sheiba, L., Dea, J., Xu, J., Hayes, L., Prochazka, F., and Parker, T., Ideal protein elasticity: The elastin models, in *Elastomeric Proteins*, Shewry, P.R., Tatham, A.S., and Bailey, A.J. (Eds.), Cambridge University Press, Cambridge, 2002, Chapter 4.

37. Keeley, F.W., Bellingham, C.M., and Woodhouse, K.A., Elastin as a self-organising biomaterial: Use of recombinantly expressed human elastin polypeptides as a model system for investigations of structure and self-assembly of elastin, *Philos. Trans. R. Soc. Lond. B Biol. Sci.*, 357, 185–189, 2002.

38. Neff, D., Frazier, S.F., Quimby, L., Wang, R.T., and Zill, S., Identification of resilin in the leg of cockroach, *Periplaneta americana*: Confirmation by a simple method using pH dependence of UV fluorescence, *Arthropod Str. Dev.*, 29, 75–83, 2000.

39. Bochicchio, B. and Tamburro, A.M., Polyproline II structure in proteins: Identification by chiroptical spectroscopies, stability, and functions, *Chirality*, 14(10), 782–792, 2002.

40. Grinberg, V.Y., Grinberg, N.V., Burova, T.V., Dalgalarrondo, M., and Haertle, T., Ethanol-induced conformational transitions in holo-α-lactalbumin: Spectral and calorimetric studies, *Biopolymers*, 46(4), 253–265, 1998.

41. Riesen, R. and Widmann, G., Protein characterization by DSC, *Thermochimica Acta*, 226, 275–279, 1993.

42. Vrhovski, B., Jensen, S., and Weiss, A.S., Coacervation characteristics of recombinant human tropoelastin, *Europ. J. Biochem.*, 250, 92–98 1997.

43. Bellingham, C.M., Lillie, M.A., Gosline, J.M., Wright, G.M., Starcher, B.C., Bailey, A.J., Woodhouse, K.A., and Keeley, F.W., Recombinant human elastin polypeptides self-assemble into biomaterials with elastin-like properties, *Biopolymers*, 70(4), 445–455, 2003.

44. Wright, C.F., Teichmann, S.A., Clarke, J., and Dobson, C.M., The importance of sequence diversity in the aggregation and evolution of proteins, *Nature*, 438(7069), 878–881, 2005.

45. Amini, F., Denison, C., Lin, H.J., Kuo, L., and Kodadek, T., Using oxidative crosslinking and proximity labelling to quantitatively characterize protein-protein and protein-peptide complexes, *Chem. Biol.*, 10(11), 1115–1127, 2003.

46. Cowie, J.M.G., *Polymers: Chemistry and Physics of Modern Materials*, Int. Textbook, Aylesbury, 1973.

47. Aaron, B.B. and Gosline, J.M., Elastin as a random-network elastomer: A mechanical and optical analysis of single elastin fibers, *Biopolymers*, 20, 1247–1260, 1980.

48. Bryant, S.J. and Anseth, K.A., The effects of scaffold thickness on tissue engineered cartilage in photocrosslinked poly(ethylene oxide) hydrogels, *Biomaterials*, 22, 619–626, 2001.

49. Shekibi, Y., Nairn, K., Bastow, T.J., and Hill, A.J., The states of water in a protein based hydrogel, Internal CSIRO report.

50. Lehmann, F.O. and Dickinson, M.H., The production of elevated flight force compromises maneuverability in the fruit fly *Drosophila melanogaster*, *J. Exp. Biol.*, 204, 627–635, 2001.

51. Magonov, S. and Chernoff, D.A., Atomic force microscopy, in *Comprehensive Desk Reference of Polymer Characterization and Analysis*, Brady, R.F., Jr. (Ed.), Oxford University Press, New York, 2003, Chapter 19.

52. Weisenhorn, A.L., Khorsandi, M., Kasas, S., Gotzos, V., and Butt, H.J., Deformation and height anomaly of soft surfaces studied with AFM, *Nanotechnology*, 4, 106–113, 1993.

53. Maxwell, J.M. and Huson, M.G., Scanning probe microscopy examination of the surface properties of keratin fibres, *Micron*, 36, 127–136, 2005.

54. A-Hassan, E., Heinz, W.F., Antonik, M.D., D'Costa, N.P., Nageswaran, S., Schoenenberger, C.-A., and Hoh, J.H., Relative microelastic mapping of living cells by atomic force microscopy, *Biophys. J.*, 74(3), 1564–1578, 1998.

55. Briscoe, B.J., Fiori, L., and Pelillo, E., Nano-indentation of polymeric surfaces, *J. Phys. D: Appl. Phys.*, 31, 2395–2405, 1998.

56. Huson, M.G. and Maxwell, J.M., The measurement of resilience with a scanning probe microscope, *Polym. Test.*, 25(1), 2–11, 2006.

57. Denny, M.W., Silks—their properties and function, in *The Mechanical Properties of Biological Materials*, Vincent, J.F.V. and Currey, J.D. (Eds.), Symposia of the Society for Experimental Biology; no. 34, Cambridge University Press, Cambridge 1980, 247–272.

58. Treloar, L.R.G., *The Physics of Rubber Elasticity*, Clarendon Press, Oxford, 1975.

10 Smart Elastomers

Dipak K. Setua

CONTENTS

10.1 INTRODUCTION

Of late, considerable convergence is taking place between many fields of science and engineering which were otherwise separate or distinct. Emergence of new technologies in every field and their impact of conglomeration have resulted in the development of smart materials and structures.

As a consequence of the above, the structural complexity of materials has increased tremendously. The smallest division of materials and devices, which were otherwise known to be of microsizes, has reached nanostates. Conventional materials have found a passage from functional materials to smart and subsequently to intelligent materials encompassing information science into materials structure. Today it is one of the major technological challenges to imbibe intelligent functions into material level, i.e., to impart capability into materials to sense changes in the environmental stimulus (light, temperature, pH, magnetic and electric field, etc.), to react and integrate the sensed information, and also to judge and adopt appropriate measures to nullify the effect of these environmental disturbances.

10.2 SMART MATERIAL

Smart materials or systems are systems that integrate the functions of sensing, actuation, logic, and control to respond adaptively to changes in their condition on the environment to which they are exposed, in a useful usually repetitive manner, and are evaluated in terms of changes in material properties, geometry, or mechanical or electromagnetic response. Smartness describes the presence of self-adaptability, self-sensing, memory, and multiple functionality in the structure, and its self-repairing characteristics to withstand sudden changes. Smart materials and systems are those which reproduce biological functions. A comparison of smart materials with other conventional systems is given in Table 10.1.

TABLE 10.1

Properties of Smart and Other Conventional Materials

Category	Fundamental Material Characteristics	Fundamental System Behaviors
Traditional materials, natural materials (stone, wood), fabricated materials (steel, aluminum)	Materials have given properties	Materials have no or limited intrinsic active response capability
High-performance materials, polymers, composites	Material properties are designed for specific purposes	Have good performance properties
Smart materials	Properties are designed to respond intelligently to varying external conditions or stimuli	Smart materials have active responses to external stimuli and can serve as sensors and actuators

10.3 SMART STRUCTURE

Smart or adaptive structures are a class of advanced structures with integrated sensors, actuators, and controls which allow it to adaptively change or respond to external conditions (Figure 10.1). Examples are buildings, bridges, and roadways that can sensor, mitigate, and control damage, e.g., aircraft that can actively minimize a structure-borne noise in the interior.

10.4 INTELLIGENT STRUCTURE

Intelligent structures are smart structures that have the added capability of learning and adapting rather than simply responding in a programmed manner, and this is usually accomplished by inclusion of Artificial Neural Network (ANN) into the structure (Figure 10.2).

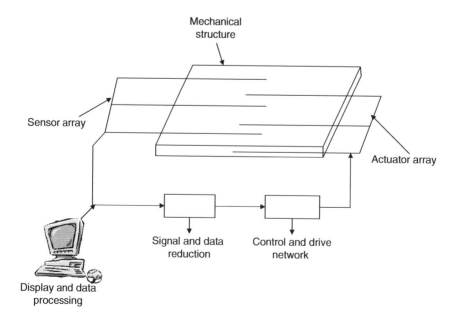

FIGURE 10.1 Conceptual diagram of smart structure.

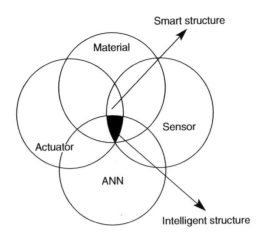

FIGURE 10.2 Venn diagram showing relationship between smart and intelligent structure.

10.5 WHAT ARE "SMART POLYMERS?"

Smart polymers are very fascinating materials that show distinct responses to differences or variations in the surrounding environment (such as thermal gradient, change in pH, and light).

They are called "intelligent" because we can in fact utilize their distinctive properties for a wide range of applications. Various types of smart polymers are shown in Figure 10.3.

Actuator materials possess the capability to convert input energy to mechanical output energy and their performance is typically measured by a number of basic parameters, e.g., strain, actuation stress, energy density (mechanical energy per stroke per unit volume or mass of the material), response time, and theoretical efficiency (the efficiency obtained with the use of an ideal electronics). Actuator materials find wide usage in drive mechanisms, where electroactive ceramics (EAC) and shape memory alloys (SMAs) are the leading alternatives. In this category of materials electroactive polymers (EAPs) offer promises for some unique performance and operational characeristics which are not found in other actuator technologies (Table 10.2).

EAPs are classified into two major groups: ionic EAPs (activated by an electrically driven diffusion of ions and molecules) and electronic EAPs (activated by an electrical field). Table 10.3 provides a summary of the mechanical performance, advantages, and cautions due to competitiveness of this class of materials with other technologies. It is worth mentioning that the fundamental mechanisms providing actuation for EAP actuators are yet to be completely understood and the robustness and stability of these polymers matching against that of natural muscle need to be accomplished. A subclass of EAPs, represented by dielectric elastomers, are able to be constituted as deformable actuators to generate high active strains and stresses, barely minimum response time but with high reliability, storage stability, and of course wide availability of materials at a lower cost.

10.6 DIELECTRIC ELASTOMER ACTUATOR

Dielectric elastomers are insulating, rubber-like structures, capable of undergoing reversible length change to a large extent and produce usable works. A dielectric elastomer actuator (DEA) works like an electrical motor by virtue of linear motion rather than rotation.

A DEA is basically a compliant capacitor where an incompressible, yet highly deformable, dielectric elastomeric material is sandwiched between two complaint electrodes. The electrodes are designed to be able to comply with the deformations of the elastomer and are generally made of a conducting material such as a colloidal carbon in a polymer binder, graphite spray, thickened electrolyte solution, etc. Dielectric elastomer films can be fabricated by conventional

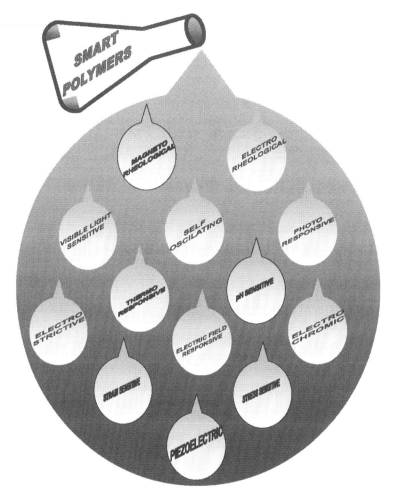

FIGURE 10.3 Types of smart polymers.

TABLE 10.2
Comparison of the Properties of EAP, EAC, and SMA

Property	Electroactive Polymers (EAPs)	Electroactive Ceramics (EACs)	Shape Memory Alloys (SMAs)
Actuation strain (%)	>10	0.1–0.3	<8
Force (MPa)	0.1–3	30–40	about 700
Reaction speed	μs to s	μs to s	s to min
Density (g/cc)	1–2.5	6–8	5–6
Drive voltage	2–7 V/10–100 V/μm	50–800 V	NA
Consumed power	mW	W	W
Fracture toughness	Resilient, elastic	Fragile	Elastics

Source: Bar-Cohen, Y. et al., Low-mass muscle actuators using electroactive polymers (EAP), *SPIE Conference on Smart Materials Technologies*, San Diego, California, March 1998, SPIE Vol. 3324, 0277-786X/98.

TABLE 10.3

Properties of Commonly Available EAP Actuators

Materials	Performance	Advantages	Critical Issues
Gels	Stress: 0.1–0.3 Mpa Strain: +1000%	Shows large volume change due to exposure to light or pH as stimuli	Slow response times, transport without liquid is critical, higher work capacity is also needed
Ionic polymers	Stress: 0.1–1 Mpa Strain: 1%–10%	Muscle-like behavior, easily processed/formed, large deformation possible	Poor understanding of processing techniques, higher speed and work capability required
Conducting polymers	Stress: 450 Mpa Strain: 1%–10%	Well understood mechanisms, reasonable deformation and forces demonstrated	Difficult to process, ability to construct devices is limited
Electrostrictive polymers	Stress: 0.2–2 MPa Strain: 10%–30%	Small strains, but large forces, easy to make into acoustic actuators	Elimination of hysteresis effects and determination of processing properties needed

Source: Wax, S.G. and Sands, R.R., Electroactive polymer actuator and devices, *Proceedings of the SPIE Conference on Electroactive Polymer Actuators and Devices*, Newport Beach, California, SPIE Vol. 3669, 0277-786X/1999.

polymer-processing techniques, e.g., spin coating or casting, and the electrodes are then created on top of the film on both sides via spraying, sputter coating, screen printing, or photolithography.

A comparative evaluation of different electrode materials on electromechanical performance of dielectric elastomer planer actuators has been given by Carpi et al.[1] The actuators were made of four different types of compliant electrode materials (thickened electrolyte solution, graphite spray, carbon grease, and graphite powder) and four different values of transverse prestress (19.6, 29.4, 39.2, and 49.0 kPa). Subsequently, a novel helical configuration of contractile linear actuator based on dielectrical elastomer (silicone rubber) which is suitable for generation of electrically driven axial contractions and radial expansions, maximum axial contraction strain of -5% at about $14V \ \mu m^{-1}$, has been designed and evaluated.[2] Molecular origins of high field electrostrictive properties of thermoplastic elastomers based on polyurethane (PU), which may be considered to be block copolymers consisting of alternating PU (hard) and polyol (soft) segments for plausible applications as acoustic projectors and transducers, have been studied by Zhenyi et al.[3] The advantages of graft elastomer (comprising two components: a flexible backbone with a grafted crystalline group) have successfully been utilized by Su et al.[4] in the construction of an efficient actuator to support tele-robotic requirements for NASA missions. The new graft elastomer has shown to demonstrate higher actuation strain, high mechanical energy density, tailorable molecular compositions, composition adjustment, and excellent processability compared to single-component dielectric elastomer, e.g., PU. The construction of ion-implanted DEA to avoid undesired stiffening of the elastomeric membrane by deposition of metal electrodes on the polymer for formation of compliant electrodes has been described by Dubois et al.[5] The authors have claimed to devise a novel method to microfabricate compliant electrodes by using implantation of metal ions into polymers to make the elastomer locally conductive without significantly increasing its stiffness. By implantation of specific micrometer-sized areas, it is possible to individually address many independent large displacements of DEAs on a single chip, thus allowing for many complex actuation schemes. The authors have measured a 110 μm vertical displacement, about eight times larger than a metal-evaporated compliant electrode, of a 35 μm thick ion-implanted polydimethyl siloxane (PDMS) square membrane of 1 mm^2 bonded to a silicone chip.

Actuators based on dielectric elastomer technology operate on a simple principle as shown in Figure 10.4. When an electric field is applied to the electrodes, positive charges appear on one

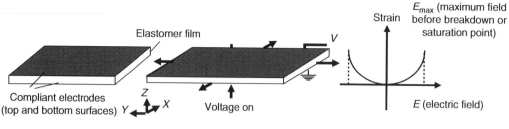

FIGURE 10.4 Principle of operation of dielectric elastomer.

electrode and negative charges on the other. This gives rise to Coulomb forces between the opposite charges, thus generating a pressure which is known as the Maxwell stress. The stress forces the electrode to move closer, thereby squeezing the elastomer. As the elastomer is thinned, it elongates in the directions perpendicular to the applied force. The net volume change of the elastomer is small because of its high bulk modulus. Therefore, the electrodes must be compliant to allow the film to strain. Exploiting this simple principle, involving the material's design and processing, actuator geometries are designed.

The effective pressure produced by the electrodes on the elastomeric film is a function of applied voltage and can be derived by using an electrostatic model proposed by Pelrine et al.[6] This pressure P is given by

$$P = \varepsilon_r \varepsilon_o E^2 = \varepsilon_r \varepsilon_o (V/t)^2 \tag{10.1}$$

where
ε_r is the permittivity of free space (8.85×10^{-12} F/m)
ε_o is the relative permittivity of the elastomer
E is the applied electric field
V is the applied voltage
t is the film thickness

The amount of strain in a particular direction, e.g., in the x direction, is given by

$$S_x = \varepsilon_r \varepsilon_o V^2 / 2Yt^2 \tag{10.2}$$

where Y is the Young's modulus. With a measurement of stress versus strain value and Poisson's ratio of the elastomer, the deformation in the direction of all three axes can be determined. In fact, the strain also depends on the boundary conditions applied and for low strains ($\leq 20\%$ with a constant Young's modulus of the material) the thickness strain s_z is given by[7]

$$s_z = -p/Y \tag{10.3}$$

The magnitude of Coulomb's force due to attraction between the positive and negative charges on the electrodes of the dielectric elastomer is given by the following equation:[8]

$$F = q_1 q_2 / 4\pi \varepsilon r^2 \tag{10.4}$$

where
q_1 and q_2 are the electric charges generated on the electrodes
ε being the permittivity of free space
r the distance between the two electrodes

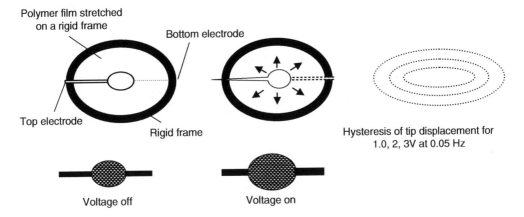

Voltage off · Voltage on

Hysteresis of tip displacement for 1.0, 2, 3V at 0.05 Hz

FIGURE 10.5 Dielectric elastomer actuator (DEA) demonstrated to expand and relax in a circular strain test.

The Maxwell's stress associated with the force is perpendicular to the applied force and is given by

$$T_{xy} = 1/4[E_xE_y + B_xB_y - 1/2(E^2 + B^2)\delta_{xy}] \tag{10.5}$$

where
  E_x and E_y are the electric fields in x and y directions
  B_x and B_y are the magnetic fields in the x and y directions
  δ = Kronecker delta, $\delta_{xy} = 0$ for $x \neq y$, $\delta_{xy} = 1$ for $x = y$

Method to test the performance of dielectric elastomer is a challenging task due to large strains associated with the flexibility of the material and also occurrence of chances of edge arcing or tearing during deformation. Figure 10.5 illustrates the principle of stretching of the film on a frame. The electrode is patterned as in a circle in the middle of the stretched film. When voltage is applied, the circle expands and the strain can be measured optically. The technique also allows us to test various prestrains, which is advantageous to increase the performance of many dielectric elastomers.

Table 10.4 summarizes the consistent responses, per Equation 10.1, of three elastomers that have been found to produce significantly large strain response.[9]

Arora[10] has studied the importance of prestrain level of dielectric elastomers of silicone and PU, and a parabolic relationship has been established between the applied electric field, the axial and radial

TABLE 10.4
Strain Response Characteristics of Dielectric Elastomers

Elastomer Type	Relative Thickness Strain (%)	Relative Area Strain (%)	Test Field Strength (MV/m)	Pressure (MPa)	Estimated Elastic Energy (MJ/m³)
		Circular Response			
Silicone A	48	93	110	0.3	0.098
Silicone B	39	64	350	3.0	0.75
Acrylic	61	158	412	16.2	3.4
		Linear Response			
Silicone A	54	117	128	0.4	0.16
Silicone B	39	63	181	0.8	0.2
Acrylic	68	215	239	2.4	1.36

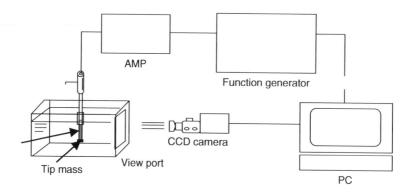

FIGURE 10.6 Test setup for strain measurement.

actuation strains, and actuation-blocking force (pressure). Silicone-based uniaxially prestrained prototype fiber actuators showed better actuation strain than PU. Kofod[11] and Kofod et al.[12] have obtained a parabolic relationship of blocking force with applied voltage, breakdown field strength, and effect of isotropic prestrain properties for optimum dimension of a dielectric silicone actuator made of 3M VHB 4910 (a commercial glue having higher dielectric constant than ordinary silicone). Prestrain provides advantages in two ways: First, it reduces the thickness of the elastomers by virtue of constancy of volume of the dielectric elastomers during actuation and the decrease in thickness results in the development of higher actuation strain and pressure at a specified voltage. Second, the prestrain also results in a change of modulus of the elastomer along the direction of prestrain. Theortical background on electrostrictive response of silicone and acrylic elastomers which can yield strain in excess of 100% has been given by Yamwong et al.,[13] Davies et al.,[14] and Krakovsky et al.[15]

Figure 10.6 gives a schematic view of the test setup of the strain measurement.[16,17] The boundary conditions in this stretched film method are modeled by finite element analysis with nonlinear material properties.

Table 10.5 gives a preliminary comparison of performance of dielectric elastomer with other actuator technologies. The reported values are collected from different sources and are not based on uniform measurement standards.[9]

10.7 ACTUATOR CONFIGURATIONS AND SOME POTENTIAL APPLICATIONS

Various types of DEAs have been reported in the literature to produce different kinds of motion and for a variety of functions.[9,18–24] Planar DEAs are fabricated by having one or more polymer films (single- or multilayered) placed between compliant electrodes. The design of DEA with a film of dielectric material between two electrodes (unimorph) or with two dielectric elastomer films with an interspaced electrode (bimorph) is shown in Figure 10.7.

Su et al.[4] have described the functions, e.g., bending and driving frequency of both unimorph and bimorph DEA actuators made of graft elastomer. The actuators were also made to function even at a low temperature of −50°C. Carpi et al.[1] have compared the effect of several parameters, namely, isotonic (at constant load) transverse strain, isometric (at constant length) transverse stress, and the driving current of dielectric elastomer planner actuator, for obtaining best values of strain, stress, and efficiency (i.e., amount of charge stored by the unit electrode area).

Diaphragm type of actuators (Figure 10.8) is made with a prestrained dielectric elastomer film placed over an opening of a rigid frame. The diaphragm is biased to actuate in one direction (up or down) rather than simply wrinkling randomly when voltage is applied. These types of actuators offer promises for making them suitable for pumps or loudspeakers and are also well suited for noise reduction in the interior cabin of aircraft for passenger/crew comfort, and enhancing weapon accuracy and component life by reducing structural vibrations, etc.

TABLE 10.5

Comparison of Properties of Dielectric Elastomers with Other Electroactive Materials

Materials	Maximum Strain (%)	Maximum Pressure (MPa)	Specific Elastic Energy Density (J/g)	Elastic Energy Density (J/cm³)	Maximum Efficiency (%)
Elastomers					
Polyurethane	100	4	2.0	0.10	<10
Acrylic	215	16	3.4	3.4	60–80
Silicone	63	3	0.75	0.75	90
Fluroelastomer	8	0.20	—	0.008	—
Polybutadiene	12	0.19	—	0.11	—
Natural rubber	11	0.094	—	0.0052	—
Elastrostrictor polymer (PVDF-TrFE)	4.3	43	0.49	0.3	80
Piezoelectrics					
Ceramic (PZT)	0.2	110	0.013	0.1	>90
Single crystal (PZN-PT)	1.7	131	0.13	1.0	>90
Polymer (PVDF)	0.1	4.8	0.0013	0.0024	80
Electrochemo-mechanical conducting polymer (Polyaniline)	10	450	23	23	<1%
Mechano-chemical polymer/gels	>40	0.3	0.06	0.06	30
Magnetostrictive (Terfenol-D)	0.2	70	0.0027	0.025	60
Shape memory alloy (TiNi)	>5	>200	>15	>100	<10
Shape memory polymer	100	4	2	2	<10

Source: Liu, C. et al., Electro-statically stricted polymers (ESSP), *SPIE Conference on Electroactive Polymer Actuator and Devices*, Newport Beach, California, March 1999, SPIE Vol. 3669, 0277-786X/99; Kornbluh, R. et al., *Application of Dielectric Elastomer EAP Actuators*, SPIE—the International Society for Optical Engineering, Bellingham, Washington, 2001, Chapter 16.

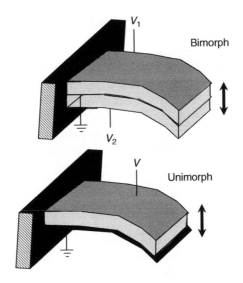

FIGURE 10.7 Planar actuator.

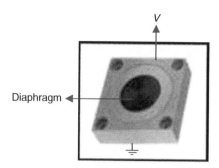

FIGURE 10.8 Diaphragm actuator.

Extended types of linear actuators (Figure 10.9) can also help control air or water flow over the surfaces of airplanes, ships, etc. They can easily change the shape or contour of surfaces of these systems, thus effecting lift and drag produced for active military camouflage applications. Some other exciting applications[25] include (a) inflatable spacecraft which can be launched in a packed form, (b) controllable lightweight large-aperture mirrors for remote-sensing applications as in space shuttles, (c) improved aerodynamics in flight by producing controlled twist in aircraft wings or helicopter rotor blades, and (d) smart windshield wipers.

By coating the compliant electrodes with a thicker but softer layer of polymer gel, the gel can spread out along with the expanding film during actuation but bunches at points where the film compresses.[26] If the polymers are imprinted with patterns of electrodes or shades of dots in a variety of shapes, these features can be raised or lowered to fabricate an active camouflage fabric which can change its reflectance for any defence systems and soldiers.

Another type of linear configuration known as "bow-tie" is shown in Figure 10.10. The actuator is constructed using dielectric elastomer film having the shape of a bow-tie with two compliant electrodes configured on its two surfaces. Application of the electric field results in planar actuation which because of the bow-tie shape is translated into linear motion. Typical applications of these types of actuators comprise a hexapod robot, to mimic the motion of insects like walking, to manufacture various animated devices like face, eyes, skin, etc., or the design of micro-air vehicle/ornithopter.

DEA configurations having cylindrical geometry are of two types: rolled (Figure 10.11) and tubular (Figure 10.12).

Rolled DEAs, also called spring rolls, are prepared by rolling up two layers of dielectric elastomer sheet laminated on both sides with flexible electrodes into the form of a compact cylinder. Sometimes, the sheets are also wrapped around a complex helical ring that holds a high circumferential prestrain on the films to enhance device performance. These actuators have one degree of freedom (1 DOF) and primarily function as a linear actuator. They can serve in many applications such as robotic arms,[27] prosthetic mechanism to powering microfluid pumps or lab-on-a-chip devices. Roll DEAs with higher degrees of freedom, have also been reported.[28] These are achieved by patterning the electrodes on the dielectric elastomer film before rolling such that

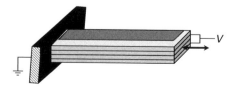

FIGURE 10.9 Extender actuator.

FIGURE 10.10 Bow-tie actuator.

after rolling the electrodes align radially on multiple circumferential spans of the rolls. Pei et al.[29] could fabricate a "Sushi Roll" with eight degrees of freedom (8 DOF) by putting four two degrees of freedom (2 DOF) DEA rolls in series and using a stretched acrylic film (3M VHB 4910). It has also been predicted that specially patterning the electrodes onto the dielectric elastomer film in such a fashion could split the roll lengthwise. If only the left half receives voltage and expands, and the right half does not move and inhibits the resulting motion, this causes the device to bend. Applications of bending rolls also include snake-like robots, manipulators, steerable catheters and endoscopes, and pointing mechanism of antennas.

Another way for fabricating cylindrical DEAs is to use elastomer tubes (Figure 10.12) instead of a dielectric film. Carpi et al.[2,22] have reported on the studies of cylindrical actuators using silicon tubes with compliant electrodes on the inner and outer walls of the tube.

The interest in the field of DEA has grown continuously during the last decade. Search for best materials (high dielectric constant, high breakdown voltage, good mechanical properties, resistance

FIGURE 10.11 Roll.

FIGURE 10.12 Tube.

to harsh environment conditions, etc.) is still an ongoing activity worldwide. Most of the published literature in DEA reports on the use of commercial materials like VHB 4910 from 3M, the silicone elastomers Sylgard 184 from DOW Corning, and Elastosil from Wacker. DEAs have one common drawback that the driving voltage is very high between 3 and 10 kV depending, however, on the polymer breakdown field and thickness of the film. Kofod[11] has mentioned about the emergence of a switch mode amplifier of small size (2.5 mm³) which is able to convert 10 to 1000 V allowing the possibility of incorporating low- to high-voltage conversion inside the DEA. Also, transistors are now available which work at 1 kV and, therefore, there are no constraints concerning the availability of electronics. Actually, working with high voltages sometimes proves to be useful as small currents are required which reduce the loss due to resistance of connecting cables and compliant electrodes. If the applications still necessitate requirement of high voltages like toys and orthopedic devices, it is possible to seal the high voltage within the actuator itself and they can be put to use for any desired operations.

In the process of conversion of electrical to mechanical energy by expanding area in dielectric elastomers, the phenomenon of stretching causes separation of like charges and hence reduces electrical energy. The reduction in electrical energy is balanced by an increase in elastic mechanical energy or mechanical work output. The two modes of conversion in DEAs (compared to only one in conventional air-gap electrostatic actuators)[26] are possible because elastomers can change shape only by maintaining a substantially constant volume (i.e., stretching in area is mechanically coupled to contraction in thickness and vice versa).

Conventional actuator applications, and the potential advantages gained by replacing existing technologies with DEA, are given in Table 10.6.

10.8   DIELECTRIC ELASTOMER GENERATORS

Dielectric elastomers also have the advantages of converting mechanical to electrical energy in a generator mode.[18] The authors have given some potential generator applications and relative advantage/disadvantage of dielectric elastomer with other competing technologies (Table 10.7).

In this mode the electrical charge is placed on the film in the stretched state (high capacitance). When the film is allowed to contract (low capacitance), the elastic stresses in the film work against the electric field pressure and thus increase the electrical energy. Figure 10.13 explains the basic mechanism.

TABLE 10.6
Potential Applications of Dielectric Elastomer as Actuator

Type of Application	Competing Technology	Potential Advantage
Linear actuators	Electromagnetics	Greater variety of shapes and sizes, performance similarity to pneumatics, greater controllability, and lower speed operation
Voice coils	Electromagnetics, electrostatics (air gap)	Higher strain capability, similar specific work output and force/weight ratio, but have limitations due to significant current leakage and thermal dissipation of load by pull-in instability
Robotic, animatronic, or prosthetic actuators	Electromagnetics (motors with gears and linkages)	Easy configuration, ability to combine actuation, sensing and structure in a single material; also similarity with pneumatics with lower cost, simplicity and greater controllability
Micro-Electro Mechanical Systems (MEMS)	Electrostatics, piezoelectrics	Higher energy and power density, simpler design, imperviousness to dust and grits, ability to combine actuation, Venn structure possible
Smart skin, aperture, large array	Piezoelectrics, shape memory alloys	Enable new functionality including modular electromagnetic properties, controlled surface texture, stealth properties, drag reduction and heat transfer capacity, rugged, etc. More cost-effective than silicon or other piezoelectrics

Source: Pelrine, R. et al., in SPIE, smart structures and materials, 2001, Electroactive polymer actuators and devices, Y. Bar-Cohen (Ed.), *Proceedings of SPIE*, Vol. 4329, 0277-786X/2001.

TABLE 10.7
Potential Applications of Dielectric Elastomer as Generator

Generator Type	Competing Technology	Potential Advantage
Engine-driven generators	Electromagnetics	Higher energy density, lower cost, good low speed performance, but the electronic cost is high and size of device is very small. Suitable electronics required for large engine applications
Shoe generators	Electromagnetics, piezoelectrics	Low cost, good load matching to eliminate mechanical complexity, lightweight, demonstrated 0.28 J energy/stroke in heel-strike device but electronics more complex than electromagnetics
Parasitic energy harvesting for remote sensors	Electromagnetics, piezoelectrics	Good load matching, remote sensing can eliminate wires, reduce cost for a number of applications but availability of power sources is limited
Wave energy	Electromagnetics	Good matching to load, low cost. But suitable coating/protective layer for water compatibility is required

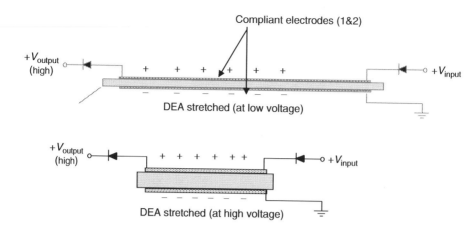

FIGURE 10.13 Basic mechanism of dielectric elastomer actuator (DEA) generator. (From Kornbluh, R., Power from plastic: How electroactive polymer "artificial muscles" will improve portable power generator in the 21st century military, Presented at *TRI-Service Power Expo*, Norfolk, Virginia, July 2003. With permission.)

Energy involved due to the change in capacitance is given by

$$E = 1/2Q_1Q_2(1/C_f - 1/C_i)$$ (10.6)

where

E is the energy for this configuration
$Q_1 = Q_2$ is the magnitude of charge on electrode 1 and 2
C_f is the final capacitance
C_i is the initial capacitance

One major application of these dielectric elastomer generators along with the right electronic device is in the area of "heel-strike generator"[30] and their recent progresses;[31] capturing the compression energy of a shoe heel when it strikes the ground during walking or running is a good way to generate portable electrical power (Figure 10.14).

Energy from the heel strike is "free"—it would otherwise be dissipated as heat. The heel-strike generator effectively couples the compression of the heel to the deformation of an array of multi-

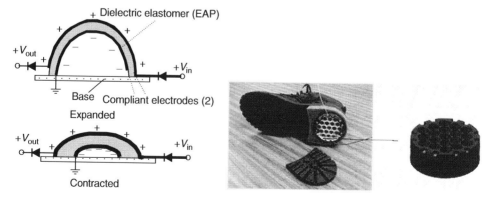

FIGURE 10.14 Dielectic elastomer powered shoe generator. (From Kornbluh, R., Power from plastic: How electroactive polymer "artificial muscles" will improve portable power generator in the 21st century military, Presented at *TRI-Service Power Expo*, Norfolk, Virginia, July 2003. With permission.)

layered diaphragms and it does not put an additional burden on the wearer. Energy converted per step can be up to 1 J/gm of DEA and the power generated (both feet) during walking is 1–10 W. The amount of polymers needed to convert 5 J is less than 50 g compared to an electromagnetic or piezoelectric device weighing more than 10 times. It is also a portable energy source for the soldiers in a battlefield to power their electronic devices, e.g., cell phone, computer, or radio set, without the use of batteries. The distinct advantages are dismounting of battery recharger, for enhanced performance of soldiers (reduced injury, improved comfort, more efficient load-carrying capacity in high altitude, and extreme cold climatic conditions). The multifunctional footwear, with a power device concealed in a boot, provides best operation for personal navigation system, medical status monitor, foot warmer, weight and space savings of night-vision devices, sensors, magnetometer, etc. Thus, DEA could fulfill a vision of "soldier as a system."

Kymissis et al.[32] have examined the possibility of generation of electrical power "parasitically" from devices built in a shoe, a wearable subsystem for the soldier. Merits of three different types of piezoelectric devices are compared. They are a unimorph strip piezoceramic composite, a stave of multilayer laminate of PVDF foil, and a shoe-mounted rotary magnetic generator as a part of technology demonstration; a piezoelectric embedded shoe has also been postulated to periodically broadcast a digital radio frequency identification (RFID) signal as the wearer walks.

Other possible applications of smart elastomers are in the area of "polymer engine" which can produce maximum power density (4 W/g) and output both in terms of electrical and mechanical power without any noise. These features are superior compared to conventional electrical generator, fuel cell, and conventional IC engine. Many DoD applications (e.g., robotics, MAV) require both mechanical and electrical (hybrid) power, and polymer engine can eliminate entire transducer steps and can also save engine parts, weight, and is more efficient.

In addition to "heel-strike generator" and "polymer engine" other inexpensive possibilities of power generation due to human motion using dielectric elastomer are depicted in Figure 10.15.

FIGURE 10.15 Harvesting energy from human movement.

FIGURE 10.16 Wind power.

The expected power levels from these types of generators are relatively small but sufficient to serve as a charging system for portable phones, radios, or musical devices.

Other efforts which could be made are to put these dielectric elastomer generators on the suspensions of automobiles to aid in the electricity use during operation. This would make things easier on the charging system of the automobile which could possibly save fuel consumption.

Dielectric elastomer's electrical energy generation scheme is also well suited for applications where electrical power can be produced from relatively large motions produced by high wind velocity at hilltops in high altitude/glacier region giving rise to wind-power generators (Figure 10.16). Alternately, currents of ocean waves, another environmental source of energy, can also be transformed into useful power generation by coupling with dielectric elastomers (Figure 10.17).

Prahlad et al.[33] have described the configuration of dielectric elastomer generators for conversion of wind or wave energy to produce localized electrical power and also suggested the relevance of this work to rural or other remote areas to fulfill their power requirement where existence of a reliable and centralized power still remained a distant possibility.

FIGURE 10.17 Wave power.

10.9 MAGNETORHEOLOGICAL ELASTOMER

The combination of polymers with nanoparticles displays novel properties and can be nurtured for evolving smart technological applications. One example of this new class of composites is magnetoreological (MR) elastomer. They originate from the concept of MR fluids which are stable colloidal dispersion of polymeric or metallic particles in nonconducting fluids such as heptanes or silicon oils. When subjected to magnetic field, MR fluids experience reversible changes in the rheological properties such as viscosity, plasticity, and elasticity. When these fluids are made to solidify by chemical methods in the presence of a magnetic field, another class of smart materials known as magnetorheological (MR) elastomer will result. These materials, which are composites consisting of small nanosized (~10 nm) magnetic particles of either ceramics, pure metals (cobalt, iron), or intermetallics (rare earth, e.g., hexagonal nickel–zinc ferrites) dispersed in an elastomer resin, e.g., silicone with density $N \sim 10^{23}/m^3$, can open up possibilities for new technological applications. The device consists of three basic components: (a) stable magnetic particles, (b) surfactant, e.g., a thin ceramic layer to mediate a uniform dispersion of these particles, and (c) an adequate liquid elastomer. Wetting and macroscopic interaction of magnetic particles with elastomers play a crucial role in determining stability and properties of an MR product. The response to an applied magnetic field is achieved by embedding magnetic particles into a cross-linked rubber matrix. During cross-linking or vulcanization, each particle in the MR elastomer is held in position until magnetic or mechanical perturbations introduce changes in the configuration of the embedded particles. The particles cannot leave the polymer matrix and thus the forces acting on the particles under external magnetic field are transmitted directly to the polymer chains resulting in either locomotion or deformation.[34] Shape distortion occurs instantaneously and also disappears rapidly when external magnetic field is applied or withdrawn. Extensive deformation, tunable elastic modulus, and shorter response time open opportunities for application of these materials as actuators, vibration-damping devices,[35,36] active mounts, and shock absorbers. Jon et al.[37] have determined Young's modulus, magneto-coupling factor, and axial strain coefficient of MR elastomers from various electroacoustic, magnetostrictive, and mechanical modeling systems. Although the density and Young's modulus obtained for MR elastomer were found to be less than those of Terfenol-D, the other parameters such as magneto-mechanical coupling factor and dynamic strain coefficient are comparable to Terfenol-D. Clarke et al.[38] have utilized a high-intensity x-ray synchrotron radiation to make novel measurements of real-time particle dynamics in MR elastomers. Small-angle scattering of transversely coherent x-rays, a time-dependent interference pattern known as speckle, was obtained. MR elastomers were excited by cycling an applied magnetic flux density (1 Tesla) and the scattered speckle patterns in millisecond timescale were imaged in a charge-coupled device (CCD) camera. These measurements demonstrate the feasibility to probe the relaxation dynamics in bulk for optically opaque samples. Banks et al.[39] have discussed issues related to modeling of nonlinearities and hysteresis arising in a class of MR-based elastomers, and substantial experimental and computational validation for the quasi-static and dynamic models are presented in the context of simple elongation of a field rubber-like rod.

10.10 ELECTROVISCOELASTIC ELASTOMERS

Similar to the principle of MR elastomers, the electroviscoelastic elastomer composites containing a polar phase can also be constituted by applying an external electric field, in place of a magnetic field during the time of cross-linkling or vulcanization of the matrix, i.e., elastomer phase. The applied electric field orients and fixes the polar phase into a solid state. The shear modulus is tunable and variable with the external field density.[40] Relative efficacy of MR elastomer and electroviscoelastic elastomer on the extent of change of shear modulus has been reported by Zrinyi et al.[41]

10.11 LIQUID CRYSTALLINE ELASTOMERS

Liquid crystalline elastomers (LCEs) are composite systems where side chains of a crystalline polymer are cross-linked. Their mesogenic domains can be ordered nematically and undergo a phase transition to a disordered state at a temperature well above the glass-transition temperature (T_g) of the polymer. Although the phase transition is thermally driven, LCEs demonstrate electrical conductivity and thus can be electrically stimulated.[42] Ratna[43] has reported contractions of nearly 30% due to the phase transition of acrylate-based LCEs.

10.12 SUMMARY AND CONCLUSION

The interest in rubber-based products has, in recent years, extended far beyond the traditional uses in tires, seals, and passive damping devices. Most recently, the interest in rubber-like smart or active material devices has motivated a number of efforts on both basic and applied aspects of the scientific engineering research required for a smart elastomer technology. Many exciting applications have been demonstrated worldwide and many premiere academic institutions across the globe, namely, Georgia Tech., Penn State, Colorado, North Carolina State (United States), University of Stuttgart, Aachen (Germany), UMIST (United Kingdom), are working on smart materials and smart gel for drug delivery, artificial muscle, display devices, reversible separation process, etc.

REFERENCES

1. Carpi, F., Chiarelli, P., Mazzoldi, A., and De Rossi, D., Electromechanical characterisation of dielectric elastomer planar actuators: Comparative evaluation of different electrode materials and different counterloads, *Sensors Actuators*, A107, 85, 2003.
2. Carpi, F., Migliore, A., Serra, G., and De Rossi, D., Helical dielectric elastomer actuators, *J. Smart Mater. Struct.*, 14, 1210, 2005.
3. Zhenyi, M.A., Scheinbeim, J.I., Lee, J.W., and Newman, B.A., High field electrostrictive response of polymers, *J. Polym. Sci., Part B, Polym. Phoi.*, 32, 2721, 1994.
4. Su, J., Harrison, J.S., St. Clair, T.L., Bar-Cohen, Y., and Leary, S., Electrostrictive graft elastomers and applications, demonstrated on the website http://www.teccenter.org/electroactivepolymers/assets/documents/all_ege_techpapers/MRS99_PA.DOC; Su, J., Costen, R.C., and Harrison, J.S., Performance evaluation of bending actuators made from electrostrictive graft elastomers, smart structures and materials 2002, in *Electroactive Polymer Actuators and Devices*, Y. Bar-Cohen (Ed.), Proceedings of SPIE, Vol. 4695, 0277–786X/2002.
5. Dubois, P., Rosset, S., Koster, S., Buforn, J.M., Stauffer, J., Mikhailov, S., Dadras, M., Rooij, Nico- F. de., and Shea, H., Microactuators based on ion-implanted dielectric electroactive polymer membranes (EAP), Presented at *13th International Conference on Solid-State Sensors, Actuators and Microsystems*, Seoul, Korea, June 5–9, 2005, 2048.
6. Pelrine, R., Kornbluh, R., and Joseph, J.P., Electrostriction of polymer dielectrics with compliant electrodes as a means of actuation, *J. Sensors Actuators, A: Phy.*, 64, 77, 1998.
7. Pelrine, R., Kornbluh, R., and Kofod, G., High strain actuator materials based on dielectric elastomers, *J. Adv. Mater.*, 12, 1223, 2000.
8. Scarborough, D., Dielectric elastomers, demonstrated on the website http://www.engr.siu.edu/mech/faculty/hippo/ME465SP05ScarboughDielectricElastomersorElectroactivePolymers.doc
9. Kornbluh, R., Pelrine, R., Pei, Q., and Shastri, S.V. *Electroactive Polymer (EAP) Actuators as Artificial Muscles.... Reality, Potential and Challenges*, First edition, SPIE—the International Society for Optical Engineering, Bellingham, Washington, 2001, Chapter 16.
10. Arora, S., Development of dielectric elastomer based prototype fiber actuator, Masters thesis submitted to North Carolina State University, July 2005.
11. Kofod, G., Dielectric elastomer actuator, Ph.D thesis submitted to Technical University of Denmark, Sept. 2001.

12. Kofod, G., Kornbluh, R., Pelrine, R., and Sommer-Larsen, P., Actuation response of polyacrylate dielectric elastomers, smart structures and materials 2001, in Electroactive polymer actuators and devices, Y. Bar-Cohen (Ed.), *Proceedings of SPIE*, Vol. 4329, 0277-786X/2001.

13. Yamwong, T., Voice, A.M., and Davies, G.R., Electrostrictive response of an ideal polar rubber, *J. Appl. Phy.*, 91(3), 1472, 2002.

14. Davies, G.R., Yamwong, T., and Voice, A.M., Smart structures and materials 2002, in Electroactive polymer actuators and devices, Y. Bar-Cohen (Ed.), *Proceedings of SPIE*, Vol. 4695, 0277-786X/2002.

15. Krakovsky, I., Romijn, T., and A. Posthuma de Boer, A few remarks on the electrostriction of elastomers, *J. Appl. Phy.*, 85(1), 628, 1999.

16. Toth, L.A. and Goldenburg, A.A., Control system design for a dielectric elastomer actuator: The sensory subsystem, smart structures and materials 2002, in Electroactive polymer actuators and devices, Y. Bar-Cohen (Ed.), *Proceedings of SPIE*, Vol. 4695, 0277-786X/2002.

17. Jeon, J.W., Park, K.C., An, S., Nam, J D., Choi, H., Kim, H., and Tak, Y., Electrostrictive polymer actuators and their control systems, Smart structures and materials, 2001, in Electroactive polymer actuators and devices, Y. Bar-Cohen (Ed.), *Proceedings of SPIE*, Vol. 4329, 0277-786X/2001.

18. Pelrine, R., Sommer-Larsen, P., Kornbluh, R., Heydt, R., Kofod, G., Pei, Q., and Gravesen, P., Applications of dielectric elastomer actuators, in SPIE, smart structures and materials, 2001, Electroactive polymer actuators and devices, Y. Bar-Cohen (Ed.), *Proceedings of SPIE*, Vol. 4329, 0277-786X/2001.

19. Pei, Q., Pelrine, R., Stanford, S., Kornbluh, R., and Rosenthan, M., Electroelastomer rolls and their application for biomimetic walking robots, *Synthetic Metals*, 135–136, 129, 2003.

20. Wingert, A., Litcher, M.D., and Dubowsky, S., On the kinematics of parallel mechanisms with bi-stable polymer actuators, demonstrated on the website http://scripts.mit.edu/~robots/robots/publications/papers/2002_06_Win_LicDub.pdf

21. Wax, S.G. and Sands, R.R., Electroactive polymer actuator and devices, *Proceedings of the SPIE Conference on Electroactive Polymer Actuators and Devices*, Newport Beach, California, SPIE Vol. 3669, 0277-786X/1999.

22. Carpi, F. and Rossi, D.D., Dielectric elastomer cylindrical actuators: Electromechanical modeling and experimental evaluation, *Mater. Sci. Eng.*, C24, 555, 2004.

23. Akers, J. and Drozd, D., Design of elastomeric piezoelectric ceramic smart structures for use as tuned resonators, *J. Rowan Eng./Mech.*, ME-S02–S09, 9(1–5), 2002.

24. Larsen, P.S., Artificial muscles, demonstrated on the website http://www.risoe.dk/fys-artmus/MIC-ARTI.pdf

25. Kornbluh, R.D., Flamm, D.S., Vujkovic-Civijin, P., Pelrine, R.E., and Huestis, D., Large light weight mirrors control by dielectric elastomer artificial muscle, AAS 200 meeting, Albuquerque, NM, June 2002, Paper no. 63–06.

26. Ashley, S., Artificial muscles, *J. Sci. Am.*, Oct., 53–60, 2003.

27. Bar-Cohen, Y., Xue, T., Shahinpoor, M., Simpson, J.O., and Smith, J., Flexible, low-mass robotic arm actuated by electroactive polymers, *Proceedings of SPIE 5th Annual International Symposium on Smart Structures and Materials*, March 1998, San Diego, CA, Paper no. 3329–07.

28. O' Halloran, A. and O' Malley, F., Dielectric elastomer actuators in the development of mechanotronic muscle, NUI, Galway faculty of engineering research day, 2004.

29. Pei, Q., Rosenthal, M., Prahlad, H., Pelrine, R., and Stanford, S. Multiple degrees of freedom electroelastomer roll actuators, *Smart Mater. Struct.*, 13, N86–N92, 2004.

30. Kornbluh, R., Power from plastic: How electroactive polymer "artificial muscle" will improve portable power generation in the 21st century military, Presented at *TRI-Service Power Expo*, Norfolk, VA, July 2003.

31. Pelrine, R. and Kornbluh, R., Recent progress in heel strike generators using electroactive polymer, demonstrated on the website http://www54.homepage.villanova.edu/michael.erwin/Research%20Papers/5sri.pdf

32. Kymissis, J., Kendall, C., Paradiso, J., and Greshenfeld, N., Parasitic power harvesting in shoes, Presented at *2nd IEEE International Conference on Wearable Computing*, Paper no. 2, Aug. 1998.

33. Prahlad, H., Kornbluh, R., Pelrine, R., Stanford, S., Eckerle, J., and Oh, S. Polymer power: Dielectric elastomers and their applications in distributed actuation and power generation, *Proceedings of ISSS International Conference on Smart Materials, Structures and Systems*, Bangalore, India, July 28–30, 2005, SA-13, pp. 100–107.

34. Zrinyi, M., New generation of smart elastomers, demonstrated on the website http://academic.sun.ac.za/unesco/Conferences/Conference2002/Zrinyi%20(3).pdf

35. Kallio, M., *The Elastic and Damping Properties of Magnetorheological Elastomers*, Espoo 2005, VTT Publications 565, pp. 1–146, 2005.

36. Davis, L.C., Model of magnetorheological elastomers, *J. Appl. Phy.*, 85(6), 3348–3351, 1999.

37. Jon, R., Suresh, G., and Natrajan, V., Studies on magnetostrictive properties of a magnetorheological elastomer, *Proceedings of ISSS International Conference on Smart Materials, Structures and Systems*, Bangalore, India, July 28–30, 2005, SA-122.

38. Clarke, R., Schlotter, W.F., Cionca, C., Paruchuri, S.S., Cunningham, J.B., Dufresne, E.M., Dierker, S.B., and Arms, D.A., Dynamics of nanomagnetic MR elastomers, demonstrated on the website http://www.mhatt.aps.anl.gov/research/publications/abstracts/APS2000clarker1.pdf

39. Banks, H.T., Pinter, G.A., and Potter, L.K., Modeling of nonlinear hysteresis in elastomers, demonstrated on the website http://www.ncsu.edu/crsc/reports/ftp/pdf/crsc-tr99–09.pdf

40. Shiga, T., Deformation and viscoelastic behavior of polymer gels in an electric fields, *Adv. Polym. Sci.*, 134, 131, 1997.

41. Zrinyi, M., Szabo, D., and Feher, J., Comparative studies of electro- and magnetic field sensitive polymer gel, *Proceedings of SPIE 6th Annual International Symposium of Smart Structures and Materials, EAPAD Conf.*, 3669, pp. 406–413, 1999.

42. Shahinpoor, M., Elastically-activated artificial muscles made with liquid crystal elastomers, *Proceedings of SPIE 7th Annual International Symposium of Smart Structures and Materials, EAPAD Conf.*, 3987, pp. 187–192, 2000.

43. Ratna, B.R., Liquid crystalline elastomers as artificial muscles: Role of side chain-backbone coupling. *Proceedings of SPIE 8th Annual International Symposium of Smart Structures and Materials, EAPAD Conf.*, 4329, 2001.

11 Recent Developments in Rubber–Rubber and Rubber–Plastics Blends

Duryodhan Mangaraj and Alice Bope Parsons

CONTENTS

11.1 RECENT WORK ON RUBBER–RUBBER BLENDS

Blending of polymers, especially elastomers, is often used to meet the performance and processing requirements of industrial products. Mangaraj has recently reviewed various aspects of elastomer blends including the thermodynamics of polymer–polymer miscibility, compatibilization of immiscible blends, distribution of fillers and processing ingredients in different phases of the blend, analytical techniques used to characterize rubber blends, and their application in the production of industrial rubber goods [1]. Subsequently, Mangaraj discussed the role of blending rubber with plastics for recycling of rubber waste [2] and the role of devulcanization, pre- and *in situ* grafting, reactive processing, etc. in achieving compatibilization [3]. Earlier, the subject had been reviewed by Corrish in Science and Technology of Elastomers (ed. Elrich) [4] and by Roland in HandBook of Elastomers (ed. Bhowmick and Stephens) [5]. In this chapter, recent work on rubber–rubber and rubber–plastics will be reviewed with an objective not only to update the information but also to appreciate the new developments in the science and technology of polymer blending.

Recent work on thermoplastic vulcanizates (TPVs) will not be included in this chapter since it is being reviewed elsewhere in the book. Abbreviations for some rubbers and accelerators will be used throughout in place of their full names as shown in Table 11.1. Acronyms for other polymers and additives will be provided in the text as required. A short discussion of polymer miscibility and compatibilization of polymer blends will be provided for better appreciation of the subject.

TABLE 11.1

Elastomer and Accelerator Acronyms

Material	Acronym
Acrylonitrile–butadiene rubber	NBR
Bromo-butyl rubber	BIIR
Butyl rubber	IIR
Chloro-butyl rubber	CIIR
Chloroprene rubber	CR
Ethylene–propylene–diene monomer rubber	EPDM
Ethylene propylene rubber	EPM
Natural rubber	NR
Polybutadiene rubber	BR
Polyisoprene	IR
Polyvinyl chloride	PVC
Styrene–butadiene rubber	SBR
Dibenzothiazyl disulfide	MBTS
Tetramethyl thiuram disulfide	TMTD

11.2 POLYMER–POLYMER MISCIBILITY AND COMPATIBILIZATION OF POLYMER BLENDS

11.2.1 MISCIBILITY

In general, the two components of a binary mixture or blend mix with each other if, in the process, Gibbs free energy of the system decreases. The change in free energy (ΔG_m) either becomes zero or negative. Free energy change due to mixing is a sum of entropy change and enthalpy change in mixing and is given by the relation

$$-\Delta G_m = -\Delta S_m + \Delta H_m \tag{11.1}$$

According to statistical thermodynamics, change in entropy is given by the change in the number of configurations that the system assumes compared to the sum of the configurations of the two components in the pure state and it is always positive. The change in enthalpy, however, can be zero, positive, or negative depending on the interaction between the molecules of the two components. It will be zero if there is no interaction, positive if they have repelling interaction, and negative if they have attractive interaction. Since polymers including elastomers are very large molecules, the change in the number of configurations on mixing is very small. Therefore, in order for two polymeric components to be miscible, the enthalpy of interaction has to be very small, zero, or negative.

Flory–Huggins model for polymer solutions, based on statistical thermodynamics, is often used for illustrating the behavior of polymer blends [6,7]. The expression for the free energy change

$$\frac{\Delta G_m}{V} = kT\left[\left(\frac{\phi_1}{V_1}\ln\phi_1 + \frac{\phi_2}{V_2}\ln\phi_2\right) + \frac{\chi_1\phi_1\phi_2}{V_1}\right] \tag{11.2}$$

contains two terms, the entropy change terms within parentheses and the enthalpy change terms outside the parentheses. Here ϕ_1 and ϕ_2 are volume fractions of the two polymeric components, V_1 and V_2, their molar volumes, and χ the interaction parameter. Since both V_1 and V_2 (polymer molar volumes) are very large, the combinatorial entropy change (ΔS_m) on mixing (the term within parentheses in Equation 11.2) is very small. Hence, the sign and the magnitude of ΔG_m largely depend on ΔH_m, represented by the last term in Equation 11.1. ΔG_m has to be very small or negative, for a binary polymer blend to be miscible. Flory–Huggins model does not provide a theoretical method for predicting the value of χ either for polymer solution or for polymer blends. Very limited data are available for polymer–polymer interaction parameters. It may be noted that χ for miscible polymer–polymer systems is five to ten times smaller than for polymer–solvent systems [1].

Earlier, Hildebrand and Scott had developed a theoretical expression for regular solutions of nonelectrolytes [8]. A regular solution is defined as one where $\Delta V_m = 0$, ΔH_m is positive, and $\Delta S_m = n_1 \ln X_1 + n_2 \ln X_2$ (i.e., only combinatorial entropy) where n_1 and n_2 are number of moles and X_1 and X_2 are mole fractions of the two components. Assuming that a solution process involves replacing similar molecules with dissimilar ones, they developed a relationship, between ΔH_m and the square root of the energy of vaporization, usually known as the solubility parameter δ. This is given by the following equation:

$$\Delta H_m / V_1 = K(\delta_1 - \delta_2)^2 \phi_1 \phi_2 \qquad (11.3)$$

In this equation, V_1 is the average molar volume of the two solvents; K is a constant close to 1; δ_1, ϕ_1 and δ_2, ϕ_2 are solubility parameter and volume fraction of component 1 and 2, respectively. This term is quite similar to the last term in Equation 11.2 that represents ΔH_m for binary polymer blends. Both are energy terms. As a result of this similarity, there have been frequent attempts in the past to replace $kT\chi_{12}$ of Equation 11.1, with a function of $(\delta_1 - \delta_2)^2$ and to estimate ΔH_m from a knowledge of the solubility parameters of component polymers. The advantage of using the solubility parameter (ϕ) concept in predicting polymer–polymer miscibility is that, for most polymers, it is either known or can be estimated by measuring solubility, swelling, or viscosity in a number of solvents or it can be estimated using group contribution approach from the knowledge of the chemical structure and molar volume of their repeating units. It has been suggested that for two polymers to be miscible their chemical structures must be similar to each other and the difference in their solubility parameters should be less than 0.1 $(\text{cal/cc})^{1/2}$. On the other hand, if the blend components have groups of opposite polarity and have interaction with each other, the enthalpy change on mixing becomes negative, thereby reducing their free energy change and enhancing their miscibility [1].

11.2.2 COMPATIBILIZATION

Since most polymers, including elastomers, are immiscible with each other, their blends undergo phase separation with poor adhesion between the matrix and dispersed phase. The properties of such blends are often poorer than the individual components. At the same time, it is often desired to combine the process and performance characteristics of two or more polymers, to develop industrially useful products. This is accomplished by compatibilizing the blend, either by adding a third component, called compatibilizer, or by chemically or mechanically enhancing the interaction of the two-component polymers. The ultimate objective is to develop a morphology that will allow smooth stress transfer from one phase to the other and allow the product to resist failure under multiple stresses. In case of elastomer blends, compatibilization is especially useful to aid uniform distribution of fillers, curatives, and plasticizers to obtain a morphologically and mechanically sound product. Compatibilization of elastomeric blends is accomplished in two ways, mechanically and chemically.

Mechanical compatibilization is accomplished by reducing the size of the dispersed phase. The latter is determined by the balance between drop breakup and coalescence process, which in turn is governed by the type and severity of the stress, interfacial tension between the two phases, and the rheological characteristics of the components [9]. The need to reduce potential energy initiates the agglomeration process, which is less severe if the interfacial tension is small. Addition of a small amount of compatibilizer acts like a solid emulsifier and stabilizes the droplets, thereby reducing the dispersed phase size. It has been observed and theoretically established that better dispersion is achieved when both components have similar viscosity. The essential condition for forming co-continuous phase is

$$\eta_1 \phi_2 / \eta_2 \phi_1 = 1 \tag{11.4}$$

where η_1, η_2 and ϕ_1, ϕ_2 are the viscosity and weight fraction of each component in the blend [10]. Co-continuous phase provides the special morphology, where the two phases behave in tandem, and exhibits the best combination of properties of the two components. The cooling rate of the blend also influences the particle size. Whereas rapid cooling provides smaller particles, slower cooling allows ripening (agglomeration) and generates large particles [11]. In the case of elastomers, the initial molecular weights are reduced by mechanochemical chain scission, by repeated cutting of the rubber layers and by using peptizers.

Chemical compatibilization is carried out in both nonreactive and reactive mode [12]. In nonreactive mode, an external polymeric material such as a copolymer, preferably a block copolymer having components similar to the blend components, is added to the blend. Both diblock and triblock copolymers have been used for compatibilization of polymer blends. Random and graft copolymers have also been used to that effect. The essential function of a compatibilizer is to wet the interface between the two phases. Block and graft copolymers achieve this by spreading at the interface and mixing with both phases through their component parts, which are similar to one phase or the other. In reactive mode, block and graft polymers formed *in situ* during mixing of the two components and additives assist in compatibilizing the blend. In the case of elastomer blends, compatibilization is also achieved by co-vulcanization between the matrix and the dispersed phase. The theoretical and practical aspects of nonreactive compatibilization have been recently discussed by Asaletha et al. in their study on compatibilization of blends of natural rubber (NR) and polystyrene (PS) using a graft copolymer of styrene onto NR as compatibilizer [13]. They found that the dispersed phase size decreased and the mechanical properties improved with increasing amounts of copolymer compatibilizer. However, these improvements leveled off at a critical copolymer volume fraction ϕ_c, which is sufficient to saturate the interphase surfaces. They found that the critical surface area Σ_c of the copolymer required to saturate the interface increases as the molecular weight of the blend components decreases, and that it is greater the higher the molecular weight of the copolymer compatibilizer. Further, better compatibilization is achieved as the molecular weight of the compatibilizer increases and as the chemical structures of the blocks resemble more the molecular structure of the blend components. Hence, optimization of molecular weight and chemical composition of the compatibilizer are desirable for developing a sound commercial blend. In earlier work, Reiss and coworkers had shown that for polyisoprene–PS blend, block copolymers provide better compatibilization than graft copolymers and solubilization of compatibilizer by phases takes place when the molecular weight of the blend components is comparable or smaller than the molecular weight of the corresponding block in the compatibilizer [14]. Teysie and coworkers, who examined the compatibilizing action of many copolymers, concluded that the structure and the molecular weight of the copolymer control the efficiency of compatibilization and that tapered block copolymers are more effective as compatibilizers than linear block copolymers [15]. Leibler et al. who have developed the thermodynamic basis for copolymer compatibilization, suggest that reduction of interfacial tension takes place due to adsorption of the copolymer at the interface and that an asymmetric copolymer is less efficient than a symmetric one [16]. Kawazura et al. have carried out a systematic investigation of the effect of diblock copolymer addition on the compatibility of rubber blends of NR–styrene–butadiene rubber (SBR) and IIR/BR and have found that addition of a small amount of diblock copolymer produced a finer morphology, increased tensile strength and abrasion resistance, and increased phase structure index, as well as interphase adhesion [17].

Reactive compatibilization is carried out during the blending process by adding a reactive material, either as a blend component or as a reactive third component. The classical example of reactive compatibilization is DuPont's production of super-tough nylon by blending nylon with maleic anhydride-grafted ethylene–propylene–diene monomer (EPDM; MA-g-EPDM). Scott and Macosko have carried out an in-depth investigation of MA-g-EPDM–nylon 6 blend and found that the chemical coupling between the two components occurs as indicated by the increase in torque and temperature and by the co-continuous morphology exhibited by the blend [18].

Reactive compatibilization is also carried out by adding a monomer which in the presence of a catalyst can react with one or both phases providing a graft copolymer *in situ* that acts as a compatibilizer. Beaty and coworkers added methyl methacrylate and peroxide to waste plastics (containing polyethylene [PE], polypropylene [PP], PS, and poly(ethylene terephthalate) [PET]). The graft copolymer formed *in situ* homogenized the blend very effectively [19].

Reactive compatibilization can also be accomplished by co-vulcanization at the interface of the component particles resulting in obliteration of phase boundary. For example, when *cis*-polybutadiene is blended with SBR (23.5% styrene), the two glass transition temperatures merge into one after vulcanization. Co-vulcanization may take place in two steps, namely generation of a block or graft copolymer during vulcanization at the phase interface and compatibilization of the components by thickening of the interface. However, this can only happen if the temperature of co-vulcanization is above the order–disorder transition and is between the upper and lower critical solution temperature (LCST) of the blend [20].

Substantial work has been done on self-cross-linking rubber blends where two elastomers with opposite ionic charges interact with each other and form cross-links and in the process provide high level of compatibility.

In summary, it may be stated that compatibilization is important between component phases in plastics and rubber blends. It can be achieved by adding an external compatibilizer such as a block or graft copolymer or by using reactive compatibilization. The latter can be carried out by functionalizing one of the blend components or by reacting with a functional chemical during blending, by surface activation, and by using interphase cross-linking reaction. The best approach to compatibilization for a blend should take into account the potential reaction at the phase boundary, the blend composition, cost, processing window, and availability of suitable reactive additive.

11.2.2.1   Blends of Saturated Hydrocarbon Elastomers (Uncompatibilized Blends)

Saturated hydrocarbon elastomers such as ethylene–propylene copolymers (EPM), polyisobutylene (PIB), and their derivatives containing small amounts of unsaturation such as EPDM and isobutylene–isoprene rubber (IIR) are widely used in rubber industry due to their light weight, high strength, low cost, low dielectric constant, and amenability to cross-linking. Their mutual miscibility is taken for granted because they all contain $-CH_2$ groups. On the other hand, they do not have any mutual interactive groups to promote mutual interaction and miscibility. Recent studies have indicated that mutual solubility is highly dependent on chemical architecture, especially the degree, length, and location of side groups and molecular weight. Conventional techniques for locating the phase boundary are hard to use to study their mutual solubility and phase diagram. Major advances in this area have been achieved by using small-angle neutron scattering (SANS) of deuterium-labeled polymers and measuring pressure–volume–temperature (PVT) relations.

Lohse et al. have summarized the results of recent work in this area [21]. The focus of the work is obtaining the interaction parameter χ of the Flory–Huggins–Stavermann equation for the free energy of mixing per unit volume for a polymer blend. For two polymers to be miscible, the interaction parameter has to be very small, of the order of 0.01. The interaction density coefficient $X = (\chi/V)RT$, a more relevant term, is directly measured by SANS using random phase approximation study. It may be related to the square of the Hildebrand solubility parameter (δ) difference which is an established criterion for polymer–polymer miscibility:

$$X \approx (ó_1 - ó_2)^2 \tag{11.5}$$

The solubility parameter is close to the square root of internal pressure, a thermodynamic term that is directly obtained from PVT measurement of the blends.

In addition, light-scattering method has been used for studying the phase diagram of polyolefin polymers such as PIB. Scattering increases with phase separation. Much light-scattering work has been done on blends of various types of PE as well as on mixtures of PP and EP copolymers. Deuterium labeling also enables getting phase diagram of polyolefin blends. PVT measurements have been used to measure internal pressure for a number of polyolefin blends. Relative solubility parameter values obtained by the PVT method agree closely with those obtained by the SANS method.

Polyolefin blends exhibit a wide range of phase diagrams, like polar polymers, even in the absence of specific interaction. Most of them exhibit upper critical solution temperature (UCST), i.e., where the interaction parameter χ decreases with increase in temperature. Some of the systems also exhibit LCST. Some polyolefin blends may exhibit both of these behaviors. Most of these systems follow solubility parameter-based correlations. The interaction energy obtained experimentally by SANS measurement agrees with those predicted from solubility parameter differences. Table 11.2 gives relative solubility parameter data (relative to the $ó$ of hydrogenated polybutadiene, $(HP_3 970)$) for a series of polyolefin polymers. It may be mentioned that group contribution method is not appropriate for estimating the solubility parameter of polyolefins.

A large number of polyolefin blends mix in a regular fashion, i.e., their interaction energy is close to the geometric mean of their component interaction energies. However, as molecular weight increases, the interaction energy becomes very small. Another interesting feature is the linear correlation between the difference in solubility parameters of the two components and packing length, p, i.e., the volume occupied by a chain divided by its radius of gyration considered as chain thickness. Table 11.3 shows irregular contribution to interaction strength for a series of polyolefin blends. Maximum irregularity appears to be with blend of PP–PEP and PP–hhPP.

Most polyolefin blends that mix irregularly have at least one homopolymer component. Regularity in backbone structure seems to ordain irregularity in mixing. Some have studied the miscibility of ethylene–octene (EO) copolymer with other EO copolymers having different octene content. The interaction strength model predicts that whereas this copolymer will be miscible with the EO copolymer containing 14.6% EO, it will be incompatible with EO copolymers with higher octene content. Results of experimental study agree with this prediction. Similar prediction of PP blends with PIB was found to have good agreement with experimental results.

It appears that the main determinant in polyolefin miscibility is the way the chains pack together, a feature that controls the intermolecular interaction between molecules. The solubility parameter approach is predictive in most cases and can be useful in designing polyolefin blends.

11.2.2.2 Compatibilized Blends

Pandey et al. have used ultrasonic velocity measurement to study compatibility of EPDM and acrylonitrile–butadiene rubber (NBR) blends at various blend ratios and in the presence of compatibilizers, namely chloro-sulfonated polyethylene (CSM) and chlorinated polyethylene (CM) [22]. They used an ultrasonic interferometer to measure sound velocity in solutions of the rubbers and their blends. A plot of ultrasonic velocity versus composition of the blends is given in Figure 11.1. Whereas the solution of the neat blends exhibits a wavy curve (with rise and fall), the curves for blends with compatibilizers (CSM and CM) are linear. They resemble the curves for free energy change versus composition, where sinusoidal curves in the middle represent immiscibility and upper and lower curves stand for miscibility. Similar curves are obtained for solutions containing 2 and 5 wt% of the blends. These results were confirmed by measurements with atomic force microscopy (AFM) and dynamic mechanical analysis as shown in Figures 11.2 and 11.3. Substantial earlier work on binary and ternary blends, particularly using EPDM and nitrile rubber, has been reported.

TABLE 11.2
Relative Solubility Parameters for Selected Polyolefins[a]

	$\delta - \delta_{ref}$ $[(J/cm^3)^{1/2}]$			
Polymer	27°C	83°C	118°C	167°C
PE polyethylene			1.60	1.48
EO22 ethylene–octene copolymer			1.38	1.28
EB22 ethylene–butene copolymer			1.32	1.23
HPB25 hydrogenated polybutadiene, model ethylen–butane copolymer, x wt% butane			1.31	1.23
EB31 ethylene–butene copolymer			1.21	1.14
HPB32 hydrogenated polybutadiene, model ethylene–butane copolymer, x wt% butene		1.28	1.22	1.15
EH37 ethylene–hexene copolymer			1.09	1.05
HPB38 hydrogenated polybutadiene, model ethylene–butane copolymer, x wt% butene	1.23	1.16	1.11	1.05
EH43 ethylene–hexene copolymer			1.02	0.97
HPB52 hydrogenated polybutadiene, model ethylene–butane copolymer, x wt% butene	1.01	0.95	0.91	0.86
EP53 ethylene–propylene copolymer			1.00	0.95
PEP hydrogenated polyisoprene, model alternating ethylene–propylene copolymer	0.92	0.90	0.89	0.88
hhPP hydrogenated poly(2,3-dimethyl butadiene), model head-to-head polypropylene	0.82	0.76	0.72	0.67
PEB hydrogenated poly(ethyl butadiene), model alternating ethylene–butene copolymer	0.72	0.71	0.69	0.66
HPB66 hydrogenated polybutadiene, model ethylene–butane copolymer, x wt% butene	0.73	0.69	0.66	0.62
HPB78 hydrogenated polybutadiene, model ethylene–butane copolymer, x wt% butene	0.49	0.46	0.44	0.41
PP hydrogenated poly(3-methyl 1.3 pentadiene), model head-to-tail polypropylene	0.22	0.25	0.26	0.25
HPB88 hydrogenated polybutadiene, model ethylene–butane copolymer, x wt% butene	0.23	0.21	0.20	0.18
HPB97 hydrogenated polybutadiene, model ethylene–butane copolymer, x wt% butene	0.00	0.00	0.00	0.00

Source: Lohse, D.J., Garner, R.T., Graessley, W.W., and Krishnamoorti, R. in Miscibility of Blends of Saturated Hydrocarbon Elastomers. Rubber Division, Proceedings of the American Chemical Society, Nashville, TN, Sept. 29–Oct. 2, 1998, Paper No. 33, 1–14.

[a] All values are with respect to that of reference polymer hydrogenated polybutadiene (HPB97), model ethylene–butene copolymer, x wt% butene.

Hussein et al. have studied the compatibilization of blends containing NBR and hydrogenated nitrile butadiene rubber (HNBR), having similar nitrile content [23]. The incentive is to obtain a blend which can replace more costly HNBR without much loss in performance. They prepared blends of NBR (Nipol-DN4555) having a 45% ACN content with HNBR (Zetpol-2010L) having ACN content of approximately 36% at different blend ratios in a Haake poly-drive blender at around 144°C. Specimens of various blend mixtures were molded at 150°C in a Carver press. Mechanical, rheological, and thermal properties were measured. The dynamic viscosity of the blends showed the log additivity rule while the storage modulus showed synergistic effects. The frequency sweep measurements of G' and viscosity suggested an emulsion-type morphology, where the

TABLE 11.3

Examples of Irregular Contribution to Interaction Strength for Polyolefin Blends

Blend	T (°C)	X (J/cm^3)	$(\delta_1-\delta_2)^2$ (J/cm^3)	X_{irr} (J/cm^3)
PP–HPB78	51	0.0643	0.0676	−0.0033
	83	0.0457	0.0441	−0.0016
	121	0.0300	0.0324	−0.0024
	167	0.0248	0.0256	−0.0008
PP–HPB97	27	0.1483	0.0484	−0.0909
	51	0.1181	0.0484	−0.0697
	83	0.0881	0.0625	−0.0255
	121	0.0648	0.0676	−0.0071
PP–PEP	27	0.1038	0.4900	−0.3862
	52	0.1233	0.4761	−0.3528
	83	0.1475	0.4225	−0.2750
	121	0.1647	0.3969	−0.2322
	167	0.1772	0.3969	−0.2197
PP–hhPP	27	0.0883	0.3600	−0.2717
	51	0.0780	0.3249	−0.2469
	83	0.0690	0.2601	−0.1911
	121	0.0634	0.2116	−0.1482
HhPP–PEP	27	0.0285	0.0100	−0.0185
	51	0.0187	0.0144	−0.0043
	83	0.0110	0.0196	−0.0086
	121	0.0134	0.0289	−0.0155
	167	0.0229	0.0441	−0.0212

Source: Lohse, D.J., Garner, R.T., Graessley, W.W., and Krishnamoorti, R. in Miscibility of Blends of Saturated Hydrocarbon Elastomers. Rubber Division, Proceedings of the American Chemical Society, Nashville, TN, Sept. 29–Oct. 2, 1998, Paper No. 33, 1–14.

droplets of NBR in HNBR matrix are very small, indicating good mechanical compatibility. All blend compositions exhibited two distinct glass transition temperatures, close to the T_g of the individual polymers, characteristic of an immiscible blend. Tensile strength and modulus of all blends followed the additivity rule. HNBR- and HNBR-rich blends exhibited strain hardening, suggesting induced crystallization. In conclusion, it was found that whereas blends of HNBR with NBR are thermodynamically immiscible, they are technologically compatible. This is evident from scanning electron micrographs shown for 10:90, 50:50, and 90:10 blends in Figure 11.4. The 10:90 blend shows uniform dispersion of the two components, with a few clusters of NBR. The 90:10 blend shows very uniform dispersion representative of an emulsion-type morphology. The 50:50 blend exhibits a co-continuous morphology.

Whereas fluoroelastomers based on tetrafluoro-ethylene/propylene/vinylidene fluoride are highly oil-resistant and strong, their low-temperature flexibility is limited. Silicone rubber, on the other hand, excels in high-temperature stability, low-temperature flexibility, chemical resistance, weatherability, electrical performance, and sealing capability. Hence, blends of silicone rubber with fluoroelastomer have the potential for providing useful performance and potential for replacing very costly fluorosilicone rubber. Ghosh and De have studied morphology as well as performance properties of elastomer blends based on fluoroelastomer (Aflas 200) and silicone rubber (SE 0075) [24]. They prepared five different compounds containing 0%–100% of the two

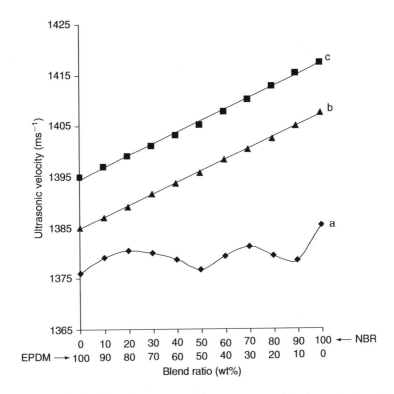

FIGURE 11.1 Ultrasonic velocity versus acrylonitrile–butadiene rubber/ethylene–propylene–diene monomer (NBR–EPDM) blend composition (a) no compatibilizer, (b) with chloro-sulfonated polyethylene (CSM), and (c) with chlorinated polyethylene (CM). (From Pandey, K.N., Setua, D.K., and Mathur, G.N., *Polym. Eng. Sci.*, 45, 1265, 2005.)

elastomers along with dicumyl peroxide (DCP) as the curing agent and triallyoxy cyanurate and calcium hydroxide as additives. Fluorosilicone rubber was also compounded in the similar manner. The compounds were mixed in a Barbender plasticorder at 80°C, sheeted out in a two-roll mill, and cured at 170°C for cure times as determined by the Monsanto rheometer. They measured a variety of blend properties including die swell and surface roughness of the extruded specimens, morphology

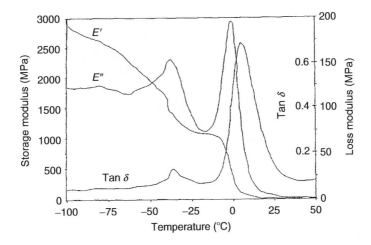

FIGURE 11.2 Dynamic mechanical analysis plots of NBR/EPDM vulcanizate of mix A without a compatibilizer. (From Pandey, K.N., Setua, D.K., and Mathur, G.N., *Polym. Eng. Sci.*, 45, 1265, 2005.)

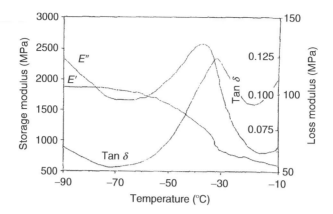

FIGURE 11.3 Dynamic mechanical analysis plots of acrylonitrile–butadiene rubber/ethylene–propylene–diene monomer (NBR/EPDM) vulcanizate of mix B containing chlorinated polyethylene (CM). (From Pandey, K.N., Setua, D.K., and Mathur, G.N., *Polym. Eng. Sci.*, 45, 1265, 2005.)

of the blends by scanning electron microscopy (SEM) and atomic force microscopy (AFM), surface energy by contact angle meter, thermal properties by TGA, limiting oxygen index, mechanical properties, and hot oil aging resistance. They found that the blends exhibit shear viscosity close to that of silicone rubber, especially at 50% and 75% silicone rubber, and a little smaller than silicone

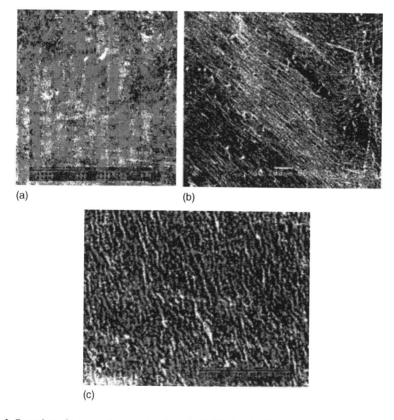

FIGURE 11.4 Scanning electron micrographs for (a) 10:90 blends, (b) 50:50 blends, and (c) 90:10 blends. (From Hussein, I.A., Chaudhry, R.A., and Abu Sharkh, B.F., *Polym. Eng. Sci.*, 44, 2346, 2004.)

TABLE 11.4

Synergistic Improvement in Modulus, Tear Strength, and Hardness of Blends

Mechanical Properties	Silicone Rubber/Fluororubber, by Weight				
	0:100	25:75	50:50	75:25	100:0
Tensile strength (MPa)	6.3	6.6	6.6	6.6	7.9
Ultimate elongation (%)	834	549	302	164	212
Modulus at 100% elongation (MPa)	0.8	1.8	3.5	4.6	4.1
Modulus at 200% elongation (MPa)	1.3	2.8	5.5	—	7.6
Tear strength (kN/m)	23.2	28.4	28.3	26.5	24.1
Tension set at 100% elongation (%)	8	8	8	6	4
Hysteresis loss (J/m^2 × 10^{-6})	0.010	0.031	0.077	0.095	0.060
Hardness (Shore A)	40	52	70	76	74

Source: Ghosh, A. and De, S.K., *Rubber Chem. Technol.*, 77, 856, 2004.

rubber at higher shear rate. The die swell of the blends is much higher than the values calculated on the basis of the additivity rule. This has been attributed to the fact that in the blend, the fluororubber is dispersed in the continuous matrix of silicone rubber, and the higher the volume fraction of fluororubber, the higher is the amount of stored elastic energy during capillary flow, and greater the deviation from the additivity rule.

The blends show synergistic improvement in modulus, tear strength, and hardness, and marginal decrease in tensile strength as seen in Table 11.4. This could be due to the formation of interfacial cross-linking between the two elastomers. On the whole, the two elastomers are technologically compatible. This is evident from SEM images. Figure 11.5 shows where the low-viscous silicone

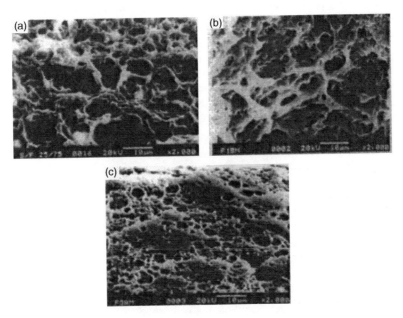

FIGURE 11.5 Low-viscous silicone rubber forms continuous matrix with fluororubber particles. (From Ghosh, A. and De, S.K., *Rubber Chem. Technol.*, 77, 856, 2004.)

rubber forms the continuous matrix containing spherical and/or elliptical fluororubber particles. The surface of the blends shows granular morphology where the low-viscous silicone rubber flows over the fluororubber matrix and tends to cover the particles, generally unreacted $Ca(OH)_2$ and CaO, on the fluororubber surface. The 50:50 blend, however, shows a distinct surface morphology characteristic of extreme polarity difference between the blend components.

The blends, irrespective of the concentration of fluororubber, show surface energy lower than neat rubbers. This is attributed to the migration of silicone rubber to the surface. The presence of silicone rubber on the surface of the blends also contributes to their lower limited oxygen index compared to that of fluoroelastomers.

The initial decomposition temperature corresponding to 3% weight loss for the blends is marginally higher than either of the blend components, possibly due to interphase cross-linking. Blending of silicone rubber with fluororubber results in the merging of the decomposition stages. The blend exhibits three-stage decomposition after the mass conversion of 5% and is essentially controlled by the fluororubber. The decomposition of the fluororubber and the blends is catalyzed by the evolution of hydrogen fluoride (HF) during the degradation of fluororubber. The residue left after thermal degradation of the blend is less than expected since HF reacts with silicon dioxide filler generating volatile silicon hexafluoride.

Ghosh and De have also studied hot oil aging of the blends and blend components in ASTM oil #1 at 150°C for 210 h. The 50:50 blend exhibits greater retention of tensile strength than the blend components, possibly due to interphase bonding. The elongation at break on hot oil aging of the fluoroelastomer and the blends remains almost intact, whereas that for silicone rubber degrades. The degrees of swelling of the blends in hot oil are in between those for the individual components. Hardness changes on hot oil aging are more or less the same for individual elastomers and their 50:50 blend.

Based on the extensive work, Ghosh and De have concluded that fluoroelastomer and silicone rubber form technologically compatible blends of micro-heterogeneous structure with thermal stability between those for the blend components. The blend could be used as a replacement for fluorosilicone rubber.

Blends of NR and polybutadiene rubber (BR) are often used in manufacturing tires. Of particular interest is the preferential location of carbon black in one phase or the other and the effect of this maldistribution on the properties. There have been contradictory suggestions regarding the uneven distribution of carbon black and its effect on performance. Clarke et al. have carried out a systematic investigation on the subject [25]. They used two modes of mixing, namely making master batches of individual rubbers with carbon black and blending them together and mixing all the ingredients in one step. The rapid decrease in domain size with mixing time in the first mode implies a high level of compatibility between the two elastomers, suggesting that addition of a compatibilizer is not needed for enhancing compatibility. Figure 11.6 shows the light micrographs of 50:50 NR–BR–carbon black master batch blends after 30, 60, and 90 s, supporting the rapid increase in compatibility. The physical properties are close to those predicted by applying simple mixing rules. Subsequently, the authors adopted single-stage mixing of all the components together. Four different mixing cycles were used to obtain uniform distribution. Usually, a high degree of blending is possible if the viscosities of the two components are similar. Since elastomers are shear-thinning in their behavior, high shear can be used to bring the viscosity of the highly viscous component down closer to that of the other (Method 1). Pre-warming the BR phase in an oven before mixing (Method 2) and premixing carbon black to the low-viscous NR phase are other ways of matching the viscosity of the two components (Method 3). Another way of matching the viscosity is adding the filler to the low-viscous component for a time before adding the high-viscous component (Method 4). The authors used these four different ways to make the compounds and cured them. All the compounds were found qualitatively similar. The fact that there was no visible difference in blend morphology implied that the domain size is less than 1 μm, probably due to a high degree of compatibility between the phases. It was suggested that the order of addition of the

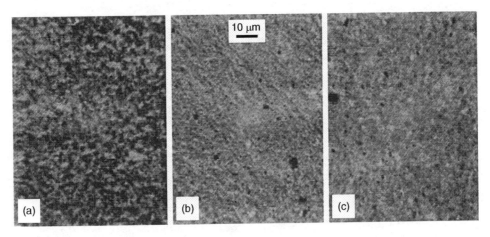

FIGURE 11.6 Light micrographs of 50:50 NR–BR–carbon black blends: (a) 30 s; (b) 60 s; (c) 120 s. (From Clarke, J., Clarke, B., and Freakley, P.K., *Rubber Chem. Technol.*, 74, 1, 2001.)

ingredients to the mixer did not affect the blend morphology, the degree of filler dispersion, or the distribution of filler between the phases to an extent to cause difference in physical properties. The difference between the single-stage and master batch mixing arises from the difference in the state of the filler distribution, which would need to be improved in order to obtain higher physicals. In conclusion, the authors observed that fine morphology is most quickly attained by blending master batches of similar viscosity. NR–BR blends made of master batches of different viscosities can have small domain sizes, due to high degree of compatibility between NR and BR and due to co-vulcanization at the interface. Hence, strength properties of different NR–BR blends can be predicted using simple mixing rules.

The emerging requirements in the automotive industry for higher service temperatures and longer warranty periods need power transmission belts to withstand higher use temperature for longer periods. This can be achieved by blending ethylene–propylene monomer (EPM) rubber or EPDM with chloroprene (chlorinated rubber, CR). However, the blends are incompatible. Arjunan has studied the technological compatibilization of CR with EPM and EPDM using maleated and epoxidized EPDM [26]. He used a number of maleated EPM Exxelor modifiers (VA 1801, 1822, and 18030) and maleated EPDM, Royaltuf 465A. It was anticipated that whereas structural similarity of the modifiers will provide compatibility with EP rubbers, specific interaction between reactive groups with chloroprene will enhance compatibility with the latter. He melt blended EPDM–CR–compatibilizers in 70:30:10 ratio. Initially, he studied the melt viscosity versus shear rate. Previous experimental studies indicated that for good compatibility and drop breakup, the viscosity ratio between the two polymer melts should stay between 3.5 and 5×10^{-5}. Initial studies suggested that EP-g-MA 1001 appeared to provide good compatibility. Subsequently, belt compounds were formulated using the blends in the above-mentioned ratio. Their ozone resistance, heat resistance, dynamic cut-growth resistance, and physical properties were evaluated following standard procedures and are given in Table 11.5. Similar work was carried out using EPDM-g-epoxy and sulfonated EPDM. It was found that the physical properties, particularly cut-growth resistance, heat, and ozone resistance, improved substantially by adding 5–10 parts of compatibilizer to the blend.

Evidence of chemical interaction between the rubbers and compatibilizers was demonstrated by extracting the blends with chloroform at room temperature and examining both soluble and insoluble fractions with Fourier transform infrared (FTIR) spectrometry. The weight of the insoluble fraction of the compatibilized melt blend was more than that in the uncompatibilized blend indicating the formation of (EP-g-MA)-g-CR due to reaction between MA and allylic chlorine of CR. The compounds containing epoxidized EPDM additive were examined by both optical and

TABLE 11.5

Evaluation of Ozone Resistance, Heat Resistance, and Dynamic Cut-Growth Resistance

Components/Properties	Blend # 1	2	3	4	5
Neoprene	100	70	70	70	70
EPDM (V 7000)	—	30	—	20	25
EPDM-g-Epoxy	—	—	30	10	—
EPDM-g-SO$_3$	—	—	—	—	5
Tensile (MPa)	13.2	10.1	15.1	10.3	11.9
Elongation (%)	736	657	664	622	646
Heat aged, 48 h, 148°C	—	—	—	—	—
Tensile (MPa)	4.8	6.2	6.3	8.1	8.5
Elongation (%)	278	399	384	448	409
Dynamic ozone resistance	144	168	500	336	>500

Source: Arjunan, P., Technological Compatibilization of Dissimilar Elastomer Blends: Part 1. Neoprene and Ethylene–co-Propylene Rubber Blends for Power Transmission Belt Application. Rubber Division, Proceedings of the American Chemical Society, Nashville, TN, Sept. 29–Oct. 2, 1998, Paper No. 52, 1–28.

scanning electron microscope. Analysis of the data indicated that the ternary blend CR–EPDM–EPDM-g-epoxy has reduced particle size compared to the binary blend CR–EPDM.

Croft has discussed the design of elastomer blends for the manufacture of wire-reinforced hoses and belting [27]. In addition to heat and oil resistance, the rubber compound used for this application should have good adhesion at the rubber–wire interface. Further, most hose compounds need to have good ozone resistance, low-temperature flexibility, excellent abrasion resistance, and flame resistance. These are achieved by using blends of chlorinated rubbers, nitrile rubber, and SBR. The level of chlorine in CR and nitrile content in NBR has to be adjusted to provide required low-temperature flexibility. The Moony viscosity of the compounds needs to be considered to provide good extrusion characteristics. Finally, reinforced hose products need to have high tensile strength, high modulus at low elongation, and high hardness (75 + durometer). Addition of CR to NBR helps adhesion and the relationship of CR to adhesion is practically linear. The same is true for PVC–NBR blends and their alloys with SBR.

For making compatible blends, the polymers should have comparable polarities and viscosities. The oil needs to be selected properly so that its solubility parameter is close to those for blend components. The cure system should be efficient for all constituent rubbers and the filler system needs to be appropriate. Finally, cost consideration should be taken into account to provide a commercially viable product.

Based on these considerations, Croft prepared six formulations containing various combinations of NBR and NBR/PVC with CR and SBR and measured their oil, heat and ozone resistance, physical properties, and adhesion characteristics. Whereas the physicals are satisfactory for all compounds, formulations based on NBR, NBR/PVC with CR performed better on heat and oil aging than the compounds containing SBR as shown in Tables 11.6 and 11.7. However, the adhesion is better with the latter compounds. It has been suggested that cuprous sulfide formed on the wire surface interacts with the double bond in SBR to provide the improvement in adhesion.

11.2.2.3   Compatibilization by Co-Vulcanization

As mentioned earlier, co-vulcanization has been effectively used for compatibilization of otherwise immiscible rubber blends. The phenomenon is much more complex. Changes in curative

TABLE 11.6
Formulations Based on NBR, NBR/PVC, CR, and SBR

Component	Formulations					
	1	**2**	**3**	**4**	**5**	**6**
NBR	100	85	70	—	—	—
NBR/PVC	—	—	—	100	85	70
CR	—	15	30	—	—	—
SBR	—	—	—	—	15	30
N762 Black	75	75	75	75	75	75
DOA (aliphatic ester)	20	20	20	20	20	20
Stearic acid	1	1	1	1	1	1
Zinc oxide	5	5	5	5	5	5
TBBS accelerator	1.5	1.5	1.5	1.5	1.5	1.5
Sulfur	2.8	2.8	2.8	2.8	2.8	2.8

Source: Croft, T.C., Elastomeric Compounds Utilizing Polymer Blends for Improved Wire Adhesion. Rubber Division, Proceedings of the American Chemical Society, Nashville, TN, Sept. 29–Oct. 2, 1998, Paper No. 63.

distribution, effect of curative distribution on tensile properties, and effect of temperature both on curative migration and cross-linking also influence the ultimate properties of the blend.

Blends of carboxylated nitrile rubber (XNBR) with EPDM are likely to provide an attractive combination of properties including oil resistance, heat and ozone resistance, high tensile strength, modulus, and hardness. However, the polar curing ingredients often diffuse from the nonpolar to polar component, thereby producing cure rate mismatch and inferior properties. Three different measures have been used to overcome the cure rate mismatch [29]:

(a) Suppressing the cure of the more reactive component (XNBR)
(b) Activating the less reactive component
(c) Blocking the diffusive path by the use of suitable additives

TABLE 11.7
Heat, Oil, and Ozone Aging Data of Blends of NBR, NBR–PVC, CR, and SBR

	1	**2**	**3**	**4**	**5**	**6**
Heat Aging, 70 h @ 212F						
Hardness change; pts	+4	+5	+5	+7	+13	+18
Elongation change; %	−18	−16	−18	−28	−38	−49
Tensile change; %	−5	−5	−4	−14	−21	−30
Oil Aging, 70 h @ 212F						
Hardness change; pts	0	−2	−4	−7	−17	−22
Elongation change; %	−3	+8	+17	−6	−24	−36
Tensile change; %	−8	−12	−23	−19	−31	−31
Volume change; %	−5	+4	+12	+4	+18	+18
Ozone aging, 70 h @100 pphm	Pass	Pass	Pass	Pass	Fail	Fail

Source: Croft, T.C., Elastomeric Compounds Utilizing Polymer Blends for Improved Wire Adhesion. Rubber Division, Proceedings of the American Chemical Society, Nashville, TN, Sept. 29–Oct. 2, 1998, Paper No. 63.

In the first process, EPDM was converted into a cure retarder and blended with NR. The resulting blend provided improved tensile strength, hardness, and modulus, even though cross-link density was not adequate. In the second route, EPDM was either halogenated, activated by maleation, partially pre-vulcanized, or grafted with accelerators or polydienes before blending with other rubbers. In the third process, the maldistribution of sulfur in a blend of nitrile rubber with EPDM is suppressed by replacing zinc oxide with lead tetroxide (Pb_3O_4) to provide a blend with good physicals.

Hopper has used the first methodology in preparing a blend of EPDM with NR with improved modulus, hardness, and tensile strength [28]. However, the cross-link density was low. The activation route was carried out by halogenation of EPDM by pre-vulcanization and grafting of accelerators to EPDM before blending with high diene rubbers. Grafting of EPDM with maleic anhydride not only improves its polarity but also makes it suitable for reactive compatiblization. The third route was used in blending EPDM with nitrile rubber where zinc oxide was replaced by lead oxide (Pb_3O_4).

Recently, Naskar et al. have used a fourth method, i.e., use of a novel cross-linking agent, bis(di-isopropyl) thio-phosphoryl disulfide, which reacts with both rubbers, thereby linking the two phases intimately and providing a robust blend [29]. The reaction scheme is given in Figure 11.7.

Vulcanization was carried out in one and two stages with and without carbon black. In one-stage vulcanization, all the ingredients were mixed and the compound was vulcanized in the mold for optimum cure time as shown by Monsanto rheometer curves. In the two-stage vulcanization process, all the ingredients except XNBR were mixed and heated for a short time, followed by mixing with required amount of XNBR and heating for rest of the time required for optimum vulcanization.

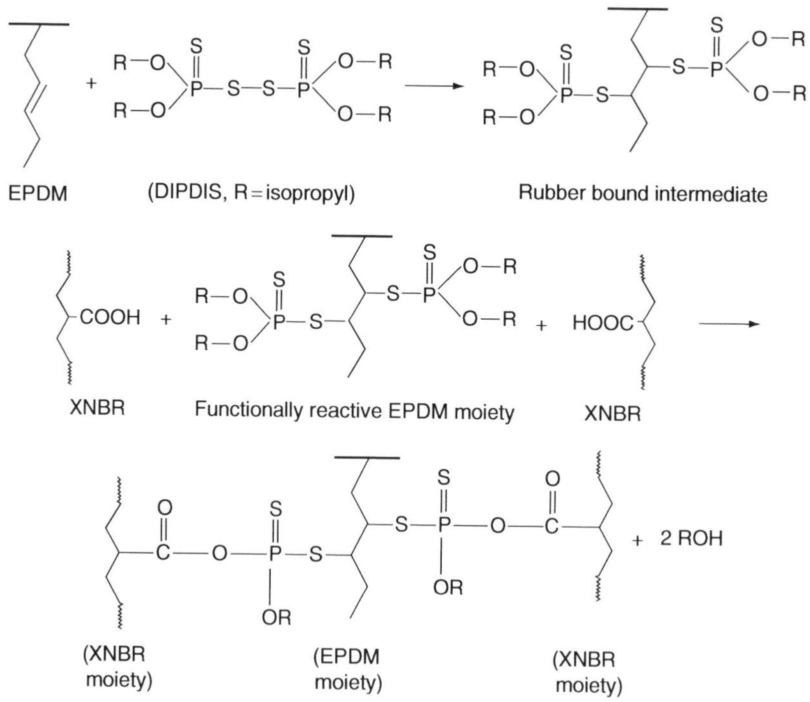

FIGURE 11.7 Inter-rubber bonding scheme for blending ethylene–propylene–diene monomer (EPDM) with nitrile rubber with a cross-linking agent. (From Naskar, M., Debnath, S.C., and Basu, D.K., *Rubber Chem. Technol.*, 75, 309, 2002.)

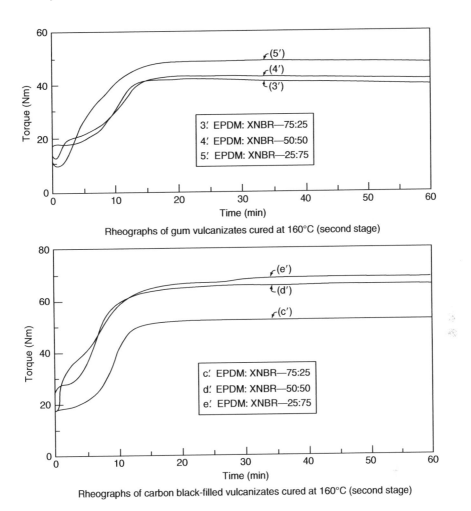

FIGURE 11.8 Torque time curves for two-step reaction. (From Naskar, M., Debnath, S.C., and Basu, D.K., *Rubber Chem. Technol.*, 75, 309, 2002.)

Both cure characteristics and physical properties were measured for the blends. Torque time curves in Figure 11.8 showed two distinct steps—the first one representing the reaction between carboxyl groups of XNBR and the ester groups of DIDPS/ZnO, and the second one, the reaction between the sulfide groups of DIDPS and EPDM. Although incorporation of carbon black increases the steepness of the curves due to enhanced rate and extent of cross-linking reaction, the nature of the cure curves remains the same. Physical properties of the blends shown in Table 11.8 increase with increase in XNBR concentration as expected of a compatible blend. Further, the rate and extent of reaction increase with increase in XNBR concentration. The negative slope of Krauss plot and higher tensile strength of the blends also support the occurrence of interfacial bonding. Electron micrographs of the tensile fractured samples in Figure 11.9 show that the texture of XNBR-rich vulcanizates is more compact than those for EPDM-rich vulcanizates and two-stage vulcanization is better for providing compatibilization and compact texture than one-stage vulcanization.

BR is incompatible with CR because of the large difference in polarity. Mingyi et al. have studied their compatibilization in the presence of a styrene–butadiene–styrene block (SBS) copolymer [30]. The blends were prepared by blending the two rubbers separately at approximately 75°C. The master batch of the BR compound contained 1 part of sulfur, 2 parts of copolymer, 1.5 parts of TMTD, 5 parts

TABLE 11.8
Physical Properties Increase with Increase in XNBR
Concentration in XNBR–EPDM Blends

Property	Mix		
	3′	4′	5′
200% Modulus (MPa)	1.73	2.10	2.15
	(4.99)[a]	(8.30)	(10.60)
Tensile strength (MPa)	7.30	12.70	18.00
	(11.30)	(20.00)	(27.30)
Elongation at break (%)	445	485	505
	(445)	(405)	(400)
Hardness (Shore A)	60	65	70
	(72)	(78)	(82)
Swelling index (Q)	1.96	1.21	0.91
Cross-linking value (1/Q)	0.51	0.82	1.47
AGED IN AIR, 72 h @ 100°C			
200% Modulus (MPa)	2.10	2.19	2.57
	(6.75)	(11.20)	(15.40)
Tensile strength (MPa)	6.30	9.00	10.60
	(11.60)	(17.90)	(24.22)
Elongation at break (%)	355	390	400
	(340)	(300)	(300)
Hardness (Shore A)	60	66	71
	(71)	(78)	(83)

Source: Naskar, M., Debnath, S.C., and Basu, D.K., *Rubber Chem. Technol.*,
              75, 309, 2002.

[a]   Values in parenthesis are the results of filled vulcanizates containing 40 phr
      black and 5 phr DOP.

of zinc oxide, and 1 part of stearic acid. The CR compound included 5 parts of zinc oxide, 4 parts of magnesium oxide, and 1 part of stearic acid. The master batches were then blended together with different amounts of SBS block copolymers. Physical properties were measured for specimens cured at 143°C for 20 min. Figure 11.10 presents tensile strength versus SBS content. It is evident that addition of SBS improves tensile strength significantly and the improvement continues up to approximately 5% of SBS. Further, addition of SBS decreases the tensile strength.

It is suggested that at higher SBS content, the copolymer forms a third phase and, as such, the cross-link density decreases and the copolymer ceases to act as a compatibilizer. Figure 11.11, an optical micrograph, shows that in BR–CR blend, large and irregular CR particles are dispersed in continuous BR phase. Addition of 1% SBS greatly reduces the particle size and homogenizes their distribution. However, when 5% or more of SBS is added, the effect on particle size reduction is very small.

The infrared spectra demonstrate that there is a strong interaction between CR and the PS segment of SBS and between SBS and BR. Scanning electron micrographs of the fracture surfaces further support the conclusion. The fracture surface of 70:30 BR/CR shows layer-shaped morphology and 30:70 blends show hillock-shaped protrusions. Addition of 5% SBS reduces the particle size, makes the particles more spherical, and enhances uniform distribution. This provides further evidence of the compatibilizing effect of SBS.

Brominated poly(isobutylene-co-4-methyl styrene) (BIMS) is often blended with general-purpose rubbers (NR, BR, SBR, etc.) for producing articles such as tire-forming bladders, treads, and side walls.

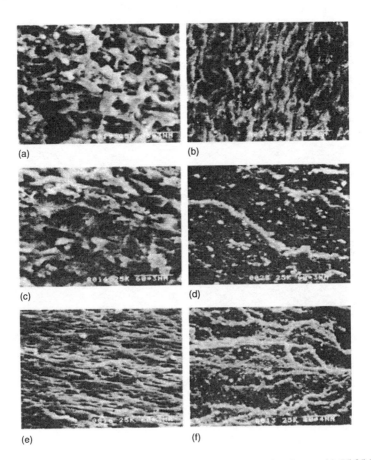

FIGURE 11.9 Tensile fractured samples show the texture of XNBR vulcanizates: (a) 75:25 EPDM/XNBR; (b) 75:25 EPDM/XNBR; (c) 50:50 EPDM/XNBR (one stage); (d) 50:50 EPDM/XNBR (two stage); (e) 25:75 EPDM/XNBR (one stage); (f) 25:75 EPDM/XNBR (two stage). (From Naskar, M., Debnath, S.C., and Basu, D.K. *Rubber Chem. Technol.*, 75, 309, 2002.)

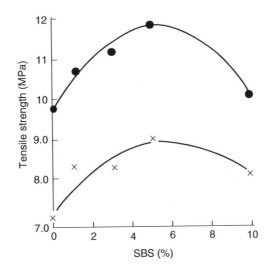

FIGURE 11.10 Effect of SBS contents on the tensile strength of blends: ●---30:70 BR/CR; ×---50:50 BR/CR. (From Mingyi, L., Hua, Z., and Jianfeng, L.J., *Appl. Polym. Sci.*, 71, 215, 1999.)

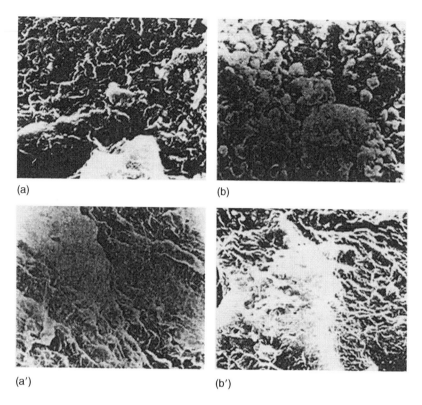

(a)   (b)

(a′)   (b′)

FIGURE 11.11 Micrographs of fracture surface of poly-butadiene rubber–chloroprene rubber (BR–CR) blends with various amounts of SBS. 70:30 BR/CR (a, b) and 30:70 BR/CR (a′, b′) with various amounts of block copolymer SBS: (a, a′) 0%; (b) 10%; (b′) 5%. (From Mingyi, L., Hua, Z., and Jianfeng, L.J. *Appl. Polym. Sci.*, 71, 215, 1999.)

It is important to know the phase behavior, i.e., the region of miscibility and kinetics of de-mixing if any. The thermodynamic interaction and phase behavior of elastomer blends based on BIMS and SBR have been investigated by Raboney et al. using SANS and cloud point measurement [31]. Flory–Huggins interaction parameter χ for the blend was estimated from a single-phase blend of low molecular analogs of the two polymers. Analysis of the results using random copolymer formalism shows that the repulsive interaction between the isobutylene and 4-methyl styrene groups enhances the compatibility between BIMS and polybutadiene. Maximum compatibility occurs around 50 mol% of 4-methyl styrene. Interaction energy density $\Lambda = (\chi/v)RT$ (where χ is the interaction parameter and v is the reference volume of the segment) plotted against 4-methyl styrene content in Figure 11.12 supports the above conclusion. Similar studies can be used to find optimum volume percent of methyl styrene in the copolymer that can form a miscible blend with other rubbers.

11.2.2.4 Compatibilization by Reactive Blending

When two polymers interact or react with each other, they are likely to provide a compatible, even a miscible, blend. Epoxidized natural rubber (ENR) interacts with chloro-sulfonated polyethylene (Hypalon) and polyvinyl chloride (PVC) forming partially miscible and miscible blends, respectively, due to the reaction between chlorosulfonic acid group and chlorine with epoxy group of ENR. Chiu et al. have studied the blends of chlorinated polyethylene (CR) with ENR at blend ratios of 75:25, 50:50, and 25:75, as well as pure rubbers using sulfur (S_8), 2-mercapto-benzothiazole, and 2-benzothiazole disulfide as vulcanizing agents [32]. They have studied Mooney viscosity, scorch

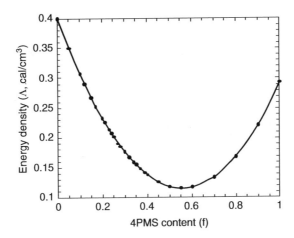

FIGURE 11.12 Interaction energy density versus 4-methyl styrene content. (From Raboney, M., Garner, R.T., Elspass, C.W., and Peiffer, D.G., Phase Behavior of Brominated Poly(Isobutylene–co-4-Methylstyrene)/ General Purpose Rubber Blends. Rubber Division, Proceedings of the American Chemical Society, Nashville, TN, Sept. 29–Oct. 2, 1998, Paper No. 36.)

time, curing characteristics, and dynamic mechanical properties. Mooney viscosity was the lowest at 75:25 CR/ENR blend ratio which represents the highest extent of plasticization and miscibility. The cure rate of ENR is much higher than that of CR. This may be due to the opening of epoxy group of ENR by sulfonic acid formed during vulcanization. On the other hand, zinc chloride formed *in situ* by the interaction of zinc oxide and CR during processing of CR retards the cure rate of the latter. However, the cure rate of the blend is enhanced by the presence of ENR and the increase is proportional to ENR content. This effect reaches maximum at 100°C. Further, the rheographs show that the torque increases sharply at the beginning and then slowly after scorch time.

This is due to the reaction between the epoxy group of ENR and chlorine of CR forming additional cross-links. The torque is the highest at 75:25 ratio where the system is highly compatible. The optimum cure time t_{c90} was the highest at ENR/CR ratio 25:75. It decreased with an increase in ENR content. The authors studied the dynamic mechanical properties of the blends and pure elastomers at three different frequencies at different temperatures. They found that T_g increases with ENR content and with frequency. They also estimated activation energy for molecular motion (E_a) leading to glass transition, using the relation $\ln f = \ln A - E_a/RT$. The authors found that activation energy increases with ENR content, reaching maximum at 75% ENR content.

Polychloroprene (CR) is often blended with polyisoprene (IR) to provide superior resistance to oxidation, oils, greases, ozone, sunlight, and combustion. Its high polarity gives a greater probability for bonding to a wide range of substrates. Polyisoprene, on the other hand, provides better elasticity, high tensile strength, tack, and lower hysteresis. It crystallizes under strain, compression, and induced orientation which enhances its strength. A blend therefore provides better balance of properties.

Kardan has studied the contribution of each component in cohesive and adhesive strength and overall durability of a blend of IR with CR [33]. Blends of IR (Natsyn 22000) with different types of CR (Neoprene W) were prepared by mill compounding and were dissolved in toluene. Each solution was applied to steel substrates and the latter were laminated by pressure. Lap shear of each laminate was measured after allowing 48 h to age. To measure cohesive forces, the adhesives were applied to steel substrates pre-coated with a primer similar to CR and then laminated. Lap shear strength of the laminates was measured. Special infrared spectroscopy was used to monitor the structural changes and molecular orientation to the metallic substrates. Figure 11.13 gives the lap shear versus CR content for the laminated steel and primed steel substrates representing cohesive and adhesive

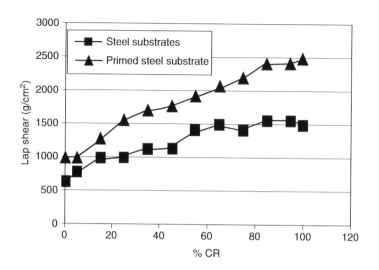

FIGURE 11.13 Lap shear versus % chloroprene rubber (CR) for CR–IR blends laminated between two steel substrates. (From Kardan, M., *Rubber Chem. Technol.*, 74, 614, 2001.)

strength. The abrupt change in the adhesive strength at 75:25 and cohesive strength at 50:50 ratios is quite obvious. This suggests that NR levels can be increased to as high as 50% without losing adhesive strength. The results of infrared analysis show that the orientation of the methylene chains remains parallel to the substrate surface until CR in the blend reaches the 25% level. The parallel orientation helps adhesion of the polymer to the metallic substrate. The molecular conformation of the polymer changes with increased amount of CR as evidenced by the increase in the ratio of the peaks at 1433 and 1450 cm^{-1} indicating different degrees of adhesion to the steel substrate. Whereas the cohesive forces in the blend reach maximum at 20:80 ratio for IR/CR, good adhesion to steel substrate increases with CR content and approaches maximum at 50:50 ratio. It is likely that the orientation and crystallization that contribute to adhesive forces are different from those in the bulk compared to those at the surface. FTIR spectra also lead to the conclusion that, at higher ratios of CR/IR, the conformation of the rubbers in the blend is largely *trans* and the orientation of $=$CH groups perpendicular to the substrate promotes adhesion.

Dikland and Van Duin have examined the miscibility regime of gum stock EPDM–EPM blends, both theoretically and experimentally [34]. EPDM grades differ from each other based on their ethylene content, molecular weight, crystallinity, etc. Very often two grades of EPDM are blended to provide either process or property advantage. They have estimated solubility parameters ($\acute{o}$) of EPDM, using Synthia module of Biosym software. The Flory interaction parameter of the two different EPM grades was calculated using the equation

$$X_{12} = [V_e(\acute{o}_1 - \acute{o}_2)^2]/RT \tag{11.6}$$

where

V_e is the ethylene reference molar volume (assumed as 33.0 cc)
$\acute{o}$ is the solubility parameter of the two polymers based on their composition
R is the gas constant
T is the absolute temperature

The effective interaction parameter of two random EPM polymers is given by

$$X_{eff} = X_{1,2}(f_1 - f_2)^2 \tag{11.7}$$

where f is the ethylene volume fraction for each polymer. The critical interaction parameter above which the blend is unstable and susceptible to phase separation is given by

$$X_{crit} = 2/r_w \tag{11.8}$$

where r_w is the weight average degree of polymerization (molecular weight divided by reference volume of ethylene (33cc)). It was assumed that the contribution of the third monomer, namely DCPD or norborene, may be added to the solubility parameter of ethylene fraction and the amount may be added to the ethylene content. They found the critical solubility parameter to be 0.0004, 0.00028, and 0.000215 for the range of PE molecular weights considered. The effective interaction parameter was calculated as a function of chemical composition and molecular weight for two EPM grades. Contours inside which the component polymers are miscible and beyond which they are immiscible were calculated. Based on these calculations, it was found out that two EPM–EPDM copolymers are compatible when their difference in ethylene content is less than 12%. Subsequently, the authors prepared a large number of EPM–EPDM blends, both by mechanical mixing and by dissolution and precipitation method. Scanning electron micrographs confirmed the above conclusion. Table 11.9 gives the blend composition, blend ratio, and both maximum and minimum domain sizes for different blends.

Schuster et al. have investigated filler partition and the role of interphase in rubber blends by dynamic mechanical analysis [35]. According to the authors, the volume fraction and the thickness of the interphase, a_{iph}, are inversely proportional to the square root of the Flory interaction parameter or the difference in the solubility parameter of the two components. At a constant volume fraction of the blend constituents, the thickness of the interphase increases with declining incompatibility of the system. Since neighbored phases with different filler content will have strong differences in mechanical response showing high stress gradient in the contact region, the amount of filler located as reinforcing agent in the interphase will play a substantial role in defining the mechanical strength of the blend. Due to the fact that both the blend ratio and filler loading lead to signal alteration in the loss modulus in the glass transition region, a specific fitting procedure may be used to obtain quantitative information about the size of the interface, the filler content located in the region, and the contribution of polymer–filler interaction in the partition of carbon black in distinct polymer phases. They studied carbon black distribution in four blends comprising NR, BR, EPDM, and SBR. Table 11.10 gives the blend composition, solubility parameter difference between the two components, interphase signal, and interphase thickness for the rubber blends. Table 11.11 gives carbon black partition coefficient from the analysis of $G''(t)$ signals.

The important yet unexpected result is that in NR–s-SBR (solution) blends, carbon black preferably locates in the interphase, especially when the rubber–filler interaction is similar for both polymers. In this case, the carbon black volume fraction is 0.6 for the interphase, 0.24 for s-SBR phase, and only 0.09 in the NR phase. The higher amount in SBR phase could be due to the presence of aromatic structure both in the black and the rubber. Further, carbon black is less compatible with NR–cis-1,4 BR blend than NR–s-SBR blend because of the crystallization tendency of the former blend. There is a preferential partition of carbon black in favor of cis-1,4 BR, a significant lower partition coefficient compared to NR–s-SBR. Further, it was observed that the partition coefficient decreases with increased filler loading. In the EPDM–BR blend, the partition coefficient is as large as 3 in favor of BR.

It was concluded that the filler partition and the contribution of the interphase thickness in rubber blends can be quantitatively estimated by dynamic mechanical analysis and good fitting results can be obtained by using modified "spline fit functions." The volume fraction and thickness of the interphase decrease in accordance with the intensity of intermolecular interaction.

Blending NBR with ENR improves physical and mechanical properties without affecting its oil resistance. Such blends can be used for making per-evaporation membranes. Mathai et al. have studied the transport properties of aromatic solvents through membranes prepared from 50:50 NBR–ENR blends [36]. Transport experiments were carried out by immersing circular specimens in the desired

TABLE 11.9

Blend Compositions, Blend Ratios, and Max/Min Domain Size of Blend

EPM Type	EPDM Type	Blend Ratio (w/w)	Minimum Domain Size (μm)	Maximum Domain Size (μm)
EPM-1	EPDM-7	50:50	0.2×1^{a}	0.2×3
EPM-1	EPDM-7	50:50	0.5^{a}	2
EPM-3	EPDM-7	50:50	Homogeneous	
EPM-3	EPDM-7	50:50	Homogeneous	
EPM-2	EPDM-2	50:50	$0.1^{b,c}$	10×1
EPM-2	EPDM-2	50:50	Homogeneous	
EPM-2	EPDM-1	50:50	$0.1^{b,c}$	15×0.5
EPM-2	EPDM-4	50:50	$0.1^{b,c}$	15×1.5
EPM-2	EPDM-5	50:50	0.05^{b}	5×0.3
EPM-2	EPDM-6	50:50	$0.1^{b,c}$	20×4
EPM-2	EPDM-6	50:50	0.1^{b}	2×0.3
EPM-3	EPDM-2	50:50	$0.1^{b,c}$	10×1
EPM-3	EPDM-2	50:50	$0.1^{b,c}$	1×0.5
EPM-3	EPDM-1	50:50	0.1^{b}	2×0.5
EPM-3	EPDM-4	50:50	$0.1^{b,c}$	15×1
EPM-3	EPDM-5	50:50	$0.1^{b,c}$	5×0.5
EPM-3	EPDM-6	50:50	0.1^{b}	0.5×0.5
EPM-3	EPDM-6	50:50	0.1^{b}	0.3×0.5
EPM-2	EPDM-2	70:30	Homogeneous	
EPM-2	EPDM-3	70:30	Homogeneous	
EPM-2	EPDM-4	70:30	<0.05	<0.05
EPM-2	EPDM-5	70:30	0.05	0.5
EPM-2	EPDM-6	70:30	0.05	0.5
EPM-3	EPDM-2	70:30	0.2	9
EPM-3	EPDM-3	70:30	0.5	8
EPM-3	EPDM-4	70:30	0.3	4
EPM-3	EPDM-5	70:30	0.1	1.5
EPM-3	EPDM-6	70:30	0.05	1.0
EPM-3	EPDM-6	70:30	Homogeneous	

Source: Dikland, H.G. and Van Duin, M., *Rubber Chem. Technol.* 76, 495, 2003.

[a] Co-continuous, although EPM sometimes tends to disperse.

[b] No good distinction between dispersed phase and matrix; only data for EPDM phase are given.

[c] Strongly elongated domains.

solvent and measuring their weight gain at time intervals until equilibrium swelling was attained. Results were expressed as moles of solvent (Q_t) per 100 g of the polymer sample. It is obvious that Q_t decreases with increase in NBR content, due to the greater polarity of the latter. Further Mathai and Singh found that Q_t decreases with increasing size of the solvent. Ficks law of diffusion suggests that

$$Q_t/Q_\infty = 4(Dt/\amalg h^2)^{1/2} \tag{11.9}$$

where
  D is diffusivity
  t is sorption time
  h is thickness

TABLE 11.10

Blend Composition, Solubility Parameter Difference between the Two Components, Interphase Signal (ϕ_{IPH}), and Interphase Thickness (a_{IPH})

Rubber Blend	Blend Ratio	$\Delta\delta$	ϕ_{IPH}	a_{IPH}
EPDM/cis-1,4-BR	70:30	1.22	1.9	3.5
NR/e-SBR (17:25)	20:80	0.96	3.8	5.4
NR/s-SBR (50:25)	20:80	0.77	9.4	13
	30:70	0.77	11.0	—
	40:60	0.77	12.0	—
NR/cis-1,4-BR	40:60	0.41	4.3	5.8
	60:40	0.41	4.9	—
	70:30	0.41	4.6	—

Source: Schuster, R.H., Meier, J., and Kluppel, M., The Role of Interphase in Filler Partition in Rubber Blends, 156th Meeting of the Rubber Division, American Chemical Society, Orlando, FL, 1999, Paper No. 60.

Hence, D can be estimated from the initial slopes Q_t/Q_∞ against square root of time. Figure 11.14 gives a plot of D for three solvents, namely benzene, toluene, and xylene, against volume fraction ENR. In blends with different fractions of ENR, diffusivity is the highest with benzene, followed by toluene and xylene, confirming that D decreases with increase in the molar volume and the solubility parameter difference of the penetrant and the polymeric substrate. Further, it is the highest for 50:50 composition. Diffusivity is also affected by the morphology and, to a large extent, controlled by the continuous phase.

Mathai and Singh have estimated the permeability coefficient P, using the formula $P = kD$ where k is the partition coefficient and D is the diffusivity. They have used both parallel and series models to calculate P. The experimental values are always greater than measured values. The poor agreement between the experimental and calculated values is attributed to the polar–polar interaction between the epoxy group and nitrile group.

TABLE 11.11

Carbon Black Partition Coefficient from the Analysis of $G''(t)$ Signals

Blend System (A/B)	Blend Ratio	Filler Loading (phr)	Mixing Procedure	C_B/C_A
NR/s-SBR (50:25)	40:60	10	CB to rubber blend	2.6
	40:60	30	CB to rubber blend	2.6
	40:60	60	CB to rubber blend	2.7
NR/cis-1,4-BR	70:30	30 N 220	CB to rubber blend	1.7
	70:30	60 N 220	CB to rubber blend	1.1
	50:50	30 N 220	CB to rubber blend	2.6
	50:50	60 N 220	CB to rubber blend	1.5
EPDM/cis-1,4-BR	—	24 N 550	CB/EPDM + BR	3.2
		24 N 550	CB/BR + EPDM	7.7
		24 N 550	CB to rubber blend	3.2

Source: Schuster, R.H., Meier, J., and Kluppel, M., The Role of Interphase in Filler Partition in Rubber Blends, 156th Meeting of the Rubber Division, American Chemical Society, Orlando, FL, 1999, Paper No. 60.

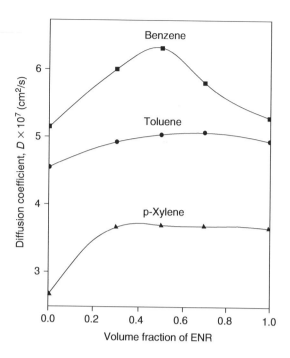

FIGURE 11.14 Variation of *D* with volume fraction of ENR in different solvents at 28°C. (From Mathai, A.E., Singh, R.P., and Thomas, S., *Polym. Eng. Sci.*, 43, 704, 2003.)

Trans-polyoctylene rubber (TOR) is a low-molecular weight hydrocarbon rubber, which can act as a compatibilizer in the blends of saturated and unsaturated rubber. Chang et al. have studied the effects of adding TOR to a blend of NR and EPDM [37]. The formulations used in this study are given in Table 11.12. It was found that the mixing torque decreased on addition of TOR but did not change much beyond 5 phr addition. The mixing temperature also decreased by about 10°C indicating that TOR acts as a plasticizer and processing aid, reduces its melt viscosity, and decreases energy consumption for mixing process. The rate of cross-linking reaction also decreases, thereby increasing scorch and time for optimum cure. However, cross-link density measured by gel content did not change appreciably. This indicated that TOR having a lower degree of unsaturation

TABLE 11.12
Formulations of NR/EPDM–TOR Blends

	Compound No. (phr)			
	1	**2**	**3**	**4**
NR	70	70	70	70
EPDM	30	30	30	30
TOR	0	5	10	20
Zinc oxide	5.00	5.25	5.50	6.00
Stearic acid	1.50	1.58	1.5	1.80
Sulfur	2.00	2.10	2.20	2.40
CBS	1.00	1.05	1.10	1.20

Source: Chang, Y.-W., Shin, Y.-S., Chun, H., and Nah, C., *J. Appl. Polym. Sci.*, 73, 749, 1999.

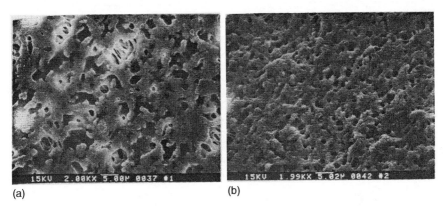

(a)   (b)

FIGURE 11.15 Diffused particles in natural rubber/ethylene–propylene monomer/*trans*-polyoctylene rubber (NR–EPM–TOR) blend (a) are much smaller than in blends without TOR (b). (From Chang, Y.-W., Shin, Y.-S., Chun, H., and Nah, C., *J. Appl. Polym. Sci.*, 73, 749, 1999.)

contributes to a very small extent (2%) to cross-link density. Scanning electron micrographs revealed that the dispersed particles in NR/EPDM–TOR blend are much smaller (approximately 1 μm) compared to the ones (about 10 μm) in blends without TOR as shown in Figure 11.15.

It is suggested that TOR having lower viscosity locates at the boundary of NR/EPDM layers, thereby increasing their interfacial strength. This was confirmed by measuring the adhesive strength (G_a) between NR and EPDM sheets with and without TOR as shown in Table 11.13. It is speculated that TOR is co-vulcanized with component elastomers, thereby increasing compatibility.

Table 11.14 gives critical stress, critical strain, and critical stored energy of NR–EPDM blends for initiation of ozone cracking. All the properties show an increase on addition of TOR, especially the critical stored energy. Stored energy is a strong indicator of ozone resistance and shows an increase by about 80% on addition of 20 parts of TOR to the compound. This was confirmed by SEM pictures of surface ozone cracks and the results of dynamic mechanical moduli and tan ∂ measurement.

In rubber–rubber blends, maldistribution of curatives such as sulfur, accelerator, and activator due to their difference in solubility and diffusivity leads to uneven distribution. Blending strategies such as adding the accelerators to each rubber followed by blending the two mixed batches has been found to be more effective than blending the curatives into both rubbers in a single step. Selection of

---

TABLE 11.13
Adhesive Strength between NR and EPDM Sheets with/without TOR

Sheet 1[a]	Sheet 2[a]	G_a (J/m^2)
NR	EPDM	279
NR	EPDM/TOR (90/10)	1274
NR/TOR (90/10)	EPDM	1176
NR/TOR (90/10)	EPDM/TOR (90/10)	1489

Source: Chang, Y.-W., Shin, Y.-S., Chun, H., and Nah, C., *J. Appl. Polym. Sci.*, 73, 749, 1999.

[a] Values in parenthesis are the results of filled vulcanizates containing 40 phr black and 5 phr DOP.

TABLE 11.14

Critical Stress, Critical Strain, and Critical Stored Energy

	Compound No.			
	1	**2**	**3**	**4**
Critical stress, σ_c (MPa)	0.142	0.180	0.210	0.240
Critical strain, $\dot{\varepsilon}_c$ (%)	10.15	11.55	12.55	11.25
Critical stored elastic energy density, W_e (kJ/m³)	7.350	10.00	12.65	13.06

Source: Chang, Y.-W., Shin, Y.-S., Chun, H., and Nah, C., *J. Appl. Polym. Sci.*, 73, 749, 1999.

the proper curing system also obviates the maldistribution of cross-link density. Curing temperature plays an important role in the distribution of cross-links. For example, in an NR–BR blend, curing at 130°C leads to higher cross-link density in NR phase than curing at 140°C where the opposite happens. In an EPDM–NR blend the cross-link density is lower in EPDM phase, but it is remarkably increased by grafting EPDM with maleic anhydride.

Wootthikanokkhan and Clythong have studied the effects of additive distribution and its effect on the distribution of cross-link density in NR–acrylic rubber (AR) blends [38]. The formulations of the four blends is given in Table 11.15.

Masticated NR was mixed with curative ingredients and the mixed batches were cured for the optimum cure time as determined by cure meter. The specimens were subjected to differential swelling in cyclohexane and acetonitrile, the former being a good solvent for NR and very poor solvent for AR and the latter being a good solvent for AR and poor solvent for NR. Volume fraction (V_r) of each rubber in the vulcanized blend was estimated from the extracted specimens. The results presented in Table 11.16 indicate that zinc diethyl thiocarbamate (ZDEC) provided better cure than either mercapto-benxo-thiazole (MBT) or diphenyl guanidine (DPG). This was also obvious from the magnitude of mechanical properties of the blends. In other words, the total number of mono- and disulfide linkages is higher in ZDEC-cured systems.

TABLE 11.15

Blend Formulations of Natural Rubber and Acrylic Rubber

	Content (phr)		
Materials	**Recipe 1**	**Recipe 2**	**Recipe 3**
NR	50	50	50
ACM	50	50	50
MBT	2.5	—	—
ZDEC	—	2.5	—
DPG	—	—	2.5
Sulfur	2.5	2.5	2.5
Sodium stearate	12.5	12.5	12.5
Zinc oxide	5.0	5.0	5.0
Stearic acid	1.0	1.0	1.0

Source: Wootthikanokkhan, J. and Clythong, N., *Rubber Chem. Technol.*, 76, 1116, 2003.

TABLE 11.16
Results from Differential Swelling Test of Various Rubber Blends

Accelerator	Curing Temperature (°C)	V_r in Cyclohexane	V_r in Acetonitrile
MBT	170	0.229 (±0.002)	0.288 (±0.005)
ZDEC	170	0.287 (±0.002)	0.340 (±0.005)
DPG	17	0.187 (±0.0025)	0.274 (±0.003)
MBT	150	0.228 (±0.002)	0.304 (±0.001)
ZDEC	150	0.293 (±0.005)	0.356 (±0.002)
DPG	150	0.190 (±0.00)	0.262 (±0.005)

Source: Wootthikanokkhan, J. and Clythong, N., *Rubber Chem. Technol.*, 76, 1116, 2003.

The authors measured proton magnetic resonance spectra of each specimen and found that the peak width (V_r) or H% value at 4.15 ppm of S/ZDEC-cured specimens was higher than that cured by MBT or DPG, regardless of curing temperature. It was concluded that the mobility of the component rubber molecules in S/ZDEC system was lower than in MBT- and DPG-cured specimen leading to a tighter cure with the former system as indicated by the V_r peak width values. Higher chain mobility in MBT and DPG systems contributed to higher tensile strength. The lower tensile properties of ZDEC-cured blends may be ascribed to formation of a larger fraction of mono- and disulfidic cross-links than polysulfidic cross-links which provide better mechanical properties.

The difference in degree of cure of the blends by different curatives has also been explained on the basis of changes in curative distribution with accelerator types and the effect of cure temperature. The tensile properties of the blend cured by S/ZDEC at 170°C were significantly lower and modulus was higher than those cured by S/MBT and S/DPG as shown in Table 11.17. Lowering of cure temperature by 20°C significantly improved these properties. However, the standard deviation in the results increased limiting the potential for any solid conclusion.

11.3 REACTIVE BLENDING

Elastomer blends in which the components react (reactive blending) with each other provide the best route to obtain a homogeneous product with improved physicals. The negative free energy

TABLE 11.17
Tensile Properties of Various NR Vulcanizates

Curing System	Sulfur Content (phr)	Strength (MPa)	Strain (%)	300% Modulus (MPa)
S/DPG	1.0	3.27 (±0.28)	1955 (±26.0)	0.115 (±0.04)
	1.5	4.59 (±0.38)	1964 (±107.0)	0.178 (±0.01)
	2.0	5.36 (±0.43)	1965 (±162.0)	0.186 (±0.01)
S/ZDEC	1.0	17.48 (±0.71)	2424 (±40.0)	0.337 (±0.005)
	1.5	18.59 (±2.13)	2114 (±198.0)	0.210 (±0.00)
	2.0	23.94 (±0.94)	1990 (±23.0)	0.278 (±0.00)
S/MBT	1.5	8.26 (±1.72)	2238 (±113.0)	0.217 (±0.008)
	2.0	14.63 (±2.00)	2470 (±85.0)	0.267 (±0.013)
	2.5	13.32 (±0.99)	2287 (±69.0)	0.302 (±0.028)

Source: Wootthikanokkhan, J. and Clythong, N., *Rubber Chem. Technol.*, 76, 1116, 2003.

generated due to chemical reaction facilitates the blending process. Nando has used this principle to develop blends between ethylene copolymers such as ethylene acrylic acid (EAA) and ethylene methacrylic acid (EMA) with poly-dimethyl siloxane (PDMS), ENR, ethylene–propylene diene rubber, and thermoplastic polyurethane (TPU) [39]. The blends were prepared at 70:30, 50:50, and 30:70 ratios using a Brabender plasticorder at 180°C and 100 rpm rotor speed for 10 min. They were characterized by FTIR, UV, and NMR spectroscopy as well as by the study of their mechanical properties, dynamic mechanical behavior, and glass transition temperature.

The gradual reduction in the absorption ratio of peaks at 1597 cm^{-1} in IR spectra and that of $\acute{o}$ at 127 and 132 ppm in the solid state NMR spectra (representing —Si—CH=CH$_2$ carbon atoms) with increased blending temperature proved that chemical reaction takes place between EMA and PDMS in the absence of any catalyst to form a graft copolymer of EMA and PDMS in an EMA/PDMS blend. It is postulated that during the reaction, EMA forms a radical after losing its tertiary hydrogen atom and the latter reacts with PDMS forming a graft copolymer. This is supported by the fact that all the blends exhibit a single glass transition temperature having values between those for the base polymers. The synergism in physical–mechanical properties such as modulus, tensile strength, and elongation at break further confirms this conclusion.

Blends of EMA and TPU also form homogeneous blends due to the interaction of the carbonyl group in EMA with the —NH group in the thermoplastic polyurethane. These blends exhibit a single T_g having values in between those for the base polymers. A clear shift of the T_g of the hard segment has been attributed to the breaking of self-associated hydrogen bonds of TPU during processing at 180°C and subsequent formation of intermolecular hydrogen bonds with carbonyl groups of EMA. The T_g of the soft segment shifts towards a higher temperature for all the blend compositions studied due to the dipolar interaction between carbonyl groups of EMA and –C–O–C group of polytetramethylene group of the TPU.

Nando has also investigated reactive blending of ENR with EAA copolymer. Homogeneous blends of ENR and EAA are formed due to the interaction of either the epoxy group or the hydroxyl group of ENR with the carbonyl group of EAA. The extent of esterification is low because EAA contains only 2.4 mol% of acrylic acid. Blends of the two polymers at different proportions have also been examined. Gradual increase in ester peak area and concurrent decrease in acid as well as epoxy/alcoholic peak area with increase in temperature from 120°C to 150°C imply that the esterification reaction is enhanced by increase in temperature. DSC studies with 50:50 blend showed a single T_g at -21°C. This is lower than T_g calculated by Fox equation. The positive shift was attributed to the occurrence of chemical reaction instead of homogeneous mixing. The miscibility of the binary blend has also been examined by using a software package developed by Graf et al. The results show the acrylic acid content of the blend increases the area of miscibility window even at low copolymer content. The interaction parameter χ, calculated with software using a reference volume of 100 cc/mole, was found close to 3.41×10^{-4} which is much below the critical value (0.002) for miscibility. Mechanical properties such as tensile strength, modulus, and hardness showed gross synergism. However, the elongation at break showed some dilution effect.

Blends of EMA copolymer and EPDM containing vinyl norborene as a third monomer were also investigated. Blending was carried out at 180°C at a rotor speed of 100 rpm. After the reaction, the blends were quenched on the cold rolls and were sheeted out. They were examined by IR spectra. The reduction of peak area related to unsaturation indicated a progressive loss of EPDM due to reaction with EMA. The extent of reaction depended on the utilization of unsaturation which is estimated to be 14% for EMA/EPDM at a 70:30 ratio and 53% at a 50:50 blend ratio. The tensile properties exhibit synergism as the EMA proportions change from 0% to 50%.

Kumar et al. have studied the milling behavior of brominated isobutylene-co-p-methyl styrene (BIMS) and its blends with EPDM [40]. Using the theoretical model of Tokita, they have tried to optimize the mill parameters. They measured the critical nip-gap at which a front-to-back roll transition takes place. They concluded that addition of different fillers reduced the critical nip-gap for BIMS as compared to the gum polymer. The critical nip-gap is lower for systems containing coarser carbon black (N330) and silica compared to systems containing finer carbon black (N550). The critical nip-gap decreased and smoothness increased with addition of EPDM up to 50 parts. Systems containing silica filler of smaller size showed a lower value of critical nip-gap. Beyond 50 parts of EPDM, the filled bands showed bagging characteristics. For blends containing 30 parts of N330 carbon black, the critical nip-gap and the mill band formation index increased on increasing the friction ratio. However, the mill band formation index N_1 remained constant. The tendency for an F–B transition decreased on increasing the loading of N330 carbon black and it increased with an increase in the loading of the processing oil.

Unlike a plastic blend where the properties largely depend on the properties of the individual component and the compatibilizer used, those of a rubber blend depend on the solubility and diffusivity of the curatives, reaction rates, scorch time, etc. Figure 11.16 gives relative cure rate and scorch time for a number of accelerators. Hence, in designing a rubber blend, all these parameters have to be taken into consideration in order to obtain good properties along with good processability.

Since EPDM is a cheaper elastomer, it is often added to SBR to reduce cost. Zhao et al. have studied the effects of curative and accelerator concentration as well as the effect of mixing on the properties of a blend compound containing 70 parts of SBR and 30 parts of EPDM [41]. Table 11.18 gives the general formulation of the blend compounds.

Table 11.19 gives the cure characteristics of the blends for high and low peroxide, sulfur, and co-agent (peroxide + sulfur) combinations. They found that the maximum and minimum torques

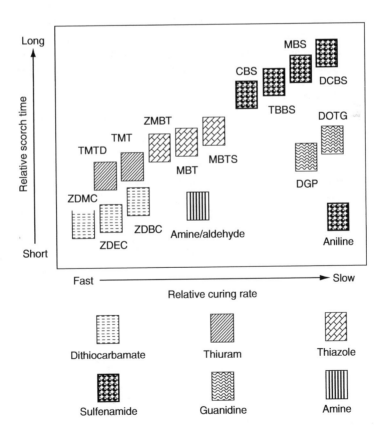

FIGURE 11.16 Schematic diagram of cure rate versus scorch time for various commercial accelerators. (From Zhao, J., Ghebremeskel, G.N., and Peasley, J., A Study of EPDM/SBR Blends, presented at a meeting of the Rubber Division, American Chemical Society, Chicago, IL, April 1999, Paper No. 16.)

TABLE 11.18

General Formulation of EPDM–SBR Blends

Sample #	P (phr)	S (phr)	C (phr)
EPDM	70	70	70
SBR1502	30	60	30
Carbon black (N330)	80	80	80
Sunpar 2280	50	50	50
ZnO	5	5	5
Stearic acid	—	1	1
Sulfur	—	Variable	Variable
TMTD	—	Variable	Variable
MBT	—	Variable	Variable
DCP	Variable	—	Variable

Source: Zhao, J., Ghebremeskel, G.N., and Peasley, J., A Study of EPDM/SBR Blends, presented at a meeting of the Rubber Division, American Chemical Society, Chicago, IL, April 1999, Paper No. 16.

with high and low levels of peroxide (HP), sulfur (HS), and co-agent (HC) are almost similar although high peroxide system is associated with higher scorch time and lower cure rate.

Tensile strength of the co-agent-cured systems, HC and LC, are the highest followed by HP- and LP-cured systems. The sulfur-cured systems gave the minimum strength. There was no significant difference in the elongation at break. The 300% modulus is higher for both the HC- and LC-cured systems. This shows that addition of a small amount of sulfur as co-agent to peroxide cure system for EPDM–SBR blends has a very positive influence on the physical properties due to formation of additional cross-links in SBR and co-vulcanization at the interface. Subsequently the authors cured a blend of 70 parts of SBR and 30 parts of EPDM with high and low levels of coagent cure systems. The physical and mechanical properties are given in Table 11.20. Replacement of 30 parts of SBR by EPDM showed only 18% decrease in tensile strength and 15% decrease in ultimate elongation. The compression set of the blends at room temperature was higher than that at 100°C, due to the dominance of physical stress relaxation. Whereas the physical properties did not deteriorate much

TABLE 11.19

Cure Characteristics of the Blends for High and Low Peroxide (P), Sulfur(S), and Co-Agent C (Peroxide + Sulfur)

Sample #	HP	HC	HS	LP	LC	LS
Max. torque (dNm)	36.7	36.02	36.7	26.7	33.9	25.9
Min. torque (dNm)	7.1	6.01	7.5	5.9	6.7	6.2
Delta torque (dNm)	29.58	30.01	29.13	20.80	27.24	19.68
Scorch time (min)	1.76	1.72	1.57	2.46	2.05	1.90
T50 (min)	6.36	3.79	2.72	7.12	5.03	3.10
T90 (min)	17.90	12.26	7.23	18.17	15.42	7.62
Cure rate index (1/min)	6.20	9.49	17.67	6.37	7.48	17.48

Source: Zhao, J., Ghebremeskel, G.N., and Peasley, J., A Study of EPDM/SBR Blends, presented at a meeting of the Rubber Division, American Chemical Society, Chicago, IL, April 1999, Paper No. 16.

TABLE 11.20

Mechanical and Physical Properties of EPDM and EPDM–SBR Blends Cured with Peroxide and Sulfur Co-Agent Cure System

Sample #	EPDM1	EPDM/SBR1	EPDM2	EPDM/SBR2
Cure System		High Level		Low Level
Stress–Strain				
Tensile strength (MPa)	23.7	19.2	22.2	18.4
Elongation at break (MPa)	391	338	529	443
100% Modulus (MPa)	3.03	4.15	2.54	3.04
200% Modulus (MPa)	9.21	10.5	6.34	7.49
300% Modulus (MPa)	17.3	17.1	11.6	12.6
Compound Mooney				
Mooney (1 + 4 min) 100°C	57.3	53.9	65.4	56.8
Hardness				
Shore A	70	72	72	71
Die C—Tear				
Tear strength (kN/m)	49.2	44.3	54.8	44.9
DIN Abrasion				
Vol. loss (mm^3)	79.9	79.9	80.1	86
Zwick Rebound				
23°C (%)	54.1	47.5	52.3	48.3
70°C (%)	62.3	54.3	56.1	50.9
Goodrich Flexometer				
Delta T (C)	47.5	59.7	70.8	80.2
Permanent set (%)	2.1	2.8	6.6	6.3
Compression Set Method B				
23°C for 70 h (%)	35.8	31.9	42.4	37.3
100°C for 70 h (%)	20.5	20.9	30.6	27.3

Source: Zhao, J., Ghebremeskel, G.N., and Peasley, J., A Study of EPDM/SBR Blends, presented at a meeting of the Rubber Division, American Chemical Society, Chicago, IL, April 1999, Paper No. 16.

on heat aging the compound at 100°C, heat aging at 140°C resulted in significant decrease of the physicals. The positive benefit on adding 30 parts of EPDM to SBR is the lowering of the Mooney by 4–10 points which is desirable for the production of extruded compounds. The authors have also examined the use of accelerators including thiuram and thiazole compounds as a function of temperature and have measured ozone resistance under dynamic conditions.

11.3.1 Rubber–Plastics Blends

Elastomers are often blended with plastics either to improve the impact resistance or to develop new materials having both plastic and elastic behavior. When the elastomer in the blend is dynamically vulcanized, the product is called a thermoplastics vulcanizate (TPV). Blends with unvulcanized rubber phase are usually known as thermoplastic elastomers. TPVs are discussed in another section of this book. This section will deal with recent developments in rubber–plastic blends.

Many of the commercially important plastics such as polystyrene, polyamide, polyester, polycarbonate, polysulfone, polyphenylene oxide alloys, epoxy, and phenolics lack good impact resistance. This is improved by blending them with a rubbery material. Compatibilization of the rubber and plastic phases is very important to achieve stress transfer from the hard to the ductile phase. A number of texts such as "Toughened Plastics" [42], "Rubber Modified Thermoset Resins" [43], and "Rubber Toughened Plastics" [44] present details of toughening mechanisms, materials used, and level of toughening achieved as a function of rubber type and content. Earlier, Mangaraj et al. reviewed past work on impact modification, especially the role of compatibilization in impact modification of plastics by blending with elastomers [45].

When a polymeric material is subjected to impact, the plastic matrix absorbs most of the energy, until the stress reaches a critical value, σ_c, when fracture growth takes place. It is postulated that in a rubber-toughened plastic the rubber particles undergo stretching and form a large number of micro-cracks instead of a large crack and thereby absorb the energy at the crack tip. Localized deformation on these sites creates micro-voids (crazing) and/or shear bands since crazing creates new surfaces. At higher stress, the fibrillar structure breaks down and a true crack forms. Even at this stage, rubber particles dissipate some of the stress through shear banding, thereby delaying the failure. Hence, toughening is best carried out by adding adequate amount of a low modulus (compared to the matrix) material having good adhesion to the matrix and optimal cross-link density. The extent of crazing depends on the number of rubber particles per unit volume of the blend and the latter depends on the volume fraction of the rubber in the blend. However, at higher volume fraction of the rubber, phase inversion takes place and the plastic forms the dispersed phase in a continuous rubber matrix as seen in high-impact PS (HIPS). Adhesion between the plastic matrix and rubber particles is important for stress transfer from glassy to the rubber matrix and initiation of crazing. For example, addition of ethylene-graft–styrene copolymer is essential to improve the impact strength of polystyrene and polyethylene blend. The particle size and degree of cross-linking of the rubbery phase also influence the impact strength.

In summary, the attributes of the elastomer that contribute to the enhanced impact strength of a plastic in plastic rubber blend include the type of rubber, plastic to rubber ratio, particle size, particle size distribution, cross-link density, and degree of grafting, if any. Molecular weight and molecular weight distribution of the plastic also exert some influence. For example, for high-impact PS, the optimal molecular weight of PS is between 170,000 and 220,000. The dispersity index M_w/M_n is kept at 2.5 to 3.0, the particle size at 2.5 μm and the swelling index, a good measure of cross-link density, is kept between 9 and 12.

Processing parameters also exert substantial influence on impact properties of the blend. For example, particle size increases with increasing speed of the agitator, leveling out at some high speed. It decreases with increase in the viscosity ratio of the component phases (becomes minimum at viscosity ratio) and the interfacial tension between the two phases. The latter can be adjusted by adding a surfactant. Block and graft copolymers often act as surfactant cum compatibilizer. However, in some rubber–plastic blends, the mutual interaction between the two components provides compatibilization such as in making high-impact nylon by blending nylon with maleated EPDM. Table 11.21 provides a list of rubber–plastic blends with improved impact resistance as reported in the literature.

Proper understanding of the toughening mechanism in rubber-toughened plastics requires a side-by-side comparison of the fracture behavior of the neat materials with and without toughening agent, over the full range of crack velocity, i.e., from the threshold stress which just maintains crack propagation up to the limiting crack velocity conditions. However, the toughening material can change other material properties including onset of yield in the material under stress. As a result, the crack velocity versus stress curve cannot be obtained by experiment over the full range of crack velocity. Dear has shown that computer simulations combined with experimental data at early stages of the failure process can provide meaningful and informative comparisons between the crack velocity versus stress curves for the material with and without toughening agents over the full crack velocity range [46]. This is possible by hosting a dynamic model in a personal computer which, with the help of a model based on mass-spring network, can build up stress and provide dynamic transfer of the strain energy to the tip of a propagating crack. The conditions can be dynamically varied for different toughness and response rate

TABLE 11.21

Selected Rubber–Plastic Blends with Improved Impact Resistance from Technical Literature

Impact Modifier	Compatibilizer	Property Enhancements
Polystyrene and Styrene Copolymer Blends		
Ethylene–propylene–diene rubber (EPDM)	EPDM-g-styrene and EPDM-g-methyl acrylate	Improved impact strength, decrease in tensile strength, consequence of better compatibility, improved adhesion of graft copolymers with PS phase
ABS (particle size 100–500A, 50–400A)	Two types of ABS (matrix and bulk polymerized)	Improves impact and glass transition temperature
SBR or SIR	Hydroboration, hydrolysis forming OH group—reacting with carboxyl containing polyolefin	High-impact tensile HDT
Graft copolymer of styrene with butadiene	SIS, SBS triblock copolymer	Higher mechanical strength
Polysiloxane	Styrene–maleic anhydride–siloxane copolymer	Improved impact resistance
Polyolefin Blends (PE,PP) with Impact Modifiers		
EPDM		Better miscibility, smaller particle size, higher rubber content, higher impact
SBR		Imcompatible, larger particle size, poor mechanical properties
EPDM, filler	Polypropylene-g-acrylic acid	Higher impact strength
Ethylene–propylene rubber	Functionalized EPDM	Improved cold impact toughness, EPDM content <30% desirable
ABS or ASA (acrylonitrile–styrene–acrylate)	ASA graft copolymer	Improved mechanical properties
Carboxylated polyolefin elastomer (SEBS-MA)	Styrene–maleic anhydride copolymer	Improved impact properties at low temperature and reduced water absorption

Polymer	Impact Modifier	Property Enhancements
Polyphenylene Oxide Blends with Impact Modifiers		
High-impact polystyrene (HIPS)	Styrene–butadiene rubber (SBR)	SBR enhances impact strength to a larger extent than MBS due to better interfacial interaction (SCM), decrease in flex strength and modulus
HIPS	Methyl methacrylate–butadiene–styrene–terpolymer (MBS)	
SBR modified	Polybutadiene-g-styrene copolymer	Reproducible mechanical/tensile strength, and good processability
Polystyrene (HIPS)	Polybutadiene–styrene blend	
PBT/PET	SIS, SBS triblock copolymers	High-impact strength, solvent resistance, and improved thermal stability
SEBS copolymer	Polyamide/polyolefin modified with dicarboxylate (propylene–maleic anhydride)	Improved impact strength, heat resistance, rigidity, and appearance
SBR	SBR–MA copolymer	Superior mechanical properties
EPDM-g-glycidyl methacrylate	Polycarbonate–polystyrene blend along with poly(alkylene-dicarboxylate) such as SMA	High-impact strength
SEBS block copolymer; polypropylene	SEBS copolymer for toughening blends of PPO with nylon and polyolefin (proprietary compatibilizer)	High-impact resistance

Source: Mangaraj, D., Markham, R., and Parsons, A., Compatibilization of polymer blends, Battelle Multiclient Report, 1994.

of the simulated materials. The advantage of the modeling arrangement is that it can be programmed to provide different simulation data quickly for comparison with experimental data. The authors have used modeling and computer simulations along with experimental results to predict the fracture behavior of a number of high-impact materials such as high-impact PS, styrene acrylonitrile copolymer, and acrylo-nitrile–butadiene–styrene (ABS) materials.

Recently, there has been considerable interest in soft-touch materials composed of thermoplastic and rubber blends. These materials when over-molded on hard or stiff structures such as hand tools, domestic appliances, writing gadgets, cellular phones, photographic equipments, etc. provide soft and comfortable feeling to the hand.

Such soft-touch materials are usually TPVs or thermoplastic elastomers (TPEs) which combine the moldability of thermoplastics in the melt state with elasticity, lower hardness, fracture resistance, and surface characteristics of elastomers. However, plastics and elastomers respond differently to mechanical stress. Hence, both rheological behavior and mechanical strength will to a large extent depend on the morphology of the blend which may change with change in the composition.

Cook et al. have done a systematic investigation of blends composed of PP and custom-synthesized SBR in which PP fractions varied from 20 to 45 wt% [47]. In addition, they also studied similar blends containing four commercial SBR materials. A commercial PP was used for blending. The elastomeric components were not vulcanized in the blend in order to study the effect of microstructure on the properties. Blends of PP–SBR were prepared using a co-rotating twin screw extruder using two different screw speeds, 70 rpm to study the effect of PP and 130 rpm for studying the effect of SBR on the blend properties. Tensile specimens were prepared by ram injection molding under conditions simulating industrial manufacturing process.

The microstructure of the blend was characterized using scanning probe microscopy of cryo-microtomed sections. The area fraction of PP was determined digitally by converting sections of the images into black and white regions. Rheology was measured by using a capillary rheometer at 200°C. The mechanical properties were determined using a servo-hydraulic test frame and wedge action grips.

The microstructure appeared well mixed although co-continuity of the phases was not obvious. The blends appeared to have a continuous PP phase containing extended, yet isolated, SBR components as shown in Figure 11.17. It appeared to be similar to the microstructure of the TPV-based on nylon and EPDM. The presence of entrapped air or mutual dissolution was not observed. As the fraction of PP increased, the microstructures became clustered into larger PP and SBR single phases, with lower SBR–PP interface area. Both the materials were shear thinning. There is a large decrease in the viscosity of the composites at small shear rate. The viscosity values of the phases followed the equation

$$\frac{\phi_1\,(\eta_1)}{\phi_2\,(\eta_2)} = 1 \tag{11.10}$$

(where ϕ and η are the fraction and viscosity for each phrase) showing PP forms the continuous phase. Further, the viscosity appeared to be more sensitive to SBR microstructure. Decrease in SBR molecular weight or use of a branched SBR decreased viscosity significantly.

The blend exhibited reversible strain only up to 2%. Below 20%, the deformation appeared to be homogeneous. Beyond this region, the stress derivative with strain decreased significantly. The decrease in stress derivative was strain rate-sensitive. The stress–strain behavior for the blends containing different SBRs are given in Figure 11.18. Both modulus and yield stress increased with increase in PP content. Neither modulus nor yield stress varied strongly with SBR type. On the other hand, all the failure properties increased with increase in PP content.

Figure 11.19 illustrates the dependence of the physical properties on PP content and SBR microstructure. All of them appeared to be highly dependent on PP content over the full composition range and the microstructure was invariant with composition. The microstructure had continuous PP and isolated SBR phases. The large decrease of viscosity and the lack of porosity in the microstructure suggest that the low viscosity PP successfully penetrated between the SBR fragments

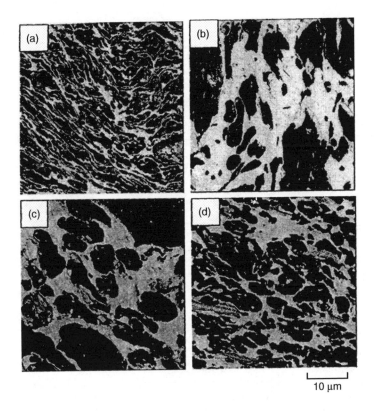

FIGURE 11.17 Microstructure of polypropylene/styrene–butadiene rubber (PP–SBR) blends, PP shown light. (From Cook, R.F., Koester, K.J., Macosko, C.W., and Ajbani, M., *Polym. Eng. Sci.*, 45, 1487, 2005.)

and coated the SBR fragments leading to continuous PP structures. Both melt viscosity and elastic modulus also depend on PP volume fraction. The elastic properties after molding largely depend on the low modulus SBR phase that carries most of the elastic strain.

One of the key parameters in morphology development in polymer blends is viscosity. It has been reported that in many cases, the viscosity of the blend is lower than that for the component. It has been suggested that systems which are more incompatible show a larger negative viscosity deviation. The polymer chains are less entangled at the interface than in the bulk. When shear stress is imposed, the resulting shear rate is higher at the interface than in the bulk of the components leading to the phenomenon of interfacial slip. Addition of a reactive compatibilizer is likely to reduce the drop in interfacial viscosity. Van Puyvelde et al. have studied the effect of reactive compatibilization on the interfacial slip of nylon/maleated ethylene–propylene rubber (EPR) blends by using co-extruded multilayers [48]. The latter generates a large but well-defined interfacial area providing a well-defined sample geometry for studying the effect of reactive compatibilization on interfacial slip.

Blends of EPR and maleated EPR (Exxon Exxelor) and nylon 6 (DSM Akulon) were prepared using a Haake Rheocord mixer at six different concentrations of compatibilizer. The viscosities were measured with capillary rheometer. Phase morphology was studied by SEM using an image analysis software. Multilayer sheets, about 1 mm thick and 40 mm wide, were prepared with alternating layers of EPR and nylon 6. In this set-up, two immiscible polymer melts are combined into two layers which subsequently enter into multiplication die, thereby largely increasing the number of layers.

Figure 11.20 shows the SEM image of the morphology of 30/70 nylon–EPR blends at two different compatibilizer concentrations. It is apparent that the addition of compatibilizer significantly reduces the size of the dispersed phase. The addition of 10% compatibilizer is sufficient to

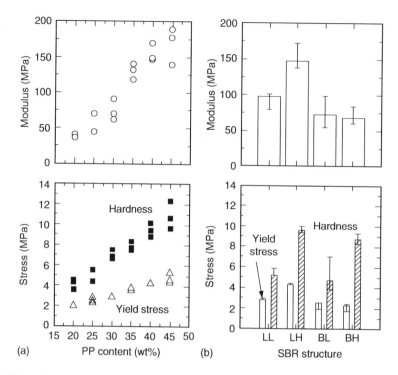

FIGURE 11.18 Tensile stress–strain responses of polypropylene/styrene–butadiene rubber (PP–SBR) blends at several ratios (where LL is linear low molecular weight; LH is linear high molecular weight; BL is branched low molecular weight; and BH is branched high molecular weight). (From Cook, R.F., Koester, K.J., Macosko, C.W., and Ajbani, M., *Polym. Eng. Sci.*, 45, 1487, 2005.)

attain critical micelle concentration to produce maximum reduction in the size of the dispersed phase. The presence of compatibilizer increases the viscosity of the blend following a log additivity rule and reaches maximum at 1% of the compatibilizer. It is predicted that the graft copolymer enhances interfacial friction, thus reducing the interfacial slip. Addition of compatibilizers beyond 1% reduces the particle size but does not change the viscosity at a particular shear rate.

For uncompatibilized blends, the viscosity decreases with increasing number of layers in a multilayer constant stress rheometer. In the present case, the viscosity rather increases due to *in situ* reaction between nylon and maleated EPDM as illustrated in Figure 11.21.

Mascia et al. have explored the possibility of improving solvent resistance of fluorosilicone rubber by blending it with polyvinylidene fluoride using diallylcyanurate and diallylphthalate (DAP) as reactive compatibilizers [49]. Copolymers are used instead of PVDF homo-polymer to enhance the possibility of forming a co-continuous phase in the blend. They postulated that the mixture of the two additives would not only increase the entropy of mixing but would also provide negative enthalpy of mixing by grafting to the polymer chains.

Silastic LS 420, possessing approximately 0.6%–0.9% pendant vinyl groups, was blended with Kynar 7201, a vinylidene fluoride copolymer with tetrafluoroethylene (Atochem), in the presence of triallylisocyanurate (TAIC) and DAP containing a small amount of benzoyl peroxide in the DAP fraction.

Initially, the co-agents were mixed with PVDF and FMVQ separately and the mixtures were subjected to mild irradiation. Solubility tests indicated no cross-linking during this operation. The polymeric components were then mixed in the presence of CaO/MgO in a Brabender plasticorder at a rotor speed of 60 rpm at 160°C. Subsequently, the temperature was lowered to 130°C and a 0.2% benzoyl peroxide paste was added. Mixing was continued for 10 more minutes. Cure characteristics

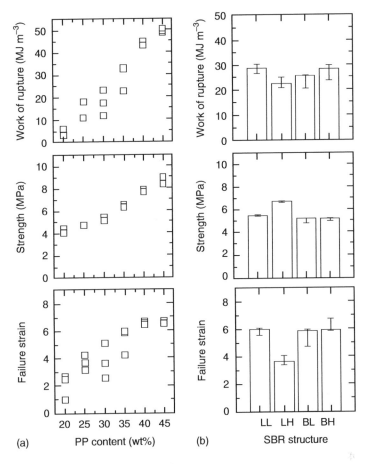

FIGURE 11.19 Physical properties depend on polypropylene (PP) content (a) and SBR microstructure (b). (From Cook, R.F., Koester, K.J., Macosko, C.W., and Ajbani, M., *Polym. Eng. Sci.*, 45, 1487, 2005.)

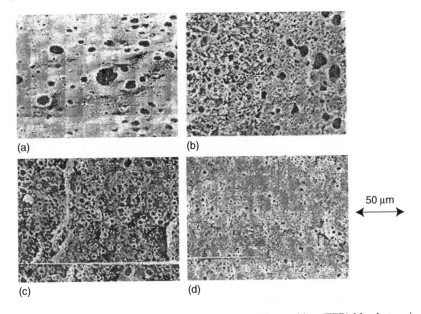

FIGURE 11.20 SEM images of a 30:70 nylon 6–ethylene–propylene rubber (EPR) blend at various compatibilizer concentrations (a) 0%, (b) 2.5%, (c) 5%, (d) 10%. (From Van Puyvelde, P., Oommen, Z., and Koets, P., *Polym. Eng. Sci.*, 43, 71, 2003.)

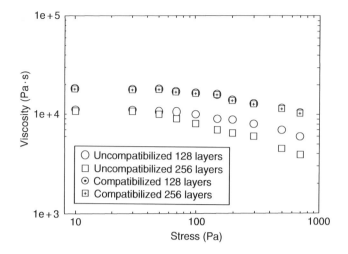

FIGURE 11.21 Viscosity of 30:70 nylon–ethylene–propylene rubber (EPR) blend at various shear rates and various compatibilizer concentrations. (From Van Puyvelde, P., Oommen, Z., and Koets, P., *Polym. Eng. Sci.*, 43, 71, 2003.)

of the resulting blend were examined by Monsanto rheometer. Gel content, thermal characteristics, dynamic mechanical behavior, morphology, and solvent resistance were also measured.

The results indicated that addition of DAP not only lowered the viscosity, but also enhanced the cure rate by TAIC. Further, the mixed co-agent system provided a synergistic effect on the state of cure. Table 11.22 provides the composition and the properties of the blends. The reduction in PVDF melting point and the increase in tan *ó* peak are rather small due to a very low degree of cross-linking. SEM images confirm the development of co-continuous phase in the compatibilized blend. The fact that the reinforcement provided by PVDF extends even beyond its melting point suggests

TABLE 11.22

Comparison of Reinforcing Efficiency of VDF-TFE Copolymers Relative to Silica Flour in Fluorosilicone Elastomers

	S_1	S_{10}
Composition		
Fluorosilicone gum	60	—
Silica reinforced fluorosilicone gum (15%)	—	60
Co-irradiated VDF-TFE/TAIC/DAP (60:5:5) mixture	35	35
TAIC (co-agent)	2.5	2.5
DAP (co-agent)	2.5	2.5
BPO paste (grafting peroxide)	0.2	0.2
MgO CaO (1:1) (acid absorber)	4.0	4.0
DMDBPH-3 (cross-linking peroxide)	1.0	1.0
Properties of Vulcanizate		
Gel content (% w/w)	40.2	67.1
Solvent uptake (%)	62.3	51.7
Tensile strength (MPa)	5.0	5.2
Elongation at break (%)	61.4	37.5

Source: Mascia, L., Pak, S.H., and Caporiccio, G., *Eur. Polym. J.*, 31, 459, 1995.

that the network exists through both phases. It was noted that tensile strength, gel content, and solvent resistance increase and elongation at break decreased with increase in PVDF content. Addition of silica does not appreciably improve any of the properties. It was concluded that PVDF can be blended with fluorosilicone elastomers to improve its solvent resistance, and multi-functional cross-linking agents such as TAIC and DAP can be used, preferably in a mixture, to compatibilize fluorosilicone rubber with PVDF.

Yanez-Flores et al. have studied the shear properties of blends of PE with polyisoprene rubber (Guayule rubber) [50]. The blends were prepared using a cam-type mixer at 50 rpm for 10 min at 140°C. The blend compositions ranged from 10% to 70% rubber content.

They found that viscosity of the blend decreases with increasing shear and PE content as shown in Figure 11.22. Further, the power law index increases with increase in high-density polyethylene (HDPE) content indicating Newtonian behavior for the blends. In general, viscosity of the blends decreases as the frequency increases. At low rubber content where HDPE predominates, variation of complex viscosity with shear rate is similar at different blend ratios NR/HDPE. However, at higher rubber content, the effect of different polyisoprenes becomes apparent. Viscosities are particularly lower for blends with Guayule rubber. The same type of behavior is seen for the variation of G' and tan δ. At higher rubber content, the effect of different polyisoprenes becomes evident suggesting that while using the same processing conditions, Hevea blends would be more prone to present elastic effects such as melt fracture. A Cole–Cole plot (i.e., plot of G' and G'' against composition), which is often used to indicate compatibility, shows that the blends are compatible only for compositions of 10%–30% rubber. The 50% and 70% rubbers give two different curves, exhibiting incompatibility.

Research concerning nylon–elastomer blends has mostly focused on the improvement of mechanical and thermal properties. Their dynamic mechanical properties are quite important both for processing and engineering applications. Wang and Zheng have studied the influence of grafting on the dynamic mechanical properties of a blend based on nylon 1212 and a graft

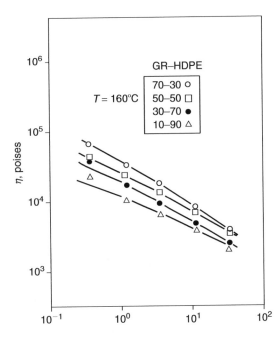

FIGURE 11.22 Variation of melt viscosity with shear rate for GR high-density polyethylene (HDPE). (From Yanez-Flores, G., Ramos De-Valle, L.F., and Rodriguez-Fernandez, O.S., Proceedings of the Society of Plastics Engineers Annual Technical Conference, 1997, p. 381.)

copolymer, namely maleated styrene-*b* (ethylene–co–butylene)–*b*-styrene triblock copolymer (SEBS-g-MA) [51]. The grafting ratio was 1.84% and PS end blocks were 30%. Blends of nylon 1212 with SEBS and SEBS-g-MA including 1 wt% antioxidant were prepared in a twin screw extruder at 210°C. Test specimens were prepared by compression molding at 180°C at 50 MPa pressure. Melt rheological tests were carried out using an ARES rheometer in parallel plates oscillatory mode using 3% strain amplitude. The dynamic storage modulus curves for the blends fall in between those for the pure components almost following a linear mixing rule. However, whereas the virgin nylon obeys the linear viscoelasticity model—i.e., G' is proportional to log frequency—the blends do not follow this proportionality rule. The G' values not only increase with increase in SEBS content, but even the slopes at low frequency are higher indicating that the nylon is still the continuous phase in the blend. The plots of storage modulus against frequency for blends containing maleated SEBS show that their dynamic mechanical behavior is quite different from those of SEBS-based blends. The values of G' at all compositions are higher than those for both virgin nylon and SEBS-g-MA, implying chemical reaction between the two components leading to the formation of a homogeneous blend. The reaction is presumed to be between the –NH groups of nylon and maleic acid groups of the grafted SEBS. The interaction stabilizes the interface leading to enhancement of interfacial viscosity and interphase adhesion. Whereas the SEMs of nylon–SEBS blends exhibit droplet-type matrix, those for nylon–SEBS-g-MA blends exhibit long strands along with some droplets.

Substantial work has been carried out on the mechanism of impact modification of plastics by blending with rubber, particularly on systems such as HIPS, acrylonitrile–styrene–butadiene terpolymer (ABS), polyphenylene oxide (PPO)–HIPS blends, and polyamide–EPR rubber blends. However, the relation between morphology, the state of the interface, and mechanical properties is still not clear. Anh and Vu-Khanh have made a detailed analysis of the fracture and yielding behavior of a blend of PS and EPR containing SEBS triblock copolymer as compatibilizer [52]. They have analyzed the time–temperature dependence of the brittle–ductile transition in terms of molecular relaxation process. PS and EPR (54% ethylene) were blended in a twin screw extruder operating at 100 rpm and 200°C. The interfacial agent SEBS triblock copolymer was added at 0%, 2.5%, 5%, 10%, 15%, 20%, and 30% levels along with the antioxidant Irganox 1010 at a 1% level. The granules were molded into 6 mm thick plates. Three-point bending impact tests were carried out on precut specimens at various temperatures. Tensile tests were carried out at various crosshead speeds from 1 mm/min to 200 mm/min at temperatures from 25°C to 70°C. Compression tests were carried out over four decades of strain rate at temperatures between −75°C and 100°C. The initial and final cross-sectional area for both tensile and compression tests were estimated at the yield point.

Earlier work had shown that lower-molecular weight compatibilizers have a more significant effect on reducing the ductile–brittle transition than higher-molecular weight ones. Further, the brittle–ductile temperature, T_{bd}, increases with the increase in loading velocity.

The time–temperature correlation implies the existence of an activation process. In other words, there is an Arrhenius-type relation between strain rate and temperature. The activation energy ΔH or the energy barrier controlling the time–temperature dependence of the fracture process can be estimated by measuring failure strain rate as a function of temperature. It was found earlier that the interfacial agent or compatibilizer reduced the strain energy barrier of the blend. The authors in this study used Charpy impact tests at five different loading speeds for blends containing three different concentrations of the compatibilizer at four different temperatures. Plotting of ln strain rate against reciprocal of temperature gave linear plots. The energy barrier ΔH controlling the time–temperature dependence of the fracture behavior was estimated from the slope. Table 11.23 provides the values of ΔH and constant A of the Arrhenius relation for different concentrations (indicated by parts) of compatibilizer, namely triblock copolymer K1 and K2. K1 (Kraton 1651) has a molecular weight of 174,000 and K2 (Kraton 1652) has a molecular weight of 50,000. The above results show that K2, the interfacial agent with lower molecular weight, is much more efficient in reducing brittle–ductile transition temperature than K1.

TABLE 11.23

Values of ΔH and A of Eyring Equation for Blends Modified with the Interfacial Agents K1 and K2

Blends	ΔH (kJ/mol)	Constant A (1/s)
K1-2.5	71.87	2.18E + 13
K1-10	52.08	1.84E + 10
K1-20	50.56	1.77E + 10
K2-2.5	45.21	5.82E + 09
K2-15	38.37	5.02E + 08

Source: Anh, T.H., and Vu-Khanh, T., *Polym. Eng. Sci.*, 41, 2073, 2001.

The authors have also analyzed the temperature dependence of yielding behavior and have estimated activation volume V^* and activation energy ΔH by analyzing temperature and loading speed dependence as shown in Table 11.24. The activation energy for the yielding process at the same composition is two- to threefold higher than that of the energy barrier controlling the brittle–ductile fracture transition. This may be due to the difference in the temperature ranges involved in the test. The activation volume may be considered as a region containing a certain number of segments which simultaneously get activated and lead to yield deformation. The results in Table 11.25 suggest that an increase in compatibilizer content would produce a larger portion of the molecular segments activated in the yielding process causing a relative increase in free volume. This contributes to an increase in activation volume and a decrease in activation energy with an increase in compatibilizer (K1 and K2) content. Subsequently, the authors have used a modified Eyring equation to analyze the results of failure under compression. The variation of activation energy and activation volume for failure under compression for different concentrations of the high-molecular weight block copolymers followed the same trend as yielding behavior considered above. The authors conclude that adding an interfacial agent (compatibilizer) in a blend lowers the brittle–ductile transition temperature. This is because the transition is an activated process and the addition of the agent lowers the activation energy and increases the activation volume. The greater this effect the lower the molecular weight of the compatibilizer.

Akhtar has studied the morphology and physical properties of NR and high-density polyethylene blends prepared in Brabender plasticorder at 150°C at a rotor speed of 60 rpm [53]. Films were molded between two chromium plates at a pressure of 0.34 MPa. The films along with mold were

TABLE 11.24

Value of Activation Volume (V^*) Per Jumping Segment and Activation Energy (ΔH) for Different Interfacial Agent Contents

	High-MW Interfacial Agent K1				Low-MW Interfacial Agent K2				
	K1-5	K1-15	K1-20	K1-30	K2-2.5	K2-5	K2-10	K2-15	K2-30
V^* (nm³)	1.426	1.527	1.533	1.601	1.269	1.295	1.318	1.360	1.436
ΔH (kJ/mol)	145	1333.1	129	134	128	124.5	119	115.4	107.2

Source: Anh, T.H. and Vu-Khanh, T., *Polym. Eng. Sci.*, 41, 2073, 2001.

TABLE 11.25

Activation Volume and Free Energy Calculated from Experimental Data

Blends	ΔH_α (kJ/mol)	V_α* (nm³/ Segment)	ΔH_β (kJ/mol)	V_β* (nm³/ Segment)
K1-2.5	145	1.43	66	0.89
K1-10	138.5	1.46	61.1	0.91
K1-20	133.5	1.58	58	1.13

Source: Anh, T.H. and Vu-Khanh, T., *Polym. Eng. Sci.*, 41, 2073, 2001.

cooled in ice, water at 25°C, cooling air at 30°C, and in press at 150°C. The properties are given in Table 11.26. As the rubber content in the blend increases, tensile strength and tangent modulus decrease along with an increase in the elongation at break. Both tensile strength and modulus of PE increase slightly as the rate of cooling decreases. The tensile strength of the blends decreases to a small extent on rapid cooling. On the other hand, the blends show a sharp drop in tensile strength when cooled in a press. Elongation at break of the blends is not affected by the method of cooling. However, when rapidly cooled in a press, the blends exhibit poor properties. In the ice-cooled PE sample, the cold drawability of PE is very high. However, when it is blended with rubber, it does not show yielding but undergoes abrupt failure. The ice-cooled samples exhibit high tear strength.

TABLE 11.26

Formulations and Properties of Blends of NR and HDPE

	NR	HDPE	Cooling Method[a]	Young's Modulus (MPa)	Tensile Strength (MPa)	Elongation at Break (%)	Tear Strength (kN/m)	Crystallinity (%)
PE	0	100	1	25.6	28.7	460	186.0	59.4
			2	27.7	29.2	270	153.5	59.5
			3	28.5	30.4	55	21.7	64.3
			4	34.1	34.1	14	21.0	60.2
A	30	70	1	12.1	12.9	4.	76.7	—
			2	11.7	12.0	45	77.3	—
			3	12.5	12.8	40	76.8	—
			4	9.3	9.3	6	13.0	—
B	50	50	1	6.2	9.1	420	43.6	—
			2	5.2	8.0	440	45.6	—
			3	5.2	8.0	350	46.0	—
			4	4.7	5.3	35	9.4	—
C	70	30	1	1.3	6.4	610	19.7	—
			2	1.0	5.6	635	20.6	—
			3	1.0	4.8	580	19.2	—
			4	0.8	2.1	390	14.0	—

Source: Akhtar, S., *Rubber Chem. Technol.*, 61, 599, 1988.

[a] Cooling method: (1) quenched in ice (−10°C); (2) quenched in water (25°C); (3) cooled in air (30°C); (4) cooled in press under pressure (0.34 MPa) at 150°C to 25°C.

The blends prepared by cooling in ice, water, and air are not very different. However, all the press-cooled samples show an abrupt reduction in strength.

Morphology of the blends was studied by both optical microscopy and SEM. It was found that HDPE forms a continuous phase and rubber is dispersed as distinct domains. The 50:50 blend shows finer particle dispersion than other blends. In 25:75 blend both HDPE and rubber form the continuous layer. The morphology is independent of the method of preparation.

Epoxy resins are often used as polymer matrices for high-performance composites for the manufacture of unidirectional tapes and woven fabric prepegs. Since temperature affects both viscous flow and cross-linking process and low-temperature cure leads to low glass transition temperature, careful control of processing temperature is essential. Tetra-functional epoxy with amine hardeners provides high cross-link density and high T_g. Therefore, matrix formulations are often made of tetraglycidyl diaminomethane (TGDDM) and diphenyl sulfone (DDS). Elastomers are often used to improve the impact resistance of the epoxy resin system either by blending or pre-reacting. The rubber is generally soluble in the epoxy precursors but may precipitate into a separate phase during the curing of epoxy resin. The compatibility between the two polymers is greatly improved by pre-reacting the rubber with the epoxy resin. Dispenza et al. have made an in-depth study of reactive blending of rubber with epoxy resins by pre-reacting the epoxy resins based on TGDDM and DDS with a rubbery terpolymer made of a functionalized acryionitrile/butadiene/methacrylic acid (Nipol 1472) [54].

Three different blending techniques were used. In the first process (PWE), the epoxy resin was pre-reacted with Nipol 1472 in the presence of ETPI catalyst. The contents were gently heated to reflux for 2 h and then allowed to cool to room temperature when other epoxy oligomers and the curing agent were added. The contents were stirred until a homogeneous mixture was obtained. The second blend (PNE) was a repeat of the first without ETPI catalyst. In the third system, the rubber pre-dissolved in acetone along with all the precursors used in the first and second process and the epoxy were charged into a flask and stirred at room temperature until a clear mixture was obtained. After each of the above operations, DDS curing agent and difunctional epoxy resins were added. Base epoxy systems with and without ETPI catalyst were used as reference. Kinetic and rheological studies were carried with uncured samples. Whereas PWE and PNE systems appeared homogeneous after solvent removal, the blended system remained turbid. Neat resin panels were prepared by curing at 180°C in appropriate mold and the specimens were used for dielectrical, morphological, and mechanical characterization. The specimens were characterized by DSC, potentiometric titration, chemo-rheological studies, dielectric spectroscopy, and mechanical testing. Morphological studies were carried out on fractured samples using SEM.

The neat blend containing both the epoxy system and the rubber appeared turbid even after heating to 180°C. Homogeneous mixtures were obtained when the rubber and epoxy monomer were pre-reacted at 60°C in the presence of ETPI catalyst both in PWE and PNE systems. The clarity was retained even after curing the specimens. The authors concluded that the carboxylic group present in the rubber reacted with epoxy groups to form ester linkages. The reaction was catalyzed by ETPI leading to almost 100% conversion in less than 2 h. Without the catalyst, the reaction occurs but it is not completed even after 5 h. When curing of the samples at 180°C was monitored by DSC measurement of exothermic heat, it was found the rate of cure of PNE system is between the PWE system and the simple blend as shown in Figure 11.23. Further, the cure curve for the epoxy with curing agent but without rubber is close to PWE and the one without curing agent is close to that of PNE system. These results were also confirmed by chemo-rheological studies at three different temperatures, 140°C, 160°C, and 180°C. The storage modulus G' and the loss modulus G'' for the PNE and blended systems cross at a certain curing time while no crossover was observed for PWE system. While crossover was repeated at 160°C, no crossover was seen at 180°C. It was concluded that the reaction between the rubber and the epoxy in the presence of ETPI in PNE system leads to development of elasticity at low temperature. The enhancement of elasticity at high temperature is due to the development of an epoxy-amino network and is not dependent on the presence of ETPI catalyst.

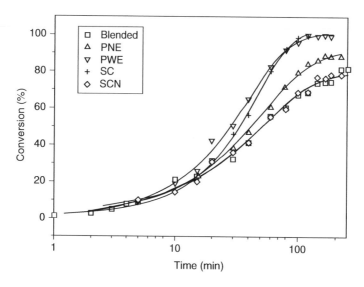

FIGURE 11.23 Advancement of the cure reactions for different base epoxy and rubber–epoxy systems. (From Dispenza, C., Carter, J.T., McGrail, P.T., and Spadaro, G., *Polym. Eng. Sci.*, 41, 1483, 2001.)

Gel time values of the three systems measured as abrupt change in the slope of $G'(t)$ under isothermal curing conditions show that gelation occurs earlier in PWE system at all temperatures considered as shown in Table 11.27. ETPI behaves like a catalyst for the primary epoxy-amino reaction which dominates the cure until vitrification occurs. Dynamic mechanical analysis and dielectric spectroscopic analysis carried out by the authors also confirm the above conclusions.

Morphology of the cured samples was analyzed by SEM of the fractured samples etched with tetrohydrofuran (THF), which is a solvent for the rubber. Figure 11.24 shows the fracture surfaces of the PWE and PNE specimens. Whereas the PWE fracture surface presents an essentially homogeneous surface with only a few small voids present, small yet uniformly distributed cavities are seen in PNE samples. The PWE morphology is consistent with the high degree of intermolecular link between rubber and epoxy macromolecules. The PNE morphology indicates incomplete reaction between epoxy and rubber.

Table 11.28 gives the yield stress (E) under compression, fracture strength (K_{ic}), fracture toughness (G_{ic}), and ductility factor (DF) obtained from three-point bend measurement in an opening mode.

The results indicate that the system containing ETPI has higher compressive yield stress and greater stiffness resulting in greater toughness as shown by the ductility factor. It was concluded that

TABLE 11.27

Gel Times Measured from Onset of the $G'(t)$ Slope Abrupt Change

Cure Temperature (°C)	PWE	PNE	Blended
180	21	27	24
160	51	54	53
140	135	140	137

Source: Dispenza, C., Carter, J.T., McGrait, P.T., and Spadaro, G., *Polym. Eng. Sci.*, 41, 1483, 2001.

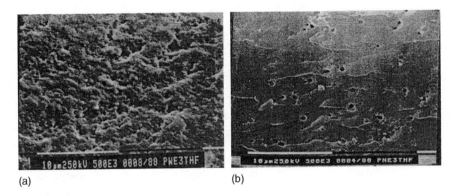

(a)   (b)

FIGURE 11.24 Fracture surfaces of resin systems (a) PWE and (b) PNE. (From Dispenza, C., Carter, J.T., McGrail, P.T., and Spadaro, G., *Polym. Eng. Sci.*, 41, 1483, 2001.)

the incorporation of a high-molecular weight rubber into the epoxy system is helpful in improving toughness and processing. Use of a rubber that can react with the epoxy system and use of a catalyst that enhances the speed of reaction between the rubber and epoxy can accelerate the curing of epoxy resin. In this case, the complete reaction between the epoxy moiety and carboxyl group led to the formation of a three-dimensional epoxy rubber network swollen by unreacted monomers, which controlled the resin flow over a wide temperature range and prevented coarse phase separation upon cure. The interpenetrating network formed upon cure provides better mechanical properties.

NR and polymethyl methacrylate (PMMA) blends can provide thermoplastic materials for fabrication of automotive components. However, the blends have a sharp interface due to poor physical and chemical interactions at the phase boundary. Such interfacial incompatibility is greatly reduced by graft polymerization of suitable monomers onto the surface. Grafting of NR has been carried out using a variety of monomers including methyl methacrylate, styrene, glycidyl methacrylate (GMA), and *N*-vinyl pyrrolidone. It has been found that the presence of styrene greatly increases the grafting of GMA to EPR and PP. In addition, GMA and GMA copolymers have also been used as compatibilizers in a number of polymer blends. Recently, Suriyachi et al. have studied grafting of GMA/styrene to NR latex and use of graft copolymer as a compatibilizer for NR–PMMA blends [55]. Grafting was carried out in a four-necked glass reactor using NR latex (60% dry rubber content) in the presence of sodium dodecylsulfate as emulsifier, isopropanol as stabilizer, and potassium hydroxide as an aid to buffer the pH. A mixture of styrene/GMA in 70:30 ratio was used for grafting. The NR latex was allowed to be swollen with the monomer for 1 h

TABLE 11.28
Mechanical Response of Neat Resin Systems

Neat Resin System	PWE	PNE
Yield stress (MPa)	158	147
Fracture strength (MPa m$^{-3/2}$)	1.398	1.004
Fracture toughness (KJ m^{-2})	0.614	0.375
Calc Mod (Gpa)	3.16	2.7
Ductility factor (mm)	0.13	0.07

Source: Dispenza, C., Carter, J.T., McGrail, P.T., and Spadaro, G., *Polym. Eng. Sci.*, 41, 1483, 2001.

before adding the catalyst and the reaction was allowed to proceed for 10 h under continuous stirring. At the end, the graft polymer was washed and dried in vacuum. Ungrafted NR was removed by Soxhlet extraction with petroleum ether and the residue was dried under vacuum for 24 h. The presence of free polymers and copolymers such as PGMA, PS, and PS/GMA was determined by extracting with acetone. Grafted polymers included NR-g-PS, NR-g-GMA, and NR-g-PS/GMA. Grafting efficiency was calculated using the equation

$$\text{Grafting efficiency}(\%) = \frac{(\text{weight of polymers grafted} \times 100)}{\text{total weight of the polymers formed}}. \quad (11.11)$$

The authors have characterized the graft polymer by solvent extraction, transmission electron microscopy, dynamic mechanical analysis, mechanical testing (including measurement of tensile, tear, and impact strength), and morphology by SEM. The reaction scheme is given in Figure 11.25.

It is postulated that addition of tetramethylene pentamine to cumene hydroperoxide accelerates the decomposition of the latter. Addition of a second monomer, styrene, eliminates the probability of steric hindrance in the graft polymerization of GMA. The epoxide group in GMA is hydrolyzed to give $-\text{OH}$ groups. The hydroxyl radicals then attach to the double bond on the NR backbone to give secondary and tertiary macro-radicals. The latter take part in the grafting reaction. The reactivity ratio of styrene/GMA suggests that styrene first reacts with NR macro-radical and the resulting macro-radical then reacts with GMA. The resulting copolymers were characterized by FTIR and NMR spectroscopy.

The conversion and grafting efficiency increase with increase in initiator concentration at different temperatures. At higher initiator concentrations, large numbers of radical chains are formed providing the possibility of their recombination in preference to graft reaction. As a result, the grafting efficiency increases up to 2.5% initiator concentration and then levels off. The grafting efficiency is highest around 2.5 phr of initiator over the temperature range of 50°C–70°C. The highest yield of graft copolymer was found around 60°C. It was postulated that at higher temperature, rapid formation of radicals produces a high concentration of radicals, which either go into recombination or other types of side reaction instead of grafting. As the monomer concentration

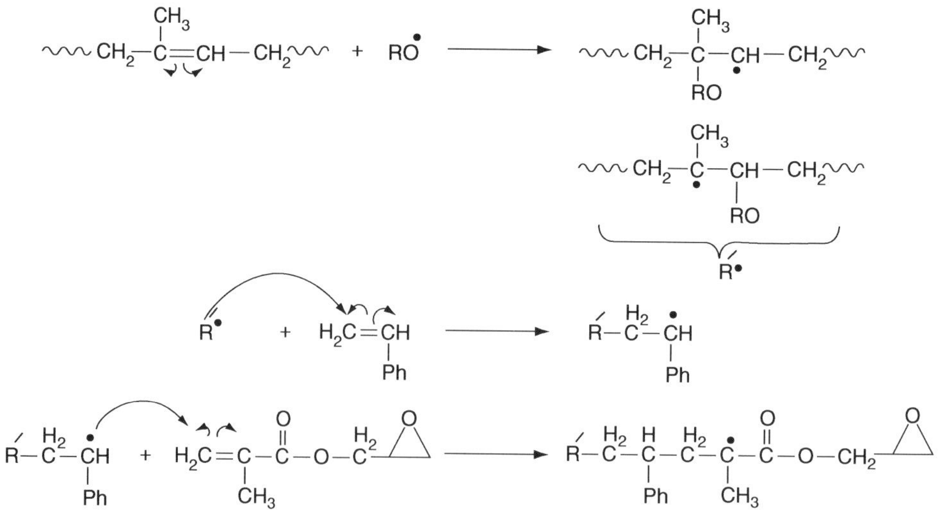

FIGURE 11.25 Reaction scheme for styrene-assisted free radical grafting of glycidyl methacrylate onto natural rubber. (From Suriyachi, P., Kiatkamjornwong, S., and Prasassarkich, P., *Rubber Chem. Technol.*, 77, 914, 2004.)

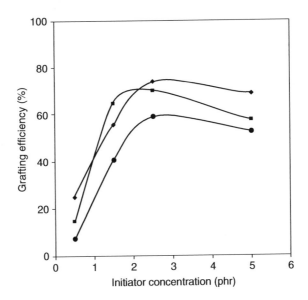

FIGURE 11.26 Effect of initiator concentration on grafting efficiency at 50°C (◆); 60°C (■); 70°C (●), monomer = 100 phr, time = 10 h. (From Suriyachi, P., Kiatkamjornwong, S., and Prasassarkich, P., *Rubber Chem. Technol.*, 77, 914, 2004.)

increases, both percent conversion and grafting efficiency increase and then decrease slightly. The grafting efficiency and percent conversion reach maximum at 100 phr monomer concentration. Figure 11.26 provides a plot of grafting efficiency against initiator concentration. Figure 11.27 plots the change in the extent of graft polymerization as a function of monomer concentration.

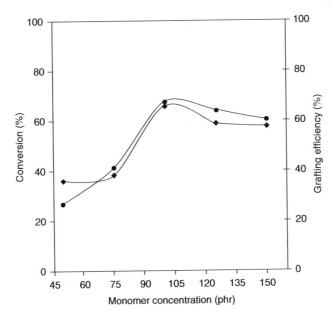

FIGURE 11.27 Effect of monomer concentration on conversion (◆) and grafting efficiency (●), initiator = 2.5 phr, T = 60°C, time = 10 h. (From Suriyachi, P., Kiatkamjornwong, S., and Prasassarkich, P., *Rubber Chem. Technol.*, 77, 914, 2004.)

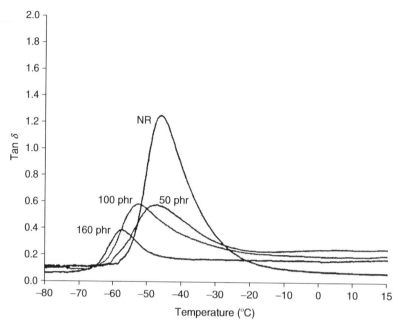

FIGURE 11.28 Dependence of glass transition temperature and tan δ on monomer concentration. (From Suriyachi, P., Kiatkamjornwong, S., and Prasassarkich, P., *Rubber Chem. Technol.*, 77, 914, 2004.)

It is suggested that the grafting reaction is a core-shell-type emulsion polymerization in which NR forms the core and styrene/GMA grafts onto the NR core forming a shell, which thickens and even forms nodules with time. During the first 4–8 h period of grafting, the core-shell structure is rather compact, after which the surface becomes smooth. Dynamic mechanical analysis of the graft polymers was used to estimate tan *ó* peak as a function of monomer concentration in Figure 11.28. The peak becomes narrower with decreasing monomer concentration as the rubber chain gets more immobilized due to the higher level of grafting.

Subsequently, NR was blended with PMMA along with various amounts of graft copolymer (prepared in the presence of 100 phr of styrene and GMA for 10 h at 60°C) in 70:30 and 50:50 ratios. Similar blends were prepared without graft copolymer. Tensile strength of the blends without graft copolymer was very low. Blends with graft copolymer exhibited higher tensile strength and elongation at break. The improvement leveled off at 10 phr of graft copolymer for 70:30 blend and at 5 phr at for 50:50 blend. Both tensile strength and elongation at break of the former were found to be higher than those of the latter. It appeared that in the former blend, PMMA was dispersed in the continuous rubber phase, whereas in the latter both the phases influenced the behavior under tensile stress. The decrease in tensile strength and elongation at break was ascribed to the brittle nature of PMMA. Figure 11.29 provides tensile strength and elongation at break for the blends at different levels of graft polymer additive.

Tear strength also increases with increase in graft copolymer content but levels off at 10 and 5 phr of the graft copolymer for 70:30 and 50:50 blends. It has been suggested that addition of the graft copolymer reduces the size of the dispersed phase and improves interfacial adhesion. This, in effect, resists crack propagation, thereby improving tear strength. The leveling off effect for both graft polymers at 10 and 5 phr levels is ascribed to the saturation of the interface with compatibilizer. Hardness which reflects resistance to indentation also follows the same trend.

Examination of the fracture by SEM shows that although there is no interphase adhesion in uncompatibilized blends, adhesion between the phases increased and dispersed domains decreased

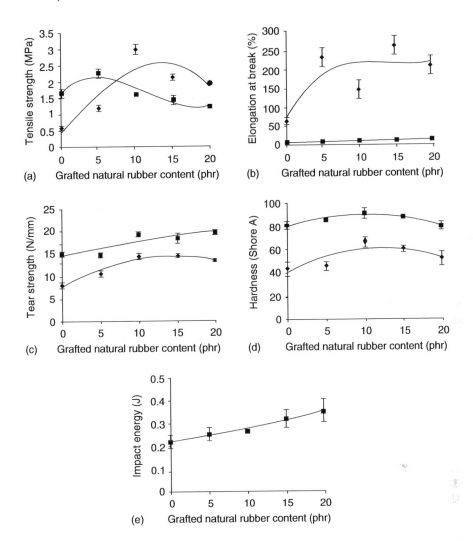

FIGURE 11.29 Effect of grafted natural rubber content on the mechanical properties of STR5L–PMMA blends at ratios of 50:50 (■); and 70:30 (◆). (From Suriyachi, P., Kiatkamjornwong, S., and Prasassarkich, P., *Rubber Chem. Technol.*, 77, 914, 2004.)

in size in the compatibilized blends. The interphase adhesion was ascribed to the presence of methyl methacrylate and rubber moieties in the compatibilizer and to the interaction of the COOH groups and OH groups formed by the hydrolysis of ester groups in PMMA and glycidyl group in GMA.

11.4 SUMMARY AND CONCLUSION

Blending of elastomers and that of elastomer with plastics is being practiced to meet property and processing requirements of new products. Recent work is more focused on understanding the science underlying the incompatibility and using suitable methods, both reactive and nonreactive, to compatibilize the blends. Techniques such as reactive blending, grafting of selected monomers prior to blending, use of special curative and co-vulcanizing agents, and compatibilizers have been used to achieve compatibility and to improve the engineering properties of the blends.

A variety of instrumental techniques including SANS, proton magnetic resonance spectroscopy, scanning electron miscroscopy, and AFM along with infrared spectroscopy are being used to characterize the interfacial interaction and its effect on static and dynamic properties of the blends. There is a serious trend in interpreting the results in the general framework of thermodynamics of polymer–polymer miscibility. Blends of elastomers with both thermoplastics and thermosets are being developed to obtain toughened materials with soft touch and thermoplastic processability. The overall result appears to provide a sound scientific base for formulating rubber compounds using two or more elastomers as well as for combining elastomers with plastics to obtain a desired portfolio of properties and processability.

REFERENCES

1. Mangaraj, D. *Rubber Chem. Technol.* 75(3), 2002, 365; Mangaraj, D. Rubber Blends. Chapter 3, *Elastomer Technology*, Special Topics (eds. K. Baranwal and H. Stephens), Rubber Division, American Chemical Society, 2003.
2. Mangaraj, D. Rubber Recycling by Blending with Plastics. Chapter 7, *Rubber Recycling* (ed. S.K. De), CRC Press, West Palm Beach, FL, 2005.
3. Mangaraj, D. Role of Compatibilization in Rubber Recycling by Blending with Plastics, presented at 166th American Chemical Society Rubber Division Meeting. Columbus, OH, 2005, Paper No. 27.
4. Corrish, P.J. *Science and Technology of Rubber* (ed. F.R. Elrich), Academic Press, New York, 1978, p. 489.
5. Roland, C.M. *Handbook of Elastomers* (eds. A.K. Bhowmick and H.L. Stevens, 2nd ed.), Marcel Dekker, New York, 2001, pp. 197, 227.
6. Flory, P.J. *J. Chem. Phys.* 10, 51, 1992; *Principles of Polymer Chemistry*, Cornell University Press, Ithaca, NY, 1962.
7. Huggins, M.L. *J. Am. Chem. Soc.* 64, 1712, 1942.
8. Hildebrand, J.H. and Scott, R.L. *Solubility of Nonelectrolytes*, Dover, New York, 1964.
9. Paul, D.R. and Barlow, J.W. *J. Macromol. Sci. Rev. Macromol. Chem.* 18, 109, 1980.
10. Sperling, L.H. Multi-Component Polymers. Chapter 2, *Advanced Chemical Series No. 211* (eds. D.R. Paul and L.H. Sperling), American Chemical Society, Washington, DC, 1986, pp. 21–56.
11. Patterson, D.L. and Robard, A. *Macromolecules*, 11(4), 690, 1978.
12. Koningsveld, R. and Staverman, A.J. *J. Polym. Sci.*, Part A 2(6), 305, 1968.
13. Asaletha, R., Thomas, S., and Kumaran, H. *J. Rubber Chem. Technol.* 68, 671, 1996.
14. Reiss, G., Kohler, J., et al. *Macromol. Chem.* 101, 58, 1967.
15. Teysie, P., Fayat, R., and Jerome, R. *Polym. Eng. Sci.* 27, 328, 1987.
16. Leibler, L., Noolandi, J., et al. *Makromol. Chem. Symp.* 16, 17, 1988.
17. Kawazura, K., Kawaza, M., et al. American Chemical Society Rubber Division Meeting, Chicago, IL, 1994.
18. Scott, C.E. and Macasko, C.W. Reactive Processing. International Polymer Processing Expo, 1995.
19. Beaty, C. Paper presented at American Chemical Society Annual Meeting, New York, 1992.
20. Antony, P. and Dey, S.K. *Rubber Chem. Technol.* 72, 449, 1999.
21. Lohse, D.J., Garner, R.T., Graessley, W.W., and Krishnamoorti, R. Miscibility of Blends of Saturated Hydrocarbon Elastomers. Rubber Division, the American Chemical Society, Nashville, Tennessee, Sept. 29–Oct. 2, 1998, Paper No. 33, 1–14.
22. Pandey, K.N., Setua, D.K., and Mathur, G.N. Ultrasonic Technique, Modulated DSC, Dynamic Mechanical Analysis, and Atomic Force Microscopy, *Polym. Eng. Sci.* 45(9), 1265–1276, September 2005.
23. Hussein, I.A., Chaudhry, R.A., and Abu Sharkh, B.F. Study of the Miscibility and Mechanical Properties of NBR/HNBR Blends, *Polym. Eng. Sci.* 44(12), 2346–2352, December 2004.
24. Ghosh, A. and De, S.K. Dependence of Physical Properties and Processing Behavior of Blends of Silicone Rubber and Fluororubber on Blend Morphology. *Rubber Chem. Technol.* 77(5), 856–872, November/December 2004.
25. Clarke, J., Clarke, B., and Freakley, P.K. Relationships between Mixing Method, Microstructure, and Strength of NR:BR Blends. *Rubber Chem. Technol.* 74(1), 1–15, March/April 2001.

26. Arjunan, P. Technological Compatibilization of Dissimilar Elastomer Blends: Part 1. Neoprene and Ethylene–co-Propylene Rubber Blends for Power Transmission Belt Application. Rubber Division, Proceedings of the American Chemical Society, Nashville, TN, Sept. 29–Oct. 2, 1998, Paper No. 52, 1–28.

27. Croft, T.C. Elastomeric Compounds Utilizing Polymer Blends for Improved Wire Adhesion. Rubber Division, Proceedings of the American Chemical Society, Nashville, TN, Sept. 29–Oct. 2, 1998, Paper No. 63.

28. Hopper, R.I. *Rubber Chem. Technol.* 49, 341, 1976.

29. Naskar, M., Debnath, S.C., and Basu, D.K. Effect of Bis (Diisopropyl) Thiophosphoryl Disulfide on the Co-Vulcanization of Carboxylic–Acrylonitrile–Butadiene Rubber and Ethylene–Propylene–Diene Rubber Blends. *Rubber Chem. Technol.* 75(3), 309–322, July/August 2002.

30. Mingyi, L., Hua, Z., and Jianfeng, L. *J. Appl. Polym. Sci.*, 71, 215–220, 1999.

31. Raboney, M., Garner, R.T., Elspass, C.W., and Peiffer, D.G. Phase Behavior of Brominated Poly (Isobutylene-co-4-Methylstyrene)/General Purpose Rubber Blends. Rubber Division, Proceedings of the American Chemical Society, Nashville, TN, Sept. 29–Oct. 2, 1998, Paper No. 36.

32. Chiu, H.-T., Tsai, P.-A., and Cheng, T.-C. *J. Mater. Eng. Perform.* 15(1), 81–87, February 2006.

33. Kardan, M. *Rubber Chem. Technol.* 74(4), 614–621, September/October 2001.

34. Dikland, H.G. and Van Duin, M. Miscibility of EPM–EPDM Blends, *Rubber Chem. Technol.* 76(2), 495–506, May/June 2003.

35. Schuster, R.H., Meier, J., and Kluppel, M. The Role of Interphase in Filler Partition in Rubber Blends, 156th Meeting of the Rubber Division, American Chemical Society, Orlando FL, September 1999, Paper No. 60.

36. Mathai, A.E., Singh, R.P., and Thomas, S. Transport of Aromatic Solvents through Nitrile Rubber/Epoxidized Natural Rubber Blend Membranes, *Polym. Eng. Sci.* 43(3), 704–712, March 2003.

37. Chang, Y.-W., Shin, Y.-S., Chun, H., and Nah, C. Effects of *trans*-Polyoctylene Rubber (TOR) on the Properties of NR/EPDM Blends, *J. Appl. Polym. Sci.* 73, 749–756, 1999.

38. Wootthikanokkhan, J. and Clythong, N. *Rubber Chem. Technol.* 76, 1116–1127, 2003.

39. Nando, G.B. Reactive Miscible Blends from Ethylene Copolymers and Specialty Rubbers, 156th Meeting of the Rubber Division, American Chemical Society, Orlando FL, September 1999, Paper No. 55.

40. Kumar, B., De, P.P., De, S.K., Bhowmick, A.K., Majumdar, S., and Peiffer, D.G. Influence of Fillers and Oil on Mill Processability of Brominated Isobutylene-co-Paramethylstyrene and its Blends with EPDM, *Polym. Eng. Sci.* 41(12), 2266–2280, December 2001.

41. Zhao, J., Ghebremeskel, G.N., and Peasley, J. A Study of EPDM/SBR Blends, presented at a meeting of the Rubber Division, American Chemical Society, Chicago IL, April 1999, Paper No.16.

42. Bucknall, C.B. *Toughened Plastics*. Applied Science Publishers, London, 1977.

43. Riew, C.K. and Gillham, J.K. Rubber Modified Thermoset Resins. *Advances in Chemistry*, 208, American Chemical Society, Washington, DC, 1984.

44. Riew, C.K. (ed.). Rubber Toughened Plastics. *Advances in Chemistry*, 222, American Chemical Society, Washington, DC, 1989.

45. Mangaraj, D., Markham, R., and Parsons, A. Compatibilization of Polymer Blends. Battelle Multiclient Report, 1994.

46. Dear, J.P. Combined Modeling and Experimental Studies of Rubber Toughening in Polymers, *J. Mater. Sci.* 38, 891–900, 2004.

47. Cook, R.F., Koester, K.J., Macosko, C.W., and Ajbani, M. Rheological and Mechanical Behavior of Blends of Styrene–Butadiene Rubber with Polypropylene, *Polym. Eng. Sci.* 45(11), 1487–1497, 2005.

48. Van Puyvelde, P., Oommen, Z., and Koets, P. Effect of Reactive Compatibilization on the Interfacial Slip in Nylon-6/EPR Blends, *Polym. Eng. Sci.* 43(1), 71–77, January 2003.

49. Mascia, L., Pak, S.H., and Caporiccio, G. Properties Enhancement of Fluorosilicone Elastomers with Compatibilized Crystalline Vinylidene Fluoride Polymers, *Eur. Polym. J.* 31(5), 459–465, 1995.

50. Yanez-Flores, G., Ramos De-Valle, L.F., and Rodriguez-Fernandez, O.S. Proceedings of the Society of Plastics Engineers Annual Technical Conference, 1997, p. 381.

51. Wang, W. and Zheng, Q. The Dynamic Rheological Behavior and Morphology of Nylon/Elastomers Blends, *J. Mater. Sci. Lett.* 40, 2005.

52. Anh, T.H. and Vu-Khanh, T. Fracture and Yielding Behaviors of Polystyrene/Ethylene–Propylene Rubber Blends: Effects of Interfacial Agents, *Polym. Eng. Sci.* 41(12), 2073–2081, December 2001.

53. Akhtar, S. Morphology and Physical Properties of Thin Films of Thermoplastic Elastomers from Blends of Natural Rubber and Polyethylene, *Rubber Chem. Technol.* 61, 599–583, 1988.
54. Dispenza, C., Carter, J.T., McGrail, P.T., and Spadaro, G. Reactive Blending of Functionalized Acrylic Rubbers, and Epoxy Resins. *Polym. Eng. Sci.* 41(9), 1483–1496, September 2001.
55. Suriyachi, P., Kiatkamjornwong, S., and Prasassarkich, P. Natural Rubber-g-Glycidyl Methacrylate/Styrene as a Compatibilizer in Natural Rubber/PMMA Blends, *Rubber Chem. Technol.* 77(5), 914–930, November/December 2004.

12 Fiber-Reinforced Elastomers

R.S. Rajeev

CONTENTS

12.1 INTRODUCTION

A composite material is defined as a macroscopic combination of two or more distinct materials having a recognizable interface between them. However, because composites are commonly used for their structural properties, the definition can be restricted to include only those materials that contain reinforcement, such as fibers or particles, supported by a binder or matrix material. Thus, composites typically have a fiber or particle phase that is stiffer and stronger than the continuous matrix phase. In order to provide reinforcement, there generally must be substantial volume fraction (~10% or more) of the reinforcing agent. In composites, the combining of two or more existing materials is done mainly by physical means. A true composite might be considered to have a matrix surrounding the reinforcing material in which the two phases act together to produce characteristics not attainable by either constituent acting alone.

Almost all high-strength or high-stiffness materials fail because of the propagation of flaws. A fiber of such a material is inherently stronger than the bulk form because the size of the flaw is limited by the small diameter of the fiber. In addition, if equal volumes of fibrous and bulk material are compared, it is found that even if a flaw does produce failure in a fiber, it will not propagate to fail the entire assemblage of fibers as would happen in the bulk material. Furthermore, preferred orientation may be used to increase the lengthwise modulus, and perhaps strength, well above isotropic values. Since this material is also lightweight, there is a tremendous potential advantage in strength/weight and/or stiffness/weight ratios over conventional materials. The desirable fiber properties can be translated into practical application when the fibers are embedded in a matrix that binds them together, transfers load to and between the fibers, and protects them from environments and handling.

High-performance rigid composites made from glass, graphite, Kevlar, boron, or silicone carbide fibers in polymeric matrices have been studied extensively because of their application in aerospace and space vehicle technology [1–3]. The reinforcement of rubbers using particulate fillers such as carbon black or precipitated silica has also been studied at length [4,5]. However, studies on composites based on fiber and rubber matrices need to be strengthened. The fiber-reinforced rubber composites are characterized by the extremely low stiffness of the rubber matrix compared to that of the reinforcing cords. Both continuous and discontinuous fibers are used to reinforce the rubber matrix; the most significant example for the former is the use of fiber-reinforced rubber in pneumatic tires. The discontinuous fiber-reinforced rubber composites are usually referred to as short fiber–rubber composites. This chapter explains some of the short and long fiber-reinforced rubber composites with special emphasis on their characterization, properties, and applications.

12.2   SHORT FIBER–RUBBER COMPOSITES

The term "short fiber" means that the fibers in the composites have a critical length which is neither too high to allow individual fibers to entangle with each other, nor too low for the fibers to lose their fibrous characteristics. The term "composite" signifies that the two main constituents, i.e., the short fibers and the rubber matrix remain recognizable in the designed material. When used properly, a degree of reinforcement can be generated from short fibers, which is sufficient for many applications. Proper utilization comprises: preservation of high aspect ratio in the fiber, control of fiber directionality to optimally reinforce the fabricated part, generation of a strong interface through physicochemical bonding, establishment of a high state of dispersion, and optimum formulation of the rubber compound itself to accommodate processing and facilitate stress transfer while as much flexibility as possible is maintained to preserve dynamic properties. Moreover, short fibers provide dimensional stability during fabrication and in extreme service environments (i.e., high temperature, solvent contact, etc.) by restricting the matrix distortion. A particular benefit is improved creep resistance. Other nonstructural benefits include improved tear and impact strength by blunting or diverting growing crack tips.

Because short fibers can be incorporated directly into the rubber compound along with other additives, the resulting composites are amenable to the standard rubber-processing steps of extrusion, calendering, and the various types of molding operations (i.e., compression, injection, and transfer). Economical high-volume outputs are thus feasible. This is in contrast to the slower processes required for incorporating and placing continuous fibers. The penalty is a sacrifice in reinforcing strength with discontinuous fibers, although they give better strength and modulus compared to particulate fillers. De and White [6] reviewed short fiber–polymer composites covering thermoplastic, thermoset, thermoplastic elastomer (TPE), and rubber matrices. Goettler and Cole [7] also reviewed the various aspects of short fiber reinforcement of rubber. The mechanics of short fiber reinforcement of rubbers based on constitutive equations and transformation laws was reviewed by Abrate [8].

12.2.1   CONSTITUENTS OF SHORT FIBER–RUBBER COMPOSITES

12.2.1.1   Rubber Matrix

Fibers find application essentially in all conventional rubber compounds. The functions of the rubber matrix are to support and protect the fibers, the principal load-carrying agent, and to provide a means of distributing the load among and transmitting it between the fibers without itself being fractured. The load transfer mechanism in short and long fibers is different. When a short fiber breaks, the load from one side of the broken fiber is first transferred to the matrix and subsequently to the other side of the broken fiber and to adjacent fibers. In short fiber–rubber composites, the shearing stress in the matrix contributes to load transfer. Typically, the matrix has a considerably lower density, stiffness (modulus), and strength than those of the reinforcing fiber material, but the combination of the two main constituents (matrix and fiber) produces high strength and stiffness, while still possessing a relatively low density.

In one of the first reports on fiber reinforcement of rubber, natural rubber (NR) was used by Collier [9] as the rubber matrix, which was reinforced using short cotton fibers. Some of the most commonly used rubber matrices for fiber reinforcement are NR, ethylene–propylene–diene monomer (EPDM) rubber, styrene–butadiene rubber (SBR), polychloroprene rubber, and nitrile rubber [10–13]. These rubbers were reinforced using short and long fibers including jute, silk, and rayon [14–16].

Rubber–rubber and rubber–plastics blends are also used as matrices for fiber reinforcement. Boustany et al. [17] used NR–SBR blend as the matrix in the short fiber reinforcement using cellulose fiber. Zhang et al. [18] prepared short sisal fiber-reinforced epoxidized NR (ENR)/polyvinyl chloride (PVC) blend and found that hydrochloric acid was generated from PVC in the composite when heated above 250°C. Arroyo and Bell [19] studied the effect of short aramid fibers on the mechanical behavior of isotactic polypropylene (PP) and EPDM rubber and their blends. Sreeja and Kutty [20] prepared NR—whole tire reclaim—short nylon fiber composites and studied the effect of urethane-based bonding agent on the cure characteristics and mechanical properties. Some of the specialty elastomers which have been used as matrices for short fiber reinforcement are silicone rubber, fluoroelastomers, ethylene vinyl acetate (EVA) rubber, and polyurethane elastomer. Zebarjad et al. [21] observed that the modification of PP with a combination of ethylene/propylene rubber and glass fiber could be used for improving the mechanical properties of the plastics. This is because, the reduction in the stiffness and strength due to the presence of rubber particles are compensated by the addition of short glass fibers.

Short fiber reinforcement of TPEs has recently opened up a new era in the field of polymer technology. Vajrasthira et al. [22] studied the fiber–matrix interactions in short aramid fiber-reinforced thermoplastic polyurethane (TPU) composites. Campbell and Goettler [23] reported the reinforcement of TPE matrix by Santoweb fibers, whereas Akhtar et al. [24] reported the reinforcement of a TPE matrix by short silk fiber. The reinforcement of thermoplastic co-polyester and TPU by short aramid fiber was reported by Watson and Frances [25]. Roy and coworkers [26–28] studied the rheological, hysteresis, mechanical, and dynamic mechanical behavior of short carbon fiber-filled styrene–isoprene–styrene (SIS) block copolymers and TPEs derived from NR and high-density polyethylene (HDPE) blends.

Various specialty elastomers have also been used as matrices for short fiber reinforcement. Some of the examples are described below:

Silicone rubber: Warrick et al. [29] reviewed the reinforcement of silicone elastomers with various types of discontinuous fibers, some of which were generated *in situ* by graft polymerization. Polyester, cellulose and carbon fibers were also used to reinforce silicone rubber [30,31].

Fluoroelastomers: Novikova et al. [32] reported improved physico-mechanical properties of fluoro rubbers by reinforcement with chopped polyamide fibers. Other fiber reinforcements are covered by Grinblat et al. [33]. Watson and Francis [34] described the use of aramid (Kevlar) as short fiber reinforcement for vulcanized fluoroelastomer along with polychloroprene rubber and a co-polyester TPE in terms of improvement in the wear properties of the composites. Rubber diaphragms, made up of fluorosilicone rubbers, can be reinforced using aramid fiber in order to impart better mechanical properties to the composite, though surface modification of the fiber is needed to improve the adhesion between fluorosilicone rubber and the fiber [35]. Bhattacharya et al. [36] studied the crack growth resistance of fluoroelastomer vulcanizates filled with Kevlar fiber.

Ethylene–vinyl acetate: Fetterman [37] reinforced compounded ethylene–vinyl acetate (EVA) copolymer by using short fibers and found that silane coupling agents were effective at establishing improved fiber–matrix adhesion. Das et al. [38] prepared carbon fiber-filled conductive composites based on EVA and studied the electromagnetic interference shielding effectiveness of the composites.

Polyurethane elastomer: Isham [39] and Leis [40] described how glass fiber reinforcement could broaden the performance spectrum in reaction–injection–molding (RIM) urethanes as well as other elastomeric systems applied to automotive fascia. Marzocchi [41] developed glass fiber-reinforced polyurethane rubber. Short fiber reinforcement of polyurethane rubber was reported by Suhara et al. [42] and Pan and Watt [43]. Bicerano and Brewbaker [44] reported that when a small amount of cellulose fiber having high aspect ratio was added to polyurethane rubber matrix, there was a significant increase in Young's modulus and tensile strength.

Thermoplastic elastomers: Kane [45] studied the physico-mechanical reinforcement of a butanediol-poly(tetramethylene glycol–terephthalic acid) TPE by short glass fibers. Kutty and Nando studied the reinforcement of TPU using Kevlar fibers [46]. López-Manchado and Arroyo [47] prepared ternary composites based on isotactic PP, ethylene–octene copolymer (EOC) TPE and poly(ethylene-terephthalate) (PET) textile fibers. The effect of matrix composition and fiber content on the final properties of the composite was investigated. They observed that PET fibers act as an effective reinforcing agent, giving rise to a sensible improvement in the tensile and flexural behavior, mainly in matrices with high copolymer percentages. They also observed that the analyzed mechanical properties depend more on the matrix composition than on the fiber percentage. The reinforcement of thermoplastic co-polyester and TPU by short aramid fiber was reported by Watson and Frances [25]. Amornsakchai et al. [48] reinforced styrene–ethylene–butylene–styrene (SEBS) TPE using aramid fiber. Increase in fiber loading caused a linear increase in the modulus of the composites whereas the tensile strength was found to be decreasing with fiber loading. Hydrolyzing the fiber using NaOH and use of a compatibilizer, maleic anhydride-grafted SEBS (MA-g-SEBS) found improving the tensile strength of the composites. Thermoplastic elastomeric matrices like styrene–butadiene–styrene (SBS), SEBS, and Hytrel were also reinforced using various short fibers including carbon, glass, and polyester [49–52].

Liquid rubbers: In order to improve the flexibility of short glass fiber-reinforced epoxy composites, Kaynak et al. [53] modified the epoxy resin matrix with hydroxyl-terminated polybutadiene (HTPB) liquid rubber. A silane coupling agent was also used to improve the interfacial adhesion between glass fibers and epoxy matrix. However, Humpidge et al. [54] reported some unique processing problems for the resulting pasty mixtures when short textile fibers were incorporated in a liquid rubber medium.

12.2.1.2   Fiber Reinforcement

The degree of short fiber reinforcement of an elastomeric matrix is governed largely by the following characteristic properties of the fiber: (a) aspect ratio, (b) adhesion to the matrix, (c) dispersion in the matrix, and (d) processibility and flexibility (which ensures minimum fiber breakage). Due to the high viscosity of the rubber compounds, the short fibers are buckled and broken by the high shear stress experienced during mixing. The length distribution of short PET, Nylon 6, aramid, rayon, carbon, and glass fibers, after mixing with polychloroprene rubber is shown in Figure 12.1 [55]. When PET fibers retained their original length of 6 mm (I), nylon, aramid, and rayon fibers were buckled or broken and gave rise to a broad distribution in a range of shorter lengths (II). The lengths of carbon and glass fibers were reduced to about 150 μm (III).

The aspect ratio, i.e., the length/diameter ratio (L/D) of the fibers is a major parameter that controls the fiber dispersion, fiber–matrix adhesion and optimum performance of short fiber–rubber

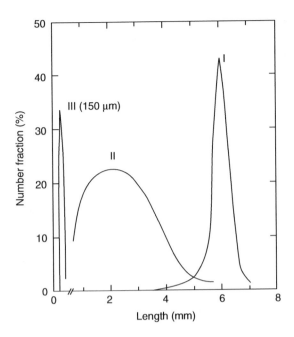

FIGURE 12.1 Length distribution of short fibers after mixing. (From Ashida, M., *Short Fibre–Polymer Composites*, De, S.K. and White, J.R. (Eds.), Woodhead Publishing Ltd., Cambridge, U.K., 1996. With permission.)

composites [56]. If the aspect ratio of the fiber is lower than the critical aspect ratio, insufficient stress will be transferred and the reinforcement will be insufficient. Studies suggest that an aspect ratio in the range 100–200 is essential for high-performance fiber–rubber composites [17,56–58]; however, the value depends on the fiber and the matrix. Chakraborty et al. [59] observed that an aspect ratio of 12 provided optimum reinforcement in the case of jute fiber–carboxylated nitrile rubber (XNBR) composites whereas Murthy and De [60] reported that an aspect ratio of 15 in the case of short jute fiber–NR system and 28 in the case of short jute fiber–SBR system [61] were sufficient for good reinforcement of the composition. Compared to glass and jute fibers, breakage of silk fiber in rubber matrix was lower. Setua and coworkers reported an average aspect ratio of 85 for short silk fiber when mixed with polychloroprene rubber [62] while in nitrile rubber-based composites, silk fiber gave an aspect ratio of 77 [63]. Rajeev et al. [64] found that the average aspect ratio of the melamine fibers in EPDM rubber–melamine fiber composition was 250 after Brabender mixing. The fibers broke down further during milling in a two-roll rubber mixing mill with the aspect ratio dropping to 11.

As far as physical length is concerned, short fibers fall into one of the following categories, shown in order of descending length: staple, pulp, whiskers, fibrils, fibrids, and nano. There is no well-defined demarcation line when one category ends and the other begins. Staples are usually more than 5 mm long; fibrils and fibrids are microscopic. Rigid straight inorganic fibers, usually with relatively low aspect ratio are called whiskers. Fibrils are diminutive staple with random length and cross-section distribution and are usually the by-product of some mechanical abrasive force. Fibrids are similar but smaller than fibrils and are usually produced by precipitation under high shear. Pulp material is usually produced by some sort of grinding action of regular staple material, which will cause randomization and reduction in the average staple length. Nanofibers are the latest addition in short fibers. Nanofiber-reinforced rubber composites, in general, come under the category of rubber nanocomposites.

A brief description of the common fibers used for the reinforcement of rubber is given below.

12.2.1.2.1 Natural Fibers

The advantage of natural fiber reinforcement of rubbers over synthetic fiber reinforcement is the lower modulus of the former, which will reduce the modulus gradation between the fiber and the matrix. In addition, the large number of hydroxyl groups on the surface of the cellulosic fibers may help in good bonding between the fiber and the matrix. However, the low availability and difficulties in processing of these fibers suitable for incorporation into the rubber matrix are the main disadvantages. Nunes and Visconte [65] published a review on the natural fiber reinforcement of rubbers with emphasis on processing and application. In a series of reports on natural fiber-reinforced rubber composites, Ismail and coworkers [66–68] added bamboo fiber, oil palm fiber, and rubber wood fiber to different rubber matrices and studied the effect of bonding agent, fiber loading and silane coupling agent on the mechanical properties, and cure characteristics of the compositions. Sisal, coir, coconut, and pineapple leaf fibers were also used to reinforce various elastomeric matrices [69–71]. Bhattacharyya et al. [72] showed that the addition of pineapple leaf fiber to NR increased hardness, compression set, tear resistance, and Mooney viscosity but decreased elongation at break, mill shrinkage, and Mooney scorch time of the composites.

12.2.1.2.2 High-Performance Synthetic Fibers

The reinforcement of rubbers using nylon, rayon, vinyl, and polyester fibers was reported by various authors [10,58,73–75]. Because of the design flexibility and suitable end-use applications, high-performance fibers such as glass, carbon, and aramid also find extensive applications in short fiber-reinforced rubbers. A brief description of some of the major high-performance fibers commonly used in short fiber–rubber composites is given below:

Glass fiber: High initial aspect ratio can be obtained with glass fibers, but brittleness causes breakage of the fibers during processing. Czarnecki and White [76] reported the mechanism of glass fiber breakage and severity of breakage with time of mixing. Gregg [77] reported that the tensile strength of glass fiber–rubber composites decreased with increasing atmospheric humidity during the glass fiber storage. Marzocchi [78] developed glass fiber–elastomer composites with improved strength and wear resistance to be used in auto tires, V-belts, and conveyor belts. Derringer [79] reported the advantages of short glass fiber-reinforced rubber composites, which included high modulus, high resilience, and low creep. Murty and coworkers studied the extent of fiber–matrix adhesion and physical properties of short glass fiber-reinforced NR and SBR composites [61,80,81]. Murty added up to 75 phr of short glass fiber to NR matrix and found that composites showed anisotropy in technical properties only when added at >25 phr [80]. Lee [82] studied the mechanical properties and morphology of short glass fiber-reinforced polychloroprene composites. He developed a thermodynamic mixing rule and a set of constitutive equations for predicting the anisotropic stress–strain properties.

Carbon fiber: Carbon fibers are characterized by a combination of lightweight, high strength, and high stiffness. The earliest commercially available carbon fibers were produced by thermal decomposition of rayon precursor materials. Although a few carbon fibers are still made from these materials, rayon has been largely supplanted as a precursor by polyacrylonitrile (PAN). PAN precursors produce much more economical fibers because the carbon yield is higher and because PAN-based fibers do not intrinsically require a final high-temperature "graphitization" step [83].

Though carbon fibers are extensively used in polymer composites, its application in rubber matrices is limited to specific end use, mainly in electrically conductive composites. An exhaustive review on the electrically conductive rubber and plastic composites with carbon particles or conductive fibers including carbon fiber was published by Jana et al. [84]. Yagi and coworkers [85] were granted a European Patent for the invention of highly conductive carbon fibers for rubber and plastics composites. They dispersed fibers in the rubber matrix by kneading. Jana and coworkers [86,87] studied the electrical conductivity of randomly distributed carbon fiber–polymer composites. It was found that the composite containing a high fiber aspect ratio, prepared by the "cement" mixing method needed a lower critical concentration of the fiber for electrical conductivity than the composite with low fiber aspect ratio prepared by the "mill" mixing method [86]. Jana et al. [87] also studied the effect of carbon fiber concentration, fiber aspect ratio (L/D), and sample thickness on the electromagnetic interference shielding effectiveness of short carbon fiber-filled polychloroprene rubber composites vulcanized by barium ferrite. Das et al. [38] studied the electromagnetic interference shielding effectiveness of EVA-based conductive composites containing short carbon fibers. They observed that increasing the fiber loading increases the shielding effectiveness of the composites. Compared to short carbon fiber-carbon black blend composites or 100% pure carbon black filled composites, 100% pure short carbon fiber filled composites showed a better shielding effect. The effects of short carbon fibers on the anisotropic swelling, mechanical, and electrical properties of radiation-cured SBR composites were studied by Abdel-Aziz et al. [88].

Aramid fiber: Aramid fiber was introduced in the 1970s when Kwolek, Blades, and colleagues discovered nematic solutions of poly(*p*-benzamide) and poly(*p*-phenylene terephthalamide) that were processable into fibers with a highly oriented, extended chain configuration [89–91]. The first commercial *p*-aramid fiber (Kevlar) was introduced in 1971 [92]. Since then, because of its high-temperature resistance coupled with chemical resistance and good mechanical properties, Kevlar fiber has been used in elastomers. Kevlar fiber imparts toughness, impact resistance, light weight, dimensional stability, and vibration damping characteristics to the composites. The processing parameters of aramid fiber-reinforced NR, SBR, nitrile rubber, and polychloroprene rubber were reported by O'Connor [58] and Moghe [93]. Moghe [93] also found that the use of a mixture of short cellulose and aramid fibers at low volume fractions improved the modulus of the composites based on NR, polychloroprene rubber, and SBR. According to O'Connor [58], aramid fiber tends to clump together and does not disperse easily in the rubber matrix. Foldi [94] pointed out the advantages and limitations of aramid fiber reinforcement of elastomers. The technical advantages are high initial tensile modulus, resistance to tear and wear, good compressive response, and good bending stiffness whereas the technical limitations are low-break elongation, high compression set, high hardness, and high anisotropy (which is advantageous in some applications where high strength in one direction and sufficient flexibility in the other direction are needed). The technical barriers are high heat buildup and uneven fiber dispersion, which results in sites of weakness and early failure. Sunan et al. [95] compared the reinforcing effect of as received and hydrolyzed Kevlar fiber-reinforced Santoprene TPE composite. López-Manchado and Arroyo [96] observed that apart from increasing the mechanical properties like tensile strength, modulus, tear, and abrasion resistance, aramid fibers decrease the time to reach optimum cure and cross-link density of NR, EPDM rubber, and SBR.

Several authors report the use of aramid fibers in pulp form. A review on the development of aramid fiber pulp to provide a viable alternative to asbestos fibers for plastics and rubber reinforcement in many end-use applications, e.g., friction brakes, clutches, gaskets, sealants, caulking, and adhesives was given by Hoiness and Frances [97]. Aramid fiber is the only fiber so far which can match the asbestos's performance as friction product [98]. An overview on reinforcing rubber compounds with *p*-aramid (Twaron) fibrillated short fibers and with *p*-aramid adhesive-activated chopped fibers was published by van der Pol [99]. He described the reinforcement of NR–polybutadiene rubber blend, polychloroprene, and EPDM rubbers using short *p*-aramid fibers [100]. Rijpkema added aramid fiber and aramid pulp to tire tread compounds based on NR and BR to reduce the rolling resistance of tire [101]. Schuler [102] reported that when Kevlar fibers were broken under very high shear, the fibers fibrillated to produce a tough pulp-like material with high surface area. Such fibrillated fibers would provide improved dispersion in the rubber matrix.

Other high-performance fibers: HDPE fibers are marketed under the trade name "Spectra" by Allied Signal and "Dy-neema" by Toyobo. HDPE fibers offer strength similar to that of *p*-aramids. Light in weight, the fiber has a specific gravity of only 0.97. HDPE fibers have very good abrasion resistance and excellent chemical and electrical resistance [103]. However, so far no report is available on its applicability as a reinforcing fiber for rubber, mainly because of its difficulty in adhering to the matrix. In addition, these fibers have low melting point and so their continuous operation temperature is relatively low (121°C). Similarly, the high-performance, thermoplastic, multifilament yarn, Vectran, introduced by Celanese Corporation also finds no application in rubber so far. This fiber is melt-spun from liquid crystal polymer and is five times stronger than steel and 10 times stronger than aluminum. Guillot got a U.S. patent for developing an insulation material for rocket motors based on Vectran [104].

Polybenzimidazole (PBI) fiber, manufactured by Celanese Corporation is an organic fiber with excellent thermal-resistant properties. Like aramid fibers, PBI fibers do not burn in air and do not melt or drop. High limiting oxygen index (LOI) coupled with good chemical resistance and good moisture regain make PBI an excellent fiber for fire-blocking end uses such as safety and protective clothing and flame-retardant fabrics. Junior et al. [105] developed rocket motor insulators based on PBI fiber-reinforced elastomeric composition. They found that the insulator composition developed by them has better erosion resistance than aramid fiber-reinforced compositions. PBI fibers were blended with aramid fiber and used in phosphonitrilic elastomeric compositions to develop rocket motor insulators [106]. Poly(p-phenylene-2,6-benzobisoxazole) (PBO) fiber, manufactured by Toyobo, Japan, under the trade name Zylon is another new entrant to the high-performance organic fiber market. Zylon has outstanding thermal properties and almost twice the tensile strength of conventional p-aramid fibers. Though its high modulus makes it an excellent candidate for reinforced composites coupled with its high LOI, so far no report is available on the reinforcement of elastomers using this fiber.

Melamine fiber: Melamine fiber is one of the recent generation high-performance fibers and is recommended for heat- and flame-resistance applications [107,108]. It is the newest fiber to be fully commercialized. Commercial preparation of this fiber is based on BASF's patented melamine technology and the fiber is marketed under the trade name Basofil. Based on melamine chemistry, Basofil offers a high operating temperature and a high LOI. These fibers combine the fire protection and heat stability properties along with good chemical, hydrolysis, and ultraviolet resistance. Like aramid fibers, these fibers also have no specific melting point and undergo decomposition above 350°C. The initial reports on melamine fiber appeared in 1974 when Japanese patents were granted to Nihongi and Yasuhira for the invention of fire-resistant melamine resin fibers by wet spinning [109,110]. They prepared these fibers by spinning a mixture of poly(vinyl alcohol), N-methylolmelamine or methylated methylolmelamine, and H_3BO_3 into an aqueous solution containing NaOH. The fiber was then treated with an aqueous solution containing H_2SO_4 and Na_2SO_4 and stretched 60% at 80°C in a bath containing Na_2SO_4, washed with water, dried using air at 170°C to give the fiber with strength 2.39 g/denier and 16% elongation.

Melamine fiber is manufactured by the following process [111]. N-methylol-2,4,6-tri-amino-s-triazine) is first produced by reacting 1 mole of melamine with 3 moles formaldehyde and the pH of the resulting mixture is adjusted to 8–9 with an alkali such as sodium carbonate or sodium bicarbonate. The reaction is conducted in a solvent, the preferred solvent being water. The reaction mixture is heated to a temperature of 75°C–85°C for a period of 1 h. Thus, the precondensate of N-methylol melamine having a degree of hydroxymethylation of 2.7 and present in the form of an aqueous solution having a concentration of 56 wt% is produced. The precondensate is either recovered as a powder followed by dissolution in water to obtain the spinning solution, or is used directly as the spinning solution and then spun into filament form and subsequently cured to form the fibers. The curing operation is performed at a temperature of 170°C–320°C where the precondensate initially formed continues, by a condensation reaction, with heating to become an insoluble and infusible three-dimensional (3D) melamine–aldehyde polymer. If the spinning temperature is high enough, the curing operation takes place simultaneously with spinning.

Melamine fiber is mainly used in heat- and flame-resistant applications, especially in the manufacture of protective clothing for the iron, steel, and automobile industries, in aircraft and other public transport systems, and in a large number of consumer goods and textiles [108]. Compared to fibers with comparable properties, its low cost is an advantage and has resulted in its being evaluated in a number of areas. Because of its variable denier and low tensile strength, melamine fiber is sometimes blended with stronger fibers such as aramid. Compared to other high-performance fibers like aramid, melamine fibers are having high elongation at break (18%) which may help in developing flexible rubber–fiber composites. The physical form of these fibers simplifies the mixing of the fiber with the rubber in two-roll mill or internal mixers.

Microscopy, x-ray diffraction (XRD), x-ray photoelectron spectroscopy (XPS), and thermogravimetry (TG) were used to characterize melamine fiber [112]. Figure 12.2A shows the scanning electron microscopy (SEM) photomicrograph of the melamine fiber filaments at a magnification of 150. The melamine fiber filaments are of variable length and narrow diameter compared to the conventional short fibers. The filaments, which are in the entangled form, show a diameter variation from 12 to 16 μm. The filament surface contains striations, which are wavy in nature, running parallel to the horizontal axis of the filament. These striations are believed to be the die lines, which occur at the time of manufacturing of the filaments. These types of striation markings due to drawing are commonly observed on the surfaces of metallic filaments [113]. The filament also contains pits on the surface as shown in Figure 12.2B. These pits are supposed to have occurred during the manufacturing of the fibers and may aid in better adhesion between the fiber and the matrix through mechanical interlocking. Figure 12.2C is the SEM micrograph of the end of a

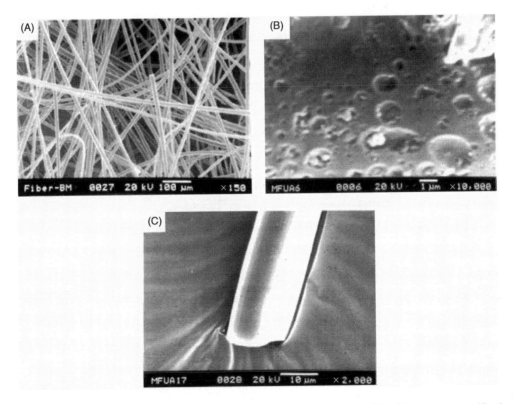

FIGURE 12.2 Scanning electron microscopy (SEM) images of: A, melamine fiber filaments at magnification 150×; B, a single melamine fiber filament at magnification 10,000 showing pits on the surface and C, the end of a melamine fiber filament. (From Rajeev, R.S., Bhowmick, A.K., De, S.K., Gong, B., and Bandyopadhyay, S., *J. Adh. Sc.Technol.*, 16, 1957, 2002. With permission.)

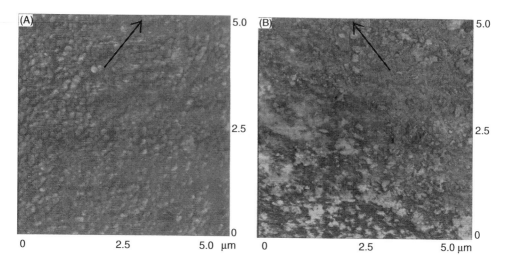

FIGURE 12.3 Atomic force microscopy (AFM) phase images of A, unaged and B, aged melamine fiber filaments at a scan area of 5×5 μm. Arrows indicate direction of major fiber axis. (From Rajeev, R.S., Bhowmick, A.K., De, S.K., Gong, B., and Bandyopadhyay, S., *J. Adh. Sci. Technol.*, 16, 1957, 2002. With permission.)

filament where it can be seen that the filaments do not have a circular cross-section. Rather, the cross-section is flattened and bean shaped, which provides them a larger surface area, which is advantageous when melamine fibers are mixed with other fibers [107]. This type of cross-section may also cause improved adhesion with the matrix.

Figure 12.3A shows the tapping mode atomic force microscopy (AFM) image of a single melamine filament surface, which reveals the surface roughness of the fiber. Surface roughness is quantified using roughness analysis in AFM. Ageing of the fibers at 150°C for 7 days was found to increase the average surface roughness (from 7.25 to 12.61 nm) which may be helpful in better fiber–matrix interaction. The change in surface roughness due to ageing is evident in the AFM image given in Figure 12.3B. The XPS analysis of the filaments (Figure 12.4) shows that apart from

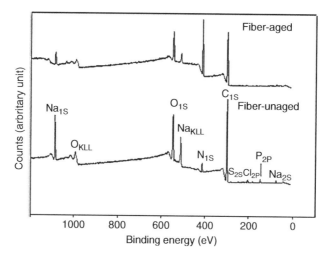

FIGURE 12.4 Wide scan x-ray photoelectron spectroscopy (XPS) images of unaged and aged melamine fiber filaments. (From Rajeev, R.S., Bhowmick, A.K., De, S.K., Gong, B., and Bandyopadhyay, S., *J. Adh. Sci. Technol.*, 16, 1957, 2002. With permission.)

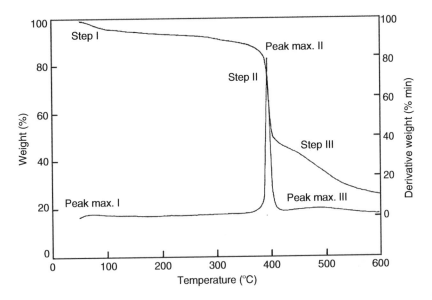

FIGURE 12.5 Thermogravimetry (TG) and derivative thermogravimetry (DTG) curves of melamine fiber filaments [114]. (From Rajeev, R.S., De, S.K., Bhowmick, A.K., and John, B., *Polym. Deg. Stabl.*, 79, 449, 2003. With permission.)

the elements carbon, nitrogen, and oxygen, which are the constituents of melamine formaldehyde polymer, the fiber surface also contains an appreciable amount of sodium along with a very low concentration of sulfur, chlorine, and phosphorous. The surface chemical composition is changed due to ageing, which is evident from the XPS spectra. Ageing causes a reduction in the hydroxyl groups and an increase in the carbonyl groups, showing that the fiber becomes less hygroscopic due to ageing. The authors confirmed the reduction of surface hydroxyl groups due to ageing by analyzing the infrared spectra of the unaged and aged fibers.

The thermogravimetric analysis (TGA) of melamine fiber is given in Figure 12.5 and the results of the thermograms are summarized in Table 12.1 [114]. The TG curve of the melamine fiber shows that three major degradation steps characterize the decomposition, the first step occurring at a temperature around 50°C. The second degradation step (starting at a temperature around 356°C) is the fastest compared to the other degradation steps. The third step starts at a temperature around

TABLE 12.1

Analysis of the Thermogram of Melamine Fiber in Nitrogen Atmosphere, at a Heating Rate of 10°C/min[a]

Onset A_f^a (°C)	Onset B_f^a (°C)	Onset C_f^a (°C)	Weight Loss in the Major Degradation Steps (%)			Peak Maximum Temperatures (T_{max}) (°C)			Rate of Decomposition (dw/dt) at T_{max} (%/min)			Char Residue at 600°C (%)
			I	II	III	T_{max1f}	T_{max2f}	T_{max3f}	T_{max1f}	T_{max2f}	T_{max3f}	
50	356	393	4.4	35.3	31.0	55	373	475	1.7	80.1	3.3	25.0

Source: Rajeev, R.S., De, S.K., Bhowmick, A.K., and John, B., *Polym. Deg. Stabl.*, 79, 449, 2003. With permission. Copyright Elsevier, 2003.

[a] See Figure 12.5 for the identification of the degradation steps.

393°C. The previous studies of thermal decomposition of melamine formaldehyde resin show that the degradation of the resin takes place in multiple steps [115,116]. However, the temperature corresponding to the onsets are slightly different, which may be due to the difference in molecular weight and crystallinity of the materials. The derivative TG curve of melamine fiber shows mainly three peaks. Hirata and Inoue [117] analyzed the pyrolysis products in each degradation step. According to them, the initial degradation step is the elimination of free side groups such as methylol groups. The sharp degradation step and the next prominent step are associated with the process of elimination of triazine rings. The highest rate of weight loss is given by the second step. Melamine fiber-reinforced rubber composites have been reported by Rajeev et al. [64,118,119]. They found out the effect of bonding system, fiber loading, and ageing on the properties of melamine fiber-reinforced EPDM [64], maleated EPDM [118], and nitrile rubber [119] composites where resorcinol, hexamethylene tetramine and hydrated silica-based dry bonding system was used as the bonding agent.

Nanofibers: Rubber composites reinforced by using nanofibers like carbon nanofiber and carbon nanotubes (CNTs) generally come under the category of rubber nanocomposites. CNTs and vapor-grown carbon nanofibers (VGCNFs) are the two important nanofibers used for reinforcing rubber matrices. The diameter of VGCNF is higher (of the order of 150 nm) than that of CNT (of the order of 20 nm). Nanofibers are highly graphitic single fibers with length up to a few hundred microns. There are mainly two types of CNTs; single-walled CNT (SWNT), and multiwalled CNT (MWNT). MWCNTs were observed by Ijima in 1991 [120]. CNTs are known for their excellent thermal conductivity twice as high as that of diamond, thermal stability up to around 2800°C, high strength, and electrical properties with current carrying capacity 1000 times higher than that of copper wires. Because of their exceptional properties, CNTs find applications in field emission (FE) displays [121], scanning probe microscopy tips [122], and micro-electronic devices [123]. They also found reinforcing various polymer matrices. A review of CNTs and their composites have been given by Thostenson et al. [124]. Theoretical and experimental studies of CNTs have shown that they have an elastic modulus greater than 1 TPa, which is close to that of diamond (1.2 TPa). In order to fully utilize the potential of CNTs as reinforcing agents for rubber matrices, their properties and interaction with the rubber matrix need to be fully understood.

12.2.2 FIBER–RUBBER ADHESION

Proper reinforcement of rubber matrix using fillers can be achieved only if there exists adequate adhesion between the filler and the rubber. Rubber–rubber adhesion and rubber–filler adhesion both without and with adhesion promoters have been studied extensively [125–127]. Fiber–matrix adhesion in short fiber–rubber composites is always a field of extensive research. If the fibers are not bonded properly with the rubber matrix, fibers will slide past each other under tension deforming the matrix, thereby reducing the strength properties. In the case of short fiber-reinforced rubber composites, loads are not directly applied to the fibers, but are applied to the matrix. To obtain a high-performance composite, the load must be effectively transferred to the fibers, which is possible only when the fiber–matrix interphase is sufficiently strong. In addition, the adhesion between the fiber and the matrix should be such that the failure occurs in the matrix rather than at the interphase [92].

Naturally occurring fibers such as cotton, cellulose, etc., have short whiskers protruding from the surface, which help to give a physical bond when mixed with rubber. Glass, nylon, polyester, and rayon have smooth surfaces and adhesion of these fibers to the rubber matrix is comparatively poor. In addition, these synthetic fibers have chemically unreactive surfaces, which must be treated to enable a bond to form with the rubber. In general, the fibers are dipped in adhesives in the latex form and this technology is the most common one used for continuous fibers. The adhesion between elastomers and fibers was discussed by Kubo [128]. Hisaki et al. [129] and Kubo [130] proposed a three-layer model composed of fibers, adhesive agents, and elastomers to explain the adhesion between elastomers and fibers. According to them, the mechanism of bonding forces between the fiber layer and the adhesive agent layer varies with each fiber whereas the bonding force between the adhesive agent layer and the elastomer layer comes from the vulcanization of the elastomer layer. The unsaturated bonds in the bonding system are bonded by a cross-linking reaction with the unsaturated bonds in the elastomer. The cross-linking reaction is similar to that of the conventional elastomer compounds consisting of various compounding ingredients, such as sulfur, accelerators, etc. These compounding ingredients migrate into the unsaturated bonds in the bonding system by the heat of the vulcanization process, and the cross-linking reaction occurs between the bonding system and elastomer. Sarkar et al. [131] studied the adhesion between rubber and fabric in heat-resistant conveyer belts based on SBR and chlorobutyl rubber.

Compared to adhesive dipping, the use of a tricomponent dry bonding system consisting of resorcinol (or a resorcinol derivative), hexamethylene tetramine (hexamine), and fine-particle hydrated silica (abbreviated as RHH) to achieve adhesion between short fiber and the rubber matrix is easy. This is because, the constituents of the dry bonding system can be added to the rubber matrix like any other compounding ingredient and extra processes like dipping and drying can be avoided. The history of dry bonding system dates back to 1967, when Dunnom [132] observed a marked difference in the adhesion level by adding silica to a compound containing resorcinol and hexa. Goettler and Shen [57], O'Connor [58], Derringer [79], and Murty and De [133] described the various aspects of short fiber adhesion to rubber in the presence of the dry bonding system. Setua and De [63] found that in short silk fiber-reinforced nitrile rubber composites, the adhesion between the fiber and the matrix was complete only when all the three components of the dry bonding system were present together. They also reported that a properly bonded fiber-filled matrix showed good ageing resistance. Ramayya et al. [73] replaced silica by carbon black in the tricomponent dry bonding system to study the adhesion between rayon fiber and NR and reported that the highest degree of bonding was achieved when silica was replaced with carbon black. Murty and De [60] made similar observations. However, Murty et al. [134] observed that the presence of carbon black did not increase the level of adhesion between short jute fiber and NR. Setua and Dutta [62] used three different dry bonding systems including RHH to study the reinforcement of polychloroprene rubber by short silk fiber. Akhtar et al. [24] reported that although the properties of short silk fiber-reinforced TPEs showed improvement on the addition of the dry bonding system, the comparatively long curing time required for the full development of adhesion between the fiber and the matrix was a major disadvantage associated with the incorporation of the bonding system. Application of a dry bonding system in short glass, carbon, cellulose, polyamide and polyester filled SBR shows that the reinforcement of the rubber matrix using these short fibers depend on the fiber type and the adhesive system [10]. The presence of the dry bonding system in these composites is found to improve the fiber–matrix interaction. The adhesion between recent generation high-performance fibers like melamine fiber and rubber matrix is also improved by using dry bonding system [64,118,119]. Studies on the adhesion of some important high-performance fibers to rubber matrices are described below:

Adhesion of glass fiber to rubber matrices: There are several reports available on the adhesion of rubber to glass fibers and most of them deal with dipping of glass fibers in the adhesive solution or application of silane coupling agent. Marzocchi [135–137] suggested different methods for improving the adhesion between glass fibers and elastomeric matrices mainly by dipping in the bonding solution and by using silane coupling agents. He developed methods for the preparation of adherent–elastomer–glass fiber composites with improved mechanical properties by assembling glass fiber bundles (coated with an organosilane and a polyurethane rubber solution) with a polyester urethane rubber [137]. Uffner [138] improved the bonding of glass fibers to elastomers by coating the glass fibers or impregnating bundles of glass fibers with a blend of lattices of acrylonitrile–butadiene–styrene copolymer, vinylchloride–vinylidene chloride copolymer and butadiene–styrene–vinylpyridine copolymer. The use of silane coupling agents for achieving better rubber–fiber adhesion is well known [139]. Murty and coworkers used RHH dry bonding system to improve the adhesion between short glass fiber and SBR [61] and short glass fiber and NR [81].

Adhesion of carbon fiber to rubber matrices: Though there are several reports available on short carbon fiber-reinforced rubber composites, only a few reports deal with the incorporation of a dry bonding system for the improvement of fiber–rubber adhesion. O'Conner [58] reported that RHH system gave best results with short carbon fiber-reinforced NR composites. In radiation vulcanized short carbon fiber-reinforced SBR, Abdel-Aziz and coworkers [88] observed that there was a good adhesion between carbon fiber and SBR matrix and that the fracture mode of the tensile fracture surfaces depended mainly on the fiber–rubber adhesion. Surface treatment is the common method adapted to achieve better interfacial adhesion between carbon fiber and rubber or thermoplastic elastomeric matrix. Ahmad et al. [140] used surface treated carbon fiber to improve the adhesion between thermoplastic NR–HDPE blend and carbon fiber. It was observed that the mechanical properties of surface treated thermoplastic NR composite increased significantly with the fiber concentration up to 20% as compared to untreated fiber sample. Oxidation was the method adopted by Ibarra and coworkers to achieve improved adhesion between carbon fiber and TPEs [141]. Fiber oxidation gave rise to composites with improved mechanical properties and the improvement depended on the degree of oxidation [142]. Carbon fiber oxidation with nitric acid caused an increase in the functional surface groups, which in turn enhanced the capacity of the fiber to interact with the thermoplastic elastomeric SBS copolymer matrix [143].

Adhesion of aramid fiber to rubber matrices: Though aramid fibers have polar groups that facilitate good adhesion, their very high crystallinity and steric hindrance due to the presence of bulky groups make it difficult for them to achieve appreciable level of adhesion by means of the dry bonding system or resorcinol–formaldehyde–latex (RFL) dip. Therefore, a two-step process is adopted. First an epoxy resin layer that has good adhesion to aramids is provided, which is followed by the RFL dip [92]. However, Doherty et al. [144] suggested that fluorination was an alternative method for activating aramid fiber surfaces, rendering the fibers more receptive to a resorcinol–formaldehyde latex dip treatment and ultimately leading to rubber composites possessing good fiber–matrix adhesion. Surface modification of aramid fibers by plasma treatment for better fiber–rubber adhesion was also reported [21]. The bond strength between a poly(*m*-aramid) fabric and fluorosilicone rubber was improved through N_2 plasma treatment of the fabric in combination with a silane coupling agent [21]. The effectiveness of the treatment is dependent on the gas type, the plasma input power, and the time of exposure. Pretreatment of the fibers using liquor ammonia will facilitate the ability of aramid fibers to react with the treating agents [145]. Rebouillat [146] replaced 0.25 to 50% of the amine sites of the molecular chains on the surface of the aramid fibers by nitrobenzyl groups or nitrostilbene groups to improve the adhesion of the modified and dipped aramid fibers with hydrogenated nitrile rubbers (HNBRs). Kutty and Nando [147] studied the self-adhesion behavior of TPU with itself and its composite with short Kevlar fibers with respect to contact time, temperature, pressure, and fiber loading. Hepburn and Aziz [148] showed that good adhesion could be obtained between aramid fibers and NR by incorporating into the rubber either a diurethane vulcanizing system or a blocked diisocyanate. Aramid fibers can be bonded effectively to rubber matrices using the dry bonding system. O'Connor [58] reported that hexamethoxymethyl-melamine improved the adhesion of Kevlar fibers to rubber matrix. Recently ring-opening meta-thesis polymerization (ROMP) was used to investigate adhesion of multifilament fibers (polyester, nylon, Kevlar) to NR through the creation of ROMP polymer coatings on the fiber surfaces followed by encapsulation in pre-cured elastomer [149].

Adhesion of melamine fiber to rubber matrices: Rajeev et al. found that the dry bonding system consisting of resorcinol, hexamethylene tetramine and hydrated silica effectively reinforces EPDM [64], maleated EPDM [118], and nitrile rubber matrices [119]. They observed increase in tensile

TABLE 12.2

Mechanical Properties and Cure Rate Index of the Mixes—Role of Dry Bonding System on the Reinforcement of Ethylene–Propylene–Diene Monomer (EPDM) Rubber with Melamine Fiber[a]

Compositions	A	B	C	D
Resorcinol/hexamine/silica loading (phr)	0/0/0	5/3/15	0/0/0	5/3/15
Fiber loading (phr)	0	0	30	30
Tensile strength (MPa)	1.5	4.2	1.6	5.9
Elongation at break (%)	141	250	124	197
Stress at 100% elongation (MPa)	1.2	1.6	1.5	4.8
Hardness (Shore A)	47	53	61	66
Cure rate index (% min^{-1})	13.2	7.8	14.7	8.5

Source: Rajeev, R.S., Bhowmick, A.K., De, S.K., Kao, G.J.P., and Bandyopadhyay, S., *Polym. Compos.*, 23, 574, 2002. With permission.

[a] All compositions contain EPDM, 100 phr; zinc oxide, 5 phr; stearic acid, 1 phr; antioxidant, 1 phr; 2-mercaptobenzothiazole (accelerator), 1.5 phr; tetramethyl thiuram disulfide (accelerator), 1 phr; and sulfur, 1.5 phr.

strength, stress at 100% elongation and hardness for the EPDM rubber–melamine fiber composites containing the dry bonding system as shown in Table 12.2. Resorcinol, hexamine, and silica were taken in the concentration 5, 3, and 15 phr, respectively. From the cure rate index (given by $100/(t_{90} - t_{s2})$ where t_{90} is the optimum cure time at 150°C and t_{s2} is the rheometer scorch time at 150°C) of the compositions given in Table 12.2, it can be observed that the addition of dry bonding system to the gum compound in the absence of fiber (that is composition B) causes substantial reduction in the cure rate. This is because of the presence of resorcinol and silica in the dry bonding system, which may retard the vulcanization rate [133]. Addition of 30 phr melamine fiber to the gum compound (i.e., composition C) causes a marginal increase in the cure rate compared to that of the gum compound presumably due to the presence of amino groups on the fiber surface. When 30 phr fiber is added to the composition containing dry bonding system, the cure rate of the resultant composition D is again lower than that of the fiber-filled composition without dry bonding system (i.e., composition C). This is due to the effect of silica, resorcinol, and hexamine on the cure characteristics of EPDM rubber [133]. From the cure characteristics, it can also be assumed that melamine fiber itself has a role in the sulfur curing process of EPDM rubber.

The polarity of the matrix helps to improve the adhesion between melamine fiber and rubber because of the polar–polar interaction between the fiber and the matrix [118]. The presence of fiber in the absence of bonding system showed only marginal improvement in tensile strength (from 1.5 to 1.6 MPa) in the case of EPDM rubber–melamine fiber composites. However, maleated EPDM rubber–melamine fiber composites showed more than 50% improvement in tensile strength in the absence of the bonding system [118]. This is because of the presence of maleic anhydride groups, which imparts polarity to the rubber.

The stress–strain curves of maleated EPDM rubber–melamine fiber composites in the absence and in the presence of dry bonding system are shown in Figure 12.6. The curves show that incorporation of melamine fiber into the maleated EPDM rubber matrix (composite B) causes an increase in modulus compared to the compound A, which is the gum compound containing no fiber and bonding system. The role of the dry bonding system in improving the adhesion between the fiber and the matrix is evident on comparing the tensile strength and Young's modulus of the fiber–rubber composites (i.e., composite B containing no dry bonding system, but fiber and composite D

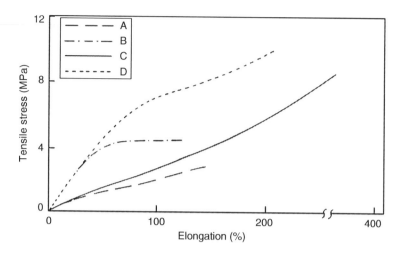

FIGURE 12.6 Stress–strain curves of maleated ethylene–propylene–diene monomer (EPDM) rubber–melamine fiber composites [118]. A, gum compound; B, composite containing 30 phr melamine fiber, but no dry bonding system; C, compound containing only dry bonding system but no fiber and D, composite containing both dry bonding system and 30 phr melamine fiber. (From Rajeev, R.S., Bhowmick, A.K., De, S.K., and Bandyopadhyay, S., *J. Appl. Polym. Sci.*, 89, 1211, 2003. With permission.)

containing both dry bonding system and fiber). Incorporation of the dry bonding system to the matrix containing 30 phr of melamine fiber (composite D) causes more than 150% improvement in tensile strength. Young's modulus also shows considerable improvement with the incorporation of the dry bonding system, so are hardness and elongation at break. The composite B, which contains no bonding system but 30 phr fiber, also shows an improvement in tensile strength and modulus compared to the gum compound A. Such an improvement in properties was not observed in the case of the corresponding EPDM rubber-based composites [64]. The polarity of maleated EPDM rubber is found to be the reason for the improved strength. The authors confirmed the improved adhesion between the fiber and the matrix in the presence of the dry bonding system by comparing the linear swelling ratio of the composites without and with dry bonding system. Even with 10 wt% melamine fiber, nitrile rubber, which is more polar than maleated EPDM rubber, showed more than 30% improvement in tensile strength in the absence of bonding system [119]. Presence of dry bonding system further improved the mechanical properties of such composites.

12.2.3 DETERMINATION OF ADHESION LEVEL AND OPTIMIZATION

Techniques like XPS, auger electron spectroscopy (AES), and energy dispersive x-ray (EDX) spectroscopy were used to study the adhesion between continuous fibers like steel cord and rubber [150,151]. New test methods were also proposed [152]. The main problem in studying the adhesion in the case of short fiber reinforcement is that it cannot be measured quantitatively. However, one can assess the adhesion level qualitatively. The rheometric curves and the shape of stress–strain curves offer a good indication of adhesion. The Monsanto rheometric cure curves of EPDM rubber–melamine fiber composites are compared with that of unfilled rubber in Figure 12.7, which further explains the role of dry bonding system in improving the fiber–matrix adhesion [64]. The gum compound A has the lowest maximum torque. Addition of fiber alone (composition B) increases the cure rate and maximum torque. However, addition of the dry bonding system provides the highest maximum torque (composition C), though the cure rate decreases compared to the other two compositions. This supports the hypothesis that the interaction between the fiber and the matrix in the presence of the dry bonding system takes place during curing when resorcinol–formaldehyde resin is formed, which migrates to the rubber–fiber interface, resulting in an efficient bond between

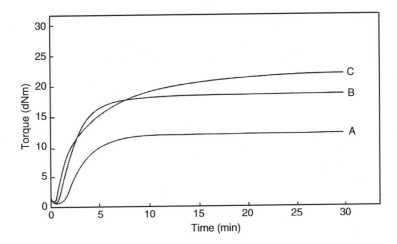

FIGURE 12.7 Monsanto rheometric curves of ethylene–propylene–diene monomer (EPDM) rubber–melamine fiber composites [64]. A, gum compound; B, compound containing 30 phr melamine fiber but no dry bonding system and C, compound containing both dry bonding system and 30 phr melamine fiber. (From Rajeev, R.S., Bhowmick, A.K., De, S.K., Kao, G.J.P., and Bandyopadhyay, S., *Polym. Compos.*, 23, 574, 2002. With permission.)

the two components. Silica, in fact, retards the vulcanization until sufficient amount of resin is formed by the reaction between resorcinol and formaldehyde donor [153].

The optimum level of the constituents of the RHH bonding system is determined by varying their concentration in the rubber–fiber compound. In EPDM rubber–melamine fiber composites [64], in order to determine the optimum concentration of the RHH bonding system, resorcinol, and hexamine were taken in three different proportions (i.e., 5/3, 10/6, and 15/9 parts by weight), keeping the fiber and silica loadings constant (15 and 30 phr, respectively). It was found that when the concentrations of resorcinol and hexamine were 10 and 6 phr, respectively (composite C), the composite provided maximum tensile strength and stress at 100% elongation for 30 phr melamine fiber loading as given in Table 12.3. After optimizing the concentration of resorcinol and hexamine, the concentration of silica was also optimized by determining the mechanical properties of compositions with different silica loadings (10, 15, and 20 phr) at a constant resorcinol–hexamine concentration (i.e., 10 and 6 parts by weight) and fiber concentration (30 phr). Analysis of the physical properties revealed that the composite containing no silica has lower tensile strength, and stress at 100% elongation as compared to those containing silica. The above observation confirms that presence of silica is essential to improve the adhesion between the rubber and the fiber in the presence of resorcinol and hexamine. However, an increase in silica concentration above 15 phr showed a reduction in tensile strength, elongation at break and modulus. The type of matrix influences the optimization of the constituents of the dry bonding system. Rajeev et al. found that the optimum concentration of bonding system is 10/6/5 resorcinol/hexamine/silica for nitrile rubber–melamine fiber composites [119] instead of the 10/6/15 concentration, which was found optimum for EPDM [64] and maleated EPDM [118] rubber-based compositions.

Dynamic mechanical analysis is also used to evaluate the fiber–rubber adhesion in short fiber–rubber composites. Figure 12.8 shows the effect of the dry bonding system on the dynamic mechanical properties of nitrile rubber–melamine fiber composites [119]. Since majority of short fibers may tend align parallel to the milling direction during the processing of the composites, the fiber orientation plays an important role in the mechanical, dynamic mechanical, and swelling characteristics of the composite. Here the test specimens were cut in the direction parallel to the milling direction (longitudinal direction), a direction to which more number of fibers are aligned. The role of the dry boding system in improving the fiber–matrix adhesion is evident on comparing

TABLE 12.3

Optimization of the Concentration of the Constituents of the Dry Bonding System in Ethylene–Propylene–Diene Monomer (EPDM) Rubber–Melamine Fiber Composites[a]

Compositions	A	B	C	D
Resorcinol/hexamine/silica loading (phr)	10/6/15	5/3/15	10/6/15	15/9/15
Fiber loading (phr)	0	30	30	30
Tensile strength (MPa)	5.5	5.9	6.9	6.6
Elongation at break (%)	362	197	154	218
Stress at 100% elongation (MPa)	1.7	4.8	6.0	4.6
Hardness (Shore A)	53	66	67	69

Source: Rajeev, R.S., Bhowmick, A.K., De, S.K., Kao, G.J.P., and Bandyopadhyay, S., *Polym. Compos.*, 23, 574, 2002. With permission.

[a] All compositions contain EPDM, 100 phr; zinc oxide, 5 phr; stearic acid, 1 phr; antioxidant, 1 phr; 2-mercaptobenzothiazole (accelerator), 1.5 phr; tetramethyl thiuram disulfide (accelerator), 1 phr; and sulfur, 1.5 phr.

the storage modulus and tan δ curves of the fiber–rubber composite containing no bonding system (designated as B) with that containing resorcinol, hexamine, and silica in the concentration 10, 6, and 5 phr, respectively (designated as D). Two other compounds, gum compound (designated as A)

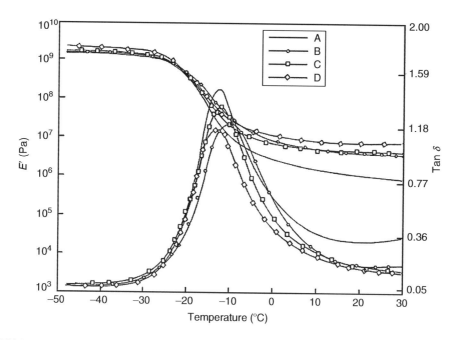

FIGURE 12.8 Plots of storage modulus and tan δ versus temperature of the nitrile rubber–melamine fiber composites [119]. A, gum compound; B, composite containing 10 phr melamine fiber but no dry bonding system; C, compound containing dry bonding system but no fiber and D, composite containing both dry bonding system and 10 phr melamine fiber. (From Rajeev, R.S., Bhowmick, A.K., De, S.K., and Bandyopadhyay, S., *J. Appl. Polym. Sci.*, 90, 544, 2003. With permission.)

and compound containing dry bonding system but no fiber (designated as C) are also given for comparison. Among the four compounds studied, the highest dynamic modulus and the lowest tan δ are shown by the fiber–rubber composite containing dry bonding system (composite D). It can also be noted that when only fiber is present in the matrix (composite B) there is a positive shift in tan δ peak temperature. Moreover, fiber causes greater reduction in tan δ peak height than the dry bonding system. This shows that there is a strong fiber–rubber bonding (even in the absence of the dry bonding system), which is, however, facilitated by the dry bonding system. This fiber–rubber bonding was also resulted in high Mooney viscosity and low Mooney scorch time for the compound B. Accordingly, the storage modulus (E') at glass/rubber transition temperature (T_g) follows the order $C < B < D$. In EPDM rubber–melamine fiber composites also [64], lower tan δ peak height was displayed by the composite containing the dry bonding system owing to better fiber–matrix adhesion, though no shift was observed in the T_g of the rubber.

Swelling characteristics were also used to determine the adhesion [59,61,64,118,119]. However, this method is erroneous since the restriction in swelling due to the presence of fibers and that due to adhesion cannot be separated out [133]. The adhesion level was also assessed by the study of fracture surfaces using SEM technique [59]. This method was extensively used by several authors to assess the adhesion between various short fibers and elastomeric matrices [24,61,63,64,118,119,134]. Sawyer and Grubb [154] explained the application of optical microscopy, SEM and transmission electron microcopy (TEM) in understanding the adhesion between fiber and polymer matrix in detail. Epstein and Shishoo [155] developed a method to measure the adhesion between fibers and fast-curing elastomer, in droplet form, using the microbond pull–put method. The SEM photomicrographs showing the fracture surface morphology of melamine fiber-reinforced nitrile rubber composite in the absence and in the presence of dry bonding system is given in Figure 12.9A and B [119]. Figure 12.9A is the SEM photomicrograph of the tensile fracture surface of the composite containing no dry bonding system. The smooth fracture surface and debonding of the fibers from the matrix show weak adhesion between the fibers and the matrix. However, when the dry bonding system is present, the morphology of the fracture surface is changed with fracture occurring in more than one plane between the fibers and the matrix as is evident in Figure 12.9B. Here fracture surface is rough compared to that of the composite containing no dry bonding system and the failure mode involves fiber breakage rather than fiber pullout. The strong attachment of the fibers on the matrix is visible here.

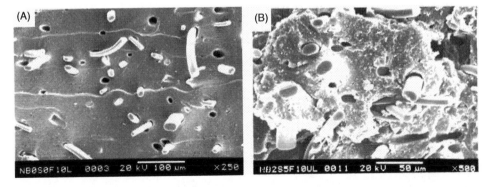

FIGURE 12.9 Scanning electron microscopy (SEM) photomicrographs of the tensile fracture surfaces of nitrile rubber–melamine fiber composites: A, composite with 10 phr fiber containing no dry bonding system; and B, composite containing both dry bonding system and 10 phr melamine fiber (in both the cases, the test specimens were cut in the direction parallel to the milling direction). (From Rajeev, R.S., Bhowmick, A.K., De, S.K., and Bandyopadhyay, S., *J. Appl. Polym. Sci.*, 90, 544, 2003. With permission.)

Stress-relaxation studies were also used as an indirect method to assess adhesion between short fiber and rubber matrix [14,143,156]. The suggestion that composites with good bonding between the fiber and the matrix have a faster stress relaxation rate than composites with poor bonding between the fiber and the matrix is supported by measurements on NR filled with untreated cellulose fibers and with allyl methacrylate-grafted cellulose fibers. Flink and Stenberg [156] explained this observation in terms of plots of relaxation modulus versus time. However, Bhagawan et al. [14] found that in short jute fiber-reinforced nitrile rubber composites, those containing bonding agent exhibited slower relaxation than those without bonding system. When bonding agent was absent, the rate of stress relaxation was dependent on strain level.

Formation of a strong interface between the fiber and the matrix is essential for high-performance composites. The stress-transfer capacity of carbon fiber–SBR interfaces was analyzed by Nardin et al. [157] in terms of interfacial shear strength. The measurement was done by means of a fragmentation test on single-fiber composites. For all the cases studied, the experimental values of the interfacial shear strength were largely higher than theoretically expected. Such a result was explained by the existence, near the fiber surface, of an interfacial layer in which the polymer chain mobility was greatly reduced. Schultz and Nardin [158] described the influence of interphase formation on the stress-transfer capacity of the fiber–matrix interface in a single-fiber composite.

Though SEM is still widely being used as the microscopic tool for the interphase analysis in fiber–rubber composites, AFM is one of the latest microscopic techniques employed for the quantification of fiber–rubber interphase. Tapping mode AFM studies showed the formation of fiber–matrix interphase in short melamine fiber-filled EPDM rubber composites [159]. The role of the dry bonding system in improving the fiber–matrix interaction is evident from the AFM images. Figure 12.10A through C gives the tapping mode AFM images of the section analyzes of the

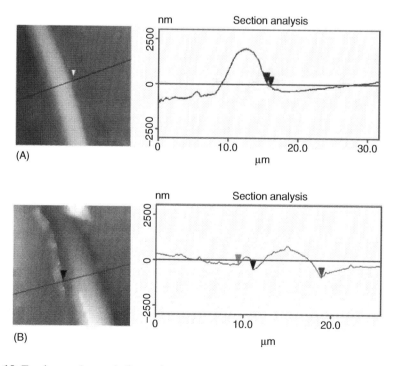

FIGURE 12.10 Tapping mode atomic force microscopy (AFM) images of the section analyzes of ethylene–propylene–diene monomer (EPDM) rubber–melamine fiber composites. A, composite containing no dry bonding system; B, composite containing resorcinol, hexamine, and silica in the concentrations 5, 3, and 15 phr, respectively.

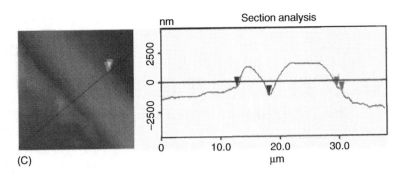

FIGURE 12.10 (continued) C, composite containing resorcinol, hexamine, and silica in the concentra-tions 10, 6, and 15 phr, respectively. (From Rajeev, R.S., De, S.K., Bhowmick, A.K., Kao, G.J.P., and Bandyopadhyay, S., *J. Mater. Sci.*, 36, 2621, 2001. With permission.)

composites with resorcinol, hexamine, and silica in the concentrations, 0/0/0, 5/3/15, and 10/6/15 phr, respectively. It can be observed that in the absence of the dry bonding system, the width of the interphase (characterized by the horizontal distance between the markers in the right hand side of the filament in the AFM image) is only 0.59 μm (Figure 12.10A). The fiber–matrix interphase is very smooth, showing little interaction between the fiber and the matrix. When resorcinol and hexamine are added in the concentrations, 5 and 3 phr, respectively, the width of the interphase (horizontal distance between the markers in the left hand side of the filament in the AFM image) is increased to 1.79 μm (Figure 12.10B). Here the morphology of the interphase is changed. When the concentration of resorcinol and hexamine is increased to 10 and 6 phr respectively, the width of the interphase (horizontal distance between the markers in the left hand side of the filament in the AFM image) is further increased to 5.39 μm (Figure 12.10C). This shows that as the concentration of the constituents of the dry bonding system is optimized, the interphase thickness is increased, leading to better fiber–matrix adhesion. The mechanical properties of the composites were found confirming the above observation.

Ageing of the composites improves the fiber–matrix adhesion especially if a heat-resistant fiber like melamine fiber is used. When short melamine fiber-filled EPDM rubber composites were aged in an air-circulating ageing oven, a better fiber–matrix adhesion was observed from SEM and AFM images, swelling studies and from mechanical property measurements [64]. The SEM fractographs of the unaged and aged composites are shown in Figure 12.11A and B, respectively. The fracto-graph of the unaged composite shows free ends of the broken fibers along with dewetting at the fiber–matrix interphase, though some fibers still remaining embedded in the matrix. These features are absent in the fractograph of the aged composite Figure 12.11B where the fiber ends and the fiber–matrix interphase are covered by the bonding system, causing higher extent of the wetting of the fiber surface by the bonding system and enhancement in the fiber–matrix adhesion. Figure 12.11A also shows that before ageing, the failure occurs primarily by the fiber pullout from the matrix, followed by the matrix failure. However, after ageing at 150°C for 48 h in air-circulating ageing oven, failure occurs through breakage of the fibers, most of which are strongly bonded to the matrix (Figure 12.11B). The authors also found that in the absence of the bonding system, there was no change in the fracture mode even after ageing [64]. This fortifies the observation that sharp improvement in adhesion between the fiber and the matrix during ageing takes place only through the bonding system.

12.2.4 PROCESSING ASPECTS OF SHORT FIBER–RUBBER COMPOSITES

Short fiber incorporated rubber compounds can be processed in a way similar to that of conventional rubber compounds. Proper dispersion of the fiber in the rubber matrix is the primary objective.

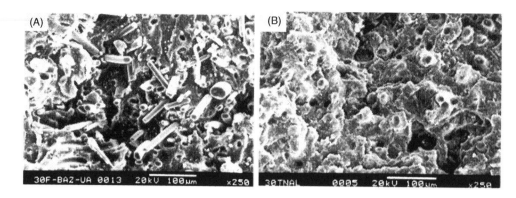

FIGURE 12.11 Scanning electron microscopy (SEM) photomicrographs of the tensile fracture surface of the ethylene–propylene–diene monomer (EPDM) rubber–melamine fiber composites. A, before ageing; and B, after ageing at 150°C for 48 h. Test specimen is cut in the direction parallel to the milling direction. (From Rajeev, R.S., Bhowmick, A.K., De, S.K., Kao, G.J.P., and Bandyopadhyay, S., *Polym. Compos.*, 23, 574, 2002. With permission.)

The various processing aspects of short fiber–rubber composites like fiber breakage, fiber aspect ratio, fiber dispersion, and fiber orientation were described in detail in several reports [56,57,93,133]. Jana and coworkers [86,160] used cement method of mixing of short carbon fibers with rubbers including polychloroprene rubber. Masticated CR was put in chloroform in order to form a paste. Short carbon fiber of about 6 mm length was dispersed into the paste by stirring. The fiber-filled paste was dried at room temperature to constant weight. The dried fiber–rubber mix was passed through a two-roll rubber mill for uniform fiber dispersion. They found that the cement method of fiber–rubber mixing caused less fiber breakage during processing as compared to conventional mill mixing method and produced rubber composites with higher hardness and modulus. Goettler and Cole [7] suggested that a finer state of final dispersion of the fibers could be achieved if the fibers were first wetted out by forming a highly concentrated mixture with a coating containing a resinous or polymeric binder. The resulting ease of dispersion and resistance to breakage for these "treated" fibers depend critically on the rheology of the formulation. This method is similar to that proposed by Jana and coworkers [86,160]. Goettler and Cole [7] put forward various standards for finding out the dispersion of wood cellulose fiber in rubber matrices. According to them, the dispersion of fiber in the rubber matrix can be affected by the factors like mixing protocol, mixing time, and rubber matrix viscosity. The Banbury fill factor and mixing conditions can affect the state of fiber dispersion as well as the degree of fiber damage, which in turn alter the mechanical properties of the vulcanized composite. The mixing action is improved when the Banbury chamber is less full as long as there is sufficient charge to keep pressure against the ram. Upside-down and two-pass mixing protocols provide improved dispersion over the standard mix. In conclusion, single-charge, single-pass mixing without premastication of the rubber and a smaller charge size are recommended for improved mixing. Fiber dispersion is also affected by mixing time. A longer mixing time and generation of greater energy through increasing mixing speed or the viscosity of the rubber stock are favorable for improved fiber dispersion. Increasing the stock viscosity increases the power input, which has two beneficial effects on the mixing process: total energy input builds faster with time, and higher stresses are generated to disperse the fibers more easily from their highly concentrated initial state into the final composite. Goettler and Cole [7] suggested that the above standards are applicable to other discontinuous fibers like rayon, nylon, or acrylic fibers.

A dimensionless dispersion number, based on the probable number of passes through the shear zone for an ingredient to be dispersed, was developed for scale-up of mixing of short fiber composites in an internal mixer by Shen et al. [161]. The dispersion and orientation of aramid fibers in a chloroprene rubber vulcanizate were investigated by Wada et al. [162] using x-ray photographs and an image analyzer. They studied the variation of tensile strength and elongation at break with fiber dispersion and variation of dynamic modulus and swelling with fiber orientation. While mixing short fiber–rubber compounds using conventional rubber mixing equipment, it should be noted that the presence of the fiber reinforcements does cause a higher rate of heat generation. To avoid scorching of the batch, it may therefore be necessary to reduce the batch size by about 10% and run the mixer at a lower speed.

Processing of short fiber–rubber composites is always associated with certain extent of fiber breakage and fiber orientation. Fiber length in a composite is critical; it should not be too long or the fibers will be entangled with each other causing problems of dispersion, on the other hand very small length of fibers does not offer sufficient stress-transfer area to achieve any significant reinforcement and the fibers thus become ineffective. The importance of fiber length and its influence on the properties of the composites were reported by many authors [17,58]. The breakage of jute, silk, and glass fibers in NR, carboxylated nitrile rubber, and nitrile rubber is reported by several authors [16,59–61]. The severity of fiber breakage depends mainly on the type of the fiber and its initial aspect ratio. For example, brittle fibers such as glass or carbon possess low bending strength and hence their breakage is severe compared to ductile fibers such as cellulose, which are more flexible and are resistant to bending. For each type of fiber, depending on its resistance to bending, no further breakage can occur below a certain aspect ratio. There exists a "critical fiber length" below which no effective stress transfer is possible. This concept can be used to predict the strength of the composite [133]. Though the fiber ends have very little influence on the reinforcing mechanism of long fiber-reinforced composites, in short fiber–rubber composites, they play a significant role in the determination of composite properties. Broutman and Aggarwal [163] analyzed the mechanism of stress transfer between the matrix and fibers of uniform radius and length and proposed a critical fiber length l_c as

$$\frac{l_c}{d} = \frac{\sigma_{fu}}{2\tau_y}$$ (12.1)

where
$\quad$ d is the fiber diameter
$\quad$ σ_{fu} is the ultimate film strength
$\quad$ τ_y is the matrix yield stress in shear

When different fibers having different diameters are compared, aspect ratio is the important factor, not the fiber length.

Campbell [164] provided a detailed description of orientation of short fibers in rubber matrices. Fiber orientation takes place mainly when the composite undergoes shear flow, the type of which is determined by the processing techniques such as milling, extrusion, or calendaring. The nature of the flow such as convergent, divergent, shear, or elongational also affects the fiber orientation as explained by Goettler and coworkers [165]. In mill processing, the number of passes, nip gap, mill roll temperature, friction ratio of the mill, and mill roll speed affect the fiber orientation. There is an optimum number of passes, which is different for different fiber–rubber compositions. The orientation of fiber causes anisotropy in mechanical properties, viscoelastic characteristics, heat buildup, and in damping [63]. Mechanical properties are greatly influenced by fiber orientation [64,93,118,119]. Goettler and Cole [7] described the recent methods in analyzing the fiber orientation in short fiber–rubber composites and pointed out the factors affecting the dispersion and breakage of short fibers during compounding. According to them, measurement of modulus, solvent swelling, ultimate tensile properties, tear strength, and morphological analysis are the methods for analyzing the fiber orientation. Green strength measurements were also done to evaluate the fiber orientation [64]. Presence of fiber would increase the green strength of the compounds along with anisotropy. Mixing of the fiber–rubber compound in the two-roll mill will always induce preferential orientation of the fibers in the milling direction though it is virtually impossible to obtain complete fiber orientation in a given direction. The extent of this orientation can be determined by comparison of physical properties of the composites with fibers in the direction parallel to the milling direction (longitudinal direction, L) and the same in the direction perpendicular to the milling direction (transverse direction, T) [166]. From the difference between the green strength in the L and T directions, expressed as ratios of improvement over the unreinforced control, the extent of orientation can be estimated. The stress–strain curves of the uncured EPDM rubber–melamine fiber compounds show a higher modulus in the longitudinal direction than in the transverse direction as shown in the Figure 12.12 [64]. It is evident that the incorporation of fibers causes considerable increase in the green strength of the compounds, which is a processing advantage and useful in the product building operation. Murty [133] described the relationship between extent of fiber orientation and composite modulus. The extent of fiber orientation is also determined using swelling studies and mechanical property determination of the cured composites. The modulus under small deformation, i.e., modulus at 5% elongation gives a clear understanding about the preferential fiber orientation. Tensile strength may not give actual fiber orientation as the fibers may align themselves under large deformations particularly in the case of composites filled with small fraction of fiber. Anisotropy in properties due to fiber orientation is important in product design. In V-belts, the transverse orientation of fibers allows sufficient flexibility in the axial direction which is a design requirement. In composites in which the fibers are randomly oriented in planes parallel to the surface, swelling takes place only in the thickness direction. Therefore, oil seals made of them tighten after swelling [56].

Dispersion of nanofibers like VGCNF and CNTs in rubber matrices is difficult because of the strong van der Waals force of attraction between the fibers. Currently there are four commonly used methods for incorporating CNTs into a polymer matrix: direct mixing, *in situ* polymerization, solution method, and melt processing. Liu et al. [167] dispersed CNTs in silicone rubber matrix by grinding method by grinding the mixture taken on the surface of a smooth glass board using the flat surface of a cylinder. Frogley et al. [168] dispersed CNTs in silicone rubber matrix by

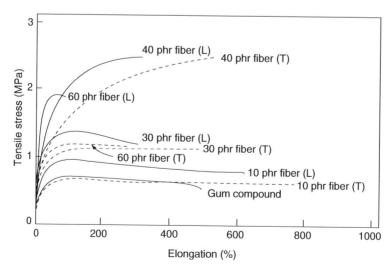

FIGURE 12.12 Green strength of the ethylene–propylene–diene monomer (EPDM) rubber–melamine fiber compounds measured in the longitudinal (L) and transverse (T) directions, showing anisotropy. (From Rajeev, R.S., Bhowmick, A.K., De, S.K., Kao, G.J.P., and Bandyopadhyay, S., *Polym. Compos.*, 23, 574, 2002. With permission.)

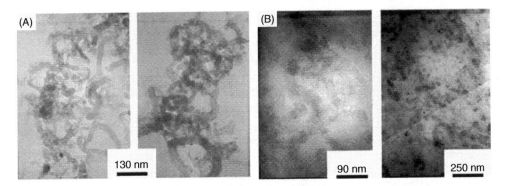

FIGURE 12.13 Transmission electron microscopy (TEM) images of carbon nanotube–hydrogenated nitrile rubber (CNT–HNBR) composites with 25 phr of CNTs prepared by two dispersion methods. A, direct mixing method (left two figures); and B, ultrasonic method (right two figures). (From Yue, D., Liu, Y., Shen, Z., and Zhang, L., *J. Mater. Sci.*, 41, 2541, 2006. With permission.)

first dissolving silicone rubber in toluene, and then adding ultrasonically dispersed CNT in toluene to the rubber solution. It is further sonicated to achieve effective dispersion of CNT in the matrix. Toluene was removed by evaporation. Yue et al. [169] used two different dispersion methods to prepare the nanocomposites of HNBR matrix and CNT. In one method, CNTs were directly mixed with HNBR in a two-roll mixing mill. In the second method, liquid HNBR was firstly dissolved in acetone, subsequently, the surface modified CNTs were added into the solution and then the ultrasonic dispersion was used to have a proper dispersion of CNT in the rubber matrix. The morphology of CNTs and their dispersion in HNBR matrix was studied by using TEM. On comparing the TEM images given in Figure 12.13A and B, it can be seen that CNTs disperse very well in HNBR when the ultrasonic predispersion technique is employed for the preparation of the composites. Kim et al. [170] fabricated aligned MWNT filled polychloroprene rubber composites. In this case, the homogeneity was achieved by mixing in an internal mixer and the CNTs were aligned by passing through a two-roll mixing mill by controlling the mixing conditions like nip gap and mixing time. Figure 12.14A and B shows the FE–SEM images of the fracture surface of the composites. The CNTs are found to disperse well in the matrix along the aligned directions.

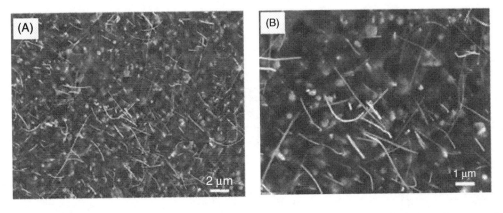

FIGURE 12.14 Field emission–scanning electron microscopy (FE–SEM) images of 30 phr carbon nanotube (CNT)–polychloroprene rubber composites in A, longitudinal; and B, transverse directions showing the dispersion of CNTs along the aligned directions. (From Kim, Y.A., Hayashi, T., Endo, M., Gotoh, Y., Wada, N., and Seiyama, J., *Scripta Materialia*, 54, 31, 2006. With permission.)

No aggregation of CNTs in the rubber matrix could be observed, indicating good dispersion of the nanotubes.

12.2.5 PROPERTIES OF SHORT FIBER–RUBBER COMPOSITES

Mooney viscosity still remains the most commonly applied tool in the rubber industry when characterizing polymer flow behavior, although it must be recognized that its low shear rate is not the most adequate one to predict rubber behavior at the shear rates occurring during typical production. Capillary rheometers are utilized to study the processing characteristics including die swell. Ibarra [171] studied the properties of uncured EPDM rubber–short polyester fiber using Mooney viscometer and Moving Die Rheometer. The variation of Mooney viscosity with volume fraction of fiber (V_f) in short melamine fiber-filled EPDM rubber composite follows the equation [64]:

$$\frac{M_f}{M_g} = 1.06 \left\{ 1 - 0.02 \left(\left(\frac{l_f}{d_f}\right) V_f - \left(\frac{l_f}{d_f}\right)^2 V_f^2 \right) \right\} \tag{12.2}$$

where

M_f and M_g stand for the Mooney viscosity of the fiber-filled compounds and gum compound (i.e., without fiber), respectively
l_f is the length of the fiber in the compound
d_f is the diameter of the fibers in the compound
l_f/d_f gives the fiber aspect ratio

In general, Mooney viscosity increases with increasing fiber loading as shown in Figure 12.15 [64].
Determination of mechanical properties like tensile strength, tear strength, modulus, and elongation at break are the most common methods adopted to determine the cured properties of short fiber–rubber composites. Murty and De [133] discussed the technical properties of short fiber–rubber composites whereas Abrate [8] reviewed the mechanism of short fiber reinforcement of rubber. Fiber concentration in the matrix plays an important role in the optimization of the required

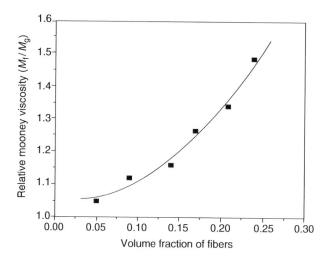

FIGURE 12.15 Plot of relative Mooney viscosity at 120°C versus volume fraction of the fiber for ethylene–propylene–diene monomer (EPDM) rubber–melamine fiber compounds. (From Rajeev, R.S., Bhowmick, A.K., De, S.K., Kao, G.J.P., and Bandyopadhyay, S., *Polym. Compos.*, 23, 574, 2002. With permission.)

properties. When fibers are added to an elastomeric compound, as with any polymer, the tensile strength first drops due to a dilution effect, even when the fibers are strongly bonded to the matrix. Depending upon the nature of the fiber, elastomer, bonding level, and state of dispersion of the fibers, recovery of the matrix strength occurs at concentrations higher than around 10–15 vol% loading. Park and Kim [172] found that the critical fiber loading with regard to tensile strength of the aramid fiber-reinforced rubber composites varied with the gum strength of the rubber. Arroyo and Bell [19] showed that at optimum concentration of PP and EPDM rubber, incorporation of more than 10% aramid fiber to the PP–EPDM rubber blend improved the impact strength of the composite. However, Watson and Frances [25] reported that only <10% (wt) fiber loading was sufficient in short aramid fiber-reinforced TPEs to achieve significant improvement in wear resistance and other properties like low elongation and extension resistance. It is generally found that the incorporation of short fibers at low (<5%) loadings can elevate the tear strength of a composite above that of its unreinforced rubber matrix [173]. At higher concentrations, the strain amplification in the matrix between the closely packed fibers can promote tearing parallel to the fiber direction, thus reducing the tear strength. In many cases, the fiber seems to have no effect on tear strength [81], and the tear strength is mainly elastomer dependent [174]. Rajeev et al. [64] added up to 60 wt% melamine fiber in EPDM rubber matrix and found that 40 wt% is the optimum fiber loading for optimum mechanical properties including tensile strength. They also found that the optimum fiber concentration for optimum tensile strength depends on factors like nature of the matrix (polar or nonpolar) and cross-linking system used (sulfur, ionic, and mixed cross-linking involving both sulfur and ionic cross-linking in the same coimpound). For maleated EPDM rubber-based composites cured by using sulfur cross-linking system, there was a reduction in tensile strength after 10 phr melamine fiber loading [118]. When the maleated EPDM rubber was cured by using ionic cross-linking, presence of fibers reduce the strength of the composite. A fiber loading of 10 phr was found the optimum for EPDM rubber–nitrile rubber composites also [119].

It is observed that fibers and constituents of dry bonding system sometimes hinder the formation of ionic cross-links in short fiber–rubber composites cured by using ionic or mixed cross-linking mechanism which involves both sulfur/accelerator and ionic cross-linking (involving zinc oxide and zinc stearate) in the same compound [118]. Dynamic mechanical analysis is a useful tool in observing such phenomena. Figure 12.16 gives the tan δ curves of the maleated EPDM rubber–melamine fiber compounds cured by using the mixed cross-linking system.

The tan δ measurement is made above 50°C. Here A is the gum compound in which both fiber and the dry bonding system are absent; B is the composite filled with 30 phr melamine fiber, which contains no dry bonding system; and C is the compound which contains the dry bonding system in the ratio of 10:6:15 phr resorcinol/hexa/silica, but no fiber. D is the fiber–rubber composite containing both the dry bonding system in the same ratio as that of C and 30 phr melamine fiber. It can be seen that the unfilled compound A displays a clear ionic transition around 130°C. Incorporation of the dry bonding system into the gum compound (i.e., compound C) reduces the tan δ peak height and the reduction is more prominent when fiber is present in the matrix (composites B and D). It was further observed that composites cured by using only ionic cross-linking mechanism were having lower mechanical properties in presence of the fiber and the dry bonding system compared to those cured by using sulfur/accelerator curing system or mixed cross-linking system. This phenomenon can also be attributed to the hindrance of formation of ionic cross-links by the fibers and the constituents of the dry bonding system. It has been reported that the increase in the tan δ peak height at ionic transition can be used to assess the degree of reinforcement due to ionic aggregates, while the lowering of the tan δ peak height is ascribed to the plasticization of the ionic aggregates [175].

The behavior of the composites under tension can be explained based on the stress-transfer theory of Cox et al. [176]. The analysis of mechanics of short fiber–rubber composites is difficult than that for continuous fiber–rubber composites. This is because of the fact that fiber end effects are important in short fiber reinforcement which are absent in the case of continuous fiber

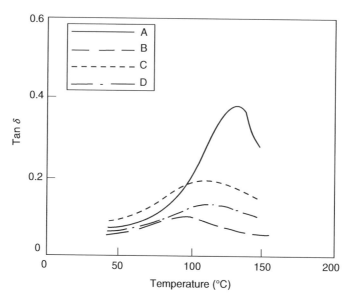

FIGURE 12.16 Plots of tan δ versus temperature of maleated ethylene–propylene–diene monomer (EPDM) rubber–melamine fiber composites cured by using mixed cross-linking system. (From Rajeev, R.S., Bhowmick, A.K., De, S.K., and Bandyopadhyay, S., *J. Appl. Polym. Sci.*, 89, 1211, 2003. With permission.)

reinforcement. There is nonuniformity in the stress transfer between the fiber and the matrix. A perfect fiber orientation is not possible with short fiber reinforcement unlike in the case of continuous fiber reinforcement. In short fiber composites, the efficiency of fiber reinforcement depends on the maximum tensile stress that can be transferred to the fiber by the shearing mechanism between the fibers and the matrix. Since the matrix has a lower modulus, the longitudinal strain in the matrix is higher than that in adjacent fibers. Assuming a perfect bond between the fiber and the matrix, the difference in longitudinal strains creates a shear stress distribution across the fiber–matrix interface. In this condition, maximum reinforcement would be achieved when the fibers are long enough so that maximum stress transfer occurs. Table 12.4 gives the calculated

TABLE 12.4

Calculated Minimum Aspect Ratio of Melamine Fibers in Ethylene–Propylene–Diene Monomer (EPDM) Rubber–Melamine Fiber Composites

Compositions	Fiber Loading (phr)	Volume Fraction of Fibers	Minimum Aspect Ratio Calculated Using Equation 12.3
$EB_2Si_{15}F_{10}$	10	0.05	115
$EB_2Si_{15}F_{20}$	20	0.09	95
$EB_2Si_{15}F_{30}$	30	0.14	80
$EB_2Si_{15}F_{40}$	40	0.17	74
$EB_2Si_{15}F_{50}$	50	0.21	67
$EB_2Si_{15}F_{60}$	60	0.24	63

Source: Rajeev, R.S., Bhowmick, A.K., De, S.K., Kao, G.J.P., and Bandyopadhyay, S., *Polym. Compos.*, 23, 574, 2002. With permission.

minimum aspect ratio in short melamine fiber-reinforced EPDM rubber composites, which is required to obtain complete stress transfer [64]. The calculation is based on the equation proposed by Rosen [177], given by:

$$l_f/d_f = \left\{ (1/2)(E_f/G_m) \left[\frac{\left(1 - V_f^{1/2}\right)}{V_f^{1/2}} \right] \right\}^{1/2}$$
(12.3)

where
l_f/d_f is the fiber aspect ratio (l_f is the length and d_f is the diameter of the fiber)
E_f is fiber modulus
G_m is shear modulus of the matrix
V_f is the fiber volume fraction

The values in Table 12.4 show that minimum aspect ratio decreases with increasing volume fraction of the fibers and the calculated values of aspect ratio for each volume fraction of fiber is greater than the observed aspect ratio of melamine fiber in the composite for all the volume fraction of fibers, i.e., 11. From the table, it can be seen that the composite with a fiber loading of 10 phr (i.e., composite $EB_2Si_{15}F_{10}$) gives the maximum difference between the minimum aspect ratio and the observed aspect ratio. As fiber loading increases, this difference decreases. Therefore, it can be assumed that up to the fiber loading of 40 phr, this reduction in difference between the calculated and observed fiber aspect ratios is one of the reasons for the increasing composite strength with increasing fiber loading. At higher fiber loadings (>40 phr), the matrix cannot effectively wet all the fibers and the strength decreases.

Short fiber reinforcement generally has a negative effect on flex fatigue, particularly at high fiber loading and high strain [57]. Heat build of these composites is also higher. The mechanical loss near the fiber–matrix interface accounts for the higher heat buildup [62,63]. Generally, an increase in fiber loading increases the heat buildup. Kwon et al. [178] found that the fatigue resistance of composites of NR with short aramid, nylon, and polyester fibers, useful in tires, was optimum with fiber length of 0.20 in. and fiber content of 6 phr. The deformation form of the interfacial regions in polychloroprene rubber composites with treated short polyester fibers was studied by Mashimo et al. [179] by examining the stress decay and the surface temperature distribution during fatigue. When the composites extended in the fiber direction, the stress decreased and the surface temperature rose due to failure of the interfacial regions between fibers and the matrix until 5000 cycles. The effect of short carbon fiber on the anisotropy in hysteresis loss and tension set of two TPEs, based on NR–HDPE blend and SIS block copolymer were studied by Roy et al. [180]. The composites based on NR–HDPE blend showed anisotropy in both hysteresis loss and tension set, whereas SIS showed lack of anisotropy in both cases. Viscoelastic properties of short fiber–rubber composites were also reported [49,51,74]. Generally, creep can be reduced greatly by the addition of short fibers to a rubber matrix. Bhattacharya et al. [36] found that Kevlar fiber-reinforced fluoroelastomer vulcanizates showed higher tearing energy than both carbon black-filled sample and the control compound. However, under dynamic fatigue conditions, the black-filled sample was found stronger.

The modulus of the composites can be theoretically calculated using the well-known Halpin–Tsai equation [181], given by:

$$E_L = E_m \left\{ \frac{1 + 2(l_f/d_f)\eta_L V_f}{1 - \eta_L V_f} \right\}$$
(12.4)

$$E_T = E_m \left\{ \frac{1 + 2\eta_T V_f}{1 - \eta_T V_f} \right\}$$
(12.5)

where

$$\eta_L = \frac{(E_f/E_m) - 1}{(E_f/E_m) + 2(l_f/d_f)} \qquad (12.6)$$

$$\text{and } \eta_T = \frac{(E_f/E_m) - 1}{(E_f/E_m) + 2} \qquad (12.7)$$

where

E_L and E_T are the Young's modulus of the composite in the longitudinal and transverse directions respectively

E_m, Young's modulus of the matrix

l_f, length of the fiber

d_f, diameter of the fiber

V_f, volume fraction of the fibers in the composite

E_f, Young's modulus of the fiber

The Young's modulus values calculated using the Halpin–Tsai equation are compared with the observed values in Figure 12.17A and B for the unaged and the aged EPDM rubber–melamine fiber composites [64]. The observed values in the longitudinal and transverse directions vary exponentially with increasing volume fraction of fibers and the effect is pronounced in the transverse direction. The observed composite modulus in the longitudinal direction is less than the calculated value at all fiber loadings, both before and after ageing. Observed modulus values of the aged composites lie closer to the calculated values. However, in the transverse direction, the trend is just the reverse. Here the observed modulus values are higher than the calculated values and the aged composites show a larger deviation from the calculated modulus values. The authors assume that the opposite trend in deviation between the calculated and the observed moduli in the case of longitudinally and transversely oriented fibers is due to some degree of randomness in fiber orientation in the composites. In the case of composites with longitudinally oriented fibers, a portion of the fibers remains transversely oriented, causing negative deviation in experimental moduli from the theoretical values. Similarly, in the case of composites with transversely oriented fibers, a portion of the fibers remains longitudinally oriented, causing the positive deviation in the observed moduli from the calculated values. It is believed that the physical form of the melamine fibers does not allow perfect orientation of the fibers during processing. The considerable breakage of the fibers during mill mixing also contributes to the deviation in the observed Young's modulus values from the theoretically calculated values.

12.2.6 MORPHOLOGICAL CHARACTERIZATION OF SHORT FIBER–RUBBER COMPOSITES

SEM is the most widely used imaging technique for the study of both short- and continuous-fiber composites. Murty and coworkers [61,81] observed the failure surfaces of short glass fiber-reinforced rubber composites using SEM. They examined tensile, tear, flexing, and abrasion failure modes of the composites and found that different modes generate typical fracture surfaces depending on the nature of the test. According to them, failure occurred primarily at the fiber–matrix interface, except in the case of abrasion where fiber breakage was prominent. SEM can also be used to understand the fiber orientation in the composites. Setua and Dutta [62] observed different fracture modes depending on the fiber orientation in the SEM photomicrograph of the tear-fractured surfaces of short silk fiber-reinforced polychloroprene rubber composites. In short fiber–rubber composites, the proper adhesion between the fiber and the matrix causes less pulling out of the fiber from the matrix and therefore, a higher extent of fiber breakage, which can be observed in the SEM photomicrographs of the tensile fracture surfaces of the composites [63].

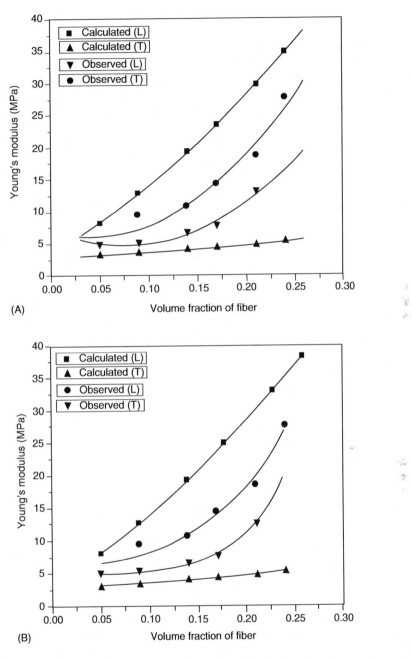

FIGURE 12.17 Young's modulus versus volume fraction of fibers for A, the unaged composites; and B, the composites aged at 150°C for 48 h of ethylene–propylene–diene monomer (EPDM) rubber–melamine fiber composites. "L" indicates test specimens cut in the direction parallel to the milling direction (longitudinal) and "T" indicates test specimens cut in the direction perpendicular (transverse) to the milling direction. (From Rajeev, R.S., Bhowmick, A.K., De, S.K., Kao, G.J.P., and Bandyopadhyay, S., *Polym. Compos.*, 23, 574, 2002. With permission.)

Microscopic techniques are extensively used to study the surface morphology of reinforcing fibers. The characterization of microstructure of polymer fibers provides an insight into structure–property relationship of the fiber. Microscopy techniques have been employed for the characterization of polymeric fibers to study features such as fiber shape, diameter, crystal size, voids, and molecular orientation [154]. In order to, understand the relation between the structure and properties of any polymeric material including fibers, their morphology should be clearly understood [182]. XRD, optical microscopy, and TEM along with a range of other analytical tools are commonly employed to determine the structure of polymeric fibers. The surface and bulk analysis of melamine fibers using different characterization techniques are already explained [112].

The latest microscopic technique to study the surface morphology of fibers and fiber-reinforced composites is AFM [183]. AFM is widely used in the field of polymer, because properties of polymeric materials are mainly governed by their composition, morphology, and interfacial structure. In a review, Tsukruk [184] illustrated the basic principles and results of AFM studies on polymer surfaces. Hautojarvi and Leijala [185] investigated the surface morphology of melt-spun PP filaments using the AFM and found that during stretching, there was a gradual transformation of surface morphology from a spherulitic morphology to fibrillar morphology. Rebouillat and coworkers [186] employed AFM to study the surface structure of Kevlar fibers. The surface structure of polyurethane fiber obtained by using electrospinning was analyzed by Demir et al. [187] by using optical microscopy, AFM, and SEM. AFM was used to study the surface morphologies of unaged and aged melamine fiber filaments by Rajeev et al. [112]. The development of AFM also enables one to measure the adhesion and adhesive forces, cross-link density, and surface modulus in nanometer level using techniques like phase imaging, force modulation, and generation of force curves [188,189]. Rajeev et al. studied the surface morphology and fiber–matrix interphase of short melamine fiber-reinforced EPDM [159], maleated EPDM [118], and nitrile [119] rubber composites using AFM.

12.2.7 APPLICATION OF SHORT FIBER–RUBBER COMPOSITES

Though short fiber-reinforced rubber composites find application in hose, belt, tires, and automotives [57,98,133,164] recent attention has been focused on the suitability of such composites in high-performance applications. One of the most important recent applications of short fiber–rubber composite is as thermal insulators where the material will protect the metallic casing by undergoing a process called ablation, which is described in a broad sense as the sacrificial removal of material to protect structures subjected to high rates of heat transfer [190]. Fiber-reinforced polymer composites are potential ablative materials because of their high specific heat, low thermal conductivity, and ability of the fiber to retain the char formed during ablation [191–194].

A solid propellant rocket motor consists of a solid propellant placed inside a metallic or composite casing which is protected from the hot combustion gases by means of insulators, liners, and inhibitors. The solid propellant on combustion is decomposed into low molecular weight gaseous products liberating very high temperature, of the order of 2500–3500 K. The chamber is to be protected from these hot combustion gases for its smooth functioning. This thermal protection is achieved by insulating the inner surface of the metallic case by suitable materials. Organic compounds are considered to be the most suitable material for rocket motor insulators because of the formation of surface char layer, which plays a predominant role in the absorption of heat because of its high heat capacity. The high heat capacity of the char reduces the heating rate to the surface of the material to be protected and coupled with a high surface emittance dissipate a large fraction of the incident heat [195]. Also, the insulator should be resilient enough to withstand the high thermal stresses generated during the burning of the propellant and it should be able to tolerate the thermal expansions and contractions at the time of launching of the rocket so that it will not debond from the metallic casing and propellant. All these requirements show that vulcanized elastomers are an ideal choice as the matrix material for solid rocket motor insulators.

If the char formed by the burning of the elastomer is not strong, it will easily erode due to mechanical shear forces, thermal stresses, and internal pressure generated by the hot volatile gases [98]. Therefore, rubber should be properly reinforced so that the char can withstand both mechanical forces and high temperature. Particulate and fibrous fillers can be used to reinforce the rubber. According to Fabrizi and La Motta [196], compared to particulate filled rubber compounds, the fiber-filled rubber composites produce a thicker char so that the thermal protection is more effective. In addition, the fibers will aid in retaining the char generated during combustion. Also, fibrous fillers counteract the erosion of char by withstanding both mechanical forces and high temperature. The principal prerequisites for a reinforcing fiber in ablative applications are (i) good thermal resistance; (ii) substantial mechanical properties at very low and very high temperatures; (iii) low specific gravity; (iv) high reinforcing efficiency; (v) no physiological hazards; and (vi) no smoke generation [98]. Thermal resistance implies high or no melting point. Good mechanical properties are not only needed at high temperatures (during combustion) but also at very low temperatures depending on where the rocket is going to be stored prior to firing. Low thermal conductivity is needed so that it should support the thermal insulation provided by the char. Lower specific gravity results in higher payloads and lower launch costs. The inhalation hazards such as that presented by asbestos during manufacturing and firing must be avoided. In military rockets not only the propellant, but also the insulator (both elastomer and reinforcing fiber) have to be smokeless.

Choice of fiber for rocket motor insulation composition depends on its mechanical and thermal properties as well as density. Originally, asbestos filled with phenolic resins and elastomers were used as insulators [98]. Later, asbestos loaded with nitrile rubber was selected as insulator because of its better ablative properties and easy processablity. The different processing aspects as well as thermal and ablative properties of insulator compound based on EPDM rubber filled with asbestos fiber, cork, and iron oxide were reported by Deuri et al. [197–200]. However, because of the health hazards posed by asbestos fibers, studies were initiated for the replacement of asbestos fibers in high-temperature applications [201] and asbestos was replaced by silica in the insulator compounds. Polyolefines fibers have low density and high specific strength. However, these fibers are not suitable at high temperatures due to their reduction in strength and thermal shrinkage. Polyamide fibers are also not recommended because of their comparatively high density and thermoplasticity. Poor adhesion to matrix and low oxidation resistance are the problems associated with carbon fibers. Glass fiber is also not suitable because of their high density, poor adhesion to matrix, and brittleness. Because of its good thermal stability, lack of a specific melting point and high strength, aramid fibers can be used as a reinforcing fiber in rocket motor insulator compounds. Fitch et al. [192] developed 7% loaded and 11% loaded Kevlar pulp filled with EPDM rubber as a replacement for asbestos and carbon fiber-filled EPDM rubber composites and found that Kevlar fiber-filled EPDM rubber composites can be used as a substrate to the carbon fiber-filled EPDM rubber for cost, processing, case bonding, and structural considerations. Herring [202] developed erosion-resistant elastomeric compositions based on EPDM rubber, aramid fiber, aramid pulp, and particulate filler. Polyamide fiber-reinforced EPDM rubber molding for rocket motor case insulator was patented by Okamoto et al. [203]. A low-density thermoplastic elastomeric ablative insulation for rocket motors based on maleated EPDM and fillers like ammonium polyphosphates, phosphazenes, and chopped fibers was developed by Guillot [204]. Parry [205] developed EPDM rubber-based thermal insulation materials filled with oxidized PAN fibers. Other fibers which have been tried in the rocket motor insulator compound recipe are PBI fiber and polybisbenzoxazole fiber. Rubber composites based on PBI fibers are also used as high-temperature insulators [105]. The latest rocket motor insulation composition based on short fibers and rubber is melamine fiber-reinforced rubber composites [114]. Because of its substantially high elongation at break (18%), lower modulus, easy availability and suitable physical form compared to other high-performance fibers, it is assumed that the melamine fiber can be easily incorporated into rubber matrices with adequate dispersion of the fiber in the matrix for the preparation of solid rocket motor insulator compositions. The rubber composites based on melamine fiber are found to have very good thermal erosion resistance when measured by using a plasma arc jet. Even for a nonpolar rubber like EPDM rubber, the thermal erosion resistance was found to be of the order of 0.20 mm/s. Nitrile rubber–melamine fiber composites also display very good ablative

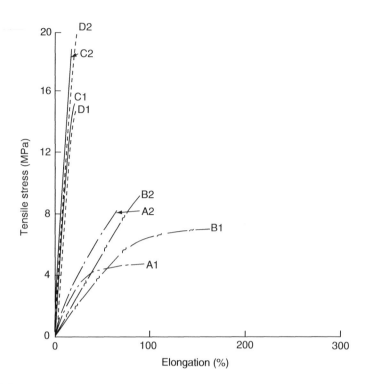

FIGURE 12.18 Stress–strain curves of rubber–fiber composites developed for solid rocket motor insulator; A, ethylene–propylene–diene monomer (EPDM) rubber–carbon fiber composites B, EPDM rubber–melamine fiber composites; C, EPDM rubber–aramid fiber composites and D, EPDM rubber–aramid pulp composites. 1 and 2 stands for unaged and aged composites respectively. Carbon fiber- and melamine fiber-reinforced composites contain resorcinol, hexamine, and silica in the concentrations 10, 6 and 15, respectively and aramid fiber- and aramid pulp-based composites contain resorcinol, hexamine, and silica in the concentrations 5, 3 and 15, respectively. (From Rajeev, R.S., Bhowmick, A.K., De, S.K., and John, B., Internal communication, Rubber Technology Center, Indian Institute of Technology, Kharagpur, India, 2002.)

properties with thermal erosion rate in the order or 0.17 mm/s. Stress–strain curves of the unaged and aged rocket motor insulator compositions based on EPDM rubber and melamine, carbon, aramid fiber, and aramid pulp are shown in Figure 12.18 [206]. Though the highest strength and modulus are displayed by aramid-based composites, the flexibility of the EPDM rubber–melamine fiber and EPDM rubber–carbon fiber composites are better than that of the aramid-based composites. EPDM rubber–melamine fiber composites retain their flexibility and strength even after ageing at 150°C for 48 h.

12.3 CONTINUOUS FIBER-REINFORCED ELASTOMERS

Continuous fiber-reinforced elastomers or cord–rubber composites are being used in aircrafts, automotives, conveyer belts, and hoses. Tires and hoses in automotives and landing gears in aircrafts are some of the examples of applications of such composites. Like short fiber–rubber composites, continuous fiber–rubber composites are also characterized by the extremely low stiffness of the rubber matrix compared to that of the reinforcing cords. Pneumatic tires are subjected to extreme loading conditions of high cyclic frequency and large deflections. One of the major reasons of failure of these composites is the debonding at the interphase between the matrix and fiber. Fatigue and pressure causes stresses at the interphase and heat generation which may ultimately lead to the failure of the composite. The geometry of the cord fibers also contribute to failures. A detailed description of cord–rubber composites with special emphasis on automobile tires has been given by Chawla [207].

12.3.1 Fibrous Reinforcement for Cord–Rubber Composites

Nylon, polyester, and aramid are the three major fibers used for reinforcing cord–rubber composites like tires. Though nylon has good fatigue resistance and high strength, which make its suitable for bias truck and aircraft tires, its lower T_g, lower modulus and poor dimensional stability make it unattractive for reinforcing radial tires [207]. Flat spotting, which is temporary or permanent flattening of the tire due to a long stop of the vehicle, is a problem associated with nylon fabrics. Because of its high modulus, high T_g and good dimensional stability, polyester fabrics are used for radial tire reinforcement. Polyester fabrics are also known for their high speed performance and moisture resistance. Aramid fibers are used mainly in belt reinforcement because of their poor fatigue strength though these fibers are known for their high strength, modulus and dimensional stability.

Steel cords are used as belt reinforcement for radial tires. The bead wire of all types of tires and the carcass of off-the-road and light truck tires are also made of steel cords. Two-ply steel cord–rubber composite laminate are being used in air spring construction. In general, high carbon steel with a carbon content of around 0.7% is used for the preparation of steel cords for tire. Steel cords are manufactured by passing the steel wire through a die for numerous times. By repeatedly drawing the wire through the die, they reach the required cord diameter. During the process, heat treatment of the wires to prevent the steel from hardening and brass coating with copper and zinc in the ratio 60:40 or 70:30 are also carried out. Brass coating will improve the adhesion between the steel cord and rubber and prevents the cord from rusting. Steel cords are characterized by their high tensile strength and high modulus of elasticity compared to other fibers, as shown in Figure 12.19 [208].

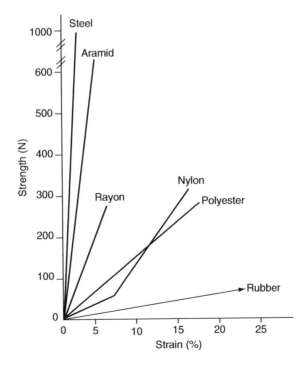

FIGURE 12.19 Stress–strain curves of some important tire-cords compared with that of rubber. (From Toyo Tire Talk, Vol. 11. Toyo Corporation, Japan. With permission.)

12.3.2 ADHESION OF FABRICS TO RUBBER

The most common adhesive system used for bonding continuous fibers and fabrics to rubber is resorcinol–formaldehyde latex (RFL) system. In general, RFL system is a water-based material. Different lattices including nitrile and SBR are used as the latex for the adhesive system. 2-Vinylpyridine–butadiene–styrene is the common latex used in the adhesive recipe. RFL system is widely being used in tires, diaphragms, power transmission belts, hoses, and conveyor belts because of its dynamic properties, adhesion, heat resistance, and the capacity to bond a wide range of fabrics and rubbers.

Adhesion of some of the high-performance fabrics to rubber is a challenging problem. Fabrics like aramid and polyester, due to the lack of active sites available on the surface, may have very little bonding to rubber in the absence of a bonding system. For the same reason, bonding of adhesion systems like RFL to these fabrics is also difficult. If the fiber is polar in nature, as in the case of nylon, hydrogen bonding is possible between the NH groups in the fiber and RF components of the RFL system. However, for polyester, the lack of functional groups in the fiber that promote hydrogen bonding makes it difficult to have proper bonding between the fabric and the adhesive system. In the case of aramid, the hydrogen bonding groups (NH groups) are sterically hindered due to the presence of bulky aromatic groups and are not available for hydrogen bonding. In order to have a chemical reaction between the fabric and the bonding system, the fabric surface should be very clean and there should be contact between the fabric and the adhesive. However, the presence of spin finish in many fabrics hinders proper chemical bonding between the fabric and the adhesive system. If fabric is relatively open, the adhesive system can penetrate into the fabric, facilitating mechanical bonding. The bonding system can also be thermodynamically bonded to the fabric by adhesive causing the swelling of the top layer of the fabric and mechanically locking into the fiber layer. For an effective bonding between the fabric and adhesive system, the adhesive should wet the fabric surface, which is governed by the energy difference between the fiber and the adhesive. The surface energy of the fiber must be greater than or equal to the surface energy of the adhesive in order to have a proper wetting.

Since steel does not have the required level of adhesion to rubber, adhesion promoters are used. A resin former (resorcinol–hexamethylene tetramine–silica) and cobalt-based organometallic promoter, either individually or jointly are used in the rubber compound in order to improve the adhesion between steel cord and rubber. The brass coating given in steel not only protects it from rusting, but also improves the adhesion between the rubber and the steel fabric. The copper in the brass cross-links with the sulfur in the rubber during curing process to form copper sulfide. However, the brass-coated steel cords are prone to sever electrochemical corrosion in silane and humid environment, which weakens the reinforcing effect of the cord and lead to the loss of cord-rubber adhesive strength. Chandra et al. [150] studied the influence of cations and anions of the adhesion promoters on the interface between brass-coated steel cord and NR skim compounds by employing AES and XPS. Cobalt stearate, cobalt boroacylate, and nickel boroacylate have been used as organometallic adhesion promoters. In this study, first tire cord adhesion test specimens were prepared for pull-out test. To study the nature of the interface by XPS, the failed metal surface after the pull-out test was dipped in xylene for 12 h at room temperature. Then, it was subjected to light mechanical abrasion by rubbing gently with a tissue paper to remove the swollen rubber vulcanizate from the steel cord. All the samples were etched for 2, 7, and 15 min at 7 kV with 50 μA target current using argon gas at 25°C. They observed that the atomic concentration of sulfur and carbon decreased with etching time whereas the oxygen concentration increased marginally in the initial state and later decreased after 15 min etching. The copper and zinc concentrations found increased significantly at higher etching times. Among the various organometallic adhesion promoters, the atomic concentration of zinc and sulfur were the highest for the system containing cobalt boroacylate.

12.3.3 ANALYSIS OF CORD–RUBBER COMPOSITES

Both static and dynamic tests are employed to evaluate the adhesion strength of cord–rubber composites. The major static tests used in tire industry are H-adhesion, 90/180° peel test, tire cord adhesion test (TCAT) and co-axial shear pull-out test (CSPT). Although these methods are helpful in evaluating cord–rubber adhesion strength, they are not sensitive to subtle changes. Moreover, the nature of the tests is not simulating the service condition of the tire. As a result, the data obtained in the laboratory are not an indication of the real performance properties of the product. Therefore, the dynamic tests are of importance and are more meaningful as far as prediction of durability of tire is concerned.

The first dynamic adhesion testing technique was introduced in 1940 [209]. With the rapid growth of radial tires, there has been much interest in dynamic adhesion testing of steel cord and a number of dynamic tests have been developed for measuring rubber to cord adhesion. Most of the dynamic test methods can be classified into four major categories on the basis of the mode of deformation, e.g., tension, compression, shear and flexing. Dynamic deformation in tension is the most common one and thus exploited in most of the existing methods. Popular dynamic test techniques are Buist method [210], Iyenger method [211], Voracheck method [212], and Khromove method [213].

Since some of the dynamic test techniques are highly time-consuming, costly, and need extra capital investment, Chandra et al. [152] developed a new dynamic adhesion test technique for evaluating steel cord–rubber adhesion. It involves the tension mode of deformation and can be used to calculate the energy required to rupture the unit area of the interface apart from calculating the adhesion strength. According to the authors, this test can be performed in Monsanto fatigue-to-failure tester (FTFT) type equipment.

The TCAT specimen selected as the test piece for the development of the dynamic adhesion testing technique for steel wire is shown in Figure 12.20. Both sides of the specimen were held by using a specially designed holder and then mounted into the FIFT machine in the unstrained condition by placing the holders between the dynamic and static beams. The samples were then subjected to repeated strain cycles and the number of cycles elapsed were automatically recorded. The samples were subjected to a specified strain for a quarter of each cycle followed by relaxation in another quarter. In the remaining half cycle, the samples were allowed to recover to the original length. The maximum extension imposed was varied over a wide range by changing the drive cams. Cord–rubber adhesion was evaluated by subjecting the TCAT specimen to cyclic extension and

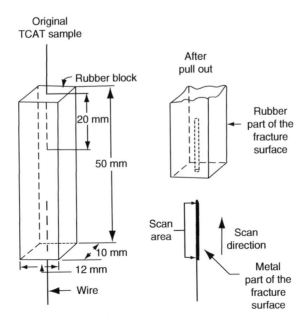

FIGURE 12.20 Tire cord adhesion test specimens and scan area of fracture surfaces. (From Chandra, A.K., Mukhopadhyay, R., and Bhowmick, A.K., *Polym. Test.*, 15, 13, 1996. With permission.)

conducting a pull-out test. The number of cycles required to pull out the cord during cycling was also used as a measure of adhesion under dynamic condition.

Jamshidi et al. [214] studied the cord–rubber interface of nylon 6, nylon 66, and PET cords and NR–SBR blends at elevated temperature using H-pull out test. In all cord–rubber systems, they observed decrease in adhesion strength with increase in temperature. Pidaparti [215] used a 3D beam finite element model to investigate the mechanical behavior of two-ply cord–rubber composite laminates under combined tension and torsional loading. Cord–rubber composites exhibit coupled axial bending-twisting behavior under different loading conditions owing to the nonlinear geometric and material properties. A number of finite element analysis methods are available for analyzing the mechanical behavior of cord–rubber composites. According to Turner and Ford [216], the interply shear strains, which occur when a two-ply composite is subjected to simple tension, are responsible for the fatigue-induced delamination in such composites. The results of various finite element analyzes on cord–rubber composites indicate that coupled bending shear and twisting shear deformations are important for a realistic analysis of the behavior of two-ply cord–rubber composite structures [215]. The complexity of the cord construction also needs to be taken into account for such analyzes as the cord itself is not a single filament, but many filaments which are twisted together.

Fiber–matrix debonding, matrix cracking, fiber fracture, and delamination are some of the failure modes observed in cord–rubber composites. Breidenbach and Lake [217] studied the mechanics of propagation of interplay cracks in an angle-plied cord–rubber composite laminate. The specimens were partitioned into three deformation regions: a central region where deformations were relatively uniform and approximately obeyed a pantographic model, and two regions along the free edges where deformations varied in a complex manner. In these edge regions, shear strains up to 1000% can occur due to an overall extension of roughly 5%. These high shear strains created stress concentrations at the edge and eventually lead to the initiation of penny-shaped cracks at the cord ends. It was found that the interply cracks propagated until delamination aroused and led to laminate's failure. Fatigue of cord–rubber composites has been studied by Pidaparti and May [218] and Song [219].

12.4 SUMMARY

Some of the important aspects of fiber–rubber composites are explained with special emphasis on development and characterization. A description on the constituents of fiber–rubber composites helps in understanding the type of fibers and rubber matrices used for their preparation. Almost all types of natural and synthetic rubbers are used for the development of short fiber–rubber composites whereas the types of rubbers used for long fiber–rubber composites are limited. Natural fibers like cotton cellulose, synthetic fibers like nylon, polyester and high-performance fibers like aramid, carbon, and melamine fibers are some of the reinforcing fibers used for the preparation of short fiber–rubber composites. An understanding of the surface and bulk properties of the reinforcing fiber is important in developing fiber–rubber composites. Various fiber characterization methods are described in detail. Proper adhesion between the fiber and the rubber matrix is one of the most important criteria for achieving desired properties in fiber–rubber composites. The different types of interactions between the fibers (both short and long) and rubber matrices and the methods to improve the fiber–rubber adhesion are explained. The fiber breakage, fiber aspect ratio, fiber dispersion, and fiber orientation are some of the processing aspects of short fiber–rubber composites which affect the performance of the composite. Apart from applications in conveyer belts, power transmission belts, V-belts, hoses, and friction products, short fiber–rubber composites are also used in high-performance applications like solid rocket motor insulators. Unlike short fibers, the fiber ends do not play an important role in determining the ultimate properties of long fiber-reinforced rubber composites. Though continuous fiber-reinforced rubber display better mechanical properties compared to that of short fiber–rubber composites, such materials are generally confined to product in which the property benefits outweigh the cost penalty. Some of the major rubber products which utilize continuous fiber reinforcement are tires, conveyer belts and automotive hoses. Highly parallel fiber orientation is one of the major reasons for the improved mechanical properties of continuous fiber-reinforced rubber composites, which is absent in the case of short fiber reinforcement where the fiber orientation in many cases is random. Nylon, polyester, and aramid are the three major fibers used for reinforcing cord–rubber composites like tires. The mechanism of reinforcement in such composites is different from that of short fiber–rubber composites mainly because of the fiber end effects, nonuniform stress transfer and random orientation in short fiber reinforcement.

ACKNOWLEDGMENT

The author wishes to acknowledge all the researchers whose works have been referred in this chapter. He would like to thank Dr. Yutaka Sato of Japan Aerospace Exploration Agency (JAXA), Tokyo, for permitting him to prepare the manuscript using the facilities at JAXA along with the assigned research activities. The author would also like to express his gratitude to Dr. A.M. Shanmugharaj of Kyung Hee University, Republic of Korea, and Mrs. Supriya Radhakrishnan of Rubber Research Institute of India, Kottayam, for arranging some of the reference materials for the preparation of the manuscript.

ABBREVIATIONS

AFM	Atomic force microscopy
CNT	Carbon nanotube
DTG	Derivative thermogravimetry
ENR	Epoxidized natural rubber
EPDM	Ethylene–propylene–diene monomer
EVA	Ethylene vinyl acetate
FE	Field emission
HDPE	High-density polyethylene
HNBR	Hydrogenated nitrile rubber
LOI	Limiting oxygen index
NR	Natural rubber
PAN	Polyacrylonitrile
PBI	Polybenzimidazole
PBO	Poly(p-phenylene 2,6-benzobisoxazole)
PET	Poly(ethylene terephthalate)
phr	Parts per hundred rubber
PP	Polypropylene
PVC	Poly(vinyl chloride)
RFL	Resorcinol–formaldehyde latex
RHH	Resorcinol–hexamethylene tetramine–hydrated silica
SBR	Styrene–butadiene rubber
SEM	Scanning electron microscopy
TEM	Transmission electron microscopy
TG	Thermogravimetry
T_g	Glass-to-rubber transition temperature
TGA	Thermogravimetric analysis
TPE	Thermoplastic elastomer
TPU	Thermoplastic polyurethane
VGCNF	Vapor-grown carbon nanofiber
XNBR	Carboxylated nitrile rubber
XPS	X-ray photoelectron spectroscopy
XRD	X-ray diffraction

REFERENCES

1. Woods, D.W. and Ward, I.M., *J. Mater. Sci.*, **29**, 2572, 1994.
2. Sohn, M.S., Hu, X.Z., and Kim, J.K., *Polym. Polym. Compos.*, **9**, 157, 2001.
3. Hiley, M.J., *Plast. Rubber Compos.*, **28**, 210, 1999.
4. Neogi, C., Basu, S.P., and Bhowmick, A.K., *J. Mater. Sci.*, **25**, 3524, 1990.
5. Waddell, W.H. and Evans, L.R., *Rubber Chem. Technol.*, **69**, 377, 1996.
6. De, S.K. and White, J.R. (Eds.), *Short Fiber–Polymer Composites*, Woodhead Publishing, Cambridge, 1996.
7. Goettler, L.A. and Cole, W.F., *Handbook of Elastomers*, 2nd ed., Marcel Dekker, New York, 2001, p. 241.
8. Abrate, S., *Rubber Chem. Technol.*, **59**, 384, 1986.
9. Collier, A.T., U.S. 1000781, August 15, 1911.
10. Rueda, L.I., Anton, C.C., and Rodriguez, M.C.T., *Polym. Compos.*, **9**, 198, 1988.
11. Ibarra, L., *J. Appl. Polym. Sci.*, **54**, 1721, 1994.
12. Pecorini, T.J. and Hertzberg, R.W., *Polym. Compos.*, **15**, 174, 1994.
13. Nunes, R.C.R. and Affonso, J.E.S., *Kautsch. Gum. Kunstst.*, **52**, 787, 1999.
14. Bhagawan, S.S., Tripathy, D.K., and De, S.K., *J. Appl. Polym. Sci.*, **33**, 1623, 1987.
15. Setua, D.K. and De, S.K., *J. Mater. Sci.*, **20**, 2653, 1985.
16. Setua, D.K. and De, S.K., *Rubber Chem. Technol.*, **56**, 808, 1983.
17. Boustany, K. and Arnold, R.L., *J. Elastomer Plast.*, **82**, 160, 1976.
18. Zhang, B.L., Chen, M., and Liu, H.L., *J. Rubber Res.*, **4**, 82, 2001.
19. Arroyo, M. and Bell, M., *J. Appl. Polym. Sci.*, **83**, 2474, 2002.
20. Sreeja, T.D. and Kutty, S.K.N., *Polym. Plast. Technol. Eng.*, **41**, 77, 2002.
21. Zebarjad, S.M., Bagheri, R., and Lazzeri, A., *Plast. Rubber Compos.*, **30**, 370, 2001.
22. Vajrasthira, C., Amornsakchai, T., and Bualek-Limcharoen, S., *J. Appl. Polym. Sci.*, **87**, 1059, 2003.
23. Campbell, J.M. and Goettler, L.A., *Proc. of PRI National Conference on Short-Fiber Reinforced Thermoplastics*, Brunel University, Uxbridge, Middlesex, United Kingdom. Paper 14, 1985.
24. Akhtar, S., De, P.P., and De, S.K., *J. Appl. Polym. Sci.*, **32**, 5123, 1986.
25. Watson, K.R. and Frances, A., *Amer. Chem. Soc. Rubber Div., Meeting*, Cleveland, OH, Oct. 6–9, 1987.
26. Roy, D., Bhowmick, A.K., and De, S.K., *Polym. Eng. Sci.*, **32**, 971, 1992.
27. Roy, D. and Gupta, B.R., *J. Appl. Polym. Sci.*, **49**, 1475, 1993.
28. Roy, D., Bhatttacharya, A.K., and Gupta, B.R., *J. Elastomer. Plast.*, **25**, 46, 1993.
29. Warrick, E.L., Pierce, O.R., Polmanteer, K.E., and Saam, J.C., *Rubber Chem. Technol.*, **52**, 437, 1979.
30. Sieron, J.K., *Rubber World*, **148**, 50, 1963.
31. Sau, K.P., Khastgir, D., and Chaki, T.K., *Angew. Makromol. Chem.*, **258**, 11, 1998.
32. Novikova, L.A., Kolesnikova, M.N., and Tolstukhina, F.S., *Kauch. Rezina.*, **6**, 19, 1978.
33. Grinblat, M.P., Lundstrem, A.M., Levit, R.M., and Veselinova, N.M., *Int. Polym. Sci. Technol.*, **1**, T136, 1974.
34. Watson, K.R. and Francis, A., *Compos. Polym.*, **1**, 187, 1988.
35. Zhang, Z., Liang, H., Hou, X., and Yu, Y., *J. Adhesion Sci. Technol.*, **15**, 809, 2001.
36. Bhattacharya, S.K., Bhowmick, A.K., and Singh, R.K., *J. Mater. Sci.*, **30**, 243, 1995.
37. Fatterman, M.O., *J. Elastomer. Plast.*, **9**, 226, 1977.
38. Das, N.C., Chaki, T.K., Khastgir, D., and Chakraborty, A., *J. Appl. Polym. Sci.*, **80**, 1601, 2001.
39. Isham, A.B., Glass Fiber-Reinforced Elastomers for Automotive Applications—A Comparison of RIM Urethanes and Alternate Material Systems, *Soc. Autom. Engi. Meet.*, Preprint No. 760333, Feb. 1976.
40. Leis, D.G., Reinforced RIM Urethane Elastomer—A New Material of the Automotive Industry, *34th SPI Reinf. Plast./Compos. Conf.*, p. 11A, 1979.
41. Marzocchi, A., U.S. 72–221680, January 28, 1972.
42. Suhara, F., Kutty, S.K.N., and Nando, G.B., *Plast. Rubber Compos.*, **24**, 37, 1995.
43. Pan, R. and Watt, D.F., *Polym. Compos.*, **17**, 780, 1996.
44. Bicerano, J. and Brewbaker, J.L., *J. Chem. Soc. Fara. Trans.*, **91**, 2507, 1995.
45. Kane, R.P., *J. Elastomer. Plast.*, **9**, 226, 1977.
46. Kutty, S.K.N. and Nando, G.B., *Plast Rubber Compos. Proc. Appl.*, **19**, 105, 1993.
47. López-Manchado, M.A. and Arroyo, M., *Polymer*, **42**, 6557, 2001.
48. Amornsakchai, T., Sinpatanapan, B., Bualek-Limcharoen, S., and Meesiri, W., *Polymer*, **40**, 2993, 1999.

49. Ibarra, L. and Panos, D., *J. Appl. Polym. Sci.*, **67**, 1819, 1998.
50. Kim, J.K., *J. Appl. Polym. Sci.*, **61**, 431, 1996.
51. Guo, W. and Ashida, M., *J. Appl. Polym. Sci.*, **50**, 1435, 1993.
52. Guo, W. and Ashida, M., *J. Appl. Polym. Sci.*, **49**, 1081, 1993.
53. Kaynak, C., Arikan A., and Tincer, T., *Ann. Tech. Conf., Soc. Plast. Eng.*, Vol. 2, 2002.
54. Humpidge, R.T., Matthews, D., Morrell, S.H., and Pyne, J.R., *Rubber Chem. Technol.*, **46**, 148, 1973.
55. Ashida, M., Composites of polychloroprene rubber with short fibres of poly(ethylene terephthalate) and nylon, in *Short Fiber–Polymer Composites*, De, S.K. and White, J.R. (Eds.), Woodhead Publishing, Cambridge, 1996, Chapter 5.
56. Nando, G.B. and Gupta, B.R., Short fibre-thermoplastic elastomer composites, in *Short Fiber–Polymer Composites*, De, S.K. and White, J.R. (Eds.), Woodhead Publishing, Cambridge, 1996, Chapter 4.
57. Goettler, L.A. and Shen, K.S., *Rubber Chem. Technol.*, **56**, 619, 1983.
58. O'Connor, J.E., *Rubber Chem. Technol.*, **50**, 945, 1977.
59. Chakraborty, S.K., Setua, D.K., and De, S.K., *Rubber Chem. Technol.*, **55**, 1286, 1982.
60. Murthy, V.M. and De, S.K., *J. Appl. Polym. Sci.*, **27**, 4611, 1982.
61. Murthy, V.M. and De, S.K., *J. Appl. Polym. Sci.*, **29**, 1355, 1984.
62. Setua, D.K. and Dutta, B., *J. Appl. Polym. Sci.*, **29**, 3097, 1984.
63. Setua, D.K. and De, S.K., *J. Mater. Sci.*, **19**, 983, 1984.
64. Rajeev, R.S., Bhowmick, A.K., De, S.K., Kao, G.J.P., and Bandyopadhyay, S., *Polym. Compos.*, **23**, 574, 2002.
65. Nunes, R.C.R. and Visconte, L.L.Y., in *Natural Polymers and Agrofibers Based Composites*, Frollini, E., Leao, A.L., and Capparelli, M.L.H. (Eds.), Embrapa Instrumentacao Agropecuaria, Sao Carlos, Brazil, 2000, p. 135.
66. Ismail, H., Shuhelmy, S., and Edyham, M.R., *Eur. Polym. J.*, **38**, 39, 2002.
67. Ismail, H., Rosnah, N., and Rozman, H.D., *Eur. Polym. J.*, **33**, 1231, 1997.
68. Ismail, H. and Rozman, H.D., *Int. J. Polym. Mater.*, **41**, 325, 1998.
69. Zhang, B., Chen, M., and Liu, H.L., *J. Rubber Res.*, **4**, 82, 2001.
70. Mothe, C.G. and de Araujo, C.R., *Thermochim. Acta.*, **357–358**, 321, 2000.
71. Geethamma, V.G., Mathew, K.T., Lakshminarayanan, R., and Thomas, S., *Polymer*, **39**, 1483, 1998.
72. Bhattacharyya, T.B., Biswas, A.K., Chatterjee, J., and Pramanick, D., *Plast. Rubber Process. Appl.*, **6**, 119, 1986.
73. Ramayya, A.P., Chakraborty, S.K., and De, S.K., *J. Appl. Polym. Sci.*, **29**, 1911, 1984.
74. Ashida, M., Noguchi, T., and Mashimo, S., *J. Appl. Polym. Sci.*, **29**, 661, 1984.
75. Goettler, L.A. and Swiderski, Z., in *Composite Applications, The Role of Matrix, Fiber and Interface*, Vigo, T.L. and Kenzing, B.T. (Eds.), VCH Publishers, New York, 1992, p. 333.
76. Czarnecki, L. and White, J.L., *J. Appl. Polym. Sci.*, **25**, 1217, 1980.
77. Gregg, R.A., *Rubber Chem. Technol.*, **45**, 49, 1972.
78. Marzocchi, A., S. African Patent No. ZA 6805086, June 3, 1969.
79. Derringer, G.C., *J. Elastoplast.*, **3**, 230, 1971.
80. Murty, V.M., *Int. J. Polym. Mater.*, **10**, 149, 1983.
81. Murty, V.M., Bhowmick, A.K., and De, S.K., *J. Mater. Sci.*, **17**, 709, 1982.
82. Lee, M.C.H., *J. Polym. Eng.*, **8**, 257, 1988.
83. Riggs, D.M., Shuford, R.J., and Lewis, R.W., in *Handbook of Composite Materials*, Lubin, G. (Ed.), Van Nostrand Reinhold, 1982, p. 196.
84. Jana, P.B., Mallick, A.K., and De, S.K., Electrically conductive rubber and plastic composites with carbon particles or conductive fibres, in *Short Fiber–Polymer Composites*, De, S.K. and White, J.R. (Eds.), Woodhead Publishing, Cambridge, 1996, Chapter 7.
85. Yagi, K., Kanno, T., and Inada, T., Eur. Patent No. EP 296613, December 28, 1988.
86. Jana, P.B., Chaudhury, S., Pal, A.K., and De, S.K., *Rubber Chem. Technol.*, **65**, 7 1993.
87. Jana, P.B., Mallick, A.K., and De, S.K., *J. Mater. Sci.*, **28**, 2097, 1993.
88. Abdel–Aziz, M.M., Youssef, H.A., Miligy, A.A., Yoshii, F., and Makuuchi, K., *Polym. Polym. Compos.*, **4**, 259, 1996.
89. Kwolek, S.L., U.S. 3600350, 1971, 3671542, 1972.
90. Blades, H., U.S. 3869429, 1975, 3869430, 1975.
91. Kwolek, S.L., Morgan, P.W., Schaefgen, J.R., and Gulrich, L.W., *Macromolecules*, **10**, 1390, 1977.

92. Gabara, V., High performance fibres 1: aramid fibres, in *Synthetic Fiber Materials*, Brody, H. (Ed.), Longman, Harlow, United Kingdom, 1994, Chapter 9.
93. Moghe, S.R., *Rubber Chem. Technol.*, **49**, 1160, 1976.
94. Foldi, A.P., *Rubber World*, **196**, 22, 1987.
95. Saikrasun, S., Amornsakchai, T., Sirisinha, C., Meesiri, W., and Bualek-Limcharoen, S., *Polymer*, **40**, 6437, 1999.
96. López-Manchado, M.A. and Arroyo, M., *J. Appl. Polym. Sci.*, **23**, 666, 2004.
97. Hoiness, D.E. and Frances, A., *Int. SAMPE Symp. Exhib. Adv. Mater. Technol.*, **32**, 1173, 1987.
98. Foldi, A.P., Design and applications of short fibre reinforced rubbers, in *Short Fiber–Polymer Composites*, De, S.K. and White, J.R. (Eds.), Woodhead Publishing, Cambridge, 1996, Chapter 9.
99. van der Pol, J.F., *Kautsch. Gummi, Kunstst.*, **48**, 799, 1995.
100. van der Pol, J.F., *Rubber World*, **210**, 32, 1994.
101. Rijpkema, B., *Kautsch. Gummi, Kunstst.*, **47**, 748, 1994.
102. Schuler, T.F., *Kautsch. Gummi, Kunstst.*, **45**, 548, 1992.
103. Reinhart, T.J., in *Engineered Materials Handbook, Vol. 1. Composites*, ASM International, Ohio, 1987, p. 31.
104. Guillot, D.G., U.S. 5352312, October 4, 1994.
105. Junior, K.E., Byrd, J.D., and Hightower. Jr. J.O., U.S. 4600732 July 15, 1986.
106. Chang, S.C., U.S. 5024860, June 18, 1991.
107. Hopen, T.J., *Microscope*, **48**, 107, 2000.
108. Technical Literature, BASOFIL fiber, BASF Corporation, Enka, NC, 1999.
109. Nihongi, Y. and Yasuhira, N., Japanese. Patent No. JP 49109625, October 18, 1974.
110. Nihongi, Y. and Yasuhira, N., Japanese Patent No. JP 49109628, October 18, 1974.
111. Nihongi, Y. and Yasuhira, N., U.S. 4088620, May 9, 1978.
112. Rajeev, R.S., Bhowmick, A.K., De, S.K., Gong, B., and Bandyopadhyay, S., *J. Adhes. Sci. Technol.*, **16**, 1957, 2002.
113. Chawla, K.K., in *Fibrous Materials*, Cambridge University Press, Cambridge, 1998, Chapter 2.
114. Rajeev, R.S., De, S.K., Bhowmick, A.K., and John, B., *Polym. Deg. Stabil.*, **79**, 449, 2003.
115. Devallencourt, C., Saiter, J.M., Fafet, A., and Ubrich, E., *Thermochim. Acta*, **259**, 143, 1995.
116. Rong, M. and Li, X., *J. Appl. Polym. Sci.*, **68**, 293, 1998.
117. Hirata T. and Inoue A., *Polymeric Materials Encyclopedia*, Vol. 6, Salamone, J.C. (Ed.), CRC Press, New York, 1996, p. 4051.
118. Rajeev, R.S., Bhowmick, A.K., De, S.K., and Bandyopadhyay, S., *J. Appl. Polym. Sci.*, **89**, 1211, 2003.
119. Rajeev, R.S., Bhowmick, A.K., De, S.K., and Bandyopadhyay, S., *J. Appl. Polym. Sci.*, **90**, 544, 2003.
120. Iijima, S., *Nature*, **354**, 56, 1991.
121. Fan, S., Chapline, M.G., Franklin, N.R., Tombler, T.W., Cassell, A.M., and Dai, H., *Science*, **283**, 512, 1999.
122. Wong, S.S., Joselevich, E., Woolley, A.T., Cheung, C.L., and Lieber, C.M., *Nature*, **394**, 52, 1998.
123. Rueckes, T., Kim, K., Joselevich, E., Tseng, G.Y, Cheung, C.-L., and Lieber, C.M., *Science*, **289**, 94, 2000.
124. Thostenson, E.T., Ren, Z., and Chou, T.W., *Compos. Sci. Technol.*, **61**, 1899, 2001.
125. Bhowmick, A.K. and Gent, A.N., *Rubber Chem. Technol.*, **57**, 216, 1984.
126. Bhowmick, A.K., De, P.P., and Bhattacharyya, A.K., *Polym. Eng. Sci.*, **27**, 1195, 1987.
127. Bhowmick, A.K. and Inoue, T., *J. Adhesion.*, **59**, 265, 1996.
128. Kubo, Y., in *Elastomer Technology Handbook*, Cheremisinoff, N.P. (Ed.), CRC Press, Boca Raton, FL, 1993, p. 857.
129. Hisaki, H. and Mori, O., *J. Sci. Rubber Ind. Jpn.*, **64**, 33, 1991.
130. Kubo, Y., *Rubber Chem. Technol.*, **64**, 8, 1991.
131. Sarkar, P.P., Ghosh, S.K., Gupta, B.R., and Bhowmick, A.K., *Int. J. Adhes.*, **9**, 26, 1989.
132. Dunnom, D.D., Hi-Sil Bulletin, PPG Ind. Inc. No. 35, 1967.
133. Murty, V.M. and De, S.K., *Polym. Eng. Rev.*, **4**, 313, 1984.
134. Murty, V.M., De, S.K., Bhagawan, S.S., Sivaramakrishnan, R., and Athithan, S.K., *J. Appl. Polym. Sci.*, **28**, 3485, 1983.
135. Marzocchi, A., U.S. 3508950, April 28, 1970.
136. Marzocchi, A., U.S. 3509012, April 28, 1970.
137. Marzocchi, A., U.S. 3796627, March 12, 1974.

138. Uffner, W.E., U.S. 3755009, August 28, 1973.
139. Oyama, M., Mori, O., and Sugi, N., Eur. Patent No. EP 194678, September 17, 1986.
140. Ahmad, S.H., Abdullah, I., Mohamed, N.H., and Satapa, M.N., *J. Phy. Sci.*, **10**, 137, 1999.
141. Ibarra, L., Macias, A., and Palma, E., *Kautsch. Gumm. Kunstst.*, **48**, 180, 1995.
142. Ibarra, L. and Ponos, D., *Polym. Int.*, **43**, 251, 1997.
143. Ibarra, L., Macias, A., and Palma, E., *J. Appl. Polym. Sci.*, **61**, 2447, 1996.
144. Doherty, M.A., Rijpkema, B., and Weening, W., *Rubber World*, **212**, 21, 1995.
145. Nechwatal, A., Reussmann, Th., and Hausprun, Ch., *Gummi. Fasern. Kunstst.*, **54**, 527, 2000.
146. Rebouillat, S., U.S. 6045907, April 4, 2000.
147. Kutty, S.K.N. and Nando, G.B., *J. Adhes. Sci. Technol.*, **7**, 105, 1993.
148. Hepburn, C. and Aziz, Y.B., *Int. J. Adhes.*, **5**, 153, 1985.
149. Caster, K.C. and Walls, R.D., *Adv. Synth. Catal.*, **344**, 764, 2002.
150. Chandra, A.K., Mukhopadhyay, R., Konar, J., Ghosh, T.B., and Bhowmick, A.K., *J. Mater. Sci.*, **31**, 2667, 1996.
151. Chandra, A.K., Biswas, A., Mukhopadhyay, R., and Bhowmick, A.K., *J. Adhes. Sci. Technol.*, **10**, 431, 1996.
152. Chandra, A.K., Mukhopadhyay, R., and Bhowmick, A.K., *Polym. Test.*, **15**, 13, 1996.
153. Dutta, B., in *Rubber Products Manufacturing Technology*, Bhowmick, A.K., Hall, M.M., and Benarey, H.A. (Eds.), Marcel Dekker, New York, 1994, p. 478.
154. Sawyer, L.C. and Grubb, D.T., *Polymer Microscopy*, Chapman & Hall, London, 1987, p. 219.
155. Epstein, M. and Shishoo, R.L., *J. Appl. Polym. Sci.*, **50**, 863, 1993.
156. Flink, P. and Stenberg, B., *Br. Polym. J.*, **22**, 193, 1990.
157. Nardin, M., El Maliki, A., and Schultz, J., *J. Adhes.*, **40**, 93, 1993.
158. Schultz, J. and Nardin, M., *J. Adhes.*, **45**, 59, 1994.
159. Rajeev, R.S., De, S.K., Bhowmick, A.K., Kao, G.J.P., and Bandyopadhyay, S., *J. Mater. Sci.*, **36**, 2621, 2001.
160. Jana, P.B. and De, S.K., *Plast. Rubber. Compos. Proc. Appl.*, **17**, 43, 1992.
161. Shen, K.S. and Rains, R.K., *Rubber Chem. Technol.*, **52**, 764, 1979.
162. Wada, N., Fukunaga, K., and Uchiyama, Y., *Kautsch. Gummi, Kunstst.*, **12**, 1142, 1991.
163. Broutman, L.J. and Aggarwal, B.D., in *Analysis and Performance of Fiber Composites*, SPE/Wiley, New York, 1980.
164. Campbell, J.M., *Prog. Rubber Technol.*, **41**, 43, 1978.
165. Goettler, L.A., Lambright, A.J., Leib, R.I., and DiMauro, P.J., *Rubber Chem. Technol.*, **54**, 273, 1981.
166. Foldi, A.P., *Rubber Chem. Technol.*, **49**, 379, 1976.
167. Liu, C.H., Huang, H., Wu, Y., and Fan, S.S., *Appl. Phy. Lett.*, **84**, 2004, 4248.
168. Frogley, M.D., Ravich, D., and Wagner, H.D., *Compos. Sci. Technol.*, **63**, 1647, 2003.
169. Yue, D., Liu, Y., Shen, Z., and Zhang, L., *J. Mater. Sci.*, **41**, 2541, 2006.
170. Kim, Y.A., Hayashi, T., Endo, M., Gotoh, Y., Wada, N., and Seiyama, J., *Scripta Materialia*, **54**, 31, 2006.
171. Ibarra, L., *Kautsch. Gum. Kunstst.*, **47**, 578, 1994.
172. Park, C.Y. and Kim, B.K., *Pollimo*, **14**, 197, 1990.
173. Beatty, J.R. and Hamed, P., *Elastomerics*, **110**, 27, 1978.
174. Sheeler, J.W., *J. Elastomer. Plast.*, **9**, 267, 1977.
175. Kurian, T., Khastgir, D., De, P.P., Thripathy, D.K., De, S.K., and Peiffer, D.G., *Polymer*, **37**, 4865, 1996.
176. Cox, H.L., *Brit.J. Appl. Phys.*, **3**, 72, 1952.
177. Rosen, B.W. and Dow, N.F., in *Fracture*, Vol. 7, Leibowitz (Ed.), Academic Press, New York, 1972, p. 612.
178. Kwon, Y.D., Beringer, C.W., Feldstein, M.A., and Prevorsek, D.O., *Rubber World*, **202**, 29, 1990.
179. Mashimo, S., Nakajima, N., Noguchi, T., Yamaguchi, Y., and Ashida, M., *Rubber World*, **200**, 28, 1989.
180. Roy, D., De, S.K., and Gupta, B.R., *J. Mater. Sci.*, **29**, 4113, 1994.
181. Halpin, J.C. and Tsai, S.C., Effects of environmental factors on composite materials, *AFML*-TR-67, 1969.
182. Young, R.J., *Introduction to Polymers*, Chapman & Hall, London, 1981.
183. Binnig, G., Quate, C.F., and Gerber, C.H., *Phys. Rev. Lett.*, **56**, 930, 1986.
184. Tsukruk, V.V., *Rubber Chem. Tech.*, **70**, 430, 1997.

185. Hautojarvi, J. and Leijala, A., *J. Appl. Polym. Sci.*, **74**, 1242, 1999.
186. Rebouillat, S., Peng, J.C.M., and Donnet, J.B., *Polymer*, **40**, 7341, 1999.
187. Demir, M.M., Yilgor, I., Yilgor, E., and Erman, B., *Polymer*, **43**, 3303, 2002.
188. Cappella, B., Sturm, H., and Schulz, E., *J. Adhes. Sci. Technol.*, **16**, 921, 2002.
189. Mareanukroh, M., Hamed, G.R., and Eby, R.K., *Rubber Chem. Technol.*, **69**, 801, 1996.
190. Vojvodich, N.S., in *Ablative Plastics*, D'Alelio, G.F. and Parker, J.A. (Eds.), Marcel Dekker, New York, 1971, p. 41.
191. Nelson, D.S. and Prince, A.S., in AIAA-94-3184, 30th AIAA/ASME/SAE/ASEE Joint Propulsion Conf. Indianapolis, June 27–29, 1994.
192. Fitch, V. and Eddy, N., in AIAA-97-2992, 33rd AIAA/ASME/SAE/ASEE Joint Propulsion Conf. Seattle, June 6–9, 1997.
193. Graham, M., Levi, L., and Clarke, B., U.S. 5821284, October 13, 1998.
194. Weisshaus, H., Engleberg, I., and Rafae, H.I., *J. Adv. Mater.*, **28**, 16, 1997.
195. Schmidt, D.L. in *Ablative Plastics*, D'Alelio, G.F. and Parker, J.A. (Eds.), Marcel Dekker, New York, 1971, p.4.
196. Fabrizi, M. and La Motta, G., in AIAA 96-2903, 32nd AIAA/ASME/SAE/ASEE Joint Propulsion Conf., Lake Buena Vista, FL, July 1–3, 1996.
197. Deuri, A.S. and Bhowmick, A.K., *Polym. Degrad. Stabil.*, **16**, 221, 1986.
198. Deuri, A.S., De, P.P., Bhowmick, A.K., and De, S.K., *Polym. Degrad. Stabil.*, **20**, 135, 1988.
199. Deuri, A.S., Bhowmick, A.K., Ghosh, R., John, B., Sriram, T., and De, S.K., *Polym. Degrad. Stabil.*, **21**, 21, 1988.
200. Deuri, A.S., Bhowmick, A.K., John, B., and Ram, T.S., *J. Mater. Sci. Lett.*, **6**, 1117, 1987.
201. Tam, W.F.S. and Bell, M., in AIAA-93–2211, 29th AIAA/ASME/SAE/ASEE Joint Propulsion Conf. and Exhibit, Monterey, CA, June 28–30, 1993.
202. Herring, L.G., U.S. 4878431, November 7, 1989.
203. Okamoto, H., Nakamura, K., and Yokoo, Y., Jap. Patent No. JP 03021596, January 30, 1991.
204. Guillot, D.G., U.S. 5498649, March 12, 1996.
205. Parry, M.J., U.K. Patent No. GB 2295396, May 29, 1996.
206. Rajeev, R.S., Bhowmick, A.K., De, S.K., and John, B., Internal communication, Rubber Technology Center, Indian Institute of Technology, Kharagpur, India, 2002.
207. Chawla, S.K., Rubber composites, in *Synthetic Fiber Materials*, Brody, H. (Ed.), Longman, Harlow, United Kingdom, 1994, chap. 8.
208. Toyo Tire Talk, Vol. 11, www.toyojapan.com/tires/pdf/TTT_11.pdf.
209. American Society for the Testing and Materials, Standards, ASTM D430–40.
210. Buist, J.M. and Naunton, W.J.S., *Trans. ZRZ*, **25**, 378, 1950.
211. Iyengar, Y., *Rubber World*, **148**(6), 39, 1963.
212. Vorachek, J.J., *108th Meeting, ACS-Rubber Division*, Paper 60, Oct., 1975.
213. Khromov, M.K., *Soviet Rubber Technol.*, **21**, 25, 1962.
214. Jamshidi, M., Afshar, F., Mohammadi, N., and Pourmahdian, S., *Appl. Surf. Sci.*, **249**, 208, 2005.
215. Pidaparti, R.M.V., *Composites Part B*, **28**, 433, 1997.
216. Turner, J.L. and Ford, J.L., *Rubber Chem. Technol.*, **55**, 1079, 1982.
217. Breidenbach, R.F. and Lake, G.J., *Rubber Chem. Technol.*, **52**, 96, 1979.
218. Pidaparti, R.M.V. and May, A.W., *Compos. Struct.*, **54**, 459, 2001.
219. Song, J., PhD Thesis, College of Engineering, Pennsylvania State University, 2004.

13 A Chemical Modification Approach for Improving the Oil Resistance of Ethylene–Propylene Copolymers

Herman Dikland and Martin van Duin

CONTENTS

13.1 INTRODUCTION

Polymerization of ethylene with propylene yields random amorphous or slightly crystalline copolymers (EPM), which have a glass transition temperature (T_g) of approximately $-55°C$, and, therefore, are rubbery at room temperature [1–3]. Ethylene–propylene monomer (EPM) must be compounded and cross-linked to achieve a good balance of engineering properties. Traditionally, rubber products are cross-linked with activated sulfur systems, and, therefore, require unsaturated hydrocarbon structures. Since EPM lacks such unsaturated hydrocarbon structures, terpolymers of ethylene, propylene, and nonconjugated dienes (EPDM) have been developed. Because of their fully saturated main-chain, EPDM rubbers have a much better resistance against oxygen, ozone, heat, irradiation, and chemicals when compared to main-chain unsaturated polydiene rubbers, such as natural rubber (NR), polybutadiene (BR), styrene–butadiene copolymer (SBR), polychloroprene (CR), and acrylonitrile–butadiene copolymer (NBR). This makes EPDM the material of choice for many rubber outdoor applications and for rubber applications that are required to withstand high temperatures. Examples of traditional EPDM applications include automotive sealing systems, building and construction profiles and sheeting, cable and wire insulation and jacketing, tubes and hoses, and a wide variety of other automotive, industrial, and domestic applications. Recently, EPDM is applied in thermoplastic vulcanizates, water pipe seals, and grass infill for artificial soccer pitches. EPDM is currently produced at approximately 1000 kt per year globally and is considered the most important rubber product in nontire applications.

The production and consumption of EPDM would have been significantly higher if the product would offer better resistance to apolar media, such as oil, gasoline, and grease. A low-cost, general-purpose elastomer that has an oil resistance comparable with CR or better as well as the weatherability and heat resistance of EPDM would make a valuable contribution to the portfolio of synthetic rubber materials. High-cost products with such a set of properties are available in the marketplace. Polyepichlorohydrin (ECO), acrylic-based elastomers (ACM), and hydrogenated NBR (HNBR) are some of the products available. The recent development of a new generation of HNBR products by DSM [4,5] has certainly contributed to a more attractive product offering in this respect. However, the cost of ECO, ACM, and HNBR is often still prohibitive for large-scale applications. Moreover, in some cases these high-cost products have additional debits, such as poor processability or poor low-temperature performance.

The objective of recent DSM studies was to develop new EPM-based elastomers that have improved oil resistance. The idea was to develop such products by chemical modification of EPM copolymers using highly polar graft monomers, such as maleic anhydride (MA), and, optionally, by reacting these EPM-g-MA polymers with other chemicals as a way to (cross-link and) further enhance the polarity of the products. It is expected that the enhanced polarity will eventually lead to improved oil resistance of the final (cross-linked) products. It is noted that the EPM copolymers with extremely high MA graft levels as employed in this study are not commercially available.

In order to validate and assess the feasibility of this approach of improving the oil resistance of EPDM, first a correlation between the oil resistance and the polarity of rubber products was established (Section 13.3), which was subsequently used as a guideline for the experimental studies. Since most of our experimental work is starting from EPM-g-MA, the mechanism of producing EPM-g-MA and the corresponding structure of EPM-g-MA that is obtained are discussed separately (Section 13.4). The main part of the overview deals with several experimental routes pursuing the preparation of modified and cross-linked EPM-g-MA and their effect on oil resistance (Section 13.5).

13.2 OIL RESISTANCE OF RUBBERS

The oil resistance of a rubber is one of the key characteristics in the final selection of the rubber to be used, especially in automotive, industrial, and seal applications. Oil resistance is the degree of swelling in a particular family of solvents, i.e., hydrocarbons, and is governed by the competition between the driving force towards "dissolution of the rubber in the low-molecular-weight oil," i.e., swelling, and the elastic force which increases upon oil uptake. The oil resistance is usually expressed as the amount of oil that is absorbed under specified conditions: a low oil swell agrees with a high oil resistance and vice versa. The oil resistance of rubbers is strongly dependent on the chemical composition of the rubber. In addition, the presence of crystallinity of the polymer, the level and nature of other compound ingredients (especially fillers and oils), the degree of cross-linking, the nature of the oil, and the temperature determine the oil swell properties.

In Table 13.1 an overview of the resistance against ASTM #3 oil (70 h at 150°C) and fuel C (70 h at 150°C) for a large number of industrially applied rubbers is presented [6] (additional data are taken from reference [7]). Rubbers such as EPDM, polyisoprene (IR), butyl rubber (IIR), NR, and BR have a high oil swell, due to their hydrocarbon structure, which is very similar to the structure of the oil itself. Rubbers such as CR, NBR, HNBR, and ECO have a much lower oil swell, because the polarity of these rubbers is higher. The oil resistance can also be improved by lowering

TABLE 13.1

Calculated Solubility Parameter and Experimental Oil Swell for a Variety of Rubbers

Rubber[a]	Solubility Parameter[b] $(J/cm^3)^{1/2}$	Oil Swell ASTM Oil #3 (70 h at 150°C) (%) [6]	Oil Swell Fuel C (70 h at 25°C) (%) [7]
ACM	18.6	25	65
BR	16.6	>140[c]	
ECO	20.7	5	10
CR	19.8	80[d]	
EPDM	15.8	>140[c]	
EVM45	17.9	80	
FKM	14.4	2	5
HNBR17	17.8	53[e]	
HNBR25	18.6	28[e]	
HNBR36	19.7	23[e]	
HNBR42	20.3		45
HNBR44	20.6	9[e]	65
IIR	14.7	>140[c]	
IR	16.6	>140[c]	
NBR20	18.3	25[d]	45
NBR30	19.3	10[d]	35
NBR40	20.3	5[d]	25
NR	16.6	>140[c]	
QM	12.8	50	
SBR25	17.3	>140[c]	

[a] Numeric values following the abbreviations (with reference to list of abbreviations) for EVM, (H) NBR, and SBR indicate the polar monomer content (vinyl acetate, acrylonitrile, and styrene, respectively) in wt%; values for NBR are estimates.

[b] Calculated data using Coleman and Painter's procedure [9].

[c] At 70°C.

[d] At 100°C.

[e] 72 h.

the polarity of the rubber, compared with the swelling agent, such as in silicon rubber (QM) and fluoro elastomers (FKM). The solubility parameter (δ), i.e., the square root of the cohesive energy density, is a straightforward measure for the polarity of a polymer. It should be noted that it is a rather simplistic measure, combining van der Waals dipolar and hydrogen-bonding interactions into one single parameter. The value of δ is dependent on the experimental method and the conditions used for its determination and on the presence of other compound ingredients, which explains the rather large scatter in experimental data. For example, for BR values of δ are listed between 14.7 and 17.6 $(J/cm^3)^{1/2}$ and for PS between 15.6 and 21.1 $(J/cm^3)^{1/2}$ [8]. In addition, for many polymers δ has not been determined experimentally. Therefore, a more consistent set of data was obtained by calculation of δ.

The value of δ of the various rubbers in Table 13.1 has been calculated using the group contribution method developed by Coleman et al. [9]. Figure 13.1 provides a plot of the oil swell in ASTM oil #3 (70 h at 150°C) versus these calculated δ values. The highest degree of swelling is obtained for those rubbers which have a δ which is close to that of the oil ($\delta = 15.7$ $(J/cm^3)^{1/2}$), i.e., BR, EPDM, IIR, IR, and NR. Rubbers with notably higher or lower δ than the oil have a lower oil swell. On the right hand side of the maximum in Figure 13.1 there is a good correlation between the experimental oil

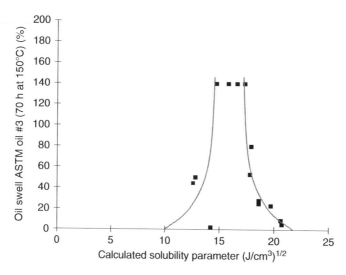

FIGURE 13.1 Oil swell in ASTM oil #3 (70 h at 150°C) for a variety of rubbers versus calculated solubility parameter.

swell and the calculated δ, especially considering that the oil swell data are "typical" values corresponding to rubber vulcanizates with different degrees of cross-linking and different types and levels of fillers and plasticizers. On the left-hand side of the maximum in oil swell in Figure 13.1 only data points for FKM and QM are shown, which do confirm a decrease in oil swell but with a rather poor correlation.

The correlation in Figure 13.1 allows the prediction of δ required for a cross-linked rubber with a desired oil resistance. For example, the structure of EPDM has to be modified in such a way that a δ outside the range of 13.5–18.0 $(J/cm^3)^{1/2}$ is achieved in order to have an oil swell similar to HNBR20 (hydrogenated nitrile rubber with 20% acrylonitrile).

In a second series of calculations, properties of modified EPM, including δ and T_g, have been generated using the Synthia software [10]. Calculations have been performed for EPM copolymers (50/50 [w/w] E/P) randomly grafted with single monomer units of a polar monomer X (EPM-g-X). It was found that grafting of 26 single styrene monomer units on 100 C-atoms of EPM results in a calculated δ of 18.0 $(J/cm^3)^{1/2}$. Grafting of a similar amount of the more polar monomer MA yields a δ of 20.8 $(J/cm^3)^{1/2}$, whereas grafting of maleimide, an even more polar and H-bonding monomer, yields a δ of 27.0 $(J/cm^3)^{1/2}$. As expected, the higher the polarity of the monomer (styrene < MA < maleimide), the larger the increase in polarity of EPM-g-X and the smaller the amount of polar monomer required to reach a certain oil resistance. The δ of HNBR20 can be reached by grafting EPM with 26, 7, and 4 monomer units of styrene, MA, and maleimide, respectively, per 100 C-atoms in EPM.

The glass transition temperature has been included in the calculations, because grafting of the (polar) monomer results in a reduction of the chain flexibility, as was witnessed by an increase of the molecular weight between entanglements, and in an increase of the dipolar interactions between the polymer chains. Starting at a calculated T_g value for EPM of $-58°C$ (close to the experimental value of $-55°C$), T_g increases to $+43°C$, $-2°C$, and $-30°C$ upon grafting of 26, 7, and 4 monomeric units of styrene, MA, and maleimide, respectively, on 100 C-atoms in EPM. The calculated T_g value for EPM-g-X, where X is styrene, is very high and thus, this modified EPM will not be rubber-like at room temperature. Figure 13.2 shows a plot of the calculated values of δ versus T_g for EPM-g-X with varying levels of grafted X. The "window" required for rubbers with δ below 13.5 or above 18.0 $(J/cm^3)^{1/2}$ and T_g below $-20°C$, i.e., for rubbers which have an oil resistance similar to HNBR20 and are rubbery at room temperature, is also shown. It is evident that neither styrene nor the more polar MA can be used as polar monomers for the modification of EPM in order to obtain an oil-resistant rubber. For the highly polar maleimide grafting levels above 11 wt% are

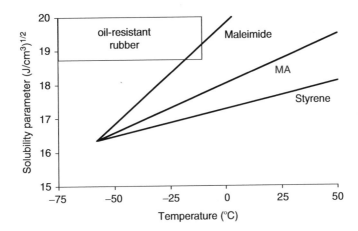

FIGURE 13.2 Calculated relation between the solubility parameter and glass transition temperature (T_g) for a variety of ethylene–propylene copolymers (EPMs) grafted with polar monomers; the window for rubbers with an oil resistance similar to or better than hydrogenated acrylonitrile–butadiene copolymer (NBR) (20 wt% acrylonitrile) is also shown.

needed to achieve HNBR20-like characteristics. This shows that EPM-g-MA is a suitable starting material for achieving our goal of an EPM-based oil-resistant rubber, although further chemical modification is needed.

13.3 ETHYLENE–PROPYLENE COPOLYMERS GRAFTED WITH MALEIC ANHYDRIDE

As much of the practical work discussed in the latter sections start from EPM-g-MA, a short survey on its preparation and its structure is deemed relevant at this stage. EPM-g-MA is an excellent polymer for our purpose, since the process of MA grafting provides favorable kinetics, grafted MA adds significant polarity and the cyclic anhydride graft has a very versatile reactivity. It readily reacts with amines and metal salts and also shows reactivity with amides, water, alcohols, epoxides, isocyanates, and oxazolines [11,12]. As a result, the polarity can be further increased by subsequent derivatization. Compatibilization with other plastics or rubbers can be achieved in many cases; the most prominent example of the latter being polyamide toughening [13,14]. Finally, it is mentioned that the grafted anhydride group can promote adhesion to polar substrates, such as metals, glass, inorganic fillers, and wood [11].

Grafting of MA onto EPM is usually carried out via free-radical chemistry [15,16] in solution in a stirred reactor or in the melt in thermoplastic mixing equipment, such as internal mixers or extruders, the latter process often being referred to as reactive extrusion (REX) [17]. Grafting of MA in solution has the advantage of the relatively high solubility of MA in the solvent, probably resulting in more homogenous products. The low reaction temperature limits the extent to which side reactions occur. In the case of solution processes, grafting may also occur on the solvent molecules. In this process the MA-grafted solvent has to be removed (and recovered) from the final product. The main advantage of REX over grafting in solution is that no solvent is involved, which is beneficial from both a cost (no recycling of solvent) and an environmental perspective. Other advantages of REX are the flexible operation and the ease of development and scale-up.

A general reaction mechanism for the grafting of MA onto EPM is given in Figure 13.3 [15,16]. Free-radical grafting of MA starts with the decomposition of the radical initiator, usually a peroxide [15,18]. The peroxide decomposes at elevated temperatures into the corresponding oxy radicals, which may further degrade to alkyl radicals and ketones. These oxy and alkyl radicals abstract

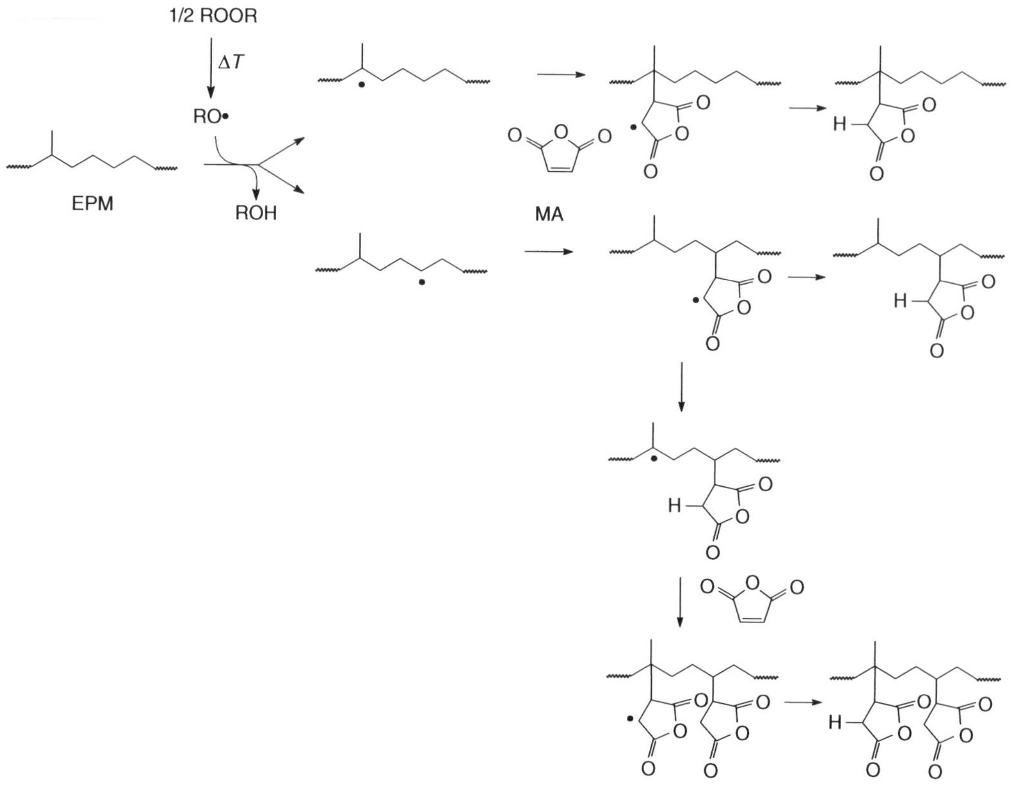

FIGURE 13.3 Mechanism of (multiple) grafting of maleic anhydride onto methine and methylene units of ethylene–propylene copolymer (EPM).

hydrogen atoms from the EPM backbone. The probability for hydrogen abstraction from the methine CH, methylene CH_2, and methyl CH_3 units is given by the product of their abundance and their intrinsic reactivity (selectivity), which decreases in the series methine > methylene > methyl (the relative reactivities at temperatures below 100°C are 2–4 > 1 > 0.1, respectively [18]; note that the differences are not large) and the concentration of the corresponding hydrogen atoms. Hydrogen abstraction from EPM occurs mainly at methine and methylene units and the higher the polymer ethylene content, the more methylene-derived, secondary radicals are formed [19,20].

After hydrogen abstraction from the EPM chain, an MA molecule is grafted onto the macroradical. The addition of the nucleophilic alkyl radical to the electron-poor unsaturation of MA is favorable [21]. It is well known that MA by itself has a very low tendency to homopolymerize [11] and has a low ceiling temperature of about 150°C [22], which can be understood in terms of the unfavorable 1,2-disubstitution of the unsaturation [23]. Once the first MA monomer has grafted onto the EPM chain, propagation will not or hardly take place. Instead, fast hydrogen transfer will yield monomeric graft structures [24]. Hydrogen transfer from abundant EPM chains to the succinyl radical yields saturated MA grafts and a new macroradical, which in its turn may react with another MA, etc. Both inter- and intramolecular hydrogen transfer may occur, the latter resulting in MA grafts in close proximity of each other along the chain [22,25,26]. For a particular EPM-g-MA grade ($M_w = 40$ kg/mol; 2.0 wt% MA) about 70% of the grafted MA molecules is present in such "bunch of grapes" structures [27]. If the amount of MA added to the polymer melt exceeds the solubility limit, droplets of molten MA will be present, resulting in very high, local MA concentrations, which will enhance the formation of multiple MA grafts.

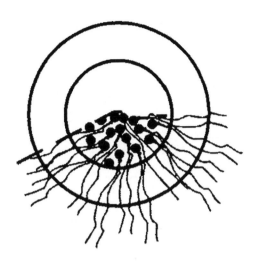

FIGURE 13.4 Schematic representation of maleic anhydride graft-rich clusters in maleated ethylene–propylene copolymers (EPMs).

For ethylene-rich EPM copolymers termination seems to proceed via combination and almost never via disproportionation, resulting in long-chain branching and cross-linking as side reactions [28]. It is noted that side reactions may occur during free-radical grafting of MA onto EPM. For instance, at higher propylene contents and/or higher reaction temperatures, EPM chain scission will occur. Finally, degradation reactions involving MA (grafts) may occur, resulting in yellow- or even brown-colored products.

Apart from the molecular structure and the topology of EPM-g-MA, the supramolecular structure of EPM-g-MA is also of interest. The formation of MA graft-rich clusters (several tens of nanometer in diameter) involving grafted MA units from different polymer chains (Figure 13.4) has been demonstrated in the solid state [29,30] and in solution [31,32]. Cluster formation of EPM-g-MA in apolar solvents is used in the application of EPM-g-MA as oil additive with dispersant properties [33]. The driving force for the formation of such supramolecular structures is the strong dipolar attraction between the polar MA grafts and the strong repulsion between the polar MA graft units and the apolar EPM matrix. When the MA grafts are hydrolyzed upon reaction with water into the corresponding diacids, cluster formation will even be enhanced due to strong hydrogen bonding. Cluster formation is limited during processing at elevated temperatures or even suppressed in polar solvents such as tetrahydrofuran (THF).

It is concluded that EPM-g-MA is not fully homogeneous, as the MA grafts are not randomly distributed along the EPM polymer chain and clustering of grafted MA units belonging to different chains occurs. The calculations of δ and T_g presented in Section 13.3 are only valid for homogeneous EPM graft copolymers. Heterogeneous graft copolymers will have a lower bulk δ and, thus, a lower oil resistance is expected. However, the latter may be compensated to some extent by the extra "cross-linking" via the formation of intermolecular clusters. Therefore, the values of δ should be seen as a first approximation of the polarity and the oil resistance of modified and cross-linked EPM.

13.4 EXPERIMENTAL VALIDATION OF CONCEPTS

13.4.1 CROSS-LINKING OF MALEATED POLYBUTADIENE

For an initial validation of the model presented in Section 13.3, a number of low-molecular-weight MA-grafted BR rubbers with MA graft levels ranging from 5 to 25 wt%, that are commercially available from Sartomer/Cray Valley, were selected. The values of δ were calculated and converted

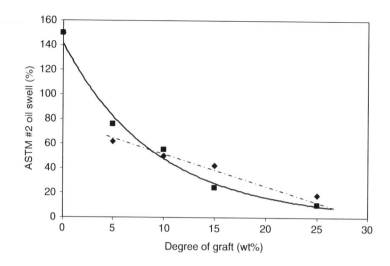

FIGURE 13.5 Measured (■) and calculated (◆) oil swell of maleated polybutadienes with varying degrees of grafting.

into oil swell data, using Figure 13.1. The samples were cured with 1 wt% of dicumyl peroxide (DCP) by compression moulding in a hot press for 20 min at 180°C. Next, the oil swell of the samples was measured according to ISO 1817–1985 (E) (Figure 13.5).

A fair match between the experimental and the measured results was obtained, indicating that the calculations can indeed be used for screening purposes. Using the approach as outlined in Section 13.3, one can now design a variety of modified EPM-g-MA structures and calculate δ and T_g of the copolymers. Based on the calculated data we were able to select the most promising cases and validate them experimentally.

13.4.2 Reaction of Maleated EPM with Alcohols and Amines

The reaction of EPM-g-MA with primary aliphatic amines results in amide acids, which at elevated temperature convert to imides [12] (Figure 13.6). The reaction of EPM-g-MA with alcohols yields

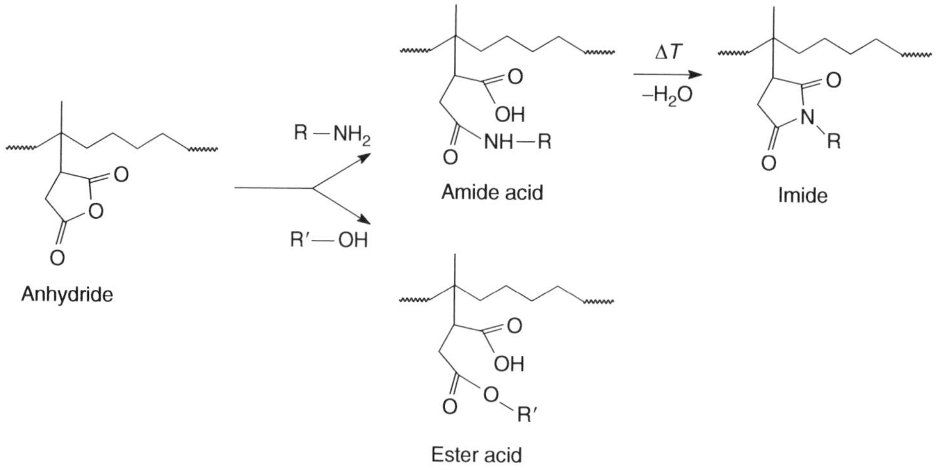

FIGURE 13.6 Reaction of maleated ethylene–propylene copolymer (EPM) with primary amines and alcohols.

ester acids (Figure 13.6). The reactions of EPM-g-MA with either amines or alcohols are also referred to as "capping" reactions.

A range of amines and alcohols has been selected for calculating δ and T_g of fully capped EPM-g-MA as a function of the MA graft content. Examples of these amines and alcohols included ammonia, 2-aminoethanol, 2-(methylamino)ethanol, diethanolamine, diisopropanolamine, urea, glycine, 1,4-diaminobutane, and the amide acid of ammonia-capped MA. Typical results are presented in Figure 13.7.

The target area for rubbers with a δ higher than 18.0 $(J/cm^3)^{1/2}$ and a T_g below $-20°C$ is indicated. Based on calculated results it should be possible to prepare grafted and fully capped EPM copolymers that fulfill the requirement for δ. In Table 13.2 the minimum amount of MA that must be grafted in order to fulfill this δ requirement is presented, including the corresponding T_g.

From Figure 13.7 it can be seen that capping of the maleated EPM is required in order to achieve the target δ while maintaining a low T_g for the modified rubber. Capping with ammonia to the corresponding imide provides a noticeable increase of polarity, also resulting from the contribution of hydrogen bonds. In the case of capping with diethanolamine very high calculated values for δ were obtained, resulting from the contribution of hydrogen bond formation of the pending alcohol and amide acid moieties.

The alcohols and the amines referred to in Table 13.2 were used for reaction with EPM-g-MA to validate the calculations. For all syntheses an experimental, very low-molecular weight (M_n of approximately 3.0 kg/mol), amorphous EPM copolymer was used that had been grafted with 13 wt % of MA. Capping reactions were performed in 10 wt% THF solutions allowing the amine or alcohol to react for 3 h at room temperature. Experiments were carried out with either stoichiometric amounts based on amines or with 0.5 or 0.33 mol equivalents of the di- and tri-functional chemicals, respectively, enabling all amine and alcohol groups to react and contribute to cross-linking during the final curing stage. Full conversion of the anhydride to the amide acid and minor conversion ($<10\%$) to ester acids were evident from Fourier transform infrared (FTIR) spectroscopy analysis. Subsequently, 1 wt% of DCP was admixed to the solution and the solvent was evaporated to obtain a polymer film of approximately 250 μm thickness. The products were cured by compression moulding for 10 min at 200°C. Virtually all amide acid and alcohol groups were converted to maleimides and ester acids, respectively, during the curing process. Finally, the cured polymer films were swollen for 72 h at 100°C in ASTM #2 oil and the swelling ratios were determined.

The results in Table 13.2 indicate that the degree of swell decreases in the order of products capped with ammonia, 2-aminoethanol, diisopropanolamine, and diethanolamine, which is in line

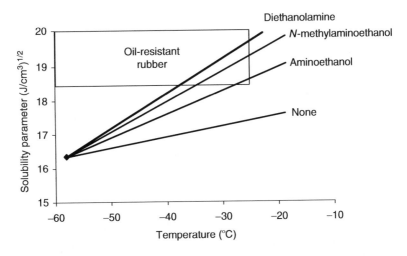

FIGURE 13.7 Calculated values for the solubility parameter and glass transition temperature (T_g) of a variety of designed ethylene–propylene graft copolymers.

TABLE 13.2

Calculated Degree of Maleic Anhydride Grafting Required to Obtain a Solubility Parameter of 18.0 $(J/cm^3)^{1/2}$, Assuming Full Conversion of the Grafted Anhydride with the Amines Listed, and Corresponding Glass Transition Temperatures (T_g)

Graft Monomer	Grafted MA (wt%)	T_g (°C)	Oil Swell in ASTM #2 Oil (%)[a]
None	>25	17	n.m.[b]
Ammonia	11.2	−22	163
2-Aminoethanol	10.9	−26	100
2-(Methylamino)ethanol	10.5	−35	n.m.[b]
Diisopropanolamine	8.9	−30	85
Diethanolamine	6.9	−41	79

[a] Oil swell for EPM grafted with maleic anhydride (EPM-g-MA) (13 wt% grafted MA) capped with stoichiometric amount of amine.
[b] Not measured.

with the expected trend (note that data for oil swell in Table 13.2 refer to stoichiometric additions of the amine). Unmodified EPM of comparable molecular weight that had been cured with 1 phr of DCP provided an oil swell of 373%. Nonetheless, the oil swell ratios that have experimentally been determined were significantly higher than expected. Partially cured CR, which was used as a reference product, still provided lower oil swell. One reason may be the lower cross-link densities that were apparent for the grafted and capped products versus the CR reference. Most likely, 1 wt% of DCP is not sufficient to fully cross-link the very low-molecular-weight EPM. As a result, the cross-link density is not high enough to exploit the full potential of the increased polarity. Another likely explanation for the relatively high oil swell obtained may be that the very high polarity of the grafted and capped structures enhances the formation of polar clusters in the rubber matrix. In this case the calculations will not apply, since a homogeneous distribution of grafted and capped sites throughout the product has been assumed. Additional work is necessary to evaluate the value of this concept for improving the oil resistance of EPM copolymers.

13.4.3 Peroxide Cross-Linking of EPM Using Polar Co-Agents

It was recognized that the grafting of high amounts of MA, followed by capping the corresponding EPM-g-MA with amines or alcohols and subsequent compounding, processing, and (peroxide) cross-linking, is quite impractical to achieve our goal. In an alternative approach we started from straight EPDM polymer, followed by processing and peroxide curing in the presence of high amounts of polar co-agents, i.e., *m*-phenylenedimaleimide, di(maleamic acid)s or trimethylolpropane trimethacrylate (TMPTMA) (Figure 13.8). It is conceivable that after cross-linking structures are obtained, comparable to those explored in Section 13.4.2.

In order to test this concept a series of compounds was prepared in a 5 L Shaw Intermix (rubber internal mixer, Mark IV, K1) with EPDM (Keltan 720 *ex*-DSM elastomers; an amorphous EPDM containing 4.5 wt% of dicyclopentadiene and having a Mooney viscosity ML(1 + 4) 125°C of 64 MU; 100 phr), N550 carbon black (50 phr), diisododecyl phthalate (10 phr), stearic acid (2 phr), and 1,3-bis(*tert*-butylperoxy-isopropyl)benzene (Perkadox 14/40 MB *ex* Akzo Nobel: 40% active material; 6 or 10 phr). A polar co-agent (15 phr) was admixed to the masterbatch on an open mill and compounds were cured for 20 min at 180°C in a rheometer (MDR2000, Alpha Technologies). The maximum torque difference obtained in the rheometer experiments was used as a measure of

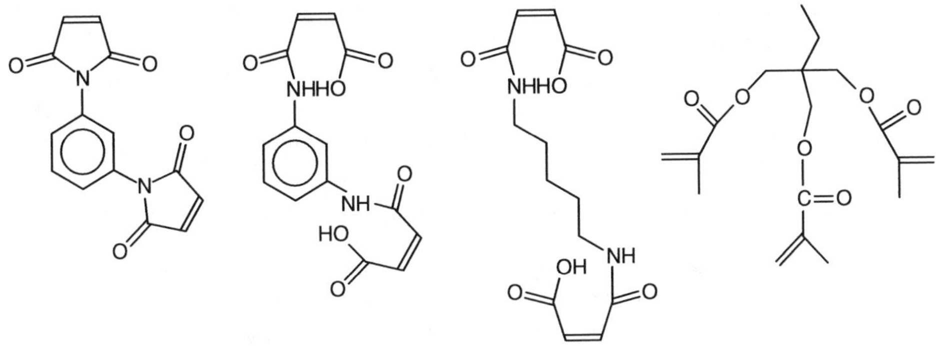

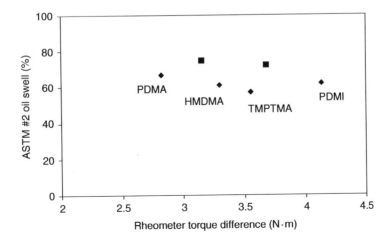

FIGURE 13.8 Co-agents used. From left to right *m*-phenylenedimaleimide (PDMI), *m*-phenylenedi(maleamic acid) (PDMA), 1,6-hexamethylenedi(maleamic acid) (HMDMA), and trimethylolpropane trimethacrylate (TMPTMA).

cross-link density [34]. Parts of the cross-linked products obtained were swollen until equilibrium in ASTM #2 oil for 72 h at 100°C. The four different co-agents, as presented in Figure 13.8, were evaluated. Two reference compounds were included in which no co-agent had been admixed. By the use of two different peroxide levels, i.e., 6 or 10 phr, different degrees of cure have been obtained. The results of the experiments are presented in Figure 13.9 by means of plotting the oil swell as a function of the rheometer torque differences obtained for the various compounds.

The results in Figure 13.9 indicate that adding these polar co-agents provide a modest improvement of oil swell properties. The cross-linked products obtained are most likely highly heterogeneous in nature. The small fraction of co-agent molecules that dissolve will increase the bulk polarity of the EPM phase and will improve the oil swell properties accordingly. However, the major part of the polar co-agent will not dissolve and will form phase-separated domains in the EPM matrix. These domains will co-cure during the peroxide cure process, rendering a network structure with chemically bound domains [35–37]. They are expected to contribute to some reduction of the oil swell, based on the hydrodynamic volume of these polar domains. Additionally, an improvement of oil resistance results from an increase of the cross-link density.

FIGURE 13.9 Effect of polar co-agents on oil swell properties of peroxide-cured ethylene–propylene–dicyclopentadiene terpolymer.

These explanations are supported by a study in which it was concluded that the presence of co-agents in peroxide-cured EPDM does not significantly affect the polymer–(swelling agent) interaction parameter [38]. This shows that the bulk polarity of the rubber is hardly affected, probably because the polar co-agent is present in the form of insoluble domains, therefore only affecting the swelling behavior of the cross-linked products by negligible hydrodynamic effects.

13.4.4 CROSS-LINKING OF MALEATED EPM WITH PROTECTED DIAMINE

A combination of the concepts presented in the previous subsections is found in curing EPM-g-MA with aliphatic diamines. When taking into account the relatively homogeneous distribution of the reactive anhydride sites at elevated temperatures (read: cure conditions), the formation of relatively homogeneous vulcanizates is expected. However, this concept proved impractical, because aliphatic diamines react extremely fast with anhydrides already under ambient conditions, making the mixing and processing of such compounds impossible due to premature curing of the compounds. Protected diamines, such as carbamate salts, that release the curing diamine only at high temperatures, could overcome this problem. For instance, 4,4′-methylenedicyclohexylamine carbamate (Diak No. 4 *ex* DuPont) is known to release the reactive diamine by decarboxylation at practical rates at temperatures starting from 200°C (Figure 13.10) [39].

Compounds based on EPM-g-MA (13 wt% MA) were mixed with equimolar amounts of Diak No. 4, according to the solution procedure described in the previous subsection, and compounds were cured for 10 min at 250°C. The oil swell of 27% in ASTM #2 oil can be regarded as a very good result, in the light of the very low-molecular weight of the starting polymer ($M_n = 3.0$ kg/mol). Comparing the results of this experiment with those reported in the previous subsection, it is concluded that in this case indeed a relatively homogeneous distribution of the polar moieties is achieved. This concept seems to allow an oil swell performance typical for relatively polar elastomers such as CR.

13.4.5 NEUTRALIZATION OF MALEATED EPM WITH ZINC ACETATE

In another approach, neutralization of maleated EPM copolymers was pursued to improve the oil resistance. EPM copolymers ($M_n = 10$ kg/mol) were grafted with different MA levels (0–8.5 wt%) and were neutralized quantitatively in a THF solution with zinc acetate. These products have been prepared in the framework of a more general study on zinc ionomers (Figure 13.11) and full experimental details are given elsewhere [29]. As expected, the oil swell is reduced with increasing degree of grafting (Figure 13.12), however, the effects are marginal. This result is explained by the strong phase separation in these ionomers.

13.4.6 GRAFTING OF POLYAMIDE-6 BRANCHES ONTO MALEATED EPM

In other experiments EPM-g-MA was reacted with polyamide-6 (PA-6). Here the idea is that the cyclic anhydride will react with the amino end groups of PA-6 to form comblike polymers with a flexible EPM backbone and crystallizable PA-6 arms. In case the number of PA-6 amino end groups is too low for full conversion of the grafted anhydride groups, a transamidation reaction may occur at sufficiently high temperatures, thus lowering the average molecular weight of the PA-6 arms

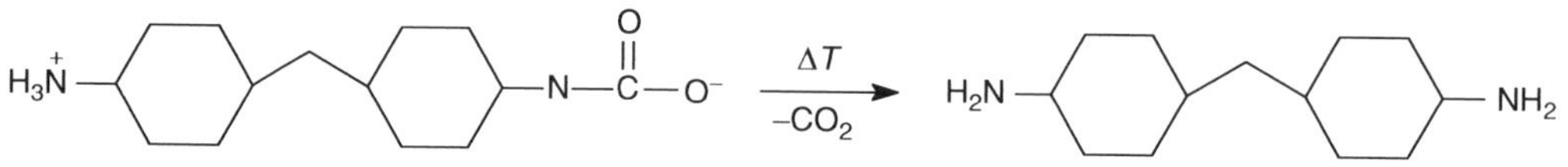

FIGURE 13.10 Decarboxylation of 4,4′-methylenedicyclohexylamine carbamate to yield the reactive diamine.

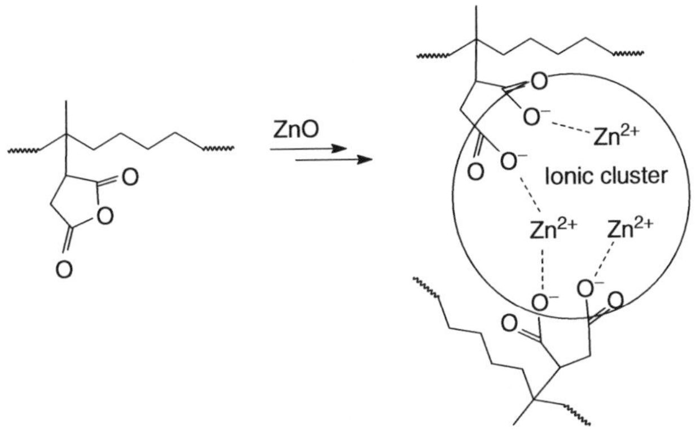

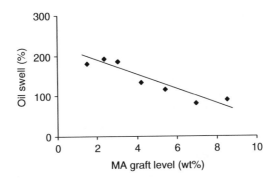

FIGURE 13.11 Schematic representation of ionomers prepared by neutralization of maleated ethylene–propylene copolymer (EPM) with zinc acetate.

[14,40]. The chemistry is similar to polyamide toughening with EPM-g-MA rubber, however, here the phases are reversed by adding relatively small amounts of PA-6 to EPM-g-MA. In addition to introducing polar PA-6 grafts, the grafted PA-6 arms are phase separated and crystallize on cooling (Figure 13.13), yielding physical cross-linking and room temperature elasticity, similar to that in segmented block copolymers as used in thermoplastic elastomer applications.

EPM-g-MA of various molecular weight and degrees of grafting was blended with low-molecular weight PA-6 variations in a so-called midi-extruder (15 cm^3) developed at DSM Research. The time (10 and 30 min) and temperature (250°C, 280°C, and 310°C) were varied, while the rotational speed of the extruder was fixed at 120 rpm. Reactively extruded products were quenched on exiting the die, thus preventing oxidation of the product. Samples were stored under nitrogen awaiting further analysis. The conversion of the grafted MA was quantified by FTIR, taking the relative decrease of the anhydride absorbance at 1860 cm^{-1} against an internal standard (CH$_2$ rock vibration at approximately 720 cm^{-1}) as a measure for the conversion. The amounts of unreacted PA-6 and unreacted EPM-g-MA were quantified by extraction using formic acid and heptane, respectively. Furthermore, differential scanning calorimetry (DSC) heating and cooling curves were obtained from selected samples, in order to assess the crystallization behavior of the products, using a Mettler 821 calorimeter (10°C/min). Finally, physical properties were measured

FIGURE 13.12 Swell in ASTM #2 oil as a function of MA graft level in zinc acetate neutralized maleated ethylene–propylene copolymer (EPM).

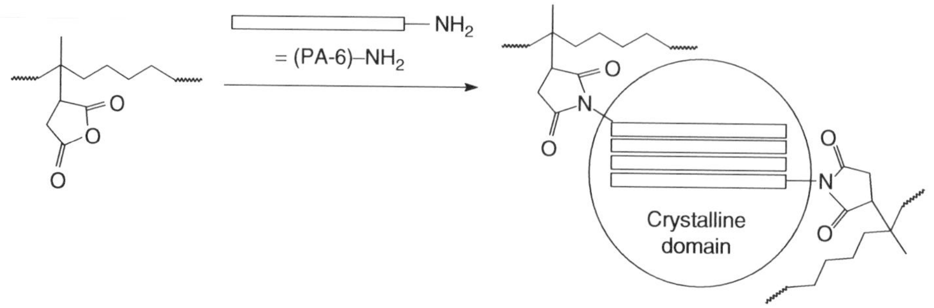

FIGURE 13.13 Schematic representation of thermoplastic elastomers obtained through reaction of maleated ethylene–propylene copolymer (EPM) with polyamide-6 (PA-6).

on compression-moulded samples according to standard procedures, viz. hardness Sh A: DIN 53505; compression set: ISO 815 type B; stress–strain behavior: ISO 37.

The conversion of the reaction of the EPM-g-MA with PA-6 varied from 50% up to 95% in all experiments. High conversions were found for reaction products of low-molecular-weight PA-6, for blends that contained relatively large amounts of PA-6, when using EPM-g-MA with a low degree of grafting, and when applying higher reaction temperatures and longer reaction times. The trends of all the influences investigated were as expected. Use of PA-6, which was not dried prior to REX, provided even higher conversions when compared to predried products. This suggests that the hydrolysis of the PA-6 chains promotes the imidization reaction. Therefore, it is concluded that the amount of free amine groups is the key driver in obtaining high grafting efficiencies. The amounts of extractable PA-6 and extractable EPM-g-MA were negligible in most cases. This is in agreement with the high imidization conversions and also suggests that an intricate product morphology has been obtained that is difficult to penetrate by either highly polar or apolar extraction media.

A remarkable reduction in PA-6 crystallinity was noticed after reactive blending for most of the products. In one particular case the PA-6 crystallinity was reduced from 47% to 7%. For most of the blends prepared, T_c of the PA-6 phase could not be observed. These results suggest that the PA-6 phase is present in a sub-micrometer dispersion in which crystallization of PA-6 is strongly suppressed. Similar results have been found for blends of PA-6 with LLDPE-g-(diethyl succinate) or styrene–MA copolymer [14,41]. In order to promote the crystallization of the PA-6 grafts, a number of experiments were carried out in which 0.1 wt% of a nucleating agent was added during the REX step. No significant effect on the crystallization behavior could be detected with DSC measurements. Therefore, it is concluded that the crystallization of the PA-6 domains is limited by the rate of crystallization rather than by crystal nucleus formation.

In Table 13.3 some typical rubber properties are presented along with the definition of the starting materials and the corresponding REX conditions. It is noted that the hardness of the products varies over a wide range, which results most likely from the fact that the nature of the EPM-g-MAs used differs, either being amorphous (samples 2 and 5) or semicrystalline (samples 1, 3, and 4). As expected, the tensile strength, the modulus, and the elongation at break are significantly better for the products based on crystalline EPM-g-MA. The elastic recovery was better for the materials starting from amorphous EPM-g-MA. In general, it can be concluded that the combination of elastic and engineering properties shows clear promise for these products for their use as thermoplastic elastomers, especially when considering that the engineering properties can be further enhanced by extending the REX formulation with for instance processing oils, (mineral) fillers, and antioxidants. The addition of free PA-6 after full imidization could be an avenue to further improve the overall balance of properties.

TABLE 13.3

Properties of Maleated Ethylene–Propylene Copolymer (EPM) Blended with Polyamide-6 (PA-6)

Sample	EPM–PA-6 Blend Ratio (w/w)	Mixing Time and Temperature (min/°C)	M_n of EPM-g-MA (kg/mol)	Degree of Grafting (wt%)	M_n of PA-6 (kg/mol)	MA Graft Conversion (%)	Comment
1	83/17	10/250	100	0.7	1.5	90	Crystalline EPM
2	83/17	10/250	40	2.4	1.5	76	Amorphous EPM
3	71/29	10/310	100	0.7	19	88	Crystalline EPM
4	71/29	10/310	100	0.7	19	90	Crystalline EPM; 0.1 wt% nucleating agent
5	48/52	10/310	40	2.7	19	81	Amorphous EPM

Sample	Hardness (°Sh A)	Tensile Strength (MPa)	Modulus 100% (MPa)	Elongation at Break (%)	Compression Set 22 h/23°C (%)	Compression Set 22 h/70°C (%)	Compression Set 22 h/100°C (%)
1	87	5.2	3.4	355	54	83	72
2	38	1.9	1.2	268	45	75	81
3	88	6.5	3.9	379	62	85	81
4	89	6.6	4.0	425	57	84	83
5	56	3.0	n.a.	98	19	45	55

A relatively low oil swell in ASTM#2 oil of 34% has been established for sample 2, indicating that an oil resistance comparable with CR-based vulcanizates can be obtained.

13.5 CONCLUSIONS

The concept of improving the oil resistance of apolar rubbers by introducing polar moieties along the chain has been validated by peroxide cross-linking of MA-grafted BR. However, capping of MA-grafted EPM copolymers with polar amines and alcohols does not result in the expected improvement of the oil resistance. The same can be concluded for the use of highly polar co-agents during peroxide cure of EPM and neutralization of EPM-g-MA with zinc acetate. This shows that obtaining a homogeneous vulcanizate in which the polar sites are evenly distributed in the polymer matrix is important in improving the oil swell properties. The curing of highly grafted EPM copolymers with a carbamate-protected diamine does show promise: the oil swell obtained in this way was close to that of a CR-based vulcanizate. Grafting of PA-6 onto EPM-g-MA results in a morphology with PA-6 domains dispersed in an EPM matrix, providing TPE-like properties. For this type of new product, oil swell characteristics comparable with CR-based vulcanizates are attainable.

ACKNOWLEDGMENTS

The authors acknowledge their DSM Research colleagues Michel van Boggelen, Marco Driessen, Dave Hofkens, Patric Meessen, and Ger Rademakers for their contributions to the various studies and the joy of working together in the Modified Polyolefins research group. Alexander Stroeks of DSM Research is thanked for calculating the solubility parameters. Rozenne Sourdrille, Christelle Noirtin, and Gerard Bosscher are acknowledged for their contributions in parts of the studies during their traineeships at DSM Research B.V.

ABBREVIATIONS

δ	Solubility parameter
ACM	Acrylic rubber
BR	Polybutadiene = butadiene rubber
CR	Polychloroprene = chloroprene rubber
DCP	Dicumyl peroxide
DSC	Differential scanning calorimetry
ECO	Polyepichlorohydrin = epichlorohydrin rubber
EPDM	Ethylene–propylene–diene monomer
EPM	Ethylene–propylene copolymer
EPM-g-MA	EPM grafted with maleic anhydride
EVM	Ethylene–vinylacetate copolymer
FKM	Fluoro elastomer
IIR	Butyl rubber
FTIR	Fourier transform infrared
IR	Polyisoprene = isoprene rubber
HMDMA	1,6-Hexamethylenedi(maleamic acid)
HNBR	Hydrogenated nitrile rubber
MA	Maleic anhydride
NBR	Acrylonitrile–butadiene copolymer = nitrile rubber
NMR	Nuclear magnetic resonance
NR	Natural rubber
PA-6	Polyamide-6
PDMA	*m*-Phenylenedi(maleamic acid)
PDMI	*m*-Phenylenedimaleimide
QM	Silicon rubber
REX	Reactive extrusion
SBR	Styrene–butadiene copolymer
TMPTMA	Trimethylolpropane trimethacrylate
T_g	Glass transition temperature
THF	Tetrahydrofuran

REFERENCES

1. Elastomere auf Basis Äthylen-Propylen, Beitrage zur Technologie des Kautschuks und verwandter Stoffe, Grünes Buch Nr. 39, Witschaftsverband der deutschen Kautschukindustrie (WdK), 1979.
2. Baldwin, F.B. and Verstrate, G., *Rubber Chem. Technol.*, 45, 709, 1972.
3. Blow, C.M. and Hepburn, C., *Rubber Technology and Manufacture*, 2nd ed., Butterworth Science, London, 1982.
4. Belt, J.W., Vermeulen, J.A.A., and Köstermann, M., US Patent 6,521,695 to DSM N.V., 2003.
5. Belt, J.J.W., Vermeulen, A.A., Singha, N.K., Aagaard, O.M., and Köstermann, M., US Patent 6,552,132 to DSM N.V., 2003.
6. Hofmann, W., *Rubber Technology Handbook*, Hanser Publishers, Munich, 1989, 162.
7. Kube, O., *Kautsch. Gummi Kunstst.* 51, 242, 1998.
8. Brandrup, J. and Immergut, E.H., *Polymer Handbook*, 3rd ed., Wiley, New York, 1989.
9. Coleman, M.M., Graf, J.F., and Painter, P.C., *Specific Interactions and the Miscibility of Polymer Blends*, Technomic Publishing, Lancaster, 1991.
10. Synthia software module, Accelrys (formerly Molecular Simulations Inc.).
11. Trivedi, B.C. and Culbertson, B.M., *Maleic Anhydride*, Plenum Press, New York, 1982.
12. Rätzsch, M., *Prog. Polym. Sci.*, 13, 277, 1988.
13. Koning, C., Van Duin, M., Pagnoulle, C., and Jeromé, R., *Prog. Polym. Sci.*, 23, 707, 1998.

14. Van Duin, M. and Borggreve, R., Blends of polyamides and maleic anhydride containing polymers: Interfacial chemistry and properties, in *Reactive Modifiers for Polymers*, Al-Malaika, S. (Ed.), Blackie Academic & Professional, London, 1997.
15. Moad, G., *Prog. Polym. Sci.*, 24, 81, 1999.
16. Van Duin, M., *Recent Res. Devel. Macromol.*, 7, 1, 2003.
17. Xanthos, M. (Ed.), *Reactive Extrusion: Principles and Practice*, Hanser Publishers, Munich, 1992.
18. Kochi, J.K. (Ed.), *Free Radicals*, Wiley, New York, 1973.
19. Van Duin, M., and Coussens, B., (Re)evaluation of the Importance of Hydrogen Abstraction during Radical Grafting of Polyolefins, Polymer Processing Society 11, Stuttgart, 1995.
20. Camara, S., Gilbert, B., Meier, R.J., Van Duin, M., and Whitwood, A.C., *Org. Biomol. Chem.*, 1181, 1, 2003.
21. Fossey, J., Lefort, D., and Sorba, J., *Free Radicals in Organic Chemistry*, Wiley, Chichester 1995.
22. Russell, K.E., *J. Polym. Sci., Part A: Polym. Chem.*, 33, 555, 1995.
23. Moad, G. and Solomon, D.H., *The Chemistry of Free Radical Polymerization*, Pergamon, Oxford, 1995.
24. Heinen, W., Rosenmöller, C.H., Wenzel, C.B., De Groot, H.J.M., Lugtenburg, J., and Van Duin, M., *Macromolecules*, 26, 1151, 1996.
25. Ranganathan, S., Baker, W.E., Russell, K.E., and Whitney, R.A., *J. Polym. Sci., Part A: Polym. Chem.*, 37, 3817, 1999.
26. Sipos, A., McCarthy, J., and Russel, K.E., *J. Polym. Sci., Part A: Polym. Chem.*, 27, 3353, 1989.
27. Vangani, V., Drage, J., Mehta, J., Mathew, A.K., and Duhamel, J., *J. Phys. Chem. B*, 105, 4827, 2001.
28. Machado, A.V., Covas, J.A., and Van Duin, M., *Polymer*, 42, 3649, 2001.
29. Wouters, M.E.L., Litvinov, V.M., Binsbergen, F.L., Goossens, J.G.P., Van Duin, M., and Dikland, H.G., *Macromolecules*, 36, 1147, 2003.
30. Yang, J., Laurion, T., Jao, T., and Fendler, J.H., *J. Phys. Chem.*, 98, 9391, 1994.
31. Nemeth, S., Jao, T., and Fendler, J.H., *Macromolecules*, 27, 5449, 1994.
32. Vangani, V., Duhamel, J., Nemeth, S., and Jao, T., *Macromolecules*, 32, 2845, 1999.
33. Li, S. et al., Proceeding of the 5th World Surfactant Congress, Florence, Italy, 955, 2000.
34. Litvinov, V. and Van Duin, M., *Kautsch. Gummi Kunstst.*, 55, 460, 2002.
35. Dikland, H.G., Van der Does, L., and Bantjes, A., *Rubber Chem. Technol.* 66, 196, 1993.
36. Dikland, H.G., Hulskotte, R.J.M., Van der Does, L., and Bantjes, A., *Kautsch. Gummi Kunstst.*, 46, 608, 1993.
37. Dikland, H.G., Ruardy, T., Van der Does, L., and Bantjes, A., *Rubber Chem. Technol.*, 66, 693, 1993.
38. Dikland, H.G., Hulskotte, R.J.M., Van der Does, L., and Bantjes, A., *Polym. Bull.*, 30, 477, 1993.
39. A Review of Fast-Cure Systems for VAMAC® Elastomers, Technical Bulletin E.I. DuPont de Nemours and Company, Inc., Document reference VAM020507, 2002.
40. Van Duin, M., Aussems, M., and Borggreve, R.J.M., *J. Polym. Sci., Part A: Polym. Chem.*, 36, 179, 1998.
41. Passaglia, E., *Polym. Adv. Technol.*, 9, 273, 1998.

Section III

Rubber Ingredients

14 Rubber-Curing Systems

Rabin N. Datta

CONTENTS

14.1 INTRODUCTION

Cross-linking or curing, i.e., forming covalent, hydrogen, or other bonds between polymer molecules, is a technique used very widely to alter polymer properties. The first commercial method of cross-linking has been attributed to Charles Goodyear [1] in 1839. His process, heating rubber with sulfur, was first successfully used in Springfield, Massachusetts, in 1841. Thomas Hancock used essentially the same process about a year later in England. Heating natural rubber (NR) with sulfur resulted in improved physical properties. However, the vulcanization time was still too long (>5 h) and vulcanizates suffer from disadvantages, e.g., aging properties.

Since these early days, the process and the resulting vulcanized articles have been greatly improved. In addition to NR, many synthetic rubbers have been introduced over the years. Furthermore, many substances other than sulfur have been introduced as components of curing (vulcanization) systems.

The accelerated sulfur vulcanization of general-purpose diene rubbers (e.g., NR, styrene–butadiene rubber [SBR], and butadiene rubber [BR]) by sulfur in the presence of organic accelerators and other rubbers, which are vulcanized by closely related technology (e.g., ethylene–propylene–diene monomer [EPDM] rubber, butyl rubber [IIR], halobutyl rubber [XIIR], nitrile rubber [NBR]) comprises more than 90% of all vulcanizations.

14.2 CURING SYSTEMS

Curing systems can be classified into four categories. Guidelines and examples of selecting different curing systems in cross-linking are reviewed in this section.

14.2.1 SULFUR-CURING SYSTEMS

Initially, vulcanization was accomplished by heating elemental sulfur at a concentration of 8 parts per 100 parts of rubber (phr) for 5 h at 140°C. The addition of zinc oxide reduced the time to 3 h. Accelerator in concentrations as low as 0.5 phr have since reduced time to 1–3 min. As a result, elastomer vulcanization by sulfur without accelerator is no longer of commercial significance. An exception is the use of about 30 or more phr of sulfur, with little or no accelerator, to produce molded products of hard rubber called ebonite.

Organic chemical accelerators were not used until 1906, 65 years after the Goodyear–Hancook development of unaccelerated vulcanization (Figure 14.1), when the effect of aniline on sulfur vulcanization was discovered by Oenslayer [3].

Aniline, however, is too toxic for use in rubber products. Its less toxic reaction product with carbondisulfide, thiocarbanilide, was introduced as an accelerator in 1907. Further developments led to guanidine accelerator [4]. Reaction products formed between carbon disulfide and aliphatic amines (dithiocarbamates) were first used as accelerators in 1919 [5]. These were and still are the most active accelerators in respect to both cross-linking rates and extent of cross-link formation. However, most dithiocarbamates accelerators give little or no scorch resistance and therefore cannot be used in all applications.

The first delayed action accelerators were introduced in 1925 with the development of 2-mercaptobenzothiazole (MBT) and 2-mercaptobenzothiazole disulfide (or 2,2′-dithiobisbenzothiazole) (MBTS) [6–8]. Even more delayed action and yet faster-curing vulcanization became possible in 1937 with the introduction of the first commercial benzothiazolesulfenamide accelerator [9–10]. Further progress was made in 1968 with the introduction [11] of prevulcanization inhibitor (PVI), N-cyclohexylthiophthalimide (CTP), which can be used in small concentrations together with benzothiazole sulfenamide accelerators. The history of the progress toward faster vulcanization with better control of premature vulcanization or scorch is illustrated in Figures 14.1 through 14.3.

Accelerated sulfur vulcanization is the most widely used method. This method is useful to vulcanize NR, SBR, BR, IIR, NBR, chloroprene rubber (CR), XIIR, and EPDM rubber. The reactive moiety present in all these rubbers is

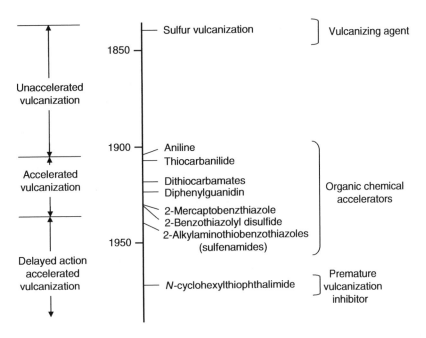

FIGURE 14.1 The history of vulcanization. (From A.Y. Coran, *Chem. Tech.*, 23, 106, 1983.)

Functionally, accelerators are classified as primary or secondary. Primary accelerators provide considerable scorch delay, medium fast cure, and good modulus development. Secondary accelerators, on the other hand, are usually scorchy and provide very fast cure. There are a wide variety of accelerators available to the compounder; including accelerator blends this number well over 100. In order to rationalize the extensive range of materials it is useful to classify them in terms of their generic chemical structure listed below and shown in Figure 14.4.

Class 1	Sulfenamide
Class 2	Thiazoles
Class 3	Guanidines
Class 4	Thiurams
Class 5	Dithiocarbamates
Class 6	Dithiophosphates

Primary:

The structures of commonly used accelerators are given in Table 14.1.

Table 14.2 provides comparisons of the different classes of accelerators based on their rates of vulcanization. The secondary accelerators are seldom used alone, but are generally found in combination with primary accelerators to gain faster cures.

The comparisons between accelerator classes are shown in Figure 14.5. Figures 14.6 and 14.7 show comparisons of primary accelerators CBS, TBBS, MBS, and MBTS in NR and SBR, respectively. Major differences in scorch safety, cure rate, and modulus development are observed.

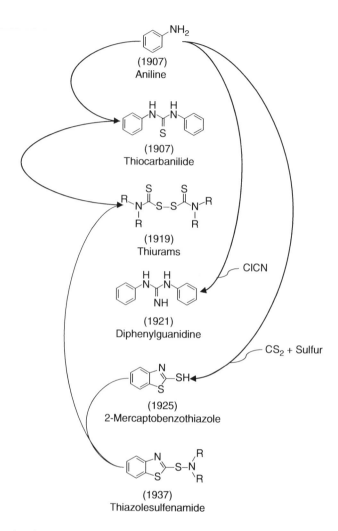

FIGURE 14.2 The chemistry of accelerator synthesis. Approximate year of commercial introduction given in parentheses. (a) Rubber + sulfur (1884); (b) Aniline (1907); (c) Diphenylguanidine (1921); (d), Thiurams (1919); (e) 2-Mercaptobenzothiazole (1925); (f) MBTS (1925); (g) Benzothiazolesulfenamides (1937); (h) Benzothiazolesulfenamide + PVI (1968). (From A.Y. Coran, *Chem. Tech.*, 23, 106, 1983.)

A large number of secondary accelerators can be used with primary accelerators, thereby providing a great deal of flexibility in processing and curing properties. Tables 14.3 and 14.4 examine [12] some of the more commonly used secondary accelerators and their effect with sulfenamide accelerator, TBBS. Seven different secondary accelerators are evaluated with TBBS in both NR and SBR compounds with MBTS/DPG as the control. The secondary accelerators also change the network structures. The polysulfidic cross-links are converted to monosulfidic exhibiting heat stability.

Aside from the sulfur, the sulfur bearing compounds that can liberate sulfur at the vulcanization temperature can be used as vulcanizing agents. A few sulfur donors are given in Table 14.5, which include some compounds like dithiodimorpholine (DTDM), which can directly substitute sulfur. Others, like tetramethylthiuramdisulfide (TMTD), can act simultaneously as vulcanization accelerators. The amount of active sulfur, as shown in Table 14.5, is also different for each compound. Sulfur donors may be used when high amount of sulfur is not tolerated in the

Cross-links
D
G
H
E
F
C
B
A
0
10
20
30
40
50
60
70
80
Time, min at 140°C

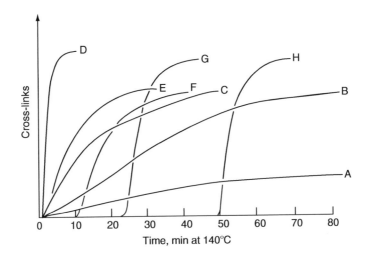

FIGURE 14.3 Development of accelerated sulfur vulcanization of natural rubber (NR). (From A.Y. Coran, *Chem. Tech.*, 23, 106, 1983.)

compounding recipe—for example, the high-temperature vulcanization of rubber. They find their use in efficient vulcanizing (EV) and semi-EV systems.

A sulfur donor system in NR (DTDM, 1.5; CBS, 1.5; TMTD, 0.5) yields a stable network structure with a contribution of 80% mono- and disulfidic cross-links at the optimum cure at 143°C or 183°C curing [13].

Cis-isoprene rubber cured with bis(diisopropyl)thiophosphoryl disulfide (DIPDIS) shows results at 160°C, producing a predominantly monosulfidic network structure [14]. Similar work on heat-resistant network structures has been carried out on other synthetic rubbers. For example, a sulfur-less system using 1 phr TBBS, 2.0 phr DTDM, and 0.4 phr TMTD in SBR gives the best aging resistance [15].

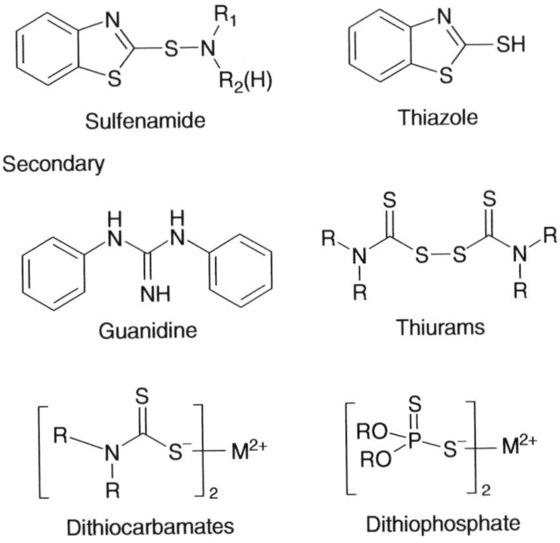

Sulfenamide
Thiazole
Secondary
Guanidine
Thiurams
Dithiocarbamates
Dithiophosphate

FIGURE 14.4 Chemical structure of accelerators.

TABLE 14.1

Accelerators for Sulfur Vulcanization

Compound	Abbreviation	Structure
Benzothiazole		
2-Mercaptobenzothiazole	MBT	
2,2′-Dithiobenzothiazole	MBTS	
Benzothizolesulfenamide		
N-cyclohexylbenzothiazole-2-sulfenamide	CBS	
N-butylbenzothiazole-2-sulfenamide	TBBS	
2-Morpholinothio benzothiazole	MBS	
N-dicyclohexylbenzothiazole-2-sulfenamide	DCBS	
Dithiocarbamate		
Tetramethylthiuram monosulfide	TMTM	
Tetramethylthiuram disulfide	TMTD	
Tetraethylthiuram disulfide	TETD	

TABLE 14.1 *(continued)*
Accelerators for Sulfur Vulcanization

Compound	Abbreviation	Structure
Tetrabenzylthiuram disulfide	TBzTD	
Zinc dimethyldithiocarbamate	ZDMC	
Zinc diethyldithiocarbamate	ZDEC	
Zinc dibenzyldithiocarbamate	ZBEC	
Amines Diphenyl guanidine	DPG	
Di-*o*-tolylguanidine	DOTG	

TABLE 14.2
Comparison of Different Classes of Accelerators

Class	Vulcanization Rate	Acronyms
Aldehyde-amine	Slow	
Guanidines	Medium	DPG, DOTG
Thiazoles	Semi-fast	MBT, MBTS
Sulfenamides	Fast-delayed action	CBS, TBBS, DCBS, MBS
Sulfenimides	Fast-delayed action	TBSI
Dithiophosphates	Fast	ZBPD
Thiurams	Very fast	TMTD, TETD, TBzTD
Dithiocarbamates	Very fast	ZDMC, ZDBC, ZBEC

Source: To, B.H., in *Rubber Technology*, Hanser Verlag, Munich, Germany, 2001.

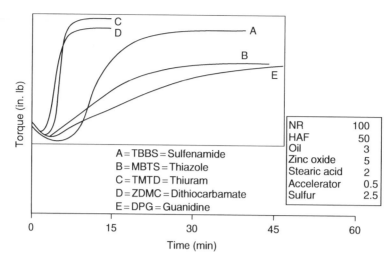

FIGURE 14.5 Comparison of accelerator classes in natural rubber (NR). (From B.H. To, *Rubber Technology*, Hanser Verlag, Munich, Germany, 2001.)

Although ultra accelerators or sulfur donors can be used together with primary accelerator (such as sulfenamide, TBBS) to improve cure rate as well as the heat resistance [16–18], their use is restricted because of the associated nitrosamine issue [19]. Accelerators derived from secondary amines, for example, MBS, TMTD, TETD, TMTM, and OTOS fall into this category. The combination of sulfenamide, such as CBS or TBBS, and a thiuram, such as TMTD or TETD, shows high-cure rates but suffers from the adverse effects on scorch resistance and vulcanizate dynamic property [20]. Additionally as previously mentioned, the use of TMTD or Tetra-ethylthiuram disulfide (TETD) or N-oxidiethylene dithiocarbamyl-N'-oxidiethylene sulfenamide (OTOS) or 4,4'-Dithiodimorpholine (DTDM) is undesirable [21] due to concerns over carcinogenic nature of the N-nitrosamines formed from the parent amines. The solution to this originated by introduction of nitrosamine safe ultra accelerator such as TBzTD [22,23].

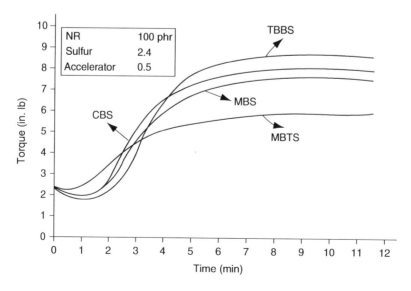

FIGURE 14.6 Comparison of primary accelerators in natural rubber (NR). (From B.H. To, *Rubber Technology*, Hanser Verlag, Munich, Germany, 2001.)

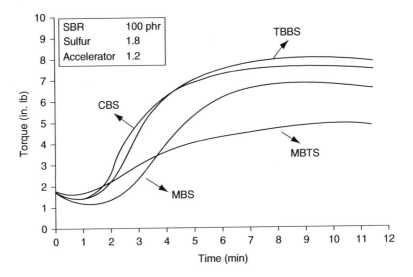

FIGURE 14.7 Comparison of primary accelerators in styrene–butadiene rubber (SBR). (From B.H. To, *Rubber Technology*, Hanser Verlag, Munich, Germany, 2001.)

Unlike TMTD, TBzTD is unique since use of small amount of TBzTD (0.1–0.2 phr) with sulfenamide system does not influence the processing characteristics while improving cure rate and dynamic properties. A comparative data are tabulated in Tables 14.6 through 14.10. Details are reported by Datta et al. [24].

TABLE 14.3
Comparison of Secondary Accelerators in NR

Stocks	1	2	3	4	5	6	7	8
Sulfur	2.5	2.5	2.5	2.5	2.5	2.5	2.5	2.5
MBTS	1.2	0	0	0	0	0	0	0
DPG	0.4	0	0	0	0	0	0	0
TBBS	0	0.6	0.6	0.6	0.6	0.6	0.6	0.6
TMTD	0	0.4	0	0	0	0	0	0
TMTM	0	0	0.4	0	0	0	0	0
TETD	0	0	0	0.4	0	0	0	0
ZDMC	0	0	0	0	0.4	0	0	0
ZDEC	0	0	0	0	0	0.4	0	0
ZDBC	0	0	0	0	0	0	0.4	0
DOTG	0	0	0	0	0	0	0	0.4
Mooney scorch at 120°C, t_5, min	7.2	16.8	21.5	23.5	13.7	16.7	20.2	21.7
Rheometer at 145°C, t_{90}, min	9.2	7.5	9.5	9.8	7.0	7.8	9.3	16.5
Stress–strain data (cure: 145°C/t_{90})								
100% Modulus, MPa	2.6	3.4	3.5	3.0	3.1	2.9	2.8	2.8
Elongation at break,%	550	450	430	480	470	500	510	510

Source: To, B.H., in *Rubber Technology*, Hanser Verlag, Munich, Germany, 2001.

NR, 100; FEF, Black 40; Aromatic oil, 10.0; Zinc oxide, 5.0; Stearic acid, 1.5; 6PPD, 2.0.

TABLE 14.4

Comparison of Secondary Accelerators in Styrene–Butadiene Rubber (SBR)

Stocks	1	2	3	4	5	6	7
Sulfur	1.8	1.8	1.8	1.8	1.8	1.8	1.8
MBTS	1.2	0	0	0	0	0	0
DPG	0.4	0	0	0	0	0	0
TBBS	0	0.5	0.5	0.5	0.5	0.5	0.5
TMTD	0	0.3	0	0	0	0	0
TMTM	0	0	0.3	0	0	0	0
TETD	0	0	0	0.3	0	0	0
ZDMC	0	0	0	0	0.3	0	0
ZDBC	0	0	0	0	0	0.3	0
ZBPD	0	0	0	0	0	0	0.3
Mooney scorch at 135°C, t_5, min	10.4	12.3	22.0	14.5	13.2	18.7	24.4
Rheometer at 160°C, t_{90}, min	9.2	7.4	9.3	8.6	8.7	12.0	21.3
Stress–strain data (cure: 160°C/t_{90})							
100% Modulus, MPa	2.0	2.1	2.1	2.0	1.9	1.9	1.6
Elongation at break, %	450	430	420	430	470	470	500

Source: To, B.H., in *Rubber Technology*, Hanser Verlag, Munich, Germany, 2001.

SBR 1500, 100; N-330, 50; Aromatic oil, 10; Zinc oxide, 4; Stearic acid, 2; 6PPD, 2

14.2.2 CURES FOR SPECIALITY ELASTOMERS

The curing systems used to vulcanize specialty elastomers such as EPDM, CR, IIR, and NBR are different than those used to cure NR, SBR, BR, and its blends. The former elastomers are less unsaturated and therefore need high ratio of accelerator to sulfur.

TABLE 14.5
Sulfur Donors

Material	Structure	Active Sulfur (%)	M.P. (°C)
TMTD		13.3	124
DTDM		13.6	125
MBSS		28.4	129
CLD		28.8	131

TABLE 14.5 *(continued)*
Sulfur Donors

Material	Structure	Active Sulfur (%)	M.P. (°C)
DPTT		16.6	120
OTOS		12.9	136
TBzTD[a]		5.8	124

[a] TBzTD acts as sulfur donor only at very high concentration.

14.2.2.1 Cure System for EPDM

EPDM vulcanizates exhibit some unique properties such as ozone, heat, light, weathering, and chemical resistance [25]. Because of this attractive combination of properties, EPDM has taken over a wide variety of applications. EPDM has relatively low unsaturation and therefore requires complex cure systems to achieve the desired properties. Nearly every conceivable combination of curing ingredients has been evaluated in various EPDM polymers over the years [26], the following systems are described as summarized in Table 14.11.

TABLE 14.6
Formulations

Ingredients/Mixes	1	2	3	4	5
NR SMR CV	100	100	100	100	100
N-330	50	50	50	50	50
Zinc oxide	5	5	5	5	5
Stearic acid	2	2	2	2	2
Santoflex 6PPD	2	2	2	2	2
Santocure CBS	0.6	0.6	0.6	0.6	0.6
Perkacit TBzTD	—	0.1	0.2	—	—
Perkacit TMTD	—	—	—	0.1	0.2
Sulfur	2.3	2.3	2.3	2.3	2.3

TABLE 14.7
Processing and Cure Characteristics

Properties	1 Control	2 0.1 phr TBzTD	3 0.2 phr TBzTD	4 0.1 phr TMTD	5 0.2 phr TMTD
Processing data					
Mooney viscosity ML(1 + 4) 100°C, MU	51	52	52	53	51
Mooney scorch, 121°C, t_5, min	30	30	27	22	18
Rheometer at 150°C					
ML, Nm	0.20	0.20	0.20	0.21	0.20
MH-ML, Nm	1.66	1.82	1.90	1.85	2.03
t_{s2}, min	3.8	3.7	3.4	3.0	2.5
t_{90}, min	10.0	7.3	5.8	6.2	4.6
Cure rate, $t_{90} - t_{s2}$	6.2	3.6	2.4	3.2	2.1

TABLE 14.8
Vulcanizate Properties (Cure: 150°C/t_{90})

Properties	1 Control	2 0.1 phr TBzTD	3 0.2 phr TBzTD	4 0.1 phr TMTD	5 0.2 phr TMTD
Hardness, IRHD	72	74	75	74	75
M100%, MPa	2.9	3.6	3.8	3.8	4.2
M300%, MPa	14.0	15.7	16.4	16.0	17.5
Tensile strength, MPa	27.7	28.1	28.7	29.0	28.3
Elongation at break, %	530	500	505	525	480
Compression set, %					
(100°C/72 h)	55	53	50	49	45
(70°C/24 h)	37	32	29	33	29
Fatigue to failure, kc (100% ext.)	143	142	142	122	122
Abrasion loss, mm^3	104	101	94	93	91

TABLE 14.9
Distribution of Cross-Link Types (Cure: 150°C/t_{90})

Cross-Link Concentration[a]	1 Control	2 0.1 phr TBzTD	3 0.2 phr TBzTD	4 0.1 phr TMTD	5 0.2 phr TMTD
Total	5.20	5.41	5.62	5.86	6.25
Poly-S	4.15	2.34	2.41	2.30	2.33
Di-S	0.88	0.78	0.60	0.58	0.55
Mono-S	0.17	2.29	2.61	2.98	3.37

[a] Expressed in gram mole/gram rubber hydrocarbon ×105.

TABLE 14.10
Vulcanizate Viscoelastic Properties (Cure: 150°C/t_{90})

Properties	1 Control	2 0.1 phr TBzTD	3 0.2 phr TBzTD	4 0.1 phr TMTD	5 0.2 phr TMTD
Storage mod. (E′) MPa	7.9	8.6	8.9	8.7	9.0
Loss mod. (E″) MPa	1.13	1.19	1.21	1.17	1.12
Tan δ	0.143	0.139	0.137	0.135	0.125

Temp: 60°C, Frequency: 15Hz and Strain: 2%.

In all cases, nitrosamines (NAs) free or safe alternatives are looked for. For system 1, the following alternative was suggested:

System 1	NA-Free Alternative
S, 1.5	S, 1.3
MBT, 0.5	MBT, 0.75
TMTD, 1.5	CBS, 3.8

The relevant properties are listed in Table 14.12.
Package 2, the "Triple 8" system can be replaced by NA-safe alternative as shown below. The properties are listed in Table 14.13.

TABLE 14.11
Cure Systems in Ethylene–Propylene Diene Monomer (EPDM)

Systems, phr	Merits	Drawbacks
System 1 S, 1.5 TMTD, 1.5 MBT, 0.5	Low cost	Bloom
System 2 (Triple 8) S, 2.0 MBT, 1.5 TEDC, 0.8 DPTT, 0.8 TMTD, 0.8	Excellent physical properties and fast cure	Scorchy and expensive
System 3 S, 0.5 ZDBC, 3.0 ZDMC, 3.0 DTDM, 2.0 TMTD, 3.0	Excellent compression set and good heat aging resistance	Bloom and very high cost

(*continued*)

TABLE 14.11 *(continued)*
Cure Systems in Ethylene–Propylene Diene Monomer (EPDM)

Systems, phr	Merits	Drawbacks
System 4		
S, 2.0	Nonblooming	Cure relatively slow and
MBTS, 1.5		worse compression set
ZDBC, 2.5		
TMTD, 0.8		
System 5		
"2121 system"	Fast cure and good	Bloom
ZBPD, 2.0	physical properties	
TMTD, 1.0		
TBBS, 2.0		
S, 1.0		

System 2, Triple 8	NA-Free Alternative
S, 2.0	S, 1.5
MBT, 1.5	ZMBT, 2.0
TDEC, 0.8	ZBEC, 0.5
DPTT, 0.8	ZBPD (75), 2.0
TMTD, 0.8	

The package 5 (2121 system) can be replaced by NA-safe alternative as described below:

TABLE 14.12
Properties

Properties	Low Cost	Alternative
Mooney scorch at 121°C		
t_5, min	12	19
t_{35}, min	15	29
Compression set, %		
72 h/23°C	6	8
72 h/70°C	24	23
72 h/100°C	63	65

Base formulation: Keltan 314, 25; Keltan 4903F, 75; N-539, 110; N-772, 60; Whiting, 30; Sunpar 2280, 85; Escorez 1102, 5; Struktol WB 212, 2; Wax 4000, 3; Caloxol W5, 6.4; Stearic acid, 1.5; Zinc oxide, 5.

Package 2, the "Triple 8" system can be replaced by NA-safe alternative as shown below. The properties are listed in Table 14.13.

TABLE 14.13

Properties

Properties	Triple 8	Alternative
Mooney scorch at 121°C		
t_5, min	5	5
t_{35}, min	8	8
Compression set, %		
72 h/23°C	5	9
72 h/70°C	33	35
72 h/100°C	70	80

Package 5, 2121 System	NA-Free Alternative
ZBPD (75), 2.0	ZBPD (75), 2.5
TMTD, 1.0	0
TBBS, 2.0	TBBS, 2.0
S, 1.0	S, 1.2

Recent developments on EPDM formulation is based on DTDM, TMTD, ZDBC, DPTT called "2828" system. The system comprises

Curatives	phr
DTDM	2.0
TMTD	0.8
ZDBC	2.0
DPTT	0.8

The beauty of this system is that it has low compression set and is a nonbloom. The data are listed in Tables 14.14 and 14.15.

14.2.2.2 Cure Systems for Nitrile Rubber

Cure systems for NBR are somewhat analogous to those of SBR except magnesium carbonate treated sulfur is usually used to aid its dispersion into the polymer [20]. Common accelerator systems include thiazoles, thiouam, thiazole/thiuram, or sulfenamide/thiuram types. Examples of these systems are shown in Table 14.16.

- Magnesium carbanate-treated sulfur
- Recipe: Medium acrylonitrile NBR, 100; N-550, 40; N-770, 40; Plasticizer (DOP), 15; Zinc oxide, 5; Stearic acid, 1; TMQ, 1; 6PPD, 2

As operating requirements for NBR become more stringent, improved aging and set properties become important. In order to address those requirements, formulations with sulfur donor systems (as a partial or total replacement of rhombic sulfur) are proposed. They are summarized in

TABLE 14.14

Curing Systems for Low Compression Set

Curatives	1	2	3	4
Sulfur	0.5	0.5	0	0.5
DTDM	2.0	1.7	2.0	0
TMTD	3.0	2.5	0.8	1.0
ZDMC	3.0	2.5	0	0
ZDBC	3.0	2.5	2.0	0
DPTT	0	0	0.8	0
CBS	0	0	0	2.0
ZBPD (75)	0	0	0	3.2
Mooney viscosity at 135°C	20.0	21.4	21.2	20.5
ML(1 + 4), MU t_5, min	14.2	13.7	16.4	14.2

Recipe: Vistalon 3708, 100; N-550, 50; N-762, 150; Circosol 4240, 120; Stearic acid, 1.0; Zinc oxide, 5.0; TMQ, 2.0.

Tables 14.17 and 14.18. The advantages of these systems are improved set and aging resistance while adequate processing safety and fast cures are maintained.

14.2.2.3 Cure Systems for Polychloroprene

Chlorobutadiene or chloroprene rubbers (CRs), also called neoprene rubbers, are usually vulcanized by the action of metal oxides. The cross-linking agent is usually zinc oxide in combination with magnesium oxide [27]. CR can be vulcanized in the presence of zinc oxide alone, but magnesium

TABLE 14.15

Properties

Properties	1	2	3	4
Rheometer, 160°C				
Delta S, Nm	2.19	2.13	1.76	1.41
t_{s2}, min	5.0	4.8	6.7	5.2
t_{90}, min	11.2	11.2	15.2	11.2
Stress–strain properties (Cure: 160°C/t_{90})				
Hardness, Shore A	70	69	70	66
M100, MPa	3.6	2.7	2.3	1.9
M300, MPa	6.8	6.4	5.7	4.7
Tensile strength, MPa	8.8	8.3	7.8	6.5
EB, %	550	545	670	640
Aged 70 h/121°C				
Hardness, Shore A	72	73	70	71
M100, MPa	3.4	3.2	2.5	2.8
M300, MPa	8.2	7.8	6.5	6.7
Tensile strength, MPa	9.1	8.9	7.9	7.8
EB, %	435	425	560	450
Compression set, 70 h/121°C, %	57	58	60	76

TABLE 14.16
High Sulfur-Cure Systems for Nitrile Rubber

Curatives/Stock	1	2	3
MC treated sulfur[a]	1.5	1.5	1.5
TMTM	0.4	0	0
MBTS	0	1.5	0
TBBS	0	0	1.2
TMTD	0	0	0.1
Processing and curing properties			
Mooney scorch at 121°C, t_5, min	6.8	8.1	5.7
Rheometer at 160°C, t_{90}, min	8.7	15.2	4.7
Physical properties (Cure: 160°C/t_{90})			
Hardness, Shore A	73	71	75
Modulus, 100%, MPa	4.2	3.6	5.0
Tensile strength, MPa	16.3	16.2	17.3
Elongation at break, %	380	475	355
Heat aging at 100°C/72 h			
Hardness, Shore A	80	78	82
Tensile Retention, %	85	68	57
Compression set, 100°C/22 h, %	31	50	55

[a] Magnesium carbonate.

oxide is necessary to confer scorch resistance. The reaction may involve the allylic chlorine atom, which is the result of the small amount of 1,2-polymerization [28].

TABLE 14.17
High Sulfur- versus Low Sulfur-Cure Systems for Nitrile Rubber

Curatives/Stock	1	2
MC sulfur	1.5	0.3
MBT	1.5	0
TBBS	0	1.0
TMTD	0	1.0
Mooney scorch at 135°C		
T_5, min	8.1	8.1
Rheometer at 160°C		
T_{90}, min	15.2	10.2
Physical properties (Cure: 160°C/t_{90})		
Hardness, Shore A	71	69
Modulus, 100%, MPa	3.6	3.1
Tensile strength, MPa	16.2	15.1
Elongation at break, %	475	485
Heat aging at 100°C/70 h		
Hardness, Shore A	78	74
Tensile retention, %	68	89
Compression set, 100°C/22 h, %	50	24

TABLE 14.18

Sulfur-Less Cure Systems for Nitrile Rubber

Curatives/Stock	1	2	3	4
MC sulfur	1.5	0	0	0
MBTS	1.5	0	0	0
TBBS	0	1.0	3.0	2.0
TMTD	0	1.0	3.0	2.0
DTDM	0	1.0	1.0	2.0
Mooney scorch at 135°C				
t_5, min	8.1	10.7	7.0	7.9
Rheometer at 160°C				
t_{90}, min	15.2	14.7	12.5	13.3
Physical properties (Cure: 160°C/t_{90})				
Hardness, Shore A	71	68	71	73
Modulus, 100%, MPa	3.6	3.1	4.3	5.7
Tensile strength, MPa	16.2	15.4	16.3	16.7
Elongation at break, %	475	485	360	290
Heat aging at 100°C/70 h				
Hardness, Shore A	78	73	75	78
Tensile retention, %	68	87	89	83
Compression set, 100°C/22 h, %	50	22	13	12

Most accelerators used in the accelerated sulfur vulcanization of other high diene rubbers are not applicable to the metal oxide vulcanization of CR. An exception is the use of so-called mixed-curing system for CR, in which metal oxide and accelerated sulfur vulcanization are combined. Along with the metal oxides, TMTD, DOTG, and sulfur are used. This is a good method to obtain high resilience and dimensional stability.

The accelerator that has been widely used with metal oxide cures is ethylene thiurea (ETU) or 2-mercaptoimidazoline. Further extensive use of ETU in vulcanization of CR is restricted because of suspected carcinogen. The related compound, thiocarbanalide, used formerly as an accelerator for sulfur vulcanization, has been revived for CR vulcanization; other substitute for ETU has been proposed [29,30].

The following mechanism for ETU acceleration has been proposed [31], Figure 14.8.

Recipes for metal oxide vulcanization are given in Table 14.19. In one case, calcium stearate was used instead of magnesium oxide to obtain better aging characteristic [32].

14.2.2.4   Cure Systems for Butyl and Halobutyl Rubber

Isobutylene-based elastomers include IIR, the copolymer of isobutylene and isoprene, halogenated IIR, star-branched versions of these polymers, and the terpolymer isobutylene–*p*-methylene styrene–bromo-*p*-methyl styrene (BIMS). A number of recent reviews on isobutylene-based elastomers are available [33–35].

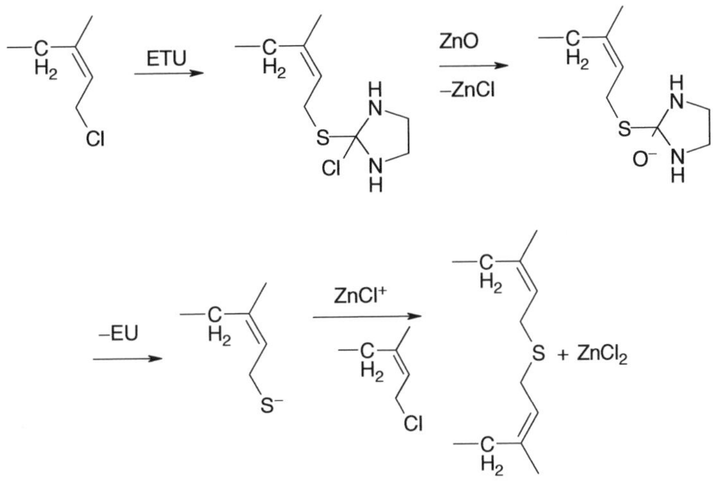

FIGURE 14.8 Chloroprene rubber (CR) cross-linking.

Polyisobutylene and IIR have chemical resistance expected of saturated hydrocarbons. Oxidative degradation is slow and the material may be further protected by antioxidants, for example, hindered phenols.

In IIR, the hydrocarbon atom positioned alpha to the C–C double bond permit vulcanization into a cross-linked network with sulfur and organic accelerators [36]. The low degree of unsaturation requires the use of ultra accelerators, such as thiuram or dithiocarbamates. Phenolic resins, bisazoformates [37], and quinone derivatives can also be employed. Vulcanization introduces a chemical cross-link every 250-carbon atom along with the polymer chain, producing a molecular network. The number of sulfur atoms per cross-link is between one and four or more [38]. IIR, apart from others goes to tire tuber application. A typical formulation is shown in Table 14.20. Recent investigation suggested [33] the use of antireversion chemical, 1,3-bis(citraconimidomethyl) benzene (Perkalink 900) for improving heat resistance. Comparative data are reported [39] and summarized in Tables 14.21 through 14.23. Figure 14.9 demonstrates the effect of Perkalink 900.

Sulfur cross-links have limited stability at elevated temperatures and can rearrange to form new cross-links. These results in poor permanent set and creep for vulcanizates when exposed for long periods of time at high temperatures. Resin cure systems provide C–C cross-links and heat stability. Alkyl phenol–formaldehyde derivatives are usually employed for tire bladder application. Typical vulcanization system is shown in Table 14.24. The properties are summarized in Tables 14.25 and 14.26.

TABLE 14.19

Vulcanization Systems for Chloroprene Rubber (CR), phr

Curatives/Stock	1	2	3
Zinc oxide	5	5	5
Magnesium oxide	4	0	4
Calcium sterate	0	5.5	0
Stearic acid	0	0	1
TMTM	0	0	1
DOTG	0	0	1
ETU	0.5	0.5	0
Sulfur	0	0	1

TABLE 14.20
Model Formulation (Inner Tubes)

Ingredients/Mixes No.	01	02
IIR Polysar 301	100	100
N-660	70	70
Zinc oxide	5	5
Stearic acid	1	1
Par. Oil Sunpar 2280	28	28
Perkacit MBTS	0.5	0.5
Perkacit TMTD	1.0	1.0
Sulfur	2.0	2.0
Perkalink 900	—	0.75

TABLE 14.21
Cure Data of the Mixes at 170°C

Properties/Mixes No.	01	02
Delta S, Nm	0.78	0.95
Scorch safety, t_{s2}, min	2.6	2.3
Opt. Cure time, t_{90}, min	9.5	11.9
Tan δ ($30'-t_{90}$)	0.112	0.056
ML(1 + 4) at 100°C, MU	55	53

TABLE 14.22
Properties (Cure 170°C/t_{90} and 30′, in the Parentheses)

Properties	01	01 (Aged)	02	02 (Aged)
Hardness, Shore A	47 (45)	50 (48)	48 (48)	50 (50)
Modulus, MPa				
50%	0.94 (0.87)	0.82 (0.72)	0.94 (0.90)	0.90 (0.88)
100%	1.62 (1.5)	1.27 (1.01)	1.60 (1.58)	1.63 (1.60)
300%	4.9 (4.6)	4.2 (3.2)	5.2 (5.2)	4.8 (4.7)
Tensile strength, MPa	11.2 (10.0)	8.8 (6.2)	11.6 (11.0)	11.1 (9.5)
Elongation at break, %	630 (615)	590 (530)	640 (625)	620 (510)
Tension set, %	7.1	12.5	5.7	6.2

TABLE 14.23
Air Permeability in mL (STP)/m_2 in 24 h at 1 bar (Cure: 170°C/30′)

Samples	1 (Control)	2 (+Pk 900)
A1	87	74
A2	84	90
A3	87	91
Thickness, mm	0.53	0.55
Average	86	85

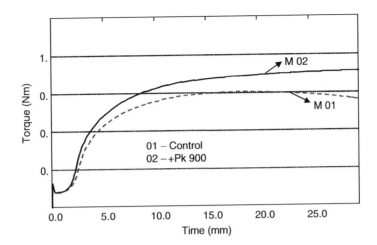

FIGURE 14.9 Cure characteristics of the mixes 01 (Control) and 02 (+Perkalink 900) at 170°C.

XIIR vulcanizates are generally acoustic loss materials having essentially the same physical and dynamic mechanical properties as regular IIE. They have the advantage of rapid rate of cure with reduced curative levels, cure compatibility with other polymers, and good cure adhesion to themselves and other elastomers [40–44]. This combination of properties makes them potentially superior base elastomers for the manufacture of a variety of mechanical goods ranging from heavy-duty shock absorbers to smaller sound and vibration damping mounting, hoses, belting, proofed goods, and many other tire and nontire applications. One of the primary applications for XIIR is in the "inner liners" for tires.

Type and level of a suitable chemical cross-linking system for compounding must be selected very carefully to achieve the desired combination of properties and service life. Several cross-linking systems are available, one of them is cross-linking via bis-maleimides (HVA-2) [45]. Unfortunately cross-linking with HVA-2 does not generate sufficient advantage with respect to heat generation and set characteristics. The use of antireversion agent, 1,3-bis(citraconimidomethyl) benzene (Perkalink 900) is explored in this application to bring the desired advantage. The formulations are provided in Table 14.27. The cure data and physicomechanical properties are tabulated in Table 14.28. The data clearly show that Perkalink 900 acts as an efficient cross-linker in XIIR. This can be explained via the reaction mechanism as shown in Figure 14.10.

It can be concluded from this study that Perkalink 900 can be used as a cross-linker in XIIR and could provide the additional advantages such as, better high-temperature compression set and lower heat built up in Goodrich Flexometer test over HVA-2.

TABLE 14.24
Model Formulations (Bladder)

Ingredients/Mixes No.	03	04	05
Butyl 268	100	100	100
Neoprene W	5	5	5
N-330	50	50	50
Castor oil	5	5	5
Perkalink 900	0	0.5	0.75
Zinc oxide	5	5	5
Resin SP 1045	10	10	10

TABLE 14.25

Cure Data of the Mixes Obtained at 190°C

Properties/Mixes No.	03	04	05
Delta S, Nm	1.1	1.0	1.2
ML, Nm	0.25	0.24	0.24
t_{s2}, min	1.7	1.7	1.7
t_{90}, min	27.4	26.9	26.5
Mooney scorch, t_5, min at 150°C	13.7	14.1	14.3

14.2.3 PEROXIDE CURE SYSTEMS

Cross-linking with peroxides has been known since 1915 when Ostromyslenski [46] disclosed that NR could be transformed into a cross-linked state with dibenzoyl peroxide. However, little interest in peroxide cross-linking evolved until the development of fully saturated ethylene–propylene copolymers in the early 1970s.

The use of peroxides for the cross-linking of elastomers is limited to those that are stable during storage, safe to handle during processing but, on the other hand, decompose sufficiently fast at cure temperatures. In order to meet these requirements peroxides containing tertiary carbon atoms are most suitable, whilst peroxy groups bonded to primary and secondary carbon atoms are less stable. Organic peroxides that are suitable for cross-linking elastomers are as follows (Figure 14.11).

In addition to the symmetrical peroxides, asymmetrical peroxides are also in use, as for example, *tert*-butyl perbenzoate, *tert*-butylcumyl peroxide, and some polymeric peroxides [47].

A further limitation with regard to the suitability of peroxides concerns the efficiency of cross-linking. Higher efficiencies are observed for those peroxides that form one of the following radicals during homolytic decomposition [48] (Figure 14.12).

TABLE 14.26

Physical Properties (Cure: 190°C/t_{90} + 2′)

Properties/Mixes No.	03	04	05
At RT			
Hardness, Shore A	65	65	65
100% Modulus, MPa	1.7	1.6	1.7
300% Modulus, MPa	5.2	5.0	5.5
Tensile strength, MPa	13.0	13.2	13.5
Elongation at break, %	670	680	660
300% Tension set, %	12.5	10.0	9.1
At 190°C			
100% Modulus, MPa	1.7	1.6	1.8
300% Modulus, MPa	5.3	4.8	5.8
Tensile strength, MPa	6.1	7.0	7.7
Elongation at break, %	350	370	380
Aged 2d/177°C; pulled at 190°C			
100% Modulus, MPa	3.1	3.2	3.4
Tensile strength, MPa	4.2	5.0	6.2
Elongation at break, %	210	220	230
200% Tension set, %	28	20	18

TABLE 14.27
Model Formulation (Inner Liners)

Ingredients/Mixes No.	06	07	08	09
CIIR Polysar 1240	100	100	100	100
N-550	55	55	55	65
Zinc oxide	5	5	5	5
Stearic acid	1	1	1	1
Par. oil	10	10	10	10
HVA-2	0	3	0	0
Perkalink 900	0	0	2	2

The thermal stability of peroxides can be expressed in terms of their half-life ($t_{1/2}$). Half-life values can be estimated in solution utilizing the technique of differential thermal analysis. These values, or more precisely the temperatures at which their half-life is equivalent, provide an indication of practical vulcanization temperatures [49] (Table 14.29).

There are a number of advantages, listed below, associated with the peroxide vulcanization of elastomers:

- Scorch-free storage of compounds
- Possibility to apply high vulcanization temperatures without reversion
- Simple compound formulation
- Low compression set even at high-cure temperatures
- Good electrical properties of vulcanizates
- Good high-temperature vulcanizate stability
- No discoloration of compounds

TABLE 14.28
Properties of the Vulcanizates (Cure: 160°C/t_{90})

Properties/Mixes No.	06	07	08	09
Cure data at 160°C				
Delta S, Nm	0.5	1.5	1.1	1.5
t_{s2}, min	4.2	2.2	4.0	2.6
t_{90}, min	9	16	26	18
Cure: 160°C/t_{90}				
Hardness, IRHD	50	69	62	71
M100, MPa	1.5	4.5	2.8	5.3
M300, MPa	3.9	11.5	9.4	11.5
Tensile, MPa	12.2	12.1	14.1	13.5
Elongation at break, %	500	210	310	260
Tear, kN/m	34	20	24	28
Compression set, %				
72 h/RT	nm	4.2	nm	4.5
24 h/100°C	nm	11	nm	6
Heat buildup at 100°C, ΔT in 25'	60	30	18	20

nm signifies "not measured."

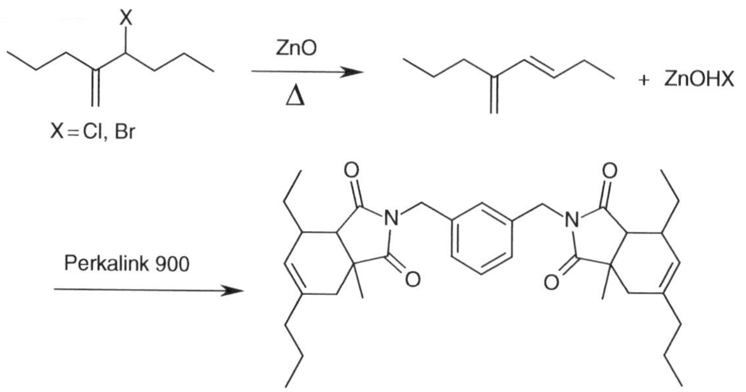

FIGURE 14.10 Reaction of Perkalink 900 with halobutyl rubber (XIIR) in the presence of zinc oxide.

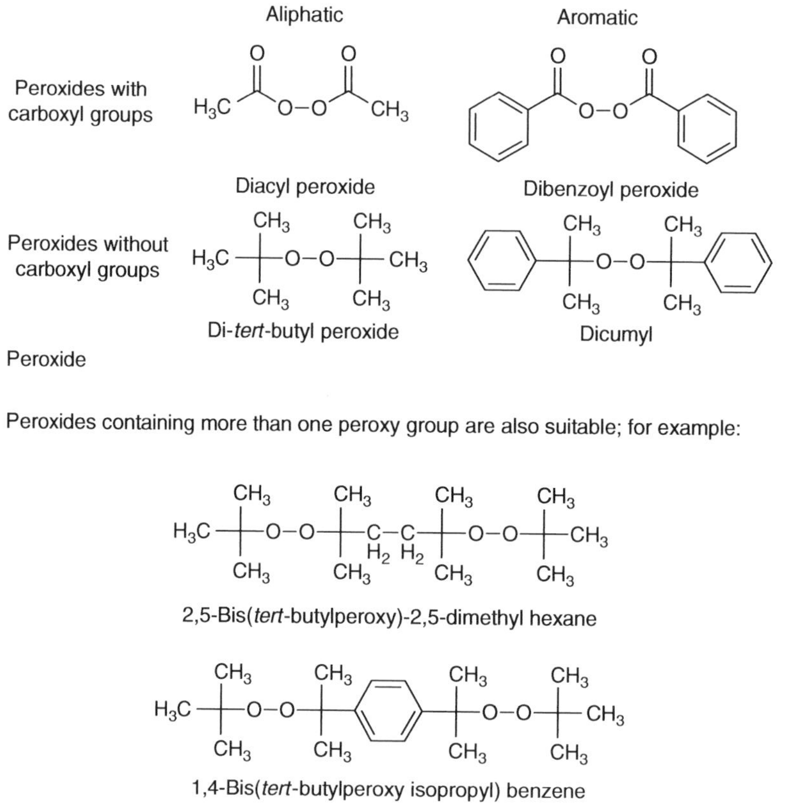

FIGURE 14.11 Some examples of typical peroxides.

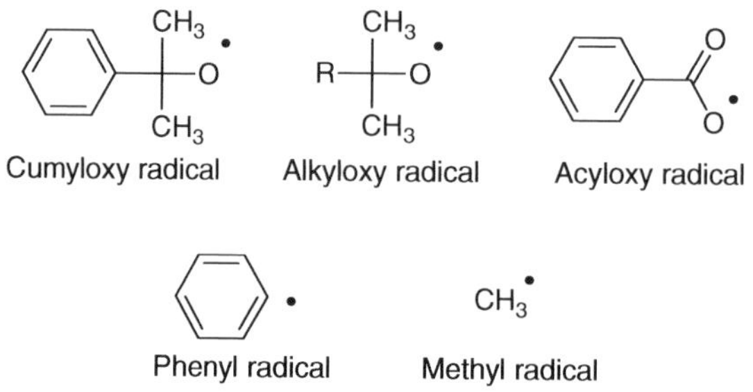

FIGURE 14.12 Radicals formed during the homolytic cleavage of peroxides that are most effective in producing cross-links.

TABLE 14.29

Typical Cross-Linking Temperatures of Cross-Linking Peroxides Based on Their Half-Life

Class	Example	Temp. (°C) $T^{1/2} = 6$ min	Typical Cross-Linking Temp. (°C)
Dialkyl peroxide	2,5-Bis(*tert*-butylperoxy) 2,5-Dimethyl-3-hexyne	173	190
	2,5-Bis(*tert*-butylperoxy) 2,5-Dimethyl hexane	159	180
Alkyl–aralkyl peroxide	Bis(*tert*-butylperoxy isopropyl) benzene	160	180
	Tert-butylcumyl peroxide	160	180
Diaralkyl peroxide	Dicumyl peroxide	155	170
Peroxy ketals	Butyl 4,4-bis (*tert*-butyl peroxy) valerate	143	160
	1,1-Bis(*tert*-butyl peroxy) 3,3,5-Trimethyl cyclohexane	129	150
Peroxy ester	*Tert*-butylperoxy benzoate	146	140

There are, however, some drawbacks compared to sulfur vulcanization:

- Limited compounding flexibility due to the reaction of peroxides with other compounding ingredients; for example, with antioxidants, plasticizers, and resins
- Sensitivity of vulcanization reactions to oxygen
- Lack of flexibility in regulating scorch and optimum cure time
- Inferior tensile, tear, and flex properties
- Inferior abrasion resistance
- Frequently disturbing odors of peroxide decomposition products
- Generally higher cost

A major variety of polymers can be cross-linked by peroxides but the reaction rates and mechanism of different polymers with peroxides vary considerably. Some ploymers are readily cross-linked by peroxides while others suffer degradation. Table 14.30 provides a list of some of the polymers that can and cannot be cured with peroxides.

Of the polymers that can be cross-linked, the cross-link efficiency varies considerably. In general the relative efficiency of peroxides vulcanization of polymers [50,51] follows as:

$$BR > NRSBR > NBR > CR > EPDM$$

Peroxides cross-linking of the more highly unsaturated polymers is more efficient due to higher concentration of allylic hydrogen. These are readily abstracted and efficiently converted to cross-link.

Peroxides vulcanization of EPDM is growing in popularity because of enhanced aging resistance. A comparison of sulfur- and peroxide-cure system is shown in Table 14.31 [53].

Apart from peroxide types and amount of peroxide incorporated in compounds, the efficiency of cross-linking depends on co-agents. The commercially notably ones are

- Maleimide type: N,N'-phenyl maleimide
- Allylic type: Triallyl cyanurate (TAC)

TABLE 14.30

Polymer Classification in Peroxide Vulcanization

Polymers That Can Be Effectively Cross-Linked by Peroxides	Polymers That Cannot Be Effectively Cross-Linked with Peroxides
Natural rubber	Polyisobutylene rubber
Styrene–butadiene rubber	Butyl rubber
Polybutadiene	Halobutyl rubber
Polyisoprene	Polyepichlorohydrin
Nitrile rubber	Polypropylene
Halogenated nitrile rubber	Polypropylene oxide
Ethylene–propylene rubber	
EPDM	
Ethylene–vinyl accetate	
Acrylonitrile–butadiene styrene	
Silicones	
Flurocarbon elastomers	
Acrylic elastomers	
Polyurethanes	
Polyethylene	
Chlorinated polyethylene	
Poly(vinyl chloride)	
Chlorosulfonated polyethylene	

Source: Dluzheski, P.R., *Rubber Chem. Technol.*, 74, 451, 2001.

- Triallyl isocyanurate (TAIC)
- Methacrylate type: Zinc dimethacrylate
- Acrylate type: Ethylene glycol diacrylate
- Zinc diacrylate
- Polymeric co-agents: Liquid 1,2-polybutadiene resin

Excellent ozone and weathering resistance, good heat and chemical resistance, good low temperature flexibility, and outstanding electric properties, make EPDM rubber preferred for a great number of specific applications. For many years peroxide cured EPDM-based compounds have been applied, e.g., for window seals, automotive hoses, steam hoses, conveyer belts, roof sheeting, tank lining, roll coverings, mouldings, and last but not least, for electrical insulation and jacketing compounds. A couple of examples are mentioned [47] as shown in Tables 14.32 and 14.33.

14.2.4 SULFUR-FREE CURING SYSTEMS

Some special vulcanizing agents can vulcanize diene rubbers such as NR, SBR, and BR. They are described in the following sections.

14.2.4.1 Phenolic Curatives, Benzoquinone Derivatives, and Bis-Maleimides

Diene rubbers can be vulcanized by the action of phenolic compounds like phenol-formaldehyde resin (5–10 phr). Resin-cured NR offers good set properties and low hysteresis [54].

Resin curing of SBR and BR imparts excellent cut growth and abrasion resistance. Resin-cured NBR shows high fatigue life and high relaxation, while resin-cured IIR shows outstanding ozone and age resistance [55].

TABLE 14.31

Comparison of Properties of Sulfur- and Peroxide-Cured Ethylene–Propylene–Diene Rubber (EPDM)

Ingredients	Sulfur Cure	Di-Cup Cure	Vul-Cup Cure
Nordel 1040	100	100	100
HAF black	50	50	50
Zinc oxide	5	5	5
Stearic acid	1	0	0
Sulfur	1.5	0	0
TMTM	1.5	0	0
MBT	0.5	0	0
DI-Cup 40KE	0	6.6	0
Vul-Cup 40KE	0	0	4.1
AgeRite Resin D	0.5	0.5	0.5
Cure temp., °C	160	171	177
Cure time, min	20	20	20
Original properties			
M100, MPa	2.2	1.8	1.9
M200, MPa	6.1	4.9	5.3
Tensile, MPa	17.9	17.2	16.7
EB, %	400	375	375
Hardness, Shore A	68	62	60
Aged properties (70 h/150°C)			
M100, MPa	5.4	1.7	2.1
M200, MPa	12.8	4.7	5.2
Tensile, MPa	15.0	17.4	16.7
EB, %	220	400	350
Hardness, Shore A	78	58	60
Compression set, % (70 h/150°C)	77	21	19

Source: Allen, R.D., *Gummi Asbestos, Kunststoffe*, 36, 534, 1983.

TABLE 14.32

Pilot Recipe for Automotive Radiator Hose (Steam Cure: 180°C/20′)

Properties

Hardness, Shore A	60
Tensile strength, MPa	13.3
Modulus, 100%, MPa	3.0
EB, %	340
Change after aging, %	
168 h/140°C	−4
168 h/150°C	−12
168 h/160°C	−21
Extractables, 168 h/100°C, water glycol (g/100 g rubber)	0.05

Compound composition: Vistalon 7500, 100; N-550, 100; Flexon 815, 45; TMQ, 1; ZMMBI, 3; Co-agent EDMA, 0.5, Perkadox 14–40, 7.

TABLE 14.33
Pilot Recipe for Injection Molded Article

Properties	01	02
BHT (phr)	0	0.4
Scorch time, t_2, min		
(120°C)	5.8	9.1
(150°C)	2.1	2.5
Cure time, t_{90}, min (150°C)	8.4	9.6
Torque increase, Nm (150°C)	4.4	4.0
Mold temperature, °C	180	200
Cure time, min	3.0	0.5
Hardness, Shore A	67	65
Tensile strength, MPa	12.1	9.0
Elongation at break, %	250	230
Compression set, 24 h/100°C, %	26	22

Keltan 578, 100, N-770, 70; N-550, 70; Sunpar 150, 70; Stearic acid, 0.5; Perkalink 400, 1; BHT, as indicated; Trigonox 2940, 11.4.

A high diene rubber can also be vulcanized by the action of a dinitrosobenzene, made *in situ* by the oxidation of a quinonediooxime (Figure 14.13) [56–60] incorporated into the rubber together with the vulcanizing agent lead peroxide.

Another vulcanizing agent for diene rubbers is *m*-phenylenebismaleimide. A catalytic free-radical source such as dicumyl peroxide or benzothiazyldisulfide (MBTS) is commonly used to initiate the reaction [61]. Phenolic curatives, benzoquinonedioxime, and *m*-phenylenebismaleimide are particularly useful where thermal stability is required.

14.2.4.2   Vulcanization by Triazine Accelerators

Logothetis [62] describes the use of triazine accelerators in the vulcanization of nitrile and fluoro elastomers. The triazine is more effective than the thiazole accelerators and produces highly reversion-resistant vulcanizates.

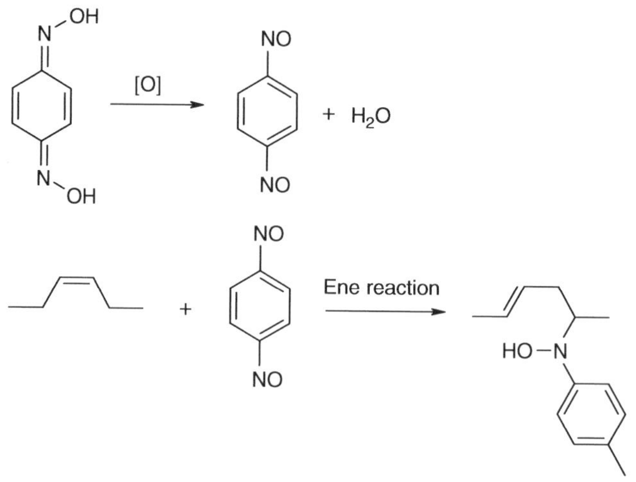

FIGURE 14.13 Vulcanization by benzoquinonedioxime.

14.2.4.3 Urethane Cross-Linkers

NR can be cross-linked by a blocked diphenyl methanes diisocyanate to produce urethane cross-links. The cross-linking agent dissociates into two quinonedioxime molecules and one diphenyl methane diisocyanate. The quinone reacts with the rubber via a nitroso group and forms cross-links via diisocyanato group. The performance of this system in NR is characterized by excellent age resistance and outstanding reversion resistance.

Further variation of the structure of nitroso compound and diisocyanate yielded NOVOR 924. An NR vulcanizate containing NOVOR 924 is more reversion-resistant than any EV system [63].

14.2.4.4 Other Cross-Linking Agents

There exist a considerable number of compounds containing labile chlorine which bring about sulfur-less vulcanization at levels of approximately 3 phr [64] as basic chemicals such as lead oxides and amines are needed. Additionally, it may be assumed that diene rubbers are cross-linked by such systems through the formation of C–C links; this would mean, initially, hydrogen chloride is split off and later neutralized by the base. Examples of chemicals that act in the manner are listed below:

- 2,4-Dichlorobenzotrichloride [65]
- Hexachloro-*p*-xylene [66]
- Chloranil [67]
- Bis-chloromethylxylene [68]
- Benzene disulfochloride [69]
- Hexachlorocyclopentadiene [70]
- Trichloromethane sulfochloride [70]
- Trichloromelamene [70]

Apart from the list above some N-bearing molecules are described in the literature [64].

14.2.5 New Developments

Maintaining properties and performance throughout a rubber product's service life is directly related to maintaining the integrity of the vulcanizate structure under both thermal and thermal oxidative conditions. Historically, this can be achieved by reducing the sulfur content in the cross-links by using efficient or semi-efficient cure systems. However, as with many changes in rubber compounding, there is a trade-off, which, in this case, is a reduction in performance in dynamic fatigue and tear resistance. Two recently developed materials have allowed compounders to forget this compromise, namely hexamethylene-1,6-bisthiosulfate (HTS) [71–73], a postvulcanization stabilizer, and 1,3-bis(citraconimidomethyl) benzene (BCI-MX, Perkalink 900) [74–86], an antireversion agent. The structures of HTS and BCI-MX are shown in Figure 14.14.

14.3 SOME PRACTICAL EXAMPLES WITH VARYING CURE SYSTEMS

Good compound means formulas are developed that are environmentally safe, factory processable, provide a satisfactory service life, and are cost-competitive to other compounds used in the same applications. Costs are always a major concern and constantly increasing environmental safety regulations must not be overlooked. In this particular section, focus has been made on cure systems and formulations of practical interest. This provides a guideline for compounding in various applications.

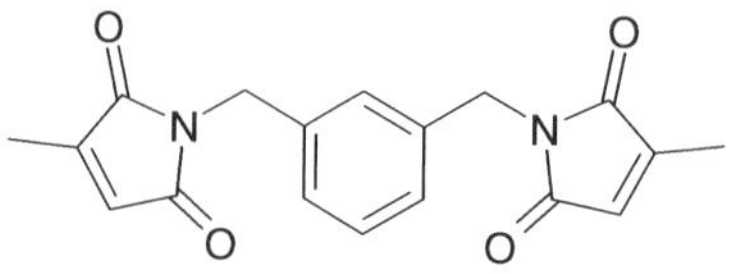

FIGURE 14.14 Structures of hexamethylene-1,6-bisthiosulfate (HTS) and 1,3-bis(citraconimidomethyl) benzene (BCI-MX).

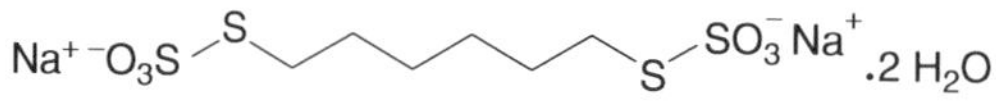

14.3.1 Tires

14.3.1.1 Tread

The tread is probably the most critical component of the tire determining the final performance. This is also the thickest component of the tire and it contributes most of the energy losses that in turn will cause a rise in the tire's running temperature and an increase in fuel consumption for the vehicle. Tread is also responsible for the safety component of the tire and its surface is designed to provide better grip in all conditions of dry, wet, ice, or snow but with minimum noise generation. Trying to balance the three main apparently conflicting needs of wear, wet grip, and rolling resistance, together with many other performance requirement leads to a wide range of tread formulations covering several natural and synthetic rubbers combined with different ratios with alternative filler types. The cure system should be CV/SEV.

14.3.1.1.1 Truck Tread (Radial)
Typical formulation is shown in Table 14.34.

For improving heat resistance, the use of Perkalink 900 is recommended [78]. The recommended loading for the above cure system should be 0.5 phr.

14.3.1.1.2 Truck Tread (Bias)
Typical formulation is shown in Table 14.35.

14.3.1.1.3 Passenger Tread
Typical formulation is shown in Table 14.36.

TABLE 14.34
Radial Truck Tread

Ingredients	phr
NR	100
N-234	55
Zinc oxide	5
Stearic acid	2
Aromatic oil	8
6PPD	2
TMQ	1.5
TBBS	1.5
PVI	0.1
Sulfur	1.5

TABLE 14.35
Bias Truck Tread (Lug)

Ingredients	phr
NR, RSS	100
Renocit 11	0.1
N-330	50
Aromatic oil	8
Zinc oxide	5
Stearic acid	2
TMQ	1
6PPD	2
MBS	0.6
PVI	0.2
Sulfur	2.3
Perkalink 900	0.5

For possibilities to raise the curing temperature, the use of antireversion chemical, Perkalink 900 is recommended [87]. The amount of Perkalink 900 required for above cure system is <0.5 phr.

14.3.1.1.4 Passenger Tread (Silica)

Silica as a reinforcing filler is being used more extensively in the tire industry to provide improved tear resistance and decreased rolling resistance. In order to maximize these benefits the silane coupling agent bis(3-triethoxysilylpropyl) tetrasulfide (TESPT) is often employed, functioning by chemically bonding the filler particles to the elastomer network. In addition, this coupling agent may act as a delayed action sulfur donor, providing some degree of reversion resistance via the formation of sulfidic cross-links of low sulfur rank.

With the advent of "Green tire" the use of silica has significantly increased. The typical "Green tire" formulation is shown in Table 14.37.

In order to achieve high-temperature cure possibilities, the use of Perkalink 900 is reported [86].

14.3.1.1.5 OTR Tread

The OTR compounds suffer from worse chipping and chunking resistance. In order to improve this unwanted phenomenon, the use of silica together with black is recommended. A typical formulation is shown in Table 14.38.

TABLE 14.36
Conventional Passenger Tread

Ingredients	phr
SBR 1712	82.5
BR CB 29	55
N-220	70
Zinc oxide	3
Stearic acid	2
6PPD	1
TMQ	2
MC wax	3
TBBS	1
Sulfur	2

TABLE 14.37
Green Tire Formulation—Passenger Tread

Ingredients	phr
SBR Cariflex S1215	75
BR Buna CB 10	25
Silica, Perkacil KS 408	80
Zinc oxide	3
Stearic acid	2
Aromatic oil	34
Coupling agent, TESPT	6.4
Santoflex 6PPD	2
Wax PEG 4000	3
Perkacit TBBS	1
Perkacit DPG	2
Sulfur	2

In order to improve reversion resistance, the use of Perkalink 900 is recommended. The highlights of performance enhancement are document by Datta and Hondeveld [86].

14.3.1.1.6   Aircraft Tread
A typical formulation for aircraft tread recipe is shown in Table 14.39.

14.3.1.2   Tread Base or Subtread

An NR-rich undertread layer can enhance the adhesion between belt or cap-ply and tread whilst a thicker subtread compound may be included to offer some additional benefits of low hysteresis for car tires and low heat generation for truck tires within the bulk of a thick section. The cure system needs better flexibility and low heat generation. Typically the cure system will be based on CV/SEV.

Tread base is generally having a composition as depicted in Table 14.40.

TABLE 14.38
Model OTR Formulation

Ingredients/Mixes	A (Blank)
NR SMR 10	100
Carbon black N-220	40
Silica, Perkasil KS 408	20
Zinc oxide	5
Stearic acid	2
Aromatic oil	3
Resin Cumarone	3
Resin Colophony	0
Coupling agent, TESPT	0
TMQ	1.5
6PPD	2.5
Wax Sunolite 240	1.0
CBS	1.4
DPG	1
Sulfur	1.4

TABLE 14.39
Aircraft Tread

Ingredients/Mixes	A (Blank)
NR	100
N-330	50
Zinc oxide	5
Stearic acid	2
6PPD	1.1
TMQ	1.0
Resorcinol	1.5
MC wax	1.0
Aromatic oil	3.0
HMT Cohedur H 30	0.8
DCBS	1.1
PVI	0.2
IS OT 20	2.2
Perkalink 900	0.5

14.3.1.3 Belts

In radial tires, sets of belts or breakers made from brass-coated steel cords are layered at alternate bias angles to provide a trellising effect, to stiffen the area under the tread and also to prevent growth under inflation or high-speed rotation. These belts provide a rigid support to the tread offering a more controlled contact with the road. They therefore play an important role in the wear of tires and the handling and stability of vehicles. The design of the belt package should give sufficient smooth ride with minimum energy loss, but with sufficient stiffness to prevent undue movement within the contact area causing irregular or rapid tire wear.

The cure package is composed of slow accelerator (DCBS or TBSI) with high level of sulfur for improving bond strength between rubber and brass layer.

Typical formulation is shown in Table 14.41.

Addition of Duralink HTS improves adhesion characteristics under aging environments [85]. Formulations without H/R bonding systems are also available [85].

TABLE 14.40
Tread Base

Ingredients	phr
NR	100
N-550	45
Zinc oxide	5
Stearic acid	2
Aromatic oil	10
6PPD	2.5
Microcrystalline wax	2
CBS	0.9
PVI	0.1
Sulfur	1.5
Perkalink 900	0.5

TABLE 14.41

Belts

Ingredients	phr
NR	100
N-326	45
Silica	15
Zinc oxide	8
Stearic acid	1.2
Tackifier SP 1068	2
Cobalt salt (NAPCO 105)	0.75
Resorcinol	2
HMMM	3
TMQ	1
6PPD	1
DCBS	1
PVI	0.1
OT 20	5
Duralink HTS	1.5

Sidewall

Ingredients	phr
NR	50
BR	50
N-550	50
Naphthenic oil	6
Zinc oxide	4
Stearic acid	1
TMQ	1.5
6PPD	2.5
MC wax	2.0
CBS	1.0
Sulfur	1.2

14.3.1.4   Sidewall

The sidewall rubber provides protection for the body plies and in the case of heavy-duty truck tires may be significant gauge in areas where abrasion or damage from kerb strikes is a likely occurrence. Sidewall compounds cover the thinnest part of the tire where most flexing occurs as the tire deflects. They therefore need to have a high degree of flex resistance, and good dynamic properties. Sidewall compounds contain additional ingredients to prevent oxidative or ozone attack since this part of the tire will be particularly exposed to the sun and the elements.

Typical formulation is shown in Table 14.41.

14.3.1.5   Carcass

The carcass or body ply of the tire is made up of fabric yarns, typically of steel, nylon, rayon, or polyester, twisted into parallel weft-less cord layers known as plies. These plies are loaded with NR-based compound loaded with adhesion promoters to generate a bond between the cord surface and other tire components.

TABLE 14.42
Truck Radial Carcass

Ingredients	phr
NR SMR L	100
N-339	55
Aromatic oil	3.0
Zinc oxide	8.0
Stearic acid	0.5
TMQ	1.0
6PPD	2.0
Wingstay L	1.0
NAPCO105(10% Co)	0.63
D-HTS	1.5
DCBS	1.1
Insoluble sulfur OT 20	5.0

The cure system of choice for a carcass compound is based on scorch-resistant cure package. Care must also be taken that the necessary high loadings of sulfur or other additives do not bloom to the surface of the calendared sheet since good tack levels must be maintained throughout the building stage. For this reason, thermally stable, insoluble sulfur, and bloom-resistant adhesion additives are required.

A typical formulation for a truck radial carcass compound is shown in Table 14.42.

Passenger carcass formulation is shown in Table 14.43.

14.3.1.6 Bead

The bead is usually wound hoop of high strength monofilament steel wire coated with rubber, providing the tire with a secure fitment to the wheel rim such that it does not move or dislodge as the vehicle undergoes severe maneuvers.

The cure package is shown in Table 14.44.

TABLE 14.43
Passenger Carcass

Ingredients	phr
NR	60
BR	20
SBR	27.5
N-660	40
Zinc oxide	3
Stearic acid	2.0
Processing oil	13
TMQ	0.5
Octylated diphenylamine	1
Resin	1.0
MBTS	0.1
TBBS	0.7
IS OT 20	2.8
PVI	0.1

TABLE 14.44
Bead

Ingredients	phr
NR	100
N-330	65
Zinc oxide	10
Stearic acid	2
Aromatic oil	8
TMQ	1
Resorcinol	3.5
HMMM (0.65%)	3
PVI	0.1
CBS	1.0
IS OT 20	6.5

14.3.1.7 Apex

The apex or filler-insert components provide the gradual shape and stiffness reduction from the rigid bead coil to the flexible mid-sidewall of the tire. These components need to be very hard to provide good vehicle handling and reduce the risk of flexural fatigue at component endings. High loadings of filler or reinforcing resins make such component difficult to process and notoriously dry, lacking in tack. Tackifying additives and process aids may therefore be required to help in the tire-manufacturing process.

The cure system is shown in formulation as tabulated in Table 14.45.

14.3.1.8 Cap-Ply

Cap-plies are more commonly used on high-performance car tires. Having a circumferential cord direction they provide an additional contractive force. They are also found more often now in normal-performance tires where they can act as a barrier layer between the tread and the casing to restrict migration of the chemicals from the tread into the belt.

A typical formulation of a cap-ply (passenger) is shown in Table 14.46.

TABLE 14.45
Apex

Ingredients	phr
NR	100
N-351	55
Zinc oxide	10
Stearic acid	2
SP 6700 Resin	2
Phenol-formaldehyde resin	10
Bonding agent	2
6PPD	2
TMQ	1
TBBS	0.6
TBzTD	0.2
PVI	0.2
IS OT 20	5.0

TABLE 14.46
Model Formulation—Cap-Ply
(Passenger)

Ingredients	phr
NR	80
BR	20
Mineral oil	7
N-326	55
Silica	7.5
DTPD	1
TMQ	1
Resorcinol	1.25
HMMM	1.5
CBS	0.4
DCBS	1.0
ZnO	8
Zinc stearate	1.0
IS OT 20	4.5

14.3.1.9 Inner Liner

The inner liner forms the vital internal membrane which holds the inflation medium at an elevated pressure within the structure of the tire. In early days the liner was a separate tube of natural or butyl, or more particularly, XIIR compound as an integral part of the tire structure. Adhesion levels of butyl compounds can be critically low requiring an insulating or barrier layer of an NR compound to act as an interface between the liner and the casing.

A typical formulation of an Inner liner is documented in Table 14.47.

14.3.2 INDUSTRIAL RUBBER PRODUCTS

This section provides some laboratory formulation and test data illustrating the effects of the antireversion agent applied in compounds typical of some industrial rubber products (IRPs) (Table 14.48). A more extensive list of potential applications is given at the end of this section. Details are reported by Datta and Ingham [88].

TABLE 14.47
Inner Liner

Ingredients	phr
Chlorobutyl 1065	100
N-660	60
Naphthenic oil	8
Stearic acid	2
Phenolic resin tackifier	4
Homogenizers	7
Magnesium oxide	0.15
Zinc oxide	3
Sulfur	0.5
MBTS	1.5

TABLE 14.48

Examples of Industrial Rubber Product Applications

Application	Polymer
Conveyor belt cover	NR
Engine mount	NR
Tank pad	NR/SBR/BR blend
Oil seal	NBR

14.3.2.1   Conveyor Belt Cover—NR

The key performance properties of conveyor belts, particularly belt cover compounds, are flex resistance, abrasion resistance, and low heat buildup. The effect of 1,3-bis(citraconimidomethyl) benzene on these key properties, in addition to cure characteristics and tensile properties, has been evaluated in a typical, NR-based belt cover formulation listed in Table 14.49.

Cure characteristics are shown in Figure 14.15; the antireversion agent has no effect on scorch resistance and time to optimum cure. The beneficial effect becomes apparent when reversion occurs, as observed in the control compound. The compound containing the antireversion agent on the other hand maintains a torque level close to the maximum.

The antireversion effect is also evident in the stability of vulcanizate properties following overcure or air aging at 100°C. Tensile, flex/cut growth characteristics, and abrasion resistance are superior for the compound containing the antireversion agent (Figures 14.16 through 14.18).

A major benefit imparted by the antireversion agent concerns heat buildup under dynamic conditions as encountered during the service of conveyor belts. Both at optimum cure and overcure the antireversion agent is extremely effective in reducing heat buildup determined by the Goodrich Flexometer test (Figure 14.19). The control compound, containing no antireversion agent, exhibits a significantly higher heat buildup. This suggests that rolling energy losses can be reduced through the use of Perkalink 900, an important consideration in today's energy-conscious environment.

14.3.2.2   Engine Mount—NR

The essential performance properties for an engine mount are low heat buildup and low dynamic compression set. These have been determined for a typical engine mount compound based on NR, with and without Perkalink 900—Table 14.50.

TABLE 14.49

Conveyor Belt Cover Formulation

Ingredients	Control	Perkalink 900
NR	100	100
N330	45	45
Aromatic Oil	4	4
ZnO	5	5
Stearic Acid	2	2
6PPD	1	1
CBS	0.5	0.5
Sulfur	2.5	2.5
Perkalink 900	—	1

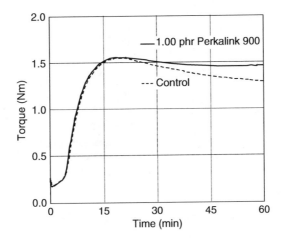

FIGURE 14.15 Cure characteristics at 150°C.

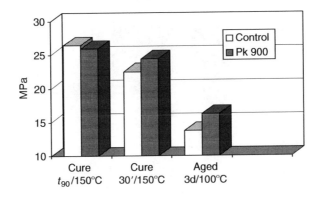

FIGURE 14.16 Tensile strength.

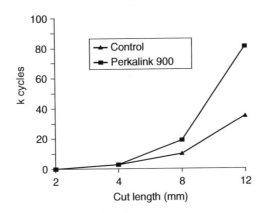

FIGURE 14.17 De Mattia flex/cut growth—cure 30′/150°C.

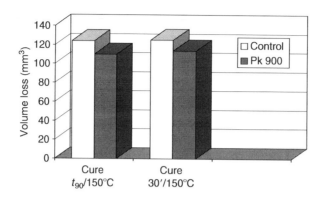

FIGURE 14.18 DIN abrasion loss.

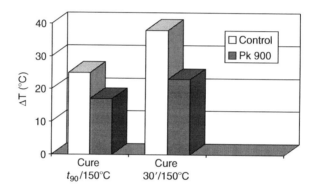

FIGURE 14.19 Goodrich flexometer heat buildup at 100°C after 30 min (load 108 N, stroke 4.45 mm, frequency 30 Hz).

TABLE 14.50
Engine Mount Formulation

Ingredients	Control	Perkalink 900
NR	100	100
N-660	42	42
Zinc oxide	5	5
Stearic acid	2	2
Oil	9	9
MC Wax	2	2
6PPD	2	2
TMQ	2	2
Resin	2	2
Santocure CBS	1.5	1.5
Perkacit TMTD	0.2	0.2
Sulfur	2.25	2.25
Perkalink 900	—	0.75

TABLE 14.51

Cure Characteristics at 150°C (170°C) and Mooney Scorch

	Control	Perkalink 900
ML, Nm	0.17 (0.16)	0.16 (0.15)
Delta torque, Nm	1.49 (1.35)	1.43 (1.34)
t_{s2}, min	4.1 (1.1)	4.1 (1.1)
t_{90}, min	8.7 (2.5)	8.4 (2.4)
Mooney scorch, t_5, 121°C	25.6	25.4

Compared to the previous example, the engine mount formulation contains a cure package tending to an SEV system; this will provide a greater degree of reversion resistance compared to a CV cure system. Nonetheless, the antireversion agent is still able to provide significant, additional benefit.

The cure data show that Perkalink 900 does not affect scorch time or time to optimum cure (Table 14.51). Also, tensile data following overcure (at 150°C and 170°C) and aging at 100°C indicate a trend of improved strength characteristics (Figure 14.20).

The important vulcanizate properties demanded by this application, low heat buildup, and low dynamic set have been determined in the Goodrich Flexometer test. The compound containing the antireversion agent exhibits a marked decrease in heat buildup and dynamic permanent set (Figures 14.21 and 14.22).

Even at the increased cure temperature of 170°C the compound containing the antireversion agent is little changed with regard to heat buildup and dynamic set compared to the lower cure temperature. The control compound on the other hand exhibits both increased heat buildup and dynamic set at the higher cure temperature. This suggests that increasing the cure temperature as a means of increasing productivity would have no detrimental effect on compound performance in this application.

14.3.2.3 Tank Pad—NR–SBR–BR Blend

Perkalink 900 is also active in compounds based on blends of NR with the synthetic elastomers SBR and BR. An evaluation in a tank pad formulation has provided evidence of reduced heat buildup on overcure; overcure is a common problem in the manufacture of these relatively bulky components. The control and test formulations, in which two levels of the antireversion agent have been evaluated, are listed in Table 14.52. Cure characteristics are given in Table 14.53.

Although little advantage is observed with regard to vulcanizate physical properties (Table 14.54), a benefit is apparent in terms of reduced heat buildup (Figure 14.23). In addition, the antireversion agent provides greater thermal stability in terms of blow-out resistance

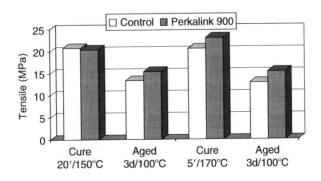

FIGURE 14.20 Tensile strength.

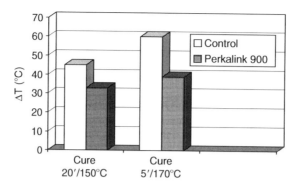

FIGURE 14.21 Goodrich flexometer heat buildup at 100°C after 30 min (load 108 N, stroke 4.45 mm, frequency 30 Hz).

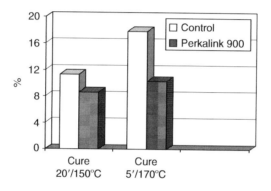

FIGURE 14.22 Goodrich flexometer dynamic permanent set at 100°C after 30 min (load 108 N, stroke 4.45 mm, frequency 30 Hz).

TABLE 14.52
Tank Pad Formulation

Ingredients	1	2	3
NR	60	60	60
SBR 1500	20	20	20
BR Buna CB10	20	20	20
N-220	70	70	70
Zinc oxide	4	4	4
Stearic acid	2	2	2
Oil	40	40	40
Santoflex 6PPD	1	1	1
MC wax	3	3	3
Santocure TBBS	1	1	1
Perkacit MBTS	2	2	2
Santogard PVI	0.15	0.15	0.15
Sulfur	2	2	2
Perkalink 900	—	0.5	0.75

TABLE 14.53
Cure Characteristics at 155°C

	1	2	3
ML, Nm	0.18	0.18	0.18
Delta Torque, Nm	1.37	1.33	1.31
t_{s2}, min	3.4	3.6	3.7
t_{90}, min	7.2	7.3	7.3

(Figure 14.24). This observation is of important practical significance since it is not uncommon for tank pads to blow out during service.

14.3.2.4 Oil Seal—NBR

Perkalink 900 has been evaluated in an NBR-based oil seal formulation in which the key properties of interest are compression set and oil resistance. Test formulations are listed in Table 14.55 and cure characteristics in Table 14.56.

On extended overcure, which can be considered as an anaerobic aging process, the compound containing the antireversion agent exhibits a higher tensile strength as shown in Table 14.57 together with additional physical properties.

Both dynamic permanent set and oil swell resistance are improved on addition of Perkalink 900, indicative of the greater thermal and mechanochemical strength of the cross-links formed from Perkalink 900 compared to sulfidic cross-links. Data are shown in Figures 14.25 and 14.26.

Perkalink 900 has been evaluated in some typical IRP applications providing a benefit with regard to reversion resistance, both during cure and following aging. This improvement in reversion resistance leads to a better retention of vulcanizate properties such as tensile strength, cut growth resistance, and abrasion resistance. Furthermore, heat buildup is significantly reduced in the presence of the antireversion agent.

In addition to the compounds evaluated, further potential applications include:

Bridgebearings
Air springs
Bushings
Dock fenders
Rollers
Mill liners
Membranes
O-Rings
Bellows
Curing bladders

TABLE 14.54
Vulcanizate Properties—Cure 30′/155°C

Properties	1	2	3
Modulus 300%, MPa	9.9	9.9	9.9
Tensile, MPa	17.4	17.8	18.0
Elongation, %	475	470	470
Tear strength, kN/m	67	75	72
Abrasion loss, mm^3	88	75	72

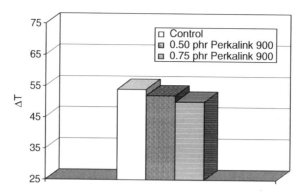

FIGURE 14.23 Goodrich flexometer heat buildup at 100°C after 60 min (load 108 N, stroke 4.45 mm, frequency 30 Hz).

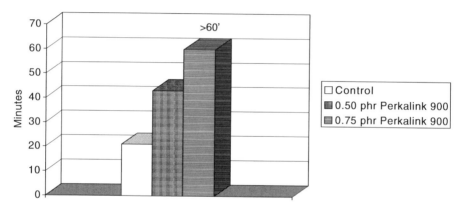

FIGURE 14.24 Goodrich flexometer blowout at 100°C (load 216 N, stroke 4.45 mm, frequency 30 Hz, test time 60').

TABLE 14.55
NBR-Based Oil Seal Formulation

Ingredients	Control	Perkalink 900
NBR	100	100
N-550	60	60
ZnO	5	5
Stearic acid	1	1
DOP	15	15
TMQ	2	2
MBTS	1	1
TMTD	0.1	0.1
Sulfur	1.5	1.5
Perkalink 900	—	0.75

TABLE 14.56
Cure Characteristics at 170°C

	Control	Perkalink 900
Delta Torque, Nm	1.36	1.39
t_{s2}, min	0.9	0.9
t_{90}, min	1.9	2.1

TABLE 14.57
Physical Properties—Cure 5′/170°C (and 120′/170°C)

Properties	Control	Perkalink 900
Modulus 300%, MPa	12.9 (13.6)	13.3 (14.0)
Tensile strength, MPa	19.0 (18.2)	20.4 (21.1)
Elongation, %	530 (470)	560 (470)
Tear strength, kN/m	62 (55)	65 (60)

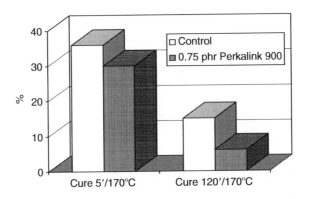

FIGURE 14.25 Goodrich flexometer dynamic permanent set at 100°C after 120 min (load 108 N, stroke 4.45 mm, frequency 30 Hz).

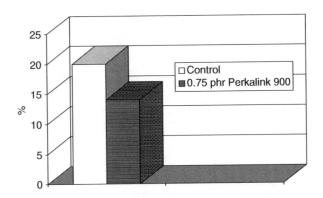

FIGURE 14.26 Swelling in ASTM oil #3, 168 h at 150°C—cure 5′/170°C.

14.4   CONCLUSIONS

Cure systems play an important role in achieving the performance requirements of rubber articles. Furthermore, these performance requirements are becoming more and more demanding due to ever-increasing severity conditions, health and safety concerns, and quality expectations.

Today, more and more know how to design a cure system capable of meet these demanding requirements. Some new materials allow compounder to reformulate a cure system capable of providing improved performance. Antireversion chemicals constitute a class of such materials, once again emphasizing the need to meet the service conditions. Health and safety issues, as for example, that concerning *N*-nitrosamines, have also lead to the introduction of new rubber additives.

Cure systems, however, paint only part of the picture with regard to the manufacture of rubber articles; polymers and fillers are equally important in meeting performance requirements. In addition, compound mixing and processing play an important role in achieving final vulcanizate properties.

REFERENCES

1. C. Goodyear, US Patent 3,633; 1944.
2. A.Y. Coran, *Chem. Tech.* 23, 106, 1983.
3. L. Bateman, C.G. Moore, M. Porter, and B. Saville, in *The Chemistry and Physics of Rubber like Substances*, L. Bateman, Ed., Wiley, New York, 1963, Chapter 19.
4. U.S. Patent 1,411,231 (March 28, 1922), M. Weiss (to Davon Chemical Corp.).
5. U.S. Patent 1,343,224 (June 15, 1920), S. Molony (to Michigan Chemical Company).
6. U.S. Patent 1,371,662-4 (March 15, 1921), C. Bedford (to Goodyear Tire & Rubber Company).
7. U.S. Patent 1,544,687 (July 7, 1925), L. Sebrell and C. Bedford (to Goodyear Tire & Rubber Company).
8. G. Bruni and E. Romani, *Indian Rubber J.*, 62, 63, 1921.
9. U.S. Patent 1,942,790 (Jan. 9, 1934), E. Zaucker, M. Bogemann, and L. Orthner (to I.G. Farben AG).
10. U.S. Patent 2,100,692 (Nov. 30, 1937), M.W. Harman (to Monsanto Chemical Company).
11. U.S. Patent 3,546,185 (Dec. 8, 1970), A.Y. Coran and J.E. Kerwood (to Monsanto Company).
12. B.H. To, *Rubber Technology*, Hanser Verlag, Munich 2001, Chapter 15.
13. A.N. Stepkin, F.S. Tolstukhina, A.A. Lapshova, and A.G. Fedorov, *Kautschuk Rezina*, 34(9), 15, 1975.
14. J.G. Pimblott, G. Scott, and J.E. Stuckey, *J. Appl. Polym. Sci.*, 19, 865, 1975.
15. M.C. Kirkham, Current Status of Elastomer Vulcanization, in *Progress of Rubber Technology*, Vol. 41, Plastics and Rubber Institute, London, 1978.
16. J.F. Krymowski and R.D. Taylor, *Rubber Chem. Technol.*, 50, 671, 1977.
17. B. Adhikari, D. Pal, D.K. Basu, and A.K. Chaudhuri, *Rubber Chem. Technol.*, 56, 327, 1983.
18. M.P. Ferrandino, J.A. Sanders, and S.W. Hong, Paper presented at 147th meeting of the Rubber Division, ACS, Philadelphia, Pennsylvania, May 2–5, 1995.
19. L. Lohwasser, *Kautschuk Gummi Kunststoffe*, 42, 22, 1989.
20. R.M. Russell, T.D. Skinner, and A.A. Watson, *Rubber Chem. Technol.*, 42, 418, 1969.
21. R.N. Datta and F.A.A. Ingham, *Kautschuk Gummi Kunststoffe*, 52, 758, 1999.
22. D. Seeberger, *Flexsys Technical Bulletin*, 2AC.0.12.1 A/09-92.
23. D. Seeberger and G. Raabe, *Kautschuk Gummi Kunststoffe*, 42, 27, 1989.
24. R. Datta, B. Oude Egbrink, F. Ingham, and T. Mori, *Kautschuk Gummi Kunststoffe*, 54, 612, 2001.
25. M. Morton, *Rubber Technology*, 3rd ed, Chapter 9, p. 271.
26. B.H. To, Rubber Technology, J.S. Dick, Ed., Hanser Publishers, Chapter 16.
27. A.G. Stevenson, G. Alliger, and I.J. Sjothun, Eds., *Vulcanization of Elastomers*, Reinhold, New York, 1964, p. 271.
28. A.Y. Coran, *Encyclopedia of Polymer Science and Engineering*, 17, 666, 1989.
29. K. Mori and Y. Nakamura, *Rubber Chem. Technol.*, 57, 34, 1984.
30. H. Kato and H. Jujita, *Rubber Chem. Technol.*, 55, 949, 1982.
31. R. Pariser, *Kunststoffe*, 50, 623, 1960.
32. R.O. Becker, *Rubber Chem. Technol.*, 37, 76, 1964.

33. G.J. Wilson, Butyl and the Halobutyls, Paper presented to Educational Symposium at the meeting of the Rubber Division, ACS, Houston, TX, Oct. 24, 1983.
34. E.N. Kresge and H.C. Wang, Butyl Rubber, in *Kirkothmer Encyclopedia of Chemical Technology*, 4th ed., M. Howe-Grant, Ed., Wiley, New York 1993, Vol. 8, p. 934.
35. D. Kruse and J.V. Fusu, *Rubber and Plastics News 1993 Technical Year Book*, May 1994, p. 10.
36. A.Y. Coran in *Science and Technology of Rubber*, F.R. Eirich, Ed., Academic Press, Orlando, 1978.
37. D.S. Breslow, W.D. Willis, and L.O. Amberg, *Rubber Chem. Technol.*, 43, 605, 1970.
38. R.L. Zapp, R.A. Decker, M.S. Dyroll, and H.A. Rayner, *J. Polym. Sci.*, 6, 31, 1951.
39. R.N. Datta and A.G. Talma, *Kautschuk Gummi Kunststoffe*, 54, 372, 2001.
40. J. Timmer and W.J. Edwards, *Rubber Chem. Technol.*, 52, 319, 1979.
41. L.D. Hertz Jr. in *Handbook of Elastomers*, A.K. Bhoumick and H.L. Stephen, Eds., Marcel Dekker, New York, 1988, p. 445.
42. M.L. Studebaker and J.R. Beatty, *Rubber Chem. Technol.*, 47, 803, 1974.
43. R.N. Capps, *Rubber Chem. Technol.*, 59, 103, 1986.
44. A.R. Payne and R.E. Whittaker, *J. Appl. Polym. Sci.*, 16, 1191, 1972.
45. R. Vukov and G.J. Wilson, Paper presented at the meeting of the ACS Rubber Division, Denver, Colorado, Oct. 23–26, 1984, Paper no. 73.
46. I.I. Ostromysklenki, *Russian Phys. Chem. Soc. J.*, 47, 1467, 1915.
47. R.D. Allen, Gummi Asbestos, *Kunststoffe*, 36, 534, 1983.
48. R.C. Keller, *Rubber Chem. Technol.*, 61, 238, 1988.
49. W.C. Endstra and C.T.J. Wressmann, in *Elastomer Technology Handbook*, N.P. Cheremisinoff and Pn. Cheremisinoff, Eds., CRC Press, New Jersey, 1993, Chapter 12.
50. L. Ganzales, A. Rodriguez, and A. Macros, *Recent Res. Dev. Polym. Sci.*, 2, 485, 1998.
51. J. Johansson, Peroxide Dispersions and Their Applications, Paper presented at the Southern Rubber Group, Inc., Winter Technical Session, March 2000.
52. P.R. Dluzheski, Rubber Chem. Technol., 74, 451, 2001.
53. W.C. Endstra and C.T.J. Wressmann, Peroxide Crosslinking of EPDM Rubbers, in *Elastomer Technology Handbook*, N.P. Cheremisinoff and Pn. Cheremisinoff, Eds., CRC Press, New Jersey, 1993.
54. A.J. Wildschut, *Rubber Chem. Technol.*, 19, 86, 1946.
55. M.C. Kirkham, Current Status of Elatomer Vulcanization, in *Progress of Rubber Technology*, Vol. 41, Plastics and Rubber Institute, London, 1978.
56. J. Rehner and P.J. Flory, *Rubber Chem. Technol.*, 19, 900, 1946.
57. R.F. Martell and D.E. Smith, *Rubber Chem. Technol.*, 35, 141, 1962.
58. A.B. Sullivan, *J. Org. Chem.*, 31, 2811, 1966.
59. C.S.L. Baker, D. Barnard, and M. Porter, *Rubber Chem. Technol.*, 35, 141, 1962.
60. L.M. Gan and C.H. Chew, *Rubber Chem. Technol.*, 56, 883, 1983.
61. P. Kovacic and P.W. Hein, *Rubber Chem. Technol.*, 35, 528, 1962.
62. A.Y. Logothetis, Paper presented at the IRC, Kyoto, Japan, p. 73, 1985.
63. C.S.L. Baker, Non-Sulfur Vulcanization, in *Natural Rubber Science and Technology*, A. Rober, Eds., Oxford Science, London, 1988.
64. Th. Kempermann, *Rubber Chem. Technol.*, 61, 426, 1988.
65. Th. Kempermann and A.G. Bayer, Internal report, AN 594.
66. V.A. Shershnev, V.V. Glushko, B.A. Dogadkin, and A.I. Kurmeeva, *Kaut. Rezina*, 31, 20, 1972.
67. P.A. Vinogradev, L.P. Kastorski, K.E. Gavshinova, and N.G. Areskena, *Vysokomol. Soedin. Ser, B* 13, 692, 1971.
68. V.I. Salnikova, A.A. Chekonova, N.D. Zakharov, G.I. Kastrykine, and L.T. Litvinova, *Izv. Vyssh. Uchebn. Zaved., Khim. Khim. Tekhnol*, 21, 433, 1978.
69. A.A. Chekanova, G.I. Kostrygina, N.D. Zakharov, and E.V. Degtyarev, *Izv. Vyssh. Uchebn. Zaved., Khim. Khim. Tekhnol*, 23, 619, 1980.
70. W. Hofamann, *Vulkanisation und Vulkanizationshilfsmittel*, Geutner, Stuttgart, 1965, p. 335.
71. P.G. Moniotte, EP 70143 A1, 1982.
72. J.M. Delseth, D.E. Maier, and P.G. Moniotte, EP 109955 A2, 1983.
73. R.N. Datta and W.F. Helt, *Rubber World*, 216(5), 24, 1997.
74. R.N. Datta and F.A.A. Ingham, *Indian Rubber J.*, 8, 52, 1994.
75. A.J. Hogt, A. Talma, R.F. De Block, and R.N. Datta, EP 0555288 B1, 1996.

76. R.N. Datta et al., *Rubber World*, 212, 24, 1995.
77. R.N. Datta, A.G. Talma, and J.C. Wagenmakers, *Kautchuk Gummi Kunststoffe*, 49, 671, 1996.
78. R.N. Datta et al., *Rubber Chem. Technol.*, 69, 727, 1996.
79. R.N. Datta et al., *Rubber Chem. Technol.*, 70, 129, 1997.
80. R.N. Datta, D. Seeberger, A.G. Talma, and J.C. Wagenmakers, *Gummi Fasern Kunststoffe*, 49, 892, 1996.
81. R.N. Datta and J.C. Wagenmakers, *Kautchuk Gummi Kunststoffe*, 50, 274, 1997.
82. R.N. Datta and W.F. Helt, *Rubber World*, 216, 24, 1997.
83. R.N. Datta and J.C. Wagenmakers, *J. Polym. Mater.*, 15, 379, 1998.
84. R.N. Datta and F.A.A. Ingham, *Kautchuk Gummi Kunststoffe*, 51, 662, 1998.
85. R.N. Datta and F.A.A. Ingham, *Kautchuk Gummi Kunststoffe*, 52, 322, 1999.
86. R.N. Datta and M.J. Hondeveld, *Kautchuk Gummi Kunststoffe*, 54, 308, 2001.
87. R.N. Datta and F.A.A. Ingham, *Kautchuk Gummi Kunststoffe*, 52, 758, 1999.
88. R.N. Datta and F.A. Ingham, *Gummi Fasern Kunststoffe*, 55, 225, 2002.

15 Degradation and Protection

Rabin N. Datta and Nico M. Huntink

CONTENTS

15.1 INTRODUCTION

The relatively low-molecular weight (MW) antioxidants have undergone an evolutionary change towards higher-MW products with the objective to achieve permanence in the rubber polymer, without loss of antioxidant activity. In the last two decades, several approaches have been evaluated in order to achieve this objective: attachment of hydrocarbon chains to conventional antioxidants in order to increase the MW and compatibility with the rubber matrix; oligomeric or polymeric antioxidants; and polymer-bound or covulcanizable antioxidants. The disadvantage of polymer-bound antioxidants was tackled by grafting antioxidants on low-MW polysiloxanes, which are compatible with many polymers. New developments on antiozonants have focused on nonstaining and slow migrating products, which last longer in rubber compounds. Several new types of nonstaining antiozonants have been developed, but none of them appeared to be as efficient as the chemically substituted *p*-phenylenediamines. The most prevalent approach to achieve nonstaining ozone protection of rubber compounds is to use an inherently ozone-resistant, saturated backbone polymer in blends with a diene rubber. The disadvantage of this approach, however, is the complicated mixing procedure needed to ensure that the required small polymer domain size is achieved.

Rubber compounds can be degraded by reactions with oxygen, ozone, light, metal ions, and heat. Antidegradants protect rubber against aerobic aging (oxygen) and ozone attack. They are of prime importance and play a vital role in rubber products to maintain the properties at service conditions. Protection of rubbers or stabilization of cross-linked networks against anaerobic aging can be achieved via other approaches: employing an EV-curing system, application of 1,3-bis(citraconimidomethyl) benzene, hexamethylene-1,6-bis(thiosulfate) disodium salt dihydrate, hexamethylene-1,6-bis(dibenzyl-thiuram disulfide), Zn-soaps, etc. [1].

Degradation by oxygen and ozone proceeds via different chemical mechanisms and results in different effects on the physical properties of rubber [2–13]. Ozone degradation results in discoloration and eventual cracking of samples. Ozone degradation is primarily a surface phenomenon. Oxygen degradation results in hardening or softening (depending on the base polymer) throughout the rubber article. For example, vulcanizates that are based on natural rubber (NR), polyisoprene rubber (IR), and butyl rubber (IIR) preferably undergo cleavage reactions during the oxidation process; they generally become softer. During progressive aging, a cross-linking mechanism starts to dominate again: completely oxidized NR is usually hard and brittle. On the other hand, vulcanizates obtained from styrene–butadiene rubber (SBR), nitrile butadiene rubber (NBR), chloroprene rubber (CR), ethylene–propylene–diene monomer (EPDM) rubber, etc. undergo cyclization and cross-linking reactions that lead to hardening of the aged part. When completely oxidized, these vulcanizates are turned into hard and brittle products. Rubbers that do not contain C=C unsaturation, such as acrylic rubber (ACM), chlorinated polyethylene (CM), chlorosulfonated polyethylene (CSM), polychloromethyloxiran (CO), ethylene–ethyl acrylate copolymer (EAM), epichlorohydrin rubber (ECO), ethylene propylene rubber (EPM), ethylene–vinylacetate copolymer (EVM), rubbers with fluoro and fluoralkyl or fluoralkoxy substituent groups on the polymer chain (FKM), silicone rubber (Q), and others are much less sensitive to oxidation than diene rubbers.

Although conventional antidegradants such as (N-1,3-dimethylbutyl)-N′-phenyl-p-phenylene-diamine (6PPD) and N-isopropyl-N′-phenyl-p-phenylenediamine (IPPD) provide protection against oxidation and ozonation, the protection lasts only for a short term; therefore, there is a trend and demand for longer-lasting and nonstaining products. Longer-term protection requires a different class of antidegradants. Long-lasting antioxidants must be polymer-bound or must have a lower volatility and leachability than conventional antioxidants, whereas long-lasting antiozonants must have a lower migration rate than the conventional antiozonants.

The purpose of this chapter is to review the developments on long-term protection of rubbers against aerobic aging, especially on long-term (more than five years, depending on service conditions) protection against ozone. Although numerous reviews of antioxidant and antiozonant aspects have been published [2–13], in most cases they only cover one element (i.e., fracture and fatigue in SBR and butadiene rubber (BR) vulcanizates [5]; the black sidewall surface discoloration and nonstaining technology [6]; ozonolysis of NR [11]; etc.), bound antioxidants, migration of the total field. In this chapter attempts will be made to summarize most developments with emphasis on long-term antioxidant as well as antiozonant protection.

15.2 OXIDATION AND ANTIOXIDANT CHEMISTRY

15.2.1 INTRODUCTION

The changes in properties observed on aging of different elastomers and their vulcanizates, and of many other polymeric materials, are well known. Antiozonants and antioxidants are employed to limit these changes. However, the most effective antioxidant for one material may be ineffective, or even harmful in another material or under different conditions. A rubber compounder must be aware of the effect of oxygen attack on rubber and should know how to compound for oxygen resistance.

15.2.2 MECHANISM OF RUBBER OXIDATION

The oxidation of polymers is most commonly depicted in terms of the kinetic scheme developed by Bolland [14]. The scheme is summarized in Figure 15.1. The key to the process is the initial formation of a free-radical species. At high temperatures and at large shear forces, it is likely that free-radical formation takes place by cleavage of C–C and C–H bonds.

Many elastomers are already observed to oxidize at moderate temperatures (below 60°C), where the energetics would not favor cleavage of C–C and C–H bonds. Thus, several studies have been conducted to determine whether trace impurities present in the polymer systems could account for the relative ease of oxidation. Two separate studies concluded that traces of peroxide were present in the polymer and that initiation occurred at low temperatures due to the relatively easy homolysis of these peroxides into free radicals [15,16]. Due to the high reactivity of free radicals, only trace amounts of these peroxides need to be present to provide initiation of the oxidative chain process. On the other hand, mechanical shear during processing and bale compaction and localized heat during the drying and packaging of the raw polymer are the most important causes of C–C and C–H bond cleavage. The resultant free radicals react with oxygen to form the peroxides responsible for degradation.

The oxidation of hydrocarbon polymers resembles the oxidation of low-molecular weight (MW) hydrocarbons, with the polymer having its own internal source of peroxide initiators present. By making the assumption that peroxides are present in even the most carefully prepared raw rubber, the ease of oxidation of rubber at low to moderate temperatures can be understood. Therefore, it is extremely important to compound rubber for extended oxidation resistance through the use of protective additives and to be aware of pro-oxidant impurities present in the rubber or the rubber compound.

Probably an important pro-oxidant for all rubbers is ultraviolet (UV) light. Blake and Bruce [17] performed a study of the oxygen absorption rates of unvulcanized NR rubbers under exposure to UV light. It was observed that exposure to light caused dramatic increases in the oxygen absorption

Initiation

$$RH \xrightarrow[\text{Shear}]{\Delta T} R\bullet + H\bullet$$

$$R-R \xrightarrow[\text{Shear}]{\Delta T} 2 R\bullet$$

$$R\bullet + O_2 \longrightarrow ROO\bullet$$

Propagation

$$ROO\bullet + RH \longrightarrow R\bullet + ROOH$$

$$ROOH \longrightarrow RO\bullet + OH\bullet$$

$$ROOH + RH \longrightarrow ROH + R\bullet + H_2O$$

$$RO\bullet + RH \longrightarrow ROH + R\bullet$$

$$OH\bullet + RH \longrightarrow HOH + R\bullet$$

Termination

$$ROO\bullet + R\bullet \longrightarrow ROOR$$

$$RO\bullet + R\bullet \longrightarrow ROR$$

$$R\bullet + R\bullet \longrightarrow RR$$

FIGURE 15.1 Bolland oxidation mechanism. (From Bolland, J.L., *Quartenary Rev. Chem. Soc.*, 3, 1, 1949.)

TABLE 15.1

Oxidation of Natural Rubber (NR) Pale Crepe at 46°C, Accelerated by Ultraviolet (UV) Light

Additive	Absorption of O_2 (cm^3/h)
None	0.067
2% Sulfur	0.028
2% Benzidine	0.014
2% Hydroquinone	0.014
2% Phenyl-β-naphtylamine	0.076
5% Zinc oxide	0.010
1% P-33 carbon black[a]	0.018

Source: From Blake, J.T. and Bruce, P.L., *Ind. Eng. Chem.*, 33, 1198, 1941.

[a] P-33 is a fine thermal black, ASTM nomenclature N880.

rate of NR. They studied the oxygen absorption rates of NR with various compounding additives. A summary of their results is given in Table 15.1. This table shows that phenyl-β-naphtylamine, an additive previously used for prevention of rubber oxidation (hardly used anymore because of toxicity reasons) can operate as a pro-oxidant under exposure to UV light. Fillers like zinc oxide, titanium dioxide, whiting, and especially carbon black lowered the rate of oxygen absorption of NR with exposure to UV light. This was attributed to the ability to make the compound opaque, thus limiting the penetration of UV light into the test films of NR. In the case of benzidine and hydroquinone, the effects were attributed to the ability of these materials to preferentially absorb the harmful UV light. Thus, it is very important to consider the pro-oxidant behavior of UV light when compounding rubbers for extended life.

The rate of peroxide decomposition and the resultant rate of oxidation are markedly increased by the presence of ions of metals such as iron, copper, manganese, and cobalt [13]. This catalytic decomposition is based on a redox mechanism, as in Figure 15.2. Consequently, it is important to control and limit the amounts of metal impurities in raw rubber. The influence of antioxidants against these rubber poisons depends at least partially on a complex formation (chelation) of the damaging ion. In favor of this theory is the fact that simple chelating agents that have no aging-protective activity, like ethylene diamine tetracetic acid (EDTA), act as copper protectors.

The rather simple sequence of reactions described in Figure 15.1 is complicated by other reactions, when oxidizable impurities or compounding ingredients are present. There are also the secondary processes whereby peroxides and free radicals undergo reactions leading to chain scission as well as cross-linking reactions. These reactions are closely related to the primary oxidation process, so that for a given type of polymer or vulcanizate the degree of deterioration of physical properties is generally proportional to the extent of oxidation.

15.2.3 STABILIZATION MECHANISM OF ANTIOXIDANTS

Complete inhibition of oxidation is seldom obtained in elastomers by addition of antioxidants or stabilizers. What is usually observed is an extended period of retarded oxidation in the presence of

$$ROOH + Fe^{2+} \longrightarrow RO\bullet + Fe^{3+} + OH^-$$
$$ROOH + Fe^{3+} \longrightarrow ROO\bullet + Fe^{2+} + H^+$$

FIGURE 15.2 Decomposition of peroxides by ions of metals (Redox mechanism).

the antioxidant. It has been demonstrated that during this period the rate of oxidation decreases with inhibitor concentration until the optimum concentration is reached and then increases again. The rate of the retarded reaction is affected by changes in oxygen concentration [18], in contrast to the uninhibited reaction, which proceeds at the same rate in oxygen or in air. These and other differences observed in the presence of oxidation inhibitors reflect significant changes in initiation and propagation, as well as in termination reactions.

It is important to recognize that different types of inhibitors often function by different mechanisms, and that a given antioxidant may react in more than one way. Thus, a material that acts as an antioxidant under one set of conditions may become a pro-oxidant in another situation. The search for possible synergistic combinations of antioxidants can be conducted more logically and efficiently if we seek to combine the effects of different modes of action. Five general modes of oxidation inhibition are commonly recognized:

1. *Metal deactivators*—Organic compounds capable of forming coordination complexes with metals are known to be useful in inhibiting metal-activated oxidation. These compounds have multiple coordination sites and are capable of forming cyclic structures, which "cage" the pro-oxidant metal ions. EDTA and its various salts are examples of this type of metal chelating compounds.
2. *Light absorbers*—These chemicals protect from photo-oxidation by absorbing the UV light energy, which would otherwise initiate oxidation, either by decomposing a peroxide or by sensitizing the oxidizable material to oxygen attack. The absorbed energy must be disposed of by processes which do not produce activated sites or free radicals. Fillers which impart opacity to the compound (e.g., carbon black, zinc oxide) tend to stabilize rubbers against UV-catalyzed oxidation.
3. *Peroxide decomposers*—These function by reacting with the initiating peroxides to form nonradical products. Presumably mercaptans, thiophenols, and other organic sulfur compounds function in this way [19]. It has been suggested that zinc dialkyldithiocarbamates function as peroxide decomposers, thus giving rubber compounds good initial oxidative stability.
4. *Free-radical chain stoppers*—These chemicals interact with chain propagating $RO_2\bullet$ radicals to form inactive products.
5. *Inhibitor regenerators*—These react with intermediates or products formed in the chain-stopping reaction so as to regenerate the original inhibitor or form another product capable of functioning as an antioxidant.

Termination of propagating radicals during the oxidative chain reaction is believed to be the dominant mechanism by which amine and phenolic antioxidants operate. The mechanism proposed to account for this behavior is given in Figures 15.3 and 15.4. The deactivation of R• via chain-breaking electron acceptors (CBA) is demonstrated for a hindered amine light stabilizer (HALS). The mechanism involves reaction of the HALS with a hydroperoxide, resulting in the formation of a stable nitroxyl radical that traps a hydrocarbon radical or abstracts a labile hydrogen from a hydrocarbon radical under formation of stable products. The hydroxylamine (CB-AH) formed via this mechanism can be used for the stabilization of peroxide radicals.

R• that is not fully deactivated via the mechanism described in Figure 15.3 reacts with oxygen resulting in a peroxide radical. These peroxy radicals abstract a labile hydrogen from primary stabilizers like hindered phenols or secondary amines, resulting in less active hydroperoxides and preventing hydrogen abstraction from the polymer chain. The resulting antioxidant radical is more stable than the initial peroxy radical and terminates by reaction with another radical in the system. This mechanism was proposed by Shelton [20], who demonstrated that replacement of the reactive hydrogen in aromatic amine antioxidants by deuterium results in a slower abstraction of deuterium by peroxy radicals and therefore in a less effective antioxidant. It has also been proposed that

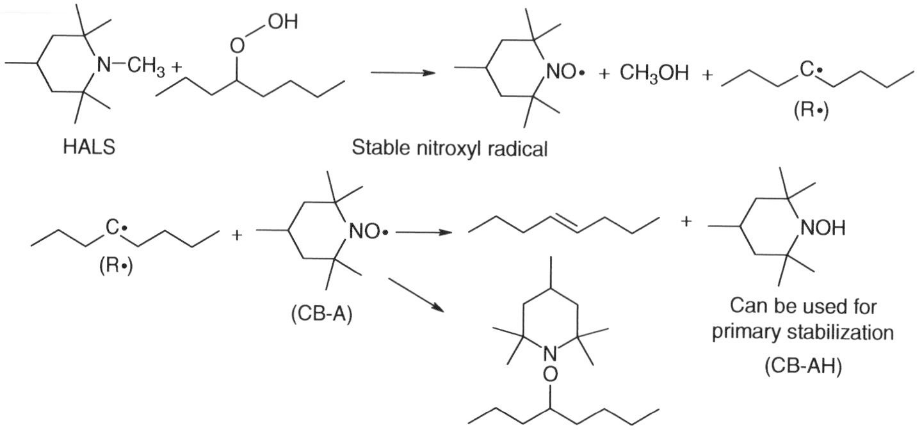

FIGURE 15.3 Deactivation of R• via chain-breaking electron acceptors (CB-A).

aromatic compounds such as phenols and aromatic amines can form π-electron complexes with peroxy radicals, which terminate to form stable products [15]. It appears that direct hydrogen absorption, π-electron complex formation, or both describe the antioxidant action of most amine and phenolic antioxidants. It is important that the level of antioxidant be kept at the optimum, since excess antioxidant can result in a pro-oxidant effect ($A-H + O_2 \rightarrow AOOH$).

The mechanism of secondary stabilization by antioxidants is demonstrated in Figure 15.5. *Tris*-nonylphenyl phosphites, derived from PCl_3 and various alcohols, and thio-compounds are active as a secondary stabilizer [21]. They are used to decompose peroxides into non-free-radical products, presumably by a polar mechanism. The secondary antioxidant is reacting with the hydroperoxide resulting in an oxidized antioxidant and an alcohol. The thio-compounds can react with two hydroperoxide molecules.

15.2.4 Methods of Studying the Oxidation Resistance of Rubber

The most common test used to study the oxidation resistance of rubber compounds involves the accelerated aging of tensile dumbbell samples in an oxygen-containing atmosphere. Brown et al. [22] recently reviewed long-term and accelerated aging test procedures. The ASTM practices (D 454 (09.01); D 865 (09.01); D 2000 (09.01, 09.02); D3137 (09.01); D 572 (09.01); D 3676 (09.02); D 380 (09.02)) for these tests clearly state that they are accelerated tests and should be used for relative comparisons of various compounds and that the tests may not correlate to actual long-term

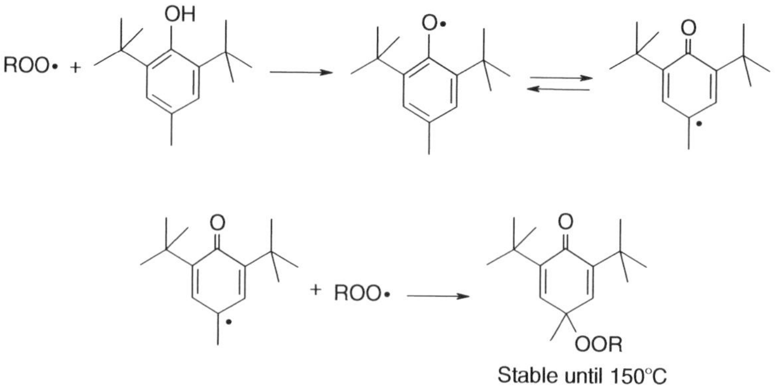

FIGURE 15.4 Primary stabilization via radical scavenging by hindered phenolics.

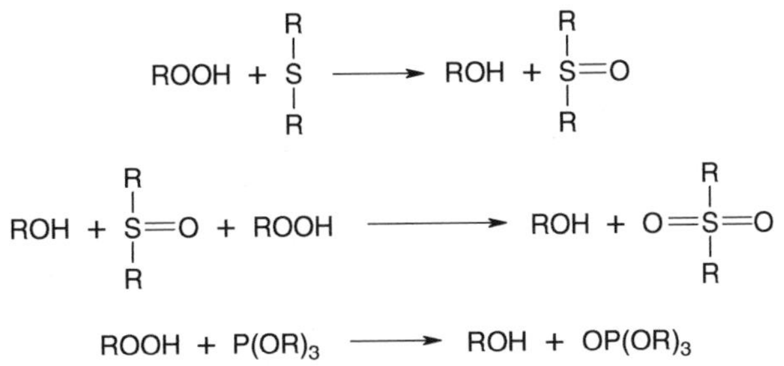

$$\text{ROOH} + \underset{R}{\overset{R}{S}} \longrightarrow \text{ROH} + \underset{R}{\overset{R}{S}}=O$$

$$\text{ROH} + \underset{R}{\overset{R}{S}}=O + \text{ROOH} \longrightarrow \text{ROH} + O=\underset{R}{\overset{R}{S}}=O$$

$$\text{ROOH} + P(OR)_3 \longrightarrow \text{ROH} + OP(OR)_3$$

FIGURE 15.5 Secondary stabilization by phosphates and thio-compounds.

aging behavior. However, these tests are useful in evaluating aging-resistant compounds and various antioxidant packages. The resistance of a compound to oxidation is generally measured by the percentage change in the various physical properties (e.g., tensile strength, elongation at break, hardness, modulus). For an elastomer which reacts with oxygen, resulting in cross-linking (generally butadiene-based elastomers such as BR, SBR, NBR), the accelerated tests result in increases in tensile modulus and hardness with a corresponding decrease in ultimate elongation. For an elastomer which reacts with oxygen resulting in chain scission (generally isoprene-based elastomers such as NR and IR), the accelerated aging tests result in decreases in tensile modulus and hardness with either increasing or decreasing ultimate elongation, depending on the extent of degradation [23]. The most effective antioxidant package for a given elastomer compound gives the smallest changes in physical properties during an accelerated aging test.

Thermoanalytical techniques such as differential scanning calorimetry (DSC) and thermogravimetric analysis (TGA) have also been widely used to study rubber oxidation [24–27]. The oxidative stability of rubbers and the effectiveness of various antioxidants can be evaluated with DSC based on the heat change (oxidation exotherm) during oxidation, the activation energy of oxidation, the isothermal induction time, the onset temperature of oxidation, and the oxidation peak temperature.

Spectroscopic techniques as 13C-NMR [28], ESR [29], pyrolysis-GC/MS, and pyrolysis-Fourier transform infrared (FTIR) [30], x-ray diffraction [31], and SEM [32] techniques are also used to study rubber oxidation.

15.3 OZONE AND ANTIOZONANT CHEMISTRY

15.3.1 INTRODUCTION

Layer and Lattimer [7] and Bailey [33] gave the historical background regarding protection of rubber against ozone.

As early as 1885 Thomson [34] observed that stretched rubber cracked on aging. In the early 1920s, a number of investigators studied this phenomenon in more detail. They found that cracks occurred only in stretched rubber, formed in a direction perpendicular to the elongation, and grew most rapidly at an elongation of about 10% [35,36]. Fabry and Buisson [37] observed crack formation in the presence of ozone, but questioned the influence of this ozone. Ozone was believed to be present only in the upper atmosphere and not at those places where rubber is commonly used. By 1935, analytical techniques had developed sufficiently to be able to measure the trace amounts of ozone, parts per hundred million (pphm), that were present in the troposphere [38]. Even so, these trace amounts were felt to be too insignificant to be the cause of severe damage. Therefore, other factors responsible for cracking were sought. Sunlight seemed to be an indispensable factor; hence names like "suncracking" and "sunchecking" were frequently used to describe this phenomenon. Direct sunlight, however, was not necessary, since cracking occurred equally well on the shady side as well as on the sunny side of the rubber [39]. Also dust was thought to be responsible for cracking.

Dust, once settled on the rubber and activated by sunlight, would give off oxidizing moieties and crack the rubber [39,40] Today, we know that only a few pphm of ozone in our atmosphere can cause severe cracking of rubber and that sunlight is responsible for its formation.

Ozone in the atmosphere is formed by the chemical reaction of atomic and molecular oxygen:

$$O + O_2 \rightarrow O_3$$

At high altitudes, the oxygen atoms are generated by the photolysis of molecular oxygen by the far UV light of the sun. In the troposphere, where only longer wavelength UV light exists, photolysis of nitrogen dioxide is the major source of oxygen atoms [41]:

$$NO_2 + h\nu \rightarrow NO + O$$

The nitric oxide produced in this reaction reacts with ozone to regenerate oxygen and nitrogen dioxide:

$$NO + O_3 \rightarrow NO_2 + O_2$$

An equilibrium is established that gives rise to a so-called photostationary-state relation, which depends on the relative rates of the above reactions:

$$[O_3] = j[NO_2]/k[NO]$$

where
  j = reaction rate of the formation of O_3
  k = reaction rate of the decomposition of O_3

Based solely on this relationship, it has been predicted that the ozone concentration should be about 2 pphm at solar noon in the United States. Indeed [7], in unpolluted environments, ozone concentrations are usually in the range of 2–5 pphm. However, in polluted urban areas, ozone concentrations can be as high as 50 pphm. Peroxy radicals formed from hydrocarbon emissions cause this enhanced ozone concentration. These radicals oxidize nitric oxide to nitrogen dioxide, thereby shifting the above steady-state relationship to higher ozone levels.

Since ozone is generated by photolytic reactions, anything which affects available sunlight will affect the ozone concentration. Consequently, ozone levels are the highest in the summer months, when the days are longer and the sun is more intense [42]. Similarly, ozone levels are highest near midday and decrease almost to zero at night [43]. Temperature has little effect on ozone formation.

The ozone-cracking problem was first taken seriously by the U.S. Government in the early 1950s. On reactivating military vehicles, moth-balled since World War II, it was found that tires were severely cracked and useless. Government-sponsored research projects rapidly led to the discovery of *p*-phenylenediamine antiozonants. Since then, these original antiozonants have been displaced by longer-lasting *p*-phenylenediamine derivatives.

15.3.2   Mechanism of Ozone Attack on Elastomers

Ozone cracking is an electrophilic reaction and starts with the attack of ozone at a location where the electron density is high [44]. In this respect unsaturated organic compounds are highly reactive with ozone. The reaction of ozone is a bimolecular reaction where one molecule of ozone reacts with one double bond of the rubber, as can be seen in Figure 15.6. The first step is a direct 1,3-dipolar addition of the ozone to the double bond to form a primary ozonide (I), or molozinide, which is only detectable at very low temperatures. At room temperature, these ozonides cleave as soon as they are formed to give an aldehyde or ketone and a zwitterion (carbonyl oxide). Cleavage occurs in the direction, which

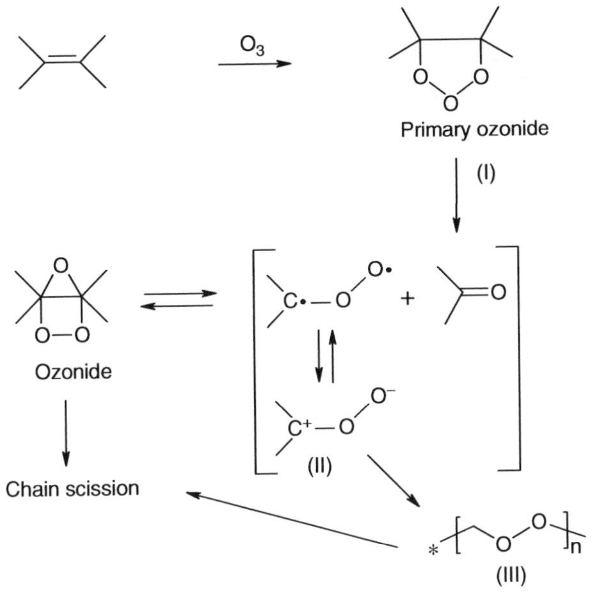

FIGURE 15.6 Ozone attack on double bonds.

favors the formation of the most stable zwitterion (II). Thus, electron-donating groups, such as the methyl group in NR, are predominately attached to the zwitterion, while electron-withdrawing groups, such as the chlorine in chloroprene rubber, are found on the aldehyde [45]. Normally, in solution, the aldehyde and zwitterion fragments recombine to form an ozonide, but higher-MW polymeric peroxides (III) can also be formed by a combination of zwitterions. The presence of water increases the rate of chain cleavage, which is probably related to the formation of hydroperoxides. The same chemistry occurs on ozonation of rubber, in solution and in the solid state [46].

Due to the retractive forces in stretched rubber, the aldehyde and zwitterion fragments are separated at the molecular-relaxation rate. Therefore, the ozonides and peroxides form at sites remote from the initial cleavage, and underlying rubber chains are exposed to ozone. These unstable ozonides and polymeric peroxides cleave to a variety of oxygenated products, such as acids, esters, ketones, and aldehydes, and also expose new rubber chains to the effects of ozone. The net result is that when rubber chains are cleaved, they retract in the direction of the stress and expose underlying unsaturation. Continuation of this process results in the formation of the characteristic ozone cracks. It should be noted that in the case of butadiene rubbers a small amount of cross-linking occurs during ozonation. This is considered to be due to the reaction between the biradical of the carbonyl oxide and the double bonds of the butadiene rubber [47].

The reaction of ozone with olefinic compounds is very rapid. Substituents on the double bond, which donate electrons, increase the rate of reaction, while electron-withdrawing substituents slow the reaction down. Thus, the rate of reaction with ozone decreases as follows: polyisoprene > polybutadiene > polychloroprene [48]. The effect of substituents on the double bond is clearly demonstrated in Tables 15.2 and 15.3. Rubbers that contain only pendant double bonds such as EPDM do not cleave since the double bond is not in the polymer backbone.

Although the cracking of rubbers is related to the reaction of ozone on the double bond, it must be mentioned that ozone reacts also with sulfur cross-links. These reactions, however, are much slower. The reaction of ozone with di- and polysulfides is at least 50 times slower than the corresponding reaction with olefins [49].

$$RSSSR + O_3 \rightarrow SO_2 + RSO_2 - O - SO_2R(+H_2O) \rightarrow 2\ RSO_2H$$

TABLE 15.2

Relative Second-Order Rate Constants for Ozonations of Selected Olefins in CCl$_4$ at Room Temperature

Olefins	Reaction Rate K_{rel} (L/mol s^{-1})
Cl$_2$C $=$ CCl$_2$	1.0
ClH $=$ CCl$_2$	3.6
H$_2$C $=$ CCl$_2$	22.1
cis-ClCH $=$ CHCl	35.7
trans-ClCH $=$ CHCl	591
H$_2$C $=$ CHCl	1180
H$_2$C $=$ CH$_2$	25,000
H$_2$C $=$ CHPr	81,000
H$_2$C $=$ CMe$_2$	97,000
cis-MeCH $=$ CHMe	163,000
Me$_2$C $=$ CHMe	167,000
Me$_2$C $=$ CMe$_2$	200,000
1,3-Butadiene	74,000
Styrene	103,000

Source: From Saito, Y., *Nippon Gomu Kyokaishi*, 5, 284, 1995.

Unstretched rubber reacts with ozone until all of the surface double bonds are consumed, and then the reaction stops [50]. The reaction is fast in the beginning, the rate progressively decreases while the available unsaturation is depleted and ultimately stops. During this reaction, a gray film, or frosting, forms on the surface of the rubber, but no cracks are noticed. The thickness of this film of ozonized rubber is estimated to be 10–40 molecular layers (60–240 Å) thick, based on the measurements of the ozone absorbed by unstretched rubber [51,52]. Disrupting this film by stretching brings new unsaturation to the surface and allows more ozone to be absorbed.

TABLE 15.3

Relative Second-Order Rate Constants for Ozonations of Different Unsaturated Rubbers in CCl$_4$ at Room Temperature

Rubbers	Reaction Rate K_{rel} (L/mol s^{-1})
CR	1.0
BR	1.5
SBR	1.5
IR	3.5

Source: From Layer, R.W. and Lattimer, R.P., *Rubber Chem. Technol.*, 63, 426, 1990; Bailey, P.S., in *Ozonation in Organic Chemistry*, Academic Press, New York, 1978; Razumovskii, S.D., Podmasteriev, V.V., and Zaikov, G.E., *Polymer Degrad. Stab.*, 20, 37, 1988.

Cracks are only observed when the rubber is stretched above a critical elongation. Two factors determine cracking under static conditions: the critical stress necessary for cracks to form and the rate of crack growth. It was established that all rubbers require the same critical stored energy for cracking to occur [53]. This energy is thought to be the energy necessary to separate the two surfaces of a growing crack. Thus depending on the stiffness of the polymer, cracks are formed above a certain elongation. Cracks will form and grow only if the ozonized surface products are moved aside to expose underlying unsaturation. Energy of some form is required to accomplish this. Under static conditions, this is equal to the critical stored energy. Under dynamic conditions, flexing by itself supplies the energy to disturb the surface and no critical energy is required.

The rate of crack growth depends on the polymer and is directly proportional to the ozone concentration. The rate of crack growth is independent of the applied stress as long as it exceeds the critical value. The rate of crack growth also depends on the mobility of the underlying chain segments of rubber, which is necessary to untangle and position double bonds for further attack by ozone. Consequently, anything that will increase the mobility of the rubber chains will increase the rate of crack growth. For example, the slow crack growth rate in IIR becomes equal to that of NR and SBR when sufficient plasticizer is added or when the temperature is raised [54]. Conversely, decreasing chain mobility diminishes the crack growth rate. For this reason, increasing the cross-linking density in some cases decreases mobility and reduces the rate of crack growth.

15.3.3 MECHANISM OF ANTIOZONANTS

Rubbers can be protected against ozone by use of chemical antiozonants and via several physical methods. The chemical antiozonants protect rubber under both static and dynamic conditions, whereas the physical methods are more related towards protection under static conditions.

15.3.3.1 Protection against Ozone under Static Conditions

There are several physical methods that can be used to protect rubber against ozone. They are wrapping, covering, or coating the rubber surface [55]. This can be accomplished by adding waxes to the rubber and/or adding an ozone-resistant polymer which increases the critical stress. Waxes are the most important in this respect. Two types of waxes are used to protect rubber against ozone, paraffinic, and microcrystalline. Paraffinic waxes are predominantly straight-chain hydrocarbons of relatively low MW of about 350–420. They are highly crystalline due to their linear structure and form large crystals having a melting range from 38°C to 74°C. Microcrystalline waxes are obtained from higher molecular petroleum residuals and have higher MWs than the paraffinic waxes, ranging from 490 to 800. In contrast to the paraffinic waxes, microcrystalline waxes are predominantly branched, and therefore form smaller, more irregular crystals which melt from about 57°C to 100°C. Waxes exert their protection by blooming to the surface to form a film of hydrocarbons which is impermeable to ozone. Protection is only obtained when the film is thick enough to provide a barrier to the ozone. Thus, the thicker the film, the better the protection. The obtained thickness of the bloom layer depends both on the solubility and the diffusion rate of the wax, which depend on the temperature. Bloom occurs whenever the solubility of the wax in the rubber is exceeded. Therefore, at temperatures lower than 40°C, the smaller and more soluble paraffinic waxes provide the best protection. Lowering the temperature reduces the solubility of the paraffinic waxes and increases the thickness of their bloom. Yet, their small size allows them to migrate rapidly to the surface, in spite of lower temperatures. Conversely, as the temperature increases, the high solubility of the paraffinic waxes becomes a disadvantage. They become too soluble in the rubber and do not form a thick enough protective bloom. Microcrystalline waxes perform better at higher temperatures, since higher temperatures increase their rate of migration to the surface and this allows more wax to be incorporated into the rubber. Therefore, blends of paraffinic and microcrystalline waxes are commonly used to guarantee protection over the widest possible temperature range [56]. Combinations of waxes and chemical antiozonants show synergistic improvement in ozone resistance [57]. The presence of the antiozonant results in a thicker bloom layer.

Another way to protect rubber against ozone is to add an ozone-resistant polymer (i.e., EPM, EPDM, halobutyl, polyethylene, polyvinyl acetate, etc.) to the rubber. Microscopic studies of these mixtures show that the added polymer exists as a separate, dispersed phase [58]. Consequently, as a crack grows in the rubber, it encounters a domain of the added polymer, which reduces the stress at the crack tip. This raises the critical stress required for cracking to occur, and crack growth ceases. Under dynamic conditions, where almost no critical stress is required, these polymer blends do not completely prevent cracking. In this case, they function by reducing the segmental mobility of the rubber chains and this slows the rate of crack growth. This method is effective when the polymer is added at a level between 20% and 50%. Higher levels do not result in further improvement of the ozone resistance [59]. At lower levels, propagation cracks circumvent the stress-relieving domains or will not reduce segmental mobility sufficiently. This method of protecting rubber against ozone is used on a limited basis, since vulcanizates of these blended rubbers frequently exhibit poorer properties. However, it is the only effective nondiscoloring method of protecting rubber under dynamic conditions.

15.3.3.2 Protection against Ozone under Dynamic Conditions

Under dynamic conditions, i.e., under cyclic deformations (stretching and compression), the physical methods to protect against ozone are no longer valid.

Chemical antiozonants have been developed to protect rubber against ozone under such dynamic conditions. Several mechanisms have been proposed to explain how chemical antiozonants protect rubber. The scavenging mechanism, the protective film mechanism, or a combination of both are nowadays the most accepted mechanisms.

The scavenging mechanism states that antiozonants function by migrating towards the surface of the rubber and, due to their exceptional reactivity towards ozone, scavenge the ozone before it can react with the rubber [60]. The scavenging mechanism is based on the fact that all antiozonants react much more rapidly with ozone than do the double bonds of the rubber molecules. This fact distinguishes antiozonants from antioxidants.

Studies of the reaction rates of various substituted paraphenylene diamines (PPDAs) towards ozone show that their reactivities are directly related to the electron density on the nitrogens due to the different substituents. Reactivity decreases in the following order: N,N,N',N'-tetraalkyl $> N,N,$ N'-trialkyl $> N,N'$-dialkyl $> N$-alkyl-N'-aryl $> N,N'$-diaryl [61]. It should be noted that their ease of oxidation decreases in the same order. As expected, PPDAs substituted by normal, secondary, and tertiary alkyl groups all exhibit essentially the same reaction rates. Only the initial reaction of the antiozonant with the ozone is rapid; the resulting ozonized products always react much more slowly. Thus, the number of moles of ozone absorbed by a compound is not necessarily an indication of its effectiveness. It is only the rate of reaction which is important.

By itself, the scavenging mechanism suffers from a number of shortcomings. According to this mechanism, the antiozonant must rapidly migrate to the surface of the rubber in order to scavenge the ozone. However, calculations show that the rate of diffusion of antiozonants to the rubber surface is too slow to scavenge all the available ozone [62]. Many compounds, which react very rapidly with ozone and therefore should be excellent scavengers, are not effective. A good example is the poor activity of N,N'-di-n-octyl-PPDA (DnOPPD) compared to the excellent activity of its *sec*-octyl isomer, DOPPD [13]. Since these isomers have the same reactivity towards ozone, the same solubility in rubber, the same MW (and diffusion rates) and melting points (both are liquids), and the difference in their antiozonant activities must reside in the nature of their ozonized products.

The protective film mechanism states that the rapid reaction of ozone with the antiozonant produces a film on the surface of the rubber, which prevents attack on the rubber, like waxes do [63].

This mechanism is based on the fact that the ozone uptake of elongated rubber containing a substituted p-phenylene diamine type of antiozonant is very fast initially and then decreases rather rapidly with time and eventually stops almost completely. The film has been studied spectroscopically and shown to consist of unreacted antiozonant and its ozonized products, but no ozonized rubber is involved [64]. Since these ozonized products are polar, they have poor solubility in the rubber and accumulate on the surface.

Currently, the most accepted mechanism of antiozonant action is a combination of the scavenging and the protective film formation. Based on this mechanism, one concludes that the higher critical elongation exhibited by DOPPD is due to the nature of the protective film which forms while scavenging ozone. The only way a film or coating can increase critical stress is, if it completely prevents ozone from reaching the surface. Only a continuous flexible film can do this. For example, wax forms such a protective but nonflexible film and increases the critical elongation [65]. A continuous flexible film also explains why DOPPD does not increase the critical elongation under dynamic conditions. In this case, flexing would disrupt the continuity of the film and destroy its ability to completely coat the rubber surface, just as flexing destroys the effectiveness of waxes. It also explains why DOPPD does not increase the critical elongation in NBR [66]. In NBR, very little DOPPD is found on the surface. Consequently, any film, which forms on the surface, is too thin to be effective. The difference in the amount of DOPPD on the surface of NBR compared to SBR is attributed to the higher solubility of DOPPD in NBR.

The effect of ozone and DOPPD concentrations on critical stress can be explained by considering the factors involved in film formation and destruction. At a fixed ozone concentration, increasing the concentration of DOPPD will increase the critical elongation because the equilibrium concentration of DOPPD on the surface of the rubber increases with loading. This results in the formation of a thicker, more durable and flexible film. The higher equilibrium surface concentration of DOPPD, lying just below the film, also guarantees that any of the film destroyed by ozone will be efficiently repaired before cracks can form. On the other hand, increasing the ozone concentration at a fixed DOPPD level decreases the critical stress because the film reacts and is too rapidly destroyed by ozone, to be repaired. Thus, the critical elongation will be that point, where the ozone concentration destroys the film more quickly than that it can be repaired. At very high ozone levels, this barrier is so quickly destroyed that the critical stress is the same as the value for an unprotected stock. Since IPPD does not increase critical elongation, its reaction products with ozone must form a barrier which contains many flaws. Indeed, IPPD is known to give a powdery bloom. However, combining IPPD with waxes results in a dramatic increase in the critical stress. This has been attributed to the ability of IPPD to facilitate wax migration and increase the thickness and continuity of the wax bloom [67].

15.3.3.3   Protection against Ozone by Substituted PPDs

The most effective antiozonants are the substituted PPDs. Their mechanism of protection against ozone is based on the "scavenger-protective film" mechanism [68–70]. The reaction of ozone with the antiozonant is much faster than the reaction with the $C=C$ bond of the rubber on the rubber surface [56]. The rubber is protected from the ozone attack till the surface antiozonant is depleted. As the antiozonant is continuously consumed through its reaction with ozone at the rubber surface, diffusion of the antiozonant from the inner parts to the surface replenishes the surface concentration to provide the continuous protection against ozone. A thin flexible film developed from the antiozonant/ozone reaction products on the rubber surface also offers protection.

In a PPD molecule, the aryl alkyl-substituted NH group is more reactive towards ozone than the bisaryl-substituted NH group owing to the higher charge density on the N-atom of the aryl alkyl-substitute [71]. This correlates very well with the literature report that aryl alkyl-PPD (e.g., 6PPD) produced nitrone, while the bisalkyl-PPD such as 77PD (N,N'-bis(1,4-dimethylpentyl)-p-phenylenediamine) produced dinitrone instead [68,69]. Apparently, the stabilizing effect of the N-aryl

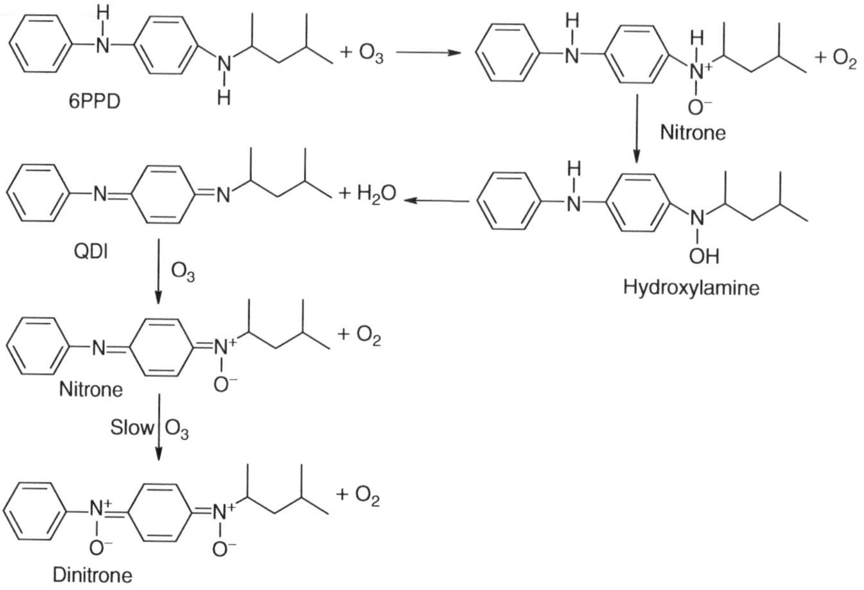

FIGURE 15.7 Ozonation mechanism for aryl alkyl-PPDs. (From Hong, S.W. and Lin, C.-Y., *Rubber World*, 36, August 2000.)

group on the nitrone retards further reaction of the nitrone with ozone. A simplified reaction mechanism for the aryl alkyl-PPDs such as 6PPD is depicted in Figure 15.7.

15.3.3.4 Methods of Studying the Ozone Resistance of Rubber

Since ozone attack on rubber is essentially a surface phenomenon, the test methods involve exposure of the rubber samples under static and/or dynamic strain, in a closed chamber at a constant temperature, to an atmosphere containing a given concentration of ozone. Cured test pieces are examined periodically for cracking.

The length and amount of cracks is assessed according to the Bayer method [72,73]. The ISO standard ozone test conditions involve a test temperature of $40°C \pm 1°C$ and an ozone level of 50 ± 5 pphm, with a test duration of 72 h. Testing is done under static [72] and/or dynamic strain [73]. These are accelerated tests and should be used for the relative comparison of compounds, rather than for the prediction of long-term service life. The method is rather complicated and demands a long duration of ozone exposure. Therefore, in some cases the rate constants of the antiozonants reaction with ozone in solution are used instead to evaluate the efficiency of different antiozonants [74].

The loss of antiozonants, either in a chemical or physical manner, appears to be the limiting factor in providing long-term protection of rubber products. That is why for new antiozonants not only the efficiency of the antiozonants must be evaluated, but one also has to watch other properties which influence their protective functions in an indifferent manner. For example, the molecule's mobility, its ability to migrate, is one of the parameters determining the efficiency of antiozonant action. Determination of the mobility kinetics of antiozonants can be done with a gravimetric method elaborated by Kavun et al. [75]. This method was used to determine the diffusion coefficient of several substituted PPDs, in different rubbers and at different temperatures [76]. The diffusion coefficients were calculated using the classical diffusion theory: Table 15.4. The diffusion coefficients increase with increasing temperature and with decreased compatibility with the rubber.

TABLE 15.4

Diffusion Coefficients for *N*-Isopropyl-*N'*-Phenyl-*p*-Phenylenediamine (IPPD), *N*-(1,3-Dimethylbutyl)-*N'*-Phenyl-*p*-Phenylenediamine (6PPD), and *N*-(1-Phenylethyl)-*N'*-Phenyl-*p*-Phenylenediamine (SPPD) (See Figure 15.14), in Different Rubbers and at Different Temperatures

Rubber	Temperature (°C)	D (cm^2/s)		
		IPPD	6PPD	SPPD
NR/BR	10	1.16E–8	7.82E–9	6.56E–9
	25	2.99E–8	1.92E–8	1.54E–8
	38	6.89E–8	4.55E–8	3.58E–8
	62	1.88E–7	1.47E–7	1.20E–7
	85	3.51E–7	2.79E–7	2.17E–7
NR	10	3.40E–9	1.70E–9	1.30E–9
	38	2.56E–8	1.39E–8	1.11E–8
	62	1.19E–7	7.05E–8	6.05E–8
	85	3.11E–7	2.34E–7	1.66E–7
SBR1500	38	1.03E–8	6.13E–9	4.62E–9
	62	4.28E–8	3.05E–8	2.47E–8
	85	1.36E–7	9.72E–8	6.02E–8
BR	38	1.32E–7	8.56E–8	6.79E–8
	62	2.71E–7	1.99E–7	1.64E–7

Source:　From Lehocky, P., Syrovy, L., and Kavun, S.M., RubberChem'01, Brussels, 2001, Paper 18.

The lower diffusion coefficient observed for *N*-(1-phenylethyl)-*N'*-phenyl-*p*-phenylenediamine (SPPD) compared to that of IPPD and 6PPD was explained by an increased MW and/or increased compatibility with the rubbers.

15.4　MECHANISM OF PROTECTION AGAINST FLEX CRACKING

Flex cracking, the occurrence and growth of cracks in the surface of rubber when repeatedly submitted to a deformation cycle, is determined by fatigue testing. Fatiguing of rubbers at room temperature is a degradation process caused by repeated mechanical stress under limited access of oxygen. The mechanical deformation stress is believed to generate macroalkyl radicals (R•). A small fraction of the macroalkyl radicals reacts with oxygen to form alkylperoxy radicals, still leaving a high concentration of the macroalkyl radicals. Consequently, removal of the macroalkyl radicals in a catalytic process constitutes a prevailing antifatigue process [9]. On the other hand, the macroalkyl radicals are rapidly converted to the alkylperoxy radicals under air–oven heat aging. The auto-oxidation propagated by the alkylperoxy radicals thus dominates the degradation process. Therefore, removal of the alkylperoxy radicals becomes the primary function of an antioxidant.

It has been shown that diarylamines are good antifatigue agents and that diarylamine nitroxyl radicals are even more effective than the parent amines. The antifatigue mechanism of the amine antidegradants, shown in Figure 15.8, has been proposed where the formation of the intermediate nitroxyl radicals plays an active role [77]. Generation of nitroxyl radicals from the free amines is first depicted in Figure 15.8. In the fatigue process, macroalkyl radicals are generated and subsequently removed by reaction with these nitroxyl radicals. The resulting hydroxylamine can be reoxidized by alkylperoxy radicals to regenerate the nitroxyl radicals in an auto-oxidation chain-breaking process.

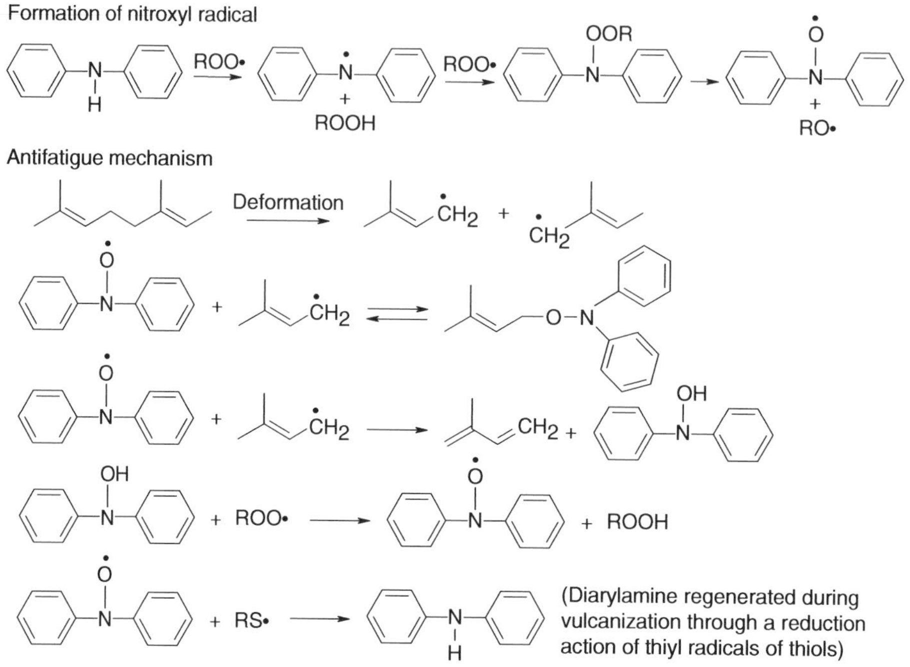

FIGURE 15.8 Antifatigue mechanism of diarylamines.

The nitroxyl radicals can be partially converted back to the free diarylamine during vulcanization through the reductive action of thiyl radicals of thiols. The free diarylamine thus regenerated, would repeat the reaction described in Figure 15.8 to form more nitroxyl radicals.

15.5 TRENDS TOWARD LONG-LASTING ANTIDEGRADANTS

15.5.1 INTRODUCTION

Antidegradants are very important compounding additives in their role to economically maintain rubber properties at service conditions. Although conventional antidegradants such as 6PPD and IPPD provide protection against oxygen and ozone, this protection is of short duration (1–5 years, depending on the service conditions). Producers of rubber chemicals are focussing on new developments, addressing longer and better protection of rubber products [78]. Therefore, several new types of long-lasting antioxidants and long-lasting antiozonants (5–15 years, depending on the service conditions) have been developed over the last two decades.

15.5.2 LONG-LASTING ANTIOXIDANTS

To limit the thermal oxidative deterioration of elastomers and their vulcanizates during storage, processing, and use, different systems of antioxidants are used. The activity of the antioxidants depends on their ability to trap peroxy and hydroperoxy radicals and their catalytic action in hydroperoxide decomposition. Their compatibility with the polymers also plays a major role. Moreover, it is very important to limit antioxidant loss by extraction (leaching) or by volatilization. Food packaging and medical devices are areas in which additive migration or extraction is of major concern. Contact with oils or fats could conceivably lead to ingestion of mobile polymer stabilizers. In an effort to address this concern, the U.S. Food and Drug Administration (FDA) has set a code of regulations governing the use of additives in food contact applications [79]. These regulations contain a list of acceptable polymer additives and dose limits for polymers, which may be used for

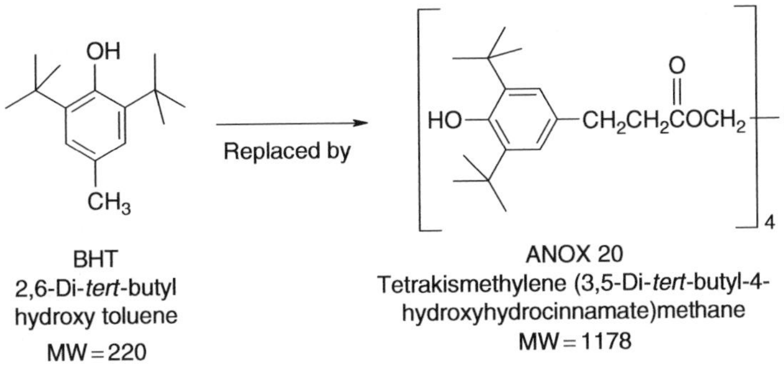

FIGURE 15.9 Replacement of low-molecular weight phenolic AO$_x$ by high-molecular weight product.

specific food contact. Inclusion of a particular compound in this list depends both on specific extractability and toxicological factors. Obviously, polymer-bound stabilizers cannot be extracted and would therefore prevent inadvertent food contamination. An additional consideration is the effect of additive migration on surface properties. As additives migrate or bloom to the polymer surface, the ability to seal or coat the surface may deteriorate. This affects coating adhesion and lamination peel strength.

The above-described issues are the reasons for an increased interest in the synthesis of new antioxidants with the possibility to graft to the polymer backbone or to form polymeric or oligomeric antidegradants. In the last two decades, several approaches have been evaluated in order to develop such new antioxidants:

- Attachment of hydrocarbon chains to conventional antioxidants in order to increase the MW and compatibility with polymers [80]
- Polymeric or oligomeric antioxidants [81]
- Polymer-bound or covulcanizable antioxidants [82–84]
- Binding of several functional groups onto a single platform [85]

Examination of the history of antioxidants such as hindered phenols and amines shows a move from low-MW products to higher-MW products. Specifically, polymer industries have abandoned the use of, e.g., butylated hydroxy toluene (BHT) in favor of tetrakismethylene (3,5-di-*t*-butyl-4-hydroxydrocinnamate)methane (see Figure 15.9). Likewise, polymeric HALS, like poly-methylpropyl-3-oxy-(4(2,2,6,6-tetramethyl)piperidinyl) siloxane, replaced the low-MW hindered amine Lowilite 77 (see Figure 15.10). The next obvious step was to produce a new class of stabilizers,

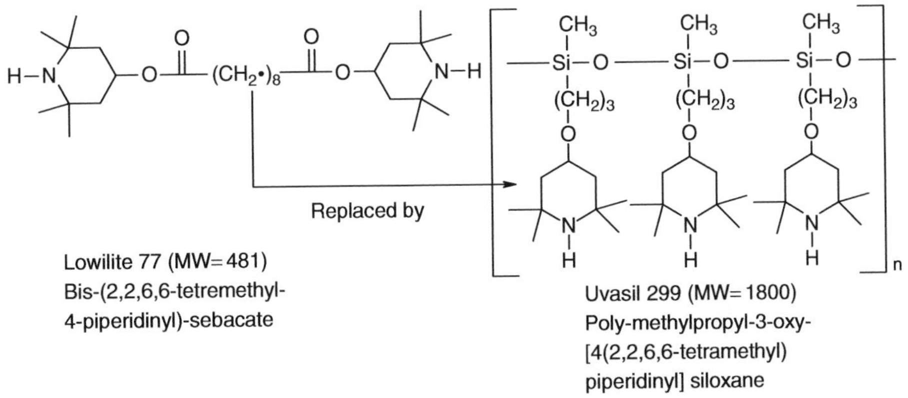

FIGURE 15.10 Replacement of low-molecular weight HALS by high-molecular weight product.

which are chemically bound to the polymer chain. This approach has had varying degrees of success. While the extraction resistance of the bound stabilizers was significantly improved, performance suffered greatly. Because degradation processes may occur in localized portions of the bulk of the polymer, mobility of the stabilizer plays a key role in antioxidant activity.

The antioxidative activities of polymeric antioxidants prepared from Verona oil and the conventional phenolic antioxidant 3-(3,5-di-*tert*-butyl-4-hydroxyphenyl)propionic acid (DTBH), chemically grafted to polystyrene and polyurethanes, is similar and in some cases even better than that of the corresponding low-MW phenolic antioxidants [81].

Several ways of obtaining polymer-bound antioxidants have been described. Roos and D'Amico reported polymerizable *p*-phenylene diamine antioxidants [86,87]. Cain et al. [88] reported the "ene" addition of nitrosophenols or aniline derivatives to produce polymer-bound stabilizers. The most versatile method of preparation of bound antioxidant is by the direct reaction of a conventional antioxidant with a polymer. Sirimevan Kularatne and Scott [89] have demonstrated that simple hindered phenols, which contain a methyl group in the *o*- or *p*-position, can react with NR in the presence of oxidizing free radicals to yield polymer-bound antioxidants. The antioxidants like styrenated phenol, diphenylamine, etc. bound to hydroxyl-terminated liquid NR by modified Friedel-Craft's reactions were found to be effective in improving the aging resistance [90]. PPDs bound to NR showed improved aging resistance compared to conventional PPDs, but as expected a worse ozone resistance because the bounded antidegradants cannot migrate to the surface [82]. Quinone diimines (QDIs) have been reported as bound antioxidants and diffusable antiozonants. During vulcanization, part of the QDI is grafted to the polymer backbone and acts as a bound antioxidant, whereas the other part is reduced to PPD and is active as diffusable an antiozonant [83].

The protection efficiency of antioxidant couples consisting of a classical compound (disub-stituted *p*-phenylenediamines and dihydroquinoline derivatives) and compounds with a disulfide bridge, resulting from diamine and phenolic structures, was reported by Meghea and Giurginca [84]. Antioxidants containing a disulfide bridge are able to graft onto the elastomer chain during processing and curing, leading to a level of protection superior to the classical antidegradants.

One of the latest developments in stabilizers is the polysiloxanes, which provide flexible, versatile backbones for a variety of classes of polymer stabilizers. The siloxanes appear to be good backbones, because they are rather inexpensive, easily functionalized, have a high level of functionalizable sites, good compatibility with many polymers, and excellent thermo and photolytic stability [85]. Hindered amines, hindered phenolics, and metal deactivators have been grafted onto the polysiloxanes. The low extractability of siloxane-based additives was further enhanced by the inclusion of graftable pendant groups onto the polysiloxane backbone [91]. The grafted stabilizers maintain their activity due to the flexible siloxane platform. This was seen as a limitation of monomeric stabilizers, which have been grafted onto the polymer matrix and thus are not mobile at all.

15.5.3 Long-Lasting Antiozonants

There is a clear demand for long-lasting antiozonants (two or three times longer-lasting than conventional antiozonants as IPPD and 6PPD) and for nonstaining and nondiscoloring antiozonants for better appearance products such as tire sidewalls. The functional classes of antiozonants include substituted monophenols, hindered bisphenols and thiobisphenols, substituted hydroquinones, organic phosphites, and thioesters [2]. Triphenyl phosphine, substituted thioureas and isothioureas, thiosemicarbazides, esters of dithiocarbamates, lactams, and olefinic and enamine compounds are reported as nonstaining antiozonants [92,93]. Approaches to completely replace the PPD's with a nondiscoloring antiozonant have had only limited success, leading to the development of new classes of nonstaining antiozonants.

Warrach and Tsou [115] reported that bis-(1,2,3,6-tetrahydrobenzaldehyde)-pentaerythrityl acetal provides superior ozone protection for polychloroprene, butyl rubber, chlorobutyl, and

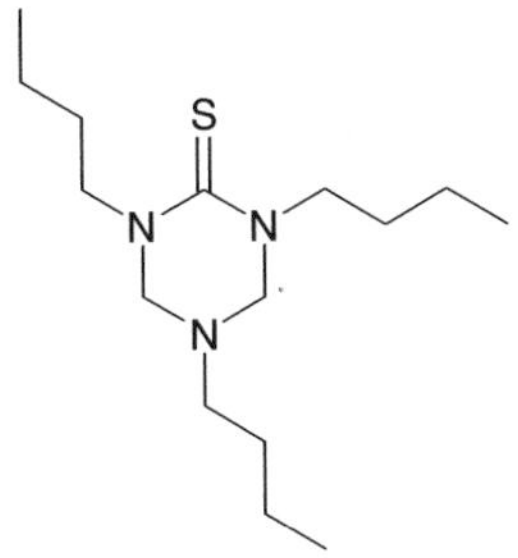

FIGURE 15.11 Tetrahydro-1,3,5-tri-*n*-butyl-(*S*)-triazinethione.

bromobutyl rubbers relative to *p*-phenylenediamine antiozonants, without discoloring the rubber or staining white-painted steel test panels. However, the ozone resistance of diene elastomers (NR, polyisoprene, SBR, polybutadiene, nitrile rubber) or inherently ozone-resistant elastomers (ethylene–propylene copolymers, ethylene–propylene–diene terpolymers, chlorosulfonated polyethylene, ethylene vinylacetate) is not improved by this compound.

Rollick et al. [92–94] reported a new class of antiozonants for rubber that do not discolor upon exposure to oxygen, ozone, or UV light, namely triazinethiones. Only changes in substitution on the nitrogen adjacent to the thiocarbonyl group affected their antiozonant efficiency. Accelerated weatherometer aging of a titanium dioxide/treated-clay filled SBR compound showed their nondiscoloring nature. Use of 4 phr tetrahydro-1,3,5-(*n*)-butyl(*S*)-triazinethione (see Figure 15.11) resulted essentially in no change in color, whereas use of only 1phr of *N*-(1,3-dimethylbutyl)-*N'*-phenyl-para-phenylenediamine significantly discolored the rubber compared to the control, see Table 15.5. The triazinethiones provided a significant degree of ozone protection to an

TABLE 15.5
Weatherometer-Induced Discoloration of 6PPD- and TBTT-Containing Styrene–Butadiene Rubber (SBR)

Antiozonant[a]	Hours	L	A	B	ΔE^b
None	24	89.89	0	7.78	
	48	89.61	−0.10	10.19	
	96	89.72	−0.10	11.90	
1 phr 6PPD[c]	24	34.21	3.69	8.21	55.80
	48	40.25	2.47	7.93	49.48
	96	45.93	1.86	8.40	43.97
4 phr TBTT[d]	24	86.66	0.80	10.28	4.09
	48	87.81	0.81	10.99	2.17
	96	87.52	1.01	11.49	2.50

Source: From Rollick, K.L., Gillick, J.G., Bush, J.L., and Kuczkowski, J.A., *ACS*, Cincinnati, OH, October 18–21, 1988, Paper 52.

[a] Formulation: 100 phr SBR 1502, 30 phr titanium dioxide, 30 phr mercaptosilinated clay, 10 phr ZnO, 5 phr naphthenic oil, 2 phr stearic acid, 2 phr sulfur, 0.25 phr TMTD (tetramethyl thiuram disulfide).

[b] $\Delta E = \sqrt{((\Delta L)^2 + (\Delta a)^2 + (\Delta b)^2)}$ = change in whiteness (ΔL), hue (Δa) and chroma (Δb) upon aging.

[c] *N*-(1,3-dimethylbutyl)-*N'*-phenyl-para-phenylenediamine.

[d] Tetrahydro-1,3,5-tri-(*n*)-butyl-(*S*)-triazinethione.

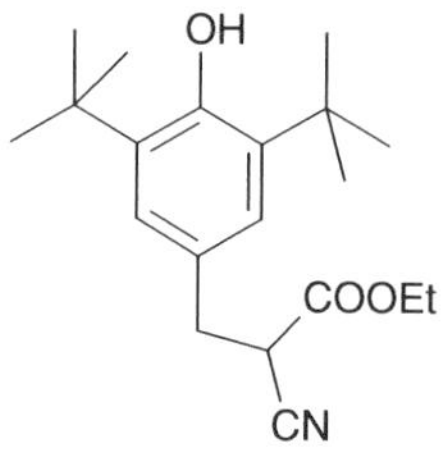

FIGURE 15.12 3,5-di-*tert*-butyl-4-hydroxybenzyl cyanoacetate.

NR–butadiene rubber black sidewall compound. They are reported as particularly valuable for use in light-colored stocks.

Ivan et al. [95] reported that 3,5-di-*tert*-butyl-4-hydroxybenzylcyanoacetate is a nonstaining antiozonant that affords similar protection to NR and to *cis*-polyisoprene compounds such as *N*-isopropyl-*N'*-phenyl-para-phenylenediamine does (see Figure 15.12). This product lasts longer than conventional nonstaining antiozonants, like the styrenated phenols.

Wheeler [96] described a new class of nonstaining antiozonants, namely the *tris-N*-substituted-triazines. 2,4,6-*tris*-(*N*-1,4-dimethylpentyl-para-phenylenediamino)-1,3,5-triazine (see Figure 15.13) gave excellent ozone resistance in an NR–butadiene rubber compound when compared to *N*-(1,3-dimethylbutyl)-*N'*-phenyl-para-phenylenediamine, but without contact, migration, or diffusion staining (see Table 15.6). Hong [97] reported equal dynamic ozone performance of this triazine antiozonant to the PPDs in both NR and butadiene rubber compounds. Birdsall et al. [98] described that the triazine antiozonant formed a discoloring bloom on an NR–butadiene rubber compound, but that the bloom was minimal when compounded at 2 phr or lower levels. When used in combination with a PPD, better ozone protection is obtained compared to using the triazine antiozonant alone, at the same total level of antiozonant. The combination of PPD and triazine antiozonant provides longer-term protection [99].

Lehocky et al. [76] reported the migration rates of IPPD, 6PPD, and SPPD (see Figure 15.14), determined in different polymers and at different temperatures. SPPD showed the lowest migration rate and is therefore expected to last longest in rubber compounds, see Table 15.4. However, the importance of the migration rate should not be overestimated, as the value is not sufficient to determine the effect and the efficiency of the antiozonant.

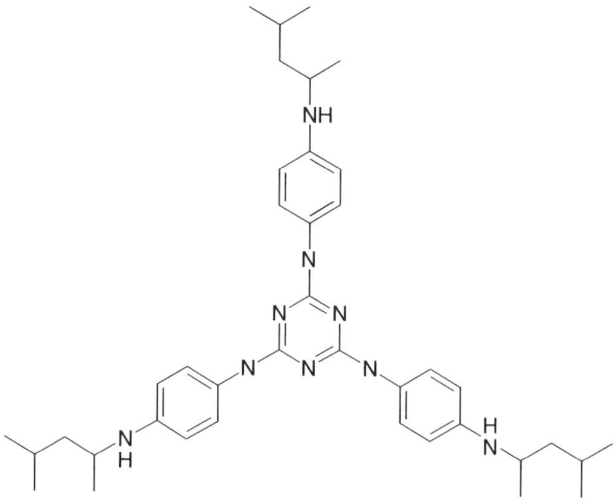

FIGURE 15.13 2,4,6-*tris*-(*N*-1,4-dimethylpentyl-*p*-phenylenediamino)-1,3,5-triazine (TAPTD).

TABLE 15.6
Staining Experiments[a] of the Triazine Antiozonant

Method	Antiozonant	L
A. Contact stain	Blank	87.10
After 96 h	TAPDT[b]	83.77
	HPPD[c]	65.59
B. Migration stain	Blank	86.89
After 96 h	TAPDT[b]	87.53
	HPPD[c]	77.79
C. Diffusion stain, exposed to sunlamp	Blank	88.10
4 h at 328 K	TAPDT[b]	82.42
	HPPD[c]	32.65

Source: From Wheeler, E.L., *ACS*, Detroit, MI, October 17, 1989, Paper 73.

[a] Tests were carried out as designated in ASTM Method D-925–83 related to staining of surfaces by contact, migration, or by diffusion. Hunter color values were measured on the L-scale. On this scale 100 is white and 0 is black.

[b] *Tris*-(N-1,4-dimethylpentyl-*p*-phenylenediamino)-1,3,5-triazine.

[c] N-(1,3-dimethylbutyl)-N'-phenyl-*p*-phenylenediamine.

The most prevalent approach to achieve long-lasting and nonstaining ozone protection of rubber compounds is to use an inherently ozone-resistant, saturated backbone polymer in blends with a diene rubber. The ozone-resistant polymer must be used in sufficient concentration (minimum 25 phr) and must also be sufficiently dispersed to form domains that effectively block the continuous propagation of an ozone-initiated crack through the diene rubber phase within the compound. Elastomers such as ethylene–propylene–diene terpolymers, halogenated butyl rubbers, or brominated isobutylene-co-para-methylstyrene elastomers have been proposed in combination with NR and/or butadiene rubber.

Ogawa et al. [100,101] reported the use of various EPDM polymers in blends with NR in black sidewall formulations. Laboratory testing showed improved resistance to crack growth and thermal aging.

Hong [102] reported that a polymer blend of 60 phr of NR and 40 phr of EPDM rubber afforded the best protection of a black sidewall compound to ozone attack. Use of a higher-MW EPDM rubber gave good flex fatigue-to-failure and adhesion to both carcass and tread compounds. TAPDT mixed with the NR to form a masterbatch followed by blending with the EPDM rubber and other ingredients, afforded the most effective processing in order to protect the NR phase. Compounds containing this NR–EPDM rubber blend (60/40) with 2.4 phr of the triazine antiozonant passed all requirements for the tire black sidewall.

Sumner and Fries [103] reported that ozone resistance depended on the level of the EPDM rubber. When using 40 phr of EPDM rubber in the compound there is no cracking throughout the

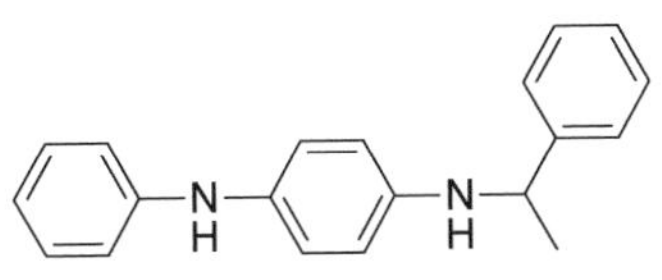

FIGURE 15.14 Structure of SPPD (N-1-phenylethyl)-N'-phenyl-*p*-phenylenediamine).

life of the black sidewall. Ozone resistance also depends on proper mixing of the EPDM rubber with NR in order to achieve a polymer domain size of less than 1 μm. Otherwise cracking can be severe. The combination of high-MW and high ethylidene norbornene (ENB) content afforded good adhesion to highly unsaturated polymers. The adhesion mechanism involves the creation of radicals when long chains of EPDM rubber and NR are broken down by shearing and mechanical work. Grafting between the two elastomers is believed to occur. The graft polymer is thought to act as compatibilizer. The NR–EPDM rubber compound does not rely on migration of antidegradants to achieve ozone resistance and therefore does not stain the sidewall. Appearance is excellent throughout the service life of the tire. However, at the current stage of development, NR–EPDM rubber sidewall compounds are difficult to mix, too expensive, result in an increased rolling resistance, and have a reduced tack compared to NR–butadiene rubber sidewall compounds. Related work carried out by Polysar in this field is described in detail [104–114].

As a general view on antidegradant activity with respect to type, structures, etc., authors would like to refer to the references described below. Antidegradants are divided into staining and nonstaining products, with or without fatigue, ozone, and oxygen protection. The most commonly used antidegradants for general-purpose rubbers were reported by Huntink et al. [116].

REFERENCES

1. R.N. Datta and N.M. Huntink, RubberChem'01, Brussels, 2001, Paper 5; R.N. Datta, *Progress in Rubber, Plastics and Recycling Technology*, 2003, **19**(3), 143.
2. J.R. Davies, *Plastics and Rubber International*, 1986, **11**(1), 16.
3. R.W. Keller, *Rubber Chemistry and Technology*, 1985, **58**, 637.
4. B.L. Stuck, Technical Report, Sovereign Chemical Company, Sept. 17, 1995.
5. J. Zhao and G.N. Ghebremeskel, *Rubber Chemistry and Technology*, 2001, **74**, 409.
6. W.H. Waddel, *Rubber Chemistry and Technology*, 1998, **71**, 590.
7. R.W. Layer and R.P. Lattimer, *Rubber Chemistry and Technology*, 1990, **63**, 426.
8. D. Erhardt, *International Polymer Science and Technology*, 1998, **25**, 7, 11.
9. S.K. Rakovski and D.R. Cherneva, *International Journal of Polymer Materials*, 1990, **14**, 21.
10. G. Scott, *Rubber Chemistry and Technology*, 1985, **58**, 269.
11. S.S. Solanky and R.P. Singh, *Progress in Rubber Plastics Technology*, 2001, **17**(1), 13.
12. R.L. Gray and R.E. Lee, *Chemistry and Technology of Polymer Additives*, 1999, **21**, 21.
13. J.C. Ambelang, R.H. Kline, O.M. Lorenz, C.R. Parks, and C. Wadelin, *Rubber Chemistry and Technology*, 1963, **36**, 1497.
14. J.L. Bolland, *Quartenary Reviews Chemistry Society*, 1949, **3**, 1.
15. J.R. Shelton and D.N. Vincent, *Journal of the American Chemistry Society*, 1963, **85**, 2433.
16. L. Bateman, M. Cain, T. Colclough, and J.I. Cunneen, *Journal of the Chemical Society*, 1962, 3570.
17. J.T. Blake and P.L. Bruce, *Industrial and Engineering Chemistry*, 1941, **33**, 1198.
18. J.R. Shelton, *Rubber Chemistry and Technology*, 1957, **30**, 1270.
19. W.L. Hawkins and M.A. Worthington, *Journal of Polymeric Science*, Part A, *Polymer Chemistry*, 1963, **1**, 3489.
20. J.R. Shelton, *Journal of Applied Polymer Science*, 1959, **2**, 345.
21. J.S. Dick, *Rubber Technology: Compounding and Testing for Performance*, Hanser Publishers, Munich, 2001, 453.
22. R.P. Brown, M.J. Forrest, and G. Soulagnet, *Rapra Review Reports*, 2000, **10**, 2.
23. A.N. Gent, *Journal of Applied Polymer Science*, 1962, **6**, 497.
24. D.J. Burlett, *Rubber Chemistry and Technology*, 1999, **72**, 165.
25. N.C. Billingham, D.C. Bott, and A.S. Manke, *Developments in Polymer Degradation*, Applied Science Publishers, London, 1981, Chapter 3.
26. D.I. Marshall, E. George, J.M. Turnipseed, and J.L. Glenn, *Polymer Engineering and Science*, 1973, **13**, 415.
27. H.E. Bair, *Polymer Engineering and Science*, 1973, **13**, 435.
28. D.D. Parker and J.L. Koenig, ACS, *Polymer Materials Science and Engineering*, 1997, **76**, Conference proceedings, 193.

29. K. Kaluder, M. Andreis, T. Marinovic, and Z. Veksli, *Journal of Elastomers and Plastics*, 1997, **29**(4), 270.

30. G.N. Ghebremeskel, J.K. Sekinger, J.L. Hoffpauir, and C. Hendrix, *Rubber Chemistry and Technology*, 1996, **69**, 874.

31. H. Lavabratt, E. Ostman, S. Persson, B. Stenberg, and A.B. Skega, *Journal of Applied Polymer Science*, 1992, **44**(1), 83.

32. D.K. Setua, *Polymer Degradation and Stability*, 1985, **12**(2), 169.

33. P.S. Bailey, *Ozonation in Organic Chemistry*, Volume 39.1, Academic Press, New York, 1978, 3.

34. J.J. Thomson, *Journal of the Society of the Chemical Industry*, 1885, **4**, 710.

35. N.A. Shepard, S. Krall, and H.L. Morris, *Industrial and Engineering Chemistry*, 1926, **18**, 615.

36. I. Williams, *Industrial and Engineering Chemistry*, 1926, **18**, 367.

37. C. Fabry and H. Buisson, *Journal of Physics and Radium*, 1921, **2**(6), 197.

38. C. Fabry and H. Buisson, *Compte Rendu*, 1931, **192**, 457.

39. A. van Rossem and H.W. Talen, *Kautschuk Gummi Kunststoffe*, 1931, **7**, 115.

40. C. Dufraisse, *Rubber Chemistry and Technology*, 1933, **6**, 157.

41. J.H. Seinfeld, *Science*, 1989, **243**, 745.

42. D. Bruck, *Kautschuk Gummi Kunststoffe*, 1989, **42**(9), 76.

43. A.J. Haagen-Smit, M.F. Brunelle, and J.W. Haagen-Smit, *Rubber Chemistry and Technology*, 1959, **32**, 1134.

44. Y. Saito, *Nippon Gomu Kyokaishi*, 1995, **5**, 284.

45. D. Bruck, H. Konigshofen, and L. Ruetz, *Rubber Chemistry and Technology*, 1985, **58**, 728.

46. P.S. Bailey, *Ozonation in Organic Chemistry*, Volume I, Academic Press, New York, 1978, 25.

47. K.W. Ho, *Journal of Polymer Science*, 1986, **24**, 2467.

48. S.D. Razumovskii, V.V. Podmasteriev, and G.E. Zaikov, *Polymer Degradation and Stability*, 1988, **20**, 37.

49. D. Barnard, *Journal of the Chemical Society*, 1957, **79**, 4547.

50. F.R. Erickson, R.A. Berntsen, E.L. Hill, and P. Kusy, *Rubber Chemistry and Technology*, 1959, **32**, 1062.

51. E.H. Andrews and M. Braden, *Journal of Polymer Science*, 1961, **55**, 787.

52. J.H. Gilbert, in Proceedings of the Rubber Technology, 4th Conference (T.H. Messenger ed.), Editorial Institute of the Rubber Industry, London, 1963, 696.

53. M. Braden and A.N. Gent, *Journal of Applied Polymer Science*, 1960, **3**, 90.

54. M. Braden and A.N. Gent, *Rubber Chemistry and Technology*, 1962, **35**, 200.

55. A. Cottin and G. Peyron, 2001, WO 200123464-A1, Michelin.

56. F. Cataldo, *Polymer Degradation and Stability*, 2001, **72**, 287.

57. D.A. Lederer and M. Fath, *Rubber Chemistry and Technology*, 1981, **54**, 415.

58. E.H. Andrews, *Journal of Applied Polymer Science*, 1966, **10**, 47.

59. E.Z. Levit, T.E. Ognevskaya, B.N. Dedusenko, and D.B. Boguslovskii, *Kauchuk Rezina*, 1979, **5**, 14.

60. S.D. Razumovskii and L.S. Batashova, *Rubber Chemistry and Technology*, 1970, **43**, 1340.

61. M. Braden, *Journal of Applied Polymer Science*, 1962, **6**, 86.

62. H.W. Engels, H. Hammer, D. Brück, and W. Redetzky, *Rubber Chemistry and Technology*, 1989, **62**, 609.

63. W. Hofmann, *Rubber Technology Handbook*, Hanser Publishers, Munich, 1989, 273.

64. J.C. Andries, C.K. Rhee, R.W. Smith, D.B. Ross, and H.E. Diem, *Rubber Chemistry and Technology*, 1979, **52**, 823.

65. P.M. Lewes, *Polymer Degradation and Stability*, 1986, **15**, 33.

66. E.H. Andrews and M. Braden, *Journal of Applied Polymer Science*, 1963, **7**, 1003.

67. P.M. Mavrina, L.G. Angert, I.G. Anisimov, and A.V. Melikova, *Soviet Rubber Technology*, 1972, **31**(12), 18.

68. R.P. Latimer, E.R. Hooser, R.W. Layer, and C.K. Rhee, *Rubber Chemistry and Technology*, 1980, **53**, 1170.

69. R.P. Latimer, E.R. Hooser, R.W. Layer, and C.K. Rhee, *Rubber Chemistry and Technology*, 1983, **56**, 431.

70. S.W. Hong and C.-Y. Lin, *Rubber World*, August 2000, **222**, 36.

71. S.W. Hong, P.K. Greene, and C.-Y. Lin, ACS Rubber Division 155th Conference, Chicago, IL, April 13–16, 1999, Paper no. 65.

72. ISO 1431-1 '89, Resistance to ozone under static strain, 1989.
73. ISO 1431-2 '82, Resistance to ozone under dynamic strain, 1982.
74. M.P. Anachkov, S.K. Rakovsky, and S.D. Razumovskii, *Journal of International Polymeric Materials*, 1990, **14**, 79.
75. S.M. Kavun, Yu.M. Genkina, and V.S. Filippov, *Kauchuk Rezina*, 1995, **6**, 10.
76. P. Lehocky, L. Syrovy, and S.M. Kavun, RubberChem'01, Brussels, 2001, Paper 18.
77. H.S. Dweik and G. Scott, *Rubber Chemistry and Technology*, 1984, **57**, 735.
78. S.W. Hong, *Elastomer*, 1999, **34**(2), 156.
79. US Food and Drug Administration, Code of Federal Regulations 21, 1995.
80. G. Scott, *ACS Symposium Series*, 280 (P. Klemschuk, ed.), ACS, Washington DC, 1985, 173.
81. R.K. Kimwomi, G. Kossmehl, E.B. Zeinalov, P.M. Gitu, and B.P. Bhatt, *Macromolecular Chemistry and Physics*, 2001, **202**(13), 2790.
82. S. Avirah, M.I. Geetha, and R. Joseph, *Kautschuk Gummi und Kunststoffe*, 1996, **49**(12), 831.
83. F. Ignatz-Hoover, O. Maender, and R. Lohr, *Rubber World*, 1998, **218**(2), 38.
84. A. Meghea and M. Giurginca, *Polymer Degradation and Stability*, 2001, **73**, 481.
85. R.L. Gray, R.E. Lee, and C. Neri, *Polyolefins X*, International Conference, Houston, TX, 1997, 599.
86. E. Roos (To Bayer AG), 1965, U.S. 3, 211, 793.
87. J.J. D'Amico and S.T. Webster (to Monsanto Co.), 1972, U.S. 3, 668, 245.
88. M.E. Cain, G.T. Knight, P.M. Lewis, and B. Saville, *Journal of the Rubber Research Institute of Malaysia*, 1969, **22**, 289.
89. K.W. Sirimevan Kularatne and G. Scott, *European Polymer Journal*, 1978, **4**, 835.
90. S. Avirah and R. Joseph, *Angewante Makromoleculaire Chemistry*, 1991, **1**, 193.
91. F. Gratani, MBS Conference, Zurich, October 1993.
92. K.L. Rollick, J.G. Gillick, J.L. Bush, and J.A. Kuczkowski, ACS, Cincinnati, OH, October 18–21, 1988, Paper 52.
93. K. Fujiwara and T. Ueno, *International Polymeric Science and Technology*, 1991, **18**, 36.
94. K.L. Rollick, J.G. Gillick, and J.A. Kuczkowski (to The Goodyear Tire & Rubber Co.), May 28, 1991, U.S. 5, 019, 611.
95. G. Ivan, M. Giurginca, and J.M. Herdan, *Journal of Polymeric Materials*, 1992, **18**, 87.
96. E.L. Wheeler, ACS, Detroit, MI, October 17, 1989, Paper 73.
97. S.W. Hong, ACS, Detroit, MI, October 17, 1989, Paper 72.
98. D.A. Birdsall, S.W. Hong, and D.J. Hajdasz, Tyretech '91 Conference Proceedings, Berlin, Germany, October 24–25, 1991, 108.
99. S.W. Hong and M.P. Ferrandio, ACS, Cleveland, OH, October 1997.
100. M. Ogawa, Y. Shiomura, and T. Takizawa (to Bridgestone Corporation), January 31, 1989, U.S. 4, 801, 641.
101. M. Ogawa, Y. Shiomura, and T. Takizawa (to Bridgestone Corporation), December 12, 1989, U.S. 4, 886, 850.
102. S.W. Hong, *Rubber Plastic News*, 1989, December 25, 14.
103. A.J.M. Sumner and H. Fries, *Kautschuk Gummi Kunststoffe*, 1992, **45**, 558.
104. G.W. Marwede, B. Stollfuss, and A.J.M. *Sumner, Kautschuk Gummi Kunststoffe*, 1993, **46**, 380.
105. G.W. Marwede, B. Stollfuss, and A.J.M. *Sumner, Plastics and Rubber Weekly*, February 1993, **1471**, 10.
106. W. Hellens, ACS, Nashville, November 3–6, 1992, Paper 23.
107. A.J.M. Sumner, H. Fries, and Tyretech '91, Berlin, October 24–25, 1991, 102.
108. A.J.M. Sumner, ACS, Detroit, Michigan, October 17–20, 1989, Paper 37.
109. W. Hellens, ACS, Detroit, Michigan, October 17–20, 1989, Paper 40.
110. W. Hellens, D.C. Edwards, and Z.J. Lobos, *Rubber Plastic News*, 1990, October 1, 61.
111. J.R. Dunn and D.C. Edwards, *Industria Della Gomma*, January 1986, **30**(1), 18.
112. W. Hellens, S.A.H. Mohammed, and R. Hallman, to Polysar Ltd., February 24, 1987, U.S. 4, 645, 793.
113. D.C. Edwards and J.A. Crossman, to Polysar Ltd., May 13, 1986, U.S. 4, 588, 780.
114. G.C. Blackshaw and I.M. Kristensen, *Journal of Elastomers and Plastics*, 1975, **7**(3), 215.
115. W. Warrach and D. Tsou, *Rubber Plastic News*, 1984, June 4, 18.
116. N.M. Huntink, R.N. Datta, and J.W.M. Noordermeer, *Rubber Chemistry and Technology*, 2004, **77**(3), 476.

16 Q-Flex QDI Quinone Diimine Antidegradant—Improved Mixing Chemistry Resulting in a Better Balance of Productivity and Performance

Frederick Ignatz-Hoover and Byron To

CONTENTS

16.1 INTRODUCTION

Quinone diimine antidegradant (Q-Flex QDI) reacts faster than conventional antidegradants to stabilize broken chains and minimize the oxidative chain degradation that occurs during mixing of natural rubber (NR) compounds in intensive mixing equipment. The quinone diimine reacts by adding to the radical chain end. This reaction not only produces a polymer-bound PPD moiety, but also prevents the recombination of the broken chains. These two benefits accelerate viscosity reduction and improve polymer–filler interaction. Target viscosities are achieved faster, improving productivity and significantly reducing processing costs. Even with shorter mixing cycles, the improved polymer–filler interaction results in final products having comparable or improved performance characteristics. Thus, the balance of productivity and performance improves by mixing NR compounds in the presence of Q-Flex QDI antidegradant.

16.2   REVIEW OF MECHANICAL DEGRADATION OF NR

The molecular weight of natural rubber (NR) can be altered by mechanical means. Mastication of NR results in a reduction of molecular weight.[1] In the distant past, NR was commonly subjected to mechanical degradation on open mills for extended periods of time to improve processability.

The mechanical reduction in molecular weight originates from two mechanisms having inverse temperature coefficients. The inverse temperature dependence of the two mechanisms results in minimum chain cleavage when mixing occurs between about 110°C and 130°C.[2] As temperatures decrease, lower than about 110°C, the rate of mechanically induced chain cleavage increases. As the mixing temperature increases above about 130°C, the rate of chain cleavage also increases. The mechanism of this process is less understood, but has been shown to be fundamentally a stress-enhanced oxidative mechanism.

Mixing or milling under an inert atmosphere produces less reduction in viscosity as opposed to mixing in air.[3] Mixing at higher partial pressures of oxygen enhances the molecular weight reduction especially at higher temperatures. In the absence of oxygen, radical chain ends formed from the shear-induced cleavage terminate by combination. In the presence of oxygen, the radical chain end reacts with molecular oxygen then subsequently forms polymeric peroxy-radicals which terminate by hydrogen abstraction. Termination reactions by such radical capture or stabilization reactions produce higher concentrations of shorter chains faster. Thus, the rate of reduction of molecular weight is accelerated to the extent that the mechanism of termination changes from recombination to disproportionation, hydrogen abstraction or radical capture, or other radical stabilization mechanisms.

Bueche[3] developed the quantitative theoretical treatment demonstrating that the high-molecular weight chains have the highest probability of cleavage. Bueche showed that the rate of reduction of molecular weight is proportional to the force acting on a bond centrally located in the polymer chain.

$$N \sim N_o \exp(-P\omega t)$$

where

   N is the number of unbroken chains at time t
   N_o is the original number of chains under consideration
   P is the probability of a chain breaking

   Further, employing the Boltzmann distribution:

$$P \sim K \exp[-(E - F_o\delta)]$$

where

   E is the energy needed to break a bond
   F_o is the force acting on the bond located at the center of the chain under consideration
   δ is the distance necessary to stretch the bond to break

   The force acting on a centrally located bond is proportional to the molecular weight of the chain containing that bond. Simplifying the relation developed by Bueche, the force, F_o, is proportional to the molecular weight, shear rate, and viscosity:

$$F_o \sim \eta\gamma M^2/M_t$$

where η is the viscosity, γ is the shear rate, M is the molecular weight of the polymer molecule supporting the force F_o upon a bond located at its center, M_t is the average of the weight and Z average molecular weights of the bulk polymer subjected to the shear field. Thus, chains of high-molecular weight experience the highest force and have the highest probability of cleavage.

The probability that a chain will break at its midpoint increases with increasing viscosity, shear rate, and the square of the molecular weight of the chain under consideration.

Low-molecular weight chains do not experience enough shear force to induce scission. Watson et al.[4] demonstrated (by the intrinsic viscosity characterization of masticated NR) that the limiting molecular weight for the shear-induced degradation is in the order of $0.7–1.0 \times 10^5$. Frenkel[5] independently speculated that shear-induced cleavage occurs near the midpoint of the polymer chain.

The shear-induced cleavage at the midpoint of a polymer chain, as agreed to by Watson et al.[6] persists in both the cold and hot mastication processes. Yet unexpectedly, as demonstrated by mastication under an inert atmosphere, the initiation of high-temperature shear-induced chain scission requires oxidative activation (i.e., the presence of oxygen or other radical acceptors or promoters). However and most importantly, while the initial high-temperature oxidatively induced scission reactions have a strong mechanical activation component, subsequent chain-propagating oxidative scission reactions are random by nature.[6] It is these random scissions which produce short chains resulting in diluent effects and dangling chain ends which detract from the mechanical properties of networks.

Since pure shear-induced scission occurs at or near the midpoint of high-molecular weight chains, the population of low-molecular weight molecules is not significantly altered solely by mastication (i.e., oxidative mechanisms being absent). Consequently, pure mechanical degradation of molecular weight produces nominal effects on such dynamic mechanical properties as G'' and tan δ of the final vulcanizates. On the other hand, oxidatively induced random scission results in an increase in the population of low-molecular weight chains and thus contributes to a degradation of dynamic mechanical properties (i.e., increases in both G'' and tan δ.)

In the presence of oxygen, the radical chain ends react with molecular oxygen forming polymeric peroxy-radicals, which terminate by hydrogen abstraction forming polymeric hydroperoxides. Oxidatively catalyzed shear-induced cleavage becomes "autoxidative" at high temperatures, leading to the rapid reductions in viscosity observed in high-temperature mixing. Thus, newly formed polymeric hydroperoxides decompose generating alkoxy- and hydroxyl-radicals. These radicals react with molecular oxygen to form additional polymeric hydroperoxides which in turn decompose to continue the cycle, i.e., the definition of "autoxidation".

16.3 REACTION OF QUINONE DIIMINE WITH RADICALS

Several studies characterizing the reactions of alkenyl radicals with quinone diimines and quinoneimines were published in the late 1970s.[7] Quinone diimines react with allylic radicals yielding both the reduced PPD and the alkylated product. In these experiments 2-methyl-2-pentene served as a model olefin (model for NR). Samples of the olefin and quinoneimines or quinone diimine were heated to 140°C. Isolation and analysis of products demonstrated that 40%–70% of the imine or diimine was reduced to the corresponding PPD, while 20%–50% was isolated as the alkylated product. This alkylation reaction (via an allylic radical) represents the pathway to the formation of rubber-bound antidegradant.[16]

The addition of a 2-methyl-2-penten-4-yl radical to the QDI (based on *p*-phenylene diamines [PPDs] thus producing the corresponding PPD radical) is highly exothermic. The reaction not only stabilizes the relatively unstable alkenyl radical, but also results in the aromatization of the diiminocylcohexadienyl ring. The enthalpy of reaction for this reaction is calculated (using MOPAC/AM1 Hamiltonian[8]) to be about −40 kcal/mol.

The results of the modeling study of the chain transfer chemistry have been published elsewhere.[9] Quinone diimines are predicted to be more than two orders of magnitude more reactive toward free radicals than the corresponding PPD. The reactivity of a radical with another molecule should be related to the Lowest Unoccupied Molecular Orbital (LUMO) energy of that molecule. The reaction of a radical with a PPD differs from the reaction of a radical with QDI.

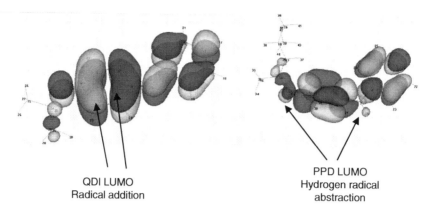

QDI LUMO
Radical addition

PPD LUMO
Hydrogen radical
abstraction

FIGURE 16.1 Lowest unoccupied molecular orbital (LUMO) representations of QDI and *p*-phenylene diamine (PPD).

PPDs donate hydrogen atoms while QDI reacts generally by addition to the radical. Examining the LUMO orbitals of PPD and QDI rationalizes one source of the large difference in reactivity. The reaction site of the PPD will be the LUMO of the H atom attached to the N atoms. While for QDI the delocalized LUMO in the conjugated diimine ring is the reaction site. These are shown in Figure 16.1.

QDI has significantly greater probability to react since the LUMO surface area is so much larger than that of PPD. The energy of aromatization (discussed above) also serves as a driving force for this reaction. Thus, QDI having a larger reaction surface area, and gaining energy of stabilization through the aromatization process, provides for higher reaction rate with radicals compared to the PPD molecules.

16.3.1 Comparison of Mixing NR—QDI versus Peptizers

RSS#2 was mixed with N-234 carbon black without additive or in the presence of quinonediime, or with a peptizer (dibenzamidodiphenyl disulfide, activated with a metal chelate).[10] In the case of the peptizer, rubber and carbon black were mixed for 30 s before the addition of carbon black. In the case of the quinone diimine, carbon black and QDI were added 15 s after addition of rubber to the mixer.

Two sets of mixes were performed, a low shear–low oxidation condition (LTM, low temperature mixing) and a high shear–highly oxidative mix condition (HTM, high-temperature mixing). In the first case the mixing was done at low speeds and the batches were dropped when the internal temperature probe in the mixer reached 230°F (LTM). In the second case, the mixer was operated at high speeds and the batches were dropped at 330°F (HTM). As discussed above the peptizers have two functions: The first is to prevent recombination of radical polymer chain ends formed during mixing. The second function is to promote the formation of radicals during the mix. Again in terms of mixing conditions defined by Busse, maintaining the mix at approximately 200°F–240°F will provide for a minimal amount of oxidative and shear-induced scission. Consequently, LTM conditions should lead to minimal reductions in viscosity due to oxidative chain scission. The LTM condition was designed to observe the effect of the QDI or peptizer under these milder mixing conditions. The HTM condition is designed to observe the effect of the additives under more severe oxidative mixing conditions.

The change in Mooney viscosity is plotted as a function of turns of the mixer (i.e., rpm × cummulative minutes of mixing). The results are presented in Figures 16.2 and 16.3. Under the LTM conditions little discrimination is observed between triplicate batches of the control (no additive) and the batches containing disulfidic peptizer. However, even under the LTM condition,

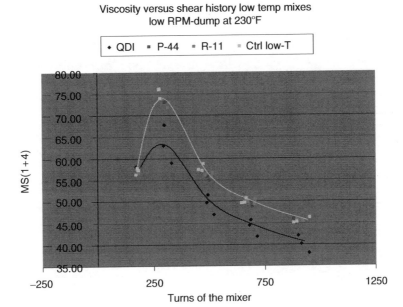

FIGURE 16.2 Mild-mixing conditions do not discriminate between peptizers (P-44 and R-11) and the control compound. While QDI-containing compounds show significant viscosity reduction.

the QDI-containing compound shows significant viscosity reduction as a function of rotations of the mixer.

Goodrich heat buildup shows some discrimination between the QDI-containing compounds and the others under both mild and oxidative mixing conditions. Figure 16.4 shows the delta temperature

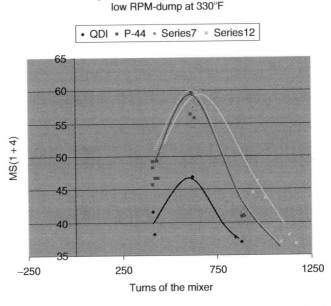

FIGURE 16.3 Under oxidative mixing conditions peptizers begin to demonstrate the chemical effectiveness at promoting oxidatively induced radical degradation. QDI-containing compounds show large viscosity reductions in the early stages of mixing.

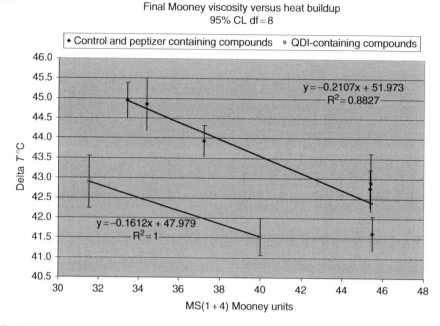

FIGURE 16.4 A general trend toward higher heat buildup as a function of lower Mooney viscosity is observed.

observed during Goodrich Flexometer testing as a function of mix viscosity. The compounds containing the disulfidic peptizers and the control show increased heat buildup as a function of decreasing Mooney viscosity. In QDI-containing compounds a trend to higher heat buildup with lower viscosity is also observed. However, the slope and intercept values of regression line for QDI are lower than for those of the control or peptized compound.

16.3.2 COMPARISONS IN MIXING NR—QDI VERSUS PREMASTICATION (WITHOUT PEPTIZER)

NR, RSS#2, was premasticated with 5 phr of N-234 in a "OO" Banbury. Four batches were mixed until the thermocouple probe in the mixer registered 320°F.[11] The average batch temperature measured by direct measurement of the batch after mixing was 350°F. Compounds were prepared from nonpremasticated NR, premasticated NR, and both nonpremasticated NR and premasticated NR containing QDI. The results of the mulitstage mix in terms of viscosity per mixing step is shown in Figure 16.5.

The viscosity of the final mix shown in Figure 16.5 for the compound containing QDI without premastication is virtually the same as the compound produced using premasticated NR without QDI.

A similar experiment was conducted using N-299 carbon black. In this case the premastication was limited to 3 min of mixing time. The average batch temperature measured after this mixing operation was 309°F. Each experiment was performed in duplicate; the average of two mixes is shown in Figure 16.6. The viscosity of the final control compound was similar to that of the premasticated rubber.

16.3.3 TWO PASS MIXING—ROTOR CONFIGURATIONS AND ANTIDEGRADANT
IN MASTERBATCH

A series of mixing experiments[12] were conducted using two pass mixing protocols rather than the multipass processes as described in the previous references. Five grades of carbon black (N-121,

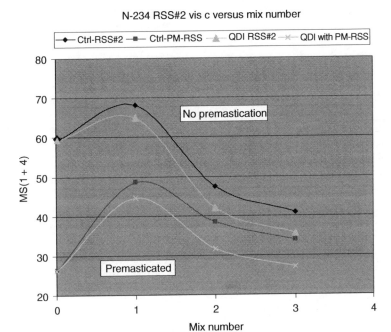

FIGURE 16.5 Viscosity changes per mixing step demonstrate QDI reduces the viscosity of natural rubber (NR) compounds. (Note that the viscosities reported are MS(1 + 4) or small rotor viscosities.)

N-330, N-550, N-660, and N-774), and NR (RSS) were mixed without antidegradant or with antidegradant (PPD or quinone diimine) using various rotor designs. The rotors used in the study include the 2 Wing Standard, 4 Wing H swirl, Kobelco's patented 6 Wing Various Clearance Mixing Technology (VCMT), and Intermeshing rotors. Comparison of viscosity and physical

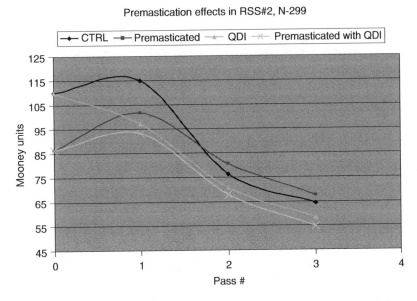

FIGURE 16.6 Viscosity of RSS#2 mixes using N-299 carbon black. QDI gave consistently lower viscosities compared to the control compounds.

property characteristics demonstrate that, on balance, cost savings are realized while producing compounds exhibiting equal or improved performance characteristics.

16.3.3.1 Mooney Viscosity

The Mooney viscosities of the final compounds were determined using the small rotor. Values are reported as MS(1 + 4). The final Mooney viscosity can be explained very well using three variables; the iodine number of the carbon black, the unit work input, kWh/lb, and the mode and type of antidegradant. The discrete variable, antidegradant, was treated as a continuous variable by assigning the values of 0, 1, and 2 to the conditions of PPD, Ctrl, and QDI. The Ctrl and PPD case both employed PPD antidegradant but the PPD case has PPD added in the master batch while the Ctrl case has the PPD added in the final stage. The Ctrl case is a master batch mixed with no added antidegradant.

For the data the squared correlation coefficient was 0.93 with a root mean square error of 2.2. The graph of predicted versus actual observed MS(1 + 4) along with the summary of fit statistics and parameter estimates is shown in Figure 16.7.

Two data points in the graph lie significantly off the line and represent the N-330 carbon black mixed with QDI. These results are exceptionally good and follow from principles expected. Viscosity is known to increase as the surface area of the carbon black increases (the particle size

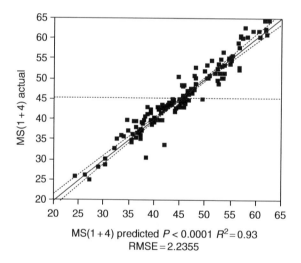

FIGURE 16.7 Graph of the predicted vs observed mooney viscosity.

Summary of Fit

R^2	0.933684
R^2 adj	0.932293
Root mean square error	2.235457
Mean of response	45.35735
Observations (or sum wt)	147

Parameter Estimates

Term	Estimate	Std Error	t Ratio	Prob > |t|
Intercept	52.686438	1.025494	51.38	<0.0001
I2No	0.183554	0.005405	33.96	<0.0001
kWh/lb	−107.3101	6.519149	−16.46	<0.0001
AD	−4.514523	0.227415	−19.85	<0.0001

decreases.) The work input into mix will reduce the viscosity as the carbon black becomes better dispersed and the molecular weight of the polymer decreases. Finally, the drop in viscosity is highest in the QDI case. The lack of any antidegradant in the masterbatch provides viscosities intermediate to the PPD case and the QDI case.

16.3.3.2   Unit Power kWh/lb

The unit power to mix can be generally described as a function of the rotor surface area, the RPM used during mixing, the time required to reach dump temperature for the first pass mix, and a treatment for the type of antidegradant used. Again the discrete variable of antidegradant was treated continuously by assigning the values of 0, 1, and 2 to the conditions of Ctrl, PPD, and QDI.

For the 150 data points the squared correlation coefficient was 0.955 with a root mean square error of 0.0062. The graph of predicted versus actual observed kWh/lb along with the summary of fit statistics and parameter estimates is shown in Figure 16.8.

These results fit with the general premises of mixing. Motor power requirements will increase with increasing time and RPM. In addition the increase in surface area of the rotor (i.e., that area

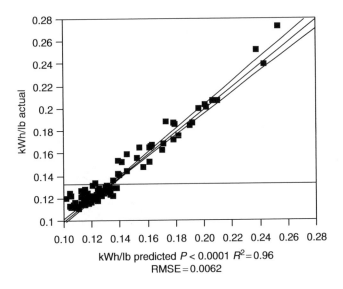

FIGURE 16.8 Graph of the predicted vs observed specific mixing power consumption.

Summary of Fit

R^2	0.955003
R^2 adj	0.953736
Root mean square error	0.006177
Mean of response	0.132641
Observations (or sum wt)	147

Parameter Estimates

Term	Estimate	Std Error	t Ratio	Prob > |t|
Intercept	−0.174306	0.009342	−18.66	<0.0001
DA	−0.003182	0.000621	−5.12	<0.0001
SA	9.8843e-7	2.682e-8	36.85	<0.0001
RPM	0.000639	0.000157	4.07	<0.0001
t_1	0.0003306	0.000009	35.74	<0.0001

useful for dispersive mixing) will also increase the energy demand on the system. The parameter of the antidegradant is also expected by the fact that we know the QDI will reduce the molecular weight faster and or improve dispersion more than either the PPD or the Ctrl case (as seen in the viscosity model above) thus reducing the energy demand in the mix.

16.3.4 POLYMER–FILLER INTERACTIONS

The formation of PPD groups on the polymer backbone provides a mechanism to improve the polymer–filler interactions. The nitrogen–hydrogen bonds are capable of hydrogen bonding with polar groups on the surface of the filler. This enhanced interaction provides for somewhat unique dynamic mechanical properties. Under ideal conditions rolling resistance improves when QDI is used in the mix. Also, abrasion characteristics are maintained and in some cases even modest improvements occur.

Chain segments between chain ends and network junction points are not load-bearing segments. Hamed[13] concludes that dangling chain ends adsorbed onto filler surfaces could " ... bear load ... by slipping (reversibly) along the filler surface." He further points out that as the chain-end is " ... 'pulled' along the filler surface ... molecular 'friction' ... is sufficiently high ... (That) ... more energy is expended ... than would be required to simply rupture the chain" thus improving network durability.

Storage and loss moduli measured at high strain quantify network strength and to some extent the adsorption-slipping process described by Hamed. Adsorption energy increases when polar end-groups (e.g., PPD moieties provided by Quinone diimine (QDI)-modified NR) can interact with polar substituents on the surface of the filler. Numerous examples in the literature demonstrate that increasing hysteresis (measured at high strain) contributes to improvements in wear, fatigue, and tear characteristics of elastomers.[14] Thus, at constant energy input with more energy dissipated through loss mechanisms at high strain, less energy is available to tear or abrade the elastomers. Measuring the loss modulus at high strain contributes to understanding the fundamentals of wear and assessing wear characteristics of various compounds.

The observed effect with QDI is more than an end-group effect. Analysis shows that roughly half of the QDI placed in a formulation will be bound into the polymer matrix. Thus, stoichiometry dictates that when compounding with 2 phr of QDI in an NR formulation 5–10 PPD moieties will be attached to each NR chain. Given that each hydrogen bond is on the order of 3–4 kcal/mol, the total increase in polymer–filler interaction will be on the order of 30–60 kcal/mol. The increase in interaction of the modified polymer with the filler is manifested in the increase in G'' especially at shear strains above 25%. (At strains lower than 10%–25%, the loss mechanisms are dominated by filler networking commonly referred to as the Payne effect.) Thus, strains above 25% represent long-range motions of the polymer chains as they are uncoiled and stretched. It is inevitable then that interactions with the filler will enhance the frictional forces of the chains as they slide past filler particles. The net result is an increase in the loss modulus G''. These effects are demonstrated in the measurements of G'' as a function of strain in the Figure 16.9.

In addition to increases in high-strain loss modulus, reductions in low-strain loss modulus are also observed. This may be attributed to the improvements in polymer–filler interactions which may reduce the amount of filler networking occurring in the compound. The low-strain losses are dominated by disruptions in the filler–filler network, the Payne effect.

According to the previous mechanism as discussed in the literature, increased loss modulus at high strain should yield improvements in abrasion resistance. Indeed in laboratory PICO abrasion experiments this result is supported. Example results are given in Table 16.1.

Over the last few years at the Flexsys laboratories in Akron, OH, data has been collected on both RPA and PICO abrasion as discussed above. In total more than 500 compounds have been tested. RPA loss modulus strain sweeps and PICO abrasion results have been collected and compiled into a database. The diversity of this database is rather broad. Many formulations based on NR,

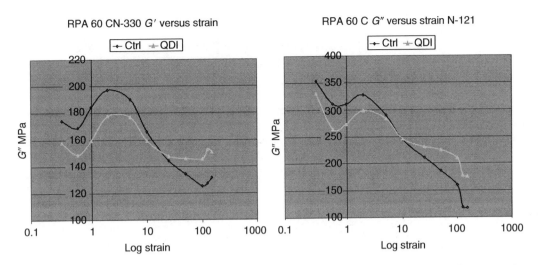

FIGURE 16.9 Lower moduli, both G′ and G″ at low strain and higher moduli at high strain support the improvement in polymer to filler interaction.

NR/butadiene rubber (BR), styrene–butadiene rubber (SBR), and SBR/BR are included. In addition, mixing conditions and procedures have varied considerably. Cure chemistries and filler type and loading are all varied. Carbon black types include N-110, N-121, N-220, N-234, N-326, N-347, N-550, and N-650 at various loading. Some compounds also include highly loaded silica filler or high black-loaded compounds with low loading of silica (at 5 and 10 phr). The compounds which are highly silica-filled compositions represent low rolling resistance OE-type formulations.

Linear regression analysis was performed on the relation of $G''(\varepsilon)$ versus PICO abrasion index. Figure 16.10 plots the correlation coefficient as a function of strain employed in the measurement of loss modulus. The regression results show poor correlation at low strain with increasing correlations at higher strains. These correlations were performed on 189 data points.

The following Figure 16.11 shows the actual correlation of G'' (100%) strain for 492 data points. The correlation coefficient has dropped slightly from the data set containing 189 data points but the relationship is still very highly significant, R^2 is nearly 0.79.

16.4 COMPOUND VISCOSITY—A BALANCING ACT OF PROPERTIES

Increasing the bound rubber content increases the effective volume fraction of filler by intimately bonding polymer to the filler. This polymer is no longer available to contribute to viscous flow. As a consequence, the viscosity of the compound increases.

Free radicals generated in butadiene-based elastomers promote cross-linking or growth of the molecular weight. These reactions lead to the increase in molecular weight or gelation. This is clear

TABLE 16.1

Pico Abrasion Data for Carbon Black Filled NR Compounds with or without QDI

	Control	QDI-Containing Compound
N-330	126	147
N-121	149	179

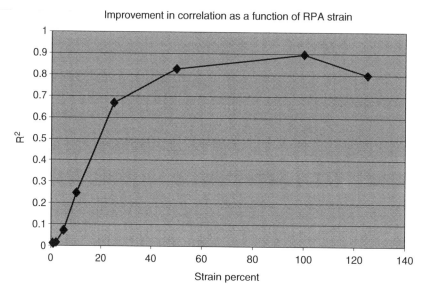

FIGURE 16.10 Correlation of G'' at various strains to PICO abrasion results.

from peroxide vulcanization chemistry where the kinetic chain length for the free-radical reaction in butadiene-based elastomers is between 1–2 orders of magnitude greater than in isoprene-based elastomers. Patterson et al.[15] report that the cross-linking efficiency in the butadiene-based elastomers can be 10–50 times more efficient than in isoprene-based rubber where one cross-link forms for each mole of peroxide.

This effect is very important when working in blended compounds of NR blended with diene elastomer. The mechanical degradation of the NR produces broken chains with free-radical end

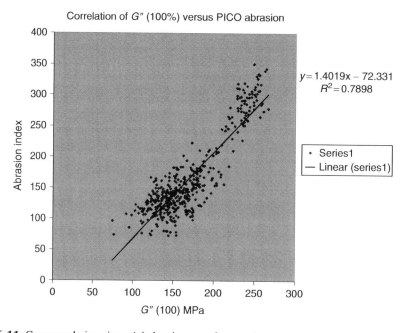

FIGURE 16.11 Compound viscosity—A balancing act of properties.

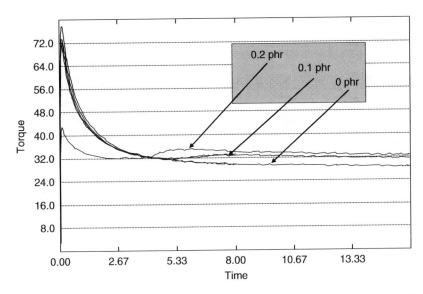

FIGURE 16.12 Mooney viscosity at 175°C of butadiene elastomer mill mixed with 50 phr of N-234 10 phr of naphthenic oil with the labeled loading of a commercial peptizer (diphenyl dibenzamido-disulfide).

groups. As discussed earlier, capturing or stabilizing the radical ends before they can recombine accelerates the molecular weight reduction of the NR. Peptizers promote these free-radical degradation reactions. However, when used in the presence of butadiene elastomers, the viscosity may actually increase. The effects of a peptizer on butadiene elastomers at a typically high mixing temperature were studied by examining the change in Mooney viscosity of a compound at high temperature. Figure 16.12 shows results of a 175°C Mooney viscosity measurement of butadiene compound mixed with peptizer (diphenyl dibenzamido-disulfide) at low temperature (70°C.) The molecular weight-building free-radical chemistry is clearly underway after about 3–7 min at this temperature.

Previous studies have demonstrated that QDI improves the formation of bound rubber.[16] From the previous discussion of free-radical chemistry, the formation of bound rubber in butadiene elastomer compounds would be expected to occur at a higher rate than in NR.

In fact viscosity reduction of diene elastomer blends with QDI show an optimum based on mixer discharge temperature. Figure 16.13 shows the results of two experiments done on the 16 L scale. The viscosities represent the viscosity of the fourth stage of a multistage mixing experiment. These compounds were 60/40 blends of either BR or SBR with NR and contained 50 phr of N-234 carbon black.

The compounds containing QDI (red lines) exhibit much lower viscosity than either the control compound (yellow lines) or the compound containing the peptizer (blue lines). The increase in the viscosity as high discharge temperatures are reached is attributed to an increase in bound rubber content.[17]

16.5 CONCLUSION

Consistent with historical results, the loss modulus at high strain correlates well with laboratory abrasion results. The best correlations occur at high strains, i.e., on the order of 50%–125%. Abrasion losses are considered to be predominantly high-strain events. Thus, the ability of a compound to dissipate energy at high strain will improve the toughness or abrasion resistance of the compound. However, it is important in the tire industry that this increase in hysteresis at high

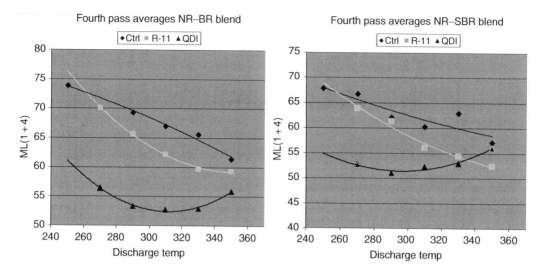

FIGURE 16.13 Fourth pass viscosity of a multistage mixing experiment of butadiene rubber–natural rubber (BR–NR) and styrene–butadiene rubber (SBR)–NR blends (60/40) with 50 phr of N-234 carbon black.

strain does not come at the expense of increasing hysteresis at low strain which of course would contribute to increased rolling resistance and reduced fuel economy.

Quinone diimines are capable of reacting rapidly with radicals formed during intensive mixing. The product, a polymer-bound PPD moiety, provides a polar functionality which is capable of improving polymer–filler interactions. In general the improvements can result in modest reductions in tangent delta (rolling resistance) and modest improvements in abrasion resistance.

The improvement in PICO abrasion observed by use of QDI can be considerable.[18] Improvements in treadwear have been confirmed in road testing.

ACKNOWLEDGMENTS

The authors wish to thank contributions made by R. Jorkasky, at KSBI in Hudson, OH, and Donald Peloso, Mike Merzweiler and Alan Hutson at Flexsys America LP for contributions to this chapter.

REFERENCES

1. Staudinger, *Ann.*, **468**, 1, 1928; *Kolloid-Z.*, **51**, 71, 1930; *Gummi-Ztg.*, **43**, 1928.
2. W.F. Busse, *Ind. and Eng. Chem.*, **24**, 2, 1932; W.F. Busse, E.N. Cunningham, *Rubber Chem. Technol.*, **12**, 2, 1932.
3. F. Bueche, *J. Appl. Poly. Sci.*, **4**, 101–106, 1960.
4. D.J. Angier, W.T. Chambers, and W.F. Watson, *J. Poly. Sci.*, **25**(109), 129–128, 1957; *J. Appl. Poly. Sci.*, **1**, 245–249, 1959.
5. Frenkel, *Acta Physiochem.* USSR, **19**, 51, 1944.
6. W.F. Busse, *Ind. Eng. Chem.*, **24**, 2, 1932; W.F. Busse and E.N. Cunningham, *Rubber Chem. Technol.*, **12**, 2, 1932.
7. I.R. Gelling and G.T. Knight, *Plastics Rubber: Process.*, **2**(3), 83–88, 1977.
8. F. Ignatz-Hoover, unpublished results.
9. F. Ignatz-Hoover et al., *J. Chem. Inf. Comput. Sci.*, **41**(2), 295–299, 2001; F. Ignatz-Hoover and B. To, Paper No. 11, Rubber Division, ACS Spring Meeting, April, 2003.
10. F. Ignatz-Hoover and B. To, Paper No. 11, Rubber Division, ACS Spring Meeting, April, 2003.
11. Ibid.

12. F. Ignatz-Hoover, B. To, and R. Jorkasky, Paper No. 10, Rubber Division, ACS Fall Technical Meeting, October 2005.
13. H. Gary, *Rubber Chem. Technol.*, **67**, 529–536, 1996.
14. F. Ignatz-Hoover, Paper No. 46, Rubber Division, ACS, Fall Technical Meeting, October 2004., and references therein.
15. D.J. Patterson, J.L. Koenig, and J.R. Shelton, *Rubber Chem. Technol.*, **56**, 971–994, 1983.
16. R. Lamba, F. Ignatz-Hoover, US Pat No. 6533859, March 18, 2003; R. Datta, P. Ebel, and F. Ignatz-Hoover, Gummi Fasern, Kunstoffe, **53**(7), 457–463, 2000.
17. F. Ignatz-Hoover, B.H. To, and R. Jorkasky, Paper number 114, "Q-Flex QDI, Quinone Diimine Antidegradant – Evidence of Competitive Chemistries During Mixing—Part II – Protecting Polymers During Mixing – Continued Developments in Diene/NR Blend Compounds", Fall Technical Meeting, Rubber Division, ACS, October, 2008.
18. S. Laube, S. Monthey, and M.-J. Wang, Chapter 12 Compounding with Carbon Black and Oil, in *Rubber Technology Compounding for Testing and Performance*, D. John ed., Hanser Publishers, Munich, 2001; F. Ignatz-Hoover, B. To, and R. Jorkasky, Effect of rotor type and antidegradant in mixing efficiency, Rubber Division, ACS Fall Technical Meeting, Paper No. 10, October 2005.

17 Silica Fillers for Elastomer Reinforcement

Doug J. Kohls, Dale W. Schaefer, Raissa Kosso, and Ephraim Feinblum

CONTENTS

17.1 INTRODUCTION

This article summarizes recent work on the structure of precipitated silica used in the reinforcement of elastomers. Silica has a unique morphology, consisting of multiple structural levels that can be controlled through processing. The ability to control and characterize the multiple structures of precipitated silica is an example of morphological engineering for reinforcement applications. In this summary of some recent research efforts using precipitated silica, small-angle scattering techniques are described and their usefulness for determining the morphology of silica in terms of primary particles, aggregates, and agglomerates are discussed. The structure of several different precipitated silica powders is shown as well as the mechanical properties of elastomers reinforced with these silica particles.

The study of the mechanical properties of filled elastomer systems is a challenging and exciting topic for both fundamental science and industrial application. It is known that the addition of hard particulates to a soft elastomer matrix results in properties that do not follow a straightforward rule of mixtures. Research efforts in this area have shown that the properties of filled elastomers are influenced by the nature of both the filler and the matrix, as well as the interactions between them. Several articles have reviewed the influence of fillers like silica and carbon black on the reinforcement of elastomers.[1–4] In general, the structure–property relationships developed for filled elastomers have evolved into the following major areas: Filler structure, hydrodynamic reinforcement, and interactions between fillers and elastomers.

In the rubber industry there is specific interest in understanding how fillers reinforce elastomers. There is a wealth of knowledge regarding the synthesis of these fillers, the mechanisms that govern the synthesis, and the characterization of the resulting filler structures.[5–9] There is interest in using fillers that are both highly dispersive and highly reinforcing as evidenced by new fillers that have been marketed over the last decade.[10–13] Different approaches have been used in improving end-use properties of filled elastomers that include refinement of particle structure, filler surface chemistry, and elastomer chemistry.[1,2,4,14] With regard to fillers, the goal has been to have a filler that is highly structured, weakly agglomerated, and that has improved surface compatibility with the elastomer.[13,14]

Presented here is the use of small-angle scattering for determining the morphology of several different silica samples and the mechanical testing of filled elastomers. The primary particle and aggregate structures are studied using ultra small-angle x-ray scattering (USAXS) while the agglomerate structures are studied using small-angle light scattering. These studies quantify the complex silica structure in each of the structural regimes.

17.2 SMALL-ANGLE SCATTERING

Particles are characterized by a variety of techniques to determine particle size, shape, and surface structure. Microscopy allows for a quick visualization of the particles and can give information on a broad range of sizes, from nanometer to millimeter particle diameters. Another approach to particle characterization uses the scattering of light, x-rays, and neutrons. Small-angle scattering is useful for examining intermediate sizes, nanometer–micrometer, of the particles and it is possible to probe the internal structure of the sample. Small-angle scattering techniques are well known and established for characterizing the morphology of micron and submicron polymeric and ceramic materials.[15-24]

17.2.1 SCATTERING EXPERIMENTS

A scattering experiment involves a source of monochromatic radiation that interacts with electrons of the sample and a detector that measures the scattered radiation. Figure 17.1 shows a schematic of a scattering experiment.

The scattering angle θ, is the angle between the transmitted beam and the scattered beam. The scattered intensities are measured at different angles for the given sample and a plot of intensity verses angle (θ) or scattering vector (q) is made. The scattering vector q is related to the scattering angle by

$$q = \frac{4\pi \sin(\theta/2)}{\lambda} \qquad (17.1)$$

where

θ is the scattering angle
λ is the wavelength of the x-ray
q is inversely related to size and has the units of 1/size

Figure 17.2 shows a combined small-angle light and x-ray curve for a precipitated silica sample, A1. For particle characterization, the range of q typically probed falls between $10^{-4} < q < 1 \text{ Å}^{-1}$ for x-rays and $10^{-6} < q < 10^{-4} \text{ Å}^{-1}$ for light.[24] The q range shown corresponds to the complex

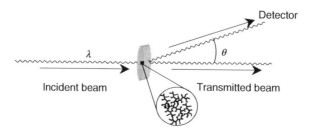

FIGURE 17.1 Schematic of a small-angle scattering experiment showing the incident x-ray striking a sample with the main "transmitted beam" passing straight through the sample. X-rays are also scattered from the sample at an angle θ. The expanded region shows cartoon of an aggregated structure that is within the sample.

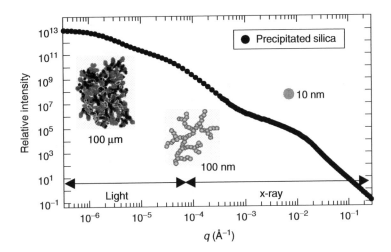

FIGURE 17.2 Combined plot of small-angle x-ray and light scattering for a precipitated silica sample. The silica sample is Dimosil 288, produced by Dimona Silica Ltd.

structures of fillers that are made up of primary particles at the smallest-size scale (high q) and agglomerates at larger sizes (low q).

By measuring the scattered intensity as a function of q over multiple decades in q, information from multiple structural levels can be obtained. Two distinct scattering features distinguish each structural level, a power–law region and a "knee" region in a log–log plot of $I(q)$ versus q. An example of power–law scattering is shown in Figure 17.2. For particles with a smooth surface the scattering intensity varies as[25]

$$I(q) \approx B_P q^{-4} \quad \text{where } B = 2\pi N(\Delta\rho)^2 S \tag{17.2}$$

where
  ρ is the scattering length density of the particle
  S is the total area of the boundary between the two scattering phases
  N is the number density of particles within the scattering volume[26]

Equation 17.2 is known as Porod's[27] law and is typical of scattering from a smooth interface. When the scattering is from the interface of a sphere, the scattering intensity decays as $I \sim q^{-4}$. The Guinier[28] region is noted by a change from power–law scattering, to a bend or "knee" as the intensity changes with q. The location of this knee in q relates to the radius of gyration of the particle. The generic relationship is that $I \sim (R_g)^2$ and Guinier's law is given as[28]

$$I(q) \approx G \exp\left(-\frac{1}{3}q^2 R_g^2\right) \quad \text{where } G = N(\Delta\rho)^2 V^2 \tag{17.3}$$

where
  ρ_0 is the average scattering length density of the particle
  V is the volume of the particle
  R_g is the radius of gyration

The Porod and Guinier regions are shown in Figure 17.3 for a precipitated silica sample. The plot shows two Guinier regions and three power–law regions. The Guinier at higher $q(R_{g1})$, $q \sim 0.04$ Å^{-1}, is the R_g of the primary particle with a value of 66 Å. The Guinier region at lower $q(R_{g2})$, $q \sim 0.004$

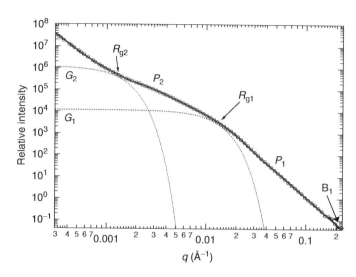

FIGURE 17.3 Log–log plot of I versus q for precipitated silica particles. The plot shows two distinct regions of power–law decay noted as P (Porod's law) and Guinier scattering, noted by R_g. The subscripts indicate the structural level associated with the scattering curve. The Guinier "knee" at higher q indicates the radius of gyration (R_{g1}) of the primary particle. At lower q the second "knee" indicates the size for the aggregates (R_{g2}). The solid line is the fit line obtained from using a unified model for the entire scattering curve. The two dashed lines highlight the Guinier fits for each structure level.

Å^{-1}, is the radius of gyration for the silica aggregates. The first power–law region (P_1), slope $= -4$, indicates scattering from the surface of the primary particles. The second power–law region (P_2) is from the organization of the particles into aggregates. This is the power–law region that can give information about the structure of the aggregates.

17.2.2 SCATTERING DATA ANALYSIS

Power–law scattering features will be discussed in relation to mass–fractal scaling laws. Fractal scaling concepts used to interpret the power–law decay are well published in the literature.[19–21,29–31] The slope of the power–law decay of intensity with q gives fractal dimensions that are related to the surface of particles as well as the arrangement of primary particles into aggregates. While there are other methods for determining the fractal dimension of aggregated particles such as microscopy (TEM, SEM), the primary tool used in measuring the fractal dimension in this work is small-angle x-ray scattering.

The scattering data collected were analyzed using a Unified Fit Model, established by Beaucage.[20,32] The model fits the power–law and Guinier regions of the scattering plot of I versus q, like the one shown in Figure 17.3. The model is capable of fitting multiple levels observed in some scattering curves where more than one R_g and power–law scattering are present. The model was used to obtain the values for R_{g1}, R_{g2}, and the power–law values displayed in the figure.

The power, [P], in the fractal power–law regime gives as the fractal dimension, d_f. $P = -df$ for each level of the fit, the parameters obtained using the unified model are: G, R_g, B, and P. P is the exponent of the power–law decay. When more than one level is fitted, numbered subscripts are used to indicate the level—i.e., G_1—level 1 Guinier pre-factor. The scattering analysis in the studies summarized here uses two-level fits, as they apply to scattering from the primary particles (level 1) and the aggregates (level 2).

For the analysis of primary particles it is possible to calculate the spherical diameter for a particle from R_g described above as $R = (5/3)^{1/2} R_g$ or $d \sim 2.6\ R_g$. It is also possible to calculate diameter for a particle through the volume/surface ratio, which is called the "Sauter mean diameter." With the parameters obtained from the unified fit this diameter, $d_{v/s}$ is calculated using the Porod invariant, Q. If the scattered intensity is integrated over all scattering angles for a particular structure, primary particles in this case, the Porod invariant related to the scattering power of the structure is calculated as:

$$Q = \int_0^\infty q^2 I(q) dq = 2\pi^2 N (\Delta\rho)^2 V \tag{17.4}$$

The value for $d_{v/s}$ is then calculated for spheres as:

$$d_{v/s} = \frac{6Q}{\pi B_1} = \frac{6 \cdot 2\pi^2 N (\Delta\rho)^2 V}{\pi \cdot 2\pi N (\Delta\rho)^2 S} = \frac{6V}{S} \tag{17.5}$$

Calculating the particle diameter through the volume/surface ratio (ratio of the third moment to the second moment) is the same moment for particle size calculation measured by nitrogen adsorption (BET). The close correlation between $d_{v/s}$ and d_{BET} was demonstrated by Kammler and Beaucage.[18,33]

Aggregates can be described by the number of primary particles that they contain. This description is termed the degree of aggregation, z. Beaucage has shown several ways for calculating the average number of primary particles which make up an aggregate from small-angle scattering data.[15,16,18] For aggregates two ratios are expressed that give the degree of aggregation, z. The value for z can be expressed as a ratio of the R_g values obtained from a two-level unified fit or a ratio of the G values from a two-level fit and they express different orders of moments.[15,18] However, certain assumptions are made when calculating z by a ratio of R_g. The ratio is scaled to the mass–fractal dimension, d_f, and it is assumed that R_g is a measure of size for both the primary particle and the aggregate, with the distribution of those sizes being narrow.[16] The degree of aggregation may also be calculated by the ratio of the Guinier pre-factors G, from a two-level unified fit where G_1 is the Guinier pre-factor for the primary particles and G_2 is the Guinier pre-factor for the aggregates. The calculation for z by the ratio of the Guinier pre-factors is given by:

$$z = \frac{G_2}{G_1} \tag{17.6}$$

17.3 SMALL-ANGLE X-RAY STUDIES ON SILICA POWDERS

A series of precipitated silica samples were studied using small-angle scattering and elastomer compounds were made to test the reinforcing effects of the silica. The small-angle x-ray scattering data presented in this paper were obtained using the ultra small-angle scattering instrument at the Advanced Photon Source (beam line 32-ID-B), Argonne National Laboratory. The small-angle scattering results for three different silica samples are plotted in Figure 17.4. Four levels of structure are present that related to the primary particle, aggregate, and two agglomerate structures. The samples shown in Figure 17.4 include A1-Dimosil 288 (Dimona Silica), A2—Ultrasil 7005 (Degussa), and A3—Zeosil 1165 (Rhodia).

The multi-level features of the silica scattering curves shown in Figure 17.4 are tabulated in Table 17.1. The values in the table are obtained using the scattering analysis presented in the previous section. The A1 sample shows larger R_g for the primary particle and aggregate.

The A2 and A3 samples showed similar sizes and dimensions at each level. The A1 sample has a larger primary particle size, indicated by an R_{g1} of 126 Å, while the A2 and A3 samples showed smaller primary particles with R_{g1} values of 85 Å for A2 and 96 Å for A3.

The four fit parameters for the first two structural levels are shown because it is possible to describe the aggregates of these samples in more detail using the Guinier and power–law

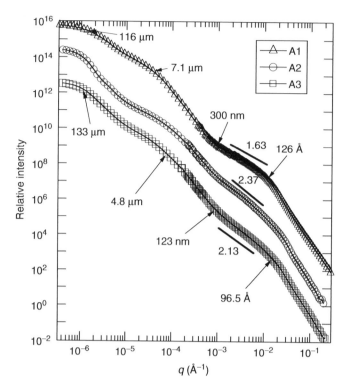

FIGURE 17.4 USAXS and SALS results for samples A1, A2, and A3 described in the text. Each sample shows four structural levels with the R_g for some of the levels indicated in the graph. The power–law value for the second level, corresponding to the mass–fractal dimension, d_f, is also indicated.

pre-factors. The Guinier pre-factor, G, from the primary particle and aggregate fits are used to calculate the degree of aggregation, z, as described by Beaucage.[16] The degree of aggregation is calculated as the ratio of the Guinier pre-factor of the aggregates to the pre-factor of the primary particles. The degree of aggregation represents the number of primary particles that make up the aggregates. This parameter gives additional information about the morphology of the aggregates beyond the simple measure of size that is obtained through the radius of gyration. The A1 sample

TABLE 17.1

Unified Fit Parameters Obtained for the Three Silica Samples Shown in Figure 17.4

Level 1	A1	A2	A3
R_g (Å)	126.4	85.2	96.5
$-|P|$	4	4	4
G	2659	592	1028
B	2.87×10^{-4}	7.36×10^{-5}	1.02×10^{-4}

Level 2	A1	A2	A3
R_g (Å)	295	123	125
$-|P|$	1.63	2.37	2.13
G	1.97×10^6	5.41×10^5	5.87×10^5
B	11.9	0.07	0.30

TABLE 17.2

Some Physical and Chemical Properties of Three Silica Samples

Measurement	A1	A2	A3
BET surface (sm^2/g)	148	174	164
CTAB surface (sm^2/g)	144	165	158
DBP adsorption (g/100 g)	234	228	236
DBP/BET	1.58	1.31	1.44
OH-groups, OH/nm^2	4.02	2.86	3.31

Note: BET = Brunaues, Emmett, and Teller
CTAB = Cetultrimethylammonium bromid

showed a degree of aggregation of 742 while A2 and A3 showed degree of aggregation values of 914 and 571, respectively.

Table 17.1 also displays the mass–fractal dimension, d_f, given by the power–law value, P, for the second level (aggregate). The $d_f = 1.63$ for A1 is less than that, and is less than 2.37 for A2 and 2.1 for A3. The lower value of d_f could indicate a more open structure of the A1 sample,[34] but it could also arise from less interpenetration of the silica aggregates occurring during drying.[35]

Table 17.2 presents results of physical–chemical testing of the silica powders. The A1 sample has the lowest Cetultrimethyammonium bromid (CTAB) and Brunaues, Emmett, and Teller (BET) surface area, higher structure (DBP) and more silanol (–OH) groups on surface per unit area.

A second study was performed in which the reactor conditions were manipulated to make silica powders with a range of structural sizes. The powders from this study are labeled with the letter "B." Figure 17.5 shows the small-angle x-ray scattering results for the B-series powders and the unified fit parameters for the scattering curves are shown in Table 17.3. The scattering results show a range of primary particle sizes, R_{g1}, from 96.6 to 3500 Å. The aggregate size, R_{g2}, had values ranging from 958 to 38,400 Å. Following the distribution in particle and aggregate sizes, the degree of aggregation, z, showed a range of 92–606 particles per aggregate. Rubber compounds containing the B-series powders should show variations in mechanical properties. The test results are discussed in the following section.

For the B-series of samples, the goal was to synthesize silica containing various particle sizes. The scattering curves for this series of samples are shown in Figure 17.5. Sample B1 had the smallest primary particle size followed by sample B3, with R_g values of 96.6 Å for sample B1, and 108.8 Å for B3. Both of these samples are close in size to samples A2 ($R_g = 85.2$ Å) and A3 ($R_g = 96.5$ Å). The B1 through B8 samples showed a range of sizes with B8 being much larger while the rest of the samples were between 100 and 460 nm. The aggregate sizes for B1 and B3 samples varied slightly from the aggregate size of the A2 and A3. The aggregate size for B1 was $R_g = 1100$ Å, and for B3 the aggregate size was $R_g = 1590$ Å. There is also similarity in the degree of aggregation for B1, B3, A2, and A3 (Table 17.3). No trend in the degree of aggregation with particle size is observed for the B1 through B8 samples as samples with similar primary particle sizes have different aggregation numbers.

17.4 MECHANICAL TESTING OF FILLED ELASTOMERS

Rubber compounds were made using A1 and A2 shown in Table 17.2 and using a model tire recipe shown in Table 17.4. The ingredients listed in Table 17.4 are in grams per 100 g of rubber (PHR). The mixing procedure used for the samples presented in Table 17.5. Both cured and un-cured compounds were tested and the mechanical testing results are shown in Table 17.6.

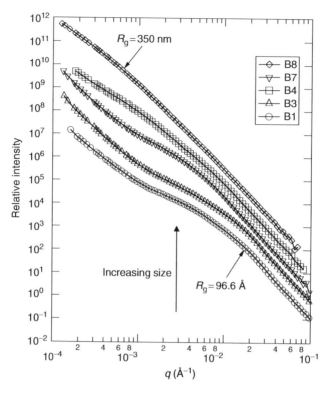

FIGURE 17.5 Silica samples B1, B3, B4, B7, and B8. The samples at the bottom of the figure have the smallest primary particle size. The increasing size is revealed by a shift of the "knee" of the curve to smaller q.

Mooney Viscometer studies at 100°C and 120°C show lower viscosity of the A1-filled gums. Lower viscosity compounds require less energy input for extrusion. Comparative results on the dynamic mechanical properties, measured using a rubber process analyzer (RPA), show that at the

TABLE 17.3
Unified Fit Parameters Obtained for Eight Precipitated Samples Provided by Dimona Silica Ltd.

Level 1	B1	B2	B3	B4	B5	B6	B7	B8
R_g (Å)	96.6	166	108	464	214	143	240	3500
$-\|P\|$	4	4	4	4	4	4	4	3.73
G	701	4290	1720	42900	7760	2230	7740	1.77×10^8
B	7.3×10^{-5}	3.8×10^{-5}	1.1×10^{-4}	1.1×10^{-5}	2.4×10^{-5}	3.3×10^{-5}	1.9×10^{-5}	1.7×10^{-5}

Level 2	B1	B2	B3	B4	B5	B6	B7	B8[a]
R_g (Å)	1100	958	1590	2340	1030	992	1840	38400
$-\|P\|$	2.40	2.85	2.11	3.37	3.20	2.84	2.51	2.87
G	4.3×10^5	6.5×10^5	1.2×10^6	4.6×10^6	9.7×10^5	6.5×10^5	1.8×10^6	1.6×10^{10}
B	8.7×10^{-2}	9.5×10^{-3}	5.2×10^{-1}	1.6×10^{-4}	1.4×10^{-3}	8.6×10^{-3}	4.5×10^{-2}	1.7×10^{-5}
z	606	151	716	108	125	293	227	92

[a] Level 2 fit values for sample B8 were obtained from light scattering data.

TABLE 17.4
Rubber Compound Formulation

Ingredient	phr
SBR	120
BR	25
Silica filler	80
Silica coupling agent	6.3
Stearic acid	2.5
Mineral oil	3.5
Zinc oxide active	2.5
DPG	2
CBS	1.7
Sulfur	2.1

0.5%-strain the dynamic modulus of the A1 compound is about 15% higher than the reference compound. This difference is an indication of higher filler–filler interaction of the A1 particles.

The reinforcing properties measured as M300/M100 are similar for the two vulcanized samples. However, the cured A1 compounds show lower ultimate elongation at break, higher curing rate (T90), and shorter breakdown time (T2), which are the signs of lower adsorption of accelerators on the A1 particles surface.

"Energy loss under low frequency conditions" is often used as a key factor for evaluating elastomer compositions. In this respect, the rolling resistance and skidding properties are usually evaluated by tan delta (tan δ) at 60°C and 0°C, where the 60°C value correlates with low frequency loss and 0°C correlates with skid resistance (high frequency loss).

TABLE 17.5
Mixing Procedure

First mixing stage
Chamber load coefficient, 0.75
Initial chamber temperature, 60°C
Initial rotor speed, 60 rpm
Load rubber, close the ram, 0–20 s
Open the ram, load 2/3 silica, 2/3 of aromatic oil, stearic acid, 50–70 s
Open the ram, load 1/3 of silica, 1/3 aromatic oil, 1/3 TESPD, 130–150 s
Open the ram, shape off the remaining silica, close the ram
Increase the rotor speed up to 80 rpm, 210–220 s
Unload the compound at the temperature 160°C, 260–300 s

Mixing stage 2 is repeated three times
Initial chamber temperature, 60°C
Initial rotor speed, 60 rpm
Load the compound, 0–30 s
Unload the compound when temperature reaches 150°C or mixing time 150 s is reached

Third mixing stage
Initial chamber temperature, 50°C
Initial rotor speed, 40 rpm
Load the compound, 0–30 s
Open the ram, load sulfur and accelerants, close the ram, 50–70 s
Unload the compound at the temperature 100°C–110°C, 160–180 s

TABLE 17.6
Rubber Testing

	A1	A2
Rheometer MDR 2000, 160°C		
Torque: max, f/d	16.52	17.39
Torque: min, f/d	1.95	2.30
ΔM, f/d	14.6	15.0
t_2 (s)	308	393
t_{90} (s)	751	1078
Mooney viscosity, 100°C	69.6	71.1
Mooney scorch, 120°C t_{35} (s)	2158.8	2879.8
Minimum viscosity	28	31.2
Extrusion:		
Density (g/cm³)	1.163	1.162
Conveyor energy (KNm/cm)	19.85	20.86
Conveyor volume (cm³/s)	0.41	0.42
Storage modulus, RPA, 60°C, G', ε—0.5% (kPa)	603.3	524.9
Tan δ, RPA, 60°C, ε—0.5%	0.322	0.360
M300%/M100%	5.27	5.33
Tensile strength (MPa)	17.1	17.7
Ultimate elongation (%)	348	410
Tear resistance (kN/m)	8.6	7.9
Hardness Shore A	61.0	59.0
Rebound, 23°C (%)	35.7	33.5
Wear resistance:		
Loss in weight (mg)	108	124
Abrasion (mm³)	87	103
Density (g/cm³)	1.178	1.180
Compression set, 70 yo°C (%)	12.28	10.61
Friction coefficient (pendulum method):		
Wet skidding	60. 6	57.3
Dry skidding	99.3	98.2
Dynamic tests:		
Tan δ at 0°C	0.41	0.35
Tan δ at 60°C	0.095	0.098

The operational-factor tests performed with A1 filled compounds in comparison with A2 filled compounds revealed higher predicted wear resistance and wet traction with the same level of rolling resistance. At the same time the graph of tan δ versus temperature in Figure 17.6 shows that tan δ is higher than that of the A2 compound at low temperatures (–20°C).

The B-series of silica samples were also blended with rubber and the compound formulation is shown in Table 17.6. The uncured gums were then tested according to ISO 5794-2 1998. The uncured samples were tested using a Mooney viscometer and an RPA, which measures the dynamic mechanical properties as the samples cure. Figure 17.7 shows the results of these two tests for the Mooney viscosity at 100°C, storage modulus, loss modulus, and tan δ.

In all of the rheometer testing of the uncured compounds, the commercial silica AZ showed the highest values with the B1 and B3 samples having the highest values among the B-series silica samples. The Mooney viscosity at 100°C increases as the number of particles in the aggregates increases. The same compounds were cured and tested, measuring tensile properties, tear resistance,

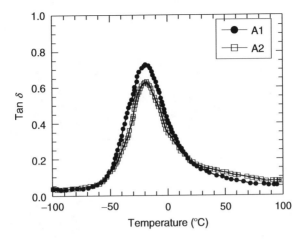

FIGURE 17.6 Plot of tan δ versus temperature for A1- and A2-cured compounds measured using a rubber process analyzer (RPA).

hardness, and wear resistance. The values from these tests are shown in Figure 17.7. The mechanical testing on the vulcanized compounds is shown in Figure 17.8.

The mechanical test data show an increase in ultimate elongation for several of the B1 through B8. Samples B1, B3, and B6 in particular showed similar tensile strength and elongation as the A2 and A3 samples. The tear resistance values were also similar for the B1 and B3 samples compared to A2 and A3 samples.

The stress–strain curves for the two commercial samples and the B1 through B4 samples are plotted in Figure 17.9. The samples B1, B3, and B6 are plotted with the commercial silica samples, showing the similarity in the stress–strain behavior (Figure 17.9a). As the primary particle size and

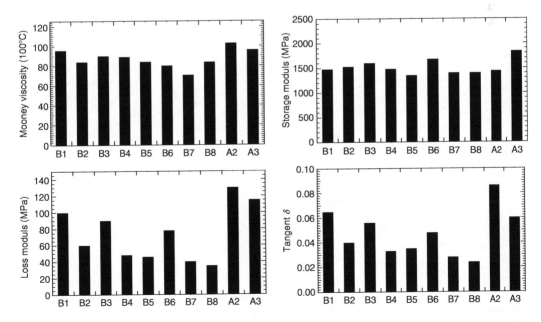

FIGURE 17.7 Results for the B1 through B8 and A2 and A3 silica samples showing the Mooney viscosity at 100°C, and the dynamic mechanical properties measured by rubber process analyzer (RPA) during curing of the compounds.

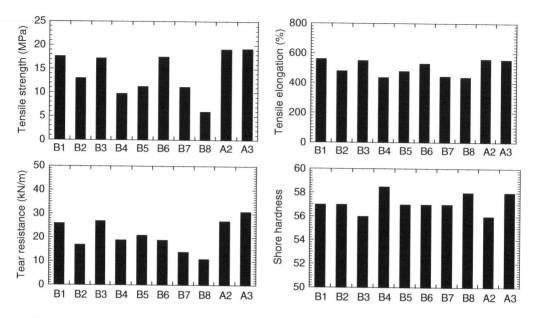

FIGURE 17.8 Mechanical testing results for cured compounds showing tensile, hardness, and tear properties.

degree of aggregation decreases for the B-series samples (Table 17.3), the tensile elongation and ultimate stress decrease as well.

17.5 CONCLUSION

The purpose of the summary of recent research was to demonstrate the application of small-angle scattering to study silica. Using small-angle x-ray scattering it was possible to measure the morphological characteristic of primary particles, aggregates and agglomerates. In addition to primary particle size, the degree of aggregation is also a parameter controllable by synthetic

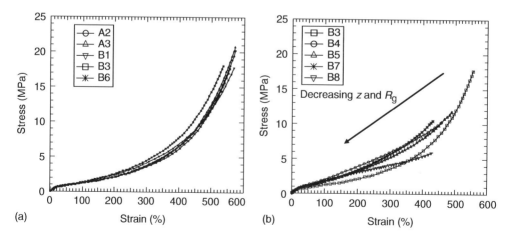

FIGURE 17.9 Stress–strain curves for the eight B-series silica samples and the A2 and A3 silica samples. B1, B3, B6, and the two commercial silica compounds. (a), The effect of particle size and the degree of aggregation on the stress–strain properties for the B-series samples is shown (b), with the elongation and ultimate stress decreasing with decreasing particle size and degree of aggregation.

protocol. Structural information for several types of precipitated silica showed that a range of morphologies is present among the samples. The relative mechanical properties of the rubber compounds showed that the mechanical properties vary, with the aggregation being the most important morphological parameter with the aggregate structure.

REFERENCES

1. Donnet, J.B., *Rubber Chem. Technol.* **71**, 323, 1998.
2. Heinrich, G., Kluppel, M., and Vilgis, T.A., *Curr. Opin. Solid State Mater. Sci.* **6**, 195, 2002.
3. Kluppel, M., *Adv. Polym. Sci.* **164**, 1, 2003.
4. Kohls, D.J. and Beaucage, G., *Curr. Opin. Solid State Mater. Sci.* **6**, 183, 2002.
5. Barthel, H., Rosch, L., and Weis, J., *Fumed Silica-Production, Properties, and Applications, in Organosilicon Chemistry II: From Molecules to Materials*, VCH: Weinheim, Germany, 1996; Vol. II, 761.
6. Brinker, J.C. and Scherer, G.W., *Sol–Gel Science*, Academic Press: San Diego, 1990, 908.
7. Donnet, J.B., Roop, C.B., and Wang, M.J., *Carbon Black: Science and Technology*, Marcel Dekker: New York, 1993, 461.
8. Friedlander, S.K., *Smoke Dust and Haze*, Wiley: New York, 2000, 175.
9. Iller, R.K., *The Chemistry of Silica*, Wiley: New York, 1979.
10. Krivak, T.G., Okel, T.A., and Wagner, M.P. (to PPG Industries, Inc.), U.S. Patent No. 5,094,829, 1990.
11. Feinblum, E. (to R&D Silicate Products Ltd.), U.S. Patent No. 5,302,364, 1992.
12. Esch, Heinze, Gorl, U., Udo, Kuhlmann, Robert, Rausch and Ralf (to Degussa AG), U.S. Patent No. 6,977,065, 2005.
13. Bice, J.E., Kellring, S.D., and Okel, T.A. (to PPG Industries Ohio, Inc.), U.S. Patent No. 7,015,271, 2006.
14. Mark, J.E., Erman, B., and Frederick, R.E., *Science and Technology of Rubber*, Elsevier/Academic Press: Burlington, MA, 2005, 743.
15. Beaucage, G., *J. Appl. Crystallogr.* **29**, 134, 1996.
16. Beaucage, G., *Phy. Rev. E* **70**, 2004.
17. Beaucage, G., Kammler, H.K., Mueller, R., Strobel, R., Agashe, N., Pratsinis, S.E., and Narayanan, T., *Nat. Mater.* **3**, 370, 2004.
18. Beaucage, G., Kammler, H.K., and Pratsinis, S.E., *J. Appl. Crystallogr.* **37**, 523, 2004.
19. Beaucage, G., Rane, S., Schaefer, D.W., Long, G., and Fischer, D., *J. Polymer Sci. Part B-Polymer Phys.* **37**, 1105, 1999.
20. Beaucage, G. and Schaefer, D.W., *J. Non-Cryst. Solids* **172**, 797, 1994.
21. Hyeon-Lee, J., Beaucage, G., Pratsinis, S.E., and Vemury, S., *Langmuir* **14**, 5751, 1998.
22. Hyeon-Lee, J., Beaucage, G., and Pratsinis, S.E., *Chem. Mater.* **9**, 2400, 1997.
23. Schaefer, D.W. and Chen, C.Y., *Rubber Chem. Technol.* **75**, 773, 2002.
24. Schaefer, D.W., Rieker, T., Agamalian, M., Lin, J. S., Fischer, D., Sukumaran, S., Chen, C.Y., Beaucage, G., Herd, C., and Ivie, J., *J. Appl. Crystallogr.* **33**, 587, 2000.
25. Porod, G., *Kolloid-Z* **124**, 83, 1951.
26. Roe, R.J., *X-Ray and Neutron Scattering from Polymers*, Academic Press: New York, 2001.
27. Porod, G., *Small-Angle X-Ray Scattering*, Academic Press: London, 1982.
28. Guinier, A. and Fournet, G., *Small-Angle Scattering of X-Rays*, Wiley: New York, 1955.
29. Schaefer, D.W. and Keefer, K.D., *Phys. Rev. Lett.* **56**, 2199, 1986.
30. Schaefer, D.W., Martin, J.E., and Keefer, K.D., *J. Phys.* **46**, 127, 1985.
31. Schaefer, D.W., Martin, J.E., Wiltzius, P., and Cannell, D.S., *Phys. Rev. Lett.* **52**, 2371, 1984.
32. Beaucage, G., *J. Appl. Crystallogr.* **28**, 717, 1995.
33. Kammler, H.K., Beaucage, G., Mueller, R., and Pratsinis, S.E., *Langmuir* **20**, 1915, 2004.
34. Meakin, P., *Adv. Colloid. Interface Sci.* **4**, 299, 1988.
35. Suryawanshi, C.N., Pakdel, P., and Schaefer, D.W., *J. Appl. Crystallogr.* **36**, 573, 2003.

18 Mechanism of the Carbon Black Reinforcement of Rubbers

Yoshihide Fukahori

CONTENTS

18.1   INTRODUCTION

The reinforcement of rubbers with carbon black is definitely one of the most important subjects in rubber science and technology. The addition of carbon black greatly increases modulus, tensile strength, tear strength, fatigue resistance, and wear resistance of unfilled rubbers. These improvements are called "carbon black reinforcement" and have been widely discussed for a long time. All phenomena given above, of course, resulted only from filling carbon black into rubbers, thus these phenomena are basically connected to each other and not generated independently. Therefore, any consideration concerning the mechanics or mechanism for a phenomenon included in carbon black reinforcement should be connected to other phenomena totally and systematically. That is, the explanation for one phenomenon must be consistent with other phenomena in carbon black reinforcement.

The above phenomena in carbon black reinforcement for rubbers are produced as a combined contribution of the four fundamental criteria, i.e., (1) the large stress increase (or modulus increase) at small extension, (2) the very large stress increase over middle to large extension, (3) the great tensile strength, i.e., 10–15 times larger than that of unfilled rubber, as shown in Figure 18.1, and (4) the dramatic stress softening and energy dissipation called the Mullins effect. Therefore, any viable theory or model for carbon black reinforcement must answer to the following four questions with a unified and synthetic concept:

1. Why does the stress increase greatly with filler content at a small strain?
2. Why does the above stress increase become larger and larger with increasing strain amplitude, as called the stress upturn?
3. Why is the filled system able to support such a very large stress before rupture? In other words, what structure is newly produced in the system to support the large stress concentration?
4. Why does the stress decrease so abruptly and vastly in unloading and why is the stress restored gradually during long periods at room temperature and more quickly at high temperature?

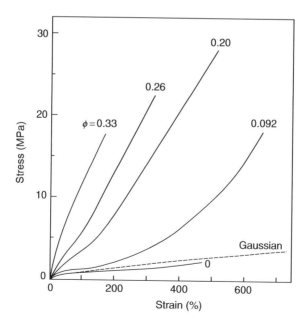

FIGURE 18.1 Stress–strain relation of HAF carbon black-filled styrene–butadiene rubber (SBR). (From Chikaraishi, T., Koubunshi Gakkkai (Japan), No 25, Polymer Free Discussion, 1987.)

Although many interface models have been given so far, they are too qualitative and we can hardly connect them to the mechanics and mechanism of carbon black reinforcement of rubbers. On the other hand, many kinds of theories have also been proposed to explain the phenomena, but most of them deal only with a part of the phenomena and they could not totally answer the above four questions. The author has proposed a new interface model and theory to understand the mechanics and mechanism of carbon black reinforcement of rubbers based on the finite element method (FEM) stress analysis of the filled system, in journals and a book.[1-5] In the new model and theory, the importance of carbon gel (bound rubber) in carbon black reinforcement of rubbers is emphasized repeatedly. Actually, it is not too much to say that the existence of bound rubber and its changeable and deformable characters depending on the magnitude of extension are the essence of carbon black reinforcement of rubbers.

Thus, we came to the following conclusion. The existence of carbon gel (bound rubber) and its changeable and deformable characters generate the carbon black reinforcement. Under large extension, the inner GH layer in bound rubber fixes molecules to the surface of a carbon particle and the outer SH layer changes its characters from soft uncross-linked amorphous molecules to extended and oriented ones depending on the magnitude of extension. Thus, there appears the super-network of strands of oriented molecules interconnected at carbon particles. This super-network generates the stress upturn in the stress–strain relation and the great tensile strength for carbon-black-filled rubber. The compatibility of molecular slippage and stress upturn can be realized, when the molecular length between adjacent carbon particles is regulated by sliding of molecules passing through entanglement points in the SH layer and stopping of slippage at the GH layer. The very large hysterisis loop in carbon-black-filled rubber results from the buckling of the super-network in unloading which corresponds to nearly a half of the hysteresis loop of the system.

The purpose of this report is to bring the author's model and theory for carbon black reinforcement of rubbers to a conclusion, with additional experiments and discussion. This research consists of three papers (Part 1, Part 2, and Part 3), where the reinforcement of elastomers is generalized with the universal and common concept. Now, preceding the detailed discussion, we would like to discuss the previous concept generally accepted for carbon black reinforcement of rubbers and the author's new model and theory.

18.2 HISTORICAL BACKGROUNDS

18.2.1 Previous General Concepts for Carbon Black Reinforcement

Figure 18.1 is the typical stress–strain curves of the filled rubber (SBR filled with fine carbon black, HAF),[6] φ the volume fraction of carbon black, showing the above three criteria from 1 to 3. The most characteristic point in stress–strain relation of the filled rubber is first, that the stress increase becomes larger and larger as extension increases (called the stress upturn), in addition to the initial stress (modulus) increase at small extension. Second, the tensile strength is 10–15 times larger than that of the unfilled rubber vulcanizate whose strength is in the order of 2 or 3 MPa ($\varphi = 0$ in Figure 18.1). Moreover, the tensile strain is also quite large, compared with the unfilled rubber of the same modulus, as shown in Figure 18.1.

These typical phenomena in the filled rubber are in a striking contrast to those given in the unfilled rubber vulcanizate of the same high modulus, produced by adding much more sulfur to rubber. In such a system, the tensile strength will be 3–4 MPa at most, being 1/10 that of the filled rubbers, as shown in Figure 18.2[7] and strain at break may be 1/5–1/10 that of the filled rubber of the same modulus. Thus, we can clearly say that it is impossible to get such a great tensile strength and tensile strain for the unfilled cross-linked rubber, even if we choose the best cross-link density for the system.

One of the general understanding about the carbon reinforcement of rubbers, which is vaguely understood but widely accepted, is to consider such a system where well-dispersed and discontinuously connected carbon particles are strongly adhered to the matrix cross-linked rubber,

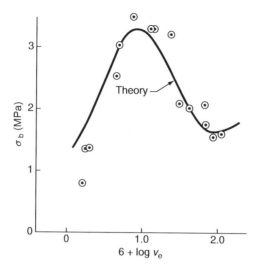

FIGURE 18.2 Tensile strength of styrene–butadiene rubber (SBR) as a function of network chain density. (From Bueche, F. and Dudek, T.J., *Rubber Chem. Tech.*, 36, 1, 1963.)

as schematically shown in Figure 18.3. In this model, a large number of cross-linked rubber molecules are concentrated on the surface of carbon particles, which makes a much higher cross-linked part there, compared with the average cross-links in matrix rubber. We can, accordingly explain the modulus increase of the system about question 1 using the model Figure 18.3, in combination with the Guth equation.[8]

However, in the model Figure 18.3, we cannot give any reasonable answers to questions 2 and 3. It is well established that the matrix cross-linked rubber in the filled system is almost the same as the unfilled cross-linked rubber. It means that the stress–strain relation of the matrix rubber in the filled

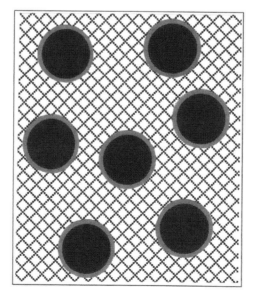

FIGURE 18.3 Schematic representation of carbon particles surrounded by densely cross-linked molecules (chemical).

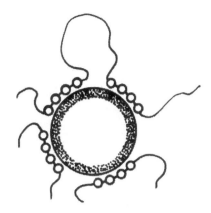

FIGURE 18.4 Schematic representation of carbon particles surrounded by molecules adsorbed on the surface of the particle (physical). (From Fukahori, Y. and Seki, W., *Polymer*, 33, 1058, 1992.)

system will be represented by the Gaussian equation up to about 500%, as shown in Figure 18.1 ($\varphi = 0$), where there is no upturn of the stress. In addition, the tensile strength of the matrix rubber in the filled rubber (unfilled cross-linked rubber) is 3–4 MPa at most. Therefore, the matrix rubber in the filled system, in which the maximum stress concentration appears, cannot support the very large stress in the system (30 MPa). Thus, the stress–strain curve of the system given in Figure 18.3 does not give the stress upturn and its tensile stress will be 3–4 MPa at most.

As is well known, of course, the model Figure 18.3 was rejected considering the stress reduction and its recoverable character in the Mullins effect,[9] concerning question 4. So far as we employ the model Figure 18.3, the stress decrease in unloading should be attributed to the breakage of the cross-linked molecules. It means that the Mullins effect must be unrecoverable in the model Figure 18.3, which is, of course against the experimental facts. Thus, instead of Figure 18.3, another model shown in Figure 18.4[10–13] was presented. In Figure 18.4, rubber molecules are adsorbed to the surface of carbon particles and slide freely on the surface under tension. The model can rather reasonably explain the stress reduction in unloading process and the stress recovery during long periods. Instead, however, this simple sliding model is evidently not available for the great stress increase at large extension, because the sliding molecules cannot generate any bigger stress than the critical one that the molecules begin to slide. This paradox is absolutely the essential question for the carbon black reinforcement of rubbers. Anyway, the simple model given in Figure 18.4 is not available to the explanation of the phenomena (1)–(3), and of course, cannot answer questions 1–3.

18.2.2 AUTHOR'S NEW INTERFACE MODEL AND CONCEPT FOR CARBON BLACK REINFORCEMENT OF RUBBERS

18.2.2.1 Interface Model

It is well known that in carbon-black-filled rubbers, carbon blacks disperse not as a separated particle but as an aggregate consisting of 5–10 particles together in matrix rubber. However, when we consider a mechanical model for the system, we can regard an aggregate as separated particles, as shown in Figure 18.5, although some consideration may be necessary for the diameter of the particle. The quite important observation for the carbon particle in carbon-filled rubbers is that the particle is surrounded by carbon gel (i.e., bound rubber). Bound rubber, produced in the mixing process of carbon black and rubber, consists mainly of two polymer phases of different molecular mobility, shown by the pulsed nuclear magnetic resonance (NMR) measurements.[14–17] In one phase, molecular mobility is strictly constrained, i.e., in the glassy state. In another phase, molecules are a little more constrained than those in unfilled rubber vulcanizate. Although a few interface

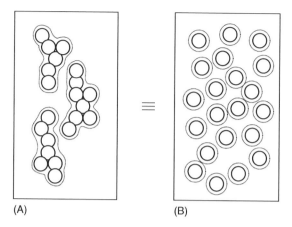

FIGURE 18.5 Equivalent models in carbon black-filled rubber: aggregates (A), separated particles (B).

models have been proposed from the viewpoint of molecular mobility so far,[17–19] they have not referred to the mechanical behavior of filled rubber.

The new interface model and the concept for the carbon black reinforcement proposed by the author fundamentally combine the structure of the carbon gel (bound rubber) with the mechanical behavior of the filled system, based on the stress analysis (FEM). As shown in Figure 18.6, the new model has a double-layer structure of bound rubber, consisting of the inner polymer layer of the glassy state (glassy hard or GH layer) and the outer polymer layer (sticky hard or SH layer). Molecular motion is strictly constrained in the GH layer and considerably constrained in the SH layer compared with unfilled rubber vulcanizate. Figure 18.7 is the more detailed representation to show molecular packing in both layers according to their molecular mobility estimated from the pulsed-NMR measurement.

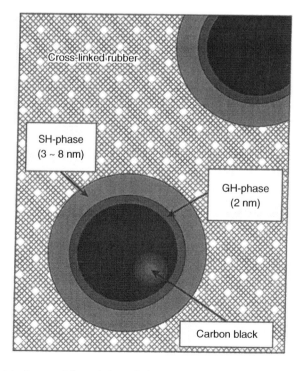

FIGURE 18.6 A new interface model consisting of glassy hard (GH) layer and sticky hard (SH) layer.

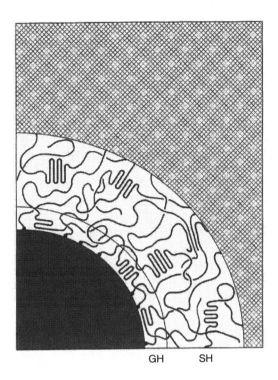

GH   SH

FIGURE 18.7 Detailed structure of Figure 18.6.

Due to the great activity of the surface of carbon particle, several parts in a molecular chain seem to be strongly adhered to the surface of the particle in the GH layer, whether physical or chemical. On the other hand, the fact that the SH layer is also insoluble in good solvents means that a part or several parts of a molecule within the SH layer is also firmly connected to the surface of the carbon particle through the GH layer, or in another case, it is tightly entangled with the molecules extending from the GH layer. In other words, the GH layer plays a role to fix several parts of a molecule in the SH layer to the surface of carbon particles. On the other hand, the SH layer seems to be strongly adhered to the surrounding matrix rubber vucanizate. It seems to be performed by chemical cross-linking or strong entanglement between molecules at the boundary surface of the SH layer and the matrix rubber. Because of the strong interaction between the SH layer and the matrix rubber, the eyeball separation is scarcely observed at the boundary even under large extension in the case of fine carbon-filled rubber vulcanizate.

The thickness is 2 nm for the GH layer and 3–8 nm for the SH layer, therefore the total thickness of the GH and SH layers is 5–10 nm. Generally, the thickness of the SH layer is smaller in fine carbon-filled rubber than in the coarse one. In the case of the fine carbon black like HAF, its diameter being about 30 nm, the thickness is 2 nm for the GH layer and 4–5 nm for the SH layer, for example. When the volume fraction of HAF carbon is 20% in SBR, the total thickness of both layers (6–7 nm) corresponds to the volume fraction of 30%–35% to the total rubber. The 2 nm thickness of the GH layer is a little less than 10% of the diameter of a fine carbon black (20–30 nm), but it is only 1% of that of a coarse carbon black (100–200 nm).

18.2.2.2 Mechanism of the Modulus Increase, the Stress Upturn, and the Great Tensile Strength

Thus, in the coarse carbon black-filled rubber, the modulus increases according to the Guth equation with filler content, because the contribution from the GH layer (1% of the diameter) will disappear as an experimental error. However, in the case of fine carbon black-filled rubber, since the contribution from the GH layer is nearly 10% of the diameter of the carbon particle, the modulus becomes larger by the increased volume fraction than that estimated from the Guth equation,[20] because we can roughly regard the modulus of the GH layer as that of a carbon particle in the calculation. This is the answer to question 1 by the new model.

As mentioned above, the GH layer plays a role only to increase the effective diameter of a carbon particle, thus the contribution of the GH layer to the stress increase of the system is kept constant over all strain amplitude. This means that the GH layer does not relate to the stress upturn at large extension, instead the SH layer must play a very important role at large extension. Here the author introduces the quite important assumption for the new model that there are very few or no cross-links in the SH layer. This assumption will be reasonably understood considering that a big sulfur particle is present outside the bound rubber during mixing. Moreover, the melted sulfur seems to be quite difficult to dissolve into the bound rubber in vulcanization at high temperature, because the bound rubber is insoluble even in good solvents, much less in such a poor solvent like the melted sulfur, which will be discussed in detail later. Anyway, if molecules within the SH layer are not cross-linked with each other, they will move and slide freely under large extension.

When carbon black (fine carbon black) is dispersed homogeneously and its volume fraction φ is less than nearly 0.2, the SH layers of adjacent carbon particles are still separated from each other in matrix rubber. In this case, the molecules in the SH layer slide, orient, and extend along the extension direction and finally produce strands of oriented molecules under large extension, as shown in Figure 18.8. In this situation, as the molecules inside the SH layer are extended much more

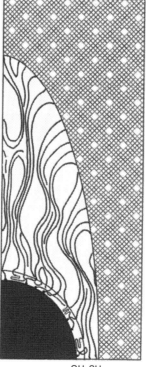

GH SH

FIGURE 18.8 Formation of the strands of oriented molecules in the sticky hard (SH) layer under large extension.

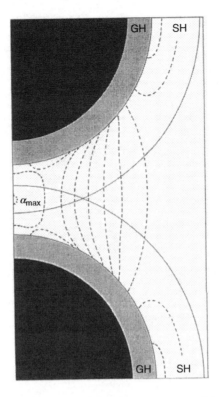

FIGURE 18.9 Overlapped sticky hard (SH) layers and the contour map of the stress concentration.

than molecules in matrix rubber, the strain at break of filled rubber becomes greater than that of unfilled rubber, as shown in Figure 18.1. On the other hand, since this phenomenon is similar to the reinforcement of rubber with some kind of short fibers oriented to the extension direction, the tensile stress will be much larger than that of unfilled rubber.

Figure 18.9 shows the contour map of the stress concentration factor α generated between two adjacent rigid particles, when the volume fraction of particles φ is larger than 0.2–0.25. It is clearly shown that the maximum stress concentration appears at the center between two rigid particles, moreover, the strong stress field like the strand of stress connects both rigid particles. In Figure 18.9, there is also shown that the SH layers surrounding the adjacent carbon particles are overlapped and adhered each other, making a continuous layer. Therefore, under large extension, the molecules in the SH layer sandwiched between two adjacent carbon particles are strongly extended by the separation of the carbon particles accompanied with molecular sliding and orientation, thus it constructs the network of the strands of oriented molecules interconnected by carbon particles, as shown in Figure 18.10.

This superstructure is expected to be quite strong to adequately support the large stress concentration around carbon particles. This is the most essential and important point in the reinforcement of rubber under large extension by filling of carbon black. The structural changing in the SH layer, from the initial soft gel (but harder than matrix rubber) to the hardened and strengthened superstructure, produces the typical feature of the stress–strain relation of the filled rubber, i.e., the abrupt upturn of stress and the very large tensile strength. Of course, these continuous phenomena in the SH layer can be realized just under the condition that there are very few or no cross-links in the SH layer. This is the answer to questions 2 and 3 from the viewpoint of the new model and concept.

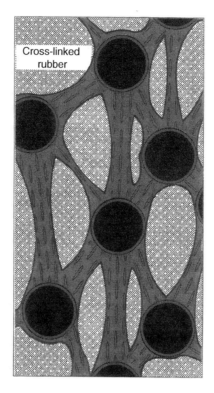

FIGURE 18.10 Super-network of the strands of oriented molecules interconnected at carbon particles.

18.2.2.3 Mechanism of Mullins Effect

Now, we consider the mechanism of stress softening of carbon black-filled rubber using the new model and concept. Firstly, we must have an insight into the buckling of the super-network of the strands of oriented molecules interconnected by carbon particles in the first unloading, as shown in Figure 18.11. Since this behavior takes place in just the beginning of the unloading, the stress decreases very rapidly, much faster than usual stress relaxation due to viscoelastic molecular motion. In the unloading process of filled rubber, the buckled bundles can hardly hold the stress of the system, and instead, the matrix cross-linked rubber mainly supports the stress. Therefore, the degree of the stress reduction in the unloading will be much faster and larger than that generated in usual stress relaxation, as shown in Figure 18.12. Thus, although the hysteresis loop is round at small extension, it becomes very sharp at large extension. This difference depends on whether it includes the buckling of the super-network or not. Anyway, we can easily understand this phenomenon to consider that the extended chewing gum buckles and abruptly loses the shrinking force when it is unloaded.

In the second loading, since the main load-bearing part is also the matrix cross-linked rubber, the stress–strain curve is similar to that of the first unloading. Therefore, the stress–strain curves in the second loading are quite similar in unfilled and filled rubbers of any filler contents and filler species, resulting in the unique master curve as indicated by Harwood et al.[21] When the extension exceeds the strain level of the first loading, since the buckled bundles return to the first extended state again, the stress increases again depending on the original stress–strain relation (Figure 18.12). The entropic and shrinking force of the matrix rubber always works to the buckled bundles, in addition the extended and buckled molecules will considerably relax and become much higher entropic state after unloading. As a result, during long periods or at higher temperature, the residual strain will recover to its origin by almost 100% and the stress also will considerably recover to its original level. These are the answers to question 4 in utilizing the new model and concept.

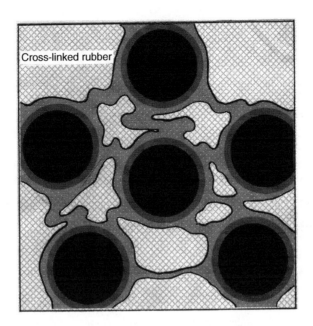

FIGURE 18.11 Buckling of the strands and the super-network.

18.3 EXPERIMENTS

18.3.1 EVALUATION OF CARBON GEL

The specimen was prepared by the following method. After mixing HAF carbon black (50 phr) with natural rubber (NR) in a laboratory mixer, carbon gel was extracted from unvulcanized mixture as an insoluble material for toluene for 48 h at room temperature and dried in a vacuum oven for 24 h at 70°C. We made the specimen as a thin sheet of the carbon gel (including carbon black) by pressing the extracted carbon gel at 90°C. The cured specimen was given by adding sulfur (1.5 phr) to the unvulcanized mixture and vulcanized for 30 min at 145°C. The dynamic viscoelastic measurement was performed with Rheometer under the condition of 0.1% strain and 15 Hz over temperatures.

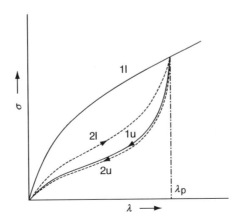

FIGURE 18.12 Mullins effect (schematic): loading (first,1l; second, 2l) and unloading (first, 1u; second, 2u).

18.3.2 EFFECTS OF THE RAZOR SLITS CUT ON THE SPECIMEN ON FRACTURE

We prepared a strip-type specimen (100 × 50 × 2 mm) from the rubber sheet of SBR filled with HAH carbon black (50 phr), vulcanized for 30 min at 155°C. On the specimen, a slit of different lengths ($s_1 = 30$ mm, $s_2 = 20$ mm, $s_3 = 10$ mm) parallel to the extension direction and a notch of different lengths (2 or 5 mm) at the center of the side surface of the specimen perpendicular to the extension direction were made by razor-cutting (see the inserted figure in Figure 18.14). The distance δ between slits and between the slit s_1 and the tip of notch was 1 and 3 mm, respectively. The no-slit specimen means that it only includes a notch, without slits.

18.3.3 EFFECT OF THE IRREGULARITY INTRODUCED IN REGULAR NETWORK

A plane string net for gardening consisting of square lattices was used. We prepared two types of specimens, one of regular lattice and another of the irregular one in which three strings were added to the regular lattice, as irregular knots. Both specimens were extended uniaxially on a plate.

18.4   RESULTS

18.4.1 EVALUATION OF CARBON GEL

Figure 18.13 shows the storage modulus G' of carbon gel (including carbon black) and the filled rubber vulcanizate plotted against temperature from −80°C to 100°C. Comparing both materials, it is clearly shown that the temperature dependence of G', i.e., the negative slope against temperature, is almost the same in both materials. This means that the negative slope of carbon gel due to its uncross-linked state is also left inside the filled rubber vulcanizate as it was, which gives the energetic property to the filled rubber vucanizate, and will be discussed later.

As a next step, we can roughly estimate the modulus of the gel (without carbon black) from Figure 18.13. Although the modulus of carbon gel is 9–10 times larger than that of the filled NR vulcanizate at 20°C, we must consider the difference of the content of carbon black in both materials. Although the volume fraction of carbon black is 0.20 (corresponding to 50 phr) in the filled NR vulcanizate, since the gel (without carbon black) is about 45% of the total NR, the volume fraction of carbon black in the carbon gel is 0.36. On the other hand, if we compare the modulus of

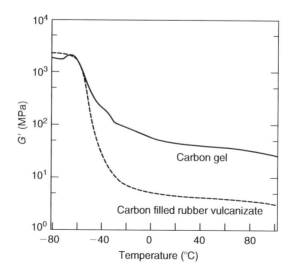

FIGURE 18.13 Storage modulus G' against temperature for carbon gel and the filled rubber vulcanizate.

the filled NR vulcanizate in the case of $\varphi = 0.2$ and $\varphi = 0.36$ from the stress–strain curves shown in Figure 18.1, it is roughly estimated that the modulus of the filled NR vulcanizate is 4–5 times larger in the case of $\varphi = 0.36$ than that in $\varphi = 0.2$. As a result, we can roughly say that the modulus of carbon gel is twice that of the filled rubber when the carbon content φ is 0.36 in both materials. Of course this difference results from the difference of the modulus of the gel and the matrix rubber vulcanizate. More accurately, we must consider that if the filled rubber vulcanizate includes the gel (45% of the total rubber) inside the system, then the modulus of the gel will be a little higher than twice. However, at the present stage, we would like to say that the modulus of the gel (i.e., GH layer in the new model) is roughly two times larger than that of the rubber vulcanizate (in the case of sulfur content of 1.5 phr). Moreover, although this is the case of the carbon black-filled NR, we can expect the same result for the carbon black-filled styrene–butadiene rubber (SBR).

This conclusion is supported by the experimental result[14–17] given by the pulsed-NMR measurement that the spin–spin relaxation time T_2 is considerably shorter for the gel than that for the matrix rubber vulcanizate, which of course, indicates that the modulus is considerably higher for the gel than for the matrix rubber. More quantitatively, Maebayashi et al.[22] measured the acoustic velocity of carbon gel by acoustic analysis and concluded that the compression modulus of the gel is about twice that of matrix rubber. Thus, at present, we can conclude that the SH layer, of course without cross-linking, is about two times harder than matrix cross-linked rubber in the filled system.

18.4.2 Effect of the Razor Slits Cut Parallel to the Extension Direction on Fracture

Figure 18.14 shows the stress–strain relation of various types of specimen with and without slits and of the different notch length and distance δ, such conditions being listed in Figure 18.14. Although the specimen with a notch, of course, shows the much lower tensile properties than that without a notch, the specimen with slits gives the higher tensile properties, such as tensile strength, strain at break, and fracture energy, compared with the specimen without slits. This means that the slits cut in front of the crack (notch) parallel to the extension direction disorganize the stress field generated at the crack front and change the propagation direction of the crack.

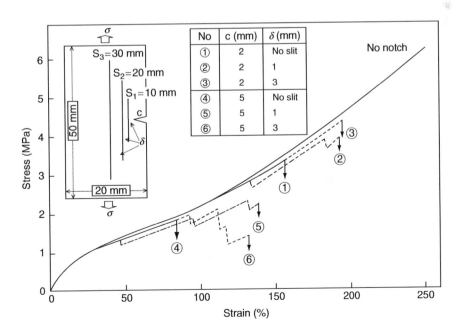

FIGURE 18.14 Effect of the slit and the notch of different length on the stress–strain relation.

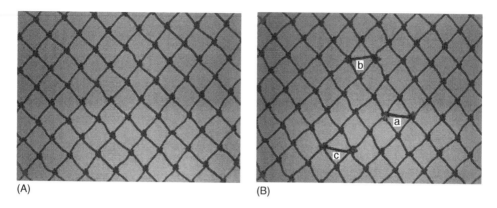

(A)   (B)

FIGURE 18.15 Plane string net of regular lattice (A) and irregular lattice indicated by a, b, c (B).

This changing of the path of crack propagation gives the material the better resistant characters for fracture. The same type of phenomenon occurs in the materials filled with fibers. Fibers play the same role more clearly on reinforcing the materials and make the tensile strength much higher. Thus, surely we can expect that the strand of molecules and the super-networks work in the same manner in the carbon black-filled rubber, which will be discussed in detail later.

18.4.3 EFFECT OF THE IRREGULARITY INTRODUCED IN REGULAR NETWORK

Figure 18.15 shows the plane string nets of regular lattice (Figure 18.15A) and irregular lattice (Figure 18.15B). In the net of irregular lattice, three strings of the same length (a, b, c) as the string between adjacent knots are securely tied to the regular knots, as shown in Figure 18.15B. When the specimens are extended horizontally (indicated by an arrow), the net of regular lattice deforms homogeneously and produces the regularly extended lattice, as shown in Figure 18.16A. On the other hand, in the net of irregular lattice, only three additional interconnected strings disturb the regular orientation of the net, the irregularity spreading over the area surrounding three interconnected points, as shown by the ellipse of broken line in Figure 18.16B.

This phenomenon gives us the very important information when we consider the orientation and extensibility of cross-linked molecules under extension. Generally, for the modeling or calculation for the stress–strain relation of cross-linked molecules, the perfect regular network is adopted a

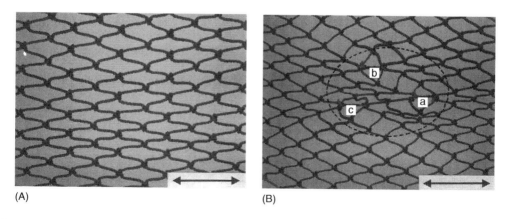

(A)   (B)

FIGURE 18.16 As Figure 18.15, but under uniaxial extension; the ellipse of broken line shows the irregular area.

priori, in the cross-link density, the molecular length between cross-links and the number of molecules consisting of a cross-link junction, for example. However, in real cross-linked rubber, as is well known, there exist many kinds of irregularities due to the very wide distribution in these cross-link properties. Therefore, we can surely understand that in such real cross-linked rubbers, the homogeneous extension and orientation of molecules must be quite difficult, which will be discussed in detail later.

18.5 DISCUSSION

18.5.1 CHARACTERISTICS OF THE SH LAYER

As shown in Section 18.2.2, the most important and essential point concerning the carbon black reinforcement of rubbers is the existence of the carbon gel (bound rubber). More exactly, almost all of the mechanical phenomena in the reinforcement by carbon black results from a series of responses of the changeable and deformable characteristics of the SH layer depending on the amplitude of extension. That is, in the SH layer under large extension, molecular movements of sliding and orientation produce the super-network of the strand of extended molecules, which supports the very large stress concentration and generates the stress upturn and the great tensile strength. Thus, first of all, the characterization of the structure and properties of the SH layer is important.

In the new interface model, the assumption that there is hardly any or no chemical cross-links in the SH layer only makes the above changeable and deformable characteristics of the SH layer possible. Gent[23] showed that in the experiment of the swelling of the carbon black-filled rubber to *n*-heptane, the swelling of the carbon gel was considerably lower than that of the matrix rubber vulcanizate and such a characteristic had been produced before vulcanization. This phenomenon tells us that the swelling of the carbon gel by the poor solvent must be quite low, so much poorer, than the melted sulfur. Very recently, Dohi et al.[24] observed the existence of sulfurs in the FT carbon-filled SBR after vulcanization, using energy-filtering TEM and visualized them as an elemental map. They showed that sulfurs were selectively adsorbed on the surface of ZnO particles and not on the surface of carbon particles. This seems to be the direct corroborative evidence to support the above assumption.

Of course, as we discussed earlier, if the SH layer is cross-linked by sulfur, just like Figure 18.3, we cannot give any reasonable explanation to questions 2 through 4. From the above total consideration, it is safe to bet that the SH layer is in the uncross-linked state. Of course, we must not forget that the uncross-linked SH layer is about two times harder than the matrix rubber vulcanizate. It means that since the modulus of the cross-linked matrix rubber is usually 5–10 times larger than that of uncross-linked rubber, the modulus of the SH layer is 10–20 times higher than that of general uncross-linked rubber. That is, the SH layer seems to be similar to thermoplastics at high temperature, and its high molecular interaction generates the large energy loss in sliding of molecules under large extension or cyclic deformation, as Mullins effect.

18.5.2 MECHANICS OF THE STRESS UPTURN AT A LARGE EXTENSION

As shown in Figure 18.1, in the stress–strain curve of the real unfilled SBR vulcanizate, the stress upturn does not appear and as a result, tensile strength and strain at break are only about 2 MPa and 400%–500%, respectively. Nevertheless, the stress–strain curve of the SBR vulcanizate filled with carbon black shows the clear stress upturn and its tensile stress becomes 30 MPa. This discrepancy between both vulcanizates is actually the essential point to understand the mechanism and mechanics of the carbon black reinforcement of rubber.

Now, we consider this phenomenon from the viewpoint of the non-Gaussian behavior of the network chain. As is well known, when we assume the idealized molecular network consisting

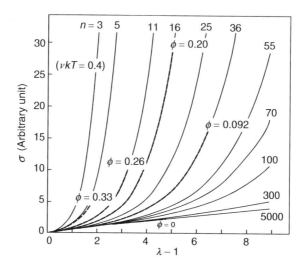

FIGURE 18.17 Stress–strain curves calculated for Equation 18.1; the thin filled lines correspond to the various n values and thick broken lines to the various φ values.

of the cross-links of four-chain model, the stress upturn, i.e., the very large stress increase of the inverse S shape under large extension shown in Figure 18.1, can be reasonably represented by the non-Gaussian network theory. When the molecular motion is constrained within the limited molecular length between cross-links, the strong upturn in the stress–strain curve at high extensions is the direct consequence of the limited extensibility of the chains. In Figure 18.17, a series of thin lines show the stress–strain relations corresponding to various n values in Equation 18.1 for the non-Gaussian networks, given by Kuhn and Grün.[38]

$$\sigma = (vkT/3)\,n^{1/2}[L^{-1}\,(\lambda/n^{1/2}) - \lambda^{-2/3}\,L^{-1}(1/(\lambda n)^{1/2})] \tag{18.1}$$

where
σ is the stress
λ is the extension ratio
L^{-1} is the inverse Langevin function
v and n are the number of network chains per unit volume and the number of the segments in a network chain, respectively
k and T are the Boltzmann's constant and the absolute temperature

Figure 18.17 shows that the characteristics of the stress–strain curve depend mainly on the value of n; the smaller the n value, the more rapid the upturn. Anyway, this non-Gaussian treatment indicates that if the rubber has the idealized molecular network structure in the system, the stress–strain relation will show the inverse S shape. However, the real rubber vulcanizate (SBR) that does not crystallize under extension at room temperature and other rubbers (NR, IR, and BR at high temperature) do not show the stress upturn at all, and as a result, their tensile strength and strain at break are all 2–3 MPa and 400%–500%. It means that the stress–strain relation of the real (noncrystallizing) rubber vulcanizate obeys the Gaussian rather than the non-Gaussian theory.

Nevertheless, it is obviously shown in Figure 18.1 that the stress–strain curve of the filled rubber gives the clear stress upturn, thus its tensile strength becomes 30 MPa. Therefore, the fundamental question is what happens or what structure is produced in the carbon black-filled rubber under large extension, which newly generates the stress upturn. In the case of the fine carbon black-filled system, when carbon blacks are dispersed ideally, the carbon gel makes the continuous phase at the

TABLE 18.1

Relation between φ and n

φ	δ/r_0	n
0	∞	∞
0.092	1.40	36
0.20	0.62	16
0.26	0.43	11
0.33	0.25	6

carbon content φ of about 0.23–0.25. In the continuous carbon gel phase, carbon particles are connected with each other by the mutual bonding of the uncross-linked SH layers. In such an uncross-linked system, the limited length for molecular motion should first be the distance between carbon particles and second the distance between molecular entanglements, just like the distance between cross-links. Thus, in the carbon-filled rubber, the distance between the carbon particles will be the fundamental unit length of molecular motion.

Now, we show the relation between the ratio of δ to r_0, δ/r_0 and the volume fraction of carbon black φ in Table 18.1, when the diameter of the hard particle (including carbon black, the GH layer and a little more contribution from the cross-links at the surface of particle) is r_0 and the distance between the hard particles is δ. In the carbon black-filled rubber ($\varphi \geqq 0.23$–0.25), the fact that the stress of the filled system is 10–15 times larger than that of the unfilled rubber as shown in Figure 18.1 indicates that more than 90% of the stress of the system is supported by the super-network and the remainder of the stress results from the matrix rubber. In the present calculation, however, we can ignore the contribution from the matrix rubber.

Now, we tentatively assume that the stress–strain curve of the case ($\varphi = 0.2$, $\delta/r_0 = 0.62$) in Figure 18.1 just corresponds to the stress–strain curve of the other case ($n = 16$) in Figure 18.17. Then, we can give the n value to other compositions of different carbon black contents, in consideration that δ/r_0 is proportional to n, the values being written in Table 18.1. The thick broken lines drawn in Figure 18.17 show the stress–strain curves of the reduced n values corresponding to the carbon black contents indicated in Table 18.1, in which the strain at break for the stress–strain curve is adjusted to the same one as given in Figure 18.1. From the similarity of the both stress–strain curves shown in Figures 18.1 and 18.17 (thick broken lines) for all carbon contents, we can say that, fundamentally, the stress–strain curves of the carbon black-filled rubber are governed by the extensibility of molecules between adjacent carbon particles. Of course, since the thick broken lines only correspond to the molecular extensibility, the contribution of filling of carbon black must be added to the lines, according to the Guth equation. For this reason, the stress and tensile strength in the broken thick line have the tendency to be considerably lower, as carbon content increases (e.g., $\varphi = 0.26$ and 0.33). Anyway, however, it must be essentially true that the stress upturn in carbon black-filled rubber results from the limited extensibility of the molecular chains in the SH layer between adjacent carbon particles. Recently, similar views have been discussed elsewhere.[25,26]

Incidentally, as is well known, in the unvulcanized state of the filled rubber, the stress upturn does not appear even in the carbon content of $\varphi = 0.2$–0.25. For understanding this phenomenon, we must consider the discontinuity between the SH layers, in addition to the very low modulus of the unvulcanized rubber. Actually, as we discussed earlier, carbon blacks disperse as an aggregate of carbon particles, then the continuity of carbon gels consisting of aggregates must be much poorer than the theoretical calculation based on the perfect dispersion of carbon particles. In this case, the stress cannot be transmitted from carbon gel to carbon gel through such a very soft medium, and as a result, the stress–strain curve of the system is rather similar to the characteristics of the unvulcanized matrix rubber. Of course, when the carbon content is more than $\varphi = 0.3$, for example, the stress–strain curve begins to generate the stress upturn even for unvulcanized filled rubber.

18.5.3 MECHANISM OF COMPATIBILITY OF MOLECULAR SLIPPAGE AND STRESS UPTURN IN CARBON BLACK-FILLED RUBBERS

The basic premise for any theory of reinforcement is based on the assumption that fillers are perfectly bonded to the surrounding matrix, thus the external force is wholly transmitted to the matrix and other particles through the interface and finally to the whole system. This procedure generates the large stress increase of the system. On the other hand, in carbon black-filled rubber, it has been pointed out that there must be grand scaled sliding of molecules under large extension, resulting in large stress relaxation and creep and great stress softening and recovery of stress, called the Mullins effect.

Moreover the stress lowers with increasing temperature in carbon-filled rubber, as shown in Figure 18.18,[13] whereas stress increases with temperature in the case of unfilled rubber vulcanizate. In addition, even at high temperature, the stress–strain curve of the carbon-filled rubber shows the inverse S letter shape with the stress upturn, indicating that it still behaves as the non-Gaussian based on the entropic contribution. Of course, the rubber elasticity theory, whether Gaussian or non-Gaussian, indicates that stress increases proportional to the absolute temperature due to entropy contribution, as given in Equation 18.1. This means that the molecules in carbon black-filled rubber are not perfectly adhered to the surface of carbon particles (as shown in Figure 18.4), but slide freely on the surface of carbon particles. As mentioned earlier, the compatibility of the above phenomena seems to be the paradoxical problem, thus the next essential question in the carbon-filled rubber is why both the stress upturn and the molecular sliding occur at the same time, without contradiction.

Now, we consider the interface of carbon black-filled rubber from the new interface model point of view. As we discussed before, the bonding is almost perfect in the three interfaces, between

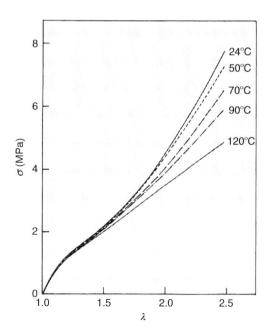

FIGURE 18.18 Temperature dependence of the stress–strain relation in carbon black-filled natural rubber (NR). (From Fukahori, Y. and Seki, W., *Polymer*, 33, 1058, 1992.)

carbon particle and the GH layer, the GH layer and the SH layer, and the SH layer and matrix rubber. On the other hand, it is quite reasonable to consider that the molecular sliding takes place in the SH layer, because we cannot expect such a large-scale sliding of molecules in the GH layer and in the cross-linked matrix rubber at all. Therefore, the point is how a molecule slides, extends, orients, and makes a bundle of molecules in the SH layer sandwiched between the adjacent rigid particles.

Here we make the following assumption for the behavior of an uncross-linked molecule in the SH layer: (1) each molecule freely slides passing through many entangled points, (2) each molecule is bonded and fixed to the GH layer at several points in a molecule, and (3) by sliding through the entangled points, each molecule regulates the molecular length to make a uniform length between adjacent particles. Figure 18.19 shows the schematic representation of the sliding and orientation of a molecule regulated under the above conditions, where points A, B, and C correspond to the position to fix a molecule to the GH layer and a', b', c', and d' to the entangled points where molecules slide freely. The sliding and regulation of a molecule is quite similar to playing the cat's cradle, in which we slide a string and regulate the length of the string equally between the fingers of both hands. The great importance in the cat's cradle is that both ends of the string are tied securely, which generates the strong force between the fingers. If they are not tied, of curse, the string will get away from the fingers and cannot produce the resistant force any more.

In the model Figure 18.19, the points A, B, and C correspond to the knots in the string and play a role to stop the sliding of the molecule between the adjacent carbon particles.

That is, in the model Figure 18.19 (i.e., in the new interface model), there are two points of the opposite contribution. The uncross-linked SH layer promotes the sliding of molecules passing

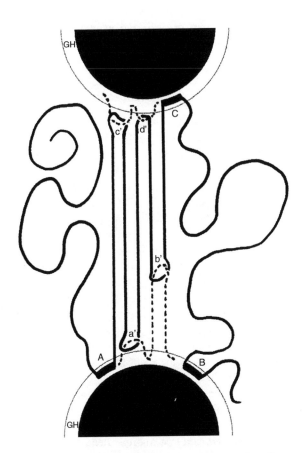

FIGURE 18.19 Regulation and orientation of molecular length between the adjacent carbon particles.

through entangled points under large extension and the GH layer stops the molecular sliding within the limited distance, which generates the stress of the system. Although the sliding of molecule itself does not increase the stress, it makes the regulation of the molecular length possible between the adjacent carbon particles. Thus, the sliding prepares the almost equal molecular length fitting exactly to the distance δ, which realizes the equal extensibility of all molecules and generates the great stress increase. Very recently, Shinohara et al.[27] observed a clear pattern showing the orientation of molecules in carbon-filled SBR with a wide-angle x-ray measurement, whereas no such pattern was observed in unfilled SBR. This obviously indicates that the uncross-linked state in the SH layer is a necessary condition for the regulation and orientation of molecules, and as a result, the formation of the strand of molecules.

Now, from a different point of view, we consider how difficult the orientation of molecules is in real cross-linked rubber. As we discussed before, the real cross-linked rubber includes the wide distribution in cross-link density, the molecular length between cross-links and the number of molecules consisting a cross-link junction, which work as irregularities in the homogeneous system. Figure 18.16 tells us that these irregularities strongly disturb the homogeneous deformation of the system and make the molecular regulation impossible. The stress–strain curves shown in Figure 18.17 are calculated under the condition that all of the molecular lengths (or the number of segments n) between cross-links are equal. Therefore, in the real cross-linked rubber, only a part of molecular length between cross-links must be n, in which the stress will increase little by little, and not abruptly.

Moreover, we must pay attention to the points that in the cross-linked rubber, the cross-link stops the sliding of molecules and has its own excluded volume. Generally, in the calculation of the stress–strain relation, the four-chain model is used for the cross-link junction and recently the eight-chain model[28] is considered to be more realistic and available. Thus, it is quite reasonable to consider that the bulky excluded volume that a cross-link junction possesses must be a real obstacle for the orientation of molecules, just like the case observed in Figure 18.16B.

18.5.4 Mechanics of Generation of Great Tensile Properties in Carbon-Filled Rubber

The fact that the tensile strength of the carbon-filled SBR is 30 MPa means that the strength of any area in the filled rubber is larger than 30 MPa, whereas that of the matrix rubber (unfilled SBR vulcanizate) is only 2–3 MPa. This means that the matrix cross-linked rubber cannot support such great stress at all, even in the case of very high cross-link density, as shown in Figure 18.2. Therefore, the question will be what structure is newly constructed in the system and supports such a large stress under extension. In the new model proposed here, the superstructure consisting of the network of the strands of oriented molecules interconnected at carbon particles supports the great stress as shown in Figure 18.10. That is, the carbon-filled rubber consists of the inhomogeneous and bi-continuous structure produced by the super-networks and the matrix rubber vulcanizate, where the super-network supports more than 90% of the tensile strength of the system.

The heterogeneity in the structure of the filled rubber has been discussed for a long time. In the tearing of the filled rubbers,[29–31] for example, fracture does not proceed smoothly but with the stick-slip motion and gives the fracture surface of knotty-tear or stick-slip tear. The tearing energy is much larger in the filled rubber than in the unfilled rubber, in particular, it is much larger in the direction parallel than perpendicular to the extension direction in the filled rubber. For this anisotropic fracture, it is considered that something like fibers should be produced in front of the crack, perpendicular to the crack propagation (i.e., parallel to the extension) under large extension. Generally, fibers change the propagation direction and moreover stop the propagation of the crack. However, since this improvement in tearing energy of carbon black-filled rubber is so different in direction, it is not necessarily accepted as a true improvement, but the apparent one depending on the direction.

Now, we return to the new model and consider the experimental results given in Figure 18.14, where even the slits (long breaks) play a role to disorganize the stress field and generate the resistant properties to fracture. Therefore, we can surely image that the existence of the super-network is closely related to the anisotropic tearing behavior, in addition to the stress upturn and the great tensile strength in carbon black-filled rubber. Although a crack propagates with accelerated force proportional to $c^{1/2}$ (where c is crack length) in homogeneous system, if even a slit exists in front of the crack, crack propagation is interrupted and changes its propagation path to a different direction, which produces the higher fracture properties, as shown in Figure 18.14. Therefore, in the system including the super-network of the strand of molecules, the crack propagation seems to be markedly delayed, because the strength of the strand seems to be almost the same as that of plastic fiber.

It had been observed at the SEM photographs[32–35] that carbon particles were surrounded with some materials slightly different from matrix rubber. The materials were vaguely believed to be carbon gel and have the relation to the above knotty-tear of the carbon filled rubber. Recently, Reichert and Göritz[26] showed much clearer photographs to indicate some oriented materials interconnecting with carbon particles, which appeared parallel to the extension direction under large extension. Göritz's photograph may depict the super-networks of molecular strands.

18.5.5 Mechanism of the Great Hysteresis Energy of Carbon Black-Filled Rubber

It has often been pointed out for a long time that the hysteresis energy given from the hysteresis loop under large extension is too big compared with the viscoelastic dissipation energy. For example, the hysteresis loop given from the stress relaxation is only 20%–30% of that from the stress–strain curve, when both measurements are performed at the same relaxation time and the same extension.[36]

As is well known, there is a close relation between the input energy at break, U_b and the hysteresis energy at break, H_b, for many kinds of filled and unfilled rubber vulcanizates, given by the following empirical Equation 18.2:[37]

$$U_b = KH_b^{2/3} = K^3 h_b^2 \tag{18.2}$$

where

K is a constant
$h_b (= H_b/U_b)$ is hysteresis ratio at break

Equation 18.2 reveals the importance of energy dissipation in the fracture of rubbery materials, that is, the strength of rubbers, whether filled or unfilled, is governed by the hysteresis energy dissipated in deforming the system.

Figure 18.20[37] shows the above relation between H_b and U_b for filled (HAF carbon black) and unfilled SBR, where it is clearly seen that Equation 18.2 is suited well for these materials. However, we must take notice of other important information in Figure 18.20 for understanding the effect of filling of carbon black. That is, to get the same energy (worked down) to break U_b, much larger hysteresis energy H_b is necessary for filled rubber than for unfilled rubber. In comparing unfilled with filled rubber in Figure 18.20, the value of H_b is 1.7 times larger for the filled SBR than the unfilled one, when the carbon black content is 60 phr, for example. Therefore, the relation between H_b and U_b for the 60 phr filled SBR must be represented by Equation 18.3, instead of Equation 18.2,

$$U_b = [K^3/1.7](h_b')^2 \tag{18.3}$$

where h_b' is the hysteresis ratio for the 60 phr filled SBR. More generally, it is written for the filled rubber of carbon content φ, as Equation 18.4,

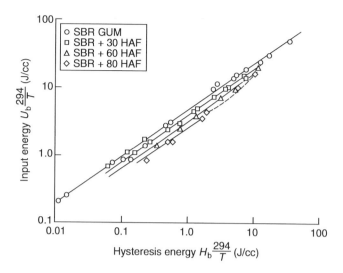

FIGURE 18.20 Relation between input energy and hysteresis energy at break for HAF carbon-filled and unfilled styrene–butadiene rubber (SBR). (From Payne, A.R., *J. Polymer, Sci.*, 48, 169, 1974.)

$$U_b = [K^3/f(\varphi)](h_b(\varphi))^2 \tag{18.4}$$

where $f(\varphi)$ and $h_{b(\varphi)}$ are functions of φ, both increasing with φ.

This means that in the case of carbon black-filled rubber, hysteresis energy given from the hysteresis loop under large extension (at breaking strain) must include some other component which has no connection with fracture. In other words, the hysteresis energy at breaking strain includes some additional "apparent hysteresis energy" depending on the filler content. In the filled SBR (60 phr), the apparent hysteresis energy is about 41% $[=(1.7-1.0)/1.7]$ of the total hysteresis energy. The next question, then, is what this apparent hysteresis energy is, i.e., what mechanism is working for the energy dissipation in the filled rubber.

Now, we return to the author's new model and concept. One of the characteristics of the model for filled rubber is that buckling of the network of strands of oriented molecules interconnected by carbon particles takes place in the unloading process under large extension, as shown in Figure 18.11. Generally, although this behavior produces a large hysteresis loop, it scarcely generates the energy dissipation, because it is only the bending of the strand of molecules and is scarcely accompanied by the frictional behavior of molecules. For this reason, we can surely say that the apparent hysteresis energy included in the hysteresis loop results from the buckling of the super-network. The magnitude of the apparent hysteresis energy, although it depends on the filler content and filler species, will roughly be a half of the hysteresis energy given from the hysteresis loop at breaking strain.

18.6 CONCLUSIONS

1. It is not too much to say that the existence of bound rubber (SH layer) and its changeable and deformable characters depending on the magnitude of extension are the essence of carbon black reinforcement of rubbers.
2. Assumption in the new model that there is hardly any or no chemical cross-links in the SH layer, only makes the changeable and deformable characteristics of the SH layer possible. This assumption has almost been verified experimentally.

3. Super-network of the strands of oriented molecules interconnected at carbon particles produced under large extension supports the increasing stress (stress upturn) and the great stress of the system at break.

4. Stress upturn in the stress–strain relation of carbon black-filled rubbers can be reasonably revealed in terms of the non-Gaussian treatment, by regarding the distance between adjacent carbon particles as the distance between cross-links in the theory.

5. Compatibility of molecular slippage and stress upturn in carbon black-filled rubbers can be realized, when the molecular length between adjacent carbon particles is regulated by sliding of molecules passing through entanglement points in the SH layer and molecular sliding is stopped at the GH layer.

6. Hysteresis energy given from the hysteresis loop at break in carbon black-filled rubbers includes the apparent hysteresis energy whose fraction is roughly a half of the total one, which results from the buckling of the super-network.

REFERENCES

1. Fukahori, Y. *The Dynamics of Polymers*, Gihoudo Press, 2000, p. 329.
2. Fukahori, Y. *Rubber Chem. Tech.*, **76**, 548, 2003.
3. Fukahori, Y. *Nippon Gomu Kyoukaishi (Japan)*, **77**, 180, 2004.
4. Fukahori, Y. *Nippon Gomu Kyoukaishi (Japan)*, **77**, 317, 2004.
5. Fukahori, Y. *J. Appl. Polym. Sci.*, **95**, 60, 2005.
6. Chikaraishi, T. *Koubunshi Gakkkai(Japan)*, No. 25, Polymer Free Discussion, 1987.
7. Bueche, F. and Dudek, T.J. *Rubber Chem. Tech.*, **36**, 1, 1963.
8. Guth, E. *J. Appl. Phys.*, **16**, 20, 1945.
9. Mullins, L. *Rubber Chem. Tech.*, **21**, 281, 1948; ibid., **23**, 733, 1950.
10. Houwink, *Rubber Chem. Tech.*, **29**, 888, 1956.
11. Dannenberg, E.M. *Tran. Inst. Rubber Ind.*, **42**, T26, 1966.
12. Boonstra, B.B. *J. Appl. Polym. Sci.*, **11**, 389, 1967.
13. Fukahori, Y. and Seki, W. *Polymer*, **33**, 1058, 1992.
14. Fujiwara, S. and Fujimoto, K. *Rubber Chem. Tech.*, **44**, 1273, 1971.
15. Kaufman, S., Slichter, W.P., and Davis, D.O. *J. Polymer Sci.*, A2, **8**, 829, 1971.
16. Nhishi, T. *J. Polymer Sci., Polym. Phys. Ed.*, **12**, 685, 1974.
17. O'Brien, J., Casshell, E, Wardell, G.E., and McBrierty, V.J. *Macromolecules*, **9**, 653, 1976.
18. Gessler, A.M. *Rubber Age*, **101**(12), 54, 1969.
19. Fujimoto, K. *Nippon Gomu Kyoukaishi (Japan)*, **37**, 602, 1964.
20. Fukahori, Y. and Seki, W. *J. Mater. Sci.*, **28**, 4471, 1993.
21. Harwood, J.A.C., Mullins, L., and Payne, A.R. *Rubber Chem. Tech.*, **39**, 814, 1966.
22. Maebayashi, M., et al. Preprints of No. 17 Elastomer Forum in Kobe, 2004, p. 8.
23. Gent, A.N. *Rubber Chem. Tech.*, **76**, 517, 2003.
24. Dohi, H., et al. Preprints of No. 17 Elastomer Forum in Kobe, 2004, p. 86.
25. Dietrich, J., Ortmann, R., and Bonart, R. *Makromol. Chem.*, **187**, 327, 1988.
26. Reichert, W.F. and Göritz, D. *Polymer*, **34**, 1216, 1993.
27. Shinohara, Y., et al. Preprints of No. 17 Elastomer Forum in Kobe, 2004, p. 24.
28. Arruda, E.M. and Boyce, M.C. *J. Mech. Phys. Solids*, **41**, 389, 1993.
29. Greensmith, H.W. *J. Polymer Sci.*, **21**, 175, 1956.
30. Andrews, E.H. *J. Appl. Phys.*, **32**, 542, 1961.
31. Glucklich, J. and Landel, R.F. *J. Appl. Polym. Sci.*, **20**, 121, 1976.
32. Hess, W.M. and Ford, F.R. *Rubber Chem. Tech.*, **36**, 1175, 1963.
33. Kraus, G. *Rubber Chem. Tech.*, **46**, 653, 1973.
34. Ban, L.L., Hess, W.M., and Papazian, I.A. *Rubber Chem. Tech.*, **47**, 858, 1984.
35. Dannenberg, E.M. *Rubber Chem. Tech.*, **59**, 512, 1986.
36. Kawabata, S., Yamashita, Y., and Ohyama, H. *Rubber Chem. Tech.*, **68**, 311, 1995.
37. Payne, A.R. *J. Polymer, Sci., Symposium*, **48**, 169, 1974.
38. Kuhn, W. and Grün, F. *Kolloid-Z*, **101**, 248, 1942.

Section IV

New Characterization Techniques

19 Visualization of Nano-Filler Dispersion and Morphology in Rubbery Matrix by 3D-TEM

Shinzo Kohjiya, Yuko Ikeda, and Atsushi Kato

CONTENTS

19.1 INTRODUCTION

Generally speaking, commercial rubber products are manufactured as a composite from a rubber and a nano-filler, which is in a group of fillers of nanometer size (mainly, carbon black and particulate silica). For an example, a pneumatic tire for heavy-duty usages such as aircrafts and heavyweight trucks is made from natural rubber (NR) and carbon black and/or silica. Their reinforcing ability onto rubbers makes them an indispensable component in the rubber products [1,2].

In material design of rubbery nano-composites, the importance of dispersion of the nano-fillers in a rubbery matrix has been well recognized, since the early stage of rubber industries. Their dispersion and the resulting morphology are reasonably assumed to be the most important factor in determining the mechanical properties of the rubbery nano-composites, i.e., filled rubber vulcanizates [2]. So far, morphology of nano-fillers in rubbery matrix has been estimated, as shown in Figure 19.1, from various evidences based on the past experiences. Nano-fillers are gathered together to form aggregates, which may further associate to give agglomerates. So far, the most useful technique for the elucidation of their morphology has been transmission electron microscopy (TEM) [3].

TEM is still the most powerful technique to elucidate the dispersion of nano-filler in rubbery matrix. However, the conventional TEM projects three-dimensional (3D) body onto two-dimensional (2D) (x, y) plane, hence the structural information on the thickness direction (z-axis) is only obtained as an accumulated one. This lack of z-axis structure poses tricky problems in estimating 3D structure in the sample to result in more or less misleading interpretations of the structure. How to elucidate the dispersion of nano-fillers in 3D space from 2D images has not been solved until the advent of 3D-TEM technique, which combines TEM and computerized tomography technique to afford 3D structural images, incidentally called "electrontomography".

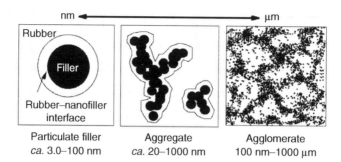

FIGURE 19.1 Morphology of nano-filler in rubbery matrix: Nano-particles are aggregated, and the aggregates also associate to give filler agglomerate in rubber. (From Kohjiya, S., Kato, A., Suda, T., Shimanuki, J., and Ikeda, Y., *Polymer*, 47, 3298, 2006. With permission.)

3D-TEM was introduced into rubber technology field in 2004 by the authors of this chapter [4]. In a recent review of 3D-TEM applied in materials science field [5], only two papers were cited on polymeric samples—one on block copolymers [6] and the other on rubbery composites with conventional and *in situ* silica [4]. Starting from the latter, 3D-TEM measurements have been carried out on rubbery nano-composites [7–16], and this recent and very important topic is described in this review.

3D-TEM gave 3D images of nano-filler dispersion in NR, which clearly indicated aggregates and agglomerates of carbon black leading to a kind of network structure in NR vulcanizates. That is, filled rubbers may have double networks, one of rubber by covalent bonding and the other of nano-filler by physical interaction. The revealed 3D network structure was in conformity with many physical properties, e.g., percolation behavior of electron conductivity.

This powerful technique is expected to increase its importance, but a careful choice of sample is essential when we consider the necessity for a good contrast in TEM observations and damages of the sample by electron-beam irradiation. Generally speaking, electron staining of the sample may be avoided, because it may change or modify it, possibly leading to a misleading structural conclusion on the original sample.

19.2 PROCEDURES FOR 3D-TEM MEASUREMENTS

3D-TEM is occasionally called "electron tomography," and is an imaging technique combining TEM and computerized tomography which enables us to visualize the nanometer-scale structure in 3D space [17]. In this measurement, TEM, which usually has resolution much better than 1 nm [18], has to be equipped with a special sample holder to be rotated typically from $+65°C$ to $-65°C$, and at each $2°C$ TEM image (a slice image) is obtained. All the slice images are subject to processing by computerized tomography for the reconstruction of a 3D image of nanometer resolution [4,5,8,17,19]. Figure 19.2 shows a generalized procedure of 3D-TEM measurement.

The instrument used in this study was Tecnai G2 F20 (FEI Co.). The 3D images were obtained by the bright-field method at 200 kV without any staining of the samples. Since three components which are matrix NR, carbon black (HAF grade), and silica (conventional VN-3 or that prepared by *in situ* technique) were all amorphous, the samples here were not diffracting, for which the bright-field electron tomography is suitable. For crystalline (i.e., diffracting) samples, high-angle annular dark-field (HAADF) imaging by scanning transmission electron microscopy (STEM) [20,21] is expected to be effective, but it is still at a developmental stage. It is noted that the NR vulcanizates were subjected in advance to extraction by organic solvents to remove Zn compounds. They were derived from ZnO (one of the additives for sulfur curing), and scattered the electron beam to result in smeared TEM images [8]. After this treatment, a thin sample for TEM (of ca. 200 nm thickness)

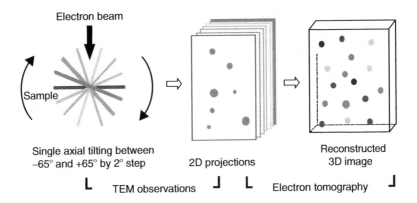

FIGURE 19.2 Procedure for three-dimensional-transmission electron microscopic (3D-TEM) observation, composed of TEM measurements and computerized tomography to reconstruct a 3D image. (From Kohjiya, S., Kato, A., Shimanuki, J., Hasegawa, T., and Ikeda, Y., *Polymer*, 46, 4440, 2005. With permission.)

was prepared by an ultramicrotome using a cryosystem in liquid nitrogen. The slice images were aligned by IMOD [22], and visualized using AMIRA [23]. For 3D-TEM, this step of computerized tomography is equally important as TEM measurements. Several software are needed for the processing other than IMOD and AMIRA.

This powerful technique for nano-structural elucidation was first developed in bioscience field [24,25], and is now more or less conventional in structural biology [17,26]. In the rubber field, the morphology of nano-fillers has been studied by conventional TEM [3,27] to enjoy subnanometer resolution. Early attempts to obtain 3D images on carbon-black-filled rubber are also known [28,29], but they did not succeed in reconstructing 3D images, due to the limited range of sample rotation. Since commercial rubber products are prepared as composites with inorganic nano-filler, i.e., they are a kind of nano-composites, the importance of elucidating 3D dispersion of nano-filler in rubbery matrix cannot be overestimated. So far, carbon black has been widely used as a reinforcer in the rubber industry [1,2,9,27]. Silica, on the other hand, is rapidly increasing its utilization [2,30,31]. Nano-structural elucidation of these fillers in rubbery matrix would afford information of much value for the establishment of structure–properties relationship, and is expected to contribute to the design of nano-composites as well as developing new reinforcing fillers other than carbon black.

19.3 CONVENTIONAL AND *IN SITU* SILICA IN RUBBERY MATRIX

Silica is unique among nonblack fillers. Its reinforcing ability is comparable to that of carbon black, especially when mixed with a suitable coupling agent, and its transparency affords many products. Additionally, it is chemically synthesized, which means that a wide range of silica (in terms of diameter, surface area, or surface activity) may be produced depending on the reaction routes and reaction conditions.

Among many kinds of reinforcing fillers for rubber, *in situ* silica has been focused [31–36]. It is produced at place by the sol–gel reaction [37], i.e., a sheet of rubber vulcanizate was swollen with tetraethoxysilane (TEOS), and the swollen sample was soaked in aqueous *n*-butylamine solution at 50°C to conduct the sol–gel reaction of TEOS. After the drying of the sheet, the nano-composite from rubber and *in situ* silica was obtained. This *in situ* silica filling was also conducted in uncured NR matrix to give *in situ* silica filled NR, which can be subjected to the conventional milling with crosslinking reagents and heat-pressing. As a reference, conventional silica (VN-3) filled rubber composite was prepared using conventional techniques.

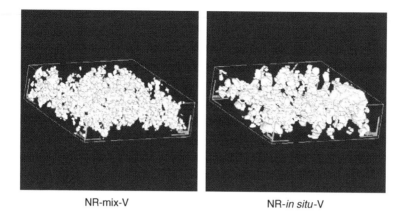

NR-mix-V　　　　　　　　　　　　NR-*in situ*-V

FIGURE 19.3 Monochromatic volume rendered views of the reconstructed 3D image of silica dispersion in natural rubber (NR) vulcanizate. The frame is shown in reconstructed perspective geometry (length and width, 630 nm; thickness, 181 nm). The green bar for each direction shows the length of 100 nm. For NR-mix-V and NR-*in situ*-V, see text. (From Kohjiya, S., Kato, A., Shimanuki, J., Hasegawa, T., and Ikeda, Y., *Polymer*, 46, 4440, 2005. With permission.)

Figures 19.3 and 19.4 show reconstructed 3D images of the conventional-silica-filled NR vulcanizate (NR-mix-V) and the *in situ* silica-filled one (NR-*in situ*-V) [4,8]. The content of silica was 33 phr, and all were aggregated, i.e., practically no isolated silica particles were observed in 3D images. Figure 19.3 shows monochromatic 3D images. To make it easy to recognize aggregates, an aggregate in which the particles are in contact is painted in color to produce Figure 19.4, in which each aggregate is easily recognized. Due to the larger size of the *in situ* silica as shown in Figure 19.3, the aggregate from *in situ* silica is also larger than that from VN-3 (though the number of particles in an aggregate is smaller in NR-*in situ*-V). However, it is noted that the size of *in situ* silica was determined by the reaction conditions of the sol–gel reaction used, and this statement is valid on

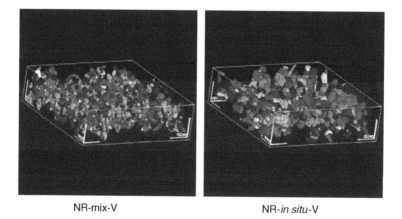

NR-mix-V　　　　　　　　　　　　NR-*in situ*-V

FIGURE 19.4 Colored volume rendered views of the reconstructed 3D image of silica dispersion in natural rubber (NR) vulcanizate. Each aggregate is isolated from the neighbors by coloring. The frame is shown in reconstructed perspective geometry (length and width, 630 nm; thickness, 181 nm). The white bar for each direction shows the length of 100 nm. For NR-mix-V and NR-*in situ*-V, see text. (From Kohjiya, S., Kato, A., Shimanuki, J., Hasegawa, T., and Ikeda, Y., *Polymer*, 46, 4440, 2005. With permission.)

the present *in situ* silica. 3D-TEM images of much smaller *in situ* silica particles may display a different feature. In the near future, this will be reported.

19.4 CARBON BLACK IN RUBBERY MATRIX

Carbon black has been extensively used in rubber industries for 100 years. Around 90% of its consumption is for tires [38]. In the rubber technology involving tires, the most important issue for many years has been "which kind of, and how much black is to be incorporated, followed by how to mix the black with rubber, under what conditions for a specific product?" In other words, how to disperse carbon black in rubbery matrix is assumed to determine the mechanical properties of final rubber products [2,3,9,14,27–29,38]. Carbon black for rubber industries are produced by the oil furnace method, and are of 10–50 nm particle diameter. They form aggregates during their manufacturing, and assumed to form agglomerates in the final rubber vulcanizates (see Figure 19.1). However, there have been no methods to quantify the 3D dispersion of carbon black in rubbery matrix. It is quite common to practice many trials to determine the optimal experimental parameters of filler usage and its mixing conditions that vast amounts of money and time are spent for the R & D of a new rubber product. In a sense, to observe the dispersion and morphology of carbon black in 3D space is the ultimate goal for eliminating the laborious work on fillers, and rubber technologists have been dreaming of the images.

Now, the dream has become a reality. Figure 19.5 shows 3D-TEM images of three NR vulcanizates, cured by sulfur/N-cyclohexyl-2-benzothiazolyl-sulfenamide (CBS) curing system, CB-10, CB-40, and CB-80 containing 10, 40, and 80 phr high-abrasion furnace (HAF) carbon black. In the black and white images the contrast was reversed, and the white is identified as the carbon black. Association to aggregates is observed even in CB-10 (volume fraction is 0.0498).

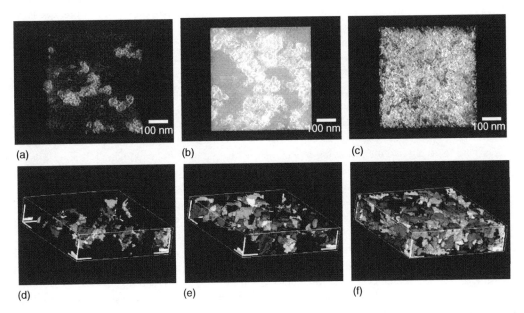

FIGURE 19.5 Reconstructed three-dimensional-transmission electron microscopic (3D-TEM) images of carbon-black-loaded natural rubber (NR) in black and white (upper) and in multicolored (lower) displays. (a) and (d) CB-10; (b) and (e) CB-40; (c) and (f) CB-80. The white bar of each direction shows the distance of 100 nm. (From Kohjiya, S., Kato, A., Suda, T., Shimanuki, J., and Ikeda, Y., *Polymer*, 47, 3298, 2006. With permission.)

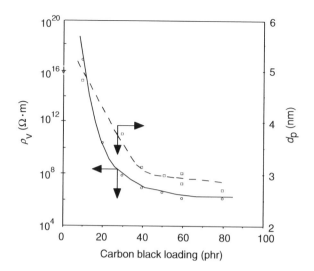

FIGURE 19.6 Dependency on carbon black loading of perimeter distance between the nearest aggregates obtained from three-dimensional-transmission electron microscopic (3D-TEM) images (d_p) and volume resistivity at room temperature (ρ_V). (From Kohjiya, S., Kato, A., Suda, T., Shimanuki, J., and Ikeda, Y., *Polymer*, 47, 3298, 2006. With permission.)

This is reasonable, because HAF carbon is aggregated during its production by the oil furnace method. For the easy recognition of the association, each aggregate is painted by a different color from the neighboring ones to result in the multicolored images in Figure 19.5. It is recognizable that all particles are in aggregation, practically leaving no isolated particles. The aggregate is the fundamental entity of carbon black existence in NR matrix.

From the 3D images shown in Figures 19.3–19.5, all structural parameters can be known on the dispersion of nanoparticles. However, for each calculation a good software is necessary, which usually costs much. As an interesting parameter, the distance between the perimeters of neighboring aggregates (d_p) is calculated from the 3D images of carbon black in NR. The results are plotted against loading amount of carbon black in Figure 19.6. The results of d_p show rapid decrease by the increase of carbon black followed by much less decrease upon further increase of the loading. It is reasonable to observe the decrease of d_p on the increase of carbon black. However, the dramatic decrease up to 40 phr and the following much smaller change have to be explained. It is well known that carbon black, especially so-called reinforcing filler such as intermediate super abrasion furnace (ISAF)- and HAF-grade carbon black, is accompanied by so-called bound rubber or carbon gel as shown in Figure 19.1. The bound rubber is assumed to be in strong chemisorption state, and is differentiated from the surrounding rubbery matrix by various techniques [3,9,12–16,27,38]. The behavior of d_p, shown in Figure 19.6, suggests that the two aggregates cannot be nearer than ca. 3 nm. Therefore, 3 nm may be the thickness of the bound rubber, which seems to be a reasonable value compared with the reported ones [38,39]. In other words, every aggregate is surface-covered by immobilized rubber molecules the thickness of which is ca. 3 nm, and the aggregates are not in direct contact but with 3 nm distance. This is the first observation of the bound rubber by TEM techniques.

19.5 VISUALIZATION OF FILLER NETWORKS OF CARBON BLACK

As physical properties of the carbon-black-loaded vulcanizates, their volume resistivity (ρ_V) were measured and plotted in Figure 19.6. Interestingly, the behavior is very similar with that of d_p.

This result of resistivity or electron conductivity, which is the inverse of resistivity, is well known and is called "percolation" [40]. Dramatic increase of conductivity is ascribable to network formation of the electron-conductive path.

What is the meaning of similarity of the behaviors of d_p and ρ_V? It strongly suggests that electron has jumped over the distance of 3 nm between the carbon aggregates to result in the conductivity of $10^{-6}\,\mathrm{S\,m^{-1}}$ at the carbon black (HAF grade) loading over 40 phr. At the nanometer level, electron can jump over the insulator distance of ca. 3 nm, which may be one of the quantum effects observed at nm level. The similar dependencies of two independent quantities, i.e., the conductivity and the distance between the neighboring aggregates (as presented in Figure 19.6) strongly indicate the formation of network structure, which is percolation in terms of conductivity or gelation in terms of association of carbon black at around 40 phr loading of carbon black. In other words, the agglomeration of carbon black aggregates with ca. 3 nm distance has resulted in a network morphology to form electron-conductive path to give rise to a conductive path of network structure.

For visualization of the networks, a kind of skeletonization is carried out connecting the centers of gravity of aggregates which are within 3 nm distance using homemade software. The procedure used here is very simple compared with the reported methods [41,42]. The skeleton obtained by connecting the centers of gravity is only an approximated one. However, the essential feature of conducting networks is retained. Results of the visualization are shown in Figure 19.7. These figures are the first nano-filler networks visualized in nanometer scale based on TEM observations. The networks are formed due to the agglomeration of carbon aggregates in NR matrix. It is noted that even in CB-10 (before the percolation threshold) the formation of networks (as pregels) is observed. This suggests a strong tendency of association of carbon black aggregates in rubbery matrix. Increasing the amount of carbon black has brought about the extension of the conductive network structure to the whole, which is percolation or gelation. This picture is against the so far common understanding that the networks of carbon black are formed only when conductivity threshold is reached [43]. The expanded image of CB-40 (Figure 19.7d) suggests much complicated network structure, but branching points and terminating chains (dangling chains) are recognized in the image. Quantitative analysis of the network structures is currently under investigation, and the result will be reported in the near future.

The morphology of the agglomerates has been problematic, although some forms of network-like structures have been assumed on the basis of percolation behavior of conductivity and some mechanical properties, e.g., the Payne effect. These network structures are assumed to be determining the electrical and mechanical properties of the carbon-black-filled vulcanizates. In tire industries also, it plays an important role for the macroscopic properties of soft nano-composites, e.g., tear,

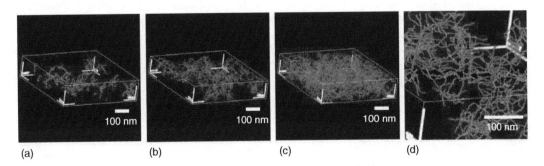

(a)   (b)   (c)   (d)

FIGURE 19.7 Visualization of network structures of carbon black in natural rubber (NR) matrix by skeletonization of the three-dimensional-transmission electron microscopic (3D-TEM) images. (a) CB-10; (b) and (d), CB-40; (c) CB-80. The image in (d) shows the expanded one from the network skeleton of CB-40. The white bar for each direction shows the distance of 100 nm. (From Kohjiya, S., Kato, A., Suda, T., Shimanuki, J., and Ikeda, Y., *Polymer*, 47, 3298, 2006. With permission.)

abrasion, and wear properties of tread rubbers in pneumatic tires. Their design in terms of reinforcing fillers has been dependent upon the trial-and-error method because of the difficulty in controlling the morphology of the carbon black agglomerates. Therefore, the structural information obtained from the visualized network structure will be of utmost importance for improving the material design of nano-composites, which should be based on scientifically elucidated structures in 3D space.

19.6   FUTURE PROSPECTS OF 3D-TEM

It is very probable that nano-fillers are associating under lots of circumstances, and in rubbery matrix they tend to further agglomerate to form network structures. This tendency may be based on the nature of nanometer size of the particles, hence huge surface area, and it is vital to take this into account in the design of nano-composites for various novel applications as well as conventional ones including tires. 3D-TEM is playing an invaluable role in elucidating the dispersion and morphology of nanofiller in 3D space.

However, there are a few limitations of 3D-TEM in using it on polymer systems. The first one is the damage by irradiation of electron beam on matrix polymers which are very sensitive to irradiation [16,18,44]. This is to be noted before subjecting a sample to TEM observation. Fortunately, NR is reported to suffer the least change by irradiation [45], because cross-linking reaction and degradation of the polymer chains occurs evenly, resulting in apparent constancy of the network structures. This point enabled us to subject NR–carbon black nano-composites to 3D-TEM measurements. In fact, in order to check the reproducibility of 3D-TEM measurements, a few samples were subjected to the same measurements after the first measurements, and exactly the same results were obtained. This suggests enough reproducibility of the 3D-TEM measurement on NR–carbon black system.

The second is the lack of contrast of polymer systems in TEM observation. In cases of rubber–inorganic nano-filler composites, the contrast between the two is more than enough without any staining of the sample. However, in many polymer systems, e.g., in block copolymers staining by a heavy metal (Os or Ru) is a must to obtain enough contrast between hard and soft domains under TEM observations. The staining gives rise to many changes in structure and properties of the sample. For example, the volume of the stained domains is surely changed, which is the most basic quantity in the evaluation of 3D structures. So far, staining by a heavy metal has been conveniently used in TEM observation of polymers, because it is beneficial not only to improve the contrast but also to improve resistance against the damage by electron. However, much care has to be paid in 3D-TEM measurements on stained samples. The results are on the modified sample by the staining, not on the original polymeric sample. To avoid these difficulties, use of phase-contrast TEM for 3D imaging is a promising technique [46,47]. It uses all the irradiated energy for forming the image, different from conventional bright-field TEM or HAADF, which is a dark-field technique using only the diffracted electron beam. Therefore, phase mode is much better, especially for electron-sensitive organic samples.

In spite of these limitations, use of 3D-TEM will dramatically increase in rubber science and technology. It is affording so valuable an information which is vital to the design of nano-composites, including tires, that 3D-TEM is finding an ever-increasing usage in rubber industries.

REFERENCES

1. Roberts, A.D. ed., *Natural Rubber Science and Technology*, Oxford University Press, Oxford, 1988.
2. Kraus, G. ed., *Reinforcement of Elastomers*, Interscience, New York, 1965.
3. Hess, W.M., in *Reinforcement of Elastomers*, ed. G. Kraus, Interscience, New York, 1965, Chapter 6.
4. Ikeda, Y., Kato, A., Shimanuki, J., and Kohjiya, S., *Macromol. Rapid Commun.*, **25**, 1186, 2004.
5. Weyland, M. and Midgley, P.A., *Mater. Today*, **7**(12), 32, 2004.

6. Yamauchi, K., Takahashi, K., Hasegawa, H., Iatrou, H., Hadjichristidis, N., Kaneko, T., Nishikawa, Y., Jinnai, H., Matsui, T., Nishioka, H., Shimizu, M., and Furukawa, H., *Macromolecules*, **36**, 6962, 2003.
7. Kohjiya, S., Kato, A., Shimanuki, J., Hasegawa, T., and Ikeda, Y., Paper presented at ACS Rubber Div. 165th Spring Technical Meeting, May 2004, Grand Rapids, MI, Paper XIII.
8. Kohjiya, S., Kato, A., Shimanuki, J., Hasegawa, T., and Ikeda, Y., *Polymer*, **46**, 4440, 2005.
9. Kohjiya, S., Kato, A., Shimanuki, J., Hasegawa, T., and Ikeda, Y., *J. Mater. Sci.*, **40**, 2553, 2005.
10. Kohjiya, S., Kato, A., Shimanuki, J., Ikeda, Y., Tosaka, M., Poompradub, S., Toki, S., and Hsiao, B.S., Paper presented at ACS Rubber Div. 167th Spring Technical Meeting, May 2005, San Antonio, TX, Paper 33.
11. Ikeda, Y., *Sen'i Gakkaisi*, **61**, 34, 2005.
12. Kato, A., Ikeda, Y., and Kohjiya, S., *Nippon Gomu Kyokaishi*, **78**, 180, 2005.
13. Kohjiya, S. and Kato, A., *Kobunshi Ronbunshu*, **62**, 467, 2005.
14. Kohjiya, S., Kato, A., Suda, T., Shimanuki, J., and Ikeda, Y., *Polymer*, **47**, 3298, 2006.
15. Kohjiya, S., *Mater. Sci. Forum*, **514–516**, 353, 2006.
16. Kato, A., Kohjiya, S., and Ikeda, Y., *Kobunshi*, **55**, 616, 2006.
17. Frank, J., *Electron Tomography—Three-dimensional Imaging with the Transmission Electron Microscope*, Plenum Press, New York, 1992.
18. Tsuji, M. and Kohjiya, S., *Prog. Polym. Sci.*, **20**, 259, 1995.
19. de Jong, K.P. and Koster, A.J., *Chem. Phys. Chem.*, **3**, 776, 2002.
20. Midgley, P.A., Weyland, M., Thomas, J.M., and Johnson, B.F.G., *Chem. Commun.*, 907, 2000.
21. Weyland, M., Midgley, P.A., and Thomas, J.M., *J. Phys. Chem. B*, **105**, 7882, 2001.
22. The IMOD homepage. Available at http://bio3d.colorado.edu/imod/index.html. Boulder Lab. for 3D Electron Microscopy of Cells.
23. The TGS homepage. Available at httm://tga.com.
24. DeRosier, D. and Klug, A., *Nature*, **217**, 130, 1968.
25. Hoppe, W., Gassman, J., Hunsmann, N., Schramm, H.J., and Sturr, M., *Hoppe-Seyler's Z. Physiol. Chem.*, **355**, 1483, 1974.
26. Mastronarde, D.N., *J. Struct. Biol.*, **120**, 343, 1997.
27. Hess, W.M. and Herd, C.R. in *Carbon Black*, eds. J.-B. Donnet, R.C. Bansal, and M.-J. Wang, Marcel Dekker, New York, 1993, Chapter 3.
28. Herd, C.R., McDonald, G.C., and Hess, W.M., *Rubber Chem. Technol.*, **65**, 107, 1992.
29. Gruber, C.T., Zerda, W., and Gerspacher, M., *Rubber Chem. Technol.*, **67**, 280, 1994.
30. Wolff, S., *Rubber Chem. Technol.*, **69**, 325, 1996.
31. Kohjiya, S. and Ikeda, Y., *Rubber. Chem. Technol.*, **73**, 534, 2000.
32. Mark, J.E. and Pan, S.-J., *Makromol. Chem. Rapid Commun.*, **3**, 681, 1982.
33. Erman, B. and Mark, J.E., *Structure and Properties of Rubberlike Networks*, Oxford University Press, 1997, Chapter 16.
34. Kohjiya, S. and Yamashita, S., *Appl. Polym. Symp.*, **50**, 213, 1992.
35. Ikeda, Y., Tanaka, A., and Kohjiya, S., *J. Mater. Chem.*, **7**, 455, 1497, 1997.
36. Hashim, A.S., Azahari, B., Ikeda, Y., and Kohjiya, S., *Rubber Chem. Technol.*, **71**, 289, 1998.
37. Brinker, C.J. and Scherer, G.W., *Sol–Gel Science*, Academic Press, New York, 1990.
38. Donnet, J.-B., Bansal, R.C., and Wang, M.-J., eds., *Carbon Black*, Marcel Dekker, New York, 1993.
39. Fujiwara, S. and Fujimoto, K., *Rubber Chem. Technol.*, **44**, 1273, 1971.
40. Probst, N., in *Carbon Black*, eds., Donnet, J.-B., Bansal, R.C., and Wang, M.-J., Marcel Dekker, New York, 1993, Chapter 8.
41. Lindquist, W.B., Venkatarangan, A., Dunsmuir, J., and Wong, T.-F., *J. Geophys. Res. Solid Earth*, **105**, 21509, 2000.
42. Nishikawa, Y., Jinnai, H., and Hasegawa, H., *Kobunshi Ronbunshu*, **58**, 13, 2001.
43. Rwei, S.-P., Ku, F.-H., and Cheng, K.-C., *Colloid Polym. Sci.*, **280**, 1110, 2002.
44. Tsuji, M., Fujita, M., and Kohjiya, S., *Nihon Reoroji Gakkaishi*, **25**, 235, 1997.
45. Kohjiya, S., Matsumura, Y., Yamashita, S., Yamaoka, H., and Matsuyama, T., *J. Mater. Sci. Lett.*, **3**, 121, 1984.
46. Danev, R. and Nagayama, K., *Ultramicroscopy*, **88**, 243, 2001.
47. Tosaka, M., Kohjiya, S., Radsin, D., and Nagayama, K., *Polym. Prepr., Japan*, **54**, 776, 2005.

20 Scanning Probe Microscopy of Elastomers and Rubbery Materials

Sergei N. Magonov and Natalya A. Yerina

CONTENTS

20.1 INTRODUCTION

Modern science and technology are focused on a design of novel materials with well-defined end properties, a miniaturization of functional structures, and exhibit a growing interest to bio-related materials. The practical realization of these tasks demands joint efforts of synthesis and processing, and both should be assisted by appropriate characterization methods. Particular attention is paid to polymer and rubbery materials, which play increasingly important roles in many industrial applications. This chapter will describe several aspects of characterization of rubbery materials by atomic force microscopy (AFM). The method provides a visualization of surface structures and an access to their local properties at the micron and submicron scales. The introductory part includes a brief description of rubbery materials and AFM. Experimental procedures and practical results obtained on a variety of samples will be outlined afterwards. Despite the evident progress achieved in AFM of polymers and rubbers, in conclusion, we will specify a number of challenging issues awaiting their solution.

20.1.1 RUBBERY MATERIALS

Elastomers and rubbery compounds are materials characterized by high degree of flexibility, excellent dynamic properties with a low hysteresis loss, and reliable behavior at low temperatures [1]. These materials of natural and synthetic origin are based on different chemical compounds: polybutene, acrylics, styrene–butadiene and ethylene–propylene copolymers, fluoropolymers, silicones, polyurethanes, etc. Rubbery-like behavior is best manifested when a 3D cross-linked network is formed in these polymers by different means (curing agents, heating, irradiation, etc.). In modern industry (automotive, aerospace, civil engineering, medical, etc.), most of mentioned elastomers are rarely used as individual materials. For many practical applications, rubber materials contain a significant amount of various compounds, such as antioxidants, plasticizers, organic and inorganic fillers, pigments, and chopped fibers [2]. For example, compositions with glass and aramid fibers provide improved strength and/or stiffness of the materials. Graphite flakes are introduced as strengthening filler, while MoS_2 flakes play lubricant role. Silica (Si) and carbon black (CB) particles are used for material strengthening, and CB can also manage the electric properties; some specific inorganic fillers are added to make the flame-retardant materials. Blending of elastomers with different plastics is commonly used for preparation of a suite of materials targeting such properties as hardness, toughness, impact resistance, thermal behavior, electric properties, and also to facilitate the processing procedures and to reduce the overall cost. The design of multicomponent materials with required properties is the main task of modern plastic and rubber industries, and knowledge of structure–properties relationship is indispensable to optimize whole manufacturing cycle from blending to processing. Due to the complex composition, the use of many characterization methods is needed for a good understanding of the effect of chemical and physical structures on the functional properties (mechanical, thermal, electrical, optical, etc.) of rubbery materials [3].

20.1.2 ATOMIC FORCE MICROSCOPY

In recent years, AFM has been recognized as a useful characterization technique for a wide diversity of polymeric materials [4,5]. In this method, imaging is performed on scales from hundreds of microns down to nanometers, and surface structures as small as lamellae and single macromolecules are resolved. Although AFM was introduced for high-resolution imaging of surfaces, it has appeared to be much more useful than anticipated by enabling a compositional mapping of heterogeneous materials and providing an access to different sample properties (mechanical, electric, magnetic, etc.). The compositional mapping is based on recognition of specific structural features and shapes of components (spherulites, lamellae, latex spheres, nanoparticles, etc.) and different image contrast that reflects local mechanical, adhesive, or electromagnetic properties of the constituents. It so happened that the stiffness of AFM probes is well suited for demarcation of mechanical properties of soft materials (polymers, rubbers, biomaterials, etc.) with elastic moduli in the range from a few MPa to 10 GPa. Therefore, AFM applications to multicomponent polymer materials such as polymer blends, composites, and block copolymers are particularly invaluable. In this function, AFM complements optical and electron microscopy and expands morphological studies to different environments (vacuum, air, liquids) and various temperatures. Visualization of bulk structure and composition of materials is important but not the ultimate goal of the characterization. Also, it is highly desirable to get quantitative information about material properties on the micron and submicron scales. First of all, this is related to measurements of local mechanical properties (elastic modulus, adhesion, viscoelasticity, friction, etc.) and definite progress has been achieved in AFM-based nanoindentation [6,7]. Quantitative evaluation of electric (conductivity, capacitance, dopant concentration) and magnetic properties at small scales is a more challenging problem, a solution of which will be welcomed by researchers working in semiconductor industry.

For the last 10 years, AFM applications to polymer and rubbery materials have been developed in a number of directions. High-resolution imaging of individual polymer molecules and studies of their conformation, mobility, and self-organization on different substrates attract increasing number of academic researchers [8]. AFM has become a common method for visualization of self-assembly structures of block copolymers with spacings in the 6–100 nm range [9]. Diversity of these structures motivates researchers to use them as nanoscale masks and templates for lithography. Morphology and nanostructure of semicrystalline polymers are clearly distinguishable in AFM images primarily due to differences in stiffness of amorphous and lamellar regions. Studies of these materials are often conducted at different temperatures to follow up structural changes associated with melting and crystallization [10]. The changes could be monitored even at the submicron scale as shown by the visualization of chain unfolding during annealing and melting of single crystals of polyethylene (PE) [11]. AFM examination of polymer blends and multicomponent systems including rubbery materials is primarily focused on compositional mapping of their constituents [12]. A practical realization of this goal demands a development of experimental procedures for recording images with high contrast of the materials' components and for finding correspondences between the image contrast and the nature of components. Several experimentally found correlations for rubbery materials have been reported [13–15]. A better understanding of tip–sample force interactions, in which elastic, viscoelastic, and adhesive responses are mixed with effects of tip geometry and various experimental parameters (probe stiffness, imaging environment, etc.), is necessary for rational analysis of AFM data, and such work is in progress.

20.2 EXPERIMENTAL AFM PROCEDURES

The basics of AFM and its applications to different materials have been described in a number of books and reviews [16–18]. Here we only emphasize on features related to the use of AFM in studies of soft materials. An implementation of tip–sample force as the probing interaction raises a question whether the imaging induces a sample deformation and its possible damage. The answer to this question is multifaceted. On one side, an operation at minimal tip–sample forces helps avoiding sample deformation and its damage, and thus favors a nondestructing and precise surface profiling. In addition, a tip contact area is minimized and that facilitates high-resolution imaging. On the other hand, imaging at elevated tip–sample forces can be useful for detecting sample regions or sample components that differ in mechanical properties. It will be desirable to operate at forces that induce only elastic deformation of a sample. Such operation could be easier to achieve in tapping mode [19] than in the contact mode. In the contact mode, a probe (a cantilever with tip attached to its free end) scans a sample surface being in permanent contact with it and the cantilever deflection is employed for measurements of tip–sample forces. Strong lateral forces leading to a damage of soft materials characterize this mode. In tapping mode, the probe oscillating at high frequencies (20–400 kHz) is rastered over the sample surface and it comes only into an intermittent contact with the sample. This mode is most often applied in studies of soft and rubbery materials; therefore, we illustrate the operation of this mode at different levels of tip–sample interactions with a couple of examples. Several basic questions of local mechanical measurements using force curves and AFM-based nanoindentation will be presented afterwards. In addition, we shortly describe sample preparation, which is another important component of AFM studies of rubbery materials.

It is worth to note that the authors' experience has been accumulated working with scanning probe microscopes MultiMode and Dimension5000 (both products of Digital Instruments/Veeco Instruments) but most of the results and conclusions are also relevant for practical work with scanning probe microscopes of other manufacturers.

20.2.1 OPTIMIZATION OF IMAGING IN TAPPING MODE

A change in the amplitude of oscillating AFM probe is chosen for a control of tip–sample force interactions in tapping mode. At the beginning, an operator adjusts the piezo-drive of the probe to its resonant frequency and chooses initial amplitude (A_0) and set-point amplitude (A_{sp}). The latter is smaller than A_0 on an amount of dissipation happening during tip–sample interaction. Low-force imaging or *light* tapping takes place when the applied probe has low stiffness, small A_0, and A_{sp} most close to A_0. All these parameters have practical limitations. Stiffness of the probes cannot be very small because surface adhesion and capillary forces, which are in place during imaging at ambient conditions, might restrict a stable tapping regime. At ambient conditions, a minimal stiffness of the probe suitable for tapping depends on sample nature. It can be as small as 0.1–0.2 N m^{-1} for imaging of surfaces with low adhesion and minimal capillary effects. The probes with stiffness in the 1–40 N m^{-1} are the ones most often applied in AFM of polymers in tapping mode. A magnitude of A_0 is limited by the instrument sensitivity, and smallest amplitude for commercial instruments is typically around 5 nm. A choice of A_{sp} depends on a number of instrumental parameters (feedback gains, scanning rate, thermal drift, etc.) as well as on a sample nature (energy of adhesion, hydrophilicity, surface roughness, etc.). Practically, one can perform imaging with A_{sp}/A_0 ratio higher than 0.9 [12].

During imaging in tapping mode a distance between a cantilever's averaged position, which stays practically constant at the level of the probe base, and a sample is changing to maintain A_{sp} chosen by an operator. Therefore, surface corrugations of a homogeneous sample are reproduced in height image, which presents vertical (z-) displacements of a piezo-scanner (carrying the sample or the probe) needed to keep the probe amplitude at the A_{sp} level. In addition to the height image, an operator can simultaneously record an amplitude (or error) image and/or phase image. The amplitude image represents a map of differences between the cantilever amplitude and A_{sp} before this error signal is used to adjust the z-displacement of a piezo-scanner annihilating this difference. Therefore, the amplitude image enhances edges of topographic features, thus helping their visualization. The phase image represents differences between a phase of the piezo-driver of the cantilever and a periodical signal of a photodetector reflecting oscillations of the probe interacting with a sample. Similar to the amplitude image, the phase image of a homogeneous sample picks mostly edges of surface structures. Height and phase images recorded during studies of multicomponent samples are more complicated because they present topographic and compositional information often convoluted with each other. Therefore, imaging at different tip–sample forces is necessary for elucidation of these contributions. Two examples shown in Figure 20.1 illustrate these issues. A surface of Xenoy—impact-resistive plastic material, which is a blend of polycarbonate (PC), poly (butyleneterephtalate) (PBET), and rubber inclusions [20], was imaged at different force conditions and height, and phase images are presented in Figure 20.1a. This surface was prepared at $-100°C$ using a cryo-ultramicrotome. In *light* tapping, the height image reveals a relatively smooth topography of the sample with surface corrugations in the 0–50 nm range (Figure 20.1a, *top*). A number of bright spots of ca. 30–50 nm in size are also seen in this area. These spots correlate in size with rubber particles present in this material. The phase image emphasizes some topography features, whereas traces of several domains of hundreds of nanometers in size are barely distinguished. With the decrease of A_{sp} (increase in imaging force), these domains become more pronounced, especially in the phase image (Figure 20.1a, *middle*), where they are seen darker than surroundings. These domains contain most of the rubber particles. The composition of the blend is best seen in the phase image (Figure 20.1a, *bottom*), which was recorded in *hard* tapping. In the same image, the domains and the incorporated rubber particles are seen as pronounced bright patterns. The observed phase contrast is most likely related to the differences of the mechanical properties of components. It is rather difficult to assign bright domains either to PC or PBET components and this is a general problem of recognition of individual components in phase images.

Assignment of phase contrast in tapping mode is not straightforward even taking in account a known relationship [21] between the phase shift (ϕ) and energy dissipation of the probe $(P_{tip}) - P_{tip} \sim [(A_0/A_{sp}) \sin\phi - 1]$. This relationship gives a general physical meaning to the phase shift but does not help in specifying what particular sample properties (elasticity, adhesion, etc.) are presented by its changes. Therefore, the problem of deciphering contributions of different properties to the energy dissipated by the probe remains opened. Furthermore, the phase images of polymer blends can show

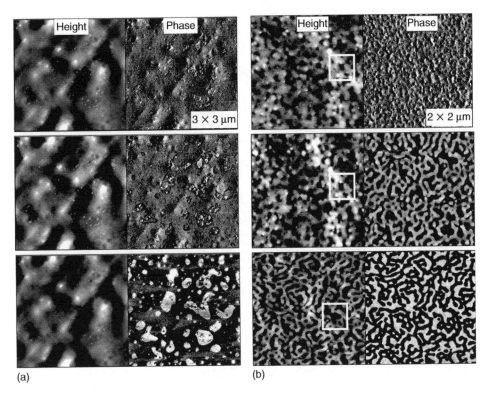

FIGURE 20.1 Height and phase images of polymer blend Xenoy (a) and triblock copolymer poly(styrene–isobutylene–styrene) (SIBS) (b) at different imaging conditions. Images at the top were obtained at $A_0 = 10$ nm, $A_{sp} = 8$ nm; images in the middle were obtained at $A_0 = 10$ nm, $A_{sp} = 7$ nm; and images at the bottom were obtained at $A_0 = 10$ nm, $A_{sp} = 4$ nm. An etched silicon (Si) probe with stiffness of 40 N m^{-1} was applied in these measurements. White squares in the height images (b) indicate the same surface location, which is slightly shifted due to the thermal drift.

a more pronounced contrast between different components than dissipation images constructed using the above-mentioned relationship between P_{tip} and sinϕ [22]. An empirical approach as an alternative to theoretical interpretation of the phase images can be helpful in studies of heterogeneous samples. It might be useful to have heterogeneous materials with known ratios of components and explore phase behavior for such systems at different A_0 and A_{sp}. For some systems, such as PE blends and elastomer systems with fillers, such knowledge has been already received [5,13]. For example, by choosing large A_0 (80–120 nm) and $A_{sp} = 0.4 A_0$ one can get a consistent correlation between brightness of phase contrast and stiffness of components in semicrystalline PE sample and blends of PE with different densities. In phase images, which are recorded at such conditions, high-density regions appear brighter than low-density and this is primarily related to differences of their mechanical properties. Differences in adhesive properties of components contribute to phase contrast more substantially when imaging is performed at low A_0.

Height and phase images of triblock copolymer of poly(styrene–isobutylene–styrene) (SIBS), which were obtained at different tip-forces, are shown in Figure 20.1b. The images at the top, which were obtained in *light* tapping, reflect surface topography (height image) and edges of the surface grainy structures (phase image). As the tip-force increased, first the phase images and then even the height images were changing. The height image in the middle shows the same pattern as the one at the top. Yet, the phase image has changed drastically and it shows a microphase-separation pattern common for block copolymers. Both height and phase images at the bottom exhibit the same structures but the contrast of the phase image is opposite to that of the height image and to that of the phase image at the middle of Figure 20.1b. It is also clear that the height image at the bottom does not reflect surface topography and is strongly "contaminated" by the sample heterogeneity. Therefore, surface topography of heterogeneous samples should be measured in light-tapping conditions to avoid a false interpretation of the height images [23]. In AFM studies of block copolymers and polymer blends with rubbery components one should be aware of the fact that sample surfaces are enriched in low-surface energy components. When this component is rubbery, then at elevated forces a probe penetrates the top surface layer and reveals structures and compositions underneath the surface. This was demonstrated in case of SIBS and other similar block copolymers, e.g., triblock copolymer of polystyrene and polybutadiene (SBS) [24].

20.2.2 PROBING OF LOCAL MECHANICAL PROPERTIES

Since the first AFM applications, researchers have examined so-called force curves. In the contact mode, these are deflection-versus-distance (DvZ) curves, as seen in Figure 20.2a. Initially, DvZ curves were employed to check whether a particular deflection set point used for imaging corresponds to a net repulsive or net attractive force [25]. This curve can also be obtained in tapping mode

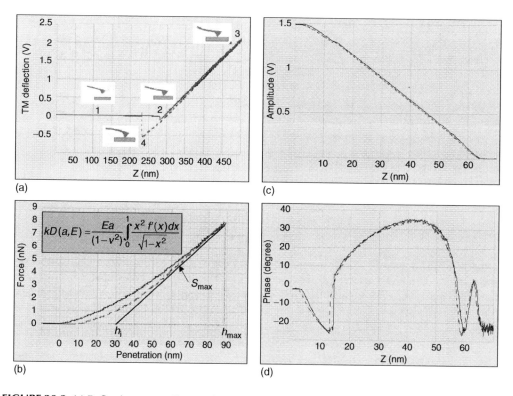

FIGURE 20.2 (a) Deflection-versus-distance (DvZ) force curve obtained in contact or tapping modes. Sketches of the probe bending in different locations (1–4) of the force curve are shown in the inserts. A solid line marks the curve recorded during tip–sample approach. The curve recorded during a tip–sample retract is marked by a dashed line. (b) Typical load-versus-penetration (LvP) curve for atomic force microscopy (AFM)-based nanoindentation. The insert shows Sneddon's formula, which describes how a load (a product of the probe stiffness k and deflection D) with reduced elastic modulus E, contact area a, sample Poisson ratio η and geometry of tip shape (under the integral). Slope is taken at the maximum of unloading curve S_{max}; maximal and remaining penetrations—h_i and h_{max} are also indicated. (c) Amplitude-versus-distance (AvZ) curve obtained in tapping mode. (d) Phase-versus-distance (PvZ) curve recorded in tapping mode.

where the deflection is called tapping mode deflection. As the probe base approaches the sample (part 1), the cantilever, first, bends toward the sample due to attractive forces (point 2). On further approach (part 2 and 3), the tip contacts the sample, and the cantilever bends upward because of repulsion. When the probe base travels in the opposite direction, the cantilever deflection decreases and changes to a downward bend, because of the tip being held by the attractive forces of probe–sample adhesion and capillary effect (point 4). Finally, the tip leaves the sample when the spring force of the cantilever overcomes the attractive forces; the cantilever deflection is essentially zero beyond this point. This result is manifested in a hysteresis—a triangular part of the force curve recorded during retraction. A slope of the DvZ curve in the repulsive force region can be related to sample stiffness if the probe is strong enough to deform it. Therefore, DvZ curves are widely employed for local mechanical properties in the AFM-based indentation, which is also called nanoindentation when probing proceeds at the submicron scale and utilizes forces in the nano-Newton range.

Practically, AFM-based nanoindentation is performed as follows. First, a sample surface is examined in tapping mode and an area of interest is chosen. Then, the operation is switched to measurements of DvZ curves in a predetermined force range. A number of indents are typically made in matrix-like or linear-like manner with perturbed locations spaced far enough from each other to avoid punching of predeformed places. After such indenting procedure, the same area is imaged in tapping mode and the leftover indents (an evidence of plastic deformation) can be visualized. In a separate experiment, an optical sensitivity of a particular probe was determined by measuring DvZ traces on rigid Si or sapphire substrates. This allows a transformation of DvZ curves into load-versus-penetration (LvP) dependencies. The hypothetical LvP curve, which is shown in Figure 20.2b, consists of loading (*solid line*) and unloading (*dashed line*) parts. When loading and unloading curves do not coincide with each other, then deformation deviates from elastic and plastic cases and viscoelastic effects should be taken into account. A quantitative analysis is typically made using the Sneddon model [26] or its offspring—Oliver–Pharr model [27], which is applied depending on the type of deformation: elastic or plastic. In case of Oliver–Pharr model, the analysis is performed only on unloading curves with additional use of such parameters as stiffness S_{max} at the maximal load, maximal and remaining tip penetrations, h_{max} and h_i, respectively.

The main Sneddon equation, which relates applied force and penetration of an axisymmetric tip, is shown in Figure 20.2b. At the left part, a product of the probe stiffness—k and the cantilever deflection, D, presents the force. The right part of the equation includes the reduced elastic modulus—E, a contact area—a, Poisson ratio of a sample—η, as well as the integral describing the tip geometry. Therefore, the calculation of elastic modulus requires knowledge of the cantilever spring constant and the tip shape. Thermal tune method [28] is one of the possible ways to determine the cantilever spring constant, which practically should be varied in a broad range (0.5–700 N m^{-1}) to match stiffness of a particular sample or its components. Tips of different nature (Si uncoated and with different hard coating, single diamond probes) are used for indentation. A choice of tip shape depends on many factors. Sharper tips with apex diameter below 15 nm provide better imaging and higher lateral resolution in indentation maps, yet a quantification of their shape demands a use of transmission electron microscope (TEM). These tips are also subjected to intense wear and, therefore, they should be used very carefully. A shape of rounded tips with apex diameter of several tens of nanometers can be estimated from scanning electron microscopic (SEM) micrographs [29]. These tips are more stable toward wear but have some limitations of imaging and resolution. In general, these probes are more suitable for AFM-based quantitative nanoindentation.

In tapping mode, the AvZ and PvZ curves (Figure 20.2c and d) can be considered as force curves. In contrast to the contact mode, where the cantilever deflection defines tip–sample force, the cantilever amplitude in tapping mode could not be considered as a measure of the tip-force. In this mode, a probe is brought into oscillation at its resonant frequency by a piezo-element. As a sample is moved from its rest position toward the oscillating probe (as it takes place in MultiMode scanning probe microscope), the amplitude is gradually reduced (see the AvZ curve in Figure 20.2c).

The amplitude drops to zero when the sample is moved from the point of first contact on a distance of the half of the full amplitude of the free-oscillating probe. From this point, a further motion of the sample will cause the cantilever bending upward, similar to what occurred in the contact mode. If the sample motion is reversed the amplitude increases as shown by a dashed curve in Figure 20.2c.

The *PvZ* curve shows how phase shift (i.e., a difference between phases of free-oscillating probe and the same probe interacting with a sample) changes with a separation between the probe base and the sample. The phase of the tapping probe, which is recorded simultaneously with amplitude, lowers initially but then reverses the direction and becomes positive with respect to the phase of the free-oscillating probe (Figure 20.2d). This corresponds to the transition from the overall attractive force regime to overall repulsive force regime. Imaging at A_{sp} corresponding to phase changes, which are negative with respect to the phase of the free-oscillating probe, proceeds at net attractive forces. Tip–sample forces are net repulsive when the phase changes are positive. At larger z-values, the phase reverses its increase and drops to negative values. Note that in a given example of DvZ and PvZ curves there is no hysteresis.

So far, AvZ and PvZ curves are rarely used for quantitative evaluation of nanomechanical properties. This situation will definitely be changed in the future. It is well established that a sample deformation, which takes place during tapping, is smaller than that in contact mode. Respectively, in tapping mode, the tip–sample contact area is smaller and this means a higher spatial and depth resolution for nanomechanical studies. Therefore, this approach will be invaluable for studies of nanomaterials and interfaces in heterogeneous polymer systems. The interplay between the experiment and theory, which is inevitable for understanding nature of amplitude and phase changes and retrieval of load–penetration dependencies needed for quantitative mechanical analysis, is underway [22].

20.2.3 Sample Preparation

Sample preparation is an important issue of AFM of polymers. High-resolution imaging is better achieved on relatively smooth surfaces with corrugations in the range of few nanometers. Many industrial polymer films and coatings as well as layers on smooth substrates made by spin-casting, dipping, and vapor deposition can be examined directly without any sample preparation. Surfaces of polymer block samples (pellets, plates, rods, etc.), which are prepared by extrusion, compression molding, or injection molding, could have rather corrugated surfaces not suitable for AFM imaging. In case of thermoplastic materials, they can be prepared for AFM investigation by melt pressing between flat plates with minimal surface corrugations and followed by cooling at room temperature (RT) and removal of one of the plates. For completely rubbery materials, this approach might not work because thermosets (cross-linked, 3D networks) do not allow materials to be melted and remolded. If surface structure of polymer materials is not a primary interest for a researcher, whose task is to investigate bulk morphology of polymer material, the only way to start an access of AFM probe to the bulk structure is cutting with an ultramicrotome. For rubbery materials or composites, where rubber is at least one of the constituents, an ultramicrotomy preparation should be performed at temperatures below glass transition temperature (T_g) of the most soft material component, and for rubbers it is usually well below RT. Therefore cryo-ultramicrotome, which is routinely employed for preparation of ultrathin polymer slices for TEM analysis, becomes an important tool in research polymer laboratories and it can be efficiently applied for sample preparation in AFM [30]. It is worth noting that time-consuming preparation of ultrathin slices of polymer material required for TEM is eliminated in AFM. Polymer blocks with a smoothly cut surface, which are best prepared with a diamond knife at optimal temperature, exhibits nanometer-scale corrugations needed for AFM imaging.

The preparation of polymer surfaces with minimal corrugations needs good practical experience. Proper choice of cutting speed and temperature helps in avoiding the chattering marks, which worsen the surface quality. Chattering features are clearly seen as arrays of horizontal lines in the height image (Figure 20.3a) of surface of a blend of polystyrene and tetramethyl bisphenol A

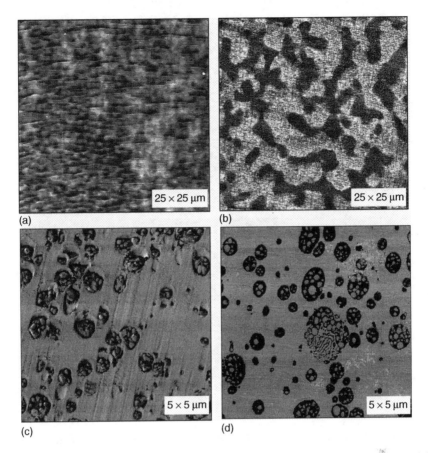

FIGURE 20.3 (a, b) Height images of microtomed surfaces of polymer blend of tetramethyl bisphenol A polycarbonate and polystyrene showing an influence of cutting speed of microtomy on surface quality. (a) Surface with chattering marks made by cutting at 1 mm s^{-1} speed is shown in (a) surface prepared by cutting at 0.25 mm s^{-1} speed is shown in (b). (c, d) Phase images of surfaces of high-impact polystyrene prepared by cutting at room temperature (RT) and at $T = -60°C$.

polycarbonate, which was ultramicrotomed at 1 mm sec^{-1} cutting speed. A slower cutting leads to smoother surface, most useful for AFM imaging and visualization of individual blend components (Figure 20.3b). Differences of surface quality of samples of high-impact polystyrene, which were cut at RT and at −60°C are evident from a comparison of phase images in Figure 20.3c and d. Cutting at RT resulted in a partial removal of soft copolymer inclusions out of rigid matrix, a smearing of their fine structures, and numerous scratch lines (Figure 20.3c). Optimization of cutting temperature of the sample was essential for distinctive visualization of the blend morphology (Figure 20.3d).

It is rather important that microtomed rubbery samples should be examined immediately after their preparation. Otherwise, surface topography and composition might change due to diffusion of low-molecular weight species and constituents with low surface energy. The images of the blend of butyl and natural elastomers, which were recorded on a freshly cut sample (Figure 20.4a) and at the same sample location after 12 h, are shown in Figure 20.4b. As one sees, height images reveal similar topography patterns whereas phase images indicate that sample surface morphology has drastically changed. This observation demonstrates also a possible use of AFM for monitoring of diffusion of low-molecular weight sample species from its bulk to the surface as well as for an environmental control because an airborne contamination can appear in a while on freshly prepared surfaces.

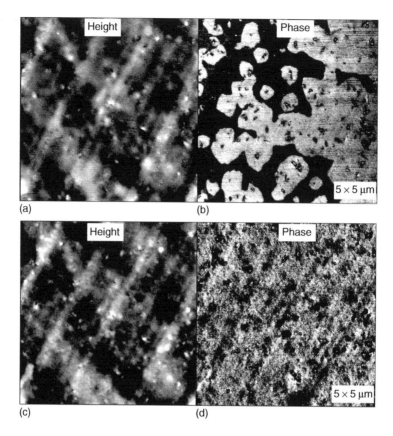

FIGURE 20.4 (a, b) Height and phase images of rubber blend of natural rubber (NR) and butylene rubber (BR) (50:50), which were taken immediately after cryo-ultramicrotomy. (c, d) Height and phase images of the same sample location as in (a, b) taken 12 h later.

20.3 PRACTICAL EXAMPLES OF AFM OF RUBBERY MATERIALS

In this part, we will discuss AFM images and nanomechanical data obtained in studies of natural and synthetic rubbers, thermoplastic elastomers (TPE), and their vulcanized counterparts—thermoplastic vulcanizates (TPV).

20.3.1 INDIVIDUAL RUBBERS

Natural and synthetic rubbers are the basic components of many industrial materials. Natural rubber (NR), which is mostly polymerized isoprene, is the ideal polymer for a variety of dynamic and static engineering applications. It also has excellent low-temperature properties ($T_g = -70°C$). Heterogeneities of NR are caused by different impurities and variations of *cis* and *trans* double bonds resulting from methods of polymerization of natural latex. Synthetic rubbers are made from a variety of monomers (isoprene, 1,3-butadiene, chloroprene, isobutylene, etc.), which can also be mixed and copolymerized in desirable proportions [1–3]. This process is applied for manufacturing of rubbers with a wide range of physical, mechanical, and chemical properties. A controlled synthetic design, which produces a required ratio of *cis* and *trans* double bonds, is often assisted by different additives that allow further optimization of material properties. Morphologies of NR, butylene rubber (BR), and random styrene–butadiene copolymer (SBR), which were chosen as the examples of synthetic rubbers, are best presented in phase images (Figure 20.5a–c). The images show locations with

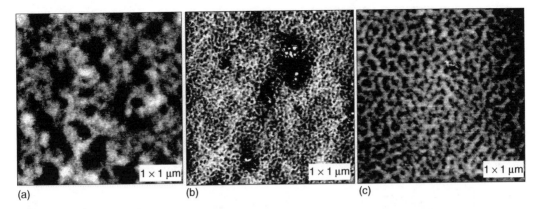

FIGURE 20.5 Phase images of neat rubbers: (a) natural rubber (NR), (b) butylene rubber (BR), and (c) styrene–butadiene copolymer (SBR).

different contrasts that can be assigned to various heterogeneities common for the elastomers which are most likely related to density variations present in these materials. A couple of hard particles, which exhibit bright contrast, are distinguished in the phase image in Figure 20.5b. Variations of the contrast in SBR image (Figure 20.5c), which shows morphological features at the larger scale than those in BR, most likely reflect a microphase separation of hard and soft blocks.

Ethylene–propylene–diene terpolymer (EPDM) is another example of synthetic rubber. The ethylene and propylene monomers are combined to form a chemically saturated, stable polymer backbone providing excellent heat, oxidation, ozone, and weather aging. A third (ethylidene norbornene), nonconjugated diene monomer can be terpolymerized in a controlled manner to maintain a saturated backbone and place the reactive unsaturation in a side chain available for vulcanization or polymer-modification chemistry. Physical properties of EPDM are largely controlled by polymer composition, the way the monomers are combined, and by molecular weight: decreasing ethylene content reduces crystallinity and associated properties such as hardness and modulus. Amorphous or low crystalline grades have excellent low temperature flexibility with T_g at $-60°C$. These polymers respond well to high filler and plasticizer (oil) loading. Durability and heat-aging resistance can be achieved with properly selected curing conditions and agents: up to 130°C with sulfur-acceleration compounds and up to 160°C with peroxide-based system.

Structural changes of EDPM caused by vulcanization with sulfur are shown in AFM phase images at the 30-micron and 5-micron scales (Figure 20.6). Morphology of an unvulcanized elastomer (Figure 20.6a and d) is characterized by small-scale phase-contrast variations without any pronounced morphological features in contrast to those observed in the rubbers shown in Figure 20.5. Vulcanization of EPDM with sulfur was accompanied by structural changes distinguished as bright rounded patches in the phase images in Figure 20.6b and e. These spots, which are of few microns in diameter, can be assigned to denser rubber structures formed during the cross-linking process. As a curative concentration increased to 2 phr, excessive sulfur is crystallized and its individual crystals are recognized as bright platelets in different sample regions (Figure 20.6c and f). Much smaller bright particles as those indicated with arrows in Figure 20.6f can represent small sulfur crystals. In other surface areas, a phase contrast is more homogeneous than in the less cross-linked material seen in Figure 20.6b and e.

Mixing of rubbers with oil is often used in industry for modification of its rheological properties and cost reduction. Morphology of oil-loaded rubbers can be easily revealed in AFM images because oil is much softer and should be recognized in phase images. Indeed, the image of noncured EPDM filled with 50% oil in Figure 20.7a shows a pattern with a binary contrast, in which dark inclusions can be assigned to oil locations. Vulcanization induces essential morphological changes

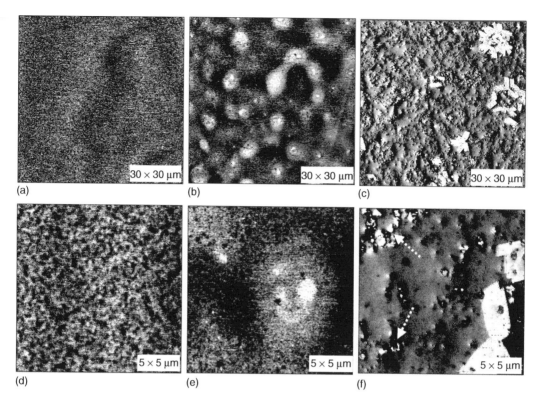

FIGURE 20.6 Phase images of ethylene–propylene–diene terpolymer (EPDM) samples at different scales. Images of the unvulcanized sample are shown in (a, d) and images of samples, which were cross-linked with different amount of sulfur curative—1 phr—in (b, e) and 2 phr—in (c, f). White arrows in (f) most likely indicate locations with small sulfur crystals.

leading to grainy structures whose dimensions increase with the degree of cross-linking. Estimates show that as sulfur content raises from 0.5 to 1.0 and to 1.5 phr, an averaged lateral size of grains changes from 50 to 84 to 146 nm, respectively (Figure 20.7b–d).

Characterization of rubber materials, whose morphology is best presented by phase images, can further benefit from local nanomechanical measurements at different locations and components. *LvP* curves were obtained on unvulcanized and vulcanized samples of neat EPDM rubbers (Figure 20.8a and b) and these rubbers were loaded with oil (Figure 20.8c and d). These measurements were performed with probes having stiffness 40 N m^{-1} and a rounded tip apex of 80 nm in diameter. The loading and unloading *LvP* curves do not coincide with each other and this hysteresis is most pronounced on noncured samples. The quantitative analysis of the unloading curves of all the samples was performed with Sneddon model modified for plastic deformation [31]. As expected, the elastic modulus of cross-linked material is higher than its precursor: 5.6 MPa versus 3.6 MPa for neat EPDM, and 2.5 MPa versus 1.9 MPa for oil-filled elastomer. It is worth to note that the obtained quantitative values of elastic moduli are overestimated because the Sneddon model does not consider the adhesion of the probe to the sample. Indeed, elastic moduli, which were obtained in the elongation measurements of macroscopic samples for both types of the material, are 1.5 times smaller than obtained through nanoindentation [32]. A combination of Sneddon with Johnson–Kandall–Roberts (JKR) and Derjaguin–Müller–Toporov (DMT) models account for adhesion contribution, and will definitely lead to more reliable quantitative results for AFM-based nanoindentation [33].

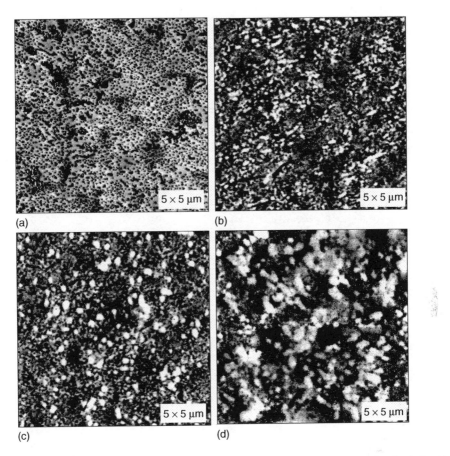

FIGURE 20.7 Phase images of ethylene–propylene–diene terpolymer (EPDM) samples loaded with oil (50 wt%). Image in (a) was obtained on the unvulcanized sample and images in (b, c, d) on samples cross-linked with different amounts of sulfur curative: 0.5, 1.0, 1.5 phr, respectively.

20.3.2 THERMOPLASTIC ELASTOMERS

Complexity of manufacturing procedures for thermosetting plastics, which include the vulcanized rubbers, motivated a search for new materials with rubbery-like behavior, which could be easily processed in a similar way to thermoplastic polymers, e.g., by extrusion or injection molding. TPEs—the novel type of materials, which combine several advantages over conventional thermosets—the wide availability of different grades, including rubbery-like materials, processing ease, and speed, and possibility to be recycled. Polyurethanes (PU) were the first major representatives of TPEs, which exhibit a number of useful properties: high abrasion resistance, low-temperature flexibility, oil resistance, and load-bearing capability, which are the best of the elastomers [34]. These materials are the reaction products of a diisocyanate and long- or short-chain polyether, polyester, or caprolactone glycols. Additives to PU will further improve their dimensional stability or heat resistance, flame retardancy, weatherability, and reduce friction. These properties make the materials useful for automotive applications, different kinds of machinery, medical instrumentation, packaging, and other industries.

PU elastomers contain alternative soft and hard segments, which separate into different phases. Hard domains play a role of cross-links, whereas soft blocks provide extensibility. Therefore, morphology and properties of PU are defined by relative amount of soft and hard segments. For example, at 70% concentration of soft segments, the material is described as a rubbery matrix with

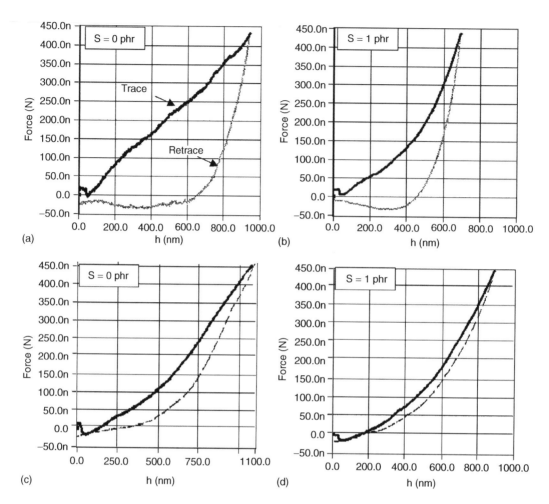

FIGURE 20.8 Load-versus-penetration (*LvP*) curves obtained during nanoindentation of ethylene–propylene–diene terpolymer (EPDM) samples. Approach curves are shown as solid line and retract curves as broken lines. The curves in (a, b) were obtained respectively on the unvulcanized and cross-linked (amount of sulfur curative was 1.0 phr) samples of neat EPDM. The curves in (c, d) were obtained respectively on the unvulcanized and cross-linked (amount of sulfur curative was 1.0 phr) samples of EPDM loaded with oil (50 wt%).

hard domains dispersed inside. At 50% concentration of soft segments, the material became much stiffer and its morphology is characterized by co-continuous phases. AFM is important characterization method of PU because it reveals fine structure of the urethane morphology without special treatment of the samples. This is demonstrated by images of PU, which were prepared from diphenylmethane diisocyanate, polypropylene oxide glycol, and butane diol (Figure 20.9) [35]. The images revealed well-defined differences of morphology and fine structure of the polymer samples with various ratios of soft and hard segments.

A large-scale phase image of the PU sample with 70 wt% of soft segments (Figure 20.9a) shows spherulites with diameters, varying from 0.8 to 7 μm, which are seen as bright rounded areas surrounded by dark rims. The spherulites are surrounded by amorphous material, which is generally seen darker than the spherulites. Also, the phase contrast of the amorphous material has two grades that indicate its heterogeneity of unknown origin. Bearing analysis showed that an area occupied by the spherulites (~28%) correlates with the ratio of the soft and hard segments. The fine structure of the spherulites is best resolved in phase image in Figure 20.9b. There is a tendency toward radial

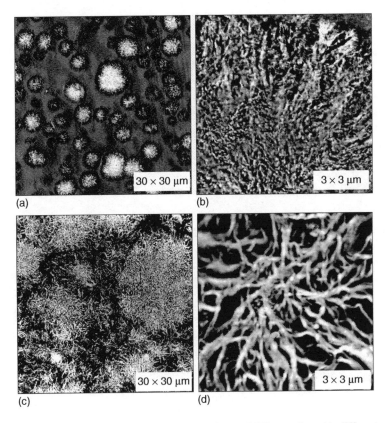

FIGURE 20.9 Phase images of cryo-ultramicrotomed surfaces of PU samples with different content of soft blocks: 70% (a–b), 50% (c–d).

growth of densely packed fibrils (~20 nm in width) from a nucleating center. The morphology of the PU sample with 50% of soft segments is revealed in the phase images in Figure 20.9c and d. In this sample the spherulites are larger, with diameters up to 20 μm. In consistency with the ratio of soft and hard segments, they occupy ~52% of surface area in Figure 20.9c. A fibrillar structure of spherulites is well resolved in phase image in Figure 20.9d. The fibrils, which have a diameter in the 50–120 nm range and length of few microns, are more densely packed in the center of spherulites.

In addition to PU, styrene block copolymers became industrially important TPEs. The block copolymers are prepared by copolymerizing two or more monomers. One of the monomers is usually a thermally stable and hard component; the other monomer is typically amorphous and rubbery segment. Varying the ratio of the monomers and the lengths of the hard and soft segments can control properties of these materials. In styrene block copolymers, hard polystyrene segments are interconnected with soft blocks of polybutadiene, polyisoprene, or ethylene–butylene. The polystyrene blocks are softening above 100°C and allow thermoplastic processing whereas soft blocks provide rubbery behavior. Such block copolymers exhibit mechanical properties as well as weather and chemical resistivity comparable to traditional rubbers. They are used in disposable medical products, food packaging, tubing, sheet, belting, mallet heads, and shoe soles. We have already demonstrated in Figure 20.1b that AFM reveals microphase separation in styrene block copolymers using different mechanical properties of individual blocks of SIBS. Another example is given in Figure 20.10, which present morphology of triblock copolymer of styrene and ethylene–butylene (SEBS) before (Figure 20.10a) and after loading with oil (Figure 20.10b). Loading of

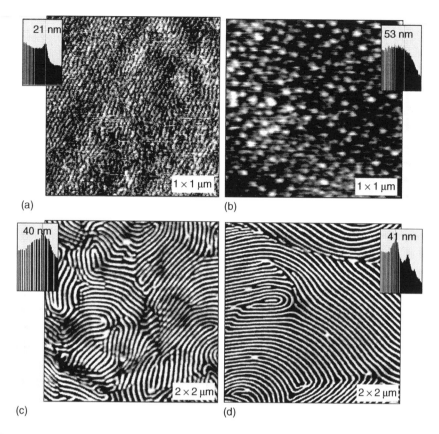

FIGURE 20.10 (a, b) Phase images of cryo-ultramicrotomed surfaces of triblock copolymer styrene and ethylene–butylene (SEBS) samples of neat material and loaded with oil (40 wt%), respectively. (c, d) Phase images of film of triblock copolymer poly(methyl methacrylate–polyisobutylene–poly(methyl methacrylate) (PMMA–PIB–PMMA) immediately after spin-casting and after 3 h annealing at 100°C, respectively. Inserts in the top left and right corners of the images show power spectra with the value structural parameter of microphase separation.

SEBS with oil leads to the transition from lamellar morphology to micellar morphology [15]. This transformation is reflected in the above images of pure SEBS and its SEBS/oil = 60:40 composition. The loading leads to an increase of the structural factor from 21 to 53 nm. Therefore, the incorporation of oil to ethylene–butylene blocks induced larger separation of micelles formed predominantly by styrene blocks.

In addition to broad industrial applications, the block copolymers attract also an increased attention of researchers dealing with nanoscale lithography. This interest is caused by well-ordered patterns in the 5–100 nm scale that are formed in these materials due to microphase separation. Such patterns are developed during high-temperature annealing as demonstrated by images in Figure 20.10c and d. The first image was obtained on a film of triblock copolymer PMMA–PIB–PMMA [(poly(methyl methacrylate)–*block*–polyisobutylene–*block* poly(methyl methacrylate)] after spin-casting procedure ((Figure 20.10c). A pattern with curved strips of different contrasts reveals a microphase separation of harder PMAA blocks and softer PIB blocks with spacing of 40 nm. The harder blocks are seen as bright strips and the softer as darker ones. The morphology is drastically changed after 3 h annealing at 100°C, which improves the microphase separation by straightening individual blocks (Figure 20.10d). From these images alone, it is not clear if the observed morphology is lamellar or cylindrical. AFM or TEM images taken on the sample cross-sections will help to differentiate these possibilities. Regular patterns (spherical, cylindrical, lamellar, etc.), which are formed during annealing of block copolymers, are defined by polymer structure and composition while the size of the nanodomains depends on the block length. Nanoscale masks and templates for lithography can be developed from various block copolymers using controlled alignment of microdomains.

20.3.3 ELASTOMER ALLOYS

Elastomer alloys are different types of polymer blends and they comprise TPVs, which are essentially a fine dispersion of highly vulcanized rubber in a continuous phase of a polyolefin. These materials are made by dynamic vulcanization [36], which includes a simultaneous mixing of components and chemical reaction of rubber cross-linking. This is one of the processes that give the material properties superior to those of the blends of the same constituents. We will consider a blend based on EPDM rubber and isotactic polypropylene (iPP)—the material similar to industrial product Santoprene to illustrate morphology of such compositions. The degree of vulcanization of the rubber and the fineness of its dispersion gives this TPV an appropriate combination of mechanical and rheological properties: high flexibility and high elongation, low residual set, and processibility common to that of thermoplastics.

AFM phase image in Figure 20.11a shows morphology of iPP and EPDM blend (40:60), which was prepared by mixing in a twin-screw extruder [37]. The brighter and darker regions of the image correspond respectively to iPP-matrix and uncured rubber domains whose size varies in the 5–7 μm range. A distribution of the components is quite homogeneous. However, after the sample was subjected to hot pressing, its morphology changes drastically because EPDM was not cross-linked (Figure 20.11b). Due to a flow of components during this process and a coalescence of small rubber domains, the size of the dark regions essentially increased. The blend with the same composition was also prepared by dynamic vulcanization. The obtained TPV is characterized by finely dispersed rubber domains (1–3 μm in size) in iPP matrix (Figure 20.11c and d) and exhibits high stability of morphology during several cycles of processing.

One of the key issues of mechanical behavior of multicomponent materials such as TPV is the structure and properties of the interface regions. The phase image in Figure 20.11d shows a part of TPV sample with few rubber domains surrounded by iPP matrix. An extended rectangle area outlined with a white dotted box includes several interfaces between rubber domains and the plastic matrix. Examination of the interfaces is a challenging task and one possible approach is AFM-based

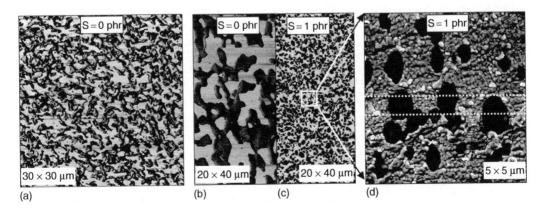

FIGURE 20.11 Phase images of polymer blend of isotactic polypropylene (iPP) and ethylene–propylene–diene terpolymer (EPDM) obtained on cryo-microtomed samples. Image in (a) shows morphology of the blend with unvulcanized EPDM obtained directly after a mixing procedure. Image in (b) shows morphology of the same blend after hot pressing. Images in (c, d) show morphology of the same blend after dynamic vulcanization (a product also known as thermoplastic vulcanization [TPV]) and hot pressing.

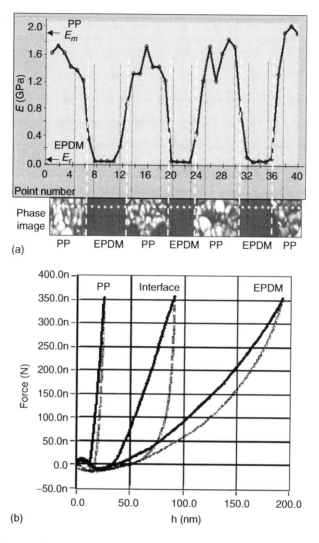

(a)

(b)

FIGURE 20.12 (a) Top part shows variations of elastic modulus profile measured in different locations of the polypropylene (PP)–ethylene–propylene–diene terpolymer (EPDM) blend. The locations are shown by white dots in the blend phase image placed at the bottom. Vertical white dashed lines show the components' borders and the elastic modulus value for this location. Vertical black dotted lines indicate the locations where elastic modulus E gradually changes between PP (E_m) and EPDM (E_r). These values are indicated with black arrows on the E axis. (b) LvP curves for PP-matrix, EPDM-domains, and one of interface locations. The approach curves are seen as solid black lines and the retract curves as gray lines.

nanoindentation. This technique was applied for the chosen area of the TPV and results of this study are summarized in Figure 20.12.

White dots, which are lined along the phase image area at the bottom of Figure 20.12a, indicate sample locations where 40 indents were made. The graph above shows a profile of elastic modulus measured in these locations using the force curves collected during nanoindentation. First of all, the modulus alteration supports our assignment of the darker areas on phase image to softer rubbery domains and brighter areas to iPP matrix. Second, the modulus changes at the interfaces between the components are not abrupt, which suggests a presence of areas mechanically different at the rubber and matrix sides of the interfaces. A different shape of LvP curves, which were recorded in these locations (Figure 20.12b), confirms the finding.

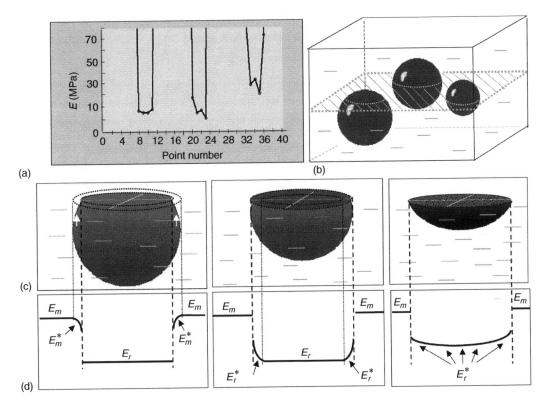

FIGURE 20.13 (a) An enlarged part of elastic modulus profile at the E_r (ethylene–propylene–diene terpolymer [EPDM]) level from Figure 20.12a. (b) A sketch illustrating a 3D model of a material having the spherical soft domains in rigid matrix. A hatched plane presents a microtomy cut. (c) Three possible geometries of cross-sections of rubbery spheres (e.g., EPDM) embedded into a rigid matrix (e.g., polypropylene [PP]). The cuts are made above the central plane (*left*), at the central plane (*middle*), and below the central plane (*right*). Dotted rings in the left and middle sketches show the matrix and sphere locations where apparent elastic moduli, E_m^* and E_r^* are expected. (d) Tentative profiles of elastic modulus change corresponding to the inclusion–matrix geometries in (c). Black arrows indicate the locations with the apparent elastic moduli E_m^* and E_r^*. Vertical dotted and dashed lines indicate the projections of such locations on the corresponding sketches in (c).

There are two issues related to these results. First of all, a magnified part of the elastic modulus graph in Figure 20.13a (black dotted rectangle from graph in Figure 20.12a) shows that the elastic modulus of rubbery domains is not the same and varies in the 7–20 MPa range. This effect might be explained by the fact that these domains can have different depths and the stiffer matrix lying underneath them influences the measurements. The sketches in Figure 20.13b and c, illustrate this situation. In a 3D model (Figure 20.13b) we assume that the rubber inclusions of a different size have a spherical shape. At least three simplified situations could be considered, depending on how a cutting plane has passed through these soft domains imbedded in a rigid matrix: along the central plane, above, or below it (Figure 20.13c). Since an under-the-surface contour of the domains varies in depth, the mechanical response at the surface locations will depend on the position of the indenter and ratio between penetration depth and subsurface thickness of the domain to be indented.

The second problem is that the experimental data indicating a presence of interfacial regions should be carefully considered, avoiding a misinterpretation. In addition to manifestation of a real interface, which is formed due to chain entanglements of the components or a presence of special additives, one should be aware about effects related to tip geometry and a complex local surrounding common to heterogeneous systems. To illustrate the possible geometrical complications during the nanoindentation, we consider two cases describing the local mechanical probing of a soft spherical inclusion in hard matrix, assuming that there is no real interface. Imagine we made $N \times N$ matrix of indents on the area covering a particle cross-section and after analysis of LvP curves a 2D map of elastic modulus was constructed. It is also assumed that tip penetrations are less than $1/10$ of the particle radius [38], otherwise the tip starts to feel the matrix lying underneath the particle in most locations.

In case 1, the ultramicrotome cut is made above the central plane of the sphere (Figure 20.13c, *left*). At the tip penetrations, which are much smaller than the depth of the soft particle, a 2D map of elastic modulus will present a true value E_r inside a whole cross-section area of the rubber domain (Figure 20.13d, *left*). As the probe is laterally moving out of the soft particle cross-section, an apparent elastic modulus (E_m^*) might be lower than true matrix modulus (E_m) due to a small thickness of matrix material tested at the immediate vicinity with the soft inclusion. At distances further away from the particle, an increase of elastic modulus E_m^* will be gradual as thickness of the matrix material underneath the tip increases until a depth of matrix substantially exceeds the tip penetration. This region with lower apparent modulus in matrix (E_m^*) is marked as a dotted ring-like zone outside the soft particle, as shown at the left in Figure 20.13c. The expected elastic modulus profile along the cross-section area is shown in Figure 20.13d (*left*).

In case 2, the cut is made at or below the central plane of the sphere (Figure 20.13c, *middle* and *right*). In this situation, the modulus 2D map will present E_m outside of the soft particle independent on a level of tip indentation (Figure 20.13d, *middle* and *right*). However, in these cases inside the soft inclusions one can expect a gradual increase of their apparent elastic modulus (E_r^*) compared to its true value E_r as the probing location shifts from the center toward the particle borders. The corresponding regions with E_r^* are expected in a dotted ring-like area within the rubber domain (Figure 20.13c, *middle*). The corresponding elastic modulus profile along the cross-section area is shown in Figure 20.13d (*middle*). At extreme case, when a thickness of the rubber inclusion becomes not much larger than the probe penetration depth (Figure 20.13c–d, both *right*), the stiff matrix underneath will cause the hardening of the soft material. Therefore, the higher E_r^* will be detected over the whole area of the rubber domain. This is demonstrated by the elastic modulus profile in Figure 20.13d (*right*).

The presented analysis shows that in absence of interface effects, one should expect gradual changes of elastic modulus on one of the sides of the border between soft inclusion and rigid matrix. In the described study of TPV, first we observed sharp changes of the phase contrast that help to distinguish the interface line between polypropylene (PP) and EPDM domains. Second, we also observed gradual changes of elastic modulus on both sides of this border. The latter observation is different from situations described in the hypothetical cases 1 and 2 and this might be considered as a proof of PP–EPDM interfaces at which mechanical response is different from properties of the main material components.

It is worth to note that nanomechanical studies of interfaces in heterogeneous polymer systems are still in infancy. The presented tentative analysis revealed only a part of hurdles in nanomechanical analysis of interfaces in complex polymer and rubbery materials. A consideration of the effects of tip dimensions, tip–sample contact area, and penetration depth will make this analysis more complicated. The fact that only a binary contrast was observed in phase images on the interface regions might be explained by a smaller tip penetration occurring in tapping mode compared to those that took place during nanoindentation experiments. At the moment it is rather difficult to determine the real dimensions of the interface. This requires higher resolution nanomechanical measurements that might be realized when AvZ curves (the attribute of tapping mode) could be employed for retrieval of quantitative mechanical data. A use of model systems such as multilayered blends, where mechanical studies at the cross-sections perpendicular to the layers are not complicated by the geometrical effects described above, can be very helpful in AFM examinations of interfaces.

20.3.4 Visualization of Fillers in Composites

Loading of rubbery materials with organic and inorganic additives and fillers is a common way of modification of their properties. Adding of oil to rubbers has been already discussed above (Figures 20.7 and 20.10). Mixing of rubber compounds with inorganic fillers is often applied for the improvement of the material performance. Silica and CB particles are employed for this purpose. Individual CB particles have dimensions of 20–40 nm as seen from AFM image in Figure 20.14a. One of the main technological challenges is obtaining fine dispersion of filler particles in high viscous rubber matrix during mixing, and this task is not trivial. Sample heterogeneities, which are most likely to represent filler aggregates at different scales up to tens of microns, are clearly distinguished in optical micrograph of a cryo-microtomed cut of rubbery compound loaded with silica (Figure 20.14b). AFM is capable of analyzing the filler distributions at smaller scales. This is illustrated by images of PU and rubber blend based on NR and SBR filled with silica (Figure 20.14c and d). In both images, the filler particles are seen as bright inclusions inside the rubber matrix. In the first image (Figure 20.14c), one sees a quite homogeneous dispersion of filler in the PU matrix. However, in case of the NR–SBR blend (Figure 20.14d) silica particles are selectively dispersed in one of the material constituents, most likely in NR.

In recent years, the interest toward nanocomposites has been intensified in response to increased efforts in nanoscience and nanotechnology. Although materials with the nanoscale-size fillers have been produced in rubber industry for many years, current design of novel nanocomposites,

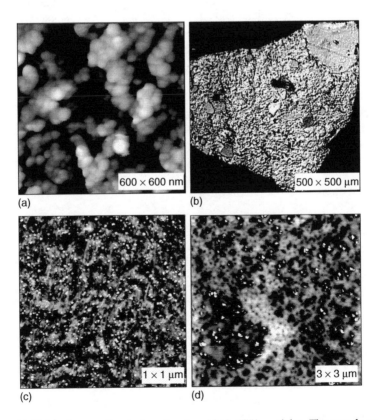

FIGURE 20.14 (a) Height image of a cluster of carbon black (CB) particles. The sample was prepared by pressing the particles into a pellet. (b) Optical micrograph of a cryo-ultramicrotome cut of a rubbery composite loaded with silica. (c, d) Phase images of a nanocomposite of polyurethane (PU) loaded with silica and a rubber blend based on natural rubber (NR) and styrene–butadiene copolymer (SBR) loaded with silica, respectively. The samples were prepared with a cryo-ultramicrotome.

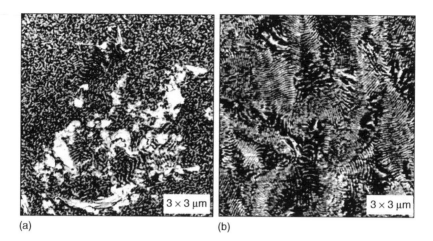

(a)   (b)

FIGURE 20.15 Phase images of samples of triblock copolymer styrene and ethylene–butylene (SEBS) filled with clay particles (5 wt%) with (a) poor mixing and exfoliation and (b) fine distribution of clay layers.

particularly with layered fillers, is focused on further improvement of mechanical, adhesive, thermal, and fire-retardant properties. One of the possible approaches is based on an exfoliation phenomenon of clay clusters into individual sheets of nanometer thickness that substantially increase a surface to volume ratio of the filler and interfacial filler–matrix interactions. The intercalation and exfoliation processes can be monitored with high-resolution TEM and AFM, the techniques complementing each other in visualization of the clay agglomerates and their individual plates. The AFM images, which were obtained on microtomed surfaces of composites of SEBS and a layered mineral fillers, are shown in Figure 20.15a and b. First of these two phase images shows a microphase-separation morphology, which is "contaminated" by a presence of submicron patches of bright contrast. These patches can be assigned to aggregates of stiff filler layers. The phase image in Figure 20.15b exhibits a more perfect block copolymer morphology that hints on successful mixing and exfoliation of the clay plates. Most likely, some of the filler layers are incorporated into lamellar morphology of SEBS as edge-on oriented sheets of size comparable to, or less than, the size of blocks of this material.

20.3.5 Conducting Species in Rubbery Materials

Compositional imaging in AFM of heterogeneous polymers is primarily based on differences of mechanical properties of components. In the cases when conducting species are present in the materials an electric force microscopy (EFM) can be applied for a detection of the related locations. In this technique, a metal-coated probe is biased at some voltage with respect to the sample and during tapping the probe resonant frequency is affected by long-range attractive forces caused by gradient of the electric field between the probe and the sample. This effect can be screened by the tip–sample mechanical interactions, which cause stronger changes of the cantilever's frequency and phase. In EFM, a so-called lift mode, or a two-pass technique is used to separate the effects of electric and mechanical forces. For each scan line the height profile is recorded during the first pass. In the second pass, along the same sample location the probe is pivoted in accordance with the just-learned height profile. This happens when the probe is lifted at 5–50 nm above the surface. During this second pass, the probe resonant frequency shifts to lower frequencies due to attractive forces caused by an electric field between a tip and sample. The frequency changes as well as related phase changes are presented in the images as dark contrast, thus pointing to conductive sample locations. This technique was used for detection of CB particles in rubber materials [39].

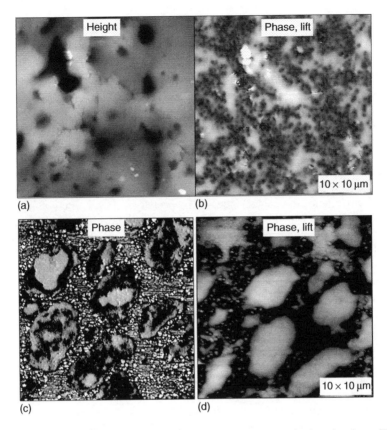

FIGURE 20.16 (a, b) Height and phase images of the surface of thermoplastic vulcanizate (TPV) based on polypropylene (PP), ethylene–propylene–diene terpolymer (EPDM), and carbon black (CB) (30:60:10) obtained in regular tapping mode The phase image was obtained in electric force microscopy (EFM) mode with lift of 20 nm and bias voltage of 10 V. (c, d) Phase images of bulk morphology of the same material obtained respectively in tapping mode and in EFM mode with lift of 50 nm and bias voltage of 10 V.

Here we illustrate the use of EFM by imaging of TPV material (PP–EPDM blend), which was specially prepared for applications (sensors, switches, and electromagnetic shields) demanding electrical conductivity [40]. This preparation was aimed on getting a percolation threshold at small CB loadings (~10 wt%) by a selective localization of the conductive particles in one of the components or at the interfaces. Surface topography of TPV sample loaded with CB is shown in Figures 20.16a and 20.16b. The height image is presented together with phase image, which was taken in the EFM mode with the probe lifted 20 nm over the surface (Figure 20.16b). Dark spots in the phase image correspond to individual CB particles and their aggregates contribute to the percolation network formed in this material. In addition to the sample surface, a distribution of the conducting regions in bulk can be examined on a freshly microtomed surface of the sample. The phase images of the same cut location were recorded in regular tapping (Figure 20.16c) and EFM modes (Figure 20.16d). The phase image in Figure 20.16c shows EPDM domains inside of PP matrix selectively filled with CB particles seen as small bright spots of harder material. The particles are not seen in the rubber domains, which exhibit only phase-contrast variations most likely related to regions of different cross-linking density. The EFM data provide important complementary information. The phase image in Figure 20.16d recorded at the lift height of 50 nm and bias voltage of 10 V exhibits dark contrast over the conducting locations of the PP filled with CB particles. There are no traces of CB particles and conducting paths in bright patches of the rubber domains. This example demonstrates that EFM can be useful in characterization of CB-filled rubbery materials.

20.4 CONCLUSIONS

Several aspects of AFM and its applications to rubbery materials were discussed and illustrated in this chapter. Despite a definite advance in this field during the last 10 years, there is plenty of room for more improvement of this method that will further broaden its use. In conclusion, we would like to mention several future capabilities.

Current instrumental developments of AFM are going toward high-resolution imaging, local probing of thermal transitions, fast scanning and compositional mapping, quantitative nanomechanical studies, and reliable operation under different liquids. AFM studies of polymer and rubber materials will benefit from these improvements. As regarding higher resolution imaging, this function might be most useful for visualization of shapes of nanoscale particles, thus helping in distinguishing fillers of different nature, e.g., silica, CB, and TiO_2 particles. An increase of scanning rate will make AFM a more efficient technique that especially is useful for combinatorial analysis of novel polymer materials in which a search for appropriate compositions includes the characterization of a large number of samples. A development of reliable AFM cells for imaging under different liquids will enhance compositional analysis of heterogeneous polymer and rubbery materials. It will be achieved by imaging under liquids, which cause selective swelling of components, thus helping their identification. For the same purpose, one can apply nanoscale thermal analysis, which allows local measurements of glass transition, melting, or other thermal phenomenon [41].

Expansion of local nanomechanical studies is expected in several directions. First of all, a quantitative analysis of nanoindentation measurements, which was partly demonstrated in the chapter, will be advanced by taking into account the adhesion and viscoelasticity contributions. DvZ curves are employed in these measurements but they can have definite limitations when high spatial and depth resolution are required for nanomechanical probing. These limitations can be overcome with AvZ curves, which are an attribute of tapping mode. There are different novel approaches of the use of AvZ curves and mechanical responses of the materials at higher harmonics [42] for nanomechanical studies. Among other modes under development, we would mention AFM-based nanoscale dynamic mechanical analysis [43], which might offer unique opportunities for mechanical measurements at high frequencies (up to 100 kHz).

The above-mentioned AFM capabilities will enhance characterization of soft materials at the nanometer scale and will make this method invaluable for researchers working in academia and industry.

ACKNOWLEDGMENTS

The authors are thankful to Dr. P. Sadhukhan (Bridgestone/Firestone Research, Akron, OH) for different rubber samples; Dr. T. Medintseva (the Semenov's Institute of Chemical Physics, Moscow, Russia) for TPV samples; Dr. N. Dutta (University of South Australia, Adelaide, Australia) for the samples of SEBS filled with oil; Dr. Z. Petrovich (Kansas Polymer Research Center, Pittsburg State University, Pittsburg, KA) for PU samples, and NIST/ATP Award #70NANB4H3055 for financial support.

REFERENCES

1. Ciesielski, A., *Introduction to Rubber Technology*, Rapra Technology, Shawbury, United Kingdom, 1999.
2. Sperling, L.H., *Polymeric Multicomponent Materials*, Wiley, New York, 1999.
3. Litvinov, V.M. and De, P.P., Eds., *Spectroscopy of Rubber and Rubbery Materials*, Rapra Technology, Shawbury, United Kingdom, 2002.
4. Ratner, B. and Tsukruk, V., Eds., *Scanning Probe Microscopy of Polymers*, ASC Symposium Series 694, Washington DC, 1998.
5. Magonov, S., Atomic Force Microscopy in Analysis of Polymers, in *Encyclopedia of Analytical Chemistry*, Meyers, R.A., Ed., Wiley, New York, 2000.

6. Van Landingham, M.R., A review of instrumented indentation, *J. Res. Nat. Inst. Stand. Technol.*, 108, 249, 2003.
7. Wahl, K.J., Asif, S.A.S., Greenwood, J.A., and Johnson, K.L., Oscillating adhesive contacts between micron-scale tips and compliant polymers, *J. Colloid Interface Sci.*, 296, 178, 2006.
8. Sheiko, S.S. and Möller, M., Visualization of individual molecules—an intriguing way to discover new functional properties, *Chem. Rev.*, 101, 4099, 2001.
9. Segalman, R.A., Schaefer, K.E., Fredrickson, G.H., Kramer, E.J., and Magonov, S.N., Topographic templating of islands and holes in highly asymmetric block copolymer films, *Macromolecules*, 36, 4498, 2003.
10. Ivanov, D.A. and Magonov, S.N., Atomic Force Microscopy Studies of Semicrystalline Polymers at Variable Temperature, in *Polymer Crystallization: Observations, Concepts and Interpretations*, Sommer, J.-U. and Reiter, G., Eds., Springer, Heidelberg, Germany, 2003, chap. 7.
11. Magonov, S.N., Yerina, N.A., Godovsky, Y.K., and Reneker, D.H., Annealing and recrystallization of single crystals of polyethylene on graphite: An atomic force microscopy study, *J. Macromol. Sci. Part B Phys.*, 45, 169, 2006.
12. Chernoff, D. and Magonov, S., Atomic Force Microscopy of Polymer Characterization and Analysis, in *Comprehensive Desk Reference*, Brady, Jr., R.F., Ed., ACS, Oxford University Press, Oxford, United Kingdom, 2003, chap. 19.
13. Galuska, A.A., Poulter, R.R., and McElrath, K.O., Force modulation AFM of elastomer blends: Morphology, fillers and cross-linking, *Surf. Interface Anal.*, 25, 418, 1997.
14. Bar, G., Ganter, M., Brandsch, R., Delineau, L., and Whangbo, M.-H., Examination of butadiene/styrene-co-butadiene rubber blends by tapping mode atomic force microscopy. Importance of the indentation depth and reduced tip–sample energy dissipation in tapping mode atomic force microscopy study of elastomers, *Langmuir*, 16, 5702, 2000.
15. Yerina, N. and Magonov, S., Atomic force microscopy in analysis of rubber materials, *Rubber Chem. Technol.*, 76, 846, 2003.
16. Bonnell, D., *Scanning Probe Microscopy and Spectroscopy: Theory, Techniques, and Applications*, 2nd ed., Wiley-VCH, New York, 2000.
17. Meyer, E., Hug, H.J., and Bennewitz, R., *Scanning Probe Microscopy: The Lab on a Tip*, Springer, Heidelberg, Germany, 2001.
18. Vancso, G.J., Hillborg, H., and Schönherr, R., Chemical composition of polymer surfaces imaged by atomic force microscopy and complimentary approaches, *Adv. Polym. Sci.*, 182, 55, 2005.
19. Zhong, Q., Innis, D., Kjoller, K., and Elings, V., Fractured polymer/silica fiber surface studied by tapping mode atomic force microscopy, *Surf. Sci. Lett.*, 290, 688, 1993.
20. Birley, A.W. and Chen, X.Y., A preliminary study of blends of bisphenol A polycarbonate and poly (ethylene terephtalate), *Brit. Polym. J.*, 17, 297, 1985.
21. Cleveland, J.P., Anczykowski, B., Schmid, A.E., and Elings, V.B., Energy dissipation in tapping-mode atomic force microscopy, *Appl. Phys. Lett.*, 72, 2613, 1998.
22. Belikov, S., Erina, N., and Magonov, S., Interplay between an experiment and theory in probing mechanical properties and phase imaging of heterogeneous polymer materials, *J. Phys., Conference Series*, 61, 765, 2007.
23. Ebenstein, Y., Nahum, E., and Banin, U., Tapping mode atomic force microscopy for nanoparticle sizing: Tip–sample interaction effects, *Nano Lett.*, 2, 945, 2002.
24. Magonov, S.N., Elings, V., Cleveland, J., Denley, D., and Whangbo, M.-H., Tapping-mode atomic force microscopy study of the near-surface composition of a styrene–butadiene–styrene triblock copolymer film, *Surf. Sci.*, 389, 201, 1997.
25. Weisenhorn, A.L., Maivald, P., Butt, H.-J., and Hansma, P.K., Measuring adhesion, attraction, and repulsion between surfaces in liquids with AFM, *Phys. Rev. B*, **45**, 11226, 1992.
26. Sneddon, I., The relation between load and penetration in the axisymmetric Boussinesq problem for a punch of arbitrary profile, *Int. J. Eng. Sci.*, 3, 47, 1965.
27. Oliver, W.C. and Pharr, G.M., An improved technique for determining hardness and elastic modulus using load and displacement sensing indentation experiments, *J. Mater. Res.*, 7, 1564, 1992.
28. Hutter, J.L. and Bechhoeter, J., Calibration of atomic force microscopy tips, *Rev. Sci. Instrum.*, 64, 1868, 1993.
29. AFM probes with tip having rounded apex whose diameter is in the 50–150 nm are available from two probe manufacturers (Team Nanotec, http://www.team-nanotec.de/) and Nanoworld (http://www.nanoworld.com).

30. Sawyer, L.C. and Grubb, D.T., *Polymer Microscopy*, 2nd ed., Chapman & Hall, London, 1996.
31. Belikov, S., Magonov, S., Erina, N., Huang, L., Su, C., Rice, A., Meyer, C., Prater, C., Ginzburg, V., Meyers, G., McIntyre, R., and Lakrout, H., Theoretical modeling and implementation of elastic modulus measurements at the nanoscale using AFM, *J. Phys., Conference Series*, 61, 1303, 2007.
32. Medintseva, T.I., Dreval, V.E., Erina, N.A., and Prut, E.V., Rheological properties thermoplastic elastomers based on isotactic polypropylene with an ethylene-propylene-diene terpolymer, *Polym. Sci.* A, 45, 2032, 2003.
33. Belikov, S. et al. in preparation.
34. Hepburn, C., *Polyurethane Elastomers*, 2nd ed., Elsevier, London, 1991.
35. Petrović, Z.S. et al. Effect of silica nanoparticles on morphology of segmented polyurethanes, *Polymer*, 45, 4285, 2004.
36. Coran, A.Y. and Patel, R. Rubber-thermoplastic compositions. Part I. EPDM-polypropylene thermoplastic vulcanizates, *Rubber Chem. Technol.*, 5, 141, 1980.
37. Prut, E.V., Zelenetskii, A.N., Chepel, L.M., Erina, N.A., Dubnikova, I.L., and Novikov, D.D., Thermoplastic elastomer composition and the way of its manufacturing, Russian Patent 206927/B.I., 32, 1996.
38. ASTM. *Standard Test Method for Vickers Hardness of Metallic Materials*, 1987.
39. Viswanathan, R. and Heaney, M.B., Direct imaging of the percolation network in a three-dimensional disordered conductor-insulator composite, *Phys. Rev. Lett.*, 75, 4433, 1995.
40. Prut, E.V. and Erina, N.A., Dynamic vulcanization and thermoplastic elastomers, *Macromol. Symp.*, 170, 73, 2001.
41. Price, D.M., Reading, M., Caswell, A., Hammiche, A., and Pollock, H.M., Micro-thermal analysis: A new form of analytical microscopy, *Microsc. Anal.*, 65, 17, 1998.
42. Sahin, O., Magonov, S., Su, C., Quate, C., and Solgard, O., An atomic force microscope tip designed to measure time-varying nanomechanical forces, *Nature Nanotechnology*, 336, 1037, 2007.
43. Su, C. and Magonov, S., System for wide frequency dynamic nanomechanical analysis, US Patent 7055378, 6, 2006.

21 Recent Developments in Rubber Research Using Atomic Force Microscopy

Ken Nakajima and Toshio Nishi

CONTENTS

21.1 INTRODUCTION

Let us open the first pages of the books on polymer or rubber physics.[1–3] These books usually begin with the explanation of the origin of polymer chains' elasticity. They do not originate from energetic elasticity mainly possessed by solid-state materials such as a metallic spring but from so-called entropic elasticity. The entropic elasticity is the consequence of an essential property of polymer chains, i.e., a single polymer chain being composed of a large number of unit constituents. Thus, even a single polymer chain can be treated by statistical mechanics. The calculation with a boundary condition that these constituents are bonded to each other to form a chain structure gives the final formula of entropic elasticity. This is the story, developed during the era when a great advocate of polymer science, professor H. Staudinger, proposed that a single polymer takes the form of a linear string or chain molecule.

The entropic elasticity of a single polymer chain is treated similarly to the statistical mechanical property of gas molecules. Now, let us compare gas molecules to children moving around freely

FIGURE 21.1 Drawing to help explain entropic elasticity. (From Saito, N., *Polymer Physics* [in Japanese], Syokabo, Tokyo, Japan, 1967.)

(with random speed and direction) in a schoolyard even though this situation is two-dimensional. Then, in the case of a single polymer chain, the children must move hand in hand as in a schematic illustration in a famous textbook of polymer physics (Figure 21.1).[1] This is the only restraint condition for the children who become constituents of a polymer chain. If two adult men take the hands of the children at each end of the chain and they stand at fixed points, they must feel a very strong force. This is analogous to the restoring force of rubber, entropic elasticity. Higher temperature is equivalent to the children's cheerfulness. In fact, the entropic elasticity for a single polymer chain is calculated to be proportional to absolute temperature using a statistical mechanical treatment,[2]

$$F = \frac{3k_{\mathrm{B}}T}{na^2}x \tag{21.1}$$

where
k_{B} is the Boltzmann constant
n and a are the number and size of the constituents (monomers, children), respectively

The front factor of x can be treated as the spring constant of rubbery elasticity, which obeys Hooke's law.

Some readers might be suspicious of the above description. Polymers are very small chain molecules and are under intense thermal motion. By using statistical mechanics, the spring constant of a single polymer chain can be calculated. Furthermore, this spring is not like a metallic spring but is entropic. However, who has observed it directly? Has anyone experienced pulling a single polymer chain as in the case of macroscopic rubber? The actual situation is that nobody has seen this theoretically equipped situation of the microscopic world. In any case, molecular-level rubber elasticity has been "a house of cards" based on the single polymer chain entropic elasticity without any experimental evidence.

The fact that rubber shows rubber elasticity was discovered more than 100 years earlier than professor H. Staudinger's proposal. The memory effect acquired by vulcanization, so-called Gough–Joule effect, and its thermodynamic explanation were the great achievements in the nineteenth century. As seen in many textbooks,[1–3] this thermodynamic approach was the easiest one to gain consistency between ever-performed experiments and theory. In fact, thermodynamics of rubbery material can be treated in parallel with thermodynamics of gas. One could show experimentally that

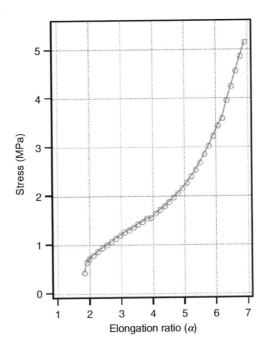

FIGURE 21.2 Macroscopic stress–strain (S–S) curve of natural rubber (NR) vulcanizate.

rubber elasticity was mainly dominated by entropic elasticity, which at present all the rubber researchers should know. However, the situation is almost the same with the case of a single polymer chain; nobody has seen the movement of chain molecules in rubbery materials as in the case of gas molecules.

The large deformability as shown in Figure 21.2, one of the main features of rubber, can be discussed in the category of continuum mechanics, which itself is complete theoretical framework. However, in the textbooks on rubber, we have to explain this feature with molecular theory. This would be the statistical mechanics of network structure where we encounter another serious pitfall and this is what we are concerned with in this chapter: the assumption of affine deformation. The assumption is the core idea that appeared both in Gaussian network that treats infinitesimal deformation and in Mooney–Rivlin equation that treats large deformation. The microscopic deformation of a single polymer chain must be proportional to the macroscopic rubber deformation. However, the assumption is merely hypothesis and there is no experimental support. In summary, the theory of rubbery materials is built like "a *two-storied* house of cards," without any experimental evidence on a single polymer chain entropic elasticity and affine deformation.

In this chapter, we describe the experimental methods that can be used to answer these serious questions; one is whether a single polymer chain really has entropic elasticity. Another is whether an elongated rubber shows affine deformation or not. For the former case, we introduce the technique named "single-molecule force spectroscopy" or "nanofishing," where a single polystyrene chain, as an example, is stretched at both ends in a solvent. For the latter, an example will be shown where an elongated natural rubber (NR) exhibits nonaffine deformation at least at its surface. Both experiments were realized by atomic force microscope (AFM)[4] that was invented in the late twentieth century. It has been possible to perform a real experiment for the conceptual experiment that appeared in the famous textbook written by professor P.G. de Gennes.[5] In addition, a section will be dedicated to the Young's modulus mapping of carbon-black-reinforced NR that also constructs one of the important fields of rubber industry.

21.2 ENTROPIC ELASTICITY OF A SINGLE POLYMER CHAIN

21.2.1 BRIEF INTRODUCTION

The use of AFM has enabled us to visualize a small world consisting of atoms and molecules. Similar to reading Braille by tracing with a finger or tracking grooves with a needle in a phonograph record, AFM uses a very sharp probe tip (ultimately consisting of a single atom) to trace the surface of an object. For the reader's imagination, think of a scale with a length of 10 cm. A sharp spike of 5 mm long is attached at the end of the scale. Take another end and push the spike onto a desk. The stronger the force becomes, the larger the deflection of the scale. Then, move the assembly of the scale and the spike on the desk. The deflection changes in accordance with the surface topography. Raster scans over the two-dimensional surface thus reproduce a surface topographic image, which is the principle of AFM. By scaling down by a thousand times, the scale and the spike correspond to a cantilever and probe of AFM. The tip radius of the spike (10 μm) then corresponds to that of the probe (10 nm). If the AFM system detects a deflection of 10 nm (10 μm on the macroscopic scale), it becomes possible to detect a weak force. Since typical values of the spring constants of cantilevers become about 1 N m^{-1} or even less, the resolution of force detection reaches 1 nN.

By using AFM, we are able to visualize a single polymer chain, as many researchers have already reported. In practice, a single polymer chain looks like a piece of string.[6] It does not resemble a metallic spring at all. The recent study has revealed the dynamic movements of such strings.[7] Furthermore, dozens of researchers have placed their attention on the expanded capability of AFM as single-molecule force spectroscopy. With this technology, the relationship between force and deformation of a single molecule sandwiched between a substrate and an AFM probe has been measured. A technique "nanofishing" introduced in this section comes under this category. Now, in the twenty-first century, we have obtained a novel method for experimentally verifying the statistical mechanics of a single polymer chain, which is indeed the basis of polymer physics. As in the case of stress–strain (S–S) curves of rubber, we can now investigate such curves for single polymer chains.

21.2.2 STATIC NANOFISHING

According to the literature, trials on stretching single polymer chains by means of AFM have been extensively performed by many researchers as schematically shown in Figure 21.3. However, these trials have been especially dedicated to studies of "single protein unfolding" events,[8–11] physisorbed macromolecules.[12–15] In the former case, proteins with modified ends were sandwiched via specific chemical bonds between a substrate and a cantilever and could be successfully stretched in the course of force–distance curve measurement by AFM. Those successes were attributed to two principal factors, the unique three-dimensional structures of proteins and the genetic engineering techniques for modifying the ends of proteins. Thus, picking points were controllably fixed. From a comparison of such cases, synthesized polymers have many disadvantages. For instance, it is difficult to attach any reactive groups at their ends in general. Thus, the force–distance curve data obtained in uncontrolled experiments contain many complicated factors such as that the adsorptions occur not only at the tails, but also at the loops. It is very difficult to judge whether the stretching event is really for a single polymer chain. The polydispersity also makes the situation very complicated. As described, however, successful nanofishing results on polystyrene (PS) chains with chemically active termini have been reported and, thus, several statistical analyses could be conducted to investigate the physical properties of a single polymer chain.[16,17]

A thiol-terminated PS was used as a sample in the experiment. It was based on a living polymerized carboxyl-terminated PS with $M_n = 93,800$ and $M_w = 100,400$. The polydispersity was $M_w/M_n = 1.07$. The degree of polymerization was about 900 and, thus, its contour length was about 220 nm. The thiol groups were substituted for the carboxylic ends using 1,10-decanedithiol by means of thiolester bonding, anticipating the preferential interaction between

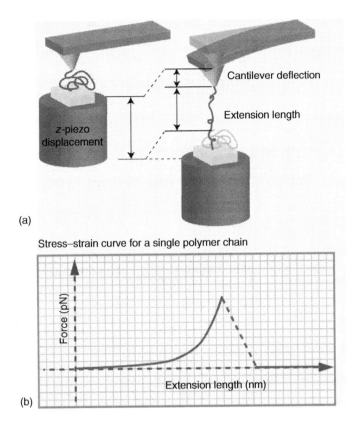

(a)

Stress–strain curve for a single polymer chain

(b)

FIGURE 21.3 Concept of nanofishing. (a) Schematic drawing and (b) force–extension (stress–strain) curve.

thiol and gold. No dimer formation was indicated from the result of gel permeation chromatography (GPC). The polymer was dissolved in a Θ solvent, cyclohexane, at 20 μM mL^{-1}, and 10 μL of the solution was cast on a Au(111) substrate. After 5 min of incubation, the surface was rinsed with pure solvent to remove weakly physisorbed molecules. These sample preparation procedures brought about well-dispersed (checked by AFM imaging, data not shown) chemisorption of polymer chains that possibly took on a random-coil form just like advertising balloons, which assured that the subsequent nanofishing measurements were on a single-molecule basis. The apparent size of the random-coil obtained by AFM imaging (data not shown), W, was about 40 nm. Then, by considering probe-shape convolution effect,[18] the deconvoluted size R_s could be calculated, with the formula $W \approx 4\sqrt{R_s R_t}$, to be about 5 nm, assuming a probe radius, R_t, of 20 nm. The value was almost comparable with the radius of a single polymer chain. Indeed, the radius of a polymer random-coil with $n = 900$ and $a = 0.25$ becomes $\sqrt{n}a = 7.5$ nm. It was very rare to fish "two fish" at the same time. At this stage, it was speculated that, except for their modified termini, the interaction between a gold substrate and polymer chains would be weak enough for them not to strongly adsorb with collapsed or extended forms. Thus, this was the case of the adsorption of polymer chains to a repulsive wall.[5]

In addition to the use of cyclohexane, dimethylformamide (DMF) was used as a good solvent. So as to pick up the thiol-modified terminal, gold-coated cantilevers were used. The nominal values of their spring constant k_1 were 30 or 110 pN nm^{-1}. A typical force–extension curve measured in cyclohexane is shown in Figure 21.4.[16] The solvent temperature was kept at about 35°C, which corresponded to its Θ temperature for PS chains. Thus, a chain should behave as an ideal chain. The slope at the lowest extension limit (dashed line in Figure 21.4) was 1.20×10^{-4} N m^{-1}.

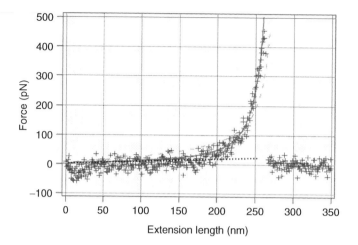

FIGURE 21.4 Nanofishing of a single polystyrene (PS) chain in cyclohexane. The solvent temperature was about 35°C (Θ temperature). A cantilever with a 110 pN nm^{-1} spring constant was used. The worm-like chain (WLC), solid line, and the freely jointed chain (FJC), dashed line models were used to obtain fitting curves. (From Nakajima, K., Watabe, H., and Nishi, T., *Polymer*, 47, 2505, 2006.)

The value should be that of single polymer chain elasticity caused by entropic contribution. At first glance, the force data fluctuated a great deal. However, this fluctuation was due to the thermal noise imposed on the cantilever. A simple estimation told us that the root-mean-square (RMS) noise in the force signal ($\Delta F \sim 15.6$ pN) for an extension length from 300 to 350 nm was almost comparable with the thermal noise, $\Delta F = (k_1 k_B T)^{1/2} \sim 21.6$ pN.

As for further analysis, curve fitting against the worm-like chain (WLC) model was conducted and indicated as a solid line in Figure 21.4. The model describes single polymer chain mechanics ranging from random-coil to fully extended forms, as follows:

$$F(x) = \frac{k_B T}{l_p} \left[\frac{1}{4(1 - x/L)^2} + \frac{x}{L} - \frac{1}{4} \right] \tag{21.2}$$

where

x is the extension length at an external load of F
l_p and L are persistence and contour lengths, respectively.

The fitting results were $l_p = 0.31 \pm 0.01$ nm and $L = 284.5 \pm 0.8$ nm. The persistence length almost corresponded to the length of a single monomer. The degree of polymerization was then calculated from these two intrinsic values to be about 918, which was in good agreement with the expected value from the synthesis information. Actually, a histogram of contour length made by dozens of successful fishing results showed a single peak with good coincidence with the GPC result. Another estimation was made with the following equation derived from Equation 21.2 for condition $x \ll L$:

$$\left. \frac{dF}{dx} \right|_{x \to 0} = \frac{3 k_B T}{2 l_p L} \tag{21.3}$$

The substitution of l_p and L resulted in a spring constant of 0.71×10^{-4} N m^{-1} for single polymer chain elasticity, which was almost equivalent to the experimentally obtained slope value,

1.20×10^{-4} N m^{-1}, and the value obtained by the simplest form, 1.45×10^{-4} N m^{-1} ($n = 918$ and $a = 0.31$ nm for Equation 21.1). These comparisons implied that the measurements were consistent with the theoretical predictions. The deviation between the rupture length of 260.9 nm and the fitted-contour length indicated that the polymer chain was not fully stretched at the rupture event. The reason for this was that the rupture event was a stochastic process and was dependent on many factors such as pulling speed, bond strength, and temperature. The validity of the freely jointed (FJC) model (dashed line) was also checked:

$$\frac{F(x)l_K}{k_B T} = L^{-1}\left(\frac{x}{L}\right), \quad L(y) = \coth y - \frac{1}{y} \tag{21.4}$$

where
   l_K is the Kuhn length
   $L(y)$ is the Langevin function

   It was obvious that the FJC model did not represent the experimental result well. Thus, nanofishing could be used to judge the ever-present basic theories of polymer physics.

   Further examinations by changing solvent temperature and solvent type proved to be useful in order to reveal statistical mechanical properties such as the second virial coefficient on a single polymer chain basis.[16,17] Such properties have normally been investigated by light scattering or osmotic pressure as averaged information from an ensemble of many polymer chains. Detailed discussions are given in the literature and readers are strongly recommended to refer them, while differences in solvent qualities cause differences in the conformations of single polymer chains, resulting in observable changes in force–extension curves. The results seemed to be explained well by Flory's lattice model, at least as the first approximation.[19]

   In a poor solvent (cyclohexane at 5°C), a polymer chain takes on a condensed globular state because constituent molecules are repulsed by the solvent molecules. Nanofishing of this chain revealed a perfectly different force–extension curve, as shown in Figure 21.5. It was observed that constant force continued from about 30 to 130 nm after nonspecific adsorption between a

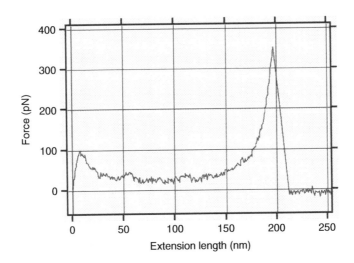

FIGURE 21.5 Nanofishing of a single polystyrene (PS) chain in cyclohexane at 5°C (poor region). A cantilever with a 30 pN nm^{-1} spring constant was used. The coil–strand coexistence was revealed in the extension range of 30–130 nm.

substrate and a probe. This phenomenon could be related to the extraction of monomers one by one from a collapsed globule into a stretched strand, which was numerically predicted by Wittkop et al.[20] Using Monte Carlo simulation, they examined the deformation behavior of a single polymer chain below the Θ temperature, thus the poor region. This type of measurement is not possible by ensemble-averaged measurements, and nanofishing is the only tool that can be used.* As seen, nanofishing seems to have a very promising future. However, there are several drawbacks to be solved. For example, it is difficult to define the transverse section perpendicular to the extension direction and, therefore, stress that is defined as force divided by section area. It is also difficult to discuss Poisson's ratio, though these definitions themselves may be unnecessary. If an ultimate goal is the measurement of single polymer chain elasticity inside a practical rubbery material, the current status that the measurement must be performed in liquid condition would be a severe problem together with the above-mentioned difficulty of uncontrolled physisorption-based experiments.

Before ending this section, we would like to introduce an experiment that checks the validity of the hypothesis of whether nanofishing is performed in a quasi-static condition. The force–extension curve in Figure 21.4 was taken at a pulling speed of 2 μm s^{-1}. Under this condition, no change in force was observed if extension was stopped at a certain point before rupture. Thus, it could be tentatively regarded that the condition was near equilibrium. In contrast, the curve in Figure 21.6 gave an apparent deviation from the former result.[21] In this experiment, a single polymer chain was exposed to a sudden, stepwise stress-increase of maximum 150 pN with the speed of 500 pN ms^{-1} four times before rupture as in the case of the macroscopic creep measurement. In this case, the pulling speed corresponded to about 17 μm s^{-1}, 8.5 times faster than the former measurement. A sudden increase in force with almost "no" deformation of the polymer chain was observed for each part. It was speculated that this phenomenon was due to topological constraint, i.e., internal entanglement of a polymer chain. The relaxation of this constraint was also found in each successive

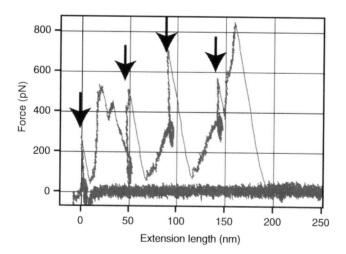

FIGURE 21.6 Nanofishing with a very fast pulling speed (about 17 μm s^{-1}) for a single polystyrene chain in dimethylformamide (DMF). A cantilever with a 30 pN nm^{-1} spring constant was used.

---

* There should be another explanation of plateau region; a chain collapsed and strongly adsorbed on the substrate. In this case, the detachment of monomer one by one from the substrate corresponded to the observation. Further studies are required to elucidate the point. However, this example was probably not the case because the substrate was repulsive wall for PS chains.

response. Further studies are necessary to explain the observed result. Nevertheless, we could conclude from this experiment that the result obtained in Figure 21.4 was for a quasi-static condition because the curve merely shows mild, gradual changes in contrast to that in Figure 21.6. In addition, we should note that the curve in Figure 21.6 might have a certain relationship with the hysteresis observed in the S–S curve for macroscopic rubbery materials since we never observe any hysteresis for quasi-static nanofishing.

21.2.3 DYNAMIC NANOFISHING

In the previous section, we showed that static nanofishing merely gave static information about polymer chains. Therefore, viscoelasticity, which is a quite important property of polymers, seldom appears. Why do polymer solutions have viscosity? How does an elastomeric isolator absorb the energy created by earthquakes? It is possible to ascribe these qualities to the viscosity of polymers. Incidentally, there is a model to describe polymeric viscoelasticity, which is composed of a single spring and a single dashpot (device composed of piston, cylinder, and viscous fluid, which gives resistance proportional to velocity). Let us say that the model is an extremely simplified one that can reproduce only a single relaxation event. Where are the spring and dashpot? What does a dashpot look like? These are frequently asked questions by many students. The students do not understand that the model is just a phenomenological one. As explained, a single polymer chain does not necessarily take the form of spring. In the same way, we do not find dashpot structures on a microscopic scale. The phenomena modeled by dashpots are several frictional contributions such as internal friction between neighboring segments (monomers) during rotational movement with respect to bonding axis, the friction between segments and solvent molecules, and the friction between segments during cooperative movement in polymer "spaghetti." These origins have their own time constants. Thus, the use of the phenomenological model requires thinking about the kinds of interactions that take place. In any case, the frictional or energy-dissipative processes are of great importance in applying polymers to realistic industrial applications. Thus, the origins must be explained experimentally. Here, we introduce a potential technique of revealing monomer–solvent friction for a single polymer chain.[22]

The experimental setup was almost same as that for the static measurement, while the cantilever was now vibrated at its resonant frequency in liquid ($f_1 = \omega_1/2\pi = 9.03$ kHz). A gold-coated cantilever was again used and its spring constant was experimentally determined by the thermal noise method ($k_1 = 29.6$ pN nm^{-1}).[23] The RMS oscillation amplitude was $A = 6.0$ nm when free oscillation (no interaction with a polymer or a substrate) was reached. In the case of successful nanofishing, the RMS amplitude decreased as in the case of the tapping-mode operation. The phase shift between input and output signals also deviated from 90°, which was the value for free resonance. The change in boundary condition by polymer chain attachment caused a change in the free oscillation of the cantilever.

To obtain information about the mechanical properties of a single polymer chain, the separation-dependent RMS amplitude and phase shift changes (data not shown) were analyzed based on the phenomenological model depicted in Figure 21.7. Assuming sinusoidal drive of the cantilever base $z_d(t) = A_d e^{i\omega t}$ with drive amplitude A_d and drive frequency ω, the deflection of the cantilever as a sinusoidal response can be written as $z(t) = A e^{i(\omega t + \theta)}$ with amplitude A and phase shift θ. The cantilever is represented by an effective tip mass together with a spring and dashpot connected in parallel (Voigt element), which are characterized by mass m, spring constant k_1, and damping coefficient η_1. Here, the damping is due to the viscous drag by the surrounding medium and the intrinsic damping caused by the deflection of the cantilever beam. A polymer chain is also described as a Voigt element, which has a spring k_2 and dashpot η_2 connected in parallel; the mass of a molecule is so small compared to the tip mass that it is ignored. The two parameters, k_2 and η_2, are what we are interested in.

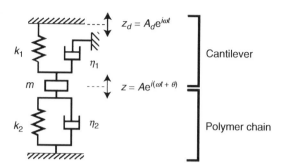

FIGURE 21.7 A double-Voigt model depicting measurement of dynamic nanofishing.

Analysis started from the following equation of motion:[24]

$$m\ddot{z}(t) + (\eta_1 + \eta_2)\dot{z}(t) + (k_1 + k_2)z(t) = k_1 z_d(t) \tag{21.5}$$

Parameters η_1 and m were obtained from the spring constant of the cantilever $k_1 (= 29.6 \text{ pN nm}^{-1})$, resonant frequency $\omega_1 (= 2\pi f_1 = 21.2 \text{ kHz})$, and quality factor $Q (= 25)$, as follows:

$$m = \frac{k_1 \sqrt{Q^2 - 1}}{Q\omega_1^2}, \quad \eta_1 = \frac{k_1}{Q\omega_1} \tag{21.6}$$

Note that these values were determined during the cantilever-tuning process and were, therefore, independent of the stretching event. Although detailed calculation is omitted, k_2 and η_2 could be calculated from the solution of Equation 21.5:

$$k_2 = \frac{A_1}{A}\left\{\left(-m\omega_1^2 + k_1\right)\cos\Delta + \eta_1\omega_1\sin\Delta\right\} - \left(-m\omega_1^2 + k_1\right) \tag{21.7}$$

$$\eta_2 = \frac{A_1}{A\omega_1}\left\{\eta_1\omega_1\cos\Delta - \left(-m\omega_1^2 + k_1\right)\sin\Delta\right\} - \eta_1 \tag{21.8}$$

where

A_1 is the oscillation amplitude without the polymer chain
Δ is the difference in phase shift with (θ) and without (θ_1) the polymer chain $(\Delta = \theta_1 - \theta)$

Amplitude and phase shift in the approaching process are regarded as A_1 and θ_1.

Figure 21.8 shows the obtained result. The stiffness of a single chain k_2 (Figure 21.8a) increased abruptly with stretching. This is commonly observed in conventional simple stretching experiments on polymers, as in Figure 21.4, i.e., an almost constant value for spring constant at low extension and an apparent increase due to the fully stretched effect, caused not only by entropic contribution, but also by enthalpic contribution such as bond angle constraint. The average value of k_2 in the low extension region was $2.86 \times 10^{-5} \text{ N m}^{-1}$. A new result, only available by dynamic nanofishing for the frictional coefficient, which might be related to viscosity, was obtained as shown in Figure 21.8b. At first glance, almost zero viscosity was observed in the low extension region. However, viscosity had a certain value in this region. From Figure 21.8b, viscosity η_2 in the region (30–150 nm) averaged $2.62 \times 10^{-9} \text{ kg s}^{-1}$. Note that this value is the first estimation of viscosity on a single polymer chain basis in low extension region.

It was also found that the viscosity also increased in the same manner as in k_2. There have been many theoretical descriptions for polymer–solvent friction.[1,5,19] As an extreme case, a single

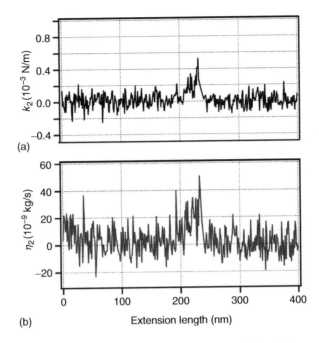

FIGURE 21.8 The behavior of (a) chain stiffness, k_2 and (b) viscosity of the chain, η_2 against the extension. The values of k_2 and η_2 were calculated by a double-Voigt model.

polymer chain is treated as a rigid sphere with a certain hydrodynamic radius. The radius must be correlated to Flory's radius, R_F.[25]

$$f_{\text{Kirkwood}} = 6\pi\eta_s R_F \frac{d\mathbf{r}}{dt} \tag{21.9}$$

where η_s is solvent viscosity. In contrast, there is a theory called the free-draining model, where all the constituents of a single polymer chain are exposed in a solvent flow.[26]

$$f_{\text{Rouse}} = \zeta_s n \frac{d\mathbf{r}}{dt} \tag{21.10}$$

where ζ_s and n are the frictional coefficient between a constituent monomer and a solvent molecule and the number of monomers, respectively. It seemed that the experimental result contradicted the assumption of the free-draining molecule, which ought to be treated as the first approximation of polymer chains interacting with solvent molecules because the increase in viscosity in the higher extension region was never explained by this simple additivity. The increase might be related to an increase in the number of constituent molecules exposed to solvent friction. Further studies with improved signal/noise ratios are necessary to explain this point including comparison with more sophisticated theories that can treat solvent flow inside a random-coil and an extended chain form.[5]

Up to here, the discussion has seemed to assume that viscosity η_2 determined by dynamic nanofishing is due to solvent friction, an idea that is not groundless. To verify this assumption, the dependence of η_2 on solvent temperature and quality was investigated. There is a well-accepted concept to describe polymer solution viscosity, intrinsic viscosity $[\eta]$. Values have been accumulated for different polymer species, solvent qualities, and temperatures by conventional techniques such as light scattering and osmotic pressure.

$$[\eta] = \lim_{c \to 0} \frac{\eta_{sp}}{c} = K M^{\alpha} \tag{21.11}$$

As is seen, $[\eta]$ is the value of specific viscosity, $\eta_{sp} = (\eta - \eta_s)/\eta_s$, is extrapolated to an infinitely dilute condition, where η, c, and M are solution viscosity, concentration, and molecular weight, respectively. K and α are constants depending on polymer species, solvent quality, or temperature.[27] Although the units of $[\eta]$ and of η_2 were different, it was considered that the origin of both parameters was the same to each other. This was because both parameters represented viscosity in an infinite dilute polymer solution. Thus, the coefficient of the viscosity–molecular weight relationship was chosen as the property for comparison. Assuming that there exists a constant K' satisfying the following relationship:

$$\eta_2 = K' M^{\alpha} \tag{21.12}$$

To compute constant K' from the experimental data, molecular weight M was calculated from the contour length obtained by WLC fitting against the force–extension curve (data not shown, however, obtained together with the curves in Figure 21.8), and a constant α was determined by reference data.[27] The temperature dependence of the viscosity constant K' and corresponding reference data K are shown in Figure 21.9. Viscosity constants K' and K were divided by those in Θ solvent K'_Θ and K_Θ for scaling, respectively. Experimental and reference data showed good agreement with each other. This agreement supports the idea that the viscosity measured here clearly depends on solvent conditions as well as intrinsic viscosity. Taking the increase in viscosity in the high extension region into account, it would be concluded that viscosity under 10 kHz perturbation was attributed to monomer–solvent friction. The investigation of molecular weight dependence will also help in checking the validity of the analysis. The realization of such an experiment is expected.

To end the section, it is worth mentioning the possible future direction of nanofishing experiments. The leading person in this field, professor H.E. Gaub and coworkers realized a kind of heat engine using a photosensitive single polymer chain.[15] They used the *cis–trans* isomerization of an azobenzene polymer to convert photochemical energy into mechanical energy. Of course, industrial application of this technique is not simple, while the research gives us a kind of dream that nanofishing will open a promising future. In the past, studies on a single polymer chain had been

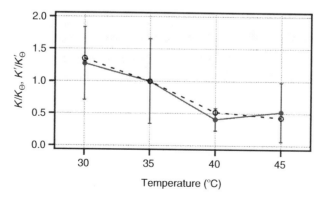

FIGURE 21.9 Temperature dependence of the viscosity constant K' (solid) and K (dashed). Both constants are divided by those in Θ solvent for scaling.

preceded by theoretical work. However, we believe that in the twenty-first century, experimental studies will take the initiative in this field.

21.3 NONAFFINE DEFORMATION OF ELONGATED NATURAL RUBBER

21.3.1 BRIEF INTRODUCTION

NR has a long-standing history. It was put to practical use long before the invention of the first synthesized elastomer, and even at present it is difficult to replace NR with other synthesized elastomers due to its superior characteristics. Thus, the science of NR has a very unusual existence among many other scientific fields, and the phenomena occurring inside NR would not be as simple as other rubbery materials. Among many scientific studies, NR vulcanizates under elongation have been an intense research interest since the 1940s. Indeed, the theory of rubber elasticity and the relationship between structure and property in polymer science were originally developed from this subject.[3,19] However, the exact behavior of NR under elongation on the microscopic scale is still not clear. In classical theories, it was assumed that NR obeys affine deformation, i.e., the deformation of polymer chains is proportional to macroscopic elongation. The assumption has been prevalently accepted as the basic formula describing the connection between the entropic elasticity of single polymer chains and macroscopic rubber elasticity for long time. On the other hand, it is interesting to investigate this quite simple assumption from a microscopic perspective.

Recently, there has been much progress in both experimental and theoretical studies. For example, Lapra et al.[28] reported that the strain undergone by a rubber matrix is highly variable and depends directly on local filler fraction, as determined by the AFM imaging of filled rubber. Murakami et al.[29] showed that during uniaxial deformation, up to 75% of polymer chain segments remain in the unoriented state even at large strains, as determined by in situ synchrotron x-ray diffraction analysis. Rubinstein and Panyukov[30] put forward a nonaffine tube model, which is an improvement of Edwards' tube model.[31] Marrucci et al.[32] suggested a new deformation tensor that considers the role of force balance in the nodes of polymer networks. Recent computer simulations like molecular dynamics will also answer the question. Despite such progress, on the other hand, there is no direct experimental evidence to show that the assumption of affine deformation is utterly incorrect.

In this section, we show the morphological changes of stretched NR without filler by AFM. Two-dimensional mappings of topography and elasticity for elongated NR will be given to confirm the breakdown of the long-believed assumption of affine deformation.

21.3.2 EXPERIMENTAL

Table 21.1 shows the recipe for obtaining NR specimen without filler. The specimen was cured at 145°C for 30 min. The cross-link density was measured by allowing it to swell in toluene and calculated to be 1.47×10^5 mol cm^{-3} using the Flory–Rehner equation.[33] The average number of

TABLE 21.1
Compound Recipe of Natural Rubber (NR) Vulcanizate without Filler

Code	NR	ZnO	Stearic Acid	Sulfur	Vulcanization Accelerator[a]	Antioxidant[b]
NR	100	5	2.0	0.5	0.5	0.5

[a] 2-Mercaptobenzothiazole.
[b] Santoflex 13 (Monsanto).

monomer units between cross-links was 450. The specimen was immersed in acetone for one day to extract impurities blooming onto surface and dried in vacuo for two days before measurement. Then, the specimen was cooled at liquid-nitrogen temperature and was broken to obtain the surface for observation. The uniaxial elongation was imposed on the sample. The extension ratios, α, were 1.0, 2.0, 3.0, 4.0, 5.0, and 7.0 for comparison (not all the corresponding data will be shown). The macroscopic S–S curve is shown in Figure 21.2.

After the elongation, two methods of AFM observation were performed. The conventional tapping-mode AFM experiments were performed under ambient condition to investigate morphological changes. The cantilever used was one whose spring constant and resonance frequency were $42\ \mathrm{N\ m^{-1}}$ and 300 kHz. In another operational mode, i.e., the force–volume (FV) measurement, was performed in distilled water to obtain Young's modulus mapping. The cantilever with the spring constant of $3.0\ \mathrm{N\ m^{-1}}$ was used. The detailed explanation of FV mode is given in Section 21.3.3.

21.3.3 RECONSTRUCTION OF REAL HEIGHT AND ELASTIC MODULUS IMAGES

AFM is widely used in the world as an imaging tool for elastic samples like polymers and biomaterials.[34,35] Contact and tapping-mode operations are known as major imaging modes. AFM probes are scanned over their surfaces with mechanical contact in both modes. Thus, it has been said among researchers that topographic images from both modes are affected by the deformation of the sample itself due to contact or tapping forces. The users might have also been able to qualitatively understand the influence of a contact or tapping forces for obtained topographic images in the past. However, the imaging with constant force condition never results in quantitative estimation under such influences.

Now, here will be demonstrated the quantitative method to obtain accurate topographic images of elastic samples together with Young's modulus distribution images. Our interest in this book, especially, exists in rubbery materials that are commonly difficult for AFM to deal with. The final goal is to understand peculiar properties of rubbery materials, i.e., mechanical and rheological properties at nanoscale (nanomechanics and nanorheology[36]). The value of a cantilever spring constant is an important factor to detect mechanical properties from a sample surface. If the spring constant is very small, the cantilever approaching the surface cannot deform the sample. If the spring constant is very large, instead, the cantilever can deform the sample, without any deflection. Therefore, necessary information about the sample is lacking. Thus, we need to choose a cantilever with an appropriate spring constant.

To obtain the mapping of the local mechanical properties of rubbery materials, FV measurement would be the most appropriate method. In this mode, force–distance curve data are recorded until a given cantilever deflection value (trigger set point) is attained for 64×64 points over a two-dimensional surface. At the same time, z-piezo displacement corresponding to the trigger set-point deflection was recorded to build a topographic image. The topographic image taken in this mode is basically same from that by conventional contact mode if contact force set point and trigger set point are identical.* If all the points over the surface are stiff enough, the set of recorded displacements represents the topographic feature (real height) for the sample. However, if the surface deforms as shown in Figure 21.10a, it is no more valid to regard an obtained data as the real topographic information. It will be later demonstrated that "real height image" can be reconstructed even for such elastic samples from the information obtained from FV measurement.

At present, there are a variety of theoretical models to describe the mechanical contact between the two bodies under external load and many of such theories have been used to analyze force–distance curves.[37] Among them, Hertz theory[38] has been the most widely used because of its

---

* Because modes of operation are different, obtained topographic images may not resemble each other if the sample deformation caused by frictional or adhesive interaction become dominant.

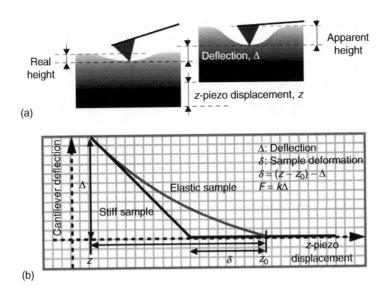

(a)

(b)

FIGURE 21.10 (a) The schematic drawing of the sample deformation for an elastic sample. (b) The comparison between force–distance curves for stiff and elastic samples.

simplicity, where the absence of adhesion is assumed. One can adopt some theories such as JKR model[39] in order to treat adhesive interaction.[40] However, Hertz contact is chosen here for the study to make our discussion clear. For the purpose of increasing the validity of Hertz model, our attention is only placed on approaching the process in a liquid circumstance.

Force curve gives the relationship between the z-piezo displacement and the cantilever deflection as shown in Figure 21.10b. When a cantilever approaches to a stiff sample surface, cantilever deflection, Δ, is equal to the z-piezo displacement, $z - z_0$. The value of z_0 is defined as the position where the tip–sample contact is realized. On the other hand, z-piezo displacement becomes larger to achieve the preset trigger value (set point) of the cantilever deflection in the case of an elastic sample due to the deformation of the sample itself. In other words, we can obtain information about a sample deformation, δ, from the force–distance curve of the elastic surface by the following relationship:

$$z - z_0 = \Delta + \delta \tag{21.13}$$

$$F = k\Delta \tag{21.14}$$

where F and k are the loading force and the cantilever spring constant, respectively. The Young's modulus can be obtained from the force–distance curve by analyzing the sample deformation under external loads. In the case of a cantilever having a stiff and conical tip, Hertz model predicts the following equation:

$$F = \frac{2E \tan \theta}{\pi(1 - \nu^2)} \delta^2 \tag{21.15}$$

where θ is the half-angle of the conical tip (35°). Here the Poisson's ratio, ν, was fixed at 0.5 for simplification. In the case of spherical tip, the following equation is used in place of Equation 21.15:

$$F = \frac{4}{3} \frac{E\sqrt{R}}{1 - \nu^2} \delta^{\frac{3}{2}} \tag{21.16}$$

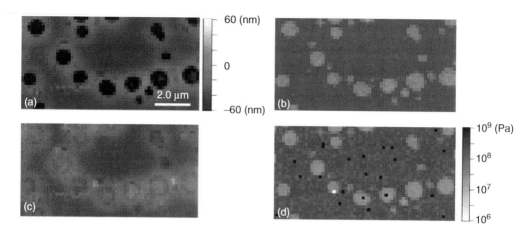

FIGURE 21.11 The result of a force–volume measurement on polystyrene (PS)/polyisobutylene (PIB) 9:1 immiscible blend. (a) The force–volume (apparent) height image, (b) the sample deformation image, (c) the reconstructed real height image, and (d) the Young's modulus distribution image. The scan size was 10 μm. Trigger threshold was 150 nm. A force–distance curve was incorporated for each 64 × 64 pixels. A cantilever with 0.58 nN nm^{-1} spring constant was used.

where R is the radius of the tip. For more details, please refer to the original paper,[34] where the procedure to determine z_0 is also described.

As an example, the result on polystyrene (PS)/polyisobutylene (PIB) 9:1 immiscible blend is shown in Figure 21.11. Their glass transition temperatures, T_g, were 100°C and −76°C, respectively. Thus, PS was in a glassy state and PIB was in rubbery state at room temperature. Their bulk Young's modulus was about 3 GPa and 3 MPa, respectively. Because of an asymmetric blend ratio, PIB-rich phase appeared as island parts in PS-rich matrix. In the apparent height image (Figure 21.11a), all the PIB-rich phases were observed as depressions. The image, on the other hand, resembled the height image obtained by contact mode (data not shown) where all the PIB-rich phases were also depressions. Guided by the idea shown in Figure 21.10b and Equation 21.13, it was possible to plot a "sample deformation image" as shown in Figure 21.11b together with a "real height image" in Figure 21.11c. As seen in this image, depressions were not the "real" depressions but rather protrusions. In short, obtained FV–height image was an artifact caused by the very low elastic modulus of PIB-rich phase.

Next, the Young's modulus was calculated by the curve fitting of a set of force–distance curves. The experimental data were fairly fitted to Equation 21.15 for PIB-rich phase. The calculated value was in the order of 10 MPa. Because the stiffness of PS-rich phase was sufficiently high, the part could not be deformed like PIB-rich one for the cantilever used here ($k = 0.58$ N m^{-1}). Then, the curve fitting the Hertz model often failed. Therefore, such parts were automatically excluded by judging a mean square-root error. The mapping of the Young's modulus distribution is shown in Figure 21.11d. Noncalculated parts are shown by the dark spots. We would like to claim that Young's modulus images must be compared with the "real" height images if one wants to make a clear correlation in-between. Otherwise, the simple correlation between "apparent" height and Young's modulus images misleads to wrong interpretations.

21.3.4 RESULTS AND DISCUSSION

The uniaxially elongated NR vulcanizate with $\alpha = 4.0$ formed line structures with several micrometers in width along the strain direction, as shown in Figure 21.12. It was evident from the series of observations with different α that there was a strong dependence of α on line-width change (data not

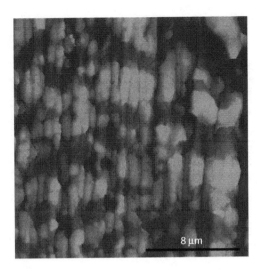

FIGURE 21.12 Tapping-mode topographic image of uniaxially elongated NR vulcanizate. The elongation ratio was about 4. The scan size was 20 μm.

shown).[21,41] That is, the larger the α was, the thinner the line structure became until its width reaches several micrometers ($\alpha < 3$). For $\alpha > 3$, on the other hand, the line width was almost unchanged. What was changed? As seen in the figure, there was inhomogeneity within a line structure. In other words, higher and lower parts appeared repeatedly. By the comparison between $\alpha = 3$ (corresponding topographic image is not shown) and $\alpha = 4$, it was observed that lower parts were elongated more than higher parts. It led us to the conclusion that the phenomenon would be related to the breakdown of affine deformation hypothesis at least for the structures in the range of micrometers. This interesting observation had to be investigated in more detail. The phase images were taken with topographic images, which implied that the higher and lower parts seemed to have different mechanical properties. However, this was merely qualitative information and, thus, the analysis with FV mode was essential.

The results of the FV-mode AFM imaging are shown in Figure 21.13.[21] Figure 21.13a is the "real" height image and Figure 21.13b is the Young's modulus image. The elongation ratio was the same as that in Figure 21.12. At this elongation ratio, the macroscopic modulus was about a few MPa as indicated in Figure 21.2, while much wider distribution of modulus was observed. The comparison between Figure 21.13a and b told us that the higher topographic parts were harder and lower parts were softer within a line structure. Considering this locality of modulus with above-mentioned locality of elongation ratio, i.e., the fact that the softer components were much easily elongated than the harder components, it was concluded that the long-believed assumption of affine deformation is not necessarily true. The reason for this double locality might be caused by an inhomogeneous cross-link. It is the future work to elucidate the conclusion with similar measurements for different cross-link density and different types of elastomeric materials. There might be a possibility that the phenomenon is specific only for NR and then this explains the superior characteristics of NR.

The most interesting observation was the very wide distribution of Young's modulus. Even the order of 100 MPa regions were spatially distributed for macroscopically soft condition ($\alpha = 4$). The regions with such high modulus did not participate in the determination of macroscopic modulus, did they? For the purpose of verifying the idea, histograms of modulus were plotted from the corresponding Young's modulus images for $\alpha = 4$ and 5. The result is shown in Figure 21.14. As seen, the lowest peak (4–6 MPa) appeared at $\alpha = 4$ was missing at $\alpha = 5$. Though the slight difference existed between the peak value and the macroscopic modulus at the same elongation

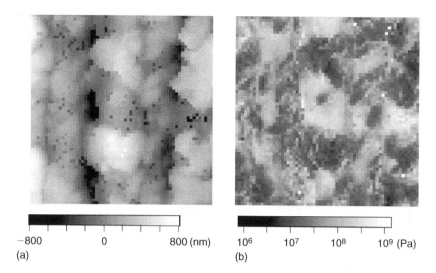

(a)   (b)

FIGURE 21.13 (a) The real height and (b) Young's modulus images of elongated NR vulcanizate. The elongation ratio was the same as that in Figure 21.12. The scan size was 5 μm. The trigger threshold was 100 nm. A cantilever with 3.0 nN nm^{-1} spring constant was used.

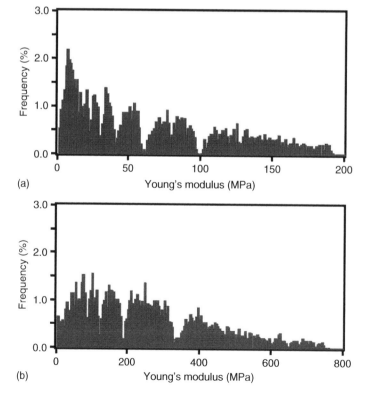

(a)

(b)

FIGURE 21.14 The comparison of Young's modulus distribution between (a) $\alpha = 4$ and (b) $\alpha = 5$. The original Young's modulus image was for (a) Figure 21.13, while that for (b) was not shown.

ratio, this was due to several reasons; uncertainty of the spring constant and the shape of the probe tip. In addition, more essential difference was the direction of modulus measurement; macroscopic one was of course done in the direction of elongation, while the one made by AFM was conducted perpendicular to this direction. However, this difference is not crucial. The emphasis must be placed on the following fact. The elongation ratio of $\alpha = 4$ is inside the plateau region of macroscopic S–S curve and the sudden increase of stress is observed at around $\alpha = 5$. Thus, contrary to the preexisting idea, it might be concluded tentatively that the plateau region observed in macroscopic S–S curve could be attributed to the remaining of the local region that had sufficiently low modulus even after stretched at the elongation ratio of around 4.

Much closer look of Figure 21.14 revealed that both distributions consisted of three peaks. These peaks in the ascending order of modulus corresponded to topographically lower, interfacial, and higher parts. More interestingly, the peak of the lowest modulus could also be divided into three peaks. It seemed there was a certain hierarchy. This hierarchy would be somewhat connected to structural hierarchy. By paying attention to the domain inside a line structure, there should exist another distribution of modulus, like in a *matryoshka* doll. With this hierarchic relationship, nonaffine deformation would continuously be valid until much smaller scale. There is a possibility of a certain correlation between this hierarchy and the specific feature of NR. Unfortunately, it is of great difficulty to visualize a single polymer chain inside rubber with the current status of arts. Therefore, we cannot discuss the affine relation between micrometer-scale deformation and that for a single polymer chain. We have many things to do. However, information based on real experimental support would complement incomplete theories, being the bridge between microscopic and macroscopic phenomena.

21.4 INTERFACIAL REGION OF NATURAL RUBBER REINFORCED BY CARBON BLACK

21.4.1 Brief Introduction

Recent demands for polymeric materials request them to be multifunctional and high performance. Therefore, the research and development of composite materials have become more important because single-polymeric materials can never satisfy such requests. Especially, nanocomposite materials where nanoscale fillers are incorporated with polymeric materials draw much more attention, which accelerates the development of evaluation techniques that have nanometer-scale resolution.[42] To date, transmission electron microscopy (TEM) has been widely used for this purpose, while the technique never catches mechanical information of such materials in general. The realization of much-higher-performance materials requires the evaluation technique that enables us to investigate morphological and mechanical properties at the same time. AFM must be an appropriate candidate because it has almost comparable resolution with TEM. Furthermore, mechanical properties can be readily obtained by AFM due to the fact that the sharp probe tip attached to soft cantilever directly touches the surface of materials in question. Therefore, many of polymer researchers have started to use this novel technique.[43] In this section, we introduce the results using the method described in Section 21.3.3 on CB-reinforced NR.

It is an unfortunate fact that several preexisting theories have tried to explain complicated mechanical phenomena of CB-reinforced rubbery materials but they have not been so successful.[44–48] However, a recent report might have a capability of explaining them collectively,[49] when the author accepted the existence of the component whose molecular mobility is different from that of matrix rubber component in addition to the existence of well-known bound rubber component. The report described that this new component might be the most important factor to determine the reinforcement. These rubber components have been verified by spin–spin relaxation time T_2 by pulsed nuclear magnetic resonance (NMR) technique,[50–53] while the information obtained by NMR is qualitative and averaged over the sample and, therefore, lacking in the spatial information. The direct correlation between such components and mechanical properties is not straightforward as well. In this section, we introduce an example of the examination of the force–distance curves on CB-reinforced NR to build a Young's modulus distribution map, resulting in the verification of the existence of new phase that has quantitatively different mechanical property from matrix rubber phase in real space.

21.4.2 EXPERIMENTAL

NR with standard recipe with 10 phr CB (NR10) was prepared as the sample. The compound recipe is shown in Table 21.2. The sectioned surface by cryo-microtome was observed by AFM. The cantilever used in this study was made of Si_3N_4. The adhesion between probe tip and sample makes the situation complicated and it becomes impossible to apply mathematical analysis with the assumption of Hertzian contact in order to estimate Young's modulus from force–distance curve. Thus, all the experiments were performed in distilled water. The selection of cantilever is another important factor to discuss the quantitative value of Young's modulus. The spring constant of 0.12 N m^{-1} (nominal) was used, which was appropriate to deform at rubbery regions. The FV technique was employed as explained in Section 21.3.3. The maximum load was defined as the load corresponding to the set-point deflection.

21.4.3 RESULTS AND DISCUSSION

Figure 21.15 shows the FV results on NR10 sample.[54] The maximum load was set to 6.0 nN. The comparison between the apparent height image obtained directly from the measurement and the reconstructed height image indicated the weaker contrast for real height image due to the larger compensation of deformation at rubbery region. The diameter of spherical structures observed in the real height image was about 100 nm. Because the diameter of CB itself was about 30 nm, the structure was attributed as so-called aggregate. Thus, chain structures of aggregates could be assigned to the secondary cohesive structure, agglomerate. By comparing the real height image with Young's modulus image, it was judged that CB region had higher Young's modulus. Figure 21.16 was taken at the same location with lower maximum load, 1.2 nN. In order to show the reliability of our reconstruction technique, line profiles along solid lines in Figures 21.15 and 21.16 are shown in Figure 21.17. Figure 21.17a was for apparent height images. It was seen that the shape of CB region (*right side*) was almost comparable to each other, while rubbery region (*left side*) became deeper for the profile taken from Figure 21.15a. This was undoubtedly due to the larger maximum load (6.0 nN). On the other hand, line profiles for real height images in Figure 21.17b had a good agreement with each other even at the rubbery region because the images (Figures 21.15b and 21.16b) contained the compensation of sample deformation for given loads. Thus, it could be concluded that the reconstruction procedure was valid enough to claim that the real height images represented the real topographic feature free from sample deformation.

Next, the details of Young's modulus mapping were investigated as shown in Figure 21.15c. The distribution of Young's modulus was divided into three regions: three representative points are indicated by open circles and the corresponding force–deformation curves are shown in Figure 21.18 ((a) for the upper open circle, (b) for the middle one, and (c) for the lower one).

TABLE 21.2
Compound Recipe of Carbon Black (CB)-Reinforced Natural Rubber (NR) Vulcanizate

Code	NR	CB	ZnO	Stearic Acid	Sulfur	Vulcanization Accelerator	Antioxidant
NR10	100	10	5	2	0.5	0.5	0.5

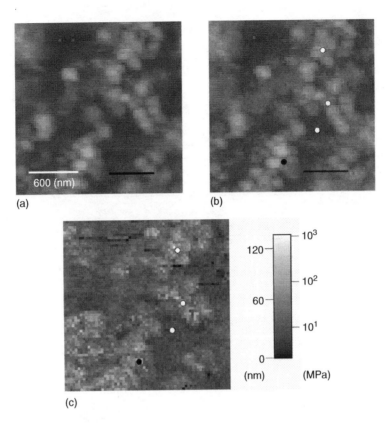

(a)   (b)

(c)

FIGURE 21.15 The results of force–volume measurement on NR10 sample. (a) Apparent height image, (b) reconstructed real height image, and (c) Young's modulus distribution image. The maximum load was set to 6.0 nN.

The curve fitting against Hertzian contact are also superimposed on each curve using the Equation 21.15. Here, the assumption was put that $\nu = 0.5$ for simplicity. Figure 21.18c (lower circle) had Young's modulus of 7.4 ± 0.1 MPa, typical for Young's modulus of NR bulk. Therefore, the region was attributed as rubbery region. The fitting error was so small that the accuracy was sufficiently guaranteed. Figure 21.18a gave Young's modulus of 1.01 ± 0.03 GPa. Although the fitting accuracy was not so low, the values vary widely from place to place, from several 100 MPa to several GPa. The region was at first attributed as stiff CB region. However, it was difficult to presume that CB

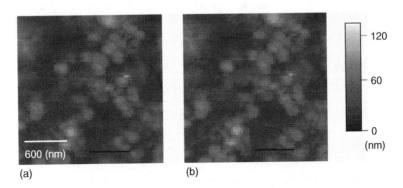

(a)   (b)

FIGURE 21.16 The results of force–volume measurement of the same location with Figure 21.15, (a) apparent height image and (b) the corresponding real height image. The maximum load was changed to 1.2 nN.

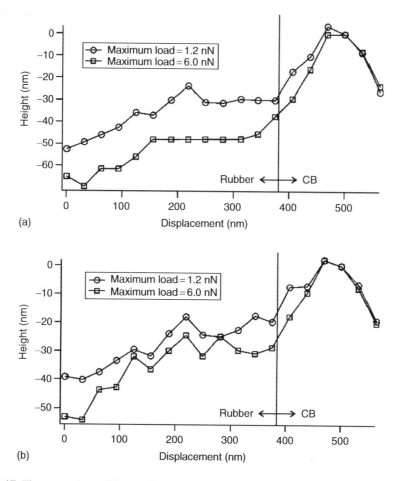

FIGURE 21.17 The comparison of line profiles along solid lines shown in each image of Figures 21.16 and 21.17. (a) Apparent and (b) the corresponding real height profiles.

particles themselves were deformed. A further conjecture must be given: the first simple, but difficult to give an actual proof, idea was that the rubber surrounding CB particles was deformed. In this case, we do not know real mechanical properties of stiff materials floating on a soft material. However, there was still some doubt whether any deformation could be detected by pushing a single CB aggregate because it was a constituent of larger agglomerate. Another possibility was that there existed a harder layer around CB aggregate, bound rubber whose existence was fully recognized by pulsed NMR study.[52] For more precise discussion, however, further study is indispensable. Thus, let us simply call the region as CB region in this section.

In addition to rubbery and CB regions, another region was found as shown in Figure 21.18b. The Young's modulus of this region was 57.3 ± 0.8 MPa, stiffer than rubbery region but softer than CB region. It was impossible to explain this value simply from the consideration of constituents. Let us call this region the intermediate region. For the purpose of the detailed investigation, the force–deformation curve was examined at the interface between rubbery and intermediate regions in Figure 21.19. As easily judged by this figure, it was impossible to give a simple curve fitting because there was an inflection point at around the sample deformation of 15 nm. Thus, the fitting in front and behind the point was tried. Near the surface the portion of 0–15 nm, the Young's modulus of 4.1 ± 0.1 MPa was obtained, which was almost the same as that of rubbery region. On the other hand, the deeper portion over 15 nm gave 76.3 ± 0.1 MPa, almost

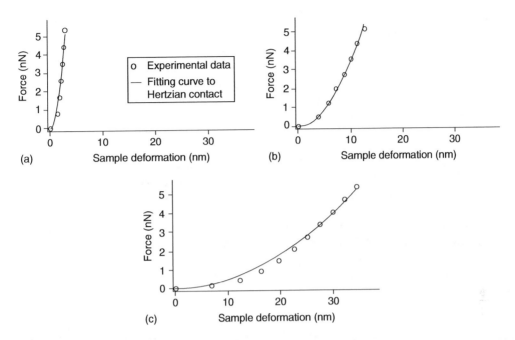

FIGURE 21.18 Force–deformation curves of local points indicated by open circles in Figure 21.15c. The curve fitting against Hertz model are superimposed on each curve. (a) Carbon black (CB) region (upper circle), 1.01 ± 0.03 GPa, (b) interfacial region (middle circle), 57.3 ± 0.8 MPa, and (c) rubber region (lower circle), 7.4 ± 0.1 MPa.

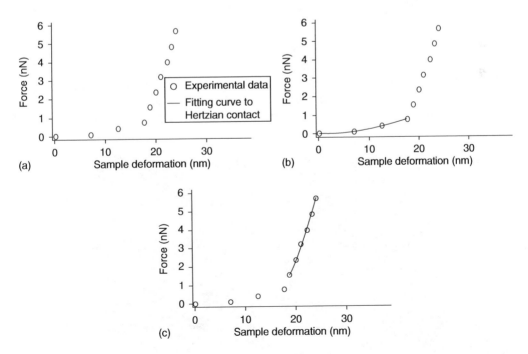

FIGURE 21.19 (a) The force–deformation curve at a point between interfacial and carbon black (CB) regions as indicated by a filled circle in Figure 21.15c. The Young's modulus calculation is conducted by dividing a curve into two parts, (b) the curve fitting within the first part, 4.1 ± 0.1 MPa, and (c) the curve fitting within the second part, 76.3 ± 0.1 MPa.

equal to intermediate region. It was, therefore, concluded that there existed rubbery region near the surface and intermediate region beneath it at this point. It was also found that another interesting force–deformation curve at the interface existed between CB and intermediate regions (data not shown), where the curve was divided into two portions, which had Young's modulus of CB and intermediate regions. Furthermore, any force–deformation curve directly connecting CB and rubbery region was never observed. Judging from the fact and the existence of these specific force–distance curves with inflection points, it was concluded that CB regions were always surrounded by intermediate regions whose Young's modulus was higher than rubbery matrix region.

It is of great importance to select an appropriate cantilever for this type of experiment. AFM users usually measure so-called sensitivity, which is the calibration value between actual cantilever deflection and detector signal. It depends on a fine adjustment of detector system and, therefore, the calibration must be performed every time. If the system is not stable enough, this sensitivity fluctuates much. In this study, the sensitivity was 54.8 nm V^{-1}. However, there is usually around 2% inaccuracy in this value. This error can be one of the largely contributing errors in analyzing force–distance curves. In practice, 1 nm V^{-1} deviation in sensitively resulted in 5% and 12% errors for rubbery and intermediate regions. In other words, the accuracy of Young's modulus in this study has such quantitative level. However, an error value exceeded 70% for CB region. This is because the cantilever used in this study has a spring constant of 0.12 N m^{-1}, too small to extract information from hard CB region. Since many types of nanocomposite materials should have similar widely distributed Young's modulus, further study is necessary to extend the applicability of the method introduced in this chapter.

As mentioned in Section 21.4.1, the existence of intermediate phase with slightly stiffer modulus than that of rubber matrix was reported,[49] which was determined by finite element calculation. The author reported that there are two phases around CB—one is almost comparable with bound rubber, 2 nm-thick glassy phase (GH-phase), another is 10 nm-thick uncross-linked one (SH-phase). The intermediate regions in this study were usually observed around CB regions and the Young's modulus of this region was higher as well. Thus, there is a possibility that SH-phase was directly observed in real space for the first time.

Tapping-mode phase contrast imaging technique has been extensively used to investigate reinforced elastomers. However, no report has been made which shows the existence of SH-phase. In practice, we could only distinguish a small amount of portion as different regions in phase contrast image as shown in Figure 21.20. The amount was almost equal to the amount of CB region

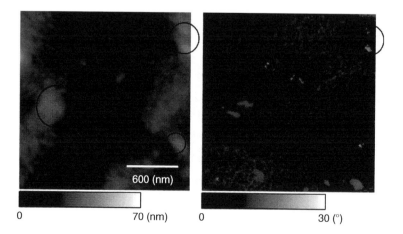

FIGURE 21.20 Tapping-mode atomic force microscopic (AFM) height image (*left*) and phase image (*right*) of NR10 sample. The structures indicated by circles are considered to be carbon black (CB) fillers.

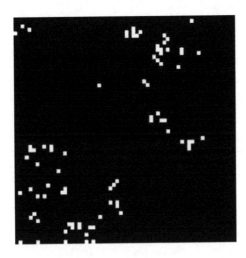

FIGURE 21.21 The binarized image of Young's modulus distribution in Figure 21.15c. The threshold value was 100 MPa.

in Figure 21.15c which was assured by the binarized data of Figure 21.15c at 100 MPa (Figure 21.21). Why it is not possible to differentiate intermediate region from rubbery region in tapping-mode? We speculate the reason as follows: mechanical properties of these regions are not so different. Then tapping with about 300 kHz never tells us about any subtle difference because the frequency is too high. We ought not to forget about the viscoelastic nature of rubbery materials. The speed of force–distance curve was corresponding to around 5 Hz excitation. As a result, nanomechanical mapping incorporated with FV technique is powerful to study nanocomposite materials, especially their interfacial structures.

21.5 SUMMARY

In this chapter, AFM *palpation* was introduced to verify the entropic elasticity of a single polymer chain and affine deformation hypothesis, both of which are the fundamental subject of rubber physics. The method was also applied to CB-reinforced NR which is one of the most important product from the industrial viewpoint. The current status of arts for the method is still unsophisticated. It would be rather said that we are now in the same stage as the ancients who acquired "fire." However, we believe that here is the clue for the conversion of rubber science from theory-guided science into experiment-guided science. AFM is not merely high-resolution microscopy, but a doctor in the twenty-first century who can palpate materials at nanometer scale.

REFERENCES

1. Saito, N., *Polymer Physics* (in Japanese), Syokabo, Tokyo, Japan (1967).
2. Kubo, R., *Rubber Elasticity/The reprint of the first edition* (in Japanese), Syokabo, Tokyo, Japan (1996).
3. Treloar, L.R.G., *The Physics of Rubber Elasticity*, Oxford University Press, Oxford, United Kingdom (1975).
4. Binnig, G., Quate, C.F., Gerber, C.H., Weibel, E., *Phys. Rev. Lett.*, **56**, 930 (1986).
5. de Gennes, P.G., *Scaling Concepts in Polymer Physics*, Cornell University Press, New York (1979).
6. Kumaki, J., Nishikawa, Y., Hashimoto, T., *J. Am. Chem. Soc.*, **118**, 3321 (1996).
7. Ando, T., Kodera, N., Naito, Y., Kinoshita, T., Furuta, K., Toyoshima, Y., *Chemphyschem*, **4**, 1196 (2003).
8. Mitsui, K., Hara, M., Ikai, A., *FEBS Lett.*, **385**, 29 (1996).
9. Rief, M., Oesterhelt, F., Fernandez, J.M., Gaub, H.E., *Science*, **276**, 1109 (1997).

10. Dietz, H., Rief, M., *PNAS*, **101**, 16192 (2004).

11. Wang, T., Sakai, Y., Nakajima, K., Miyawaki, A., Ito, K., Hara, M., *Colloid Surf.*, **B40**, 183 (2005).

12. Cortiz, C., Hadziioannou, G., *Macromolecules*, **32**, 780 (1999).

13. Li, H., Zhang, W., Xu, W., Zhang, X., *Macromolecules*, **33**, 465 (2000).

14. Haupt, B.J., Senden, T.J., Sevick, E.M., *Langmuir*, **18**, 2174 (2002).

15. Holland, N.B., Hugel, T., Neuert, G., Scholz, A.C., Renner, C., Oesterhelt, D., Moroder, L., Seitz, M., Gaub, H.E., *Macromolecules*, **36**, 2015 (2003).

16. Nakajima, K., Watabe, H., Nishi, T., *Polymer*, **47**, 2505 (2006).

17. Nakajima, K., Watabe, H., Nishi, T., *Kautsch. Gummi Kunstst.*, **5–06**, 256 (2006)

18. Vesenka, J., Guthold, M., Tang, C.L., Keller, D., Delaine, E., Bustamante, C., *Ultramicroscopy*, **42–44**, 1243 (1992).

19. Flory, P.J., *Principles of Polymer Chemistry*, Cornell University Press, New York (1953).

20. Wittkop, M., Kreitmeier, S., Goeritz, D., *Phys. Rev. E*, **53**, 838 (1996).

21. Nakajima, K., Watabe, H., Ohno, N., Nagayama, S., Watanabe, K., Nishi, T., *J. Soc. Rubber Ind., Jpn.* (in Japanese), **79**, 466 (2006).

22. Watabe, H., Nakajima, K., Sakai, Y., Nishi, T., *Macromolecules*, **39**, 5921 (2006).

23. Hutter, J.L., Bechhoefer, J., *Rev. Sci. Instrum.*, **64**, 1868 (1993).

24. Anczykowski, B., Gotsmann, B., Fuchs, H., Cleveland, J.P., Elings, V.B., *Appl. Surf. Sci.*, **140**, 376 (1999).

25. Kirkwood, J.G., Riseman, J., *J. Chem. Phys.*, **16**, 565 (1948).

26. Kramers, H.A., *J. Chem. Phys.*, **14**, 415 (1946).

27. Brandrup, J., Immergut, E.H., *Polymer Handbook*, 3rd ed., Wiley, New York (1989).

28. Lapra, A., Clement, F., Bokobza, L., Monnerie, L., *Rubber Chem. Technol.*, **76**, 60 (2003).

29. Murakami, S., Senno, K., Toki, S., Kohjiya, S., *Polymer*, **43**, 2117 (2002).

30. Rubinstein, M., Panyukov, S., *Macromolecules*, **35**, 6670 (2002).

31. Doi, M., Edwards, S.F., *The Theory of Polymer Dynamics*, Clarendon Press, Oxford, United Kingdom (1986).

32. Marrucci, G., Greco, F., Ianniruberto, G., *Macromol. Symp.*, **158**, 57 (2000).

33. Flory, P.J., Rehner, J., *J. Chem. Phys.*, **11**, 521 (1943).

34. Nukaga, H., Fujinami, S., Watabe, H., Nakajima, K., Nishi, T., *Jpn. J. Appl. Phys.*, **44**, 5425 (2005).

35. Nakajima, K., Fujinami, S., Nukaga, H., Watabe, H., Kitano, H., Ono, N., Endoh, K., Kaneko, M., Nishi, T., *Jpn. J. Polym. Sci. Technol.* (in Japanese), **62**, 476 (2005).

36. Nakajima, K., Yamaguchi, H., Lee, J.C., Kageshima, M., Ikehara, T., Nishi, T., *Jpn. J. Appl. Phys.*, **36**, 3850 (1997).

37. García, R., Pérez, R., *Surf. Sci. Rep.*, **47**, 197 (2002).

38. Landau, L., Lifchitz, E., *Theory of Elasticity*, Mir, Moscow, Russia (1967).

39. Johnson, K.L., Kendall, K., Roberts, A.D., *Proc. R. Soc. Lond.*, **A324**, 301 (1971).

40. Sun, Y., Akhremitchev, B., Walker, G.C., *Langmuir*, **20**, 5837 (2004).

41. Watabe, H., Komura, M., Nakajima, K., Nishi, T., *Jpn. J. Appl. Phys.*, **44**, 5393 (2005).

42. Nishi, T., Nakajima, K., *Polymeric Nano-Materials* (in Japanese), edited by Society of Polymer Science, Kyoritsu, Tokyo, Japan (2005).

43. Magonov, S.N., Whangbo, M.H., Surface Analysis with STM and AFM, VCH, Weinheim, Germany (1996).

44. Fukahori, Y., *J. Soc. Rubber Ind., Jpn.* (in Japanese), **76**, 460 (2003).

45. Fukahori, Y., *J. Soc. Rubber Ind., Jpn.* (in Japanese), **77**, 18 (2004).

46. Fukahori, Y., *J. Soc. Rubber Ind., Jpn.* (in Japanese), **77**, 103 (2004).

47. Fukahori, Y., *J. Soc. Rubber Ind., Jpn.* (in Japanese), **77**, 180 (2004).

48. Fukahori, Y., *J. Soc. Rubber Ind., Jpn.* (in Japanese), **77**, 317 (2004).

49. Fukahori, Y., *J. App. Polym. Sci.*, **95**, 60 (2005).

50. Fujiwara, S., Fujimoto, K., *Rubber Chem. Tech.*, **44**, 1273 (1971).

51. Kaufman, S., Slichter, W.P., Davis, D.O., *J. Polym. Sci.*, **A2**, 8, 829 (1971).

52. Nishi, T., *J. Polym. Sci., Polym. Phys. Ed.*, **12**, 685 (1974).

53. O'Brien, J., Cashell, E., Wardell, G.E., McBrierty, V.J., *Macromolecules*, **9**, 653 (1976).

54. Nukaga, H., Fujinami, S., Watabe, H., Nakajima, K., Nishi, T., *J. Soc. Rubber Ind., Jpn.* (in Japanese), 79, 509 (2006).

Section V

Physics and Engineering

22 Reinforced Elastomers: From Molecular Physics to Industrial Applications

Gert Heinrich, Manfred Klüppel, and Thomas A. Vilgis

CONTENTS

22.1 INTRODUCTION

Rubber elasticity has a long-standing history. Ancient Mesoamerican people were processing rubber by 1600 BC [1], which predated development of the vulcanization process by 3500 years. They made solid rubber balls, solid and hollow rubber human figurines, wide rubber bands to haft stone ax heads to wooden handles, and other items.

The use of fillers—especially, carbon black—together with accelerated sulfur vulcanization, has remained the fundamental technique for achieving the incredible range of mechanical properties required for a great variety of modern rubber products. Increased reinforcement of the rubber material has been defined as increased stiffness, modulus, rupture energy, tear strength, tensile strength, cracking resistance, fatigue resistance, and abrasion resistance [2]. Accordingly, a practical definition of reinforcement is the improvement in the service life of rubber articles that fail in a variety of ways, one of the most important being rupture failure accelerated by fatigue processes, such as occurs during the wear of a tire tread.

Before dealing with reinforcement of elastomers we have to introduce the basic molecular features of rubber elasticity. Then, we introduce—step-by-step—additional components into the model which consider the influence of reinforcing disordered solid fillers like carbon black or silica within a rubbery matrix. At this point, we will pay special attention to the incorporation of several additional kinds of complex interactions which then come into play: polymer–filler and filler–filler interactions. We demonstrate how a model of reinforced elastomers in its present state allows a thorough description of the large-strain materials behavior of reinforced rubbers in several fields of technical applications. In this way we present a thoroughgoing line from molecular mechanisms to industrial applications of reinforced elastomers.

22.2 STATISTICAL MECHANICS OF UNFILLED AND FILLED POLYMER NETWORKS

The classical theories of rubber elasticity are based on the hypothetical phantom-like chain which may pass freely through its neighbors and itself. In a real chain, the volume of a segment is excluded to other segments belonging either to the same chain or to others in the network. Consequently, the uncrossability of chain contours by those occupying the same volume becomes an important factor. Departures from phantom-like behavior due to entanglements are introduced in several recent models of rubber elasticity. Among these, the constrained junction (CJ) model [3,4] and the slip-link (SL) model [5,6] or tube model [7] have shown remarkable success in describing the behavior of networks in the presence of entanglements [7–11].

As Goodyear discovered, when he first vulcanized rubber in 1839, a viscous liquid of macro-molecules becomes solid when a sufficient number of permanent cross-links are introduced. In common with three-dimensional simple atomic solids, this equilibrium rigidity is a consequence of the spontaneous breakdown of translational symmetry, i.e., monomers no longer explore the space and, instead, become localized and fluctuate around preferred positions. In contrast, the formation of a crystalline macromolecular solid is frustrated by the random locations of the permanent cross-links, together with the impenetrability of the chains. As a result, rigidity is acquired through the formation of one of many equilibrium amorphous solid states. In the amorphous solid state of rubber, the monomers are localized at random spatial positions about which they exhibit Gaussian fluctuations characterized by a single-length scale, say ξ, that is finite, whereas in the liquid state ξ is infinite. During the vulcanization, the system undergoes a continuous transition from the liquid state, in which $\xi^{-1} = 0$, to the solid state, in which ξ^{-1} grows continuously from zero, as the mean number of cross-links M_c is increased beyond a critical value M_{crit}. For $0 \leq (M_c - M_{crit})/M_{crit} \ll 1$ one finds $\xi^{-1} \sim (M_c/M_{crit} - 1)^{1/2}$ [12].

It is beyond our control how the cross-links are spaced along the polymer chains during the vulcanization process. This extraordinary important fact demands a generalization of the Gibbs formula in statistical mechanics for amorphous materials that have fixed constraints of which the exact topology is unknown. Details of a modified Gibbs formula of polymer networks can be found in the pioneering paper of Deam and Edwards [13].

As known, the free energy of an uncross-linked polymer melt can be calculated by the usual Gibbs formula,

$$F = -k_B T \cdot \log Z \tag{22.1}$$

where the partition function

$$Z = \int d\{\mathbf{R}\} \exp\left(-\frac{H(\{\mathbf{R}\})}{k_B T}\right) \tag{22.2}$$

and $\{\mathbf{R}\}$ being abstract variables characterizing the Hamiltonian H. This is a typical "annealed" thermodynamic problem. However, in the "quenched" problem of a permanently cross-linked network we have two types of variables, $\{\mathbf{R}\}$ and $\{\mathbf{r}\}$, where $\{\mathbf{R}\}$ are free degrees of freedom and $\{\mathbf{r}\}$ are the quenched ones (cross-links). In this case, Equation 22.1 is no longer applicable, and a new thermodynamic formulation has to be employed. The free energy of a sample created with a certain set of frozen variables $\{\mathbf{r}\}$ is

$$\widetilde{F}\{\mathbf{r}\} = -k_B T \cdot \log Z(\{\mathbf{r}\}) = -k_B T \cdot \log \int d\{\mathbf{R}\} \exp\left(-\frac{H(\{\mathbf{R}\},\{\mathbf{r}\})}{k_B T}\right) \tag{22.3}$$

The macroscopic observable free energy is the average overall possibilities of arranging $\{\mathbf{r}\}$; i.e.,

$$F = \int dp(\{\mathbf{r}\})\widetilde{F}\{\mathbf{r}\} = -k_B T < \log Z(\{\mathbf{r}\}) > \tag{22.4}$$

where the probability distribution is

$$p\{\mathbf{r}\} = No^{-1} \int d\{\mathbf{R}\} \exp\left(-\frac{H(\{\mathbf{R}\},\{\mathbf{r}\})}{k_B T}\right) \tag{22.5}$$

with *No* being a normalization such that

$$\int p(\{\mathbf{r}\})d\{\mathbf{r}\} = 1 \tag{22.6}$$

Equation 22.4 suggests that the averages of Z have to be done after deforming the rubber, since the answer of the system on a deformation state, say $\boldsymbol{\lambda}$, depends on the configuration $\{\mathbf{r}\}$. This is now a non-Gibbsian way of performing the average. Mathematically, the average of the logarithm is difficult to perform and a trick can be used [14]:

$$< \log Z > = \lim_{n \to 0} \frac{d}{dn} < Z^n > \tag{22.7}$$

Obviously it is much easier to perform averaging of a power instead of the logarithm. Replicating the system (Hamiltonian H) n-times allows rewriting $Z^n(\{\mathbf{r}\})$ in terms of multiple integrals (replica method):

$$Z^n = \int \prod_{\alpha=1}^{n} d\{\mathbf{R}^{(\alpha)}\} \exp\left(-\frac{1}{k_B T} \sum_{\alpha=1}^{n} H^{(\alpha)}(\{\mathbf{R}^{(\alpha)}\},\{\mathbf{r}\})\right) \tag{22.8}$$

In the deformed state, the variables in the Hamiltonian change from $(\{\mathbf{R}^{(\alpha)}\}, \{\mathbf{r}\})$ to $(\{\mathbf{R}^{(\alpha)}\}, \{\boldsymbol{\lambda}\mathbf{r}\})$. However, the distribution $p(\{\mathbf{r}\})$ of finding the topology $\{\mathbf{r}\}$ depends solely on how the material is made instantaneously at thermal equilibrium (i.e., at constant temperature T, pressure p, etc.); i.e., $p(\{\mathbf{r}\})$ does not depend on the external deformation tensor $\boldsymbol{\lambda}$. Then, the final answer for the free energy of the deformed network is

$$F = -k_B T \lim_{n \to 0} \frac{d}{dn} \log Z(n) \tag{22.9}$$

$$Z(n) = \int d\{\mathbf{R}^{(0)}\} \prod_{\alpha=1}^{n} \int_{(\lambda)} d\{\mathbf{R}^{(\alpha)}\} \exp\left(-\frac{1}{k_B T} \sum_{\alpha=0}^{n} H^{(\alpha)}\right) \tag{22.10}$$

where $\alpha = 0$, is attributed to the undeformed system, and $\alpha = 1, \ldots, n$ to the replicated deformed networks. The λ assigned integral $\int_{(\lambda)}$ denotes the integration under the new conditions of strain.

$\alpha=0 \qquad \alpha=1 \qquad \ldots \qquad \alpha=n$

So far, we have not introduced a specific model of the polymer network chains. This problem can be rigorously solved for cross-linked polymer networks consisting of phantom chains [13], or even in the more general case of filled networks where the chains interact, additionally, with spherical hard filler particles [15].

In the case of filled rubbers, the network is represented by a huge chain of contour length $L = NL'$, where N is the number of primary chains of length L'. The observable (macroscopic) free energy is given by Equation 22.9, where $Z(n)$ is the replicated partition function that is estimated by functional integration ($\delta \mathbf{R}^{(\alpha)}(s)$) over the continuous chain conformations:

$$
Z(n) = \int\int \cdots \int D_w\left[\mathbf{R}^{(\alpha)}(s)\right] \left\{ \left[\int_0^L ds_1 \int_0^L ds_2 \prod_{\alpha=0}^n \delta\left(\mathbf{R}^{(\alpha)}(s_1) - \mathbf{R}^{(\alpha)}(s_2)\right) \right]^{M_c} \right.
$$
$$
\left. \times \left[\int_0^L ds_1 \int_0^L ds_2 \cdots \int_0^L ds_{f_F/2} \prod_{\alpha=0}^n \int_{V^{(\alpha)}} d\mathbf{R}_c^{(\alpha)} \prod_{i=1}^{f_F/2} g_i^{(\alpha)}\left(\mathbf{R}_i^{(\alpha)}(s_i) - \mathbf{R}_c^{(\alpha)}\right) \right]^{M_f} \right\} \quad (22.11)
$$

Here, α is the replica index ($0 \leq \alpha \leq n$), $V^{(\alpha \neq 0)}$ is the deformed sample volume, and $V^{(\alpha=0)}$ is the volume of the sample in the undeformed reference state. The partition function given in Equation 22.11 accounts for the fact that the probability distribution of the network is given by the constraint averaged over the Gaussian chain configurations. The corresponding probability is given by the Wiener measure (see, e.g., 16,17), which, in $3(n+1)$ dimensions in replica supervector space, takes the form

$$
D_w\left[\mathbf{R}^{(\alpha)}(s)\right] = \exp\left\{ -\frac{3}{2b} \sum_{\alpha=0}^n \int_0^L ds \left(\frac{\partial \mathbf{R}^\alpha(s)}{\partial s}\right)^2 \right\} \prod_{\alpha=0}^n \delta \mathbf{R}^{(\alpha)}(s) \quad (22.12)
$$

Here, b denotes the Kuhn's statistical segment length. The network is represented by a huge chain internally cross-linked at M_c cross-linking points where it touches and at the surfaces of M_f filler particles. The point-like local cross-link constraints are easy to handle and can be represented by the term

$$
\left[\int_0^L ds_1 \int_0^L ds_2 \, \delta\left(\mathbf{R}(s_1) - \mathbf{R}(s_2)\right) \right]^{M_c} \quad (22.13)
$$

corresponding to the diagram

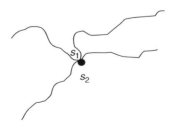

The δ-function makes sure that if two segments s_1 and s_2 meet on the huge network chain they can form a permanent constraint $\mathbf{R}(s_1) = \mathbf{R}(s_2)$. Hence, this process will produce a network junction of functionality $f_N = 4$, usually realized as sulfur bridges in technical elastomers like, for example, tire treads.

The constraints due to M_f filler particles are somewhat more complicated to mimic analytically. A crude model is to assume solid spheres of an average functionality $f_F \gg f_N$ $(=4)$ [15]. If a filler fixes segment s_i to segment s_j, another couple between the vector positions $\mathbf{R}(s_i)$, $\mathbf{R}(s_j)$ and the center of the particle, $\mathbf{R}_c$, has been produced. Graphically the following diagrams can represent it:

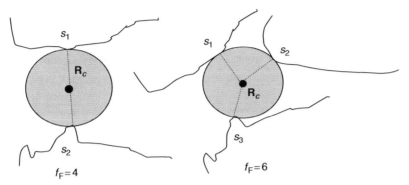

etc.

These "filler constraints" can be represented by a Gaussian potential

$$g_i(\mathbf{R}_i(s_i) - \mathbf{R}_c) = \prod_{\mu=x,y,z} \left(\frac{2}{\pi \varepsilon_\mu^2}\right)^2 \exp\left\{-\frac{2}{\pi \varepsilon_\mu^2} \left(R_{i\mu}(s_i) - R_{c\mu}\right)^2\right\} \tag{22.14}$$

where we have assumed that the average Cartesian component of the distance between particle center $\mathbf{R}_c$ and vector position $\mathbf{R}_i(s_i)$ is given by $\varepsilon_\mu/2$. The quantity ε_μ is a typical linear extension of a filler particle.

Equations 22.3–22.14 represent the simplest formulation of filled phantom polymer networks. Clearly, specific features of the fractal filler structures of carbon black, etc., are totally neglected. However, the model uses chain variables $\mathbf{R}(s)$ directly. It assumes the chains are Gaussian; the cross-links and filler particles are placed in position randomly and instantaneously and are thereafter permanent. Additionally, constraints arising from entanglements and packing effects can be introduced using the mean field approach of harmonic tube constraints [15].

To handle the constraint term mathematically, it is usually expressed in terms of a fugacity or a chemical potential via the identity

$$X^N = \frac{1}{2\pi i} \oint_C \frac{d\mu N!}{\mu^{N+1}} e^{\mu X} \tag{22.15}$$

where the contour C encloses the origin of the complex plane. The generalized partition function then takes the form

$$Z(n) = \int \prod_{\alpha=0}^n \delta \mathbf{R}^{(\alpha)}(s) \frac{1}{2\pi i} \oint_C \frac{d\mu_c M_c!}{\mu_c^{M_c}} \frac{1}{2\pi i} \oint_C \frac{d\mu_f M_f!}{\mu_f^{M_f}} \exp\left\{-\frac{3}{2b} \sum_{\alpha=0}^n \int_0^L ds \left(\frac{\partial \mathbf{R}^\alpha(s)}{\partial s}\right)^2 + Q^*\right\}$$

$$\tag{22.16}$$

$$Q^* = \mu_c \int_0^L ds_1 \int_0^L ds_2 \prod_{\alpha=0}^n \delta\left(\mathbf{R}^{(\alpha)}(s_1) - \mathbf{R}^{(\alpha)}(s_2)\right) + \mu_f \int_0^L ds_1 \cdots \int_0^L ds_k \prod_{\alpha=0}^n \int_{V^{(\alpha)}}$$

$$\times d\mathbf{R}_c^{(\varepsilon)} \prod_{i=1}^k g_i^{(\alpha)}\left(\mathbf{R}_i^{(\alpha)}(s_i) - \mathbf{R}_c^{(\alpha)}\right) \qquad (22.17)$$

where $k = f_F/2$.

The evaluation of $Z(n)$ for filled rubbers has been done [15] along the line of Edwards and Deam [13,18] for unfilled networks, who used a Feynman variational principle [19]. The rigorous derivation of the elastic free energy for the filled rubber problem leads to the following expression [15]:

$$F = \frac{1}{2} k_B T \left(M_c + g_f M_f\right) \sum_\mu \lambda_\mu^2 + \text{terms independent of deformation} \qquad (22.18)$$

where $g_f \equiv k - 1$.

Equation 22.18 is nothing but the neo-Hookean free energy of a phantom network consisting of $(M_c + g_f M_f) \approx (M_c + k M_f)$ (for k large) cross-links. The corresponding elastic modulus $G_C = k_B T(M_c + g_f M_f)$ considers the constraints due to interchain junctions. The deformation-independent terms in Equation 22.18 contain two localization parameters of the polymer–polymer junctions and polymer–filler couples, respectively. Both quantities have been introduced as the strength of harmonic trial potentials, which simulate the conformational restrictions due to the cross-links. The inverse of such a localization parameter defines the mean distance in which the cross-links are localized; i.e., it is proportional to the root mean square of the radius of gyration of a network chain between two neighbored cross-links.

The presented scheme offers several extensions. For example, the model gives a clear route for an additional inclusion of entanglement constraints and packing effects [15]. Again, this can be realized with the successful mean field models based on the conformational tube picture [7,9] where the chains do not have free access to the total space between the cross-links but are trapped in a cage due to the additional topological restrictions, as visualized in the cartoon.

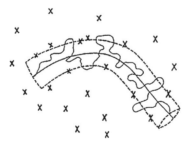

This leads to an additional deformation dependent term in Equation 22.18, which arises from the deformation dependence of the local conformational tube diameter due to the presence of entanglements. Rigorous calculations [20] or simple scaling arguments based on an affine-coupling hypothesis of network strands between successive entanglements [21] yield a nonaffine deformation law of the lateral tube dimension. In addition, further extension of this approach has been realized by consideration of limited chain extensibility of the network chains in the filled rubber matrix [22,23]. Then, the free energy density of the nonaffine tube model of rubber elasticity reads:

$$F_R(\lambda_\mu) = \frac{G_c}{2}\left\{\frac{\left(\sum_{\mu=1}^3 \lambda_\mu^2 - 3\right)\left(1 - \frac{T_e}{n_e}\right)}{1 - \frac{T_e}{n_e}\left(\sum_{\mu=1}^3 \lambda_\mu^2 - 3\right)} + \ln\left[1 - \frac{T_e}{n_e}\left(\sum_{\mu=1}^3 \lambda_\mu^2 - 3\right)\right]\right\}$$
$$+ 2G_e\left(\sum_{\mu=1}^3 \lambda_\mu^{-1} - 3\right) \tag{22.19}$$

Here, n_e is the number of statistical chain segments between two successive entanglements and T_e is the trapping factor $(0 < T_e < 1)$, characterizing the portion of elastically active entanglements. Accordingly, the finite extensibility is considered to be determined mainly by the physical network of trapped entanglements that depends on the amount of cross-linking via the trapping factor T_e [30]. The first bracket term of Equation 22.19 considers the constraints due to interchain junctions, with an elastic modulus G_c proportional to the number of network junctions per unit volume. The second addend is the result of tube constraints, whereby G_e is proportional to the entanglement density of the rubber. The parenthetical expression in the first addend considers the finite chain extensibility of polymer networks by referring to a proposal of Edwards and Vilgis [9]. For the limiting case $n_e/T_e = \Sigma\lambda_\mu^2 - 3$, a singularity is obtained for the free energy F_R, indicating the maximum strain of the network, which is reached when the chains between successive trapped entanglements are fully stretched out. In the limit $n_e \to \infty$ the original Gaussian formulation of the nonaffine tube model is recovered.

It is important to note here that the presence of rigid filler clusters, with bonds in the virgin, unbroken state of the sample, gives rise to hydrodynamic reinforcement of the rubber matrix. This must be specified by the strain amplification factor X, which relates the external strain ε_μ of the sample to the internal strain ratio λ_μ of the rubber matrix:

$$\lambda_\mu = 1 + X\varepsilon_\mu \ (\mu = 1, 2, 3) \tag{22.20}$$

The precise nature of the strain amplification factor X will be introduced in Section 22.4 about the nature of rigid filler clusters within a soft rubbery matrix. Additionally, we have to take into account the presence of filler-induced hysteresis which will be described by an anisotropic-free energy density, considering the cyclic breakdown and reaggregation of the residual fraction of more fragile filler clusters with already broken filler–filler bonds. Accordingly, the total free energy density of filler-reinforced rubber will consist of two contributions, Equation 22.19 (together with Equation 22.20) plus another contribution, say $F_A(\varepsilon_\mu)$, which considers the energy stored in the substantially strained fragile, but soft, filler clusters:

$$F(\varepsilon_\mu) = (1 - \Phi_{\text{eff}})F_R(\varepsilon_\mu) + \Phi_{\text{eff}}F_A(\varepsilon_\mu) \tag{22.21}$$

where Φ_{eff} is the effective filler-volume fraction. The expression for $F_A(\varepsilon_\mu)$ will be discussed in the following chapters.

22.3   FILLER NETWORKING IN ELASTOMER COMPOSITES

Filler networking in elastomer composites can be analyzed by applying transmission electron microscopy (TEM)-, flocculation-, and dielectric investigations. This provides information on the fractal nature of filler networks as well as the morphology of filler–filler bonds. From Figure 22.1 it becomes obvious that TEM analysis gives a limited microscopic picture of the filler-network morphology. This is mainly due to the spatial interpenetration of neighboring flocculated filler clusters.

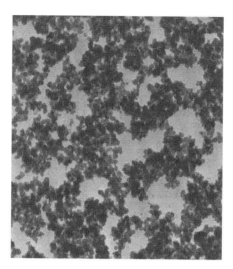

FIGURE 22.1 Transmission electron microscopy (TEM) micrograph of a carbon black network obtained from an ultrathin cut of a filled rubber sample.

Flocculation studies, considering the small-strain mechanical response of the uncross-linked composites during heat treatment (annealing), demonstrate that a relative movement of the particles takes place that depends on particle size, molar mass of the polymer, as well as polymer–filler and filler–filler interactions (Figure 22.2). This provides strong experimental evidence for a kinetic cluster–cluster aggregation (CCA) mechanism of filler particles in the rubber matrix to form a filler network [24].

The ac conductivity in the high-frequency regime is related to an anomalous diffusion mechanism of charge carriers on fractal carbon black clusters, implying a power law behavior of the conductivity with frequency. This scaling behavior of the conductivity is observed in many carbon-black-filled elastomer systems. It confirms the fractal nature of filler networks in elastomers below a certain length scale, though it gives no definite information on the particular network structure [24–26].

From the dielectric investigations it becomes obvious that charge transport above the percolation threshold is limited by a hopping or tunneling mechanism of charge carriers over small gaps of order 1 nm between adjacent carbon black particles (compare Figure 22.3). From this finding and

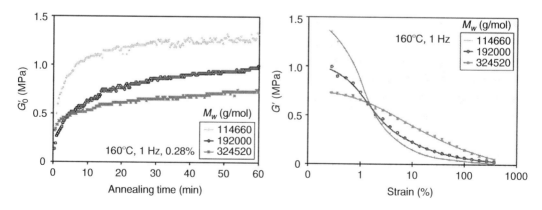

FIGURE 22.2 Flocculation behavior of the small-strain modulus at 160°C of uncross-linked solution-based styrene–butadiene rubber (S-SBR) composites of various molar mass with 50 phr N234, as indicated (*left*) and strain dependence of the annealed samples after 60 min (*right*). (From Klüppel, M. and Heinrich, G., *Kautschuk, Gummi, Kunststoffe*, 58, 217, 2005. With permission.)

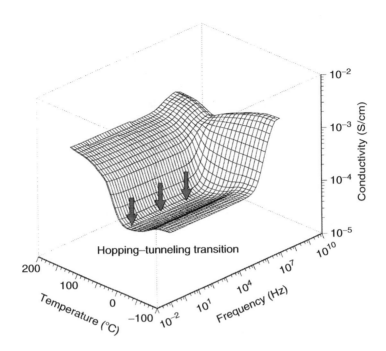

FIGURE 22.3 AC conductivity of natural rubber (NR) samples with 80 phr N339. At low temperature, conduction is due to tunneling over nanoscopic gaps between adjacent carbon black particles. At higher temperatures $(T > 20°C)$, a transition of the conductivity mechanism from tunneling to thermally activated hopping of charge carriers is observed. (From Klüppel, M. and Heinrich, G., *Kautschuk, Gummi, Kunststoffe*, 58, 217, 2005. With permission.)

the observed dependency of the flocculation dynamics on the molar mass or the amount of bound rubber (Figure 22.2), a model of filler–filler bonds can be developed that is schematically depicted in Figure 22.4.

The model of filler–filler bonds introduced above relates the mechanical stiffness of filler–filler bonds to the remaining gap size between the filler particles that develop during annealing (and cross-linking) of filled rubbers. In this model, stress between adjacent filler particles in a filler cluster is assumed to be transmitted by nanoscopic, flexible bridges of glassy polymer, implying that a high flexibility and strength of filler clusters in elastomers is reached. This picture of filler–filler bonds allows for a qualitative explanation of the observed flocculation effects by referring to the amount of bound rubber and its impact on the stiffness and strength of filler–filler bonds [24].

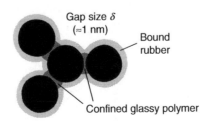

FIGURE 22.4 Schematic view considering the structure of filler–filler bonds in a bulk rubber matrix. The gap size of neighboring filler particles with confined glassy polymer and the bound rubber layer, governing the stiffness and strength of filler–filler bonds, is indicated. (In general, the black disks represent primary carbon-black aggregates.) (From Klüppel, M. and Heinrich, G., *Kautschuk, Gummi, Kunststoffe*, 58, 217, 2005. With permission.)

22.4 REINFORCEMENT OF ELASTOMERS BY FRACTAL FILLER STRUCTURES

The flocculation results presented in Figure 22.2 give strong evidence that kinetically aggregated filler clusters or networks are formed in elastomer composites, as shown in Figure 22.5.

Based on the investigations of filler-network morphology, a micromechanical model of rubber reinforcement by flexible filler clusters has been developed [24–26]. This model of rubber reinforcement refers to the kinetic CCA approach of filler networking in elastomers, which represents a reasonable theoretical basis for understanding the linear viscoelastic properties of reinforced rubbers. According to the CCA model, filler networks consist of a space-filling configuration of CCA clusters with characteristic mass fractal dimension $d_f \approx 1.8$. A schematic view of this structure is shown on the right-hand side of Figure 22.5 ($\Phi > \Phi^*$). The mechanical response of filler networks at small strain depends purely on the fractal connectivity of the CCA clusters. It can be evaluated by referring to the Kantor–Webman model of flexible chain aggregates [27]. For the small-strain modulus a power–law behavior with filler concentration is predicted. The evaluated exponent 3.5 is in good agreement with the experimental data of Payne [28] for carbon-black-filled butyl rubber. The universal scaling behavior of the small-strain modulus is demonstrated in Figures 22.6 and 22.7.

The consideration of flexible chains of filler particles, approximating the elastically effective backbone of the filler clusters, allows for a micromechanical description of the elastic properties of tender CCA clusters in elastomers. The main contribution of the elastically stored energy in the strained filler clusters results from the bending–twisting deformation of filler–filler bonds, considered by an elastic constant $\bar{G}$. The predicted power–law behavior of the small-strain modulus of filler-reinforced rubbers is confirmed by a variety of experimental data shown in Figure 22.7, including carbon black and silica-filled rubbers as well as composites with microgels [25,26,29]. Further evidence for the impact of filler networking on the mechanical properties is found at medium and large strain, where pronounced differences in filler–filler interaction, as obtained, for example, by graphitization of carbon blacks, result in a largely different stress–strain behavior. Accordingly, the observed crossover of the storage modulus and the steeper stress–strain curves of solution-based styrene–butadiene rubber (S-SBR) composites with the original highly active black N220 as compared to the deactivated black N220g can be related to characteristic differences in the stiffness and strength of filler–filler bonds [24]. This is depicted in Figure 22.8.

For a consideration of filler-network breakdown at increasing strain, the failure properties of filler–filler bonds and filler clusters have to be evaluated in dependence of cluster size. This allows for a micromechanical description of tender but fragile filler clusters in the stress field of a strained rubber matrix. A schematic view of the mechanical equivalence between a CCA–filler cluster and a series of soft and hard springs is presented in Figure 22.9. The two springs with force constants k_s

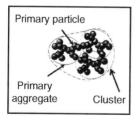
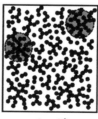

$\Phi < \Phi^*$ $\Phi > \Phi^*$

FIGURE 22.5 Schematic view of kinetically aggregated filler clusters in rubber below and above the gel point Φ^*. The left side characterizes the local structure of carbon black clusters, built by primary particles and primary aggregates. (Every black disk in the center figure [$\Phi < \Phi^*$] and on the right-hand side ($\Phi > \Phi^*$) represents a primary aggregate.) (From Klüppel, M. and Heinrich, G., *Kautschuk, Gummi, Kunststoffe*, 58, 217, 2005. With permission.)

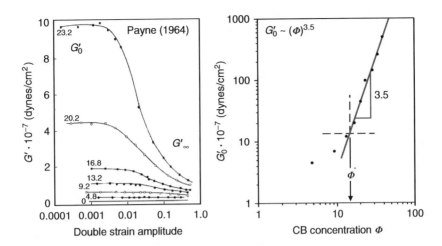

FIGURE 22.6 Payne effect of butyl composites with various amounts Φ of N330, as indicated (*left*) [28]. Scaling behavior of the small-strain modulus of the same composites (*right*). The obtained exponent 3.5 confirms the cluster–cluster aggregation model. (From Klüppel, M. and Heinrich, G., *Kautschuk, Gummi, Kunststoffe*, 58, 217, 2005. With permission.)

and k_b correspond to bending–twisting and tension deformations of the filler–filler bonds with elastic constants $\bar{G}$ and Q, respectively.

The dynamic flocculation model of stress softening and hysteresis assumes that the breakdown of filler clusters during the first deformation of the virgin samples is totally reversible, though the initial virgin state of filler–filler bonds is not recovered. This implies that, on the one side, the fraction of rigid filler clusters decreases with increasing prestrain, leading to the pronounced stress softening after the first deformation cycle. On the other side, the fraction of fragile filler clusters increases with increasing prestrain, which impacts the filler-induced hysteresis [24]. A schematic representation of the decomposition of filler clusters in rigid and fragile units for preconditioned samples is shown in Figure 22.10.

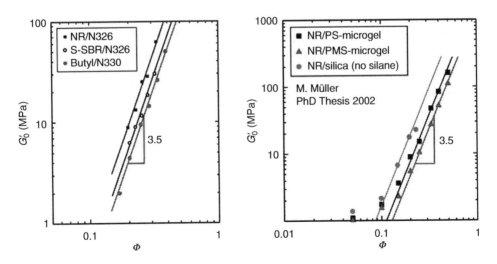

FIGURE 22.7 Scaling behavior of the small-strain modulus of carbon black composites (*left*) and microgel or silica composites (*right*). In all cases an exponent close to 3.5 is found, indicating the universal character of the cluster–cluster aggregation model. (From Klüppel, M. and Heinrich, G., *Kautschuk, Gummi, Kunststoffe*, 58, 217, 2005. With permission.)

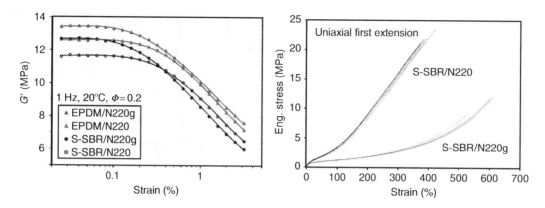

FIGURE 22.8 Payne effect (*left*) and stress–strain curves (*right*) of solution-based styrene–butadiene rubber (S-SBR) composites with fixed amounts of carbon black N220 and graphitized black N220g, as indicated. (From Klüppel, M. and Heinrich, G., *Kautschuk, Gummi, Kunststoffe*, 58, 217, 2005. With permission.)

By assuming a specific cluster size distribution $\phi(\xi_\mu)$ in reinforced rubbers, the constitutive material model of filler-reinforced rubbers can now be derived [24,30]. This model is based on the nonaffine tube model of rubber elasticity, including hydrodynamic amplification of the rubber matrix by a fraction of rigid filler clusters with filler–filler bonds in the unbroken, virgin state as reflected in Equations 22.18–22.21. The filler-induced hysteresis will then be described by an anisotropic-free energy density (Equation 22.21), considering the cyclic breakdown and reaggregation of the residual fraction of more fragile filler clusters with already broken filler–filler bonds.

In the case of a preconditioned sample and for strains smaller than the previous straining ($\varepsilon_\mu < \varepsilon_{\mu,\max}$), the strain amplification factor X in Equation 22.20 is independent of strain and determined by $\varepsilon_{\mu,\max}$ ($X = X(\varepsilon_{\mu,\max})$). For the first deformation of virgin samples it depends on the external strain ($X = X(\varepsilon_\mu)$). By applying a relation derived by Huber and Vilgis [31,32] for the strain amplification factor of overlapping fractal clusters, $X(\varepsilon_{\mu,\max})$ or $X(\varepsilon_\mu)$ can be evaluated by averaging over the size distribution of rigid clusters in all space directions. In the case of preconditioned samples this yields

$$X(\varepsilon_{\mu,\max}) = 1 + c\,\Phi_{\mathrm{eff}}^{\frac{2}{3-d_f}} \sum_{\mu=1}^{3} \frac{1}{d} \left\{ \int_0^{\xi_{\mu,\min}} \left(\frac{\xi_\mu'}{d}\right)^{d_w - d_f} \phi\left(\xi_\mu'\right) d\xi_\mu' + \int_{\xi_{\mu,\min}}^{\infty} \phi\left(\xi_\mu'\right) d\xi_\mu' \right\} \qquad (22.22)$$

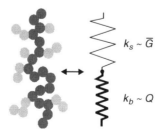

FIGURE 22.9 Schematic view demonstrating the mechanical equivalence between a filler cluster and a series of soft and stiff molecular springs, representing bending–twisting and tension deformation of filler–filler bonds, respectively. (From Klüppel, M. and Heinrich, G., *Kautschuk, Gummi, Kunststoffe*, 58, 217, 2005. With permission.)

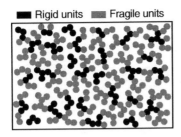 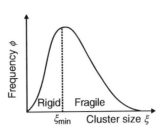

FIGURE 22.10 Schematic view of the decomposition of filler clusters in rigid and fragile units for preconditioned samples. The right side shows the cluster size distribution with the prestrain-dependent boundary size ξ_{min}. (From Klüppel, M. and Heinrich, G., *Kautschuk, Gummi, Kunststoffe*, 58, 217, 2005. With permission.)

where c is a constant of order one and $\Phi_{eff} = \Phi/\Phi_p$ is the effective filler-volume fraction, with Φ being the filler-volume fraction and Φ_p the solid fraction of possibly structured filler particles, e.g., carbon black or silica. ξ_μ is the cluster size, d is the particle size, $\phi(\xi_\mu)$ is the normalized size distribution, $d_f \approx 1.8$ is the mass fractal dimension, and $d_w \approx 3.1$ [26] is the anomalous diffusion exponent on fractal CCA clusters.

The second addend in Equation 22.21 considers the energy stored in the substantially strained fragile but soft filler clusters:

$$F_A(\varepsilon_\mu) = \sum_\mu^{\partial\varepsilon_\mu/\partial t > 0} \frac{1}{2d} \int_{\xi_{\mu,min}}^{\xi_\mu(\varepsilon_\mu)} G_A\left(\xi'_\mu\right) \varepsilon^2_{A,\mu}\left(\xi'_\mu,\varepsilon_\mu\right) \phi\left(\xi'_\mu\right) d\xi'_\mu \qquad (22.23)$$

Here, G_A is the elastic modulus and $\varepsilon_{A,\mu}$ is the strain of the fragile filler clusters in spatial direction $\mu(=x, y, z)$. The dependence of these quantities on cluster size ξ_μ and external strain ε_μ is specified in [24].

It is demonstrated in Figure 22.11 that the quasi-static stress–strain cycles at different prestrains of silica-filled rubbers can be well described in the scope of the above-mentioned dynamic flocculation model of stress softening and filler-induced hysteresis up to large strain. Thereby, the size distribution $\phi(\xi_\mu)$ has been chosen as an isotropic logarithmic normal distribution ($\phi(\xi_1) = \phi(\xi_2) = \phi(\xi_3)$):

$$\phi(x_1) = \frac{\exp\left(-\dfrac{\ln(x_1/<x_1>)^2}{2b^2}\right)}{\sqrt{\pi/2}bx_1} \qquad (22.24)$$

with mean cluster size $<x_1> \equiv \xi_o/d$ and distribution width b. A similar good agreement between experimental data and simulation as shown in Figure 22.11 can be obtained for carbon-black-filled elastomers and for biaxial tension data [24]. The obtained microscopic material parameter appears reasonable, providing information on the mean cluster size ξ_o and distribution width b, the tensile strength of filler–filler bonds $Q\varepsilon_b/d^3$, and the polymer network chain density ($\nu_c \sim G_c$).

The model describes the characteristic stress softening via the prestrain-dependent amplification factor X in Equation 22.22. It also considers the hysteresis behavior of reinforced rubbers, since the sum in Equation 22.23 has taken over the stretching directions with $\partial\varepsilon/\partial t > 0$, only, implying that up and down cycles are described differently. An example showing a fit of various hysteresis cycles of silica-filled ethylene–propylene–diene monomer (EPDM) rubber in the medium-strain regime up to 50% is depicted in Figure 22.12. It must be noted that the topological constraint modulus G_e has

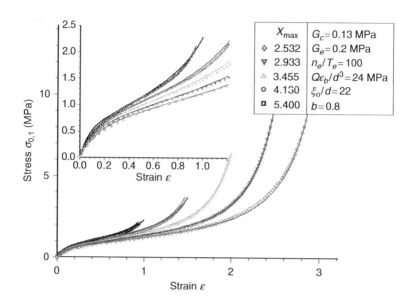

FIGURE 22.11 Uniaxial stress–strain data (up-cycles) of solution-based styrene–butadiene rubber (S-SBR) samples with 60 phr silica at different prestrains $\varepsilon_{max} = 100\%$, 150%, 200%, 250%, and 300% (symbols) and fittings (lines) with the stress-softening model Equations 22.19–22.24. The fitting parameters are indicated. The insert shows a magnification of the small-strain data. (From Klüppel, M. and Heinrich, G., *Kautschuk, Gummi, Kunststoffe*, 58, 217, 2005. With permission.)

been prefixed in the simulations shown in Figures 22.11 and 22.12 and was not used as a fitting parameter. The chosen values $G_e = 0.2$ MPa and $G_e = 0.6$ MPa correspond to the independently measured entanglement densities of the S-SBR matrix and EPDM rubber matrix, respectively.

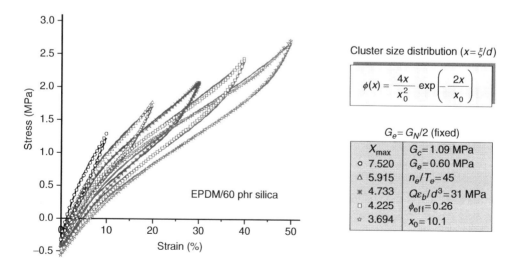

FIGURE 22.12 Uniaxial stress–strain cycles of ethylene–propylene–diene monomer (EPDM) samples with 60 phr silica at different prestrains $\varepsilon_{max} = 10\%$, 20%, 30%, 40%, and 50% (symbols) and fittings (lines) with the stress-softening model Equations 22.19–22.24. The fitting parameters are indicated. The assumed cluster-size distribution is also shown, which differs from the one in Equation 22.24. (From Klüppel, M. and Heinrich, G., *Kautschuk, Gummi, Kunststoffe*, 58, 217, 2005. With permission.)

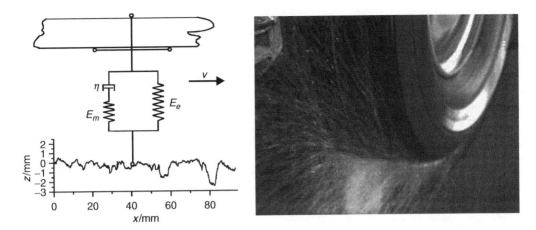

FIGURE 22.13 Simple analytical model of hysteresis friction of rubber moving over a rough road profile (*left*). Tire on a wet road track, where hysteresis energy losses dominate the traction behavior (*right*). (From Klüppel, M. and Heinrich, G., *Kautschuk, Gummi, Kunststoffe*, 58, 217, 2005. With permission.)

In particular it can be shown that the dynamic flocculation model of stress softening and hysteresis fulfils a "plausibility criterion," important, e.g., for finite element (FE) applications. Accordingly, any deformation mode can be predicted based solely on uniaxial stress–strain measurements, which can be carried out relatively easily. From the simulations of stress–strain cycles at medium and large strain it can be concluded that the model of cluster breakdown and reaggregation for prestrained samples represents a fundamental micromechanical basis for the description of nonlinear viscoelasticity of filler-reinforced rubbers. Thereby, the mechanisms of energy storage and dissipation are traced back to the elastic response of tender but fragile filler clusters [24].

22.5 INDUSTRIAL APPLICATIONS

The developed micromechanical model of reinforcement by active fillers allows for a better control of material properties and a more fundamental engineering praxis in rubber industry. In particular,

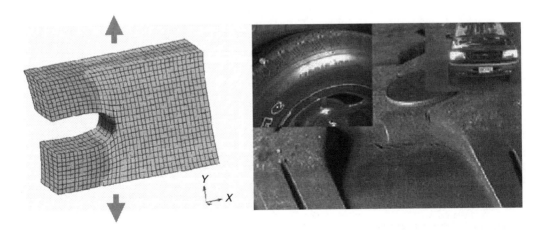

FIGURE 22.14 Finite element (FE) simulation of stress distribution in a strained rubber block, indicating high stress values in the cut area where crack propagation takes place (*left*). Photograph showing the result of undesired cuts in a tire tread (*right*). (From Klüppel, M. and Heinrich, G., *Kautschuk, Gummi, Kunststoffe*, 58, 217, 2005. With permission.)

FE simulations of the stress–strain properties of filler-reinforced elastomers are an important tool for predicting the service live performance of rubber goods. Typical examples are the evaluation of rolling resistance of tires due to hysteresis energy losses, mainly in the tire tread or the adjustment of engine mounts in automotive applications.

Another topic of interest in rubber industry is the wet traction behavior of tires, which is strongly affected by the hysteresis energy losses during slide over rough surfaces. A simple linear viscoelastic model of hysteresis friction is shown in Figure 22.13, where the friction force is estimated via an integration over different deformations on a broad frequency scale, as given by the road-roughness profile on various length scales [33]. A more fundamental model has to take into account the strongly nonlinear deformation cycles and stress-softening effects as shown, e.g., in Figure 22.11. The analytical description of this behavior can serve as an input for computer simulations of several processes involving elastomer friction and contact mechanics of soft matter with rough rigid surfaces.

Another important aspect of rubber engineering is the prediction of cut growth rates and service lifetime of elastomer goods. A worst case scenario is the appearance of macroscopic cuts in tires as shown on the right-hand side of Figure 22.14. The avoidance of such catastrophic damage pictures, resulting, for example, from large deformations on bad roads, requires the control of stresses in the tire under service conditions. FE simulation can serve as a tool for estimating the stress distribution in tires and the maximum stress values in critical areas, where crack propagation will appear. However, the prediction of cut propagation requires precise material laws, since pronounced stress softening takes place close to the cut tip and anisotropy effects play a certain role. The proposed micromechanical material law provides a physical basis for an improvement of FE simulations and advancement of its application in engineering praxis. However, a direct inclusion of this material law into fracture mechanical concepts of highly extensible elastomers is still extremely difficult. Furthermore, it is unclear whether for engineering applications the use of such a complicated material law is necessary, or not. This question is still under discussion.

REFERENCES

1. Hosler, D., Burkett, S.L., and Tarkanian, M.J., *Science*, **284**, 1988–1991, 1999.
2. Dannenberg, E.M., *Rubber Chem. Technol.*, **48**, 410–443, 1975.
3. Ronca, G. and Allegra, G.J., *J. Chem. Phys.*, **63**, 4990, 1975.
4. Flory, P.J. and Erman, B., *Macromolecules*, **15**, 800, 1982.
5. Ball, R.C., Doi, M., Edwards, S.F., and Warner, M., *Polymer*, **22**, 1010, 1981.
6. Edwards, S.F. and Vilgis, T.A., *Polymer*, **27**, 483, 1986.
7. Heinrich, G., Straube, E., and Helmis, G., *Adv. Polym. Sci.*, **85**, 33, 1988.
8. Erman, B. and Mark, J.E., *Structures and Properties of Rubberlike Networks*, Oxford University Press, New York and Oxford, 1997.
9. Edwards, S.F. and Vilgis, T.A., *Rep. Prog. Phys.*, **51**, 243, 1988.
10. Erman, B. and Mark, J.E., *Annu. Rev. Phys. Chem.*, **40**, 351, 1989.
11. Thirion, P. and Weil, T., *Polymer*, **25**, 609, 1983.
12. Goldbart, P.M. and Zippelius, A., *Phys. Rev. Lett.*, **71**, 2256, 1993.
13. Deam, R.T. and Edwards, S.F., *Phil. Trans. R. Soc.*, **11**, 317, 1976.
14. Edwards, S.F., *Polymer Networks*, Chrompft, A. and Newman, S. (Eds.), Plenum Press, New York, 1971.
15. Heinrich, G. and Vilgis, T.A., *Macromolecules*, **26**, 1109, 1993.
16. Albeverio, S.A. and Høegh-Kröhn, R.J., *Mathematical Theory of Feynman Path Integrals*, Lecture Notes in Mathematics 523, Springer, Berlin, Heidelberg, and New York, 1976.
17. Grosche, C. and Steiner, F., *Handbook of Feynman Path Integrals*, Springer Tracts in Modern Physics, Springer, Berlin, Heidelberg, and New York, 1998.
18. Deam, R.T., *Molecular Fluids*, Balian, R. and Weill, G. (Eds.), p. 209, Gordon and Breach, New York, 1976.
19. Feynman, R.P. and Hibbs, A.R., *Quantum Mechanics and Path Integrals*, McGraw Hill, New York, 1965.
20. Heinrich, G. and Straube, E., *Acta Polym.*, **35**, 115, 1984.

21. Heinrich, G. and Kaliske, M., *Comp. Theor. Polym. Sci.*, **7**, 227, 1997.
22. Kaliske, M. and Heinrich, G., *Rubber Chem. Technol.*, **72**, 602, 2000.
23. Kaliske, M. and Heinrich, G., *Kautsch. Gummi Kunstst.*, **53**, 110, 2000.
24. Klüppel, M., *Adv. Polym. Sci.*, **164**, 1, 2003.
25. Klüppel, M. and Heinrich, G., *Rubber Chem. Technol.*, **68**, 623, 1995.
26. Klüppel, M., Schuster, R.H., and Heinrich, G., *Rubber Chem. Technol.*, **70**, 243, 1997.
27. Kantor, Y. and Webman, I., *Phys. Rev. Lett.*, **52**, 1891, 1984.
28. Payne, A.R., *J. Appl. Polym. Sci.*, **8**, 266, 1964.
29. Heinrich, G. and Klüppel, M., *Adv. Polym. Sci.*, **160**, 1, 2002.
30. Klüppel, M. and Schramm, J., *Macromol. Theory Simul.*, **9**, 742, 2000.
31. Huber, G. and Vilgis, T.A., *Euro. Phys. J.*, **B3**, 217, 1998.
32. Huber, G. and Vilgis, T.A., *Kautsch. Gummi Kunstst.*, **52**, 102, 1999.
33. Klüppel, M. and Heinrich, G., *Rubber Chem. Technol.*, **73**, 578, 2000.

23 Effects of Time, Temperature, and Fluids on the Long-Term Service Durability of Elastomers and Elastomeric Components

Robert P. Campion, Des (C.J.) Derham, Glyn J. Morgan, and Michael V. Lewan

CONTENTS

23.1 INTRODUCTION

The established durability of elastomers (rubbers) in tire manufacture—the main industry of elastomer usage in volume terms—is the result of the tire industry continually researching many aspects of fatigue, abrasive wear, rubber-to-rubber adhesion, rubber-to-cord adhesion, and so on. Manufacturing procedures have improved so that these consolidated integral components arising out of multilayered constructions assembled before vulcanization are of high quality. Tires normally contain very few flaws that might eventually lead to fatigue crack-growth problems. Related effects of oxygen and sometimes ozone from the air have been minimized by employing suitable low-permeability linings and recipes containing antidegradant ingredients. An awareness of maximum temperatures reached during service (perhaps a few degrees over 100°C near the breaker edge) enables estimates of likely service life to be made once appropriate accelerated laboratory testing has been conducted; validation is then by vehicular testing.

However, durability issues remain of concern in other application areas, where elastomers again perform critical duties, even if their volume usage is much less than for tire manufacture. A major example is given by oil and gas production (especially offshore), which is annually a worldwide multibillion dollar business that involves much hardware (oil rigs, drilling rigs and ships, well and wellhead equipment, numerous ancillaries, etc.). Although the vast bulk of material used is metallic, at the scale involved, the quantity of elastomeric material employed is still significant—and its *rôle* in the harsh environments associated with gas or oil extraction is critical, if acceptable service lives are to be achieved by the various components. Elastomers of many shapes and sizes must resist extremes of fluid type, temperature, and pressure in harsh environments and locations where replacement is essentially impossible. Rubber-based hoses or subsea-flexible pipes are used to transport crude oil, gas, etc. Multilayered rubber- or metal-bonded "flexelements" accommodate the massive loads and varied motions associated with tension-leg oil platforms—these arising from tides and wind; "flexjoints" perform a similar role along some pipelines while being hollow to allow transport of fluid through them. Finally, in numerically the most widespread usage, elastomeric seals bring about appropriate containment of fluids in many production and service areas—often under hostile conditions.

The successful service of elastomeric components depends on different overriding factors concerning deformation mode and environment.[1] These factors will vary widely for different applications, or even within one general application area. Other application industries where elastomeric components can be exposed to fluids under hostile conditions include aerospace, automotive, civil engineering, cable insulation, chemical processing, and biomedical. Moreover, even under relatively benign conditions, long-term durability issues can be problematic if the definition of service failure is strict; in one example of this in automotive service, tight regulatory controls—drawn up for environmental reasons—mean that fuel system components made from elastomers and other polymers must only allow each day a minuscule escape of fuel by permeation to the atmosphere.*

With such a broad subject area, to meet the objectives of the present publication, the general approach for this chapter has been to cover as wide a range as feasible with, where possible, emphasis made on relationships of specific effects with actual service applications. The chapter illustrates various effects of time, temperature, and fluids on long-term service durability of a range of elastomers, mainly by employing various data obtained by Materials Engineering Research Laboratories (MERL) personnel over recent years.

23.2   INITIAL DESIGN CONSIDERATIONS

To introduce this section, it should be observed that the behavior of elastomers as a class stems from their unique position amongst materials of being amorphous polymers. Hence, their associated fine-scale morphological structural details govern their overall behavior, whether in their response to applied forces or to immersion in a fluid. Their molecules are long, possessing molecular weights of perhaps 5×10^4–5×10^5, are extendable, and exist in an intertwined fashion—the whole exhibiting a dynamic ever-changing myriad of conformations according to kinetic theory. Hence, some free space exists between molecular "chains" to enable applied forces to cause bulk deformations, and which allows fluid ingress providing thermodynamic considerations are favorable.

Different characteristics of the chemical structure of different elastomers cause variations in their individual responses to these external influences. The design engineer deals with behavior occurring as a result of this general structural nature.

23.2.1   Large Components

Most mechanical and civil engineering applications involving elastomers use the elastomer in compression and/or shear. In compression, a parameter known as "shape factor" (S—the ratio of one loaded area to the total force-free area) is required as well as the material modulus to predict the stress versus strain properties. In most cases, elastomer components are bonded to metal-constraining plates, so that the shape factor S remains essentially constant during and after compression. For example, the compression modulus E_c for a squat block will be

$$E_c \approx 5GS^2 \tag{23.1}$$

where G is the shear modulus of the elastomer. Barring chemical attack, or significant fluid ingress (swelling), G and S will remain constant, so that the stiffness of the component is for practical purposes also constant through the service life. For some critical components, for example, flexelements (large supporting bearings on some oil rigs), possible fluid ingress and attack must be taken into account at the design stage. Relaxation processes ("stress relaxation"—loss of stress with time in a deformed component; or "creep"—continuing deformation with time under load) are only rarely important in large components. Only where other factors of construction necessitate close tolerances, would relaxation effects have to be taken into account.

Elastomer types used successfully in these areas are natural rubber (NR), polychloroprene (CR), and nitrile rubber (NBR), and hydrogenated nitrile rubber (HNBR), where oil resistance is also required.

23.2.2   Small to Medium Components

By far the largest group of small to medium elastomer components comprises seals and gaskets. Relaxation phenomena, which would result in loss of sealing ability, can become important.

---

* Recent legislation sets "near-zero" limits for evaporative emission globally (EURO II, III/2000, and CARB).

The generally small size of seals means that they are much more vulnerable to attack by fluids (liquid or gas) than larger components. They must be resistant to the fluids they seal and to environmental factors. Hence, seals are normally resistant to oils and/or aerobic heat to a greater or lesser extent, typical elastomers employed being nitriles, hydrogenated nitriles, fluoroelastomers (including copolymers), silicones, fluorosilicones, and perfluoroelastomers. In addition, seals of ethylene propylene diene monomer (EPDM) elastomer, although oleophilic, are produced for use in sealing against water or polar solvents.

23.2.2.1 Time-Dependent Changes for O-Ring Seals

Seals can present a much more complex design problem than large mechanical components, as they

- Are either totally or predominantly unbonded
- Can change their shape substantially
  - During deformation
  - During service if higher pressures are involved
  - Over time as they acquire compression set
- Are usually relatively small in cross section, so that fluids (liquids or gases) can penetrate them, significantly or completely, by dissolution and diffusion

Thus, from the second point, even if the physical properties of a seal material remained totally constant during the service life, the stiffness characteristics of the seal would alter because the shape factor changes—see Figure 23.1.

In seals which are not pressure-energized, sealing force is provided by the residual stress in the compressed seal, so that stress relaxation becomes an important factor. Even for more common situations using pressure-energized seals, residual stress is required at initial start-up—or after a

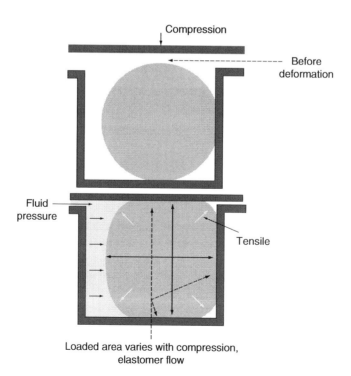

FIGURE 23.1 O-ring seal section; how shape factor S can vary in different circumstances.

shutdown—in order for the system to hold on to initially introduced fluid so that it can then develop sufficient pressure for the main energization of the seal. A further issue arises because elastomeric seals are frequently employed in steel housings. The coefficient of thermal expansion of elastomers is some ten times that of steel—a substantial mismatch. Thus, on heating, a seal may expand to overfill its housing, and on cooling, a seal may shrink away from its housing, leaving a leakage path. An additional point of which the designer must be aware is that the origin of elasticity is thermodynamic, so that the stress in a constrained elastomer is directly proportional to absolute temperature—quite independently of any dimensional changes. Thus, the sealing force when the temperature is lowered will reduce by the additive effects of contraction and thermodynamic stress reduction.

To put in context the above-mentioned complexities, let us consider two real-life scenarios where, ideally, we wish to quantify the drop in sealing force:

1. A seal operating at ambient temperature in a normal temperate climate undergoes a period at much lower temperature due to a cold spell.
2. A piece of equipment operating at high temperature is shut down.

For a while, these seals experience a much lower temperature than usual. Will they seal again when the equipment is restarted?

We could attempt to calculate the loss in stress associated with each of these temperature drops by calculations using either a mathematical model or finite element analysis. We would need the following information as input:

1. Modulus of the elastomer and how this changes with temperature
2. Thermal coefficient of expansion–contraction of the elastomer
3. Shape of the compressed seal in its "set" configuration
4. Initial sealing force
5. Residual sealing force at the beginning of the event
6. Coefficient of expansion–contraction of the housing
7. Behavior of any back-up sealing components
8. Any contribution from swelling by a fluid
9. Quantification of the thermodynamic component of stress
10. Temperature profile of the event

Items 1 and 2 are experimentally measurable, but it should be borne in mind that highly heat-resistant seals may come at, or near, their glass transition temperature (T_g) during a cooling event, and the coefficient of thermal expansion changes in this region.

For item 3, if an assumption is made that the seal completely fills its housing, the bulk modulus is the operative one. If the seal does not completely fill the housing, the shape factor effects apply. This is a very tricky area as Young's modulus and bulk modulus differ by at least two orders of magnitude.

The "initial" sealing force, item 4, should be roughly known from initial design studies, but it may be subject to considerable uncertainty and must be based on an assumption of the worst combination of seal–housing tolerance variables. The residual sealing force may be estimated if there is knowledge of the stress-relaxation rate of the elastomer, using

$$L_t = L_0\left(1 - B\log\left(\frac{t}{t_0}\right)\right) \qquad (23.2)$$

where L_0 represents initial load and L_t that after time t, and B is the fractional rate of physical-stress relaxation per decade; however, even then, because of the remaining parameters shown on the

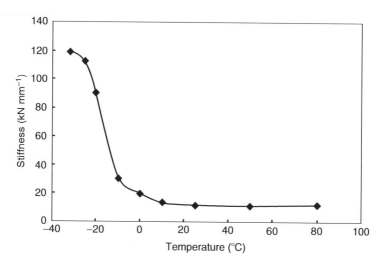

FIGURE 23.2 Seal stiffness versus temperature for a fluoroelastomer.

above-mentioned list which affect sealing force, an estimate based on stress-relaxation rate alone may be very naïve. A further complication in modeling may be provided by lack of knowledge of any backup components in the sealing system. Simplified models exist which purport to predict long-term sealing ability—they should be treated with great caution in critical applications (see below).

Regarding temperature effects, Figure 23.2 shows how in practice the stiffness of a small fluoroelastomer O-ring seal in a cup housing, i.e., unconstrained at its internal face, changed with temperature, using actual experimental measurements. The T_g for the material was about $-18°C$, but cooling-induced stiffening is seen to begin well above this temperature. Even at 10°C above the T_g, the stiffness has doubled. At the T_g the stiffness has increased by 250%. This increased stiffness is important in practice because in this particular experimental setup, it was estimated that, at $-18°C$, the seal would only need to decrease in height by 0.003 mm (just over a tenth of 1%) to reduce the sealing force by 50% based on thermal contraction alone and ignoring other factors—not a happy situation.

It is unlikely that the entire calculation of sealing loss could be done theoretically even using finite element methods, as there is such a large number of parameters, and only some of them are known at all accurately. Derham[2] has shown that for an O-ring in a similar configuration to that described here, the loss of sealing force accompanying a temperature drop from 80°C to 23°C was significant, being in practice twice as great (~50%) as the best estimate that could be made from simple theoretical considerations (~25%). This discrepancy was mainly attributed to the change in ring shape, and therefore stiffness, after it had been under compression for some time.

The unfortunate conclusion must therefore be that, although calculated predictions of sealing force retention may have some value in some situations, where a temperature drop is concerned, only experience, or experimental testing, is likely to provide anything near the correct answer.

23.2.2.2 Time-Dependent Changes Induced by Liquid Swelling

A more predictable situation exists for an O-ring seal housed as above when contacted and swollen by a liquid solvent. Generally a small amount of positive swell is beneficial to sealing, and *it is believed that many seals would not function for long without some swelling.* Excessive swelling, however, weakens the seal and leads to total failure by fracture or "extrusion."

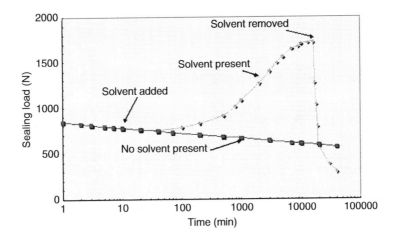

FIGURE 23.3 EPDM O-ring seal; its stress relaxation when dry (linear) and contacting solvent.

Figure 23.3 shows how the sealing force in an EPDM O-ring is influenced by the presence of a hydrocarbon solvent. Normally, when dry, the force decays with time as shown by the straight line, with stress relaxation proportional to logarithmic time. When a solvent is added, however, the swelling force overcomes the stress relaxation and the sealing force increases as shown. In this experiment, after a time the solvent was removed. The sealing force then fell and reached a lower value than would have been the case had the solvent not been introduced. This shows that the solvent has "damaged" the seal—possibly by extracting some soluble constituent from it.

From these results the component of the force which is directly due to swelling can be obtained. It is found that this increase in force is proportional to the swelling so that an equation relating sealing force increase with root time, analogous in form to that used for standard mass-absorption measurements used to determine diffusion coefficient D values, can be developed. For one EPDM elastomer O-ring in a hydrocarbon solvent—an oleophilic elastomer selected deliberately to emphasize this effect—the D value calculated, thus, was essentially equal to the value obtained by standard mass-absorption testing, leading to the conclusion that the increase in sealing force is directly proportional to the amount of solvent absorbed by the seal. To confirm this, Derham developed the expression further, to arrive at

$$\text{Total sealing force} = L_0 \left(1 - B \log\left(\frac{t}{t_0}\right)\right) + \left(\frac{2L_\infty}{h}\right)\left(\frac{D(t - t_0)}{\pi}\right)^{1/2} \tag{23.3}$$

where
  L_∞ is the final load attained at equilibrium swelling
  h is the thickness of the O-ring when initially flattened by compression into its housing

Agreement between experiment and theory was excellent and is described by Derham.[2] Clearly this approach is only appropriate where the seal is not constrained to a point where bulk compression begins.

In addition to the physical stress relaxation effect, chemical changes to the elastomer (e.g., by oxidation) can also increase relaxation rate. We then have

$$L_t = L_0 \left(1 - B \log\left(\frac{t}{t_0}\right)\right) + C(t - t_0) \tag{23.4}$$

where

  B is the fractional physical stress-relaxation rate per logarithmic decade of time

  C is the fractional chemical rate of stress relaxation per unit time

The accurate application of this approach to the long-term performance of big elastomeric-structural bearings used in the construction of a large building in London, U.K., is well documented.[3,4]

As already noted, stress relaxation is closely related to creep—that is, the increase in deformation which occurs when a material is held at constant stress. Another somewhat similar term "extrusion" is also used in oil-field sealing, pipeline, and other circles—here not as a manufacturing term, but applying when high-pressure fluid forces elastomer (usually as a seal material) into gaps between adjacent rigid surfaces (e.g., between a seal housing and the mating metal surface).

23.3   OTHER EFFECTS OF TIME ALONE

23.3.1   Accelerated Techniques

Many factors can contribute to reducing testing durations compared with anticipated component service life. However, in the simplest case where conditions of strain, pressure, and temperature are the same for both service and testing, test acceleration can arise simply because there are situations where test simulations can be continual in the laboratory, whereas in practice actual events occur only every few days or weeks, as illustrated below.

23.3.1.1   Reduction of Cycle Times for Elastomer Sleeves for Offshore Emergency Valve

In an established offshore safety application, a hollow profiled elastomeric sleeve is used at near-ambient temperatures to operate rapidly under certain emergency conditions to allow the sudden passage of copious volumes of water. The sleeve experiences a significant biaxial expansion due to this sudden passage of pressurized water, until the expansion is stopped by the sleeve coming up against a metallic grid structure. Although the relevant operational emergencies are very rare, it is critical that the sleeve will operate successfully if one should happen. Hence, the offshore operator performs a check expansion test after every few weeks of service; failure for this test would be by the expansion not happening properly or by the sleeve bursting to lose the pressurized water. While this series of check tests is deemed essential to reassure all concerned that the sleeve will operate if need be, for the sleeve elastomer it poses an extra problem, as a fatigue element is introduced. To accommodate this, the valve manufacturer developed a service-realistic laboratory fatigue test in which typical sleeves are cycled until they either exhibit premature burst (fatigue-induced) or (ideally) reach a specified number of cycles at well over the maximum number of check expansion tests predicted for the specified sleeve service life. The manufacturer's testing is performed at the same conditions of strain, temperature, and pressure as applied to service—here, testing-time acceleration is not produced by increasing the severity of conditions applied, but merely by eliminating the time periods between expansion checks that occur in practice.

In this case, the manufacturer's synthetic elastomeric compound failed the fatigue requirements, the sleeves bursting prematurely. After a review and initial laboratory testing on material mechanical properties considered relevant (especially tear strength), MERL developed a stiff NR elastomer compound leading to the best balance with cycling test fatigue resistance for the application; after subsequent sleeve fatigue testing, the NR sleeves then produced were predicted to operate for the full design service life while accommodating the periodic check tests throughout.

23.3.1.2   Reduction of Cycle Times for Seal Stacks for Deep Offshore Oil Wells

The same basic philosophy as above was applied[1,5] to the development of a high-pressure high-temperature (HPHT) dynamic service-realistic test facility for seals. This facility provided the means of testing actual seal stacks under the very severe conditions which apply in the deepest oil wells in offshore oil production. These stacks function in service at the lowest point in the well, in practice a number of them being positioned together in a continuous line inside the polished bore of the production tube, to prevent escape of the well fluids. The situation is dynamic but of very low frequency, because thermal contractions–expansions of the tubing, or fluid-pressure fluctuations, can mean that the seals move considerable distances inside the bore over a period, with some reciprocations.

To simulate this action, the test rig comprises a full-scale polished bore pressure vessel and a centrally located reciprocating mandrel with two seal-stack locations containing the high-pressure fluid zone between them—all mounted horizontally for convenience. MERL's procedures set out to test the seal stacks to the full operating conditions in terms of distance traveled during the reciprocations. At test temperatures up to 200°C, test pressure (up to 100 MPa, 15,000 psi) is applied to one side of the seals only, any leakage outside the stacks being measured as pressure increase there for gas, or as volume collected at the end of test for liquid. Once again, the delay which exists in service between expansion–contraction events means that real-time testing at a reasonable testing-cycle rate provides an acceleration; replicates of actual seals for use in service can be tested for leakage at the actual HPHT conditions while moving for the actual distance they will be forced to move in service. In this way, several designs of seals have been shown to be suitable for service in these critical stacks, including some new ones.

23.4 FLUID ENVIRONMENTS—GENERAL

23.4.1 FLUID CONTAINMENT

For successful fluid containment by an elastomeric component, it is obvious that all escape possibilities must be eliminated or minimized. Possible mechanisms applying for such fluid escape are

1. In a sealing situation, leakage *past* the elastomer, outside and around its surface, usually due to imperfections in the housing; this applies especially in a dynamic arrangement if the sealing stress diminishes due to relaxation, wear, and/or thermal contraction
2. Movement through local physical flaws arising, e.g., in hoses, from
   - Discontinuities (e.g., imperfectly bonded interfaces or rubber processing flaws)
   - Cracks, splits, and other fractures, perhaps caused from an excess of absorption—with ensuing swelling and weakening—or from "explosive decompression" (ED) fracturing in high-pressure gas situations
3. Permeation—the background ever-present contribution to leakage (albeit small) through elastomers, involving all elastomer surfaces exposed to fluid, all external free surfaces, and the intervening bulk

Aspects of point 1 have already been included in Sections 23.2.2.1 and 23.2.2.2. Regarding point 2, a new diagnostic test for discontinuities is outlined in Section 23.5.2, absorption in Section 23.4.2, and ED failures in Sections 23.5.1.1 and 23.5.1.2. As to point 3, the ubiquitous phenomenon of permeation, which occurs wherever fluid containment by elastomers exists, is examined below.

In addition, any chemical aging (Section 23.5.3) could affect these mechanisms; for instance, if an elastomer surface is degraded by contact with hostile chemicals, access to the interior may be facilitated and its permeation characteristics would change accordingly.

The question of relative rates of seal leakage and permeation (cf. points 1 and 3) has been considered by MERL.[6] For the sealing of HPHT (5000 psi, ca. 345 bar, or 34.5 MPa; 100°C) fluids with chevron seal-stack systems used at the bottom of oil wells in intermittently dynamic conditions, high-pressure permeation and seal-leakage tests using the equipment outlined in Section 23.3.1.2 have been conducted; it has been shown that methane gas permeation rate was only ca. 1% of the rate of static leakage, when that occurred, and was at an even lower proportion of the leakage rates applying to dynamic situations.

In a different application area, a dynamic test was designed at MERL some years ago to provide a realistic simulation of automobile air-conditioning units in service (Harrison and Stevenson).[7] When conducting long-term tests, leaks of hydroflurocarbon (HFC) gas were observed over a limited period at the beginning of testing before the elastomer seals being used had been energized sufficiently to function correctly. Separate permeation tests at the same conditions were conducted through sheet membrane samples of the same elastomer; permeation rates were, at most, 3% of measured leakage rates.

It might be thought as a consequence of measurements such as these that leakage factors are the main issues in fuel containment. However, although obviously important, in some cases a leak might occur only at intermittent intervals, and the associated problem might well be easily resolvable by component replacement. In contrast, the relevance of permeation to fluid containment is its continuous nature—its rate may be low, but it occurs all the time that fluid is contacting elastomer. Hence, this phenomenon is now considered in association with related processes absorption, adsorption, and diffusion.

23.4.2 ABSORPTION, DIFFUSION, AND PERMEATION

Any polymer contains some inner free space "*free volume*" distributed in a dynamic manner between its molecular chains (see Section 23.2). When it is exposed to a fluid (liquid or gas) the physical possibility exists for fluid absorption by the polymer, if the fluid molecules or atoms are small enough to fit into local regions of this distributed space during kinetic movements. As this happens, subsequent kinetic chain motion must allow for the newly absorbed fluid molecules and, hence, the polymer's overall volume will adjust accordingly; this action will coincide with the formation of more free space around these fluid molecules—so the polymer will swell a little. This process will be continued until an equilibrium is reached ("equilibrium swelling"), by which time the extent of swelling can be considerable. The amount of fluid taken up and the rate at which this happens are both important, and are discussed in this and following sections.

Elastomers as a class contain considerably more free volume than other polymers, and hence they can swell more. Thermodynamic Gibbs free-energy considerations influence the degree of fluid absorption by the elastomer; associated enthalpic effects arise from the chemical nature of the elastomer, and entropic effects from the elastomer's morphological structure and size of fluid molecules. Chemical potential (μ), the free-energy change for a single species with all other factors remaining constant, drives the fluid into the elastomer in the first place. However, μ is conveniently quantified by other parameters for different situations—usually by concentration, which itself normally depends on vapor pressure (liquid) or actual pressure (gas).

The molecular process termed "permeation" is the most fundamental physical means by which fluid can pass right through an elastomer or other polymer. Absorption and permeation, besides being interrelated, both depend on two other phenomena associated with fluid–polymer interactions, as follows:

Adsorption: The first stage, applying when a polymer contacts a fluid, is that the latter adsorbs into the polymer surface, reaching equilibrium solubility here.

Diffusion: Molecules of a fluid already inside a polymer at a high-concentration region compared with surrounding regions will diffuse over a finite time away from the high concentration until an equilibrium situation is achieved. If the high concentration is at the surface, diffusion occurs into the bulk. The diffusant molecules move stepwise into free volume holes as they form according to kinetic theory.

Thus we also have

Absorption: When an elastomer is submerged in a fluid, the latter is absorbed ("taken up" by the elastomer and distributed throughout it) due to a combination of the two above effects—i.e., the continuous processes of adsorption followed by diffusion into the bulk until equilibrium. For practical reasons, absorption tests, involving mass uptake, are mainly restricted to immersion in liquids—although exceptions to this exist,[8] these applying to high-pressure gas absorption.

Permeation: When a fluid contacts one side of an elastomer membrane, it can permeate right through the membrane, escaping on the far side. The process again combines adsorption and diffusion as above, but with the additional process eventually of evaporation—treated mathematically as "negative adsorption." (Permeation could also be viewed as combining "one-way absorption" and evaporation.) Wherever these conditions for permeation exist the phenomenon occurs, whatever the shape of the elastomer barrier—but the associated mathematics becomes complex for irregular barrier shapes.

At constant conditions, different fluids will diffuse at different rates into a particular elastomer (with their rates raised proportionally by increasing the exposed area), and each will reach the "far" elastomer-sample surface proportionally more rapidly with decreasing specimen thickness. Small molecules usually diffuse through an elastomer more readily than larger molecules, so that, as viscosity rises, diffusion rate decreases. One fluid is likely to diffuse at different rates through different elastomers. Permeation rates are generally fast for gases and slow for liquids (and fast for elastomers and slow for thermoplastics and thermosets).

Hence we have in sequence for permeation:

1. Adsorption of a fluid quantified by the surface solubility coefficient s according to Henry's law

$$c = sP \tag{23.5}$$

   where
   c is concentration and
   P is pressure (gas) or vapor pressure (liquid) of the applied fluid

2. Diffusion of the fluid into the bulk. Rates of diffusion are governed by Fick's laws,[9] which involve concentration gradient and are quantified by the diffusion coefficient D; these are differential equations that can be integrated to meet many kinds of boundary conditions applying to different diffusive processes.[10,11]
3. Desorption (evaporation) of the fluid when it reaches the opposite surface of the polymer, again involving Henry's law and now with the value of P applying to this surface.

For gas and vapor systems, by combining the laws of sorption and diffusion in the sequence (1)–(3), general permeation equations are obtained. For sheet membrane samples of polymers above T_g, if the definition is made that permeation coefficient $Q = Ds$,

$$\text{Permeation rate} = QA\left(\frac{P_1 - P_2}{h}\right) \tag{23.6}$$

where
A and h are sample area and thickness, respectively
$(P_1 - P_2)$ is the (vapor) pressure difference across the membrane

Other equations exist for other geometries, e.g., for pipes:

$$\text{Permeation rate} = \frac{2\pi LQ(P_1 - P_2)}{\ln\left(\dfrac{r_2}{r_1}\right)} \tag{23.7}$$

where

  L is pipe length

  r_2 and r_1 are outer and inner radii, respectively

and for cones

$$\text{Permeation rate} = \frac{\pi r_1 r_2 Q (P_1 - P_2)}{h} \tag{23.8}$$

where r_1 and r_2 are now the major and minor radii, respectively.

For more complex shapes, numerical solutions may be needed; coefficients obtained by testing of simple samples can be applied to realistic geometries for components of the same material.

For a gas, the magnitude of Q expresses its permeation rate (standard temperature and pressure [STP] volume per time) for passing through an elastomer cube of side unity, driven by a pressure of 1 atmosphere. Units in commonplace usage, referring to a cube of sides 1 cm, are $cm^2 \, s^{-1} \, atm^{-1}$. For liquids, mass per time is a more relevant rate, with equivalent units for Q being $g \, cm^{-1} \, s^{-1} \, atm^{-1}$ (dimensionally linked with the volume units via $M/22{,}400 \, cm^{-3}$ with molecular weight M in grams).

23.4.2.1 Temperature Effects

Increasing temperature permits greater thermal motion of diffusant and elastomer chains, thereby easing the passage of diffusant, and increasing rates; Arrhenius-type expressions apply to the diffusion coefficient applying at each temperature,[11] so that plots of the logarithm of D versus reciprocal temperature (K) are linear. A similar linear relationship also exists for solubility coefficient s at different temperatures; because $Q = Ds$, the same approach applies to permeation coefficient Q as well.

In practical situations, if permeation through a hose lining, for example, is being considered, and the hose is exposed externally to a cold environment (such as the sea) while hot fluid passes through the hose (e.g., hot oil in the offshore production industry), a temperature gradient $T_1 - T_2$ will apply across the lining. To cover this situation, for a representative permeation coefficient $Q_{\Delta T}$, Campion and Thomas developed from Fick's laws

$$Q_{\Delta T} = \frac{D_0 E_D (T_1 - T_2) s_1}{R T_1^2 \left(e^{\left(\left(\frac{E_D}{R T_1} \right) \left(2 - \frac{T_2}{T_1} \right) \right)} - e^{\left(\frac{E_D}{R T_1} \right)} \right)} \tag{23.9}$$

R is the gas constant; D_0 and activation energy E_D are constants derived from an Arrhenius plot for diffusion coefficients applying at different temperatures, and solubility coefficient s_1 was obtained from a separate permeation test at $T_1 K$. Suitable testing using a specially constructed permeation cell water-cooled at one end provided good validation data.[12]

23.4.3 Liquid Absorption by Elastomers and Swelling

23.4.3.1 Solubility Parameter

Individual liquids and elastomers each possess their own "solubility parameter," δ.[13] This is a thermodynamic property which is related to the energy of attraction between molecules. In its simplest form,* an elastomer will possess a drive to absorb a liquid of similar δ, and be swollen by it. As the difference between the solubility parameter values of species increases, so their affinity for each other decreases. The commonest units for δ in the literature are $(cal \, cm^{-3})^{1/2}$; to convert values thus to MPa, multiply by 2.05.

---

* Additional contributions to δ from, for instance, hydrogen-bonding forces, well studied by Hansen, C.M. (*J. Paint Technol.*, 39(505), 105, 1967), have not been relevant to MERL's general usage of this parameter.

Thermodynamically, intermixing as above arises from achieving a negative free energy from the process; in turn, this arises by balancing a contribution from enthalpy (heat content) with another from entropy (involving material structural aspects). The solubility parameter reflects the enthalpic term.

Solubility parameter spectroscopy (SPS)[4] is a means of determining the δ value of an elastomer, and furthermore gives a better indication of the δ range across which swelling might occur for an elastomer–liquid system—frequently more useful when selecting materials for an application involving a known liquid. For SPS, replicate elastomer samples are immersed in a series of carefully selected liquids or miscible liquid mixtures, each of known δ, and a plot (spectrum) of equilibrium mass uptake (see Section 23.4.4.1.) versus liquid solubility parameter is thus developed. The δ value coinciding with the maximum of the plot is taken as the δ value for the elastomer. SPS spectra obtained by Morgan and Campion for four common oil-field elastomers are shown in Figure 23.4; an added benefit from such plots is that the width of its solubility parameter peak indicates the δ range across which substantial swelling of an elastomer can occur in low viscosity solvents. As viscosity increases when using other solvents, peak width and height are reduced.

Intermediate liquid δ values are obtained by mixing liquids of known solubility parameter; SPS makes use of this. The δ value of the mixture is equal to the volume-weighted sum of the individual component liquid δ values. Thus, the mass uptake of a miscible liquid mixture by an elastomer may be very much greater than the swelling which would occur in the presence of either one of the constituent liquids alone. The mixture could of course comprise more than two liquid components, and an analogous situation would apply; MERL have applied this approach for the offshore oil-production industry to allow realistic hydrocarbon model oils to be developed,[14] basically by mixing one simple aliphatic (paraffinic) hydrocarbon, one naphthenic, and one aromatic to proportions that meet two criteria, namely, that

1. Mixture possesses a δ value very near to that of the crude oil being simulated.
2. Its aliphatic/naphthenic/aromatic proportioning is very near to that of the crude oil.

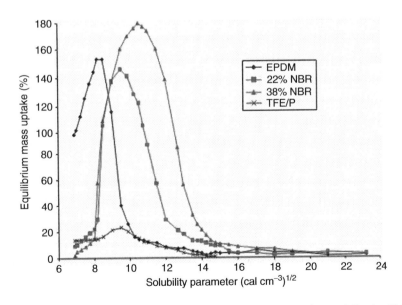

FIGURE 23.4 Solubility parameter spectra for elastomers ethylene propylene, nitrile (at 22% and 38% acrylonitrile content) and tetrafluoroethylene propylene copolymer.

The hydrocarbons selected were heptane, cyclohexane, and toluene; one such mixture is now specified within the fluids for liquid aging assessments according to Norwegian NORSOK M-710 Rev. 2 standard for qualification of offshore duty elastomer seals.[15]

Campion and Morgan developed the technique of *reverse solubility parameter spectroscopy* (RSPS)[4] to determine the δ value for a liquid (e.g., an oil), from a series of swelling measurements using a range of elastomers of known δ. In this way, δ for crude Brent oil from the North Sea has been found to be 8.2 $(cal\ cm^{-3})^{1/2}$.

A small selection of solubility parameter values in $(cal\ cm^{-3})^{1/2}$ for elastomers and liquids are

- From Figure 23.4, for EPDM 8.25; 22% NBR 9.3; 38% NBR 10.5; TFE/P 9.5
- From the same work, 38% HNBR 9.6; ECO 10.7; FKM A 10.9; FKM GFLT 10.7
- Liquids[16]: hexane 7.33; heptane 7.48; isooctane 6.9; octane 7.6; decane 7.77; cyclohexane 8.25; toluene 8.97; acetone 9.74; ethanol 12.97; methanol 14.5; ethylene glycol 14.5; water 23.2

δ values can vary slightly depending on method of measurement or estimation; within one type of elastomer, differences of comonomer proportions, compound recipes, etc., can have an effect. Good sources of δ values exist.[16,17]

Volume swell can be quantified during absorption measurements by weighing in water as well as in air, to apply the Archimedes approach.

23.4.3.2 Material Structural Factors Which Can Influence Liquid Uptake Amounts

Material structural aspects, thermodynamically associated with the entropic contribution to the free-energy equation, can sometimes be used to exert a degree of "control" over the level of liquid absorbed by sealing elastomers. If the δ difference between liquid and polymer is significantly large, negligible swelling will occur in any case; if δ values are similar, swelling may occur, unless prevented by structural aspects, such as

- Glass transition temperature (T_g): Swelling will be reduced in an elastomer with a high T_g, that is, in an elastomer with relatively little free volume available to absorb liquid.
- Cross-linking: Since cross-links provide constraints against chain portions being moved apart by incoming liquid molecules, swelling is reduced by high cross-linked levels.
- Filler: Swelling is reduced by the use of high filler loadings, with less elastomer then available to absorb fluid.

In addition, the following points can be relevant:

- Increasing liquid viscosity lowers rate of absorption and equilibrium mass absorbed.
- During liquid immersion, soluble non-bound ingredients in an elastomer can be removed by leaching—usually (but not always) a small effect.
- Increasing temperature may increase swelling by a modest amount.
- If two immiscible liquids A and B (i.e., possessing very different δ values) form two layers when brought together, and an elastomer of similar δ to A is completely immersed in the (denser) B layer (schematically in Figure 23.5), nevertheless, the elastomer will eventually swell as if immersed completely in A. This arises because each liquid of an immiscible mixture still dissolves a minute amount of the other. At equilibrium, the chemical potential μ of A will be the same, whether as pure liquid, dissolved in B, or dissolved in the elastomer. At the same temperature, the same μ would apply for the elastomer immersed directly in A. However, the *kinetics* of absorption will be different, being much slower than

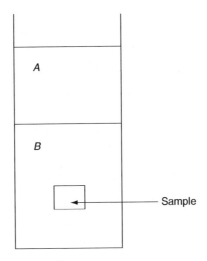

FIGURE 23.5 Immiscible liquids with elastomer sample immersed in lower layer.

in the direct case for a static immiscible mixture, or rather more rapid than this if the immiscible layers are continually agitated. Morgan and Campion have observed such behavior for EPDM elastomer with *A* being a hydrocarbon solvent and *B* being water.

General points of fluid kinetics are now discussed.

23.4.4 DIFFUSION AND PERMEATION TESTING

23.4.4.1 Kinetics of Liquid Diffusion Tests

Diffusion rates for liquids in an elastomer are easily measured by absorption (immersion) testing, a simple process as indicated in Figure 23.6. An initially weighed immersed sheet sample of elastomer is removed from the liquid periodically, rapidly dabbed with tissue paper, reweighed, and replaced. A plot of mass increase versus root time is drawn (also see Figure 23.6), root time being chosen due to the form of appropriate solutions of Fick's laws.

A solution of Fick's 2nd Law often used for absorption testing is

$$\frac{m_t}{m_\infty} = \left(\frac{2}{h'}\right)\sqrt{\left(\frac{Dt}{\pi}\right)} \qquad (23.10)$$

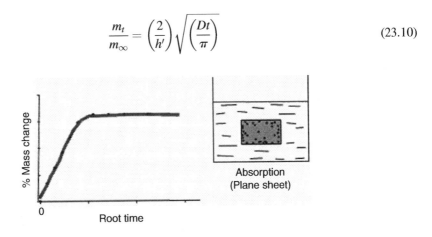

FIGURE 23.6 Immersion testing outline—an absorption plot.

where m_t is mass change at time t, m_∞ is mass change at equilibrium, D is diffusion coefficient, h' is sheet sample half-thickness—written this way because in absorption situations liquid enters a sheet sample from opposite directions, via the two large faces, so that each route effectively only needs to "fill" half the sample. D in effect is the diffusion rate across a unit cube of the elastomer. A dimensional analysis of the above equation reveals that the area term cancels out within this particular expression and its derivatives, so that area is not a factor when obtaining D by means of such equations when using absorption tests.

The horizontal portion of the plot in Figure 23.6 represents the equilibrium mass uptake—the amount absorbed at equilibrium whose magnitude is influenced by the solubility parameter δ (Section 23.4.3.1). In reality, coefficient D quite often varies with concentration, so that the overall plot takes on a "sigmoid" shape (see Section 23.4.4.3), although a horizontal portion is still usually achieved eventually; if it is not, the elastomer is possibly a two-phase material (a blend), with one phase much slower at absorbing the incoming liquid.

MERL's interest in employing these tests has included (1) the quantifying of D to enable certain predictions to be made (see Section 23.4.6), (2) testing at high temperature and pressures to simulate service realism (see Section 23.5.1.3), and (3) to provide data for use in estimating permeation rates while not needing to conduct actual permeation tests (see Section 23.4.4.3).

23.4.4.2 Liquid Permeation Testing

Water permeability is tested by transmission permeation tests, standard test ASTM E 96 being widely used. The water which permeates is absorbed in a desiccant and its weight—or alternatively, the loss of weight of the water supply from a container (or "permeation cup") sealed by the sheet test sample—is determined. The same procedure is applied for the transmission of volatile liquids in standard ISO 6179 (BS 903:A46)—see Figure 23.7.

A plot of permeant mass loss versus time is developed, at equilibrium being of the form shown in Figure 23.8, which illustrates data from an actual test at 23°C of hexane permeating through tetrafluoroethylene–propylene copolymer (popularly termed TFE/P, better FEP). A main reason for including this plot is that it demonstrates clearly that the permeation rate is unaffected by the phase of the fluid contacting the sample. Here, contact was by vapor when the sealed cup was upright as shown in Figure 23.7 and liquid, when the cup was inverted (changeovers happening several times). This observation occurs because the ultimate driving force for permeation, the chemical potential μ, is represented for the vapor by vapor pressure: when the cup is inverted, the vapor now away from the sample remains at the same μ value—and because the system is in equilibrium, this same μ value must apply to the liquid phase as well.

In Figure 23.8, the gradient of the plot at steady-state conditions gives a permeation rate which will provide coefficient Q using the general permeation equation (Section 23.4.2).

Finally, one variation on the weight-loss method—"Fuel hose testing; modified GM 9061 'fill and plug' method"—applies for fuel line aspects of fuel containment—here the sample can be from an actual hose or its liner.

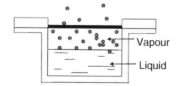

FIGURE 23.7 Permeation cup sealed by sheet sample with hydrocarbon liquid inside.

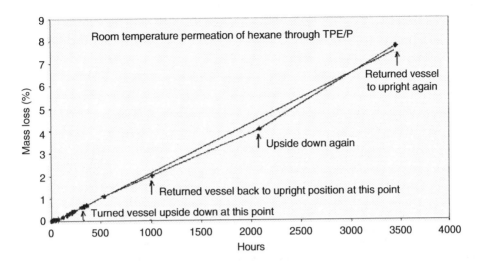

FIGURE 23.8 Mass loss versus time for liquid transmission test.

23.4.4.3 Estimations of Liquid-Permeation Rates from Absorption Testing

The interrelationships of the various coefficients associated with fluid uptake (Section 23.4.2) mean that it should be possible to estimate a rate for one of the uptake phenomena from test data for another of them. Campion proposed using this approach to estimate permeation coefficient Q from solubility coefficient s. The form of a liquid absorption plot (Figure 23.6, Section 23.4.4.1) is such that s should be obtainable from it, and inspection showed that this link was via Henry's law with concentration corrected by the polymer density ρ. The following expression was derived[18] for s:

$$s = \left(\frac{m_\infty}{m_0}\right)\frac{\rho}{P} \qquad (23.11)$$

where

m_0 is initial polymer mass

m_∞ is mass change (i.e., mass of liquid absorbed) at equilibrium

p is vapor pressure of the liquid, obtainable from standard tables

Then $Q = Ds$ (Section 23.4.2), with diffusion coefficient D also measurable by absorption testing, so that Q can be estimated. Permeation rate is then obtainable for a particular sample shape at specified conditions by using the appropriate permeation equation.

In initial tests to check the approach,[19] sheet samples of thermoplastic ethylene tetra-fluoroethylene (ETFE) were used, with a permeation cup control test yielding a Q value of 7.7×10^{-10} g cm^{-1} s^{-1} atm^{-1}, and the proposed absorption approach giving 5.8×10^{-10} g cm^{-1} s^{-1} atm^{-1}. The agreement was promising, with reasons for the measured difference of 1.9×10^{-10} g cm^{-1} s^{-1} atm^{-1} thought to include possible concentration effects, and the fact that sample thicknesses were different for the two tests—so that crystallinity levels could have also differed.

In a more extensive exercise on this topic, Smith and Campion[20] first examined several polymer–hexane systems. Where a concentration-dependent coefficient occured, the sloping portion of the resulting sigmoidal absorption plot was broken down into sections as indicated schematically in Figure 23.9, and separate values of D were calculated for sections $0 \rightarrow A$, $A \rightarrow B$, $B \rightarrow C$, $C \rightarrow D$, and $D \rightarrow E$ (where E marks the steepest gradient applying). A time-averaging process was applied to these values of D to provide an appropriately weighted average D value for fluid diffusing through the full range of concentrations present—as indeed must happen during the permeation process.

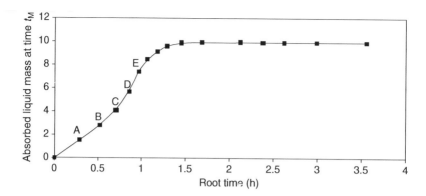

FIGURE 23.9 Schematic sigmoidal absorption plot for situation where D is concentration-dependent.

Secondly, for predicting permeation, Table 23.1 shows[20]—again for convenience using a thermoplastic, medium-density polyethylene (MDPE)—the estimated permeation coefficient result from a 40°C hexane absorption test, and the comparison with the result from a control permeation cup test. This process was actually repeated for several polymers, including two elastomers. The table also shows the result from a limited study about a further stage of prediction, where short-time absorption testing was performed at several higher temperatures leading to Q values each time, so that Q for 40°C could then be estimated using an Arrhenius extrapolation (Section 23.4.2.1). To demonstrate how the Q values might be utilized, calculated permeation rates through a sheet membrane were obtained from them by using relevant permeation equation 23.6.

All correlations were good in this study, thus validating the approach.

23.4.4.4 Gas Diffusion and Permeation Testing

For gases, both permeation and diffusion data are best measured by permeation tests, many different types been described elsewhere.[18,21] The same sheet membrane permeation test can quantify permeation coefficient Q, diffusion coefficient D, solubility coefficient s, and concentration c. The membrane, of known area and thickness, must be completely sealed to separate the high-pressure (initial) region from that containing the permeated gas; it may need an open-grid support to withstand the pressure. The permeant must be suitably detected and quantified (e.g., by pressure or volume buildup, infrared (IR) spectroscopy, ultraviolet (UV), gas chromatography, etc.).

A plot of permeant build-up versus time is then developed, of the form shown in Figure 23.10 for volume build-up. The gradient of the plot at steady-state conditions gives a permeation rate which will give Q using the general permeation equation (Section 23.6). Also, after Daynes, and as discussed elsewhere.[11]

TABLE 23.1

Hexane Permeation Predictions for Medium-Density Polyethylene (MDPE) Sheet Sample of Thickness 1.57 mm at 40°C

MDPE	Permeation Coefficient Q $\times 10^{-8}$ (g s^{-1} cm^{-1} atm^{-1})	Permeation Rate through the sheet $\times 10^{-5}$ (g h^{-1} cm^{-2})
Control permeation cup test at 40°C	2.5	22
Absorption test conducted at 40°C	2.4	21
Prediction–absorption at elevated temperatures	2.8	24

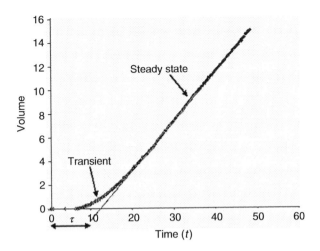

FIGURE 23.10 Permeated volume versus time for gas permeation test.

$$D = \frac{h^2}{6\tau} \tag{23.12}$$

where

τ is a time lag term, easily read from the test
h is membrane thickness

Commonly of units $cm^2\,s^{-1}$, in magnitude D is the diffusion rate across the 1 cm-sided polymer cube considered earlier for Q (Section 23.4.2). Coefficient s is then calculable because Q is defined as the product Ds, and from Henry's law (Section 23.4.2) concentration c can then be obtained.

Morgan has conducted such testing at moderate applied pressures, employing MERL's permeant-pressure buildup test.[22] He showed that a large-particle-size or low-structure carbon black particulate filler can be incorporated at high loadings into a bromobutyl tire inner liner compound to maintain levels of general mechanical properties, while reducing air permeation rate compared with those for other compounds containing established black types.

23.4.4.5 High-Pressure Gas Permeation Tests

MERL has conducted numerous high-pressure gas permeation tests on many gases and elastomers using its own design of test equipment. A general outline of the permeation system is shown schematically in Figure 23.11, the low-pressure transducer being sealed into a small fixed-volume reservoir to catch permeant after permeation. Figure 23.12 shows various parts of the permeation cell, with the sample area for testing being seen as a central disc. O-rings are used to seal a steel holder into the test cell, with the elastomer being earlier molded and bonded into the holder during its curing process. The flash elastomer from this process, seen in one photograph, lies outside an O-ring located in the circular housing groove, so that test results are unaffected.

The sample is assembled into the permeation cell after first having had its thickness accurately measured, with the sample supported by a stainless steel porous sinter (not visible in the photographs). The test cell is heated to test temperature by suitably controlled band-heaters, test temperature being measured by a thermocouple located close to the sample. Test gases are boosted to test pressure using a gas intensifier operated by pressurized air.

When testing, the permeation rate is initially measured as change in pressure (P_2) in the low-pressure reservoir versus time, to give a plot of the same form as shown in Figure 23.10, but with pressure rather than volume being plotted. To convert to volume, the following expression (referring strictly to an ideal gas) is applied to obtain rate dq/dt from the slope of the plots:

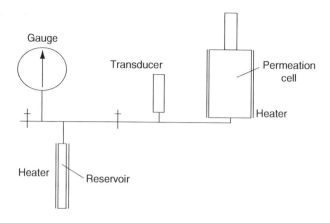

FIGURE 23.11 Schematic of high-pressure gas permeation testing system.

$$\frac{dq}{dt} = \frac{\left(\frac{\delta P_2}{\delta t}\right)_v (273 V_{LP})}{T_{LP}} \tag{23.13}$$

where V_{LP} and T_{LP} are volume and temperature (K) of the low-pressure side of the cell, P_2 is in atmospheres, and permeation rate derives from volume dq of test gas at STP permeating per time dt through the test sample.

During this operation, the equilibrium gradient is read off after a delay of at least three times greater than the time lag (Section 23.4.4.4) to avoid any influence of the transient stage on steady-state equilibrium. Permeation coefficient Q is then calculated by inserting this permeation rate and dimensional and pressure details into the general permeation equation 23.6 for a sheet sample (Section 23.4.2). The presence of an *in situ* micrometer to measure at intervals changes in sample thickness during the test period compared to the initial value leads to a more accurate calculation of Q.

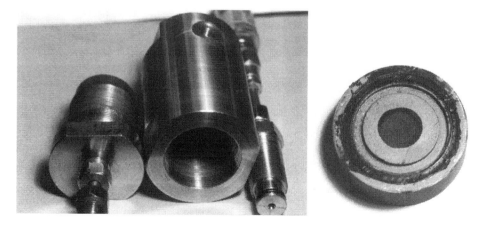

FIGURE 23.12 High-pressure permeation cell parts and molded/cured sample.

Following tests made with this equipment, publications[12,23] show values for Q and related coefficients for many high-pressure gases and polymers indicating (cf. Section 23.5.1.4) how

- Elastomers can be compacted by the hydrostatic effects of the high-pressure gas, concurrently with the gas entering the elastomer as part of the permeation process
- Following this, elastomers can be swollen by some high-pressure gases (especially CO_2); as the densities of these gases approach liquid-like levels, at appropriate temperatures they become supercritical fluids which possess a solubility parameter magnitudes that, however, are highly dependent on temperature and pressure

Both of these effects cause a reduction in Q values compared with low-pressure measurements; as time progresses, this reduction can increase.

Values from high-pressure permeation tests can give useful information regarding the selecting of elastomer types to withstand potential explosive (rapid gas) decompression damage (Sections 23.5.1.1 and 23.5.1.2).

23.4.5 General Comments on Permeation Testing

23.4.5.1 Practical

Regarding high-pressure gas permeation testing, the presence of a stainless steel sinter to support the sample against the applied gas pressure is normally considered essential. However, using thermoplastic rather than elastomer samples because of the inherent stiffness of the former, Morgan and Campion have shown that the presence of a sinter reduced measured rates from their correct values because the steel around the sinter's pores, which contacts the rear sample face, in so doing reduced the effective area of this face. To overcome this problem for testing at a moderately elevated pressure of 70 atm (bars) a sample made in the shape of a truncated cone was employed, an added advantage being the effective elimination of edge effects outside the seals usually used within the permeation cell—also usually a source of error. The conical sample is adhered within an accurately profiled sample holder which is inserted and sealed into the pressure cell such that the high pressure will be applied to the sample base (major radius); a three-dimensional "wedge effect" thus applies.

When testing, the permeation equation shown for conical samples in Section 23.4.2 is used to provide permeation coefficient Q. Measured values showed that the more normal methods employed for high-pressure permeation testing at MERL and elsewhere, using a sample support sinter, are likely to give underestimates of true permeation rates by a factor of about three.[12]

The rationale applies equally to elastomer testing, but test development is still required due to the lower stiffnesses involved.

23.4.5.2 Theoretical

For single-component systems, permeation testing can be divided into four broad categories:

1. **Gases**
   Pressure-dependent; permeation rate and all coefficients can be obtained from one test, usually relatively short.
2. **Vapor or mixed vapor/liquid below saturated vapor pressure (svp)**
   Vapour-pressure-dependent up to svp, supplying the same data as in (1) per test; tests likely to be medium-term to give sufficient permeant for reliable measurement. Can reduce vapor pressure (e.g., by using a carrier gas) on low-pressure side to maximize permeation by keeping P_2 near-zero.

3. **Mixed vapor/liquid at svp**
   Effectively pressure-independent*; same other comments as in (2).
4. **Liquid (vapor-free)**
   Essentially pressure-independent*; only permeation rate and D readily obtainable; medium-long tests (again can be shortened by reducing vapor pressure on low pressure side).

All the data from testing to these categories can be used to compare fluids and/or polymers and to make service predictions.

23.4.6 PREDICTING FROM COEFFICIENTS

Q provides insight on how much fluid passes right through a polymer at steady-state conditions whereas D is used to estimate the time to breakthrough and s governs the amount of fluid initially available for migration. Predictions stemming from Q should be applied to the correct service geometry (see Section 23.4.2). Since Arrhenius-type expressions apply to all three coefficients,[11] accelerated testing at elevated temperatures can be undertaken. A plot of log [Q or D or s] against reciprocal absolute temperature can be extrapolated to a lower service temperature, to allow predictions to be made.

23.5 FLUID ENVIRONMENTS—SPECIAL SITUATIONS

23.5.1 OTHER PHENOMENA AT HIGH FLUID PRESSURES

23.5.1.1 Explosive (Rapid Gas) Decompression and Doming

Arising from Henry's law (Section 23.4.2), a piece of rubber (elastomer) can have many times its own volume of gas dissolved in it when exposed to the gas under pressure. Just as carbon dioxide dissolved in lemonade appears as bubbles when the pressure is released by opening the cap, so gas dissolved under pressure in rubber will come out of solution when the pressure is released—e.g., by shutting down a pumping system. Carbon dioxide bubbles in lemonade can be seen to grow with time as they rise to the surface. This is because, with time, more gas is coming out of solution. However, bubbles inside an elastomer are trapped—and as they grow and coalesce, they may possess enough strength eventually to tear the rubber. In extreme cases the rubber can essentially explode—hence the term "explosive decompression."

23.5.1.1.1 Half-Bubbles or Domes at Elastomer–Substrate Interfaces
If an elastomer is bonded to a substrate such as steel, it is usual for the bond to have small areas of imperfection where the adhesive or the chemical preparation of the surface is defective. Such areas are known as "holidays." In high-pressure gas environments, these holidays form nucleation sites for the growth of half-bubbles or "domes," under conditions where gas has been dissolved in the elastomer and the pressure has subsequently been reduced. Gas collecting at the imperfection at the interface will inflate the rubber layer, and "domes" will show as bumps on the surface of the rubber-coating layer—just as a paint layer bubbles up in domes when the wood underneath gives off moisture or solvents in particular areas.

Some of the mathematics of dome growth in an elastomer layer bonded to a substrate has been published.[24] Clearly, from the point of view of potential damage, the conditions for domes to

---

* If applied pressure P is increased from condition 1 to 2, then $\ln(p_1/p_2) = V_m(P_1 - P_2)/RT$, where molar volume is V_m, the gas constant is R and temperature is T. From this, e.g., for water, a 1000-fold increase in P only approximately doubles saturated vapor pressure p. For hydrocarbons, p could be doubled by a lower pressure increase, in the order of 150 times or so; however for moderate pressures, a tenfold increase in P even here only increases p by some 5%. Hence, for most practical situations, vapor pressure of a liquid can be considered as independent of applied pressure. Vapor-free liquid may need chemical potential represented differently (possibly by work done).

continue to grow are critical. There are three forces which must balance—the elastic force due to extension of the rubber as it forms a dome, the gas pressure inside the dome, and the bond strength of the rubber to the substrate which is resisting the growth of the dome. For example, we have:

$$P_G = \frac{2PV_G}{\pi hs\left(f + \dfrac{h}{2}\right)^2} \tag{23.14}$$

which relates how the gas pressure P_G to which the elastomer has been exposed is related to the pressure P and volume V_G of the gas in the dome, where f is the flaw diameter, h the thickness of the rubber layer, and s the solubility coefficient of the gas in the rubber. In determining whether detachment will occur, the bond strength of the rubber-to-substrate bond will be required, but space does not permit further treatment here—see reference above. The phenomenon can be of great importance in lined metal pipes, in pumps—and in hoses made from multiple layers of (usually) different rubbers and reinforcements, where doming becomes layer separation.

23.5.1.1.2 Explosive Decompression in Solid Elastomer Sections—The Nature of ED Damage

Internal bubble growth occurs because flaws at an elastomer–substrate interface act as nucleation sites where gas which is coming out of solution in the rubber can collect, following a reduction in pressure. If there are flaw sites, based on energy considerations, gas will take the easiest option—which is to collect at an already preexisting bubble and cause it to grow.

It used to be believed that ED would only occur if there were preexisting flaws in the rubber—and attempts were made to purify rubber to make it more ED "resistant." This proved futile and it is now known that the "flaws" can be at the near molecular level, and are ubiquitous. Using CO_2, Derham and Thomson[24] have shown that even in rubber so pure that it is glass-clear optically transparent, thousands of bubbles can nucleate in a few cubic centimeters. Due again to energy considerations, these immediately begin to coalesce into a smaller number of larger bubbles. This process will go on with larger bubbles gobbling up smaller bubbles (see Figure 23.13), until there are a very few large bubbles—or sufficient time has elapsed for dissolved gas to have diffused out of the elastomer at its free surfaces.

Derham has shown that, not surprisingly, the predominant ED mechanism leading to bubble growth in soft (transparent) elastomer is fracture or tear of the rubber, bubbles evolving from

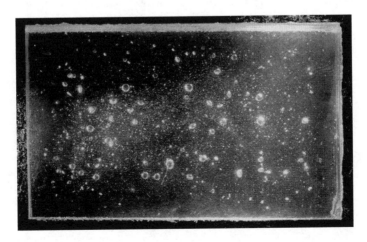

FIGURE 23.13 Explosive decompression (ED) in soft transparent elastomer; bubbles of carbon dioxide (CO_2) several days after decompression.

ellipsoids to predominantly spherical shape, by tearing along a very well-defined "equator." The surfaces of bubbles have been found to show growth rings (very much like tree rings) which indicate that normal stick–slip rubber tear is occurring. These rings have been seen on both hard and soft rubbers. It is thus clear that tear strength, and ways to resist tear, are key to ED resistance.

The amount of gas which has been dissolved in the elastomer will determine the speed of the ED process. In cases where the gas pressure has been, say, many tens of megapascals (MPa), and the pressure has been reduced to atmospheric in minutes rather than hours, the ED, once initiated, can occur catastrophically in seconds. Where it is practically possible, ED damage can be avoided by reducing the pressure slowly.

23.5.1.1.3 Explosive Decompression in Solid Elastomer Sections—Predominantly O-ring and Other Seals

In service situations it is not possible to devote long periods, possibly days, to the depressurization stage—and in other operational circumstances depressurization may be accidental and, therefore, uncontrolled. To combat such events, "engineering" elastomers as used for sealing in offshore oil and gas production are designed to be of high modulus, with high loadings of particulate filler, and being tightly cured. This target is often achieved by employing peroxide cure systems—which, however, tend to be associated with low tear-strength levels. To explain this apparent paradox between this approach and Derham's conclusions, one review of the ED topic[25] includes an expression developed by Gent and Tomkins showing how pressure within an inflated bubble relates directly to modulus for the same bubble size. Hence a low pressure is sufficient for a bubble to reach a critical size at which rupture commences for the soft elastomer—this pressure arising by gas coming out of solution in the elastomer and building up pressure in the bubble. This situation would be reached relatively early on in the decompression process, so that tearing would be occurring for most of the decompression process—to accord with Derham's observations. With the "hard" engineering elastomer seals, the initial bubble-growth stage is extended to such a degree (especially if the modulus is further increased by support from the surrounding housing groove—compare shape factor effects in Section 23.2) that critical bubble sizes for rupture are not in practice reached for moderate service application conditions.

In fact, ED damage does occur in offshore gas production at more severe conditions; rupture occurs, leading to blisters and/or splits in the elastomer (see Figure 23.14). To resist ED damage from a material viewpoint, it is traditionally difficult to develop recipes that yield both high modulus and tear-strength magnitudes. However, a MERL approach sought to achieve this goal by a less traditional way—by developing a rubber with altered morphology. Campion and Thomson[26] have developed a material with increased energy dissipation, so that it is much more resistant to developing cracks, while still retaining a high modulus to inhibit initial bubble growth.

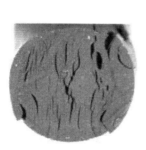

FIGURE 23.14 Explosive decompression (ED) damage in a fluoroelastomer O-ring seal section (*left*) and miscellaneous samples of a nonfluorinated oil-field elastomer (*right*) obtained using pressure vessels such as those shown in Figure 23.15.

The modified rubber is referred to as biphasic single-component elastomer (BPSCE). Here, a special cure system permits a high degree of control to be exercised over a deliberate nano-structural heterogeneity of cross-linked HNBR elastomer. The resulting BPSCEs are characterized by superior resistance to ED, and very high tear strengths—typical values for which greatly exceed those of most conventionally formulated rubbers. BPSCE, although based on a single elastomer, is a two component system comprising a rigid skeletal element for stiffness, seamlessly interspersed with a yielding phase which acts to inhibit crack growth.

Returning to the cause of fracturing by ED, although there is ample evidence above to show that the presence of flaws is not *necessary* to bring about ED damage, Campion has shown in filled sealing elastomers that micro-discontinuities present due to processing considerations do assist in promoting failure. By molding samples in nonstandard ways, it was possible to induce molding flow to traverse sheet samples—rather than follow the plane of the sheet as is more usual—and in subsequent ED testing, failures followed this crossways direction, rather than occurring in the more normal planar direction as shown in Figure 23.14.

Other means of reducing the risk of ED damage, not based on the material, also exist. All factors are outlined in Section 23.5.1.2.

23.5.1.2   Strategies to Increase Resistance to ED Damage

Table 23.2 shows a summary of possible ways whereby ED damage can be reduced.

Further to aspects discussed in Section 23.5.1.2, the simplest strategy, if it is practical, is slow decompression. If the pressure can be reduced sufficiently slowly, the gas which is dissolved in the elastomer can diffuse out of the rubber without building up sufficient internal pressure to cause damage. The rate of decompression required may be roughly calculated if the diffusion coefficient of the gas and its solubility in the elastomer are known—but in practice it will probably require a laboratory simulation.

Housing design can provide a good opportunity for reducing greatly the occurrence of ED damage. Elastomers have a very high bulk modulus: at mechanical pressures of 70 MPa (10,000 psi) a seal will only compress about 3%. Analogously, a bubble of gas trying to grow inside a rubber section will be fighting against the high bulk modulus if the seal is essentially fully constrained. Under these circumstances the destructive bubbles of gas simply cannot grow. In design, therefore, if tighter tolerances can be achieved so that a slightly softer seal can be made to fit its housing

---

TABLE 23.2
Ways to Lessen Risk of Explosive Decompression (ED) Damage for Sealing Elastomers in Service

Variable	"Ideal" Level	Comment
Material modulus	HIGH	Governs initial bubble inflation
Tear strength	HIGH	To prevent subsequent rupture; not usually compatible with high modulus
Gas pressure	LOW	Above 7–14 MPa (1000–2000 psi), risk of ED greatly increases
Gas concentration	LOW	Depends on gas, elastomer, increasing temperature
Diffusion coefficient	HIGH	Depends on gas, elastomer, increases with temperature
Temperature	PROBABLY LOW	Gases more soluble and escape by diffusion slower, but mechanical properties high
Decompression rate	LOW	If very slow, ED will not occur; often impractical in service
Seal cross section	SMALL	ED damage most likely in center of seal (for e.g., diffusion reasons)
Seal constraint (groove fill)	HIGH	Inhibits seal expanding and hence blister/rupture formation when decompressed
Decompression cycles	FEW	Risk of damage increases with cycle number

tightly, and extrusion can be avoided by design, it will be possible to have a seal which is both more effective and which is resistant to ED.

It is necessary to have a healthy scepticism of most elastomers which are described as "ED resistant" unless their recipes are known to have been designed with regard to covering all relevant factors. How much they will resist will depend very much on the conditions—and conventionally formulated elastomers have simply been designed (usually primarily by increasing hardness) to be *relatively* ED-resistant. With all elastomers, resistance will depend on numerous factors including characteristics of the elastomer itself, the housing design, the contact fluids and their temperature, pressure, and solubility. The morphology of bubble formation, i.e., large single bubbles or a tendency towards larger numbers of smaller bubbles persisting, is also a factor. Some deductions can be made by theoretical models, but in the present state of the art, so many factors are interacting that prototype testing under simulated real conditions will be necessary, certainly in safety-critical applications.

Many subtleties associated with ED, for instance, accompanying thermodynamic cooling issues, failure processes, and effects of localized stresses, are discussed in detail in the extensive review on this topic by Briscoe et al.[27] Other workers have observed similar fracture effects arising from rapid temperature increases while maintaining pressure[28,29]; the connection with ED is via Henry's law linking dissolved gas concentration and solubility coefficient, and the fact that solubility coefficient *decreases* (in an Arrhenius fashion, as it happens) for readily condensable (i.e., less volatile) gases when temperature increases.[11]

23.5.1.3 Liquid Absorption at High Pressure

The general approach employed for standard absorption (liquid mass uptake) testing outlined in Section 23.4.4.1 can also be used for high-pressure testing, but including the additional and significant step of conducting all exposures (immersions) in liquid contained in a pressure cell (see Figure 23.15). Liquid pressure is quantified by a pressure transducer, and temperatures by thermocouples, all internally located. A suitably linked pressure pump and liquid reservoir ensures that sufficient liquid is present during the exposures. The drawback is that, for each weighing, pressure and temperature must be reduced and the cell opened to access the samples; for this reason, many samples are included per test. These processes are reversed as rapidly as possible to continue the test.

FIGURE 23.15 Typical pressure cells as used for high-pressure absorption testing (before insulation applied).

TABLE 23.3

Some Pressure Effects on Liquid Diffusion Coefficients and Equilibrium Masses Absorbed

Elastomer	Solvent	Applied Pressure P (atm)	Equilibrium Mass Uptake m_∞ (%)	Diffusion Coefficient D $\times 10^{-8}$ (cm^2 s^{-1})
FEP[a]	Toluene	1	27	12
		172	23	16
		345	21	12
EPDM[b]	Toluene	1	134	36
		172	114	39
		345	116	43
FKM[c]	Methanol	1	33	4
		172	35	4
NBR[d] ("nitrile")	Methanol	1	10	13
		172	9	7

[a] Tetrafluoroethylene propylene copolymer.
[b] Ethylene propylene diene terpolymer.
[c] Fluoroelastomer copolymer.
[d] 36/64 Acrylonitrile butadiene copolymer.

By such means, Morgan and Campion have found that pressurizing solvents during absorption tests for several elastomers caused some effects, but these were not great even at pressures of hundreds of atmospheres (see Table 23.3).

Values of D for the pressurized tests may contain some errors because of the times needed to remove samples from the cells. It is therefore likely that D is not affected greatly by increases in the applied pressure of the liquid being absorbed. Although differences in equilibrium mass uptake m_∞ values at the various pressures were again not great, there may be a tendency for a slight decrease to occur with increasing pressure; if so, this could arise of hydrostatic compaction of the elastomer by the pressurized liquid (see Section 23.5.1.4 for related measurements for high-pressure gases).

23.5.1.4   Measurements of High-Pressure Gas-Induced Compaction and Swelling

Based on the high-pressure equipment shown in Figure 23.12, Lewan and Campion developed a pressure vessel for making accurate measurements of thickness changes experienced by an elastomer when exposed to high-pressure gases; as the sample is molded and bonded into a shallow steel rectangular-sectioned round dish, these thickness changes also represent volume changes. In one test, an EPDM elastomer was pressurized by an inert gas and thickness changes measured across a regime of temperature and associated pressure changes; the technique used to make these measurements (replaced by an linear variable differential transformer [LVDT] system in current versions), and the full data, are described elsewhere.[1]

To illustrate the degree of volume change that occurred, Figure 23.16 outlines the behavior at ambient temperature for 40 h following the start of test. The initial increase in pressure to 8300 psi/58 MPa gave a decrease in thickness (volume) of 14%. Presumably internal lubrication by gas molecules led to increased compaction relative to mechanical pressurization (Section 23.5.1.2). With the pressure staying constant, the sample subsequently began to swell (exhibiting a typical diffusion-type profile) until essentially a constant thickness was reached, when the system was thought to be in equilibrium; it was observed after the 25–40 h (overnight) stage that the temperature had fallen by 3°C, to explain the slight negative gradient. The swelling was about a 30% reduction of the compacted value.

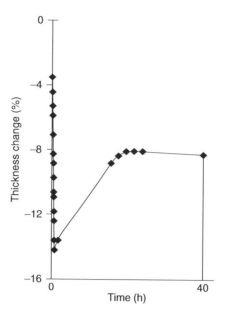

FIGURE 23.16 Thickness (volume) changes from exposing ethylene propylene diene monomer (EPDM) to 8300 psi inert gas pressures.

The work shows that compaction and swelling effects should be included in assessments when sealing high-pressure oil-field fluids. Housing and/or seal design should be such that these effects are minimized.

23.5.2 DISCONTINUITIES WITHIN THE MOLDED ELASTOMER BULK

Durability of elastomer articles can depend on fatigue issues. If flaws are present in a component due, for example, to incomplete manufacturing processes, fatigue life could well be reduced accordingly. In an extensive exercise examining relevant factors for an elastomer diaphragm in widespread usage, it was found that a combination of appropriate peel testing and diagnostic gas

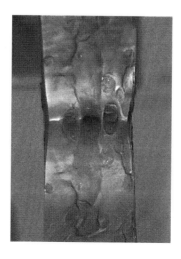

FIGURE 23.17 Styrene–butadiene copolymer (SBR) peel adhesion test piece after first applying a gas decompression (GD) procedure and then peel testing.

decompression (GD) exposures could indicate whether such flaws would be present. The GD technique arose from applying the usually deleterious phenomenon of explosive (rapid gas) decompression (Section 23.5.1.1) to advantage. For this, Lewan employed CO_2 gas but only under mild pressure, short exposure time, and moderate decompression rate (Lewan, M.V. and Campion, R.P., unpublished work).

In more detail, if two identical elastomer sheets are pressed and held together, and cured, they will adhere during this process if the base elastomeric compound is tacky (possesses a high degree of autohesion) when uncured. The tackiness ensures that the surfaces remain in intimate contact, and interdiffusion of curatives is then possible during early stages of cure, leading later on to cross-links being formed across the interface, to "cement" the good bonding. When the elastomer used was a NR recipe, its high tack level and subsequent high cured rubber-to-rubber adhesion level meant that peel-adhesion test values were very high, with test pieces tearing across tabs ("cohesive failure") rather than along the original interface. In addition, a similar result was obtained for replicate peel test pieces first exposed to the gas-decompression procedure before performing the peel testing—there was no "memory interface" remaining from the assembly stage.

When an EPDM elastomer was examined, its tack was so low that no permanent cured rubber-to-rubber bond formed and the peel sample separated cleanly during test, at very low separation forces. No GD stage was necessary.

With an elastomer of moderate tack—a styrene–butadiene copolymer (SBR)—the peel adhesion separating force measured was high, approaching the NR result. Cohesive failure occurred. But when replicate samples were exposed to the GD procedure and peel testing then applied, peel forces were reduced by 80%, and separation was at the original interface which could be clearly observed (Figure 23.17). Besides a few ED fractures, the exposed surface showed other regions which clearly had not been well bonded, despite the high-peel forces obtained for those other replicate test pieces which had not been exposed to gas before peel testing. A memory interface—which if present in analogous engineering articles could give rise to future fatigue problems in service—had been present, although peel testing alone had not shown this up.

These observations indicate that peel testing can provide high test values for interfaces which are not completely bonded, presumably providing that some regions of the interface *are* well bonded—these shield the worse regions, perhaps helped by geometrical factors applying during testing. The work also shows how a combination of techniques can be used to resolve such a point—one which was difficult to resolve by either technique alone. An analogous approach could be applied to an appropriate manufactured component if desired.

23.5.3 Summary of Chemical Aging Issues

Clearly, a complete examination of elastomer durability according to this chapter's title should include a full assessment of chemical issues involving elastomers, but these are so many and so diverse that space does not nearly permit. Instead, a brief comment only is made. Much of MERL's recent interest in chemical behavior has involved the assessment of the stability or otherwise of a wide range of elastomers—many rated as fluid- and temperature-resistant ones—when exposed to oil-field chemicals in pressure cells of the type shown in Figure 23.15. Thomson[30] has described how hydrogen sulphide (a product of "sour" oil wells) can chemically degrade elastomers, measuring changes in property levels, and providing many data to follow the associated kinetics. He concluded that, at high temperatures, a perfluoroelastomer, a tetrafluoroethylene–propylene copolymer, and a particular base-resistant fluoroelastomer offered the best resistance. In other work, Thomson and others[31] have examined reinjection issues of acid gas being returned to a well—effects on elastomers again being paramount.

Application of mechanical stresses, implicitly altering chemical potentials, can also effect further changes, as illustrated by Morgan et al.[32] for a fluoroelastomer in aqueous amine environment at several temperatures; this study indicated that a balance existed between the amine attack which brought about extra cross-linking and a physical process (a water uptake by certain filler particles), and that the addition of stress accelerates the chemical process.

An important issue regarding chemical durability is knowing when deterioration has progressed sufficiently for the elastomer no longer to be able to function as designed. Thus, failure criteria must be specified. These can be many and varied, according to the service application; in a review[1] in 2003, a dozen or so of these are listed, and a description of how the Arrhenius approach can be used to apply a chosen failure criterion is given. This approach translates data from (short) accelerated tests at high temperatures to service temperature; extrapolation becomes possible to give the time at service temperature to reach the failure criterion previously selected. The rationale and caveats are outlined, but the approach is one worth applying where possible.

23.6 CONCLUSIONS

The many and various uses of elastomers all arise from the flexible nature of this versatile class of polymer. Against the background of continuous kinetic motion, this flexibility arises for reasons associated with the general molecular structure of elastomers in general, and from particular structural and morphological features which exist for individual elastomers to lead to the different characteristics they possess. But, paradoxically, the degree of openness of the structure also provides the deterioration modes that can in time lead to failure if material selection has been wrong, if service forces applied should alter, or if exposure conditions change.

The intent of this chapter has been to outline how appropriate laboratory testing, under service-realistic conditions where possible, can contribute to the understanding of mechanisms of elastomer or component deterioration which can apply for a variety of situations. For example, both established and novel means of following certain mechanisms and determining their rates have been discussed, optimum elastomer characteristics and conditions to be avoided where possible for fracturing phenomena such as ED have been listed, and guidelines provided on how correct design might make due allowances for deterioration which follows known rules, such as creep or stress relaxation, to yield acceptable service lifetimes. Some case studies have been included.

What is certain is that elastomers can and do perform critical duties in engineering applications which function for many years in conditions that can be severe or hostile. The basic requirement is to select the right elastomer for the job—which in turn can mean to carry out the most appropriate design, test, and validation work before putting a component into service.

ACKNOWLEDGMENTS

The authors are grateful to MERL for permission to write this chapter, and acknowledge the work performed by colleagues Dr. Barry Thomson and Dr. Nickie Smith.

REFERENCES

1. Campion, R.P., *Rubber Chem. Technol.* (*Rubber Rev.*), 76(3), 719, 2003.
2. Derham, C.J., *Plast., Rubber Compos. Process. Appl.*, 26(3), 129, 1997.
3. Derham, C.J. and Waller, R.A., *The Consulting Engineer*, 39, 49, 1975.
4. Stevenson, A. and Campion, R.P., Durability, in *Engineering with Rubber*, Gent, A.N. (Ed.), Hanser-Gardner, Cincinnati, OH; also Carl Hanser, Munich, Germany, 2nd ed., 2001, chap. 7.
5. Shepherd, R., Stevenson, A., and Abrams, P.I., *Proc. Conf. NACE Expo'97*, New Orleans, Louisiana, NACE, Houston, TX, May 1997, paper 87.
6. Campion, R.P., Shepherd, R., and Priest, A.M., *Proc. 12th OMAE Conf.*, Glasgow, United Kingdom, III-A, Materials Engineering, ASME, New York, 1993, 399.
7. Harrison, C. and Stevenson, A., *Proc. Conf. SAE Congress*, Detroit, MI, SAE, Warrendale, PA, 1998, 981083.

8. Examples involve (1) measurements of changes in mass-dependent resonant frequency of a beam attached to the elastomer sample in pressurized CO_2 (Briscoe, B.J., Liatsis, D., and Mahgerefteh, H., *Proc. Conf. Diffusion in Polymers*, PRI, Reading, United Kingdom, March 1988, paper 17) and (2) weighing samples using high-pressure microbalance within a cell containing pressurized test gas (von Solms, N. et al, *J. Appl. Polym. Sci.*, 91, 1476, 2004), also with data relayed outside a cell by a magnetic system (von Schnitzler, J. and Eggers, R., *J. Supercrit. Fluids*, 16, 81–92, 1999).

9. Fick, A., *Annln. Phys.*, 94, 59, 1855.

10. Crank, J., *The Mathematics of Diffusion*, 2nd ed., Oxford: Oxford University Press, 1979.

11. van Amerongen, G.J., *Rubber Chem. Technol. (Rubber Rev.)*, 37(5), 1065, 1964.

12. Campion, R.P. and Morgan, G.J., *Proc. 4th Conf. Oilfield Engineering with Polymers*, London, MERL, Hitchin, United Kingdom, 2003, paper 11.

13. Hildebrand, J.H., Prausnitz, J.M., and Scott, R.L., *Regular and Related Solutions*, Van Nostrand Reinhold, New York, 1970.

14. Campion, R.P., *Proc. Conf. Polymer Testing*, Shawbury, UK, Rapra, Shawbury, United Kingdom, 1996, paper 14.

15. Norwegian standard NORSOK M-710 Rev. 2, Qualification of non-metallic sealing materials and manufacturers, October 2001.

16. Allen, G., Gee, G., and Nicholson, J.P., *Polymer*, 1, 56, 1960.

17. Sheehan, C.J. and Bisio, A.L., *Rubber Chem. Technol.*, 39(1), 149, 1966.

18. Campion, R.P., *Permeation through Polymers for Process Industry Applications*, MTI Publication No. 53, Materials Technology Institute of the Chemical Process Industries, St Louis, MO; distributed by Elsevier Science, Amsterdam, The Netherlands, 2000.

19. Campion, R.P. and Thomson, B., *Proc. 1st Conf. Polymers in Automotive Fuel Containment*, Birmingham, UK, Rapra, Shawbury, United Kingdom, 2000.

20. Smith, N.L. and Campion, R.P., *Proc. 2nd Conf. Polymers in Automotive Fuel Containment*, Hanover, Germany, Rapra, Shawbury, United Kingdom, 2005.

21. Lomax, M.L., *Permeability Review*, Members Reports, Rapra, Shawbury, United Kingdom, 1979, No. 33.

22. Hardy, D., et al., *Proc. Conf. Rubber Chem. '99*, Antwerp, Belgium, Rapra, Shawbury, United Kingdom, 1999, paper 2.

23. Campion, R.P. and Morgan, G.J., *Plastics, Rubber & Composites Processing & Applications*, 17, 51, 1992.

24. Derham, C.J. and Thomson, B. *Proc. 4th Conf. Oilfield Engineering with Polymers*, London, MERL, Hitchin, United Kingdom, 2003, paper 1.

25. Campion, R.P., *Cell. Polym.*, 9(3), 206, 1990.

26. Campion, R.P. and Thomson, B., *Offshore Research Focus 114*, Offshore Supplies Office (United Kingdom), October 1996, 5.

27. Briscoe, B.J., Savvas, T., and Kelly, C.T., *Rubber Chem. Technol.*, 67, 384, 1994.

28. George, A.F., *Proc. 10th Intl. Conf. on Fluid Sealing*, Innsbruck, Austria, BHRA, Cranfield, United Kingdom, 1984, D2.

29. George, A.F., Sully, S., and Davies, O.M., *Fluid Sealing*, 1997, 437.

30. Thomson, B., *Proc. 5th Conf. Oilfield Engineering with Polymers*, London, MERL, Hitchin, United Kingdom, 2006, paper 4.

31. Martin, T., Abrams, P.I., Harris, R., and Thomson, B., *Proc. 5th Conf. Oilfield Engineering with Polymers*, London, MERL, Hitchin, United Kingdom, 2006, paper 5.

32. Morgan, G.M., Campion, R.P., and Derham, C.J., *Proc. Meeting Rubber Division ACS* Chicago, IL, 1999, paper 56.

24 Extrapolating the Viscoelastic Response of Rubber

C.M. Roland

CONTENTS

24.1 INTRODUCTION

By definition, the term rubber comprises any flexible-chain, amorphous polymer above its glass transition temperature. However, T_g is rate-dependent and at sufficiently high rates rubbers can exhibit glassy behavior. This effect is exploited in various applications of elastomers, such as tire treads and acoustic tiles, and thus requires accurate characterization. Commercially available dynamic mechanical spectrometers are generally limited to low frequencies (<100 s^{-1}) and low strains. Servohydraulic test systems can achieve relatively high crosshead speeds, but in practice have limitations on their attainable strain rates [1,2]. Because of these limitations, various groups have resorted to design of their own equipment to measure the mechanical properties of rubber at high rates.

If the material characterization is limited to the linear viscoelastic response, dielectric spectroscopy [3,4] can be employed to supplement or even replace mechanical measurements. The local segmental dynamics in polymers having a dipole moment transverse to the chain axis give rise to the α-dielectric relaxation. Chain modes, corresponding to Rouse and entangled dynamics, require a dipole parallel to the chain to be dielectrically active. The dielectric α-relaxation time is smaller than the corresponding mechanical τ; however, the T- [5–9] and P-dependences [10,11] of the two quantities are the same. Dielectric spectroscopy can be carried out routinely over 11 decades, down to $\sim 10^{-7}$ s, with direct measurements of the sample capacitance. This dynamic range can be extended to ~ 18 decades [12], although corrections for contributions from connectors and cabling are required. In addition to its broader frequency range, the other obvious advantage of dielectric spectroscopy over mechanical is the former's absence of moving parts. This means that hydrostatic pressure can be easily applied, enabling the P-dependence of τ to be measured [13–15]. Few mechanical studies of the effect of pressure have been carried out. These are generally limited to studying the viscosity, η, of relatively low-molecular weight polymers [16–18] or the dynamic modulus of polymer melts in a rheometer pressurized with a gas [19–21]. Since gases generally dissolve in rubber to varying degrees, the pressurizing medium may also plasticize the sample. There is a scarcity of data, but the proportionality between mechanical relaxation times (or η) and dielectric τ appears to be maintained under pressure [10,11].

Since we are interested in this chapter in analyzing the T- and P-dependences of polymer viscoelasticity, our emphasis is on dielectric relaxation results. We focus on the means to extrapolate data measured at low strain rates and ambient pressures to higher rates and pressures. The usual practice is to invoke the time–temperature superposition principle with a similar approach for extrapolation to elevated pressures [22]. The limitations of conventional t–T superpositioning will be discussed. A newly developed thermodynamic scaling procedure, based on consideration of the intermolecular repulsive potential, is presented. Applications and limitations of this scaling procedure are described.

24.2 FREE-VOLUME MODELS OF POLYMER DYNAMICS

According to free-volume interpretations, the rate of molecular motions is governed entirely by the available unoccupied space ("free volume"). Early studies of molecular liquids led to the Doolittle equation, relating the viscosity to the fractional free volume, f [23,24]

$$\log \eta(T) = \log A_{\mathrm{D}} + B_{\mathrm{D}}/f(T) \tag{24.1}$$

where A_{D} and B_{D} are constants. However, fits of Equation 24.1 to experimental data show deviations beyond any experimental error [25–27]. By assuming that free volume expands linearly with temperature, the Williams–Landel–Ferry (WLF) equation [22,28]

$$\log \eta(T) = \log \eta(T_{\mathrm{R}}) - \frac{c_1(T - T_{\mathrm{R}})}{c_2 + T - T_{\mathrm{R}}} \tag{24.2}$$

is obtained, where T_{R} is an arbitrary reference temperature. The WLF equation is popular for fitting experimental data, including relaxation times (τ replacing η) since η and τ are proportional to each other except near the glass transition [29–31]. When the actual magnitudes of η or τ are not known, t–T shift factors of viscoelastic quantities are analyzed. In the original theory the WLF constants c_1 and c_2 were identified with functions of f and its change with temperature, with the approximate "universal values" of $c_1 = 8.86$ and $c_2 = 101.6$ [28]. An alternative equation for the T-dependence of relaxation times is the Vogel–Fulcher (VF) equation [22]

$$\log \tau(T) = \log \tau_{\mathrm{VF}} + \frac{b_{\mathrm{VF}}}{T - T_{\mathrm{V}}} \tag{24.3}$$

in which b_{VF} and τ_{VF} are constants. At the Vogel temperature, T_{V}, the dynamics would diverge if not for intervention of the glass transition. T_{V} is often taken to be the Kauzmann temperature at which the extrapolated liquid entropy equals the crystal entropy, ostensibly in violation of the third law of thermodynamics ("Kauzmann paradox") [32]. Although having different origins, Equations 24.2 and 24.3 are mathematically equivalent: $b_{\mathrm{VF}} = c_1/c_2$ and $T_{\mathrm{V}} = T_{\mathrm{R}} - c_2$ [33].

A more rigorous free-volume treatment is due to Cohen and Grest (CG) [34,35], according to which the material is comprised of liquid and solid-like cells. The former have free volume, but mobility requires continuity of the local empty space. The temperature dependence of the relaxation times according to the CG model is

$$\log \tau(T) = A_{\mathrm{CG}} + \frac{B_{\mathrm{CG}}}{T - T_{\mathrm{CG}} + \left[(T - T_{\mathrm{CG}})^2 + C_{\mathrm{CG}}T\right]^{1/2}} \tag{24.4}$$

with A_{CG}, B_{CG}, C_{CG}, and T_{CG} material constants. T_{CG} corresponds to the percolation temperature at which continuity of the liquid-like cells is attained; that is, at T_{CG} each liquid-like molecule has at least two near neighbors that are also liquid-like. Since Equation 24.3 has one parameter more than the WLF or VF equations, it more accurately fits experimental data [34,36,37]. However, over a sufficiently broad temperature range, the T-dependence of τ (and η) changes, requiring two WLF or VF equations [38,39]. Equation 24.4, on the other hand, can describe data over the entire measurable range, with the change in the temperature dependence of $\tau(T)$, usually identified by taking the temperature derivative of $\tau(T)$ [38–40], occuring at the percolation temperature T_{CG} [41].

To account for the variation of the dynamics with pressure, the free volume is allowed to compress with P, but differently than the total compressibility of the material [22]. One consequent problem is that fitting data can lead to the unphysical result that the free volume is less compressible than the occupied volume [42]. The CG model has been modified with an additional parameter to describe $\tau(P)$ [34,35]; however, the resulting expression does not accurately fit data obtained at high pressure [41,43,44]. Beyond describing experimental results, the CG fit parameters yield free volumes that are inconsistent with the unoccupied volume deduced from cell models [41]. More generally, a free-volume approach to dynamics is at odds with the experimental result that relaxation in polymers is to a significant degree a thermally activated process [14,15,45].

24.3 TIME–TEMPERATURE SUPERPOSITIONING

The dubious underpinnings of free-volume models do not undermine the use of free-volume-derived equations to fit data. Indeed, the VF (Equation 24.3), which is mathematically equivalent to the WLF equation, can be obtained from an entropy model such as the Adam-Gibbs theory [46] or by assuming activated transport [47]. Since the principle of time–temperature superpositioning was drawn from free-volume ideas [48], however, its validity must be established empirically. And it has been shown for every polymer tested to date [49] that breakdown to t–T superpositioning occurs when both the polymeric chain modes (corresponding to the lower frequency end of the transition zone and terminating for entangled polymers at the rubbery plateau) and the local segmental modes (which segue into the glass transition) contribute to the measured response. For example, this breakdown has been demonstrated for polystyrene [50], polyisobutylene [51,52], atactic polypropylene [53], polyvinylacetate [54], polymethylmethacrylate [55], and 1,4-polyisoprene [56]. The thermorheological complexity of polymers in the softening zone does not invalidate the use of t–T superpositioning in all cases. For example, rheological properties, for which the underlying molecular mechanisms have the same temperature dependence, should superpose. However, when high-temperature measurements are used to obtain terminal (or chain) relaxation times and these are combined with low-temperature measurements of segmental relaxation times, the extrapolation of the combined shift factors can be in error by orders of magnitude. For this reason t–T superpositioning is unreliable for determination of strain rates relevant for very high-frequency applications of rubber, such as impact resistance.

24.4 THERMODYNAMIC SCALING OF POLYMER DYNAMICS

A more realistic alternative to free-volume models is to interpret relaxation and related properties of polymers in terms of the interactions among segments. This is especially true as the glass transition is approached from the liquid or rubbery state. In this "supercooled" regime the molecular mobility is governed by (infrequent) jumps over barriers that are large compared to the available thermal energy [57–59]. However, due to the cooperativity inherent to vitrifying materials, each molecule or polymer segment is involved in competing interactions with its neighbors. Identifying the nature of the intermolecular potential governing these interactions is central to understanding polymer dynamics; however, the extreme complexity of the multidimensional potential energy landscape requires simplifying assumptions. A general potential that only accounts for two-body interactions is the Lennard–Jones (LJ) function, which expresses the potential energy as

$$U(r) = 4\varepsilon[(\sigma/r)^{3\gamma} - (\sigma/r)^6] \tag{24.5}$$

where
    r is the separation distance and
    ε, a microscopic energy
    σ, a molecular size, are constants

    The parameter γ in the exponent accounts for the steepness of the repulsive interaction and this repulsive exponent varies among materials [60,61]. Repulsive forces are strong and short-ranged (typically $\gamma \geq 4$) in comparison with the attractions. Since the force on any molecule is the vector sum of the contributions from all other molecules, and in condensed matter each molecule (or polymer segment) has many neighbors, the attractive forces tend to cancel locally [62]. This leads to the classic van der Waals picture of liquids, in which the arrangement of molecules (liquid structure) is governed by steric effects and local packing, with the attractions manifested as a uniform background pressure [63–65]. This suggests that in analyzing local properties the attractive term may be neglected, since attractions exert only a mean-field, density-dependent pressure, serving to cohere the material; that is, the attractive forces from the many neighbors of a given molecule or polymer segment essentially cancel [66]. This simplification is also consistent with the fact that in liquids, the static structure factor at intermediate and large wave vectors is sensitive only to the repulsive part of the potential [67,68].

    Dropping the attractive term in Equation 24.5 gives the inverse power–law (IPL) repulsive potential [69–72]

$$U(r) = Ar^{-3\gamma} \tag{24.6}$$

in which the coefficient $A = 4\varepsilon\sigma^{3\gamma}$. If the IPL accurately represents the local interaction energy between segments, then thermodynamic properties, such as the total energy, volume, and entropy, can be expressed as unique functions of $r^{3\gamma}$ or V^γ [70]. The assumption that complex intermolecular interactions can be represented by a spherically symmetric, two-body repulsive potential is only an approximation; this is especially true for polymers. The neglect of the longer-ranged attractive forces limits consideration to local properties; for example, Equation 24.6 does not yield an accurate equation of state (EOS) [73]. With these constraints we expect that local segmental relaxation times can be expressed in terms of the variable T^1V^γ (note $V^\gamma \sim r^{3\gamma}$). For the limiting case of hard spheres (volume-dominated dynamics) $\gamma = \infty$, while for thermally activated relaxation $\gamma = 0$.

    To test this idea we analyze local segmental relaxation times measured dielectrically as a function of temperature and pressure for polymethyltolylsiloxane (PMTS) [74]. The data are shown in Figure 24.1, and then replotted versus specific volume in Figure 24.2. Clearly, the local segmental relaxation times are not uniquely defined by V. However, when plotted in the form suggested by Equation 24.6, as a function of TV^γ (Figure 24.3), the data collapse onto a single master curve for $\gamma = 5$. This exponent (which is independent of T, V, and P) is determined empirically as the value yielding superposition of the τ_α. In Figure 24.4 are collected relaxation times for 11 polymers, also plotted versus TV^γ. The magnitude of γ yielding superposition for each material ranges from 1.9 for 1,4-polyisoprene to 5.6 for polymethylphenylsiloxane [15,75]; that is, γ is a material-specific constant, whose magnitude reflects the effective steepness of the intermolecular repulsive potential. Scaling plots of the viscosity for two polymers of relatively low-molecular weight (~10^4 g/mol) are displayed in Figure 24.5 [11]; these data also conform to the thermodynamic scaling.

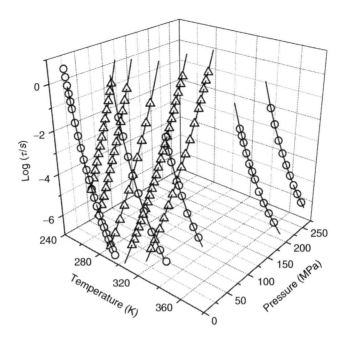

FIGURE 24.1 Local segmental relaxation times for polymethyltolylsiloxane (PMTS) measured dielectrically as a function of temperature at constant pressure (circles) and as a function of pressure at fixed temperature (triangles). (From Paluch, M., Pawlus, S., and Roland, C.M., *Macromolecules*, 35, 7338, 2002.)

Generally, the values of the scaling exponent are smaller for polymers than for molecular liquids, for which $3.2 \leq \gamma \leq 8.5$. A larger γ, or steeper repulsive potential, implies greater influence of "jamming" on the dynamics. The smaller exponent found for polymers in comparison with small-molecule liquids means that volume effects are weaker for polymers, which is ironic given their central role in the historical development of free-volume models. The reason why γ is smaller

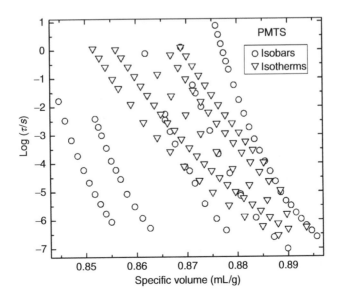

FIGURE 24.2 Local segmental relaxation times for polymethyltolylsiloxane (PMTS) from Figure 24.1 replotted as a function of specific volume.

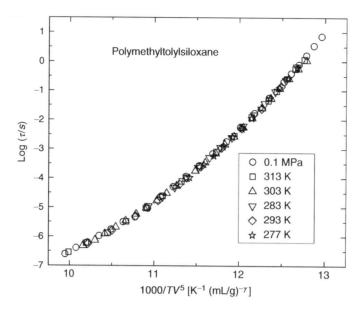

FIGURE 24.3 Local segmental relaxation times for polymethyltolylsiloxane (PMTS) plotted as a function of TV^γ. (From Casalini, R. and Roland, C.M., *Coll. Polym. Sci.*, 283, 107, 2004.)

is the relative insensitivity of the chain backbone bonds to pressure. This is seen in the small changes in chain end-to-end distance with pressure [76] and in the small contribution of these chain configurational changes to the compressibility of polymers [77].

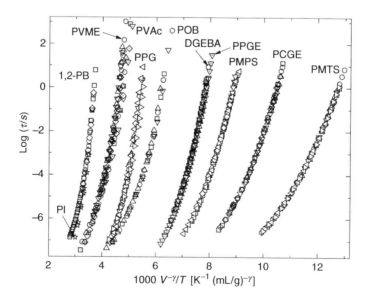

FIGURE 24.4 Master curves of the local segmental relaxation times for: 1,4-polyisoprene ($\gamma = 3.0$); 1,2-polybutadiene ($\gamma = 1.9$); polyvinylmethylether ($\gamma = 2.55$); polyvinylacetate ($\gamma = 2.6$); polypropylene glycol ($\gamma = 2.5$); polyoxybutylene ($\gamma = 2.8$); poly(phenyl glycidyl ether)-co-formaldehyde ($\gamma = 3.5$); polymethylphenylsiloxane ($\gamma = 5.6$); poly[(o-cresyl glycidyl ether)-co-formaldehyde] ($\gamma = 3.3$); and polymethyltolylsiloxane (PMTS) ($\gamma = 5.0$) [15 and references therein]. Each symbol for a given material represents a different condition of T and P.

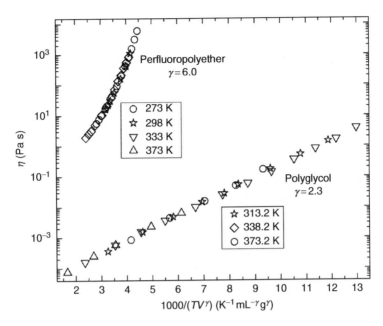

FIGURE 24.5 Master curves of the zero-shear rate viscosities of perfluoropolyether (Fomblin Z25, ($M_w = 9.5$ kg/mol)) and a copolymer of ethylene glycol and propylene glycol ($M_w = 12.5$ kg/mol). (From Roland, C.M., Bair, S., and Casalini, R., *J. Chem. Phys.*, 125, 124508, 2006.)

From the empirically demonstrated scaling, the relaxation times (or viscosity) are known to depend only on the product

$$\tau = \Im(TV^\gamma) \tag{24.7}$$

where $\Im$ is an unknown function. This means that if dynamic measurements are carried out at ambient pressure, in principle τ or η are known at all pressures, providing γ is known. In Section 24.5, we describe how the scaling exponent can be determined directly from PVT data, without the need to superposition experimental $\tau(T,V)$. Accordingly, relaxation times obtained at ambient pressure can be extended to elevated pressures (or to isochoric conditions) by utilizing the scaling relationship.

24.5 DETERMINATION OF THE SCALING EXPONENT FROM PVT DATA

To obtain the scaling exponent from EOS data, we consider a conventional metric of the relative effects of volume and temperature on the dynamics, the ratio of the apparent activation energy at constant volume, $E_V(T,V) = R \left. \dfrac{d \ln \tau}{dT^{-1}} \right|_V$ to the apparent activation enthalpy at constant pressure, $H_P(T,P) = R \left. \dfrac{d \ln \tau}{dT^{-1}} \right|_P$ [78]. This ratio varies between 0 and 1, smaller values reflecting volume-dominated relaxation, and a value near unity indicating that temperature is the dominant control variable. From the empirical scaling, the relaxation times (or viscosity) are functions of TV^γ. Taking the respective derivatives of Equation 24.7 gives

$$E_V = R \frac{d\Im}{dT^{-1}}(-T^2)V^\gamma \tag{24.8}$$

and

$$H_P = R \frac{d\Im}{dT^{-1}} (-T^2) V^\gamma \left(1 + T\gamma V^{\gamma-1} \frac{dV}{dT}\bigg|_P \right) \tag{24.9}$$

The ratio is then

$$E_V / H_P = (1 + \gamma \alpha_P T)^{-1} \tag{24.10}$$

where α_P is the isobaric thermal expansion coefficient. Parenthetically, we note that the product $\alpha_P T$ at T_g has a near universal value for polymers; that is, 0.18 ± 0.02 (the "Boyer–Spencer rule") [75,79].

We next obtain an expression for the activation energy ratio in terms of EOS properties. Following Naoki [80], the activation energy can be expressed as

$$E_V = H_P - RT^2 \frac{dP}{dT}\bigg|_V \frac{d\ln\tau}{dP}\bigg|_T \tag{24.11}$$

From Euler's chain rule

$$\frac{d\ln\tau}{dP}\bigg|_T = -\frac{d\ln\tau}{dT}\bigg|_P \frac{dT}{dP}\bigg|_\tau \tag{24.12}$$

Where the subscript τ implies the derivative is taken at constant value of the relaxation time. With the definition of H_P gives

$$E_V / H_P = 1 - \frac{dP}{dT}\bigg|_V \frac{dT}{dP}\bigg|_\tau \tag{24.13}$$

Since the glass transition corresponds to a constant value of the relaxation time [15], $dT/dP|_\tau$ is just the pressure coefficient of T_g. Comparing Equations 24.10 and 24.13, we see that the scaling exponent is related to quantities—thermal pressure coefficient, thermal expansion coefficient, T_g, and its pressure coefficient—that can all be determined from PVT measurements

$$\gamma = \left(1 - \left[1 - \frac{\partial P}{\partial T}\bigg|_V \frac{dT_g}{dP} \right]^{-1} \right) \bigg/ (\alpha_P T_g) \tag{24.14}$$

An alternative, simpler expression can be obtained as follows. Using the chain rule, the temperature derivative of the volume for constant value of the relaxation time is expressed as

$$\frac{dV}{dT}\bigg|_\tau = \frac{dV}{dT}\bigg|_P + \frac{dV}{dP}\bigg|_T \frac{dP}{dT}\bigg|_\tau \tag{24.15}$$

Dividing both sides by the temperature derivative of the volume at constant pressure gives

$$\frac{dV}{dT}\bigg|_\tau \bigg/ \frac{dV}{dT}\bigg|_P = 1 + \left(\frac{dV}{dP}\bigg|_T \bigg/ \frac{dV}{dT}\bigg|_P \right) \frac{dP}{dT}\bigg|_\tau \tag{24.16}$$

Again using Euler's chain rule, the value in brackets equals the negative of the isochoric derivative of temperature with respect to pressure. Thus, Equation 24.16 simplifies to

$$\frac{\alpha_\tau}{\alpha_P} = 1 - \frac{dT}{dP}\bigg|_V \frac{dP}{dT}\bigg|_\tau \tag{24.17}$$

where α_τ is the thermal expansion coefficient at constant τ. Equating Equations 24.13 and 24.17 gives

$$E_V/H_P = \left(1 - \frac{\alpha_P}{\alpha_\tau}\right)^{-1} \tag{24.18}$$

which substituted into Equation 24.10 yields a simple expression for the scaling exponent in terms of the isochronic thermal expansion coefficient

$$\gamma = -\frac{1}{T\alpha_\tau(T)} \tag{24.19}$$

The denominator in Equation 24.19 is evaluated along the glass transition line, $T_g(P)$, so that

$$\gamma = -\frac{V_g}{T_g}\frac{dT_g}{dV_g} \tag{24.20}$$

We illustrate this for lightly cross-linked 1,2-polybutadiene [81]. From the values obtained using the data in Figure 24.6, α_τ at $T_g \left(\equiv V_g^{-1}\frac{dV_g}{dT_g}\right) = -2.62 \times 10^{-3}\text{K}^{-1}$ and $T_g = 273.1$ K at 0.1 MPa, Equation 24.20 gives $\gamma = 1.4 \pm 0.1$. The relaxation times measured for this material at $P = 0.1$ MPa [82] in combination with the EOS for the rubber [81] enable the value of TV^γ associated with each

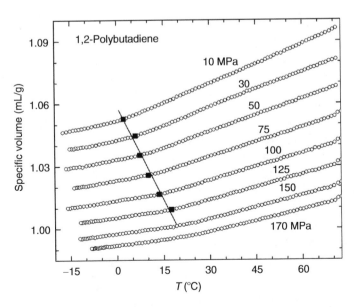

FIGURE 24.6 Specific volume as a function of temperature at the indicated pressures for 1,2-PB network (molecular weight between cross-links ~8 kg/mol) [81]. The solid squares denote the glass transition defined from the intersection of the glassy and liquid data.

(T, 0.1 MPa) condition to be calculated. τ is then known for any P and T corresponding to an ambient pressure value of TV^γ. This method is quite powerful because it allows determination of relaxation times for any thermodynamic condition (T, P, or V) provided only that the same value of τ had been measured at ambient pressure.

24.6 DETERMINATION OF THE SCALING EXPONENT FROM THE GRÜNEISEN PARAMETER

The method described above, using PVT data rather than empirical scaling of relaxation times to determine γ, relies on no assumptions other than the experimental fact that τ (or η) is uniquely defined by TV^γ. However, even though the EOS for the rubbery state must be known in order to convert measured $\tau(T,P)$ to $\tau(T,V)$, the method requires that the PVT measurements extend below the glass transition in order to evaluate α_τ at T_g (Equation 24.20). This is a serious limitation for the study of rubbers, which have low glass transition temperatures. An alternative method avoids this by taking advantage of the underlying basis for the thermodynamic scaling, e.g., Equation 24.6, to deduce the value of γ. To connect the exponent in the IPL to the scaling exponent we make use of the Grüneisen parameter, γ_G, a measure of the anharmonicity of lattice vibrations, formally defined in terms of the volume dependence of the phonon frequency, ω [83]

$$\gamma_G = -\frac{d \ln \omega}{d \ln V} = \frac{r \, d\omega}{3\omega \, dr} \tag{24.21}$$

The force constant for the vibrations, $K \equiv \frac{d^2 U}{dr^2}$, in the harmonic approximation is proportional to ω^2. Differentiating Equation 24.6 to obtain K and substituting into Equation 24.21, the result is [84,85]

$$\gamma_G = \frac{1}{2}\gamma + \frac{1}{3} \tag{24.22}$$

The γ in Equation 24.22 is from the IPL; thus, the mode Grüneisen parameter reflects the steepness of the intermolecular potential. If Equation 24.6 applies for the local segmental dynamics, the scaling exponent can be calculated using Equation 24.22. (Note that a similar result, $\gamma_G \sim \gamma$, can be obtained from an entropy model of the dynamics [86].) The limitations of this approach are twofold: (i) the mode Grüneisen parameter varies with T, whereas the scaling exponent is a constant; (ii) it is not obvious that a parameter characterizing vibrational motions should relate to relaxation dynamics, notwithstanding their mutual dependence on the intermolecular potential. Nevertheless, around T_g literature values of γ_G fall in the range between 1 and 4 [87,88], from which Equation 24.22 gives γ between 2 and 9, roughly consistent with experimental values.

An alternative definition of the Grüneisen parameter is in terms of thermodynamic properties [87]

$$\gamma_G = \frac{V\alpha_P}{C_V\kappa_T} \tag{24.23}$$

in which C_V is the isochoric heat capacity and κ_T is the isothermal compressibility. Table 24.1 shows γ_G for two polymers and three molecular liquids and compares γ calculated from Equation 24.22 to the value of γ determined by superpositioning experimental $\tau(T,V)$ data [86]. The latter are larger by about a factor of 2. At least some of this discrepancy is due to contributions to the heat capacity and thermal expansivity in Equation 24.23, such as vibrations and secondary relaxation processes, that are unrelated to relaxation. An approximate correction would be to subtract from α_P and C_V the corresponding values for the crystal, assuming the latter are known.

TABLE 24.1
Comparison of Experimental and Calculated Scaling Exponents

	γ (Equation 24.7)	γ_G (Equation 24.23)	γ (Equation 24.22)
Polyvinylacetate	2.5	0.7	0.7
Polymethylmethacrylate	1.25	0.7	0.7
Propylene carbonate	3.7	1.4	2.1
o-Terphenyl	4	1.2	1.7
Phenyl salicylate	5.2	1.9	3.1

24.7 EXTENSION OF THERMODYNAMIC SCALING TO CHAIN DYNAMICS

The scaling results above all pertain to local segmental relaxation, with the exception of the viscosity data in Figure 24.5. Higher temperature and lower times involve the chain dynamics, described, for example, by Rouse and reptation models [22,89]. These chain modes, as discussed above, have different T- and P-dependences than local segmental relaxation.

In Figures 24.7 through 24.9 are shown the volume dependences of the local and global relaxation times for 1,4-polyisoprene [90], polypropylene glycol [91], and polyoxybutylene [76]. For either mode, volume does not uniquely define the relaxation times, as the curves for different

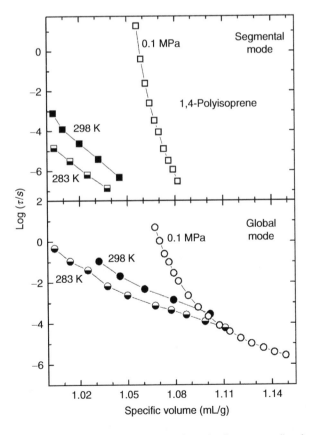

FIGURE 24.7 Local segmental (upper panel) and normal mode (lower panel) relaxation times for 1,4-polyisoprene ($M_w = 11$ kg/mol) [90], plotted versus specific volume.

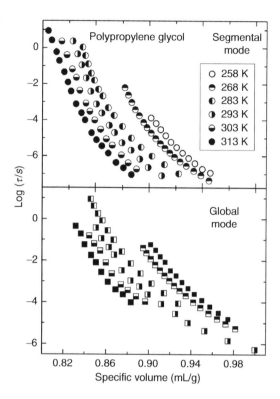

FIGURE 24.8 Local segmental (upper panel) and normal mode (lower panel) relaxation times for polypropylene glycol ($M_w = 4$ kg/mol) [91], plotted versus specific volume.

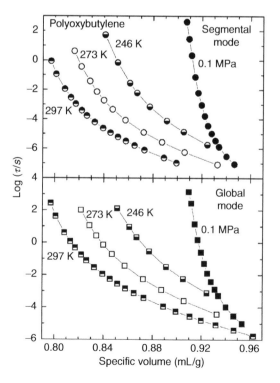

FIGURE 24.9 Local segmental (upper panel) and normal mode (lower panel) relaxation times for polyoxybutylene ($M_w = 5.3$ kg/mol) [76], plotted versus specific volume.

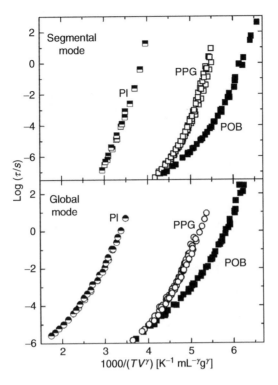

FIGURE 24.10 Dielectric relaxation times from Figures 24.7 through 24.9 plotted versus TV^γ, with mode independent $\gamma = 3.0$ (1,4-polyisoprene), $= 2.5$ (polypropylene glycol), and $= 2.65$ (polyoxybutylene).

conditions fail to superpose. It can also be noted that the V-dependences of the local segmental relaxation times are stronger than the for the normal mode τ. When these data are plotted versus TV^γ (Figure 24.10), master curves are obtained for both modes with the same value of γ [92]; that is, the same γ gives superpositioning of τ for local segmental and global chain relaxation. Their differing T- and V-dependences are manifested in the difference between the two relaxation times in the functional form of their TV^γ dependences (i.e., the $\Im$ in Equation 24.7) [93]. The segmental relaxation times change more with changes in TV^γ and this difference is more pronounced at longer τ. Such behavior is inconsistent with the Rouse and reptation models [22,89], which posit a single friction factor as governing the dependence of the dynamics on thermodynamic variables.

24.8 SUMMARY

Although time–temperature superpositioning is widely practiced, extrapolations based on it will be in substantial error when both local segmental relaxation and the chain dynamics contribute to the measured viscoelastic properties. Moreover, the original basis for the time–temperature superposition principle that free volume governs the mobility of polymers is incorrect. A more accurate method is the thermodynamic scaling of viscosities and relaxation times based on the product variable TV^γ, with the scaling exponent a material constant. Although the idea for this scaling derives from using an IPL to approximate the intermolecular potential, it is an experimental fact that τ and η are single functions of TV^γ, demonstrated for more than a dozen polymers and about 40 molecular liquids. The scaling exponent can be determined directly from PVT measurements, without recourse to superpositioning of experimental data. Accordingly, relaxation times and viscosities measured at ambient pressure can be extended to elevated pressures (or to isochoric conditions) by utilizing the scaling relationship. The only requirement is knowledge of the EOS, which is obtained from the same PVT data yielding the scaling exponent.

ACKNOWLEDGMENT

This work was supported by the Office of Naval Research.

REFERENCES

1. Bergstrom, J.S. and Boyce, M.C. *J. Mech. Phys. Solids* 46, 931, 1998.
2. Arruda, E., Wang, Y., and Przbylo, P. *Rubber Chem. Technol.* 74, 560, 2001.
3. Kremer F. and Schonhals, A., Eds., *Broadband Dielectric Spectroscopy*, Springer, Berlin, 2003.
4. McCrum, N.G., Read, B.E., and Williams, G. *Anelastic and Dielectric Effects in Polymer Solids*, Wiley, London, 1967.
5. Boese, D., Momper, B., Meier, G., Kremer, F., Hagenah, I.U., and Fischer, E.W. *Macromolecules* 22, 4416, 1989.
6. Colmenero, J., Alegria, A., Alberdi, J.M., Alvarez, F., and Frick, B. *Phys. Rev. B* 44, 7321, 1991.
7. Colmenero, J., Alegria, A., Santangelo, P.G., Ngai, K.L., and Roland, C.M. *Macromolecules* 27, 407, 1994.
8. Paluch, M., Roland, C.M., Gapinski, J., and Patkowski, A. *J. Chem. Phys.* 118, 3177, 2003.
9. Paluch, M., Roland, C.M., and Best, A. *J. Chem. Phys.* 117, 1188, 2002.
10. Bair, S. and Winer, W.O. *J. Lubr. Technol.* 102, 229, 1980.
11. Roland, C.M., Bair, S., and Casalini, R. *J. Chem. Phys.* 125, 124508, 2006.
12. Schneider, U., Lunkenheimer, P., Brand, R., and Loidl, A. *Phys. Rev. E* 59, 6924, 1999.
13. Skorodumov, V.F. and Godovskii, Y.K. *Polym. Sci.* 35, 562, 1993.
14. Floudas, G. Chapter 8 in ref. 3.
15. Roland, C.M., Hensel-Bielowka, S., Paluch, M., and Casalini, R. *Rep. Prog. Phys.* 68, 1405, 2005.
16. Hellwege, K.H., Knappe, W., Paul, F., and Semjonow, V. *Rheol. Acta* 6, 165, 1967.
17. Fillers, R.W. and Tchoegl, N.W. *Trans. Soc. Rheol.* 21, 51, 1977.
18. Bair, S. *J. Eng. Tribol.* 216, 1, 2002.
19. Han, C.D. and Ma, C. *J. Appl. Polym. Sci.* 28, 851, 1983.
20. Gerhardt, L.J., Manke, C.W., and Gulari, E. *J. Polym. Sci. Polym. Phys.* 35, 523, 1997.
21. Royer, J.R., Gay, Y.J., Desimone, J.M., and Khan, S.A. *J. Polym. Sci. Polym. Phys.* 38, 3168, 2000.
22. Ferry, J.D. *Viscoelastic Properties of Polymers*, 3rd ed., Wiley, New York, 1980.
23. Doolittle, A.K. and Doolittle, D.B. *J. Appl. Phys.* 28, 901, 1957.
24. Cohen, M.H. and Turnbull, D. *J. Chem. Phys.* 31, 1164, 1959.
25. Corezzi, S., Rolla, P.A., Paluch, M., Ziolo, J., and Fioretto, D. *Phys. Rev. E* 60, 4444, 1999.
26. Paluch, M. *J. Chem. Phys.* 115, 10029, 2001.
27. Schug, K.U., King, H.E., and Böhmer, R. *J. Chem. Phys.* 109, 1472, 1998.
28. Williams, M.L., Landel, R.F., and Ferry, J.D. *J. Am. Chem. Soc.* 77, 3701, 1955.
29. Chang, I. and Sillescu, H. *J. Phys. Chem. B* 101, 8794, 1997.
30. Rössler, E. *Phys. Rev. Lett.* 65, 1595, 1990.
31. Fischer, E.W., Donth, E., and Steffen, W. *Phys. Rev. Lett.* 68, 2344, 1992.
32. Angel, C.A. *Pure Appl. Chem.* 63, 1387, 1991.
33. Zallen, R. *The Physics of Amorphous Solids*, Wiley, New York, 1981.
34. Cohen, M.H. and Grest, G.S. *Phys. Rev. B* 20, 1077, 1979; Grest, G.S. and Cohen, M.H. *Adv. Chem. Phys.* 48, 455, 1981.
35. Grest, G.S. and Cohen, M.H. *Phys. Rev. B* 21, 4113, 1980; Cohen, M.H. and Grest, G.S. *J. Non-Cryst. Solids* 61/62, 749, 1984.
36. Cummins, H.Z., Li, G., Hwang, Y.H., Shen, Q., Du, W.M., Hernandez, J., and Tao, N.J. *Z. Phys. B* 103, 501, 1979.
37. Schneider, U., Lunkenheimer, P., Brand, R., and Loidl, A. *Phys. Rev. B* 59, 6924, 1999.
38. Stickel, F., Fischer, E.W., and Richert, R. *J. Chem. Phys.* 102, 6251, 1995; 104, 2043, 1996.
39. Hansen, C., Stickel, F., Berger, F.T., Richert, R., and Fischer, E.W. *J. Chem. Phys.* 107, 1086, 1997; Hansen, C., Stickel, F., Richert, R., and Fischer, E.W. *J. Chem. Phys.* 108, 6408, 1998.
40. Ngai, K.L. and Roland, C.M. *Polymer* 43, 567, 2002.
41. Paluch, M., Casalini, R., and Roland, C.M. *Phys. Rev. E* 67, 021508, 2003.

42. Bendler, J.T., Fontanella, J.J., Shlesinger, M.F., and Wintersgill, M.C. *Electrochim. Acta* 49, 5249, 2004.
43. Corezzi, S., Capaccioli, S., Casalini, R., Fioretto, D., Paluch, M., and Rolla, P.A. *Chem. Phys. Lett.* 320, 113, 2000.
44. Comez, L., Corezzi, S., Fioretto, D., Kriegs, H., Best, A., and Steffen, W. *Phys. Rev. E* 70, 011504, 2004.
45. Roland, C.M., Paluch, M., Pakula, T., and Casalini, R. *Philos. Mag.* 84, 1573, 2004.
46. Adam, G. and Gibbs, J.H. *J. Chem. Phys.* 43, 139, 1965.
47. Goldstein, M. *J. Chem. Phys.* 51, 3728, 1969.
48. Marvin, R.S. in "Proceedings of the Second International Congress on Rheology," Vol. 6, W. Harrison, Ed., Butterworths, London, 1953, p. 156.
49. Ngai, K.L. and Plazek, D.J. *Rubber Chem. Technol.* 68, 376, 1995.
50. Plazek, D.J. *J. Phys. Chem.* 69, 3480, 1965.
51. Plazek, D.J., Chay, I.-C., Ngai, K.L., and Roland, C.M. *Macromolecules* 28, 6432, 1995.
52. Plazek, D.J. *J. Rheol.* 40, 987, 1996.
53. Roland, C.M., Ngai, K.L., Santangelo, P.G., Qiu, X.H., Ediger, M.D., and Plazek, D.J. *Macromolecules* 34, 6159, 2001.
54. Plazek, D.J. *Polym. J.* 12, 43, 1980.
55. Plazek, D.J., Rosner, M.J., and Plazek, D.L. *J. Polym. Sci. Polym. Phys.* 26, 473, 1988.
56. Santangelo, P.G. and Roland, C.M. *Macromolecules* 31, 3715, 1998.
57. Goldstein, M. *J. Chem. Phys.* 51, 3728, 1969.
58. Stillinger, F.H. and Weber, T.A. *Science* 225, 983, 1984.
59. Sampoli, M., Benassi, P., Eramo, R., Angelani, L., and Ruocco, G. *J. Phys. Cond. Mat.* 15, S1227, 2003.
60. Moelwyn-Hughes, E.A. *Physical Chemistry*, 2nd ed., Pergamon Press, New York, 1961.
61. Bardik, V.Y. and Sysoev, V.M. *Low Temp. Phys.* 24, 601, 1998.
62. Widom, B. *Physica A* 263, 500, 1999.
63. Widom, B. *Science* 157, 375, 1967.
64. Chandler, D., Weeks, J.D., and Andersen, H.C. *Science* 220, 4599, 1983.
65. Stillinger, H., Debenedetti, P.G., and Truskett, T.M. *J. Phys. Chem. B* 105, 11809, 2001.
66. Widom, B. *Physica A* 263, 500, 1999.
67. Weeks, J.D., Chandler, D., and Andersen, H.C. *J. Chem. Phys.* 54, 5237, 1971.
68. Tölle, A., Schober, H., Wuttke, J., Randl, O.G., and Fujara, F. *Phys. Rev. Lett.* 80, 2374, 1998.
69. Longuet-Higgins, H.C. and Widom, B. *Mol. Phys.* 8, 549, 1964.
70. Hoover, W.H. and Ross, M. *Contemp. Phys.* 12, 339, 1971.
71. Speedy, R.J. *J. Phys. Cond. Mat.* 15, S1243, 2003.
72. Shell, M.S., Debenedetti, P.G., La Nave, E., and Sciortino, F. *J. Chem. Phys.* 118, 8821, 2003.
73. Roland, C.M., Feldman, J.L., and Casalini, R. *J. Non-Cryst. Solids* 352, 4895, 2006.
74. Paluch, M., Pawlus, S., and Roland, C.M. *Macromolecules* 35, 7338, 2002.
75. Casalini, R. and Roland, C.M. *Coll. Polym. Sci.* 283, 107, 2004.
76. Casalini, R. and Roland, C.M. *Macromolecules* 38, 1779, 2005.
77. Porter, D. and Gould, P.J. Presented at the *15th U.S. National Congress on Theoretical and Applied Mechanics*, University of Colorado, Boulder, June 29, 2006.
78. Williams, G. *Dielectric Spectroscopy of Polymeric Materials*, J.P. Runt and J.J. Fitzgerald, Eds., American Chemical Society, Washington, DC, 1997, Chapter 1.
79. Van Krevelen, D.W. *Properties of Polymers*, Elsevier, New York, 1990.
80. Naoki, M., Endou, H., and Matsumoto, K. *J. Phys. Chem.* 91, 4169, 1987.
81. Roland, C.M., Roland, D.F., Wang, J., and Casalini, R. *J. Chem. Phys.* 122, 134505, 2005.
82. Roland, C.M. *Macromolecules* 27, 4242, 1994.
83. Grüneisen, E. *Ann. Phys. (Leipzig)* 39, 257, 1912.
84. Moelwyn-Hughes, K.A. *J. Phys. Coll. Chem.* 55, 1246, 1951.
85. Hook, J.R. and Hall, H.E. *Solid State Physics*, 2nd ed., Wiley, West Sussex, United Kingdom, 1991.
86. Casalini, R., Mohanty, U., and Roland, C.M. *J. Chem. Phys.* 125, 014505, 2006.
87. Hartwig, G. *Polymer Properties at Room and Cryogenic Temperatures*, Plenum Press, New York, 1994, Chapter 4.
88. Curro, J.G. *J. Chem. Phys.* 58, 374, 1973.

89. Doi, M. and Edwards, S.F. *The Theory of Polymer Dynamics*, Clarendon, Oxford, 1986.
90. Floudas, G. and Reisinger, T. *J. Chem. Phys.* 111, 5201, 1999.
91. Roland, C.M., Psurek, T., Pawlus, S., and Paluch, M. *J. Polym. Sci. Polym. Phys. Ed.* 41, 3047, 2003.
92. Roland, C.M., Casalini, R., and Paluch, M. *J. Polym. Sci. Polym. Phys.* 42, 4313, 2004.
93. Ngai, K.L., Casalini, R., and Roland, C.M. *Macromolecules* 38, 4363, 2005.

25 Recent Advances in Fatigue Life Prediction Methods for Rubber Components

Ryan J. Harbour, Ali Fatemi, and Will V. Mars

CONTENTS

25.1 INTRODUCTION

Crack nucleation and crack growth approaches for analyzing the fatigue behavior of rubber components are complementary paradigms that can be applied depending on analysis objectives. Curiously, the crack growth approach has received relatively more attention in the rubber literature over the last 50 years, and seems to have arrived to a more mature state than the nucleation approach. There remains a need for further development of the nucleation approach. Traditional equivalence parameters, which are used in theories of crack nucleation to evaluate the relative level of multiaxial loading conditions, are based on scalar magnitudes, and are not adequate due to the plane-specific nature of the nucleation phenomenon. Cracking energy density is presented as one example of a plane-specific equivalence parameter. Fatigue analysis methods must account for various effects in rubber that are observed under constant amplitude testing. These include crack closure, *R*-ratio, and loading amplitude. Finally, since most real-world components experience complex loading histories unlike the simple histories used for characterizing fatigue behavior, methods of inferring the fatigue behavior of variable amplitude signals from constant amplitude test results are required. Proposed approaches of predicting variable amplitude fatigue behavior of rubber based on constant amplitude data are presented.

25.2 FATIGUE CRACK NUCLEATION AND CRACK GROWTH

Two general approaches have emerged for analyzing the fatigue performance of a rubber component[1]: (1) The fatigue crack nucleation approach; and (2) the fatigue crack growth approach. The choice of which approach to apply depends on the goals of the investigation.

Fatigue crack nucleation refers to the development, under the action of cyclic loading, of one or more cracks in a specimen, from an initial state nominally free of preexisting cracks. Although a large number of very small crack precursors are in fact present initially, these are typically not visible to casual observers, and are not considered in calculating mechanical fields (i.e., stress and strain distributions) in the specimen. As an analysis approach, the aim is to identify the location and time at which critical fatigue cracks first appear. The approach employs a continuum parameter, defined at each point in a body, to judge the effect of applied loading on each material point. A number of parameters have been investigated and applied. Maximum principal stress and strain are examples.

Fatigue crack growth refers to the development of cracks under cyclic loading, from an initial state in which the crack (or cracks) of interest is known. To use this approach, the location, geometry, and extent of the crack must be specified. The approach employs the energy release rate (or tearing energy[2]) to judge the effect of applied loads on the crack. This concept originates from the idea that, in order for a crack to grow, mechanical energy must be converted to energy associated with a crack surface, as proposed by Griffith.[3] The energy release rate T is the change in stored mechanical energy dU per unit change in crack surface area dA:

$$T = -\frac{dU}{dA} \tag{25.1}$$

Rivlin and Thomas investigated the energy release rate for multiple specimen geometries and determined that a critical energy release rate existed for a material for which crack growth would occur when the energy release rate exceeded this critical value. They found that the critical energy release rate was independent of the specimen geometry and it was a property of the material. The approach has enjoyed a great deal of success, and has been widely applied to both strength and fatigue problems.

Much more has been published over the last 50 years concerning the crack growth approach[4] than the crack nucleation approach for rubber. Consequently, the crack growth approach enjoys a relatively more mature position than the crack nucleation approach. In recent years, the development of equivalence parameters that work in cases involving multiaxial loading[5,6] has renewed interest in the crack nucleation approach.

Sometimes, crack nucleation and crack growth approaches have been discussed as if they were competing theories. Such discussions often point out correctly that crack nucleation can be rationalized in terms of the fracture mechanics of small initial flaws preexisting in the material; for example, see Mars and Fatemi.[1] So what justifies a separate analysis approach for crack nucleation? There are two unique features of the crack nucleation problem. First, precursor cracks are very small, and occur very frequently. For practical purposes, it can be assumed that a precursor crack exists at every point of the material.[7] Second, the energy release rate of a small crack obeys a simple scaling law with respect to the size of the crack, enabling the energy release rate to be estimated from the loading state of the crack-free material.[8] So, far from being competing theories, these different analysis methods are actually complementary paradigms that make powerful and efficient tools for analyzing all stages of the fatigue failure process.

25.3   MULTIAXIAL LOADING

In order to apply the crack nucleation approach, the mechanical state of the material must be quantified at each point by a suitable parameter.[9] Traditional parameters have included, for example, the maximum principal stress[10] or strain, or the strain energy density. Maximum principal strain and stress reflect that cracks in rubber often initiate on a plane normal to the loading direction. Strain energy density has sometimes been applied as a parameter for crack nucleation due to its connection to fracture mechanics for the case of edge-cracked strips under simple tension loading.[11]

Several criticisms of these parameters have recently been pointed out.[12] First, they have no specific association with a material plane (i.e., they are scalar parameters), despite the fact that cracks are known to nucleate on specific material planes. With traditional parameters it is difficult to account for the effects of crack closure under compressive loading. Traditional parameters have not been successful at unifying experimental results for simple tension and equibiaxial tension fatigue tests. Finally, a nonproportional loading history can always be constructed for a given scalar equivalence parameter that holds constant the value of the scalar parameter, but which results in cyclic loading of material planes. For such histories, scalar parameters incorrectly predict infinite fatigue life.

Due to the plane-specific nature of crack nucleation under multiaxial tests, Mars and Fatemi proposed the cracking energy density as an equivalence parameter that represents the portion of strain energy density available to be released as crack growth on a specific material plane. The form of the cracking energy density W_c is

$$dW_c = \vec{t} \cdot d\vec{\varepsilon} \tag{25.2}$$

where $\vec{t}$ is the traction vector and $d\vec{\varepsilon}$ is the corresponding strain vector increment for the given stress state $\boldsymbol{\sigma}$ and strain state $\boldsymbol{\varepsilon}$ on a specific material plane defined by the unit normal vector $\vec{r}$:

$$\vec{t} = \boldsymbol{\sigma}\vec{r} \tag{25.3}$$

$$\vec{\varepsilon} = \boldsymbol{\varepsilon}\vec{r} \tag{25.4}$$

Cracking energy density is a robust equivalence parameter because it relates approximately to the fracture mechanical behavior of small flaws and is well defined for complex strain histories. This parameter differs from the traditional parameters by being plane-specific and able to account for crack closure when compressive stresses act on a material plane. Since the cracking energy density depends on both the strain state and the cracking plane, it is impossible to design a dynamic loading history that holds the cracking energy density constant on all of the planes.

Based on comparison of three traditional equivalence parameters with cracking energy density, the maximum principal strain corresponded the closest to the cracking energy density. Thus, Mars and Fatemi judged that the maximum principal strain is the most robust and meaningful of the traditional parameters considered in their work.

Mars and Fatemi[13] investigated the above-mentioned equivalence parameters by applying them to a series of multiaxial experiments conducted using the test specimen[14] shown in Figure 25.1. The three traditional equivalence parameters and the cracking energy density all ranked proportional axial–torsion tests between pure axial and pure torsion tests in terms of fatigue lives. Crack closure (compressive forces acting normal to the plane of crack growth closes the crack) impacted both fatigue life and the plane of crack nucleation, which is only accounted for by the cracking energy density criterion. The cracking energy density generally provided good correlation to the fatigue life data for combined axial–torsion tests while none of the scalar equivalence parameters were accurate for the entire range of tests. This result signified that the cracking energy density might be a better equivalence parameter to use for multiaxial loading histories than the traditional scalar parameters.

In a recent study, Saintier et al.[10,15] investigated the multiaxial effects on fatigue crack nucleation and growth in natural rubber. They found that the same mechanisms of decohesion and cavitation of inclusions that cause crack nucleation and crack growth in uniaxial experiments were responsible for the crack behavior in multiaxial experiments. They studied crack orientations for nonproportional multiaxial fatigue loadings and found them to be related to the direction of the maximum first principal stress of a cycle when material plane rotations are taken into account. This method accounts for material rotations in the analysis due to the displacement of planes associated with large strain conditions.

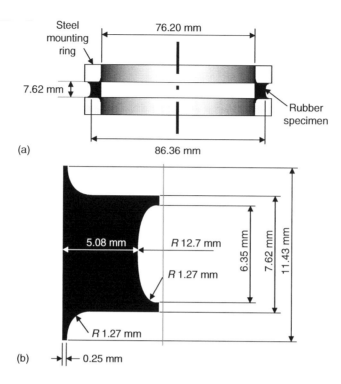

FIGURE 25.1 Multiaxial ring specimen geometry designed for axial and torsional loads: (a) general design of the specimen with a rubber ring bonded between two steel mounting rings and (b) cross-sectional geometry of the specimen. (From Mars, W.V. and Fatemi, A., *Exp. Mech.*, 44, 136, 2004.)

Saintier et al. proposed a fatigue life model for multiaxial tests in natural rubber that uses a critical plane approach. Based on the relationship between the maximum principal stress direction and crack orientations, they considered the maximum value of the first principal stress as the driving force of the damage process. The critical plane approach produced good agreement to uniaxial fatigue life results generated using Diabolo specimens with the majority of the fatigue life predictions being within a factor of 2 of the experimental lives for tests with and without R-ratios greater than 0. For some simple multiaxial tests, the critical plane approach was able to successfully predict fatigue lives as well as crack orientations. This approach was limited by its inability to capture size or gradient effects associated with most notched members when applied to specimens with smaller notch radii.

25.4 FACTORS THAT AFFECT THE FATIGUE BEHAVIOR OF RUBBER

In order to accurately model the fatigue behavior of rubber, fatigue analysis methods must account for various effects observed for rubber during constant amplitude testing. Effects associated with load level, R-ratio (ratio of minimum to maximum loading level), and crack closure are presented in this section.

25.4.1 Load Level

Load level refers to the magnitude of the loading conditions. Experimental results show that the fatigue behavior of rubber is directly linked to the level of the load. The crack growth behavior of rubber contains four distinct regions based on increasing energy release rates. These regions include an initial region of constant crack growth rate, a transition region of sharp increase in crack growth rate following the crack growth threshold T_0, a region that follows a power–law relation and a final

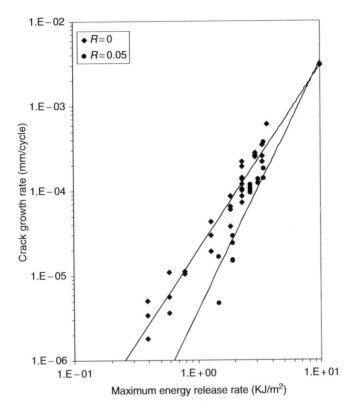

FIGURE 25.2 Constant amplitude fatigue crack growth rate results for natural rubber at R-ratios of 0 and 0.05.

region of unstable crack growth for energy release rates above the critical energy release rate T_c. Figures 25.2 and 25.3 illustrate the crack growth rates for R-ratios of 0 and 0.05 for the region of crack growth rates that follow a power–law relationship where R is defined as the ratio of the minimum energy release rate to the maximum energy release rate.

The relationship between load level and fatigue crack nucleation lives is clearly evident from the ε–N and S–N plots for the material. A sample ε–N plot for natural rubber is presented in Figure 25.4. An increase in the load level of the applied cycles results in a shorter fatigue life. Strain levels below the fatigue life threshold produce infinite fatigue lives. The relationship between the load and the fatigue life follows a linear relation when plotted on a log–log scale.

The effects of increased crack growth rates and shorter fatigue lives associated with increases in load level are in agreement when the mechanisms of crack nucleation are considered. Since crack nucleation is defined as the growth of flaws in the material until the crack reaches a critical length, shorter fatigue lives would be associated with the increased crack growth rates from higher load levels.

25.4.2 R-RATIO

The R-ratio is used to characterize constant amplitude signals as the ratio of a specified parameter at the minimum loading condition P_{min} to the parameter at the maximum loading condition P_{max} as given by

$$R = \frac{P_{min}}{P_{max}} \tag{25.5}$$

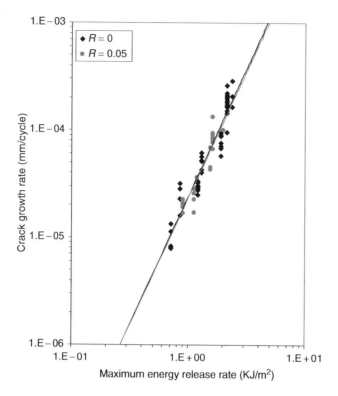

FIGURE 25.3 Constant amplitude fatigue crack growth rate results for SBR at R-ratios of 0 and 0.05.

The common parameter used to define the R-ratio for fatigue crack growth tests in rubber is the energy release rate. The effect of R-ratio on fatigue crack growth rates depends on whether the rubber strain crystallizes. In the case of a rubber that strain crystallizes, such as natural rubber, a small increase in R-ratio results in a significant drop in the fatigue crack growth rate for the same level of maximum energy release rate, as shown in Figure 25.2. This result indicates that the

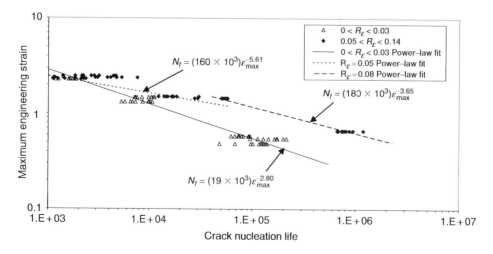

FIGURE 25.4 Typical strain-life plot for natural rubber at multiple R-ratios. (From Mars, W.V. and Fatemi, A., *Fatigue Fract. Eng. Mater. Struct.*, 26, 779, 2003.)

addition of a small minimum stress level produces a beneficial effect on the fatigue crack growth behavior, as first observed by Lindley.[16] On the other hand, the same beneficial effect is not observed in rubbers that do not strain crystallize such as styrene–butadiene rubber (SBR). The fatigue crack growth rate results for SBR shown in Figure 25.3 are independent of the R-ratio of the applied cycles.

While a single power–law relationship can model the fatigue crack growth results for a given R-ratio in either type of rubber, modeling the overall fatigue crack growth rate behavior of a material for all R-ratios requires a different approach for each material. Since the crack growth rate results for SBR are independent of R-ratio, the fatigue crack growth rate behavior can be modeled by a single power–law relationship. In order to account for the R-ratio effect in strain-crystallizing rubbers, Mars and Fatemi[17] developed a phenomenological model in the third regime of crack growth rates. The model assumes that the crack growth rate data can be modeled as a power–law relationship for each R-ratio, and that the crack rate growth data for all R-ratios converge at a point defined by the critical energy release rate T_c and maximum fatigue crack growth rate r_c, expressed by

$$\frac{da}{dN} = r_c \left(\frac{T_{max}}{T_c} \right)^{F(R)}$$

(25.6)

Here, the equation for the crack growth rate da/dN is defined in terms of the maximum energy release rate T_{max}, the power–law exponent $F(R)$, and the point at which the crack growth rate data converge. The power–law exponent $F(R)$ for the Mars–Fatemi model is of the form

$$F(R) = F_0 + F_1 R + F_2 R^2 + F_3 R^3$$

(25.7)

or alternatively an exponential form of

$$F(R) = F_0 e^{F_4 R}$$

(25.8)

where F_0, F_1, F_2, F_3, and F_4 are material constants that were used to fit the model to fatigue crack growth rate data. The advantage of the first form of $F(R)$ is that it fits the data better in some cases. The second form is easier to apply, since it has fewer parameters. Lindley estimated values of 10 kJ/m^2 for critical energy release rate and 2 for F_0. The Mars–Fatemi model represents the crack growth rates for multiple R-ratios as a single power–law relationship by defining an equivalent maximum energy release rate for an R-ratio of 0, $T_{max,0}$:

$$T_{max,0} = T_{max}^{\frac{F(R)}{F_0}} T_c^{\left(1 - \frac{F(R)}{F_0} \right)}$$

(25.9)

Similar R-ratio effects are observed during crack nucleation tests for strain-crystallizing rubbers as noted by Mars and Fatemi,[11] Cadwell et al.[18], and Fielding.[19] The results from these experiments show that R-ratios greater than 0 effectively increase the fatigue lives of strain-crystallizing rubbers. A small minimum stress level can result in longer lives as fatigue crack growth rates drop due to the R-ratio effect in strain-crystallizing rubbers as shown in Figure 25.4. The Mars–Fatemi model also is capable of modeling fatigue life curves for R-ratios greater than 0 in strain-crystallizing rubbers. It is important to note that the R-ratio is not necessarily constant in multiaxial loadings for every plane and needs to be calculated for each plane in order to properly account for the effects of R-ratio on fatigue lives.[20]

25.4.3 CRACK CLOSURE

Crack closure occurs when the loading conditions applied to the specimen produce a compressive force acting normal to the plane of crack growth. This compressive force does not contribute to the growth of the crack and can even cause retardation of crack growth in some cases. Since the work done by the compressive force is not available to be released by crack growth, it should not be considered in the fatigue analysis. When the crack is closed, only the shearing forces in the material produce crack growth if friction is small. Crack closure is of practical importance since many applications of rubber involve compressive loads. The plane-specific equivalence parameter of cracking energy density does account for crack closure, while the traditional scalar equivalence parameters previously mentioned do not account for crack closure.

25.5 VARIABLE AMPLITUDE LOADING

Constant amplitude loading conditions can be defined by a single amplitude and mean. In contrast, variable amplitude loading conditions involve a wide range of events that cannot be characterized by a single amplitude and mean. Typical constant and variable amplitude test signals are presented in Figure 25.5. Although material characterization typically focuses on constant amplitude loading conditions, such signals do not reflect the more complex loading histories experienced by components in actual service. Variable amplitude test signals are necessary to represent realistic service histories. In order to effectively predict the fatigue behavior of actual components, an understanding of the effects associated with variable amplitude loading conditions is required.

The basic steps to analyze variable amplitude loading include: (1) decomposing the test signal, by a suitable cycle counting procedure, into a list of individual constant amplitude reference events; and (2) accumulating the fatigue damage caused by each event in the list. The most common approach involves the use of the rainflow cycle counting procedure[21] along with Miner's linear damage rule[22] to analyze the variable amplitude signal.

The rubber literature contains only a handful of references that discuss the phenomenology or analysis of fatigue under conditions of variable amplitude loading. Sun et al.[23] investigated the effects of load sequencing on fatigue life, during tests designed to evaluate the applicability of Miner's linear damage rule. Mars and Fatemi[13] studied the effects of initial overload periods on fatigue life. A rainflow filtering process has been investigated by Steinwegger et al.[24] as an approach to reduce test times for both uniaxial and multiaxial test signals. While these efforts provide insights into specific variable amplitude loading cases, previous research is not available on general approaches to infer the fatigue behavior of rubber under variable amplitude loading conditions based on constant amplitude data.

A linear crack growth rate prediction model analogous to Miner's linear damage rule has been proposed to predict the fatigue crack growth rate for variable amplitude signals using constant

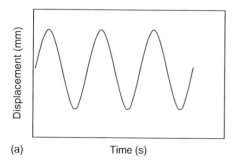

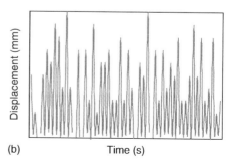

FIGURE 25.5 Sample constant amplitude signal (a) and variable amplitude signal (b).

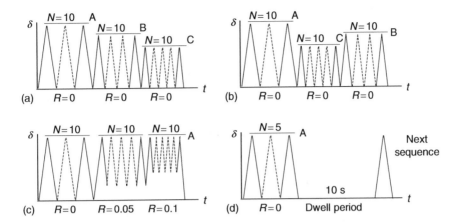

FIGURE 25.6 Examples of variable amplitude fatigue crack growth test signals applied to pure shear specimens to investigate the effects of (a) load severity, (b) load sequence, (c) R-ratio, and (d) dwell periods on crack growth rates. A, B, and C denote peak strain levels.

amplitude crack growth data as the basis for the predictions by Harbour et al.[25] The linear crack growth rate prediction model equates the total crack growth rate r for a signal as the sum of the constant amplitude crack growth rates r_i and the number of occurrences of each component event N_i.

$$r = N_1 r_1 + N_2 r_2 + N_3 r_3 + \ldots + N_i r_i \tag{25.10}$$

The linear crack growth rate model does not account for any interaction between the individual components of the test signal such as load sequencing. Variable amplitude fatigue crack growth experiments[25] were conducted on pure shear specimens for filled natural rubber and filled SBR. The tests signals consisted of repeating blocks of constant amplitude signals designed to investigate the effects of varying load level, R-ratio, and load sequencing during an experiment as illustrated in Figure 25.6. For the majority of the test signals, the fatigue crack growth rate prediction model was able to successfully predict the experimental fatigue crack growth rates to within a factor of 2. Varying load level, R-ratio, and load sequence did not have a significant effect on the crack growth rates for the tested signals. However, the periodic inclusion of a dwell period[26] at a near-zero stress level produced significantly higher crack growth rates in filled SBR. The inclusion of a 10 s dwell period between every five cycles produced crack growth rates 10 times faster than the corresponding constant amplitude crack growth rates without the dwell period. Similar results were observed in the filled natural rubber, but to a lesser extent. This dwell period effect is not predicted by any current fatigue life prediction models.

25.6 SUMMARY

This chapter summarizes some of the recent advances in fatigue life prediction methodology for rubber components. The two general approaches to analyze fatigue behavior, crack nucleation and crack growth, are presented as complementary analysis methods. In order to analyze fatigue crack nucleation under multiaxial loading conditions, it has been necessary to discard traditional scalar-based parameters in favor of place-specific parameters. Cracking energy density was presented as an example of such a parameter. Factors such as load level, R-ratio, and crack closure have been shown to have significant effects on the fatigue behavior of rubber. An approach for predicting the effects of variable amplitude loading conditions on the fatigue behavior of rubber was introduced, using a linear crack growth rate law. It was noted that, while the linear law can be applied successfully to most signals, certain signals containing a dwell period produce faster-than-expected rates of crack growth.

NOMENCLATURE

a	Crack length
A	Area
$F(R)$	Power–law exponent for the Mars–Fatemi model
N, N_i	Cycles, per block i
P_{max}, P_{min}	Maximum, minimum values of a parameter
$\vec{t}$	Traction vector
T, T_{max}, T_0	Energy release rate, maximum, threshold
T_c	Critical energy release rate
$T_{max,0}$	Equivalent maximum energy release rate
r, r_i	Crack growth rate, of block i
r_c	Critical crack growth rate
$\vec{r}$	Unit normal vector
R	Energy release rate ratio
U	Stored mechanical energy
W_c	Cracking energy density
ε	Strain tensor
$\vec{\varepsilon}$	Strain vector
σ	Stress tensor

REFERENCES

1. Mars, W.V. and Fatemi, A., A literature survey on fatigue analysis approaches for rubber, *Int. J. Fatigue*, 24, 946, 2002.
2. Rivlin, R.S. and Thomas, A.G., Rupture of rubber. I. Characteristic energy for tearing, *J. Polym. Sci.*, 10, 291, 1953.
3. Griffith, A.A., The phenomena of rupture and flow in solids, *Phil. Trans. Roy. Soc. Lond., Ser. A*, 221, 163, 1920.
4. Mars, W.V., Comparison of fatigue crack growth characterizations made at several labs, presented at *Fracture Mechanics for Elastomers: 50 Not Out, Proceedings of Anniversary Meeting*; Tun Abdul Razak Research Centre, Hertford, United Kingdom, 2008.
5. Mars, W.V., Multiaxial fatigue of rubber, Ph.D. dissertation, The University of Toledo, Toledo, OH, 2001.
6. Saintier, N., Multiaxial fatigue life of a natural rubber: crack initiation mechanisms and local fatigue life criterion, Ph.D. dissertation, Ecole des Mines de Paris, 2001.
7. Mars, W.V. et al., Fatigue life analysis of an exhaust mount, in *Constitutive Models for Rubber IV*, Austrell, K., Ed., Swets & Zeitlinger, The Netherlands, 2005, 23.
8. Mars, W.V., Heuristic approach for approximating energy release rates of small cracks under finite strain, multiaxial loading, in *Elastomers and Components—Service Life Prediction; Progress and Challenges*, Coveney, V., Ed., OCT Science, Philadelphia, 2006, 89.
9. Mars, W.V. and Fatemi, A., Multiaxial fatigue of rubber: Part I: equivalence criteria and theoretical aspects, *Fatigue Fract. Eng. Mater. Struct.*, 28, 515, 2005.
10. Saintier, N., Cailletaud, G., and Piques, R., Crack initiation and propagation under multiaxial fatigue in a natural rubber, *Int. J. Fatigue*, 28, 61, 2006.
11. Mars, W.V. and Fatemi, A., Fatigue crack nucleation and growth in filled natural rubber, *Fatigue Fract. Eng. Mater. Struct.*, 26, 779, 2003.
12. Mars, W.V. and Fatemi, A., The correlation of fatigue crack growth rates in rubber subjected to multiaxial loading using continuum mechanical parameters, *Rubber Chem. Technol.*, 80, 169, 2007.
13. Mars, W.V. and Fatemi, A., Multiaxial fatigue of rubber: Part II: experimental observations and life predictions, *Fatigue Fract. Eng. Mater. Struct.*, 28, 523, 2005.
14. Mars, W.V. and Fatemi, A., A novel specimen for investigating the mechanical behavior of elastomers under multiaxial loading conditions, *Exp. Mech.*, 44, 136, 2004.

15. Saintier, N., Cailletaud, G., and Piques, R., Crack initiation and propagation under multiaxial fatigue in a natural rubber, *Int. J. Fatigue*, 28, 530, 2006.
16. Lindley, P.B., Relation between hysteresis and the dynamic crack growth resistance of natural rubber, *Int. J. Fract.*, 9, 449, 1973.
17. Mars, W.V. and Fatemi, A., A phenomenological model for the effect of R ratio on fatigue of strain crystallizing rubbers, *Rubber Chem. Tech.*, 76, 1241, 2003.
18. Cadwell, S.M. et al., Dynamic fatigue life of rubber, *Ind. Eng. Chem., Anal. Ed.*, 12, 19, 1940 [reprinted in *Rubber Chem. Tech.*, 13, 304, 1940].
19. Fielding, J.H., Flex life and crystallization of synthetic rubber, *Ind. Eng. Chem.*, 35, 1259, 1943.
20. Mars, W.V. and Fatemi, A., Analysis of fatigue life under complex loading: revisiting Cadwell, Merill, Sloman, and Yost, to be published in *Rubber Chem. Tech.*
21. Stephens, R.I. et al., *Metal Fatigue in Engineering*, 2nd ed., Wiley, New York, 2001.
22. Miner, M.A., Cumulative damage in fatigue, *J. Appl. Mech.*, 6, 159, 1945.
23. Sun, C., Gent, A.N., and Marteny, P., Effect of fatigue step loading sequence on residual strength, *Tire Sci. Tech.*, 28, 196, 2000.
24. Steinwegger, T., Flamm, M., and Weltin, U., A methodology for test time reduction in rubber part testing, in *Constitutive Models for Rubber III*, Busfield, J.J.C. and Muhr, A.H., Eds., A.A. Balkema, Lisse, The Netherlands, 2003, 27.
25. Harbour, R.J., Fatemi, A., and Mars, W.V., Fatigue crack growth of filled rubber under constant and variable amplitude loading conditions, *Fat Fract. Eng. Mat. Struct.*, 30, 640, 2007.
26. Harbour, R., Fatemi, A., and Mars, W.V., The effect of a dwell period on fatigue crack growth rates in filled SBR, *Rubber Chem. Technol.*, 80, 838, 2007.

26 Rubber Friction and Abrasion in Relation to Tire Traction and Wear

Karl A. Grosch

CONTENTS

26.1 INTRODUCTION

Tire traction describes the force transmission between tire and road under all eventualities. It is the prerequisite for controlled steering, acceleration, and braking of self-propelled vehicles on flat tracks. It finds its upper limit in the frictional force when total sliding occurs. Two aspects have therefore to be considered: The mechanics of force transmission of elastic wheels and its relation to rubber friction.

Whenever two bodies slide against each other under friction wear is inevitably occurring.

Material is transferred from one body to the other and this process goes in both directions. Wear is therefore associated with friction.

Since the object is to achieve high traction forces, high friction is required. This suggests that wear is also inevitably high. However, wear resistance or abrasion resistance, which is understood to be wear under controlled conditions, is likely to depend on the strength of a material. Therefore, a high wear resistance may also go well together with high friction and it is the prime objective of friction and abrasion researches to try to combine high friction with low abrasion.

When forces are transmitted between tire and road the tire is deformed to some extent. It still adheres to the road in part of the contact area, but slides locally when the ratio of tangential stress to the local pressure exceeds the friction coefficient and wear occurs. It is this partial adhesion and sliding which on the one hand allows a control over the force transmission and on the other hand leads to wear of the tire.

The chapter concentrates on the aspect of force transmission during limited sliding. The friction contribution to traction and the resulting wear, both in laboratory experiments and in tire traction and wear, tries to find a correlation between laboratory experiments and road testing conditions. Basic research will be dealt with to help to understand the connection between laboratory and road test correlations.

26.2 FRICTION OF RUBBER

Rubber friction differs from that of friction between hard solids in that the friction force is a nonlinear function of the load and depends strongly on both speed and temperature, whilst in hard solids the load dependence is linear and the friction force is virtually independent of speed and temperature [1].

The friction coefficient is defined as the tangential force acting on a sliding body to the ground reaction force. For rubbers this is a function of the ground pressure. Its dependence has been discussed sufficiently in the literature where it was shown that this is important for soft rubbers on smooth surfaces [2,3], but is of little influence for tire compounds on roads which are always sufficiently rough for the load dependence to be small if not completely absent [4,5].

26.2.1 TEMPERATURE AND SPEED DEPENDENCE OF RUBBER FRICTION

26.2.1.1 WLF Transformation Applied to Rubber Friction

Rubber friction is dominated by the viscoelastic properties of the rubber. The most striking manifestation is that equivalence exists between temperature and speed. Williams et al. [6] demonstrated that this could be described by a universal function, the famous WLF transformation equation. Conceived originally for describing the viscosity of liquids [7] and later applied to the dynamic modulus of gum rubber compounds [7], tearing energy [8], and strength [9] the author has shown that it can be applied to rubber friction for both gum as well as filled rubber compounds [10,11]. Generally, the transform is carried out by obtaining the friction coefficient over a logarithmic range of speeds, with the highest speed low enough for temperature rises in the contact patch to be negligible. The experiments are then repeated for a range of temperatures, as shown in Figure 26.1(left). If the data are plotted as function of (log speed + log a_T), where log a_T is given by the WLF equation, it appears that the curves for different temperatures are segments of a single,

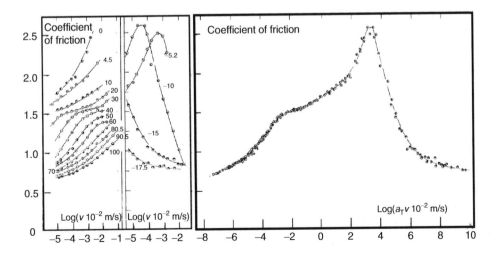

FIGURE 26.1 Experimental friction data (left) as function log speed at different temperatures and master curve (right) of an acrylate–butadiene rubber (ABR) gum compound on a clean dry silicon carbide 180 track surface referred to room temperature. (From Grosch, K.A., Sliding Friction and Abbrasion of Rubbers, PhD thesis, University of London, London, 1963.)

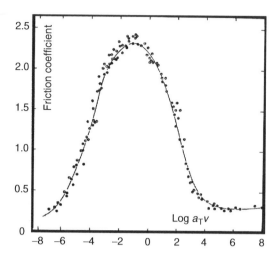

FIGURE 26.2 Master curve of the friction coefficient of an acrylate–butadiene rubber (ABR) gum compound on smooth clean dry glass, referred to room temperature. (From Grosch, K.A., *Proc. Roy. Soc.*, A 274, 21, 1963.)

so called "master curve" as shown in Figure 26.1(right) for an acrylate–butadiene rubber (ABR) gum compound on a silicon 180 track.

Master curves of the friction coefficient have been obtained for a wide range of rubber compounds on different types of tracks for dry and wet surfaces.

The shape of the maser curve not only depends on the rubber compound, but also on the surface on which it slides. On dry, clean polished glass the friction master curve for gum rubbers rises from very small values at low log $a_T v$ to a maximum which may reach friction coefficients of more than 3 and falls at high log $a_T v$ to values which are normally associated with hard materials, i.e., 0.3 shown for an ABR gum compound in Figure 26.2. If the position of the maximum on the log $a_T v$ axis for different gum rubbers is compared with that of their maximum log E'' frequency curves, a constant length $\lambda = 6 \times 10^{-9}$ m results which is of molecular dimension, indicating that this is an adhesion process [10].

On a rough silicon carbide track the friction starts at much higher values at low log $a_T v$ values than on the smooth surface. It rises and appears to pass through a maximum approximately at the same position as for the same compound on glass. It forms, however, only a small platform and begins to rise further to reach a maximum height similar to that on the glass, but at a much higher log $a_T v$. If the rough track is dusted with a fine powder, preferably with magnesium oxide, the hump disappears; the main maximum remains virtually the same in height and position as that on the clean surface, indicating that the molecular adhesion friction process is impaired.

The log $a_T v$ of the maximum on silicon carbide bears a constant relation to the frequency of the maximum loss angle of the dynamic modulus, showing that this friction process is based on a classical cyclic hysteric loss mechanism with a constant length of 5×10^{-4} m between speed and frequency which agrees with the spacing of the asperities of the track. A number of authors have published theories of this mechanism [12–14]. It is important to bear in mind that the adhesion plays a major role also in deformation friction. Purely vertical deformation leads to friction values, which are much too small to explain the magnitude of the friction coefficient [15]. The fact that the deformation friction on a dusted track is still high suggests that quite a small adhesion component is sufficient to produce large strains in the rubber as also is borne out by the presence of abrasion.

26.2.1.2 Polymer Effect on the Master Curve

The master curves of gum rubbers have all a similar shape on smooth surfaces exhibiting a single maximum of about the same height. They differ essentially in their position on the log $a_T v$ axis of

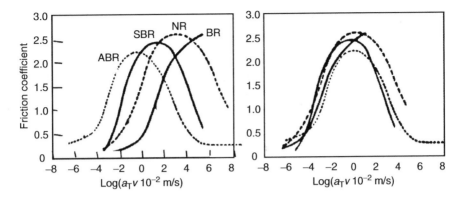

FIGURE 26.3 The position of different gum rubbers on the log $a_T v$ axis (left) referred to 20°C and (right) to their standard reference temperatures.

the master curve. If the curves are referred to a constant reference temperature, say 20°C, the position of the maximum moves further to the right the lower the glass transition temperature of the polymer. If they are referred to their standard reference temperatures or to their glass transition temperature the maxima occur almost at the same log $a_T v$ as shown in Figure 26.3. Hence the prime difference in friction behavior on smooth surfaces stems from the difference between their glass transition temperatures.

On rough surfaces the maximum depends on the position of the maximum tan δ of the polymer.

26.2.1.3 Effect of Filler and Oil Extension on the Shape of the Master Curve

Gum rubbers used above to demonstrate the influence of polymer and surfaces play no part in tires because they are too soft. Tire compounds contain invariably a considerable amount of filler predominantly carbon black but increasingly also silica. The addition of the filler does not affect the WLF transformation, but changes the shape of the master curve considerably. Figure 26.4 shows the master curves of an ABR compound (a) as gum, (b) filled with 20 pphr, and (c) with 50 parts of carbon black, respectively on smooth glass, on a dusted and clean silicon carbide track. On glass, the curves for gum and filled rubbers have a similar shape, the single maximum, however, is reduced progressively as the filler content is increased. On the two rough tracks the deformation friction maximum also decreases with increasing amount of black. On the clean rough track the deformation peak has decreased to the same level as the vestiges of the adhesion maximum so that a broad plateau appears between adhesion and deformation friction component maxima. The width of it depends on the difference between the frequencies of maximum loss modulus and loss factor.

On the dusted track the diminished adhesion friction component is clearly apparent for all three rubbers when comparing them with the master curves on the clean track The friction plateau observed for the rubbers filled with 50 pphr black which is typical for tire tread compounds is observed for most rubbers, as shown in Figure 26.5.

26.2.1.4 Friction Master Curve on Wet Tracks

If friction coefficient measurements are carried out on wet, rough surfaces at different water temperatures over a range of logarithmically spaced speeds, keeping the highest speed low, the data can be assembled into a master curve. This is shown in Figure 26.6 for an ABR gum rubber of the same formulation as used above on dry silicon carbide tracks, on an Alumina 180 grindstone surface when lubricated with distilled water and with 5% detergent added to the water, respectively. Comparison with the friction master curve obtained on the dry track shows that it displays the same features as the one on the dry surface when distilled water is used [15]. It has a peak at

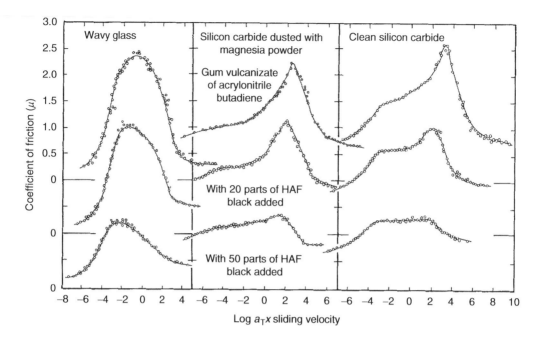

FIGURE 26.4 Master curves on smooth, wavy glass, on a silicon carbide track dusted with magnesium oxide and on a clean silicon carbide track of three acrylate–butadiene rubber (ABR) compounds as gum rubber, filled with 20 pphr carbon black and 50 pphr, respectively. (From Grosch, K.A., Sliding Friction and Abrasion of Rubbers, PhD thesis, University of London, London 1963.)

the same log $a_T v$ and the adhesion hump is also clearly visible. If a detergent is added to the water, this hump disappears. A similar effect was observed on dry tracks dusted with magnesia. Roberts [16] has shown that when a rubber sphere approaches a wet glass plate the water tends to conglomerate into globules in the contact area leaving virtually dry regions as shown in Figure 26.7(left) together with the height measurements of the water globules between rubber and glass plate. If a polar substance like soap was added to the water a very thin uniform film was maintained over the whole contact

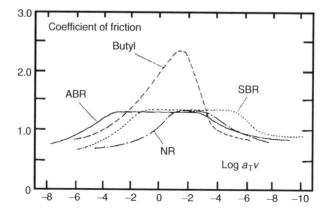

FIGURE 26.5 Master curves of different types of polymer all filled with 50 parts of carbon black referred to 20°C. (From Grosch, K.A., *Proc. Roy. Soc.*, A 274, 21, 1963, 21.)

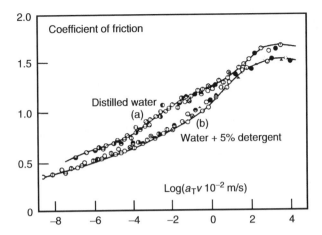

FIGURE 26.6 Friction master curve of an acrylate–butadiene rubber (ABR) gum compound on an Alumina 180 wet surface. (a) wetted with distilled water and (b) 5% detergent added to the water. (From Grosch, K.A., *The Physics of Tire Traction, Theory and Experiment*, Hayes, D.L. and Browne, A.L. (eds.), Plenum Press, New York/London, 1974.)

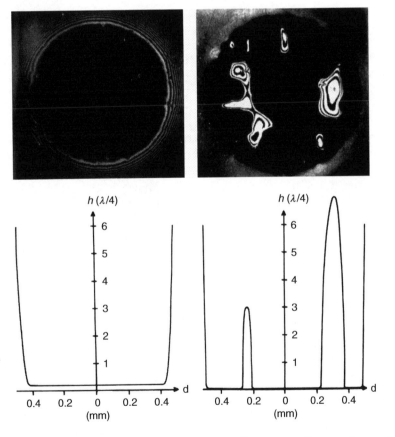

FIGURE 26.7 Newton's interference fringes and the corresponding contour diagrams showing the difference between the contact area of a rubber sphere and a glass plate when (right) wetted with distilled water and (left) with a polar substance added to the water. (From Roberts, A.D., *The Physics of Tire Traction, Theory and Experiment*, Hayes, D.L. and Browne, A.L. (eds.), Plenum Press, New York/London, 1974.)

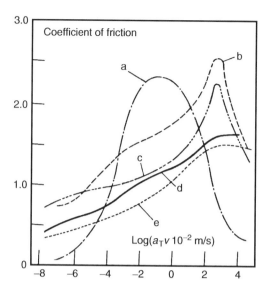

FIGURE 26.8 The friction master curves of the acrylate–butadiene rubber (ABR) gum rubber on (a) dry glass, (b) dry clean silicon carbide 180, (c) dry silicon carbide dusted with MgO powder, (d) Alumina 180 wetted with distilled water, and (e) wetted with water +5% detergent.

area to very high pressures (left of Figure 26.7). Therefore, the friction on wet tracks appears to be dry friction at a lower level. Because the water globules reduce the real contact area the frictional force is reduced accordingly, the water globules carrying some of the normal load. The total friction is then due to adhesion in the dry regions which enhances the deformation friction by producing tangential stresses in the rubber as well as by a normal deformation. Even in the presence of the polar liquid and hence presumably in the absence of dry regions friction is still very much higher than would be attributable to normal deformation so that tangential stresses must be present in the contact area. The continuous film thickness is so small—Roberts measurements indicate a thickness of about 1.5×10^{-8} m—that adhesion is still active, only its viscoelastic nature has been impaired in a similar way as is the case on dry dusted tracks. Figure 26.8 demonstrates this similarity between dry and wet friction by comparing directly the master curves of the ABR gum rubber on (a) dry glass, (b) dry clean silicon carbide, (c) dry silicon carbide dusted with magnesium oxide (MgO) powder, (d) alumina wetted with distilled water, and (e) wetted with water +5% detergent.

26.2.2 WATER LUBRICATION EFFECTS ON RUBBER FRICTION

There are two effects, which reduce the friction between rubber and a wet surface. One is the classical lubrication known from shafts rotating in sliding bearings, lubricated with oils and greases, i.e., with substances having a high viscosity. In this case the lubrication is intentional. In rubber friction on wet surfaces all efforts are being made to reduce it (lucky that water has a low viscosity). The other is due to the inertia of the liquid. When tires roll or slide on a wet road the water has to be displaced—squeezed—out of the contact area against its own inertia. This effect occurs when a tire rolls freely as well as when partial or total sliding occurs. Consider first the lubrication due to sliding: If a rectangular hard slab slides over a hard, plane, and wet surface at a velocity v and it can rotate freely around an axis parallel to its plane and normal to the sliding direction then a wedge-shaped film of water forms between the two hard surfaces separating them through a normal pressure generated by the velocity gradient between the front and the rear of the tilted plate along the sliding direction. This is the basis of sliding lubrication (for a detailed treatment consult Mitchell [17]). The minimum film thickness at the end of the slider is given by

$$h = \text{const} \left[\frac{a^2 b \eta v}{L} \right]^{1/2} \tag{26.1}$$

where

    a and b are the dimensions of the slab in and normal to the sliding direction, respectively
    η is the viscosity of the water
    v is the sliding speed
    L is the load acting on the slab

Of particular interest here is the case when a soft rubber slides over a hard smooth surface. In this case a wedge-shaped film forms even when the bodies slide nominally parallel to each other, because the rubber will deform under the hydrodynamic pressure. Roberts [16] has demonstrated this convincingly by letting a rubber sphere slide over a smooth glass surface lubricated with a silicon oil film (see Figure 26.9). The Newtonian fringes represent lines of equal height; the central contour line is shown at the bottom left. For the case of a rubber cylinder sliding on a flat hard surface the average film thickness is given by Archard and Kirk [18] as

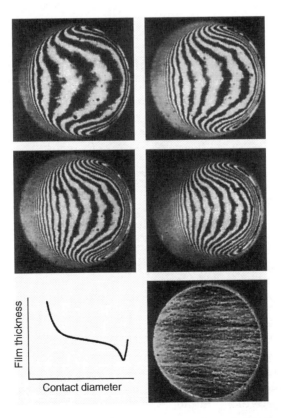

FIGURE 26.9 Newton's interference fringes formed between a glass plate lubricated with a silicon oil film and a rubber sphere sliding on it, showing the deformation of the sphere through the hydrodynamic pressure exerted on the rubber sphere. (From Roberts, A.D., *The Physics of Tire Traction, Theory and Experiment*, Hayes, D.L. and Browne, A.L. (eds.), Plenum Press, New York/London, 1974.)

$$h = 0.9 \left[\frac{\eta v R}{E} \right]^{1/2} \tag{26.2}$$

where
   R is the radius of the rubber cylinder
   E is the modulus of the rubber

For dimensional reasons a similar relation should also hold when a hard solid sphere slides on a soft rubber surface.

Assuming that the frictional force F will be proportional to the effective dry area of contact A_d

$$F = \text{const} \cdot A_d \tag{26.3}$$

and that the roughness of the hard surface—micro or otherwise—can be represented by spheres of radius r, that the thickness of the average lubricating film is h at the sliding velocity v and that h is smaller than r, the dry contact area producing through the film is

$$A_d = n\pi r^2 \left\{ 1 - \frac{h}{r} \right\} \tag{26.4}$$

where n is the number of spheres in the contact area. Using Equation 26.2 for the film thickness, the friction coefficient μ can be written as

$$\mu = \frac{F}{L} = \frac{\text{const}}{L} \left[1 - \left(\frac{a^2 b \eta v}{L r^2} \right)^{1/2} \right] \tag{26.5}$$

or with Equation 26.3

$$\mu = \frac{F}{L} = \frac{\text{const}}{L} \left[1 - 0.9 \left(\frac{R \eta v}{E r^2} \right)^{1/2} \right] \tag{26.6}$$

The speed dependence follows then

$$\mu = \mu_d \left[1 - \left(\frac{v}{v_{\text{crit}}} \right)^{1/2} \right] \tag{26.7}$$

where μ_d is the coefficient of friction on the dry track and v_{crit} is the speed at which the friction coefficient will become zero, i.e., perfect lubrication is achieved. It has a different meaning depending on which model it is based. In practice, zero friction does not happen, and a small frictional force remains because of the resistance, which the bulge of excess water creates in front of the tire or sample. Also, of course, the model simplifies the road surface structure. In reality, the road surface contains particles of different size and shape having different contact radii and different heights. In addition, often the asperities themselves have a microroughness [19–22]. When considering lubrication effects the prediction above indicates that the friction coefficient should decrease with the square root of the speed.

Practically, all data of friction measurements on wet tracks in the speed range of hydrodynamic lubrication exist as tire skid measurements. Figure 26.10 shows the results of a braking test on wet, finely structured concrete using a smooth tire and measuring the friction coefficient as function of

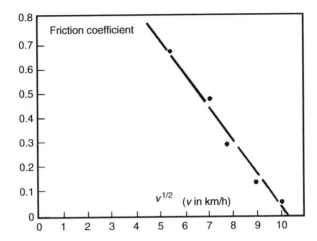

FIGURE 26.10 Locked wheel braking coefficients of a tire with a smooth tread on a wet polished concrete track as function of the square root of the speed (tire size 175 R 14, load 350 kg, inflation pressure 1.9 bar).

the speed after the wheels have been locked. Plotting the data as function of the square root of the speed results in a straight-line graph with an extrapolated value of the abscissa at zero friction coefficient of 10.2, which corresponds to a critical speed of 104 km/h. The intercept at zero gives a dry friction coefficient of 1.35, which is very close to what would be expected from laboratory measurements of tread compounds on rough surfaces. The critical speed can be related to the roughness of the track. Using Equation 26.2 with $a = 0.12$ m, $b = 0.15$ m, $L = 3500$ N, $v_{crit} = 28.9$ m/s, and $\eta = 1.01 \times 10^{-3}$ N s/m^2 the radius of the track asperities becomes $r = 0.13$ mm. If the calculation is based on Equation 26.3, using $R = 0.31$ m as the radius of the tire, $E = 24$ N/mm^2, and the viscosity as above the asperity radius $r = 0.21$ mm. Although they differ by almost a factor of about 2, both are close to reality. In any case Equation 26.6 does describe the water lubrication between rubber and rough track with two measurable quantities, i.e., the dry friction coefficient and a critical speed at which the friction becomes zero, two very useful quantities to judge the wet traction behavior of tires on a particular road surface. Notice that in this model the water film thickness does not enter the calculations.

Consider now the effect due to the inertia of the water. This plays no part in laboratory experiments with test samples; it manifests itself always in tire use. It will be discussed here to complete the discussion on models concerning the effect of water on friction. The effect is associated with aquaplaning and has been discussed extensively by Yeager [23]. Water has to be squeezed out of the contact area. For this to happen, the water film thickness has to be higher than the asperities. As a point moves along the contact area it sinks until it makes dry contact with the tips of the asperities at a point x along the contact area. This length is proportional to the sink time t_a which itself is a function of the water height above the asperities and to the speed of the vehicle v_f

$$x = v_f t_a \tag{26.8}$$

With increasing speed the point to make dry contact moves further along the contact area thereby reducing the dry contact and hence the friction. When the speed is so high or the sink time is so long that even at the end of the contact area l the water cannot be removed, all frictional contact is lost, and with it the steering and braking/accelerating capability of the vehicle. The subject has been discussed extensively in the literature (e.g., [23–27]).

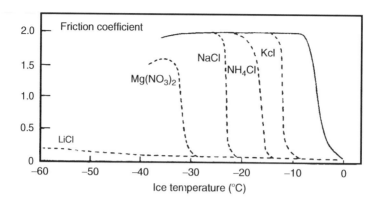

FIGURE 26.11 Frictional properties of ice when slight amounts of different salts are added to the water before freezing. (From Roberts, A.D. and Lane, J.D., *J. Phys. D*, 16, 275, 1983.)

26.2.3 FRICTION ON ICE

To establish a master curve on ice is difficult because the ice surface, being not far removed from its melting point, reacts itself very sensitively to pressure, speed, and temperature changes. Roberts and Lane [28] have shown that small additions of different salts change the ice properties at the surface drastically, as shown in Figure 26.11. This may well be a major reason for the generally observed large variability of open-air winter tests on ice. Figure 26.12 shows the friction coefficient of a natural rubber (NR) gum compound on a smooth ice track as function of the track surface temperature for three different speeds. Near the melting point of the ice the friction is low, obviously some melting in the contact area has lubricated the track. In addition, the shear strength of the ice is low so that some shearing of the ice may also take place. The friction coefficient rises sharply as the track temperature is decreased reaching similar values as on a smooth dry track. The actual height

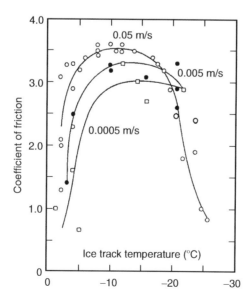

FIGURE 26.12 Friction coefficient of a natural rubber (NR) gum compound as function of the ice temperature at three different speeds (left) and friction coefficient of four different gum compounds having different glass transition temperatures as function of the ice track temperature at a constant sliding speed of 0.005 m/s. (From Heinz, M. and Grosch, K.A., ACS Spring Meeting, St Antonio, 2005.)

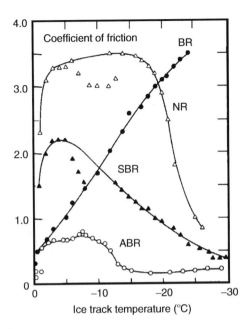

FIGURE 26.13 Friction coefficients of four different gum compounds having different glass transition temperatures as function of the ice track temperature at a constant sliding speed of 0.005 m/s. (From Heinz, M. and Grosch, K.A., ACS Spring Meeting, St Antonio, 2005.)

depends on the speed: The higher the speed, the lower the maximum. The drop in friction at $-25°C$ for the curve at the lowest speed is probably not due to compound properties. It may be that slight hoarfrost formed on the track.

Figure 26.13 shows the friction coefficient of four different gum polymers as function of the ice track temperature at a constant speed [29]. NR displays a broad maximum as already seen above. The friction coefficient of the styrene–butadiene rubber (SBR) compound rises only over a small temperature range and begins to drop already at $-5°C$ ice temperature. The coefficient of friction of butadiene rubber (BR) on the other hand rises with decreasing ice temperature over the whole experimental range, whilst that of the ABR compound remains very low. The major difference between these polymers is their glass transition temperature. The viscoelastic nature of the friction behavior becomes apparent if the above data are plotted as function of log $a_T v$. For each polymer only part of its friction master curve is produced because of the single speed and the limited temperature range. However, because of their different glass transition temperatures they occupy different ranges on the log $a_T v$ axis. If all four master curve segments are plotted as function of log $a_T v$ on one graph they form the shape of a single master curve with a single maximum, which has the same position as on glass, as shown in Figure 26.14 (compare with Figure 26.3 right). This shows clearly that the friction process on ice is the same as on smooth glass; a deformation peak is absent demonstrating clearly that the friction on ice is due to adhesion. Recently, friction coefficients were measured on the ice track disk of an LAT 100 traction and abrasion tester, comparing several compounds at different loads, speeds, and ice surface temperatures right up to $-4.5°C$, a temperature which is of great practical importance for winter conditions of many northern countries with a moderate winter climate [30]. Figure 26.15 shows the frictional force on ice (a) as function of load, at a constant speed and ice surface temperature for two compounds showing the experimental values and a fitted power function

$$F = F_0 \left[\frac{L}{L_0} \right]^n \tag{26.9}$$

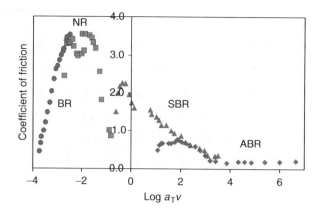

FIGURE 26.14 The friction coefficients of the four gum polymers of Figure 26.13 plotted as function of log $a_T v$. (Data from Heinz, M. and Grosch, K.A., ACS Spring Meeting, St Antonio, 2005.)

where L is the load and n is a power index, which depends on the compound and similar curves for the speed dependence at a constant load, (b) shows the fitted power functions of seven compounds, three winter and four summer compounds, left for the load and right for the speed dependence, and (c) the relative ratings with compound 1 as the reference. It is very clear that the winter tread compounds are much superior to the four tested summer tread compounds both for their load as well as for their speed dependence. The frictional force decreases with increasing speed because the ice melts more as the speed increases lubricating the track. Plotting in Figure 26.16 the friction force at a constant speed and load as function of the ice surface temperature for the seven compounds it is seen that the compound differences are maintained also over a wide range of ice surface temperatures. Table 26.1 shows friction coefficients and ratings of four tread compounds at $-4.5°C$. A is a Standard 100 OSBR summer tire compound, B is an NR + N234 traditional winter compound with a low glass transition temperature, and C is 70 BR/30 SBR + N375, a modern black-filled winter tread also with a low glass transition temperature. D has the same polymer composition, but is filled with silica with silane coupling. The difference between the winter treads and the summer compound is striking. The absolute values of the friction coefficient are very low, a dangerous situation in any case so that the advantage of a suitable tread compound is even more important.

26.2.4 FRICTION PROPERTIES OF TREAD COMPOUNDS AT HIGHER SPEEDS

In order to establish master curves of friction or side force coefficients, the speed range has to be low in order to avoid significant temperature rises in the contact area and, if the experiments are carried out on wet tracks, also lubrication effects.

Generally, however, it is necessary to allow higher speeds to study the influence of temperature rises and lubrication effects in the contact area because this influences profoundly the traction capabilities of tires.

Few experiments exist which have tried to determine the temperature rise in the contact area between rubber and a hard slider. Schallamach rubbed a thermocouple over a rubber pad at different loads and obtained considerable rises in temperature. The author had a thick rubber strip glued to a tire-testing drum and a thermocouple imbedded in a small spherical slider to which different loads could be applied. Figure 26.17 shows the temperature rise in the contact area between a dry rubber surface and the slider as function of speed and on a wet surface for two different loads [15]. On the dry surface the temperature rises rapidly with increasing speed reaching values of over $300°C$. The curves can be described exactly by a simple relation

$$\text{Temprise} = \text{const} \cdot \sqrt{v_s} \qquad (26.10)$$

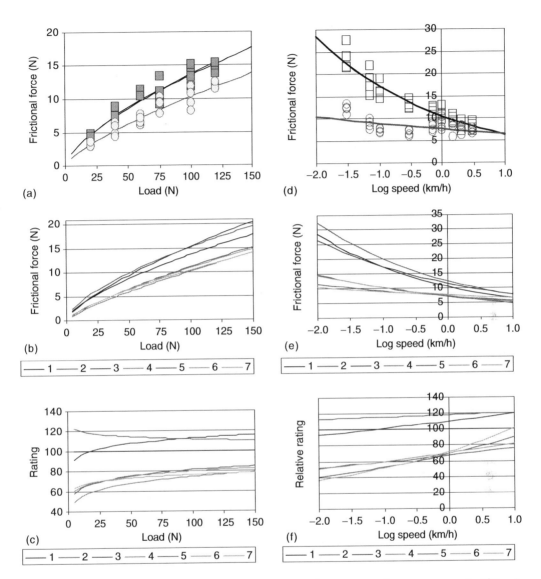

FIGURE 26.15 Friction force on ice as function of the (a) load at a constant ice temperature of $-5°C$ and a speed of 1.5 km/h, for two compounds showing the experimental points and the fitted power function (b) showing the calculated power functions of seven compounds. Their main ingredients are shown below. (c) Shows the relative ratings with compound 1 as reference. (d) to (f) Show similar graphs for the speed dependence at a load of 75 N.

where v_s is the sliding speed. The constant contains the strength of the heat source, heat conductivity, and dimensions of the heat source [30]. The rubber does not burn, although its inflammation point has been exceeded. This is due to the fact that the temperature decreases rapidly, as soon as it leaves the contact area. Because of the low heat conductivity the temperature also decreases rapidly with increasing depth.

On the wet track the temperature rises to about 100°C and then falls. This fall was accompanied by a drop of the frictional force indicating that the lubrication of the water led to an increasing separation of the two surfaces.

Figure 26.18 shows a theoretical calculation of the temperature rise along the contact area of a rubber strip under a constant load and speed over a semi-infinite dry rubber solid [30]. The figure

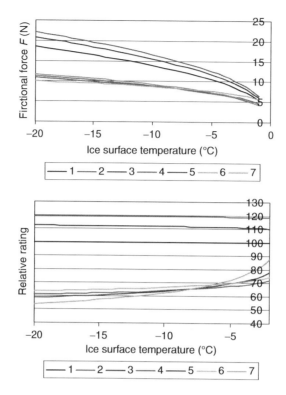

FIGURE 26.16 The frictional force and the relative rating of the seven compounds of figure 14 as a power function of the ice surface temperature at a load of 75 N and a speed of 1.5 km/h.

shows the temperature profile along the pad in the sliding direction. Heat losses towards the sides have been neglected. The maximum temperature occurs near the end of the pad and decays rapidly after the pad has passed. The temperature in the interface is, of course, the same for slider and solid semi-infinite surface. It is determined by the heat conductivities of the two surfaces and their specific heat capacities. Roberts [31] measured the decay of the surface temperature after a slider had passed over it. This agrees well with the theoretical calculations. Therefore, there can be no doubt that a considerable temperature rise occurs in the contact area between rubber sample and abrasive track.

The temperature rise in the contact area plays a major role in abrasion and tire wear. It leads to thermal degradation and aids oxidation. This will be discussed in Section 26.5. The friction is primarily influenced because the temperature rise influences the operating point $\log a_T v$ of the master curve.

TABLE 26.1
Friction Coefficients and Relative Ratings of Four Tire Tread Compounds at $-4.5°C$ Ice Surface Temperature and a Speed of 0.5 km/h

Compound	Measurement				Average	Rating
	1	2	3	4		
A	0.0528	0.0770	0.0642	0.0718	0.0665	**54.7**
B	0.1208	0.1279	0.1188	0.1182	0.1214	**100.0**
C	0.1390	0.1360	0.1370	0.1364	0.1371	**112.9**
D	0.1402	0.1375	0.1479	0.1566	0.1455	**119.8**

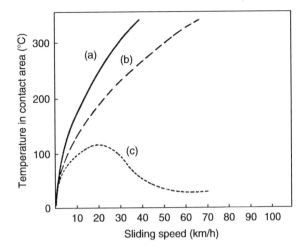

FIGURE 26.17 Temperature rise between a rubber track and a steel slider as function of speed: (a) cone tip radius 1 mm, (b) sphere 11 mm dia. and (c) cone sliding on wet track. Load in all cases 2.5 N. (From Grosch, K.A., *The Physics of Tire Traction, Theory and Experiment*, Hayes, D.L. and Browne, A.L. (eds.), Plenum Press, New York/London, 1974.)

In order to calculate the operative log $a_T v$ value a relation is required between the temperature rise in the contact area and the sliding speed. According to Carslaw and Jaeger [30] the maximum temperature rise, occurring for rubber near the end of the contact area, is given by the following relation:

$$\Delta t = Q\sqrt{\frac{2l_r}{\pi \rho c K v_s}} \tag{26.11}$$

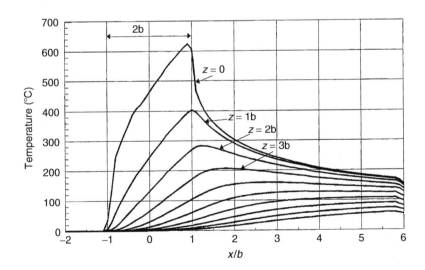

FIGURE 26.18 Theoretical temperature rise in the contact area of a pad sliding over a semi-infinite solid for different depths from the surface. Width 2b: 2 mm, speed: 3 m/s, pressure: 2 Mp, $\mu = 1$, heat conductivity: 0.15 W/m/K, heat diffusivity: 10^{-4} m^2/s.

where

  Q is the heat generated per unit area and time
  l_r is the contact length of the heat source
  ρ is the density
  c is the specific heat
  K is its heat conductivity

In this case all the heat flows into the rubber. If the heat is produced by friction, part of it will flow into the rubber and the other part into the track surface. Assuming a large heat capacity for both and neglecting heat losses to the sides, the amount of heat flowing into the rubber is given by

$$\psi = \frac{q_r}{Q} = \frac{\sqrt{\kappa_r}/K_r}{\sqrt{\kappa_r}/K_r + \sqrt{\kappa_t}/K_t} \tag{26.12}$$

where the subscripts t and r stand for track and rubber, respectively, κ is the heat diffusivity, K is the heat conductivity of the material, and q_r is the amount of heat, which flows into the rubber per unit area and time.

The temperature rise near the end of the slider will then become

$$\Delta t = \psi_r \mu p \sqrt{\frac{2 l_r v_s}{\pi \rho c K_r}} \tag{26.13}$$

This relation can be simplified to

$$\Delta t = \xi_c \mu p \cdot \sqrt{v_s} \tag{26.14}$$

which agrees well with the findings of Figure 26.17 and Equation 26.10 above.

The constant ξ_c contains the constants above. Unfortunately, it is hardly ever possible to deduce ξ_c from this formula. It has to be determined by experiment.

If the temperature is a function of speed log a_T is also a function of speed changing the operating point on the master curve. Log $a_T v$ increases with speed and decreases with increasing temperature. At very low speeds log v wins, but as the speed increases the influence of log a_T becomes larger eventually overtaking log v and log $a_T v$ passes through a maximum. Figure 26.19 shows the log $a_T v$ values as function of the sliding speed for an NR and an SBR tread compound under different track surface conditions at a constant ambient temperature. To obtain these curves the constant ξ_c had to be estimated. For the dry track conditions, this was derived from Figure 26.17. For wet conditions its value cannot exceed 100°C and was reduced accordingly and for ice it cannot exceed its melting point. In each case the range of log $a_T v$ values obtainable is limited. On dry and wet surfaces the log $a_T v$ values go through a maximum. In addition, on wet surfaces lubrication leads to a decrease of the temperature at high speeds so that the log $a_T v$ value rises again. Because of the temperature limit of 0°C on ice the curve rises continuously.

The range of log $a_T v$ values obtainable in practical tire tests is therefore limited to about 2 decades. Hence only a small part of the master curve is covered in practical traction tests. This portion, however, depends strongly on the ambient conditions. Because of the different glass transition temperatures for different polymers the operative values of log $a_T v$ also differ considerably for different compounds since they are determined by their standard reference temperature or, more basic, their glass transition temperature. Figure 26.20 indicates the regions of the operative log $a_T v$ for sliding speeds between 1 and 10 km/h (corresponding to about 10–100 km/h forward speed at an antiblocking system [ABS] braking) on the two master curves of NR and SBR gum on glass (see upper part of the diagram) under wet and icy environmental conditions, respectively.

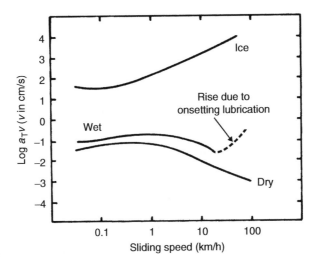

FIGURE 26.19 Log $a_T v$ values as function of speed for a natural rubber (NR) and a styrene–butadiene rubber (SBR) compound on wet and icy track surfaces.

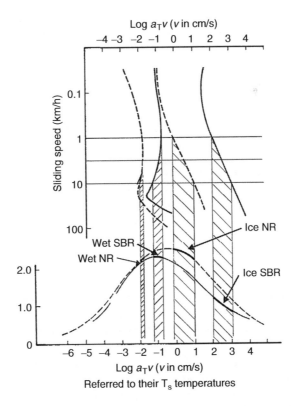

FIGURE 26.20 The log $a_T v$ speed function of the previous chart is combined with the friction master curves for a natural rubber (NR) and a styrene–butadiene rubber (SBR) gum compound on glass showing the limited range of friction values (and their position on the log $a_T v$ axis for different testing conditions) which are obtained when the sliding speed is increased.

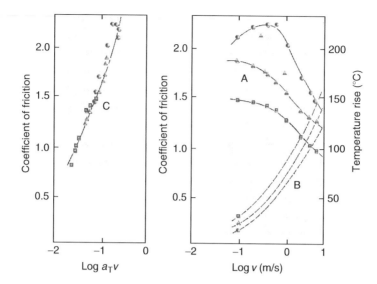

FIGURE 26.21 Coefficient of friction (A) and the temperature rise (B) in the contact area between a natural rubber (NR) gum compound and a thermocouple slider as function of speed at different loads. Also shown are the data when plotted against log $a_T v$ (C), indicating a rising portion of a single master curve which absorbs also the apparent load effect.

On wet surfaces, the SBR is better than the NR compound because their operating log $a_T v$ range is on the rising part of the master curve, whilst on ice it is on the decreasing branch, reversing their ranking, as already seen above when discussing rubber friction on ice. This is the case for using separate sets of tires for winter and summer.

The maximum of the log $a_T v$ speed curve also produces a maximum of the friction speed curve. This has nothing to do with the maximum of the friction master curve. On dry or wet surfaces the operating points lie generally on the rising part of the master curve. The maximum friction is then determined by the maximum log $a_T v$. This is demonstrated in Figure 26.21 which shows friction measurements on an NR gum compound with a slider which had a thermocouple embedded measuring both friction and the corresponding contact temperature allowing the transformation of the data into a friction (log $a_T v$) curve. The three experimental curves were obtained under different loads producing different temperature rises. It is seen that a single rising curve results from the transformation. The load dependence is revealed in this case as a disguised temperature effect.

26.3 INFLUENCE OF THE FRICTION ON THE FORCE TRANSMISSION UNDER LIMITED SLIP

Speed and direction of a self-propelled vehicle are controlled by forces between tire and road. These forces have an upper limit set by the available friction coefficient. Once the ratio between tangential traction and normal pressure exceeds this limit, local sliding occurs. It is important to remember that sliding friction has no preferred direction; once sliding sets in over the whole contact area directional control is lost as is the case when the wheels are locked. The vehicle moves to a stop in the direction in which its center of mass was moving when the wheels locked. It would appear at first sight that this would be the case whenever the circumferential speed of the tire differs from its forward speed. This would indeed be the case if the wheel and the track were infinitely stiff. For all elastic wheels the possibility exists that through deformation of the wheel only part of the contact area slides and the remainder maintains contact by adhesion and thus enables a controlled change of direction as well as longitudinal acceleration and braking. Because of the elastic nature of the contact a difference between the circumferential speed and the forward speed of the wheel occurs whenever a force is acting on the wheel.

26.3.1 SIMPLE SLIP

The difference between the circumferential and the traveling velocities is defined by the slip s_1, a vectorial kinematic quantity, as

$$\vec{s_1} = \frac{\vec{v}_f - \vec{v}_c}{|v_c|} \tag{26.15}$$

where
v_f is the traveling velocity
v_c is the circumferential velocity of the wheel

The difference between v_c and v_f is called the slip velocity v_s. When the forward velocity makes an angle θ with the plane of the wheel the slip becomes

$$s_1 = \sin\theta \tag{26.15a}$$

For circumferential slip

$$s_1 = 1 - \frac{v_s}{v_f} \tag{26.15b}$$

For braking $v_f > v_s$ is positive and has a maximum value of 1. For acceleration $v_f < v_s$ is negative and becomes infinite when the forward speed is zero, i.e., the driven wheel spins through.

Slip can be considered as the ratio of energy, which is turned into heat to the kinetic energy stored in the mass carried by the wheel. During cornering energy is lost and the forward speed is smaller than the circumferential speed. During braking the situation is similar. In both cases the slip becomes 1 when the wheel no longer turns. Accelerating energy is supplied to increase the forward speed. Hence the negative sign of the slip. If the vehicle cannot gain any kinetic energy because of spin-through the slip becomes infinite.

26.3.2 RELATION BETWEEN A FORCE ACTING ON THE WHEEL AND SIMPLE SLIP

When forces are transmitted, slip occurs because the wheel is being deformed. The relation between force and slip is one of the most important laws in tire mechanics, because it influences the all important properties of traction, durability, and tire wear. (This is not only true for tires, but also for all force transmission by adhesion friction).

Even in a homogeneous solid elastic wheel the distortion is complex and requires sophisticated methods to arrive at a precise relation between force and slip. For tires this is even more difficult because of its complex internal structure. Nevertheless, even the simplest possible model produces answers which are reasonably close to reality in describing the force–slip relation in measurable quantities. This model, called the brush model—or often also the Schallamach model [32] when it is associated with tire wear and abrasion—is based on the assumption that the wheel consists of a large, equally spaced number of identical, deformable elements (the fibers of a brush), following the linear deformation law

$$f = k_f \cdot y \tag{26.16}$$

where
  f is the force on a brush element
  y is the deformation
  k_f is the spring constant of the fiber

  This has the consequence that no distortion occurs outside the contact area.
  Consider first the case of cornering, i.e., the plane of the wheel makes an angle with its traveling velocity. This creates a force normal to the plane of the wheel referred to as side force or cornering force. In tires it serves to steer the vehicle.
  The load on the wheel produces a contact area of finite length a. The distortion caused by the load is ignored in the brush model. This means that the above relation is really a shear relation. The fibers have a large compression stiffness and a small shear stiffness, which in fact is true for rubbers. The large contact length is created by the air inflation chamber of the tire. Solid rubber wheels bulge out.
  As x increases along the contact area the deflection increases proportionally. For unit area the stiffness $k = n \times k_f$, where n is the number of fibers per unit area. Hence the distorting stress t becomes $t = k \times \tan\theta$. This situation is maintained until the limiting force set by the available friction is reached. If it is assumed that the pressure distribution along the x-axis is elliptical and the friction coefficient is a constant, sliding occurs along the curve $f(\mu p)$. A diagrammatic view of this is given in Figure 26.22. The side force is then proportional to the area under the curve. A similar diagram is also obtained for the braking or accelerating force, except that the force acts in the plane of the wheel. The force–slip relation for the brush model is given by

$$S, B, A = \frac{\mu L}{\pi} \left[q + \sin^{-1} q \right] \tag{26.17a}$$

where
  S, B, A stand for either side or braking of accelerating force
  μ is the friction coefficient

The function q is given by

$$q = \frac{2c}{1 + c^2} \tag{26.17b}$$

where c depends on the type of force considered.
  For the side force

$$c = \frac{\pi}{8} \cdot \frac{ka^2}{\mu L} \cdot \tan\theta \tag{26.17c}$$

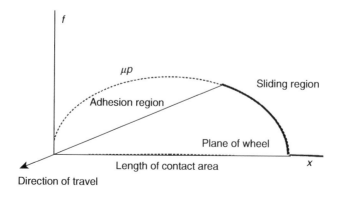

FIGURE 26.22 Diagrammatic view of the contact area for the brush wheel model under cornering.

where

    L is the load

    a is the length of the contact area

    k_s is the stiffness normal to the plane of the wheel

    μ is the friction coefficient

    θ is the slip angle, the angle which the plane of the wheel makes with the forward velocity of the wheel

At small slip angles the force–slip angle relation reduces to

$$S = \frac{k_s a^2}{2} \cdot \theta \tag{26.18a}$$

This means that at small slip angles the force is independent of load and friction coefficient and at zero slip angle the slope of the curve is given by

$$K_s = \frac{k_s a^2}{2} \tag{26.18b}$$

which is the cornering stiffness of the wheel, a measurable quantity.

Similarly, for circumferential slip

$$c = \frac{\pi}{8} \cdot \frac{k a^2}{\mu L} \cdot \frac{s_1}{1 - s_1} \tag{26.19}$$

which becomes at small slips

$$B, A = \frac{k_c a^2}{2} \cdot s_1 \tag{26.19a}$$

where

    B is the braking force

    A is the accelerating force

In this case the slope of the curve at zero slip is given by

$$K_c = \frac{k_c a^2}{2} \tag{26.19b}$$

where K_c is the circumferential slip stiffness of the wheel which has a different value from the cornering stiffness, but can also be measured. Figure 26.23 plots $F/\mu L$ as function of the quantity c together with the adhesion and the sliding contributions, making up the total force. The adhesion region dominates up to a value of c of about 0.82. At larger values the sliding region becomes more and more important. At higher c values, i.e., at a high stiffness and/or large slip, low friction coefficient and/or low load the force F tends asymptotically to μL.

26.3.3 Side/Braking Force–Slip Relation on Wet Surfaces

It was described above how the friction coefficient changes with increasing speed when a sample slides on a wet surface. The braking coefficient also changes with increasing slip at a constant

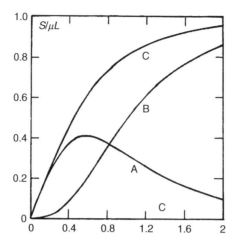

FIGURE 26.23 Side force coefficient (dimensionless quantity $S/\mu L$) as function of the quantity c (Equation 26.17c) showing the two components due to adhesion and sliding.

forward speed. This is best described using Equation 26.7 for the friction coefficient and the slip speed instead of the forward speed. Figure 26.24 shows a calculated example using the brush model and a modified friction coefficient

$$\mu = \mu_d \left[1 - \left(\frac{s_1 v_f}{v_{crit}} \right)^{\frac{1}{2}} \right] \tag{26.20}$$

where

s_1 is the braking slip

v_f is the forward speed

v_{crit} is the speed at which perfect lubrication would be achieved and the force would become zero

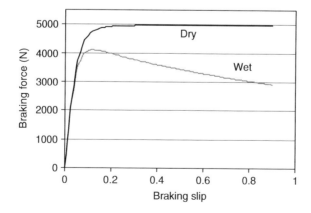

FIGURE 26.24 Braking force as function of slip using the brush model with (dry) a constant friction coefficient and (wet) a friction coefficient which depends on the slip speed because of wet lubrication. Braking stiffness 90,000 N, dry friction coefficient 1.2, load 4500 N, speed at the onset of braking 40 km/h, critical speed 210 km/h.

It is seen that the braking force passes through a maximum and decreases steadily with increasing slip. The maximum value occurs at a relatively small slip value and is much lower than the dry maximum braking force of $F = \mu L$. It is a practical experience that the braking force in a car-braking test on a wet track reaches a maximum and then drops rapidly to a steady lower value when the wheels lock. This sudden decrease in braking power occurs because the braking force decreases with increasing slip, a situation which is unstable and hence the wheel locks, and the slip virtually jumps to a 100% value instantaneously. To counteract this effect modern cars are equipped with ABS which releases the brakes for a short time as soon as the maximum is exceeded.

26.3.4 Forces under Composite Slip

The model can also be used when both slip components act simultaneously. The limiting force is still determined by the friction coefficient and pressure. The total traction at a point x in the contact area is the vector sum of the two components. This determines the point of onsetting sliding and hence the circumferential contribution is the total less the side force contribution. This is shown diagrammatically in Figure 26.25 [33].

The side force becomes

$$S = \frac{\mu L}{\pi} \cdot \frac{K_s \sin \theta}{\sqrt{(K_s^2 \sin^2 \theta + K_c^2 s^2)}} \left\{ q_{com} + \sin^{-1} q_{com} \right\} \qquad (26.21a)$$

where K_s and K_c are the cornering stiffness and the circumferential slip stiffness, respectively, and

$$q_{com} = \frac{2\sigma}{1 + \sigma^2} \qquad (26.21b)$$

and $\quad \sigma = \frac{\pi}{4\mu L} \cdot \frac{\sqrt{(K_s^2 \sin^2 \theta + K_c^2 s_{co}^2)}}{\cos \theta - s_{co}} \qquad (26.21c)$

with $s_{co} = 1 - s_1$.

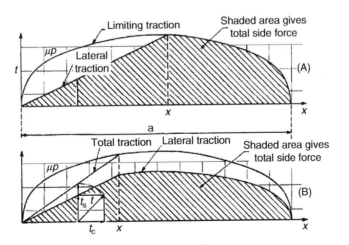

FIGURE 26.25 Diagrammatic view of the combined side and braking/accelerating force function for the brush wheel model. (From Schallamach, A. and Grosch, K.A., *Mechanics of Pneumatic Tires*, S.K. Clark (ed.), The US Department of Transportation, National Highway Safety Administration, Washington DV, p. 419.)

The braking and acceleration forces are obtained from

$$B = A = \frac{K_c s_{co}}{K_s \sin \theta} \cdot S \qquad (26.21d)$$

Because the limiting force is given by the friction force

$$F = \mu L$$

the two components cannot be increased at will. If the braking component is increased the side force component is reduced accordingly.

As discussed under wet friction, the friction coefficient decreases on wet surfaces with increasing speed. This has also an effect on the slip dependence of side force or braking force for constant forward speed, since it is the slip speed which determines the hydrodynamic pressure in the contact area and hence the friction coefficient An example calculated with the brush model and the friction coefficient of Equation 26.7 in which the forward speed is replaced by the slip speed is shown in Figure 26.26, for the braking force component as function of slip together with the side force components at different set slip angles. Even at small slip values the braking force is reduced progressively with increasing set slip angles. When the braking slip approaches 1 (locked wheel condition), the side force component tends towards zero.

If a car is braked whilst negotiating a curve, the steering force is reduced and the car suddenly understeers.

Whenever forces are transmitted and slip occurs as consequence energy is dissipated.

The general expression for the energy dissipation W in kilojoule per kilometer is

$$W = \text{force} \times \text{slip} \qquad (26.22)$$

$$\text{For cornering, this is given by } W_s = S \cdot \sin \theta \qquad (26.22a)$$

$$\text{And for circumferential slip } W_c = B \cdot s_l \qquad (26.22b)$$

The total energy loss per unit distance in a slipping wheel under composite slip is then the sum of the two components. It is partly turned into heat and is the cause of material transfer, i.e., abrasion, which itself is influenced by the raised temperature in the contact area.

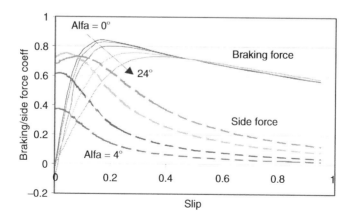

FIGURE 26.26 Braking force as function of the slip on a wet surface in the presence of side forces at different set slip angles (calculated with the brush model).

It is important to remember that the total sum of the energy is limited to $W_T = \mu L$, the product of friction coefficient and load.

26.3.5 EXPERIMENTAL VERIFICATION OF THE MODEL

Figure 26.27 shows experimental values for the side force of a bias tire as function of the variable c for different loads [32] (see also data from [34]). The solid line is the force–slip relation of the brush model. In order to make the data fit the line, the cornering stiffness, and the friction coefficient have to be determined. For this Schallamach calculated logarithmic curves of S/L vs. $\tan \theta$. Vertical and horizontal shifts produce the value for the friction coefficient and the cornering stiffness. Figure 26.28 shows similar values for a radial ply tire, obtained with a trailer equipped with a force-measuring hub. Different load and slip angle settings were used on two different wet surfaces.

Side force measurements were carried out on a laboratory abrasion tester LAT100 at different slip angles at a constant load and speed as shown in Figure 26.29. The cornering stiffness and the friction coefficient were obtained by a curve fitting method using a computer program. Since the sample runs on the flat side of a circular disk, the side force is not zero at exactly zero slip angle. Its deviation from zero was also determined by the program. The curve obtained agreed excellently with the data. A similar diagram was obtained for different loads as shown in Figure 26.30. In this case the friction coefficient depended on the load according to the relation

$$\mu = \mu_o \left[\frac{L}{L_o} \right]^{-0.1} \tag{26.23}$$

After adjusting for this effect, all data could be fitted to the tire model curve. It is seen that good agreement is reached in all cases with this simple model. Figure 26.31 shows the braking coefficient as function of the circumferential slip at a constant vehicle speed, load, and zero slip angle. The data presented here were obtained with the NHTSA Mobile Traction Laboratory (MTL). The curve was fitted to the data using the brush wheel model and a variable friction coefficient as defined in Equation 26.7 [35]. With the MTL it was also possible to establish braking force–slip curves under a set slip

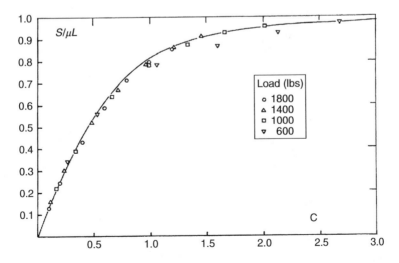

FIGURE 26.27 Side force coefficient as function of c (Equation 26.17c) for a bias tire for different loads. The solid line is the function of the brush model. (From Schallamach, A. and Grosch, K.A., *Mechanics of Pneumatic Tires*, S.K. Clark (ed.), The US Department of Transportation, National Highway Safety Administration, Washington DV, data from Nordeen, D.L. and Cortese, A.D., *Trans S.A.E.*, 72, 325, 1964.)

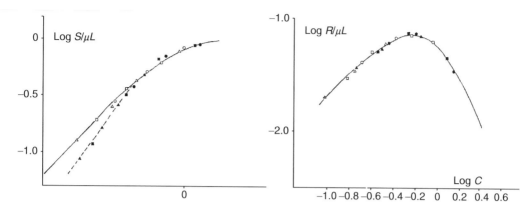

FIGURE 26.28 Side force coefficient and self-aligning torque of a radial ply tire 175 R 14 on two wet road surfaces of different friction coefficient, at three slip angles and loads as function of the quantity c (Equation 26.17c) all on log scales. The solid lines correspond to the brush model. (From Schallamach, A. and Grosch, K.A., *Mechanics of Pneumatic Tires*, S.K. Clark (ed.), The US Department of Transportation, National Highway Safety Administration, Washington DV.)

angle as shown in Figure 26.32. It was also possible to fit the curves to the data using the brush model including the decrease of the side force coefficient with increasing braking slip The model is very useful for a description of the force–slip relation using the measurable quantities cornering stiffness, longitudinal slip stiffness, and friction coefficient, including a variable friction coefficient taking into account the water lubrication. But it is of little use to the tire construction engineer since it gives no information on either the nature of the stiffness or the friction coefficient.

26.3.6 SPEED AND TEMPERATURE DEPENDENCE OF THE SIDE FORCE COEFFICIENT

Figure 26.33 shows the side force coefficient as function of log speed for different temperatures at a constant load and slip angle for a tire tread compound based on 3,4 *cis*-poly-isoprene, a polymer

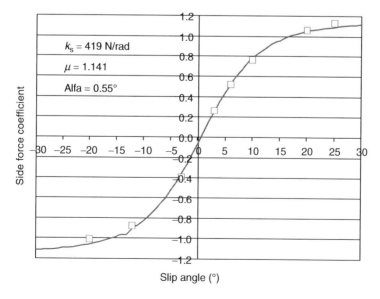

FIGURE 26.29 Side force as function of slip angle measured on the LAT 100 laboratory test equipment. Load 75 N, speed 2 km/h.

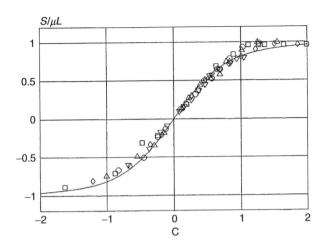

FIGURE 26.30 Side force coefficient S/L as function of slip angle for different loads. To fit the brush model curve, the friction coefficient had to be adjusted for load dependence according to $\mu = \mu_o(L/L_o)^{-0.1}$ (surface: Alumina 180, speed: 2 km/h).

with a high glass transition temperature. The data were obtained on the flat side of an Alumina 180 wet, blunt grinding stone using the LAT 100 abrasion and traction tester [36]. The temperature was regulated by the water temperature which was pumped onto the track in a closed circuit. In Figure 26.34, the data have been transformed into a master curve, shifting the data, by using the WLF equation directly with a standard reference temperature $T_s = T_g + 50$. It is seen that a single curve results which shows all the features already discussed above for the friction coefficient on dry and wet tracks. For dry friction a plateau between the adhesion and deformation friction would be expected for this type of black-filled compound. For the side force measurements there is still a deformation peak discernable, but also a clear change in curvature at the position of the adhesion maximum, expected on dry smooth surfaces. Figure 26.34 also shows the master curve obtained for the same compound on the same surface for the friction coefficient. In this case,

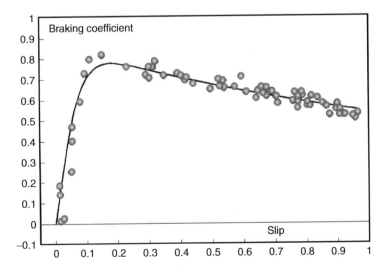

FIGURE 26.31 Braking coefficient measured on a wet test track with the Mobile Traction Laboratory (MTL) as function of slip.

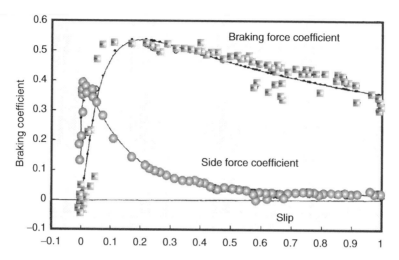

FIGURE 26.32 Braking and side force coefficient as function of the longitudinal slip for a set slip angle of 8° on wet asphalt at a constant speed of 30 mph, obtained with the Mobile Traction Laboratory (MTL) of the NHTSA. The curves were fitted using the brush model for composite slip with a variable friction coefficient.

the adhesion friction is more distinct and the deformation friction less so. The maximum of the deformation friction occurs at a lower speed than that of the side force coefficient. This is largely due to the fact that the forward speed was used for the transformation of the side force coefficient which is about 0.7 decade higher than the slip speed, which in turn is approximately equal to the sliding speed of the friction coefficient. Figure 26.35 shows the master curves of the side force coefficients of the 3,4 UR tread and an OESBR tread compound. Because of the much lower glass transition temperature of the SBR, the decrease of the side force coefficient at high log $a_T v$ values was outside the experimental range. The figure demonstrates how misleading spot measurements

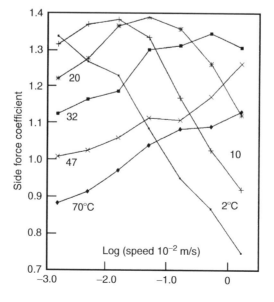

FIGURE 26.33 Side force coefficient of an 3,4 IR tread compound as function of log (speed) at different water temperatures on a wet Alumina 180 surface, of which the tips have been blunted.

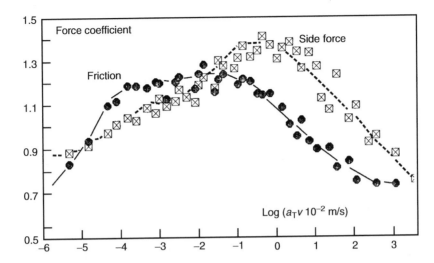

FIGURE 26.34 The data of Figure 26.32 plotted as function of log a_Tv with the standard reference temperature taken to be $T_s = T_G + 50$, also shown the friction master curve for the same compound on the same surface.

can be. Two measurements carried out at two different water temperatures under otherwise the same conditions give two totally different results, reversing the ranking of the compounds, an experience which is being made continuously in practical tire traction tests.

26.4 CORRELATION BETWEEN ROAD TEST DATA AND LABORATORY MEASUREMENTS

26.4.1 LABORATORY MEASUREMENTS

Laboratory measurements are primarily concerned with tread compound traction properties. Tread pattern and other tire parameters like cornering and longitudinal slip stiffness require still tests with tires on either large indoor machines or direct proving ground measurements.

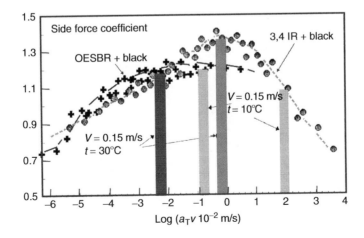

FIGURE 26.35 Master curve of the side force coefficient of two tread compounds with different glass transition temperatures on wet Alumina 180: A 3,4 IR $T_g = -21°C$ and OESBR $T_g = -46°C$. Two spot measurements at two different water temperatures show that the ranking if the two compounds reverses. (From Grosch, K.A., *Kautschuk, Gummi, Kunststoffe*, 6 m, 432, 1996.)

The basis of comparison is: Similar log $a_T v$ values and track surface structures for laboratory and road test conditions. As discussed under Section 26.3.6 side force measurements at constant slip angle and load over a suitable range of speeds and temperatures form a very good basis for comparison with road data.

The following laboratory surfaces have given high correlations with a wide range of road test data:

- Alumina of several grades of coarseness and sharpness
- Ground glass
- Stainless steel with finely lathed grooves

Because of the limited range of log $a_T v$ values encountered in road tests (see Section 26.3.5) a range of temperatures between 1°C and 50°C and a sliding speed range of 3 decades from 2 km/h downward is sufficient to cover most of the conditions of slip speed and log $a_T v$ values in the contact area. Most of the laboratory evaluations are carried out on wet surfaces, the water is used to provide the required temperatures. In order to obtain the best correlation between road test ratings at one testing condition and laboratory ratings at a given range of log $a_T v$ it is useful to fit a function to the experimental points of the master curve. A quadratic equation usually gives very good results. A more sophisticated approach which is better when minimum and maximum values of the master curve are to be included is to use the normal distribution function. This is, however, rarely necessary.

If only one speed and a range of temperatures are used the range of log $a_T v$ is more limited, but still adequate for a quick survey of compound performance. Figure 26.36 shows a data set for an oil-extended styrene–butadiene rubber (OESBR) black-filled passenger tire tread compound, obtained at different temperatures and speeds (black open squares and solid black line) and transformed into a master curve using the WLF equation, fitted with a quadratic function using the least-square method. Also shown are the points obtained at one speed and different temperatures also transformed into a master curve fitted with a quadratic function (full squares and red dotted line). The agreement is good over the range covered by both, but outside this range the two deviate considerably so that an extrapolation is not to be recommended. Table 26.2 shows the coefficients of a quadratic function for the extended experiment (black line and points) together

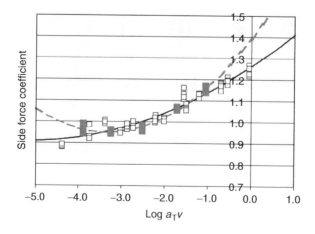

FIGURE 26.36 The side force coefficient of an OESBR black-filled tire tread compound on wet blunt Alumina 180 as function of log $a_T v$ obtained at three speeds and five temperatures (black open squares) with a quadratic equation fitted to the data (black solid line). The red marked points were obtained at one speed for five temperatures with the dotted red line the best fitting quadratic equation, indicating the risk of extrapolation with a limited set of data.

TABLE 26.2

Coefficients of a Quadratic Equation Representing Part of the Side Force—Log $a_T v$ Master Curve

| Compound | Parameter of Quadratic Equation | | | Correlation Coefficient |
	a	b_1	b_2	r
A	1.262	0.136	0.013	0.962
B	1.074	0.113	0.009	0.961
C	1.604	0.188	0.006	0.976
D	1.513	0.226	0.014	0.992
E	1.365	0.288	0.035	0.942
F	1.520	0.226	0.016	0.969

with the correlation coefficients between calculated and measured values applied to six tire tread compounds for which also road test results were available. The ratings of these compounds in relation to compound 1 are shown in Figure 26.37 as function of log $a_T v$. The reference compound 1 was an OESBR + black summer tread compound. Compound 2 was an NR winter tread and is inferior to the reference compound over the whole range of log $a_T v$, compound 3 is a modern solution SBR–BR blend with a silica filler and silane coupling. It is seen that it is better than the standard compound 1 at higher log $a_T v$ and worse at low ones. This corresponds to practical experience.

With the modern ABS braking system contact temperatures are kept low, particularly on wet surfaces so that the log $a_T v$ is relatively high and the silica-filled compound is superior to the standard black-filled one. With locked wheel braking the contact temperatures are high and log $a_T v$ values are low, and the ranking of the braking performance of these two compounds reverses.

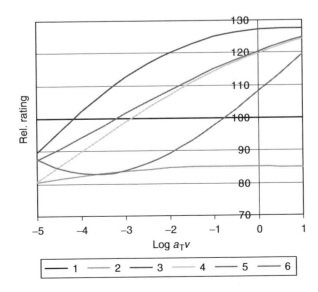

FIGURE 26.37 Compound ratings of the six compounds of Table 26.2 as function of log $a_T v$, using compound 1 as reference.

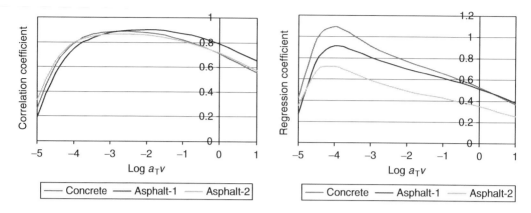

FIGURE 26.38 Correlation and regression coefficients between road test ratings obtained on three different wet road test tracks and laboratory ratings obtained on a wet Alumina 60 laboratory surface as function of log $a_T v$.

26.4.2 CORRELATION WITH ROAD TEST RATINGS OF COMPOUNDS

The six compounds mentioned above were also tested on three different wet road test surfaces. In order to obtain a quantitative assessment of the correlation of road test ratings with laboratory ratings, the road test ratings are compared with the laboratory ratings as function of the log $a_T v$ values by calculating the linear regression line and correlation coefficient between the road test ratings and the laboratory ratings over a range of log $a_T v$ values. The resulting functions for the regression and correlation coefficients are shown in Figure 26.38. The road test ratings were obtained for braking distance measurements between 90 and 10 km/h, The laboratory measurements were carried out with an LAT 100 traction and abrasion tester on Alumina 60 which is a coarse abrasive surface. Further laboratory tests on blunt Alumina 180 and on ground glass gave similar correlation results. A correlation coefficient of 0.9 is achieved over a range of log $a_T v$ between -3 and -2. In this region, the regression coefficient was about 0.8. A regression coefficient of 1 means a 1 to 1 correspondence between laboratory and road test ratings. Larger than 1 indicates a bigger spread for the road test ratings and smaller than 1 for the laboratory ratings.

26.4.3 FORCE MEASUREMENTS AS FUNCTION OF TEMPERATURE AND SPEED CONSIDERING THE TWO AS INDEPENDENT VARIABLES

Up to this point, all evaluations were based on the transformation between temperature and speed using the WLF equation. There is good reason for doing this. Even if deviation from the WLF equation occurs, there is overwhelming evidence that temperature and speed are not independent of each other. A transformation function between speed and temperature is obviously dominating the behavior. The WLF transform is therefore a good starting point. There exists certainly a very broad body of data that conforms closely to this transformation within the accuracy of careful friction or side force measurements. Occasionally, however, deviations are observed. In this case usually the speed dependence at a given temperature is less pronounced than would be expected from the transformation. One explanation could be that lubrication plays a part, despite of the low speeds used in the experiments. Since, however, the phenomenon appears also to be associated with specific compounds, a reason could be that modern compounds are often blends of different polymers, the glass transition temperatures of which are not well defined. Also the influence of modern filler systems, particular silica with silane coupling, may contribute to such deviations. Carbon black fillers generally do not influence the transformation but affect only the magnitude of

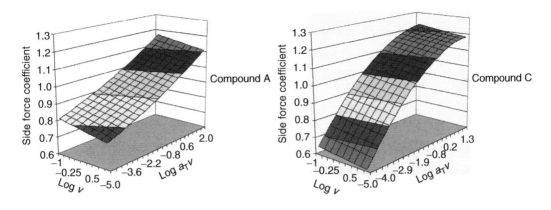

FIGURE 26.39 Side force coefficient as function of log $a_T v$ and log v for compounds 1 and 3 of Table 26.2.

the friction coefficient. This may be different for silica because of the different coupling to the polymer. At present, there is insufficient data available to verify these hypotheses.

However, a correlation with road test ratings can often be improved if an independent term of log v is added to the transformation variable log $a_T v$. The friction or side force coefficient can then be written as

$$fc = a + b_1 \cdot [\log (a_T v)] + b_2 \cdot [\log (a_T v)]^2 + b_3 \cdot \log (v) \qquad (26.24)$$

The results have to be presented as a table or as a three-dimensional graph. As an example the side force coefficients are shown in Figure 26.39 for compounds 1 and 3 of Table 26.2. It is seen that the major variation is due to the log $a_T v$ value and only a small part is due to the additional log v term. The correlation with road test ratings of the six compounds of Table 26.2 is shown in Figure 26.40, comparing the evaluation using the laboratory data sets obtained on wet blunt Alumina 180 and the road test ratings on a concrete track with stopping distance measurements between 90 and 10 km/h under ABS braking. It appears that the point of highest correlation shifts slightly with the speed and is also improved.

If the WLF transform is not obeyed rigorously, a more empirical analysis is possible, treating temperature and speed as separate variables. The temperature dependence is described by a square

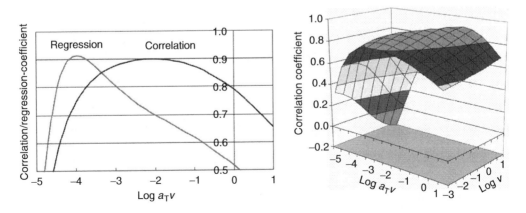

FIGURE 26.40 Correlation coefficient between road test ratings on a wet concrete track and laboratory measurements on a wet blunt Alumina 180 disk. Left as function of log $a_T v$ and right as function of log $a_T v$ and log v.

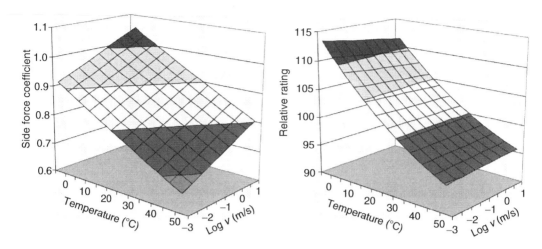

FIGURE 26.41 Side force coefficient of compound C of Table 26.4 and the rating in relation to compound A as function of temperature and log v.

law, the speed dependence by a linear relation between the friction or side force coefficient and log (speed). The addition of an interaction term, i.e., the interdependence of friction or side force on temperature and speed would amount to some kind of transformation, i.e., a treatment as already described above.

Figure 26.41 shows the side force coefficient (upper part) of compound 3 of Table 26.2 and its rating (lower) as function of temperature and log speed. The side force coefficient depends on temperature and log speed. The rating of compound 3 relative to 1 depends on temperature so strongly that a reversal in ranking occurs, the dependence on speed, however, is small.

Figure 26.42 compares the correlation coefficient between a road test on wet concrete and the laboratory side force measurements of the six compounds of Table 26.2 on blunt, wet Alumina 180 (a) with a log a_Tv–log v evaluation and (b) with a temperature–log v evaluation. It appears that the

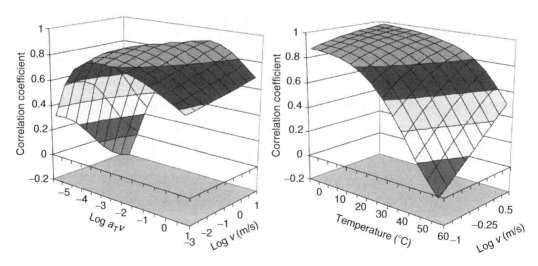

FIGURE 26.42 Comparison of the correlation coefficients between laboratory side force measurements with the six compounds of Table 26.2 on wet, blunt Alumina 180 and a concrete road test track as function of log a_Tv and log v (left) with function of temperature and log v (right).

correlation coefficient for the temperature–log v evaluation depends strongly on both temperature and speed, i.e., on the laboratory testing condition, whilst in the log $a_T v$–log v evaluation it depends on log $a_T v$ but not on log v. Both the methods of log $a_T v$–log v and temperature–log v give equivalent results and similar correlations with road test data and both are slightly more precise than the evaluation based on log $a_T v$ alone. Clearly, the direct temperature–log v evaluation is more easy to handle and probably also more easy to understand in practical routine compound evaluation and comparison. The log $a_T v$ method, however, is physically the more satisfactory method and if deviations from the master curves are observed repeatedly, then it should be worthwhile to spend some research effort on it to identify its cause. In particular, if deviations are compound specific, it may be very useful and of practical benefit to understand their origin.

26.5 LABORATORY ABRASION

26.5.1 SLIDING ABRASION

26.5.1.1 Pressure Dependence of the Sliding Abrasion

An abrasion experiment is carried out by sliding a rubber sample over a given distance and determining the volume loss by some suitable method, usually by determining the difference in weight before and after. It is always assumed that the abraded volume is proportional to the distance covered between measurements, which is reasonable if the sharpness of the track remains constant. Hence, the abraded volume is always referred to unit distance covered. This basic quantity depends on the pressure between the sliding surfaces, the temperature, speed, and the topography of the track, such as sharpness and coarseness of the asperities.

In contrast to the frictional force, the resulting abrasion is generally a nonlinear function of the pressure p

$$\text{abr} = \text{abr}_{\text{ref}} \cdot \left\{ \frac{p}{p_0} \right\}^n \tag{26.25}$$

with a distinct positive power index n. For a constant apparent area of contact, a similar relation results for the load dependence. The reference abrasion loss abr_{ref} at the reference pressure p_0 and the power index n depend both on the type of track and the rubber compound. If the abrasion is referred to the frictional energy dissipation, usually referred to as the abradability of the rubber compound, the power index would be modified, if there is a load dependence of the friction coefficient and would remain the same if there is not. If the track is smooth the abrasive loss may be so small that it deludes normal measurement, although the frictional force may be very high, whilst on rough sharp surfaces the abrasive loss may be pronounced at a moderate frictional force. It follows that stress concentrations in the contact area enhance the abrasion process at similar external forces. Figure 26.43 shows the abrasion loss of a BR tread compound on four surfaces of different sharpness, as function of the pressure both plotted on logarithmic scales. The straight-line graphs differed both for their reference loss as well as for their power index n [37]. The lowest index n was obtained on the flat side of a silicone carbide grinding wheel. The others were a tarmac road surface and two concrete floor surfaces. The power index was always significantly greater than 1. It is a general observation that the larger the power index, the blunter the surface on which the abrasion takes place. On very sharp tracks, on the other hand, the power index may become 1 as shown in Figure 26.44 [35].

26.5.1.2 Dependence of the Abrasion on the Energy Consumption in the Contact Area

A more basic approach is to consider the dependence of the abrasion loss not on pressure but the energy dissipation W caused by the frictional force F. For sliding experiments this is

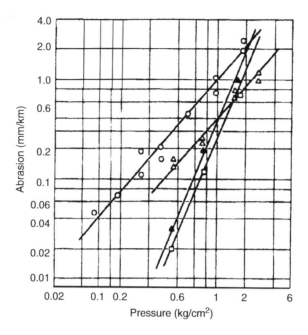

FIGURE 26.43 Abrasion loss as function of pressure for a butadiene rubber (BR) tread compound on four different abrasive surfaces. ○ tarmac, □ Akron abrasive disk, ▲ concrete I, △ concrete II. (From Grosch, K.A. and Schallamach, A., *Kautschuk, Gummi und Kunststoffe*, 22, 288, 1969.)

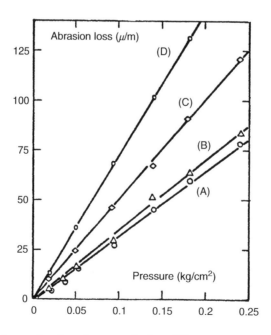

FIGURE 26.44 Abrasion loss as function of pressure for four different compounds. A: SBR + 50 HAF black, B: NR + 50 HAF black, C: NR + 50 thermal black, D: NR + 50 activated $CaCO_3$. (From Schallamach, A. and Grosch, K.A., *The Mechanics of Pneumatic Tires*, S.K. Clark (ed.), The US Department of Transportation, National Highway Safety Administration, Washington DV, p. 407.)

$$W = F \cdot s \tag{26.26}$$

where s is the distance covered. Hence the energy dissipation W per unit distance becomes equal to the frictional force. On very sharp tracks the abrasion loss is nearly proportional to the frictional force, more generally, however, it is a power function of the energy dissipation, whereby the power index depends both on the sharpness of the track as well as on the rubber compound in question. When referred to the energy dissipation, the abrasion loss is generally called the abradability of the rubber compound on the track on which it was obtained.

This behavior is similar to the cut growth and fatigue behavior of rubber compounds. The rate of the growth of a cut is a function of the tearing energy [38,39] which itself is proportional to the stored elastic energy density in the test piece. The exact value depends on the shape of the test piece.

26.5.1.3 Rate of Cut Growth: Tearing Energy Relation

Since this appears to be the basis of the abrasion process it is profitable to be familiar with its shape. Figure 26.45 shows this relation for an unfilled NR compound [40,41]. It can be divided into four regions.

Below a limiting tearing energy T_0 there occurs no mechanical tearing. Cut growth is due to ozone, which spontaneously destroys cross-links. Its rate of growth is independent of the tearing energy and depends only on the ozone concentration.

Above T_0 there is a small region for which the cut growth rate dc/dn is proportional to the tearing energy T

$$\frac{dc}{dn} = A_{cr} \cdot \{T - T_0\} \tag{26.27}$$

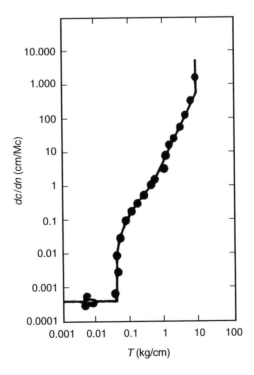

FIGURE 26.45 Rate of cut growth dc/dn as function of the tearing energy T for a natural rubber (NR) gum compound. (From Lake, G.J. and Lindley, P.B., *Rubber J.*, 146, 10, 1964.)

where A_{cr} is a material property that is influenced by temperature and oxygen. In a third region the cut growth rate can be described by a power law

$$\frac{dc}{dn} = B_{cr} T^\beta \qquad (26.28)$$

In this region the mechanical rate of growth B_{CT} is also influenced by temperature and the presence of oxygen. The power index β depends primarily on the polymer. When the tearing energy approaches a critical value T_c the rate of cut growth becomes suddenly very large, spontaneous rupture occurs. The critical tearing energy T_c is then proportional to the energy density at break.

Figure 26.46 shows the cut growth rate for six polymers as gum rubbers and filled with two levels of reinforcing carbon black [40]. They all can be represented by a power function over a considerable range of tearing energies.

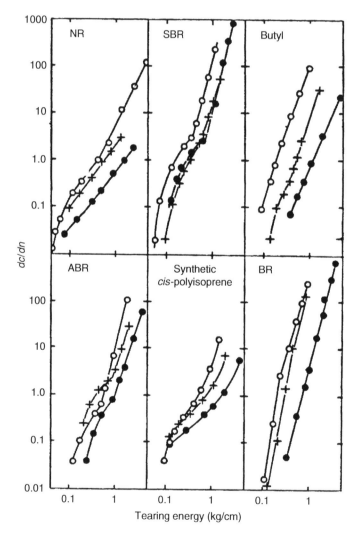

FIGURE 26.46 Rate of cut growth for six different rubbers ($\bigcirc$) gum rubber, ($+$) loaded with 20 pphr reinforcing black, and ($\bullet$) with 50 pphr reinforcing black. (From Lake, G.J. and Lindley, P.B., *Rubber J.*, 146, 10, 1964.)

Notice that the carbon black filler reduces the rate of cut growth, but has only a small effect on the power index. Generally, it appears that abrasion occurs mainly in the third region, except when the abrasive track is very sharp. In this case the number of cycles to detach a small piece of rubber becomes small and the abrasion is proportional to the reciprocal of the energy density at break of the rubber compound.

26.5.1.4 Stress and Energy Distribution in Rubber Samples Sliding on Rough Tracks

It was pointed out in Section 26.2.1 that the friction coefficient is considerably larger on a rough track than on a smooth one when the $\log a_T v$ values of the master curve are small, i.e., when the temperature is high, and the speed is low, i.e., when the viscoelastic losses are low. Moreover, the adhesion friction, which enables tangential stresses to be built up, is low.

The higher friction on the rough track can therefore only come about when elastic stored energy in the contact area is also lost in the friction process. The process can be likened to the stretching of an elastic rubber strip, fixed at one end. On releasing it, the strip contracts and all the stored elastic energy is turned into heat which raises first the temperature in the strip and is then lost to the environment. Figure 26.47 shows photos of the lines of equal stress in a transparent rubber sample under polarized light due to a wedge sliding under load over the surface. For a normal load the lines are circles with their centers in line with the normal force. When the wedge is pulled horizontally the resultant force makes an angle with the plane of the rubber sample which is given by $1/\tan \alpha = \mu$, the friction coefficient. The circles of equal stress turn through this angle and the part of the circle cut off by the surface plane of the rubber is rotated through 180°, leading to a compressive stress in front of the moving force and an extensive stress behind it. This agrees closely with calculated stress distribution due to a line force acting at an angle α to the surface of a semi-infinite solid (Figure 26.48) [44].

Figure 26.49 (upper) shows the elastic stored energy of two line forces acting at an angle α to the surface of the semi-infinite body simultaneously and Figure 26.49 (lower) the horizontal stress component. As expected both energy and stress show strong peaks at the points of contact.

The stress is in compression in front of the contact, i.e., the friction component pushes. At the end of the first contact the stress is strongly in tension, decreasing rapidly away from the contact and passes again into compression at midpoint between the two contacts. If more contact points are

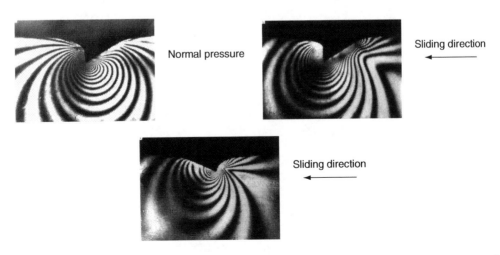

Normal pressure

Sliding direction

Sliding direction

FIGURE 26.47 Two-dimensional stress pattern in a transparent rubber block under a line force under polarized light. (From Schallamach, A., *Wear*, 13, 13, 1969.)

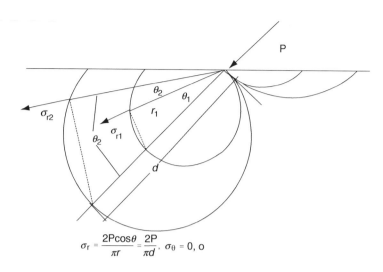

$$\sigma_r = \frac{2P\cos\theta}{\pi r} = \frac{2P}{\pi d}, \quad \sigma_\theta = 0, o$$

FIGURE 26.48 Calculated lines of equal stress for a line force made up of normal load and frictional force acting on the surface of a semi-infinite body.

considered this pattern is repeated. The stress is also limited to a small range of depths. If the distance between the two contact points is 1 mm then at a depth of 0.04 mm the horizontal stress component has dropped to about 20% to that of the value at 0.01 mm depth. The stress at the surface itself cannot be calculated because the point has an infinitely small contact area and hence the stress becomes infinitely large. In reality, the force is due to an asperity with a small but measurable tip radius, which produces a well, defined contact area particularly between a hard asperity and a soft rubber compound, keeping the contact force finite. Nevertheless, the smaller the contact radius, the higher are the stresses at the contact points. The contact radius defines the sharpness of the wedge and in a three-dimensional case the sharpness of an asperity. The stored energy between the line forces does not become zero and a tensile stress acts at the last of the line forces and a compressive stress at the front.

If these large energy concentrations meet a flaw in the rubber, the resulting tearing energy will increase the flaw at each pass until a rubber particle is detached. This is essentially the mechanical basis of the abrasion process.

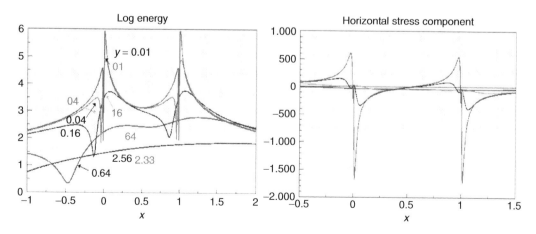

FIGURE 26.49 Calculated stored elastic energy and the horizontal stress component due to two line forces at an angle α to the plane of the rubber surface and a fixed distance x apart for different depths from the surface of a semi-infinite body.

26.5.1.5   Temperature and Speed Dependence of Sliding Abrasion

For simple unfilled, noncrystallizing rubber compounds the temperature and speed dependence of sliding abrasion on sharp tracks is dominated by the viscoelastic nature of the rubber, provided that the sliding speed is so low that temperature rises due to friction are small. Combining the two variables using the WLF transformation results in single "master" curves as shown in Figure 26.50 [43]. The abrasion loss decreases with increasing log a_Tv reaching a minimum value. A further increase of the log a_Tv leads to a sharp increase of the abrasion. This is accompanied by a drastic change of appearance of the abraded surface. This will be discussed in Section 26.5.1.6. It appears that the lowest abrasion loss occurs when the rubber has its maximum extensibility, which is reasonable because this corresponds to the highest energy density at break. The WLF transform works only well for abrasion of noncrystallizing gum rubbers.

Experiments carried out at a constant speed over a range of temperatures showed, however, that for tire tread compounds the temperature dependence of abrasion, although smaller reaches in this case, too, a minimum at a particular temperature as shown in Figure 26.51 and rises sharply with a further decrease in temperature.

Experiments measuring the energy density at a high rate of extension over a range of temperatures showed that the abrasion–temperature function for tread compounds had a similar shape as the inverse of the energy density at break temperature curve. This is also shown in Figure 26.51 which compares the abrasion–temperature function with that of the inverse of the energy at break temperature function. In order to match the two curves the rate of extension had to be 10^2 for a sliding speed of 1 cm/s.

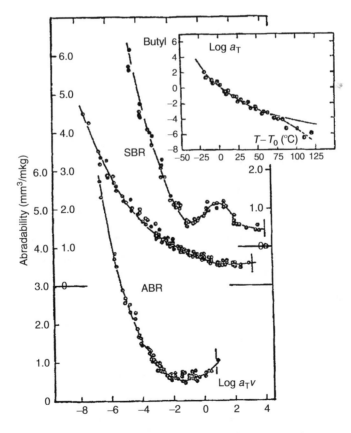

FIGURE 26.50 Sliding abradability as function of the variable log a_Tv for four noncrystallizing rubber compounds on dusted silicon carbide 180.

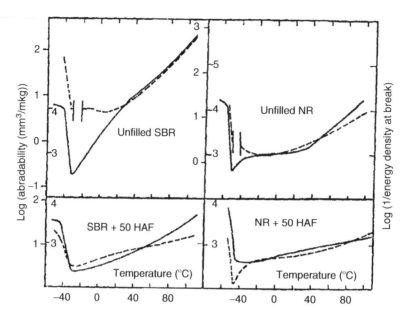

FIGURE 26.51 Brasion loss per unit energy (abradability) (—) as function of temperature for four different compounds on a silicon carbide track at a speed of 1 cm/s together with energy density measurements (—) at an extension rate of 10^2/s. (From Grosch, K.A. and Schallamach, A., *Trans IRI*, 40, T80, 1961; *Rubber Chem. Technol.*, 39, 267, 1966.

The difference between the two ordinate origins represents the log of the coefficient in the equation between abrasion resistance, defined by the energy required to remove unit volume of rubber (J/mm^3) and the energy density at break in the same units.

$$k = \frac{A_{\text{resistance}}}{U_{\text{break}}} \tag{26.29}$$

where

  $A_{\text{resistance}}$ is the inverse of the abradability defined as the abrasion loss per unit energy
  U_{break} is the energy density at break

This is about 2000 indicating that the energy required to remove unit volume of rubber even by the very sharp abrasive track is an inefficient process. A surprising result is that the tread compounds have a much higher abrasion loss than the gum rubbers, as shown in Figure 26.52. This shows the abrasion as function of temperature for three rubbers (A) SBR, (B) ABR, and (C) NR, the solid lines are for the gum rubbers, the dotted one for the same polymer, filled with 50 HAF black. The reason will become apparent when examining the appearance of the abraded surfaces.

26.5.1.6 Appearance of Abraded Rubber Surfaces

At normal temperatures the surface of a rubber sample, which has been subjected to sliding abrasion, has a ridge pattern, the direction of the ridges being normal to the sliding direction of the abrasion (see Figure 26.53a). This is generally called an abrasion or Schallamach pattern [45]. A cross-sectional view reveals that the ridges are undercut, the lips pointing against the abrasion direction (Figure 26.53b). During sliding an asperity of the track pulls the lip over and extends it until the adhesion force can no longer hold it and it snaps back. However, as the tip extends the stress at the bottom of the ridge rises and a crack can grow to a limited extent. After a number of passes of an asperity the lip has become long enough for a particle to be detached. During the sliding the pattern moves slowly in the direction of abrasion. This abrasion mechanism prevails for soft

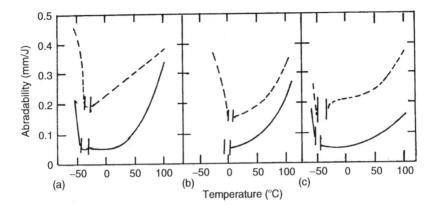

FIGURE 26.52 Sliding abrasion of three different tread compounds as function of temperature at a sliding speed of 0.01 m/s (a) styrene–butadiene rubber (SBR), (b) ANR, (c) NR, ----- tread compound, —gum compound.

rubbers with a high extensibility and the higher the extensibility, the coarser the abrasion pattern, as seen in Figure 26.54, which was obtained when carrying out the abrasion–temperature experiments described above. Although the samples were rotated through 90° at regular intervals in order to suppress their formation abrasion pattern did form. Because of the rotation of the sample they appear as nipples rather than ridges at low log $a_T v$ values (high temperatures, low sliding speeds) (upper photograph of Figure 26.54). As the minimum of the abrasion is approached, very pronounced ridges formed despite the rotation (middle photo of Figure 26.54) and the abrasion increased forming a small hump in the abrasion log $a_T v$ curves (Figure 26.50 above). At still higher log $a_T v$ values scoring marks appear (lower photo, Figure 26.54). Rubber acts now like a hard solid, in this case like a plastic material. Because of the high rate of extension both in the abrasion process and in the energy to break experiments, this temperature occurs well before the glass transition temperature is reached. The abrasion increases drastically as seen in Figures 26.51 and 26.52 above in accordance with the decreasing energy density at break.

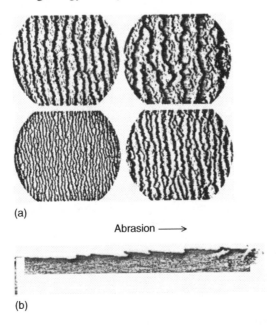

(a)

Abrasion ⟶

(b)

FIGURE 26.53 Abrasion pattern: (a) appearance for different rubbers under different testing conditions and (b) cross section through abraded samples.

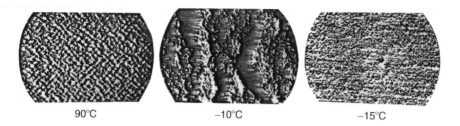

FIGURE 26.54 Abrasion pattern of the acrylate–butadiene rubber (ABR) gum rubber during the sliding abrasion experiment at different temperatures.

Tire tread compounds are filled with highly reinforcing black. This means, above all, that the compound becomes harder. Noncrystallizing rubbers also become much stronger. Their extensibility, however, is lower than that of the gum compound and accordingly the abrasion pattern is much more closely spaced and less pronounced. In many cases scoring marks are present at the same time as abrasion pattern at all temperatures as is shown in Figure 26.55. The abrasion loss is then higher for the black-filled compound than for the corresponding gum rubber, as shown in Figure 26.52. Veith [46] has interpreted the presence of coring marks in black-filled rubbers that they incorporate regions of plastic behavior. This may well be due to the bound rubber, which the strong activity of reinforcing blacks creates in its vicinity.

26.5.1.7   Abrasion by a Razor Blade

Champ et al. [47] devised a simple abrasion experiment by measuring the abrasion loss caused by a razor blade scraping under load across a rubber surface. For this an Akron abrader sample wheel was used. They used soft unfilled noncrystallizing rubber compounds based on different polymers. Under these conditions strong abrasion pattern develop.

If the increase of the crack length at the base of an abrasion pattern lip is Δc the increase in abrasion depth per pass, i.e., per cycle is $\Delta c \cdot \sin \beta$, where β is the angle which the cut growth direction makes with the surface of the sample. The rate of crack growth can be measured independently as function of the tearing energy. Assuming that the abrasion functions like the crack growth of a trouser test piece the tearing energy is given by

$$T = F/b \cdot (1' + \cos \beta) \tag{26.30}$$

where

  b is the width of the sample wheel
  F is the tangential force on the blade

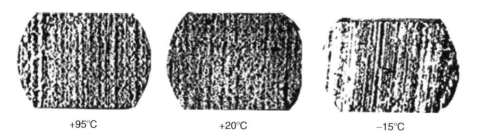

FIGURE 26.55 Abrasion surface appearance of a natural rubber (NR) black-filled tire tread compound for sliding abrasion at different temperatures.

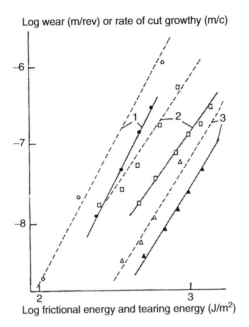

FIGURE 26.56 Log Abrasion loss by a blade (solid lines) and log cut growth rate (dashed lines) of noncrystallizing rubber compounds as function of log frictional and log tearing energy, respectively: isomerized natural rubber (NR), 2 styrene–butadiene rubber (SBR), and 3 acrylate–butadiene rubber (ABR). (From Champ, D.H., Southern, E., and Thomas, A.G., *Advances in Polymer Friction and Wear*, Lieng Huang Lee (ed.), Plenum, New York/London, 1974, p. 134.)

By using different loads, different tearing energies are produced. Figure 26.56 shows the results of their experiments for three gum rubbers. The solid lines represent the abrasion per cycle and the dashed lines the rate of cut growth.

The agreement between the two measurements is very satisfactory for the power index, and the coefficient between the two cut growth rates was about 4 depending somewhat on the compound. Whilst it works well for noncrystallizing rubbers it does not work for unfilled NR. Crystallization influences the rate of crack growth significantly whilst in the abrasion experiment it appears to be less effective so that the abrasion was much higher than expected from cut growth measurements.

Pulford and Gent [48] have extended these experiments. They obtained straight-line graphs if plotting log abrasion against log frictional energy dissipation, as shown in Figure 26.57. The slope of the graphs corresponded to those obtained from cut growth experiments for gum rubbers. For the black-filled BR, however, it was much smaller than that of the gum rubber whilst for cut growth it was similar (compare with Figure 26.42).

26.5.1.8   Effect of Smearing and Abrasion in an Inert Atmosphere

One of the most obvious laboratory abrasion effects indicating that abrasion is not only a mechanical rupture phenomenon but has also a strong chemical aspect is the so-called smearing that is observed in practically all abrasion experiments. Abrasion debris on clean surfaces tend to stick together forming rolls of abraded rubber. They are often quite large, but consist actually of a large number of very small rubber particles, which are highly degraded thermally (depolymerized) and oxidized. These rolls are sticky and protect the rubber surface from an unhindered abrasion. The abrasion loss drops and may stop altogether. In road wear this phenomenon is only observed under extreme conditions like car racing. Normally, there is sufficient dust on the road, which absorbs

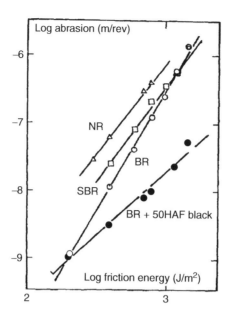

FIGURE 26.57 Log abrasion by a blade as function of log friction energy for three gum and a black-filled rubber. (From Gent, A.N. and Pulford, C.T.R., *J. Appl. Pol. Sci.*, 28, 943, 1983.)

the sticky parts; the rolls disintegrate or do not form at all. Hence, some suitable powder is applied to the track of laboratory abrasion experiments in order to absorb the smeary abrasion debris which otherwise would falsify the abrasion results. MgO has proved to be one of the most effective agents for absorbing the abrasion debris and thus preventing smearing. Schallamach [49] has carried out experiments on blunt abrasive tracks using MgO and a mixture of Fuller's earth and alumina powder.

The results are shown in Figure 26.58 for two blunt surfaces: Knurled aluminum and knurled steel. Aluminum produces a lower abrasion loss than steel. On both surfaces, however, the abrasion was much higher when MgO was used as an absorbing powder than with the mixture of Fuller's earth and alumina. The two compounds examined were an NR compound without antioxidant and one with two parts of Nonox ZA. There is a very clear difference between the two compounds indicating that oxygen plays an important role.

Abrasion experiments in nitrogen, first carried out in Russia [50], showed that abrasion was in practically all cases higher in air than in nitrogen with the exception of butyl, and the effect was larger on blunt abrasives than on sharp ones. Figure 26.59 is an abrasion time record which Schallamach obtained on an Akron grinding wheel when switching from air to nitrogen and back again. A mixture of alumina powder and Fuller's earth was fed into the nip between sample and track to counteract smearing. The sample was first run in air until a constant rate of abrasion had been reached. With the introduction of nitrogen the abrasion loss dropped rapidly, significantly more so for the unprotected compound. After air was reintroduced the abrasion rose again to approximately the former value. Each time the atmosphere was changed it took a few readings before a steady state was reached, indicating that the oxygen required a certain time to influence the surface layer to saturation.

In Figure 26.60, the difference between abrasion in air and nitrogen is plotted as function of the abrasion in air for different compounds abraded on several surfaces of different sharpness using MgO and a mixture of Fuller's earth and alumina powder, respectively, as agents to counteract

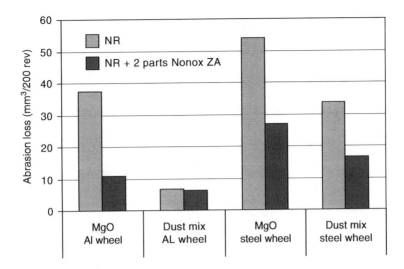

FIGURE 26.58 Abrasion of two natural rubber (NR) tread compounds, one unprotected, the other with two parts Nonox ZA on a knurled Aluminum and a knurled steel surface, respectively, in the presence of (a) magnesium oxide (MgO) powder and (b) a dust mix of Fuller's earth and alumina powder. (Deduced from Schallamach, A., *Appl. Pol. Sci.*, 12, 281, 1968.)

smearing. The powder introduced has two opposing effects on the abrasion loss in both media to differing degrees. It prevents smearing and helps in this way to increase the abrasion loss. On the other hand it can lubricate the track and protect the rubber from abrasion, reducing the abrasion loss. Thus, if smearing is small in air and the lubrication effect is significant, abrasion in nitrogen may become larger than in air resulting in a negative difference between the abrasion loss in air and nitrogen. MgO produced very large differences between abrasion in air and nitrogen when the losses in air were high, i.e., when the oxygen and smearing effect were large (unprotected NR) suggesting

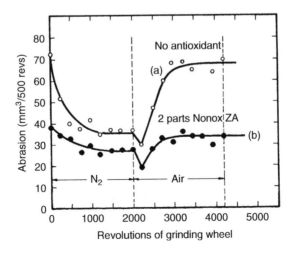

FIGURE 26.59 Time record of the abrasion loss on a standard Akron grinding wheel in nitrogen and air of a natural rubber (NR) tread compound (a) unprotected and (b) protected with an antioxidant. (From Schallamach, A., *Appl. Pol. Sci.*, 12, 281, 1968.)

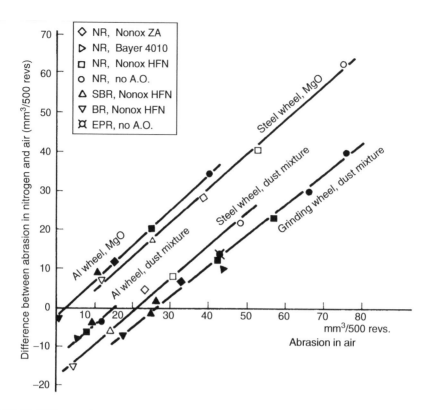

FIGURE 26.60 Difference between abrasion in air and nitrogen as function of abrasion in air for different compounds in the presence of either magnesium oxide (MgO) or the mixture of Fuller's earth and alumina. (From Schallamach, A., *Appl. Pol. Sci.*, 12, 281, 1968.)

a high effectiveness in eliminating smearing. If the compound was well protected the abrasion in air was similar to that in nitrogen, suggesting that the lubrication was also small. Thus, MgO can be seen to be a very effective agent to eliminate smearing without interfering with the abrasion process itself. On the other hand the mixture produced less abrasion in air and in some cases for well-protected rubbers the abrasion was higher in nitrogen than in air, producing a negative difference between air and nitrogen. If it assumed that the lubrication effect is about the same for both air and nitrogen, then clearly the smear prevention was poor for the mixture.

Gent and Pulford [48] noticed that black-filled compounds tend to smear more than unfilled ones, especially NR, SBR, and ethylene–propylene rubber (EPR), whilst BR and tensilized poly-propylene rubber (TPPR) showed no smearing at all producing powdery abrasion debris. This was also the case for NR in nitrogen or vacuum, showing clearly that oxidation is a major cause of smearing in air, but thermal degradation may also play a role since it is known that BR, for instance, has a very high decomposition temperature ($>500°C$) so that smearing would be expected to be less than that for NR which has a much lower thermal decomposition temperature (about 260°C). Gent and Pulford suggest three chemical contributions to the abrasion and smearing process: Thermal degradation due to local frictional heating; oxidation which is also aided by the raised temperatures; and formation of free radicals in the rupture process which then also oxidize. In all experiments, BR has the smallest smearing effect, i.e., it has the highest resistance to thermal and oxidative degradation. It is suggested that this is the reason for its well-known high-abrasion resistance rather than any special physical properties.

26.5.2 ABRASION UNDER LIMITED SLIP

26.5.2.1 Abrasion as Function of Slip and Load

Abrasion under limited slip used to be, and to some extent still is, measured either with the standard Akron abrader or the Lambourn abrader. In case of the Akron abrader the sample runs under a slip angle at a constant load against the abrasive surface of an alumina grind stone. Speed and load are fixed; the side force is not measured.

The author [51,52] conceived an extension of the Akron abrader principle by

- Measuring the side force
- Varying the slip angle over a wide range from −40° to +40°
- Having a speed range from 2 to 100 km/h, although the maximum speed is usually not necessary for abrasion, except perhaps for research purposes
- A continuous range of loads from 10 to 150 N
- Abrasive surfaces of different grain size and sharpness
- Measurements at abrasive surface temperatures between the dew point and 70°C

The instrument is now marketed as a modern fully automatic (except for mounting the sample) multiple testing tool, including also friction measurements on wet and icy surfaces Most of the following data were obtained using this instrument. The advantage of such an arrangement is the very wide range of severities (load and slip angle) and speeds which can be covered, as well as the possibility of using a considerable range of different abrasive surface structures.

Figure 26.61 shows the abrasion of an OESBR tread compound as function of load for different slip angles on a sharp Alumina 60 surface. Because of the wide range of abrasion rates for different slip angles the abrasion data were plotted on a log scale. It is seen that at the small slip angle the dependence on load is small and becomes more pronounced as the slip angle is increased. This is expected from the brush model. At small slip angle the side force is independent of the load and hence it is expected that the abrasion behave in a similar way.

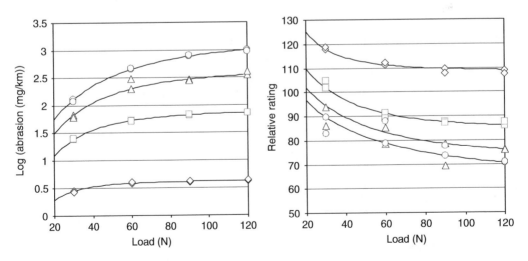

FIGURE 26.61 Log (abrasion) of an OESBR and a natural rubber (NR) tire tread compound as function of load at different slip angles at a speed of 19.2 km/h. left: Abrasion loss of the OESBR compound as function of load. Right the relative wear resistance rating of natural rubber (NR) to the OESBR as function of load for different slip angles.

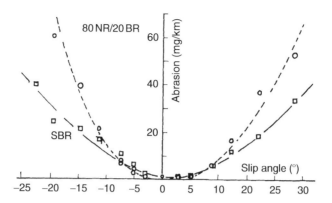

FIGURE 26.62 Abrasion of OESBR and an 80 natural rubber (NR)/20 butadiene rubber (BR) blend tire compound as function of slip angle at a load of 76 N and a speed of 19.2 km/h.

Also shown is the relative rating between the OESBR compound and an NR + black tire tread compound. At the smallest slip angle the rating of the NR is better than the OEBR but decreases with the load. As the slip angle is increased the rating reverses.

Figure 26.62 shows the abrasion as function of slip angle for an OESBR tread compound and an 80 NR/20 BR at a constant load. Clearly, the abrasion depends very strongly on the slip condition. Careful examination of the graph shows that the abrasion loss for the NR–BR blend is higher at large slip angle, but becomes better at low slip angles. This becomes more apparent if the data are plotted on a log scales as shown in Figure 26.63. Straight-line graphs are obtained with slightly different slopes for the NR-based and the SBR compound. Plotting the relative rating of the NR–BR compound in relation to the OESBR as function of log slip angle shows that the small difference between the slopes is in fact a highly significant effect when considering the relative rating (Figure 26.63 right), i.e., the practical comparison of the two compounds.

Both load and slip angle can be combined if the abrasion is considered to be a function of energy dissipation. If the side force is measured and the slip angle is known the data can be evaluated directly as function of energy. This is the case with the LAT100 and is shown in Figure 26.64 for an NR and SBR comparison on a sharp Alumina 60 and a blunted Alumina 180, respectively [54]. Again straight-line graphs are obtained if the two variables are both plotted on logarithmic scales and the two curves cross over on both surfaces. The slopes of the lines and the intercept (the average absolute abrasion loss) differ considerably for the two surfaces. The crossover points occur at low energy levels in both cases. The slopes of the lines are larger than 1, indicating that the process is more related to a fatigue failure than to a direct tearing process. This is

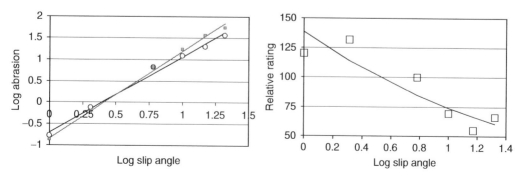

FIGURE 26.63 The abrasion data of figure 61 plotted on logarithmic scales and the relative rating of the natural rubber (NR)/BR blend in relation to the OESBR as function of log slip angle.

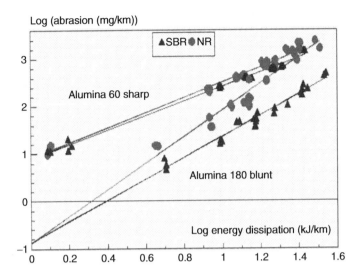

FIGURE 26.64 Log (abrasion) for two tread compounds natural rubber (NR) + black and styrene–butadiene rubber (SBR) + black on two surfaces of different sharpness Alumina 60 and Alumina 180 blunt as function of log (energy dissipation). (From Grosch, K.A. and Heinz, M., Proc. IRC 2000, Helsinki, 2000, paper 48.)

particularly so for the blunt surface. The absolute loss is distinctly smaller at low energies than on the sharp surface but at large energies the lines cross because the much larger slopes of the lines on the blunt surface lead to a larger abrasion loss at high energies than on the sharp surface. The differences of the slopes between NR and SBR are also larger on the blunt surface than on the sharp one. In fact this difference is much larger than would be expected from tire wear, hinting at the possibility that road surfaces are not blunt but act rather like a sharp surface.

26.5.2.2 Effect of Speed of the Abrasion of Slipping Wheels

If experiments are carried out as function of energy dissipation (slip angle and/or load) at different speeds and the data plotted on log scales, straight-lines graphs are still obtained but they have different slopes [35] (Figure 26.65). This is not easily explainable by a purely mechanical cut growth process. In the general abrasion process several variables are present simultaneously. In particular, as the energy dissipation in the contact area increases the temperature is going to rise since a considerable portion of the energy is being turned into heat. This would also be the case at different speeds. Surface changes may play a role and the dynamic strength may influence the abrasion, possibly through changes of the cut growth characteristic. Figure 26.66 shows the abrasion loss of three tire tread compounds as function of speed both plotted on a logarithmic scale [53]. Although the influence is much smaller than that of the energy and hence the scatter is larger, clear differences between the compounds emerge. Empirically a straight-line graph was applied to the data. Because the effect is small a large number of measurements would be required to demonstrate any significant deviation from this simple approach.

The abrasion rises with increasing speed. The slope of the straight-line differs considerably for the three compounds. The SBR has the smallest dependence on speed and the NR has the largest. This is most easily explained if it is assumed that the temperature in the contact area rises with speed and both speed and temperature together influence the abrasion speed dependence. In this range it would be expected that the abrasion rise with increasing temperature, both on viscoelastic and thermo-oxidative degradation considerations whilst in it would be expected to fall with increasing speed on viscoelastic reasoning if the temperature could be kept constant. Clearly, the thermo-oxidative effect outweighs the purely viscoelastic effect. This agrees with the finding that SBR has a

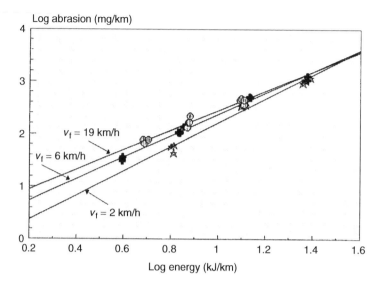

FIGURE 26.65 Log (abrasion) as function of log (energy dissipation) for a commercial tire tread compound at three different speeds. Surface: Alumina 60.

much smaller effect than NR. The SBR–BR blend filled with silica lies between these two. This is not easily explained since the author knows of no fundamental data for thermal and oxidative resistance of this type of material, or its exact viscoelastic properties.

26.5.2.3   Combining the Energy and Speed Dependence of Abrasion

Since both energy and speed give straight-line graphs when both variables are plotted on logarithmic scales these quantities may be added up. In order to allow for the fact that the slope of the log abrasion–log energy lines depend on speed and vice versa, a further term has to be included.

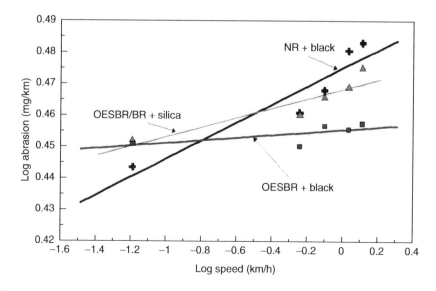

FIGURE 26.66 Log (abrasion) as function of log speed for three different tire tread compounds. Load: 76 N, slip angle: 14.6°. Surface: Alumina 60. (From Grosch, K.A. and Heinz, M., Proc. IRC 2000, Helsinki, 2000, paper 48.)

The simplest way is to introduce the product of the two variables log (speed) and log (energy), giving the following equation:

$$[\log (A)] = a + b_1[\log (W)] + b_2[\log (v)] + b_3[\log (W)] \cdot [\log (v)] \qquad (26.31)$$

where
  A is the abrasion volume loss per kilometer
  W is the energy dissipation per kilometer
  v is the forward speed of the abrasive disk in the contact area relative to the sample wheel

In order to determine the coefficients of this equation at least four different testing conditions are required, two energy levels given by two slip angle–load combinations and two speeds. The slip can also be circumferential instead of being set by a slip angle; important is that the resulting force arising from the slip is measured in order to obtain a measure of the energy dissipation. In practice, more test conditions are useful, say three energy and three speed levels and for each condition repeat measurements are required because abrasion is always subject to variation. The four coefficients are then calculated from the results using the statistical method of least-square deviations from the mean [54,55]. Table 26.3 shows the coefficients obtained in this way for four passenger tire tread compounds. The coefficient a gives the expected abrasion loss at an energy of 1 kJ/km and a speed of 1 km/h. The coefficient b_1 is the power index for the abrasion–energy relation at a constant speed and temperature. It is positive and larger than 1 for all compounds indicating that a cut growth relation is in operation. It is smaller than would be expected from cut growth experiments but larger than simple tearing.

The coefficient b_2 is related to the abrasion speed relation at constant energy and temperature. It is negative showing that at a constant temperature the abrasion would decrease with increasing speed, emphasizing the viscoelastic nature of abrasion. The coefficient b_3 is positive and can be taken as the temperature effect due to both energy and speed.

These coefficients can be used to calculate the best estimate of the abrasion loss over a wide range of energy and speed levels on the particular surface on which they were obtained. Figure 26.67 shows a scheme of testing conditions to cover a wide field of severities. This range for log energy and log speed from 0 to 1.6, respectively, has proved to be very useful. To fill it, some extrapolation from actually obtained data is required, which has to be viewed with some care. To extend the experimental conditions down to the left upper corner of the table is not practical because the very mild conditions require a long time in order to obtain a measurable abrasion loss which in addition is subject to increased variation because environmental fluctuations become very important. The four conditions at the corners are mandatory. The two further conditions at moderate speed and energy are desirable because they support the mildest condition, which takes the longest time and is subject to the largest variations. A complete scheme of nine conditions gives very reliable results.

TABLE 26.3
Coefficients of the Abrasion Equation as Function of Log Energy and Log Speed for Six Compounds Tested on Alumna 60 at Nine Testing Conditions

Compound	Coefficients				Correlation r
	a	b_1	b_2	b_3	
OESBR	0.683	1.640	−0.339	0.223	0.994
pol blend + silica I	0.583	1.706	−0.115	0.152	0.996
pol blend + silica II	0.908	1.502	−0.465	0.388	0.994
NR + black	0.690	1.636	−0.345	0.371	0.995

Energy (kJ/km)	Log energy	Speed (km/h)								
		1.00 / 0	1.58 / 0.2	2.51 / 0.4	3.98 / 0.6	6.31 / 0.8	10.00 / 1	15.85 / 1.2	25.12 / 1.4	39.81 / 1.6
1.00	0									
1.58	0.2									
2.51	0.4			▓		░			▓	
3.98	0.6									
6.31	0.8			░		▓			▓	
10.00	1									
15.85	1.2									
25.12	1.4			▓					▓	
39.81	1.6									

FIGURE 26.67 Proposed testing scheme for the evaluation of the coefficient of the abrasion equation.

The abrasion loss as log (abrasion) of log (energy) and log (speed) is best presented either in tabular form filling out the table of Figure 26.67 or as a three-dimensional graph [52] as shown in Figure 26.68. Notice that the abrasion between the mildest condition (upper left) and the most severe condition (lower right) differs by a factor of about 1000. More important for practical use is the relative rating of an experimental compound to a standard reference compound,

$$\text{Relative rating} = \frac{\text{abrasion standard compound}}{\text{abrasion experimental compound}} \cdot 100 \qquad (26.32)$$

This rates the abrasion resistance of an experimental compound in relation to a reference compound. There is no single rating but rather one, which depends on the severity of the test. This is often such that reversals in ranking occur. An example is shown in Figure 26.69 as a three-dimensional diagram. This compound (compound 2 of Table 26.3 in relation to compound 1) is better than the control at low speeds and low energies. In this case, the presentation of the ratings as a table is more useful, because it gives a detailed quantitative view of the compounds behavior. Table 26.4 shows the ratings for compounds 2 and 4 for which the abrasion coefficients are listed in Table 26.3 with compound 1 as the reference. Both compounds show reversals in ranking with changes in severity that agree closely with road wear experience. The upper compound (compound 2) which is better at

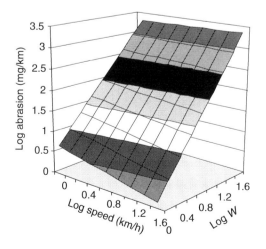

FIGURE 26.68 Log abrasion as function of log energy and log speed for a tire tread compound.

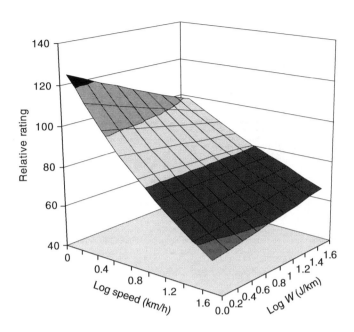

FIGURE 26.69 Elative rating of compound 2 in relation to compound 1 of Table 26.3 as function of log energy and log speed.

low energies and speeds than the reference is known to be better under driving conditions typical for the United States, i.e., speed restriction and long straight roads lead to low and moderate severity driving, whilst under European traffic conditions this type of compound is appreciated for its high wet grip and is also known to be somewhat inferior in wear resistance in relation to the reference.

The NR compound 4 is known to be better under low temperature conditions than the control, but worse under high temperature conditions. Chemically, NR has the lowest thermal stability of the polymers used for tread compounds in tire technology and it has therefore the highest temperature dependence of abrasion and wear. Thus, it is generally accepted that NR has a higher wear resistance in a moderate climate than, for instance, SBR but a much lower one in hot climates. This will be thoroughly documented below under tire wear.

In Table 26.4, NR is better whenever the sample surface temperature was near room temperature, i.e., at low speeds for all energies and at low energies of all speeds. When both were raised the sample surface temperature rose considerably and the rating of the NR deceased correspondingly.

26.5.3 CORRELATION BETWEEN LABORATORY ABRASION AND TIRE TEST COMPOUND RATINGS

If laboratory abrasion data sets as described above are available for the compound ratings obtained in a road test, they can be compared with each other for all energy and speed levels of the laboratory rating tables.

An example is shown in Table 26.5. This gives the laboratory rating as function of log energy and log speed for four passenger tire tread compounds for which road test ratings were available. They are shown on the left of each table. For compound 1 practically all laboratory ratings were less than 100 and in the tire road test the compound was also distinctly poorer than the control. Compound 3 was slightly better than the control. There is a region of higher ratings as well as one of lower ones. Hence, it is not certain whether the compound is better under all service conditions, whilst compound 4 was much better than the control over most of the testing conditions and this is reflected in the high road test rating. The compound developer gets therefore a much broader view of the compounds wear potential with this, admittedly, more time consuming laboratory abrasion method than with a single point test result.

TABLE 26.4

Relative Ratings of Compounds 2 and 4 in Relation to Compound 1 of Table 26.3 as Function of Log Energy and Log Speed in Tabular Form

SBR/BR + SI vs. OESBR Black

Log Energy	Log Speed (km/h)								
	0.0	0.2	0.4	0.6	0.8	1.0	1.2	1.4	1.6
0.0	125.9	113.6	102.4	92.4	83.3	75.2	67.8	61.2	55.2
0.2	122.1	110.9	100.7	91.4	83.0	75.3	68.4	62.1	56.4
0.4	118.5	108.3	98.9	90.4	82.6	75.5	69.0	63.1	57.6
0.6	114.9	105.7	97.2	89.4	82.3	75.7	69.6	64.0	58.9
0.8	111.5	103.2	95.6	88.5	81.9	75.9	70.2	65.0	60.2
1.0	108.1	100.8	93.9	87.5	81.6	76.0	70.9	66.0	61.5
1.2	104.9	98.4	92.3	86.6	81.2	76.2	71.5	67.1	62.9
1.4	101.8	96.1	90.7	85.7	80.9	76.4	72.1	68.1	64.3
1.6	98.7	93.8	89.2	84.8	80.6	76.6	72.8	69.2	65.7

NR + Black vs. OESBR + Black

Log Energy	Log Speed (km/h)								
	0.0	0.2	0.4	0.6	0.8	1.0	1.2	1.4	1.6
0.0	113.2	113.6	113.9	114.2	114.5	114.8	115.1	115.5	115.8
0.2	113.4	111.2	109.0	106.8	104.7	102.6	100.6	98.6	96.6
0.4	113.7	108.9	104.3	99.9	95.7	91.7	87.9	84.2	80.6
0.6	113.9	106.6	99.8	93.5	87.5	82.0	76.7	71.9	67.3
0.8	114.1	104.4	95.6	87.5	80.0	73.2	67.0	61.4	56.2
1.0	114.3	102.2	91.5	81.8	73.2	65.5	58.6	52.4	46.9
1.2	114.5	100.1	87.5	76.5	66.9	58.5	51.2	44.7	39.1
1.4	114.7	98.0	83.8	71.6	61.2	52.3	44.7	38.2	32.6
1.6	114.9	96.0	80.2	67.0	55.9	46.7	39.0	32.6	27.2

The framed cells of the laboratory abrasion table show by inspection two testing conditions for which a high correlation exists between road ratings and a single laboratory abrasion test condition. It would have been very difficult to pick one of these by chance.

By using regression analysis between the road test ratings and the ratings obtained for any one cell, i.e., any particular testing condition the correlation coefficient, the regression coefficient and the intercept of a linear regression equation can be calculated for each cell, i.e., each laboratory condition. This is shown in Table 26.6 for the data of Table 26.5. If the criterion is used that the regression coefficient should be nearly 1 and the intercept nearly zero besides a high correlation coefficient the two best testing conditions differ slightly from those by inspection. The range of a high correlation coefficient is quite large; for a 1:1 correlation, however, it is very limited.

Similar data were also obtained for truck tire compounds during an extensive survey by the European Union to improve durability and wear of re-treaded tires. An abrasion program carried out on three compounds of different polymer blends with the same carbon black gave a high correlation with the road test ratings on driven axles of different use at high energy levels. Two such correlations are shown in Figure 26.70. For seven compounds containing different types of filler with the same polymer formulation of 80 SBR/20 NR the correlation between road wear and laboratory ratings were narrower as shown in Figure 26.71. Notice that the regression coefficient

TABLE 26.5

Relative Ratings of Four Passenger Commercial Tire Tread Compounds for Which Road Test Ratings Were Available as Function of Log Energy and Log Speed

Compound 1

Road Rating	Log U	Log v								
		0.0	0.2	0.4	0.6	0.8	1.0	1.2	1.4	1.6
	0.0	51.8	57.5	63.8	70.7	78.5	87.1	96.7	107.3	119.0
	0.2	58.9	63.2	67.9	73.0	78.4	84.3	90.6	97.3	104.6
	0.4	66.9	69.6	72.4	75.4	78.4	81.6	84.9	88.3	91.9
87	0.6	76.0	76.6	77.2	77.8	78.3	78.9	79.5	80.1	80.7
	0.8	86.4	84.3	82.3	80.3	78.3	76.4	74.5	72.7	70.9
	1.0	98.3	92.8	87.7	82.8	78.2	73.9	69.8	66.0	62.3
	1.2	111.7	102.2	93.5	85.5	78.2	71.5	65.4	59.9	54.8
	1.4	127.0	112.5	99.6	88.2	78.1	69.2	61.3	54.3	48.1
	1.6	144.3	123.8	106.2	91.1	78.1	67.0	57.5	49.3	42.3

Compound 2 = Control

Road Rating	Log U	Log v								
		0.0	0.2	0.4	0.6	0.8	1.0	1.2	1.4	1.6
	0.0	100	100	100	100	100	100	100	100	100
	0.2	100	100	100	100	100	100	100	100	100
	0.4	100	100	100	100	100	100	100	100	100
100	0.6	100	100	100	100	100	100	100	100	100
	0.8	100	100	100	100	100	100	100	100	100
	1.0	100	100	100	100	100	100	100	100	100
	1.2	100	100	100	100	100	100	100	100	100
	1.4	100	100	100	100	100	100	100	100	100
	1.6	100	100	100	100	100	100	100	100	100

Compound 3

Road Rating	Log U	Log v								
		0.0	0.2	0.4	0.6	0.8	1.0	1.2	1.4	1.6
	0.0	91.9	105.0	120.1	137.3	157.0	179.5	205.2	234.7	268.3
	0.2	97.9	106.7	116.3	126.7	138.1	150.4	163.9	178.6	194.6
	0.4	104.4	108.4	112.6	116.9	121.4	126.1	130.9	135.9	141.1
107	0.6	111.3	110.2	109.0	107.9	106.8	105.6	104.5	103.5	102.4
	0.8	118.7	111.9	105.6	99.5	93.9	88.5	83.5	78.7	74.2
	1.0	126.6	113.7	102.2	91.9	82.6	74.2	66.7	59.9	53.9
	1.2	134.9	115.6	99.0	84.8	72.6	62.2	53.2	45.6	39.1
	1.4	143.9	117.4	95.8	78.2	63.8	52.1	42.5	34.7	28.3
	1.6	153.4	119.3	92.8	72.2	56.1	43.7	34.0	26.4	20.5

Compound 4

Road Rating	Log U	Log v								
		0.0	0.2	0.4	0.6	0.8	1.0	1.2	1.4	1.6
	0.0	68.1	88.0	113.7	147.0	190.0	245.5	317.3	410.1	530.0
	0.2	75.1	92.5	114.0	140.4	172.8	212.9	262.2	322.8	397.6
	0.4	82.9	97.3	114.2	134.0	157.3	184.6	216.6	254.2	298.3
170	0.6	91.5	102.3	114.4	127.9	143.1	160.0	178.9	200.1	223.8
	0.8	101.0	107.6	114.6	122.2	130.2	138.7	147.8	157.5	167.9
	1.0	111.4	113.1	114.9	116.6	118.4	120.3	122.1	124.0	125.9
	1.2	122.9	118.9	115.1	111.4	107.8	104.3	100.9	97.6	94.5
	1.4	135.6	125.1	115.3	106.3	98.0	90.4	83.4	76.9	70.9
	1.6	149.7	131.5	115.5	101.5	89.2	78.4	68.9	60.5	53.2

TABLE 26.6

Correlation, Regression Coefficients and Intercept of a Linear Relation between the Road Test Ratings from Table 26.4 and the Laboratory Ratings Obtained for Each Cell

	Log U	0.2	0.4	0.6	0.8	1.0	1.2	1.4	1.6
					Log v				
	0.2	0.283	0.623	0.824	0.911	0.949	0.965	0.973	0.978
	0.4	0.354	0.678	0.866	0.947	0.978	0.990	0.993	0.993
	0.6	0.444	0.746	0.906	0.970	0.993	0.999	0.998	0.996
Correlation coefficent	0.8	0.564	0.829	0.935	0.961	0.962	0.959	0.954	0.951
	1.0	0.698	0.925	0.923	0.877	0.839	0.812	0.793	0.782
	1.2	0.778	0.988	0.830	0.693	0.607	0.548	0.505	0.473
	1.4	0.717	0.916	0.635	0.447	0.338	0.267	0.213	0.169
	1.6	0.563	0.677	0.395	0.215	0.114	0.048	0.000	−0.037
	Log U	0.2	0.4	0.6	0.8	1.0	1.2	1.4	1.6
					Log v				
	0.2	0.55	1.03	1.02	0.81	0.60	0.45	0.34	0.26
	0.4	0.78	1.30	1.28	1.04	0.80	0.62	0.48	0.38
Regression coefficient	0.6	1.13	1.68	1.62	1.33	1.06	0.85	0.69	0.56
	0.8	1.72	2.25	2.02	1.63	1.32	1.08	0.91	0.78
	1.0	2.52	3.07	2.38	1.77	1.38	1.14	0.97	0.86
	1.2	3.04	3.94	2.41	1.51	1.08	0.83	0.68	0.59
	1.4	2.51	3.92	1.88	0.96	0.58	0.39	0.28	0.20
	1.6	1.55	2.60	1.08	0.42	0.18	0.07	0.00	−0.04
	Log U	0.2	0.4	0.6	0.8	1.0	1.2	1.4	1.6
					Log v_f				
	0.2	67	13	4	17	33	46	57	64
	0.4	43	−13	−20	−3	17	33	46	56
	0.6	6	−52	−51	−26	−2	18	33	45
Ordinate intercept	0.8	−57	−110	−87	−48	−17	6	23	35
	1.0	−148	−194	−116	−52	−11	14	31	42
	1.2	−216	−286	−114	−20	25	49	64	74
	1.4	−170	−287	−59	35	71	88	97	103
	1.6	−68	−153	17	82	103	112	116	118

was only about 0.5. This means that the laboratory experiments gave twice as high as discrimination between the compounds than the road wear data. The data are obtained on a laboratory alumina abrasive track. The question arises of how this surface corresponds to road surfaces.

26.6   TIRE WEAR

26.6.1   TIRE WEAR UNDER CONTROLLED SLIP CONDITIONS

One of the simplest instruments to measure actual tire wear under controlled load and slip is to tow a two-wheeled trailer with its wheels set at a given slip angle. Both wheels have to have tires mounted with the same construction, tread pattern, and compound. Under these conditions the tires run under the same slip angle. Such a trailer was constructed by Grosch and Schallamach [56]. The wear as function of the slip angle is shown in Figure 26.72 for three different black-filled tire tread compounds. Plotting wear and slip angle on log scales produces straight-line graphs indicating that the wear loss can be described by a power law which is in agreement with the model described above. The straight-line graphs cross, indicating reversals in ranking. The NR and SBR cross at a low slip angle. For the blend of 50 NR/50 BR this shifts towards higher slip angles. All tires developed Schallamach abrasion pattern during the experiments. The most important variable, however, was the tire temperature, measured on the surface of the tire. It increased with increasing

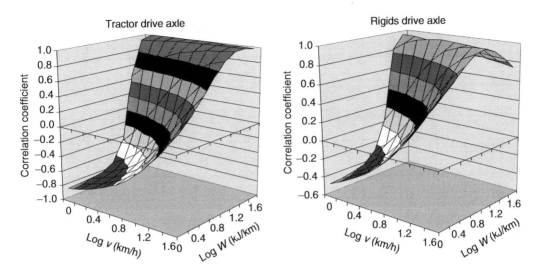

FIGURE 26.70 Correlation coefficient between road test ratings of three truck tire compounds on re-treaded tires at two different axles and laboratory ratings as function of log energy and log speed.

slip angle drastically. If the rating of NR in relation to SBR is plotted as function of the tire surface temperature, as shown in Figure 26.73, a clear function results, showing that NR is better at low tire surface temperatures, dropping rapidly as the temperature rises.

The tire surface temperature may also be caused by environmental conditions. Wear measurements with the trailer at a constant slip angle, but changing ambient temperature produced a larger temperature dependence of the wear rate for NR than for SBR as shown in Figure 26.74, indicating a crossover of the straight-line graphs at a surface temperature of about 48°C. Such temperatures are also produced in ordinary tire wear. Measurements on tires put on the towing vehicle of the trailer

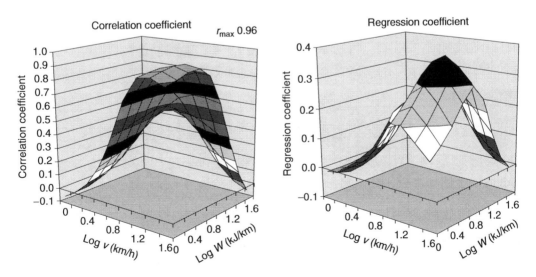

FIGURE 26.71 Correlation and regression coefficients between road test ratings of seven truck tire compounds 80 SBR/20 NR differing by type and amount of filler and laboratory abrasion ratings obtained on LAT 100 testing equipment.

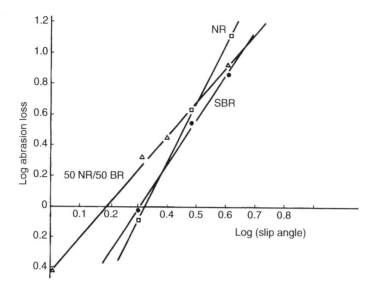

FIGURE 26.72 Log tire wear as function of log slip angle obtained with the MRPRA trailer for three tire tread compounds.

gave much lower wear rates, as expected, but the resulting wear rating of NR to SBR corresponded to that expected from the measured tire surface temperature as also shown in Figure 26.73 (open circles). The shift of the crossover point towards higher slip angles, i.e. higher surface temperatures, for the NR–BR blend indicates a higher thermal stability of the blend. All this is qualitatively in agreement with the laboratory investigations described above and underlines the important role that the temperature plays in the wear process. Its importance is not derived from the dynamic properties, but rather stems from the thermal oxidative processes and the compound resistance to them.

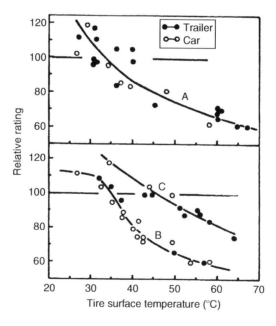

FIGURE 26.73 Relative compound rating as function of the tire surface temperature for three different comparisons obtained with the MRPRA test trailer, including also data from the towing vehicle.

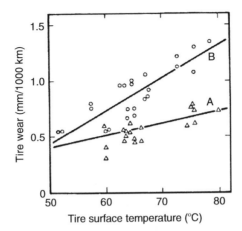

FIGURE 26.74 Abrasion of a natural rubber (NR) and a styrene–butadiene rubber (SBR) tread compound as function of the tire surface temperature brought about by ambient temperature changes at a constant slip angle measured with the MRPRA test trailer.

26.6.2 Conditions for Tire Wear in Road Tests and Normal Usage

26.6.2.1 Influence of the Road Surface

Consider first a test car, which is driven in an 8 h shift over a prescribed route. This will have a length of about 600 km. It could be selected to have special features like predominantly motorway (turnpike or freeway) or lead through winding hilly country. In either case there will inevitably be a number of changes of road surface structure resulting in an average road surface. This average may still be different for different test routes but the difference between the two averages are already likely to be smaller than the differences encountered along each route. By road surface structure is meant the structure, which is likely to influence the abrasion loss significantly, that is, its sharpness and coarseness. Moreover, the average of a test route is likely to reflect the average of a whole geographical region. Differences between the average road surfaces in highly developed countries are likely to be small whilst larger differences exist between such countries and less developed ones.

But even the structures of a chosen test route vary with the season. It is well known that road surfaces in areas of moderate climate are much sharper in winter than in summer. Even in much shorter periods the sharpness changes often drastically with the weather conditions.

Since wear results are generally obtained over a time span they are inevitably averages. This is not too serious if the ranking of test compounds is not strongly influenced by the road surface structure and in particular by the changes which occur. Nevertheless it is obvious that repeatability is limited and as the few examples given from laboratory experiments show reversals in ranking are rather the rule than the exception. Hence, a single road test result can only have a limited validity.

26.6.2.2 Tire Construction Influences

Road wear is force controlled. This is a fundamental difference to slip-controlled laboratory abrasion test machines or wear tests with a trailer as described above. In force-controlled events the abrasion loss is inversely proportional to the stiffness of the tire whilst under slip control the abrasion is proportional to its stiffness (see Equations 26.18a and 26.19a).

The effect can be seen in Figure 26.75 which shows the results obtained with the two-wheeled trailer discussed above. Tire group A was a re-treaded bias tire; group B a commercial steel-belted radial ply tire. When one tire of each group was mounted on the axle of the trailer for equal set slip angles, the direction of the tow bar adjusted itself in such a way that the tires run under the same

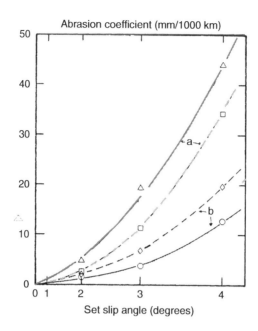

FIGURE 26.75 Comparing the rate of wear of a retread bias tire with a commercial radial ply tire (a) one tire each mounted on the trailer axle (equal force comparison) and (b) two tires of each group mounted and run consecutively (equal slip).

force, i.e., the slip angle of the stiffer tire is reduced, that of the softer (bias) is increased. Hence, a considerable difference in wear occurs for the two groups. If two tires of the same group are mounted the trailer runs now for both groups under the set slip angle. Now, the stiffer tire wears more than under the previous condition and the softer wears less. If the tires had had the same tread compound and curing conditions the wear rate would in fact have reversed. As this was not the case only a reduction in wear occurs for the one and an increase for the other.

In tire tests, therefore, a stiffer tire construction reduces the wear. The stiffer construction not only includes the carcass and belt (i.e., height to width ratio), but also the tread pattern and the stiffness of the tread compound. Therefore, a correct road test requires two identical vehicles, each equipped with a test group of four identical tires. If this is not possible at least one axle has to have identical tires. If this is not the case an average slip will balance the acting force on the axle. This will be larger for the stiffer tire than is required if both tires had the same stiffness and smaller for the softer tire. Hence, the result is falsified to the advantage of the softer tire. The same argument holds for multi-section tires.

26.6.2.3 Driving Influences

The forces acting on a car are due to acceleration and braking, wind and rolling resistance. In order to obtain a quantitative estimate of their effects, the route is divided into small sections over which it may be assumed that the force and speed conditions are constant. Consider first the cornering acceleration component. Curves will differ in radii and the speeds with which they are being negotiated. The decisive aspect is the centrifugal acceleration it produces. These accelerations can be measured. They are likely to follow a statistical distribution function symmetrical around zero. Figure 26.76a shows the measured cornering force distribution obtained in a controlled road test for passenger car tire wear. Also shown is a normal distribution adjusted for its width to fit the data. It is obvious that the real distribution is very similar to a normal one. The same holds for the fore and aft acceleration as seen in Figure 26.75b. The force due to wind resistance is a function of the speed and

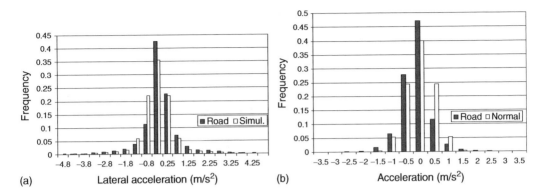

(a) Lateral acceleration (m/s²)   (b) Acceleration (m/s²)

FIGURE 26.76 Measured distribution of (a) cornering and (b) fore and aft accelerations in a controlled road wear test for passenger car tires.

the vehicle dimensions. The rolling resistance is due to the tires and the vehicle. Both are primarily functions of the load. In controlled road tests the load is usually kept constant, the speed, however, is also distributed. Figure 26.77 shows a speed distribution obtained on a controlled road test measured over 600 km. Two superposed normal distribution functions gave a good agreement with the data. For the general user the load is also not constant and can be represented by distribution functions.

26.6.3 ROAD WEAR TEST SIMULATION

Using distribution functions for accelerations, speeds, and loads it is possible to calculate the necessary slip for a large number of driving events with the help of the brush tire model, each event being defined by a cornering acceleration, a longitudinal acceleration (speed up or braking), the mass of the vehicle determined by the load and weight of the vehicle its forward speed and the forces to overcome the wind and the rolling resistances of tires and vehicle. All these are assumed to be constant for a short distance, hence, the energy dissipation in the contact area due to the resulting slip and the slip speed can be calculated. Using the laboratory abrasion equation 26.31 the volume

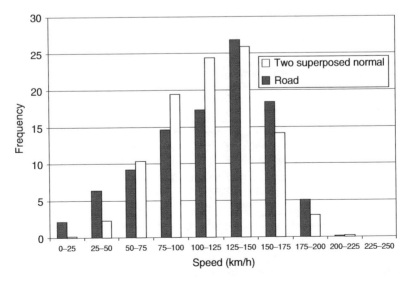

FIGURE 26.77 Speed distribution in a controlled road test for passenger car tires also shown a distribution made up of two superposed normal ones.

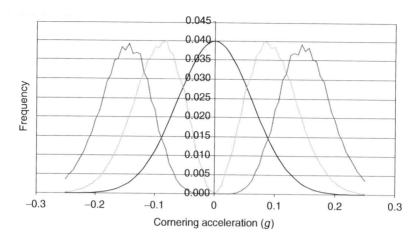

FIGURE 26.78 Theoretical acceleration (force) distribution function in a computer simulation, the resulting energy dissipation multiplied by the frequency of occurrence and the expected abrasion loss, using Equation 26.18 with parameter for a passenger tread compound.

loss for each event is obtained. With distribution functions describing the frequency of their occurrence the volume loss over the whole distance is obtained.

26.6.3.1 Force Distributions

It has already been shown that real measured distribution functions can be represented by normal distributions. Hence, in a simulation program it is useful to use such functions in absence of real functions. Normal distributions around zero are used for the acceleration components. Their width is defined by three times their standard deviations, which includes 99.9% of all events. They are a major indication of the severity of the driving style. Figure 26.78 shows a normal distribution of the cornering acceleration, the resulting energy dissipation multiplied by the frequency of occurrence, and the abrasion multiplied by the frequency, using laboratory abrasion equation 26.31 with data obtained for a test compound on the LAT 100 abrasion tester. The energy and abrasion curves were normalized so that their heights correspond approximately to that of the distribution function. The curves are shown as continuous functions for clarity, in a practical simulation program a limited number of steps have to be used.

26.6.3.2 Speed and Load Distributions

For the speed and load distributions three superposed normal distributions around three fixed mean values are used, corresponding for the speed to town, country road, and motorway (turnpike or freeway) traffic. A maximum speed fixes the total width of the curve from zero to that maximum. This corresponds to 10σ, where σ stands for standard deviation of the three superposed distributions. The three mean values are fixed at 3, 5, and 7σ. Their heights can be varied according to the frequency with which the three distributions occur, their sum has to add up to 1. A similar distribution is also used to describe the different load conditions with low, medium, and high loads. Figure 26.79 gives an example of such a triple distribution function.

With the assumption that the frequencies of the different distributions determine one event, the frequency of each event is defined by the product of the individual frequencies.

Having calculated the force for a particular event the slip is calculated using the bush model and hence the energy dissipation is obtained. Using the factors of the abrasion equation, determined with the LAT 100 on an alumina surface the abrasion loss for each event is calculated. The forces are different for a driven and a nondriven axle and accordingly different abrasion rates will result.

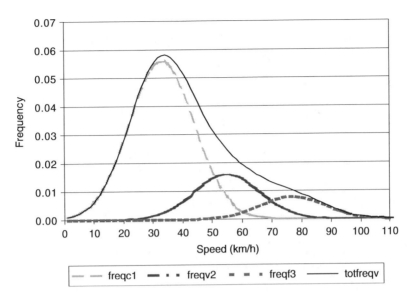

FIGURE 26.79 Theoretical triple distribution as they are used in a computer simulation for speed and load distributions (height ratios: town 0.7, country 0.2, and motorway 0.1).

The load transfer, which occurs during cornering, averages out and is therefore not included. With the data of the tires, i.e., dimension, tread width, net/gross of the tread pattern, and tread depth, expected tire lives under the above-specified conditions are calculated. This approach is highly empirical (although supported by some distribution measurements). It has the merit that with an existent set of laboratory abrasion data a large number of road test simulations with different boundary conditions can be run in a very short time and their effect on tire life and compound rating can be estimated.

26.6.3.3 Results of Road Test Simulation Calculations

Table 26.7 gives a list of the boundary conditions which define a tire wear test simulation and in fact also an actual road test. The road surface is the laboratory surface on which the abrasion data for the simulation were obtained. There is as yet no definition of a road surface and even if there were one, it would be of little use since road surface structures change frequently along the road surface as pointed out earlier.

The tire construction influences both cornering and longitudinal slip stiffness. These include the tire carcass, breaker construction, inflation pressure, and tread pattern design. However, since the two stiffness components can be measured, knowledge of the construction details is not necessary. The vehicle geometry influences the tire wear through the air resistance, which it creates, and through the load distribution between front and rear axles.

The driving parameters are determined by the maximum values of their distribution functions and in case of load and speed by the ratios between low, medium, and high loads and town, country road, and highway traffic, respectively.

The tread pattern influences the tire life through its available compound volume. Figure 26.80 shows the major influences on the tire life. The most important is clearly the influence of the driving style determined by accelerations and maximum speed and the tire construction as determined by the cornering and longitudinal slip stiffness. The figure compares two representative modern passenger tire tread compounds, an OESBR black-filled compound and a solution SBR–BR blend filled with silica. The two upper graphs show the influence of the maximum acceleration components at a constant maximum speed of the relevant distribution functions on tire life. It is clear that the tire life decreases rapidly with increasing severity. The solid lines refer to the cornering stiffness

TABLE 26.7

Listing of the Boundary Conditions for a Road Test Stimulation

| Tire size: | 205/65R 15 | | |
| Laboratory surface | Alumina 60 | | |

	Tire Construction:		**Tread Pattern**	
Tire	kso(N/rad) =	60000	Pattern width/tire width =	0.78
parameter	kco(N/sl) =	120000	Net/gros =	0.7
	Friction coefficient =	1.1	Cross-section ratio =	0.65
			Pattern depth (mm) =	8

Vehicle	Rolling resistance coefficient =	0.01
parameter	cw =	0.3
	Proj vehicle cross-section (m²) =	2.5

	Maximum corn accel. (g) =	0.15	Smallest tire load (N) =	3200
Driving	Maximum longitudinal accel. (g) =	0.15	Largest tire load (N) =	4900
parameter	Maximum vehicle speed (km/h) =	170		
		Low	Medium	High
	3-load distr. ratio =	0.2	0.5	0.3
	3-speed distr. ratios =	0.2	0.3	0.5

components of Table 26.5 whilst for the dotted lines the stiffness components have been reduced by 25%, keeping all other values the same. This reduces the tire life considerably. However, not only the tire lives are involved, the relative ratings of the two compounds are also affected. At low

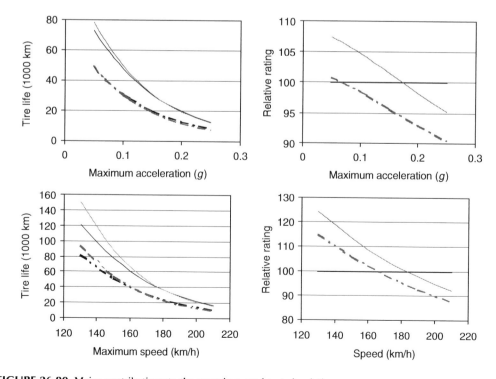

FIGURE 26.80 Major contributions to the wear in a road test simulation.

accelerations the OESBR–BR blend + silica is better than the QESBR black compound whilst the rating reverses with increasing acceleration.

The same is observed if the maximum speed of the speed distribution function is varied and the acceleration is kept constant. The tire life is drastically reduced and the rating of the silica compound reverses in relation to the OESBR as the speed is increased.

26.6.4　Correlation between Laboratory Road Test Simulation and Road Wear Test Results

26.6.4.1　Correlation with a Set of Passenger Car Tires

It has already been shown how a correlation between road test ratings and laboratory abrasion can be obtained over a range of energies and speeds (Table 26.6). Usually, a good correlation, i.e., a high correlation coefficient and a regression coefficient near 1 are obtained only over a limited range.

This can be taken a step further by comparing road test ratings with a simulated laboratory test rating. This reduces the comparison again to a single number, which is usually favored by managers and chemists. However, this number is now based on a range of boundary conditions defining the road test. This means also that some thought has to be invested when deciding on the road test conditions. In order to be able to do so this absolutely correctly the exact boundary conditions of the road test should be known. Since today it is an easy matter for tire- and vehicle-manufacturers to carry out the required measurements, such a set of boundary conditions and a comprehensive set of laboratory abrasion data could save an enormous amount of compound development time [57]. For the available road test data such information was not available. However, because a road test simulation takes only a very short time and representative boundary conditions are not difficult to guess, good correlations can be shown to exist also between laboratory road test simulation and actual abrasion road test ratings without the exact knowledge of the boundary conditions.

Table 26.8 shows three road test simulations of increasing severity by raising the maximum acceleration components in three steps. The remainder of the boundary conditions was only less known and was therefore guessed. The calculated ratings are compared with the road test ratings. At a maximum acceleration for both components of 0.35 g a very good correlation is achieved. This corresponds to a hard driving style, but one which is common for tire test drivers and the achieved tire lives are accordingly short and are in agreement with this style of driving. It is not known what they really were but they are realistic.

26.6.4.2　Correlation with Truck Tire Road Test Ratings

Principally, the simulation compares tires under equal force conditions. It calculates the resulting energy dissipation and slip speed. Using the abrasion equation 26.31 the abrasion volume loss is obtained. Generally, the program uses one cornering and one longitudinal slip stiffness, usually for the control tire for all groups. This leads to a correct prediction as long as the tire construction and tread pattern and the shear modulus of the tread are the same, which is usually the case for the former two. In most cases differences between the moduli of tread compounds are also small and hence the results are reasonably correct. Ideally, the cornering and longitudinal slip stiffness should be measured. If this is not possible the two slip stiffness components of the tire ought to be corrected for differences of shear modulus of the tread.

Considering the tire as a composite short beam under shear the following correction may be applied to the basic stiffness of the control:

$$K_x = K_o \frac{\rho(1 + \varphi)}{(1 + \varphi\rho)} \tag{26.33}$$

TABLE 26.8

Correlation between Road Test Ratings and Laboratory Road Test Simulations of a Group of Four Passenger Tires, Discussed Above for Their Laboratory Abrasion and Correlation to Road Test Ratings

Tire size	205/60R 15	

Laboratory surface:	Alumina 180

Cornering stiffness (N/rad)	45,000	Tire cross-section ratio	0.6
Circumf. slip stiffness (N/slip)	90,000	Net/gros of pattern	0.7
Friction coefficient	1	Pattern width/tire width	0.76
Tire load (N)	4500	Pattern depth (mm)	8
Rolling resistance (N)	45		
cw coefficient	0.3		
Maximum speed (km/h)	170		

Maximum cornering acceleration (g)	0.2
Maximum fore and aft acceleration (g)	0.2

Compound	Vol/km	km/mm	Tire life (km)	Rating	Rating fr, Road
1	12.2	17,563	112,404	80	87
2	9.8	21,929	140,348	100	100
3	10.9	19,617	125,549	89	107
4	5.4	39,566	253,224	180	170

Maximum cornering acceleration (g)	0.3
Maximum fore and aft acceleration (g)	0.3

Compound	Vol/km	km/mm	Tire life (km)	Rating	Rating fr, Road
1	49.0	4386	28,072	86	87
2	42.0	5120	32,766	100	100
3	41.1	5231	33,477	102	107
4	25.5	8439	54,008	165	170

Maximum cornering acceleration (g)	0.35
Maximum fore and aft acceleration (g)	0.35

Compound	Vol/km	km/mm	Tire life (km)	Rating	Rating fr, Road
1	83.7	2566	16,420	88	87
2	73.9	2906	18,600	100	100
3	69.7	3080	19,711	106	107
4	46.0	4671	29,896	161	170

where

K_o is the slip stiffness component of the control tire

φ is the ratio between the tread and carcass stiffness

ρ is the stiffness ratio of the experimental to the control tread

The latter can be estimated from the side force coefficients obtained during the abrasion experiments for a small slip angle and a high speed. They reflect directly the compound stiffness and since the dimensions are the same also the shear modulus.

For a balanced construction the ratio φ should be 1. If this is assumed to be the case the correction becomes

$$K_x = K_o \frac{2\rho}{(1 + \rho)} \tag{26.33a}$$

If $\varphi < 1$, the correction is larger, approaching ρ in the extreme case. If φ is larger than 1, the correction becomes smaller than given by Equation 26.33a, so that K_x approaches K_o.

If multi-section tires or experimental and control tires are mounted on the same axle, as is often the case for tests with truck tires on commercial fleets, the inverse of the above relation should be applied because the tires are now running under an imposed common slip. For the stiffer compound the slip is now larger and for the softer it is smaller and the stiffer compound becomes poorer and the soft compound becomes better than would be the case if whole tires of the same group had been mounted on the same axle.

This is the case for the results of the truck tire tests discussed below. Table 26.9 shows the ratings of a laboratory wear simulation and the average road test ratings obtained for the three re-tread truck tire compounds used in a European Commission project to improve retread tire quality (see Figures 26.70 and 26.71). Because on all vehicles either multi-section tires were mounted or two groups were mounted on the same axle, in each case the comparison occurred under equal slip conditions rather than equal force. The stiffness of the compounds was estimated by the measured side force coefficients obtained during the abrasion experiments for the smallest slip angle and highest speed. The simulations were carried out correcting the stiffness of the tires for shear

---

TABLE 26.9

Boundary Conditions for the Road Test Simulation Using Laboratory Abrasion Data to Achieve a Correlation with the Average Road Test Ratings Obtained with Three Truck Tire Re-Tread Compounds Mounted on Axles of Several Trucks Either as Multi-Section Tires or as Whole Tires but Two Groups on the Same Axle Together with the Simulation and Actual Ratings when Compared (a) under Equal Force, (b) under Equal Energy (This Assumes the Same Slip Stiffness for all Groups), and (c) under Equal Slip Conditions

| Tire size: | 275/80R 22.5 |
| Laboratory surface | alumina 120 |

	Tire Construction:		**Tread Pattern**	
Tire parameter	kso (N/rad) =	175,000	Pattern width/tire width =	0.82
	kco (N/sl) =	325,000	Net/gros =	0.72
	Friction coefficient =	1.1	Cross-section ratio =	0.8
			Pattern depth (mm) =	16

Vehicle parameter	Rolling resistance coefficient =	0.01
	cw=	0.9
	Proj vehicle cross-section (m²) =	5

Driving parameter	Maximum corn acceleration (g) =	0.15	Smallest tire load (N) = 13,500
	Maximum longitudinal acceleration (g) = 0.15		Largest tire load (N) = 27,500
	Maximum vehicle speed (km/h) =	80	

		Low	Medium	High
	3-load distr. ratio =	0.25	0.3	0.45
	3-speed distr. ratios =	0.3	0.4	0.3

	Comp.	Vol/km	km/mm	Tire Life (km)	Rating	Average Road Rating	Correlation Coefficient
Equal force	1	298.9	1726	24,853	100.0	100	
	2	287.7	1793	25,818	103.9	118	0.34
	3	217.5	2372	34,154	137.4	113	
Equal energy	1	298.9	1726	24,853	100.0	100	
	2	275.5	1873	26,967	108.5	118	0.648
	3	246.2	2095	30,167	121.4	113	
Equal slip	1	298.9	1726	24,853	100.0	100	
	2	260.5	1980	28,516	114.7	118	0.977
	3	277.6	1859	26,764	107.7	113	

modulus differences of the compounds according to Equation 26.33 with $\varphi = 0.3$. (a) For equal force, (b) equal energy (this assumes the same stiffness for all compounds), and (c) for equal slip conditions. In the latter case the inverse of Equation 26.33 is used.

Since no data apart from a description of the general use of the truck (mostly short-distance haulage) and the axle position on which the tires were mounted were available, again reasonable assumptions on maximum accelerations and speeds, as well as load distributions had to be made. They are listed in Table 26.9 together with the three comparisons.

The boundary conditions chosen are reasonable for the kind of application under which the tires are operated and the achieved mileages agree well with the one obtained from the simulation.

It is seen that a good agreement is only reached if equal slip conditions are assumed, with a correlation coefficient of 0.977, demonstrating also how misleading tire road test results can be if multi-section tires are used or tires of different stiffness due to either different constructions or modulus differences of the tread compounds.

26.6.5   ENERGY CONSUMPTION AND SLIP SPEEDS IN ROAD WEAR

The road test simulation program also lists the energy consumption and the slip speeds, which occur for each event taking part in the simulation. These quantities can be presented in the form of frequency distributions. The two energy distributions for a typical passenger tire and truck tire simulations are shown in Figure 26.81. Notice that they are almost identical for passenger and truck tires, although the boundary conditions are very different. The reason is twofold: First, the truck tires are much stiffer and hence the resulting slip angle for the same force is much smaller. Second, the accelerations in truck tire use are distinctly lower than for cars. Apart from engine power/kilogram mass being much lower (7–10 kW/t for trucks compared with about 70 kW/t for passenger cars), too high an acceleration could well damage the goods carried. In addition, the laboratory range of energies, marked by the blue box, is also close to the range of the energies in practical tire use. The reason for this is that much larger slip angles are used at a much smaller force.

A similar situation exists for the slip speeds (Figure 26.82). They are lower for trucks than for passenger tires but both are within the slip speed range of the laboratory abrasion tests.

Hence the extrapolation between a laboratory abrasion test and tire wear in general use is much smaller than would be expected at first sight. It is therefore not surprising that the laboratory test procedure, which admittedly is much more complex than a single point reading, gives a close correlation for ratings and indeed also for mileages.

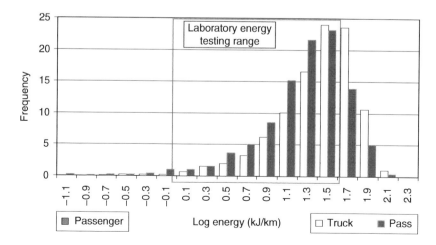

FIGURE 26.81 Energy distributions in a typical passenger car and truck tire road test simulation. Also shown is the range of the energies used in laboratory abrasion experiments.

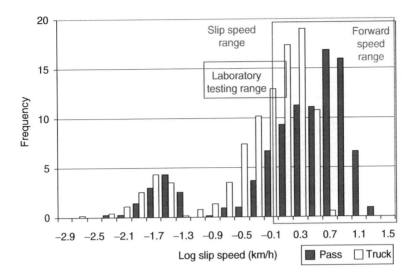

FIGURE 26.82 slip speed distributions in a typical passenger car- and truck tire road test simulation. Also shown is the range of the slip speeds used in laboratory abrasion experiments.

26.7 CONCLUSION

During the last 50 years the understanding of the very complex phenomenon of friction and abrasion of tire tread compounds and their relation to traction and wear has progressed steadily. This chapter tries to trace the main strands of this story. Some details may still be missing but the main points are now clear.

26.7.1 FRICTION AND TIRE TRACTION

The all-important difference between the friction properties of elastomers and hard solids is its strong dependence on temperature and speed, demonstrating that these materials are not only elastic, but also have a strong viscous component. Both these aspects are important to achieve a high friction capability. The most obvious effect is that temperature and speed are related through the so-called WLF transformation. For simple systems with a well-defined glass transition temperature the transform is obeyed very accurately. Even for complex polymer blends the transform dominates the behavior; deviations are quite small.

The shape of the so called master curve, however, depends strongly on the compound formulation and the type of contact surface in the friction process. Its detailed study gives an insight into the processes responsible for the frictional behavior.

Two processes contribute to the total friction: Adhesion and internal losses through cyclic deformation. Adhesion is an activated process akin to the cohesion of liquids which is demonstrated by the WLF transformation and the conversion constant between the speed of maximum friction and the relaxation time of the maximum of the relaxation spectrum of the polymer. This constant is a length of molecular dimensions. On smooth surfaces adhesion is the only process responsible for the total friction. On rough surfaces local deformation produces a cyclic process leading to internal losses. In this case, the conversion constant is a length directly linked to the asperity spacing of the track. But on rough tracks, too, adhesion plays a major role. Purely normal deformation contributes only a very limited fraction of the total friction. Tangential stresses are required to account for the large friction on rough surfaces and these are made possible by the presence of adhesion.

In the region of low viscosity the friction is higher on rough surfaces than on smooth ones whilst in the high viscosity region the reverse is the case indicating that viscosity alone is not responsible for friction. High extensibility and the loss of the stored elastic energy under high extension also contribute to friction. Fillers reduce the extensibility and molecular mobility and hence reduce the friction in the high viscosity region although in bulk measurements the filler increases the hysteresis of the compound. This is obviously due to the reduction of the real contact area due to the increased stiffness of the compound.

This basic rubber friction process is present on all surfaces, dry, wet or icy, being modified only by the external conditions. On wet surfaces this is primarily water lubrication which itself is influenced by the water depth, roughness of the road surface, and the state of the tire tread pattern. The low friction on ice near its melting point is mainly due to the properties of the ice.

A special winter compound is required because the contact temperature cannot exceed 0°C. Because of the large latent heat of melting, a film of water produced through the heat generated in the contact area is so thin that it does not affect the friction coefficient significantly. Hence, high friction coefficients on ice are possible at temperatures below about −10°C. The limitation is set by the compound properties, expressed through their glass transition temperature. Even so, it is not yet possible to construct a friction master curve from more basic measurements as, for instance, dynamic modulus or contact area measurements. The master curve is therefore at present the staring point of obtaining an insight into the traction behavior of tires.

Basically, a road traction test differs from a laboratory test only in that the temperature in the contact area is allowed to rise and is not really measurable, whilst in the laboratory the speed is kept so low that the temperature rise may be neglected.

The different surface structures of the road and laboratory surfaces appear to play a minor role.

This is not surprising since at a given speed the coarseness of a track (the average spacing of the asperities) influences the friction only on a logarithmic scale. Also the observed dependence of the friction coefficient on load of soft rubber compounds on smooth surfaces disappears for harder black or silica-filled treads compounds on rough surfaces.

A good correlation between road tests and laboratory measurements is usually obtained only over a limited range of the master curve variable log $a_T v$. Comparisons of the friction coefficients of different compounds over a wider range of log $a_T v$ values with road test ratings identify the useful ranges and give a detailed picture of its capabilities in practical tire use. Further developments of compounds can be limited to laboratory assessments over a limited range of log $a_T v$, i.e., temperatures and speeds with a very strong assurance that the road performance will reflect the laboratory findings.

Winter compounds are still best developed on an ice track because this is the easiest way to produce the necessary low temperature range.

Since modern compounds are highly sophisticated compositions of polymer types and fillers a detailed understanding of the exact shape of a master curve, its relation to the WLF transform and possible deviations from it leave still room for further research.

26.7.2 ABRASION AND TIRE WEAR

Basically abrasion is a cut growth process. This process is governed by the tearing energy as defined by Rivlin and Thomas. Schallamach was the first to conceive abrasion as a function of the energy dissipation in the contact area of rubber and track. Basic experiments demonstrated the close relation between the cut growth—tearing energy relation on the one hand and the abrasion—energy consumption relation on the other. The relation between the abrasion and the energy consumption as a power law is now firmly established and holds over a wide range of energies. Power index and constant depend on both compound and abrasive track. The prime effect is thereby due to the sharpness of the track rather than due to its coarseness. This influences both the level of abrasion as well as the power index of the abrasion–energy relation of the polymer filler system. Both processes, abrasion and cut growth, are not purely physical phenomenon but are also subject to oxidation.

For abrasion this is, however, a much more dominating process than for cut growth. The main reason is that the energy consumption in the abrasion process raises the temperature in the interface between rubber and track and thereby modifies this process. The temperature in the contact patch is a function of the power consumption and depends, therefore, also on the sliding speed. The temperature not only influences the oxidation and cut growth process, but also causes thermal degradation.

The physical and the chemical processes are so strong that they lead to different behavior of compounds under different service conditions and reversals in compound ranking are common. It is therefore, in principle, not possible to design a single laboratory abrasion test, which could reflect the practical experience of service conditions. A range of energies and speeds are a minimum requirement. Computer simulation of road test conditions has shown that both these quantities can be set in laboratory experiments in the same range as occur in practical service, increasing the likelihood of producing compound ratings which reflect the practical experience.

The question of the influence of the road surface on tire wear cannot be answered unequivocally because of the large number of different compositions, state of use, and weather influences on their abrasive power. Road surfaces are also not durable enough for laboratory use. Hence, reliance has to be placed on the correlation between laboratory results on a laboratory abrasive surface and road test experience. Alumina of different grain size (but primarily 60) has proved to be the most useful. Even its sharpness changes with time of use and disks are limited in their useful life.

The link between laboratory abrasion methods and road tests is now well established so that further research can concentrate on further elucidation of the basic underlying processes. This concerns particularly the interaction between filler systems and polymers, which is continuously expanding through both new polymers and new fillers. But now the link does not extend from a basic experiment to tire road testing. It can be limited to a laboratory abrasion testing method and suitable basic experiments.

ACKNOWLEDGMENT

This chapter is based on two chapters, which the author wrote for "The Pneumatic Tire" edited by professor A.N. Gent and professor J. Walters of the University of Akron and published by the National Highway Safety Administration of the U.S. Government who all gave their kind permission to use these two chapters when compiling this one. I am very grateful for this.

REFERENCES

1. F.B. Bowden and A.D. Tabor, *Friction and Lubrication of Solids*, Oxford University Press, London, 1954.
2. P. Thirion, *Gen. Caout.* **23**, 1946, 101.
3. A. Schallamach, *Proc. Phys. Soc.* **B65**, 1952, 657.
4. A. Schallamach, *Wear* **1**, 1958, 384.
5. K.A. Grosch, *Rubber Chem. and Tech. Rubber Rev.*, **69**, 1996, 495.
6. M.L.Williams, R.F. Landel, and J.D. Ferry, *J. Am. Chem. Soc.*, **77**, 1955, 3701.
7. See literature on visco-elasticity, for example: J.D. Ferry, The Visco-elastic Properties of Polymers, 3rd Ed, Wiley, New York, 1980.
8. L. Mullins, *Trans IRI*, **35**, 1959, 213.
9. Th.L. Smith, *J. Pol. Sci.* **32**, 1958, 99.
10. K.A. Grosch, *Proc. Roy. Soc.*, **A 274**, 1963, 21.
11. K.A. Grosch, PhD thesis. Sliding Friction and Abrasion of Rubbers, University of London, 1963.
12. F.A. Greenwood and A.D. Tabor, *Proc. Phys. Soc.*, **71**, 1958, 989.
13. F.A. Greenwood and J.B.P Williams, *Proc. Roy. Soc.*, **A 295**, 1966, 300.
14. M. Klüppel, A. Müller, A. Le Gal, and G. Heinrich, ACS, *Rubber Division Meeting San Francisco*, Spring 2003.

15. K.A. Grosch, The speed and temperature dependence of rubber friction and its bearing on the skid resistance of tires, *The Physics of Tire Traction, Theory and Experiment*. D.L. Hayes and A.L. Browne (eds.), Plenum Press, New York/London, 1974, 143.

16. A.D. Roberts, Lubrication studies of smooth rubber contacts, *The Physics of Tire Traction, Theory and Experiment*. D.L. Hayes and A.L. Browne (eds.), Plenum Press, New York/London, 1974, 143.

17. A.G.M. Mitchell, Viscosity and lubrication, *The Mechanical Properties of Fluids*. Brackie & Sons, London, 1944.

18. T. Kirk and T.F. Archard, *Proc. Roy. Soc. Lond.*, **261**, 1966, 532.

19. K.A. Grosch, *Kautschuk und Gummi*, Kunststoffe, **49**(6), 1996, 132.

20. G. Heinrich, ACS Rubber Division Spring Meeting, Montreal, 1996.

21. B. Persson, *Surf. Sci.*, **401**, 1998, 445.

22. M. Klüppel and G. Heinrich, ACS Rubber Division Meeting, Spring 1999, paper 43.

23. R.B. Yeager, Tire hydroplaning: testing, analysis and design. *The Physics of Tire Traction, Theory and Experiment*. D.L. Hayes and A.L. Browne (eds.), Plenum Press, New York/London, 1974, 143.

24. R.N.J. Saal, *J. Soc. Chem. Ind.*, **55**, 1936, 3.

25. A.L. Browne, *Tire Sci. Technol.* **1**, 1977.

26. W.B. Horne and U.T. Joyner, *SAE report*, 870 C, 1965.

27. H. Barthelt, *Automobil Technische Zeitung*, **10**, 1973, 368.

28. A.D. Roberts and J.D. Lane, *J. Phys. D*, **16**, 1983, 275.

29. M. Heinz and K.A. Grosch, ACS Spring Meeting, St Antonio, 2005.

30. H.S. Carslaw and J.C. Jaeger, *Conduction of Heat in Solids*, Oxford University Press, Oxford, 1959, p. 279.

31. A.D. Roberts, *J. Nat. Rubber Res.* **2**(4), 1987, 255.

32. A. Schallamach and D. Turner, *Wear*, **3**, 1960, 1.

33. A. Schallamach and K.A. Grosch, The Mechanics of Pneumatic Tires. S.K. Clark (ed.), The US Department of Transportation, National Highway Safety Administration Washington DV 20950, Chapter 6, p. 408.

34. D.L. Nordeen and A.D. Cortese, *Trans. SAE*, **72**, 1964, 325.

35. K.A. Grosch, *Rubber Reviews, Rubber Chem. Technol*, **69**, 1996, 496.

36. K.A. Grosch, *Kautschuk, Gummi, Kunststoffe*, **6**m, 1996, 432.

37. K.A. Grosch and A. Schallamach, *Kautschuk, Gummi und Kunststoffe*, **22**, 1969, 288.

38. R.S. Rivlin and A.G. Thomas, *J. Polym. Sci.*, **10**, 1953, 291.

39. A.G. Thomas, *J Appl. Sci.*, **3**, 1960, 168.

40. G.J. Lake and P.B. Lindley, *Rubber J.*, **146(10)**, 1964, 10.

41. P.B. Lindley and A.G. Thomas, 4th Rubber Technology Conference, London, 1962.

42. A. Schallamach, *Wear*, **13**, 1969, 13.

43. K.A. Grosch and A. Schallamach, *Trans IRI*, **41**, 1965 T 80 and *Rubber Chem. Technol.*, **39**, 1966, 287.

44. S.P. Timoshenko and J.N. Goodier, *Theory of Elasticity*, 19th ed. McGraw-Hill International Book Co., New York, 1983.

45. A. Schallamach, *Trans. Inst. Rub. Ind.*, **28**, 1952, p. 256.

46. A. Veith, *Polym. Test.* **7**, 1987 177.

47. D.H. Champ, E. Southern, and A.G. Thomas, *Advances in Polymer Friction and Wear*, Lieng Huang Lee (ed.), Plenum, New York/London, 1974, p. 134.

48. A.N. Gent and T.R. Pulford, *J. Appl. Pol. Sci.*, **28**, 1983, 943.

49. A. Schallamach, *Appl. Pol. Sci.*, **12**, 1968, 281.

50. G.J. Brodskii, Sakhuoskii, M.M. Reznikovskii, and V.F. Estratov, *Soviet Rubber Technol.*, **19**, 1960, B. 22.

51. K.A. Grosch, Kautschuk, *Gummi und Kunststoffe*, **49**(6), 1998, 432.

52. K.A. Grosch, Conference Proc. Tyretech, London, 1998, 14.6T1.

53. K.A. Grosch and M. Heinz, *Proc. IRC 2000*, Helsinki, 2000, paper 48.

54. K.A. Grosch, H. Moneypenny, and I.H. Wallace, IRC 2001 Birmingham, IOM Communications, 2001, 307.

55. K.A. Grosch, ACS Rubber Div Meeting, Fall 2001, Cleveland, OH, paper 48, pp. 33,012.

56. K.A. Grosch and A. Schallamach, *Wear*, **4**, 1961, 356.

57. D.O. Stalnaker and J.L.Turner, *Tire Sci. Technol.*, **30**(2), 2002, p. 100.

27 Improving Adhesion of Rubber

José Miguel Martín-Martínez

CONTENTS

27.1 GENERALITIES ON ADHESION OF RUBBER

The research carried out on adhesion of rubber at the Adhesion and Adhesives Laboratory of the University of Alicante (Spain) had been mainly directed to improve the upper to sole adhesion in shoe industry. Alicante region had been one of the greatest shoe manufacturers in the world and although the competitiveness of Asia and Brazil produced a tremendous decline in production, currently the local shoe industry is interested in manufacturing high added value shoes, one of the main technological efforts being directed to the complete removal of organic solvents during production. Although this aim has been commercially reached for adhesives, as several waterborne and hot-melt formulations are currently in the market, the most suitable surface treatments for rubber are still based on the use of organic solvents (mainly halogenation). On the other hand, due to the legislative pressure of the European Commission, in short the use of organic solvents will be strictly forbidden and environment-friendly surface treatments for rubbers have to be developed. In this chapter, the main efforts made during the last 15 years in the Adhesion and Adhesives Laboratory of the University of Alicante will be summarized, paying particular attention to alternative surface treatments to halogenation of rubber soles.

Adhesion of rubber is limited because of its inherent nonpolar nature and the presence of additives in formulation (processing oils, moulding agents, antiozonant waxes, vulcanization aids). Although, unvulcanized rubbers are somewhat less difficult to bond, most of the rubbers used in industry are vulcanized rubbers. To improve their adhesion, a surface treatment is always necessary.

Most of the sole materials in shoe manufacturing are made of nonvulcanized thermoplastic (TR) and sulfur-vulcanized styrene–butadiene rubbers (SBRs), and adhesion is a crucial industrial demand, as the stresses supported by shoes during use are by far more exigent than in many other rubber goods. Several strategies can be used to improve the adhesion between the upper, the adhesive, and the sole, such as surface modification of the upper and/or rubber sole (by applying an adequate surface treatments and/or primers), modification of the adhesive formulation, and modification of the design of the joint by using fracture mechanics criteria.

Styrene–butadiene–styrene (SBS) block copolymers are adequate raw materials to produce thermoplastic rubbers (TRs). SBS contains butadiene—soft and elastic—and styrene—hard and tough—domains. Because the styrene domains act as cross-links, vulcanization is not necessary to provide dimensional stability. TRs generally contain polystyrene (to impart hardness), plasticizers, fillers, and antioxidants; processing oils can also be added. Due to their nature, TR soles show low surface energy, and to reach proper adhesion a surface modification is always needed.

Although natural rubber (*crepe*) and nitrile rubbers are used, sulfur-vulcanized SBRs are the most common materials for soles. The vulcanization system contains sulfur and activators (*N*-cyclohexyl-2-benzothiazole sulphenamide, dibenzothiazyl disulfide, hexamethylene tetramine, and zinc oxide), and fillers (silica and/or carbon black, calcium carbonate) are added to control hardness and abrasion resistance. Antioxidants (zinc stearate, phenolic antioxidant) and antiozonants (microcrystalline paraffin wax) are also necessary to avoid degradation processes and early aging. In general, vulcanization is carried out in a mould within a hot plate press at about 180°C, where some chemical reactions affecting adhesion are produced. Because of the thickness of the rubber sole, the part in contact with the mould is substantially hotter than the bulk, producing over-vulcanization on the surface. Furthermore, the zinc oxide and the stearic acid react to produce zinc stearate during vulcanization, which is combined with hexamethylene tetramine to form an unstable complex. After vulcanization, contact with moisture or many solvents apparently causes the breakdown of this complex with the appearance of the zinc stearate that migrates to the SBR sole[1] which acts as an antiadherend moiety.[2] On the other hand, during cooling after vulcanization, the paraffin wax migrates to the SBR surface also giving adhesion problems.[3–6] Therefore, the causes of poor adhesion in SBR soles are quite diverse. Over-vulcanized layers are too stiff to support peel stresses. Presence of silicone release agents may reduce the bond strength of SBR sole joints.[1] The presence of antiadherend moieties on the surfaces, its low surface energy, and the migration of low-molecular moieties (antiozonant wax, processing oils) to the interface once the adhesive joint is produced are other causes of lack of adhesion in SBR.

27.2   SURFACE TREATMENTS TO IMPROVE ADHESION OF RUBBERS

The application of surface treatments to rubbers should produce improved wettability, creation of polar moieties able to react with the adhesive, cracks and heterogeneities should be formed to facilitate the mechanical interlocking with the adhesive, and an efficient removal of antiadherend moieties (zinc stearate, paraffin wax, and processing oils) have to be reached. Several types of surface preparation involving solvent wiping, mechanical and chemical treatments, and primers have been proposed to improve the adhesion of vulcanized SBR soles. However, chlorination with solutions of trichloroisocyanuric acid (TCI) in different solvents is by far the most common surface preparation for rubbers.

Solvent wiping. Rubbers tend to swell by application of solvents and the mechanical interlocking of the adhesive is favored. Although chlorinated hydrocarbon solvents are the most effective, they are toxic and cannot be used; toluene and ketones are currently the most common solvents. The treatment with solvents is effective in the removal of processing oils and plasticizers in vulcanized rubbers, but zinc stearate is not completely removed and antiozonant wax gradually migrates to the rubber/polyurethane adhesive interface.[7] Table 27.1 shows the moderate increase in adhesion produced in SBR by MEK wiping.

Roughening. Roughening of rubbers removes the overvulcanized layer and surface contaminants (oils, moulding release agents, antiadherend moieties, etc.) and creates roughness. The removal

TABLE 27.1

Peel Strength Values of Vulcanized Styrene–Butadiene Rubber (SBR) Rubber/Polyurethane Adhesive/Leather Joints

Surface Preparation	T-Peel Strength Value (kN/m)
As-received	0.1
MEK wiping	3.0
Roughening	4.2

Source: Pastor-Blas, M.M., Sánchez-Adsuar, M.S., and Martín-Martínez, J.M., *J. Adhes.*, 50, 191, 1995; Romero-Sánchez, M.D., Pastor-Blas, M.M., and Martín-Martínez, J.M., *Int. J. Adhes. Adhes.*, 21, 325, 2001.

of weak boundary layers and the creation of surface heterogeneities produces improved adhesion of rubber (Table 27.1), a greater value than that produced by solvent wiping is obtained.[2,8]

Halogenation and cyclization (treatment with sulfuric acid) are the most common surface treatments for rubbers.

27.2.1 CYCLIZATION

Cyclization is generally very effective in improving the adhesion of TR[9] and SBR[10] to polyurethane adhesives. Rubbers treated with concentrated sulfuric acid yield a cyclized layer on the surface. This layer is quite brittle, and when flexed develops microcracks, which are believed to help in subsequent bonding by favoring the mechanical interlocking of the adhesive with the rubber.

The treatment with sulfuric acid produces a noticeable decrease in contact angle (i.e., improved wettability) due to the removal of zinc stearate and the formation of polar moieties on the rubber, mainly the creation of highly conjugated C=C bonds and the sulfonation of the butadiene units (Figure 27.1), i.e., the hydrogen of C—H bond is removed and replaced by a SO_3 molecule, which is then hydrogenated to form a sulfonic acid at the site of attachment. The treatment is not restricted to the surface but also produces a bulk modification of the rubber.

An immersion time less than 1 min, the neutralization with ammonium hydroxide (it extracts the hydrogen from the sulfonic acid and leaves stabilized $SO_3^+NH_4^-$ ion pair), and the high concentration of the sulfuric acid (95 wt%) are essential to produce adequate effectiveness of the treatment.[9,10] H_2SO_4 treatment increases the T-peel strength of treated TR or SBR–polyurethane adhesive joints (Figure 27.2).

On the other hand, if the rubber contains paraffin wax, the surface treatment with sulfuric acid promotes its migration to the surface creating a weak boundary layer able to produce poor adhesion.[11] Solvent wiping with petroleum ether after surface treatment removes this antiadherent moiety but a progressive migration from the bulk to the rubber surface with time occurs.

27.2.2 HALOGENATION

The use of chlorination as surface treatment to improve the adhesion of rubbers to polyurethane adhesives was proposed in 1971. The employ of chlorination in the industry is due to its high effectiveness in improving the adhesion of several types and formulations of rubbers, it is cheap and easy to apply. Furthermore, chlorination makes the rubber surface compatible with many adhesives

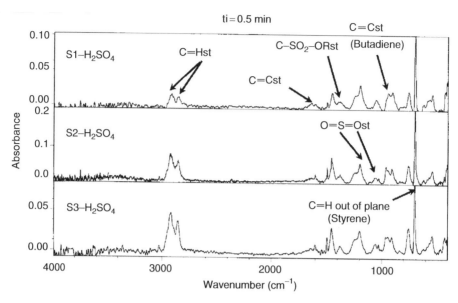

FIGURE 27.1 ATR-IR spectra of sulfuric acid-treated thermoplastic rubbers (TRs) with different styrene content. Immersion time = 0.5 min. (From Cepeda-Jiménez, C.M., Pastor-Blas, M.M., Ferrándiz-Gómez, T.P., and Martín-Martínez, J.M., *Int. J. Adhes. Adhes.*, 21, 161, 2001.)

(epoxy, polyurethane, and acrylic); removes contaminants and antiadherent moieties from the surface avoiding their migration to the rubber–adhesive interface; improved durability and aging resistance are imparted to the joints; and the treated surfaces remain reactive with the adhesive for at least three months after surface preparation.[12] Chlorination of rubber soles also shows some limitations: (i) Poor stability of the solutions (chlorine evolution with time is produced); (ii) organic solvents are necessary for being effective; (iii) treatment is not restricted to the outermost surface, but penetrates into the rubber decreasing its mechanical properties. The extent of penetration of the

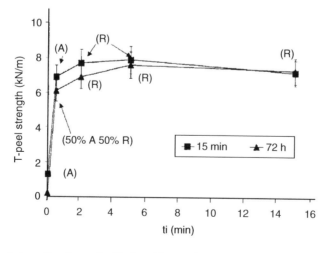

FIGURE 27.2 T-peel strength values of sulfuric acid-treated styrene–butadiene rubber (SBR)/polyurethane adhesive joints as a function of the immersion time in sulfuric acid. A = adhesion failure; R = cohesion failure in the rubber. (From Cepeda-Jiménez, C.M., Pastor-Blas, M.M., Ferrándiz-Gómez, T.P., and Martín-Martínez, J.M., *J. Adhes.*, 73, 135, 2000.)

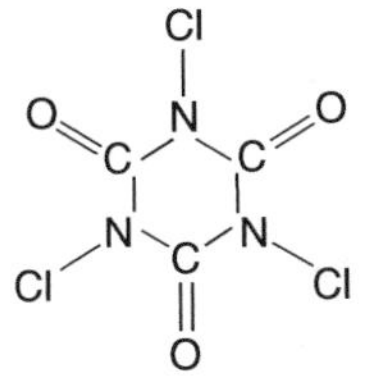

FIGURE 27.3 Chemical structure of trichloroisocyanuric acid (TCI).

chlorinating agent into the rubber depends on its concentration and rubber formulation. Oldfield and Symes[13] show that a depth between 1 and 5 μm is reached by treatment of different rubbers with chlorinating agent.

27.2.2.1 Halogenation with Organic Solvent Solutions of Trichloroisocyanuric Acid

Several chlorinating agents have been used in shoe industry for rubber sole bonding (aqueous solutions of chlorine, acidified sodium hypochlorite aqueous solutions, etc.), but the most commonly used chlorinating agent is an organic solvent (ketone or ester) solution of TCI—1,3,5-trichloro-1,3,5-triazin-2,4,6-trione—(Figure 27.3). The organic solvent in TCI solutions determines the degree of rubber wetting and furthermore, the actual chlorinating species are produced by the reaction of TCI with the organic solvent. Thus, the chlorinating species in TCI/MEK solutions are α-chloro ketones, whereas acid chlorides are formed in TCI–ethyl acetate (EA) solutions.[14]

The improved adhesion of rubber soles treated with TCI solutions are due to the contribution of several mechanisms of adhesion:

1. *Mechanical adhesion.* Cracks and pits are produced on the treated rubber surface which favor the mechanical interlocking with the adhesive. Furthermore, unreacted solid prismatic TCI crystals on the treated rubber surface can be dissolved by the organic solvent into the adhesive, favoring the reaction with the adhesive.
2. *Chemical adhesion.* Chlorination is produced in the *trans* C=C bonds of butadiene creating chlorinated hydrocarbon and other C—Cl moieties.[12,14] According to Table 27.2, C—Cl

TABLE 27.2
Atomic Percentages (at%) on As-Received and Surface-Chlorinated Styrene–Butadiene Rubber (SBR) Rubber. XPS Experiments

Element	As-Received (at%)	3 wt% TCI/MEK (at%)
C	94.9	78.0
O	3.8	10.6
Si	1.1	2.4
Zn	0.2	—
N	—	3.6
Cl	—	0.6
S	—	4.8
O/C	0.04	0.14

Source: Pastor-Blas, M.M., Ferrándiz-Gómez, T.P., and Martín-Martínez, J.M., *J. Adhes. Sci. Technol.*, 14, 561, 2000.

moieties are produced on the rubber surface, and the increase in O/C ratio and the creation of carboxylic acid moieties show that oxidation is produced by treatment with TCI solutions. Furthermore, zinc stearate and paraffin wax are removed, and isocyanuric acid formation is produced (reaction by-product of TCI).

3. *Thermodynamical adhesion.* Treatment with TCI solution improves the wettability on SBR, due to the creation of surface chemistry and roughness.

The effectiveness of the chlorination of rubbers with TCI solutions strongly depends on several experimental variables.

(a) *Application procedure of the TCI solution.* TCI solution can be applied on rubber surface by spraying, brushing, or immersion. In general, brushing provided better performance than the other application procedures (peel strength values for 2 wt% TCI solution in MEK-treated rubber/polyurethane adhesive joint of 3.6 kN/m when chlorinating agent is applied by immersion vs 12.5 kN/m when applied by brushing).[15] This is due to deeper penetration of the chlorinating agent into rubber by brushing. Furthermore, lower water contact angle values are obtained (91° by immersion vs 73° by brushing), greater removal of antiadherent moieties is reached by brushing (Figure 27.4), and higher number and deeper cracks are produced on the brushed TCI solution on rubber surface.

(b) *Time between the application of the TCI solution and the adhesive (time after halogenation treatment) on rubber surface.* Chemical reaction of chlorinating species with SBR needs some time. At least, the solvent from the chlorinating agent must be completely removed from the rubber surface, so a minimum reaction time of 10 min is necessary, although the influence of the time after TCI treatment on the rubber surface properties was recently monitored up to one year.[16] It has been shown that the increase in the time after TCI treatment produces some aging of the rubber surface.

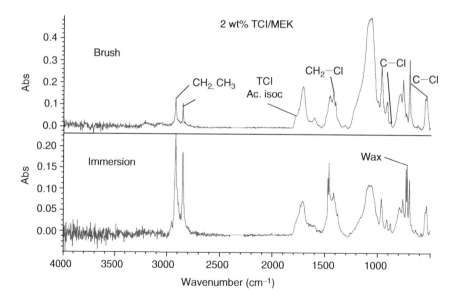

FIGURE 27.4 ATR-IR spectra of 2 wt% trichloroisocyanuric acid (TCI) solution in MEK-treated styrene–butadiene rubber (SBR). Application of the chlorinating agent by immersion or by brush. (From Romero-Sánchez, M.D., Pastor-Blas, M.M., and Martín-Martínez, J.M., *J. Adhes. Sci. Technol.*, 15, 1601, 2001.)

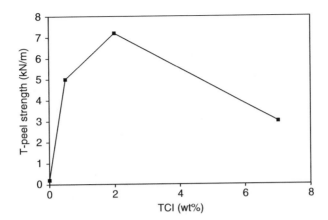

FIGURE 27.5 T-peel strength values of chlorinated styrene–butadiene rubber (SBR)/polyurethane adhesive joints as a function of the trichloroisocyanuric acid (TCI) concentration in the solution. (From Pastor-Blas, M.M., Sánchez-Adsuar, M.S., and Martín-Martínez, J.M., *J. Adhes.*, 50, 191, 1995.)

(c) *TCI concentration in the solution*. It has been shown[17] that the peel strength of some surface-chlorinated SBR or natural rubber/epoxy adhesive joint increases by increasing the TCI concentration in the solution up to a given value, however, a noticeable decrease is obtained for high TCI concentrations[4] (Figure 27.5). This has been ascribed to the formation of weak boundary layers due to mechanical and chemical degradation of the outermost rubber surface to which the failure of the joint is directed during peel tests.

(d) *Solvent in the TCI solution*. Solvent reacts with TCI to produce the actual chlorinating species which may differ in nature and halogenating ability depending on the solvent nature. Ketones and esters are the most common solvents in TCI solutions used in SBR sole bonding. MEK solutions of TCI are more effective than EA solutions in the removal of zinc stearate.[18] However, EA solutions of TCI are generally used in the commercial chlorinating agents because of its higher stability against chlorine evolution. Different esters having different evaporation rate have also been used as solvents for TCI. All these ester solutions provide good level of adhesion in SBR, although EA imparts the highest peel strength value (Figure 27.6).

(e) *Roughening before chlorination with TCI solution*. The application of TCI solution on a previously roughened SBR, enhances the effects due to chlorination because a higher extent of chemical reaction is produced.[19]

(f) *Solvent wiping of rubber before application of TCI solution*. Solvent wiping removes antiadherent moieties from the SBR surface and therefore higher degree of reaction of TCI solution can be expected. In fact, a greater increase in wettability is obtained because the higher concentration of chlorinated and oxidized moieties on the SBR surface.[20] As a consequence higher peel strength values are obtained.

27.2.2.2 Water-Based Halogenating Agents

Limitations in the VOCs emissions will certainly push the shoe industry to use alternative bonding technologies to the current solvent-based surface preparations. Several limitations are found in the removal or substitution of solvents in shoe bonding, such as the use of new equipments and machinery, the modification in the procedure to produce bonding, costs increase, and difficulty in bonding some materials without the use of solvents.

Several water-based chlorination treatments have been proposed,[21–23] as their use do not require noticeable changes with respect to the current technology in the shoe industry. Bleach has been

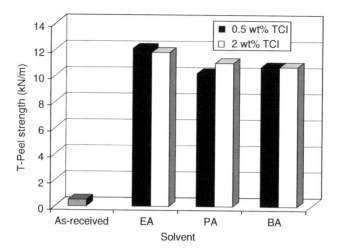

FIGURE 27.6 T-peel strength values of chlorinated styrene–butadiene rubber (SBR)/polyurethane adhesive joints as a function of the ester used to prepare the 2 wt% trichloroisocyanuric acid (TCI) solution. EA = ethyl acetate; PA = propyl acetate; BA = butyl acetate. (From Ramero-Sánchez, M.D., Pastor-Blas, M.M., and Martin-Martinez, J.M., *Int. J. Adhes. Adhes.*, 21, 325, 2001.)

successfully used to chlorinate rubber soles in the past. Recently *acidified sodium hypochlorite solutions* containing 1-octyl-2-pyrrolidone as wetting agent have been effectively used for the treatment of TR soles.[22] An increase in wettability and creation of chlorine moieties in TR are obtained but the treatment in an ultrasonic bath is mandatory to obtain adequate adhesion to waterborne polyurethane adhesive. Furthermore, neutralization with ammonium hydroxide and extensive washing of the treated rubber sole with water are necessary, followed by drying; in addition, the stability of the chlorinating solutions is relatively poor. The treatment is restricted to about 1 μm depth of the TR.

Aqueous solutions of sodium dichloroisocyanurate (DCI) have been recently used to increase the adhesion of SBR and TR soles.[21] The chemical structure of DCI (Figure 27.7) is somewhat similar to that of TCI. The surface treatment with aqueous DCI solutions modifies the surface chemistry of TR and SBR, creating C–Cl moieties and removing the zinc stearate from the SBR surface. The use of a low DCI concentration in water is less effective in modifying the TR, but is sufficient to obtain good peel strength values for SBR and waterborne polyurethane joints[21] (Figure 27.8). On the other hand, heterogeneities and cracks are created on the rubber surface (mainly on the SBR surface) that may contribute to enhance the mechanical interlocking with the adhesive.

Successful treatment of SBR and TR has also been obtained by immersion for 1 min in aqueous *N*-chloro-*p*-toluensulphonamide solutions (obtained by acidifying chloramine T solutions).[24,25]

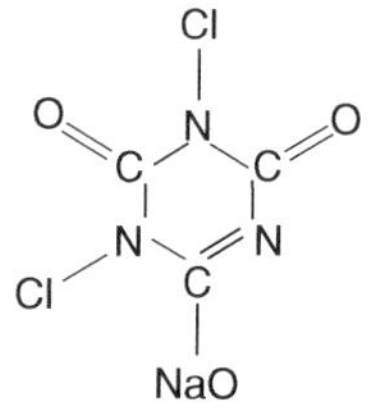

FIGURE 27.7 Chemical formula of sodium dichloroisocyanurate (DCI).

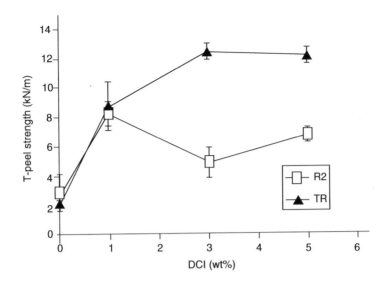

FIGURE 27.8 T-peel strength values of dichloroisocyanurate (DCI)-treated styrene–butadiene rubber (SBR) (R2 rubber) and thermoplastic rubber (TR)/waterborne polyurethane adhesive/canvas joints. Influence of the dichloroisocyanurate (DCI) concentration. (From Cepeda-Jiménez, C.M., Pastor-Blas, M.M., Martín-Martínez, J.M., and Gottschalk, P., *J. Adhes. Sci. Technol.*, 16, 257, 2002.)

Acidification of chloramine T with sulfuric acid produces the formation of dichloramine T (DCT) and hypochlorous acid (HClO), species which react with C=C bonds of the butadiene units. The effectiveness of the treatment is ascribed to the introduction of chlorine and oxygen moieties on the rubber surface. A decrease in the pH of the chloramine T aqueous solutions produced more extended surface modifications and improved adhesion properties in the joints produced with waterborne polyurethane adhesive (Figure 27.9). The adhesive strength obtained is slightly lower than that obtained for the rubber treated with 3 wt% TCI/MEK, and its increases as the pH of the chloramine T solution decreases (Figure 27.9). A cohesive failure in the rubber is generally obtained.

Recently, cryo-ground rubber tire (GRT) has been successfully treated by immersion for 20 min at room temperature in TCI solution in methanol of different concentrations (1–5 wt%).[26] The wettability and the extent of chlorination of GRT increases with increased concentrations of TCI, and a small extent of surface oxidation occurs. Optimum performance was obtained for 3 wt% TCI–methanol solutions.

27.2.3 RADIATION-BASED TREATMENTS

Several environment-friendly surface preparation for the treatment of rubber soles with radiations have been recently studied.[27–33] These treatments are clean (no chemicals or reactions by-products are produced) and fast, and furthermore online bonding at shoe factory can be produced, so the future trend in surface modification of substrates in shoe industry will be likely directed to the industrial application of those treatments. Corona discharge, low-pressure RF gas plasma, and ultraviolet (UV) treatments have been successfully used at laboratory scale to improve the adhesion of several sole materials in shoe industry. Recently, surface modification of SBR and TR by UV radiation has been industrially demonstrated in shoe industry.[27]

Corona discharge has been successfully used to improve the adhesion of TRs to polyurethane adhesive.[28,29] The treatment with corona discharge improves the wettability of TR due to the surface

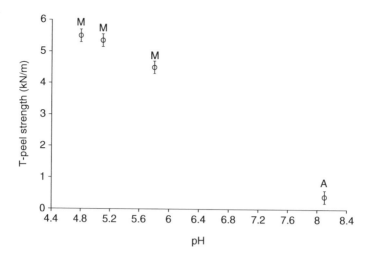

FIGURE 27.9 T-peel strength values of styrene–butadiene rubber (SBS) treated with chloramine T aqueous solutions with different pH/waterborne polyurethane adhesive/roughened leather joints, as a function of the pH value of the chloramine T aqueous solutions. A: adhesion failure to the rubber, M: cohesive failure in the rubber. (From Navarro-Bañón, M.V., Pastor-Blas, M.M., and Martín-Martínez, J.M., *Proceedings of the 27th Adhesion Society*, Wilmington, NC.)

formation of C—O, C=O, and COO$^-$ groups. Besides, surface cleaning can be produced but roughness is not created, and the adhesive strength obtained is only moderate.

Low-pressure RF gas plasma is effective to enhance the adhesion of SBR and TR.[30–32] Oxygen, nitrogen, and oxygen–nitrogen mixture plasmas have been successfully used to increase the wettability, create C—O and C=O polar moieties, and partially remove hydrocarbon moieties in SBR and TR, these effects are more marked by treating with oxygen plasma for 1 min.

Treatment of TR with *UV radiation* has been shown to be successful in increasing its adhesion to polyurethane adhesive.[33] A low-pressure mercury vapor lamp (main emission at 254 nm; power = 20 mW/cm^2) has been used. The UV treatment of TR improves the wettability, produces the formation of C—O, C=O, and COO$^-$ moieties, and ablation is also produced. The extended UV treatment produces greater surface modifications, as well as the incorporation of nitrogen moieties at the surface. Peel strength values increase after UV treatment of TR, in a greater extent by increasing the treatment time.

27.3 CARBOXYLIC ACIDS PRIMERS TO IMPROVE THE ADHESION OF RUBBER

To improve the adhesion of rubber soles containing noticeable amounts of zinc stearate on their surfaces, several primers have been suggested, such as isocyanate wiping[34] and acid solutions. Application to the surface of an acid stronger than stearic acid (such as hydrochloric acid) improves the bond strength on some rubbers, due to stearic acid formation which is less deleterious to joint strength. Lactic acid is also strong enough to remove metal soaps contamination (i.e., zinc stearate) by brushing the surface of some rubbers.

Solutions of different carboxylic acids (fumaric acid [FA], maleic acid, acrylic acid, succinic acid, and malonic acid) in ethanol have been effectively used as primers to increase the adhesion of synthetic vulcanized SBRs.[35,36] The increase in the adhesion properties of SBR treated with carboxylic acid is attributed to the elimination of zinc stearate moieties and the deposition of acid on the rubber which migrates into the solvent-borne polyurethane adhesive layer once the adhesive joint is formed. The nature of the carboxylic acid determines the rate of diffusion into the adhesive and the extent of rubber–adhesive interfacial interaction.

27.3.1 COMBINATION OF TCI AND FUMARIC ACID SOLUTIONS FOR TREATMENT OF RUBBER

The treatment of SBR with fumaric acid solutions avoids the migration of antiadherent moieties to the surface and the treatment with TCI solutions is effective to enhance the adhesion of several rubbers to polyurethane adhesive. Therefore, the combined use of mixtures of TCI and FA solutions should be more effective in improving the adhesion of difficult to bond SBR. The wettability of SBR is improved by treatment with 3 wt% TCI/EA followed by treatment with 0.5 wt% FA/EtOH (3 wt% TCI–0.5 wt% FA), with 0.5 wt% FA/EtOH followed by treatment with 3 wt% TCI/EA (0.5 wt% FA–3 wt% TCI), or with TCI + FA mixtures.[37] However, the extent of the surface modifications produced and the adhesive strength of adhesive joints are mainly due to chlorination with TCI/EA.

27.4 MECHANISMS OF ADHESION INVOLVED IN RUBBER BONDING

Considering that most of the upper materials are porous and the adhesives used in shoe bonding are applied as liquids, an acceptable penetration of the adhesive into the upper is expected and then the mechanical adhesion is generally favored. However, because the different rubber materials are nonporous and have in general a relatively low surface energy, both mechanical (e.g., roughening) and chemical (e.g., halogenation) surface preparations are necessary and enhanced chemical adhesion is required. Therefore, the main mechanisms of adhesion involved in shoe bonding are mechanical and chemical. Furthermore, in the bonding of some rubber soles the weak boundary layers produced by surface contaminants and/or by migration of antiadherent moieties to the interface must be removed before joint formation.

The improved adhesion between the surface-chlorinated layer and the polyurethane adhesive has been ascribed to the creation of hydrogen bonds between the urethane group and the chlorine groups on the surface.[38] An alternative recent explanation[39] suggests that the chlorination produces cross-linking of the rubber surface and a diffusion of TCI and chlorinated rubber through the polyurethane is produced. Besides, there is interpenetration between the chlorinated rubber and the polyurethane adhesive at the interface, which is likely responsible for the improvement of adhesion. This interface has a distinctive chemistry, which differs from the bulk rubber and the polyurethane adhesive. Furthermore, the residual unreacted chlorinating agent on the rubber surface seems to react with the polyurethane adhesive producing additional chlorinated moieties at the interface.

ACKNOWLEDGMENTS

Author thanks M.D. Romero-Sánchez and C.M. Cepeda-Jiménez for obtaining most of the experimental results included in this chapter. Author also thanks the Spanish Research Agency (MICYT, Ministerio de Educación y Ciencia) and the Valencian Community Research Agency (Conselleria de Educación y Ciencia, Generalitat Valenciana) for granting different projects in the improvement of adhesion of rubber.

REFERENCES

1. Pettit D. and Carter A.R., February 1964, Adhesion of translucent rubber soling, *SATRA Bulletin*, **11(2)**, 17–21.
2. Blackwell F.B., October 1973, Adhesion of solings, *SATRA Bulletin*, **15(22)**, 423–427.
3. Pettit D. and Carter A.R., 1964, Adhesion of translucent rubber: application of infra-red spectrometry to the problem, SATRA Research Report 165, Kettering.
4. Pastor-Blas M.M., Sánchez-Adsuar M.S., and Martín-Martínez, J.M., 1995, Weak surface boundary layers in styrene–butadiene rubber, *J. Adhes.*, **50**, 191–210.

5. Pastor-Blas M.M. and Martín-Martínez J.M., 1995, Mechanisms of formation of weak boundary layers in styrene–butadiene rubber, in *Proceedings of The International Adhesion Symposium*, H. Mizumachi (Ed.), Gordon and Breach Science Publishers, Melbourne, 215–233.

6. Bernabéu-Gonzálvez A., Pastor-Blas M.M., and Martín-Martínez J.M., 1998, Modified adhesion of rubber materials by surface migration of wax and zinc stearate, in *Proceedings of the World Polymer Congress, 37th International Symposium on Macromolecules MACRO 98*, Gold Coast, Australia, 705.

7. Romero-Sánchez M.D., Pastor-Blas M.M., and Martín-Martínez J.M., 2001, Adhesion improvement of SBR rubber by treatment with trichloroisocyanuric acid solutions in different esters, *Int. J. Adhes. Adhes.*, **21**, 325–337.

8. Romero-Sánchez M.D., Pastor-Blas M.M., and Martín-Martínez J.M., 2002, Improved peel strength in vulcanized SBR rubber roughened before chlorination with trichloroisocyanuric acid, *J. Adhes.*, **78**, 15–38.

9. Cepeda-Jiménez C.M., Pastor-Blas M.M., Ferrándiz-Gómez T.P., and Martín-Martínez J.M., 2001, Influence of the styrene content of thermoplastic styrene–butadiene rubbers in the effectiveness of the treatment with sulfuric acid, *Int. J. Adhes. Adhes.*, **21**, 161–172.

10. Cepeda-Jiménez C.M., Pastor-Blas M.M., Ferrándiz-Gómez T.P., and Martín-Martínez J.M., 2000, Surface characterization of vulcanized rubber treated with sulfuric acid and its adhesion to polyurethane adhesive, *J. Adhes.*, **73**, 135–160.

11. Cepeda-Jiménez C.M., Pastor-Blas M.M., and Martín-Martínez J.M., 2001, Weak boundary layer on vulcanized styrene–butadiene rubber treated with sulfuric acid, *J. Adhes. Sci. Technol.*, **15(11)**, 1323–1350.

12. Fernández-García J.C., Orgilés-Barceló, and A.C., Martín-Martínez J.M., 1991, Halogenation of styrene–butadiene rubber to improve its adhesion to polyurethanes, *J. Adhes. Sci Technol.*, 5, 1065–1080.

13. Oldfield D. and Symes T.E.F., 1983, Surface modification of elastomers for bonding, *J. Adhes.*, **16**, 77–96.

14. Pastor-Blas M.M., Ferrándiz-Gómez T.P., and Martín-Martínez J.M., 2000, Chlorination of vulcanized styrene–butadiene rubber using solutions of trichloroisocyanuric acid in different solvents, *J. Adhes. Sci. Technol.*, **14**, 561–581.

15. Romero-Sánchez M.D., Pastor-Blas M.M., and Martín-Martínez J.M., 2001, Chlorination of vulcanized SBR rubber by immersion or brushing in TCI solutions, *J. Adhes. Sci. Technol.*, **15**, 1601–1620.

16. Romero-Sánchez M.D., Pastor-Blas M.M., Ferrándiz-Gómez T.P., and Martín-Martínez J.M., 2001, Durability of the halogenation in synthetic rubber, *Int. J. Adhes. Adhes.*, **21**, 101–106.

17. Oldfield D. and Symes T.E.F., 1983, Surface modification of elastomers for bonding, *J. Adhes.*, **16**, 77–96.

18. Romero-Sánchez M.D., Pastor-Blas M.M., and Martín-Martínez J.M., 2003, Improved adhesion between polyurethane and SBR rubber treated with trichloroisocyanuric acid solutions containing different concentrations of chlorine, *Compos Interface*, **10(1)**, 77–94.

19. Romero-Sánchez M.D., Pastor-Blas M.M., and Martín-Martínez J.M., 2002, Improved peel strength in vulcanized SBR rubber roughened before chlorination with trichloroisocyanuric acid, *J. Adhes.*, **78**, 15–38.

20. Vélez-Pagés T., Romero-Sánchez M.D., Pastor-Blas M.M., and Martín-Martínez J.M., 2002, MEK wiping prior to chlorination to improve the adhesion of vulcanized SBR rubber containing paraffin wax, *J. Adhes. Sci. Technol.*, **16**, 1765–1780.

21. Cepeda-Jiménez C.M., Pastor-Blas M.M., Martín-Martínez J.M., and Gottschalk P., 2002, A new water-based chemical treatment based on sodium dichloroisocyanurate (DCI) for rubber soles in footwear industry, *J. Adhes. Sci. Technol.*, **16(3)**, 257–284.

22. Cepeda-Jiménez C.M., Pastor-Blas M.M., Martín-Martínez J.M., and Gottschalk P., 2003, Treatment of thermoplastic rubber with bleach as an alternative halogenation treatment in the footwear industry, *J. Adhes.*, **79(3)**, 207–237.

23. Brewis D.M. and Dahm R.H., 2003, Mechanistic studies of pretreatments for elastomers, in *Proceedings of Swiss Bonding 2003*, Zurich, 69–75.

24. Petravicius A., Rajackas V., and Kabaev M.M., 1980, Study of chemical modification of the surface of sole rubbers with *N*-halosulfamides, *Izvestiya Vysshikh Uchebnykh Zavedenii, Tekhnologiya Legkoi Promyshlennosti*, 76–79.

25. Navarro-Bañón M.V., Pastor-Blas M.M., and Martín-Martínez J.M., 15–18 February 2004, Water based halogenation for rubber materials, in *Proceedings of the 27th Adhesion Society*, Wilmington, NC.

26. Amit Kumar Naskar A.K., De S.K., and Bhowmick A.K., 2001, Surface chlorination of ground rubber tire and its characterization, *Rubber Chem. Technol.*, **74**, 645–661.

27. AS-3000 treatment system for Shoe industry 2002, CELME brochure, Alicante.

28. Romero-Sánchez M.D., Pastor-Blas M.M., Martín-Martínez J.M., Zhdan P.A., and Watts J.M., 2001, Surface modifications in a vulcanized rubber using corona discharge and ultraviolet radiation treatments, *J. Materials Sci.*, **36(24)**, 5789–5799.

29. Romero-Sánchez M.D., Pastor-Blas M.M., and Martín-Martínez J.M., 2003, Treatment of a styrene-butadiene-styrene rubber with corona discharge to improve the adhesion to polyurethane adhesive, *Int. J. Adhes. Adhes*, **23(1)**, 49–57.

30. Pastor-Blas M.M. and Martín-Martínez J.M., 2002, Different surface modifications produced by oxygen plasma and halogenation treatments on a vulcanised rubber, *J. Adhes. Sci. Technol.*, **16(4)**, 409–428.

31. Pastor-Blas M.M., Martín-Martínez J.M., and Dillard J.G., 1998, Surface characterization of synthetic vulcanized rubber treated with oxygen plasma, *Surf. Interf. Analysis*, **26**, 385–399.

32. Ortiz-Magán A.B., Pastor-Blas M.M., Ferrándiz-Gómez T.P., Morant-Zacarés C., and Martín-Martínez J.M., 2001, Surface modifications produced by N_2 and O_2 RF-plasma treatment on a synthetic vulcanised rubber, *Plasmas Polym.*, **6(1,2)**, 81–105.

33. Romero-Sánchez M.D., Pastor-Blas M.M., Martín-Martínez J.M., and Walzak M.J., 2003, UV treatment of synthetic styrene–butadiene–styrene rubber, *J. Adhes. Sci. Technol.*, **17(1)**, 25–46.

34. Carter A.R., August 1968, Isocyanate wipe, *SATRA Bulletin*, **13(8)**, 125–126.

35. Pastor-Sempere N., Fernández-García J.C., Orgilés-Barceló A.C., Torregrosa-Maciá R., and Martín-Martínez J.M., 1995, Fumaric acid as a promoter of adhesion in vulcanized synthetic rubbers, *J. Adhes.*, **50**, 25–42.

36. Pastor-Sempere N., Fernández-García J.C., Orgilés-Barceló A.C., Pastor-Blas M.M., Martín-Martínez J.M., and Dillard J.G., 1998, Surface treatment of styrene–butadiene rubber with carboxylic acid, *in First International Congress on Adhesion Science and Technology*, W.J. van Ooij and H.R. Anderson Jr. (Eds), Utrecht, VSP, 461–494.

37. Romero-Sánchez M.D. and Martín-Martínez J.M., 2003, Treatment of vulcanised styrene–butadiene rubber (SBR) with mixtures of trichloroisocyanuric acid and fumaric acid, *J. Adhes.*, **79**, 1111–1133.

38. Kinloch A.J., 1987, *Adhesion and Adhesives, Science and Technology*. Chapman & Hall, London, pp. 127–128.

39. Pastor-Blas M.M., Martín-Martínez J.M., and Boerio F.J., 2002, Mechanisms of adhesion in surface chlorinated thermoplastic rubber/thermoplastic polyurethane adhesive joints, *Rubber Chem. Technol.*, **75(5)**, 825–837.

28 Rheology of Rubber and Rubber Nanocomposites

Muthukumaraswamy Pannirselvam
and Satinath Bhattacharya

CONTENTS

28.1 INTRODUCTION

Rheology deals with the deformation and flow of any material under the influence of an applied stress. In practical applications, it is related with flow, transport, and handling any simple and complex fluids [1]. It deals with a variety of materials from elastic Hookean solids to viscous Newtonian liquid. In general, rheology is concerned with the deformation of solid materials including metals, plastics, and rubbers, and liquids such as polymer melts, slurries, and polymer solutions.

Any rheometric technique involves the simultaneous assessment of force, and deformation and/or rate as a function of temperature. Through the appropriate rheometrical equations, such basic measurements are converted into quantities of rheological interest, for instance, shear or extensional stress and rate in isothermal condition. The rheometrical equations are established by considering the test geometry and type of flow involved, with respect to several hypotheses dealing with the nature of the fluid and the boundary conditions; the fluid is generally assumed to be homogeneous and incompressible, and ideal boundaries are considered, for instance, no wall slip.

28.2 TYPES OF RUBBER: NATURAL AND SYNTHETIC RUBBER

Rubber can be classified into two types: Natural rubber (NR) and synthetic rubber.

28.2.1 Natural Rubber

NR is one of the most useful products from the fast disappearing rain forests. Indeed a major portion of pharmaceutical organic chemicals owe their origin to the vast diversity of plant life found there [2]. Polyisoprene extracted from *Hevea brazilensis* is called NR. Early rubber scientists were found among the Aztecs and Mayas of South America, who used rubber for shoe soles, and coated fabrics well over 2000 years ago. Maurice Morton's book on rubber technology mentioned that Christopher Columbus was the first European to discover NR, in the early 1490s, when he found natives in Haiti playing ball with an extract from a tree. In 1832, Goodyear purchased the claim of combining sulfur with India rubber [2]. In 1889, John Dunlop in England invented the first commercially successful pneumatic tire, which was at that time used for bicycles [3]. In 1904, Stern in England found that carbon black, blended with rubber, significantly increased a number of its mechanical properties.

28.2.2 Synthetic Rubber

In the early 1900s, research scientists were eagerly seeking rubber materials which could be manufactured artificially. Kuzma [5] notes that the Russians prepared such a rubber, known chemically as polybutadiene. In the 1930s, Germans began commercial production of synthetic rubber called Buna-S (styrene–butadiene copolymer). With the outbreak of the World War II, there was a huge shortage of NR. Styrene–butadiene rubber (SBR) was improved, then manufactured on a large scale, and later called as SBR, which today is a major material source in the rubber industry. Butyl rubber (IIR, isobutylene isoprene copolymer), Hypalon (CSM, chlorosulfonated polyethylene), and Viton (FKM, flouroelastomer), and ethylene–propylene terpolymer rubber (EPDM) are some of other synthetic rubbers. Polyurethane, discovered by Bayer in the 1950s is an important invention in the class of synthetic rubber. A significant addition to the rubber industry is a class of materials called thermoplastic elastomers (TPEs) which is gaining increasing prominence in the marketplace. They behave like rubber at room temperature, but soften like plastic when heated. When cooled down, TPEs return to their rubbery state. Table 28.1 illustrates some trade names of natural and synthetic rubber.

NR has endured all the inventions of synthetic rubbers exceptionally well and still represents nearly one-third of all rubber in the marketplace [2]. The new awareness of our environment gives NR the added advantage of being seen as a renewable source.

TABLE 28.1

Trade Names of Natural and Synthetic Rubber

Symbol	Generic Name	Some Trade Names
SBR	Styrene–butadiene rubber	Cariflex, Copo
SSBR	Solution	Duradene, Soloflex
CR	Chloroprene rubber	Neoprene, Denka
NBR	Nitrile	Nipol, Krynac, Paracril
EPDM	Ethylene propylene diene rubber	Buna EP, Royalene, Keltan
CIIR, BIIR	Butyl	Exxon butyl, Polysar Butyl
MQ	Silicone elastomers	Elastosil, Silopren
BR	Polybutadiene rubber	Budene, Taktene, Diene, Intene
ACM	Polyacrylate	HyTemp, Europrene
ECO	Epichlorohydrin ethylene oxide	Hydrin C
AU	Urethane (Ester)	Urepan, Millathane, Vibrathane
EU	Urethane (Ether)	Millathane, Adiprene

Source: Andrew, C., in *Introduction to Rubber Technology*, Knovel e-book publishers, 1999, 25–26.

28.3 THEORY BEHIND RHEOLOGICAL MEASUREMENTS

The flow behavior of a material in response to an applied stress (stress = force/area) varies depending on the composition of the material. A polymer solution or a melt consists of large molecular compounds sometimes referred to as macromolecules. These macromolecules deform under shear force, but will try to return to their original form when the stress is removed [3]. This mechanism is called viscoelastic property. Most of the polymer solutions and polymer melts exhibit viscoelastic property. Complex fluids such as mining wastes, pastes, slurries, and mineral suspensions are usually nonhomogeneous on the microscopic or colloidal state. Their flow behavior is altered with the application of shear as the "structure" of these materials is affected when shear induces alteration of the particle orientation, relative position, and particle fluid interactions. The flow behavior of a material during processing may vary significantly because the consistency and composition of the material could be drastically altered due to certain unit operations process like aeration, cooling, compounding, crystallization, heating, homogenization, and mixing.

28.4 RHEOLOGY MEASUREMENTS ON RUBBER AND RUBBER NANOCOMPOSITES

Rubber possesses many properties, that allow it to process well on rubber machinery, and that the product performs according to the different applications. The important parameters which come under the heading "Rheology" includes melting point for purity of accelerators, flow behavior, scorch time, cross-linking in rubber matrix, particle size, structure of fillers, specific gravity, Mooney viscosity, and much more. Sections 28.4.1 through 28.4.6 briefly discuss various instruments used to measure rheological properties of rubber and rubber nanocomposites.

28.4.1 MOONEY VISCOMETER

Melvin Mooney devised an instrument to measure the stiffness of uncured compounds, also known as the compound's viscosity. The unit of measurement is expressed in Mooney units. Figure 28.1 shows an instrument set up for measuring viscosity, the Mooney viscometer, in which a knurled knob (rotor) rotates (at 2 revolutions per minute) in a closed heated cavity filled with rubber [2].

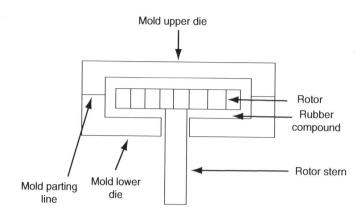

FIGURE 28.1 Principle of Mooney viscometer.

A shearing action grows between the compound and the rotor, and the resulting torque is measured in arbitrary units called Mooney units, which directly relate to torque. Normally, a preheat period is given to the elastomer following which the disk starts to rotate. An initial high viscosity is recorded which decreases to a minimum value. If the viscosity is more, then the Mooney unit (number) is more and viceversa.

28.4.2 MOONEY SCORCH

Mooney viscometer is also used to measure the time it takes, from initial exposure of the compound to a particular temperature, to the time of onset of cure at that temperature [2]. This is known as the scorch time. Scorch time is an important parameter to the rubber processor, as a short time may lead to problems of premature vulcanization. As the test is taken past the onset of cure, the rotor tears the cured rubber, and therefore this device cannot be used to investigate rheological properties after the scorch time.

28.4.3 OSCILLATING DISC CURE METER

Oscillating disc cure meter or rheometer (ODR) solves the challenge of not being able to measure after the scorch time, by converting the rotor from a rotating mode to an oscillating one [2]. The magnitude of the oscillation is measured in degrees of arc (1° and 3° are the most common), and the rate of oscillation is suggested as 1.7 Hz. The cure meter is an essential piece of equipment used extensively in the rubber laboratory. The machine plots a graph of torque vs. time for any given curing temperature.

A piece of rubber is placed on the heated rotor, and the heated top die cavity is immediately brought down on to the lower die thus filling the cavity [2]. In Figure 28.2, the curve shows an immediate initial rise in torque upon closure of the heated cavity. At the top of the first peak, the compound has not had more chance to absorb heat from its surroundings, and since viscosity is temperature-dependent it will be somewhat higher in these first few seconds. As the rubber absorbs heat from the instrument, it softens. Its temperature then stabilizes, and its viscosity has a constant value prior to the onset of cure. This assumes that it is not masked by a very short scorch time. The first important feature of the curve is the minimum viscosity of the rubber at the selected temperature and degree of oscillation, denoted by symbol M_L. After a certain time, the viscosity and torque begin to increase, indicating the initiation of curing process.

The torque continues to increase, until there is no more significant rise. At this point the compound is vulcanized, and this maximum torque value is designated by the symbol M_H. The last major piece of data to be extracted is the time it takes to complete the cure, known as the cure time ($t'x$).

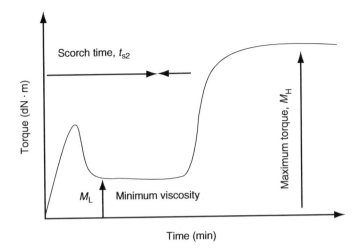

FIGURE 28.2 Standard oscillating disc curemeter curve. (Redrawn from Andrew, C., *Introduction to Rubber Technology*, Knovel e-book publishers, 1999.)

28.4.4 ROTORLESS CUREMETER OR MOVING DIE RHEOMETER

Moving die rheometer (MDR) is testing equipment which analyses the cure characteristics of rubber and monitors the processing properties and physical properties of the material. MDR measures the change in stiffness of a rubber sample, compressed between two heated platens, by an applied oscillating force. The degree of vulcanization determines the cure characteristics of the rubber. Here, the oscillating disc in the ODR is replaced with an oscillating die. A top and bottom form the cavity that holds the rubber sample. The dies permit the instrument to separate the elastic and viscous components of the compound and plot them as two separate curves (new). Although the two curves can be shown as S' (elastic stress) and S'' (viscous stress), very often it is preferred to be shown as S' and $\tan \delta$ which is the ratio of S'' and S'. Figure 28.3 shows $\tan \delta$ decreasing as the cross-linking process continues, reaching a maximum limit at completion of cure [2]. The final value of $\tan \delta$ indicated in figure is 0.22, indicating a lower damping (more elastic) material.

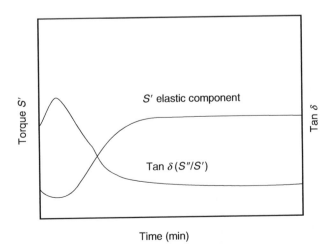

FIGURE 28.3 Moving die rheometer (MDR) curve. (Redrawn from Andrew, C., *Introduction to Rubber Technology*, Knovel e-book publishers, 1999.)

28.4.5 RUBBER PROCESS ANALYZER

Rubber process analyser (RPA) is a dynamic mechanical rheological tester designed to provide dynamic property data on rubber, polymers, and cured and uncured compounds [4]. The test chamber is a biconical cavity working under pressurized condition to offer easy sample loading. A distinct advantage of RPA over conventional dynamic rheological tester is that no special sample preparation is necessary. The measurements of the viscoelastic properties of rubber at various strains and frequencies can be experimented in one test. Therefore, it is very simple to apply RPA to measure the viscoelastic properties of rubber and rubber nanocomposites [4]. RPA offer many test techniques to cater for the special needs such as simple strain and frequency sweeps at different temperatures to more sophisticated procedures, especially designed for more complex systems. Special techniques can be designed whose basic principle is either to observe the effect of progressively higher strains on subsequent low-strain dynamic test or to study how low-strain dynamic modulus recovers with time after a large strain has been applied. In addition to the general strain sweep tests and frequency sweep tests, RPA also provides two such test procedures for uncured filled rubber compounds; the morphology recovery test (MRT) and damaged morphology recovery test (DMRT). MDT consists of a sequence of alternatively low-strain test and increasingly higher-strain test; the low-strain test measures the evolution—if any—of the dynamic properties after the application of increasingly larger-strain amplitude. DMRT procedure consists of first measuring the dynamic properties at low strain (normally 0.5°) then applying a large strain to damage the rubber–filler morphology, and subsequently repeating the low-strain measurement at various intervals.

28.4.6 CAPILLARY RHEOMETER

A capillary rheometer is another type of instrument, in which the uncured rubber is extruded through a small orifice and the change in dimensions of the extrudate is measured with a laser [2]. This instrument generates high shear rates, compared to Mooney rheometer. The capillary rheometer can thus represent flow of compounds on rubber processing machinery, such as injection molds.

28.4.7 OTHER LABORATORY TECHNIQUES

There are many other techniques available to test the rheological properties of rubber and rubber composites. Tests include volume resistivity compared with filler loadings, ozone resistance, heat aging, and swell in liquids, low-temperature brittleness, and flame propagation. Thermal analysers such as differential scanning calorimeters (DSC), thermogravimetric analysers (TGA), and thermo-mechanical analysers (TMA) are few instruments which measure how heat interacts with rubber. Dynamic mechanical thermal analysers and dynamic mechanical rheological testers are other types of instruments to measure the rheological data on rubber by applying various deformation modes at different frequencies and temperatures.

28.5 FLOW CLASSIFICATION

Flow is generally classified as shear flow and extensional flow [2]. Simple shear flow is further divided into two categories: Steady and unsteady shear flow. Extensional flow also could be steady and unsteady; however, it is very difficult to measure steady extensional flow. Unsteady flow conditions are quite often measured. Extensional flow differs from both steady and unsteady simple shear flows in that it is a shear free flow. In extensional flow, the volume of a fluid element must remain constant. Extensional flow can be visualized as occurring when a material is longitudinally stretched as, for example, in fibre spinning. When extension occurs in a single direction, the related flow is termed uniaxial extensional flow. Extension of polymers or fibers can occur in two directions simultaneously, and hence the flow is referred as biaxial extensional or planar extensional flow.

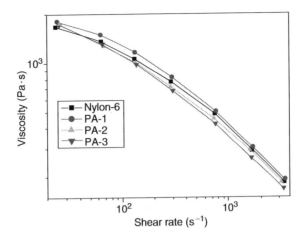

FIGURE 28.4 Shear viscosity vs shear rate from a capillary rheometer at 250°C for nylon-6, PA-1, PA-2, PA-3. (From Dong, W., Zhang, X., Liu, Y., Guia, H., Wang, Q., Gao, J., Song, Z., Lai, J., Huang, F., and Qiao, J., *Eur. Polym. J.*, 42, 2515, 2006.)

28.5.1 Measurement of Steady Shear Rheology

Most of the steady shear measurements for micro- and nanocomposites have been carried out using the rotational parallel plate and cone and plate geometries. Rotational rheometers are suitable for low- to medium-range shear rate of nanocomposites. Viscometric flow is assumed to have been generated in the fluid layer for an applied rate of shear. A variety of rheometers are available for the steady shear measurements. Dong et al. [7] experimented a novel technique to prepare nylon-6–unmodified clay–rubber nanocomposites. The rubber used in this new process is a kind of ultrafine full-vulcanized powder rubber (UFPRM) with special structure [7]. Shear viscosity of the nylon-6–UFPRM nanocomposites is shown in Figure 28.4. Three compound powders (PA-1, PA-2, and PA-3) of UFPRM–montmorillonite (UFPRM), butadiene–styrene–vinyl-pyridine (UFPRM), and silicone UFPRM (S-UFPRM) were obtained after spray-drying the three different mixtures.

Viscosity of PA-1 is higher than that of pure nylon-6. At low shear rate the viscosities of PA-2 and PA-3 are similar, but PA-3 has much lower viscosity than PA-2 at high shear rate. PA-3 shows lower viscosity than pure nylon-6, suggesting that nylon-6–UFPRM nanocomposite has better processability than pure nylon-6 in most of the applications. Stephen et al. studied the flow properties of sodium bentonite and sodium fluorohecctorite-filled NR with reference to shear rate, filler loading, and temperature. Research suggested that the viscosity of layered silicate-reinforced latex samples decreases with increase in shear rate, representing pseudoplastic nature, i.e., shear thinning behavior.

28.5.2 Measurement of Dynamic Shear Rheology

Dynamic measurement is a very useful technique for investigating the structure of delicate materials and deals with the state of the material due to unperturbed structure at small deformations. Dynamic measurements yield valuable information regarding the extent and dynamics of structure formed by particles in viscoelastic fluids. Most dynamic tests are experimented in the linear viscoelastic range of the material. This is tested by (1) dynamic strain sweep test; (2) dynamic time sweep test; and (3) dynamic frequency sweep test. Strain sweeps are conducted at different frequencies (0.1–100 rad/s) in order to determine the linear viscoelastic region of the material. The samples are subjected to a shear stress at a given frequency. Storage modulus (G') is determined as a function of strain or stress. The range in which G' remains constant gives the linear viscoelastic region for the material at the given temperature and frequency. This test indicates the region in which the deformation is small

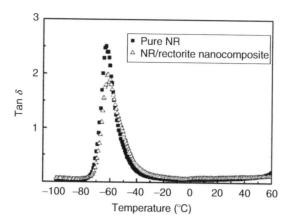

FIGURE 28.5 Tan δ vs temperature for pure natural rubber (NR) and NR/rectorite (10 phr) nanocomposite. (From Wang, Y. et al., *Eur. Polym. J.*, 41, 2776, 2005.)

enough for the modulus to be independent of deformation. Wang et al. [8] measured dynamic mechanical properties to examine the degree of filler–matrix interaction of NR–rectorite nanocomposites. The loss factor (tan δ) and the storage modulus (G') of the pure NR and the NR–rectorite nanocomposite vs. temperature are depicted in Figures 28.5 and 28.6, respectively. Nanocomposite exhibits a strong enhancement of the modulus over the temperature range studied, indicating that the elastic responses of pure NR towards deformation are strongly influenced by the presence of nano-dispersed rectorite layers.

Glass transition temperature (T_g) for pure NR is $-63.43°C$, while for the nanocomposite it increases to $-61.92°C$. NR–rectorite nanocomposite shows a higher glass transition temperature, lower tan δ peak, and slightly broader glass transition region compared to pure NR.

28.5.3 Measurement of Dynamic Mechanical Characteristics Using Rubber Process Analyser

Konecny et al. [9] suggested that there exists a linear correlation between dynamic mechanical characteristics from DMA and RPA for carbon black-filled rubber compounds. The amplitude dependencies of the analogous dynamic mechanical characteristics obtained from DMA and RPA

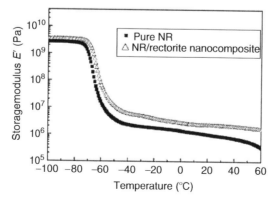

FIGURE 28.6 Storage modulus vs temperature for pure NR and NR/rectorite (10 phr) nanocomposite. (From Wang, Y. et al., *Eur. Polym. J.*, 41, 2776, 2005.)

measurements seem to be similar. The authors also studied carbon black and clay-filled rubber compounds using DMA and RPA. The author found that the presence of carbon black (filler) with higher reinforcing effect in vulcanized rubber results in higher values of modulus. The lowest values were obtained for unfilled rubber. The relationships between E'' and G'' modulus for all rubber clay composites and rubber carbon black composites are shown here. Wang et al. [10] studied that the complex viscosity, η^*, decreased with the prolongation of the aging time in the region of Newtonian flow, but in the region of non-Newtonian flow, the decrement of η^* with a rising shear rate decreased with the prolongation of the aging time. The authors also studied the aging of rubber. Zhang et al. [11] used RPA to evaluate the processing properties of the rubber compound at different strain and temperature. Figure shows the dependence of storage modulus of the compound on shear strain. At very low strain, the plateau storage modulus of the compound mounted up with the increasing filler ratio. Author also demonstrated the dependence of Mooney viscosity of the compound blend ratio. Mooney viscosity of the compound rose up straight with increasing filler ratio; because the cross-linked rubber particles greatly blocked the flow of the compound due to extremely high viscosity.

28.6 PARAMETERS MEASURED IN THE RHEOLOGICAL INSTRUMENTS

28.6.1 VISCOELASTICITY

Viscoelasticity illustrates materials that exhibit both viscous and elastic characteristics. Viscous materials like honey resist shear flow and strain linearly with time when a stress is applied. Elastic materials strain instantaneously when stretched and just as quickly return to their original state once the stress is removed. Viscoelastic materials have elements of both of these properties and, as such, exhibit time-dependent strain. Viscoelasticity is the result of the diffusion of atoms or molecules inside an amorphous material. Rubber is highly elastic, but yet a viscous material. This property can be defined by the term viscoelasticity. Viscoelasticity is a combination of two separate mechanisms occurring at the same time in rubber. A spring represents the elastic portion, and a dashpot represents the viscous component (Figure 28.7).

When a force is applied to this combination, causing a deformation, the spring and dashpot behave quite differently. The spring will store deformational energy, and then release it when the deformation is removed. When the piston in the dashpot moves, it cannot return to its original position when the force is removed. The kinetic energy that the piston has when it is moving is irreversibly converted into heat. No energy remains constant to move it back to its original position.

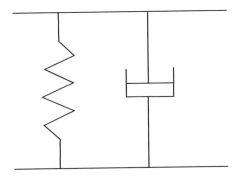

FIGURE 28.7 Viscoelastic Kelvin Model. (Redrawn from Bhattacharya, S.N., *Rheology Fundamentals and Measurements*, RMIT University, Melbourne, Australia, 2004.)

28.6.2 Viscoelasticity and Modulus

In the viscoelastic model, both the spring and the dashpot change their position, in response to an applied force. The applied force provides the stress, and the change in position of the spring, and the dashpot relates to the strain. Stress divided by strain equals modulus. The part of the modulus provided by the dashpot is called loss modulus (G'') which is related to energy lost as heat, while the spring's contribution is called storage modulus (G') which is related to the storage of converted kinetic energy as potential energy. Both G' and G'' play an important part in how rubber behaves in dynamic situations. Tan δ (ratio of G'' and G') will allow determination of the amount of damping that the rubber will provide when used as a spring. Tan δ will also be useful in determining how much heat will be produced in dynamic applications.

28.6.3 Viscoelasticity in Cyclic Deformation

A number of rubber engineering applications involve deformation of the product in a repetitious manner, called cyclic deformation. Consider the case of rubber band stretched through only a few percent, to represent the origin point for movement. The behavior of the band may be followed, moving a few percent above this point, then back through the origin to a few percent less, and finally back to its origin point. This should keep the band approximately within its mathematically linear region and will be one full cycle of movement. The curves (Figure 28.8) provide both the elastic and viscous components.

The elastic stress curve in figure perfectly follows elastic strain [2]. This constant is the elastic modulus of the material. In this idealized example, this would be equal to Young's modulus. Here at this point of maximum stretch, the viscous stress is not a maximum, it is zero. This state is called Newton's law of viscosity, which states that, viscous stress is proportional to strain rate. Rubber has some properties of a liquid. At the point when the elastic band is fully stretched and is about to return, its velocity or strain rate is zero, and therefore its viscous stress is also zero.

Figure 28.9 shows the total stress curve (a combination of the viscous and elastic), it is called viscoelastic curve. It is following, the elastic one, by a distance, equivalent to a certain number of degrees, known as delta, δ.

$$\tan\delta = \frac{G''}{G'} = \frac{S''}{S'}$$

where

  G'' is loss modulus
  G' is storage modulus
  S'' is viscous stress
  S' is elastic stress

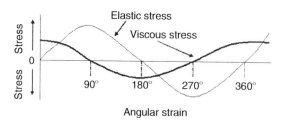

FIGURE 28.8 Idealized cyclic stress–strains, showing the viscoelastic curve split up into its two primary components, elastic and viscous. (Redrawn from Andrew, C., *Introduction to Rubber Technology*, Knovel e-book publishers, 1999.)

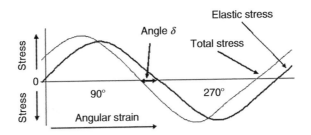

FIGURE 28.9 Idealized cyclic stress–strain curve, showing the full viscoelastic curve together with its elastic component. (Redrawn from Andrew, C., *Introduction to Rubber Technology*, Knovel e-book publishers, 1999.)

28.7 EFFECTS OF FILLERS ON RHEOLOGICAL PROPERTIES OF RUBBER AND RUBBER NANOCOMPOSITES

Rubber and rubber nanocomposites are highly complex polymer systems in which various solid and liquid ingredients are dispersed in an elastomer matrix that by nature exhibits a strong viscoelastic character. In all composites, there are relationships between the magnitude of rubber–filler interactions and severity of flow singularities. An elastomer molecule of rubber interacts with a filler particle, and then the reaction takes place at nanometre scale, in regions of the material that can be called the rubber-filled mesophase. Rubber nanocomposites exhibit very different rheological properties when compared with pure, unfilled elastomers, such as, lower extrudate swell with increasing filler content, disappearance of any linear viscoelastic region, and anisotropic effects in flow [1–6].

Relationship between the capability of certain minerals to impart reinforcement to elastomers and the development of interactions between filler particle surfaces and rubbers date back to the earliest observations in rubber technology (Figure 28.10). Particles larger than 10^3 nm do not have reinforcing capabilities or have a detrimental action, and generally increase viscosity by a mere hydrodynamic effect [33]. Reinforcement is readily obtained with sizes smaller than 100 nm, but particle structure appears as a more decisive factor. Two classes of minerals have been found to offer significant reinforcing capabilities, carbon black and silica, providing their manufacturing technology is such that quite complicated tridimensional objects involving several elementary particles of appropriate sizes are obtained.

As previously discussed in earlier topics, extensional flow occurs when the material is not in contact with solid boundaries, as is the case during drawing of filaments, film, sheets, or inflating

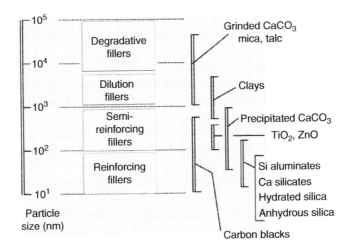

FIGURE 28.10 Classification of fillers according to average particle size. (From Leblanc, J.L., *Prog. Polym. Sci.*, 27, 627, 2002.)

bubbles. Converging flows at the inlet of die area is also extensional in nature. In extension, the material is stretched continually in a particular direction. The principal axis of strain keeps doubling in length at equal intervals of time during a steady extensional flow. The effect of fillers on the extensional flow properties of different polymers were experimented by many researchers since 1975 [14–16]. Leblanc [14] studied the effect of fillers in rheological properties of rubber composites. The report also discusses the nature of rubber–filler interactions and their effects on rheological properties of uncured materials. The concept of rubber-filled mesophase was introduced in order to underline the fundamental scaling problem that exists when attempting to relate phenomena occurring in the nanometer range to flow singularities. Many researchers have studied the rubber–filler interactions and reinforcing effect in vulcanizates [15–18]. Lobe and White [36] also studied that the extensional viscosity may tend to become constant at very low deformation rates, but become unbounded at higher and higher deformation rates. At higher filler concentrations, constant extensional viscosities were achieved with time and the values were found to decrease with increasing extensional rate.

28.7.1 ORGANOCLAYS

Carbon black used to be the most important reinforcing agent in the rubber industry. The unique black color of the rubber material, and its dependence on petroleum caused researchers to develop other satisfying reinforcing agents instead. Experiments showed that the property enhanced with reduction in particle size of fillers. It could be of great interest to use white fillers (clays) as substitutes for carbon black in rubber compounding. Clays are ultrafine in nature, ranging to nanometer size. The challenge with clay is its low surface activity. Recent researchers have made the clay attractive by treating it with long-chain amines called organoclays. Organoclays, as their name suggests, are mineral-based rheology modifiers. They are produced from clay minerals of the smectite group of clays, principally hectorite and bentonite. Clay in nature is water swellable. These clays are made of different tactoid layers (sheets of layers). In order to increase the compatibility of clay, the clay is modified with a number of different long-chain organic compounds. These compounds, usually cationic (Na^+ ions) in nature, are electrostatically bonded to the surfaces of individual clay platelets. The modification process converts the clay to a hydrophobic, solvent-dispersible ingredient that can modify the rheology of organic solvent-based systems. Bhowmick and Madhuchhanda [19] studied the phase image of the filled nanocomposites and revealed the presence of clay as better filler in the rubber matrix. Pinnavaia and Lan [20] reported the preparation of nanocomposites with a rubber–epoxy matrix obtained from diglycidyl ether of Bisphenol A (DGEBA) derivative cured with a diamine so as to reach subambient glass transition temperatures. It has been shown that depending on the alkyl chain length of modified MMT, an intercalated and partially exfoliated or a totally exfoliated nanocomposite can be obtained. The same authors [21] also studied other parameters such as the nature of alkyl ammonium cations present in the gallery and the effect of the cation exchange capacity (CEC) of the MMT when DGEBA was cured with *m*-phenylene diamine. The same kind of study was also conducted by Zilg et al. [22], who cured DGEBA with hexahydrophthalic acid anhydride in the presence of different types of clays, and also modified with a wide variety of surfactants. Sadhu and Bhowmick [23] studied the effect of MMT clay on three different grades of acrylonitrile–butadiene rubber (NBR), SBR, and polybutadiene rubber (BR). Shear viscosity decreased with increasing shear rate and incorporation of unmodified (pristine clay) and modified (organoclay) clay. The effect became more prominent with increasing polarity of the rubber. Die swell decreased with increase in clay content of rubber composite. Varghese and Karger-Kocsis [24] studied the effect of untreated clay and organophilic clay on rubber nanocomposites, and concluded that property improvements caused by the fillers in order as follows: organophilic clays > pristine synthetic-layered silicate (sodium flourohectorite) > pristine natural clay (purified sodium bentonite) > precipitated nonlayered silica (Table 28.2). Organophilic clays accelerated the sulfur-curing of NR, which was believed to occur because of a complexation reaction in which the amine groups of the clay intercalants participated.

TABLE 28.2

Rheometric Properties of Natural Rubber (NR) Mix Containing 10 phr Fillers

Parameter	Silica	Fluorohectorite	Bentonite	MMT–ODA	MMT–TMDA
Minimum torque (dNm)	0.61	0.67	0.58	1.04	1.23
Maximum torque (dNm)	7.87	8.30	7.99	8.34	8.69
Maximum–minimum torque (dNm)	7.27	7.63	7.41	7.30	7.46
Scorch time t_2 (min)	8.37	7.42	5.48	5.17	3.17
Cure time t_{90} (min)	14.71	12.99	11.77	10.12	8.54
Cure rate $t_{90} - t_2$ (min)	6.34	5.57	6.29	4.95	5.37

Source: Siby Varghese, J. and Karger–Kocsis, J., *J. Appl. Polym. Sci.*, 91, 813, 2004.

The storage modulus of different silicate-filled rubber nanocomposites as a function of temperature is shown in Figure 28.11. The MMT–ODA (Nanomer I.30P)-filled composite shows the highest storage modulus. Tan δ as a function of temperature is shown in Figure 28.12. All silicates except MMT–ODA (Nanomer I.30 P) have a pronounced peak, which elucidates the contact to the filler are low. A secondary relaxation peak for this composite in the tan δ–temperature curve is evident in Figure 28.12. This may be assigned to the rubber between the clay layers in an intercalated (partially exfoliated) and/or confined (reaggregated) system. Here, the rubber behaves differently than that of the bulk, being under high constraints. Varghese also studied the difference in the moduli for the different silicate-filled composites that are prominent at low elongations (100% and 200%). This is because the filler–polymer interactions play a critical role at low elongations, just as at high elongations the strain-induced crystallization of NR dominates. Zhang et al. [25] studied the dependence of conversion on reaction time of SBR copolymerization with different OMMT contents. Figure 28.13 shows that the addition of OMMT does not change the kinetics of copolymerization on the whole, when its content is lower than 3 wt%. Figure 28.14 also shows the dependence of the SBR composition with different OMMT contents on the conversion.

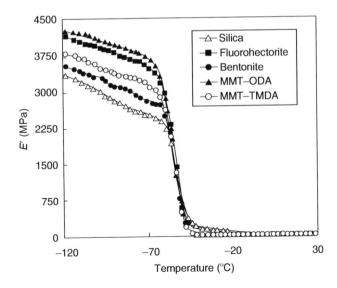

FIGURE 28.11 Storage modulus as a function of temperature for the ENR stocks containing various fillers in 10 h. (From Siby Varghese, J. and Karger–Kocsis, J., *J. Appl. Polym. Sci.*, 91, 813, 2004.)

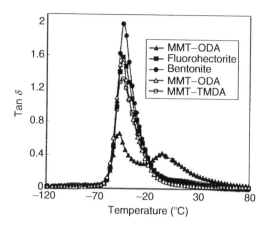

FIGURE 28.12 Mechanical loss factor as a function of temperature for the ENR stocks containing various fillers. (From Siby Varghese, J. and Karger–Kocsis, J., *J. Appl. Polym. Sci.*, 91, 813, 2004.)

For OMMT content to be 3 wt% and changing the monomer ratio of styrene and butadiene (such as 0:100, 10:90, 25:75, 40:60, 60:40, and 100:0), a series of the polymers–MMT products were prepared, and coded as S-0M3, S-10M3, S-25M3, S-40M3, S-60M3, S-100M3, respectively.

SBR–MMT nanocomposites containing 0–4 wt% OMMT were prepared, and coded as SB, S-25M1, S-25M2, S-25M2.5, S-25M3, and S-25M4, respectively. The results indicate that styrene plays an important role in the improvement of the exfoliated clay structure of the SBR–MMT nanocomposite. When styrene content is more than 25%, completely exfoliated nanocomposites can be observed. Styrene molecules have relatively stronger polarity and more easily enter into the galleries between silicate layers than butadiene and cyclohexane molecules. Expanded galleries between layers as a result of the entering of styrene can accommodate more monomers and solvent molecules to assure the randomness of this copolymer. Teh et al. [26] grafted the rubber onto MMT layers by the chemical modification technique in one step by a cation exchange process between the

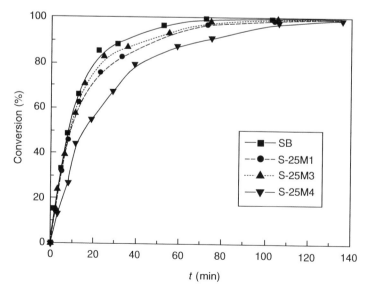

FIGURE 28.13 Dependence of conversion on reaction time of the styrene–butadiene. (From Zhang, Q. et al., *Polymer*, 46, 129, 2005.)

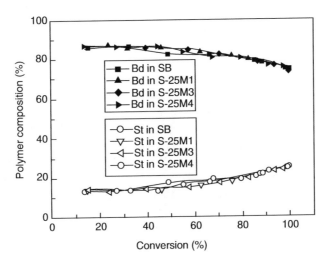

FIGURE 28.14 Dependence of the SBR composition with different OMMT contents. (From Zhang, Q. et al., *Polymer*, 46, 129, 2005.)

chains and ammonium groups of amine-terminated butadiene–acrylonitrile copolymers (ATBN). The basal spacing increases as the rubber molecules intercalate between lamellae of the clay.

The storage modulus of the different silicate-filled vulcanizates as a function of temperature is shown in Figure 28.11. The MMT–ODA-filled composite registers the highest storage modulus. After the glass transition temperature (T_g), the mixes show a similar change with the temperature. The intercalated and exfoliated silicate layers of MMT–ODA reinforce the matrix to a great extent, as reflected in the storage moduli. Clearly, all silicate versions except MMT–ODA have a pronounced T_g peak, which reflects the high mobility of the polymer chains when their contacts to the filler are low. However, in the MMT–ODA-filled composite, the T_g peak is strongly reduced, indicating a strong interaction between NR and silicate. Moreover, there is a secondary relaxation peak for this composite in the tan δ–temperature curve. This may be assigned to the rubber between the clay layers in an intercalated (partially exfoliated) and/or confined (reaggregated) system. Here, the rubber behaves differently than that of the bulk, being under high constraints.

Teh et al. [26] studied the effect of different filler loading up to 10 parts per 100 parts of rubber (phr) and observed that the tensile strength, elongation at break, and tear properties went through a maximum (at about 2 phr) as a function of the organoclay content. The hardness, moduli at 100% and 300% increased continuously with increased organoclay loading (Figure 28.15).

The researchers also studied that the addition of organoclay into the NR increases the stiffness remarkably (Figure 28.16). It has been studied that the effect of organoclay is not significant below the glass transition temperature (Figure 28.17) [28].

Teh et al. [27] compared the effects among organoclays, silica, and carbon black in rubber composite (Figure 28.18). The improvement of tensile strength, elongation at break, and tear properties in organoclay-filled composite was significantly higher compared to other fillers.

28.7.2 CARBON BLACK

Carbon black is reinforced in polymer and rubber engineering as filler since many decades. Automotive and truck tires are the best examples of exploitation of carbon black in rubber components. Wu and Wang [28] studied that the interaction between carbon black and rubber macromolecules is better than that of nanoclay and rubber macromolecules, the bound rubber content of SBR–clay nanocompound with 30 phr is still of high interest. This could be ascribed to the huge surface area of clay dispersed at nanometer level and the largest aspect ratio of silicate layers, which result in the increased silicate layer networking [29–32].

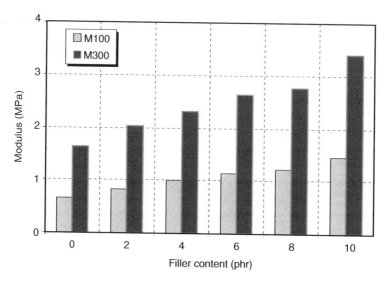

FIGURE 28.15 Modulus vs filler content of organoclay in nanocomposite. (From Teh, P.L. et al., *J. Appl. Polym. Sci.*, 100, 1083, 2006.)

Reinforcing fillers in general and high-structure carbon blacks in particular improve the extrusion characteristics of elastomers by decreasing extrudate swell. The extrudate swell decreases with increasing carbon black content [33].

28.7.3 CARBON NANOTUBES

It is well known that the particle size, structure, and surface characteristics are important parameters that determine the reinforcing ability of filler; particle size is important because a reduction in size

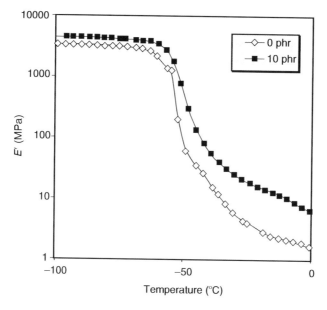

FIGURE 28.16 Storage modulus vs temperature of organoclay-loaded rubber nanocomposite. (From Teh, P.L. et al., *J. Appl. Polym. Sci.*, 100, 1083, 2006.)

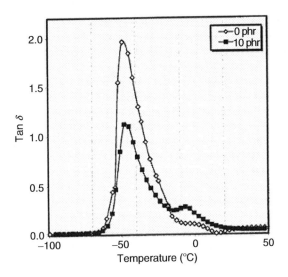

FIGURE 28.17 Curve of tan δ vs temperature. (From Teh, P.L. et al., *J. Appl. Polym. Sci.*, 100, 1083, 2006.)

provides a greater surface area. In comparison with nanotubes, the reinforcing ability of clay is poor because of its big particle size and low surface activity. On the other hand carbon nanotubes (CNTs) exhibit a significant reinforcement effect. Nazabal et al. [35] demonstrated that organoclay increased the toughness and modulus of the matrix of the polymer. It is very clear from the figure that the maximum stress of pure rubber is 0.2839 MPa. The addition of 1 wt% CNT increases the stress level from 0.2839 to 0.5641 MPa. The stress level increases further to 2.55 MPa for 10 wt% CNTs (which is

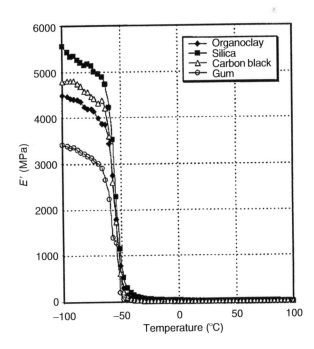

FIGURE 28.18 Storage modulus vs temperature (for different fillers). (From Teh, P.L. et al., *J. Appl. Polym. Sci.*, 94, 2438, 2004.)

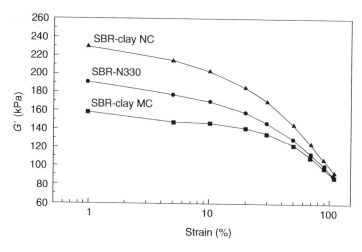

FIGURE 28.19 Relationship between G' and strain (80°C, 1 Hz, 20 phr filler). (From Wu, Y. and Wang, Y., *Comp. Sci. Technol.*, 65, 1195, 2005.)

nine times stronger than pure NR). The same researchers also found that increasing the CNTs content in rubber makes the composite stronger and tougher, but at the same time more brittle. Lobe and White [36] studied the effect of carbon black concentration on the rheological properties of polystyrene melt (Figure 28.19). Lopez-Manchado et al. [34] studied the effect of CNT in rubber composites. Dynamic mechanical analysis indicates a stronger filler–matrix interaction in the case of single-wall nanotubes (SWNTs) incorporation (Figure 28.20), showing a noticeable decrease of the height of tan δ peak, as well as shift of T_g towards higher temperatures. The storage modulus increased with incorporation of single-walled nanotube in rubber composite. In Figure 28.20, the height of tan δ peak corresponding to the glass transition temperature (T_g) is reduced. A shift of the tan δ peak towards higher temperatures by addition of the fillers is observed. Glass transition temperature increases from −49.4°C for pristine NR to −45.7°C for SWNT-reinforced rubber composite. This proves the strong interfacial action between the filler and the matrix, by incorporation of the nanotubes [37–42].

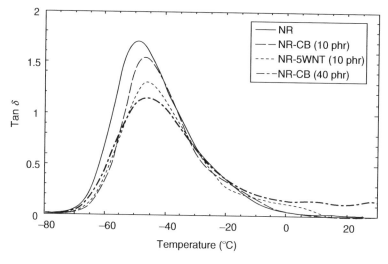

FIGURE 28.20 Curves of tan δ vs temperature for rubber and carbon nanotubes (CNTs)/rubber nanocomposite. (From Lopez-Manchado, M.A. et al., *J. Appl. Polym. Sci.*, 92, 3394, 2004.)

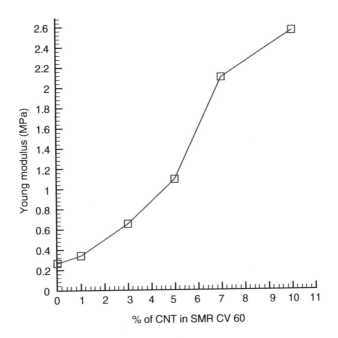

FIGURE 28.21 Young modulus of SMR CV 60 at different percentage of CNTs. (From Atieth, A. et al., *Comp. Struct.*, 75, 496, 2006.)

Figure 28.21 shows that the Young's modulus increased with increase in the amount of the CNTs. However, at 1 and 3 wt% of CNTs, the increment of the modulus is not as high as that of the tensile strength. The same value of the modulus and the tensile strength were observed at 5 wt% of CNTs.

Researchers [37] also compared the storage modulus of a 40 phr carbon black-filled compound and a 10 phr SWNT–NR nanocomposite. The different properties between carbon black- and SWNTs-filled NR nanocomposites can be explained in terms of two different filler morphology, particularly surface area, aspect ratio, and structure. It can be observed from Figure 28.22 that

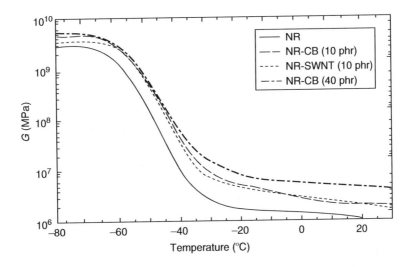

FIGURE 28.22 Storage modulus G' vs temperature of carbon nanotube (CNT)/rubber composite. (From Lopez-Manchado, M.A. et al., *J. Appl. Polym. Sci.*, 92, 3394, 2004.)

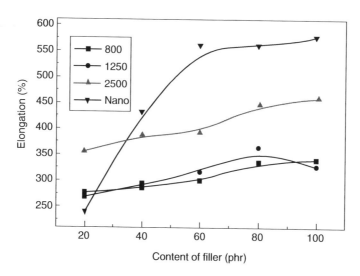

FIGURE 28.23 Effect of magnesium hydroxide with different particle sizes on the elongation at break at composites. (From Zhang, Q. et al., *J. Appl. Polym. Sci.*, 94, 2341, 2004.)

the modulus increases in the presence of both fillers, which is attributed to the hydrodynamic reinforcement upon introducing the filler. Atieh et al. [6] dispersed CNTs homogeneously in NR in an attempt to increase the mechanical properties of the nanocomposites. The properties of the nanocomposites such as tensile strength, tensile modulus, tear strength, elongation at break, and hardness were studied. Mechanical test results show an increase in the initial modulus for up to 12 times compared to that of pure NR. The study also shows that for the same filler content the effect of CNT is higher than that of carbon black. Kim and Hayashi [43] prepared rubber composite sheets filled with 5–30 wt% of highly aligned CNTs through conventional rubber technology. Selective alignment of CNTs led to enhancements in the elastic modulus, thermal, and electric conductivity compared to neat rubber sheet.

28.7.4 Magnesium Hydroxide

Magnesium hydroxide, an effective flame-retardant filler, possesses high decomposition temperature, can be used in a wider range of thermoplastics than aluminum hydroxide [44–48]. Zhang et al. [49] discussed that the properties of rubber composites improved with decreasing particle size of magnesium hydroxide powder (Figures 28.23 and 28.24). The effect of particle size on the fire resistance of composites was investigated, which also showed that the particle size of powder has an impact on the fire-resistant properties of composites. The research also mentioned the surface treatment effect on the properties of rubber nanocomposites. Three different microsize magnesium hydroxide (800, 1250, and 2500 mesh) and nanosize of magnesium hydroxide were used to prepare rubber nanocomposites. Dynamic modulus of composites does not change with increasing strain and suggests that the network structure is not destroyed. Rapid decrease of modulus is related to the breakage of the particles network. The modulus increases with increase in filler content of the composite.

It is clearly visible from Figure 28.24 that initial moduli of nanocomposites are higher than those of microcomposites at any loading, given that the surface activity and surface area between nanoparticles and microparticles are different.

Figure 28.24 also shows the Payne effect of gross rubber with magnesium hydroxide contents, indicating that there is a plateau region in the curve at a certain content, which means that the dynamic modulus of composites does not change with increasing strain and suggests that the network structure is not destroyed. The modulus increases with the filler content in the polymer

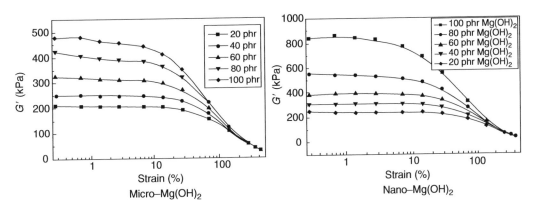

FIGURE 28.24 Plots of Payne effect of composites filled with different contents of Mg(OH)$_2$. (From Zhang, Q. et al., *J. Appl. Polym. Sci.*, 94, 2341, 2004.)

matrix. Zhang also concluded that the initial moduli of nanocomposites are higher than those of microcomposites at any loading, given that the surface activity and quantity between nanoparticles and microparticles are distinct. Zhang et al. studied that the surface activity of nanoparticles is much beneficial for the properties of composites compared with microcomposites. The disadvantage of magnesium hydroxide is that it requires large amounts to achieve the desired degree of flame retardance, which would increase the agglomeration of particles and affect the mechanical properties. The surface treatment on inorganic powder can efficiently improve the interfacial action between powder and rubber, dispersibility, and reinforcement ability of powder.

28.7.5 Calcium Carbonate

It is of great interest to use inorganic fillers such as calcium carbonate as substitute in rubber compounding. Modulus at 300% elongation increases with increase in nano-calcium carbonate loading in rubber composite, up to 8 wt% filler (Figure 28.25). Mishra and Shimpi [50] reported that the modulus also increased with addition of extender. The extender helps the nano-calcium carbonate disperse uniformly in the rubber matrix. For all nanosizes of calcium carbonate (21, 15, and 9 nm), the increase in the elongation was significantly greater than that of the commercial microcalcium carbonate

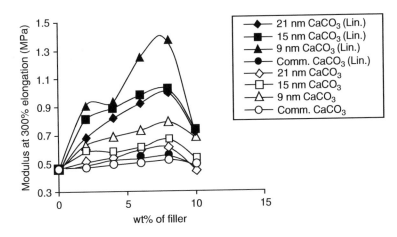

FIGURE 28.25 Modulus at 300% elongation of SBR filled with different fillers at various compositions. (From Mishra, S. and Shimpi, N.G., *J. Appl. Polym. Sci.*, 98, 2563, 2005.)

(with an extender). However, the elongation at break increased with decrease in the nanosize for all cases. Studies show that the viscosity of lattices increases as a function of weight percentage of filler. The increase in viscosity is owing to the reinforcement occurred in the system in the presence of layered silicates because of the high polymer–filler interaction. Since at each shear rate there is a structural equilibrium between the immobilized and mobilized part, the effective volume fraction is a function of shear rate. This is the reason behind the shear thinning behavior of layered silicates-reinforced latex samples. From the viscosity curve it is clear that latex nanocomposites have non-Newtonian behavior. As the shear rate increases, the viscosity of the samples decreases. The shear thinning behavior is predominant in these samples as the filler loading increases, which indicates the network formation at low shear rate and its gradual destruction at higher shear rates. Stephen [51] studied the rheological behavior of NR, carboxylated SBR lattices, and their blends with special reference to shear rate, temperature, and filler loading. It was suggested that in the presence of layered silicates, latex systems exhibited enhancement in viscosity due to the network at higher temperature, the viscosity of all systems decreased with increase in temperature. Layered silicates-reinforced latex systems showed pseudoplastic flow behavior and possess enhanced zero shear viscosity and yield stress. At a fixed extension rate, the extensional viscosity increased with increasing filler concentrations because the solid particles of calcium carbonate did not deform under stretching and hence exerted more resistance to the flow of the molten thread line with an increase in concentration. Modeling shows the effect of frictional and ductile detachments on thin skinned extension. Extension occurred above two different basal configurations: a stretchable rubber sheet and a folded, banded sheet intended to produce homogeneous and heterogeneous extensions.

28.7.6 FULLERENE

Researchers studied the effect of fullerene in rubber composites with different temperature range [52]. There was no substantial influence of fullerene on T_g, tan δ, and G-modulus within the temperature range from $-150°C$ to $-50°C$ (glassy state), and properties increase at rubbery state ($0°C–150°C$). At temperatures between $-150°C$ and $-50°C$ when rubber is rigid, G-modulus is virtually independent of the fullerene concentrations between 0.065 and 0.75 phr and a single major peak in Figure 28.26 shows that the fullerene does not influence the glass transition.

The properties are evaluated at temperatures between $-50°C$ and $250°C$ (high elastic state of rubber). For rubbers in a high elastic state, an increase in fullerene concentrations is followed by a rise in modulus. It is very lucid at temperatures between $-10°C$ and $150°C$. Most likely, it is caused

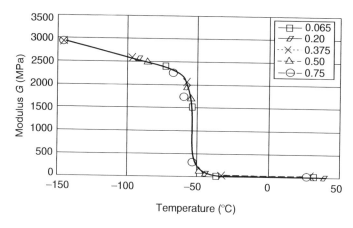

FIGURE 28.26 The dependency of modulus from DMA test at twisting on fullerene concentration for low-temperature range. (From Jurkowska, B., Jurkowski, B., Kamrowski, P., Pesetskii, S.S., Koval, V.N., Pinchuk, L.S., and Olkhov, Y.A., *J. Appl. Polym. Sci.*, 100, 390, 2006.)

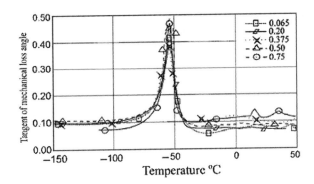

FIGURE 28.27 The dependency of tan δ from DMA test at twisting on fullerene concentration for low temperature range. (From Jurkowska, B., Jurkowski, B., Kamrowski, P., Pesetskii, S.S., Koval, V.N., Pinchuk, L.S., and Olkhov, Y.A., *J. Appl. Polym. Sci.*, 100, 390, 2006.)

by creation of additional strong physical junctions of the rubber network. Some changes in the slope of segments of $G(T)$ curves between 0°C and 150°C and lowered temperature at the beginning of creation of a second polymer network suggest the growth of degradation energies of the branching junctions as concentration of fullerene increases. The decrease of loss tangent at temperature range between 20°C and 150°C is the lowest for the higher concentration of fullerene (Figure 28.27). It evidences that part of the branching junctions in the rubber network related to the presence of fullerene has high degradation energy as it was supposed earlier.

28.8 FUTURE DEVELOPMENTS IN RUBBER NANOCOMPOSITES

There is a growing interest in the development of rubber nanocomposites because of the enhanced rheological, mechanical, gas barrier, and thermal properties that can be achieved by intercalated and exfoliated silicate layers of nanoscale thickness. Melt compounding would be used for rubber nanocomposite manufacturing in the long term. Melt compounding will be pushed forward by direct intercalation strategies, i.e., attempts will be made to produce the organoclay and disperse it in the rubber matrix at the same time. Novel organophilic silicates will be developed that are modified for good compatibility with rubbers. Surfactant or the intercalant used in the clay will be involved in the curing procedure. Organoclays, coupling agents like silanes, and titanates are also a promising strategy. Ultrasonics or high-speed mixing will be utilized to disperse the clay particles in the dissolved rubber solution. Excellent dispersion between the organic and inorganic phases can be achieved and less energy is necessary during processing to ensure product homogeneity. There may also be preparation of rubber nanocomposites by solution, latex, and melt-compounding routes. It is possible to anchor certain monomers prior to polymerization on the surface of inorganic particles to produce polymer chains chemically connected to the inorganic dispersed phase. This interaction creates a true hybrid organic–inorganic material that can be supplied as a master batch to improve properties of other commodities. Research and development activity in the field of nanoreinforcement is far more advanced with thermoplastic resins than rubbers; as the number of applications of rubber is less than that of thermoplastics especially polyolefins. More information to the structural property relationships can be expected from molecular modeling works.

28.9 CONCLUSION

In conclusion, the field of rubber and rubber nanocomposites continues to be a very fertile area of research, with many new advances in both basic and applied topics of research. The flow properties of filled rubber compounds arise from their heterogeneous nature and the strong interactions that develop between their various ingredients. Many evidences suggest that the complex nature of the processes takes place between the surface of filler particles and rubber segments. Rubber by itself is macromolecules and fillers (mostly) are nanoparticles. The interaction between the rubber and nanofillers results in long-range modifications in terms of flow mechanics. In low shear region, there is thus a rubber–filler mesophase whose dynamics reaches at best equilibrium between adsorption and desorption of rubber segments on appropriate sites on filler particles. If the temperature is changed or shear is applied, this equilibrium is likely to be displaced. The presence of nanoparticles is observed to increase the glass transition temperature. The storage modulus is usually found to increase with nanofiller concentration.

There are two important parameters of fillers, which control the mechanical, rheological, and physical properties of the rubber composite or matrix. They are average particle size and surface characteristics of the filler. Particles larger than 10^3 nm do not have reinforcing capabilities or have a detrimental action, and generally increase viscosity by a mere hydrodynamic effect. Many of the available fillers are suitable for direct usage without any treatment of the fillers. However, some of the nonpolar matrix is not compatible with filler because of the hydrophilic nature of the fillers. Basic research is still underway in areas to find a better surfactant with dual properties: chain which reacts with the nonpolar matrix, and a molecule that reacts with surface oxygen atoms of the fillers. Example of the above-mentioned situation is applicable for silica fillers in rubber matrix. Silica rubber interaction is a much more complicated reaction than that for carbon black, because the surface chemistry of silica plays a significant role. In addition to the two parameters of the fillers, previously discussed, the property of the rubber nanocomposite also depends on the chemical nature, size and structure of filler, mixing conditions, the storage time and temperature.

Rubber nanocomposites possess higher glass transition temperature, thermal stability, tensile strength, and elongation at break than virgin or pure rubber. Rubber nanocomposites reinforced with CNTs has pronounced and positive effect on mechanical and physical properties compared to NR. The selective alignment of CNTs led to enhancements in the elastic modulus, thermal conductivity, electrical conductivity, and electromagnetic shielding property compared to neat rubber sheet. NR–rectorite nanocomposite showed that the nanocomposite exhibited a higher glass transition temperature, lower tan δ peak, and slightly broader glass transition region compared with pure NR. Silicate layers with nanometer dispersion and high aspect ratio improves the tensile strength, gas permeability, and rheological properties better than micrometer clay or silica. Rubber nanocomposites exhibit outstanding mechanical performances, and improved gas barrier properties, which are likely attributed to the nanometer-scale dispersion and high aspect ratio of clay layers in rubber matrix. Rubber nanocomposite showed reduced swelling rate due to the tortuosity of the path and the reduced transport area in polymeric membrane. Gas permeability of the nanocomposite materials decreased remarkably by the presence of the high loadings of vermiculite. Although a significant amount of work has been reported on rheological properties of rubber, much research still remains in the area of rubber nanocomposites.

NOTATIONS

tan δ	loss factor
G'	storage modulus
G''	loss modulus
T_g	glass transition temperature
M_L	minimum viscosity
M_H	maximum torque value
$t'x$	cure time
S'	elastic stress
S''	viscous stress
E'	mechanical loss factor

REFERENCES

1. Bhattacharya, S.N., *Rheology Fundamentals and Measurements*, RMIT University, Melbourne, Australia, 2004.
2. Andrew, C., *Introduction to Rubber Technology*, Knovel e-book publishers, 1999, 25–26.
3. Stern, J., in *History in Rubber Technology and Manufacture*, C.M. Blow, Ed., Newnes-Butterworths, London, 1977, p. 18.
4. Pawlowski, H. and Dick, J., *Rubber World*, 206, 1992, 35–40.
5. Kuzma, L.J., *Rubber Technology*, 3rd ed., M. Morton, Ed., Van Nostrand Reinhold, New York, 1987, p. 235.
6. Atieh, M.A., Razi, A., Girun, N., Chuah, T.G., El-sading, M., and Biak, D.R.A., *Comp Struct*, 75, 2006, 496–500.
7. Dong, W., Zhang, X., Liu, Y., Guia, H., Wang, Q., Gao, J., Song, Z., Lai, J., Huang, F., and Qiao, J., *Eur Polym J*, 42, 2006, 2515–2522.
8. Wang, Y., Zhang, H., and Wu, Y., *Eur Polym J*, 41, 2005, 2276–2783.
9. Konecny, P., Cerny, M., Voldanova, J., Malac, J., and Simonik, J., *Polym Adv Technol*, 18, 2007, 122–127.
10. Wang, P., Qian, H., Yang, C., and Chen, Y., *J Appl Polym Sci*, 100, 2006, 1277–1281.
11. Zhang, L.Q., Yang, J., Lu, M., and Jia, Q., *J Appl Polym Sci*, 103, 2007, 1826–1833.
12. White, J.L., *Rubber Chem Technol*, 50, 1977, 163–185.
13. Nakajima, N. and Harrell, E.R., Analyzing steady-state flow of elastomers. *Encyclopaedia of Fluid Mechanics*, Vol. 7, Gulf Publishing, Houston, 1988, pp. 703–723, Chapter 24.
14. Leblanc, J.L., *Kautch Gummi Kunstst*, 43(10), 1990, 883–892.
15. White, J.L., Czarnecki, I., and Tanaka, H., *Rubber Chem Technol*, 53, 1980, 823–835.
16. Ide, Y. and White, J.L., *J Appl Polym Sci*, 22, 1978, 1061–1079.
17. Lockett, F.J., National Physical Laboratory report no. 25, Division of materials applications, Teddington, England, 1972.
18. Donnet, J.B. and Voet, A., *Carbon Black Physics, Chemistry and Elastomer Reinforcement*, Marcel Dekker, New York, 1976.
19. Bhowmick, A. and Madhuchhanda, M., *Polymer*, 47, 2006, 6156–6166.
20. Lan, T. and Pinnavaia, T.J., Clay-reinforced epoxynanocomposites. *Chem Mater*, 6, 1994, 2216.
21. Lan, T., Kaviratna, P.D., and Pinnavaia, T.J., *Chem Mater*, 7, 1995, 2144–2150.
22. Zilg, C., Mulhaupt, R., and Finter, J. *Macromol Chem Phys*, 200, 1999, 661–670.
23. Sadhu, S. and Bhowmick, A., *J Polym Sci Part B: Polym Phys*, 43, 2005, 1854–1864.
24. Siby Varghese, J. and Karger-Kocsis, J., *Appl Polym Sci*, 91, 2004, 813–819.
25. Zhang, L., Zhang, Z., Yang, L., and Xu, H., *Polymer*, 46, 2005, 129–136.
26. Teh, P.L., Mohd Ishak, M.A., Hashim, A.S., and Ishiaku, U.S., *J Appl Polym Sci*, 100, 2006, 1083–1092.
27. Teh, P.L., Ishak, Z.A., Hashim, A.S., Karger-Kocsis, J., and Ishiaku, U.S., *J Appl Polym Sci*, 94, 2004, 2438–2445.
28. Wu, Y. and Wang, Y., *Comp Sci Technol*, 65, 2005, 1195–1202.
29. Donnet, J.B., *Carbon Black*, Marcel Dekker, New York, 1993.
30. Leblanc, J.L., *Eur Rubber J*, 171(10), 1989, 33–35.
31. Leblanc, J.L., *Prog Rubber Plast Technol*, 10(2), 1994, 112–129.
32. Zhang, L., Wang, Y., Wang, Y., Sui, Y., and Yu, D., *J Appl Polym Sci*, 78, 2000, 1873.
33. Leblanc, J.L., *Prog Polym Sci*, 27, 2002, 627–687.
34. Lopez-Manchado, M.A., Herrero, B., and Arroyo, M., *Polym Int*, 52, 2003, 1070.
35. Gonzalez, I., Eguiazabal, J.I., and Nazabal, J., 2005, *Polymer*, 2005, vol. 46, 2978–2985.
36. Lobe, V.M. and White, J.L., *Polym Engg Sci*, 19, 1979, 617–624.
37. Kenny et al., *J Appl Polym Sci*, 92, 2004, 3394–3400.
38. George, S.C. et al., *J Appl Polym Sci*, 78, 2000, 1280.
39. Gonza'lez, L., Rodriguez, A., and Maras, A., *Recent Res Dev Polym Sci*, 2, 1998, 485.
40. Wolf, S. and Wang, M.J., *Rubber Chem Technol*, 65, 1992, 329.
41. Wolf, S., *Rubber Chem Technol*, 69, 1996, 325.
42. Abbas Bahroudi, et al., *J Struct Geol*, 25, 2003, 1401–1423.
43. Kim, Y. and Hayashi, T., *Scripta Materialia*, 54, 2006, 31–35.
44. Herbiet, R., *Polym Polym Comp*, 8, 2000, 551.

45. Hornsby, P.R. and Wang, J., *Prog Rubber Plast Technol*, 10, 1994, 204.
46. Yu, Y., Wu, Q., Ge, S., Zhou, H., and Xu, S., *Handbook of Flame Retardant Materials*, Qunzhong Press, Beijing, China, 1997.
47. Molesky, F.J., *Vinyl Addit Technol*, 3, 1995, 159.
48. Hornsby, P.R., *Fire Mater*, 18, 1994, 269.
49. Zhang, Q., Tian, M., Wu, Y., and Liqun, Z., *J Appl Polym Sci*, 94, 2004, 2341–2346.
50. Mishra, S. and Shimpi, N.G., *J Appl Polym Sci*, 98, 2005, 2563–2571.
51. Stephen, R., Varghese, S., Joseph, K., Oommen, Z., and Thomas, S., *J Membrane Sci*, 282, 2006, 162–170.
52. Jurkowska, B., Jurkowski, B., Kamrowski, P., Pesetskii, S.S., Koval, V.N., Pinchuk, L.S., and Olkhov, Y.A., *J Appl Polym Sci*, 100, 2006, 390–398.

29 Rubber–Silica Mixing

Wilma Dierkes and Jacques Noordermeer

CONTENTS

29.1 INTRODUCTION

29.1.1 SILICA TECHNOLOGY

Silica has gained commercial importance as filler in the 1970s in heavy-service tires for earth moving and mining equipment [1]. It has become a common practice to improve the cutting and chipping resistance of heavy truck tires by adding 10–25 phr of silica to the conventional carbon black-loaded tread compounds, with adjustment of the carbon black level to maintain appropriate compound hardness. The addition of silica improved (wet) traction of the truck tires, however at the expense of some increased heat buildup. The latter can be kept at an acceptable level by use of an appropriate coupling agent. Further, silica enhances the adhesion between the steel cord reinforcement and the rubber of radial tires.

Since the introduction of the "green tire," silica is also used more and more as reinforcing filler in passenger car tire applications. The increased attitude of protecting the environment gives rise to a demand for tires combining a long service life with driving safety and low fuel consumption. Furthermore, replacing carbon black by silica lifts the "magic triangle of tire technology," the compromise between rolling resistance, wet grip, and longevity of a tire, to a higher level. The replacement of carbon black by easy dispersion silica fillers, combined with special rubber types and the selection of a proper coupling agent, allows for a significantly reduced rolling resistance of tires, and as a consequence reduced fuel consumption of the vehicle, while keeping the wet traction and abrasion resistance on the same level [2]. The achievable reduction of the rolling resistance is around 20%, corresponding to fuel savings of 3%–4%.

However, the use of silica as reinforcing filler is difficult: when silica is mixed with the commonly nonpolar olefinic hydrocarbon rubbers, there will be a greater tendency towards hydrogen-bond interactions between surface silanol-groups of silica aggregates than to interactions between polar siloxane or silanol-groups on the silica surface and the rubber polymers. So mixing silica with rubber involves major problems. For this reason, there is great interest in enhancing the compatibility of hydrocarbon rubbers and precipitated silica by modification of the surface of the silica. Bifunctional organosilanes are commonly used as coupling agents to chemically modify silica surfaces in order to promote interactions with hydrocarbon rubbers. As a consequence, silica compounds need to undergo a chemical reaction between the surface silanol-groups of the filler and the alkoxy-groups of the coupling agents. During this silanization reaction ethanol is formed. This reaction is slowed down by the hold-up of the alcohol in the mixer and the compound, making the mixing process inefficient and expensive.

29.1.2 Network Formation with Silica: The Silanization Reaction

The surface of silica is covered by a layer of acidic silanol and siloxane groups. This highly polar and hydrophilic character of the filler surface results in a low compatibility with the rather apolar polymer. Besides, highly attractive forces between silica particles result in strong agglomeration forces. The formation of a hydrophobic shell around the silica particle by the silica–silane reaction prevents the formation of a filler–filler network by reduction of the specific surface energy [3].

Prior to the chemical reaction of the silane with the silanol-groups on the silica surface, the silane molecule has to make contact with the silica surface by adsorption. Then the chemical reaction of silica with an alkoxy–silyl moiety of the coupling agent takes place in a two-step, endothermic reaction. The primary step is the reaction of alkoxy-groups with silanol-groups on the silica filler surface [4]. Two possible mechanisms are reported:

(a) Direct reaction of the silanol-groups with the alkoxy-groups of the coupling agent
(b) Hydrolysis of the alkoxy-groups followed by a condensation reaction with the silanol-groups

The fact that the rate of silanization is influenced by the moisture content of the silica supports the mechanism wherein a hydrolysis step is involved. The reaction follows pseudo first-order kinetics. Figure 29.1 shows the mechanism of the primary reaction.

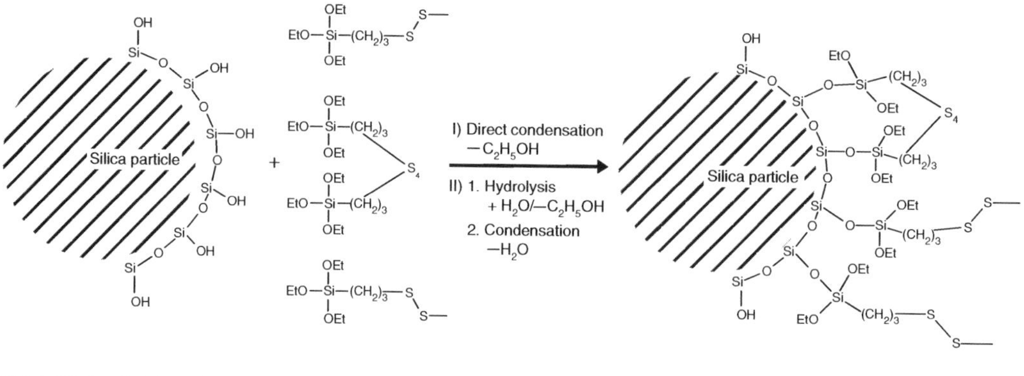

FIGURE 29.1 The primary reaction of silica with a silane. (From Hunsche, A. et al., *Kautschuk und Gummi Kunststoffe*, 50, 881, 1997.)

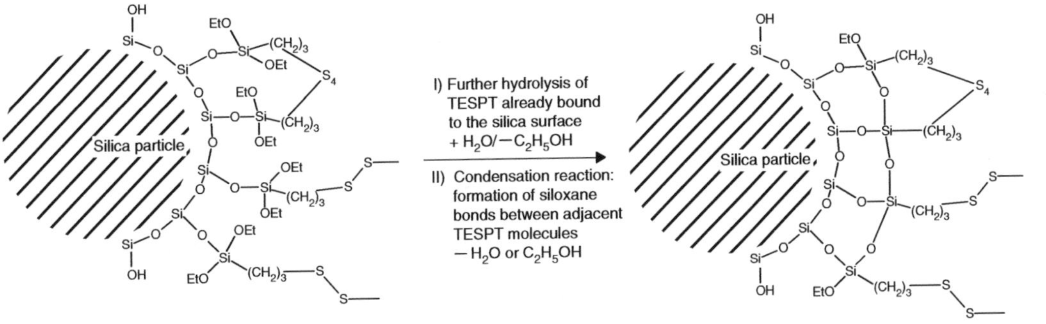

FIGURE 29.2 The secondary reaction of silica with a silane. (From Hunsche, A. et al., *Kautschuk und Gummi Kunststoffe*, 50, 881, 1997.)

The secondary reaction, depicted in Figure 29.2, is a condensation reaction between adjacent molecules of the coupling agent on the filler surface or between alkoxy-groups of the coupling agent and silanol-groups of the silica [6–8]. It is generally accepted, that again a hydrolysis step is involved in the reaction. In comparison to the primary reaction, this step is slower by a factor of approximately 10.

The coupling agents need a moiety enabling them to react with the polymer during vulcanization in order to be reinforcing. This is shown in experiments with coupling agents, which can react with the filler but not with the polymer: the reinforcing effect is reduced, as the formation of bond rubber is impossible [9]. In general, the moiety reacting with the polymer is a sulfur-group, either a poly- or disulphidic group or a blocked sulfur-group. Other functional groups used to link the coupling agents to the polymer are double bonds. These moieties need to be activated by the addition of an active sulfur-compound or by the generation of radical species, in order to obtain simultaneous cross-linking of the polymer and the coupling agent with comparable reaction rates during curing. Figure 29.3 shows a general reaction scheme for the formation of the filler–polymer bond.

The polysulfidic moieties of the silanes are unstable, and cleavage of the sulfur groups results in active sulfur species. A notorious problem with this kind of coupling agents is the balance between its reactivity towards the silica, requiring a temperature of at least 130°C to obtain an acceptable speed, and its eagerness to react with the rubber polymer, which starts to become noticeable at temperatures above 145°C. Furthermore, the primary and secondary reactions are chemical

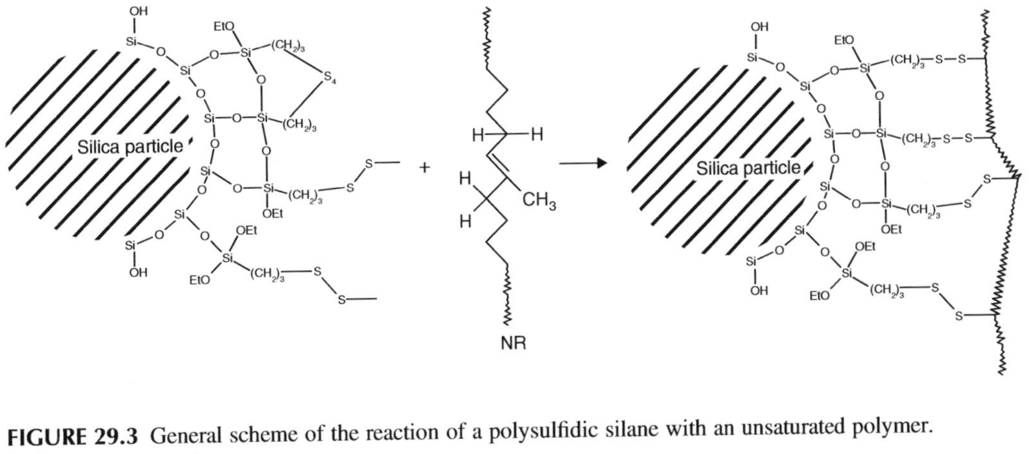

FIGURE 29.3 General scheme of the reaction of a polysulfidic silane with an unsaturated polymer.

equilibria, negatively influenced by the hold-up of ethanol in the mixer and the rubber batch. Removal of this ethanol is favored by high-temperature mixing. It leaves the rubber processor a very narrow window in the mixing operation, so as to limit the duration of the mixing cycle by quickly reaching a high temperature in the mix without the risk of creating premature scorch in the mixer [10].

29.1.3 RUBBER MIXING WITH EMPHASIS ON SILICA COMPOUNDS

Mixing of rubber compounds is a rather complicated process as components differing in structure, viscosity, and rheological behavior have to be homogenized. Physical processes, for example, the adsorption of polymers or other additives onto filler surfaces and the incorporation of polymers into voids of the filler structure, need to occur. In the case of silica, a chemical reaction between the coupling agent and the silanol-groups on the filler surface has to take place. Besides this, the mixing equipment—in general an internal mixer—is not very flexible and a short-term adjustment of the mixing conditions is difficult. Different physical mixing processes have to take place in order to guarantee a homogenous blend of all the compound ingredients on the required scale: laminar, distributive, and dispersive mixing [11–13].

(a) Extensive or laminar mixing: The primary mechanism of this mixing process is convection, associated with deformation of the matrix.
(b) Distributive mixing: The randomness of the spatial distribution of a minor constituent within a matrix is increased without change in the dimensions of this constituent. Examples are the distribution of a carbon black, silica, or accelerator masterbatch throughout a polymer matrix.
(c) Intensive or dispersive mixing: The size of clusters of the minor constituent is reduced. Examples are the breakdown of carbon black or silica agglomerates into smaller aggregates and primary particles. Filler agglomerates are broken, when the stress level within the deforming matrix exceeds the cohesive forces or surface tension forces of the agglomerates.

The most challenging part of rubber mixing is the dispersion of the filler: The filler agglomerates have to be broken into smaller particles, the aggregates, but not completely to the level of primary particles. An optimal particle size distribution has to be achieved in order to obtain the best properties of the final rubber product [14].

Figure 29.4 illustrates the transition from large agglomerates into smaller aggregates and primary particles, and gives an indication of their dimensions.

The temperature window for mixing of silica compounds is rather narrow, limited by the decreasing silanization rate and increasing risk of scorch. High temperatures improve the silanization rate due to the temperature dependence of the reaction and the enhanced rate of alcohol

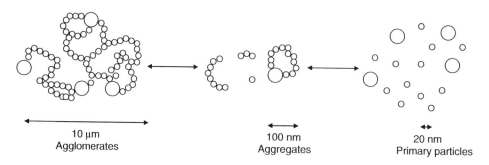

| 10 µm | 100 nm | 20 nm |
| Agglomerates | Aggregates | Primary particles |

FIGURE 29.4 Filler aggregation and dispersion. (From Li, Y. et al., *Rubber Chem. Technol.*, 67, 693, 1994.)

removal, as well as a reduced sterical hindrance of the silylpropyl-group of the coupling agent by increased thermal mobility [16]. The mixing temperature should be in the range of 145°C–155°C to achieve a good silanization and to avoid pre-cross-linking. Experience shows, that at a discharge temperature of 155°C the primary reaction is completed after approximately 8 min, when mixing is done in two separate steps, and a total mixing time of approximately 11 min are necessary to fully complete the silanization including the secondary reaction in 3–5 steps. The reaction rate is increased with increasing concentration of the silane, but finally reaches a plateau: The rate is limited by the reduced amount of accessible silanol-groups on the silica surface [17].

The silica surface is covered by silanol-groups, responsible for the strong interparticle forces. Silica particles agglomerate, are difficult to disperse, and tend to re-agglomerate after mixing [18]. One of the effects of re-agglomeration is the reduction of processability during storage: The viscosity of a compound increases during storage, with an increasing rate at higher temperatures. The viscosity increase follows a second-order kinetic law, suggesting that the process is based either on coalescence of filler-bound rubber entities or re-agglomeration of filler particles. The tendency of re-agglomeration is influenced by the water content of the silica: A higher water content results in a lower rate of viscosity increase during storage due to a shielding effect of the water molecules on the surface of the silica, which reduces the interparticle forces [19].

29.1.4 Special Characterization Methods for Silica–Rubber Coupling

29.1.4.1 Payne Effect

Above a critical filler concentration, the percolation threshold, the properties of the reinforced rubber material change drastically, because a filler–filler network is established. This results, for example, in an overproportional increase of electrical conductivity of a carbon black-filled compound. The continuous disruption and restoration of this filler network upon deformation is well visible in the so-called Payne effect [20,21], as represented in Figure 29.5. It illustrates the strain-dependence of the modulus and the strain-independent contributions to the complex shear or tensile moduli for carbon black-filled compounds and silica-filled compounds.

29.1.4.2 Hysteretic Properties: The Loss Angle δ at 60°C

The incorporation of reinforcing fillers into rubber results in most cases in an increase of the storage and loss moduli, G' and G'', and an increase in hysteresis, as quantified by the loss angle δ, where $\tan \delta$ is G''/G'. When properly dispersed and coupled to the rubber matrix via a coupling agent, as represented by a low Payne effect, silica also shows less hysteretic loss at elevated temperatures,

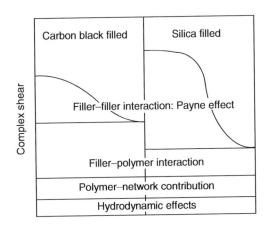

FIGURE 29.5 Effects contributing to the complex shear modulus.

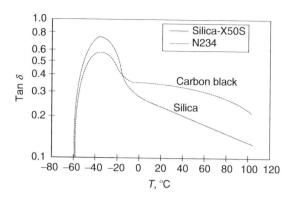

FIGURE 29.6 Temperature profile of the phase angle tan δ for a 75 phr N234 carbon black-filled versus a 75 phr silica/silane-reinforced green-tire compound. (From Wang, M.-J., *Rubber Chem. Technol.*, 71, 520, 1998.)

relative to carbon black. Figure 29.6 depicts the typical dependence of the tan δ as a function of temperature for a typical "green tire" compound containing 75 phr N234 carbon black versus 75 phr highly dispersible silica and 12 phr of a mixture of 50% bis-(triethoxysilylpropyl)tetrasulfane (TESPT) and 50% N330 carbon black [22]. In tire technology, it is common to take the tan δ in the temperature range below 0°C as indicative for winter use and from around 0°C to 30°C as representative for the performance of a tire with respect to (wet) traction or skid resistance. The temperature range between 30°C and 70°C comprises the running temperature of a tire. Under these temperature conditions the loss angle essentially is a measure for the degree of rolling resistance. Figure 29.6 clearly shows the large gain in lower tan δ-values at 60°C with silica relative to carbon black, responsible for the lower rolling resistance of silica, generally at the cost of some lower (dry) skid resistance. The loss in skid resistance of silica-reinforced tires on wet roads is compensated by the more hydrophilic nature of silica versus carbon black. Tan δ at 60°C is also often used as a measure of the reinforcing power of silica in rubber compounds, more closely related to its application in tires.

29.2 COMPOUNDING AND MIXING PRINCIPLES WITH FOCUS ON ZINC OXIDE ADDITION AND DUMP TEMPERATURE

Thurn and Wolff [23] mentioned in 1975 that zinc oxide might have a positive effect on the coupling reaction. The presence of zinc oxide in the compound caused a rapid decrease in the compound viscosity as measured by the mixing torque. Lateron Reuvekamp [14] did an elaborated investigation on the influence of zinc oxide (ZnO) on the processing behavior and final properties of silica compounds. A summary of these investigations and the consequences for compounding and mixing of silica compounds is given in the following paragraphs. From the results the conclusion can be drawn that ZnO acts as a catalyst for the primary reaction.

29.2.1 EXPERIMENTAL

Experiments were performed using a tire-tread composition as shown in Table 29.1, representing a common silica recipe corresponding to the fuel-saving green-tire technology [2,24].

The compounds were mixed in three steps: The first two steps were done in an internal mixer with a mixing chamber volume of 390 mL. The mixing procedures employed in the first two steps are indicated in Table 29.2. The starting temperature was 50°C and the cooling water was kept at a constant temperature of 50°C. The rotor speed was 100 rpm and the fill factor 66%. After every mixing step the compound was sheeted out on a 100-mL two-roll mill. The third mixing step was done on the same two-roll mill. The accelerators and sulfur were added during this step.

TABLE 29.1
Compound Composition

Component	phr
S-SBR	75.0
BR	25.0
Silica	80.0
Silane	7.0
Aromatic oil	32.5
ZnO	2.5
Stearic acid	2.5
Wax	1.5
6PPD	2.0
Sulfur	1.4
CBS	1.7
DPG	2.0
Total	233.1

The dump temperature of the compound was varied by changing the mixer's rotor speed and fill factor while keeping the other mixing conditions and the mixing time constant. Under the assumption that the final dump temperature is the main parameter influencing the degree of the silanization reaction, the effect of the presence of ZnO on the dynamic and mechanical properties of the compound was investigated. ZnO was either added on the two-roll mill or in the mixer.

The curing properties were determined with the aid of a rubber process analyzer (RPA 2000, Alpha Technologies) at a temperature of 160°C. The compounds were cured in a laboratory press at 160°C and 100 bar at a time corresponding to the t_{90} of the specific compound.

Dynamic measurements of the uncured compounds were also performed with the aid of the RPA 2000 at 100°C and a frequency of 0.5 Hz. The Payne effect was measured as the storage modulus G' at a low strain of 0.56%.

The mechanical properties of the cured compounds were determined using a tensile tester according to ISO-37.

TABLE 29.2
Mixing Protocol

Time (min/s)	Action
0.00	Open ram; add rubbers
0.20	Close ram
1.20	Open ram; add $\frac{1}{2}$ silica, $\frac{1}{2}$ silane, $\frac{1}{2}$ oil, ZnO, and stearic acid
1.50	Close ram
2.50	Open ram; add $\frac{1}{2}$ silica, $\frac{1}{2}$ silane, $\frac{1}{2}$ oil
3.30	Close ram
4.30	Open ram; sweep
4.45	Close ram
6.45	Dump
0.00	Load compound
5.00	Dump

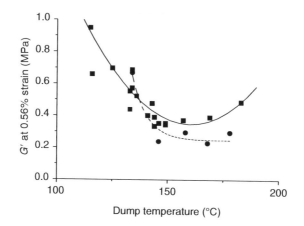

FIGURE 29.7 G' at 0.56% strain as a function of the dump temperature; ■ with ZnO; ● without ZnO.

29.2.2 RESULTS AND DISCUSSION

In Figure 29.7, the storage modulus G' decreases for a compound without ZnO as well as for a compound containing ZnO. However, the G' above 140°C stays on a low level for the compound without ZnO, while the modulus for a compound with ZnO tends to increase again. At higher strain values, the increase in G' above 150°C for the compound without ZnO is less compared to a compound with ZnO (Figure 29.8). This can be explained by the lower scorch sensitivity in the absence of ZnO and the interference of ZnO with the silanization reaction: ZnO acts as an activator for the reaction of the coupling agent with the rubber matrix.

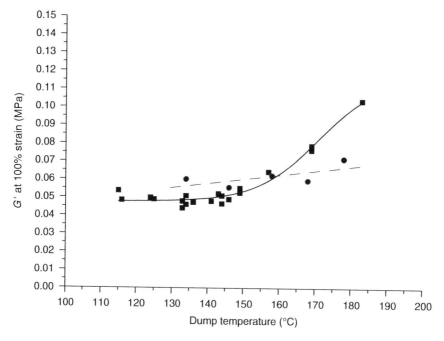

FIGURE 29.8 G' at 100% strain as a function of the dump temperature; ■ with ZnO; ● without ZnO.

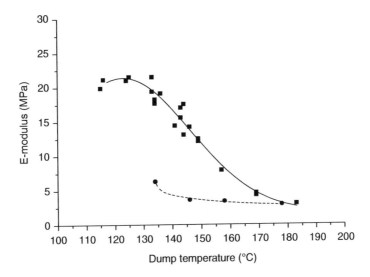

FIGURE 29.9 Modulus of elasticity as a function of the dump temperature; ■ ZnO added in the mixer; ● ZnO added on the mill.

At dump temperatures above 130°C, the Young's modulus of the vulcanized samples starts to decrease, as can be seen in Figure 29.9. As with the effects observed in the case of the storage modulus G', the decrease in the modulus of elasticity must be attributable to a reaction between the silica and the coupling agent, either the primary reaction alone or both the primary and the secondary reaction. As the Young's modulus is per definition a low-strain property, it is as sensitive to filler–filler interaction between silica aggregates as the G' at low-strain amplitudes. If ZnO is not present during mixing, but added later on the two-roll mill together with the curing agents, the Young's modulus is lower. The lower Young's modulus can be caused by a better hydrophobation reaction of the coupling agent leading to a lower filler–filler interaction.

The tensile properties of the vulcanized sample are listed in Table 29.3. The difference in properties between the compounds with ZnO added in the mixer, resp. on the mill, is rather small except for the 300% modulus. If ZnO is added in the mixer, the 300% modulus remains practically

TABLE 29.3

Mechanical Properties of the Vulcanized Compounds with ZnO Added in the Mixer, Resp. on the Two-Roll Mill at Different Dump Temperatures

Dump Temperature (°C)	Addition of ZnO	E-Modulus (MPa)	100% Modulus (MPa)	200% Modulus (MPa)	300% Modulus (MPa)	Stress at Break (MPa)	Strain at Break (%)
134	Mixer	19.2	1.9	4.8	8.8	14.7	440
145	Mixer	14.4	1.8	4.6	8.7	15.3	452
157	Mixer	7.9	1.6	4.6	9.1	15.6	438
169	Mixer	4.5	1.8	5.7	11.5	16.1	376
183	Mixer	3.0	1.8	6.8	14.3	15.1	311
134	Mill	6.3	1.4	4.1	8.4	16.2	476
146	Mill	3.6	1.5	4.8	10.1	14.9	390
158	Mill	3.4	1.5	4.7	9.6	15.4	409
168	Mill	2.9	1.5	5.2	11.4	13.8	341
178	Mill	2.9	1.5	5.1	10.6	13.2	344

constant up to a dump temperature of 150°C and then increases. The 300% modulus for the compound with ZnO added on the mill is barely influenced. The results of the dynamic measurements of unvulcanized compounds at low- and high-strain amplitudes (Figures 29.7 and 29.8) show the same trends as the stress–strain measurements of vulcanized compounds as given in Table 29.3.

ZnO influences the characteristics of a compound twofold: Primarily, ZnO influences the kinetics of the silica–silane reaction. Secondly, ZnO enhances the scorching behavior of a sulfur-containing coupling agent. This effect is clearly visible when looking at the G' at 100% strain and the 300% modulus.

29.3 INFLUENCE OF MIXING EQUIPMENT AND CONDITIONS ON THE SILANIZATION EFFICIENCY

Silica compounds are generally processed in conventional internal mixers, preferably with intermeshing rotors. These mixers are designed and optimized for carbon black-filled compounds in which mixing is based only on physical processes. When a silica–silane reinforcing system is used, additionally a chemical reaction, the silanization, occurs. One of the main influencing factors of the silanization reaction is the concentration of ethanol in the compound as well as in the mixer [25,26]. As the silanization finally reaches an equilibrium, low concentrations of ethanol in the compound are expected to enhance the reaction rate.

29.3.1 EXPERIMENTAL

In this series of experiments for the comparison of different mixer types and sizes, a pre-dispersed masterbatch was heated up to the final silanization temperature. The compound was a passenger car tire-tread masterbatch based on a blend of solution-based styrene–butadiene rubber (S-SBR) and butadiene rubber (BR) with silica and a coupling agent, either TESPT or TESPD, the disulfane variant of TESPT. Mixing and predispersion was done in a 320 L intermeshing mixer. A typical fingerprint of the mixing step is given in Figure 29.10.

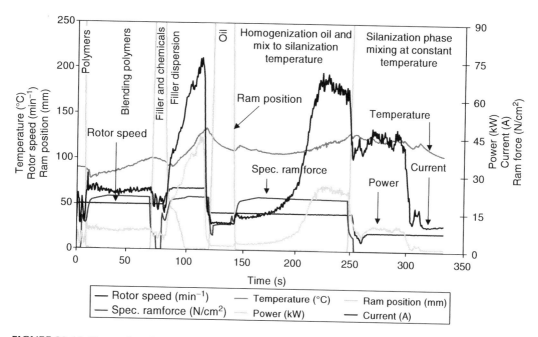

FIGURE 29.10 Fingerprint of masterbatch mixing.

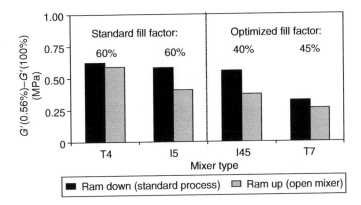

FIGURE 29.11 Effect of silanization in an open mixer on the Payne effect for different mixer types and fill factors (silanization time 150 s, temperature 145°C, T4: tangential 3.6 L, I5: intermeshing 5 L, I45: intermeshing 45 L, T7: tangential 7 L).

29.3.2 RESULTS AND DISCUSSION

One measure to improve the devolatilization of the compound is to work in an open mixer during the silanization step. In the experiments described in the following paragraphs, silanization was done pressure-less in an open mixer for 150 s at a constant temperature (145°C). The temperature was held constant during the silanization period by adjustment of the rotor speed. After 150 s of silanization the compound was discharged and further analyzed.

Figures 29.11 and 29.12 demonstrate the results of the silanization efficiency for different mixers and different fill factors under standard conditions (closed mixer) compared to working with an open mixer. The silanization efficiency is measured by viscosity and Payne effect.

A significant improvement of the silanization efficiency is found when silanization is done in an open mixer with a high fill factor (60%) as well as with a reduced optimized fill factor (40%, 45%) as indicated by a reduced Payne effect and compound viscosity. This effect is more pronounced in an intermeshing mixer compared to a tangential mixer. The results on small scale (T4, I5, T7) are verified on larger scale in the intermeshing 45-L mixer (I45). In this mixer, the increase in silanization efficiency by working with an open mixing chamber is even stronger compared to the smaller mixer types. The absolute level of silanization is independent of mixer size.

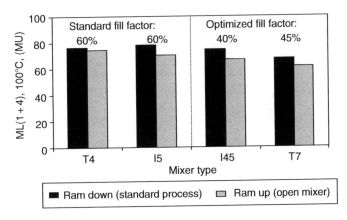

FIGURE 29.12 Effect of silanization in an open mixer on Mooney viscosity for different mixer types and fill factors (silanization time 150 s, temperature 145°C).

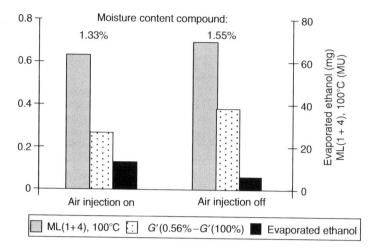

FIGURE 29.13 Influence of air injection on the silanization efficiency (mixer volume 45 L, fill factor 0.4, silanization temperature 145°C, time 150 s).

Reduction of the fill factor in order to improve the intake behavior when working in an open mixer has an additional positive effect on the silanization efficiency. A combination of both measures, silanization in an open mixer and reduction of the fill factor, leads to further improvement.

Another measure to improve the removal of ethanol is air injection into the mixer during the silanization step. Air can be injected from the bottom part of the mixer using existing valves without any special outlet for the injected air. In these experiments air injection is switched on once the compound reached the silanization temperature (145°C) and the rotor speed is adjusted in order to maintain the silanization temperature. Figure 29.13 shows the properties of this compound compared to a compound that was silanized under the same conditions except with air injection switched off. Air injection lowers the Payne effect, Mooney viscosity, and water content in the compound, and ethanol removal is more effective. All other properties are comparable to the properties of a standard silica compound.

A better cooling efficiency of the compound can be achieved by a lower cooling temperature of the mixing chamber and rotors. Tests were done in order to measure the effect of a decreasing cooling temperature on the silanization efficiency. Figure 29.14 shows the effect of this measure in a

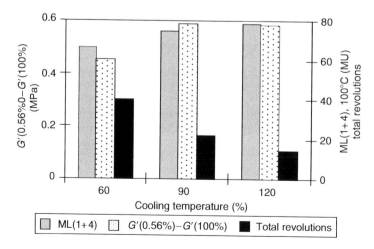

FIGURE 29.14 Influence of cooling temperature on silanization efficiency in a tangential mixer (mixer volume 3.6 L, open mixer, fill factor 0.6, silanization temperature 145°C, time 150 s).

tangential mixer: In both cases viscosity as well as Payne effect is reduced when the cooling temperature is reduced from 120°C to 60°C. At the same time the input of mechanical energy and total number of revolutions necessary to maintain the temperature level are increased.

To summarize, the following factors are found to increase the silanization efficiency:

- Mixing pressure-less in an open mixer: After mixing and dispersion, the following silanization step can be done in an open mixer. This enhances ethanol evaporation and thus increases the silanization efficiency. The best way to achieve this is working with two mixers: One standard mixer (preferably with intermeshing rotor geometry) for mixing and dispersion, followed by the silanization step done in a specially designed reactor. The role of the silanization reactor is to keep the temperature on the desired level, to create a high level of fresh surface for devolatilization of the compound and to allow the ethanol to escape out of the reactor.
- Air injection: This measure has a positive effect on the silanization efficiency without negatively influencing the properties of the material. The combination of air injection with working in an open mixer adds the positive effects of the two measures.
- Temperature control settings: Intensive cooling of the mixing chamber and the rotors increases the silanization efficiency due to a better devolatilization of ethanol from the compound.

29.3.3 Modeling Devolatilization of Ethanol in an Internal Mixer

The devolatilization of a component in an internal mixer can be described by a model based on the penetration theory [27,28]. The main characteristic of this model is the separation of the bulk of material into two parts: A layer periodically wiped onto the wall of the mixing chamber, and a pool of material rotating in front of the rotor flights, as shown in Figure 29.15. This flow pattern results in a constant exposure time of the interface between the material and the vapor phase in the void space of the internal mixer. Devolatilization occurs according to two different mechanisms: Molecular diffusion between the fluid elements in the surface layer of the wall film and the pool, and mass transport between the rubber phase and the vapor phase due to evaporation of the volatile component. As the diffusion rate of a liquid or a gas in a polymeric matrix is rather low, the main contribution to devolatilization is based on the mass transport between the surface layer of the polymeric material and the vapor phase.

Based on this model the flow rate of ethanol can be estimated by using the specific parameters of the mixing process and kinetic factors (Equation 29.1):

$$Q\frac{t}{V} = \left[2\sqrt{\frac{D}{\pi \cdot t_{e,l}}}(c - c_{eq})A_l + 2\sqrt{\frac{D}{\pi t_{e,p}}}(c - c_{eq})A_p \right]\frac{t}{V} = k_1 \cdot c \cdot t \qquad (29.1)$$

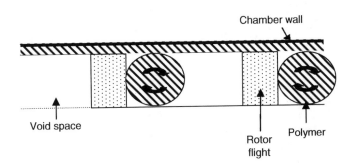

FIGURE 29.15 Flattened model description of the cross-section of a partially filled mixing chamber.

The flow rate and, as a consequence, the silanization efficiency is influenced by the following factors:

D	Diffusion coefficient of ethanol in the polymer matrix
A_l	Area of the compound layer on the mixer wall
A_p	Area of the surface of the pool material
c	Concentration of ethanol in the compound
c_{eq}	Concentration of ethanol in the compound surface layer in equilibrium with the gas phase
k_I	First-order reaction constant for the silanization reaction
Q	Volumetric flow rate of ethanol from the compound to the gas phase
t	Time
$t_{e,l}$	Exposure time of the wall layer
$t_{e,p}$	Exposure time of the pool surface
V	Volume of the compound

This model demonstrates the following factors to enhance devolatilization and hence silanization:

- High operation temperature (limited by scorch phenomena) in order to raise the speed of the silane reaction with silica and the diffusion coefficient of ethanol
- Small mixer volume
- Low fill factor of the mixer
- High rotor speed
- Long reaction/mixing time
- Low partial pressure of ethanol in the void space of the mixer

To summarize, the kinetics of the silanization reaction are strongly influenced by the efficiency of the devolatilization process. The degree of devolatilization mainly depends on processing conditions (e.g., rotor speed and fill factor), mixer design (e.g., number of rotor flights, size of the mixer), and material characteristics. The diffusion coefficient of the volatile component in the polymeric matrix is of minor influence.

29.4 SUMMARY

Mixing of silica compounds basically differs from mixing of carbon black compounds as a chemical reaction has to take place: the silanization reaction. The crucial steps in this reaction are absorption of the coupling agent onto the silica surface and the chemical reaction of the coupling agent with the silanol-groups on the filler surface during mixing, followed by the reaction of the coupling agent with the polymer during curing of the material.

The main influencing factors for the silanization reaction are:

1. Interaction with other compound ingredients, e.g., polar compounding additives such as ZnO, amines or oils
2. Kinetic hindrance of the silanization reaction due to high concentration of the reaction product, ethanol, in the compound

In the presence of ZnO, the dump temperature is a parameter of paramount importance in mixing of a silica compound containing a coupling agent with a polysulfidic moiety: The sulfur eliminated from the coupling agent causes premature scorch in the presence of ZnO. This problem is partially solved when ZnO is added in the productive mixing step on the two-roll mill.

The efficiency of the silanization reaction is increased by all measures enhancing devolatilization of ethanol from the silica compound in the mixer. One possible measure is the reduction of the ethanol concentration in the void space of the mixer, thus increasing the driving force for mass transfer across the interface between the compound and the gas phase. This can practically be implemented by working in an open mixer during the silanization step or by flushing the mixer with a strong current of air. The driving force for ethanol devolatilization can also be increased by an intensive cooling of the mixing chamber and the rotors. As a consequence, the rotor speed has to be increased in order to keep the temperature at the required level, resulting in higher silanization efficiency due to a better devolatilization of the compound.

REFERENCES

1. Wagner, M.P., *Rubber Chemistry and Technology*, 49, 703, 1976.
2. Rauline, R., European Patent Appl. 0 501227A1, 1992.
3. Wang, M.J., *Rubber Chemistry and Technology*, 71(3), 520, 1998.
4. Luginsland, H.-D., in *Proc. Rubberchem 99*, Antwerp, Belgium, Paper No. 22, 1999.
5. Hunsche, A. et al., *Kautschuk und Gummi Kunststoffe*, 50(12), 881, 1997.
6. Görl, U. and Hunsche, A., in *Proc. 150th ACS Rubber Division Meeting*, Louisville, KY, Paper No. 76, 1996.
7. Hunsche, A. et al., *Kautschuk und Gummi Kunststoffe*, 51(7–8), 525, 1998.
8. Görl, U. et al., *Rubber Chemistry and Technology*, 70(4), 608, 1997.
9. Brinke, J.W. et al., *Rubber Chemistry and Technology*, 76(1), 12, 2003.
10. Dierkes, W., *Economic mixing of silica compounds*, Print Partners Ipskamp, Enschede, The Netherlands, 2005.
11. Leblanc, J.L., *Progress and Trends in Rheology*, II, 32, 1988.
12. Johnson, P.S., *Elastomerics*, 9, January 1983.
13. Chohan, R.K. et al., *International Polymer Processing*, 2(1), 13, 1987.
14. Reuvekamp, L.A.E.M., *Reactive Mixing of Silica and Rubber for Tyres and Engine Mounts*, Twente University Press, Enschede, The Netherlands, 2003.
15. Li, Y. et al., *Rubber Chemistry and Technology*, 67, 693, 1994.
16. Patkar, S.D., Evans, L.R., and Waddel, W.H., *Rubber and Plastics News*, 237, September 1997.
17. Luginsland, H.-D., in *Proc. 11th Intern. SRC Meeting*, Púchov, 1999.
18. Wolff, S. et al., *European Rubber Journal*, 16(January), 16–19, 1994.
19. Schaal, S., Coran, A.Y., and Mowdood, S.K., *Rubber Chemistry and Technology*, 73, 240, 2001.
20. Payne, A.R. and Whittaker, R.E., *Rubber Chemistry and Technology*, 44, 440, 1971.
21. Payne, A.R., *Rubber Chemistry and Technology*, 39, 365, 1966.
22. Wang, M.-J., *Rubber Chemistry and Technology*, 71(3), 520, 1998.
23. Thurn, F. and Wolff, S., *Kautschuk und Gummi Kunststoffe*, 53, 10, 1975.
24. Leblanc, J.L. and Mongruel, A., *Progress in Rubber and Plastics Technology*, 17, 162, 2001.
25. Dierkes, W. et al., *Kautschuk Gummi Kunststoffe*, 56, 338, 2003.
26. Dierkes, W. et al., *Rubber World*, 229(6), 33, 2004.
27. Dierkes, W. and Noordermeer, J., *International Polymer Processing*, 22(3), 259, 2007.
28. Dierkes, W.K. and Noordermeer, J.W.M., *Proc. 167th ACS Rubber Division Meeting*, San Antonio, TX, Paper No. 46, 2005.

30 Stream Effects and Nonlinear Viscoelasticity in Rubber Processing Operations

Jean L. Leblanc

CONTENTS

30.1   INTRODUCTION

With respect to the history of sciences and techniques, it is a common observation that pragmatical engineering, or even craft industry, came before—often, by several centuries—the scientific understanding of material properties and/or natural phenomena exploited in industrial operations. Quite obvious for a number of old-known materials such as mineral glass and metals, this observation is particularly pertinent in rubber technology. Indeed the principles of essential operations such as mastication (Hancock—1832) and vulcanization (Goodyear—1939) were discovered long before the macromolecular nature of polymers was demonstrated (Staudinger—1922). But elastomers are practically never used alone, but as compounds with various ingredients, with filler(s), e.g., carbon black or silica, and processing oil being the major ones, and it is through more than one century of pragmatic developments (i.e., *trial-and-error*), that rubber engineers eventually mastered the preparation and processing of complicated formulations. Filled rubber compounds are highly viscous and heterogeneous systems, exhibiting a number of singular viscoelastic effects, and the rigorous approach of their flow properties is only at the beginning, namely by recognizing at last their character of "complex polymer systems". Correctly assessing the rheological properties of such materials is therefore a challenging task, in part because most of traditional rheological instruments, at first designed with respect to much simpler fluids, do not match well with the peculiar flow properties of complex polymer systems.

From a rheological point of view, rubber compounds may be viewed as complex polymer systems that consist of essentially three classes of ingredients: High-molecular weight polymer(s), powdery filler(s), and low viscosity chemicals. Such systems are of course heterogeneous by nature and exhibit a strong viscoelastic character, most of the time nonlinear in current stress and strain rate conditions. In addition, there are strong interactions between the various phases, either purely physical, or chemical, or both. This makes flow measurement on such materials a very difficult task, which requires specific instruments and adequate test protocols for obtaining significant and reproducible results.

Most rubber processing operations occur at a high rate of strain, and therefore it is essentially the nonlinear viscoelastic response of rubber compounds that is of interest. This called for developments in the rheometry of complex polymer systems and, amongst various proposals, commercial torsional dynamic rheometers, with a closed test cavity, i.e., the "Rubber Process Analyzer", RPA 2000, or the "Production Process Analyzer", PPA (Alpha Technologies, now a Dynisco company), proved to provide very reproducible and meaningful results with a number of complex polymer systems, including of course highly filled rubber compounds. In order to extend its capabilities, such an instrument was purposely modified in our laboratory with the objective to develop what has been called Fourier transform (FT) rheometry. The aim of this chapter is to thoroughly describe this test technique and to use it to study how viscoelastic properties of filled rubber compounds evolve along the processing line, owing to what we call "stream effects".

30.2 SPECIFIC ASPECTS OF RUBBER RHEOMETRY

With respect to the high rate of strain region concerned by rubber processing operation, one should at least address the flow functions, for instance the shear viscosity function, i.e., $\eta = f(\dot{\gamma}, T)$ where $\dot{\gamma}$ is the shear rate and T the temperature, and it is worth recalling here that the shear thinning behavior of most polymer materials is of course a nonlinear viscoelastic response. The shear rate range of interest is obviously the processing range, which is generally between 10 and 1000 s^{-1} (Note: certain processing operations like injection molding work in a higher shear rate range, i.e., several thousands per second, where testing difficulties are of another nature because elastic and anisotropic effects are dominant). Only the capillary rheometer has the capability to investigate the 10–1000 s^{-1} shear rate range but it is nowadays recognized that this test technique, owing to its tediousness, the long testing time needed, and the necessary operator skills for significant results, cannot be considered for routine testing on the factory floor. Furthermore, if capillary rheometry is clearly understood in terms of steady shear conditions, the information that the instrument can provide about the elastic character of tested materials is still subject to debate. There are also a number of concerns about wall slippage effects, most likely with rubber formulations owing to their small labile ingredients, which complicate the treatment of capillary rheometer results, notably with respect to antagonistic effects between compressibility and wall slip. Those aspects are beyond the scope of the present chapter but a thorough discussion can be read elsewhere.[1]

Over the twentieth century, the rubber industry has developed special rheometers, essentially factory floor instruments either for checking process regularity or for quality control purposes, for instance, the well-known Mooney rheometer (1931), the oscillating disk rheometer (1962), and the rotorless rheometer (1976). All those instruments basically perform simple drag flow measurements but they share a common feature: During the test, the sample is maintained in a closed cavity, under pressure, a practice intuitively considered essential for avoiding any wall slip effects. Indeed it has long been shown that, with rubber materials, only pressurized conditions allow reproducible data to be obtained.[2] Over the last decades, closed cavity torsional dynamic rheometers proved to be convenient instruments to assess the viscoelastic properties of rubber materials, for instance the Rubber Process Analyzer, RPA 2000 (Alpha Technologies), extensively described elsewhere[3] and whose operation in both the linear and the nonlinear viscoelastic region of simple polymer systems was fully validated.[4]

The measuring principle of torsional rheometers consists in applying a sinusoidal strain in order to assess the resulting complex stress supported by the material. When the latter is viscoelastic, there is out-phasing of the (complex) stress with respect to the applied strain. Both the measured complex stress and the phase angle give access to the viscoelastic properties of the material, in terms of elastic and viscous moduli, currently termed real and imaginary parts of the complex modulus, with respect to a standard treatment of measured quantities. The test principle of the RPA is described in Figure 30.1. The testing gap consists in a bi-conical test chamber with grooved dies to prevent slippage. The lower die can be oscillated in torsion at controlled strain and frequency. The torque measuring system is fixed on the upper die and calibrated with a torsion spring. The dies temperature control system has a resolution to the nearest 0.1°C and the test cavity is maintained closed through the application of a force of 15.7 kN, which corresponds to a pressure of around 12 MPa. The instrument is fully monitored by an external computer (PC) with the capability to combine preprogrammed test sequences in any order.

Like any dynamic strain instrument, the RPA readily measures a complex torque, $S*$ (see Figure 30.1) that gives the complex (shear) modulus $G*$ when multiplied by a shape factor $B = 2\pi R^3/3\alpha$, where R is the radius of the cavity and α the angle between the two conical dies. The error imparted by the closure of the test cavity (i.e., the sample's periphery is neither free nor spherical) is negligible for Newtonian fluids and of the order of maximum 10% in the case of viscoelastic systems, as demonstrated through numerical simulation of the actual test cavity.[4]

The RPA built-in data treatment extracts from the recorded torque signal 16 discrete values in order to calculate through a discrete FT (U.S. Patent 4,794,788) the real and imaginary components

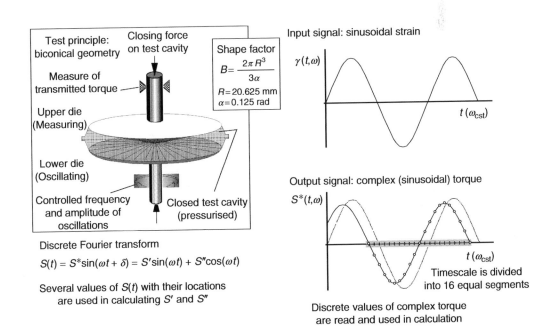

FIGURE 30.1 Testing principle and built-in data treatment of the RPA.

of the complex torque, that is the elastic S' and the viscous torque S'' (at the frequency ω and strain amplitude γ considered), respectively. The following classical viscoelastic relationships exist between the various dynamic quantities:

$$S*(\omega) = \sqrt{S'^2(\omega) + S''^2(\omega)} \quad \delta = \tan^{-1} \frac{S''}{S'}$$

$$G* = \frac{S*}{B\gamma} \quad G' = \frac{S* \cos \delta}{B\gamma} \quad G'' = \frac{S* \sin \delta}{B\gamma}$$

$$\eta'(\omega) = \frac{G''(\omega)}{\omega}: \text{ dynamic viscosity function} \quad \tan \delta(\omega) = \frac{G''(\omega)}{G'(\omega)}$$

It is clear that this data treatment is strictly valid providing the tested material exhibits linear viscoelastic behavior, i.e., that the measured torque remains always proportional to the applied strain. In other words, when the applied strain is sinusoidal, so must remain the measured torque. The RPA built-in data treatment does not check this $\gamma(\omega)/S*(\omega)$ proportionality but a strain sweep test is the usual manner to verify the strain amplitude range for constant complex torque reading at fixed frequency (and constant temperature).

At sufficiently low strain, most polymer materials exhibit a linear viscoelastic response and, once the appropriate strain amplitude has been determined through a preliminary strain sweep test, valid frequency sweep tests can be performed. Filled rubber compounds however hardly exhibit a linear viscoelastic response when submitted to harmonic strains and the current practice consists in testing such materials at the lowest permitted strain for satisfactory reproducibility; an approach that obviously provides "apparent" material properties, at best. From a fundamental point of view, for instance in terms of material sciences, such measurements have a limited meaning because theoretical relationships that relate material structure to properties have so far been established only in the linear viscoelastic domain. Nevertheless, experience proves that "apparent" test results can be well reproducible and related to a number of other viscoelastic effects, including certain processing phenomena.

When compared to standard (open cavity) cone-plate or parallel disks rheometers, closed cavity torsional rheometers such as the RPA or the PPA have unique high-strain capabilities, which prompted us to modify the instrument in order to investigate the promises of FT rheometry, as outlined a few years ago by the pioneering works of Wilhelm.[5] The technique consists of capturing strain and torque signals and in using FT calculation algorithms to resolve it into their harmonic components, as detailed below.

30.3 FLOW SINGULARITIES OF FILLED RUBBER COMPOUNDS: A BRIEF OVERVIEW

30.3.1 INTRINSIC COMPLEXITY OF RUBBER COMPOUNDS

Rubber compounds are complex polymer systems and a brief discussion on their hidden complexity is an interesting preliminary step in foreseeing the rheological test techniques (including data treatment) that can be best considered. Rubber processing is a sequence of operations, with varied complexity, whose development has been essentially pragmatic over the decades. Indeed, the objectives assigned to the rubber engineer are to eventually produce parts with appropriate design, properties, and performances, at minimal cost, with a constant quality. Whatever the concept and design of the rubber part, the first step consists in preparing "compounds", that is, formulations with a number of different ingredients, with the elastomer(s) and the filler(s) the major components (see Figure 30.2).

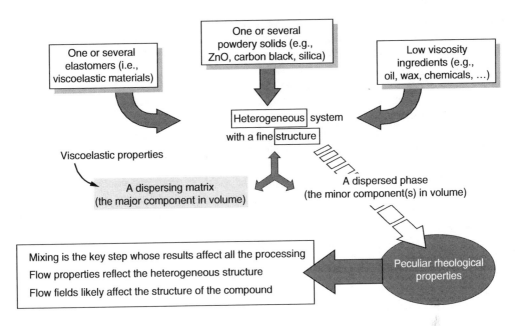

FIGURE 30.2 Intrinsic complexity of rubber compounds.

All the compounding ingredients have either limited or no mutual solubility (or compatibility) and it results in an heterogeneous system with a fine structure[6] whose importance in the final properties of cured parts is known for long. Obviously the rubber(s) and the filler(s) are the most important elements of the structure not only because they are the largest components on a weight (and volume) basis, but also because there are very strong interactions occurring between them. Contemporary research has revealed that the rubber–filler structure is playing the key role in the flow properties of uncured compounds;[7,8] but the structure is also somewhat sensitive to the flow fields in processing steps.

30.3.2 STREAM RHEOLOGICAL EFFECTS IN RUBBER PROCESSING

In rubber technology, mixing is the first step in a series of operations and obviously the most important one because there are "cascade" or "stream" effects (see Figure 30.3). Along a

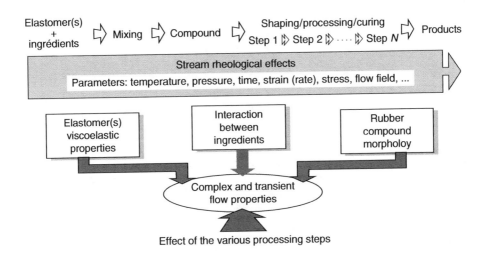

FIGURE 30.3 Stream or "cascade" effects in rubber processing.

processing line, a rubber compound experiences a large spectrum of strain, stress, and temperature conditions, generally badly documented. At a given point of the process, the material has thus an important strain–stress–temperature history that is obviously affecting its behavior during the subsequent steps. This is what we mean by the term "stream effects".

Engineers on the factory floor are well aware of such stream effects and they know very well that regular and consistent production is achieved only through the strict respect of procedures, with many controls at various steps. The key operation remains of course the mixing which, in addition to the obvious objective of achieving an optimal dispersion of all compounding ingredients, must yield compounds with stabilized rheological properties, necessary for easy processing in the subsequent operation, extrusion or molding, and of course vulcanization. Such goals are achieved through a number of steps, generally set up in a pragmatic manner, with the associated large energy consumption. A practical demonstration of the magnitude of stream effects was brought through purposely design experiments, performed within the framework of a recent European cooperative project (Brite-Euram "Prodesc"; *Process Description for Silica and Carbon Black compounds*— contract BRPR-CT98–0625). As shown in Figure 30.4, samples were taken at different positions along the compounding line of a silica-based tread band compound when preparing five different batches, and the Mooney viscosity ML(1+4) was measured at 100°C on all samples. As can be seen, viscosity data are scattered when measured at the beginning of the process, and the viscosity is at its highest level, then it reduces, as well as the scatter, until a stable rheological state is achieved. If one associates the highest batch viscosity with an incomplete dispersion of compounding ingredients and/or with still developing rubber–filler interactions, it is clearly seen that open mill and extrusion operations (steps 1 and 2) have a contribution that is nearly equivalent to internal mixing steps. In other terms, the imparted mixing energy is as important on open mill and in the extruder as in the internal mixer.

It follows of course that if something wrong happens at a given step of the compounding procedure, it will likely reflect immediately on the subsequent steps, without guarantee that the

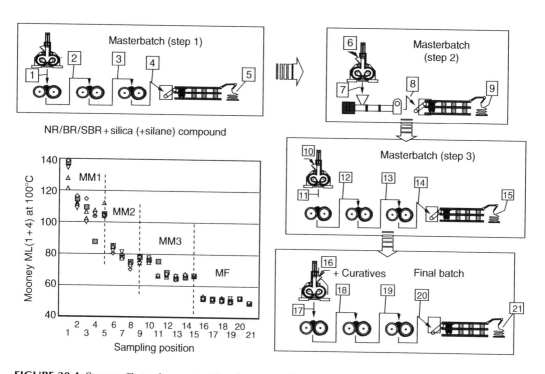

FIGURE 30.4 Stream effects along a tread band compounding line.

perturbation will be conveniently dampened down. It means also that the capability to predict (to foresee) how a rubber compound will likely behave in subsequent processing steps not only depends on a complex set of properties that one should measure on the material, but also on the exact knowledge of the flow conditions that the material will encounter. An immediate conclusion is that, with respect to the complexity of processing operations and to the presently limited understanding of the fundamental aspects of most processing situations (essentially in the nonlinear viscoelastic domain of the rubber material), it would be somewhat questionable to assign predicting capabilities to any single rate laboratory test. This is largely demonstrated by the recurrent failure of all so-called "processability tests (or testers)" that were considered over the last decades. In the author's opinion, the best one can expect is that a suitable testing procedure, providing it addresses the appropriate range of strain, stress, and temperature, might be able to somewhat anticipate the likeliness of a bad or acceptable behavior of the material. But the predicting capabilities of any approach will be limited by the very random nature of rubber compounds: indeed the rubber itself is a statistical assembly of a large number of polymer chains with various lengths (or molecular weights), the dispersion of the filler(s) and the other ingredients is by nature a stochastic (i.e., random) process and, finally, along the processing line the material experiences a complicated history of strain, rate, and temperature.

Prediction would only be possible if "trajectories" were known in advance, through the appropriate mathematical tools, as do classical mechanics. Applied to rubber processing, the concept of "trajectory" must receive obviously a very broad sense, i.e., how the material evolves along a complex space where stress, strain rate, temperature, filler dispersion, material memory, etc. are parameters. One understands intuitively that, with material that have a "statistical" nature, which are handled through processes that have also a "stochastic" nature, there are an infinity of such trajectories; and it follows that whatever is the complexity of a test on the material at a given point of the process, the results of such a test have predictive capabilities that are so far limited to our understanding of the complex rheological events and the associated interactions between compounding ingredients.

The above discussion provides valuable indication about the type of instrument and/or testing procedure one should consider. Obviously single rate instrument/method are inappropriate because, with respect to stream rheological effects observed in rubber processing, it is the knowledge of how the flow properties are likely to vary with the strain rate that is needed. Ideally how the strain and rate fields affect the rubber–filler structure would be valuable information but this is not really attainable, with the present level of science and technology—at least in a practical manner—despite encouraging progress.[9,10]

With respect to stream rheological effects in rubber processing, and despite all the restrictions discussed above, it seems nevertheless that the key information is how the nonlinear viscoelasticity is related to the processing behavior of rubber compounds. Such information can be deduced from the appropriate test procedure with the RPA, providing one considers the capabilities of the instrument to provide nonlinear viscoelastic data.

30.4 FOURIER TRANSFORM RHEOMETRY WITH AN UPDATED CLOSED CAVITY TORSIONAL RHEOMETER

30.4.1 Fourier Transform Rheometry: Principle of the Technique

FT rheometry is a powerful technique to document the nonlinear viscoelastic behavior of pure polymers as observed when performing large amplitude oscillatory strain (LAOS) experiments.[5,11] When implemented on appropriate instruments, this test technique can readily be applied on complex polymer systems, for instance, filled rubber compounds, in order to yield significant and reliable information.[12] Any simple polymer can exhibit nonlinear viscoelastic properties when submitted to sufficiently large strain; in such a case the observed behavior is so-called extrinsic nonlinear viscoelasticity owing to external factors (i.e., the applied strain). Complex polymer systems appear to be intrinsically nonlinear, since their response to applied strain has an internal origin, i.e., their morphology, and superposes to the nonlinearity associated with large strain. Filled rubber compounds are known for long to be nonlinear and FT rheometry allows a very fine characterization of their behavior, which has however to be supplemented by additional data treatment, as detailed below.

Essentially, FT rheometry consists in capturing strain and torque signals during dynamic testing and in using FT calculation algorithms to resolve it into their harmonic components. Figure 30.5 illustrates the principle of this test technique, with data gathered when submitting, at 100°C, to an harmonic strain of 22.5° amplitude at 1 Hz frequency, either a gum polybutadiene or a carbon black-filled butadiene rubber (BR) compound (i.e., NeoCis BR 40: 100; N330 carbon black: 50; zinc oxide: 5; stearic acid: 3; process oil: 5; trimethyl quinoline, polymerized: 2; iso-propyl-paraphenylenediamine: 1; compound prepared in Banbury mixer with a mixing energy of ≈ 1500 MJ/m^3). In such testing conditions, no linear viscoelastic response is obtained as demonstrated by the significant distortion of the recorded torque signals.

FT is essentially a mathematical treatment of harmonic signals that resolved the information gathered in the time domain into a representation of the measured material property in the frequency domain, as a spectrum of harmonic components. If the response of the material was strictly linear, then the torque signal would be a simple sinusoid and the torque spectrum reduced to a single peak at the applied frequency, for instance 1 Hz, in the case of the experiments displayed in the figure. A nonlinear response is thus characterized by a number of additional peaks at odd multiples of the

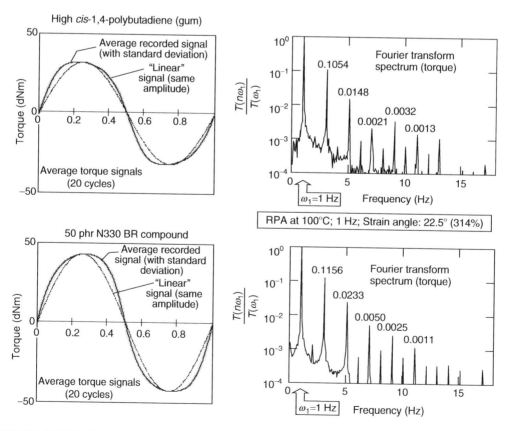

FIGURE 30.5 Fourier transform (FT) resolution of torque signals into harmonic components; notes that relative harmonic components are displayed.

applied strain frequency, as clearly shown in the right part of Figure 30.5. Wilhelm et al. demonstrated the fact that only odd harmonics are significant in terms of the material's response in theoretical considerations.[13,14] Indeed if a shear strain of maximum strain amplitude γ_0(rad) is applied at a frequency ω(rad/s) to a viscoelastic material, the strain varies with time t(s) according to $\gamma(t) = \gamma_0\sin(\omega t)$ and the shear stress response can be expected to be given by a series of odd harmonics, i.e.,

$$\sigma(t) = \sum_{j=1,3,5,\ldots}^{\infty} \sigma_j\sin(j\omega t + \delta_j) \qquad (30.1)$$

provided one assumes that, over the whole viscoelastic domain (i.e., linear and nonlinear), the viscosity function $\eta = f(\dot{\gamma})$ can be approximated by a polynomial series with respect to the shear rate. If the tested material exhibits a pure linear viscoelastic response, Equation 30.1 reduces to the first term of the series, as considered in most standard dynamic test methods.

At 314% strain, both the gum and the filled BR give clearly a nonlinear response, as shown by the distortion of the torque signal, which reflects a lack of proportionality between the applied strain (perfectly sinusoidal) and the measured torque. With a simple, homogeneous polymer material (here BR gum) nonlinearity occurs from the application of high strain. The distortion is more severe with the compound and in addition seems to affect more the right part of the half signal, an effect that is attributed to the presence of carbon particles. In other terms, there is a substantial difference between the nonlinear viscoelastic behavior of a pure, unfilled polymer and of a filled material. The former essentially exhibits nonlinearity through the application of a sufficiently large strain and we called this behavior *extrinsic nonlinear viscoelasticity* (because occurring through *external* causes, i.e., the applied strain), while the latter shows *intrinsic nonlinear viscoelasticity* (because owing to the *internal* morphology of the material). It is quite obvious that standard strain sweep experiments, while possibly detecting the occurrence of the nonlinear viscoelastic response, have absolutely no capability to distinguish extrinsic and intrinsic characters.

30.4.2 UPDATING A TORSIONAL DYNAMIC TESTER FOR FOURIER TRANSFORM RHEOMETRY

The appropriate modifications were brought to a commercial torsional rheometer, i.e., the RPA, in order to capture strain and torque signals, as schematically illustrated in Figure 30.6. Details on the modification and the measuring technique were previously reported.[15] Essentially strain and torque signals, as provided by the instrument, are collected by means of an electronic analogic-digital conversion card that reads tensions, which are proportional to the applied strain and the transmitted torque. The card has a resolution of 16 bits with a maximum sampling rate of around 205 K sample/s for two inputs of ± 10 mV to ± 10 V. Read values are converted into physical quantities (i.e., degree angle and deci Newton meter [dNm]) and recorded with a PC, through the appropriate codes written in LabView software (National Instruments). Test results consist thus of data files of actual harmonic strain and stress readings versus time. Proprietary data handling programs, written in MathCad (MathSoft Inc.), are used to perform FT calculations and other data treatments.

Extensive experiments were in fact needed before optimal test and acquisition conditions were eventually set (for details, see[12,16]). In any fixed strain and frequency conditions, data acquisition is made in order to record 10,240 points at the rate of 512 pt/s. Twenty cycles are consequently recorded at each strain step, with the immediate requirement that the instrument is set in order to apply a sufficient number of cycles (for instance, 40 cycles at 1.0 Hz, 20 cycles at 0.5 Hz; the "stability" condition with the RPA) for the steady harmonic regime to be reached. Data acquisition is activated as the set strain is reached and stable.

FT of harmonic signals yields essentially two types of information: First the main signal component, i.e., the peak in the FT spectrum that corresponds to the applied frequency (hereafter

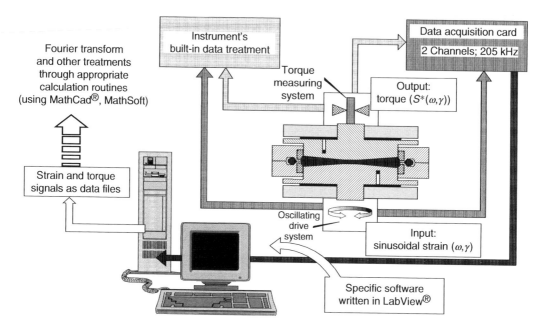

FIGURE 30.6 Updating a closed cavity torsional dynamic tester for Fourier transform (FT) rheometry.

noted either $T(1\omega)$ or $S(1\omega)$ with respect to the torque or strain signals, respectively), second the harmonics, with the third (i.e., the peak at $3\times$ the applied frequency) the most intense one. A specific calculation program, written using the FT algorithm available in MathCad 8.0 (MathSoft Inc.), is used to obtain the amplitude of the main stress and strain components (corresponding the test frequency) and the relative magnitudes (in %) of the odd-harmonic components, i.e., $I(n \times \omega_1)/I(\omega_1)$. Note that in this paper, $I(n\omega_1)/I(\omega)$ or the abridged form $I(n/1)$, is used to describe the nth relative harmonic component of any harmonic signal; $S(n\omega_1)/S(\omega_1)$ or $S(n/1)$ specifically means that a strain signal is considered; $T(n\omega_1)/T(\omega_1)$ or $T(n/1)$ is used for the torque signal. The number of data points used, the frequency resolution (Hz), the acquisition time (s), and the sampling rate (point/s) are also provided. Figure 30.7 shows the averaged torque signal recorded when submitting a gum high cis-1,4-BR (i.e., NeoCis BR40, Polimeri) to 20° strain. As can be seen, the signal is harmonic but clearly distorted in comparison with a sinusoid of same amplitude. The displayed torque signal was averaged out of 20 recorded cycles (i.e., 10,240 data points) and the standard deviation drawn as a shaded area is barely visible, which demonstrates the excellent stability of the torque response. The single FT spectrum, obtained through calculation on the last 8192 points of the recorded torque (upper left in the figure), exhibits significant 3rd and 5th harmonics with further ones becoming very small. The results of the odd-harmonic components analysis on both the torque and strain signals are displayed in the inserted table. The very low strain harmonic peaks (<0.55%) indicate the excellent quality of the applied signal at this strain angle (20°).

30.4.3 STRAIN SWEEP TEST PROTOCOLS FOR NONLINEAR VISCOELASTICITY INVESTIGATIONS

Test protocols were developed for nonlinear viscoelastic investigations, which essentially consist in performing strain sweep experiments through two subsequent runs separated by a resting period of 2 min. At least two samples of the same material are tested (more if results reveal test material heterogeneity), in such a manner that, through inversion of the strain sequences (i.e., runs 1 and 2), sample fatigue effects are detected, if any. At each strain sweep step, data acquisition is made in order to record 10,240 points at the rate of 512 pt/s. Differences are expected between runs 1 and 2 for materials exhibiting strain memory effects. With the RPA, the maximum applicable strain angle

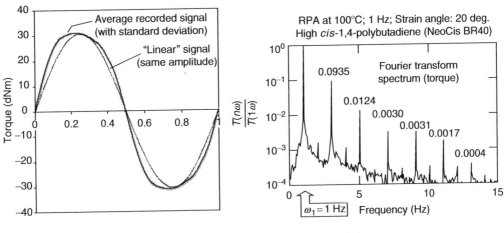

Fourier transform spectrum; odd-harmonic components analysis

Nr pts	Freq. resol	Main freq.	3rd harm.	5th harm.	7th harm.	9th harm.	11th harm.	13th harm.	15th harm.
t_{aq}	Sampling pt/s	Main torque a.u.	$T(3/1)$, %	$T(5/1)$, %	$T(7/1)$, %	$T(9/1)$, %	$T(11/1)$, %	$T(13/1)$, %	$T(15/1)$, %
		Main strain a.u.	$S(3/1)$, %	$S(5/1)$, %	$S(7/1)$, %	$S(9/1)$, %	$S(11/1)$, %	$S(13/1)$, %	$S(15/1)$, %

8.192	0.063	1	3	5	7	9	11	13	15
7.998	512	1541	9.35	1.24	0.30	0.31	0.17	0.04	0.05
		911	0.55	0.27	0.14	0.08	0.05	0.01	0.00

FIGURE 30.7 Typical (averaged) torque traces as recorded when a gum polybutadiene sample is submitted to high strain; the Fourier transform (FT) spectrum exhibits accordingly significant harmonic contributions; the inset table gives the results of the automatic analysis of torque and strain signals.

depends on the frequency, for instance around 68° ($\approx$950%) at 0.5 Hz, considerably larger than with open cavity cone-plan or parallel disks torsional rheometers.[17] Whatever the frequency, the lower-strain angle limit is 0.5° (6.98%) below which the harmonic content of the strain signal becomes so high that measured torque is excessively scattered and likely meaningless. Test protocols at 0.5 Hz were designed to probe the material's viscoelastic response within the 0.5° to 68° range, with up to 20 strain angles investigated.

Ideal dynamic testing requires that a perfect sinusoidal deformation at controlled frequency and strain be applied on the test material. Whatever are the quality design and the care in manufacturing instruments, there are always technical limits in accurately submitting test material to harmonic strain. FT analysis of the RPA strain (i.e., applied) signal allowed its quality to be precisely documented,[12] and revealed significant (i.e., larger than noise) odd-harmonics components, with obviously the 3rd the larger one, particularly at low-strain amplitude. The relative 3rd harmonic strain component, i.e., $S(3/1)$, decreases as strain amplitude increases, whatever the test conditions, and generally passes below 1% of the main component when the strain angle is higher than 1.3°–1.5° (18%–20%). In other terms, high-strain tests are performed in better-applied signal conditions than low-strain ones.

With respect to the harmonic content of the strain signal, a correction method was therefore developed for torque harmonics, based on observations made when testing an ideal elastic body, for instance, the calibration spring. $T(n\omega/1\omega)$ data are corrected according to

$$T(n\omega/1\omega)_{corr} = T(n\omega/1\omega)_{TF} - CF \times S(n\omega/1\omega)_{TF} \qquad (30.2)$$

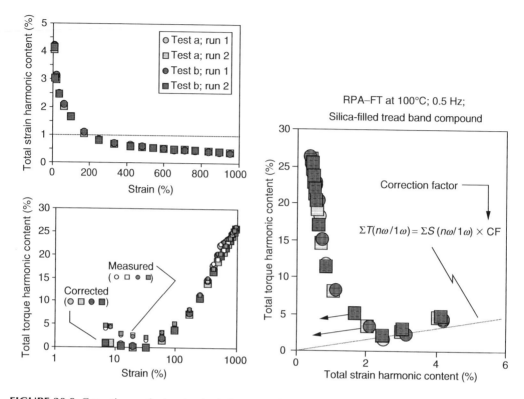

FIGURE 30.8 Correction method on total relative torque harmonic content.

where $T(n\omega/1\omega)_{TF}$ and $S(n\omega/1\omega)_{TF}$ are the nth relative harmonic components of the torque and strain signals, respectively, and CF the correction factor, as derived from a plot of $T(n\omega/1\omega)$ versus $S(n\omega/1\omega)$. A detailed demonstration of this correction method has been previously reported for the 3rd relative torque harmonic[18] and, as illustrated in Figure 30.8, the correction applies also when considering the "total torque harmonic content", TTHC, i.e., the sum $\sum T(n\omega/1\omega)$ of all the odd harmonics up to the 15th. As can be seen, at low strain, when the viscoelastic response of the material is expected to be linear, the corrected relative torque harmonics vanish, in agreement with theory.[5] As shown in the right graph of Figure 30.7, $\sum T(n\omega/1\omega)$ versus $\sum S(n\omega/1\omega)$ decreases, passes through a minimum, and appears to be bounded by a straight line whose slope provides the correction factor. The correction method is based on the simple argument that, if the applied strain were perfect, all $\sum T(n\omega/1\omega)$ data points would fall on the vertical axis.

FT analysis of torque signal allows clearly quantifying the nonlinear response of viscoelastic materials submitted to high strain, but experiments with complex polymer systems have revealed a limit of this data treatment. Whether the torque signal is distorted "on the left" or "on the right", with respect to a vertical axis drawn at the first quarter of the cycle, does not reflect in the FT spectrum. It was observed, however, that most complex polymer systems exhibit severe distortions, which sometimes affect more the right part of the half signal, when strong interactions can be suspected between components (i.e., phases) of materials. As previously mentioned, we expressed this difference between the nonlinear viscoelastic behavior of a pure, unfilled polymer and of a complex polymer material, namely filled rubber compounds, through the terms *extrinsic* and *intrinsic* nonlinear viscoelasticity. The former occurs through *external* causes, i.e., the applied strain, while the latter reflects the *internal* morphology of the material, and further complicate the response to the applied harmonic strain. As was seen in Figure 30.5, it is clear that FT analysis of torque signal, while offering an attractive quantification of the nonlinear viscoelastic response, has limited capabilities to distinguish extrinsic and intrinsic characters. Therefore, in order to supplement FT analysis, quarter cycle integration was developed as an easy data treatment technique to distinguish extrinsic and intrinsic nonlinear viscoelasticity. The ratio of the first to second quarter torque signal integration, i.e., Q1/Q2 allows clearly distinguishing between the strain amplitude effect on a pure and complex polymer materials. With the former, Q1/Q2 ratio is always higher than one and increases with strain amplitude; in such a case the torque signal is always distorted "on the left" (i.e., Q1 > Q2). In addition, with certain complex systems, Q1/Q2 varies with strain amplitude, which likely reflects changes in interactions between phases that sometimes happen to vanish at high strain, thus indicating profound change in the compound morphology.

30.4.4 Modeling the Variation of FT Analysis Results with Strain Amplitude

According to strain sweep test protocols described above, RPA–FT experiments and data treatment yield essentially two types of information, which reflects how the main torque component, i.e., $T(1\omega)$, and the relative torque harmonics vary with strain amplitude.

The ratio $T(1\omega)/\gamma$ has obviously the meaning of a complex modulus, i.e., $G^* = 12.335 \times \frac{T(1\omega)}{\gamma}$ (with G^* in kPa, $T(1\omega)$ in arbitrary unit and γ in %) and, for a material exhibiting linear viscoelasticity within the experimental window, a plot of G^* versus γ shows the most familiar picture of a plateau region at low strain, than a typical strain dependence. Such a behavior is well modeled with the following equation, in which one recognizes the mathematical form of the so-called Cross-equation for the shear viscosity function:

$$G^*(\gamma) = G_f^* + \left[\frac{G_0^* - G_f^*}{1 + (A\gamma)^B} \right] \qquad (30.3)$$

where G_0^* is the modulus in the linear region, G_f^* the final modulus, A the reverse of a critical strain for a mid-modulus value to be reached, and B a parameter describing the strain sensitivity of the material, as illustrated in Figure 30.9 with data on a pure (gum) natural rubber and a carbon black-filled ethylene–propylene–diene monomer (EPDM) compound.

Odd torque harmonics become significant as strain increases and are therefore considered as the nonlinear viscoelastic "signature" of tested materials, only available through FT rheometry. Figure 30.10 shows the typical pattern of corrected relative TTHC (i.e., $\sum T(n\omega/1\omega)$) versus strain γ that, in most cases, can be modeled with a simple equation, i.e.,

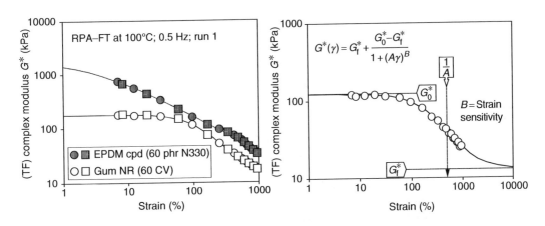

FIGURE 30.9 Modeling the strain dependence of complex modulus.

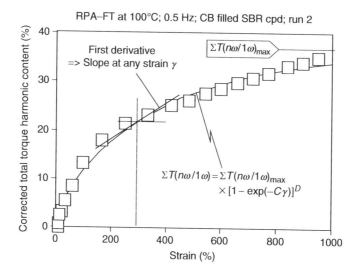

FIGURE 30.10 Modeling the variation of total torque harmonic content (TTHC) with increasing strain.

$$\text{TTHC}(\gamma) = \text{TTHC}_{max} \times \left[1 - \exp\left(-C\gamma\right)\right]^{D} \qquad (30.4)$$

where

γ is the deformation (%)
TTHC_{max} (i.e., $\sum T(n\omega/1\omega)_{max}$) a maximum plateau value at high (infinite) strain
C and D fit parameters

The curve drawn illustrates how the model fits measured data. The first derivative of Equation 30.4 allows calculating the slope at any strain. The same model can be used to fit any relative torque harmonic, for instance the 3rd one, $T(3/1)$. Note that in using Equation 30.4 to model harmonics variation with strain, one may express the deformation (or strain) γ either in degree angle or in percent. Obviously all parameters remain the same except C, whose value depends on the unit for γ. The following equality applies for the conversion: $C(\gamma,\text{deg}) = \frac{180\alpha}{100\pi} \times C(\gamma,\%)$, where $\alpha = 0.125$ rad.

In summary, the FT rheometry protocols described above and the associated data treatment yield a considerable number of information about the viscoelastic character of materials. Within less than 1 h, two samples are tested and the full data treatment is performed (using the present combination of VBA macros and MathCad routines).

30.5 NONLINEAR VISCOELASTICITY ALONG RUBBER PROCESSING LINES

30.5.1 EPDM COMPOUNDING LINE

30.5.1.1 Experimental Approach

Thanks to the courtesy of a French manufacturer of general rubber goods, a compounding line for an carbon black-filled EPDM formulation was the source of a series of samples, as indicated in Figure 30.11. For obvious reasons, limited information is given about the compound studied, which is 35% weight carbon black filled, with 19% weight processing oil and 22% weight other ingredients. Curatives addition was made on open mill, and samples were taken at eight positions along the processing line: After dumping from the internal mixer (Banbury type), before and after curatives addition on open mill, at three positions on the batch-off during cooling, before and after granulation.

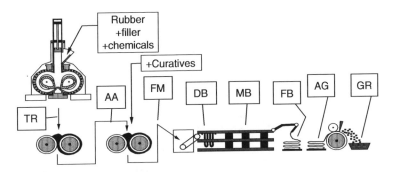

FIGURE 30.11 Sampling along a compounding line for a filled ethylene–propylene–diene monomer (EPDM) compound; curatives addition on open mill.

According to test protocols described above, RPA–FT test were performed at 100°C, 1 Hz on all samples; additional tests at 100°C, 0.5 Hz were performed on IMA_TR, IMA_FM, and IMA–AG samples. Essentially three types of data will be discussed hereafter: The complex modulus G^* (as derived from the main torque component in the FT torque spectrum), the corrected total torque harmonic component, i.e., cTTHC, and the Q1/Q2 ratio.

30.5.1.2 Results and Discussion

30.5.1.2.1 (FT) Complex Modulus versus Strain

Complex modulus G^* versus strain curves are shown in Figure 30.12, as obtained on samples taken just after the internal mixing step (TR), after curatives addition (FM), and before the granulation (AG). In all cases, the two tested samples (a & b) give identical results thus demonstrating that immediately after dumping from the internal mixer, the material has already an excellent homogeneity. One notes however several substantial differences between and after curatives addition: (i) On sample TR, test frequency has no effect on complex modulus measurement except a singular upturn appearing at high strain, depending on frequency, observed only on run 1 data; (ii) after curatives addition, the expected frequency effect is observed (i.e., 0.5 Hz tests yield higher modulus) but the effect is larger on run 1 than on run 2 data; (iii) on run 2 data, curatives addition is associated with a significant drop in modulus. Whatever the test sample, no linear region is observed within the experimental window, a net inflection is observed at around 100% strain and strain history effects are revealed through differences between run 1 and run 2 data, as expected with any highly filled compound.

Lines in Figure 30.12 were drawn with parameters obtained when fitting data with Equation 30.3. It is fairly obvious that, outside the experimental window, data would not necessarily conform to such a simple model, which in addition cannot meet the inflection at 100% strain. All results were nevertheless fitted with the model essentially because correlation coefficient were excellent, thus meaning that the essential features of G^* versus strain dependence are conveniently captured through fit parameters. Furthermore any data can be recalculated with confidence within the experimental strain range with an implicit correction for experimental scatter. Results are given in Table 30.1; note that 1/A values are given instead of A.

Data in Table 30.1 clearly show that, whatever the position of the test sample along the compounding line, there is a substantial difference between run 1 and run 2 data, particularly in what the "linear" modulus data are concerned. However, G_0^* is an extrapolated value and quite unrealistic values are obtained on certain samples, e.g., TR and AA, and it might be safer to consider modulus variations along the compounding line by using the (recalculated) complex modulus at 10% strain (Figure 30.13).

In agreement with observations previously made, recalculated $G_{10\%}^*$ data offer an overall picture of processing effects along the mixing line. As can be seen, the expected effect of test frequency is observed, except on TR sample, but the most significant observations are made on run 2 data, which

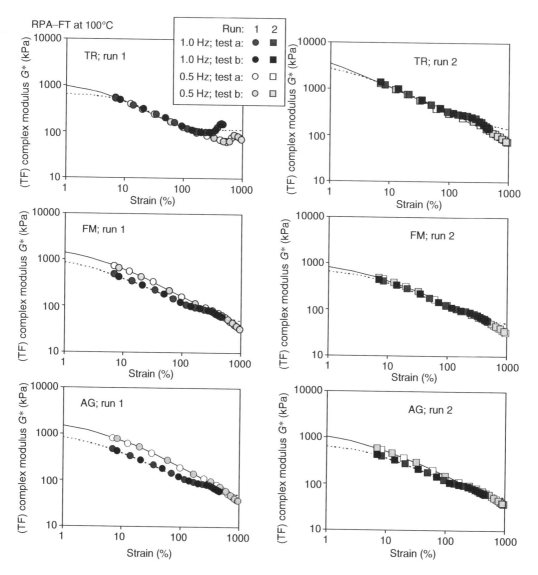

FIGURE 30.12 Ethylene–propylene–diene monomer (EPDM) compounding; complex modulus strain dependence; samples TR, FM, and AG.

converge to a stabilized level of 366.4 kPa at the end of the compounding process. The strain history effect is well reflected by differences between run 1 and run 2 data, which remain practically constant after curatives have been added. Curatives' addition is in fact associated with a huge drop in modulus (compare run 2 data of samples TR and AA with sample FM). Despite the extra mixing energy obviously associated with the milling process, it is clear that the modulus drop reflects rather a kind of plasticizing effect of curatives. Indeed there is not much difference between samples TR and AA. Cooling on open mills further stabilizes the compound, but the effect is very small. The higher modulus after granulation is due to the addition of talc, just before this last step.

Figure 30.14 shows an interesting aspect of RPA–FT experiments, i.e., the capability to quantify the strain sensitivity of materials through parameter B of fit Equation 30.3. As can be seen, curatives' addition strongly modifies this aspect of nonlinear viscoelastic behavior, with furthermore a substantial change in strain history effect. Before curatives addition, run 2 data show very lower-strain

TABLE 30.1

Ethylene–Propylene–Diene Monomer (EPDM) Compounding; RPA–FT Results at 100°C; Complex Modulus Dependence on Strain; Fit Parameters of Equation 30.3

	Test: S_Sweep_1Hz_2Runs_100			Temperature (°C):	100	
Modeling Summary Report:		(FT) G^* versus Strain%		Frequency (Hz):	1.0	
Sample	**Run (a & b)**	G_0^* kPa	G_f^* kPa	1/A %	B	r^2
TR	1	645.8	113.0	15.31	1.578	0.9894
	2	3953.0	102.4	2.86	0.776	0.9986
AA	1	600.1	104.6	16.76	1.530	0.9918
	2	3589.0	98.2	2.91	0.781	0.9988
FM	1	1158.0	32.7	3.97	0.752	0.9992
	2	756.8	39.8	9.29	0.853	0.9996
DB	1	932.9	34.8	6.33	0.800	0.9996
	2	767.7	39.2	8.64	0.837	0.9996
MB	1	1206.0	32.9	3.60	0.745	0.9996
	2	737.3	40.7	9.41	0.854	0.9996
FB	1	1315.0	29.7	2.74	0.710	0.9996
	2	726.0	40.3	9.49	0.855	0.9996
AG	1	1114.0	34.3	4.10	0.758	0.9996
	2	739.8	40.6	9.23	0.851	0.9996
GR	1	909.0	32.8	7.60	0.827	0.9997
	2	714.7	37.4	9.34	0.836	0.9996

	Test: S_Sweep_05Hz_2Runs_100			Temperature (°C):	100	
Modeling Summary Report:		(FT) G^* versus Strain%		Frequency (Hz):	0.5	
Sample	**Run (a & b)**	G_0^* kPa	G_f^* kPa	1/A %	B	r^2
TR	1	1125.0	61.6	5.72	0.958	0.9983
	2	14,110.0	10.6	0.16	0.602	0.9995
FM	1	1993.0	6.3	3.38	0.727	0.9998
	2	1066.0	9.2	5.26	0.696	0.9998
AG	1	2129.0	−6.3	3.78	0.686	0.9996
	2	1519.0	4.2	3.34	0.652	0.9997

sensitivity than run 1 data; after curatives addition, the reverse is observed but the difference is smaller. One clearly also sees that the nonlinear character is essentially stabilized at the beginning of the cooling process (position DB) since all subsequent results remain within the average values.

30.5.1.2.2   (FT) Torque Harmonics versus Strain

Odd harmonics become significant as strain increases and are therefore considered as the nonlinear viscoelastic "signature" of tested materials, available only through FT rheometry. Figure 30.15 shows the typical patterns of both the (corrected) relative 3rd torque harmonic and the TTHC, when plotted versus strain γ, as exhibited at sample positions TR (at dump), FM (after curatives addition), and AG (before granulation). Similar graphs are obtained with all the other samples. Only sample TR exhibits significant differences between runs 1 and 2 data, likely associated which an incomplete dispersion of filler. Of course the TTHC curve envelops the $T(3/1)$ curve, with the latter clearly evolving towards a plateau, in line with the modeling by Equation 30.4. There are however some additional strain effects, occurring at around 300% strain, which are obviously not addressed by

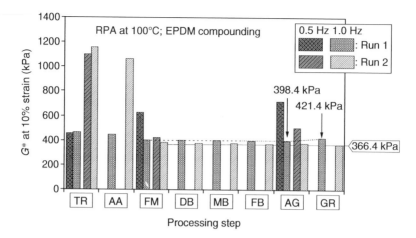

FIGURE 30.13 Ethylene–propylene–diene monomer (EPDM) compounding; variation of complex modulus at 10% strain along the compounding line; $G^*_{10\%}$ values calculated with fit parameters given in Table 30.1.

Equation 30.4, and will not be discussed here since they are likely related to formulation details, not available for this report.

Figure 30.16 shows that the nonlinear viscoelastic character is essentially not depending on the test frequency, and does evolve much along the compounding line considered. Results on sample TR suffer from some scatter and there is a substantial difference between run 1 and run 2 data, likely because mixing operations are not completed within the internal mixer. It might seem surprising that no frequency effect is observed on the TTHC, but such a behavior just conforms to the separability of strain and time effects, as postulated in advanced nonlinear viscoelastic modeling.

As shown by the dashed curves in Figure 30.16, fitting 1.0 Hz results with Equation 30.4 yields plateau values, which likely have no physical meaning, with respect to 0.5 Hz data. However, fit parameters of this equation (as given in Table 30.2) allow the slope to be calculated at any strain within the experimental window. It follows that the slope at a given strain, for instance 200%,

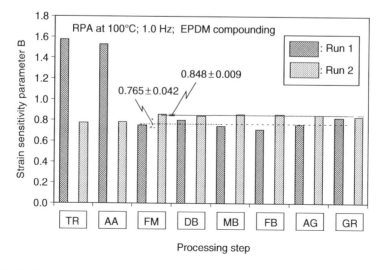

FIGURE 30.14 Ethylene–propylene–diene monomer (EPDM) compounding; variation of strain sensitivity (parameter B of Equation 30.3) of complex modulus along the compounding line.

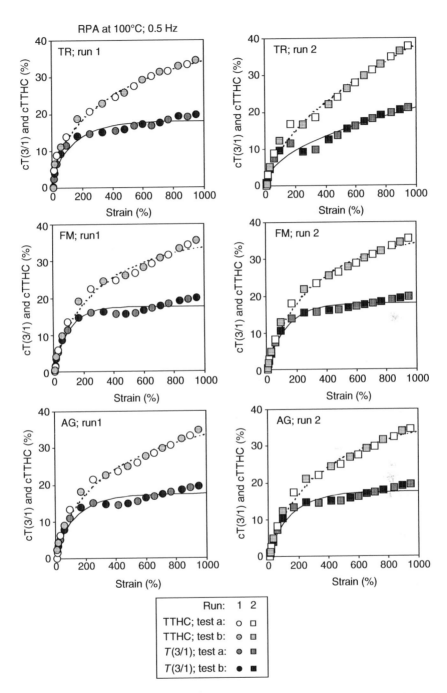

FIGURE 30.15 Ethylene–propylene–diene monomer (EPDM) compounding; corrected relative 3rd torque harmonic component and total torque harmonic content (TTHC) versus strain, at 0.5 Hz; samples TR, FM, and AG.

i.e., $\left. \dfrac{d\,TTHC}{d\gamma}\right|_{\gamma=200\%}$, clearly shows how the nonlinear character is evolving along the whole compounding line (see Figure 30.17). Up to position FM, the compound is submitted to shear on open mills and, as reflected by slopes calculated from run 1 data, the nonlinear character markedly increases. Then during cooling steps, a kind of dampening effect is observed (clearly on slopes from

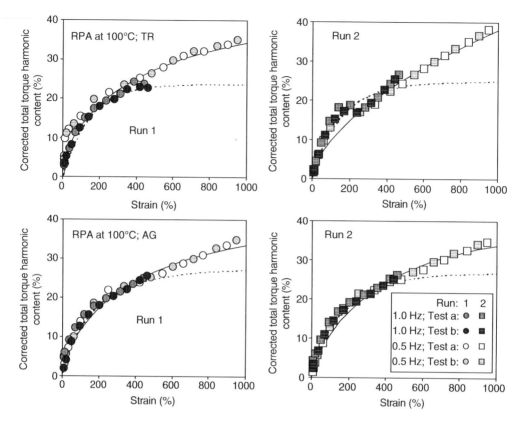

FIGURE 30.16 Ethylene–propylene–diene monomer (EPDM) compounding; corrected total torque harmonic content (TTHC) versus strain, at 1.0 and 0.5 Hz; samples TR and AG.

run 2 data) and, finally, the granulation process also increases the nonlinearity, likely because of local shear heating and the addition of talc as antisticking agent.

30.5.1.2.3 Quarter Cycle Integrations

Quarter cycle integration data are shown in Figure 30.18, in terms of Q1/Q2 ratio, for samples taken at positions TR, FM, and AG, i.e., at three critical steps of the process. As can be seen, Q1/Q2 ratio is always lower than 1, and exhibits significant variations with applied strain. Torque cycles are always distorted "on the right" which, as commented above (Section 30.4.3), results from an intrinsic nonlinear viscoelastic character of the material. In other terms, the filler level and/or the rubber–filler interactions are dominating the harmonic response of the material. Q1/Q2 < 1 means that the filler content is above the percolation level (around 13% in volume) and the peculiar variation of Q1/Q2 with strain suggests that the carbon black is a moderately reinforcing one. Indeed, up to around 200%, the rubber–filler morphology has a growing influence on the nonlinear character then, as strain is further increasing, the rubber matrix (likely the "unbound rubber"[7,8]) is playing a growing role and Q1/Q2 evolves towards 1.

Along the mixing line, significant changes occur on Q1/Q2. At dump, data are scattered and there is little, if any, effect of the test frequency. After curatives addition, the test frequency significantly affects the Q1/Q2 signature, as expected since quarter cycle integration provides all-inclusive parameters (i.e., main torque component and all harmonics). One notes also a net strain history effect, i.e., run 1 and run 2 data do not superimpose.

Except for samples at dump, a significant frequency effect is observed on Q1/Q2 versus strain graphs, in such a manner that the 0.5 Hz data envelop the 1.0 Hz ones. The higher the frequency, the

TABLE 30.2

Ethylene–Propylene–Diene Monomer (EPDM) Compounding; Rubber Process Analyzer–Fourier Transform (RPA–FT) Results at 100°C; Total Torque Harmonic Content (TTHC) versus Strain; Fit Parameters of Equation 30.4

	Test: S_Sweep_1Hz_2Runs_100			Temperature (°C):	100
Modeling Summary Report:		**cTTHC versus Strain%**		**Frequency (Hz):**	**1.0**
Sample	**Run (a & b)**	**TTHC$_{max}$, %**	**C**	**D**	**r^2**
TR	1	23.8	0.0069	0.989	0.9962
	2	25.1	0.0057	0.865	0.9787
AA	1	25.2	0.0059	1.027	0.9984
	2	25.5	0.0057	0.848	0.9812
FM	1	26.7	0.0048	1.001	0.9975
	2	26.7	0.0051	0.997	0.9983
DB	1	28.4	0.0036	0.754	0.9956
	2	25.9	0.0057	1.033	0.9967
MB	1	26.5	0.0051	0.941	0.9959
	2	26.0	0.0060	1.013	0.9965
FB	1	25.9	0.0059	0.990	0.9959
	2	27.0	0.0052	0.895	0.9965
AG	1	27.6	0.0041	0.727	0.9931
	2	27.0	0.0050	0.838	0.9954
GR	1	26.9	0.0051	0.894	0.9961
	2	27.0	0.0054	0.913	0.9965

	Test: S_Sweep_05Hz_2Runs_100			Temperature (°C):	100
Modeling Summary Report:		**cTTHC versus Strain%**		**Frequency (Hz):**	**0.5**
Sample	**Run (a & b)**	**TTHC$_{max}$, %**	**C**	**D**	**r^2**
TR	1	40.9	0.0013	0.539	0.9937
	2	91.8	0.0003	0.623	0.9936
FM	1	36.5	0.0022	0.673	0.9931
	2	37.0	0.0024	0.745	0.9963
AG	1	38.7	0.0016	0.606	0.9943
	2	37.9	0.0019	0.705	0.9954

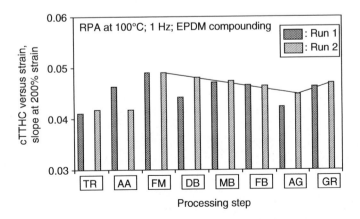

FIGURE 30.17 Ethylene–propylene–diene monomer (EPDM) compounding; corrected total torque harmonic content (TTHC) versus strain, at 1.0 Hz; slope at 200%, as calculated with first derivative of Equation 30.4 and fit parameters in Table 30.2.

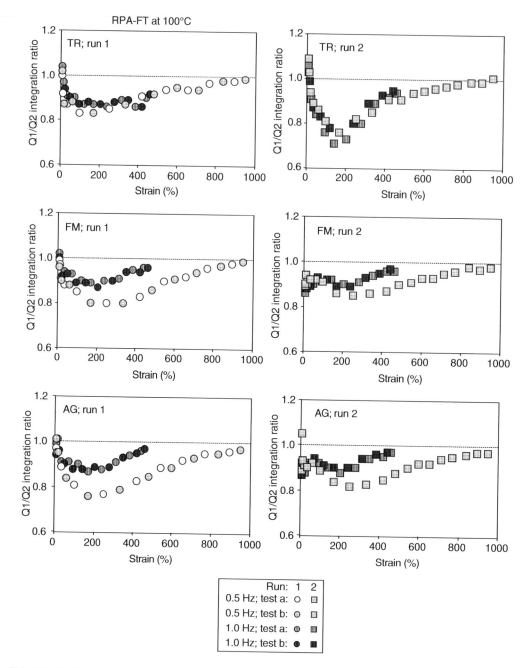

FIGURE 30.18 Ethylene–propylene–diene monomer (EPDM) compounding; quarter cycle integration ratio versus strain, at 0.5 and 1.0 Hz; samples at position TR (at dump), FM (after curatives addition), and AG (after cooling).

smaller the minimum Q1/Q2, but the smaller also the strain for Q1/Q2 ratio to be back to 1. One also notes that, at position AG (i.e., near the end of the compounding process), the difference in Q1/Q2 versus strain graphs is larger, particularly with run 1 data. Such an observation is well consistent with the meaning we give to the Q1/Q2 ratio. Indeed, at low strain, rubber–filler interactions play the major role in the nonlinear viscoelastic response but there is a maximum strain above which the rubber–filler morphology is disrupted, in such a manner that its influence on dynamic properties reduces as strain further increases. However, nothing is really "broken" within the compound upon increasing strain since the above mechanism is essentially repeated during run 2.

30.5.2 SILICA-FILLED COMPOUND MIXING LINE

30.5.2.1 Experimental Approach

Preparing silica-filled compounds is by far the most complex mixing operation in tire technology, essentially because the silanization process is carried out in internal mixer, concomitantly with the dispersion of filler particles and other compounding ingredients The order of addition of formulation ingredients, as well as the control of mix temperature variation through shear warming-up, are important aspects of the process, likely typical of the mixing line considered and an essential know-how, about which tire manufacturers remain quite discreet. Little is really known about the complex physical (i.e., filler particles dispersion and distribution) and chemical (i.e., the silane–silica reactions) events that take place in the earlier steps of such compounding operations, essentially because practical observations are really challenging, if not impossible.[19] In quite a practical manner, rubber engineers have set up relatively complicated compounding operations for silica–silane-filled formulations, involving several dumping, cooling, and "remixing" steps, in order to eventually obtain quality final batches. By courtesy of a Dutch tire manufacturer, we had the opportunity to obtain in confidence samples taken along a mixing line for a silica-filled tire tread compound (Figure 30.19). For obvious reasons, no formulation or mixing procedure details can be given in this report, but are in fact not necessary for the point we want to make.

At the positions indicated in Figure 30.19, three samples were taken at random during the preparation of four different batches. Samples were received within one week of the operations as rough plaques of up to 2 cm thickness. In order to eliminate any sample size/geometry effect on RPA–FT tests, cylindrical pieces were die cut out of the plaques, and their weight adjusted before compression molding for 5 min at 60°C in a mold mimicking the RPA test cavity geometry and volume +5%. According to the test protocol described in Section 30.4.1, RPA–FT strain sweep tests (2 tests: a & b) were performed at 100°C, 0.5 Hz on all samples. All tests were repeated three weeks later (2 tests: c & d), after storage in darkness at room temperature. No significant aging effect was

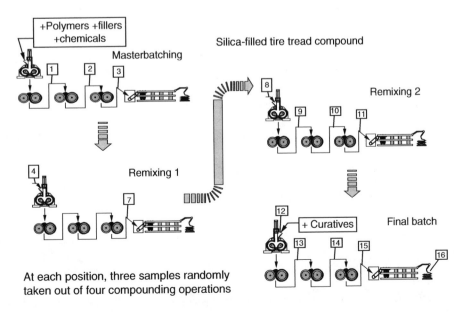

FIGURE 30.19 Sampling along a mixing line for a silica-filled compound.

noted and only modeling results of the second series of tests will be reported in the tables below, while some figures will show data from both series of experiments. We will discuss hereafter how the following data vary with applied strain: the complex modulus G^*, the corrected total torque harmonic component cTTHC, and the Q1/Q2 ratio.

30.5.2.2   Results and Discussion

30.5.2.2.1   (FT) Complex Modulus versus Strain

Figure 30.20 shows complex modulus G^* versus strain curves, as measured on samples taken at expected key positions in the process: at dump of the masterbatch (position 1), when loading for the first remixing (position 4), when loading for the second remixing (position 8, when loading for curatives addition (position 12), at the end of the line (position 16). Results for 2 × 2 tests are shown and as can be seen a large scatter is observed at dump of the masterbatch, likely reflecting a poor

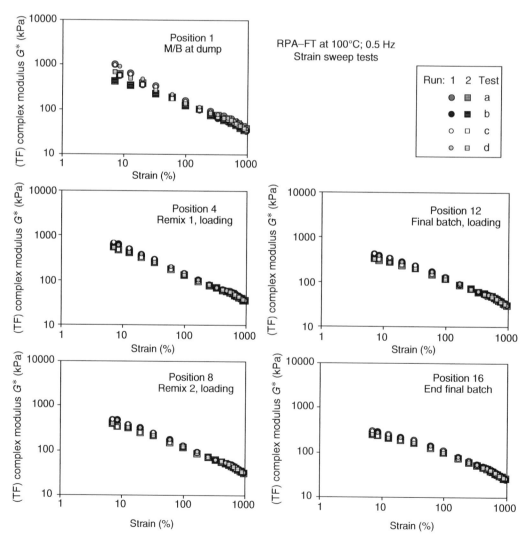

FIGURE 30.20 Mixing silica-filled compound: Complex modulus strain dependence at selected position along the mixing line.

homogeneity, which is immediately improved by passing on the three subsequent open mills. Indeed, at position 4, results appear well reproducible with a small but significant strain history effect (i.e., run 1 and run 2 data do not superimpose). At all positions, strong strain dependence is observed, with the limit of the linear region barely visible at the end of the process (position 16). Within the experimental strain window, the compound maintains a strong nonlinear viscoelastic character along the full mixing line.

Fitting data with Equation 30.3 is excellent, as reflected by correlation coefficients given in Table 30.3 with all fit parameters, and as illustrated by the curves drawn in Figure 30.21. By somewhat compensating for experimental scatter, fit curves clearly show that the nonlinear character significantly changes along the compounding line (compare graphs in Figure 30.21). One notes in Table 30.3 that negative values for G_f^* are sometimes obtained, which of course must be kept as such for data-fitting purposes, but must be discarded from discussion because they have no physical meaning.

TABLE 30.3
Mixing Silica-Filled Compound; Rubber Process Analyzer–Fourier Transform (RPA–FT) Results at 100°C; Complex Modulus Dependence on Strain; Fit Parameters of Equation 30.3

Test: S_Sweep_05Hz_2Runs_100					Temperature (°C):	100	
Modeling Summary Report:		G^* versus Strain%			**Frequency (Hz):**	**0.5**	
Sample Identification			G_0^*	G_f^*	$1/A$		
Location	**Position**	**Run (a & b)**	**kPa**	**kPa**	**%**	**B**	r^2
M/B at dump	1	1	8602.0	23.3	0.53	0.8015	0.9992
		2	3526.0	15.5	0.83	0.6990	0.9985
	2	1	2925.0	7.9	1.15	0.6721	0.9997
		2	1313.0	15.7	4.32	0.7325	0.9999
	3	1	2272.0	19.0	2.65	0.7838	0.9989
		2	2217.0	6.6	1.16	0.6290	0.9980
Remix 1 loading	4	1	3421.0	9.9	0.96	0.6847	0.9996
		2	1259.0	13.8	4.16	0.7084	0.9999
	7	1	1095.0	8.4	5.36	0.7096	0.9998
		2	723.9	4.1	7.78	0.6477	0.9999
Remix 2 loading	8	1	1068.0	8.0	5.68	0.7084	0.9998
		2	908.9	(−4.0)	4.07	0.5744	0.9983
	9	1	890.5	3.8	7.34	0.6900	0.9992
		2	661.1	(−6.0)	8.24	0.5888	0.9994
	10	1	972.8	0.4	5.81	0.6601	0.9999
		2	533.7	0.6	14.28	0.6538	0.9994
	11	1	922.1	(−0.8)	6.08	0.6495	0.9996
		2	550.2	(−2.0)	12.85	0.6394	0.9998
Final batch loading	12	1	882.4	(−3.0)	6.08	0.6277	0.9994
		2	549.7	0.0	13.02	0.6411	0.9996
	13	1	498.1	(−7.0)	14.91	0.6310	0.9998
		2	368.7	(−7.1)	22.89	0.6213	0.9999
	14	1	467.6	(−5.4)	16.31	0.6425	0.9998
		2	381.0	(−6.7)	21.13	0.6218	0.9993
	15	1	443.5	(−5.0)	18.88	0.6564	0.9998
		2	361.8	(−6.1)	24.09	0.6353	0.9999
End batch	16	1	468.7	(−6.4)	16.91	0.6394	0.9997
		2	356.7	(−5.1)	25.11	0.6422	0.9999

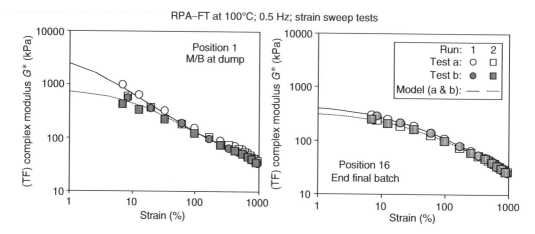

FIGURE 30.21 Mixing silica-filled compound: overall changes in nonlinear viscoelastic character along the mixing line.

A quick examination of fit parameters (Table 30.3) reveals the basic aspects of nonlinear viscoelasticity changes along the compounding line: The (extrapolated) zero strain modulus and the strain sensitivity parameter decrease, and the critical strain $1/A$ increases. But fit parameters of G^* versus γ curves allow a very fine analysis of the variation of viscoelastic parameters during mixing operations, namely by revealing key steps of the process. For instance, recalculated complex modulus at 10% strain show that expectedly intensive shearing steps in internal mixer readily improve the processing behavior, likely by contributing to the completion of the silica–silane reactions and by further homogenizing the compound, as clearly seen in the upper graph of Figure 30.22. Another interesting observation is the strong effect associated with curatives addition; not only the modulus drops by around 100 kPa, which reflects a kind of plasticizing effect of such chemicals, but also, the mid-modulus critical strain $1/A$ increases by 6%–7%. With respect to the mathematical form of Equation 30.3, the parameter $1/A$ is somewhat related with the extent of the linear viscoelastic region and, therefore, adding curatives appears as an essential step in smoothing the processability character of the compound.

30.5.2.2.2 (FT) Torque Harmonics versus Strain

Figure 30.23 shows how the TTHC versus strain γ curves vary along the compounding line, using run 1 data. Again a large scatter is observed on results obtained with samples taken at position 1, but at the subsequent sampling positions, test results appear repeatable, likely reflecting the improving homogeneity along the mixing line. Curves shown suggest the existence of a maximum TTHC at very large strain, outside however the experimental window.

One notes a singularity at around 600% strain, which is of course not considered by the simple modeling of Equation 30.4, whose results are nevertheless given in Table 30.4, since they provide a simple manner to capture the main changes in the nonlinear response of the material, as they evolve along the compounding line. When for instance, slopes $\frac{d\,\text{TTHC}}{d\gamma}\big|_{200\%}$ are calculated using the first derivative of Equation 30.4 and fit parameters given in Table 30.4, one observes a singular variation of nonlinear properties occurring during remix 1, at least when run 1 data are considered (Figure 30.24). In fact, from the earlier steps of the initial masterbatching process up to the end of remix 1, the slope increases and reaches a maximum value. Then during the subsequent steps, this nonlinear quantity decreases and levels out. A similar observation is made on run 2 data, while the maximum is smoother, and at the end of the compounding process, leveled out slopes for runs 1 and 2 reveal a significant strain history effect, as reflected by the difference between the average values calculated

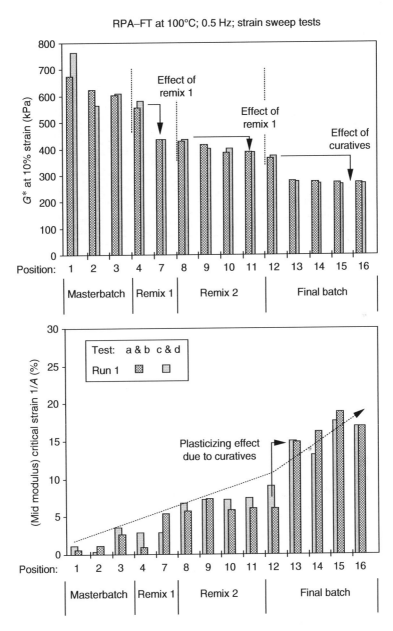

FIGURE 30.22 Mixing silica-filled compound: Complex modulus versus strain amplitude as modeled with Equation 30.3; typical changes in nonlinear viscoelastic features along the mixing line.

from results at positions 14 to 16. Nonlinear viscoelastic properties appear thus stabilized at the end of the compounding process, but still have the capability to be changed through large deformation, i.e., the so-called strain history effect.

As shown in Figure 30.23 above, the singularity at 600% in harmonics content versus γ curves tends to disappear as compounding progresses. Figure 30.25 further shows this singular event at two key positions along the compounding line: At the loading for remix 1 (position 4) and at the end of the process (position 16). Both the TTHC and the 3rd relative torque harmonic curves are drawn, in order to demonstrate that this event is not an experimental artifact. As can be seen, data from two tests (i.e., a & b, according to test protocols described in Section 30.4.3) entangle well, which further

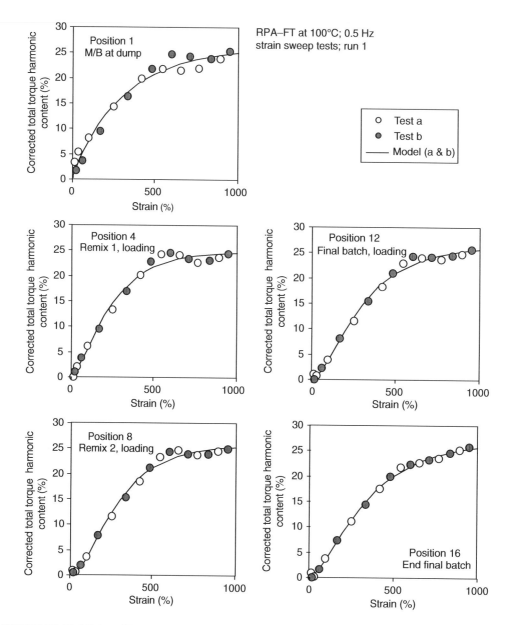

FIGURE 30.23 Mixing silica-filled compound: Total torque harmonic content (TTHC) versus strain amplitude; observed behavior at selected position along the compounding line.

supports their physical significance. A possible interpretation of this singularity would be that some physicochemical events, for instance, the silica silanization and dispersion, initiated in the earlier stages of the compounding process, need quite an extended time (and energy) to be completed. Indeed, if the silanization is not complete, the rubber–filler interphase is not stabilized and, for instance, some silanol groups might remain available that would hinder the full dispersion of silica particles. The nonlinear viscoelastic response to increasing (dynamic) strain of a bulk compound with incomplete silanization would therefore have two contributions, one from the rubber matrix plus one from the rubber–filler interphase region, until strain is large enough to dislocate the latter, thus leaving only the rubber matrix viscoelastic response. The fact that, at position 16, the

TABLE 30.4

Mixing Silica-Filled Compound; Rubber Process Analyzer–Fourier Transform (RPA–FT) Results at 100°C; Total Torque Harmonic Content (TTHC) versus Strain; Fit Parameters of Equation 30.4

		Test: S_Sweep_05Hz_2Runs_100			Temperature (°C):	100
Modeling Summary Report:		cTTHC versus Strain%			Frequency (Hz):	0.5
Sample Identification			$TTHC_{max}$			
Location	Position	Run (a & b)	%	C	D	r^2
M/B at dump	1	1	27.53	0.0023	0.655	0.9961
		2	28.70	0.0025	1.003	0.9988
	2	1	24.76	0.0040	1.261	0.9954
		2	25.52	0.0036	1.395	0.9986
	3	1	25.88	0.0035	1.042	0.9952
		2	26.92	0.0031	1.296	0.9988
Remix 1 loading	4	1	24.37	0.0045	1.434	0.9954
		2	26.24	0.0032	1.282	0.9983
	7	1	26.02	0.0043	1.618	0.9960
		2	26.37	0.0034	1.489	0.9985
Remix 2 loading	8	1	25.67	0.0049	1.993	0.9973
		2	26.75	0.0033	1.425	0.9987
	9	1	27.33	0.0038	1.514	0.9973
		2	27.81	0.0032	1.447	0.9985
	10	1	27.13	0.0038	1.544	0.9980
		2	28.07	0.0030	1.387	0.9985
	11	1	26.98	0.0040	1.589	0.9978
		2	27.93	0.0032	1.439	0.9990
Final batch loading	12	1	26.65	0.0042	1.765	0.9970
		2	27.54	0.0033	1.544	0.9986
	13	1	27.04	0.0036	1.588	0.9980
		2	27.73	0.0029	1.463	0.9991
	14	1	27.25	0.0037	1.663	0.9985
		2	28.86	0.0027	1.337	0.9992
	15	1	26.75	0.0037	1.707	0.9989
		2	28.70	0.0026	1.352	0.9987
End batch	16	1	26.99	0.0036	1.631	0.9982
		2	27.92	0.0029	1.446	0.9992

singularity is still visible on run 1 data, but not on run 2, supports this explanation. A more precise one could likely be offered with respect to formulation details, which cannot be disclosed here.

30.5.2.2.3 Quarter Cycle Integrations

Figure 30.26 shows quarter cycle integration results at five key positions of the compounding line, in terms of Q1/Q2 ratio. As can be seen, Q1/Q2 ratio is always lower than 1, and exhibits large variations with increasing strain amplitude. Such a behavior is the "signature" of intrinsically nonlinear viscoelastic materials. The morphology that results from rubber–silica interactions is the controlling factor in the harmonic response of the compound. Consequently, a large scatter is observed at position 1 because the dumped masterbatch is still far from being well dispersed and, as previously shown, the silanization process is also not yet completed. Passage on three open mills is however sufficient to improve the dispersion, as reflected by the considerably reduced scatter on data at position 4, but physicochemical events still occur in the subsequent stages. How Q1/Q2

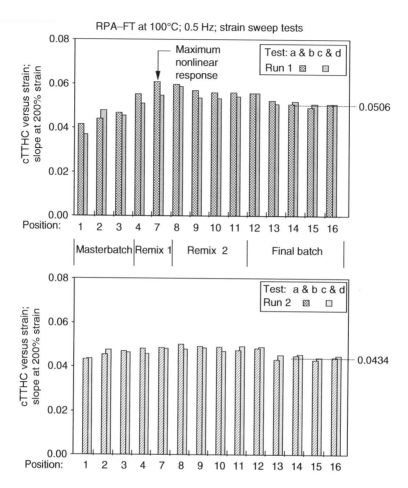

FIGURE 30.24 Mixing silica-filled compound: Total torque harmonic content (TTHC) versus strain amplitude; slope at 200% of TTHC versus γ curves varies along the compounding line.

versus strain curves evolve along the line is quite interesting since it further supports some of our arguments above regarding the still on-going silanization reactions up to the very end of the compounding process. Figure 30.25 indeed shows that Q1/Q2 versus γ curves pass through a minimum at around 600%, but the extent of this minimum decreases towards the end of the compounding line. At high (dynamic) strain amplitude, the final batch has a Q1/Q2 ratio which tends to be back to zero; in other terms, the torque signal is then nearly symmetrical, while not sinusoidal, as reflected by the important harmonic content. In (silane treated) silica-filled systems, the widely accepted chemical scheme for the silanization is a two-step process, with the second one, i.e., the strong interactions with the rubber network, occurring only during the vulcanization process, when sulfane groups are broken to form covalent bonds with rubber chains.[20] The first step concerns condensation reactions between silane's ethoxy functions and silanol groups present on the silica surface, with an alcohol as side product. It is obvious that optimal dispersion of silica particles is achieved when this primary reaction is complete and it is known that, providing mixing temperature is kept below the activation level of cross-linking reaction, the second step does not occur during compounding operations. This suggests that the lower right graph in Figure 30.25 describes the nonlinear viscoelastic behavior of a heterogeneous system consisting of fully silanated silica particles, well dispersed in a rubber matrix but not forming a strong soft network, in sharp contrast with carbon black-filled compounds.

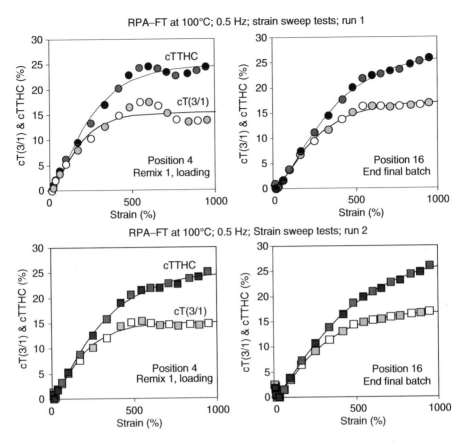

FIGURE 30.25 Mixing silica-filled compound; singularity in harmonic content versus strain amplitude curves; variation along the compounding line.

30.6 CONCLUSIONS AND PERSPECTIVES

Filled rubber compounds are complex polymer systems, which are conveniently characterized with advanced harmonic testing methods, as developed for closed cavity rheometers, through the appropriate modifications. Strain sweep test protocols, the associated data treatment and modeling allow obtaining a number of meaningful results about the nonlinear viscoelastic character of materials. Because material's response is obtained at large strain amplitude, results must be analyzed through special techniques, for instance the well-known FT. The FT spectra contain all the information available through harmonic testing and considering the main torque component and relative harmonic torque components versus the strain amplitude makes a basic analysis. Easy modeling methods give access to various parameters, which clearly reflect the many facets of nonlinear viscoelasticity of test materials. Quarter cycle integration of (average) torque signal provides additional information to FT analysis.

Rubber processing is quite a complex sequence of operations, with significant stream effects, i.e., the processing behavior at a given step is strongly depending on the previous stages. On the factory floor regular and consistent production is achieved only through the strict respect of procedures, with many controls at various steps. The key operation is of course the mixing which, in addition to the obvious objective of achieving an optimal dispersion of all formulation ingredients, must yield compounds with stabilized rheological properties, necessary for easy processing in the subsequent operations, extrusion or molding, and of course vulcanization. Such goals are achieved through a number of steps, using specific machines, with the associated

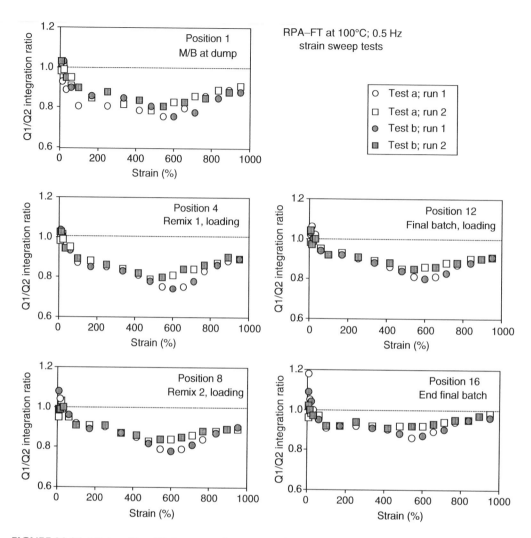

FIGURE 30.26 Mixing silica-filled compound; quarter cycle integrations of (average) torque signal; variation of Q1/Q2 ratio along the compounding line.

large energy consumption. The belief that a suitable assessment technique for nonlinear viscoelastic properties along compounding line would help understanding stream effects, was the motivation for the work reported in this chapter.

Large amplitude harmonic strain experiments were consequently performed on samples taken along industrial mixing (compounding) lines, by using a commercial torsional dynamic rheometer, conveniently updated for the "FT rheometry". Either with respect to the single-step mixing of a general purpose EPDM-based formulation or to the more sophisticated compounding of a silica-filled compound, FT rheometry proved to be a valuable technique, yielding a number of accurate results in a reproducible manner, providing the appropriate test protocols and data treatments are used. It was found that filled rubber compounds exhibit viscoelastic properties that are essentially nonlinear (at least in the most practical strain window) owing to their heterogeneity, the amplitude of the involved strain, and strong interactions occurring between the rubber matrix and the filler particles. It was also demonstrated that rheological properties are stabilized during rubber compounding operations, thanks to the sequences of processes as pragmatically optimized by factory floor engineers. In addition, nonlinear harmonic testing offers the possibility not only to precisely document how and when stabilization occurs along the compounding line, but also to reveal certain effects, observable at high dynamic strain only, which might be interpreted with respect to compound's morphology. FT rheometry is therefore a very promising test technique, worth further developments.

ACKNOWLEDGMENTS

The author would like to thank Mr. J.M. Beurrier and Mrs. I. Le Blanc (CF Gomma, Rennes, France) and Dr. G. Nijman (Vredestein AB, Enschede, the Netherlands), and their coworkers for kindly submitting the samples that were used in generating the results discussed in this chapter.

REFERENCES

1. J.L. Leblanc, Wall slip and compressibility like effects in capillary rheometry tests on complex polymer systems, *Plast. Rubber Comp.*, **30**(6), 282–293, 2001.
2. S. Montès, J.L. White, N. Nakajima, F.C. Weissert, and K. Min, An experimental study of flow and stress fields in a pressurized Mooney viscometer, *Rubber Chem. Technol.*, **61**, 698–716, 1988.
3. H. Pawlowski and J. Dick, Viscoelastic characterization of rubber with a new dynamic mechanical tester, *Rubber World*, **206**(3), 35–40, 1992.
4. J.L. Leblanc and A. Mongruel, A thorough examination of a commercial torsional dynamic rheometer with a closed oscillating cavity, *Prog. Rubber Plast. Technol.*, **17**, 162–185, 2001.
5. M. Wilhelm, Fourier-Transform rheology, *Macromol. Mater. Eng.*, **287**(2), 83–105, 2002.
6. See, J.L. Leblanc, Insight into elastomer—filler interactions and their role in the processing behaviour of rubber compounds, *Prog. Rubber Plast. Technol.*, **10/2**, 110–129, 1994, for a pictorial representation of such a morphology.
7. J.L. Leblanc and C. Barrès, *Bound Rubber: A Key Factor in Understanding the Rheological Properties of Carbon Black Filled Rubber Compounds*, Rub. Div. Mtg, ACS, Chicago, IL, April 13–16, 1999, p. 70.
8. J.L. Leblanc, Elastomer—filler interactions and the rheology of filled rubber compounds, *J. Appl. Polym. Sci.*, **78**, 1541–1550, 2000.
9. J.L. Leblanc and B. Stragliati, An extraction kinetics method to study the morphology of carbon black filled rubber compounds, *J. Appl. Polym. Sci.*, **63**, 959–970, 1997.
10. V.M. Litvinov and P.A.M. Steeman, EPDM–Carbon black interactions and the reinforcement mechanisms, as studied by low resolution ^{1}H NMR, *Macromolecules*, **32**(25), 8476–8490, 1999.
11. G. Fleury, G. Schlatter, and R. Muller, Non linear rheology for long branching characterization, comparison of two methodologies: Fourier transform rheology and relaxation, *Rheol. Acta*, **44**, 174–187, 2004.
12. J.L. Leblanc, Investigating the non-linear viscoelastic behavior of filled rubber compounds through Fourier transform rheometry, *Rubber Chem. Technol.*, **78**, 54–75, 2005.
13. M. Wilhelm, D. Maring, and H.-W. Spiess, Fourier-transform rheology, *Rheol. Acta*, **37**, 399–405, 1998.
14. M. Wilhelm, P. Reinheimer, and M. Ortseifer, High sensitivity Fourier-transform rheology, *Rheol. Acta*, **38**, 349–356, 1999.
15. J.L. Leblanc and C. de la Chapelle, Updating a torsional dynamic rheometer for Fourier transform rheometry on rubber materials, *Rubber Chem. Technol.*, **76**, 287–298, 2003.
16. J.L. Leblanc and C. de la Chapelle, Characterizing gum elastomers by Fourier transform rheometry, *Rubber Chem. Technol.*, **76**, 979–1000, 2003.
17. C. Friedrich, K. Mattes, and D. Schulze, *Non-linear Viscoelastic Properties of Polymer Melts as Analyzed by LAOS-FT Experiments*, IUPAC Macro 2004, Paris, France, July 4–9, 2004, Paper 6.1.3.
18. J.L. Leblanc, Fourier Transform rheometry: A new tool to investigate intrinsically non-linear viscoelastic materials, *Ann. Trans. Nordic Rheol. Soc.*, **13**, 3–21, 2005.
19. W. Dierkes, Economic mixing of silica-rubber compounds; Interaction between the chemistry of the silica-silane reaction and the physics of mixing, Ph.D. thesis, University of Twente, The Netherlands, 22 April 2005, ISBN 90–365–2185–8.
20. S. Wolff, Reinforcing and vulcanization effects of silane Si-69 in silica-filled compounds, *Kautsch. Gummi, Kunstst.*, **34**, 280–284, 1981.

31 Electron Beam Processing of Rubber

Rajatendu Sengupta, Indranil Banik, Papiya Sen Majumder,
V. Vijayabaskar, and Anil K. Bhowmick

CONTENTS

31.1 INTRODUCTION

Radiation processing of polymers was introduced after World War II with the development of the nuclear reactor. In the current years, various radiation sources, e.g., X-rays (soft and hard), gamma (γ) and ultraviolet (UV) rays and electron beam (EB) are being widely used.

The electromagnetic radiations can profoundly alter the molecular structure and macroscopic properties of polymeric materials. In almost all the cases, the absorbed radiation dose necessary to bring about changes in physical properties of polymers is considerably lower than that required to cause any significant change in glasses, ceramics, or metals. The high-energy irradiation of some polymers (e.g., natural rubber [NR] and polyethylene [PE]) can lead to the formation of three-dimensional network structures, which generally improve the overall physical or chemical properties of the original polymer [1,2]. A comparative study on vulcanization of rubber mixtures based on NR and polybutadiene rubber with carbon black by applying separate EB irradiation and simultaneous EB and microwave irradiation has been carried out [3]. Improvement of surface hardness by surface modification of ultrahigh molecular weight polyethylene (UHMWPE) with EB and cross-linking parameters of low-density polyethylene (LDPE) by EB irradiation with different current intensities have been studied [4,5]. On the other hand, some polymers such as polyisobutylene and polypropylene (PP), degrade on exposure to radiations [6,7]. Network structures are formed under a wide variety of radiation conditions, e.g., ionizing or nonionizing radiations, photochemical or thermochemical systems. The unique advantage of ionizing radiations is that they provide a means of generating cross-linked structure from thermoplastics and elastomers without any chemical agent and heat, thus improving the mechanical and chemical properties of the original polymer.

Both qualitative and quantitative evaluations of radiation-induced modifications (cross-linking and grafting) have been pursued between 1949 and 1964, but in recent years, radiation processing of various polymers has been commercialized. The development of the nuclear reactor after World War II and early studies of radiation effects on polymeric materials [6,8–11] culminated in the electron accelerator equipment [12]. Manufacturers continue to develop and improve the EB accelerator equipment like self-shielded systems with compactness and high throughput capability to meet the requirements of new applications [13,14]. Physical and chemical changes in some polymers and mechanisms of cross-linking and chain scission under exposure to radiation energies have been investigated first in the 1950s [15–21]. The efforts are however still continuing today.

Over the last few decades, the use of radiation sources for industrial applications has been widespread. The areas of radiation applications are as follows: (i) Wires and cables; (ii) heat shrinkable tubes and films; (iii) polymeric foam; (iv) coating on wooden panels; (v) coating on thin film–video/audio tapes; (vi) printing and lithography; (vii) degradation of polymers; (viii) irradiation of diamonds; (ix) vulcanization of rubber and rubber latex; (x) grain irradiation.

There are potential applications in the area of pollution control. These include:

(a) Treatment of sewage and sludge
(b) Removal of SO_x and NO_x from flue gases

Radiation-grafting methods in bioengineering applications, especially to produce biocompatible materials are very promising and rapidly developing areas. Synthetic polymers produced by conventional methods have several disadvantages, because they contain toxic compounds such as catalysts which can be leached into aqueous biological medium. Toxic additives can be avoided, if polymerization is initiated by radiation [22]. Chapiro [20] and Charlesby [21] are the pioneers who geared up the research on radiation synthesis for interdisciplinary fields like biomedical applications. Synthesis of several intelligent biomembranes for controlled drug releases by irradiation techniques have been recently reported [23].

This review will highlight various aspects of electron beam processing of rubber. In addition, other rubber-like materials and non-EB techniques have also been briefly discussed for comparison.

31.2 RADIATION-INDUCED MODIFICATION

31.2.1 RADIATION TYPES AND SOURCES

Cross-linking of monomer, oligomer, and polymer or degradation reactions may be induced by various radiations such as microwave, ultrasonic, γ-rays, high-energy electrons, light energies (UV), plasma discharge, etc.

1. Microwaves (having a frequency of 10^{10}–10^{12} Hz) are essentially a part of the electromagnetic spectrum sandwiched between the quasistatic regime of lower frequencies and optical regimes of higher frequency. The distinctive feature of microwave is the large depth of penetration for most materials, particularly biological tissues. The principle underlying

microwave absorption is similar to dielectric heating except that the alternating field in this case is of higher frequency. The dipoles in the dielectric material line up with the field and as the field alters, the molecules move to keep the dipoles in phase with the field. At higher frequency, the polar molecules fail to keep in phase with the electric field. As a result, a portion of the energy is converted into very fast kinetic energy, which results in the generation of heat throughout the bulk of the polar materials. For microwave heating, the material must be polar or should contain polar or conductive ingredients. Since most polymers are not polar, their dipoles are not strong enough to produce a fast rate of heating. Hence, nonpolar polymeric materials must be compounded with dipolar and electrically semiconducting materials such as metal oxides and carbon black to compensate for the lack of dipoles in the polymer compound.

2. Unlike microwave, ultrasonic waves are not electromagnetic radiations. They are essentially elastic waves created by the cyclic vibratory movement of an ultrasonic transmitter [24]. Two types of transmitters are frequently used: Piezoelectric crystals and magnetostrictive devices. Piezoelectric crystals contain electric dipoles. When voltage is applied across the crystal, the dipoles and hence, the crystals lengthen as a result of electric repulsion. Under AC voltage, the expansion and contraction of piezoelectric crystals produce alternative cycles of compression and rarefaction in any material in contact with the piezoelectric crystals. Rigid bodies like metals and ceramics reflect the ultrasonic waves, whereas low-viscosity liquids transmit them without any appreciable loss of ultrasonic energy. On the other hand, if the medium in contact is constituted of a viscoelastic material such as rubber, the ultrasonic energy is converted to heat by relaxation and frictional processes. Since the ultrasonic waves are generated throughout the bulk of the material, the heating is uniform.

3. Radioisotope sources (^{60}Co) produce γ-rays with high penetration capabilities in the energy range 1.17–1.33 MeV and are primarily used for the preservation of food, sterilization of medical devices, treatment of sewage and sludge, and cross-linking of high-performance monomer–polymer impregnated wood composites for flooring.

4. Various sources of energetic electrons such as Van de Graff generator (energy 0.9–3 MeV), linear accelerator (energy 25 MeV), dynamitron (energy 0.3–4 MeV), etc. are available for industrial applications. The EB radiations have high penetrating power and are used for curing thick polymeric articles, wires and cables, highly pigmented inks and coatings, etc.

5. UV radiation is a part of the spectrum of electromagnetic radiations. It adjoins the short wave range of the visible light and extends up to the ionizing radiations.

According to DIN 5031, section 7, the radiation in the wavelength ranging from 380 nm to 100 nm is called UV radiation. Electromagnetic waves have a double nature: They could be described either as a wave or a photon. It is well known that each radiation has a certain quantum of energy.

Consequently, radiation energy is always an integral multiple of the energy of a photon. The energy is given as

$$E = hc/\lambda \tag{31.1}$$

where

E = energy of a photon (J)
h = Planck's constant = $6.6260755 \times 10^{-34}$ J s^{-1}
c = light velocity = 299792458 m s^{-1}
λ = wavelength (m)

This equation shows that the radiation energy depends only on the wavelength. As a consequence, it can be stated that the shorter the radiation wavelength, the more is the energy contained in the radiation.

TABLE 31.1

Frequency and Wavelength of Various Radiations

Radiation	Frequency (Hz)	Wavelength (μm)
Ultraviolet	10^{17}–10^{15}	10^{-2}–1
Microwave	10^{12}–10^{10}	10^{3}–10^{5}
Electron beam	10^{21}–10^{18}	10^{-7}–10^{-4}

Source: From Bhowmick, A.K. and Mangaraj, D. in *Rubber Products Manufacturing Technology*, Bhowmick, A.K., Hall, M.M., and Benary, H.A., Eds., Marcel Dekker, New York, 1994, 315. With permission.

6. Plasma or glow discharge radiation sources fall into three categories:
   (a) Thermal plasmas are produced by gas arcs under atmospheric pressure in the 5000–50,000 K region (arc welding).
   (b) Cold plasmas, are produced by glow discharges such as those found in neon signs.
   (c) Hybrid plasmas are between cold and thermal plasmas and are defined as having numerous small thermal sparks uniformly distributed throughout large volumes of nonionized gas molecules, thus producing a relatively low average temperature of the plasma state. Corona and ozone generator discharges are associated with this type of plasma. High voltages are required to maintain its discharge characteristics (60–10,000 Hz at several thousand volts). Table 31.1 gives the frequency and wavelength of a few radiations.

31.2.2 INTERACTION OF RADIATION WITH POLYMERS

High-energy radiation interacts with a material by three principal mechanisms [25–28]:

(a) Photoelectric effect: By which a photon is absorbed and an orbital electron, having the energy of the incident radiation minus the binding energy, is ejected.
(b) Compton scattering: By which an electron is ejected, while the photon (of reduced energy) is scattered.
(c) Pair production (above 1.022 MeV): By which a positron–electron pair is produced, and where subsequent annihilation of the positron generates γ-rays of ≈0.511 MeV which may further interact by the first two mechanisms.

Compton scattering is the principal mechanism for organic polymers (R) when high energy interacts with them.

$$R \rightarrow R^+ + e^- \text{ (ionization)}$$
$$e^- + R \rightarrow R^+ + 2e^- \text{ (ionization of further molecules)}$$
$$e^- + R^+ \rightarrow R^* \text{ (coulombic attraction producing highly excited electronic states)}$$
$$R \rightarrow R^* \text{ (direct formation of electronically excited state)}$$

The excited-state molecules may either undergo radiationless decay to the ground state leading to the formal generation of heat under conditions of high radiation flux or radiative decay (i.e., phosphorescence), thereby emitting light.

Excited states can also decay by means of chemical reaction via heterolytic bond cleavage, leading to ions, or by homolytic bond cleavage generating free radicals.

$$RX^* \rightarrow R^{\bullet} + X^{\bullet}$$

Evidence indicates [28,29] that in most cases, for organic materials, the predominant intermediate in radiation chemistry is the free radical. It is only the highly localized concentrations of radicals formed by radiation, compared to those formed by other means, that can make recombination more favored compared with other possible radical reactions involving other species present in the polymer [30]. Also, the mobility of the radicals in solid polymers is much less than that of radicals in the liquid or gas phase with the result that the radical lifetimes in polymers can be very long (i.e., minutes, days, weeks, or longer at room temperature). The fate of long-lived radicals in irradiated polymers has been extensively studied by electron-spin resonance and UV spectroscopy, especially in the case of allyl or polyene radicals [30–32].

In photochemistry, the excited-state sites, produced as a result of absorption of energy by particular chromophores located at specific sites, can be expected to undergo a certain set of defined reactions. By comparison, the excited states, induced by radiation, are distributed at random throughout the molecular structure. However, because of the energy transfer processes, radiation-induced reactions are not totally random, e.g., energy can be trapped by certain functional groups, such as aromatic rings, that undergo efficient nonreactive decay to the ground state [30].

Homolytic bond cleavage from excited states in irradiated polymers [30] can lead to a pair of free radicals via bond scission, involving main chain or side-chain substituents.

In the case of radical formation by the main-chain scission of the polymer molecule, a high concentration of the radicals initiates the reactions involving the geminate pair [30].

In the case of radicals formed by the cleavage of the side groups, a small reactive radical fragment, e.g., hydrogen radical, is obtained which then leads to a series of reactions involving either the geminate pair or the generation of a second macroradical together with the formation of a stable low-molecular weight product, due to the increased mobility of the small radical fragment [30].

The above-mentioned mode of reactions changes when the irradiation is carried out in the presence of gases such as oxygen. In this case, energy transfer, the reaction of oxygen with polymer radicals [32] (leading to the formation of peroxy radicals) and other reactions may affect the type and concentration of products formed [33]. The same can be said for certain additives mixed into the elastomer for one or the other purpose.

The lifetime of the radicals, controlling the nature of reactions in polymers (e.g., cross-linking or scission), is governed by morphology of the polymer [34], temperature [35], as well as the stability of the radical [36]. Hence, radicals formed in the crystalline regions of a semicrystalline polymer have much lower mobility and longer lifetimes; such radicals can migrate slowly to the amorphous region, where they subsequently react. Radical movement in solid polymers, especially in the crystalline region occurs by subsequent hydrogen abstraction reactions [34,37] as evidenced by the isotopic exchange studies:

$$\left\{ \begin{array}{l} -CH_2-CH_2-CH_2-CH_2- \\ -CH_2-\dot{C}H-CH_2-CH_2- \end{array} \right\} \rightarrow \left\{ \begin{array}{l} -CH_2-CH_2-\dot{C}H-CH_2- \\ -CH_2-CH_2-CH_2-CH_2- \end{array} \right\}$$

The existence of closely spaced radical pairs can be identified by spin–spin interactions in organic materials irradiated at low temperature [38] and these coupled spins disappear as the temperature is raised, because of both termination and radical migration.

The time-dependent nature of migration and chemical reaction of free radicals [30] in irradiated polymers can play an important role in altering the polymer structure and properties, e.g., cross-link formation via reactive sites or chain scission, or postirradiation oxidative influences (irradiation in presence of air or oxygen).

The interaction of electromagnetic radiation with materials falls into two classes: monomers, oligomers, polymers, and additives interacting with ionizing radiation or high-energy electrons (0.1–10 MeV); and monomers, oligomers, polymers, catalysts, or additives, undergoing photochemical or thermal reaction processes (0.01–1 eV).

Reaction schemes (Schemes 31.1 through 31.3) are shown as follows [39]:

$$\text{Preformed high molecular weight polymer} \xrightarrow[\text{or high-energy electrons}]{\text{ionizing radiation}} \text{Cross-linked or degraded polymer}$$

SCHEME 31.1 Effect of high energy electrons on high molecular weight polymer.

Low-molecular weight polymers + Vinyl monomers + Multifunctional vinyl monomers —(ionizing radiation or light energy/catalyst)→ cross-linked polymer systems

Preformed polymer containing photosensitive functional groups —(light energy)→ cross-linked polymer systems

Preformed polymers containing reactive functionalities other than vinyl unsaturation (oxirane, hydroxyl, carboxyl, etc.) + multifunctional (nonvinyl substitution) cross-linking oligomers (methylol ureas, amines, etc.) —(thermal energy,)→ Cross-linked polymer system

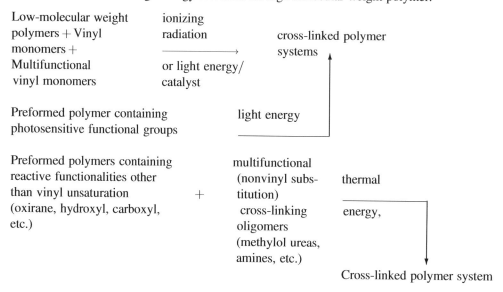

SCHEME 31.2 Formation of crosslinked polymers.

Monomers [39] are defined singly as vinyl unsaturated organic molecules with well-characterized molecular weights (100–500 g/mol).

$$CH_2{=}CHR \xrightarrow[\text{or peroxide catalyst+heat}]{\text{High-energy radiation, or photoactive catalyst and } h\upsilon,} \text{Uncross-linked linear polymers}$$

Cross-linking oligomers (mol wt ~200–1000), containing multiple unsaturation, are used to promote rapid free-radical-induced propagation reactions leading to cross-linked structures [39].

$$\begin{matrix} CH_2{=}CH \\ CH_2{=}CH \end{matrix} R{-}CH{=}CH_2 \xrightarrow[\text{Low-energy (nonionizing) radiation + catalyst}]{\text{High-energy radiation or}} \text{Cross-linked or network structures}$$

SCHEME 31.3 Generation of crosslinked and uncrosslinked polymers from monomers.

Some polymers like PE and NR get cross-linked on exposure to radiation while others like those based on vinylidene polymers, e.g., polymethylmethacrylate (PMMA), polyisobutylene, degrade. Certain other types of polymer structures (high aromatic content or thermoset) resist degradation by high-energy radiation. Coating polymers usually contain acrylic, methacrylic, or fumaric vinyl unsaturation along or attached to the backbone.

Presence of photosensitive groups is an essence for a polymer to get cross-linked by light energy (UV or visible).

From a practical stand point almost any of the lower-molecular weight vinyl monomers, cross-linking oligomers, and polymers can be blended with a high-molecular weight thermoplastic polymer to enhance cross-link density at lower dose rates [39]. The influence of various other additives on the efficiency of cross-linking of polymers will be highlighted in the subsequent sections.

31.2.3 Advantages of Electron Beam Irradiation

Some of the potential advantages of EB irradiation are as follows:

1. Radiation doses required to effect desirable changes in polymers (2–10 Mrad) can be delivered to the product in question in a matter of seconds [39] whereas, it might have taken hours to accomplish the same with γ-radiation sources of reasonable intensity [33].
2. It is possible to selectively direct the radiation beam toward any part of the product and to control its intensity by simple turning of a potentiometer. In this way, the radiation can also be turned off completely, rendering the equipment safe for maintenance and repair work. Special irradiation procedures, e.g., double side exposure, variation in angle of incidence, and special shielding techniques allow an even greater control of energy deposition far beyond anything achievable with other techniques.
3. EB processing of polymeric materials including other radiation-induced modifications require less energy "cold cure" [40]. The ratio of energy requirements for heat and radiation cure is often larger than 3, since the energy conversion and utilization efficiency of an electron accelerator is generally higher than that of a heated mould.
4. EB-induced modifications permit greater processing speed.
5. Compared to a ^{60}Co-γ source, the electron accelerator yields an accurately focusable and constant radiation. The ^{60}Co-γ source always creates a diffuse radiation, the energy of which decreases with time according to the half-life of the material [40].
6. Though the primary reactive species consisting of ionized molecules, excited states, and radicals are same for different types of incident ionizing photons or particles, the amount of energy deposition per unit length of the ionization track, and hence the density of radical pairs formed, depends on the type of radiation. This is described in terms of linear energy transfer (LET) [30]. High-energy electrons with higher LET coefficients than X-rays and γ-rays produce more densely ionized tracks and have a fairly definite maximum range of penetration in a polymer (on the order of millimeters) [39]. Chemical modifications with high and low LET effects and LET dependence on chemical reactions and change of mechanical properties have been reported [41,42]. Comparisons of dose-rate effects and LET effects on different polymer matrix have also been reported [43].
7. There is no induced radioactivity [30,44] with EB irradiation (unlike γ- or neutron irradiation) and hence EB-induced modification of materials is less hazardous compared to neutron or γ-ray-induced modifications.
8. Microwave curing of polymeric materials requires the presence of dipolar materials for effective modification to occur through dielectric heating [45]. This is not an essence in the case of EB modification of polymers which requires the presence of only labile reactive site, e.g., hydrogen in the polymeric structure.
9. EB irradiation (like the other ionizing radiation techniques) can bring about the vulcanization of saturated chemically inert polymers which cannot be achieved in the conventional thermochemical curing methods [44].
10. EB cross-linking leads to the production of homogenous massive articles, compared to the conventional curing methods, owing to high penetration of the radiation [29,39]. The vulcanization time can also be shortened considerably, since the degree of cross-linking is determined by the integral dose of radiation absorbed [30,46].

11. Compounding formulation for EB cross-linking is far less critical than for conventionally cured compounds. Even dispersion of vulcanizing agents in conventional compounds is vital for a viable product. Excessive milling or mixing will cause scorch (premature cure), and render the compound unusable. In the case of EB radiation-curable compounds, this problem does not occur [44] and the processing is similar to thermoplastic polymers.

12. Material containing conventional vulcanization system, once formed (e.g., profile) is normally cured immediately. If the product is off-size, or undercured, it is not possible to run again. In the case of EB-cross-linked equivalent, the product, in the case of being undercured, can be treated with additional dose to make up to the required level. Hence, EB-processing of polymers is expected to generate less scrap [47].

31.2.4 Parameters/Units Used in Electron Beam Radiation-Induced Modification of Materials

1. Absorbed dose: It is the amount of energy [30] imparted to the matter. Its unit is expressed in rad, or gray (Gy). The main units are interrelated by the following relation:

$$1 \text{ Gy} = 100 \text{ rad} = 1 \text{ J/kg} = 6.24 \times 10^{15} \text{ eV/g} = 10^4 \text{ erg/g}$$

2. Absorbed dose rate: This is the absorbed dose per unit time expressed in grays per unit time (kGy/s or kGy/min). Dose rate (D_R) for an electron accelerator [48] can be written in terms of beam current (I) and irradiation field area (A) as follows:

$$D_R = K \cdot I / A \tag{31.2}$$

where K is stopping power of the electrons which depends on energy of the electrons and the density of materials being irradiated.

3. Depth of EB penetration: The depth of penetration of energetic electrons into a material at normal angle of incidence is directly proportional to the energy of the electrons and inversely proportional to the density of the material [49,50]. The depth is expressed as a product of penetration distance and the density of the material (i.e., 1 g/cm^2 = 1 cm $\times$ 1 g/cm^3). The radiation energy and thus the type of electron accelerator to be used are dependent on the required penetration depth, the density of the irradiated material, and the chosen irradiation system. If one measures the density (d) in gram per cubic centimeter (g/cm^3) and the layer thickness (T) in millimeter (mm), one can determine the radiation energy (E) necessary for optimal homogeneity from [40]:

$$E = \frac{T \cdot d}{3} \text{ (MeV), for one-sided irradiation} \tag{31.3}$$

$$\text{or} \quad E = \frac{T \cdot d}{8} \text{ (MeV), for two-sided irradiation} \tag{31.4}$$

In general, single-sided radiation is suggested for the articles of 6 mm thickness having equivalent density of water, but for materials having density higher than water, double-sided irradiation is recommended.

The mode of energy dissipation by electrons in materials is either through inelastic or elastic collisions with the nuclei of the medium [28,51]. The angle of incidence of the electrons to the surface of the product affects the depth–dose distribution for materials [49].

The depth of light energy penetration for a photo cross-linking (photo-initiated free-radical propagation reaction) process can be calculated as

$$r = k'C^{1/2}e^{-1.151\epsilon cx} \tag{31.5}$$

where
    r = rate of the cross-linking reaction at thickness x
    C = concentration of the photoinitiator or light-absorbing functional group
    ϵ = extinction coefficient of the light-absorbing functional group or photoinitiator
    x = thickness of the reactive polymer layer (in cm)
    k' is a constant

4. Processing capacity: This can be readily calculated on either the basis of weight or area by the following equations:

$$\text{Capacity (Kg/h)} = 360 \times \frac{\text{Beam power (kW)}}{\text{Dose (Mrads)}} \times \text{Fractional beam utilization efficiency} \tag{31.6}$$

The efficiency ranges from 50%–70% depending upon techniques of irradiation used.

$$\text{Capacity (m}^2\text{/min)} = K \times \frac{\text{Beam current (mA)}}{\text{Dose (Mrads)}} \times \text{Fractional area irradiation efficiency} \tag{31.7}$$

K used in Equation 31.7 is a slowly changing inverse function of beam energy and ranges from 1.76 to 0.85 Mrad m^2 min^{-1} mA as the energy varies from 300 to 3000 KeV [52]. Surface dose is to be used in Equation 31.6 and average dose for Equation 31.7.

5. Irradiation-processing efficiency: The maximum irradiation efficiency [33] of a product is obtained when all of the EB power is absorbed in the product in such a manner that the required dose distribution is attained. The area efficiency can be maximized by adjusting the scanned length of the EB to match the width of the product which is conveyed past the scan horn. Maximum thickness efficiency can be effected by appropriately matching the accelerator energy to the product thickness. A lower efficiency is obtained (~50%) for sheets for optimum thickness receiving single-sided irradiation; the efficiency will be 75%–80% for double-sided irradiation at optimum thickness.

6. Dosimetry: It is necessary to verify, from time to time, the energy being received by the product. This is readily accomplished by various dosimetry procedures.

    (a) Absolute dose can be measured calorimetrically [53]. If a chemically inert substance, such as carbon or water, is irradiated, the kinetic energy of the electrons is converted to heat. The resulting temperature rise can be measured and, knowing the specific heat, the average dose can be calculated.

    (b) Of more practical use in industrial radiation processing are relative dosimeters whose response must be calibrated against a primary standard (calorimetry or ionization chamber data). Most of these are plastic films containing radiochromic dyes. These show color change upon irradiation which can be measured spectrophotometrically. The change in optical density (OD) measured at a specific wavelength is then used as a measure of dose. They have the advantages of being cheap, easy to use, physically thin, and show reasonable independence of dose rate, atmosphere, and temperature. The plastic film dosimeters [54,55] used by the industry are blue cellophane (measurements from 0.5–12 Mrad with a OD change at 650 nm), nylon, or polyvinylbutyral dyed with triphenylmethane cyanide or methoxide dyes giving response up to 20 Mrad when measured at 600 and 510 nm [56,57].

7. Product yield: The yield of an irradiation process [39] can be defined as the amount of a particular reaction (cross-linking, degradation, gas formation, formation of unsaturation, etc.) produced per 100 eV of absorbed energy. It is expressed as the G value, for example,

G(scission) $= G$(S) $=$ Number of polymer chain scissions per 100 eV absorbed. G(cross-linking) $= G$(X) $=$ Number of polymer cross-link sites per 100 eV absorbed.

The efficiency of a photochemical process [39] requiring high-intensity light sources, e.g., UV or visible light or lasers (1–10^{12} eV), is measured in terms of a quantum yield:

$$\Phi = \text{Quantum yield} = \frac{\text{Number of molecules undergoing a particular process}}{\text{Number of photons or quanta absorbed by the system}}$$

G(S) and G(X) have been estimated by quantifying the effect on molecular size distributions inferred from sedimentation velocity, gel permeation chromatography, and dynamic light-scattering measurements [58].

31.2.5   RADIATION RESISTANCE OF POLYMERS

Polymers can be radiation protected by stabilizing additives (antirads) through two principal mechanisms:

(i) Energy scavenging involving the initially formed excited states and (ii) radical scavenging by radical scavengers (antirads and/or antioxidants).

31.2.5.1   Energy Scavenging

Aromatic molecules like benzene, styrene, naphthalene, anthracene, and aromatic plasticizers serve to act as local traps for excitation energy after it is deposited in the material. For example [30], the cross-linking yield, G(X), of PE (1.0–2.0) is greater than that of polystyrene (0.035–0.05). Again, the scission yield of polyisobutylene is found to be higher than poly(α-methylstyrene). Aromatic nylons are more stable to radiation than aliphatic nylons. Among elastomers, a number of polyurethane rubbers (having high content of aromatic groups), poly(styrene-co-butadiene) rubber (with a high aromatic content) exhibit radiation resistance. Among fabrics, aromatic polyamides and poly (ethylene terephthalate) show radiation resistance that is superior to most synthetic and natural fibers. The greater the resonance energy of the aromatic additive used, the higher will be the protection to radiation. The stabilization can be affected either through physical mixing of the additive with the polymer matrix (called external protection) or chemically binding, often by copolymerization and grafting the aromatic species directly onto the polymer chain (internal protection). The former is exemplified by the radiation stabilization of PMMA by incorporation of various aromatic additives [59] and the latter by grafting styrene onto the polymer backbone [60]. The lower radiation resistance of polymer containing the bulky groups on their backbone is reported in the literature [30].

TABLE 31.2

Effect of Antirad Agents on the Scission Yield of a Natural Rubber Compound

	Chain-Scission Efficiency, $G(S)$	
Additive	N_2	Air
Control	2.7	13
2-Naphthylamine	1.6	5.6
1,4-Naphthoquinone	2	5.6
2-Naphthol	1.3	4.5

Source: From McGinnise, V.D. in *Encyclopedia of Polymer Science and Engineering*, Kroschwitz, J.I., editor-in-chief, Wiley, New York, 1986, 418. With permission.

31.2.5.2 Radical Scavenging

The antirads, including numerous commercial antioxidants, having labile hydrogens in their structure interrupt the radical chain reactions, through transfer of the labile hydrogen and yield a highly stabilized radical from them, e.g., with a hindered phenol.

$$R^\bullet \text{ (Polymer radical)} + \phi\text{-OH (a hindered phenol)} \rightarrow RH + \phi\text{-O}^\bullet \text{ (a stabilized radical)}$$

where ϕ = a bulky aromatic species.

Hindered phenols, amines, mercaptans, iodine, sulfur, and miscellaneous compounds with labile hydrogen atoms act as radical scavengers. Table 31.2 shows the protection effect of antirads on a typical component. The performance of an antirad, besides acting as an energy deactivator (aromatics) or radical scavenger, also depends on its compatibility with the host polymer and its long-term retention (possible by covalently binding with antioxidant moieties or energy scavenging aromatics) in the polymer matrix. Low-molecular weight additives incorporated in the formulation, forming the easiest and cheapest way to induce stability, can be lost (with time) by migration to sample surfaces, often followed by evaporation. Certain fillers, such as carbon black, may also enhance radiation resistance [30].

31.2.6 RADIATION-INDUCED MODIFICATION/PROCESSING OF POLYMERS

Radiation-induced modification or processing of a polymer is a relatively sophisticated method than conventional thermal and chemical processes. The radiation-induced changes in polymer materials such as plastics or elastomers provide some desirable combinations of physical and chemical properties in the end product. Radiation can be applied to various industrial processes involving polymerization, cross-linking, graft copolymerization, curing of paints and coatings, etc.

Among all the radiation sources, an EB accelerator is the most useful due to the advantages discussed in an earlier section. During radiation processing, the kinetic energy of electrons either gets converted into heat or is used for excitation of atoms or molecules. The interaction of accelerated electrons with organic systems produces very reactive, short-lived, ionic and free-radical species. The chemical changes brought about by these species are very useful in several systems, and are the basis of the growth of the electron-processing industry.

31.2.6.1 Radiation-Induced Cross-Linking of Polymer

Cross-linking of polymers is done routinely by radiation technology. The factors influencing the cross-linking or the efficiency of cross-linking of polymeric materials are:

TABLE 31.3

Approximate Cross-Link and Scission Yields of Irradiated Polymers

Polymer	G(X)	G(S)
Polyethylene	3.0	0.88
Linear low-density polyethylene	2.53	0.4
Polypropylene	0.6	1.1
Natural rubber	1.1	0.22
Polybutadiene rubber	3.8	
Poly(styrene–butadiene) copolymer	2.8	0.39
Polystyrene	0.045	<0.018
Poly(dimethyl siloxane)	2.7	<0.54
Poly(methyl methacrylates)	—	1.22–3.5
Poly(ethylene terephthalate)	0.035–0.14	0.07–0.17
Poly(vinyl chloride)	2.15	—
	0.33	0.23
Polycaprolactum (nylon-6)	0.5	0.6
Poly(hexamethylene adipamide-6,6)	0.5	0.6
Poly(vinyl acetate)	0.1–0.3	0.06–0.17
Polyisobutylene	0.05	1.5–5
Ethylene–propylene rubber	0.26–0.5	0.3–0.46
Ethylene–propylene–dicyclopentadiene terpolymer	0.91	0.29
Poly(vinylidene fluoride-co-hexafluoropropylene)	1.70	1.36
Poly(vinylidene fluoride-co-chlorotrifluoroethylene)	1.03	1.56

Source: From McGinnise, V.D. in *Encyclopedia of Polymer Science and Engineering*, Kroschwitz, J.I., editor-in-chief, Wiley, New York, 1986, 418. With permission.

1. Building blocks of a polymer must contain at least one α-hydrogen [61] and the energy of the absorbed radiation must be greater than the bond energy of the labile site. On the other hand, polymers with a high concentration of quaternary carbon atoms [30] along the chain (e.g., polyisobutylene, PMMA) primarily undergo scissions. Steric inhibition of radical–radical recombination following the chain break is the most likely explanation of this mechanism. Table 31.3 lists the cross-link and scission yields of irradiated polymers.

2. Cross-linking yield depends upon temperature, e.g., below T_g, the radiation yields of H_2 and cross-linking of PE are independent of temperature, whereas above T_g the yields of both increase steadily with increasing temperature [30]. In polybutadiene rubber, it has been observed that the cross-linking yield steadily increases with temperature [33]. It has been reported that EB irradiation at high temperature and at small dose improves certain mechanical properties like Rockwell hardness and resistance to wear for polycarbonate (PC) and polysulfone [62]. For polyisoprene rubber, lowering of temperature to 77 K had little effect on the cross-linking yield. In a given polymer, radiation cross-linking in the rubbery condition tends to exceed that in the glassy condition, which in turn generally exceeds that in the crystalline state if accessible. At temperatures above the ceiling temperatures of vinyl polymers, free-radical depolymerization occurs by irradiation [39].

   Reduction in properties of polymers under the mechanical stress in presence of radiation due to chain scission is also reported [63].

3. Cross-linking efficiency of a polymeric material also depends whether the irradiation is carried out in the presence or absence of air, e.g., oxygen has been found to increase the

rate of scission and decrease cross-linking in polyisoprene [64], polybutadiene, and copolymers of butadiene [65], PE [30] and a variety of other polymers. Oxygen stabilizes [66] the macroradicals formed on irradiation and thus prevents the radical–radical termination (through recombination).

4. Dose rate of the radiation employed, for the modification of a particular polymer, also influences the degree of cross-linking in the presence of oxygen. Lower dose rates often increase the chain-scission yields, and result in more extensive material degradation per equivalent absorbed dose [67]. Dose-rate effects are not important under inert atmosphere [25,26]. Similarly, cross-linking of polytetrafluoroethylene (PTFE), which had been recognized to be a typical chain-scission polymer, could be induced to cross-linking by irradiation at the molten state in oxygen-free atmosphere [68].

Heterogeneous oxidation of polymer samples, occurring whenever oxygen diffusion into an irradiated sample becomes a rate limiting step, can be obtained by means of microhardness measurements by plotting the depth of indentation of tiny, weighted probe as a function of position across the cross-sectional surface of an irradiated sample [69–71]. Other profiling techniques involve sectioning of degraded samples into successive slices using a vibratome and then measuring a given property for each slice, e.g., density [30], by a density gradient column, gel fraction after solvent extraction [71], IR spectroscopy [72], chemiluminscence [73], electron-spin resonance (ESR) spectroscopy [31], and gas chromatography [30]. Other material property measurements include volume resistivity, tensile strength, elongation at break, bend strength, etc.

EB irradiation of polymeric materials leads to superior properties than the γ-ray-induced modification due to the latter having lower achievable dose rate than the former. Because of the lower dose rate, oxygen has an opportunity to diffuse into the polymer and react with the free radicals generated thus causing the greater amount of chain scissions. EB radiation is so rapid that there is insufficient time for any significant amount of oxygen to diffuse into the polymer. Stabilizers (antirads) reduce the dose-rate effect [74]. Their effectiveness depends on the ability to survive irradiation and then to act as an antioxidant in the absence of radiation.

5. Efficiency of cross-linking is also governed by the morphology (degree of crystallinity or amorphous characteristics) of the polymer, e.g., for PE [39], irradiation below the melting point allows for cross-linking to occur within the amorphous part of the material (due to greater freedom of radical movement there), whereas more radicals are likely to be trapped in its crystalline phases.

The rate of γ-radiation-induced cross-linking in the crystalline and amorphous regions of a crystallizable polychloroprene has been measured by Makhlis et al. [75] who have found a considerably lower cross-link density and less degradation in the crystalline portion of the rubber. The cross-links have been postulated to be mainly intramolecular in crystalline regions and intermolecular in the amorphous phases.

The yield or efficiency of cross-linking is also dictated by the hydrostatic pressure to which the polymer is subjected during irradiation [76]. The increase in $G(X)$ is due to an enhancement of chain-reaction cross-linking furthered by a closer proximity of the polymer chains.

6. Cross-linking of polymeric materials can also be promoted by the incorporation of additives often known as cross-linking enhancers [39] or cross-link promoters [33]. They are classified as (a) indirect cross-link promoters and (b) direct cross-link promoters.

(a) *Indirect Cross-link Promoters* Chemical agents, e.g., alkyl halides, nitrous oxide, sulfur monochloride, and bases (ammonia, amines, water) do not enter directly into the cross-linking reaction, but merely enhance the production of reactive species such as free radicals [30,33] which then, through secondary reactions, lead to the formation of cross-links.

An increased proportion of cross-linking has been observed in PE, PP, and ethylene–propylene (EPs) copolymers by incorporating nitrous oxide into the polymer matrix [39].

Bases like water or ammonia [33] have been found to enhance cross-linking in polymers like PP and ethylene–propylene elastomers. The mechanism of action of various indirect cross-link promoters is reported in the literature [77–81].

(b) *Direct Cross-link Promoters* These compounds enter directly into the cross-linking reaction and become an interconnecting molecular link per se [33].

Maleimides: Alkyl and aryl maleimides in small concentrations, e.g., 5–10 wt% significantly enhance yield of cross-link for γ-irradiated (in vacuo) NR, *cis*-1,4-polyisoprene, poly(styrene-co-butadiene) rubber, and polychloroprene rubber. *N*-phenylmaleimide and *m*-phenylene dimaleimide have been found to be most effective. The solubility of the maleimides in the polymer matrix, reactivity of the double bond and the influence of substituent groups also affect the cross-link promoting ability of these promoters [82]. The mechanism for the cross-link promotion of maleimides is considered to be the copolymerization of the rubber via its unsaturations with the maleimide molecules initiated by radicals and, in particular, by allylic radicals produced during the radiolysis of the elastomer. Maleimides have also been found to increase the rate of cross-linking in saturated polymers like PE and polyvinylacetate [33].

Thiols (polymercaptans): Polyfunctional thiol compounds also enhance the rate of cross-linking in unsaturated elastomers and polymers used in the graphic arts, electronics, and coating industries [39]. It has also been found to induce cross-linking in polyisobutylene and its copolymers which normally degrade on radiation exposure. As the thiyl radicals are electrophilic [83], olefins with electron-donating substituents will react faster with thiyl radicals than those with electron-withdrawing groups. The rate of addition will of course also depend on the stability of the thiyl radical, the olefin and the radical formed by the addition reaction. The steric hindrance to the addition step and polar factors will contribute too.

Acrylic and allylic compounds: These have been reported to enhance cross-linking in polyvinylchloride [84], PE [85], PP [86], polyisobutylene, and EP copolymers [87], butadiene–styrene copolymers, and NR [88] each compounded with 50 phr of carbon black.

Efforts to achieve a retardation of cross-linking in elastomers are based on the general assumption of a radical mechanism for retardation cross-linking and the possibility of its inhibition by a deactivation of the reactive macromolecular radical [33]. These compounds generally contain one or more labile hydrogen atoms, which after, donation of this atom, will form relatively inactive radicals. Typical antirad agents are quinones, hydroquinones, and aromatic amines (phenyl and napthylamines).

31.2.6.2 Radiation-Induced Polymerization

This technique is used for the production of radiation-cured coatings, adhesives, and inks. The process is not accompanied by the release of heat, which is particularly important in the case of heat-sensitive materials, e.g., wood, cardboard, paper, plastics, etc. Various radiation sources are used for this technique, among which UV and EB are most useful.

In recent years, a growing interest has appeared in using UV radiation to induce polymerization of monomers. Because of the limited penetration of UV light into organic materials, this technology has found its major sectors of application in the hardening of thin films used either as coatings for the surface protection of various substrates or as photoresists for the production of high-resolution relief images. In light-induced polymerization or UV-curing process, the liquid monomer or resin is converted instantly into a solid polymer network by simple exposure to the radiation of an UV lamp. The advantages of an UV-radiation source are as follows:

1. No use of solvent
2. Environmentally attractive
3. No contamination of work place
4. No green house effect
5. Fast dry
6. Adhesion to various substrates
7. Press stability (printing)
8. High product quality
9. Reduction of waste
10. Rapidity of the cure, which is usually completed within a fraction of a second, even in the presence of air
11. Low overall energy consumption
12. Ability to operate at ambient temperature

The chemistry involved in UV-curable resin systems has been extensively investigated and thoroughly surveyed [88–94]. UV-radiation polymerization, is in principle, completely analogous to the conventional addition polymerization. A photoinitiator is used in UV polymerization. Its function is the same as the free-radical initiator. A conventional initiator possesses a thermally labile bond which is cleaved to form free-radical species, but the photoinitiator has a bond which breaks upon absorption of radiant energy. Benzoin ethers, benzyldialkyl ketals, benzophenone, and acetophenone derivatives are the important UV-photoinitiators [95–99].

UV-curing of acrylic–vinyl acetate copolymers in conjunction with monofunctional and trifunctional acrylate monomers (2-ethyl hexyl acrylate and trimethylolpropane triacrylate [TMPTA]) have been studied [100]. Improvement in adhesive properties of hot melts (e.g., compositions like vinyl acetate/2-ethyl hexyl acrylate or vinyl acetate/ethyl acrylate/octyl acrylate of William's plasticity numbers between 1.2–5.5), when irradiated by UV radiation in the presence of polynuclear quinones such as 1,2- and 1,4-naphthaquinone, phenanthraquinone and others, have been noted [101,102]. A further variation of the copolymerized photoinitiator approach has been presented in the use of the reaction products of polychlorophenyl glycidyl ethers (such as the pentachlorophenyl ether) and acrylic or methacrylic acids (MAs) [103]. Such polymerizable photoinitiators have been copolymerized with the usual adhesive-producing monomers, such as methyl acrylates/2-ethyl hexyl acrylates, vinyl acetate, and others to produce polymers which can be used as hot melts and irradiated by UV. The advantage of this process is said to be the constancy of adhesive properties over a wide range of cure conditions. In other words, they are difficult to overcure with the resulting loss of adhesive properties. Other investigations with polymerizable photoinitiators [104] include the use of benzoic acrylate which is copolymerized with a variety of monomers, such as 2-ethyl hexyl acrylate, butyl acrylate, tetrahydrofurfuryl acrylate, acrylamide acrylic acid (AA), and others, and then irradiated for varying periods of time. Voluminous adhesive data have been given to show improvement in adhesive properties on irradiation. Acrylated and methacrylated epoxy oligomers have been used to increase the reactivity and strength characteristics of paints. Three-dimensional membranes have been produced via radiation-induced precipitation polymerization of oligomers exposed to UV light. The radiation method ensures highly homogeneous distribution of the pore size in a preset range as compared to other membrane manufacturing techniques. The radiation-chemical technology enables to control porosity from 15% to 85% and pore size from 0.005 to 2.0 micro meter. Owing to a three-dimensional structure, the membranes acquire a number of valuable properties namely, operating temperature up to 150°C, resistance to organic solvents, acidic and low alkali solutions, etc. [105]. A coating formulation from diacrylated epoxy bisphenol-A-epichlorohydrin tripropylene glycol diacrylate (as reactive monomer), additives such as silanes and fluorochemicals has been prepared. Thin films coated on wood substrates have been exposed to UV radiation in the presence of the photoinitiator, benzyl ethyl ether. The additives enhance the adhesion of the film onto the wood substrates, but do not appear to have any effect on the hardness of the films [106].

In γ- and EB polymerization, no such photoinitiator is required unlike UV-polymerization process. The passage of energetic electrons through organic media can lead to production of free-radical species without using photoinitiators [21]. Although other types of reaction occur in EB polymerization (ion formation, rearrangement, cross-linking), free-radical reactions are considered to be the most important in polymerizations of coatings and adhesive films [107]. One of the simplest ways to prepare a pressure-sensitive adhesive is to cure a mixture of acrylic monomers by γ- and EB radiation. A mixture of an alkyl acrylate or a methacrylate with less than 10 carbon atoms in the alkyl group and 0.2 mol or less of a polar vinyl monomer (containing carboxyl, cyano, keto or amido groups) has been polymerized with ionizing radiation at a dose of about 5–7 Mrad (50—70 kGy) to produce pressure-sensitive adhesives. The peel adhesion and shear adhesion of the radiation polymerized/cured adhesives are in most cases higher than those of the conventional adhesives. The polar monomer is said to be important for attaining good adhesive properties [108–110]. An example of radiation-induced polymerization is exemplified by EB polymerization of the polyfunctional monomer tetramethylol methane tetraacrylate (TMMT) [111]. Figure 31.1a shows the variation of absorbance ratio of 1632 cm^{-1} with respect to 1470 cm^{-1} (A_{1632}/A_{1470}) of TMMT irradiated with doses of 20, 50, 100, and 200 kGy. The reduction of absorbance, due to double bonds, is also corroborated from the absorbance ratio of 2970 cm^{-1}, shown in Figure 31.1b. The decrease is sharp up to 50 kGy, beyond which the change is marginal. The monomer TMMT polymerizes through its double bonds (Scheme 31.4) in the presence of high-energy EB irradiation and forms three-dimensional network structures. As a result, it transforms from the liquid to the solid state. The residual unsaturation observed in the cross-linked material must be due to the unreacted double bonds and this may be due to the bulky structure of the monomer which shields double bonds from further reaction in the cross-linked material. It is interesting to note that the absorbance ratio of A_{1740}/A_{1470}, owing to carboxyl group, of the multifunctional monomers shown in Figure 31.1c is constant throughout the entire range of radiation dose studied, supporting the absence of oxidation of the monomer under the influence of EB. Polyacrylates, polyethers, and rubber-tackifier blends have been irradiated by EB radiation to obtain pressure-sensitive adhesive tapes with improved high-temperature shear adhesion [112,113].

A different approach, although still working with essentially "non-functional" polymers has been exemplified [114,115], in which, a 100% solid (solvent free) hot melt has been irradiated to produce pressure-sensitive adhesives with substantially improved adhesive properties. Acrylic polymers, vinyl acetate copolymers with small amounts of N,N'-dimethylaminoethyl methacrylate, diacetone acrylamide, N-vinyl pyrrolidone (NVP) or AA have been used in this study. Polyfunctional acrylates, such as trimethylolpropane trimethacrylate (TMPTMA) and thermal stabilizers can also be used.

NR, styrene–butadiene rubber (SBR), polybutadiene rubber, nitrile rubber, acrylic copolymer, ethylene–vinyl acetate (EVA) copolymer, and A–B–A type block copolymer with conjugated dienes have been used to prepare pressure-sensitive adhesives by EB radiation [116–126]. It is not necessary to heat up the sample to join the elastomeric joints. This has only been possible due to cross-linking procedure by EB irradiation [127]. Polyfunctional acrylates, tackifier resin, and other additives have also been used to improve adhesive properties. Sasaki et al. [128] have studied the EB radiation-curable pressure-sensitive adhesives from dimer acid-based polyester urethane diacry-late with various methacrylate monomers. Acrylamide has been polymerized in the intercalation space of montmorillonite using an EB. The polymerization condition has been studied using a statistical method. The product shows a good water adsorption and retention capacity [129].

31.2.6.3 Radiation-Induced Grafting

In order to modify the properties of polymeric materials or create new materials with specific properties, various techniques have been used including polymer blending, copolymerization, and

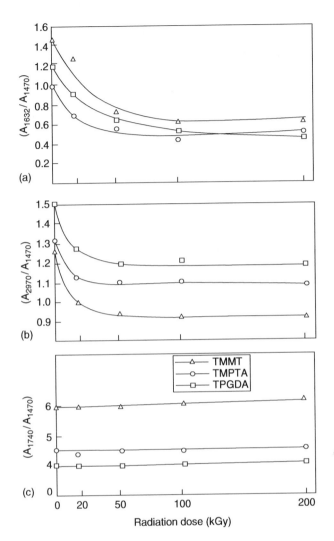

FIGURE 31.1 (a) Plot showing the variation of the absorbance ratio A_{1632}/A_{1470} of tetramethylol methane tetraacrylate (TMMT) irradiated with doses of 20, 50, 100, and 200 kGy. (b) Plot showing the variation of A_{2970}/A_{1470} of TMMT irradiated with doses of 20, 50, 100, and 200 kGy. (c) Variation of A_{1740}/A_{1470} of TMMT irradiated with doses of 20, 50, 100, and 200 kGy. (From Banik, I., PhD thesis, IIT Kharagpur, 2000.)

composite forming. Of these methods, copolymerization is unique because the monomer units of two or more different polymers are chemically bonded together. Various types of copolymers with unique properties may be formed depending on the method used and the specific monomers involved.

Random copolymers usually exhibit properties which are intermediate between those of the specific homopolymers. The fact that graft copolymers contain long sequences of two different monomer units indicates that it should be possible to select polymer combinations to give highly specific properties which are characteristics of the homopolymers involved.

A number of methods have been used to prepare graft copolymers in the past few decades including both conventional chemical and radiation-chemical methods [20,86,87]. In the latter case, graft copolymerization is usually initiated by creating active radical sites on existing polymer chains. The advantages of radiation-chemical methods are: (i) ease of preparation as compared to

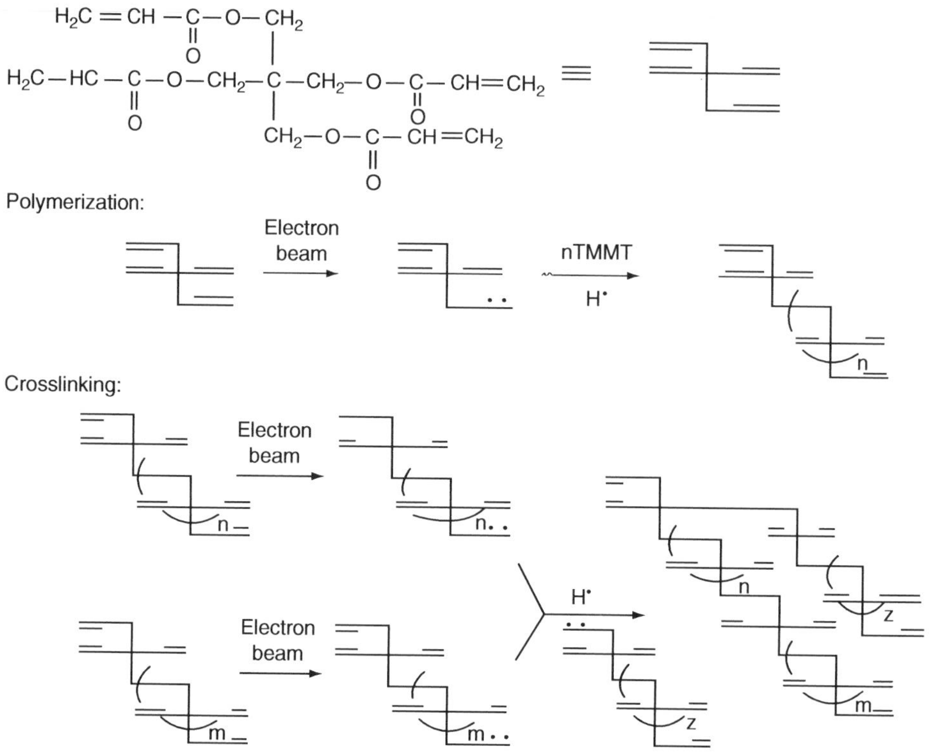

SCHEME 31.4 Polymerization and subsequent crosslinking of the monomer TMMT by high-energy EB irradiation.

conventional chemical methods, (ii) general applicability to a wide range of polymer combinations (due to the relatively unselective absorption of radiation in matter), and (iii) more efficient (and, thus more economical) energy transfer provided by radiation compared to chemical methods.

The methods of achieving graft copolymerization using radiation include:

1. Simultaneous irradiation of the backbone polymer in the presence of the monomer
2. Prior irradiation of the backbone polymer in the presence of oxygen and subsequent monomer grafting by polymeric peroxides
3. Prior irradiation of the backbone polymer in vacuum and subsequent monomer grafting by trapped radicals
4. Simultaneous irradiation of two polymers to cause inter-cross-linking

Of the various forms of radiation available, γ-rays and EB radiations (ionizing radiations) have several advantages. First, the development of high-power and dependable EB accelerators has made a wide range of dose rates easily attainable. The depth of electron penetration has also been increased with accelerated voltage. Second, the high dose rate and increased penetration depth, coupled with the energy efficiency of radiation use has made EB-processing economically competitive and commercially appealing [130].

The theory of radiation-induced grafting has received extensive treatment [21,131,132]. The typical steps involved in free-radical polymerization are also applicable to graft polymerization including initiation, propagation, and chain transfer [133]. However, the complicating role of diffusion prevents any simple correlation of individual rate constants to the overall reaction rates. Changes in temperature, for example, increase the rate of monomer diffusion and monomer reaction, but also, increase the rate of radical decay by various competing reactions. The direct effect of ionizing radiation in materials is to produce active radical sites. During the vacuum irradiation of polymers, some of the radicals become trapped and can be used to initiate the free-radical polymerization of a second monomer to produce a graft copolymer.

In PE, these trapped radicals have been identified as, mainly, alkyl and allyl radicals with the structures ($—CH_2CHCH_2—$) and ($—CH—CH=CH—$) [134,135]. In the presence of oxygen, the polymeric radicals will react to form diperoxides and hydroperoxides, as well as certain amount of less stable peroxy radicals ($—CH_2OO^•$).

Effect of additives such as mineral acids and polyfunctional monomers on radiation grafting has been discussed by Ang et al. [136]. Mineral acids and polyfunctional monomers, such as divinyl-benzene and TMPTA in small amounts (1% vol/vol), act synergistically in enhancing radiation-grafting yield for a typical system involving the copolymerization of styrene in methanol with polyolefins, e.g., PE and PP films. Effect of acids, in particular, has also been extensively investigated for a wide variety of monomers/substrates (e.g., cellulose, wool, PE, poly(vinyl chloride) (PVC), and leather by several authors [137–142]. Polyfunctional monomers, especially multifunctional acrylates and methacrylates appear to accelerate the radiation-induced grafting reaction by chain branching. These polyfunctional monomers are important, because they are used in both radiation grafting and curing reactions. The effect of propylene content on the EB-radiation grafting of AA onto ethylene–propylene rubber (EPR) to obtain hydrophilic elastomers has been studied. The degree of grafting has been found to increase with the addition of alcohol (methanol or 1-propanol) into water as solvent for grafting, and grafting occurs easily in EPR of lower propylene content of 22% [143]. Little work has been reported on radiation grafting on powdered polymer, presumably due to the difficulties involved in handling powdered polymers [144]. However, there are two areas worth reporting. Firstly, the high surface/volume ratio of small particles should have a strong influence on the radiation-grafting process, where diffusion has been reported to play a major role [145]. Secondly, graft copolymers have been shown to play an important part in the improved properties of polymer blends [146]. Small grafted particles could provide the compatibility required for improved blend properties from otherwise highly incompatible polymers. Morgan and Corelli [147] have studied the grafting of acrylonitrile (AN) to LDPE powder having an average particle size of 20 μm diameter, by EB irradiation. The irradiation has been performed in vacuum, air, and dry oxygen environment. Initiation of grafting from both alkyl and peroxy radicals at 77°C has been inferred directly from the influence of several factors, showing higher grafting at lower dose rate and higher temperature and lower grafting from vacuum irradiation. Decreased grafting from irradiation in oxygen suggests a change in grafting mechanisms unique to this system in which the large particle surface area, providing a high peroxide concentration, leads to rapid initiation and rapid termination via radical occlusion. Sidorova and Kabanov [148] and Hegaz et al. [149] have reported the effect of Mohr's salt on radiation-induced grafting of AA and acrylamide onto LDPE. It has been found that presence of Mohr's salt suppressed the homopolymerization of monomer. Similar effect of ammonium ferrous sulfate in radiation-induced grafting of 4-vinyl pyridine onto both pure and plasticized PVC has been reported [150]. The grafting yield is higher in the case of plasticized PVC than the pure one, at constant grafting condition, due to higher diffusibility of monomer into the plasticized PVC. Graft copolymerization of various monomers, e.g., styrene, methyl methacrylate (MMA), or AA on PE by ionizing radiation has been studied [151]. Reactivity ratios and comonomer compositions in the radio/chemical graft polymerization of the above monomers have shown significant difference at low radiation doses. It has been proposed that grafting with a low radiation dose occurs by a free-radical mechanism, whereas grafting with a high radiation dose proceeds with significant contribution by a cationic mechanism. Grafting of AA and MA onto PE has been carried out by EB radiation as well as γ-radiation [152]. Grafting of several monomers onto PE takes place to a higher monomer conversion during EB irradiation, whereas no noticeable grafting has been observed on PE films previously exposed to EB. Grafting of PE film with the monomers during direct γ-irradiation of the system has been unsuccessful due to homopolymer formation. The grafting effectiveness depends on irradiation modes, irradiation dose, and monomer concentration significantly. However, no change in the IR spectra of graft copolymers obtained by both irradiation methods (γ- and EB radiation) has been observed. Radio-chemical grafting of tetrafluoroethylene (TFE) onto PE by γ-irradiation has been reported by Miller and Perkowshi [153]. The grafting yield increases linearly with increasing concentration of TFE and irradiation dose. The grafting degree also increases with increasing grafting time but levels off after 24 h. The modified polymer is partially crystalline. Hegazy et al. [154] have prepared hydrophilic membranes based on graft copolymers of AA/2- and 4-vinyl pyridine and LDPE by irradiation method. Characterization of some selected properties of the grafted membrane has been studied and accordingly the possibility of its practical use in wastewater treatment from heavy and toxic metals such as Pb, Zn, Cd, Fe, etc. have been investigated. The metal uptake by the membranes has been determined using atomic absorption technique. Cation exchange membranes modified with carboxylic acid group for battery separator have been prepared by radiation-induced grafting of AA and MA on PE film [155]. The surface area, thickness, volume water uptake, ion-exchange capacity, specific electric resistance, and electrolyte flux have been evaluated after PE has been grafted with AA and MA. KOH diffusion flux has been found to increase with the degree of grafting. Electrical resistance of two cation exchange membranes modified with AA and MA has been observed to decrease rapidly with an increase in the degree of grafting.

The technique of graft copolymerization is used for the production of radiation-modified fabrics and fibers. The process consists of saturating the fabrics with vinyl monomers and then irradiating it in moist state with accelerated electrons. The fabrics thus produced have improved properties such as resistance to wrinkling and shrinkage, resistance to fire, color-fastness, good launderability, and dissipation of static charge.

Rao and Lokhande [156] have reported the modification of synthetic fibers by radiation-induced grafting reaction. Radiation grafting of AA, MA, AN or vinyl acetate on polyester or PP fibers gives better results than chemical grafting, because of prevention of homopolymerization of monomer. Grafted fibers have decreased strength, but increased hydrophilicity, improved dyeability, and increased electrical conductivity compared to nongrafted fibers. The influence of high energy (EB/γ) and UV radiation on the UHMWPE fiber has been reviewed by Chodak [157]. Cross-linking leads to improvement in properties, e.g., improvement of impact resistance at low temperatures, although lowering of strength and an increase in creep rate are observed as a result of chain scission. Cross-linking enhancers such as low-molecular weight unsaturated compounds (acetylene) are used to increase the rate of cross-linking or lower the required radiation dose. Photoinitiators like acetophenone, benzophenone have been used to effect UV-radiation cross-linking.

Radio-chemical graft copolymerization with good efficiency on halogenated polyolefins has been carried out by contacting the substrate with monomer (styrene) vapor [158,159]. Interpenetrating polymer network (IPN) could be made by grafting the monomers on preirradiated substrates [160]. Trapped radicals on samples with different storage conditions and reaction conditions are usually determined by ESR. Modification of PVC by radiation-grafting technique is an attractive method to improve the mechanical strength, printing ink adhesion, and adhesive receptance. Wang [161] has reported the radiation-induced graft copolymerization of AN onto PVC. The grafting yield in γ-induced graft copolymerization of AN onto PVC increases with increasing radiation dose. Irradiation in ambient air results in retardation of the grafting reaction and reduction of graft percentage due to breakdown of some grafting peroxide bonds at high dose. The rate of grafting increases with increasing dose rate. The γ-ray and EB-induced graft copolymerization of AA, MA, TMPTMA, and tetraethylene glycol dimethacrylate onto PVC has also been investigated [162]. The graft percentage increases with increasing dose rate and the total dose. EB irradiation at an extremely high dose stimulates the grafting rate especially that of polyfunctional monomers, because of the quick formation of a three-dimensional network formed by intra- and intermolecular cross-linking. Although the grafting results of both γ-ray and EB irradiations are similar, the latter favors the mass graft processing of PVC films. The formation of the grafted layer on PVC film surfaces induces a great improvement in tensile strength and excellent printing ink adhesion and adhesive receptance. Ion-exchange membrane has been prepared by γ-radiation-induced grafting of 4-vinyl pyridine onto PVC film [163]. The grafted film has sufficient flexibility and mechanical strength.

The γ-radiation-induced grafting of diethylene glycol dimethacrylate and its mixture with β-hydroxy ethyl methacrylate in ethanol–water systems onto silicone rubber has been reported [164]. The grafting yield increases as the radiation dose, concentration of monomer and concentration of transfer agent increase. At the same radiation dose, the degree of grafting decreases, as the dose level increases. However, at the same dose rate, the grafting level increases with radiation dose.

The modification of polymer surfaces by graft copolymerization of a monomer or monomers from active sites has been reported in numerous references [165–169]. The most common techniques are γ- and EB radiations, which generate surface radicals. Monomers can be present in gas phase (sublimed solid), in solution or as neat liquid.

Reactive species can be generated prior to monomer exposure (preirradiation grafting), during contact with monomer, or, after the polymer surface has been saturated with monomer and isolated (postirradiation grafting). The radiation-induced (γ-ray and EB) graft copolymerization of AA and vinyl acetate monomer onto PE surface has been reported [170]. The grafted sheets show excellent bonding with an epoxy adhesive and enhanced adhesion with aluminum.

Graft copolymerization is a common method for modifying polymer properties [171–173]. Radiation-induced graft copolymerization is also quite common. For example, ^{60}Co γ-radiation graft modification of EPR improves surface wettability and biocompatibility [174]. Taher et al. [175] have studied the mechanical and electrical properties, swelling behavior and gel content of the graft copolymers obtained by the radiation (using a ^{60}Co γ-source) grafting of MA onto PP films. Radiation grafting of dimethylaminoethylmethacrylate onto PP by the preirradiation method and the effect of the grafting yield on the crystallinity of the modified polymer have been determined by Bucio et al. [176]. The hydrophilicity of EVA copolymer is found to increase with the increase in grafting degree during the radiation graft copolymerization of AA onto EVA [177]. Konar and Bhowmick [178] have reported the surface properties of PE grafted with triallyl cyanurate (TAC) in the presence of EB. The changes in surface energy and thermodynamic work of adhesion of the modified PE have been determined. The surface energy increases with the grafting level and with the irradiation dose. Modification of NR by initiating the polymerization of AN in NR field latex using γ-rays was carried out by Claramma et al. [179]. Acrylamide has been graft copolymerized onto LDPE, EVA, and LDPE/EVA films using γ-radiation [180]. It has been found that such grafted films can be used for copper ion recovery from their aqueous solutions with an efficiency of up to 98%. As the main chain and branch chain are usually thermodynamically incompatible in most graft copolymers, they can be classified as multiphase polymers in the solid state.

31.2.6.4 Radiation-Induced Surface Modification

A surface is that part of an object which is in direct contact with its environment and hence, is most affected by it. The surface properties of solid organic polymers have a strong impact on many, if not most, of their applications. The properties and structure of these surfaces are, therefore, of utmost importance. The chemical structure and thermodynamic state of polymer surfaces are important factors that determine many of their practical characteristics. Examples of properties affected by polymer surface structure include adhesion, wettability, friction, coatability, permeability, dyeability, gloss, corrosion, surface electrostatic charging, cellular recognition, and biocompatibility. Interfacial characteristics of polymer systems control the domain size and the stability of polymer–polymer dispersions, adhesive strength of laminates and composites, cohesive strength of polymer blends, mechanical properties of adhesive joints, etc.

Surface modification of any material including polymers is concerned with a change of the surface properties. The surface properties can be controlled by adjusting the concentration of one component in a two-component blend or the orientation of polymer molecules near the surface [181–185].

Surface modification as a technology has been employed in various ways for many years. However, the breadth and magnitude of its applications have grown significantly during the last decade. Much of this growth has been facilitated by the development and spread of rapid and reliable surface characterization techniques. Both scientific literature and patents describe a wide range of modification techniques. Methods include chemical, photochemical, and high-energy physical techniques to modify polymer surfaces [186].

Reduced tack in rubber articles can be achieved by increasing the extent of cross-linking at the surface. The low mechanical strength of the weak boundary layer of polymer surfaces which prevents the formation of strong adhesive joints can be increased rapidly and dramatically by allowing electronically excited species of rare gases to impinge upon the surface of the polymer. As these metastable and ionic gases come in contact with the polymer surface (e.g., PE), they cause abstraction of hydrogen atoms and formation of polymer radicals at and near the surface of the polymer. The radicals formed by this process interact to form cross-links and unsaturated groups without appreciable scission of the polymer chains. The mechanical strength of the surface region is increased remarkably by the formation of a cross-linked matrix. This surface treatment technique is called CASING (Cross-linking by Activated Species of Inert Gases). CASING of polymers permits the formation of strong adhesive joints with conventional adhesives [187].

31.2.6.4.1 Physical Techniques

1. Corona Discharge: Corona treatment is widely used with plastics [188,189]. Exposure to the discharge is usually in air and at atmospheric pressure. PE treated in this way undergoes surface oxidation and some unsaturation is introduced [190]. This results in an increase in surface energy and wettability by polar adhesives, such as epoxies. Additionally, surface roughening takes place due to nonuniform degradation of the surface region [191]. This is thought to be caused by preferential attack of the more vulnerable amorphous portions compared to the crystalline regions. Enhanced joint strength with corona-treated PE is attributed both to increased surface roughness and to increased intrinsic adhesion with surface polar species, e.g., hydroxyl, aldehyde, carbonyl, and carboxyl groups [192,193]. Oxygen-containing functional groups have been confirmed by x-ray photoelectron spectroscopy (XPS) studies [192]. For a typical corona discharge treatment of PE, these functional groups range in concentration from 4×10^{-3} to about 1.4×10^{-2} groups per surface methylene unit [192].

2. Plasma Treatment: Low-pressure-activated gas plasma produced in a radio-frequency (RF) field, is employed in this method. When a polymer is exposed, active radicals and ions are created on the surface which leads to oxidation. Both inert gases (e.g., argon, helium, etc.) and reactive (e.g., oxygen) plasmas have been used apart from several other gases, although, oxygen is generally too active because of rapid and extensive degradation and ablation. Plasma treatment increases the wettability and bondability of nonpolar polyolefins as well as polar plastics such as nylon, although in the latter case, the improvement is modest [194]. Behnisch et al. [195] have investigated the surface modification of PE in remote nitrogen-, oxygen-, and hydrogen-discharge plasma. van Ooij et al. [196] have reported the surface modification of films and fibers by plasma treatments.

3. Radiation Treatment: NVP, 2-hydroxyethylmethacrylate (HEMA), and acrylamide (AAm) have been grafted to the surface of ethylene–propylene–diene monomer (EPDM) rubber vulcanizates using the radiation method (from a ^{60}Co γ source) to alter surface properties such as wettability and therefore biocompatibility [197]. Poncin-Epaillard et al. [198] have reported the modification of isotactic PP surface by EB and grafting of AA onto the activated polymer. Radiation-induced grafting of acrylamide onto PE is very important

as modified PE is of interest for various applications such as biomaterials, membranes, and coatings. Acrylamide (AAm) has been grafted on the surface of starch-filled LDPE (SLDPE) and LDPE films by γ-irradiation technique [199]. The aim of the work has been to prepare biocompatible materials by grafting a monomer onto the surface of the polymer. In order to change a hydrophobic polypeptide (poly γ-methyl L-glutamate, PMLG) into a hydrophilic polypeptide, Yue-E et al. [200] have investigated the radiation-induced grafting of a hydrogel γ-hydroxyethyl methacrylate onto PMLG membranes. This provides the basis for developing useful biocompatible and antithrombogenic materials in the biomedical field. The radiation-induced graft polymerization of methyl acrylate onto PTFE and polychlorotrifluoroethylene (PCTFE) sheets has been investigated [201] at 10°C with a 500 Gy/h ^{60}Co source. The peel strength of the grafted sheets bonded with an epoxy adhesive is much higher than chemically etched joints. PMMA and polyimide samples have been irradiated with 500 KeV carbon and hydrogen ions generated by a high-intensity pulsed power ion beam source [202] for the purpose of volatilization of surface material as a means to obtain a physically modified material surface. PE foams produced by radiation-induced cross-linking by EB show a smooth and homogeneous surface, when compared to chemical cross-linking method using peroxide as a cross-linking agent. This process fosters excellent adhesive and printability properties. Besides that, closed cells, intrinsic to these foams, imparts optimum mechanical, shock and insulation resistance, making these foams suitable for some market segments as automotive and transport, buoyancy, floatation and marine, building and insulation, packaging, domestic sports, and leisure goods [203].

4. Photochemical Methods: Literature provides examples of direct surface modification of polymers by photohalogenation in order to improve specific surface properties without affecting the bulk. Improved solvent resistance of acrylonitrile butadiene rubber (NBR) results from exposure to actinic radiation followed by contact with potassium bromide/bromate which functions as a free bromine source. Bromination of NBR increases compatibility to both aqueous- and hydrocarbon-based inks for application as a photosensitive flexographic printing plate coating [204]. Chlorination has no substantial effect. Transparent polymers such as isobutylene–isoprene rubber (IIR), NBR, EPDM rubber, PC, and neoprene have been chlorinated by irradiation in gaseous chlorine [205]. Changes in contact angles for isopropanol/water mixtures with PE, PP, and EPDM have been used to characterize the chlorinated polymers. The result is improved molecular surface layer characteristics such as resistance to dirt, solvent and penetration by toxic agents such as dichloroethylene sulfide (mustard gas). Tensile sheets of vulcanized and unvulcanized carbon black-filled rubber compounds of NR, isoprene rubber (IR), butyl rubber (BR), SBR, EPDM, etc. have been surface modified by spray depositing an alkyl halide on the substrate surface, followed by irradiation with UV light. The result is reduced air permeability, enhanced ozone resistance, and improved release properties for potential application in tires as an inner liner or in air springs [206]. Contact angle measurements in water have been used to determine the success of grafting. Polyamide and PLP fibers have been dyed by photografting an amino substituted nitroazobenzene onto the surface [207]. The dye in toluene has been syringed onto the surface and UV irradiated without a sensitizer. Surface grafting has been evaluated based upon the inability of excellent solvents to remove the dye. Polyester fabric pretreated at 120°C with *N,N*-dimethylformamide/ AA and irradiated in the presence/absence of biacetyl vapor affords grafts that significantly affect dyeing [208]. Hamilton et al. [209] have reported the surface modification of PE by solution phase photochemical grafting with AA and acrylamide to improve surface wettability. The results of x-ray photoelectron spectroscopy (XPS) and attenuated total reflectance IR spectroscopy (ATR-IR) of the modified PE surfaces are consistent with the presence of grafts of polyacrylamide or poly(acrylic acid). A water soluble polymer,

poly(vinyl pyrrolidone) has been coated onto medical grade silicone rubber and illumin-
ated with a Dymax PC—2 light welder for 3 min. The bond strength of the treated silicone
rubber to PVC, using cyanoacrylate adhesive, is increased by more than 18-fold [210].

31.2.6.4.2 Other Physical Treatments

Other direct methods of oxidation have also been used to modify polymer adherends effectively.
With flame treatment [211,212] in air, an oxidizing flame briefly impinges on the surfaces. XPS
analysis [213] has shown that amide surface groups are generated as well as typical oxidation
functionalities. Biaxially oriented films of PP and PET have been surface modified with this
treatment [214]. XPS analysis shows that the polymeric surfaces are rapidly oxidized, attaining
XPS O/C atomic ratios on PP of greater than 0.10 in less than 0.5 s. PP surface is chemically
modified to render it hydrophilic using ArF excimer laser radiation [215]. It is considered that
H atoms are selectively pulled out from the area irradiated with ArF excimer laser light and are
replaced by OH functional groups in the presence of water. In this treatment, the irradiated sample
becomes hydrophilic with enhanced adhesion properties. The treated PP and stainless steel bonded
with epoxy adhesive shows a tensile shear strength of 4.6 MPa. Polyimide films have also been surface
modified using ArF excimer laser and Xe_2^* excimer lamp irradiation [216] in an atmosphere of CF_4.
In this treatment, the contact angle of water is increased to 134° as compared to the original 58°.
CO_2 pulsed laser-induced surface grafting of acrylamide onto EPR in the presence and absence of
sensitizers, investigated by Mirzadeh et al. [217,218], produces bio- and water compatible polymers.
Water compatibility has been evaluated by measuring water drop contact angle. Breuer et al. [219]
have carried out photolytic surface modification of PP with UV laser. On the basis of such activation, a
significant enhancement of the adhesive bond strength between a polymer and an adhesive (in this case
epoxy resin) has been achieved. Pulsed UV lasers and ArF excimer laser at 1934 nm have been used for
surface modification of polymer surfaces, nanostructure fabrications on polymers and diamond
deposition using photoablated polymer plumes [220].

31.2.6.5   Effect of Irradiation on Various Polymers

Under the influence of ionizing radiation, additional chemical bonds are formed between the long
chains of polymers. As a result of such cross-linking, the materials with higher resistance to
temperature and chemicals are formed. They also display much better strength characteristics. EB
accelerator and ^{60}Co (γ-radiation source) are used on a commercial scale most frequently for
radiation-induced cross-linking of polymers for production of high-quality cables, heat shrinkable
foils, and tubings, although evidence of photocross-linking can also be found.

The radiation-treated cables find wide applications in control instrumentation of nuclear power
reactors, particle accelerators, aviation, and telephone equipments. Usually PE and PVC are
radiation cross-linked for production of such cables. The heat shrinkable foils are widely used in
packaging, electrical and electronic industries. The radiation cross-linked PE possesses the property
of elastic memory which is utilized to produce heat shrinkable products.

Unsaturation in the polymer system can dramatically affect cross-linking induced by ionizing
radiation [221–225]. Generally, vinylene (internal) and particularly, vinylidene or terminal vinyl
groups influence the ease of cross-linking of PE, like the unsaturation in NR and polybutadiene
[226–230]. Differences in cross-linking behavior among these systems arise from several factors,
including differences in polymer structure and morphology. Cross-linking in these polymers can
occur by coupling of polymer chain radicals and by involvement of the unsaturation in interchain
free-radical processes. Small amount of mineral acids, polyfunctional monomers, such as TMPTA,
TAC, *N,N'*-bis(2,4-diallyoxy-6–5triazinyl)-piperazine, AA, etc. facilitate the cross-linking of PE
[126,148,231,232]. Klier et al. [233] have reported that in EB cross-linking of PE, the addition of
1 phr ethylene glycol dimethacrylate, TMPTMA, TMPTA, tetravinylsilane, or triallylphosphate
decreases the necessary dose by 20%–30%, while 1 phr TAC reduces the necessary dose by 45%.

Maleimides and dimaleimides are known to accelerate the thermally induced peroxide cure of rubbers [234]. Their sensitizing action for radiation-induced cross-linking has been first observed by Miller and coworkers [235,236]. Experiments performed on purified NR and on a number of other elastomers show that a large number of maleimides as well as alkyl and aryl dimaleimides significantly enhance the rate of cross-linking. Among these, the most effective ones are N-phenylmaleimide and m-phenylene dimaleimide, which when added to the purified NR in an amount of approximately 5 wt%, increase the $G(X)$ value (as defined earlier) from 1.89 (value for the pure rubber) to 46 and 30, respectively. The reactivity of the double bond in the maleimide and the influence of substituent groups also affect the ability of these promoters in cross-linking. Beside unsaturated elastomers, maleimides have also been tested in a number of other polymers. In polydimethylsiloxane (PDMS), polyisobutylene, and PVC, maleimides do not affect the rate of cross-linking. In PE, the addition of 5 wt% of m-phenylene dimaleimide causes an increase of $G(X)$ from 1.8 to 7.2. An even more pronounced effect has been observed in polyvinyl acetate where the dose for gelation can be reduced by about a factor of 50.

The usefulness of polyfunctional thiol compounds as cross-linking agents for the radiation curing of unsaturated elastomers and polymers used in the graphic arts, electronics, and coating industries has been explored by a number of investigators [237–240]. They have reported that the rate of cross-linking can be considerably enhanced by only small amounts of these promoters. The basis for cross-linking in these elastomers is the addition of thiols to olefinic double bonds, a reaction which has been studied by many investigators [239–242]. It has been shown that the addition reaction proceeds mainly by a radical mechanism. However, when light and oxygen are carefully excluded and purified materials are used, the reaction may occur by an ionic process.

The acrylic compounds, maleimides, dimaleimides, and thiols are known as direct cross-link promoters, since they enter directly into the cross-linking reaction and become interconnecting molecular links [243].

Another group of compounds also can lead to the formation of cross-links. These do not enter directly into the cross-linking reaction, but merely enhance the production of reactive species such as free radicals. The most extensively studied compounds of this type are those containing halogen. A number of publications [244–251] have dealt with the cross-link promoting effect of such chemicals in various elastomers. The cross-linking promoting effect increases on passing from iodo-to-bromo-to-chloro substituted compounds and with increasing degree of halogenation. For chlorinated aliphatics, it is apparent that the sensitizing effect increases with decreasing number of carbon atoms in the molecule.

Okada [252] has demonstrated that enhanced radiation cross-linking can be achieved in PE, PP, and polyisobutylene by incorporation of nitrous oxide into the polymer matrix. Observations on PE have further shown that the increase in cross-linking is accompanied by an increase in the yield of trans-vinylene unsaturations. The cross-link enhancing effects of nitrous oxide has also been observed by Karpov [253] in EP copolymer, by Kondo and Dole [254] and Lyons and Dole [255] in PP. Lugao et al. [256] have investigated the mechanism and rheological properties of PP irradiated under various atmospheres. The study deals with the role of branching, cross-linking and degradation on melt strength properties, the mechanism and kinetics of PP with time of irradiation and the importance of double bond formation. Double bond formation is found to increase the branching and cross-linking reactions and long-chain branching is found to have a marked effect on the crystallization temperature. The same group has studied the effect of radiation dose on the isothermal and nonisothermal crystallization of LLDPE, LDPE, and HDPE by differential scanning calorimetry (DSC). The isothermal crystallization allows the observation of the changes in the crystallization rate, related to the decrease in the crystallization temperature caused by the cross-linking of the polymer. Nonisothermal crystallization shows the development of crystallites of very different sizes in the polymer.

Chodak [257] has reviewed the properties (include mechanical, processing, orientation) of radiation (UV, γ, EB) of polyolefin-based materials include PE, PP, and their blends based on ethylene–propylene and EVA copolymers. The structure and morphology of isotactic polypropylene (iPP) functionalized by EB irradiation at room temperature [258] in air have been investigated by elementary analysis, Fourier transform infrared (FTIR) spectroscopy, ESCA, and static contact-angle studies. Studies reveal the introduction of oxygen-containing groups onto iPP molecular chain after irradiation. The static contact angle of iPP decreases with increasing radiation dose due to the increase in polar component of surface energy owing to introduction of polar-oxygenated groups.

A drastic increase in the rate of cross-linking has been observed by Karkova et al. [259] in PE and PP when irradiated in the presence of sulfur monochloride.

The effect of antioxidants such as hindered phenolics, secondary amine, and thioester on the radiation cross-linking efficiency of LDPE has been reported [260]. Amount of cross-linking at a given dose decreases with all the antioxidants, the thioester being the most effective. IR absorption spectroscopy has been used to demonstrate dose-rate dependence of *trans*-vinylene unsaturation in irradiated Marlex 50 PE [261]. When the irradiated polymer is stored in vacuum a decrease is observed in *trans*-vinylene absorbance over a period of several weeks. After high dose-rate irradiation the decay is preceded by an initial increase. These phenomena have been ascribed to the reaction of "trapped" radicals.

Radiation-induced cross-linking of chlorinated and chlorosulfonated high-density PE has been reported by Korolev et al. [262]. It has been observed that the extent of cross-linking is strongly dependent on the chlorine content in the sample, chlorosulfonated PE is cross-linked more readily than the chlorinated sample in air and inert atmosphere.

It has been reported that the stability and heat storage capacity of high-density PE is improved by EB irradiation [263]. It has also been inferred from X-ray diffraction measurement that the cross-linking occurs primarily in the amorphous regions. Gel content increases with radiation dose initially, while other reactions take place in addition to cross-linking reaction at higher doses. Recently, Tikku et al. [264] and Chaki et al. [265,266] have reported the EB radiation-induced cross-linking of PE using various unsaturated monomers, e.g., TMPTMA, MMA, and TAC. The mechanical properties improve on radiation. The dielectric constant and dielectric loss decrease with radiation dose. Dynamic mechanical thermal analysis studies have revealed that the multi-functional monomers plasticize the base polymer prior to irradiation and enhance cross-linking during irradiation.

Besides PE, PVC is another versatile polymer which is used as wire and cable insulation. Ordinary plasticized PVC used in large quantities as wire and cable insulation lacks two properties, i.e., heat resistance and abrasion resistance, which are required in certain applications. For example, the insulation must be capable of withstanding temporarily the temperature of heat soldering iron and it must resist physical damage by abrasion when wires lying adjacent to it are pulled out of the wire-holding troughs. Cross-linking of PVC greatly enhances these important properties [267].

Radiation cross-linking of PVC in the absence of additives has been reported by several authors [20,21,267,268]. PVC by itself is not readily cross-linkable by EB radiation. In the study of radiation chemistry of PVC, it is known that the oxidation takes place in the presence of atmospheric oxygen during irradiation [269] and hydroperoxide groups are produced on oxidation. The formation of carbonyl group along with the elimination of HCl from PVC on irradiation in air has also been reported [270,271].

Certain polyfunctional unsaturated monomers act as radiation sensitizers. When these are added to PVC, cross-linking takes place readily, upon irradiation [272,273]. A series of monomers containing different number of unsaturated groups has been investigated to study the effect of functionality on the cross-linking of PVC. The polyfunctional methacrylates and acrylates have been found to have the greatest sensitivity towards cross-linking of PVC [274–278]. The order of gelling ability has been found to be tri > di > mono unsaturation. Bowmer et al. [279] have reported the radiation cross-linking of PVC with TMPTMA under nitrogen atmosphere at temperature from −30°C to 125°C. Before irradiation, PVC becomes plasticized by the low-molecular weight monomers. This has been reflected in lower ultimate strength and higher elongation compared to the pure PVC. Dynamic mechanical properties have also proven the plasticization of PVC by the monomer prior to irradiation. The gel content and ultimate tensile strength increase on irradiation, but the elongation at break is reduced significantly [280–282]. The kinetics and cross-linking mechanism of PVC with TMPTMA have been studied in detail [283]. The cross-linking rate has been found to be proportional to the TMPTMA concentration. Cross-linking of PVC by ionizing radiation in the presence of tetraethyleneglycol dimethacrylate has been investigated in terms of viscoelastic and calorimetric properties as well as the macroscopic morphology of the cured samples [274–277]. Discoloration and polyene formation on irradiation of PVC have been suppressed in the presence of polyfunctional monomers [284]. Mixture of PVC with monomer plus a variety of secondary additives, such as diundecyl phthalate (DUP) and dioctyl phthalate (DOP) for wire and cable insulation have been reported by several investigators [275–288]. The cross-linked materials have been characterized by different techniques [289–291]. The radiation-induced cross-linking of PVC insulation improves the life span with high resistance to abrasion cut through, crushing fire, heat from soldering iron, etc. [292–300]. Nethsinghe and Gilbert [301] have also investigated the EB modification of PVC in the presence of polyfunctional monomer TMPTMA.

EB radiation-cured PVC film surface induces a big improvement in tensile strength and excellent printing ink adhesion and adhesive receptance [302]. PVC foams containing plasticizers, thermally activated blowing agents, cross-linkers, and heat stabilizers have been manufactured by EB radiation technology [303].

EB beam irradiation technique is used for the processing of many other polymers and copolymers besides PE and PVC [304]. The radiation chemistry of PP has been reported by several authors [305–308]. It has been reported that oxygen accelerates the degradation reactions by peroxidation of the polymer chain followed by decomposition and rearrangement [309,310]. Dunn et al. [311] have studied the radical structure and its role in the oxidative degradation of γ-irradiated PP. The mechanical properties also deteriorate due to degradation. However in the presence of polyfunctional monomers, cross-linking of PP takes place [312]. Cold and heat-resistant foam has been manufactured from the blends of PP, poly(1-butene), linear LDPE and blowing agent by cross-linking with ionizing radiation and heating the cross-linked product above the decomposition temperature of the blowing agent [313].

Bartl and Seheurlen [314] have reported the comparative study of irradiation-induced cross-linking of PE and EVA copolymer. The incorporation of black and nonblack filler in EVA prior to irradiation enhances the tensile strength of the irradiated product much more than PE with same filler and radiation dose. Dobo et al. [315,316] have reported some properties of radiation cross-linked EVA. The effect of accelerated electrons on PE containing EVA copolymer has been reported by Gordiichuk and Gordienko [317]. The LDPE is cured more effectively by irradiation than higher density PE due to a lower crystallinity of the former. Datta et al. [318] have studied the EB-initiated cross-linking of a semicrystalline EVA copolymer with TMPTMA over a range of radiation doses (2–20 Mrad) and concentrations of TMPTMA (0.5–5 parts per weight). The residual concentration of carbonyl groups due to oxidation and incorporation of TMPTMA into the cross-linked EVA, as determined from IR difference spectra, increases with radiation dose and with the TMPTA level in the initial stages. The gel content has been found to increase with radiation dose. X-ray studies indicate that there are two peaks at 10.8–10.9 degree and the corresponding interplanar distances are 4.07 and 3.73 Å. With increase in the radiation dose or TMPTMA level, the crystallinity increases in the initial stage and then decreases. Figure 31.2 shows the variation of crystalline melting point, heat of fusion, and percent crystallinity of samples against radiation doses and TMPTMA levels. At higher irradiation doses and TMPTMA levels, more polar groups ($>C=0$) are formed. Higher intermolecular interactions are developed due to the presence of an increased number of polar groups, leading to the increase of percentage crystallinity. After attaining a maximum, it decreases because of the breakdown of structure as a result of chain scissions. The interplanar distance or the interchain distance of the grafted polymer has been found to remain unchanged. DSC studies also support the X-ray results. Tensile properties have been observed to

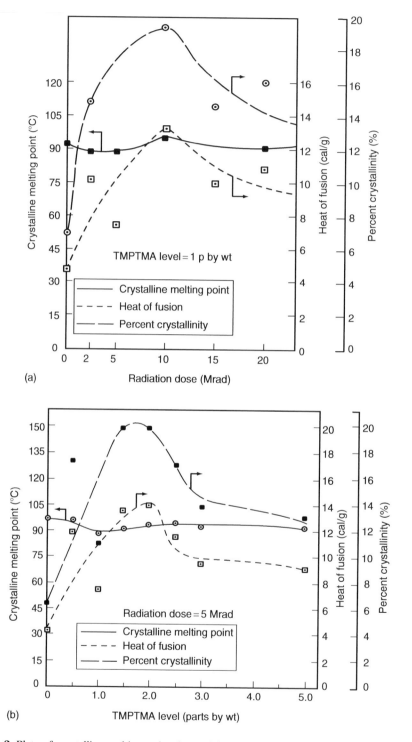

FIGURE 31.2 Plots of crystalline melting point, heat of fusion and percent crystallinity of ethylene–vinyl acetate (EVA) samples versus (a) radiation dose; (b) trimethylolpropane trimethacrylate (TMPTMA) level from differential scanning calorimetry (DSC) studies. (From Datta, S.K., Bhowmick, A.K., Chaki, T.K., Majali, A.B., and Deshpande, R.S., *Polymer*, 37, 45, 1996. With permission.)

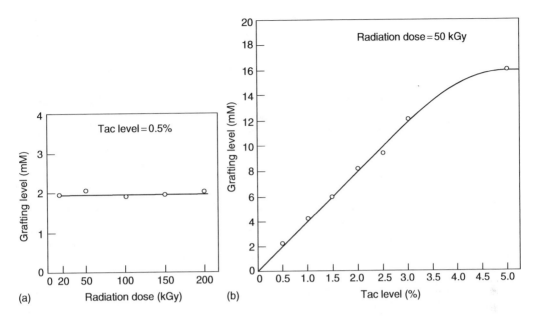

FIGURE 31.3 Plot of (a) grafting level versus irradiation dose having constant level of triallyl cyanurate (TAC), (b) grafting level versus TAC level at constant irradiation dose of 50 kGy. (From Datta, S.K., Bhowmick, A.K., and Chaki, T.K., *Radiat. Phys. Chem.*, 47, 913, 1996. With permission.)

increase with the radiation dose and TMPTA level up to 50 kGy at 1 part TMPTA level, and 1 part at a 5 Mrad dose, because of the formation of a large network structure. At higher radiation doses, the tensile properties decrease due to a breakdown of the network structure. With TAC, [319] the grafting level (Figure 31.3) has been found to remain almost constant with increase of irradiation dose, because most of the TAC is grafted at very early stage of irradiation. On the other hand, the grafting level is increased regularly with the increase of TAC level. This is due to the fact that more TAC is grafted onto EVA as irradiation generates more free radicals. The procedure for determining the grafting level can be found in the literature.

31.2.6.6 Influence of Irradiation with Specific Reference to Rubbers

31.2.6.6.1 Natural Rubber and Polyisoprene

Among the most apparent changes in microstructure on irradiation is the decrease in unsaturation with accompanying *cis–trans* isomerization [320]. Antioxidants and sulfur have been found to reduce the rate of decay of unsaturation, while thiols and other compounds lead to an increase [33]. UV-absorption spectroscopic studies [321] have indicated the formation of conjugated polyenyl structures. Kaufman and Heusinger [322] have also observed the decay in unsaturation in 1,2- and 3,4-isomers of polyisoprenes leading to cross-linking or cyclization. The yield of cross-links in 1,2- and 3,4-isomers of polyisoprenes [322] is observed to be higher than high *cis*-1,4-polyisoprenes and NR [323,324]. Oxygen has been found to increase the rate of scission and decrease cross-linking [325]. Cross-link promoters, maleimides, and a number of halogenated compounds, increase the yield of cross-links. Cross-linking in NR and 1,4-polyisoprene is believed to involve coupling of allylic free radicals produced during radiolysis as observed by ESR.

The tensile strength of NR compounds in the presence of cross-link promoters such as dichlorobenzene is increased as compared to the sulfur-accelerator and peroxide-curing systems. The retention of the maximum tensile strength at elevated temperatures is greater for radiation cured than for chemically vulcanized NR [326,327]. Also reported are a higher abrasion resistance [328] and a lower flex life in the case of radiation-cured system. Effect of phenoxy ethyl acrylate (PEA) as radiation vulcanization accelerator on radiation vulcanization of NR has been evaluated by Jayasuriya et al. [329]. However, optimum mechanical properties could be achieved only by high EB dose. The electron-beam-cured NR is reported to be superior in properties as compared to γ-radiation-modified analogue, when irradiation is carried out in air. The properties have been found to be equal in the presence of antioxidants [327,330] due to the radiation protection offered by them from oxygen and ozone. The photo-cross-linking of acrylated NR has been studied by Xuan and Decker [331]. A method of radiation graft copolymerization of *N*-butyl acrylate (NBA) on NR latex has been studied [33]. The rate of conversion has been found to increase with the increase of NBA in the latex and an irradiation dose of about 12 kGy is needed to obtain 90% conversion with 40 phr of NBA in the latex. Radiation-induced graft copolymerization of AN on NR latex has been carried out by Claramma et al. [332]. Influence of several additives like Irganox 1010 and tribasic lead sulfate on radiation cross-linking of epoxidized NR (ENR) have been studied [333]. The influence of EB irradiation and step-cross-linking on high-abrasion furnace (HAF) carbon black-loaded NR vulcanizates has been studied [334]. With irradiation, a higher cross-link density, thermal conductivity, and T_g were observed as compared to the unirradiated NR vulcanizates. However, the step-cross-linking process had slight effects on the cross-link density, T_g, and thermal properties of these composites. Radiation cross-linking of NR/montmorillonite clay nanocomposites prepared by melt mixing were studied by Sharif et al. [335]. The tensile strength and gel content of the nanocomposites showed optimum values at a range of 3.0–5.0 phr clay content.

31.2.6.6.2 Polybutadiene and Butadiene Copolymers

By far, the most predominating effect observed in the radiolysis of polybutadiene (*cis*-1,4, *trans*-1,4, and 1,2-vinyl) and its copolymers is the decay of unsaturation [336], the decrease being greater for polybutadiene irradiated in vacuum. The modes of decay of the double bonds in *cis*-1,4-polybutadiene are through *cis–trans* isomerization involving intermolecular energy transfer [28,337], and cross-linking as well as cyclization. The products formed by the destruction of the *trans* and vinyl isomers consist of conjugated double bonds [338,339] and their detection by infrared (IR) spectroscopy has been difficult owing to the overriding bands, from both cyclic structure and olefinic bonds.

The yield of cross-linking depends on the microstructure of polybutadiene and purity of the polymer as well as on whether it is irradiated in air or in vacuum. The cross-link yield, $G(X)$, has been calculated to be lowest for *trans* and highest for vinyl isomer [339]. The introduction of styrene into the butadiene chain leads to a greater reduction in the yield of cross-linking, than the physical blends of polybutadiene and polystyrene [340]. This is due to the intra- and probably also intermolecular energy transfer from the butadiene to the styrene constituent and to the radiation stability of the latter unit.

The partial replacement of styrene in a 60/40 butadiene–styrene copolymer with methylmethacrylate results in an increased rate of chain scission [340]. However, a terpolymer of MA, butadiene, and styrene shows increased response to radiation cross-linking [341] and the increase is attributed to the radiation-induced decarboxylation and subsequent intermolecular reaction of the polymeric radicals formed with unsaturations. For poly(butadiene-co-acrylonitrile) [342], $G(X)$ increases with AN content and low dose exposure. The effect of γ-irradiation on both the electrical conductance and diffusion coefficient of AN–butadiene rubber [343] mixed with different concentrations (1, 3, and 5 phr) of LDPE that is swollen in benzene, have been studied. Diffusion coefficient is found to decrease with increasing γ-irradiation dose for samples loaded with 1 and 3 phr of LDPE, while samples with 5 phr of LDPE show a significant increase of diffusion coefficient. The electrical conductance is found to be highly affected by γ-irradiation dose. A partial radiation curing has been used to decrease the melt flow index of blends containing a 1-olefin polymer, a butadiene–styrene rubber, and certain diblock or triblock copolymers [344]. Extremely fast cross-linking is observed in epoxidized *cis*-1,4-polybutadienes [345] and this is due to a very efficient chain-reaction process which gets progressively inhibited at higher doses. Polymercaptans in conjunction with haloaromatics sensitizes the cross-linking in polybutadiene and its copolymers [346]. Carbon black and silica fillers are reported to enhance cross-linking [347] becoming either chemically linked to polymer chain or resulting in a higher cross-link density near the surface of the reinforcement particle [348]. Aromatic oils have been found to enhance the yield of scission [349], while tetramethylthiuram disulphide, sulfur, diphenylguanidine, and mercaptobenzothiazole retard cross-linking. The grafting of styrene [350], mercaptans [351], carbon tetrachloride [352], and other compounds onto the backbone of polybutadiene is reported.

Optimum dose requirement for a full cure of polybutadiene and its copolymers is lower than that of the NR and a further reduction of the cure dose occurs with the addition of cross-link promoter. Solution polymerized rubber requires lower dose than emulsion polymerized rubber due to the former having lesser impurities than the latter. In the absence of oxygen, the tensile strength, hardness, resilience and permanent set of the radiation-cured rubbers are found to be equal to the chemically cured rubbers [33], while the abrasion resistance of the former is higher. The cross-linking mechanism of polybutadiene analyzed by ^{13}C nuclear magnetic resonance (NMR) technique, has been reported by O'Donnell and Whittaker [353]. The decrease in C=C bonds, formation of cross-links *cis* to *trans* isomerization during the γ-irradiation of (a) >99% and *cis*-1,4-polybutadiene and (b) 54% *trans*, 41% *cis*-1,4-polybutadiene. G(-*cis* C=C) and G(-*trans* C=C), have been found to be similar and decrease with dose from ≈40 for 0–1 MGy to 5 for 5–10 MGy. G(-double bonds) and G(cross-link) have been found to be comparable, indicating that cross-linking occurs through the double bonds. The 1,2-polybutadiene has been found to be much more sensitive to cross-linking and a value of G(C=C) = 240 is obtained at low doses. The tensile strength of radiation-cured poly(butadiene-co-acrylonitrile) rubber is also reported [354] to be equal to that of chemically cured systems. The tensile strength of the EB-radiation-cured poly(styrene-co-butadiene rubber) SBR [348] is found to be lower, and flex life higher (due to either slower crack growth or smaller rate of crack initiation) than that of sulfur-cured rubber. Also, the extensibility of the radiation-cured carbon black-filled SBR has been reported to be lower than that of the sulfur-cured system and this is believed to be due to an increased cross-link density in the immediate vicinity of the carbon black particles. The mechanical properties of SBR vulcanizates containing EB-modified surface-treated dual phase filler have been reported by Shanmugharaj and Bhowmick [355]. It was observed that the scorch time of the SBR vulcanizates increased with the irradiation of the fillers due to the introduction of quinone-type oxygen on the filler surface.

Microstructural changes on irradiation have been observed by IR and UV spectroscopy. Changes in absorption bands due to vinylidene double bonds [356,357], substituted double bonds, and ethyl and methyl groups give a measure of modifications in the presence of radiation. The ratio of the double bonds (located mainly at the end of a polymer chain) and scission is reported by some investigators [356–358] and found to be independent of temperature and dose. This is believed to be due to the reaction of the methyl radical side group with hydrogen atoms on the backbone of the parent chain.

31.2.6.6.3 *Polyisobutylene and Its Copolymers*

Degradation predominates over cross-linking in large amounts in the case of polyisobutylene and its copolymers. The yield of scission for polyisobutylene has been analyzed by various workers [200,358–360] and an enhanced scission occurs in solvents like chloroform, heptane, and carbon tetrachloride [361,362]. Unsaturated hydrocarbons like cyclohexene and diisobutylene strongly retard degradation [356,44]. The amount of scission is also lower in poly(isobutylene-co-isoprene). This is due to the energy transfer in the former case and to the simultaneous occurrence of cross-linking in the latter. For chlorinated poly(isobutylene-co-isoprene), poly(isobutylene-co-isoprene-co-divinylbenzene) [33], and dehydrohalogenated chlorobutyl rubber [33,363], rapid gelation occurs at low doses followed by predominating scission due to the gradual depletion of the reactive sites by cross-linking. The increase in the yield of cross-linking for polyisobutylene and its copolymers in the presence of cross-link promoters like polymercaptans, aromatic halides, triallyl acrylates, and dimaleimides [31,84,364–366] has also been investigated. Physical properties comparable to a conventional curing system are achieved mostly in chlorobutyl compounds containing polythiols as cross-link promoters. A very important application of EB irradiation is the radiation destruction of spent BR [367]. The radiation-degraded material was evaluated for reuse in the production of diaphragms, rubberized fabric, and roofing materials.

31.2.6.6.4   Ethylene–Propylene Copolymers

On γ-irradiation, a decay of unsaturation is observed in the microstructure of EPDM rubber containing 2–4 mol% of 5-ethylidene-2-norbornene [368]. The formation of *trans*-vinylenes and conjugated dienes has been noted in ethylene-rich EPR rubbers [369]. Gas production in EPR under the influence of γ-radiation has also been reported [370].

The cross-linking efficiency of EP copolymer has been investigated by some workers [368,371] and the yield is found to increase with increasing ethylene content. The value lies closer to PP homopolymer. The amounts of cross-linking as well as scission have been observed to increase with increasing diene content, the labile point. Cross-linking enhancers like chlorobenzene, nitrous oxide, allylacrylate, and *N*-phenylmaleimide are reported to promote cross-linking [85,370,371].

Physical properties of carbon black-filled EPR and EPDM elastomers have been found to be comparable with the sulfur-cured analogues [372]. Aromatic oils increase the optimum dose requirement for these compounds due to the reaction of the transient intermediates formed during radiolysis of the polymer with the oil as well as energy transfer which is particularly effective when the oil contains aromatic groups. The performance and oxidative stability of unfilled EPDM as well as its blend with PE [373], and the thermal stability and radiation-initiated oxidation of EPR compounds are reported by a number of workers [374,375].

The effect of propylene content on the EB radiation grafting of AA onto EPR to obtain hydrophilic elastomers has been investigated. Degree of grafting has been found to increase with the addition of alcohol (methanol and 1-propanol) into water as solvent for grafting which occurs easily in EPR of lower propylene content of 22% [376].

The degradation by γ-radiation of amorphous EP copolymers containing 23% and 36% of propylene has been correlated with the migration of radicals to the crystalline-amorphous boundaries of the polymer [377]. Hilborn and Ranby [378] have studied the photocross linking of EPDM elastomers with 1-hydroxycyclohexyl phenyl ketone as the photoinitiator with the help of the ESR technique. Cross-links are formed through recombination of allylic radicals, formed from the model EPDM compound. Rak [379] has reported the use of EPDM as a polymeric sensitizer for the radiation-induced cross-linking of PE. With the addition of 20% and 40% EPDM in PE, the EB radiation dose decreases by 30% and 50%, respectively. Blends of iPP and EPDM rubber have been treated by EB irradiation. Mutual cross-linking of the components through the interface has been detected. This inter-cross-linking causes a remarkable improvement in the mechanical properties particularly of the thermoplastic elastomer [380]. The vulcanization of EPM-EPDM rubbers by EB of varying intensity and radiation dosage has opened a new way for the production of heat, weather, and ozone-resistant articles, particularly for automotive applications. The scorch, surface roughness, and tackiness problems can be avoided by EB treatment of the green predecessors of tubes, profile or sealings which are cured afterwards by hot air [9].

Majumder and Bhowmick [381] have investigated the influence of the concentration of TMPTA on the surface properties of EPDM rubber, modified in the presence of EB. The surface energy of the TMPTA-modified EPDM rubber has been observed to increase as compared to the unmodified one. Table 31.4 displays the contact angles and the work of adhesion of water, for the control and the modified EPDM surfaces.

The decrease in the contact angle and corresponding increase in the work adhesion of the modified surfaces are due to the generation of polar carbonyl (C=O) and ether (C=O=C) groups on the surfaces formed through interaction of macroradicals on EPDM backbone with atmospheric oxygen. The results have been confirmed from IR (Figure 31.4) and X-ray photoelectron spectroscopic (XPS) observation (Table 31.5).

TABLE 31.4

Contact Angle (θ) and the Work of Adhesion (W_A) for the Control and the Modified EPDM Surfaces

Sample	θ_{water} (deg)	W_A (mJ/m^2)
Unmodified EPDM	84	80.4
EPDM + 100 kGy radiation dose + 2% TMPTA	78	85.3
EPDM + 100 kGy dose + 10% TMPTA	76	87.9
EPDM + 100 kGy + 20% TMPTA	71	91.6

Source: From Sen Majumder, P. and Bhowmick, A.K., *J. Adhesion Sci. Technol.*, 11, 1321, 1997. With permission.

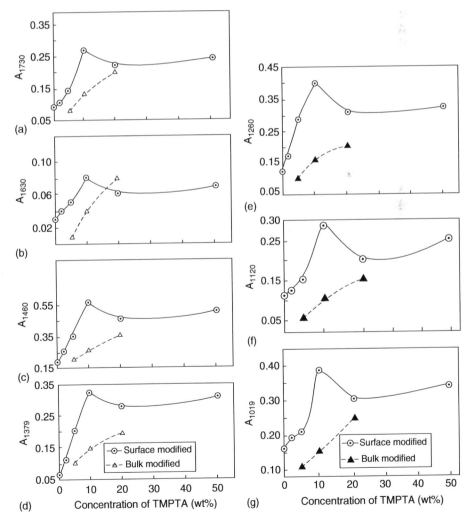

FIGURE 31.4 Absorbance of control and EPDMs modified with different trimethylolpropane triacrylate (TMPTA) levels at an irradiation dose of 100 kGy. (From Sen Majumder, P. and Bhowmick, A.K., *J. Adhesion Sci. Technol.*, 11, 1321, 1997. With permission.)

TABLE 31.5

XPS Details of C1s and O1s Spectra from Control and Grafted EPDMs

Sample Come	C1s Peak Position (eV)	C1s Area	O1s Peak Position (eV)	O1s Area	Concentration (%) C	Concentration (%) O	O/C Ratio
E$_{0/0}$, Unirradiated EPDM rubber	—	1221 (Relative area %)	—	896 (Relative area %)	79.5	20.5	0.26
	284.9	80.5	532.4	78.8			
	286.2	14.2	533.2	21.2			
	287.6	5.3					
E$_{0/100}$, EPDM surface irradiated with a dose of 100 kGy	—	1092 (Relative area %)	—	915 (Relative area %)	77.3	22.7	0.29
	285.3	76.9	532.7	79.0			
	286.6	16.5	533.3	21.0			
	288.2	6.6					
E$_{2/100}$, EPDM surface grafted with 2% TMPTA at a radiation dose of 100 kGy	—	1121 (Relative area %)	—	1121 (Relative area %)	74.0	26.0	0.35
	285.3	75.0	532.6	80.4			
	286.8	18.1	533.8	19.6			
	288.6	6.9					
E$_{10/100}$, EPDM surface grafted with 10% TMPTA at a radiation dose of 100 kGy	—	1115 (Relative area %)	—	1131 (Relative area %)	73.8	26.2	0.36
	285.9	70.0	533.5	80.5			
	287.6	18.4	534.4	19.5			
	289.6	9.1					
	292.0	2.5					
EB$_{5/100}$, EPDM rubber bulk modified with 5% TMPTA and irradiated with a dose of 100 kGy	—	886 (Relative area %)	—	883 (Relative area %)	74.1	25.9	0.35
	285.4	80.0	532.6	78.0			
	286.6	14.6	533.6	22.0			
	287.6	5.4					

Source: From Sen Majumder, P. and Bhowmick, A.K., *J. Adhesion Sci. Technol.*, 11, 1321, 1997. With permission.

IR-ATR spectra of EPDM rubber sheets modified on the surface with TMPTA solutions and irradiated at a constant dose of 100 kGy have been taken. The main peaks of interest are observed at 1730, 1630, 1460, 1379, 1260, 1120, and 1019 cm^{-1}. The peak absorbances are plotted against concentration of TMPTA in Figure 31.4a through g. It is observed that with increasing concentration of TMPTA, the peaks at 1730, 1630, 1460, and 1379 cm^{-1} steadily increase up to a certain concentration level, namely, 10% TMPTA, after which there is a slight drop in the absorbance values. These rise again at a still higher concentration of 50%, although the magnitudes of the absorbance values are lower than that obtained at 10% concentration. A similar result is obtained, when the absorbances at 1260, 1120, and 1019 cm^{-1} are analyzed with respect to the concentration of TMPTA.

During irradiation, a large number of free radicals are generated on the EPDM backbone. These radicals, being highly reactive, undergo aerial oxidation to produce carbonyl and ether linkages on the rubber surface. They may cause cross-linking as well as chain scission of the EPDM rubber. On the other hand, active free radicals are also produced in profuse numbers on the TMPTA molecule itself. With the help of these radicals, TMPTA participates in several reactions, namely, self-cross-linking, grafting, cyclization, cyclopolymerization, etc. The grafting of TMPTA onto the

EPDM backbone, together with the carbonyl and ether groups generated on the rubber itself causes the carbonyl and ether absorptions to increase. The absorption at 1630 cm^{-1} of the *trans*-vinylene bonds of TMPTA is mainly due to the grafting of TMPTA onto EPDM and partly due to the formation of vinylidene bonds at the chain ends on irradiation. The absorptions at 1460 and 1379 cm^{-1} due to >CH$_2$ scissoring vibration and —CH$_3$ stretching vibration, respectively, result from the rubber and the grafted TMPTA.

As the concentration of TMPTA increases from 0% to 10%, the concentration of the active radicals also increases, thereby increasing the grafting and cross-linking level with EPDM. However, it must be pointed out that there are competitions amongst various reactions in this short time of exposure. The ultimate surface generated is probably a function of concentration, radiation dose, the level and nature of the monomer, etc. For example, with a further rise in the concentration of TMPTA solution to 20%, the TMPTA molecules preferably engage themselves in reactions such as self-cross-linking, cyclopolymerization, etc. generating new but different molecules. This would probably lower the above absorptions. However, as the TMPTA concentration further increases to 50%, some of the molecules, once again, are grafted onto EPDM together with the self-cross-linking, cyclization reactions, etc. Thus, a balance is struck amongst the reactions and the absorptions show a relative increase, although not as much as before.

The absorbances at 1730, 1630, 1460, 1379, 1260, 1120, and 1019 cm^{-1} follow an upward trend with concentration in the case of the bulk-modified samples also (Figure 31.4a through g) in line with the gel content, due to the reasons, pointed out above. Since the surface concentration of TMPTA per unit volume of EPDM is lower in the case of bulk modification as compared to surface modification, the optimum value of the concentration of TMPTA is not observed in these plots.

XPS studies have also been carried out on the modified surfaces. Five representative samples are chosen: Control EPDM (both unirradiated $E_{0/0}$ and irradiated with a dose of 100 kGy $E_{0/100}$), surface-grafted EPDMs at 100 kGy irradiation dose using 2% ($E_{2/100}$) and 10% ($E_{10/100}$) TMPTA solution and EPDM bulk grafted with five parts TMPTA ($EB_{5/100}$). The individual peak positions and areas and the concentrations of carbon and oxygen along with the O/C ratios are reported in Table 31.5. The C1s peak appears at 284.9 eV binding energy for the control EPDM ($E_{0/0}$). In order to quantify, the C1s peaks of all the samples have been curve fitted. There are additional peaks observed on curve fitting of C1s at 286.2 eV (a shift of 1.3 eV) and 287.6 eV (a shift of 2.7 eV), which may be ascribed to C—O and C=O, respectively. On modification of the surface, there is a shift of the C1s core peak to 285.3 eV. The peaks at 286.2 and 287.6 eV also undergo a shift to higher binding energies of 286.8 and 288.6 eV, respectively for the $E_{2/100}$ sample. This shift is even further pronounced in the case of $E_{10/100}$ sample, where an additional peak is also observed at 292.0 eV, probably due to O=C—O groups or π–π* transitions. In the case of the bulk-modified sample ($EB_{5/100}$), peak shifts to higher binding energies, compared to $E_{0/0}$, are also observed. Clearly, modification in the presence of EB induces oxidation and incorporates polar functionalities in EPDM. A gradual decrease in the relative area of the C1s core peak with a corresponding increase in the area of the constituent peaks with modification also confirms this fact.

The same effect is also corroborated from the appearance of the O1s peaks where the core peak at 532.4 eV due to C—O and 533.2 eV due to C=O groups are also promoted to higher binding energies with increasing concentration of TMPTA with a simultaneous rise in the area of the peaks. For example, the oxygen concentration rises from 20.5% to 26.2% when the EPDM surface is modified with 10% TMPTA at 100 kGy radiation. The increase in the O/C ratio further supports these observations. It may be pointed out that the theoretical value of the O/C ratio for pure TMPTA is 0.53, which is much higher than the values reported in Table 31.5. The trend observed in the XPS studies is in accordance with that observed in the IR-ATR measurements (Figure 31.4) where the absorbances at 1730 cm^{-1} due to C=O and at 1260, 1120, and 1019 cm^{-1} due to C—O—C group, increase gradually with increase in the TMPTA level until 10% TMPTA.

The effect of EB-radiation dose intensity on the surface properties of surface and bulk-modified EPDM rubber have been investigated [382]. Predominant chain scission at higher radiation doses reduces the EPDM backbone to segmental moieties as a result of which a fraction of the carbonyl moieties may vaporize from the surface under the action of the high-energy electrons, thereby decreasing the carbonyl absorption. Figure 31.5, displays the highly expanded spectra for EPDM in the C1s and O1s regions. Four representative samples are shown: Unirradiated control EPDM ($E_{0/0}$) and EPDM surface grafted with 10% TMPTA solution at irradiation doses of 50, 100, and 200 kGy ($E_{10/50}$, $E_{10/100}$, and $E_{10/200}$). For the purpose of quantification, both the C1s and O1s peaks have been curve fitted. The C1s peak for the control EPDM ($E_{0/0}$) appears at 284.9 eV binding energy as shown earlier. The additional peaks observed due to curve fitting of C1s at 286.2 and 287.6 eV are due to the C—O and C=O groups, respectively. On modification of the surface in the presence of EB, there is a distinct shift of the peaks to higher binding energies. The C1s core peak is shifted to 286.3 eV, while those at 286.2 and 287.6 eV also undergo a shift to 288.6 and 290.9 eV, respectively for the $E_{10/50}$ sample. Apart from these, an additional peak is observed at 293.5 eV which is probably due to the O=C—O group. There are peak shifts observed in the case of all the surface-modified samples as compared to the control surface, although, this shift is maximum for the $E_{10/50}$ surface. A marked decrease in the relative area concentration of the C1s core peak with a corresponding increase in the concentrations of the other constituent peaks is also observed. These phenomena are the highest in the case of the $E_{10/50}$ sample. Again, modification in the presence of EB has induced oxidation and polarity in EPDM. The same effect is also prominent in the appearance of the peaks in the O1s region where the peaks at 532.4 eV due to C—O and 533.2 eV due to C=O groups for the control sample, are also promoted to higher binding energies on modification, with a simultaneous rise in the relative concentrations of the peaks. This is the maximum at 50 kGy irradiation dose.

The trend observed in the above studies is once again in accordance with that of the IR-ATR measurements where the absorbances at 1730 cm^{-1} due to C=O and at 1260, 1120, and 1019 cm^{-1} due to C—O—C exhibit maximum values at 50 kGy irradiation dose [382].

The total surface energy γ_s, together with γ_s^{LW} (Liftshitz van der Waal's component), $\gamma_s^{(+)}$ (acidic component), $\gamma_s^{(-)}$ (basic component), and γ_s^{AB} (acidic and basic component) have been determined for the modified and control EPDMs. The results are plotted against radiation dose in Figure 31.6.

The total surface energy increases with radiation dose for both the control and the TMPTA-modified surface up to a certain dose. While the maximum increase is from 46.1 to 49.7 mJ/m^2 ($\cong$8%) for the control EPDM, the TMPTA-modified surface with 50 kGy radiation dose shows a value of 60.7 mJ/m^2 (an increase of 31%). At a higher radiation level, the total surface energy, γ_s is lowered. The γ_s^{LW} value of the control and the surface-modified sample lie in the range of 29.7 to about 35.9 mJ/m^2 (Figure 31.6e). At higher radiation doses, the values are lower. Hence, the increase in total surface energy must be due to the larger contribution of the acid–base components of the surface energy. It is observed that γ_s^{AB} value for the TMPTA surface-modified sample is increased from 11.7 to 26.7 mJ/m^2 at 50 kGy radiation dose. It decreases with further dosage. The control sample shows, however, a much lower value of γ_s^{AB}. At 50 kGy radiation dose, γ_s^{AB} is 18.8 mJ/m^2. The major contribution of the γ_s^{AB} comes from the increase in $\gamma_s^{(-)}$ value with surface modification. For example, this has changed from 13.1 mJ/m^2 for $E_{10/0}$ to about 42.0 mJ/m^2 for $E_{10/50}$. The contribution of $\gamma_s^{(+)}$ is much lower. While the value of $\gamma_s^{(-)}$ for $E_{0/50}$ is 36.9 mJ/m^2, $\gamma_s^{(+)}$ is only 2.4 mJ/m^2. Similarly, $\gamma_s^{(-)}$ for $E_{10/50}$ is 42.0 mJ/m^2, whereas $\gamma_s^{(+)}$ is 4.2 mJ/m^2. The bulk-modified samples, on the other hand, register a slight increase in γ_s on irradiation. This is mainly due to the increased values of γ_s^{AB} and $\gamma_s^{(-)}$.

($\gamma_s^{AB}/\gamma_s^{LW}$) has also been calculated for all the samples (Figure 31.6f). It has been observed that for the surface-modified samples, the ratio rises from 0.3 to 0.9 with 50 kGy radiation dose. It decreases thereafter. This ratio is much higher than that obtained with the control samples (0.3–0.6).

It is also interesting to note from the IR results described earlier that the absorbance at 1730 cm^{-1} increases with irradiation dose up to 50 kGy. Similarly, A_{1260}, A_{1120}, and A_{1019} also increase with irradiation. There is also a rise in A_{1460} and A_{1379} with irradiation dose. The maximum oxygen

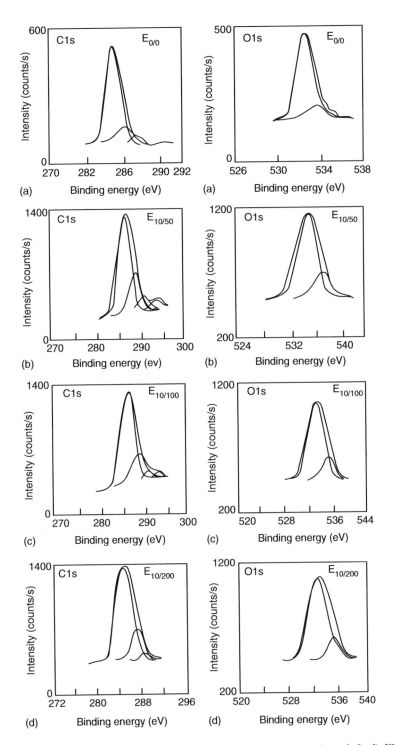

FIGURE 31.5 C1s and O1s spectra of (a) control unirradiated EPDM ($E_{0/0}$); and (b–d) EPDM surface modified with 10% trimethylolpropane triacrylate (TMPTA) at 50, 100, and 200 kGy doses. (From Sen Majumder, P. and Bhowmick, A.K., *Radiat. Phys. Chem.*, 53, 63, 1998. With permission.)

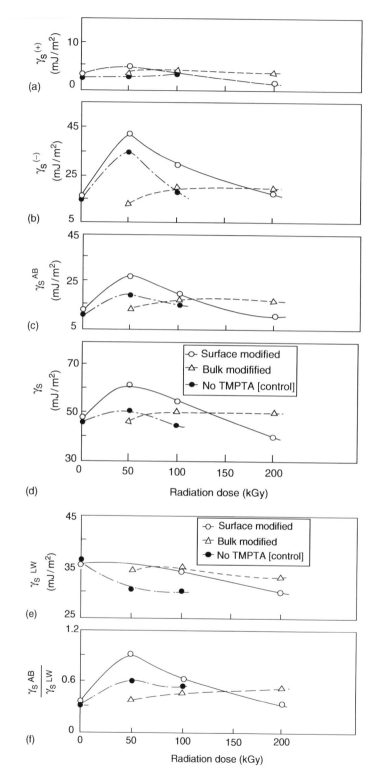

FIGURE 31.6 Change of the surface energy parameters of control and modified ethylene–propylene–diene monomers (EPDMs) with irradiation dose. (From Sen Majumder, P. and Bhowmick, A.K., *Radiat. Phys. Chem.*, 53, 63, 1998. With permission.)

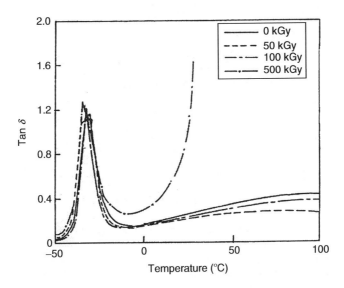

FIGURE 31.7 Representative plots showing the variation of tan δ with temperature for the control ethylene–propylene–diene monomer (EPDM) irradiated to various doses. (From Sen Majumder, P. and Bhowmick, A.K., *J. Appl. Polym. Sci.*, 77, 323, 2000. With permission.)

concentration at 50 kGy irradiation dose has also been observed from the XPS data. Hence, the changes in surface energy values must be ascribed to an interplay of various factors like grafting, cross-linking, oxidation, chain scission, disproportionation, etc.

Dynamic mechanical thermal analysis [383] of the EB-modified rubber (Figure 31.7) indicates that with increase in the radiation dose, there is a slight shift in the glass transition temperature accompanied by a marginal decrease in the (loss tangent)$_{max}$ value. These shifts are due to the increasing degree of cross-linking. In presence of TMPTA, there is a steady fall of the tan δ_{max} value from 1.15 to 0.875 with a corresponding rise in the T_g from $-32°C$ to $-27°C$ on account of structural modifications of the rubber. Among the various polyfunctional monomers used, the storage modulus and T_g of the tetracrylate-based system have been found to exhibit the maximum value owing to the highest cross-link density. At a particular radiation dose, the tensile properties have been observed to improve with increase in TMPTA level. Similar observations have been made with increase in number of unsaturations in the initial starting material (i.e., the polyfunctional monomer). The friction behavior [384] of the EPDM rubber pre-cross-linked with 0.2 phr dicumyl peroxide (DCP) and surface modified with various concentrations of TMPTA at a particular radiation dose and at different EB doses using a fixed TMPTA concentration has been studied. The dynamic coefficient of friction between EPDM and a solid aluminum surface has been shown to rise initially up to a certain level (5 wt%) after which there is a decrease. Figure 31.8 shows the variation of μ with concentration of TMPTA. At a particular radiation dose, with increase in TMPTA concentration, there is an increase in the surface energy due to increased polarity of the surface owing to the grafting of TMPTA onto EPDM. This is responsible for increasing the interfacial shear strength due to adhesion with the counterface. Thus, the friction coefficient increases. However, after a certain level, with a further rise in the modification, the gradual increase in the elastic modulus of the rubber due to more and more cross-linking gains predominance over the adhesion term. This lowers the true area of contact with the counterface (due to increase in surface hardness), thereby reducing the friction. Similar results have been obtained with variation in radiation doses. μ Reaches a maximum at a dose of 50 kGy and then decreases due to surface hardening. The adhesion behavior of EPDM rubber has also been investigated [385]. Figure 31.9 shows the effect of radiation dose on the adhesive bond strength of EPDM

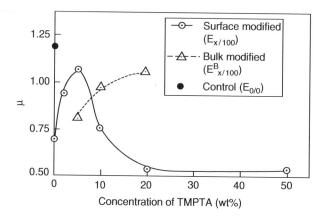

FIGURE 31.8 Change in the coefficient of friction of modified dicumyl peroxide/ethylene–propylene–diene monomer (DCPD/EPDM) with the concentration of trimethylolpropane triacrylate (TMPTA) at a fixed irradiation dose of 100 kGy. ($\odot$) Surface modified with 100 kGy, ($\triangle$) Bulk modified with 100 kGy dose, ($\bullet$) Control EPDM rubber. (From Sen Majumder, P. and Bhowmick, A.K., *Wear*, 221, 15, 1998. With permission.)

rubber with itself, NR, and aluminum. In all the cases, the strength increases initially up to about 50 kGy radiation dose after which there is a decrease. The peel behavior of the joints shows a change from a smooth failure to a stick—slip mode on irradiation.

IR-ATR absorptions at 1730, 1260, and 1019 cm^{-1} have been observed to rise up to the 50 kGy level beyond which there is a slight drop in the values. With increase in the radiation dose, the generation of active radicals on the EPDM backbone increases and so also that of the carbonyl and ether groups due to aerial oxidation of these radicals as noted in the earlier sections. However, at higher irradiation doses above 50 kGy, the chain scission of the EPDM molecules predominates as a

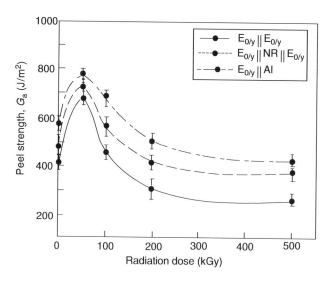

FIGURE 31.9 Change in the peel strength and radiation dose of: ethylene–propylene–diene monomer (EPDM)/EPDM joint [$E_{0/y} \parallel E_{0/y}$]; EPDM/natural rubber (NR)/EPDM joint [$E_{0/y} \parallel NR \parallel E_{0/y}$]; and EPDM/Al joint [$E_{0/y} \parallel Al$]. (From Sen Majumder, P. and Bhowmick, A.K., in *Polymer Surface Modification: Relevance to Adhesion*, Vol. II, Mittal, K.L., Ed., VSP, Utrecht, The Netherlands, 2000, 425. With permission.)

result of which the rubber chains are broken down into segmental moieties and a fraction of the polarity is lost from the surface. Thus, the absorbance values decrease at this stage. These changes give rise to changes in the surface energy values, $\gamma_s^{(+)}$ and $\gamma_s^{(-)}$, as shown in Figure 31.6. Following the above arguments, with increase in the generation of the polar groups on the surface, the intermolecular forces of attraction between the EPDM rubber surfaces increase and hence, the peel strength rises (from 415 to 682 J/m^2, a rise of about 65%). Above 50 kGy, due to a partial degradation of the rubber chains, the polarity drops and so does the adhesion strength.

For the EPDM/NR joint, the modification of the EPDM rubber increases its cure compatibility with NR. This, thus, increases with radiation dose up to 50 kGy beyond which a drop in the absorbance values due to predominant chain scission of the rubber also lowers the bond strength. Besides, interdiffusion of the rubber molecules across the interface also contributes to the formation of the bond.

In the case of EPDM against aluminum surface, the rise in the polar surface forces brings about an increase in the adsorption forces of attraction between the rubber and the metal which in turn, raises the peel strength. Again, beyond the 50 kGy level, the scission of the rubber chains brings down the polarity and the peel strength as well. It is interesting to note that the modified EPDM/Al joint displays the highest peel strength at a similar irradiation dose among the three joints studied. These results could be explained on the basis of similar arguments stated earlier.

31.2.6.6.5   Polychloroprene Rubber

The cross-linking efficiency of the more branched polychloroprene latex (mercaptan modified) has been found to be higher than the less branched sulfur-modified one. The latex dispersion is found to display higher rate of cross-linking than the coagulated and subsequently dried rubber films [386] due to higher concentration of radical in the former.

In the case of crystallizable polychloroprene, the cross-link density has been found to be lower in the crystalline region as compared to the amorphous portion of the rubber due to the lower radical mobility in the former. The cross-links are found to be intramolecular in the crystalline zone [387].

Chemicals like polyorthoaminophenol, diphenylamine in small amounts have been found to decrease the yield of cross-linking [388]. The tensile strength of the carbon black-filled polychloroprene compounds has been found to be comparable to the conventional thermally cured one. The physical properties [389] have been observed to improve on adding cross-linking promoters like N,N'-hexamethylene-bis-methacrylamide into the polymer matrix.

Radiation-induced cross-linking of polychloroprene with different polyfunctional monomers have been investigated [390]. Data on tensile strength of the polychloroprene cross-linked with (polyethyleneglycol dimethacrylate) or urethane polymer indicates the higher cross-linking efficiency of the above monomers on the polymer matrix. The degree of cross-linking has been found to be proportional to the absorbed dose and the initial polyfunctional monomer concentration.

31.2.6.6.6   Nitrile Rubber

Yasin et al. [391] not only compared the efficiency of six polyfunctional monomers as cross-linking coagents for EB cross-linking of nitrile rubber (NBR) but also showed that radiation vulcanization of NBR with diethylene glycol dimethacrylate at 20 kGy yielded results similar to that obtained from conventionally vulcanized NBR. The influence of EB irradiation as one of the mixed cross-linking systems on the structure and properties of NBR was carried out and the results were explained by IR analysis, solid-state NMR analysis, and chemical probe treatments of the irradiated vulcanizates [392]. A mechanistic path was also proposed to explain the structural changes on irradiation of the NBR vulcanizates. Vijaybaskar et al. [393] studied the EB irradiation on different grades of NBR with varying AN content and found that up to the radiation dose range of 500 kGy, the percentage increment in 100% modulus for NBR having 18% AN content was more than NBRs containing 33% and 50%, respectively. This was attributed to the higher percentage of butadiene chains in NBR with 18% AN content which accounted for the higher amount of gel formation. Further, the concept of "cross-link clustering" was found to be more in NBR with 18% AN content and was explained by the authors [393] on the basis of cyclization of butadiene chains as indicated from the difference between the spectroscopic cross-link densities and chemical cross-link densities.

31.2.6.6.7   Ethylene–Octene Copolymer Rubber

Clay fillers were surface modified with TMPTA or triethoxyvinyl silane (TEVS) followed by EB irradiation by Ray and Bhowmick [394]. Both the untreated and treated fillers were incorporated in an ethylene–octene copolymer. Mechanical, dynamic mechanical, and rheological properties of the EB-cured unfilled and filled composites were studied and a significant improvement in tensile strength, elongation at break, modulus, and tear strength was observed in the case of surface-treated clay-filled vulcanizates. Dynamic mechanical studies conducted on these systems support the above findings.

31.2.6.6.8   Polysiloxane Polymers

The gases resulting from the irradiation of PDMS have been reported in the literature [395] and consist entirely of hydrogen, methane, and ethane. The yield has been found to be proportional to the degree of cross-linking since double bonds cannot be formed.

The cross-linking of polysiloxanes has been reported extensively by some investigators [395–398]. IR measurements on PDMS have indicated the formation of three types of cross-links $>Si-Si<$, $>Si-CH_2-Si<$, and $>Si-(CH_2)_2-Si<$ formed through coupling of various radicals formed by hydrogen and methyl abstraction. The degree of cross-linking was found to increase with dose due to the availability of more radicals and the greater freedom of polymer chains to undergo translational and segmental motion.

Antirads [399] like *tert*-dodecyl mercaptan and diethylsulphide have been found to retard cross-linking in PDMS. The acid effect in mutual X-radiation grafting of AA, 4-vinyl pyridine (4-VP) and N-2-vinylpyrrolidone (N-VP) in methanol on silicone rubber [400] as a function of radiation dose has been investigated. In the case of AA radiation grafting, under a certain dose, the grafting yield shows a decrease which is proposed to be caused by its autoacceleration polymerization. In the presence of acid (0.01 mol/1 of H_2SO_4), the curve of grafting yield as a function of dose turns out to be complex. Two regions of dose have been observed: Below 6 kGy, an enhancement of grafting yield is observed, but between 6 and 16 kGy, the opposite effect is observed. The homopolymerization of AA induced by γ-irradiation has been observed to be serious at higher doses. On the other hand, Cu^{+2} ions (0.01 mol/1) have been found to diffuse into Si–rubber backbone inhibiting the radiation grafting. In methanol, H ions have been found to increase the $G(H_2)$ value. In grafting systems the H atom produced can initiate the hydrogen abstract reaction with matrix and addition reaction with monomer followed by increasing grafting probability or initiating homopolymerization of monomers. The competition of the above two reactions is related to the viscosity of grafting systems. At higher dose, the viscosity of systems becomes more serious, the H^+ ions are difficult to diffuse onto the surface of the matrix, then the homopolymerization in solution will be dominant. On the opposite the enhancement of H^+ ions to grafting reaction will be preferential. The grafting of basic monomer 4-VP in presence of acid is found to increase with radiation dose. The 4-VP system (a neutral monomer) is found to have an inhibiting effect of grafting yield. In presence of H^+, N-VP molecules react to form salt before grafting and this decreases the N-VP concentration. This accounts for the inhibition on grafting yield in acid medium. The mechanism of chain scission occurring in polysiloxane model compounds and PDMS elastomers have been studied [401]. Linear as well as cyclic structures have been detected as the scission fragments.

The physical properties of radiation and peroxide cured polysiloxanes have been compared by several investigators [402]. Vinyl-substituted (0.14 mol%) radiation-cured polysiloxane is found to have better strength properties than the chemically cured analogue. The phenyl substitution (7.5 mol%) has only marginal effect. The physical properties of radiation and thermally cured silica-filled polyvinylethyldimethylsiloxane compounds are found to be similar.

31.2.6.6.9 *Fluoropolymers*

Lipko et al. [403] have synthesized semi-II IPN fluorocarbon elastomers by the addition of small amounts of multifunctional monomer to the linear fluorocarbon polymer, followed by radiation (β and γ) polymerization *in situ*. The linear fluorocarbon binds with the cross-linked network II, producing a semi-IPN of the second kind. Due to high-energy radiation attack on the fluorocarbon backbone, a significant amount of graft copolymer is reported to be formed which improve the compatibility and yield clear and transparent elastomers. Low-energy UV (which does not promote grafting and gives inferior quality elastomers) is reported to display less optical clarity, much lower strength, and higher creep. The number average molecular weight of PTFE irradiated by a ^{60}Co-γ source in air at room temperature has been established from the experimental results of tensile creep measurements and electron microscopy by Takanega and Yamagata [404]. ESR measurements have been performed by Hedvig [405] at different stages of γ-radiation degradation of PTFE. It is found that under radiation, peroxy radicals are transformed into chain-end fluorocarbon radicals, which in turn become side-chain radicals. The influence of γ-radiation on the melting and two solid–solid transitions, occurring near 19°C and 30°C, of PTFE has been studied by Kusy and Turner [406] with the help of DSC. A continuous dispersion of all three transition temperatures, with increasing dose is observed in a first scan of highly crystalline samples at a heating rate of 20°C/min. Hegazy et al. [407] have studied the irradation grafting of AA onto poly(tetrafluoroethylene-co-hexafluoropropylene) film. The degree of grafting increases with irradiation dose (γ-ray) and monomer concentration and slightly decreases as the grafting temperature is elevated. The relationship between the grafting rate and film thickness has shown a negative first-order dependency. The synergistic effect of the cross-linking monomers—TMPTMA and triallyl isocyanurate—in the presence of EB on the degree of cross-linking of Viton GLT (a copolymer of vinylidenefluoride, perfluoromethylvinylether, and TFE) leading to an improvement in properties [408] has been reported. A mixed system of TMPTMA and triallyl isocyanurate is observed to have better cross-linking efficiency with the same EB dose than the cross-linking systems based on individual monomers. The predominance of cross-linking in TFE copolymers of perfluoro methyl vinyl ether (PMVE) over perfluro-3,6-dioxa-4-methyl-7-octene sulphonyl fluoride (PSEPVE), and ease of the grafting of mixtures of TFE and methyl perfluoro-3,6-dioxa-4-methyl-7-octenoate onto PSEPVE copolymer as compared to PMVE copolymers have been investigated by Uschold [409]. Zhong et al. [410] have carried out XPS studies on the ^{60}Co-γ-ray-induced structural changes like scission, branching, and formation of double bond in the copolymer of TFE with hexafluoropropylene. A comparative study of the effect of EB irradiation on the properties of PTFE, polyvinyl fluoride, and polyvinylidene fluoride is reported by Timmerman and Greyson [411].

Irradiation effects of excimer laser radiation and EB on PP and ethylene–tetrafluoroethylene (ETFE) copolymer films have been reported [412]. It is found that irradiation effects of UV radiation from an excimer laser on PP and ETFE greatly differ from those from an EB. For PP film, irradiation with a KrF laser and EB is found to degrade the polymer with the formation of double bonds and carbonyl groups. Golub et al. [413] have carried out a comparative study with the help of electron spectroscopy for chemical analysis (ESCA) of the films of polyvinyl fluoride, Tedlar (E.I. Du Pont de Nemours & Company, Inc., Wilmington, Delaware), TFE–hexafluoropropylene copolymer, Kapton F (E.I. Du Pont de Nemours & Company, Inc., Wilmington, Delaware), and PTFE, Teflon (E.I. Du Pont de Nemours & Company, Inc., Wilmington, Delaware), exposed to atomic oxygen (O(3P))—either in Low Earth Orbit (LEO) on the STS-8 Space Shuttle or within or downstream from a radio-frequency oxygen plasma. The major difference in surface chemistry of polyvinyl-fluoride induced by the various exposures to oxygen is a much larger uptake of oxygen when etched either in or out of the glow of oxygen plasma than when etched in LEO. In contrast, Kapton F and Teflon have been found to exhibit very little surface oxidation during any of the three different exposures to O(3P), due to the absence of hydrogen in their structures.

Swelling water uptake, electric conductivity, and transport number of the membranes are measured as a function of the ion-exchange capacity (IEC). IEC has been estimated in terms of the degree of sulfonation in the grafted film. The water absorption of the grafted membrane is found to increase with the IEC, while the specific electric resistance and the transport number decreases linearly with the IEC. The thermal and chemical stability increase with the degree of grafting. The stable membrane properties are established due to a homogenous ion-exchange group (the potassium sulfate group) distribution in the membrane. The grafted membrane has cation exchange capacity.

Momose et al. [414] have studied the electrochemical, thermal, and chemical properties of cation exchange membrane obtained through radiation grafting of α,β,β-trifluorostyrene onto poly (ethylene-co-tetrafluoroethylene) film by preirradiation method followed by sulphonation and hydrolysis of the grafted film. Zhong et al. [415] have investigated the radiation (^{60}Co-γ-ray) stability of PTFE with different crystallinity. The reduction in molecular weight and the G values for scission, $G(S)$, has been observed to be higher for the sintered sample of PTFE having a relatively low crystallinity as compared to the as-polymerized sample having a much higher crystallinity. Photoelectron spectroscopic evidence for radiation-induced cross-linking of PTFE has been established by Rye [416]. The C1s spectrum obtained after irradiation with 2 KeV electrons for all doses greater than 1 μA min/cm^2 consists of four peak spectrum identical to that previously obtained for plasma-polymerized PTFE and assigned to carbon atoms with variable numbers of bound atoms (CF$_3$, CF$_2$, CF$_1$, and CF$_0$). The CF$_1$ and CF$_0$ components increase with increasing electron dose and at high electron doses dominate the spectrum. With increasing dose the CF$_3$ component approached a constant value, while both the CF$_2$ component and the total F/C ratio decrease. These four components are those expected to result from radiation-induced cross-linking reactions of the polymer and are consistent with previous suggestions that cross-linking is the basis of radiation patterned adhesion to PTFE.

Lappan et al. [417] have reported the thermal stability and the degradation fragments of the EB-irradiated PTFE by thermogravimetric analysis coupled with mass spectrometry. The TGA results confirm the known decrease in the thermal stability of irradiated PTFE with increasing radiation dose. CO$_2$, HF, and fluorocarbon fragments are evolved on thermal degradation of the irradiated samples. CO$_2$ and HF are formed by decomposition of peroxy radicals up to 250°C. In addition, low-molecular weight fluorocarbons are desorbed from the irradiated PTFE. At temperatures above 300°C, CO$_2$ is formed by decarboxylation of radiation-induced COOH groups inside PTFE. High-molecular weight PTFE is transformed to free-flowing micropowder by treatment with EB [418]. In case of irradiation in presence of air, carboxylic groups are incorporated which rapidly hydrolyze to carboxylic groups in the surface regions due to atmospheric humidity. These polar groups reduce the hydrophobic and oleophobic properties so much that homogenous compounding with other materials become possible. In addition to PTFE, copolymers of TFE and hexafluoropropylene (HFP), poly(tetrafluoroethylene-co-perfluoropropylene) (FEP) and perfluoropropylvinyl ether (PFA) have been modified. In case of identical irradiation conditions, the concentration of carboxylic group is much higher in FEP and PFA than in PTFE, which is due to lower crystallinity of copolymers. EB irradiation of PTFE has been performed in vacuum at elevated temperature above the melting point. The changes in the chemical structure have been studied. The concentration of CF$_3$ branches has been found to be much higher as compared to room temperature irradiation. In a practical test, PTFE micropowders functionalized by EB have been compounded with epoxy resins, polyoxymethylene, and polyamides. Such compounds are characterized by very good frictional and wearing behavior in dry-running tests.

Temperature dependence (related to the temperature dependence of the conformational structure and the morphology of polymers) of the radiation effect on various fluoropolymers e.g., poly (tetrafluoroethylene-co-hexafluoropropylene), poly(tetrafluoroethylene-co-perfluoroalkylvinylether), and poly(tetrafluoroethylene-co-ethylene) copolymers has been reported by Tabata [419]. Hill et al. [420] have investigated the effect of environment and temperature on the radiolysis of FEP. While the irradiation is carried out at temperatures above the glass transition temperature of FEP, cross-linking reactions predominate over chain scission or degradation. Forsythe et al. [421]

have investigated the effect of X-irradiation temperature on the polymer properties of the fluoro-elastomer poly(tetrafluoroethylene-co-perfluoromethylvinylether). The TFE/PMVE samples have been γ-irradiated to 150 kGy at temperatures ranging from 77 to 373 K. Analysis of sol–gel behavior, tensile properties, and glass temperature indicate that cross-linking commences in the temperature range 195–263 K, at a dose of 150 kGy. The latter temperature has been found to be 13 K below T_g. Cross-linking remains relatively constant at higher temperatures. Chain-scission reactions have been found to occur well below T_g and increase at higher temperature. The optimum temperature for the radiation cross-linking of TFE/PMVE copolymer has been found out to be 263 K. The chemical and mechanical analysis of TFE (tetraflouroethylene)/(perfluoromethylviny-lether) PMVE copolymer forms an insoluble network at a dose (gelation) of 15.8 kGy. The tensile properties indicate the predominance of cross-linking with optimal elastomeric properties reached in the dose range of 120–200 kGy. FTIR spectroscopy shows the formation of new carboxylic acid end groups on irradiation. These new groups have been shown to decrease the thermo-oxidative stability of the cross-linked network as determined by thermogravimetric analysis. ESR studies on the polymer at 77 K indicate the presence of radical precursors. A G value of 1.1 has been determined for radical production at 77 K. Comparison of radical concentrations for a copolymer with a different mole-ratio of perfluoromethyl vinyl ether indicates PMVE units contribute to chain scission.

The γ-radiolysis of poly(tetrafluoroethylene-co-perfluoromethylvinylether) (TFE/PMVE) has been investigated [422] using solid-state ^{19}F and ^{13}C NMR spectroscopy. Chain-scission products identified in the polymer have been observed to be saturated chain ends $-CF_2CF_3$ ($G = 1.0$), methyl ether and groups $-CF_2OCF_3$ ($G = 0.9$), acid end groups $-CF_2COOH$ ($G = 0.5$), and a small amount of terminal unsaturation $-CF{=}CH_2$ ($G = 0.2$). The G value for main-chain scission, $G(S)$ is determined to be 1.4. Cross-linking of TFE/PMVE is found to proceed via a Y-linking mechanism. The G values for cross-linking, $G(X)$ has been observed to be 0.9. A maximum of 0.2 mol% cross-links are formed under experimental conditions.

Pacansky et al. [423] have investigated the structural changes of EB-irradiated poly(tetrafluor-oethylene/perfluoromethylvinylether) by IR spectroscopy when the copolymer is irradiated, gases are produced, mass loss is induced, and acid fluoride and groups are formed in the bulk. CF_4, COF_2, CO_2, and CF_3OCF_3 are produced with G values of 0.93, 0.31, 0.055, and 0.14, respectively. The sum of the G values for gas formation, 1.43, provides an upper limit estimate for side-chain cleavage. The G value for formation of acid fluoride in the bulk is 0.34 and provides a lower limit estimate for main-chain scission. Hence, most of the side-chain scissioning leads to reactions that do not produce acid fluoride end groups.

The γ-radiation-induced cross-linking of PTFE in the molten state under oxygen free atmosphere has been investigated by Oshima et al. [424]. The γ-radiation cross-linking of PTFE is confirmed by the improvement in yield strength and modulus, increase in transparency due to decrease in crystallinity, improvement in radiation resistance, and by nature of transition and relaxation of molecular motions. The yield of alkyl type free radicals are enhanced by cross-linking as evidenced by ESR study. The free radicals are converted effectively to peroxyradicals by reaction with oxygen at room temperature. The conversion rate of free radicals into peroxy radicals is decreased by increased cross-linking density for the film (thickness: 0.5 mm) specimen, which means oxygen diffusion is depressed by the cross-linking. When the specimen containing peroxy radicals is heated up to 373 K under vacuum, the alkyl radicals are found to appear with the decay of the peroxy radicals, and the radicals are converted to peroxy radicals by introduction of oxygen once again. It means that the peroxy radicals extract fluorine atom from the main chain of PTFE molecules to induce the alkyl radicals.

The influence of polyfunctional monomers—tripropyleneglycol diacrylate (TPGDA), TMPTA, TMPTMA, TMMT, and TAC on the structural changes of fluorocarbon terpolymer poly(vinylide-nefluoride-co-hexafluoropropylene-co-tetrafluoroethylene) has been investigated [425]. The ATR-IR studies show that the absorbance due to the double bond at 1632 cm^{-1} decreases both in the

(Figure 31.10a) polyfunctional monomer and its blends with the polymer as a result of grafting and cross-linking. The concentration of the carbonyl group (Figure 31.10a) indicated by the absorbance at 1740 cm^{-1}, increases with radiation dose but shows marginal variation at higher radiation doses. The absorbances at 1397, 1021, 672, and 504 cm^{-1}, due to C—F group (Figure 31.10b) decrease on irradiation of the mixtures of fluoroelastomer and TMPTA, indicating dehydrofluorination and scission involving C—F bond. The cross-linking involves the union of macroradicals generated through splitting off labile hydrogen on the polymer chain or addition of macroradicals across the double bonds so generated on dehydrofluorination (Scheme 31.5a). The scission reactions are shown in Scheme 31.5b. The grafting occurs through unsaturation of the polyfunctional monomers through free-radical mechanism (Scheme 31.6). The concentration of the carbonyl group increases with radiation dose due to atmospheric oxidation. TMPTA lowers the amount of oxidation, especially at relatively higher levels owing to its dual role as a plasticizer and cross-linking sensitizer. As a plasticizer TMPTA mobilizes the polymeric chain under the influence of radiation, decreasing the amount of oxidative scission and increasing the extent of radical–radical recombination. At a particular radiation dose, the behaviors of TPGDA, TMPTMA, TMMT, and TAC-based

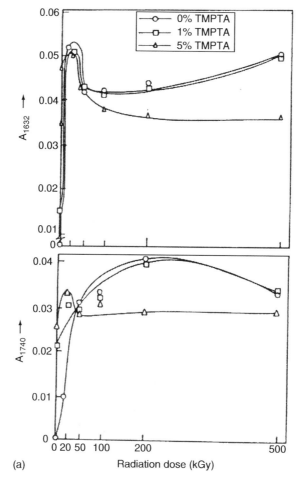

(a)

FIGURE 31.10 (a) Plot showing the variation of absorbances at 1740 and 1632 cm^{-1} of the control and the trimethylolpropane triacrylate (TMPTA) mixed rubber irradiated with 20, 50, 100, 200, and 500 kGy.

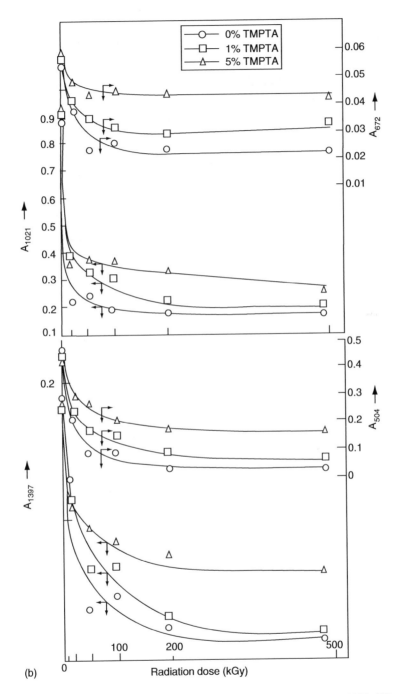

FIGURE 31.10 (continued) (b) Plot showing the variation of absorbances at 1397, 1021, 672, and 504 cm^{-1} of the control and the blends of fluororubber with trimethylolpropane triacrylate (TMPTA) irradiated with doses of 20, 50, 100, 200, and 500 kGy. (From Banik, I., Dutta, S.K., Chaki, T.K., and Bhowmick, A.K., *Polymer*, 40, 447, 1998. With permission.)

systems have been observed similar to that of TMPTA-based one. The gel content increases with radiation dose and TMPTA level (Figure 31.11a and b) due to formation of network structure. The gel fraction has been found to be lowest for system based on TPGDA (Figure 31.11c) due to its

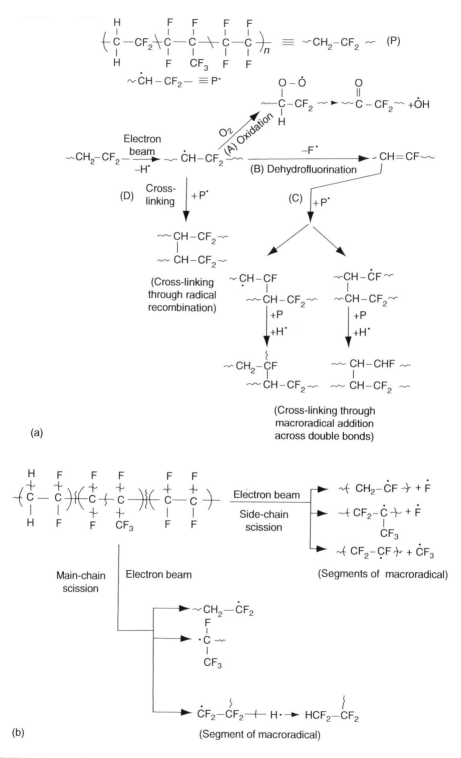

SCHEME 31.5 (a) Oxidation, dehydrofluorination and crosslinking reactions as a result of EB irradiation of poly (vinylidenefluoride-co-hexafluoropropylene-co-tetrafluoroethylene). (b) Chain scission reactions as a result of EB irradiation of poly (vinylidenefluoride-co-hexafluoropropylene-co-tetrafluoroethylene).

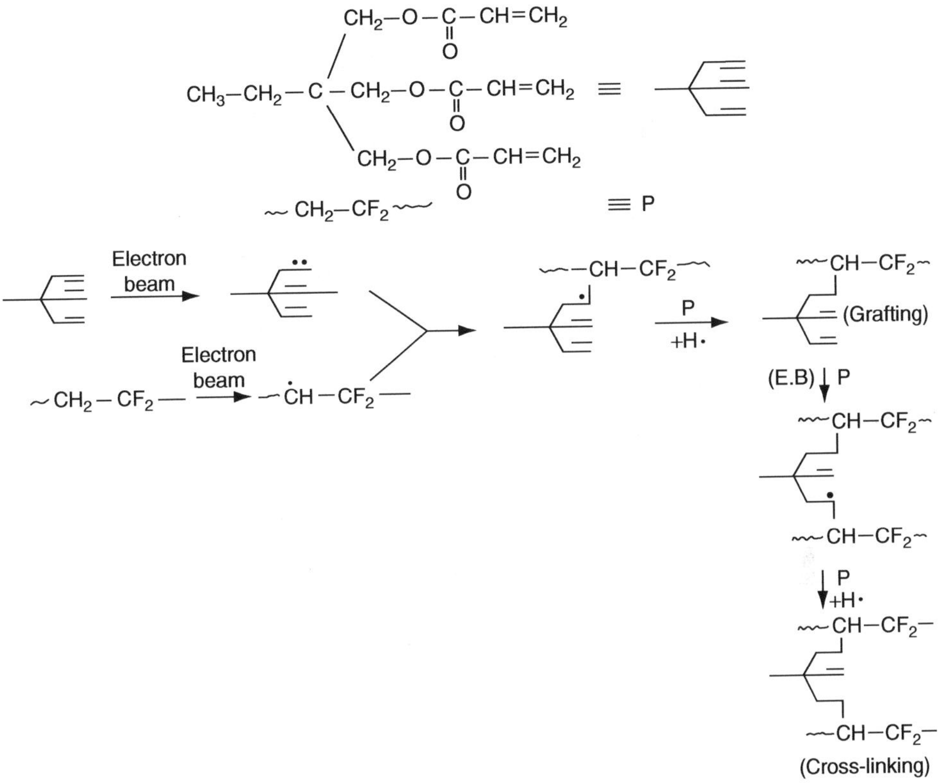

SCHEME 31.6 Grafting of TMPTA on poly (vinylidenefluoride-co-hexafluoropropylene-co-tetrafluoroethylene) by EB irradiation.

having lower number of initial unsaturation and hence lower efficiency of cross-linking. The ratio of chain scission to cross-links calculated using Charlesby–Pinner equation (Figure 31.11d) indicates a much higher cross-linking efficiency with polyfunctional monomer fluorocarbon rubber matrix as compared to the control rubber.

XPS studies have been reported to be in line with the IR observations [426]. F1s spectrum of the irradiated rubber shows a shift towards lower binding energy indicating dehydrofluorination and scission. O1s spectrum of the irradiated sample shows marginal increase in both peak area and width supporting the atmospheric oxidation. Comparative study of copolymeric poly(vinyldenefluoride-co-tetrafluoroethylene) fluorocarbon rubber over a range of radiation doses reveals an increased tendency towards oxidation and dehydrofluorination of the former owing to the increased amount of hydrogen. The gel content is also slightly lower owing to greater extent of oxidative scission. IR-observations show that magnesium oxide, which is generally added in fluorocarbon recipe, shows an increased tendency towards dehydrofluorination at a particular radiation dose. This is due to the promotion of formation of unsaturation by basic MgO through accepting the acidic HF liberated on dehydrofluorination. Mechanical and dynamic mechanical properties [427,428] indicate that with increase in radiation dose, an increase in degree of cross-linking results which leads to increase in modulus and T_g with a corresponding decrease in elongation at break, set and tan δ. Figures 31.12 and 31.13 represent dynamic mechanical and failure properties of the radiation-modified rubbers, respectively. Increase in modulus and T_g is in accordance with the increase in degree of cross-linking. The effect of multifunctional monomer has been realized only at relatively

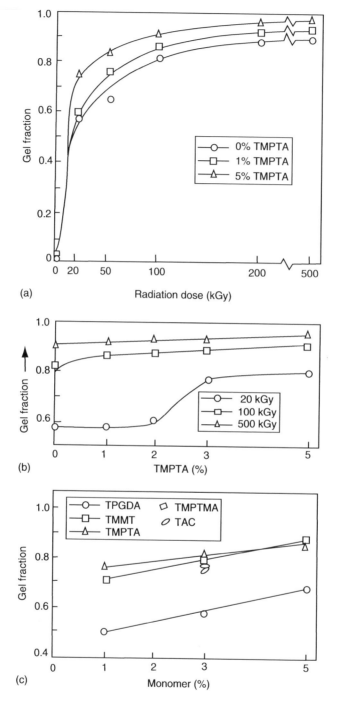

FIGURE 31.11 (a) Plot of gel fraction of 0, 1, and 5 wt% trimethylolpropane triacrylate (TMPTA) systems with radiation doses of 20, 50, 100, 200, and 500 kGy. (b) Plot showing the variation of gel fraction at radiation doses of 20, 100, and 500 kGy of 0, 1, and 5 wt% TMPTA systems. (c) Plot of gel fraction of tripropyleneglycol diacrylate (TPGDA), TMPTA, trimethylolpropane trimethacrylate (TMPTMA), tetramethylol methane tetraacrylate (TMMT), and triallyl cyanurate (TAC)-blended systems against the level of monomer at constant radiation dose (50 kGy).

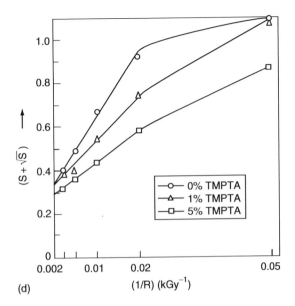

(d)

FIGURE 31.11 (continued) (d) Plot of $S + \sqrt{S}$ versus 1/R for 0, 1, and 5 wt% trimethylolpropane triacrylate (TMPTA) systems. (From Banik, I., Dutta, S.K., Chaki, T.K., and Bhowmick, A.K., *Polymer*, 40, 447, 1998. With permission.)

higher level of TMPTA where improvement in strength and failure properties is observed. The dynamic storage modulus at 50°C is higher and loss tangent becomes lower. Among the various polyfunctional monomers, TPGDA, TMPTA, and TMMT, the mechanical properties and the degree of cross-linking have been observed to be lowest for systems based on TPGDA. MgO used in the formulation leads to the improvement in mechanical properties. Increase in the cross-link generated on dehydrofluorination has been found to in line with reduced loss tangent and increased storage modulus for the MgO-filled rubber vulcanizates. Dielectric studies [429] on the EB-modified fluorocarbon rubber reveal that the dielectric constant and loss factor (Figure 31.14a and b) decreases with increase in radiation dose. The decrease in the dielectric constant and the loss

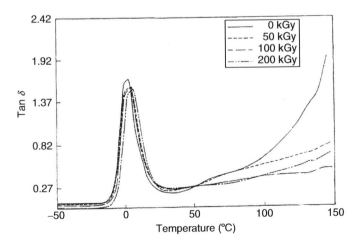

FIGURE 31.12 Tan δ of pure fluorocarbon rubber (terpolymer) irradiated to different radiation doses, viz. 0, 50, 100, and 200 kGy. (From Banik, I. and Bhowmick, A.K., *J. Appl. Polym. Sci.*, 69, 2079, 1998. With permission.)

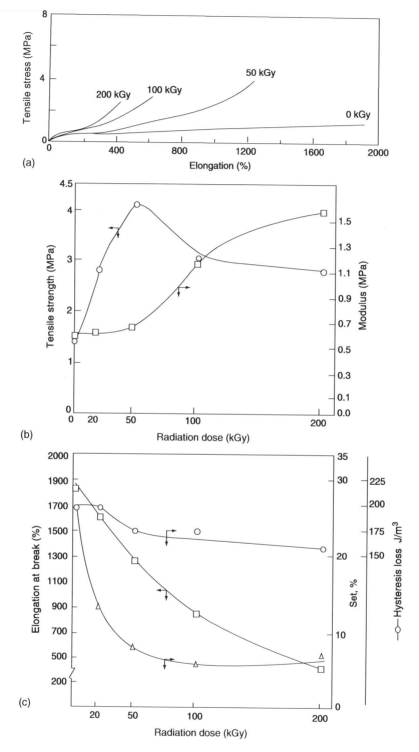

FIGURE 31.13 (a) Plot showing the stress–strain behavior of various irradiated rubbers. (b) Plot showing the variation of tensile strength and modulus of rubbers irradiated with different doses. (c) Plot showing the variation of hysteresis loss, set, and elongation at break of irradiated fluorocarbon rubbers. (From Banik, I. and Bhowmick, A.K., *Radiat. Phys. Chem.*, 54, 135, 1999. With permission.)

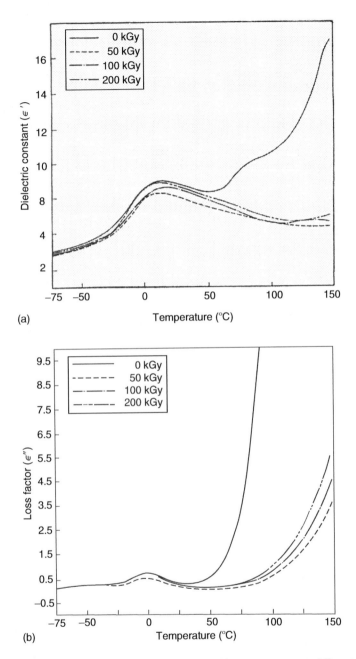

FIGURE 31.14 (a) Variation of the dielectric constant with temperature at different radiation doses. (b) Variation of the dielectric loss factor with temperature at different radiation doses. (From Banik, I., Chaki, T.K., Tikku, V.K., and Bhowmick, A.K., *Angew. Makromol. Chem.*, 263, 5, 1998. With permission.)

factor is attributed to the decrease in dipole orientation polarization and the dielectric relaxation as a result of cross-linking the polymeric chains.

Thermal degradation studies of EB-cured terpolymeric fluorocarbon rubber [430] by nonisothermal thermogravimetry in the absence and presence of cross-link promoter TMPTA reveal that thermal stability is improved on radiation and more so in the presence of TMPTA. Initial decomposition temperature, maximum decomposition temperature and the decomposition temperature at various amounts of decomposition (all giving a measure of thermal stability), all have been found to increase with increase in radiation dose (up to a certain level) and rise in TMPTA level. Decrease of the various decomposition temperatures at higher doses is related to probable scission reactions at higher doses. Decomposition of the irradiated rubber has been proposed to occur through splitting of pendent hydrogen and fluorine on the backbone of the macromolecular chain leaving a char residue. MgO stabilizes the rubber matrix more against thermal attack because of the additional cross-links formed in its presence. The effect of EB irradiation on the properties of carbon black, silica, and clay-filled fluorocarbon rubber has been studied over a range of radiation doses, loadings, and nature of the fillers [431]. Compared to the unfilled irradiated rubber, the tensile and modulus improve with a decrease in the particle size of the carbon black filler. Improved stress–strain properties of the filled samples are due to the increased rubber filler interaction (leading to the formation of strong chemicals bonds) under the influence of EB.

31.2.6.7 Radiation Effect on Thermoplastic Elastomers

The undesirable properties of thermoplastic polyurethane elastomer, i.e., softening at high temperatures and flow under pressure, which limit their use at elevated temperatures have been reduced by cross-linking with EB radiation. The cross-linked polyurethane shows good mechanical properties and also displays good resistance to aggressive chemicals, e.g., brake fluid [432–435].

Heat-resistant flexible insulated wires have been manufactured by directly coating the outer surface of the twisted conductors with fluorine containing elastomer and blend, then radio-chemically cross-linking them. The resulting wire shows no cracking when wrapped around a rod of its own diameter and heated to high temperature for several days in contrast to wire coated [436–442] with the same compound with organic peroxide and cross-linked in saturated steam [435–442]. Haruvy et al. [443] have studied the grafting of acrylamide to nylon-6 by the EB preirradiation technique. X-ray diffraction measurements of nylon-6 grafted with AA indicate that the crystallites of nylon-6 are only marginally affected by very extensive radiation grafting. Considerable stresses are built up in the samples during their grafting due to distortion of the crystallites. Fusion endotherms of the grafted samples are masked to a large extent by the stress-release exotherms. Consequently, values of heat of fusion (ΔH_f) derived from DSC measurements of the unannealed samples are not reliable. Stresses built up during grafting may be at least partially released by treating these with 65% formic acid at room temperature (annealing). The perfection of the crystallite and/or their size increases as a result of annealing, since the reflections become much sharper. The overall degree of crystallinity also seems to increase significantly as a result of such treatment. Increase in the perfection of crystallites and in the degree of crystallinity has been attributed to increased mobility of polymer chains within the solvent swollen amorphous regions, which enables rearrangement and enlargement of the adjacent crystalline phase. When crystallites become larger as a result of such changes, the entanglement of the polymer chains in the amorphous regions diminishes and the amorphous phase becomes more accessible to penetration by alien molecules.

Akhtar et al. [444] have studied the effect of γ-irradiation on NR–PE blend. The high-energy radiation at a high dose rate has been found to cause extensive cross-linking in the bulk. The rupture energy values increase subsequently in the range of 15–25 Mrad and then decrease, as the absorbed dose increases further.

The effect of ^{60}Co γ-ray irradiation on the mechanical properties, surface morphology, and fractography of blends of plasticized PVC and thermoplastic copolyester elastomer, Hytrel (E.I. Du Pont de Nemours & Company, Inc., Wilmington, Delaware), have been studied by Thomas et al. [445]. Radiation has two major effects on the blend cross-linking of the Hytrel phase and degradation of PVC phase. Both effects are found more prominent at higher radiation dose.

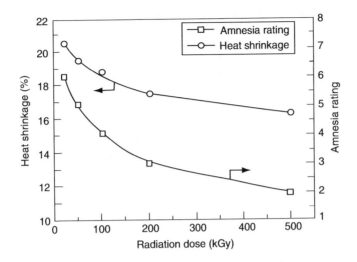

FIGURE 31.15 Variation of percent heat shrinkage and amnesia rating with radiation doses of low-density polyethylene/ethylene–vinyl acetate (LDPE/EVA) film from 50:50 blend without ditrimethylol propane tetra-crylate (DTMPTA). (From Chattopadhyay, S., Chaki, T.K., and Bhowmick, A.K., *Radiat. Phys. Chem.*, 59, 501, 2000. With permission.)

Influence of EB radiation doses, multifunctional sensitizer levels (ditrimethylol propane tetra-crylate, DTMPTA), and blend proportions on heat shrinkability of thermoplastic elastomeric films from blends of EVA copolymer and LDPE have been investigated [446]. The results are explained with the help of gel fraction and X-ray data. With the increase in radiation dose, gel fraction increases, which in turn gives rise to low values of heat shrinkage and amnesia rating (a measure of change in dimension) (Figure 31.15). Increase in the gel fraction increases the elasticity of polymer and hence facilitates the recovery, which in turn lowers the heat shrinkage and amnesia rating for a particular set of heat shrinkage condition. Reduced crystallinity at higher doses is one of the reasons for the low heat shrinkability of the irradiated samples. At a constant radiation dose and blend ratio, percent heat shrinkage is found to decrease with increase in DTMPTA level due to the increase in gel fraction and reduced crystallinity. The dynamic mechanical spectra [447] of the blends indicate both their immiscibility and two phase structure. Reprocessability studies reveal that blends irradiated at 50 kGy and below can be reprocessed. The decrease in properties after the processing cycles is probably due to the change in molecular architecture at a high temperature of 150°C. Although the processability of the compositions containing a greater amount of PE is better, they suffer from a poor permanent set. Dielectric studies [448] show that compared to the unirradiated samples, the permittivity and the loss factor decrease for the samples treated to a certain dose level beyond which there is a substantial increase. The decrease has been attributed to the increase in degree of cross-linking. Variation of the electrical properties with DTMPTA levels at a constant radiation dose clearly indicates the plasticizing effect of the monomer, apart from its function as an aid to cross-linking. Both the permittivity and the dielectric loss factor increase with an increase in the EVA content of the blends as a result of which the electrical insulating properties of the blends decrease. As the proportion of the EVA in the blend increases the overall polarization and amorphous portion of the blend also increases. This gives rise to increases in relaxation mechanisms as a result of both orientational and interfacial polarization losses. The effect of EB irradiation on the morphological, mechanical properties and heat and hot oil resistance of a thermoplastic elastomeric blend of nylon 6 and hydrogenated nitrile rubber (HNBR) has been investigated by Das et al. [449]. The irradiated blends have excellent hot oil resistance and the tensile strength of the blends does not change much after aging at 150°C for 72 h.

31.3 APPLICATIONS OF ELECTRON BEAM IRRADIATION

The EB irradiation has been successfully employed in the following areas [31,254]:

- Cross-linking of wire and cables, heat shrinkable polymeric tubes and films, and PE foam.
- Curing of coatings on wooden panels, in printing and lithography, precuring of tire components (e.g., inner liners, body plies, bead cores, tread, and sidewall) to improve the dimensional stability (green strength) in order to withstand the processing operations. The properties of a typical EB-irradiated halobutyl inner liner have been reported by Mohammed and Walker [450]. Irradiation of the compounds improves the green strength and fatigue properties, although, there is a marginal decrease in the tensile strength.
- Degradation of polymers (e.g., PTFE, organic glass, and silicone elastomers), cross-linking and grafting of plastics, rubbers, fibers, and polymer blends (as mentioned in the preceding sections), cross-linking of roofing materials, grain and food irradiation, and irradiation of diamonds and semiconductors.
- In pollution control in treatment of sewerage and sludge, removal of SO_x and NO_x from flue gases.
- In radiation sterilization of disposable supplies such as hypodermic syringes, needles, catheters, and sterile dressings in medicine.

31.4 CONCLUSION

Among the various radiation-induced modifications, the EB-processing of polymers has gained special importance as it requires less energy, is simple, fast, and versatile in application. The overall properties of EB-irradiated polymeric materials are also improved compared to those induced by other ionizing radiation.

REFERENCES

1. Bamford, C.H. and Tipper, C.F.H., *Comprehensive Chemical Kinetics*, Vol. 14, Elsevier, New York, 1975, 175.
2. Dole, M., *The Radiation Chemistry of Macromolecules*, Vol. 2, Academic Press, New York, 1972, Chapter 12.
3. Martin, D., Ighigeanu, D., Mateescu, E., Craciun, G., and Ighigeanu, A. *Radiat. Phys. Chem.*, 65, 63, 2002.
4. Tretinnikov, O.N., Ogata, S., and Ikada, Y., *Polymer*, 39, 6115, 1998.
5. Mateev, M., Karageorgiev, S., and Atanasova, B., *Radiat. Phys. Chem.*, 48, 437, 1996.
6. Davidson, W.L. and Geib, I.G., *J. Appl. Phys.*, 19, 427, 1948.
7. Aoshima, M., Jinno, T., and Sassa, T., *Kautsch. Gummi Kunstst.*, 45, 644, 1992.
8. Burr, J.G. and Garrison, W.M., US Atomic Energy Document 2078, Office of Technical Services, Oak Ridge, TN, 1948.
9. Sisman, O. and Bopp, C.D., US Atomic Energy Commission Report, ORNL-928, TISE, Office of Technical Services, Oak Ridge, TN, 1951.
10. Bopp, C.D. and Sisman, O., US Atomic Energy Commission Report, ORNL-1373, Office of Technical Services, Oak Ridge, TN, 1958.
11. Little, K., *Nature*, 170, 1075, 1952.
12. Schiller, S., Heisig, U., and Panzer, S., *Electron Beam Technology*, Wiley, New York, 1982, 30.
13. Hoshi, Y., Sakamoto, I., Mizusawa, K., and Kashiwagi, M., *Radiat. Phys. Chem.*, 46, 477, 1995.
14. Tartz, M., Hartmann, E., Lenk, M., and Mehnert, R., *Nucl. Instr. and Meth. A*, 427, 261, 1999.
15. Charlesby, A. and Ross, A., *Proc. R. Soc. London Ser. A*, 217, 122, 1953.
16. Charlesby, A. and Hancock, N.H., *Proc. R. Soc. London Ser. A*, 218, 245, 1953.
17. Charlesby, A., *Proc. R. Soc. London Ser. A*, 222, 60, 1954.

18. Lawton, E.S., Bueche, A.M., and Balwit, J.S., *Nature*, 76, 172, 1953.
19. Lawton, E.S., Balwit, J.S., and Bueche, A.M., *Ind. Eng. Chem.*, 48, 1703, 1964.
20. Chapiro, A., *Radiation Chemistry of Polymeric Systems*, Wiley, New York, 1962, 50.
21. Charlesby, A., *Atomic Radiation and Polymer*, Pergamon Press, Oxford, 1965, 65.
22. Nho, Y.C., Proceedings of the Fifth International Conference on Radiation Curing, Rad Tech Asia'95, Thailand, December 14–16, 1995, 47.
23. Kaetsu, I., Uchida, K., Sutani, K., and Sakata, S., *Radiat. Phys. Chem.*, 57, 465, 2000.
24. Bhowmick, A.K. and Mangaraj, D., in *Rubber Products Manufacturing Technology*, Bhowmick, A.K., Hall, M.M., and Benary, H.A., Eds., Marcel Dekker, New York, 1994, 315.
25. Chapiro, A., *Radiation Chemistry of Polymeric System*, Wiley, New York, 1962.
26. Makhlis, F., *Radiation Physics and Chemistry of Polymers*, Halsted Press, New York, 1975.
27. Ausloss, P., *Fundamental Processes in Radiation Chemistry*, Wiley, New York, 1998.
28. Dole, M., *The Radiation Chemistry of Macromolecules*, Vol. 1, Academic Press, New York, 1975.
29. David, D., *Comprehensive Chemical Kinetics*, Bamford, C.H. and Tiffer, C.F.H., Eds., Elsevier, Amsterdam, 1975.
30. Clough, R.L., Radiation resistant polymers, in *Encyclopedia of Polymer Science and Engineering*, Vol. 13, Kroschwitz, J.I., Ed., Wiley, New York, 1986, 669.
31. Ranby, B. and Rabek, J.F., *ESR Spectroscopy in Polymer Research*, Springer, Berlin, 1977.
32. Niki, E., Decker, C., and Mayo, F.R., *J. Polym. Sci.*, 11, 2813, 1973.
33. Bohm, G.G.A. and Tveekrem, J.O., *Rubber Chem. Technol.*, 55, 578, 1982.
34. Hori, Y., Shimada, S., and Kashiwabara, H., *Polymer*, 18, 151, 1977.
35. Wundrich, K., *J. Polym. Sci. Polym. Phys.*, 11, 1292, 1973.
36. Dole, M. and Cracco, F., *J. Phys. Chem.*, 66, 192, 1962.
37. Clough, R.L., *J. Chem. Phys.*, 87, 1588, 1987.
38. Iwasaki, M., Toriyama, K., Muto, H., and Nunome, K., *J. Chem. Phys.*, 65, 596, 1976.
39. McGinnise, V.D., Crosslinking with radiation, in *Encyclopedia of Polymer Science and Engineering*, Kroschwitz, J.I., editor-in-chief, Wiley, New York, 1986, 418.
40. Hofmann, W., *Rubber Technology Handbook*, Hanser, Munich, 1989.
41. Chapiro, A., *Nucl. Instr. and Meth. B*, 105, 5, 1995.
42. Seguchi, T., Kudoh, H., Sugimoto, M., and Hama, Y., *Nucl. Instr. and Meth. B*, 151, 154, 1999.
43. Kudoh, H., Celina, M., Malone, G.M., Kaye, R.J., Gillen, K.T., and Clough, R.L., *Radiat. Phys. Chem.*, 48, 555, 1996.
44. Dogadkin, B., Tarasava, Z.N., Mapunov, M.I., Karpov, V.L., and Kaluzen, N.A., *Koll. Zh.*, 20, 260, 1958.
45. Bhowmick, A.K. and Mangaraj, D., Vulcanization and curing techniques, in *Rubber Products Manufacturing Technology*, Bhowmick, A.K., Hall, M.M., and Benarey, H.A., Eds., Marcel Dekker, New York, 1994, 363.
46. Charlesby, A., *Nature*, A22, 60, 1954.
47. Bly, J.H., Rubber ACS Division Meeting, Pensylvania, 3–6 May, 1982.
48. Ramamurthi, S.S., Bapna, S.C., Soni, H.C., and Kotaiah, K., Proceedings of the Indo-USSR Seminar on Industrial Applications of Electron Accelerator Bhabha Atomic Research Centre, Mumbai, India, November 1–3, Vol. 2, 1982, 53.
49. Rosenstein, M. and Silverman, J., *J. Appl. Phys.*, 43, 3191, 1972.
50. Cleland, M.R. and Farrell, J.P., Proceedings of the Fourth Conference on Application of Small Accelerators, North Texas State Univ, Denton, Texas, 1976.
51. Mclaughlin, W.L., Harjtenberg, P.E., and Pedersen, W.B., *Int. J. Appl. Radiat. Isot.*, 26, 95, 1975.
52. Becker, R.C., Bly, J.H., Cleland, M.R., and Farrell, J.P., *Radiat. Phys. Chem.*, 14, 353, 1979.
53. Laughlin, J.S., in *Radiation Dosimetry*, Vol. III, Attix, F.H. and Jochilin, E., Eds., Academic Press, New York, 1969.
54. Charlesby, A. and Woods, R.J., *Int. J. Appl. Radiat. Isot.*, 14, 413, 1963.
55. Daview, I.M. and McQue, B., *Int. J. Appl. Radiat. Isot.*, 21, 283, 1970.
56. Levine, H., McLaughlin, W.L., and Miller, A., *Radiat. Phys. Chem.*, 14, 551, 1979.
57. McLaughlin, W.L., Humphreys, J.C., Radak, B.B., Miller, A., and Olejnik, J.A., *Radiat. Phys. Chem.*, 14, 535, 1979.
58. Moad, C.L. and Winzor, D.J., *Prog. Polym. Sc.*, 22, 759, 1998.
59. Wundrich, K., *J. Polym. Sci. Polym. Phys.*, 12, 201, 1974.

60. Witt, E., *J. Polym. Sci.*, 41, 507, 1959.
61. Miller, A.A., Lawton, E.J., and Balwit, J.S., *J. Polym. Sci.*, 14, 503, 1954.
62. Seguchi, T., Yagi, T., Ishikawa, S., and Sano, Y., *Radiat. Phys. Chem.*, 63, 35, 2002.
63. Clough, R.L., Gillen, K.T., Campon, J.L., Gaussens, G., Schonbacher, H., Seguchi, T., Hilski, H., and Machi, S., *Nucl. Safety*, 25, 198, 1984.
64. Kilb, R.W., *J. Phys. Chem.*, 63, 1838, 1959.
65. Makhlis, F., Gubanova, G., and Popova, V., *Vysokomol. Soedin.*, 15, 1995, 1973.
66. Arnold, P.M., Graus, G., and Anderson, R.H., *Kautsch. Gummi. Kunstst.*, 12, 27, 1959.
67. Seguchi, T., Arakawa, K., Hayakawa, N., and Machi, S., *Radiat. Phys. Chem.*, 18, 671, 1981.
68. Seguchi, T., *Radiat. Phys. Chem.*, 57, 367, 2000.
69. Clough, R.L., Gillen, K.T., and Quintana, C.A., *J. Polym. Sci. Polym. Chem.*, 23, 359, 1985.
70. Gillen, K.T., Clough, R.L., and Quintana, C.A., *Polym. Deg. Stab.*, 17, 331, 1987.
71. Gillen, K.T., Clough, R.L., and Dhooge, N.J., *Polymer*, 27, 225, 1986.
72. Giberson, R.C., *J. Phys. Chem.*, 66, 463, 1962.
73. Yoshi, F., Sasaki, T., Makuchi, K., and Tamura, N., *J. Appl. Polym. Sci.*, 31, 1343, 1986.
74. Seguchi, T., Arakawa, K., Hayakawa, N., and Machi, S., *Radiat. Phys. Chem.*, 18, 671, 1981.
75. Makhlis, F., Nikitin, Ya L., Volkova, A.K., and Tikhonova, M.N., *Vysokomol. Soedin.*, A14, 1782, 1972.
76. Sasuga, T. and Takehisa, M., *J. Macromol. Sci. Phys.*, B11, 389, 1975.
77. Okada, Y., *Adv. Chem. Ser.*, 66, 44, 1967.
78. Scholes, G. and Simic, M., *Nature*, 202, 895, 1964.
79. Kando, M. and Dole, M., *J. Phys. Chem.*, 70, 883, 1966.
80. Geymer, D.O., *Macromol. Chem.*, 100, 186, 1967.
81. Bennett, J.V., Pearson, R.W., and Mills, I.G., *Ind. Eng. Chem.*, 39, 469, 1963.
82. Miller, S.M., Roberts, R., and Vale, R.L., *J. Polym. Sci.*, 58, 737, 1962.
83. Walling, G. and Helmreich, W., *J. Am. Chem. Soc.*, 81, 1144, 1959.
84. Miller, A.A., *Ind. Eng. Chem.*, 51, 1271, 1959.
85. Lyons, B.J., *Nature*, 185, 604, 1960.
86. Odian, G. and Bernstein, B.S., *Nucleonics*, 21, 80, 1963.
87. Geymer, D.O. and Wagner, C.D., *Polym. Prepr.*, 9, 255, 1968.
88. Pappas, S.P., *UV Curing Science and Technology*, Vols. 1 and 2, Technology Marketing Corporation, Connecticut, 1978, 1985.
89. Wenting, S.G. and Koch, S.D., *UV Curing in Screen Printing for Printed Circuit and the Graphic Arts*, Technology Marketing Corporation, Norwalk, Connecticut, 981, 70.
90. Delzenno, G.A., *Makromol. Chem. Supp.*, 2, 169, 1979.
91. Green, G.E., Stark, B.P., and Zahir, S.A., *J. Macromol. Sci. Rev. Macromol. Chem.*, 21, 187, 1982.
92. Hara, K.O., *Ind. Chem. Bull.*, 189, 1982.
93. Oster, G. and Yang, N.L., *Chem. Rev.*, 68, 125, 1968.
94. Calvert, J.A. and Pitts, J.N., *Photochemistry*, Wiley, New York, 1966, 70.
95. Ledwith, A., Russell, P.J., and Sutcliffe, L.H., *J. Chem. Soc. Perkin. Trans.*, 2, 1925, 1972.
96. Ledwith, A., *J. Oil Col. Chem. Assoc.*, 59, 157, 1976.
97. Sandner, M.R. and Osborn, C.L., *Tetrahedron Lett.*, 415, 1974.
98. Sandner, M.R., Osborn, C.L., and Trecker, D.J., *J. Polym. Sci.*, 10, 3173, 1972.
99. Sandner, M.R. and Osborn, C.L., US Patent 3, 715, 923, 1973.
100. Barzynski, H., Marx, M., Storck, G., Drucheke, W., and Spoor, H., Ger. Patent 2, 357, 486, 1975.
101. Skoultchi, M.M. and Davis, I.J., US Patent 4, 069, 123, 1978.
102. Skoultchi, M.M. and Davis, I.J., Ger. Patent 2, 411, 169, 1974.
103. Pastor, S.D. and Skoultchi, M.M., US Patent 4, 052, 527, 1977.
104. Guse G., Lukat, E., and Schulte, D., Ger. Patent 2, 743, 979, 1979.
105. Shiryaeva, G.V., Proceedings of the Fifth International Conference on Radiation Curing, Rad Tech Asis'95, Thailand, December 14–16, 1995.
106. Viengkhou, V., Garnett, J.L., and Teck, Ng L., Proceedings of the Fifth International Conference on Radiation Curing, Rad Tech Asis'95, Thailand, December 14–16, 1995.
107. Tawn, A.R.H., *J. Oil Col. Chem. Assoc.*, 51, 782, 1968.
108. Fukukawa, S., Shimomura, K., Ijichi, I., Yoshikawa, N., and Murakami, T., Ger. Patent 2, 134, 468, 1972.
109. Yoshikawa, N., Shimitzu, Y., and Sunagawa, M., Jap. Kokai 75, 136, 328, 1975.

110. Nishizaki, T., Nishimura, K., Okazaki, H., and Yamano, K., Jap. Kokai 75, 64, 329, 1975.

111. Banik, I., PhD thesis, IIT Kharagpur, 2000.

112. Hendricks, J.O., US Patent 2, 956, 904, 1960.

113. Aubrey, D.W., British Patent 866, 003, 1962.

114. Christenson, R.M. and Anderson, C.C., US Patent 2, 131, 059, 1972.

115. Dowbenko, R., Christenson, R.M., Anderson, C.C., and Maska, R., *Rubber Chem. Technol.*, 46, 539, 1974.

116. Gleichenhagen, P. and Karmann, W., Ger. Patent 2, 455, 133, 1976.

117. Karmann, W. and Gleichenhagen, P., Ger. Patent 2, 350, 030, 1975.

118. Yoshikawa, N., Kamano, T., and Sunagawa, M., Jap. Kokai 51, 046, 331, 1976.

119. Vehara, K., Murakoshi, S., and Hisamtsu, H., Jap. Kokai 74, 05, 145, 1974.

120. Sasaki, T., Araki, K., Kawaguchi, T., and Ishiyama, H., Jap. Kokai 76, 26, 940, 1976.

121. Sasaki, T., Araki, K., Kawaguchi, T., and Ishiyama, H., Jap. Kokai 76, 26, 941, 1976.

122. Sasaki, T., Araki, K., Kawaguchi, T., and Ishiyama, H., Jap. Kokai 76, 26, 942, 1976.

123. Fukukawa, T., Shimomura, T., Yoshikawa, S., and Sunagawa, M., Jap. Kokai 74, 037, 692, 1975.

124. Nakata, S., Honi, H., and Ohnishi, K., Jap. Kokai 74, 029, 613, 1974.

125. Kasper, A.A., US Patent 3, 328, 914, 1967.

126. Hansen, D.R. and St. Clair, D.J., US Patent 4, 133, 731, 1979.

127. Vallat, M.F., Ruch, F., and David, M.O., *Nucl. Instr. and Meth. B*, 185, 175, 2001.

128. Sasaki, T., Takeda, S., and Shiraishi, K., Proceedings of the Sixth Japan–China Bilateral Symposium on Radiation Chemistry, Waseda University, Tokyo, Japan, November 6–11, 1994.

129. Gao, D., Heimann, R.B., Liu, Y., and Hou, J., Proceedings of the Fifth International Conference on Radiation Curing, Rad Tech Asia'95, Thailand, December 14–16, 1995.

130. Report on Survey and Workshop on Industrial Radiation Proceedings, Industrial Research Institute, Research Corporation for the Office of Cooperative Technology, National Bureau of Standards, US Department of Commerce, October 1979.

131. Odain, G. and Chandler, H.W., Radiation induced graft polymerization, in *Advances in Nuclear Science and Technology*, Vol. 1, Henley, E.S. and Kauts, H., Eds., Academic Press, New York, 1962, 65.

132. Garnett, J.L., *Radiat. Phys. Chem.*, 14, 79, 1979.

133. Billmeyer, F.W., *Textbook of Polymer Science*, Wiley, New York, 1971, 85.

134. Waterman, D.C. and Dole, M., *J. Phys. Chem.*, 74, 1906, 1970.

135. Devries, K.L., Smith, R.H., and Fanconi, B.M., *Polymer*, 21, 949, 1980.

136. Ang, C.H., Garnett, J.L., Long, M.A., and Levot, R., *Radiat. Phys. Chem*, 22, 831, 1983.

137. Zahran, A.H. and Zohdy, M.H., *J. Appl. Polym. Sci.*, 31, 1925, 1986.

138. Gupta, B.D. and Chapiro, A., *Eur. Polym. J.*, 25, 1137, 1989.

139. Stannett, V.T., *Radiat. Phys. Chem.*, 35, 82, 1990.

140. El-Assy, N.B., *J. Appl. Polym. Sci.*, 42, 885, 1991.

141. Misra, B.N., Chauhan, G.S., and Rawat, B.R., *J. Appl. Polym. Sci.*, 42, 3233, 1991.

142. Dworjanyn, P.A., Garnett, J.L., Khan, M.A., Maojun, X., Ping, Q.M., and Nho, Y.C., *Radiat. Phys. Chem.*, 42, 31, 1993.

143. Soebianto, Y.S., Yoshi, F., Makuchi, K., and Ishigaki, I., *Angew. Makromol. Chem.*, 152, 149, 1987.

144. Omichi, H. and Araki, K., *J. Polym. Sci., Polym. Chem.*, 16, 179, 1978.

145. Odian, G. and Kruse, R.L., *J. Polym. Sci. Part—C*, 22, 691, 1969.

146. Machi, S., Kamel, I., and Silverman, J., *J. Polym. Sci. Part A-1*, 8, 3329, 1970.

147. Morgan, P.W. and Corelli, J.C., *J. Appl. Polym. Sci.*, 28, 1879, 1983.

148. Sidorova, L.P. and Kabanov, V.Y., *Khim. Vys. Energ.*, 18, 36, 1984.

149. Hegaz, A.E., Dessouki, A.M., and El-Boohy, A.H., *J. Polym. Sci. Part—A, Polym. Chem.*, 24, 1933, 1986.

150. El-Dessouky, M.M. and El-Sawy, N.M., *Radiat. Phys. Chem.*, 26, 143, 1985.

151. Aliev, R.E. and Kabanov, V.Y., *Vysokomol. Soedin. Ser. B*, 26, 871, 1984.

152. Achmolowicz, T. and Wisniewski, A., *Polimery*, 30, 141, 1985.

153. Miller, J.S. and Perkowshi, J., *Polimery*, 31, 50, 1986.

154. Hegazy, E.S.A., Rehim, H.A.A.I., and Shawky, H.A., *Radiat. Phys. Chem.*, 57, 60, 2000.

155. Choi, S.H., Park, S.Y., and Nho, Y.C., *Radiat. Phys. Chem.*, 57, 179, 2000.

156. Rao, K.N. and Lokhande, H.T., *Man-Made Text.*, 30, 67, 1987.

157. Chodak, I., *Prog. Polym. Sci.*, 23, 1409, 1998.
158. Sugo, T., Okamokto, J., Fujiwara, K., and Sekiguchi, H., Jap. Kokai 89, 188, 506, 1989.
159. Sugo, T., Okamokto, J., Fujiwara, K., and Sekiguchi, H., Jap. Kokai 89, 292, 174, 1989.
160. Chen, J., Yang, L., Wu, M., Xi, Q., He, S., Li, Y., and Nho, Y.C., *Radiat. Phys. Chem.*, 59, 313, 2000.
161. Wang, U.P., *J. Ind. Irradiat. Technol.*, 1, 219, 1983.
162. Wang, U.P., *Radiat. Phys. Chem.*, 25, 491, 1985.
163. Kudryavtsev, V.N., Shapiro, A., and Endrikhovska, B.A.M., *Plast. Massay.*, 4, 6, 1984.
164. Yamakawa, S., Yamamoto, F., and Kato, Y., *Macromolecules*, 9, 754, 1976.
165. Cohn, D., Hoffman, A.S., and Ratner, B.D., *J. Appl. Polym. Sci.*, 33, 1, 1987.
166. Lenka, S., Nayak, P.L., and Mohanty, I.B., *J. Appl. Polym. Sci.*, 33, 21, 1987.
167. Rao, M.H., Rao, K.N., Lokhande, H.T., and Teli, M.D., *J. Appl. Polym. Sci.*, 33, 2707, 1987.
168. Omichi, H., Chaudhury, D., Stannett, V.T., *J. Appl. Polym. Sci.*, 32, 4827, 1986.
169. Chan, C.M., *Polymer Surface Modification and Characterization*, Hanser, Munich, 1994.
170. Schultz, J., Carre, A.C., and Mazeau, C., *Int. J. Adhesion Adhesives*, 4, 1963, 1984.
171. Sen, A.K., Mukherjee, B., Bhattacharyya, A.S., De, P.P., and Bhowmick, A.K., *J. Appl. Polym. Sci.*, 44, 1153, 1992.
172. Battaered, H.A.J. and Tregear, G.W., *Graft Copolymers*, Wiley, New York, 1967, 124.
173. Ceresa, R.J., *Block and Graft Copolymerization*, Vol. 1, Wiley, New York, 1973, 130.
174. Haddadi-asl, V. and Burford, R.P., *Radiat. Phys. Chem.*, 47, 907, 1996.
175. Taher, N.H., Hegazy, E.A., Dessouki, A.M., and El-Arnaouty, M.B., *Radiat. Phys. Chem.*, 33, 129, 1989.
176. Bucio, E., Aliev, R., and Burillo, G., *Radiat. Phys. Chem.*, 52, 193, 1998.
177. Che, J.T. and Zhang, W.X., *Radiat. Phys. Chem.*, 42, 85, 1993.
178. Konar, J. and Bhowmick, A.K., *J. Adhesion Sci. Technol.*, 8, 1169, 1994.
179. Claramma, N.M., Mathew, N.M., and Thomas, E.V., *Radiat. Phys. Chem.*, 33, 87, 1989.
180. Abdel-bary, E.M. and El-Nesr, E.M., *Radiat. Phys. Chem.*, 48, 689, 1996.
181. Schonhorn, H., *Macromolecules*, 1, 145, 1968.
182. Gagnon, D.R. and McCarthy, T.J., *J. Appl. Polym. Sci.*, 29, 4335, 1984.
183. Schmitt, R.L., Gardell, Jr. J.A., Magill, J.H., Salvati, Jr. L., and Chin, R.L., *Macromolecules*, 18, 2675, 1985.
184. Weidmaier, J.M. and Meyer, G.C., *J. Appl. Polym. Sci.*, 28, 1429, 1983.
185. Takahara, A., Tashita, J., Kajiyama, T., Takayanagi, M., and MacKnight, W.J., *Polymer*, 27, 987, 1985.
186. Waddell, W.H., Evans, L.R., Gillick, J.G., and Shuttleworth, D., *Rubber Chem. Technol.*, 65, 687, 1992.
187. Mark, H.F., Bikales, N.M., Overberger, C.G., and Menges, G., Eds., *Encyclopedia of Polymer Science and Engineering*, Vol. 4, Wiley, New York, 1988.
188. Rabe, J., Bischoff, G., and Schmidt, W.F., *J. Appl. Phys. Part–1*, 28, 518, 1989.
189. George, W.T., US Patent 3, 018, 189, 1962.
190. Rossman, K., *J. Polym. Sci.*, 19, 141, 1956.
191. Kim, C.Y. and Goring, D.A.I., *J. Appl. Polym. Sci.*, 15, 1357, 1971.
192. Briggs, D., *J. Adhesion*, 13, 287, 1982.
193. Blythe, A.R., Briggs, D., Kendall, C.R., Rance, D.G., and Zichy, V.I., *Polymer*, 19, 1273, 1978.
194. Delollis, N.J., *Rubber Chem. Technol.*, 46, 549, 1973.
195. Behnisch, J., Hollander, A., and Zimmermann, H., *J. Appl. Polym. Sci.*, 49, 117, 1993.
196. van Ooij, W.J., Zhang, N., Guo, S., and Luo, S., Paper Presented at Functional Fillers and Fibers for Plastics'98, Beijing, China, June 15–17, 1998.
197. Katbab, A.A., Burford, R.P., and Garnett, J.L., *Radiat. Phys. Chem.*, 39, 293, 1992.
198. Poncil-Epaillard, F., Chevet, B., and Brosse, J.C., *J. Appl. Polym. Sci.*, 53, 1291, 1994.
199. Bagheri, R., Naimain, F., and Sheikh, N., *Radiat. Phys. Chem.*, 49, 497, 1997.
200. Yue-E, F., Xia, Z., Xuewu, G., and Tianyi, S., *Radiat. Phys. Chem.*, 49, 589, 1997.
201. Yamakawa, S., *Macromolecules*, 12, 1222, 1979.
202. Celina, M., Kudoh, H., Renk, T.J., Gillen, K.T., and Clough, R.L., *Radiat. Phys. Chem.*, 51, 191, 1998.
203. Cardoso, E.C.L., Lugao, A.B., Andrade, E., and Silva, L.G., *Radiat. Phys. Chem.*, 52, 197, 1998.
204. Fickes, M.G. and Rakoczy, B., US Patent 4, 400, 460, 1983.
205. McGinniss, V.D., US Patent 4, 661, 534, 1987.
206. Gillick, J.G. and Waddell, W.H., US Patent 4, 824, 692, 1989.
207. Bellobono, I.R., Tolusso, F., and Selli, E., *J. Appl. Polym. Sci.*, 26, 619, 1981.

208. Siddiqui, S.A., McGee, K., Lu, W.C., Alger, K., and Nedles, H.L., *Am. Dyestuff Rep.*, 70, 20, 1981.
209. Hamilton, L.M., Green, A., Edge, S., Badyal, J.P.S., Feast, W.J., and Pacynko, W.F., *J. Appl. Polym. Sci.*, 52, 413, 1994.
210. Swanson, M.J. and Opperman, G.W., in *Polymer Surface Modification: Relevance to Adhension*, Mittal, K.L., Ed., VSP Utrecht, The Netherlands, 1995, 319.
211. Brewis, D.M. and Briggs, D., *Polymer*, 22, 7, 1981.
212. Kreidl, W.H., US Patent 2, 632, 921, 1953.
213. Briggs, D., Brewis, D.M., and Konieczko, M.B., *J. Mat. Sci.*, 14, 1344, 1979.
214. Strobel, M., Walzak, M.J., Hill, J.M., Lin, A., Karbashewski, E., and Lyons, C.S., in *Polymer Surface Modification: Relevance to Adhension*, Mittal, K.L., Ed., VSP Utrecht, The Netherlands, 1995, 233.
215. Murahara, M. and Okoshi, M., in *Polymer Surface Modification: Relevance to Adhension*, Mittal, K.L., Ed., VSP Utrecht, The Netherlands, 1995, 223.
216. Murahara, M., Ikegame, T., and Tonita, M., Paper Presented at the Second International Symposium on Polymer Surface Modification, Relevance to adhension, New Jersey, May 24–26, 1999.
217. Mirzadeh, H., Katbab, A.A., and Burford, R.P., *Radiat. Phys. Chem.*, 41, 507, 1993.
218. Mirzadeh, H., Katbab, A.A., and Burford, R.P., *Radiat. Phys. Chem.*, 42, 53, 1993.
219. Breuer, J., Metev, S., and Sepold, G., in *Polymer Surface Modification: Relevance to Adhension*, Mittal, K.L., Ed., VSP Utrecht, The Netherlands, 1995, 185.
220. Hiraoka, H., Sendova, M., Lee, C.H., Latsch, S., Wang, T.M., Sung, J., Hung, C.T., and Smith, T., Proceedings of the SPIE—The International Society for Optical Engineering, Society of photo-optical instrumentation engineering, Bellingham, WA, 1995, 260.
221. Okada, T., Mandelkern, L., and Glick, R., *J. Amer. Chem. Soc.*, 89, 4790, 1967.
222. Lyons, B.J. and Vaughn, C.R., *Adv. Chem. Ser.*, 66, 139, 1967.
223. Lyons, B.J., *J. Polym. Sci. Part-A*, 3, 777, 1965.
224. Miller, S.M., Spindler, M.W., and Vale, R.L., *J. Polym. Sci. Part-A*, 1, 2537, 1963.
225. Bernstein, B.S., Odian, G., Orban, G., and Tirelli, S., *J. Polym. Sci. Part-A*, 3, 3405, 1965.
226. Auerbach, I., *Polymer*, 8, 63, 1967.
227. Charlesby, A., *Adv. Chem. Ser.*, 66, 1, 1967.
228. Vaughan, G., Eaves, D.E., and Cooper, W., *Polymer*, 2, 235, 1961.
229. Parkinson, W.W. and Sears, W.C., *Adv. Chem. Ser.*, 66, 57, 1967.
230. Shultz, A.R., *Encycl. Polym. Sci. Tech.*, 4, 398, 1966.
231. Yoon, B.M., Pyun, H.C., Kim, T.R., and Jin, J., *Pollimo*, 7, 115, 1983.
232. Bao, W.S., Liu, W., and Shi, Z., *Sanghai Keji Daxue Xeubao*, 1, 72, 1983.
233. Klier, I., Ladyr, V., and Vokal, A., *Sb. Vys. Sk. Chem. Technol. Praze*, S11, 47, 1984.
234. Kovacic, P. and Hein, R.W., *J. Amer. Chem. Soc.*, 81, 1190, 1959.
235. Miller, S.M., Roberts, R., and Vale, R.L., *J. Polym. Sci.*, 58, 737, 1962.
236. Miller, S.M., Spindler, M.W., and Vale, R.L., *Proc. IAEA Conf. Appl. Large Rad. Sources Ind.*, 1, 329, 1963.
237. Pearson, D.S. and Shurpik, A., US Patent 3, 843, 502, 1974.
238. Zapp, R.L. and Oswald, A.A., Paper 55 Presented at the Rubber Division, ACS Meeting, Cleveland, 1975.
239. Morgan, C.R., Magnotta, F., and Ketlev, A.D., *J. Polym. Sci. Polym. Chem.*, 15, 627, 1977.
240. Morgan, C.R. and Ketlev, A.D., *J. Polym. Sci. Polym. Lett.*, 16, 75, 1978.
241. Pierson, R.M., Gibbs, W.E., Mever, G.E., Naples, F.J., Saltman, W.M., Schrock, R.W., Tewksbury, L.B., and Trick, G.S., *Rubber Plast. Age*, 38, 592,708 and 721, 1957.
242. Walling, C. and Helmreich, W., *J. Amer. Chem. Soc.*, 81, 1144, 1959.
243. Bohm, G.G.A. and Tveekrem, J.O., *Rubber Chem. Technol.*, 55, 575, 1982.
244. Chapiro, A., *J. Chem. Phys.*, 47, 747, 1950.
245. Dogadkin, B.A., Mladenov, I., and Tutorskii, I.A., *Vysokomol. Soedin.*, 2, 259, 1960.
246. Jankowski, B. and Kroh, J., *J. Appl. Polym. Sci.*, 13, 1795, 1969.
247. Jankowski, B. and Kroh, J., *J. Appl. Polym. Sci.*, 9, 1363, 1965.
248. Kozlov, V.T., Kaplunov, M.Y., Tarasova, Z.N., and Dogadkin, B.G., *Vysokomol. Soedin.*, A10, 987, 1968.
249. Kozlov, V.T., Klauzen, N.A., and Tarasova, Z.N., *Vysokomol. Soedin.*, A10, 1949, 1968.
250. Mesrobian, R.B., *Conf. Int. Geneva*, 15, 826, 1958.

251. Turner, D., *J. Polym. Sci.*, 15, 503, 1958.
252. Okada, Y., *Adv. Chem. Ser.*, 66, 44, 1967.
253. Karpov, V.L., *Vysokomol. Soedin.*, 7, 1319, 1965.
254. Kondo, M. and Dole, M., *J. Phys. Chem.*, 70, 883, 1966.
255. Lyons, B.J. and Dole, M., *J. Phys. Chem.*, 68, 526, 1964.
256. Lugao, A.B., Hutzler, B., Ojeda, J., Tokumoto, S., Siemens, R., Makuchii, K., and Villavicencio, A.L.C. H., *Radiat. Phys. Chem.*, 57, 389, 2000.
257. Chodak, I., *Prog. Polym. Sci.*, 20, 1165, 1995.
258. Guan, R., *Radiat. Phys. Chem.*, 76, 75, 2000.
259. Karkova, G.K., Kachan, A.A., and Chervyatsova, L.L., *J. Polym. Sci. Part—C*, 16, 3041, 1967.
260. Gal, O.S., Markovic, V.M., Novakovic, L.R., and Stannett, V.T., *Radiat. Phys. Chem.*, 26, 325, 1985.
261. Fydelor, P.J. and Pearson, R.W., *J. Appl. Polym. Sci.*, 5, 171, 1961.
262. Korolev, B.M., Glazunova, E.D., Korotyanski, M.A., and Ronkin, G.M., *Chem. Abstr.*, 96, 181751w, 1982.
263. Rak, D., *Sb. Vys. Sk. Chem. Technol. Praze*, S11, 185, 1984.
264. Tikku, V.K., Biswas, G., Despande, R.S., Majali, A.B., Chaki, T.K., and Bhowmick, A.K., *Radiat. Phys. Chem.*, 45, 829, 1995.
265. Chaki, T.K., Despande, R.S., Majali, A.B., Tikku, V.K., and Bhowmick, A.K., *Die Angew. Makromol. Chem.*, 61, 217, 1994.
266. Chaki, T.K., Roy, S., Despande, R.S., Majali, A.B., Tikku, V.K., and Bhowmick, A.K., *J. Appl. Polym. Sci.*, 53, 141, 1994.
267. Salmon, W.A. and Loan, L.D., *J. Appl. Polym. Sci.*, 16, 671, 1972.
268. Miller, A.A., *J. Phys. Chem.*, 63, 1755, 1959.
269. Hegazy, E.S.A., Seguchi, T., and Machi, S., *J. Appl. Polym. Sci.*, 26, 2947, 1981.
270. Pelit, J. and Zaiton, G., *OCR Acad. Sci. Ser. C*, 256, 2610, 1966.
271. Scarbrough, A.L., Keller, W.L., and Rizzo, P.W., *Nat. Bur. Stand Circ. No. 525*, 95, 1953.
272. Pinner, S.H., *Nature*, 183, 1108, 1959.
273. Miller, A.A., *Ind. Eng. Chem.*, 51, 1271, 1959.
274. Dakin, V.I., Egorova, Z.S., and Karpov, V.L., *Plast. Massy*, 40, 91, 1979.
275. Bair, H.E., Matsuo, M., Salmon, W.A., and Kwei, T.K., *Macromolecules*, 5, 114, 1972.
276. Dobo, J., *Pure Appl. Chem.*, 46, 1, 1976.
277. Daries, D.D. and Slichter, W.P., *Macromolecules*, 6, 728, 1973.
278. Szymczak, T.J. and Manson, J.A., *Mod. Plast.*, 66, 1974.
279. Bowmer, T.N., Davis, D.D., Kwei, T.K., and Vroom, W.L., *J. Appl. Polym. Sci.*, 26, 3669, 1981.
280. Dakin, V.I., Egorova, Z.S., and Karpov, V.L., *Khim. Vys. Energ.*, 11, 378, 1977.
281. Egorova, Z.S., Dakin, V.I., and Karpov, V.L., *Vysokomol. Soedin.*, A21, 2117, 1979.
282. Bohm, G.G.A., Oliver, F., and Pearson, D.S., *Soc. Plast. Eng. J.*, 27, 21, 1970.
283. Bowmer, T.N., Hellman, M.Y., and Vroom, W.I., *J. Appl. Polym. Sci.*, 28, 2083, 1983.
284. Dole, M., *The Radiation Chemistry of Macromolecules*, Academic Press, New York, 1973, 70.
285. Loan, L.D., *Radiat. Phys. Chem.*, 9, 253, 1977.
286. Bellino, R.A., *Wire J.*, 75, 1975.
287. Alexander, D.C., *Plastics and Polymers*, 195, 1975.
288. Saito, E., *Radiat. Phys. Chem.*, 9, 675, 1977.
289. Bowmer, T.N. and Vroom, W.I., *J. Appl. Polym. Sci.*, 28, 3527, 1983.
290. Bowmer, T.N. and Hellman, M.Y., *J. Appl. Polym. Sci.*, 28, 2553, 1983.
291. Zyball, A., *Kautsch. Gummi. Kunstst.*, 72, 487, 1982.
292. Pearson, D.S., *Radiat. Phys. Chem.*, 18, 89, 1981.
293. Scalco, E. and Moore, W.F., Proceedings of the Plastics and Rubber Institute International Conference on Radiation Processing for Plastics and Rubber, Brighton, June, 1981.
294. Morganstern, K.H., *Radiat. Phys. Chem.*, 18, 1, 1981.
295. Oda, E., *Radiat. Phys. Chem.*, 18, 241, 1981.
296. Bowden, M.J., *Crit. CRC. Rev. Solid State Sci.*, 8, 223, 1979.
297. Ward, J.M., *Circuits Manuf.*, 22, 78, 1982.
298. Scalco, E., and Moore, W.F., *Radiat. Phys. Chem.*, 21, 389, 1983.
299. Kudryavtsev, V.N., Shapiro, A., and Endriknovska, A.M., *Plast. Massy*, 4, 6, 1984.

300. Dessouky, M.M.E. and Ebsawy, M.N., *Radiat. Phys. Chem.*, 26, 143, 1985.

301. Nethsinghe, L.P. and Gilbert, M., *Polymer*, 29, 1935, 1988.

302. Klier, I. and Cernoch, D., *Sb. Vys. Sk. Chem. Technol. Praz Polym Chem. Vlastnostizpracov*, S12, 145, 1985.

303. Nakae, T., Okubo, S., and Shunji, K.S., Jap. Kokai 89, 185, 338, 1989.

304. Herring, C.M., *Radiat. Phys. Chem.*, 14, 55, 1979.

305. Black, R.M. and Lyons, B.J., *Nature*, 180, 1346, 1957.

306. Black, R.M. and Lyons, B.J., *Proc. Rubber Society Lond Ser. A*, 253, 322, 1959.

307. Schnabel, W. and Dole, M., *J. Phys. Chem.*, 67, 295, 1963.

308. Novikov, A.S., Karpov, V.L., Galil-oglay, F.A., Slovokhotova, N.A., and Dyumaeva, T.N., *Vysokomol. Soedin.*, 2, 485, 1960.

309. Ranby, B. and Rabek, J., *Photo Degradation, Photo Oxidation and Photo Stabilization of Polymers*, Wiley, New York, 1975, 75.

310. Ranby, B. and Rabek, J., *ESR Spectroscopy in Polymer Research*, Springer, New York, 1977, 70.

311. Dunn, T.S., Epperson, B.J., Sugg, H.W., Stannett, V.T., and Williams, J.L., *Radiat. Phys. Chem.*, 14, 625, 1979.

312. Najiri, B. and Sawasaki, T., *Radiat. Phys. Chem.*, 26, 339, 1985.

313. Suzuki, T., Nakayama, K., and Isono, M., Jap. Kokai 91, 126, 736, 1991.

314. Bartl, H. and Seheurlen, H. British Patent 901, 117, 1962.

315. Dobo, J., Forgacs, P., Somogyi, A., and Roder, M., *Proc. Tihany Symp. Radiat. Chem.*, 4, 359, 1977.

316. Dobo, J., Forgacs, P., and Somogyi, A., *Radiat. Phys. Chem.*, 18, 399, 1981.

317. Gordiichuk, T.N. and Gordienko, V.P., *Kompoz. Polim. Mater.*, 38, 33, 1988.

318. Datta, S.K., Bhowmick, A.K., Chaki, T.K., Majali, A.B., and Deshpande, R.S., *Polymer*, 37, 45, 1996.

319. Datta, S.K., Bhowmick, A.K., and Chaki, T.K., *Radiat. Phys. Chem.*, 47, 913, 1996.

320. Hayden, P., *Int. J. Appl. Radiat. Iso.*, 8, 65, 1960.

321. Evans, G., Higgins, M., and Turner, D., *J. Appl. Polym. Sci.*, 2, 340, 1959.

322. Kaufman, P. and Heusinger, H., *Makromol. Chem.*, 177, 871, 1976.

323. Flory, P.J., Rabjohn, N., and Schaffer, M.C., *J. Polym. Sci.*, 4, 435, 1949.

324. Langley, N.R. and Ferry, J.D., *Macromolecules*, 1, 353, 1968.

325. Turner, D.T., *J. Polym. Sci.*, 35, 541, 1960.

326. Adams, H. and Johnson, B.L., *Ind. Eng. Chem.*, 45, 1539, 1953.

327. Harmon, D.J., *Rubber Age*, 86, 251, 1959.

328. Sasuga, T. and Takehisa, M., *J. Macromol. Sci. Phys.*, 11, 389, 1975.

329. Jayasuriya, M.M., Makuuchi, K., and Yoshi, Y., *Eur. Polym. J.*, 37, 93, 2001.

330. Harmon, D.J., *Rubber World*, 136, 585, 1958.

331. Xuan, H.L. and Decker, C., *J. Polym. Sci. Part A. Polym. Chem.*, 31, 769, 1993.

332. Claramma, N.M., Mathew, N.M., and Thomas, E.V., *Radiat. Phys. Chem.*, 33, 87, 1989.

333. Ratnam, C.T., Nasir, M., Baharin, A., and Zaman, K., *Nucl. Instr. and Meth. B*, 171, 455, 2000.

334. Madani, M. and Badawy, M.M., *Polymers Polym. Compos.*, 13, 93, 2005.

335. Sharif, J., Zin Wan Yunus, W.M., Mohd. Dahlan, K.Z.H., and Ahmad, M.H., *Polym. Test.*, 24, 211, 2005.

336. Kuzminski, A., Nikinna, T.S., Zhuravskaya, E.V., Oksentievich, L.A., Sunitsa, I., and Vitushkin, N., Proceedings of the International Conference on Peaceful Uses of Atom Energy, Geneva, 29, 258, 1958.

337. Golub, M.A., *J. Am. Chem. Soc.*, 82, 5093, 1960.

338. Burnay, S.C., *Radiat. Phys. Chem.*, 13, 171, 1979

339. Pearson, D.S., Skutnik, B.J., and Bohm, G.G.A., *J. Polym. Sci. Polym. Phys.*, 12, 925, 1974.

340. Witt, E., *J. Polym. Sci.*, 41, 507, 1959.

341. Dogadkin, B.A., Mladenov, I., and Tutorskii, I.A., *Vysokomol. Soedin.*, 2, 259, 1960.

342. Makhlis, F., Grubanova, G., and Papouva, V., *Vysokomol. Soedin.*, 15, 1995, 1973.

343. Ateria, E., *J. Polym. Sci.*, 66, 339, 1997.

344. Bohm, G.G.A., Hamed, G.R., and Vescelius, L.E., US Patent 4, 250, 273, 1981.

345. Nanogaki, S., Morishita, H., and Saito, N., *Appl. Polym. Symp.*, 23, 117, 1974.

346. Bohm, G.G.A., in *The Radiation Chemistry of Macromolecules*, Vol. II, Dole, M., Ed., Academic Press, New York, 1972, Chapter 12.

347. Amenija, A., *J. Phys. Soc. Japan*, 17, 1245 and 1694, 1962.

348. Bohm, G.G.A., De Trano, M.N., Pearson, D.S., and Carter, D.R., *J. Appl. Polym. Sci.*, 21, 31930, 1977.

349. Anderson, H.R., *J. Appl. Polym Sci.*, 3, 316, 1960.
350. Yoshida, K., Ishigura, K., Garreau, M., and Stannet, V., *J. Macromol. Sci.*, 414, 739, 1980.
351. Okamoto, H., Adachi, S., and Twaia, T., *J. Macromol. Sci.*, A11, 1949, 1977.
352. Adam, C., Lacaste, J., and Dowphin, G., *Polymer*, 32, 317, 1991.
353. O'Donnell, J.H. and Whittaker, A.K., *J. Polym. Sci. Part A. Polym. Chem.*, 30, 185, 1992.
354. Harmon, D.J., *Rubber Age*, 86, 251, 1959.
355. Shanmugharaj, A.M. and Bhowmick, A.K., *Rubber Chem. Technol.*, 76, 299, 2003.
356. Alexander, P. and Charlesby, A., *Proc. Roy. Soc. London Ser. A*, 230, 136, 1955.
357. Alexander, P., Black, R.M., and Charlesby, A., *Proc. Roy. Soc. London Ser. A*, 232, 31, 1955.
358. Turner, D.T., in *The Chemistry and Physics of Rubber like Substances*, Bateman, L., Ed., Wiley, New York, 1963.
359. Very, G.A. and Turner, D.T., *Polymer*, 5, 589, 1964.
360. Wundrich, K., *Eur. Polym. J.*, 110, 341, 1974.
361. Henglen, A. and Schneider, G., *J. Phys. Chem.*, 19, 367, 1959.
362. Lukhovitsku, V.I., Tsingister, V.A., Sharadina, N.A., and Karpov, V.L., *Vysokomol Soedin*, 8, 1932, 1966.
363. Exxon Chemical Brochure on Conjugated Butyl Elastomer.
364. Zapp, R.L. and Oswald, A.A., Rubber ACS Meeting, Cleveland, 1975.
365. Odian, G. and Bernstein, B.S., *J. Polym. Sci. Polym. Lett.*, 2, 819, 1964.
366. Miller, S.M., Spindler, M.W., and Vale, R., *J. Polym. Sci.*, A1, 2537, 1963.
367. Telnov, A.V., Zavyalov, N.V., Khokhlov, Y.A., Sitnikov, N.P., Smetanin, M.L., Tarantasov, V.P., Shorikov, I.V., Liakumovich, A.L., and Miryasova, F.K., *Radiat. Phys. Chem.*, 63, 245, 2002.
368. Geissler, W., Zott, M., and Heusinger, H., *Macromol. Chem.*, 179, 697, 1978.
369. Dzhibgashvili, G.G., Slovokhotova, N., Leshchenko, S., and Karpov, V.L., *Vysokomol Soedin*, A13, 1087, 1971.
370. Smetanina, L.B., Leshchenko, S.S., Yegotova, Z.A., Starodubtsev, D.S., Klinshpont, E.I., Maplunov, M., and Karpov, V.L., *Vysokomol Soedin*, A12, 2401, 1970.
371. Odian, G., Lanparella, D., and Canamare, J., *J. Polym. Sci.*, C16, 3619, 1967.
372. Pearson, D.S. and Bohm G.G.A., *Rubber Chem. Technol.*, 45, 193, 1972.
373. Kammel, G. and Widenman, R., *J. Polym. Sci.*, 57, 73, 1976.
374. Decker, C., Mayo, F., and Richardson, H., *J. Polym. Sci.*, 11, 2879, 1973.
375. Makhlis, F., Nikitin, L., Biryukava, T., and Kryukova, A., *Kauch Rezina*, 4, 19, 1976.
376. Soebianto, Y.S., Yoshii, F., Makuchi, K., and Ishigaki, I., *Angew. Makromol. Chem.*, 152, 149, 1987.
377. O'donell, J.H. and Whittaker, A.K., *J. Macromol. Sci. Pure and Applied Chemistry*, 1, 29, 1992.
378. Hilborn, J. and Ranby, B., *Macromolecules*, 22, 1154, 1989.
379. Rak, D., *Sb. Vys. Sk. Chem. Technol. Praze*, S11, 185, 1984.
380. Harnischfeger, P., Kinzel, P., and Jungnickel, B.J., *Angew. Makromol. Chem.*, 175, 175, 1990.
381. Sen Majumder, P. and Bhowmick, A.K., *J. Adhesion Sci. Technol.*, 11, 1321, 1997.
382. Sen Majumder, P. and Bhowmick, A.K., *Radiat. Phys. Chem.*, 53, 63, 1998.
383. Sen Majumder, P. and Bhowmick, A.K., *J. Appl. Polym. Sci.*, 77, 323, 2000.
384. Sen Majumder, P. and Bhowmick, A.K., *Wear*, 221, 15, 1998.
385. Sen Majumder, P. and Bhowmick, A.K., in *Polymer Surface Modification: Relevance to Adhesion*, Vol. II, Mittal, K.L., Ed., VSP Utrecht, The Netherlands, 2000, 425.
386. Tarasova, Z.N., Erkina, I.A., Kazlov, V.T., Zagumennya, T.I., and Kaplunov, M., *Vysokomol Soedin*, A14, 1782, 1972.
387. Makhlis, F., Nikitin, L.Y., Volkova, A.K., and Tikhonova, M.N., *Vysokomol Soedin*, A14, 17822, 1972.
388. Ito, M., Okada, S., and Kuriyama, I., *J. Mater. Sci.*, 16, 10, 1981.
389. Miligy, A.A.E.I., Monsour, M.A., Khalifa, W.M., and Abdel Bary, M., *Elastomerics*, 111, 28, 1979.
390. Yunshu, X., Yoshii, F., and Makuchi, K., *Radiat. Phys. Chem.*, 53, 669, 1998.
391. Yasin, T., Abmed, S., Yoshii, F., and Makuuchi, K., *React. Func. Polym.*, 53, 173, 2002.
392. Vijayabaskar, V., Costa, F.R., and Bhowmick, A.K., *Rubber Chem. Technol.*, 77, 624, 2004.
393. Vijayabaskar, V., Tikku, V.K., and Bhowmick, A.K., *Radiat Phys. Chem.*, 75, 779, 2006.
394. Ray, S. and Bhowmick, A.K., *Radiat Phys. Chem.*, 65, 259, 2002.
395. Ormerod, M.G. and Charlesby, A., *Polymer*, 4, 459, 1963.
396. Charlesby, A., *Proc. Roy. Soc. London Ser. A*, 230, 120, 1955.

397. Langley, N.R. and Polmanteer, K.E., *J. Polym. Sci. Polym. Phys.*, 12, 10232, 1974.
398. Delides, C.G. and Shepherd, I.W., *Radiat Phys. Chem.*, 10, 379, 1977.
399. Miller, A.A., *J. Am. Chem. Soc.*, 83, 31, 1961.
400. Hongfei, H., Xiaohong, L., and Jilam, W., *Radiat Phys. Chem.*, 39, 513, 1992.
401. Tanny, G.B. and St. Pierre, L.E., *J. Polym. Sci. Polym. Lett.*, 9, 868, 1971.
402. Mironov, E.I., Kuziniskii, A.S., Khazen, L.Z., and Seregnina, S.V., *Kautch Rezine*, 6, 19, 1971.
403. Lipko, J.D., George, H.F., Thomas, D.A., Hargest, S.C., and Sperling, L.H., *J. Appl. Polym. Sci.*, 23, 2739, 1979.
404. Takanega, M. and Yamagata, K., *J. Appl. Polym. Sci.*, 26, 1373, 1981.
405. Hedvig, P., *J. Polym. Sci. Pt. A1 Polym. Chem.*, 7, 1145, 1969.
406. Kusy, R.P. and Turner, D.T., *J. Polym. Sci. Pt. A1: Polym. Chem.*, 10, 1745, 1972.
407. Hegazy, E.S.A., Ishigaki, I., Dessouki, A.M., Rabie, A., and Okamoto, J., *J. Appl. Polym. Sci.*, 27, 535, 1982.
408. Kaiser, R.J., Miller, G.A., Thomas, D.A., and Sperling, L.H., *J. Appl. Polym. Sci.*, 27, 957, 1982.
409. Uschold, R.E., *J. Appl. Polym. Sci.*, 29, 1335, 1984.
410. Zhong, X., Sun, J., Wang, F., and Sun, Y., *J. Appl. Polym. Sci.*, 44, 638, 1992.
411. Timmerman, R. and Greyson, W., *J. Appl. Polym. Sci.*, 6, 456, 1962.
412. Kawanishi, S., Shimizu, Y., Sugimoto, S., and Suzuki, N., *Polymer*, 32, 979, 1991.
413. Golub, M.A., Wydeven, T., and Cornia, R.D., *Polymer*, 30, 1571, 1989.
414. Mamose, T., Yashioka, H., Ishigaki, I., and Okamoto, J., *J. Appl. Polym. Sci.*, 38, 2091, 1989.
415. Zhong, X., Yu, L., Zhao, W., Zhang, Y., and Sun, J., *J. Appl. Polym. Sci.*, 48, 741, 1993.
416. Rye, R., *J. Polym. Sci. Pt. B: Polym. Phys.*, 31, 357, 1993.
417. Lappan, U., Hauss, L., Pompe, G., and Lunkwitz, K., *J. Appl. Polym. Sci.*, 66, 2287, 1997.
418. Lunkwitz, K., Lappan, U., and Lechman, D., *Radiat. Phys. Chem.*, 57, 373, 2000.
419. Tabata, Y., Proceedings of the IUPAC Sponsored 37th International Symposium on Macromolocules, Macro'98, Australia, July 12–17, 1998, 2.
420. Hill, D.J.T., Mahajerani, S., Pomery, P.J., and Whittaker, A.K., Proceedings of the IUPAC Sponsored 37th International Symposium on Macromolocules, Macro'98, Australia, July 12–17, 1998, 24.
421. Forsythe, J.S., Hill, D.J.T., Logothetis, A.L., Seguchi, T., and Whittaker, A.K., *Radiat. Phys. Chem.*, 53, 611, 1998.
422. Forsythe, J.S., Hill, D.J.T., Logothetis, A.L., Seguchi, T., and Whittaker, A.K., *Macromolecules*, 30, 8101, 1997.
423. Pacansky, J., Waltman, R.J., and Jebens, D., *Macromolecules*, 29, 7699, 1996.
424. Oshima, A., Seguchi, T., and Tabata, Y., Proceedings of the IUPAC Sponsored 37th International Symposium on Macromolocules, Macro'98, Australia, July 12–17, 1998, 56.
425. Banik, I., Dutta, S.K., Chaki, T.K., and Bhowmick, A.K., *Polymer*, 40, 447, 1998.
426. Banik, I., Bhowmick, A.K., Tikku, V.K., Majali, A.B., and Deshpande, R.S., *Radiat. Phys. Chem.*, 51, 195, 1998.
427. Banik, I. and Bhowmick, A.K., *J. Appl. Polym. Sci.*, 69, 2079, 1998.
428. Banik, I. and Bhowmick, A.K., *Radiat. Phys. Chem.*, 54, 135, 1999.
429. Banik, I., Chaki, T.K., Tikku, V.K., and Bhowmick, A.K., *Angew. Makromol. Chem.*, 263, 5, 1998.
430. Banik, I., Bhowmick, A.K., Raghavan, S.V., Majali, A.B., and Tikku, V.K., *Polym. Degrad. Stab.*, 63, 413, 1999.
431. Banik, I. and Bhowmick, A.K., *J. Appl. Polym. Sci.*, 76, 2016, 2000.
432. Electron beam-curable acrylate-modified colored polyurethanes. (Goodyear Tire and Rubber Co., USA). Jpn. Kokai Tokkyo Koho (1984), 4 pp. CODEN: JKXXAF JP 59066413 A 19840414 Showa. Patent written in Japanese. Application: JP 83-165284 19830909. Priority: US 82-419864 19820920. CAN 101:73632 AN 1984:473632 CAPLUS.
433. Assink, R.A., *J. Appl. Polym. Sci.*, 30, 2701, 1985.
434. Jonsen, B., *J. Biomater. Appl.*, 1, 502, 1987.
435. Bayer, G. and Steckenbieglen, B., *Gummi. Fasern. Kunstst.*, 44, 614, 1991.
436. Chem. Abstr., 102, 63800f, 1985.
437. Glaister, F.J., US Patent 4, 637, 955, 1987.
438. Tanaka, T. and Oe, T., Jap Kokai 90, 99, 526, 1990.
439. Shingyoehi, K., Kashiwazaki, S., Ando, Y., and Yamazaki, Y., Jap Kokai 85, 137, 952, 1985.

440. Kawashima, C., Ogasawara, S., Tanaka, I., and Koga, Y., Ger Patent 3, 524, 370, 1986.
441. Seki, I., Ando, Y., and Yagyu, H., Jap Kokai 85, 260, 636, 1985.
442. Samura, Y. and Kajiwara, K., Jap Kokai 86, 266, 439, 1986.
443. Haruvy, Y., Rajbenbach, L.A., and Jagur-Grodzinski, J., *Polymer*, 25, 1431, 1984.
444. Akhtar, S., De, P.P., and De, S.K., *J. Appl. Polym. Sci.*, 32, 4169, 1986.
445. Thomas, S., Gupta, B.R., De, S.K., and Thomas, K.T., *Radiat. Phys. Chem.*, 28, 283, 1986.
446. Chattopadhyay, S., Chaki, T.K., and Bhowmick, A.K., *Radiat. Phys. Chem.*, 59, 501, 2000.
447. Chattopadhyay, S., Chaki, T.K., and Bhowmick, A.K., *J. Appl. Polym. Sci.*, 79, 1877, 2001.
448. Chattopadhyay, S., Chaki, T.K., Khastgir, D., and Bhowmick, A.K., *Polym. Polym. Comp.*, 8, 345, 2005.
449. Das, P.K., Ambatkar, S.U., Sarma, K.S.S., Sabharwal, S., and Banerji, M.S., *Polym. Intl.*, 55, 118, 2006.
450. Mohammed, S.A.H. and Walker, J., *Rubber Chem. Technol.*, 59, 482, 1986.

Section VI

Tires

32 Tire Technology—Recent Advances and Future Trends

Arup K. Chandra

CONTENTS

32.1 INTRODUCTION

The tire can be considered today a matured and fully developed primary engineering component of surface transport vehicles. Born in 1888 in the hands of J.B. Dunlop, the tire has undergone a series of changes over the decades and graduated to a high technological product (Figure 32.1). Though the basic functions of a tire which include dry and wet grip, lower rolling resistance (RR), reliability, comfort, and aesthetics remain unchanged, the advances in the technology are continuing on many fronts to meet never-ending customer demands and to accommodate new applications.

In order to support and meet this demand an all-around development has taken place on the material front. Increasing automobile manufacturers' requirements and ever-growing customer expectations have resulted in the evolution of new product technology. As a consequence, run-flat tire, closed cellular polyurethane (PU) tire, tweel tire, and active wheel system have become a reality on the road today and indicate a big change in the years to come. The manufacturing technology

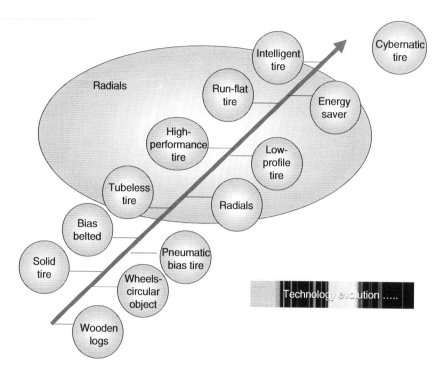

FIGURE 32.1 Trends in tire technology.

adopted by different companies differs slightly from company to company based on their technical competency; however, major players are trying to develop a modular manufacturing system.

The salient features of the recent development and future trends in all areas of tire technology, i.e., materials, technology, and manufacturing, will be covered in the following sections.

32.2 TRENDS IN MATERIALS DEVELOPMENT

In order to support and meet this demand, an all-around development has taken place on the material front too, be it an elastomer; new-generation nanofiller, surface-modified or plasma-treated filler; reinforcing materials like aramid, polyethylene naphthenate (PEN), and carbonfiber; nitrosoamine-free vulcanization and vulcanizing agents; antioxidants and antiozonents; series of post-vulcanization stabilizers; environment-friendly process oil, etc.

32.2.1 POLYMERS/ELASTOMERS

Today, world wide huge crunch of natural rubber (NR) and shooting up of NR prices are a major concern for all tire manufacturers. The worldwide shortage of NR is arising mainly due to production cut in Malaysia and the shift towards palm oil cultivation, growing usages of NR in radial tire, and increasing demand in China. Usage of more synthetic rubber and partial replacement of NR by synthetic polyisoprene is expected to rise. Even though NR is traded above US$2/kg, it is still the first choice for radial truck tire manufacturers because of its excellent physical and mechanical properties and better adhesion to steel cord.[1–4]

Styrene–butadiene rubber (SBR) among the available synthetic rubbers possesses the greatest balance of functional qualities in the widest range of applications. Today, solution SBR (S-SBR) is widely replacing emulsion SBR in passenger radial tire. S-SBR is prepared by solution polymerization using Ziegler–Natta catalyst. S-SBR has no organic fatty acid and has more control on its microstructure (tailor-made product). This results in better physical properties like lower RR and better wet grip which is the most stringent requirement for ultrahigh performance (UHP) and other tires.[5]

Butadiene rubber (BR) prepared using neodymium catalyst has a very high *cis* content, very high molecular weight, and narrow mole weight distribution. This polymer exhibits better physical properties and higher abrasion resistance but is difficult to process.[5] Recently, Bridgestone Corporation has announced the development of a synthesis technology that chemically bonds a high *cis*-butadiene rubber (*cis*-BR) with silica at its molecular chain end, resulting in a better-performing, more abrasion-resistant tire.[6] Bridgestone has generated two technologies that, when integrated, led to the development of the end-functionalized high *cis*-BR with reactivity to silica. This first creates a polymer chain of high *cis*-BR with an active site at chain end by redeveloping the polymerization catalyst, which converts monomer molecular chain to a polymer chain. The second relates to the chemical conversion of the chain-end active site to the silica-reactive functionality, by reaction with a carefully designed functionalizing agent. The company said that if the new compound, nonporous-tech (nanostructure-oriented properties control technology), is used in tire tread, it would improve flexibility in cold conditions yet retain stiffness for warm conditions.

32.2.2 Reinforcing Fillers

Carbon black is incorporated into the polymer to achieve several requirements.

The most stringent requirements are good tread wear, low RR, and superior traction. Out of these parameters, if one improves, the other will deteriorate.

These three interrelated properties are sometimes assigned to a so-called magic triangle (Figure 32.2). In recent years, the world's leading carbon black manufacturers have come with innovative grades to meet conflicting demands.[7]

In the recent past, many developments have taken place in reinforcing-filler technology. These developments include introduction and usage of improved grades of silica, developments of dual filler, nanotechnology for improved tire performance, introduction of biofillers such as corn starch in tire treads, ecofriendly tire (Goodyear), introduction of polymeric fillers and ground rubber as recycling materials and ecofriendly fillers.

Use of highly dispersible silica together with silanes and high vinyl S-SBR has met all critical magic triangle requirements like good dry and wet traction, reduction in tread wear, and low RR.

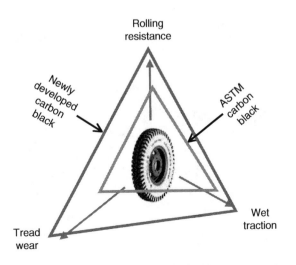

FIGURE 32.2 Magic triangle.

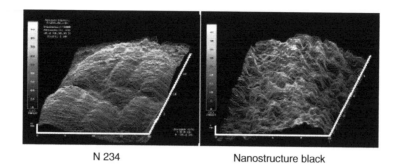

N 234   Nanostructure black

FIGURE 32.3 Surface topography of new-generation nanostructure black.

Highly dispersible, high-surface area silica offers nearly 40% reduction of RR and approximately 30% reduction of heat buildup.

Increasing price of crude oil has built up pressure on tire and automobile industry to develop low rolling-resistant tire with better traction. Combination of carbon and silica with coupling agent (dual filler technology) shows low RR with better traction and skid resistance in tire tread compound. Carbon black developed by plasma process and nanostructure black are other new significant developments in filler technology.

Carbon black–silica dual phase fillers reduce hysteresis while maintaining or improving abrasion. This system is less expensive as coupling agent requirement is less and produces semiconductive product compared to full silica–filler system. Dual phase filler is less abrasive to the processing equipment compared to the usage of silica filler alone, but use of dual phase filler increases cost of compound compared to traditional carbon black.

Carbon black by plasma process replaces incomplete combustion with direct splitting of the hydrocarbon feedstock oil using plasma into hydrogen and carbon black. It produces carbon black with predetermined characteristics. Nanostructure black is a family of new carbon blacks characterized by rough surface and enhanced filler–polymer interaction (Figures 32.3 and 32.4). It hinders the slippage of polymer molecule along the rough nanostructure surface and reduces the hysteresis significantly. This type of black ideally meets truck tire requirement as it provides improved tread wear in addition to low hysteresis.

Silica filler and clay are hydrophilic in nature while most of the polymers are hydrophobic in nature; this generates huge surface energy difference at the polymer–filler interface and results in macrophase separation. Coupling agent is found to be one of the alternative ways to improve reinforcement by silica; however, it requires very lengthy and multistage mixing technology and

Polymer–filler interaction of carbon black surface

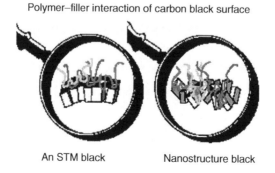

An STM black   Nanostructure black

FIGURE 32.4 Reinforcement by nanostructure black.

consumes energy. The efficiency of coupling agents is very low in the presence of other rubber chemicals. This is overcome by the recently developed functionalized filler. A functionalized filler is a new category of fillers in which surface chemistry of the filler is changed for better polymer–filler interaction. This is a suitable way to improve the degree of reinforcement and polymer–filler interaction of rubber by providing functionality to the filler surface, which is stable at normal mixing temperature but can react with the rubber matrix during vulcanization. This leads to grafting of filler particles on the rubber matrix with great improvement of physical properties. This can be achieved through chemical surface modification of carbon black.[7] The shape of the aggregate plays an important role in the distribution of filler in rubber matrix and controlled aggregate size distribution (ASD) also helps in processing.

32.2.3 NANOPARTICLES AND NANOCOMPOSITES

Development in nanocomposites is a significant development of the twenty-first century. Nowadays market value of nanocomposites is approximately US$75 million and is expected to soar to US$250 million by 2010. The rapid growth in this field will definitely change the material demand scenario for tire industry to a large extent in coming years. Polymer nanocomposites are organic–inorganic hybrid materials where the inorganic phases like nanosilica, nano-ZnO, and nano-clay are distributed in nanoscale within the organic polymer matrix. Increased surface area of nanofillers produces better interfacial polymer–nanofiller interaction. Polymer nanocomposites consist of a very high degree of reinforcement, high thermal stability, improved impermeability to gases, vapor and liquids, good optical clarity in comparison to conventionally filled composites, and flame retardancy with reduced smoke emission.[8–11]

Fillers with extremely high aspect ratios (1000–10,000) such as carbon nanotubes (Figure 32.5) have a much lower percolation threshold (lower amount is required for equivalent reinforcement).

Therefore, it might be possible to formulate a compound which both conducts and has all the other advantages of the "green" tire by the incorporation of relatively small amounts of carbon nanotubes into the silica-filled compound.

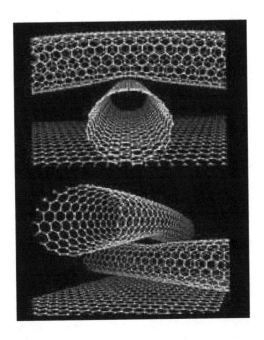

FIGURE 32.5 Carbon nanotube.

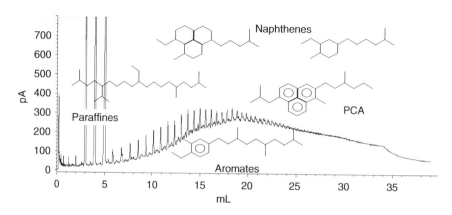

FIGURE 32.6 Constitutes of mineral oil.

32.2.4 PLASTICIERS/OILS

Mineral oils also known as extender oils comprise of a wide range of minimum 1000 different chemical components (Figure 32.6) and are used extensively for reduction of compound costs and improved processing behaviors.[12] They are also used as plastisizers for improved low temperature properties and improved rubber elasticity. Basically they are a mixture of aromatic, naphthanic, paraffinic, and polycyclic aromatic (PCA) materials. Mostly, 75% of extender oils are used in the tread, subtread, and shoulder; 10%–15% in the sidewall; approximately 5% in the inner liner; and less than 10% in the remaining parts for a typical PCR tire. In total, one passanger tire can contain up to 700 g of oil.

A lot of studies since 1970 have been done that proved the carcinogenic potential of PCAs. Concern about toxicity of highly aromatic oil led to the usage of nontoxic type of extender oil. These nontoxic oils are mostly Distillate Aromatic Extract (DAE), Treated Distillate Aromatic Extract (TDAE), and Mild Extraction Solvates (MESs). Existing extender oils can be replaced by Residual Aromatic Extract (RAE), Hydrogenated Naphthenic Oils (HNAP), and blends of highly immobile asphaltenic hydrocarbons. Low-molecular weight polymer, which acts as a process aid during mixing, processing, and gets cured along with the rubber and offers better performance, is a potential candidate that can replace the conventional process oil. Use of vegetable oils from different sources is an area where lot of developmental work is going on owing to toxicity and scarcity of mineral oil.

32.2.5 REINFORCING MATERIALS/FABRICS

There are several fibers currently used in carcass reinforcement. High modulus low shrinkage (HMLS) polyester is typically used in most radial passenger tires and offers a good mixture of properties. Nylon is used extensively in bias ply tires where strength is of paramount importance. Rayon is the reinforcement cord of choice in Europe for high-performance tires, due to its ability to maintain mechanical properties at high temperature. Aramid fiber offers UHP properties but at a high cost. This cost limits usage of aramid for tires to high-end luxury and race cars.

The performance properties of PEN present opportunities for replacement of rayon or polyamide in carcass construction. The use of PEN cord in these applications is currently being evaluated in both Asia and Europe. PEN has demonstrated acceptable flexural fatigue equivalent to polyethylene terephthalate (PET) and rayon. It has equivalent toughness to rayon, which is important for sidewall impact resistance. PEN's superior mechanical properties also afford opportunities to use less fiber in carcass construction enabling production of lighter-weight, more fuel-efficient tires.

TABLE 32.1
Physical Properties of High Strength Industrial Yarns Used in Reinforcement

Polymer	PEN	PET	Nylon	Rayon	Aramid
Density	1.36	1.39	1.14	1.52	1.44
Tenacity, g/d	10	9	9	5	23
Tensile Modulus, g/d	360	110	50	125	560
Tensile Elongation, %	8	14	15	12	4
Boiling Water Shrinkage, %	1	5	8	Decomposes	<0.1
Dry Heat Shrinkage (1 min at 177°C)	4	8	10	6	<0.1

In radial tires, the carcass provides reinforcement along the radial direction. To provide reinforcement in the rolling direction, belts are necessary.[13] Steel is the traditional material of choice but there has been a drive to replace it with lighter-weight synthetic cords. Aramids offer a synthetic alternative but their cost has been prohibited. PEN has good fatigue at low twist and superior compressive modulus to aramid. This can be translated into a significant reduction in tire weight and improvement in fuel saving. In addition, use of a polyester fiber like PEN offers greater opportunity to recycle used tires.

The cap ply is wound over the shoulders or the entire body of the belt to provide a compressive force, which resists the centrifugal forces created in the belt at high speed. Nylon 6,6 has been typically used for cap ply applications due to its high retractive force at elevated temperatures. PEN cord has been demonstrated to offer similar properties to nylon at high temperature and is an excellent replacement material. Honeywell (formerly Allied Signal) has commercialized high-tenacity/high-modulus PEN fiber under the trade name PENTEX. In January 1999, the Italian tire company, Pirelli, launched a new tire reinforced with cap made from PENTEX in its radial motorcycle tires. PEN's significantly higher modulus versus nylon or rayon (Table 32.1) gives it the dimensional stability required for these high-performance uses.

Bridgestone Corporation, the largest tire company in Japan and in the world, is also using PEN cord replacing nylon cap in its UHP tire. In addition to the dimensional stability, the rigidity that PEN brings to the tire prevents noise generated by the road surface transmitted to the car. Tests show that these PEN-containing tires reduce road noise by as much as 30%, making the tires ideally suited for high-end luxury cars, such as Lexus.

32.2.6 POST-VULCANIZATION STABILIZER

Reversion characteristics of NR are of great concern. Lot of novel chemicals have been introduced to increase the reversion resistance of NR. Examples of these are zinc soap activator (Structol-A73), silane coupling agent (Si-69), anti-reversion agent (Perkalink 900), and post-vulcanization stabilizer (Duralink HTS and Vulcuren KA 9188; Figure 32.7). These materials will enhance the life of the tire, enable the users for more retreading, and thereby reduce the material demand.[14–16]

32.3 PRODUCT

With all-around development, expectations of tire customers have also grown. Now, customers are more demanding and are looking for better mileage, lower heat buildup, better ride and handling (dry and wet traction), and an environment-friendly (i.e., low RR, reduced noise, more durable, and less polluting) product. Keeping customers' expectations in mind, major tire manufacturers have also introduced several innovative products. Development of super single tire, solid tire (closed-cellular PU tire), run-flat technology, multi-air chamber tire, active wheel system, and tweel tire are among the major pathbreaking achievements in the recent past of the tire industry.

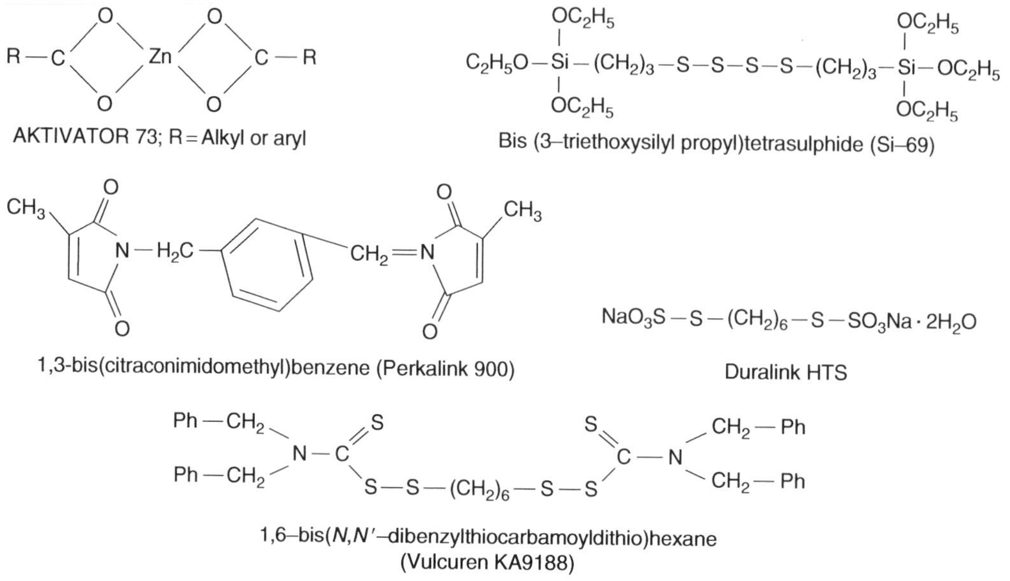

R—C(O)(O)Zn(O)(O)C—R

AKTIVATOR 73; R = Alkyl or aryl

$C_2H_5O-\underset{\underset{OC_2H_5}{|}}{\overset{\overset{OC_2H_5}{|}}{Si}}-(CH_2)_3-S-S-S-S-(CH_2)_3-\underset{\underset{OC_2H_5}{|}}{\overset{\overset{OC_2H_5}{|}}{Si}}-OC_2H_5$

Bis (3–triethoxysilyl propyl)tetrasulphide (Si–69)

1,3-bis(citraconimidomethyl)benzene (Perkalink 900)

$NaO_3S-S-(CH_2)_6-S-SO_3Na \cdot 2H_2O$

Duralink HTS

1,6–bis(N,N′–dibenzylthiocarbamoyldithio)hexane
(Vulcuren KA9188)

FIGURE 32.7 Structure of post-vulcanization stabilizers.

32.3.1 SUPER SINGLE TIRE

Wide-base super single radial truck tire replaces dual tire, increases freight volume by decreasing floor level, decreases total tire weight and RR, reduces tire waste, and reduces materials (Figure 32.8). This is made possible by replacement of twin tire by low-aspect ratio, wide-section single tire in multi-axle present-generation low-platform trailers. This significantly enhances safety, speed, and RR. Lower weight of the tire wheel assembly, compared to that of dual tire is expected to reduce fuel consumption by 3%–5%.

FIGURE 32.8 Super single tire.

32.3.2 POLYURETHANE TIRE

Microcellular tire will be one of the biggest revolutions in the field of tire technology. PU tire made of microcellular PU foam consists of hundreds of thousands of microscopic air cells, both open and closed, trapped in a dense and very tough PU rubber matrix. The result is a puncture-proof PU tire. Properties of PU rubber can be easily changed from rigid to flexible; hence, it becomes an excellent alternative to the traditional pneumatic tire. In PU tires, each microcellular cell acts as an independent cushioning agent, which will not cause flat to the whole tire. In addition, these microscopic air cells will regain their original size, shape, and form once the sharp object is removed from the tire, returning the rider to the original comfort and grip of the journey. Solid PU tire can have weight equal to or less than a pneumatic tire and tube, strength, and durability, with wear resistance exceeding that of pneumatic rubber tires by up to 300% or a life span of up to four times that of the normal rubber tire. A wide range of PU tires can be developed with pressures ranging from 30 to 175 psi, providing a riding quality and firmness normally associated with rubber pneumatic tires, can fully be molded with a thick self-skin, and fitted on standard rims with inexpensive mounting tools. Competitive pricing makes these tires affordable compared to regular rubber tires and thorn-prone tubes. In tests performed by an independent laboratory, tire performance was shown to be equal to or better than pneumatic tires. Derek Crofton (huntsman polyurethane, United States) has already used microcellular PU tires as bicycle tires. As per recently published Smithers report,[17] Amrityre Corp. had "positive results" from both independent laboratory testing and field-testing of its proposed PU tread stock for tire retreading. Microcellular PU tires are recyclable, with worn-out material being used for such applications as doormats, truck bed liners, racetracks, and nonpneumatic tires.

32.3.3 RUN-FLAT TIRE

The spare tire is a heavy and in many cases, unnecessary weight to be carried around in the vehicle. Tires are becoming more reliable and durable, but punctures can still occur. Companies have been studying techniques for eliminating the spare tire for many years. Most of the major tire companies now have products that will run for a defined safe distance and speed when the air is lost. This run-flat tire and special rim combination enables vehicle mobility in the event of a puncture or air loss to transit a significant distance at a reasonable speed before the problem is attended. Self-supporting tires with heavy sidewall inserts have to be formulated with new recipes that are stiff enough to support the load, but also resistant to the excessive temperatures that are generated as the tire deflects. Ultrahigh tensile steel cords replace or supplement the conventional polyester or rayon carcass of run-flat car tires. The downside of this insert philosophy, which is not required for a very high percentage of the tire's life and in most cases never at all, is that it is a heavy addition detracting from weight, RR, and ride comfort (Figure 32.9).

32.3.3.1 Tire Pressure Monitoring System for Run-Flat Tires

The market is still reluctant to adopt these run-flat tires until there is an integral system that warns the driver of pressure loss. Many new "smart" developments are hitting the markets that monitor tire temperature and pressure. This type of monitoring is also beneficial to fleet operators who need to measure and maintain tire-operating pressure to minimize tire wear and fuel consumption.

32.3.4 MULTI-AIR CHAMBER TIRE

Bridgestone Corporation has announced in its news release in February 2006 the development of a multi-air chamber tire, which has three separate chambers acting as pressure vessels.[18]

The all-new type of tire contour was born from extensive R&D efforts with a view to the future of tires. The inside of the tire consists of a main chamber in the center and subchambers on both

FIGURE 32.9 Cross-section of a typical run-flat tire—Michelin PAX system.

sides. The air pressure in each of the chambers can be controlled independently, making it possible to adjust the stiffness balance between the tread area and the left and right sidewalls. This ensures ideal contact in any condition, and thus a more stable and comfortable drive. It also means performance can be adjusted in line with customer needs such as comfort or maneuverability.

As the tire is divided into three parts, even if the air pressure in one of the chambers drops to 0 KPa caused by a nail puncture, for instance, the remaining chambers will be able to support the weight on the tire, so that it will be able to continue to operate for a specified distance (Figure 32.10).

32.3.5 ACTIVE WHEEL SYSTEM

In active wheel system, active wheel is equipped with wheel, tire, electric motors, and other electromechanical components and connections all packed within the circumferences of the wheel and tire. Electric motor not only runs the wheels but also slows and stops them; as a result, traditional disk or drum brakes might eventually be eliminated. By using electric motors to turn the wheels, large heavy transmission and differential become obsolete (Figure 32.11).

32.3.6 TWEEL: THE FUTURE TIRE

Michelin's new innovation tweel is a single unit consisting of four pieces: hub, PU spokes, a shear band surrounding the spokes, and tread band. Tweel's hub functions as it would in a normal wheel.

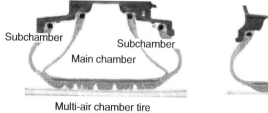
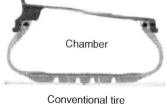

Multi-air chamber tire   Conventional tire

FIGURE 32.10 Multi-air chamber tire.

FIGURE 32.11 Active wheel system.

A rigid attachment connects the hub to the axle. PU spokes are flexible to help absorb road impacts. The shear band surrounding the spokes effectively takes the place of the air pressure, distributing the load. The tread is similar in appearance to a conventional tire[19] (Figures 32.12 and 32.13). Its high lateral stiffness improves cornering. Tweel can be engineered to give five times the lateral stiffness of a pneumatic tire without any loss in ride comfort. It is impervious to nails on the road. The tread will last two to three times as long as today's radial tires and can be retreaded again.

The heart of tweel innovation is its deceptively simple looking hub and spoke design that replaces the need for air pressure while delivering performance previously only available from pneumatic tire. The flexible spokes are fused with a flexible wheel that deforms to absorb shock and rebound with unimaginable ease. Without the air needed by conventional tires, tweel still delivers

FIGURE 32.12 Michelin tweel tire.

FIGURE 32.13 Car on tweel.

pneumatic-like performance in weight-carrying capacity, ride comfort, and the ability to envelope road hazards. In tweel, the vertical stiffness (which primarily affects ride comfort) and lateral stiffness (which affects handling and cornering) can be optimized independently of each other, enabling new applications not possible for current pneumatic tires. Tweel is in line with a long-term vision and a step towards innovations.

32.4 MANUFACTURING TECHNOLOGY

The dream of every manufacturer is to create a fully integrated factory where raw materials enter through one door and with minimal manhandling or supplementary processing finished high-quality goods leave the other end. Some industries have achieved pretty close to this ultimate dream factory concept; however, tire industry still has some way to go. However, major players are trying to develop modular manufacturing systems, e.g., C3M (Carcasse, Monofill, Moulage at Mechanique) by Michelin, MIRS (Modular Integrated Robotized System) by Pirelli, BIRD (Bridgestone Innovative Rational Development) by Bridgestone, or IMPACT (Integrated Manufacturing Precision Assembly Cellular Tech) by Goodyear.

In a modular-type factory, processes are being combined where rubber and other ingredients are put into a mixer at the start of the process and they come out as a nice tire combined with other processes at the end. Bridgestone has earmarked several million US dollars for investment in the BIRD—promotion line at Hikone. A huge bonus of this modular system is its ability to respond to flexibility of demand for tires of any size and design. The system can produce multiple sizes of tires simultaneously, which allows for making tires in smaller lots and will therefore reduce inventories. In addition, this modular system is easy to deploy globally, while its automated efficiency, coupled with fewer operations will help sustain the viability of tire making in the huge wage environments of industrial nations. Moreover, small factories with a modular system can produce 200,000–350,000 passenger car tires a year, yet occupying only a quarter to one-third of the space taken by a conventional production layout and reduced environmental impact including conservation of energy (consumed 40% less energy than conventional system).[20] Today the cost of modular equipment is high. In the long run it will reduce to the extent that even the reader can dream to be an employer rather than an employee.

32.5 FUTURE CHALLENGES IN MATERIAL REQUIREMENT

Performance requirements, environmental issues, and availability/cost of the material will mainly drive material requirement in the future. In order to face the huge tire wastage problem causing major hazards to the environment, future development in rubbery materials will be focused on development of thermoplastic polymer so that used polymer could be recovered by thermal treatment and separation, biological degradation by radiation/addition of chemical into the rubber compound that could be activated by exposure to radiation and development of biopolymer.

S-SBR with star-branched structure, alpha methyl styrene grades of SBR, and high styrene ESBR will replace conventional SBR grades. Neodium catalyst-based BR in spite of its processing difficulties but owing to better performance properties will gain foothold in the industry. Epoxized NR having much higher glass transition temperature (T_g) than NR and close to S-SBR can be the future choice for better wet grip. Synthetic IR may be a substitute for NR. PU usage in solid tires, usage of ethyleme–propylene–diene monomer (EPDM) with improved filler intake/odor-related issues ("Insite" Technology with metallocene catalysts), and increased popularity of bromobutyl in place of chloro are other major developments in elastomers to meet growing performance expectations of the customer.

In reinforcing materials double-dipped polyesters for improved tire durability, plasma-treated yarns for improved bonding in tire, and increased usage of aramid fabric as belt and application of PEN are the areas where manufacturers are showing interest. Introduction of new styles of steel wire geometry for improved rubber to metal adhesion and new steel wire coating formulations for improved rubber to metal bonding are other focused areas of development.

A major breakthrough in reinforcing materials can come with the development of nanofibers, which can be aligned, in a polymer matrix and give high strength to the polymer. It can reduce capital equipment requirement and recycling problem. Successful development of a molecular architecture where crystallization of planes within the polymer structure would give the reinforcement to a flexible amorphous phase can produce tire without fabric reinforcement.[21–23]

32.6 TECHNOLOGICAL CHALLENGE

New expressways and highways in developing countries demand more speed capability, low cost, high mileage, and energy-efficient tires. With the rapid growth of mining and infrastructure industry all over the world there is a high demand for off-the-road (OTR) tires. New tire technology has to face the demand arising out of changes in modern vehicles with faster speed, safety regulations, mechanization in agriculture and construction work, continuous crude price increase, and aggressive competition.

Technology needs to be geared up for growing radialization, which would catch up even 90%–95% in developing countries. H- and V-rated tires would emerge as brand builders; ride and handling, noise level, RR, etc. as unique selling propositions (USP); and wet skid a mandatory requirement for high-performance passenger car tires. If LT and truck radials in developing countries are to grow, mileage coupled with high performance would be the future challenge in this category.

Bias segment is to face challenges imposed by radials. This segment would demand more mileage with high load-carrying capacity at higher speed. Fuel cost would mount and cause pressure on overall cost; thus high mileage, fuel efficiency, and better retreadability will be the USP. Mechanization would increase the demand of farm and OTR tire. Emergence of high horse power (HP) tractors leading to larger-size radial tire with a demand to improve mileage and cut/stubble resistance would be the key challenges.

32.7 FUTURE DEVELOPMENTS

Lower-aspect ratio for automobile tires, higher load-carrying capacity, and reduced size for truck tires would be key sectors for development. Farm and off-road tire will face demand for higher speeds, more ride comfort, and enhanced traction. Development of advanced higher-strength reinforcement will lead to lower weight and reduced thickness for the tire.

Advancement in tire technology is expected to continue, to take advantage of emerging technologies, to meet further customer demands, and to accommodate new applications. Different tires will be introduced in the front and rear positions of automobiles as a result of the different loadings. "Unsymmetrical" tread designs will have different patterns on the outer and inner halves of the tread, thus accommodating the different contact forces on the two halves of the footprint, especially during cornering.

Reduction in tire development time and improvement in quality continue to advance by use of computers. Customer demands for enhanced performance will lead to further customizing of tire designs for particular applications. Different vehicles may require different tires, fine-tuned to the specific features of each vehicle and its suspensions. Designing "zoned" treads with different materials used in the shoulders and in the crown area to optimize the overall performance or zone tread designs with a relatively wide and deep circumferential groove in the middle of the tread to facilitate water removal and prevent aquaplaning is also being researched.[24] The development of tire that is able to run some distance after air loss will continue to be persuaded to eliminate the requirement of spare tire. Pressure-warning devices, which provide a safety alarm and may give the driver enough warning to be able to drive to the nearest aid station, are another research area.

32.8   CONCLUSION

The pneumatic tire, which already performs remarkable functions at a very modest cost, will continue to develop and further enhance its values to the customer.

This fascinating product will still continue to develop to accommodate new applications, safety, health and environment (SHE) issues, advantages of novel materials like nano-composites, plasma-surface-modified carbon black, development of computer simulation techniques, and finally to develop a cybernetic or thinking tire.

ACKNOWLEDGMENTS

The author would like to express his sincere thanks to the management of Apollo Tires Ltd. for kind permission to publish this article and Mr. Vivek Bhandari for his sincere help in compilation of this work.

REFERENCES

1. M.P. Wagner, Automobile Engineering Meeting, Society of Automotive Engineers, Detroit, MI, 14–18 May, 1973.
2. Arup K. Chandra, A. Biswas, R. Mukhopadhyay, B.R. Gupta, and Anil K. Bhowmick, *Plast. Rubber Compo. Pro. Appl.*, 22, 249, 1994.
3. Arup K. Chandra, A. Biswas, R. Mukhopadhyay, and Anil K. Bhowmick, *J. Adhesion*, 44, 177, 1994.
4. N.M. Mathew, in *Rubber Technologist Handbook*, J.R. White and S.K. De, Eds., Rapra Technology Limited, U.K., 2003.
5. D.C. Blackley, *Synthetic Rubbers: Their Chemistry and Technology*, Applied Science Publishers, London, 1983.
6. ICIS News, Houston, 21 July, 2006.
7. W.M. Hess and C.R. Herd, in *Carbon Black*, J.B. Donnet, R.C. Bansal and M.J. Wang, Eds., Marcel Dekker, New York, 1993.
8. A. Okada and A. Usuki, *Mater Sci. Eng.*, C3, 109, 1995.
9. A. Bandyopadhyay, A.K. Bhowmick, and M. De Sarkar, *J. Appl. Polym. Sci.*, 93, 2579, 2004.
10. J.W. Gilman, *Appl. Clay Sci.*, 15, 31, 1999.
11. P.B. Messersmith and E.P. Giannelis, *J. Polym Sci.*, 33, 1047 1995.
12. R.N. Datta and F.A. Ingham, in *Rubber Technologist Handbook*, J.R. White and S.K. De, Eds., Rapra Technology Limited, U.K., 2003.
13. B. Rodgers and W. Waddell, in *The Science and Technology of Rubber*, J.E. Mark, B. Erman, and F.R. Eirich, Eds., Elsevier academic Press, 2005.
14. N.R. Kumar, A.K. Chandra, and R. Mukhopadhyay, *J. Mat. Sci.*, 32, 3717, 1997.
15. P.P. Chattaraj, A.K. Chandra, R. Mukhopadhya, and Th. Abraham, *Kaut. Gummi Kuns*, 44(6), 555, 1991.

16. R.N. Dutta, A.M. Schotman, P.J.C. Van Haeren, A.J.M. Weber, and F.H. Van Wijk, *Rubber Chem. Technol.*, 69, 727, 1996.
17. The Smithers Report, Smithers Scientific Services Inc., Akron, OH, January, 2006.
18. News Release, Bridgestone, Tokyo, 28 February, 2006.
19. News Clippings, Motoring Channel Staff, Web Wombat, 11 February, 2005.
20. Dream Machines, Tire Technology International, 27 December, 2004.
21. Arup K. Chandra, New Technology Development and Implication for Rubber Demand, Paper presented at Rubber & Tire Asia/China Markets 2006—Conference, Shanghai, China, 28 February–1 March, 2006.
22. Arup K. Chandra, Tyre Technology—Recent Advances and Future Trends, (Paper no. 40) presented at the Fall 170th Technical Meeting of the Rubber Division, American Chemical Society, Cincinnati, OH, 10–12 October, 2006.
23. Vivek Bhandari and Arup K. Chandra, Trends in Tyre Materials, Manufacturing and Technology, Paper presented at "INDIA RUBBER EXPO—2007, Chennai, India, 17–20 January. 2007.
24. R.A. Ridha and W.W. Curtiss, in *Rubber Products Manufacturing Technology*, A.K. Bhowmick, M.M. Hall, and H.A. Benarey, Eds., Marcel Dekker, New York, 1994.

33 Recent Developments in Fillers for Tire Applications

Meng-Jiao Wang and Michael D. Morris

CONTENTS

33.1 INTRODUCTION

Since the invention of the pneumatic tire, fillers have played an important role in tire performance, alongside that of the polymer. The introduction of carbon black as a reinforcing filler early in the last century was possibly the most important single advance ever in tire technology. It provided about a 10-fold increase in service life compared to what was used previously. Since then, carbon black has remained the predominant reinforcing material used in tires, as well as in other rubber products, due to its outstanding reinforcing ability in a range of polymers, and relatively low cost. In the intervening years, great advances have been made in the production process, quality, and grades of carbon black available. Since the early 1990s, precipitated silica has found a significant and increasing niche as a tire filler material, particularly in tread compounds for passenger car tires, because of the wet grip and low rolling resistance it can provide.

The tire industry is by far the largest consumer for carbon black, so it should not be surprising that a major focus for research and development at carbon black producers has been understanding and improving the performance of carbon black and other fillers in tires. This chapter will describe and discuss some recent advances in the understanding and improvement of filler reinforcement of elastomers, with tire performance as the main focus.

It is well known that tires are continually being demanded to provide increased durability, better fuel economy, and improved grip, particularly in wet conditions. This triad of performance requirements can be translated to three measurable tire properties, namely, wear resistance, rolling resistance, and skid resistance, especially wet skid. In addition to these three, heat buildup is a major concern, particularly for heavier-vehicle tires. Optimizing tire performance in only one of these areas is almost always insufficient. The challenge is to provide the best balance of performance in all three dimensions, and the filler, or reinforcing agent, can play a major role in achieving that goal. Before considering the impact of fillers on the key performances of tires, it is necessary to consider filler dispersion and dispersibility in elastomers, as this is critical in achieving the full performance of any filler.

33.2 DISPERSION OF CARBON BLACK IN ELASTOMERS

33.2.1 Effect of Morphology on Dispersibility of Carbon Black

During mixing of carbon black pellets with polymer, distributive and dispersive mixing processes control the dispersion quality, after incorporation has taken place. There are various factors that influence the dispersion process, such as the mixing equipment, compound formulation, mixing sequence, and mixing conditions. It has been shown theoretically that a critical stress is required to separate an aggregate from an agglomerate, and this stress increases with decreasing particle size. Alternatively, the smallest radius to which agglomerates can be dispersed under given shear condition can be expressed by:

$$R_{\text{crit}} \geq \frac{1}{\sin \psi} \sqrt{\frac{2\,Hv}{5\pi\eta\dot{\gamma}}} \tag{33.1}$$

where

ψ is the angle which is describes the size of the aggregate relative to the agglomerate (Figure 33.1)
v is the average connection number with which an aggregate binds to the agglomerate
H is the mean force of a connection
η and $\dot{\gamma}$ are the medium viscosity and shear rate[1]

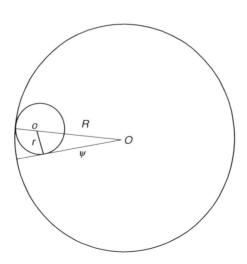

FIGURE 33.1 Schematic representation of equivalent agglomerate and aggregate.

Clearly, the smaller the aggregate, the larger will be R_{crit}. It has been shown that higher surface area carbon blacks not only have smaller primary particles, but also smaller aggregates than lower-surface area counterparts. High-area blacks should therefore give poorer dispersion than lower-surface area blacks under the same dispersion conditions.

For carbon blacks with the same size primary particles, i.e., the same surface area, high-structure blacks have more void volume, as measured by higher dibutylphthalate (DBP) absorption, and larger aggregate size. The higher void volume leads to a reduced number of connections between aggregates within an agglomerate, v, and hence lower critical stress for dispersion. Thus, high-structure blacks should be easier to disperse, or give better dispersion, than lower-structure blacks of the same surface area. In addition, the effect of structure on the dispersibility is also related to the occlusion of polymer by carbon black aggregates, which was described by Medalia in 1970.[2] When structured carbon blacks are dispersed in rubber, the rubber portion filling the internal void of the carbon black aggregates is unable to participate fully in the macrodeformation of the filled system. The partial immobilization of the occluded rubber makes it behave like filler rather than like polymer matrix. Due to this phenomenon, the effective volume of the filler, with regard to the stress–strain behavior of the filled system, is increased considerably. The viscosity of the filled matrix is then given by a modified Guth–Gold equation:

$$\eta = \eta_0 \left(1 + 2.5\phi_{eff} + 14.1\phi_{eff}^2\right) \tag{33.2}$$

where
η_o is the viscosity of the unfilled polymer
ϕ_{eff} is the effective volume fraction of filler

By relating the endpoint of crushed DBP absorption to the void space within and between equivalent spheres of aggregates, and assuming the spheres to be packed at random, Wang et al.[3] obtained the following equation for the effective volume fraction of carbon black:

$$\phi_{eff} = \phi \frac{0.0181 \cdot CDBPA + 1}{1.59} \tag{33.3}$$

where CDBPA is the compressed DBP absorption number. The higher viscosity provided by higher structure black leads to greater shear stress and therefore better dispersion than from an equivalent low-structure material.

33.2.2 Effect of Surface Energy on Dispersibility of Carbon Black

It is known that the possible interaction between two materials 1 and 2 is determined by their surface energies, which consist of two components, dispersive, γ^d and specific or polar, γ^p. When hydrogen bonding and acid–base interactions are also involved, the adhesion energy between the two materials, W_a will be[4]

$$W_a = 2(\gamma_1^d \gamma_2^d)^{1/2} + 2(\gamma_1^p \gamma_2^p)^{1/2} + W_a^h + W_a^{ab} \tag{33.4}$$

where
γ_1^d and γ_2^d are the dispersive components of surface energy of materials 1 and 2
γ_1^p and γ_2^p are the polar components
W_a^h and W_a^{ab} are the adhesion energies due to hydrogen bonding and acid–base interaction between these two materials

For filler agglomerates, the cohesive energy between aggregates, W_c, can, therefore, be estimated by

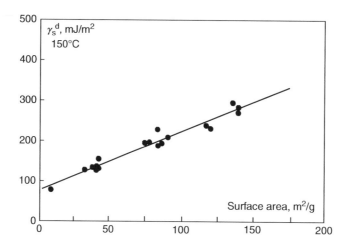

FIGURE 33.2 Dispersive component of carbon black surface energy as a function of its surface area.

$$W_c = 2\gamma_f^d + 2\gamma_f^p + W_c^h + W_c^{ab} \tag{33.5}$$

where

γ_f^d and γ_f^p are the dispersive and polar components of the filler surface energy

W_c^h and W_c^{ab} are the cohesive energies from hydrogen bonding and acid–base interactions

The surface energies of various carbon blacks have been measured by inverse gas chromatography (IGC).[5] It can be seen that the dispersive component of the surface energy of a series of carbon blacks increases with increasing surface area (Figure 33.2). Polar components of surface energy show a similar upward trend with increasing surface area.[5] The implication of these results is that based on the mean force of connection, which is related to surface energy, higher surface area blacks will be harder to disperse.

We have already shown that higher-surface area blacks are expected to be more difficult to disperse based on physical effects alone, so demonstrating the effect of surface energy of the filler on dispersibility has not been easy. One approach is to use simple heat treatment of blacks to modify the surface, without changing the morphology. When carbon black is heated in an inert atmosphere (nitrogen in this case), the γ_s^d of the carbon black increases with increasing temperature, up to 900°C, which is far below the graphitization temperature of about 1500°C (Figure 33.3). When these treated blacks (N234) are compounded into emulsion SBR at 50 phr loading, the dispersion of carbon black gets worse with increasing treatment temperature.[1] Since, at these temperatures the morphologies of the carbon black cannot be changed, the variation of dispersion of the black upon heat treatment can only be interpreted in terms of their surface characteristics.

The effects of carbon black morphology on dispersibility described above have been borne out by practical experience. Higher surface area and lower-structure blacks are known to be more difficult to disperse. Traditionally, carbon blacks with surface areas higher than 160 m²/g and CDBP lower than 60 mL/100 g cannot be sufficiently well dispersed using normal dry-mixing equipment, so they are not considered rubber grades. Figure 33.4 shows the ASTM carbon black spectrum used in the rubber industry, expressed by compressed DBPA versus surface area.

33.2.3 New Technology for Carbon Black Dispersion in Rubber—Continuous Liquid Phase Mixing

An enormous amount of work has been reported on the issue of carbon black dispersion improvement. One of the approaches has been to produce carbon black–polymer masterbatches by mixing

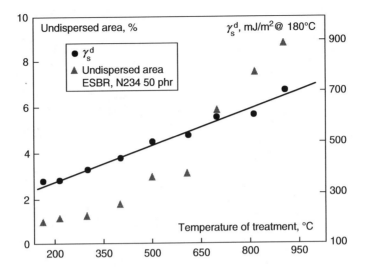

FIGURE 33.3 Dispersive component of surface energy and dispersion quality in ESBR as a function of heat treatment of N234.

polymer latex with filler slurry and then coagulating the mixture chemically. With this process, the filler dispersion can generally be improved relative to dry mixing. However, this process is relatively slow and has low productivity. In addition, polymer latexes contain nonrubber substances, which can be adsorbed on the carbon black surface, interfering with polymer–filler interaction. A newer masterbatch technology that involves rapid mechanical coagulation, has overcome most of these shortcomings.[6] Cabot Elastomer Composites (CECs) are produced by mechanically dispersing carbon black in water, then injecting this slurry into a mixer at very high speed, where it is mixed with a latex stream. Under highly energetic and turbulent conditions, the mixing and coagulation of

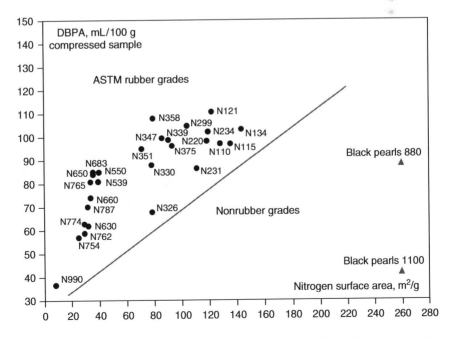

FIGURE 33.4 Surface area versus structure for rubber grades of carbon black with some nonrubber grades.

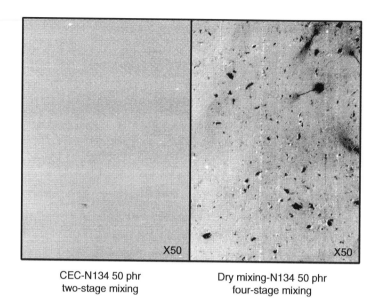

CEC-N134 50 phr
two-stage mixing

Dry mixing-N134 50 phr
four-stage mixing

FIGURE 33.5 Dispersion of carbon black N134 in Cabot Elastomer Composite (CEC) and dry-mixed compounds.

the polymer with filler is completed mechanically at room temperature in less than 0.1 s, without the aid of chemicals. The masterbatch is dewatered and dried using established processes.

The macrodispersion measured by means of optical microscopy for two carbon black–N134-filled vulcanizates, one prepared with CEC and the other using dry mixing, is shown in Figure 33.5. A two-stage mixing process was used for adding small chemicals and curatives to CEC, whereas a four-stage mixing process was used for dry mixing. Clearly, the dispersion of the filler in CEC is considerably better than that of the dry-mixed compound, even though energy input is much lower. In fact, it was shown by transmission electron microscopy that carbon black dispersion and distribution are fully completed at the very early stages of the CEC process. Since only minor mechanical energy had been input at that stage, it shows that good filler dispersion can be achieved using CEC technology without significant mechanical breakdown of the polymer molecules. For the high-surface area- and/or low-structure carbon blacks that have poor dispersibility, CEC is particularly advantageous, especially at high filler loading.[7]

Significantly, the dispersion of carbon black in the compound is independent of its morphology when CEC technology is used. Thus, the grades of carbon black that can be used for rubber reinforcement are greatly expanded to the high-surface area and low-structure blacks.[8] For example, over a normal range of loading, Black Pearls1100 (BP1100) with surface area of 260 m^2/g and CDBP of 42 mL/100 g (see Figure 33.4) has been dispersed in natural rubber (NR) with excellent quality. With this material, the vulcanizates possess not only significantly higher ultimate properties, but also higher tear strength than has ever been seen with conventional compounds (Table 33.1). More interestingly, the trade-off among tensile strength, elongation at break, and hardness can be improved, as shown in Figure 33.6.

33.3 IMPROVEMENTS IN ROLLING RESISTANCE AND HEAT BUILDUP

Both rolling resistance and heat buildup are related to hysteresis; that is the amount of energy that is converted to heat during cyclic deformation. It is well known that hysteresis of tread compounds, characterized by the loss factor, tan δ, at high temperature, is a key parameter. It not only governs heat buildup of the compounds under dynamic strain but also shows a good correlation with the

TABLE 33.1
Tear Strength of BP1100–CEC and Dry-Mixed N234-Filled Vulcanizates

| Process | Carbon Black | | | | Tear Strength | | |
	Type	Loading (phr)	Oil (phr)		Die C (kN/m)	Die B (kN/m)	Trouser (Relative) (%)
Dry mixing	N234	50	5		115	164	100
CEC	BP1100	80	0		150	220	290
CEC	BP1100	95	0		170	204	275
CEC	BP1100	80	7.5		140	236	365

rolling resistance of tires. The hysteresis is in turn determined by filler networking in the polymer matrix. Filler networking, which can also be viewed as microdispersion, is a function of both the degree or quality of initial dispersion, and the degree to which the filler flocculates in the rubber compound. Thus, the mean distance between aggregates, and the attractive force between aggregates are the key parameters controlling filler networking.[4] Improvement of filler dispersion, increase in the inter-aggregate distance, δ_{agg}, reduction in filler–filler interactions, and increase in polymer–filler interaction all favor depressing the filler networking, resulting in lower hysteresis.[4]

33.3.1 CARBON BLACK–NATURAL RUBBER MASTERBATCHES

The masterbatch technology used to make CEC has been shown in Section 33.2 to provide excellent carbon black dispersion for all grades of carbon black. It would be expected therefore, that compounds made from CEC should show lower hysteresis, higher resilience, and lower heat buildup than compounds made from the same formulation and same carbon black, by dry mixing. Furthermore, the hysteresis advantage of CEC over dry mix should increase as the surface area of the black increases, or its structure decreases. Laboratory test results and commercial trials have shown this to be true. For example, Figure 33.7 shows a comparison of tan δ_{max} at 60°C between CEC and its dry-mixed counterpart for N134 carbon black. With no oil in the compounds, the CEC compound shows lower hysteresis. As the loading of both filler and oil is increased (to keep the hardness constant), the gap between the two types of mixing process becomes larger, i.e., the advantage of CEC is even greater at higher black and oil loadings. The advantageous hysteresis performance of CEC is attributed mainly to the excellent microdispersion that is achieved. As oil is added to a conventional compound, the level of polymer–filler interaction is reduced somewhat by adsorption of oil on the carbon black surface, enabling more filler flocculation during processing and

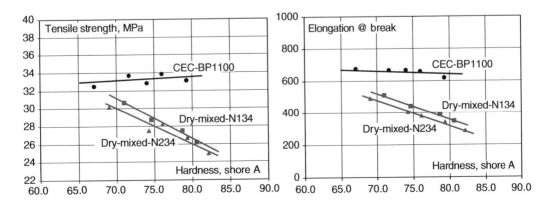

FIGURE 33.6 Improvement of the trade-off between tensile strength, elongation at break, and hardness with BP1100-filled Cabot Elastomer Composite (CEC).

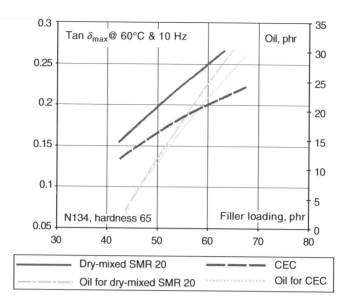

FIGURE 33.7 Effect of carbon black and oil loadings on hysteresis.

curing. In CEC compounds, strong polymer–filler interaction is already established by the time oil is added; therefore, the effect of the oil on filler flocculation is less.

33.3.2 CARBON–SILICA DUAL PHASE FILLERS

A new carbon black-based technology that has been introduced to reduce hysteresis in both car and truck tread compounds is the carbon–silica dual phase filler (CSDPF). These materials are produced using a unique co-fuming process, and are grouped into two different product categories. The Ecoblack CRX 2XXX series contains silica domains finely distributed throughout the aggregates of the filler. The CRX 4XXX series, on the other hand, has silica predominantly on the particle surface, giving higher silica surface coverage. Both types of CSDPF are characterized by lower filler–filler interactions and higher polymer–filler interactions than conventional carbon blacks. The lower filler–filler interaction of CSDPFs can be demonstrated by the strain dependence of the elastic modulus, G'. In filled rubber, the elastic modulus decreases with increasing strain amplitude, which has been termed the "Payne effect".[9] This effect is generally used as a measure of filler networking.[10] It can be seen from Figure 33.8 that both CRX 2124 and CRX 4210 gave lower Payne effects than either carbon black or silica, even though the chemical composition of these materials are intermediate between carbon black and silica. This unique behavior of the two new materials is readily explained by the fact that based on surface energy considerations, the interaction between unlike surfaces is weaker than that between the same category surfaces.[4,11] In the dual phase particles, the probability of unlike surfaces coming into contact with each other is much higher than in any single-phase filler material.

A result of weaker filler–filler interactions in the CSDPFs is that the microdispersion can be substantially improved, and the filler network is less developed. As a result, tan δ_{max} of CRX 4210 in a passenger tread compound is reduced by about 49%, compared with carbon blacks of similar morphologies (Figure 33.9). In the passenger tread compound, the hysteresis of the CRX 4210-filled compound is comparable to that of a silica-filled compound of the same polymer system, which is generally known to be better than carbon black compounds in this regard. In a typical truck tread compound, CRX 2124 had 33% lower tan δ_{max} than a comparable carbon black compound.[12] Based on the well-known correlation between rolling resistance and tan δ at high temperature, it can be expected that the fuel economy of a truck equipped with tires having tread compounds containing CRX 2124 can be significantly improved.

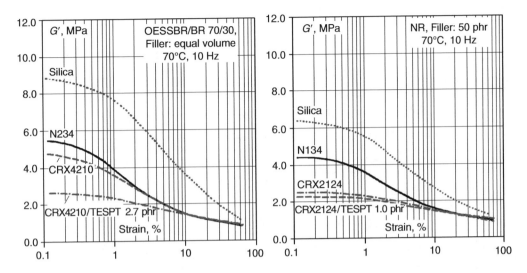

FIGURE 33.8 Strain dependence of G' for oil extended solution SBR (OESSBR)/BR and natural rubber (NR) filled with a variety of fillers.

In terms of heat buildup, CRX 2124 compounds showed lower temperatures than comparable black-filled compounds under test conditions of 20 kg static load, 80 kg dynamic load, and 15 Hz. The advantage of CRX 2124 compounds was 18°C when no coupling agent was used, and 23°C when 1 phr of TESPT was used.[12] In a passenger-tire compound, the advantage of CRX 4210 with 3.1 phr of coupling agent was 40°C. It is widely accepted that heat buildup is closely related to the service life of tires. Lower running temperatures lead to less mechano-oxidation of polymer and retard ply and shoulder separations, among other benefits.

33.3.3 CHEMICALLY MODIFIED CARBON BLACK

A new method for achieving stable attachment of organic functional groups to carbon black surface using diazonium chemistry, has been applied to the 4-aminophenyl disulfide (APDS) precursor.[13]

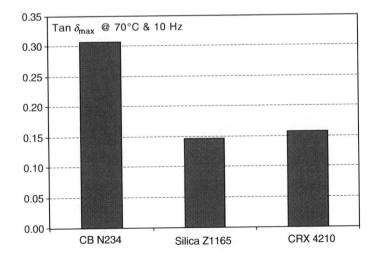

FIGURE 33.9 Comparison of tan δ_{max} for a variety of fillers in solution-based styrene–butadiene rubber/butadiene rubber/natural rubber (S-SBR/BR/NR) tread compounds for passenger tires.

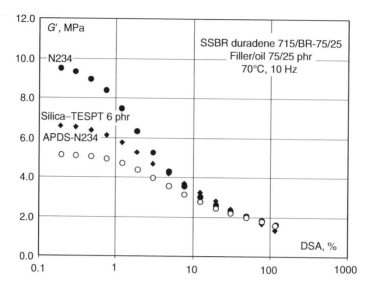

FIGURE 33.10 Strain dependence of G' for vulcanizates filled with a variety of fillers.

With APDS-modified carbon black, the dispersive component of surface energy of the carbon black is drastically reduced,[14] reducing the driving force of black agglomeration, resulting in better micro-dispersion. At the same time, the sulfide group is able to react with polymer molecules upon heating, so that a coupling reaction between polymer and filler can be achieved during vulcanization. The effectiveness of surface modification in suppression of filler networking due to both lowering filler–filler interaction, and increasing polymer–filler interaction via chemical bridges, has been verified by dynamic rheological testing.[15] These fillers were tested in typical passenger tire tread compounds alongside the virgin carbon black and silane-modified silica. The strain dependence of elastic modulus results (Figure 33.10) show that the APDS-modified carbon exhibits a lower Payne effect compared to bis(3-(triethoxysilyl)propyl)tetrasulphide (TESPT)-modified silica, which is itself much lower than that found for unmodified carbon black. A striking similarity is observed between the modified carbon black and coupled silica in the tan δ at 70°C results (Figure 33.11).

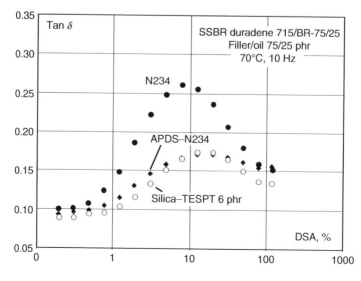

FIGURE 33.11 Strain dependence of tan δ at 70°C for vulcanizates filled with a variety of fillers.

Thus, the APDS-modified carbon black provides viscoelastic properties comparable to TESPT-modified silica, which is now well known to be better in terms of hysteresis and hence rolling resistance compared to conventional carbon black.

33.4 ADVANCES IN ABRASION RESISTANCE

The wear resistance of rubber compounds is of great practical importance for tires, but the mechanisms involved in tire wear are multiple and may vary depending on the vehicle, the driving conditions, and other extraneous variables. Many attempts have been made to develop laboratory test methods to simulate tire wear under various conditions. Whilst none can fully replicate road wear, various laboratory abrasion tests can be used to provide an indicator of wear resistance of tires under certain conditions.

It has been well established that wear resistance of filled rubber is essentially determined by filler loading, filler morphology, and polymer–filler interaction. For fillers having similar morphologies, an increase in polymer–filler interaction, either through enhancement of physical adsorption of polymer chains on the filler surface, or via creation of chemical linkages between filler and polymer, is crucial to the enhancement of wear resistance. In addition, filler dispersion is also essential as it is directly related to the contact area of polymer with filler, hence polymer–filler interaction.

33.4.1 CARBON BLACK MORPHOLOGY EFFECTS ON ABRASION RESISTANCE

For conventional carbon blacks, there is generally a trade-off between wear resistance and hysteresis. High surface area carbon blacks, which provide a large interfacial area between filler and polymer, generally show a high degree of interaction with the hydrocarbon polymer,[5,16] favoring high abrasion and wear resistance. On the other hand, high-surface area carbon blacks tend to give poorer dispersion, both on the macro- and microscales (see Section 33.2). It has already been shown that poor microdispersion leads to more developed filler networking, hence higher hysteresis. In addition, poor macrodispersion can have a large negative impact on abrasion resistance. In practice, the selection of carbon black grades for abrasion resistance is usually a compromise between maximizing surface area and surface activity for performance, and keeping the carbon black particles large enough to enable an adequate level of dispersion to be achieved.

CEC masterbatch technology can provide a way to eliminate this compromise by providing an excellent level of dispersion to all grades of carbon black. The effect of loading (carbon black N134) on abrasion resistance measured at 7% slip ratio, shows a characteristic maximum, then drops at high filler loading (Figure 33.12). The maximum for the CEC compound is higher, and occurs at a higher carbon black loading compared to an equivalent dry-mixed compounds with SMR 20. Several mechanisms may be involved in the reduction in abrasion resistance at high filler loading. These factors include reduction in rubber content, rapid increase in hardness, and deterioration of fatigue resistance. Poor dispersion of carbon black at high loading, however, plays an important role in the decline of abrasion resistance, and the excellent dispersion in the CEC compounds accounts for the better abrasion resistance at high loading. It has been found that at constant hardness, abrasion resistance follows different curves as a function of filler and oil loading (Figure 33.13). For dry-mixed compounds, abrasion resistance decreases monotonically with increasing carbon black and oil loading. In contrast, the abrasion resistance of CEC vulcanizates passes through a maximum. This can be explained by the effect of oil adsorption on polymer–filler interaction, one of the most critical parameters controlling abrasion resistance. In dry-mixed materials, oil adsorption interferes with the polymer–filler interaction, so it is observed that any addition of oil to the compound has a negative effect on abrasion resistance. The polymer–filler interaction in CEC is much less affected by the addition of oil during mixing, since the adsorption of polymer chains on carbon black surface has already been completed. It can be seen (Figure 33.13) that in CEC compounds, increased carbon black loading initially causes an increase in abrasion resistance. The effect of the increasing oil

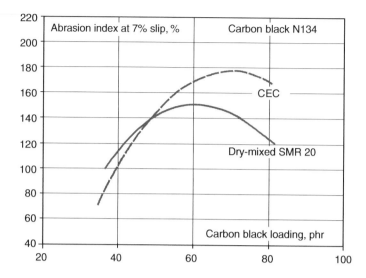

FIGURE 33.12 Effect of carbon black loading on abrasion resistance.

content, required to keep hardness constant, is not enough to offset the gain in abrasion resistance caused by the higher black loading in these compounds. Only at oil contents above about 20 phr, does performance begin to decline in this system.

33.4.2 SILICA-MODIFIED CARBON BLACK

Silica has been used increasingly as a reinforcing agent in passenger car tire tread compounds in recent years due to the advantages it provides in rolling resistance and wet traction. However, it is widely known that the abrasion resistance of silica compounds is inferior to carbon black due to the poorer polymer–filler interaction, related to surface energy.[17] Even the use of coupling agents, which provide chemical linkages between polymer molecules and the silica surface, cannot provide

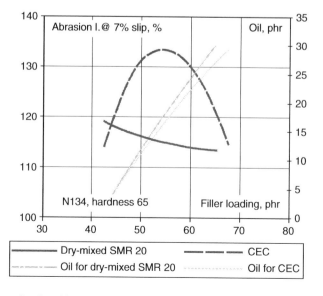

FIGURE 33.13 Effect of carbon black and oil loadings on abrasion resistance.

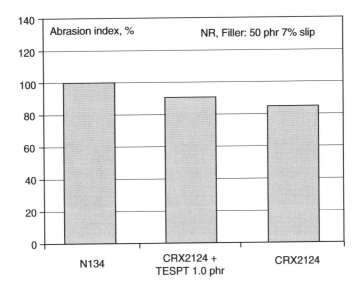

FIGURE 33.14 Comparisons of abrasion resistance of truck tread compounds with a variety of fillers. Basic formulation: NR 100, filler 50, processing oil 2.5.

the same level of polymer–filler interaction provided by carbon black. This is especially true in NR-based compounds where the reaction of the coupling agent with silanol groups on the silica surface may be inhibited by adsorbed nonrubbers.[18] Dual phase carbon–silica particles provide a solution to this problem. For example, the abrasion index of a CRX 2124-filled compound without TESPT, measured by the Cabot Abrader at 7% slip ratio, is only about 17% lower than its carbon black counterpart. By addition of 1.0 phr of TESPT, the abrasion resistance of CRX 2124 can be further improved making it closer to that of N134, one of the preferred reinforcing blacks for truck tire treads (Figure 33.14). The negative impact of the silica domains on abrasion resistance seems to be partially offset by the high-surface activity of the carbon domains on the aggregates. Evidence of the higher filler–polymer interaction of these fillers in hydrocarbon polymers can be obtained by comparing the adsorption energies of polymer analogs on the filler surfaces. The adsorption-free energies of heptane, a model compound for hydrocarbon polymers, measured by IGC, is higher on the CSDPFs than on a mixture of carbon black and silica (Figure 33.15). The higher polymer–filler interaction of CSDPFs with hydrocarbon polymers can be attributed to higher surface activity of their carbon domain, which is related to the unique microstructure of the carbon domain.[11]

33.5 DEVELOPMENTS ON WET SKID RESISTANCE

The mechanism by which rubber compounds affect the wet-skid resistance of tires is complicated. Dynamic hysteresis and modulus at low temperature has long been used as an indicator of the wet-skid resistance of rubber vulcanizates because the high-frequency nature of the strains involved in the skid process can be transposed to a low temperature test. While this generally works well under certain conditions, models based exclusively on the dynamic properties of rubber cannot explain all skid test results under all conditions. For example, silica tread compounds have been shown to give better grip than carbon black-filled compounds under wet conditions on car tires with antilock brake system (ABS)-type skidding. However, the carbon black compounds give better skid resistance under dry conditions,[1] under heavy loads such as those found in truck tires,[19] and with locked wheel skidding. Thus the same compounds, with the same dynamic properties, can give different relative traction performance, depending on the test conditions. Clearly, something more than the dynamic hysteresis of the compounds must be involved.

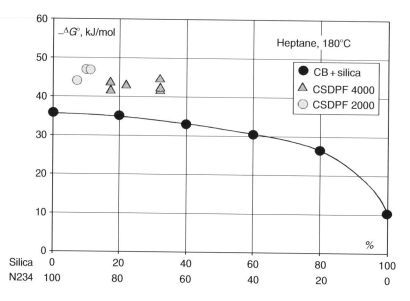

FIGURE 33.15 Adsorption energies of heptane at 180°C on a variety of fillers.

33.5.1 Three-Zone Concept of Wet Traction or Wet Skidding of Tires

A more complete model for the wet traction of tires is based on a "Three-Zone Concept" originally put forward by Gough[20] for sliding wheels and later extended by Moore[21] to cover the case of the rolling tire. This concept suggests that at speeds below the hydroplaning limit, the tire contact area can be divided longitudinally into three distinct regions (Figure 33.16). The three characteristic zones of contact may be identified as follows:

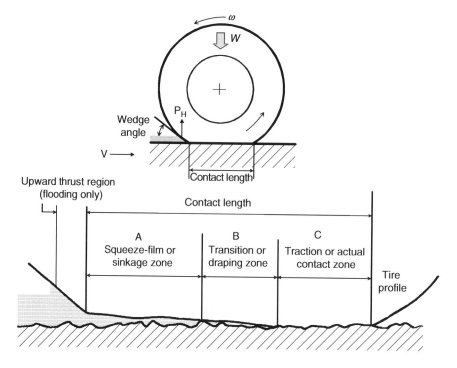

FIGURE 33.16 Three-zone model of tire contact with road surface under wet conditions.

- Squeeze-film zone or sinkage zone: In this forward region of contact area, a water wedge is formed due to the displacement inertia of the intercepted water film. The thickness of this film decreases progressively as the individual tread elements pass through the contact area, squeezing out the water film. The key mechanism operating here is elasto-hydrodynamic lubrication (EHL), but very little friction is generated in this zone.
- Transition zone: The transition zone begins when the tire elements, having penetrated the squeeze-film, commence to drape dynamically about the major asperities of the road surface, and make contact with the lesser asperities. In this zone, a progressive breakdown of the water film occurs down to thickness of a few layers of water molecules. Therefore, a mixed-lubrication regime exists, which is part hydrodynamic and part boundary. The effective friction coefficient changes widely from a very low value of viscous hydroplaning at the leading edge of the transition zone, to the friction value of boundary layer lubrication at the end edge of this zone, which is comparable to the dry friction.
- Traction zone or actual contact zone: This region is the rear part of the contact area, beginning with the end of transition zone. It is the zone where most of the traction or skid resistance is developed. Here, the lubricating water film has been totally or substantially removed, and vertical equilibrium of the tread elements on road surface has been attained. In this zone, boundary lubrication (BL) is the dominant mechanism.

The overall traction performance will be determined both by the relative size of the three zones, and the level of friction in each zone. Thus, to improve wet-skid resistance, the goals should be to increase the traction zone relative to the squeeze-film and transition zones, and to increase the level of friction in the transition and traction zones.

33.5.2 EFFECTS OF FILLERS IN THE THREE ZONES

33.5.2.1 Squeeze-Film Zone

In the squeeze zone, EHL is important. This mechanism can be considered a special case of hydrodynamic lubrication in which the elastic deformation of the surrounding solids plays a significant role in the lubrication process. Unlike hydrodynamic lubrication between rigid surfaces, the modulus of the material is an important factor in EHL, in addition to the normal load, velocity of the tangential motion, viscosity of the lubricant, and the shape of the asperities.[22] The relative size of this zone is determined primarily by the speed, road surface texture, and tire design, but also by the modulus of the rubber compound. These factors determine the rate at which the film is squeezed out, and consequently, the relative size of the squeeze zone. The higher the modulus of the materials, the higher would be the squeezing rate of lubricant film, leading to a relatively shorter squeeze zone, and therefore relatively longer transition and traction zones where almost all of the friction is generated. The result would be higher overall traction.

33.5.2.2 Transition Zone

In the transition zone, EHL is still important, but as more water is removed, EHL at the microscale (MEHL) becomes more important, and when the water layer is reduced to molecular levels, another mechanism, BL takes over. Since BL is the main mechanism by which friction is generated in the overall skidding process, any material properties which increase the proportion of BL in the transition zone relative to EHL, i.e., accelerate the transition from EHL to BL, will have an impact on overall skid performance. As discussed above, modulus is an important factor in determining the rate of water removal in EHL. For MEHL, it is the modulus on the microscale at the worn surface of the tread that is critical. There is evidence that after a certain amount of normal wear, a significant part of the surface of silica-filled compounds is bare silica, whereas in black-filled compounds, the surface is fully covered by rubber.[23] The difference in modulus between rubber and silica is very large, so even if only part of the worn surface is bare silica, it would make a significant impact on the modulus of the microsurface layer. The fact that more bare filler is exposed on the worn surface of a silica compound compared to a similar carbon black compound, leads to significantly faster water removal in the transition zone, therefore more BL in this zone and a relatively longer traction zone.

The reason why silica compounds reveal bare silica after some wear whilst black-filled compounds do not, is related to the type and levels of filler–polymer interaction in these two systems. It has been shown that carbon black has stronger interaction with hydrocarbon polymers than silica, both from energetic characteristics of the filler surfaces,[5] and from in-rubber properties.[24] The use of coupling agents such as TESPT certainly increases the level of polymer–filler interaction in silica compounds by introducing covalent bonds; however only a small proportion of the surface silanol groups are reacted. The remaining silanol groups still have a strong affinity for water. It seems possible that under conditions of wet skid, water may penetrate to the silica surface at the outer layer of the compound, and weaken the polymer–silica bonding through adsorption, and possibly hydrolysis of siloxane linkages. This would cause cavitation to occur at the silica–polymer interface, which would facilitate failure at the filler surface, rather than rupture in the polymer matrix, as is the case in black-filled compounds.

33.5.2.3   Traction Zone

When the water film is squeezed out, the thick water layer is removed and the surfaces are separated by lubricant film of only molecular dimensions. Under these conditions, which are referred to as BL conditions, the very thin film of water is bonded to the substrate by very strong molecular adhesion forces and it has obviously lost its bulk fluid properties. The bulk viscosity of the water plays little or no part in the frictional behavior, which is influenced by the nature of the underlying surface. By comparing with the friction force of an elastomer sliding on a rigid surface in a dry state, Moore was able to conclude that for an elastomer sliding on a rigid surface under BL conditions, one can expect[22]:

$$\mu_{BL} \geq \mu_{dry} \tag{33.6}$$

In contrast to a rigid–rigid system, where μ_{BL} is lower than μ_{dry}, the addition of water to an elastomer–rigid sliding system in a boundary state does not cause a reduction in the resulting sliding friction coefficient. Under boundary conditions, silica, as a rigid material, gives slightly lower friction than it does in the dry state, whereas for rubber, the friction is comparable to dry friction. Table 33.2 summarizes the friction coefficients for different materials against glass under both wet and dry conditions. It can be seen that the friction of rubber against a solid surface is much higher than that of other solids, including silica and carbon black.[23] It would therefore be expected that carbon black compounds, in which the filler is always covered by a layer of rubber, are favored in this region relative to silica compounds, even under wet-skid conditions. This can explain why truck tires, which carry a high load and therefore have a relatively large traction zone, show better traction

TABLE 33.2
Friction Coefficients of Different Materials against Glass Surface

	Dry Surface	Wet Surface
Carbon black[a]	0.16[26]	0.13[26]
Silica	1.04[27]	0.9[27]
Rubber	2.5 ~ 4[28]	Similar to dry surface—BL
Rubber		≪0.1—EHL

[a] For graphite and hard nongraphite carbon surface.

from carbon black compounds than from comparable silica compounds, even in the wet. It should be noted that in the traction zone, the deformation component of the frictional force will play an important part hence, the dynamic properties of the compounds are also important.

33.5.3 New Fillers for Improved Wet-Skid Resistance

Laboratory studies on a range of different fillers in typical passenger tread compounds have shown that the friction coefficient, measured by means of an improved British Portable Skid Tester (BPST) on wet and sandblasted glass, is closely related to the silica–polymer interfacial area in the compounds (Figure 33.17). A higher silica–polymer interfacial area gives better wet-skid resistance. Giustino and Emerson,[25] have shown that the BPST number (or friction coefficient) correlates well with peak friction coefficients obtained from the skid testing of passenger tires on a wet road. Therefore, fillers giving higher silica–polymer interfacial area in the tire tread compound should impart higher wet-skid resistance under ABS conditions where braking takes place in the peak friction regime.

CRX 4XXX, which are carbon–silica dual phase particles with high-surface silica coverage, were developed partly to provide improved wet-skid resistance compared to carbon black. The dynamic moduli of compounds filled with this material are considerably reduced compared to carbon black, and the low hysteresis at low temperature in typical passenger-tire tread compounds is very similar to that of silica compounds. Besides imparting low rolling resistance, the wet-skid resistance of CSDPF compounds, measured with BPST, is significantly higher than that of carbon black compounds (Figure 33.17). In fact, several road tests have demonstrated that for vehicles with ABS, the wet-skid resistance of CSDPF 4210 with silica-surface coverage about 55% tested on smooth pavement surface gives comparable results to silica.[1] This can be explained as follows: In the water squeeze and transition zones, where EHL and MEHL are predominant, CSDPF 4210 may not be as effective as silica but it is better than carbon black. In the traction zone where the friction coefficient is controlled by BL, it should be better than silica because 45% of the aggregate surface is covered by rubber due to the strong polymer–filler interaction. It also provides higher hysteresis and low dynamic modulus at low temperature compared to carbon black, which aids performance in the traction zone.

Figure 33.18 shows a comparison of the wet-skid resistance under different loads of carbon black-, silica-, and CRX 2XXX-filled tread compounds, using smooth sandblasted glass as the friction substrate. These results were obtained using a Grosch Abrasion and Friction Tester (GAFT). It can be seen that under high load, where BL is relatively dominant, the wet friction coefficients of

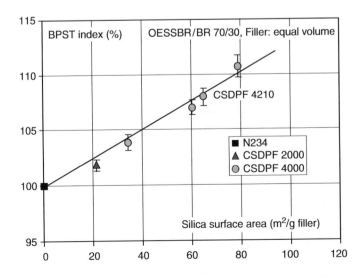

FIGURE 33.17 Effect of silica surface area on wet-skid resistance of vulcanizates filled with different fillers.

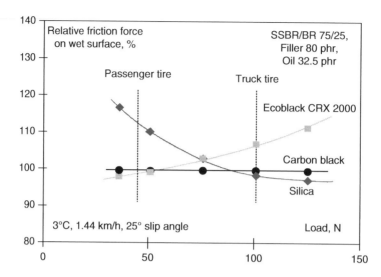

FIGURE 33.18 Wet-skid resistance measured by GAFT as a function of load for compounds with a variety of fillers.

the carbon black and silica compounds are similar, but the CRX 2XXX filler is significantly better. These load conditions correspond to those encountered by truck tires. At low loads, which are relevant to passenger car tires, and where MEHL plays an important role, silica shows a significant advantage over carbon black and the CRX 2XXX filler.

REFERENCES

1. Wang, M.-J., *Kautsch. Gummi Kunstst.*, **58**, 626, 2005.
2. Medalia, A.I., *J. Colloid Interf. Sci.*, **32**, 115, 1970.
3. Wang, M.-J., Wolff, S. and Tan, E.-H., *Rubber Chem. Technol.*, **66**, 178, 1993.
4. Wang, M.-J., *Rubber Chem. Technol.*, **71**, 520, 1998.
5. Wang, M.-J., Wolff, S., and Donnet, J.B., *Rubber Chem. Technol.*, **64**, 714, 1991.
6. Mabry, M.A., Rumpf, F.H., Podobnik, J.Z., Westveer, S.A., Morgan, A.C., Chung, B., and Andrew, M.J., US Patent 6,048,923, 2000.
7. Wang, M.-J., Wang, T., Wong, Y.L., Shell, J., and Mahmud, K., *Kautsch. Gummi Kunstst.*, **55**, 388, 2002.
8. Wang, T., Wang, M.-J., McConnell, G.A., and Reznek, S.R., WO Patent 2003/050182, 2003.
9. Payne, A.R., *J. Polym. Sci.*, **6**, 57, 1962; Payne, A.R. and Whittaker, R.E. *Rubber Chem. Technol.*, **44**, 440, 1971.
10. Wang, M.-J., *Rubber Chem. Technol.*, **72**, 430, 1999.
11. Wang, M.-J., Mahmud, K., Murphy, L.J., and Patterson, W.J., *Kautsch. Gummi Kunstst.*, **51**, 348, 1998.
12. Wang, M.-J., Zhang, P., Mahmud, K., Lanoye, T., and Vejins, V., *Tire Technology International 2002*, pp. 54–59, UK & International Press, 2002.
13. Belmont, J.A., Amici, R.M., and Galloway, C.P., US Patent 5,851,280, 1998.
14. Wang, M.-J., Paper presented at Carbon Black '05 Perspective in Asia Pacific, Oct. 16–18, 2005, Suzhou, China.
15. Tokita, N., Wang, M.-J., Chung, B., and Mahmud, K., *J. Soc. Rubber Ind., Jpn.*, **72**, 522, 1998.
16. Wolff, S., Wang, M.-J., and Tan, E.-H., *Rubber Chem. Technol.*, **66**, 163, 1993.
17. Wang, M.-J., Tu, H., Murphy, L., and Mahmud, K., *Rubber Chem. Technol.*, **73**, 666, 2000.
18. Wang, M.-J., Zhang, P., and Mahmud, K., *Rubber Chem. Technol.*, **74**, 124, 2001.
19. Wang, M.-J., Kutsovsky, Y., Zhang, P., Murphy, L.J., Laube, S., and Mahmud, K., *Rubber Chem. Technol.*, **75**, 247, 2002.
20. Gough, V.E., *Revue Generale Du Caoutchouc* (Discussion of Paper by D. Tabor) **36**, 1409, 1959.

21. Moore, D.F., *Wear*, **8**, 245, 1965.
22. Moore, D.F., *Principles and Applications of Tribology*, Chapter 7, Pergamon Press, Oxford, 1975.
23. Wang, M.-J., Effect of Filler–Elastomer Interaction on Hysteresis, Wet Friction and Abrasion of Filled Vulcanizates, Paper presented at a meeting of the Rubber Division, ACS, Akron, May 8–10, 2006.
24. Wolff, S. and Wang, M.-J., *Rubber Chem. Technol.*, **65**, 329, 1992.
25. Giustino, J.M. and Emerson, R.J., Instrumentation of the British Portable Skid Tester, Paper presented at a meeting of the Rubber Division, ACS, Toronto, Canada, May 10–12, 1983.
26. Bowden, F.P. and Tabor, D., *The Friction and Lubrication of Solids*, Chapter VIII, pp. 163–169, Clarendon Press, Oxford, 1950.
27. Sameshima, J., Akamatu, H., and Isemura, T., *Rev. Physical Chem.*, **37**, 90, 1940.
28. Grosch, K.A., *Rubber Chem. Technol.*, **69**, 495, 1996.

34 Rubber Oxidation in Tires

John M. Baldwin and David R. Bauer

CONTENTS

34.1 INTRODUCTION

While a new tire may have excellent resistance to crack initiation and propagation between the steel belts, an aged tire of the exact same construction can exhibit dramatically reduced crack growth resistance, which in some cases may contribute to tire failure. To further underscore the point, the United States National Highway Traffic Safety Administration (NHTSA) determined that tire aging was a main contributor to the Firestone Wilderness AT tire recall [1]. Figure 34.1 shows the failure rate of Wilderness AT tires as a function of tire age (determined from date of tire manufacture).

Even for the lowest-performing tires, it took a minimum of two years in the field before significant numbers of failures were reported. Based on the results of tire rubber analysis done by both Ford Motor Company and NHTSA on the recalled tire set, it was decided to embark upon a research project into understanding and quantifying the field aging of tires and then using that knowledge to develop an accelerated aging test for tires.

This chapter will explain the setup and methodology used for the field retrieval study. The physical and chemical analysis techniques used to study the oxidation of the tire rubber, along with the data analysis developed to interpret the results, will also be explained. Then, the development of an oven-aging protocol that attempts to reproduce the mechanism and rate of tire field aging will be described.

34.2 EXPERIMENTAL

34.2.1 MEASUREMENTS

Being cut from tires, the rubber samples in this work cannot always be studied using standard test procedures. The specific techniques used to measure rubber aging have been described in detail elsewhere and are summarized below [2]. The same techniques have been used to evaluate rubber aging in both field and the laboratory oven-aging studies. For reference, Figure 34.2 shows a diagram of the internal components of a radial-ply tire.

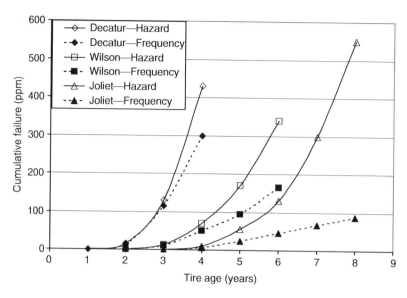

FIGURE 34.1 United States Government (NHTSA) data on recalled tires. Frequency (or hazard) analysis versus tire age.

Peel Strength. Samples were prepared by cutting 2.5″ (63.5 mm) wide radial sections, bead to bead. The sample was then sectioned into two 1.25″ (31.75 mm) radial strips, which were each cut circumferentially at the centerline of the tread resulting in four test specimens (2-SS and 2-OSS). Each sample was cut with a razor knife for a length of 1″ (25.4 mm) from the skim end of the test strip, midway between the belts, to facilitate gripping the ends in the T-2000 stress–strain tester jaws. The sides of each specimen were scored midway between the belts, to a depth of 1/8″ (3.175 mm) radially from the end of the gripping surface to the end of belt #2 in the shoulder area, providing a 1″ wide peel section. The peel test was performed at 2″/min (50.8 mm/min) at 24°C.

Tensile and Elongation. Samples of the belt wedge rubber, located between belts 1 and 2, were removed from both shoulders (serial side and opposite serial side) and buffed to a uniform thickness of 0.5–1.0 mm. Care was taken so that no significant heat was introduced to the samples by the buffing. Specimens were die-cut using an ASTM D 638 Type V dumbbell die and tested per ASTM D 412. Results obtained included modulus values at 100%, ultimate elongation, and tensile strength. Samples were tested at 20″/min (50.8 cm/min).

Cross-Link Density. Cross-link density measurements were determined on samples swollen to equilibrium in toluene after 24 h. Five specimens were tested for each sample. The volume fraction

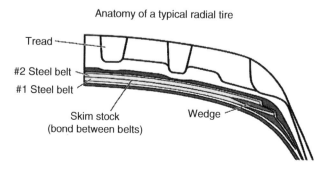

FIGURE 34.2 Diagram of steel belt portions of a tire.

TABLE 34.1

Vehicle-Class, Tire-Brand Combinations

	Large Car F	Small Car C	Truck U3855 B	Truck O3855 A	Perf Car E	SUV Minivan D
Brand A	x	x	N/A	x	x	x
Brand B	x	x	x	N/A	x	x
Brand C	x	x	x	x	N/A	x

of polymer in the swollen gel was measured at equilibrium swelling and the cross-link density determined using the Flory–Rehner equation [3]. The polymer–solvent interaction parameter χ with toluene as the swelling media for natural rubber (the skim rubber) is 0.391 [4]. The volume fraction of rubber was not corrected for the presence of fillers.

34.2.2 FIELD RETRIEVAL OF TIRES

A full description of the field retrieval design of experiment and initial results are contained in previous papers and are summarized below [5–7].

Fifteen tire brand–vehicle combinations were collected from Phoenix, AZ; Los Angeles, CA; Detroit, MI; and Hartford, CT. Six different vehicle types and three different tire manufacturers were studied as detailed in Table 34.1. Vehicle–brand combinations are denoted by the vehicle type followed by the letter denoting the specific manufacturer, e.g., SUV/Minivan-A. The geographic locations were chosen based on ambient temperature and ozone level. Both on-road and full-size spare tires were collected ranging in age from two weeks to six years old. In total, over 1500 tires were analyzed in this study. A small number of tires from Miami, Fl and Denver, CO were collected to evaluate the effects for road roughness on aging. No significant effect was observed and these tires were not included in the analysis.

In addition to the Ford field-retrieval study, NHTSA has carried out further studies of aging from tires retrieved from Phoenix [8]. Analysis in this chapter will be limited to tires SUV/Minivan-A, -B, and -C and Large Car-C from the Ford study. Tire sizes and maximum inflation pressures for these tires are reported in Table 34.2. The analysis will focus on extracting accurate rate data from the field data. Of key importance is developing an understanding regarding the sources of variability in field aging.

34.2.3 OVEN-AGING CONDITIONS

Tires were mounted and inflated to the maximum pressure listed on the sidewall prior to oven aging using either air or a 50/50 blend of N_2/O_2. In the case of tires inflated with the 50/50 blend of N_2/O_2, the atmospheric air present was not purged; the blend was added on top of it yielding a tire cavity concentration of approximately 41%–44% O_2. Two of the tires in this study (SUV/Minivan-A and SUV/Minivan-C) were aged at 50°C, 60°C, and 70°C for up to eight weeks, while the tires and

TABLE 34.2

Tire Sizes and Inflation Pressures

Vehicle Type-Brand	Tire Size	Maximum Sidewall Pressure (kPa)
SUV/Minivan-A	215/70R15	240
SUV/Minivan-B	235/75R16	300
SUV/Minivan-C	215/65R16	300
Large Car-C	225/60R16	240

Large Car-C were aged up to eight weeks at 70°C. The ovens were calibrated per ASTM E 145 with an A2LA approved, modified, method for temperature uniformity, consistency, airflow exchanges, and airflow velocity.

34.3 RESULTS AND DISCUSSION

34.3.1 FIELD TIRE STUDY

Aging of the belt skim and wedge rubber was found to be quite general. Values for peel strength and elongation to break decreased with age while the cross-link density and modulus increased [5–7]. Detailed analysis of these changes following the method of Ahagon confirmed that in all cases the changes were a result of oxidation [9–11]. The oxidation was relatively uniform across the belt. The rate of oxidation varied widely for different tire sizes and brands. Ambient temperature was found to be the only important external environmental factor in aging rate. Ozone did not affect the rate of aging. Tires in Phoenix aged ~2.5 times faster than tires in Detroit or Hartford and ~1.7 times faster than those in Los Angeles. These rates did not depend on the specific tire. Full-size spare tires aged ~70% as fast as on-road tires, again, independent of external environment, tire brand, or vehicle.

As discussed above, we have already shown that the property changes in the skim and wedge rubber from field-aged tires are consistent with oxidative degradation. The kinetics of aerobic oxidation in different rubber materials have been studied extensively by Gillen et al. through measurements of oxygen-consumption rates, modulus profile, and elongation to break over a wide range of temperature and time. These studies demonstrated that under constant, controlled conditions, the oxygen consumption rate depended strongly on rubber composition, but for a given material, was relatively constant for long periods of time [12–14]. The rate of oxygen consumption increased with increasing temperature consistent with typical oxidation activation energies. The relative rates of oxygen consumption at different temperatures were consistent with accelerated shift factors derived from property change measurements such as modulus and elongation to break implying that the property changes were controlled by the extent of oxidation over the whole temperature range. At low extents of oxidation, the increase in modulus was proportional to the extent of oxidation. At high oxidation levels, the modulus increased exponentially with oxidation. Cross-link density and modulus are strongly correlated. At relatively low oxidation levels, the number of reactive sites is large compared to the number of sites that have reacted (oxidized). Thus, one would expect that the change in cross-link density (and modulus) would be proportional to the extent of oxidation. Over the range of aging observed in our studies, we find that both the modulus and cross-link density of the skim and wedge rubber in tires can be fit to a simple linear equation in time [15].

$$A(t) = A(0)(1 + \alpha t) \qquad (34.1)$$

where the rate constant α reflects the oxidation rate and its effect on modulus or cross-link density. Thus, α will depend on the intrinsic oxidation rate of the rubber material of interest, the permeation rate of oxygen between the rubber sample and the oxygen source, and the composition factors that influence the exact products of oxidation (cross-linking vs. scission). We expect the values of α to be different for different rubber compositions in the same tire and even for the same composition in different locations within the tire. For similar locations, the values of α for modulus and cross-link density should be similar (this is the reason that we have chosen to calculate the cross-link density from the swelling ratio). It should be noted that cross-link density and modulus are affected not only by oxidation but also by state of cure. The rate of oxidation will clearly be a function of the temperature history of the tire. Figure 34.3 compares the increase in modulus of the wedge rubber from SUV/Minivan-A tires obtained from two different cities. Figure 34.3 also illustrates the shift factor analysis, which was a method developed by Gillen et al., who employed the classic

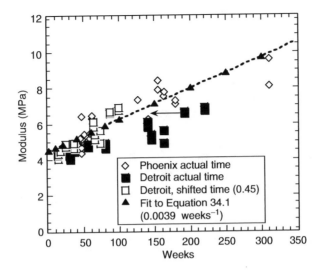

FIGURE 34.3 Shift factor approach and fit to Equation 34.1 for typical field data (modulus).

time–temperature superposition approach using strain at break data, used here to derive the relative rates of oxidation in the two cities [12].

For properties such as peel strength and elongation to break, which decrease with oxidation, a different empirical relation has to be used. We have found that elongation to break and peel strength retention data can be fit to the following expression [15]:

$$B(t) = \frac{B(0)}{(1 + \beta t)},$$ (34.2)

where the rate constant β reflects the oxidation rate and its effect on loss of elongation or peel strength. The same factors that influence α also will affect β. We have shown that Equation 34.2 correctly describes the time dependence of elongation to break down to 5% retention (nitrile rubber data, taken from Gillen et al.) [12]. Fits to Equation 34.2 and illustration of the shift factor analysis are shown for skim rubber peel strength retention in the SUV/Minivan-A tire in Figure 34.4.

The field data shown in Figures 34.3 and 34.4 also illustrate the relatively high level of variability in aging results. The sources of variability will be discussed in more detail below. In order to maximize the available data, we have chosen to combine the data from all four cities using the shift factor analysis described above. All of the changes are driven by oxidation and the relative behaviors correlate well with one another from tire to tire. In general, the proportional changes in peel strength and elongation to break are larger than those for modulus and cross-link density. In addition, peel strength and elongation to break are likely to be more closely correlated to tire tread separation performance, because these measurements reflect the ultimate properties of the rubber, rather than modulus or cross-link density. For these reasons, we will concentrate the analysis mostly on peel strength and elongation to break.

Elongation to break for the wedge rubber versus aging time are shown in Figures 34.5 through 34.8 for the four tires from the Ford field study. Skim rubber peel strength versus aging time is shown in Figures 34.9 through 34.12 for the same tires. Aging rates were determined by a least squares fit of the data to Equation 34.2. The skim rubber peel strength data is for the center of the skim. Values have also been measured for the skim rubber in the belt edge region. Aging rates for these properties for all of the tires are summarized in Table 34.3. These rates will be compared to rates measured for oven-aged tires in the following section. The aging rates for the SUV/Minivan-A tire are considerably faster than the other tires in this study.

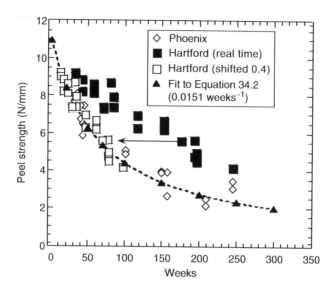

FIGURE 34.4 Shift factor approach and fit to Equation 34.2 for typical field data (peel strength).

The variability in property retention for field-tire aging shown in Figures 34.3 through 34.12 is significant. Relatively large sample sizes are necessary to determine accurate aging rates. The variability is a result of variation both in initial property value and in aging rate. Assuming that the variations in these two parameters are independent, the standard deviation expected for peel strength retention as a function of time can be approximated by the following equation [16]:

$$\sigma(t) \cong \text{Peel}(t) \times \sqrt{\left(\frac{\sigma_{\text{Peel}}}{\text{Peel}_0}\right)^2 + \left(\frac{\sigma_{\beta^t}}{1 + \beta t}\right)^2} \qquad (34.3)$$

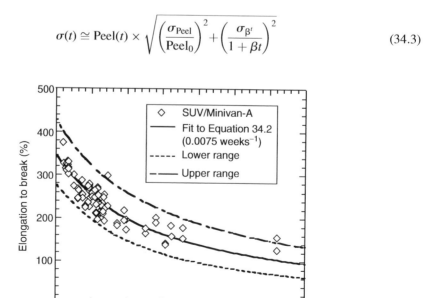

FIGURE 34.5 Elongation to break of wedge rubber versus weeks in Phoenix for SUV/Minivan-A tire.

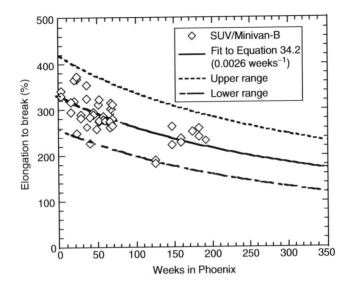

FIGURE 34.6 Elongation to break of wedge rubber versus weeks in Phoenix for SUV/Minivan-B tire.

where

σ_{Peel} is the standard deviation in initial peel strength
σ_{β} is the standard deviation in aging rate, β
Peel(t) is the value calculated from Equation 34.2

A similar dependence applies to the standard deviation for elongation to break. The upper and lower ranges shown in Figures 34.5 through 34.12 were generated using the following equations:

$$\text{Upper Range} = \text{Peel}(t) + 3\sigma(t) \qquad (34.4)$$

and

$$\text{Lower Range} = \text{Peel}(t) - 2.5\sigma(t) \qquad (34.5)$$

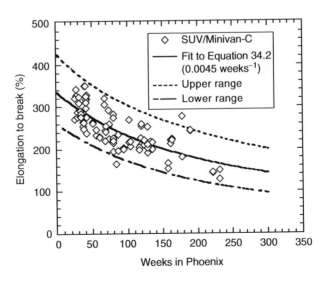

FIGURE 34.7 Elongation to break of wedge rubber versus weeks in Phoenix for SUV/Minivan-C tire.

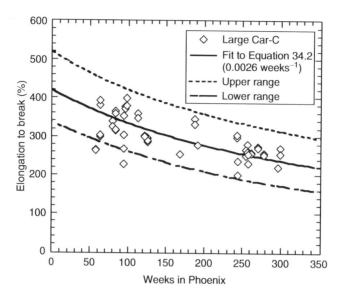

FIGURE 34.8 Elongation to break of wedge rubber versus weeks in Phoenix for Large Car-C tire.

The reason that the parameters are different in the two ranges is that Peel(t) is not a symmetric distribution, even though the initial peel and the aging rate are assumed to be symmetric. The choice of 2.5σ yields the best estimate of the typical "worst-case" tire as confirmed by a Monte Carlo simulation of 1200 tires (Figure 34.13). The values for $\sigma_{Peel}/Peel_0$ and σ_β/β that best fit the data shown in Figures 34.5 through 34.12 are 9% and 17%, respectively. These values are independent of tire type and have been found to apply reasonably well to all of the tires in the Ford field studies. A similar analysis of oven-aged samples presented in the following section yields a value for both $\sigma_{Peel}/Peel_0$ and σ_β/β of ~7%.

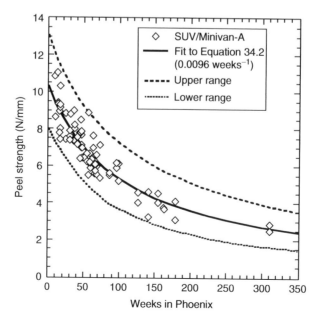

FIGURE 34.9 Peel strength retention for skim rubber (center) versus week in Phoenix for SUV/Minivan-A tire.

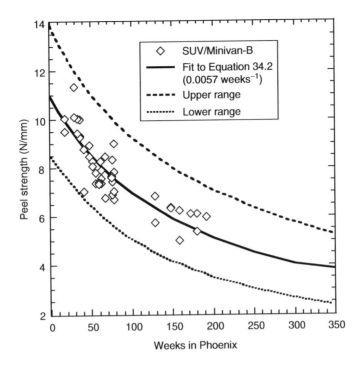

FIGURE 34.10 Peel strength retention for skim rubber (center) versus week in Phoenix for SUV/Minivan-B tire.

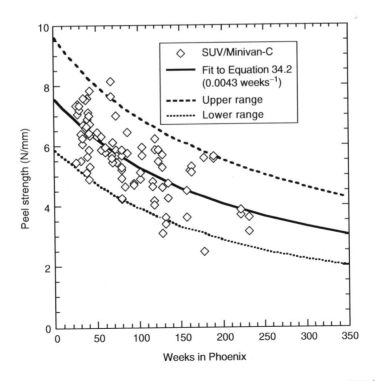

FIGURE 34.11 Peel strength retention for skim rubber (center) versus week in Phoenix for SUV/Minivan-C tire.

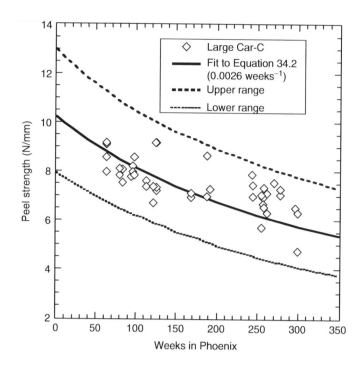

FIGURE 34.12 Peel strength retention for skim rubber (center) versus week in Phoenix for Large Car-C tire.

There are many sources that can contribute to potential variation in initial peel (or elongation to break) and aging rate. The main factors that contribute to initial property variation include measurement error and variations in compounding, molding, and cure that affect the initial properties. While these factors influence both oven and field samples, it seems likely that the variation in initial properties for the field samples will be somewhat greater than for the oven samples due to the wide time span over which the field tires were manufactured (the oven samples had identical DOT codes). Factors that contribute to σ_β include differences in inner liner thickness (which affects permeation of oxygen to the skim rubber), compounding variability (which could affect antioxidant effectiveness), and environmental variability. For the oven, the environmental variability should be small; however, for the field we expect significant environmental variability even in tires obtained from the same location. This variability can result from variations in both parking and driving habits including fraction of time parked in the shade versus the sun, road surface that the vehicle is parked or driven on (asphalt vs. concrete), inflation pressure, load, and driving mileage and other details (speed, aggressiveness, etc.). All of these factors influence the temperature history of a tire relative to the ambient air temperature. A variation of 17% corresponds to a 2°C average temperature

TABLE 34.3
Field Aging Rates (Weeks)$^{-1}$

Vehicle Type-Brand	Elongation to Break	Peel Strength Center	Peel Strength Belt Edge
SUV/Minivan-A	0.0075	0.0096	0.0120
SUV/Minivan-B	0.0026	0.0057	0.0055
SUV/Minivan-C	0.0045	0.0043	0.0047
Large Car-C	0.0026	0.0026	0.0022

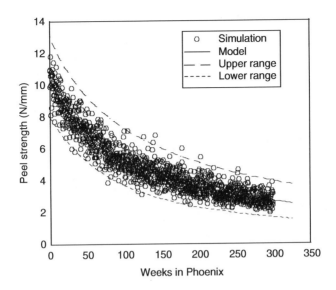

FIGURE 34.13 Monte Carlo simulation of Peel strength retention for SUV/Minivan-A tire. See Figure 34.10 for comparison.

variation. Analysis of all of the field data with respect to mileage reveals that spare tires age at roughly 70% the rate of the average on-road tire and that tires that are driven 24,000 km/year age at a rate that is 15%–20% higher than those that are driven only 12,000 km/year. Similarly, we have estimated that parking in full sun versus parking in the shade can lead to a variation in peak temperature of easily 10°C. Distributions in parking behavior could easily account for a significant fraction of the total variability. It is interesting in this regard that the variability in tire property retention for tires collected from the same vehicle is considerably smaller than the overall variability.

Irrespective of the source of the aging variability, the magnitude has a significant impact on developing a specification for tire aging. For example, when an average tire drops to 40% of its initial peel strength, the worst-case tire of the same age will have dropped to ~25% of its initial peel strength. It would be necessary to age an average tire almost twice as long to reduce the properties to this level. Since tire disablements are relatively rare events (parts per million), they likely occur only under worst-case conditions. Aging a tire in the laboratory to reproduce a worst-case tire would be extremely challenging. Fortunately, the similarity in variability across tire types and brands suggests that it should be possible to age a tire in the laboratory that has the properties of an average tire aged in the field and to require performance levels with a sufficient safety margin to insure acceptable performance of the worst-case tire.

34.3.2 Oven Aging

Elongation to break for the wedge rubber versus oven-aging time are shown in Figures 34.14 through 34.17 for the 3 SUV/Minivan tires and the Large Car tire. Skim rubber peel strength versus oven-aging time is shown in Figures 34.18 through 34.21 for the same tires. In comparison with the field data, the oven-aging data show much smaller variability. This is consistent with the argument that most of the field variability is due to variations in external environment. Aging rates were determined by a least squares fit of the data to Equation 34.2. The skim rubber peel strength data is for the center of the skim. Values have also been measured for the skim rubber in the belt edge region. The data for the different oven conditions have been shifted to 70°C, air inflation at the maximum sidewall pressure. The shift factor for 50/50 N_2/O_2 relative to air is 1.3–1.4. The shift factors for 60°C and 50°C relative to 70°C are 0.6–0.65 and 0.3–0.35, respectively. The shift factors

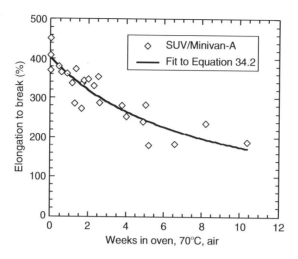

FIGURE 34.14 Elongation to break of wedge rubber versus time in oven at 70°C, air for the SUV/Minivan-A tire.

for the different variables (temperature, fill gas...) are independent of one another and are the same for the different tires. Based on the observed shift factors, we estimate that the shift factor between 65°C and 50/50 N_2/O_2 relative to 70°C and air inflation is ~1.15. That is, the higher concentration of oxygen in the 50/50 N_2/O_2 with 65°C more than compensates for the slightly higher temperature using air as the fill gas.

One key consideration in the comparison of oven and field aged tires is that they were produced at different times. The tires used in the oven-aging studies were produced 2–4 years after the newest tires in the field aging studies. If rubber compounding strategies or tire design and construction changed significantly during this period, then a comparison of aging rates may not be valid. Comparison of tire cross sections from the oven-aged and field-aged sets did not suggest any obvious changes in design and/or construction that would affect rubber aging rates. While it

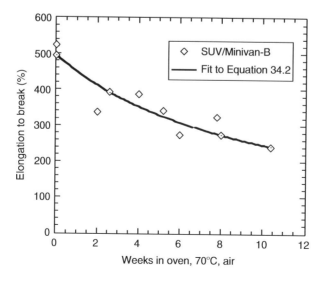

FIGURE 34.15 Elongation to break of wedge rubber versus time in oven at 70°C, air for the SUV/Minivan-B tire.

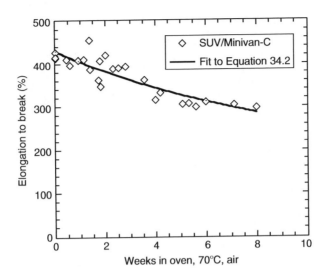

FIGURE 34.16 Elongation to break of wedge rubber versus time in oven at 70°C, air for the SUV/Minivan-C tire.

is not possible to quantify explicitly how any changes in rubber compounding might influence oxidation rates, it is possible to compare the initial properties from nominally identical tires in the oven and field studies. Another way to evaluate for possible changes in rubber oxidation chemistry is to plot properties such as strain ratio at break versus modulus at 100% elongation for the different data sets. As discussed elsewhere, log–log plots of these properties yield a straight line with a negative slope close to 0.75, which is indicative of oxidative aging. The exact value of the slope has been found to depend on rubber composition [9–11]. If the slope in the oven data set is different from that of the field data set, then it is likely there has been a change in rubber composition and that acceleration factors derived from kinetic analysis of the two data sets may be distorted by that change.

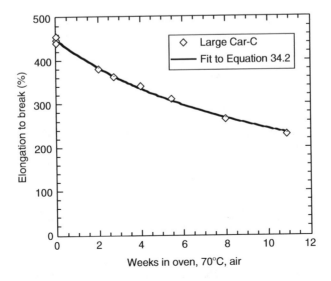

FIGURE 34.17 Elongation to break of wedge rubber versus time in oven at 70°C, air for the Large Car-C tire.

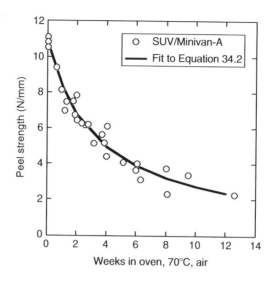

FIGURE 34.18 Peel strength of skim rubber (center) versus time in oven at 70°C, air for the SUV/Minivan-A tire.

The ratio of initial properties between the newer tires used in the oven study and the older tires used in the field study are reported in Table 34.4. The initial skim peel strengths (both center and belt edge) of the three SUV/Minivan tires are identical between the field and oven data sets. The initial cross-link density is somewhat more variable, but this may be due in part to different levels of cure. While this does not guarantee a constant composition over this time period, it does suggest that peel retention in the skim rubber is likely to yield a consistent correlation for these three tires. By contrast, the initial elongation to break and modulus values are significantly different in the two data sets for the three SUV/Minivan tires. The elongation to break is higher and the modulus is lower for

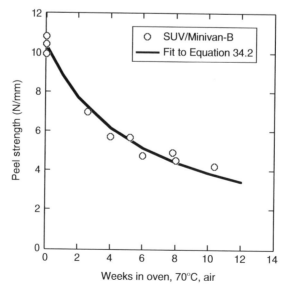

FIGURE 34.19 Peel strength of skim rubber (center) versus time in oven at 70°C, air for the SUV/Minivan-B tire.

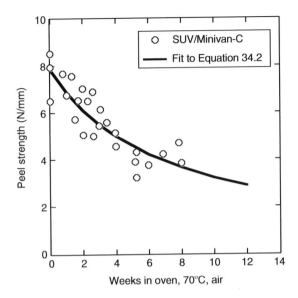

FIGURE 34.20 Peel strength of skim rubber (center) versus time in oven at 70°C, air for the SUV/Minivan-C tire.

the newer oven tire set. Further evidence that a change in composition in the wedge rubber between the two data sets is provided by the measurably different slopes in the Ahagon plot for SUV/Minivan tire "A" for the two data sets, as shown in Figure 34.22. It is impossible to predict what effect this change will have on oxidation kinetics; however, the clear evidence of different composition makes any attempt at correlation suspect. In contrast to the three SUV/Minivan tires, the large car tire shows significant differences in the initial properties in the skim. The cross-link density in the newer tires used in the oven test is higher and the peel strength lower than observed in the field set. For this tire, the initial elongations in the two sets are close, but modulus values are significantly different.

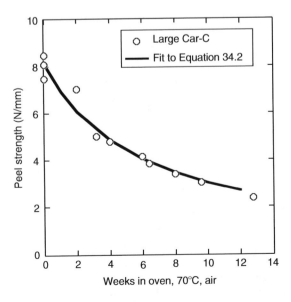

FIGURE 34.21 Peel strength of skim rubber (center) versus time in oven at 70°C, air for the Large Car-C tire.

TABLE 34.4

Ratio of Initial Properties, Oven/Field Evaluation

Vehicle Type-/Tire Brand	Skim Peel Strength	Skim Cross-Link Density	Wedge Peel Strength	Wedge Elongation to Break	Wedge Modulus
SUV/Minivan-A	1.01	1.13	1.01	1.17	0.88
SUV/Minivan-B	0.95	0.99	1.10	1.50	0.87
SUV/Minivan-C	1.01	1.10	0.96	1.28	0.87
Large Car-C	0.79	1.32	0.79	1.07	1.14

Based on the above arguments, with peel strength measurements for the three SUV/Minivan tires. We will also include the elongation to break data for the Large Car-C tire even though the difference in wedge rubber modulus between the field and oven sets makes the comparison between these data sets somewhat questionable.

The rates of property change in the oven at 70°C are compared in Table 34.5 to those determined for field data from Phoenix. The acceleration factors for peel strength retention of the skim rubber in the skim and wedge regions for the 3 SUV/Minivan tires are identical within experimental error (30 ± 2.5). The acceleration factor for the skim cross-link density, having a value of 21, is somewhat smaller but also reasonably constant. One possible explanation could be the effects of diffusion limited oxidation (DLO). It may be envisaged that in oven-aged tires, the bottom (#1) belt might be measurably more oxidized than the top (#2) belt, due to its closer proximity to the high oxygen concentration inside the tire, as demonstrated in earlier modeling work for diffusion limited oxidation in tires [12,13]. In a peel test, the ultimate value will be determined by the least extensible bonds, i.e., those nearest the bottom belt. For the cross-link

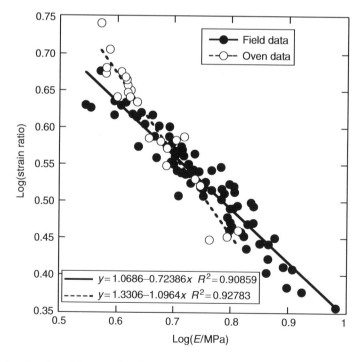

FIGURE 34.22 Log (strain ratio) versus log (modulus) for field and oven-aged SUV/Minivan-A tires. Both sets are linear indicating oxidative aging, but the slopes are different indicating a change in composition of the wedge rubber.

TABLE 34.5

Comparison of Property Change Rates, Oven/Field

A. Skim Peel Strength Retention Rate Constant (Center), β, weeks^{-1}

Vehicle Type-Tire Brand	70°C-Air	Phoenix	Acceleration Factor
SUV/Minivan-A	0.292	0.0096	30
SUV/Minivan-B	0.168	0.0057	29
SUV/Minivan-C	0.142	0.0043	33
Large Car-C	0.167	0.0026	64

B. Skim Cross-Link Density Rate Constant, α, weeks^{-1}

Vehicle Type-Tire Brand	70°C-Air	Phoenix	Acceleration Factor
SUV/Minivan-A	0.049	0.0023	21
SUV/Minivan-B	0.048	0.0021	23
SUV/Minivan-C	0.031	0.0017	18
Large Car-C	0.045	0.0012	38

C. Skim Peel Strength Retention Rate Constant (Belt Edge), β, weeks^{-1}

Vehicle Type-Tire Brand	70°C-Air	Phoenix	Acceleration Factor
SUV/Minivan-A	0.295	0.012	25
SUV/Minivan-B	0.18	0.0055	33
SUV/Minivan-C	0.14	0.0047	30
Large Car-C	0.098	0.0022	45

D. Wedge Elongation to Break Rate Constant, β, weeks^{-1}

Vehicle Type-Tire Brand	70°C-Air	Phoenix	Acceleration Factor
SUV/Minivan-A	0.130	0.0075	17
SUV/Minivan-B	0.102	0.0027	38
SUV/Minivan-C	0.063	0.0045	14
Large Car-C	0.085	0.0026	33

E. Wedge Modulus Rate Constant, α, weeks^{-1}

Vehicle Type-Tire Brand	70°C-Air	Phoenix	Acceleration Factor
SUV/Minivan-A	0.057	0.00390	15
SUV/Minivan-B	0.052	0.00158	33
SUV/Minivan-C	0.018	0.0024	8
Large Car-C	0.038	0.0019	20
Small Car-BFG	0.0655	0.0023	28

density measurement, however, the value will be the average of the cross-link density gradient, of which the values decrease from the bottom belt to the top belt, thus resulting in a lower measured overall acceleration factor. The oxidation of field tires is not nearly as limited by diffusion because of the lower temperatures; thereby the skim rubber exhibits much less of a gradient between the belts. The acceleration factor for the elongation to break and modulus for these three tire sets is highly variable. This result is totally consistent with observed initial property data that strongly suggested peel retention would yield more consistent results than elongation to break due to the likely rubber compound changes, which have affected the elongation to break properties in the wedge region. The low acceleration factors for elongation to break and modulus for two of the

SUV/Minivan tires cannot be a result of a fundamentally different temperature dependence (leading to a different acceleration factor) in the wedge regions in these tires since the measured temperature dependence between 50°C and 70°C is identical for elongation and peel in these tires. Changes in geometry can also be ruled out as an influence on the acceleration factor as the peel strength of the skim rubber in the belt edge near the wedge yields good correlation. The peel retention acceleration factor for the Large Car-C tire is much higher than for the other tires studied here. Again, this can be explained by the observed differences in initial properties that suggest direct comparisons of rates may be suspect. The tensile properties for the wedge rubber in this tire were relatively similar between the field and oven data sets. The acceleration factor for elongation to break is 33 while that for the modulus is 20. These results are reasonably consistent with the peel strength and cross-link density results for the SUV/Minivan tires. An overall acceleration factor was determined by combining the peel loss rates of the SUV/Minivan tires and the elongation to break loss rate for the Large Car-C, with the value of the acceleration factor being 31 ± 2.5. It is interesting to note that the oven to field acceleration factor estimated from the light truck tire previously studied was 34 ± 6. While this agreement must be considered somewhat fortuitous, it is nonetheless consistent.

Based on comparisons of oven and field aging rates of properties for which there is a reasonable expectation that the initial properties are constant in time over the course of the experiment, we conclude that the acceleration factor between oven aging at 70°C with air inflation at the maximum sidewall pressure and the average tire aging in Phoenix is ~31. The acceleration factor is insensitive to tire size and tire construction ranging from small car tires to at least SUV/Minivan tires and possibly even light truck tires. The fact that aging acceleration factors appear to not depend strongly on tire size may seem somewhat surprising. Further support for this result is provided by a more detailed comparison of the acceleration factors at different temperatures relative to Phoenix as shown in Figure 34.23. The temperature dependence for the different tires is very similar. The only significant discrepancy is the 60°C for the light truck tire, which has a significantly lower acceleration factor than other tires in the set at this temperature. The Phoenix point falls somewhat above the line extrapolated from the oven data. This is hardly surprising since the point is plotted at the

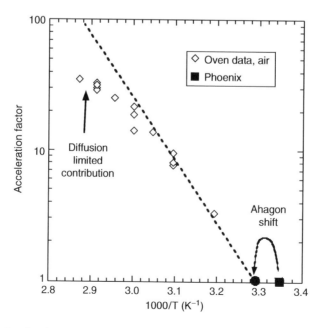

FIGURE 34.23 Acceleration factor versus temperature. Phoenix field data point is shifted 5°K based on work published by Ahagon. (From Kaidou, H. and Ahagon, A., *Rubber Chem. Technol.*, 63, 698, 1990.)

ambient air temperature. As noted in the previous paper, tire temperatures will certainly be higher than ambient due to convective heating from the road, radiative heating from the sun, and internal heating from driving use. Ahagon has estimated that this leads to an average temperature increase of 3°C–5°C, consistent with Figure 34.23 and with our variability analysis in the previous paper. Another point to make is that the data appears to curve off of the straight line over 6°C because of the effects of diffusion limited oxidation.

An acceleration factor of 31 means that reproducing the "age" of an average six-year-old tire in Phoenix would require ~10 weeks in the oven at 70°C using air as the inflation media. The time can be shortened to about 8.5 weeks using 65°C and 50/50 N_2/O_2 inflation or seven weeks using 70°C and 50/50 N_2/O_2. Other conditions can be estimated from the different shift factors.

34.4 CONCLUSIONS

Despite the significant effects of field variability on rubber aging in tires, we have shown that it is possible to derive reasonable estimates of aging rates for key rubber properties in tires. The kinetics of aging for key rubber properties determined from sets of four different tires types and brands collected from the field and from exposure in laboratory ovens are compared to show that oven aging of tires mounted on wheels and inflated can anticipate field aging results. For a given tire, field aging rates are controlled by the ambient temperature. In North America, Phoenix serves as a reasonable worst-case location. Rubber properties for which aging rates have been measured include peel strength in the skim rubber, cross-link density of the skim rubber, elongation to break of the wedge rubber, and modulus of the wedge rubber. Since the oven exposures were performed on tires that were manufactured at significantly different times from the tires in the field set, initial properties were compared to select tires and properties that were most likely to be unchanged. For three different brands of SUV/Minivan tires, the skim rubber appears to have remained consistent for the oven and field sets, but there was evidence that the wedge rubber composition had changed. By contrast, the wedge rubber for the Large Car-C tire was reasonably consistent, but the skim rubber properties were not. The acceleration factor for all of the properties whose initial values were consistent between the field and oven sets were consistent with one another. Various different conditions can be defined for oven aging of tires: mount tires on wheel, inflate tire with air or 50/50 N_2/O_2 to the maximum sidewall pressure, and expose at 65°C–70°C. Depending on the specific conditions, it is possible to produce tires in 7–10 weeks that appear to be equivalent to tires aged in Phoenix for six years independent of tire size or brand. Future work will focus on mechanical evaluations of aged tires. As discussed in the body of this chapter, a critical part of this evaluation is determining the safety margin in property retention required to insure that worst-case tires will continue to perform in an acceptable manner.

One final point concerns the subject of fill gas replenishment. It is undoubtedly apparent to the reader that as time spent in the oven increases, the concentration of oxygen in the tire cavity concomitantly decreases. There is a critical minimum concentration of oxygen required to maintain the correct reaction rate and this concentration may be reached for certain tires if the tire is not occasionally removed from the oven and refilled with the appropriate oxygenated gas. The frequency of filling is determined by such factors as inflation pressure retention (IPR), initial oxygen concentration, and oven temperature. The oven-aged tires mentioned in this paper were not replenished. The tires chosen, along with the conditions, did not lead to oxygen starvation in the times reported (based on measurement of fill gas composition as a function of oven-aging conditions). Other tires not reported here, however, were taken out to much longer aging times and the oxygen concentration did indeed fall below the critical minimum. Henceforth, to avoid starving the belt area of oxygen, a regimen of replenishment has been implemented on all future tire aging studies conducted by the authors. Tires are to be removed from the oven every two weeks, the valve core pulled, and the tire reinflated with 50/50 N_2/O_2. This regimen will ensure that the critical concentration of oxygen is always present in the tire cavity.

REFERENCES

1. NHTSA Office of Defect Investigation, Engineering Analysis Report and Initial Decision Regarding EA00–023: Firestone Wilderness AT Tires. 2001, NHTSA.
2. Baldwin, J.M. Accelerated Aging of Tires, Part I. Presented at a meeting of the Rubber Division, American Chemical Society. 2003. Cleveland, OH.
3. Flory, P.J. and J. Rehner, *Journal of Chemical Physics*, **11**, 512, 1943.
4. Till, P.H., *Journal of Polymer Science*, **24**: 301, 1957.
5. Baldwin, J.M., D.R. Bauer, and P.D. Hurley. Field Aging of Tires, Part III. Presented at a meeting of the Rubber Division, American Chemical Society. 2005. San Antonio, TX.
6. J.M. Baldwin, David R. Bauer, and Paul D. Hurley, *Rubber Chemistry and Technology*, **78**(5), 754–766, 2005.
7. Baldwin, J.M., M.A. Dawson, and P.D. Hurley. Field Aging of Tires, Part I. Presented at a meeting of the Rubber Division, American Chemical Society. 2003. Cleveland, OH.
8. NHTSA Phoenix Tire Dataset 4.0, http://www-nrd.nhtsa.dot.gov/vrtc/ca/tires.htm
9. Ahagon, A., *Rubber Chemistry and Technology*, **59**, 187, 1986.
10. Ahagon, A., M. Kida, and H. Kaidou, *Rubber Chemistry and Technology*, **63**, 683, 1990.
11. Kaidou, H. and A. Ahagon, *Rubber Chemistry and Technology*, **63**, 698, 1990.
12. Gillen, K.T., M. Celina, and R. Bernstein, Validation of improved methods for predicting long-term elastomeric seal lifetimes from compression stress-relaxation and oxygen consumption techniques. *Polymer Degradation and Stability*, **82**(1), 25–35, 2003.
13. Gillen, K.T. et al., *Trends in Polymer Science*, **5**, 250, 1997.
14. Wise, J., K.T. Gillen, and R.L. Clough, Quantitative model for the time development of diffusion-limited oxidation profiles. *Polymer*, **38**(8), 1929–1944, 1997.
15. David R. Bauer J.M. Baldwin, and Kevin R. Ellwood, *Rubber Chemistry and Technology*, 78(5), 777–792, 2005.
16. David R. Bauer J.M. Baldwin, and Kevin R. Ellwood, Correlation of Rubber Properties Between Field Aged Tires and Laboratory Aged Tires, Part II, presented at a meeting of the Rubber Division, American Chemical Society, San Antonio, TX, May 16–18, 2005.

35 Recent Developments in Rubber Mixing and Cord Calendering in Tire Production

Gerard Nijman

CONTENTS

35.1 INTRODUCTION

Sport cars with a maximum speed of higher than 270 km/h and sport utility vans are widely available nowadays. Tires for these cars have to fulfil very tough requirements regarding car handling, safety, and grip as well as slip resistance under extreme road conditions. Environmental regulations limit the freedom of possibilities in tire design. Introduction of silica tires by almost every premium tire manufacturer in the early 1990s of the last century meant a big step forwards in tire performance, especially on wet, snowy, and icy road conditions. A very good understanding of the tire manufacturing process is a precondition for a capable production of these very high performance tires. Tires are produced in four production steps. Before tire moulding and vulcanization the tire has to be built on tire building machines out of various components. Tire components are made on extrusion lines or calendering lines. The first step is the mixing of rubber compounds out of raw materials such as natural or synthetic polymers, fillers, process oils, curing systems, chemicals, etc.

This chapter deals with recent developments of rubber mixing and calendering. Rubber extrusion is discussed elsewhere in this book. Although it dominates rubber mixing in the last ten years, the mixing of silica-filled tread compounds is only slightly discussed in this chapter, since another chapter in this book totally covers the mixing of silica compounds.

35.2 INTRODUCTION ON RUBBER MIXING

For almost a century, rubber compounds are produced by internal mixers. These mixers are batch mixers. Figure 35.1 shows the concept of an internal mixer. It is characterized by a mixer entry chute through which the material can be fed. The ram or floating weight pushes the raw material towards the mixing chamber. The mixing chamber has got the shape of a horizontal "8" in which two rotors counter-rotate. At the bottom a drop door can be opened to discharge the stock into a downstream device. If both the ram and drop door are closed the mixing process is taking place in a closed mixing chamber. All the materials take part in the mixing process.

Basically two rotor types are applied in the internal mixer: The tangential rotor type and the intermeshing rotor type. The latter one interferes with the adjacent rotor and turns at the same rotor speed compulsory. In the drop door or through the side plates a thermocouple is mounted to record the temperature of the rubber.

In the mixing chamber, both rotors and the drop door are temperature controlled by means of a closed-loop water control. The positive effect of steady thermal conditions on batch to batch uniformity has been reported by Melotto.[1]

The material flow in an internal mixer is oriented vertically. On top of the internal mixer a more or less sophisticated feeding system can be found. Underneath the mixer a downstream device, such as a sheeting extruder with one or two tangential screws, or two roll mills are placed.

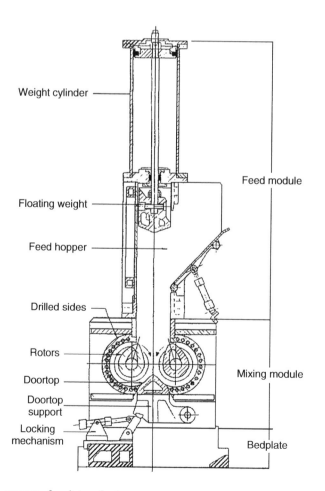

FIGURE 35.1 Basic concept of an internal mixer.

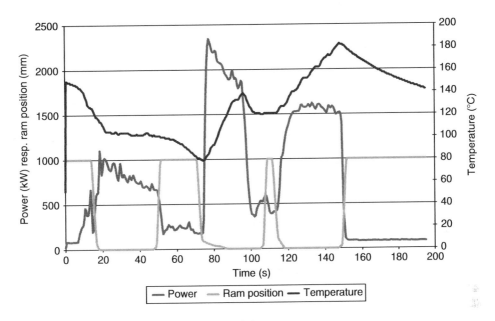

FIGURE 35.2 Typical fingerprint for masterbatch mixing.

The shape of the mixer entry chute allows the feeding of raw materials in almost any trade form. Rubber bales, the most widely used trade form for rubbers and polymers, can be fed as one piece in an industrial-sized mixer. Feeding of material is very convenient, especially for tangential type of mixers.

By recording the motor power, the temperature, and the position of the ram the progress of the mixing process can be followed. Such a diagram of a complete mixing process is called a fingerprint of the batch. The integrated power, which is the total energy input, can be added in the fingerprint. Figure 35.2 shows a typical fingerprint for a masterbatch produced on an industrial-sized mixer. It is characterized by a high temperature increase and a highly fluctuating power input. It can be clearly seen that the mixing process is highly non-isothermal. The maximum temperature determines the total mixing time.

The concept of the internal mixer is well accepted in the rubber industry, especially for the production of masterbatch compounds. Within a variety of materials the deformation of rubber compounds is relatively difficult due to the very high viscosity. Mixing devices have to be strongly powered and have to be designed in a very robust way in order to withstand the very high forces and momentum acting on the device. The most important benefits of an internal mixer are

- Compact however robust design
- Wide mixer entry chute to adopt almost any supply form of the raw materials
- Relatively low raw material prices, since special preparation and handling devices during raw material production are not needed
- Closed mixing chamber to enable a clean and dust-free rubber mixing

The high price for these benefits of the internal mixer is the batch-wise production of rubber compounds. This leads to quality variations and batch nonuniformity. The temperature increase of the compound limits the mixing process and therefore the mixing quality. Eventually the mixing quality is so poor that an extra mixing step, remilling, is needed. In most cases the curing system can only be supplied during a separate, final batch-mixing, step because curatives can only be processed below 100°C–110°C.

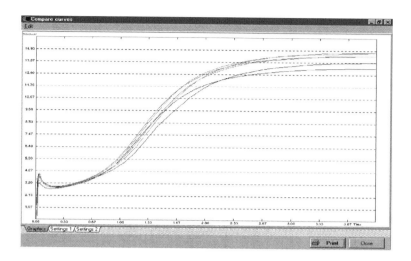

FIGURE 35.3 Rheo curves of various batches of a compound with the same recipe.

It is well known that the rheo curve, a diagram in which the curing characteristics are given, is a fairly good measure to determine the consistency and uniformity of batch mixing. For an example, Figure 35.3 shows rheo curves of a compound with the same recipe. This is generally regarded as a sufficient batch to batch consistency. Variations in curing characteristics are caused partly by variations in the mixing process. Viscosity variations and therefore variations in the compound's processability can be expected as well. This yields variations in extruded profile cross-section geometry.

The effect of the mixing behavior on the rheological properties of styrene–butadiene rubber (SBR) compounds has been reported by Leblanc.[2]

Figure 35.4 shows the concept of the internal mixer proposed by Banbury.[3] It can be seen easily that the basic concept of the internal mixer did not change since almost 90 years. Obviously the advantage of the easily feeding is preferred above the rather poor between batch consistency.

In the last 20–30 years material handling in mill rooms as well as the mixing process is highly automated. Also, downstream equipment runs in an automatic control mode.

As an example a typical layout of a mill room is depicted in Figure 35.5. Bulk materials and fluids are supplied by means of road tankers and transferred and stored in storage silos. On top of the internal mixer day bins can be found. The storage of the "daily" consumption of bulky materials is done very close to the mixer to allow short weighing and charging times and letting the materials come to the required ambient conditions. The elastomers are usually supplied in bales and have to be cut or granulated in order to meet the weighing tolerances needed for the next batch to be mixed. Charging of rubber is usually done through conveying belts.

Small bags of chemicals and other small quantities of raw materials are pre-weighed offline and can be put on the same conveyor belt.

The total volume of the raw materials is much higher than the volume of the rubber batch which is going to be produced out of it. This is because of the fact that especially fillers contain a lot of air. To avoid any material leaving unmixed out of the mixing chamber and to achieve the required mixing quality, the mixing chamber is principally underfilled. The fill factor, which is a very important mixing parameter is defined as the ratio between the volume of the batch and the gross (fluid filled) volume of the mixing chamber. In other words

$$\Phi = \frac{m}{\rho V_{\text{mixer}}} \tag{35.1}$$

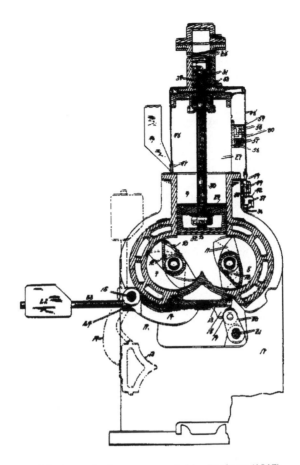

FIGURE 35.4 The concept of the internal mixer as suggested by Banbury (1917).

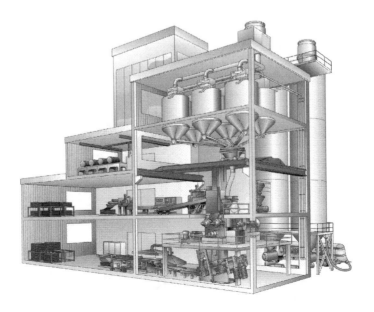

FIGURE 35.5 Typical general layout of a mixing room.

where
  Φ is the fill factor
  m is the batch weight
  ρ is the density of the compound
  V_{mixer} is the volume of the mixing chamber

Typical values for the fill factor are 0.6–0.8 depending on the compound's recipe and the type of rotors being applied.

Especially in tire manufacturing, the mixing process has been optimized with respect to productivity. High rotor speeds combined with high fill factors enable to increase the productivity and are applied widely. Various textbooks deal with the mixing process in such mixers.[4–7] Many papers have been published which deal with the mixing quality obtained by means of batch mixers.[8–10] It is well known that the compound quality is limited due to the energy balance, the flow behavior in the mixer, and batch to batch variations. By a correct compound design one could correct for the influence of the mixing process on the final properties of such a compound. However, the viscosity of the rubber compound and therefore its processability depend on the mixing process as well.[11,12] This yields the maintaining of rather big geometry tolerances on extruded profiles[13] and other products made of rubber.

Various papers deal with the flow behavior of the rubber compound in an internal mixer. Basically the mixed good is treated as a fluid in these papers.[14–18] Flow visualization studies have been published in which a laboratory mixing chamber has been made of Plexiglas and the mixing chamber is filled with a high viscous fluid. By adding tracers, the flow in an internal mixer can be visualized. Numerical simulations have been carried out by Regalia,[19] Hu,[20] Clarke,[21] Yao,[22] and Kim.[23] Although very interesting flow behaviors have been achieved by means of numerical simulation, calculation results differ significantly from observations made from industrial mixing processes, mainly due to the complex rheological behavior of a rubber compound (disregarding the fact that during at least a rather big part of the masterbatch mixing process there is not any flow of a continuum as numerical fluid flow simulation requires). A lot of free surfaces exist due to the partially filled mixing chamber. As long as free surfaces are very difficult to predict in numerical simulation, simulation and practise might differ quite tremendously.

An analysis of the flow behavior which is more or less confirmed by numerical simulations is presented by Leblanc.[14] The total flow in an internal mixer (tangential type) can be divided into a laminar, circulating flow which is pushed by the rotor tip, an axial directed flow, as an effect of the helix angle between axis and rotor wing and a leakage flow between rotor tip and mixing chamber wall (refer to Figure 35.6), or

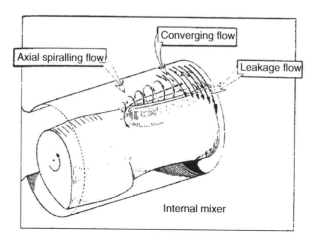

FIGURE 35.6 Schematic representation of flows in a mixing chamber of an internal mixer.

$$\varphi_{tot} = \varphi_{lam} + \varphi_{ax} + \varphi_{leak} \tag{35.2}$$

If the transport from the left rotor regime to the right rotor regime is taken into account, this equation is completed by

$$\varphi_{tot} = \varphi_{trans} + \varphi_{lam} + \varphi_{ax} + \varphi_{leak} \tag{35.3}$$

Laminar mixing is basically taking place throughout the mixing chamber. Because of the general layout the rotor design axial flow is split up at the edge of a rotor wing, favoring distributive mixing. Dispersive mixing is mainly taking place in high shear zones, e.g., between rotor tip and mixing chamber wall because for dispersive mixing a sufficient high normal or shear stress has to be applied on the particle to be dispersed. If φ_{tr} and φ_{ax} are relatively large with respect to the flow contributions, distributive mixing is favored. The narrower the clearance between rotor tip and mixing chamber wall, the higher the shear forces will be. However, φ_{leak}, the contribution of the total flow in the dispersive mixing region, will be smaller as well. The dispersion effect of an internal mixer should be understood by optimizing the chance that every part of the batch will flow over the rotor tip combined with a sufficient high shear forces in the dispersive region.

35.3 MIXING BY MEANS OF INTERNAL MIXERS

35.3.1 MIXING PROCESS IN AN INTERNAL MIXER

It is well known that the cycle time of the mixing process in an industrial-sized internal mixer is limited by the increase of the compound temperature. For example, typical mixing times on a 400 L tangential mixer with four wing rotors is of the order 200 s as can be derived from Figure 35.2. Especially during the last mixing phase the temperature increase of the compound is approximately $1°C/s$.

The cost price of rubber mixing is determined by the prices of the applied raw materials, the direct machine and labor costs, as well as overhead costs such as depreciation, maintenance, and indirect labor costs. If mixing times should be shorter or the fill factor should be higher the throughput of the mixer line would be higher; however it is well known that mostly the degree of mixing (e.g., carbon black dispersion) would be lower. Wiedmann and Schmidt[8] have shown that increasing the fill factor of a mixing process beyond the optimum value leads to a tremendous decrease of the filler dispersion quality. It can be explained by the fact that the mixing chamber is overfilled and the floating weight will not come to its closed position during the mixing cycle. It means that a part of the raw materials will remain in the entry chute and will not be mixed well into the batch. On the other hand Büker has shown that remilling of a tire tread compound leads to both better carbon black dispersion and processability of the compound on extrusion lines.[24] It is expected that reducing of mixing time and overfilling the mixing chamber will lead not only to a (slight) higher productivity but also to a poorer quality, more rejected production, and higher scrap rates.

To achieve costs improvement in the mixing operation a sort of hyperbolic relationship between productivity and mixing quality has to be taken into account. By reducing the rotor speed, for instance, a lower temperature increase can be expected, due to less friction losses, and therefore a longer mixing time and more intensive mixing. In other words, a better mixing quality leads in this case to a lower throughput.

The mixing process generally starts with the charging of (a part of) the raw materials into the mixing chamber, followed by putting the floating weight down to the mixing chamber. After following the mixing cycle procedure, finally the mixing process ends by opening the drop door and discharging the batch into the downstream equipment. For masterbatch mixing, the most important mixing cycle, is depicted in Figure 35.7. This mixing cycle is well known as the straight mixing process. Upside down mixing is in some cases preferable especially if large quantities of

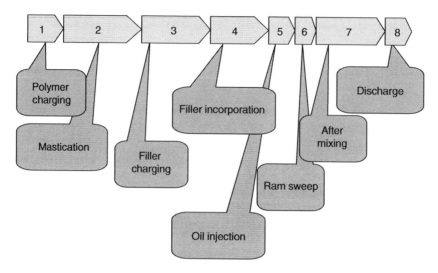

FIGURE 35.7 Masterbatch mixing procedure (straight mixing).

process oil have to be mixed into the compound. With upside down mixing the fillers are charged first, followed by process oil and polymers. This process will not be discussed in this chapter.

A typical fingerprint of the mixing process is given in Figure 35.8 in which the corresponding phases of the mixing cycle of Figure 35.7 are given as well. The fingerprint is a diagram in which the power supply, the signal of the thermocouple, the rotor speed, and the position of the floating weight is given as a function of the progressing mixing time. The phases of the mixing process can be easily recognized from this diagram. After charging the polymers, the floating weight comes down to its lowest position almost immediately, because in this stage of the mixing process the mixing chamber is rather underfilled. During the first "ram down" phase, the polymers are plasticated and blended. At the end of this phase a more or less homogeneous blend can be observed. After lifting the floating weight the fillers are added into the mixing chamber and it takes a while before the floating weight reaches its lowest position, due to the low bulk density of the fillers compared to its density. If the ram comes down almost direct after closing the mixing

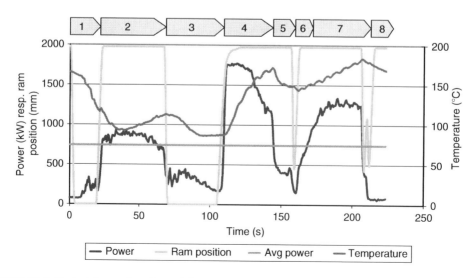

FIGURE 35.8 Typical fingerprint of a straight masterbatch mixing process.

chamber, the fill factor has been chosen too low and a less optimal mixing quality can be expected. On the other hand, if the ram does not come to its final position before the process oil is being injected, the mixing chamber is overfilled and especially the filler dispersion is expected to be poor.

After oil injection the power input drops, since the process oil initially acts as a lubricant between rubber batch and mixing chamber wall. However, after reasonable time the process oil has been incorporated, which can be seen from the increasing power supply. A cleaning action (lifting and lowering the floating weight) is followed by the last mixing phase in which ideally the power input reaches a horizontal asymptote, indicating that every material has been incorporated and filler dispersion is proceeding, according to a study by Leblanc.[14]

A typical example of a fingerprint of a final batch-mixing process is given in Figure 35.9. Remilling as well as final batch mixing start with adding the masterbatch compound followed by adding the small components (usually in a small bag). After material feeding the floating weight comes down and then the compound is mixed until a certain discharge criterion is reached. Sometimes the floating weight is lifted for a cleaning action.

Leblanc[14] also refers to a second power maximum after adding the fillers in masterbatch mixing. It principally indicates the completion of filler incorporation. In industrial mixing, however, this second power peak is often confused with an arbitrary power maximum when finally the floating weight comes to its lowest position (refer to Figure 35.10). Especially in intermeshing mixers, the interclearance region between the two rotors can cause a temporary material jam, which results into a slight lifting of the ram. As soon as the material has been taken into the region between rotors, mixing chamber, and drop door the floating weight is able to come to its lowest position.

Clearly recognizable is the big fluctuation of the power supply. During the phases in which the power input is high, mixing action takes place together with a remarkable temperature increase.

Balancing the power input with the enthalpy increase of the compound and mixer body, drive losses give the heat flow into the cooling system of the internal mixer. A typical energy balance of a complete mixing process can be found in Figure 35.11. Less than 50% of the power input can be cooled away during typical masterbatch mixing. If the rotor speed is reduced until rather low speed an (more or less) isothermal mixing process can be achieved. Isothermal mixing is favored for

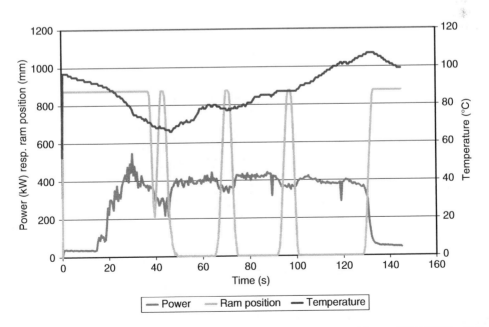

FIGURE 35.9 Typical fingerprint of a final batch-mixing process data obtained from a GK160SUK with fixed rotor speed (18/25 rpm).

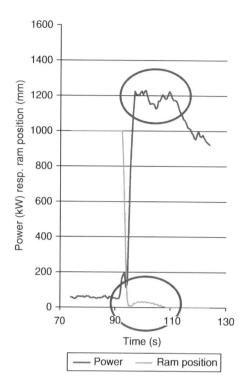

FIGURE 35.10 Illustration of the effect in the fingerprint of material jam in the interclearance zone.

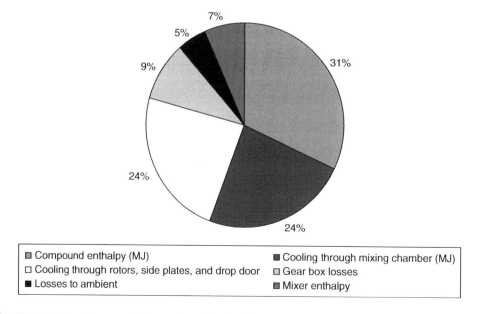

FIGURE 35.11 Typical energy balance of one batch-mixing process on a GK320E mixer (styrene–butadiene rubber/carbon black [SBR/CB] compound).

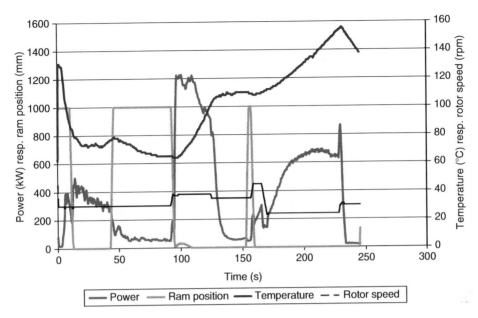

FIGURE 35.12 Typical fingerprint of a masterbatch mixing process on an intermeshing internal mixer (GK 320E (Harburg Freudenberger) with PES5 rotors; styrene–butadiene rubber/carbon black [SBR/CB] tread compound).

achieving a high dispersion quality and/or high degree of hydrophobation during silica mixing.[10,25,26]

Intermeshing mixers show a larger mixer and rotor surface–mixer volume ratio compared to a tangential mixer. It means that the cooling performance of intermeshing mixers is principally better than tangential mixers. However, charging and discharging of intermeshing mixers are poorer compared to tangential mixers. Both effects lead to a longer mixing and cycle time compared to a tangential mixing process with similar batch weight. A typical fingerprint of a masterbatch mixing cycle can be found in Figure 35.12. The differences between fingerprints of similar compounds mixed in intermeshing mixers compared to tangential mixers are characterized by

- Poorer intake of fillers
- Lower temperature increase during mixing phases and therefore a bigger cooling energy to power supply ratio
- Poorer discharge after the drop door is opened
- more generally, by a longer cycle time

35.3.2 MIXING OF SILICA COMPOUNDS

Today, modern tires have to meet very demanding handling requirements with respect to grip and skid under wet, snowy, or icy conditions. Rolling resistance shall be as low as possible and a high mileage will become more important because of environmental reasons. It is necessary to apply silica-filled tread compounds based on solution-based SBR (S-SBR) and high *cis*-butadiene rubber (*cis*-BR) polymers in high-performance tires.[27] The mixing of these compounds is complicated because the silica has to be hydrophobated by means of silane, e.g., TESPT.[28] This chemical reaction, also called the silanization reaction, lasts long with respect to mixing times and temperature may not rise too high (it is said that the reaction temperature may not exceed 160°C in order to avoid secondary reaction effects). It therefore requires one or two remilling phases to complete the silanization process. Although extensively discussed elsewhere in this book, the mixing of silica

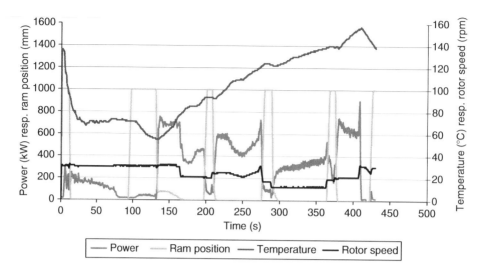

FIGURE 35.13 Typical fingerprint of a masterbatch mixing process of a solution-based styrene–butadiene rubber (S-SBR)/Silica/TESPT tread compound on a GK 320E (Harburg Freudenberger) with PES5 rotors.

compounds on industrial-sized mixers should be discussed here as well, especially in the context of the cooling performance of industrial mixers.

Silica compounds are mixed preferably in intermeshing mixers with variable and adjustable rotor speeds. Since the hydrophobation reaction is influenced by the content of ethanol (a split off product from this reaction) and by the moisture content, it is required to lift the floating weight several times during one mixing cycle. A typical masterbatch mixing cycle for S-SBR Silica silane compounds on a 320E internal mixer is given in Figure 35.13, in which it can be clearly observed that the temperature increase is limited by lowering the rotor speed.

The application of two roll mills in the downstream of such mixing lines is preferred, especially if the two roll mills are equipped with stock blenders (Figure 35.14). After discharging, the rubber batch will be sheeted on the two roll mill. By putting the sheet through the stock blender, the free surface from which the ethanol and moisture can be evaporated is rather big allowing the chemical reaction to proceed even at somewhat lower temperatures. A study by Zuuring et al.[29] has shown

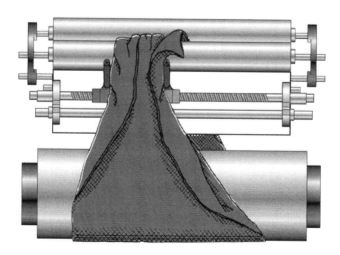

FIGURE 35.14 Two-roll mills with stock blender.

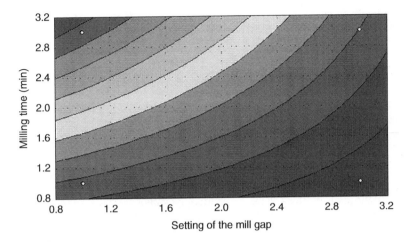

FIGURE 35.15 Influence of the open mill mixing process on the degree of silanization in a solution-based styrene–butadiene rubber (S-SBR)/Silica/TESPT compound as given in a contour plot of the Payne effect ($G'(0,56)$) according to Zuuring. With proceeding silanization the Payne effect drops tremendously. (From Zuuring, A, Diploma thesis, Saxion University, Enschede, NL, 2000.)

that a smaller mill clearance setting and a longer milling time yield better silanization conditions. This study also has shown that by proper two-roll milling after mixing in an intermeshing mixer, one remill step can be eliminated as can be recognized from Figure 35.15.

35.3.3 Cooling Efficiency of an Internal Mixer

It is interesting to study the mixing process of masterbatch mixing with remilling and final batch mixing. Within masterbatch mixing the filler incorporation phase can be compared with that after mixing phase. A typical tire tread formulation has been mixed in three steps on a Kobe Steel BB 270 tangential mixer equipped with four-flighted rotors of Mark H II. The mixing procedure for masterbatch mixing, remilling, and final batch mixing is given in Table 35.1. Fingerprints including power supply and records of the thermocouple have been analyzed as follows:

– During the black incorporation phase, the aftermixing phase, the second part of the remilling phase and the second part of the final batch mixing phase, an average power consumption has been taken

TABLE 35.1
Mixing Procedures on Kobe Steel's BB270 Mixer with 4WH Mark II Rotors

	Masterbatch Mixing	Remilling	Final Batch Mixing
Rotor speed (rpm)	60	60	30
Charging	Polymer	Masterbatch	Remilled batch
Mastication	30 s	—	—
Charging	Fillers + small components	—	Small components
Mixing	15 s	50 s	50 s
Injection	Of process oil	—	—
Oil incorporation	30 s	—	—
Ram sweep	15 s	15 s	15 s
After mixing	Until 155°C	Until 155°C	Until 105°C

- At the same phases the average temperature increase has been derived
- With the assumption that the specific heat of the compound is 1700 J/kg/K, the compound enthalpy increase per second has been estimated by means of

$$\frac{\Delta H}{\Delta t} = mc_{\mathrm{v}} \frac{\Delta T}{\Delta t} \tag{35.4}$$

in which ΔH is the enthalpy increase of the batch during time frame Δt, ΔT the corresponding temperature increase of the batch, m is the batch weight, and c_{v} is the specific heat of the compound.
- The ratio between average power and enthalpy increase per unit of time is compared between the various mixing phases.
- The area of the total cooling surface of the mixing chamber and the rotors is estimated to be 6 m^2. The temperature of the cooling water was 30°C. The heat flux can be calculated by dividing the cooling power by cooling area, or

$$\Theta_{\mathrm{cool}} = \frac{P_{\mathrm{cool}}}{A} = \frac{\Delta P_{\mathrm{input}} - \Delta H}{A \Delta t} \tag{35.5}$$

- The heat transfer coefficient from the rubber batch to the cooling water can be estimated by dividing the heat flux by the difference between the average rubber temperature and the cooling water temperature, or

$$\alpha_{\mathrm{th}} = \frac{\Theta_{\mathrm{cool}}}{T_{\mathrm{av,batch}} - T_{\mathrm{water}}} \tag{35.6}$$

Figure 35.16 shows the fingerprints of the masterbatch mixing process, remill process, and final batch-mixing process. The derivation of the data as mentioned above is clearly shown in the fingerprints. The results are summarized in Table 35.2. What can be seen is that cooling power is approximately one-third of the total power supply. Since the masterbatch cycle time is remarkably longer than the

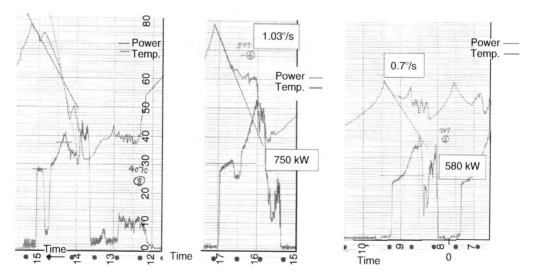

FIGURE 35.16 Fingerprint of a masterbatch (left), remill (middle) and final batch (right) mixing process of a natural rubber–styrene–butadiene rubber–butadiene rubber/carbon black (NR–SBR–BR/CB) compound, mixed on a Kobe Steel BB270 tangential mixer with 4WH mark II rotors. The temperature increase and average power during dispersion (only masterbatch) and after mixing phase are indicated in the diagram.

TABLE 35.2
Results of the Analysis of Fingerprints of Figure 35.16

		Masterbatch		Remilling	Final Mix
		Incorporation	After Mixing		
Avg. power	kW	900	700	750	580
Temp. increase	°C/s	2.15	0.9	1.03	0.7
Enthalpy increase	kW	844	353	403	267
Id. in % of avg. power	%	94	50	54	46

remilling time, it seems that the cooling effect during masterbatch mixing is poorer with respect to remilling and final batch mixing. Hence, it is found that during filler incorporation there is hardly any heat transfer from rubber to mixer. It can be explained by the fact that during filler incorporation the mixed good is not a continuum, but there are still a lot of (isolating) loose fillers available.

This study leads to interesting conclusions:

- Masterbatch rotors should be designed in order to incorporate the fillers and additives as efficient as possible.
- Remill batch and final batch rotors should be designed in order to favor distributive mixing and, hence, optimized cooling effect.
- The masterbatch mixing cycle should not be extended too long, but, in case of demanding mixing quality requirements two short mixing cycles, i.e., masterbatch mixing and remilling, might be more efficient than one extended masterbatch mixing cycle without remilling.

A similar study of a masterbatch mixing process on an intermeshing mixer (Harburg Freudenberger GK320E with PES3 rotors (former Krupp, Werner & Pfleiderer mixer concept) resulted in similar heat transfer coefficients and slightly more effective cooling due to higher cooling surface availability per unit weight of rubber batch). The mixing details can be found in Table 35.3.

In literature, some heat transfer coefficients from rubber to cooling water in an internal mixer can be found, e.g., in Refs. 30 and 31. The present heat transfer coefficients are in the same region as referred in literature. Wiedmann and Schmidt found a rather huge dependency of the heat transfer coefficient on the rotor speed. This can be explained by the fact that every time a rotor tip passe the mixing chamber wall, it leaves a thin layer of hot material to the wall, which is refreshed after the next tip passage. Knowing this, the benefits of the cooling effect of an intermeshing mixer with respect to the tangential type should be seen at the same rotor speed. In other words, the cooling performance of an intermeshing mixer at low speed might be poorer than the cooling effect of a tangential mixer at high rotor speed.

TABLE 35.3
Results of a Fingerprint Analysis of a Masterbatch and Remill Mixing Process of a Styrene–Butadiene Rubber–Carbon Black (SBR–CB) Compound on a GK320E Intermeshing Mixer with PES3 Rotors (Harburg Freudenberger)

		Masterbatch		Remilling
		Incorporation	After Mixing	
Avg. power	kW	980	520	650
Temp. increase	°C/s	1.7	0.24	0.26
Enthalpy increase	kW	673	99	103
Id. in % of avg. power	%	69	19	16

35.3.4 Mixing Process Optimization

Optimizing the mixing process can be done by studying the fingerprints. Schlag[32] has presented a study on the influence of the duration of each mixing phase on the carbon black dispersion quality of an NR/SBR/BR/N375-based tread compound. By observing the visual appearance of the compound at several phases during masterbatch mixing on a GK45E mixer he could distinguish the following appearances:

α. Start of the mixing process: Polymers are dosed in lumps.
β. Power input is going through a minimum. The structure of the blend is crumby and pigeon egg size BR lumps are still visible.
γ. The polymer blend shows a homogeneous appearance.
   a. Floating weight is down. The power input shows a peak. Carbon black is mainly loose and the incorporation into the polymers has just started.
   b. The power input goes through a minimum due to the rising of floating weight (material jam in the inter-rotor zone) A lot of loose black is visible.
   c. The floating weight has come to its lowest position. The power runs through a second peak. Carbon black dispersion is poor. Loose black visible.
   d. (power minimum).
   e. Carbon black incorporation completed. Dispersion quality is reasonable.

   In an experimental design the duration of the blending time was chosen between 2 (30 s) and 3 (60 s). After 180 s the mixing process was stopped. The discharge temperature was approximately the same (160°C); however, the quality of the compounds differs tremendously. The compound with the short blending time had a poor dispersion quality and some black was not yet incorporated. The mixing process of the compound with long blending time was completed, as shown in the power chart (Figure 35.17). In other words within the same mixing time, mixing quality can be improved by optimizing the mixing phases during one mixing cycle.

35.3.5 Recent Developments in Mixing Equipment

Although the basic concept of the internal mixer already exists for a long time, a couple of new developments in rotor designs and mixing concepts have been published.[33–40] In rotor designs two development directions can be recognized. The first direction is optimizing the cooling effect, in which the distributive mixing components of the flow (transport flow and splitting up of the axial flow) is being optimized.[37,39,40] This can be done by shortening the long wing and adding short

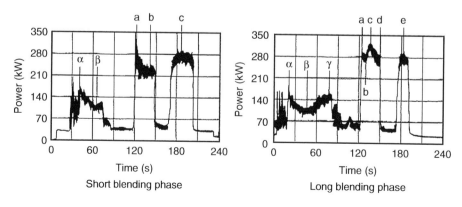

FIGURE 35.17 Power charts mode on a GK45E (PES3 rotors; Harburg Freudenberger) as a part of a design of experiments reported by Schlag. (From Schlag, L., Diploma thesis, University OOW, Emden, Germany, 1999.)

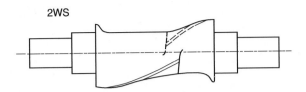

FIGURE 35.18 Example of a two-wing (tangential) rotor (2WS rotor of Kobe Steel Ltd).

wings in the middle of the rotor. The other direction is encouraging the incorporation of the material in the compound. Therefore, the material has to force to come into the laminar region and the high shear region.[33,38] A very long rotor wing or more long wings per rotor are needed to optimize the incorporation performance. A disadvantage is the relatively strong heat buildup in the compound during mixing which might lead to an additional remill step in order to fulfil the mixing quality requirement. Alternatively two roll mills in the downstream of the mixing line might solve the problem of achieving the required mixing quality.

Traditionally tangential rotors were two-wing flighted (see Figure 35.18). One wing was large, starting from the first edge and another shorter wing started from the other edge. Later on four-wing rotors were applied, by simply adding the same couple of wings 180° opposite. The incorporation and dispersion effect of such rotors is better compared to two-wing rotors[33] but also the temperature buildup is higher, which might lead to an insufficient distributive mixing performance. This disadvantage was the motivation for more sophisticated rotor designs.

Internal mixer supplier Pomini has developed the high distributive mixing (HDM) rotor, a tangential rotor with two rotor wings placed to the edges and three somewhat smaller rotor wings, placed somewhere in the middle.[37] The HDM rotor is given in Figures 35.19 and 35.20. It is reported that distributive mixing performance is better compared to traditional rotor geometries, which can be understood by analyzing the transport flow and the splitting-up effect of the axial flow.

On the other hand Harburg Freudenberger has developed the HD-SC rotor geometry and the MD-SC rotor geometry.[40] Both rotors are four-flighted (refer to Figure 35.21) but the opposite long wing starts from the opposite edge. The MD-SC rotor geometry has got four long rotor wings.

Farrel has reported the development of the ST rotor.[33,35] The ST rotor is four flighted (see Figure 35.22), and is similar to the HD-SC rotor. These rotor designs result in a more aggressive mixing performance (better incorporation performance however higher temperature buildup). The effect of heat buildup is reported to be controlled by the synchronous technology, which Farrel applies. It means that both rotors rotate with the same speed. The orientation of the rotors is determining the effect of heat buildup.[33]

Kobe Steel reported two rotor developments. The first development is an improvement of the four flight rotor by extending the long rotor wing until four times the short rotor wing (see Figure 35.23). As for the ST rotor it can be run in an even speed or with friction. The performance improvement can

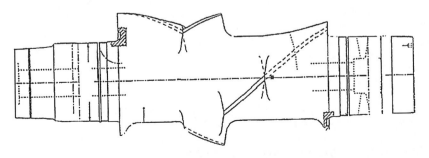

FIGURE 35.19 High distributive mixing (HDM) rotor design (tangential) of Pomini SpA.

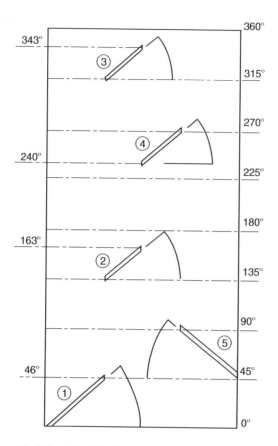

FIGURE 35.20 Lay flat model of Pomini's high distributive mixing (HDM) rotor design.

be found in a higher incorporation and better dispersion effect.[38] Another development is the six-wing rotor (refer to Figure 35.24). The reported excellent dispersion effect might be disturbed by a rather extreme temperature buildup during mixing. In order to reduce this effect the clearance between rotor tip and mixing chamber wall is divided in three sections: small clearance, moderate, and "large" clearance.[38]

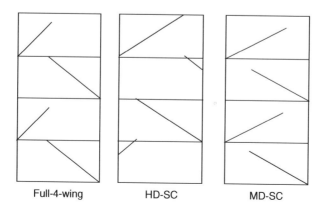

FIGURE 35.21 Maximum dispersion super cooling (MD-SC) and high dispersion super cooling (HD-SC) rotor designs of Harburg Freudenberger compared to their full four-wing rotor.

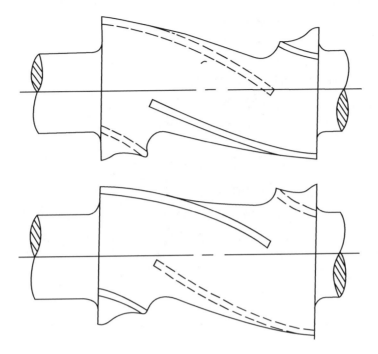

FIGURE 35.22 ST rotor design by Farrel.

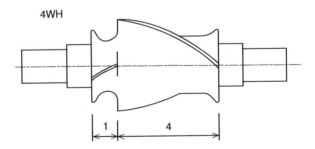

FIGURE 35.23 4WH rotor design of Kobe Steel.

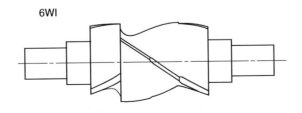

FIGURE 35.24 6WI rotor design of Kobe Steel.

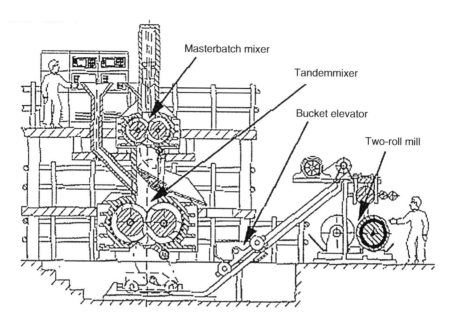

FIGURE 35.25 Tandem mixing concept by Peter. (From Peter, J. and Weckerle, G., *Kautschuk Gummi Kunst.*, 43, 606, 1990.)

Pomini reported that their VIC technology combines the fair raw material intake behavior of the tangential mixer with the excellent dispersion and cooling behavior of an intermeshing mixer.

The VIC mixer is an intermeshing mixer in which the distance between the two rotors can be varied during mixing. Power input can be controlled better, it is said.[41–44] During masterbatch mixing the interclearance section between the two rotors can be maximal. After material intake and incorporation, the interclearance section can be reduced by reducing the distance between the two rotor axes, which leads to better dispersion performance.

Peter has presented the tandem mixing concept in a series of papers.[45–50] The concept consists of an internal mixer which is placed on top of an oversized intermeshing mixer operated without a ram (see Figure 35.25). After masterbatch mixing the batch is being charged in the second mixer. Due to the size the fill factor of the lower mixer friction losses in the compound are lower than the cooling performance, allowing the compound to cool down. After cooling down until a sufficient level, the curative system can be added into the bottom mixer. If the bottom mixer should be of a tangential type the cooling effect as well as the distributive mixing performance (needed for mixing of the curative system) is poor. While the next masterbatch is mixed in the top machine, the final batch is being produced in the bottom machine. Needless to say, the tandem mixing concept can be applied for other purposes like remilling as well.

35.4   RESEARCH TREND: CONTINUOUS MIXING

One of the main disadvantages of batch mixing in internal mixers is the relatively poor batch to batch uniformity[51] as well as difficulties in material handling and their impact on compound and product quality varation.[52–57] It is well known from thermoplastics processing that continuous mixed compounds are very consistent and very narrow tolerated. However, the feeding of such devices requires free-flowing feedstock. Obviously such raw materials used to be rarely available for rubber compounds. Recently powdery rubber became available.[58,59] Research work on continuous mixing of rubber compounds has shown that the degree of mixing could be very high and its variation is remarkably low.[60] Continuous mixing of rubber compounds has gained attention recently, especially since Pirelli announced the continuous compounding mixer.[61]

In this section an overview on the available technologies for continuous mixing and to compare its benefits with traditional batch mixing is discussed. Further work needs to be done before continuous mixing is widely accepted and an outlook in the next future will be discussed as well.

In the last 20 years several attempts have been made to apply continuous mixing devices for the production of rubber compounds. The following devices have been applied with more or less success:

- Single rotor continuous mixer[62]
- Corotating twin-screw extruders (TSEs)[63]
- Counter-rotating TSE[68]
- Multiscrew extruders[69]
- Planetary gear extruders[70]

The single rotor continuous mixer basically consists of a feeding extruder and a mixing extruder and is fed with free-flowing materials. The first device is a proper preheating and densifying device, based on a single-screw extruder. It feeds its material into a distributive mixing and homogenizing device with moderate shear zones to avoid a too strong heat buildup. After that the dispersive zone can be found in which each material part is subjected to high stress events between cylinder and single rotor.[62]

Corotating TSEs are widely applied in the compounding of thermoplastics. The extruder consists of two screws which intermesh each other. The screw concept consists of modules with different screw elements. Figure 35.26 shows typical examples of screws which are applied in twin-screw extrusion. Figure 35.27 shows typical screw elements which have been used in TSE applications in rubber compounding. A few attempts to apply corotating TSEs into rubber mixing

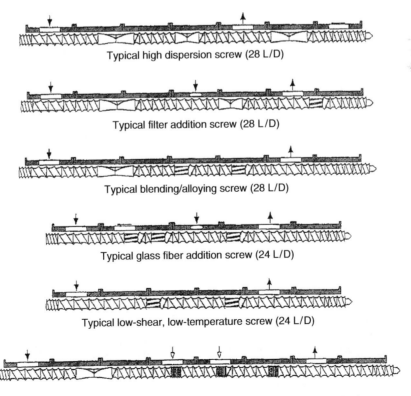

Typical high dispersion screw (28 L/D)

Typical filter addition screw (28 L/D)

Typical blending/alloying screw (28 L/D)

Typical glass fiber addition screw (24 L/D)

Typical low-shear, low-temperature screw (24 L/D)

FIGURE 35.26 Typical screw configurations for corotating twin-screw extruders (TSEs).

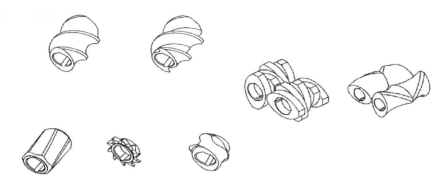

FIGURE 35.27 Typical screw elements as applied in corotating twin-screw extruders (TSEs).

were reported.[63–66] The first report deals with the mixing of rubber out of raw materials which were supplied in their normal trade form. Polymers were granulated just prior to their feeding into the extruder. The other three reports deal with the mixing of powder rubber and the main results obtained were a very high degree of dispersion and a remarkable consistency. The basic outcome of these papers is that from a technological point of view it is possible to mix rubber compounds by means of counter-rotating TSEs. Limper and Keuter proposed a model for the pressure buildup along the screw axes of a corotating TSE.[67] They assumed that such a device can be presented as a series of reaction vessels. A typical result of a model calculation is shown in Figure 35.28. It can be clearly seen that the corotating TSE is a poor pressurizer. This is well known from practise.

Counter-rotating TSEs are applied in PVC compounding and processing. This concept has been used by Farrel for rubber mixing and they call this aggregate the Farrel Continuous Mixer.[68]

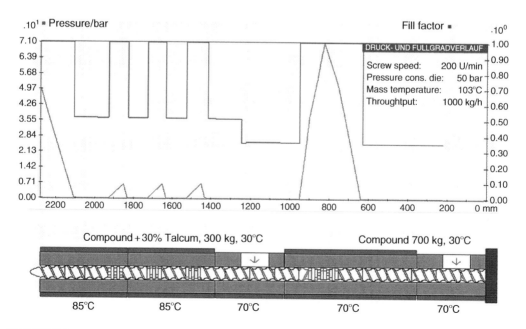

FIGURE 35.28 Typical pressure buildup along a screw axis of a corotating twin-screw extruders (TSEs).

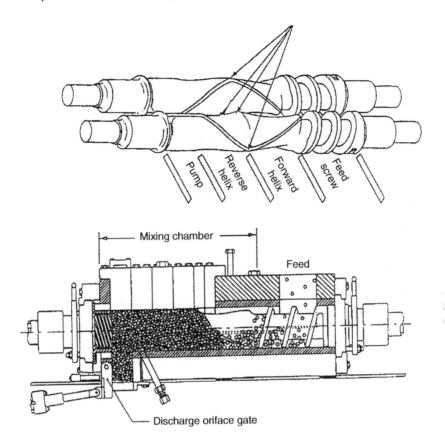

FIGURE 35.29 Farrel continuous mixer.

The concept, shown in Figure 35.29, consists of a short feeding zone, a mixing zone which looks similar to typical rotor design as applied in tangential internal mixers, and an adjustable orifice gate. The feeding of the mixer is starved. The residence time in the mixer and hence the degree of mixing are determined by the discharge orifice gate. Although high degree of mixing can be achieved, due to its low throughput rate, this concept did not find a wide application.

Multiscrew extruders like the Bühler Ring extruders are proposed to be applicable for continuous rubber mixing as well.[69] This extruder type is in fact an extension of the corotating TSE (refer to Figure 35.30). Several screws are arranged in a ring and two adjacent screws intermesh each other, creating relatively more shear zones than TSEs at similar throughput ranges. In order to achieve the same mixing degree, a shorter extruder could be applied with respect to TSE, but a homogeneous feeding seems to be less beneficial.

Planetary gear extruders are widely applied in PVC processing, since its said favorable temperature control.[70] The concept is shown in Figure 35.31 and consists of a main screw, partially responsible for the feeding and partially acting as a drive for various gears which rotate between the drive and the barrel. The pressure buildup is very poor; a gear extruder or a single-screw extruder should be added in order to pressurize the rubber. Luther reported the mixing effect of a planetary gear extruder when preparing silica compounds using E-SBR silica rubber filler composites.[71]

Among other devices the Farrel MVX extruder seems to be applied successfully at least in some niches of rubber mixing.[72] It basically consists of a single-screw venting extruder and in its upstream a sort of tangential mixing unit (refer to Figure 35.32). The screw speed together with the rotor speed determines the mixing degree. The feeding units and the reciprocating ram determine the throughput.

FIGURE 35.30 The Bühler ring extruder.

Table 35.4 contains all benefits and disadvantages of batch mixing with respect to continuous mixing. Table 35.5 contains basically the same for continuous mixing with respect to batch mixing. The major advantages of internal mixers are

1. It accepts almost any trade form.
2. It is widely applied for several decades. It is based on proven technology.

FIGURE 35.31 The planetary gear extruder (supplied by Battenfeld and Entex).

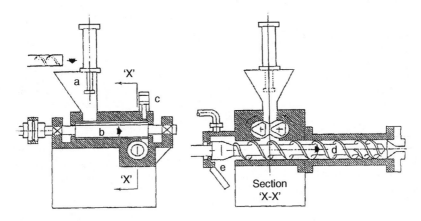

FIGURE 35.32 The Farrel MVX extruder.

The disadvantages of the internal mixers are benefits of the continuous mixer, namely

3. The absolute degree of mixing is not very high.
4. The between batch uniformity is poor, especially at high throughput requirements.

So if raw materials should be available in free-flowing form, continuous mixing could be applied in order to ensure high-quality compounds.

Another benefit of continuous mixing is the better temperature control. This enables the control of a chemical reaction, if any, and the application of thermal sensitive raw materials, which cannot be used in internal mixers.

Today it can be stated that continuous mixing can be applied for rubber compounding success-fully. The very recent announcement of the continuous compounding machine (CCM) by Pirelli shows that the benefits of continuous mixing have found evidence.[73] The basic precondition of applying continuous mixing devices is the availability of all raw materials in free-flowing form in order to achieve a controlled feeding of the aggregate. Either the polymers or rubbers have to be supplied as powders or granulates or advanced granulator techniques (including antitacking facilities) have to be applied.

Continuous mixing of rubber compounds is a very promising method in order to obtain a more consistent quality at very high levels (e.g., better dispersion and far better consistency).

However, neither are present trade forms of traditional raw materials suitable for continuous mixing processes nor have R&D activities on this subject been reported so far.

TABLE 35.4
Benefits of Internal Mixers

Benefits	Disadvantages
Accepts almost any trade form	High power peaks
Proven technology	Batch to batch variation
No need for special dosing equipment	Heat history
Both automatic and manual control	Material weight variation
Efficient dispersion effect	Dust forming by air movement
Flexibility in run length	Labor intensive
"Easy" to maintain and robust machine	Many adjustable parameters
Wide application field	High installation costs
	Complicated downstream

TABLE 35.5
Benefits of Continuous Mixing Devices

Benefits	Disadvantages
1. Steady power supply, i.e., steady process and constant heat history	1. Free-flowing raw materials only
2. Fine-tuning possible	2. Developing technology
3. Labor extensive	3. Complicated dosing machines necessary
4. Constant high degree of mixing, no between batch var.	4. Automatic control only
5. Lower erection costs	5. Beneficial for long runs only
6. Higher degree of dispersion because of lack of dead spots	6. High mass temperatures (to some extent)
7. Direct extrusion possible	7. Screw optimization for one compound
8. Direct calendering possible	

Powdery rubber is an excellent material for continuous mixing but replaces traditional raw materials as a special engineered material.

A breakthrough of application of continuous mixers in rubber mixing cannot be expected until the performance of the final product desires a substantially more consistent mixing quality.

35.5 A FUNDAMENTAL APPROACH OF THE CORD CALENDERING PROCESS

35.5.1 INTRODUCTION

Calendering is one of the oldest rubber processing technologies. It is known that coating of fabrics is done for almost 200 years. Steel cord and fabric cord topping is one of the processes in tire manufacturing. Specifications and tolerances of calendered cords are very tough. Especially the thickness variation across the calendered ply and the cord density (end count) are rather difficult to fulfill. Modern calender lines are very complicated in order to fulfill the requirements; despite that there is rather little research work done in the field of rubber calendering. The intensity of research reduced when extruders started to get more common in the 1970s. Although calenders have been applied for almost 200 years there are only a few attempts to interpret the calendering process in a technological way.

The product quality, however, is related to the calendering process. Thickness gradients across the calendered product, thickness variations along the product, as well as cord density distribution determine the quality of the product and are tremendously related to the process parameters.

Typical requirement figures of coated steel cord or fabric cord sheets as applied in the tire industry are:

Cross section
    Thickness: 0.4–2.5 mm
                ± 0.05 mm
    Width: 900–1450 mm
                ± 0.5 mm/m
    Porosity: not allowed
    Blisters: not allowed

Cord density
    Number of cords: 60–150 cords/10 cm
    Missing cords: 3 cords/10 cm for textile cord/
                          1 cord/10 cm for steel cord
    Off balance: max 0.01 mm
    Weight: ± 50 g/m^2
    Cord tension: until 20 kN per total width
    Cord tension variation: ± 250 N per total width both in calendering direction and across

The purpose of this section is to explain in a qualitative way how the product quality is related to the calendering parameters. Therefore a simplified calender model is presented. The model describes the pressure buildup in the calender nip region as a function of compound viscosity, clearance, calender line speed, rolling bank height, as well as geometrical data. The general layout of a typical steel and fabric cord calender is explained by means of the result of the presented calender model.

35.5.2 Clearance Flow between Two Rolls

The calendering process can be described by means of a combined drag and pressure flow. The rotating rolls of a calender drag the rubber through the calender nip. The clearance as well as viscoelastic properties determine the sheet thickness.

The flow in the clearance of two parallel rolls is visualized in Figure 35.33. Two adjacent and parallel placed rolls are counter-rotating. Both rolls drag the material in the converging region of the calender nip through the clearance. The flow is highly converging, which means that a high pressure will be built up. A part of the dragged material is pushed back by the pressure flow. The result is a rolling bank in the calender nip region, which can be observed quite easily. The nip pressure causes a nip force and, therefore, rolls deflection.

A first model of the calender nip flow has been presented by Ardichvilli.[74] Further on Gaskell presented a more precise and well-known model.[75] Both models are very simplified, which yields that the flow is Newtonian and isothermal, and they predict that the nip force is inversely proportional with the clearance. Since rubber materials show a shear thinning behavior Ardichvilli's model seems not to be very realistic. The purpose of this section is to present a calender nip flow model based on the power law. The model is still being considered isothermal. Such a model was first presented by McKelvey.[76]

For the present approximation of the force, acting on the roll surface, the following assumptions are made:

1. A two-roll calender with symmetrical rolls
2. Nip force deflects the rolls. No lateral roll movement
3. The compound has a uniform temperature. Viscous heating is neglected
4. Rolling bank in the nip is small compared with the roll diameter
5. The rheological behavior of the compound can be described by the well-known power law

$h(x)$ describes the distance between the surface given by $y = 0$ and the roll surface.

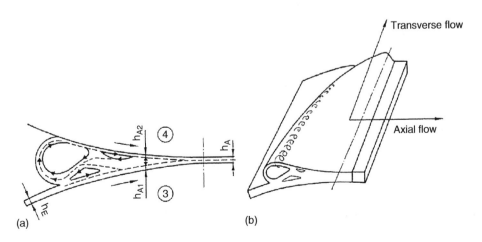

(a)   (b)

FIGURE 35.33 Illustration of the calender nip flow.

The resulting force on the calender roll can be estimated as follows:

Let $p(x)$ be the pressure as a function of the distance x from the calender nip perpendicular to the calender axes. The coordinate y describes the distance from the center of the clearance to the roll axis. The equation of motion in this case can be reduced to

$$\frac{dp}{dx} = \frac{d\tau}{dy} \tag{35.7}$$

in which τ represents the shear stress. Applying the power law as material function and two times integration with respect to y yields

$$v(y) = \frac{n}{n+1} \left(\frac{1}{m} \frac{dp}{dx} \right)^{\frac{1}{n}} \left(y^{\frac{n+1}{n}} - h^{\frac{n+1}{n}} \right) + U \tag{35.8}$$

using the boundary condition that at $y = h$, $v(h) = U$ and using symmetry at $y = 0$. The throughput through the calender clearance per width unit is given by the integral of Equation 35.8:

$$\frac{Q}{W} = \int_{-h}^{h} v(y) \, dy = \frac{2n}{n+1} \left(\frac{1}{m} \frac{dp}{dx} \right)^{\frac{1}{n}} h^{\frac{2n+1}{n}} + 2Uh \tag{35.9}$$

At an undefined position x^* the pressure gradient will vanish since obviously the pressure is maximal. The applicable distance between the two roll surfaces is $2h^*$. The throughput at that position is given by the following expression:

$$\frac{Q}{W} = 2Uh^* \tag{35.10}$$

Combining both Equations 35.9 and 35.10 yields into an explicit equation for the pressure gradient:

$$\frac{dp}{dx} = m \left(\frac{2n+1}{n} U \right)^n \frac{|h - h^*|^{n-1} (h - h^*)}{h^{2n+1}} \tag{35.11}$$

The pressure through the calender nip is determined with this equation. The distance from the calender flow centerline to each roll surface is given by

$$h = h_0 \left(1 + \frac{D}{2} - \sqrt{\frac{D^2}{4h_0^2} - \frac{x^2}{h_0^2}} \right) \tag{35.12}$$

which can be approximated by means of the binomial theorem by

$$h = h_0 \left(1 + \frac{x^2}{Dh_0} \right) \tag{35.13}$$

In fact a parabolic surface of the rolls in the flow region is assumed, which is valid as long as the rolling bank diameter is small compared with the roll diameter. This is normally the case in a calendering process.

A coordinate transformation is suggested, namely

$$a = \frac{x}{\sqrt{Dh_0}} \tag{35.14}$$

The equation for the pressure gradient (Equation 35.11) yields

$$\frac{dp}{da} = \frac{\sqrt{Dh_0}m}{h_0^{n+1}} \left(\frac{2n+1}{n} U \right)^n \left(\frac{\left| a^2 - a^{*2} \right|^{n-1} \left(a^2 - a^{*2} \right)}{\left(a^2 + 1 \right)^{2n+1}} \right) \tag{35.15}$$

The pressure buildup through the calender nip, which is the difference in pressure between the rolling bank surface and $p(x)$ is given by the integral of Equation 35.15:

$$p(a) = \frac{\sqrt{Dh_0}m}{h_0^{n+1}} \left(\frac{2n+1}{n} U \right)^n \int_{-a_H}^{a} \frac{\left| a'^2 - a^{*2} \right|^{n-1} \left(a'^2 - a^{*2} \right)}{\left(a'^2 + 1 \right)^{2n+1}} \, da' \tag{35.16}$$

in which a^* determines the position where the calendered sheet leaves one of the roll surfaces and is determined by the following equation:

$$p(a^*) = 0 = \int_{-a_H}^{a^*} \frac{\left| a'^2 - a^{*2} \right|^{n-1} \left(a'^2 - a^{*2} \right)}{\left(a'^2 + 1 \right)^{2n+1}} \, da' \tag{35.17}$$

The pressure distribution given by Equation 35.16 causes an acting force of both rolls. The magnitude of this force is given by integrating the following equation:

$$\frac{F}{W} = \int_{-H}^{x^*} p(x) \, dx \tag{35.18}$$

in other words

$$\frac{F}{W} = \frac{mD}{h_0^n} \left(\frac{2n+1}{n} U \right)^n \int_{-A_H}^{A^*} \int_{a^*}^{a} \frac{\left| a'^2 - a^{*2} \right|^{n-1} \left(a'^2 - a^{*2} \right)}{\left(a'^2 + 1 \right)^{2n+1}} \, da' \, da \tag{35.19}$$

where
  $2h_0$ represents the clearance
  D the diameter of the calender roll
  a_H the rolling bank height
  a^* the distance from the clearance behind the calender where $p = 0$
  W the total width of the calender roll
  F the resulting force on the calender roll
  n the power–law index
  m the power–law coefficient

Although Equation 35.19 cannot be solved analytically it is possible to approximate Equation 35.19 by using

$$\int f(x) \, dx \approx \left(f(x + \Delta x) + f(x) \right) \cdot \frac{\Delta x}{2} \tag{35.20}$$

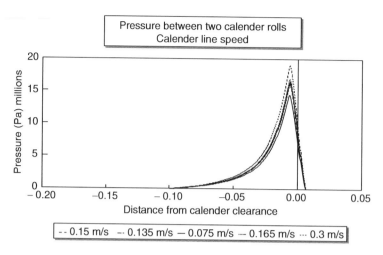

FIGURE 35.34 Pressure between two calender rolls: variation of the calender line speed.

The following diagrams show typical pressure curves in the flow region of the calender nip as a function of various parameters. The following parameters are used for the presented model calculations:

$$U = 0.15 \text{ m/s}; \ h_0 = 0.25 \text{ mm}; \ m = 100 \text{ kPa s}; \ n = 0.2; \ H = 0.1 \text{ m}$$

Figure 35.34 shows a slight dependency of the pressure buildup on the calender line speed, which equals the circumferential roll speed. The general shape of the pressure curve can be understood as follows. A converging drag flow yields a pressure buildup until a barrier has been passed. The material left (=upstream) from the pressure maximum will take part in the rolling bank flow. The material between the pressure maximum and the clearance of the calender flows by means of the drag flow and pressure flow. Each material volume element will pass the clearance. At the position where the pressure vanishes the sheet will be taken apart from one of the rolls.

Figures 35.35 through 35.37 show the dependency of the pressure buildup on the rolling bank content. The pressure curve and therefore the nip force will change dramatically with the rubber content in the nip region. A varying feeding of the calender will cause varying nip forces.

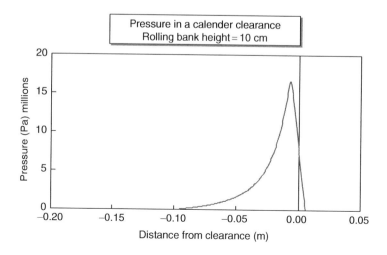

FIGURE 35.35 Pressure buildup in the nip between two calender rolls. Rolling bank height = 10 cm.

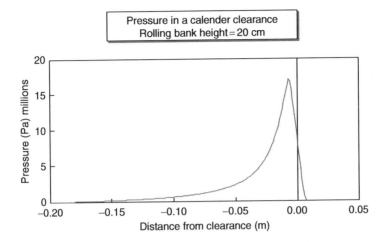

FIGURE 35.36 Pressure buildup in the nip between two calender rolls. Rolling bank height = 20 cm.

Figure 35.38 shows the influence of the clearance setting. A smaller clearance will cause a higher pressure buildup and therefore higher nip forces.

Figures 35.39 and 35.40 show a tremendous dependency on both power–law parameters. The pressure buildup and nip force very much depend on the viscous behavior of the rubber compound. Although not calculated one can simply understand that a varying feedstock temperature will cause variations in the nip force because the viscosity of rubber compounds very much depends on temperature.

35.5.3 ROLL DEFLECTION COMPENSATING METHODS

From the diagrams in Figures 35.34 through 35.40 it might become clear that all parameters will have a tremendous influence on the calendering process and therefore on the calendered product. The calender line speed seems to have the least influence on the calendering performance.

Although viscous heating has been neglected in the modeling of the clearance flow it can be easily understood that a varying compound temperature (and therefore varying viscosity) will cause

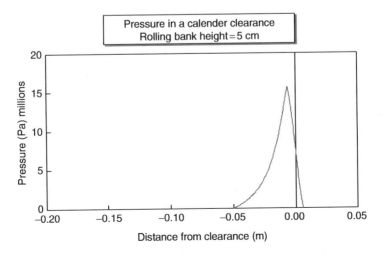

FIGURE 35.37 Pressure buildup in the nip region between two calender rolls. Rolling bank height = 5 cm.

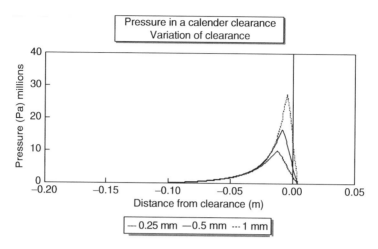

FIGURE 35.38 Pressure buildup in the nip region of two calender rolls as a function of the clearance setting.

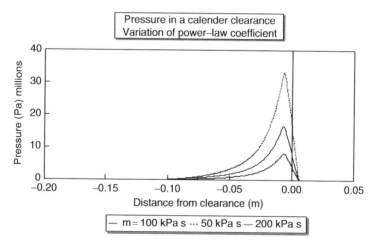

FIGURE 35.39 Pressure buildup in the nip region between two calender rolls as a function of the power–law coefficient.

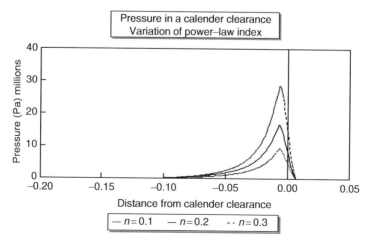

FIGURE 35.40 Pressure buildup in the nip region between two calender rolls as a function of the power–law index.

a tremendous variation of the product quality. Model predictions have shown a dramatic dependency on viscosity data.

High pressures can be found in the calender nip region. The resulting force acts on the adjacent calender rolls and causes a deflection of the rolls and therefore a thickness gradient across the calendered product. The total displacement $Y(z)$ of the roll surface caused by roll deflection due to this viscous force can be approximated by Kopsch.[77]

$$Y(z) = \frac{FW^2 z}{24EJ}\left[1 - 2\left(\frac{z}{W}\right)^2 + \left(\frac{z}{W}\right)^3 + 6\frac{b}{W} - \frac{6bz}{W^2}\right] \tag{35.21}$$

where E represents the elasticity modulus of the material of which the rolls are made (say 210 GPa); J the moment of inertia of the roll; b the distance between roll edge and bearing edge; z the coordinate parallel to the roll axis with respect to the roll edge; J is given by

$$J = \frac{\pi}{64}\left(D_o^4 - D_i^4\right) \tag{35.22}$$

in which D_o is the outer diameter and D_i is the inner diameter of the calender roll.

The enlargement of the clearance caused by roll deflection can be compensated completely by applying an external bending force on the outer ends of the roll axis.

If the acting force causes a roll deflection which exceeds a certain lateral displacement of the roll the thickness gradient across the calendered product might be beyond the required value which means that a calender shall include some special devices to compensate this roll deflection. Some well-known devices are

1. Cross axes setting
2. Counter-roll bending
3. Roll crowning

By crossing the axes of both rolls in the surface of the calendered sheet the effect of roll deflection can be compensated partially. If C_0 represents the cross axes setting displacement the resulting clearance $Y_{CA}(z)$ is given by

$$Y_{CA}(z) = \left(\sqrt{C_0^2 + (D + 2h_0)^2} - D - 2h_0\right)\left(1 - \frac{2z}{W}\right)^2 \tag{35.23}$$

in which the Pythagoras theorem has been applied.

Figure 35.41 shows the effect of the cross axes setting on the clearance along the calender roll. The effect of the deflection of both rolls on the clearance is given for the reference situation as given before. The total width W of the roll is 1600 mm and the diameter D is 650 mm (inner diameter $D_i = 350$ mm). Z is the coordinate parallel to the rolls and originates at one of the calender roll edges.

It seems that the right setting of the cross axes setting practically can compensate the deflection of the rolls due to the nip forces. A compensation of roll deflection can also be achieved by means of outer roll bending devices, which is depicted in Figure 35.42. An additional counterdirected bending force is applied on the outer ends of the rolls.

In modern calender lines both compensating devices can be found. It is common that wide calender rolls are crowned in such a way that under working conditions the crowning compensates the roll deflection due to high shear forces. Crowning specifications can be understood by means of the presented theory. Since the lowest nip forces can be achieved at high clearance settings, low line

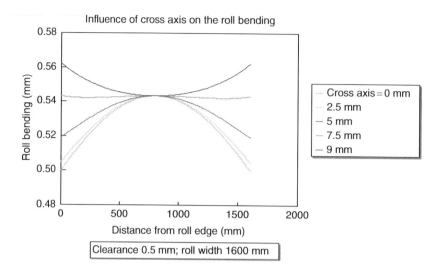

FIGURE 35.41 Effect of the cross axes setting compensation method on the clearance in the total width of the calender roll in working condition.

speeds, with a small rolling bank and by using a low viscous rubber compound roll crowning should be specified for these cases. Outer roll bending and cross axes setting devices will compensate the effect of additional nip forces.

35.5.4 A CALENDER LINE FOR COATING OF TEXTILE AND STEEL CORDS

There are various textbooks available on the calendering process,[77,78] which is referred for an extensive explanation. In this section the impact of the presented theory on the general layout of a dual-purpose calendering line for textile cord and steel cord coating is considered.

Obviously the nip forces do influence the calendering process and determine the product quality. From the theory as presented previously in this chapter it can be easily seen that a constant product quality yields a constant calendering process; constant with respect to line speed, viscosity, and therefore preheating history and rolling bank geometry, i.e., nip feeding. The following specifications for the process of steel and textile cord coating can be understood by the model in a qualitative way:

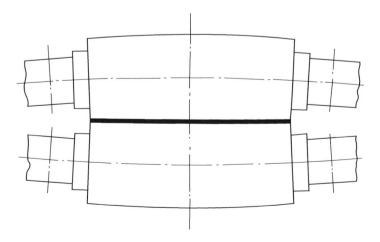

FIGURE 35.42 Illustration of the outer roll bending working principle.

1. Constant and steady nip feeding
2. Constant rubber temperature with which the calender is fed
3. Cord tension as low as possible, but sufficient to keep cord distribution
4. Adjustable in at least three steps
5. Cord density control
6. Calender to be equipped with crowned rolls, external roll bending, and adjustable cross axes setting (controlled)
7. Thickness control equipment
8. Peaks in tension to be damped
9. Batch to batch variations to be avoided
10. Major changes in calender line speed to be avoided

Figure 35.43 shows a general layout of a dual-purpose calender line for coating steel cord and textile cord. The heart of the production line is the four-roll calender in an S-configuration. Two rubber sheets are formed in the upper and lower nip. The thin sheets are guided to the middle nip and the cords are coated in the middle nip between the two rubber sheets. Generally outer roll bending is applied on rolls 2 and 3 to compensate the roll deflection caused by the nip force in nip 2. Rolls 1 and 4 can be set crossed respectively to rolls 2 and 3.

The accumulator in the upstream is designed in such a way that the calendering process can run steadily while splicing the tail end of a fabric with the lead end of the fabric of the next roll. The accumulator in the downstream takes care for a constant process during the exchange procedure at the winding station.

Beside roll deflection and therefore thickness gradients in the product high pressure in the calender nip may cause varying positions of the cords in the calendered sheet in case of cord coating.

By applying a sufficient high cord tension this can be avoided. However, cord tension might cause some elongation. The duration of application of high cord tension has to be as short as possible in order to avoid creep. The complete upstream and downstream of the calender is mainly dedicated to maintain the preset cord tension. In order to keep the cord tension as low as possible the tension can be set in 3–5 steps. One (or two) in the textile upstream, one in the calender area, and

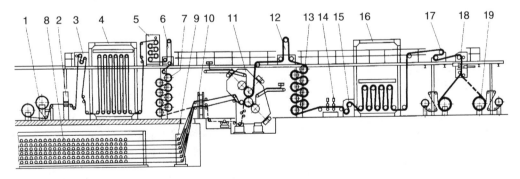

FIGURE 35.43 Example of a general layout of a dual-purpose calender line for coating textile and steel cord.

Legend
1. Textile cord let off
2. Splice press
3. Tension control unit
4. Accumulator
5. Pull unit
6. Tension control unit

7. Preheating drums
8. Creel room
9. Guiding roll for steel cord
10. Splice press
11. Four rolls S-calender
12. Tension control unit

13. Cooing drums
14. Cross yarn breaker
15. Tension control unit
16. Accumulator
17. Pull unit
18. Guiding roll
19. Wind up station

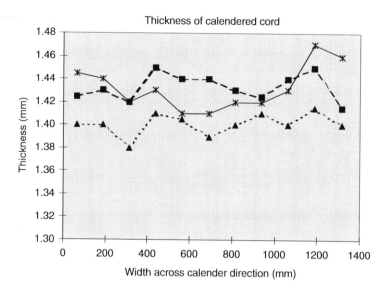

FIGURE 35.44 Typical thickness deviation across the calender direction for various products (textile cord sheet).

one (or two) in the calender downstream. In order to keep the tension of each steel cord constant (steel cord is supplied on separate bobbins, which are put on shafts in the creel room) each unwinding position for steel cord bobbins is equipped with a controlled braking system.

Only in the calender area the cord tension is high enough to resist the pressure caused by the rubber flow in the calender nip region. Therefore, a constant cord density can be maintained by means of locally high cord tension.

Figure 35.44 typically shows the capability of a modern calender line to produce a 1450 mm coated sheet without a big thickness gradient across by applying the proper cross axes settings for various products. The products fulfil the requirement mentioned earlier in this paper.

REFERENCES

1. M.M. Melotto; *Rubberworld*, **217**, 17, 1995.
2. J.L. Leblanc, C. Evo, and R. Lionnet; *Kautschuk Gummi Kunst.* **47**, 401, 1994.
3. F.H. Banbury; US Patent 1, 227, 522, 1917.
4. A. Limper, P. Barth, and F. Grajewski; Technologie der Kautschukverarbeitung; *Carl Hanser Verlag*, München, 1989.
5. Anil K. Bhowmick, Malcolm M. Hall, and Henry A. Benarey (Eds); *Rubber Products Manufacturing Technology*; Marcel Dekker; New York/Basel/Hong Kong; 1994; ISBN 0-8247-9112-6.
6. James L. White; *Rubber Processing Technology, Materials and Principals*; Hanser, Munich/Vienna/New York; 1995; ISBN 3-446-16600-9.
7. Fritz Röthemeyer/Franz Sommer; *Kautschuktechnologie*; Hanser, München/Wien; 2001; ISBN 3-446-16169-4.
8. W.M. Wiedmann and H.M. Schmidt; *Kautschuk Gummi Kunst.* **34**, 479, 1981.
9. J.L. White and J.K. Min; International Seminar on Elastomers, *J. Appl. Polym. Sci: Appl. Polym. Symp.* **44**, 59, 1989.
10. G. Nijman: Anforderungen an einem Mischsaal aus Sicht eines Reifenfertigers. DIK Symposium on Mixing, Hannover, 1993.
11. S. Ghafouri; *Rubberworld*, **222**, 31, 2000.
12. S. Schaal, A.Y. Coran, and S.K. Mowdood; *Rubber Chem. Tech.* **73**, 240, 2000.
13. G. Nijman and R. Luscalu; *Rubberworld*, **221**, 27, 1999.

14. J.L. Leblanc and R. Lionnet; *Polymer Eng. Sci.* **32**, 989, 1992.
15. K. Min and J.L. White; *Rubber Chem. Tech.* **58**, 1024, 1985.
16. A. Morikawa, J.L. White, and K. Min; *Kautschuk Gummi Kunst.* **41**, 1226, 1988.
17. A. Morikawa, K. Min, and J.L. White; *Adv. Polym. Technol.* **8**, 383, 1988.
18. P.S. Kim and J.L. White; *Rubber Chem. Tech.* **69**, 686, 1996.
19. R. Regalia and S. Sibilla; *Gummi Fasern Kunst.* **52**, 264, 1999.
20. B. Hu and J.L. White; *Rubber Chem. Tech.* **66**, 257, 1993.
21. J. Clarke, B. Clarke, and P.K. Freakley; *Rubber Chem. Tech.* **74**, 1, 2001.
22. C.-H. Yao, I. Manas-Zloczower, R. Regalia, and L. Pomini; *Rubber Chem. Tech.* **71**, 690, 1998.
23. J.K. Kim and J.L. White; *Nihon Reoroji Gakkaishi*, **17**, 203, 1989.
24. U. Büker; Diploma thesis, RWTH Aachen University, Aachen (D), 2000.
25. L.A.E.M. Reuvekamp, J.W. ten Brinke, P.J. van Swaaij, and J.W.M. Noordermeer; *Rubber Chem. Tech.* **75**, 187, 2002.
26. L.A.E.M. Reuvekamp, J.W. ten Brinke, P.J. van Swaaij, and J.W.M. Noordermeer; *Kautschuk Gummi Kunst.* **55**, 41, 2002.
27. M.P. Wagner; *Rubber Chem. Tech.* **49**, 703, 1976.
28. A.G. Degussa; Organosilanes for the rubber industry, Technical Inf. 1995.
29. A. Zuuring; Diploma thesis, Saxion University, Enschede, NL, 2000.
30. W.M. Wiedmann and H.M. Schmidt; *Rubber Chem. Tech.* **55**, 363, 1982.
31. N. Nakajima, E.R. Harrell, and D.A. Seil; *Rubber Chem. Tech.* **55**, 456, 1982.
32. L. Schlag; Diploma thesis, University OOW, Emden (D), 1999.
33. G.S. Donoian, E.L. Canedo, and L.N. Valsamis; *Rubber Chem. Tech.* **65**, 792, 1992.
34. N. Yamada, K. Takakura, and K. Inoue; Paper presented at a meeting of the ACS Rubber Division in Dallas, 2000.
35. M.A. Melotto; Paper presented at a meeting of the ACS Rubber Division, May 2000.
36. F.J. Borzenski; Paper 27C of the ITEC 1998.
37. L. Pomini; Paper presented at a meeting of the ACS Rubber Division in Cleveland, 1997.
38. R.J. Jorkasky; Paper 32B of the ITEC 2006.
39. Kobe Steel, Ltd; Introduction of new 4 wing rotor (4WN Rotor); technical inf..
40. www.harburg-freudenberger.com company brochure.
41. L. Pomini and M. Jacobi; *Kautschuk Gummi Kunst.* **54**, 684, 2001.
42. E. Sheehan and L. Pomini; *Rubberworld*, **219**, 21, 1997.
43. L. Pomini, *Kautschuk Gummi Kunst.* **47**, 865, 1994.
44. C.-H. Yao, I. Manas-Zloczower, R. Regalia, and L. Pomini; *Rubber Chem. Tech.* **71**, 690, 1998.
45. J. Peter and G. Weckerle; *Kautschuk Gummi Kunst.* **43**, 606, 1990.
46. J. Peter and G. Weckerle; *Kautschuk Gummi Kunst.* **43**, 896, 1990.
47. J. Peter, G. Weckerle, F. Johnson, and F. Thurn; *Kautschuk Gummi Kunst.* **44**, 758, 1991.
48. J. Peter and G. Weckerle; *Kautschuk Gummi Kunst.* **46**, 545, 1993.
49. J. Peter, F. Röthemeyer, and J. Jennissen; *Kautschuk Gummi Kunst.* **47**, 666, 1994.
50. F. Johnson; *Rubberworld*, **213**, 31, 1993.
51. G. Nijman; *Kautschuk Gummi Kunst.* **57**, 430, 2004.
52. H. Keuter and A. Limper; *Kautschuk Gummi Kunst.* **56**, 250, 2003.
53. H. Keuter and A. Limper; *Kautschuk Gummi Kunst.* **58**, 90, 2005.
54. H. Keuter and A. Limper; *Kautschuk Gummi Kunst.* **58**, 581, 2005.
55. H. Keuter, A. Limper, and C. Rüter; *Gummi Fasern Kunst.* **57**, 715, 2004.
56. H. Keuter, A. Limper, A. Holzmüller, and V. Hasemann; *Gummi Fasern Kunst.* **57**, 791, 2004.
57. H. Keuter, A. Limper, and U. Magnusson; *Gummi Fasern Kunst.* **58**, 27, 2005.
58. U. Görl and K.H. Nordsiek; *Kautschuk. Gummi Kunst.* **51**, 250, 1998.
59. E.T. Italiaander; *Rubber Technol. Intl.* **26**, 177, 1997.
60. U. Görl, M. Schmitt, A. Amash, and M. Bogun; *Kautschuk. Gummi Kunst.* **55**, 22, 2002.
61. Pirelli World, **32**, no. 3; 4, 2002.
62. P.K. Freakley and J.B. Fletcher; *Kautschuk Gummi Kunst.* **55**, 535, 2002.
63. G. Capelle; *Gummi Fasern Kunst.* **49**, 470, 1996.
64. R. Uphus, O. Skibba, R.H. Schuster, and U. Görl; *Kautschuk Gummi Kunst.* **53**, 279, 2000.
65. A. Amash, M. Bogun, R.H. Schuster, and U. Görl; *Kautschuk Gummi Kunst.* **55**, 367, 2002.

66. M. Bogun, F. Abraham, L. Muresan, R.H. Schuster, and H.J. Radush; *Kautschuk Gummi Kunst.* **57**, 363, 2004.
67. A. Limper and H. Keuter; Paper presented on a symposium on Mixing at the DIK German Rubber Institute; September 11, 2000 in Hannover, Germany.
68. F.J. Borzenski; Paper presented on a symposium on Continuous Mixing at the DIK German Rubber Institute; October 25 and 26, 1999 in Hannover, Germany.
69. F. Innerebner; Paper presented on a symposium on Continuous Mixing at the DIK German Rubber Institute; October 25 and 26, 1999 in Hannover, Germany.
70. M. Roth; Paper presented on a symposium on Continuous Mixing at the DIK German Rubber Institute; October 25 and 26, 1999 in Hannover, Germany.
71. S. Luther, M. Bogun, and H. Rust; *Kautschuk Gummi Kunst.* **58**, 371, 2005.
72. F.J. Borzenski; Paper presented on a symposium on Continuous Mixing at the DIK German Rubber Institute; October 25 and 26, 1999 in Hannover, Germany.
73. ERJ **184**, no. 7; 30, 2002.
74. G. Ardichvilli; *Kautschuk,* **14**, 23, 1938.
75. R.E. Gaskel; *J. Appl. Mech.* **17**, 334, 1950.
76. J.M. McKelvey; *Polymer Processing.* Wiley, New York, 1962, p. 211.
77. H. Kopsch; Kalandertechnik. *Carl Hanser Verlag*, München, Wien, 1978, p. 23.
78. G. Capelle; Calendering technology, in A.K. Bhowmick, M.M. Hall, and H.A. Benarey (ed): *Rubber Products Manufacturing Technology*, Marcel Dekker, New York/Basel/Hong Kong, 1994.

36 High-Tech Quadroplex Extrusion Technology for the Tire Industry

Hans-Georg Meyer

CONTENTS

36.1 INTRODUCTION

Revolutionary tire designs for high-tech passenger car radial (PCR) tires drives the developmental pace of the tire equipment manufactures. Triple-tread, asymetrical tread, or highly loaded silica compounds have a consequential impact on the extrusion machinery. In this context the design of the extrusion head and the extruders itself requires adaption to the new, challenging tire technology of today.

This paper discusses these recent developments and outlines how tire machinery manufacturers have developed possible solutions to adequately meet the needs of the fast-paced new tire design evolutions.

36.2 TRIPLE-TREAD, ASYMMETRICAL TREAD, CHIMNEY DESIGN, SILICA COMPOUNDS

All above tire designs have a direct impact on the configuration of a multiplex extrusion system.

The multiple tire designs require the extrusion systems to be easily adaptable in terms of extruder configuration and extrusion head design. The presentation demonstrates various examples and possibile solutions.

Against the background of increasing fuel prices, fuel-saving technologies are constantly gaining importance. Nowadays, the rolling resistance is a key factor in fuel consumption as about every fifth tank filling is used to overcome it. The formula is quite simple: the lower the rolling drag, the lower the fuel consumption.

The rolling resistance can be decisively influenced by the rubber compound the treads are made from. Tire compounds usually contain approximately 30% of reinforcing fillers to provide the rubber compound with the required properties in terms of adhesion and abrasion resistance, initial tearing, and tear growth resistance. For decades, these properties have been exclusively obtained by using industrial black especially developed for this purpose, i.e., carbon black. Today, up-to-date passenger car tires are provided with more individual features produced by an additional ingredient: silica. Thanks to the incorporation of silica, the rolling resistance of a tread can be considerably reduced, while the quality standards of the tire produced in terms of wear, abrasion, and tear resistance, for instance, remain unchanged.

36.3 SILICA-LOADED COMPOUNDS

Today, highly loaded silica compounds with relatively low carbon black content for tread profiles are becoming the standard for high-tech PCR tires.

The characteristics of silica in the extrusion process bears the following challenges:

– Considerably increased abrasion on the extrusion machinery
– Tendency to stickiness on extruder parts and flow channels
– Different behavior in the extrusion process with regard to extrusion temperature and pressure thus necessitating design changes of the extruder process section

Prior to the advent of the use of highly loaded silica compounds the wear protection of rubber extrusion screws was virtually covered by either the nitriding or the stelliting process.

Due to the aforementioned reasons the silica-loaded compounds have forced the manufacturers to research for more adequate wear and antisticking coatings.

Recent trials and long-term field tests have shown that the combined application of plasmanitriding and physical vapor deposited (PVD) hard-coating currently appears to be the best solution, both in terms of improved wear life and antisticking characteristics.

For rubber extrusion screws the plasmanitriding CrN-modified combined with PVD coating has proven to be the most effective surface treatment. The provision of a highly polished surface prior to the application of the plasmanitriding process is important.

This is why companies like Berstorff use PVD-coated screws for this purpose as they exhibit better wear protection than screws with nitrided or stellited surfaces. PVD stands for physical vapor deposition and refers to the evaporation of chrome and its accelerated application onto the surface. In combination with nitrogenous gases, the metal ions form hard nitrides that multiply the wear resistance of the screws.

The PVD process by means of electric arc evaporation has attained a high degree of importance in deposition of hard layers on an industrial scale. The hardness achieved is a minimum of 2100 HV. Compare this to the standard nitriting process which provides a hardness of about 900 HV or the stelliting process which offers about 500–600 HV. Extrusion screws treaded with this special process in the tire industry have seen a much improved wear life when processing highly filled silica compounds or very hard apex compounds.

PVD-coated screws for highly loaded silica compounds are mainly used in Berstorff Multiplex extrusion lines. With these lines, the different compounds are led together and shaped in special profile extrusion dies. Flow channels with optimized geometry ensure a uniform material flow of the individual profile components. The extrusion tools Berstorff developed for this purpose are composed of splicing bar and final die and convert the profiles into treads with extremely close dimensional tolerances.

36.4 DEMAND FOR HIGH-OUTPUT EXTRUSION SYSTEMS

The ever-increasing pressure on the tire manufacturers to compete in an environment of increased cost for raw materials and energy is demanding highly efficient tire production machinery.

This paper presents the world's most productive extrusion line for PCR tires.

The target output capability for such a multiplex extrusion line is 25,000–30,000 treads per day. This results in a rubber processing capacity of 10,000 kg/h.

The massive quadroplex extrusion head is designed to handle four extruders with two each 10 in. (250 mm) and 8 in. (200 mm) pin-barrel extruders.

Furthermore, the head has an extrusion working width of 850 mm allowing the simultaneous extrusion of two-tread profiles at a speed of about 30 m/min.

The extrusion head must be capable of holding peak stock pressures of 250 bars in every flow channel. Thus, the required increased locking forces have to be capable to exert a force of 10,000 KN per head section.

This by itself necessitated a much stiffer design of the head.

A new flow channel design and extruder positioning were part of the high-tech development.

The increased stiffness of this massive head and a vastly increased center section resulted in a total weight of 45,000 kg of the quadroplex head.

36.5 EXTRUSION HEAD ENGINEERING

Such supersize extrusion heads withstand massive forces created by the compound pressure in the flow channels, particularly when dealing with large area extrusion outlets up to 850 mm.

Correspondingly all calculations are finite element method (FEM)-based. Furthermore, the flow channel calculations are based on computer fluid dynamics (CFD) research test for the optimization of the rubber flow.

Optimization of the flow paths of the multiplex head resulted in an additional reserve of capacity. The layout of the inlet angle of the extruders and the arrangement of the flow channels were made in such a way that necessitates the pressure loss to a miniumum. The benefit of this can be clearly seen in a reduced stock temperature. This lower temperature can be utilized to increase the throughput capacity.

The final layout of the flow channels is aided by three-dimentional (3D) models which also form part of our integrated CAD/CAM system.

In addition to these special demands, the design of a modern multiple extrusion head had to meet the following requirements:

- High degree of personnel and product rationalization
- High product precision
- Minimum start-up scrap
- Minimal set-up times
- High plant availability

These criterions are also applied to the entire extrusion unit.

A further factor in the supply of new plant and equipment is the necessity to achieve short delivery and short commissioning times.

A multiplex head, as normally used in tread or sidewall extrusion, is fed by two, three, or four extruders. With the feeding units for the extruders, the resulting height of the extrusion unit is up to 9 m. The extuders are arranged on a steel structure platform, to allow access to the individual machines and the platform is designed for easy service access to all the auxiliary equipment, such as gearboxes, and motors.

The layout and design of this unit is carried out with the help of 3D computer software. The steelstructure, the extruders, and the head, can thus be viewed and any conflicts that arise can be dealt with in the early stages of the engineering. In addition, existing restrictions to the erection of the unit, such as building columns and energy supplies, can easily be integrated in this software and individual solutions can be worked out.

To keep the installation and commissioning times to a minimum, the entire unit is preassembled at the machine builder's factory. As far as possible, the necessary auxiliary equipment is also fitted. For the multiplex head, this means prefitting of the hydraulic pipework from the cylinders to the hydraulic power unit.

The electrical supply and the control lines for the hydraulic power unit are similarly prepared with Profibus connection terminals.

By these measures, it is possible to reduce installation times and ensure trouble-free commissioning of the equipment.

36.6 QUICK CHANGEOVER FOR RAPID PRODUCT CHANGES

Each head section can be opened separately. This reduces the required cleaning and saves on scrap compound when changes are made. The swing-away flow channel inserts allow fast product changes. Easy access to the flow channels and extruder screw tips allow for fast removal of compound residues, so that cleaning times are shortened and waste minimized.

If only the dimensions of the profile are to be changed for subsequent production, then only the final die plate has to be replaced. For this purpose, the head is equipped with a double comb-wedge unit, which allows the changing of the final dies without the need to exchange the splice bar. This is another decisive step towards minimal changeover times.

36.6.1 Multiplex Head Double Wedge System

A modern high-out extrusion line requires certain facilities for quick product changes.

- Double comb wedge system
- Final die change without cassette change

The extrusion tools for a multiplex head consists of the following parts:

- Cassette with electrical heaters and thermocouple
- Splice bars divided into two parts for triplex three parts for quadroplex
- Final dies

36.6.2 Extrusion Screws

The multiplex extrusion screws have to cover a variety of applications and depending on the task and the type of rubber compound the following principal screw designs are available:

- A-type screws for low output
- B-type screws for high homogenization requirements
- C-type screws for high performance
- D-type screws for ultrahigh output
- High-performance intake design

- Stellited flights
- High tensile steel
- PVD-coated screws

36.6.2.1 Extruder Barrel Section

The extruder barrel is designed as follows:

- Peripherically drilled cooling bores for intensive heating/cooling
- Scope of delivery: 100% pins, 50% plugs
- Thickwall bimetallic liner

36.6.3 GEAR BOX

The massive gear boxes of the extruder drive has the following design criterions:

- Heavy-duty design
- Force feed lubrication if required
- No sound-absorbing hood necessary to garantuee max. 80 dB (A) in 1 m distance
- All bearings with calculated lifetime of min. 40,000 h up to 100,000 h (B10)

36.6.4 EXTRUDER FEEDING SECTION

This section of the extruder for the tire industry consists of the following compounds:

- Feed roll flap hinged
- Bearings protected against rubber leakage by additionally scraper cheek
- Feed roll flap opens/closes by means of a hydraulic cylinder

The large feeding conveyors are principally designed as follows:

- Intralox conveying belts for easy maintenance and excellent belt guiding with metal detectors and chalk marking device for marking of metal-contaminated slabs
- Preinstalled to terminal box
- Pneumatic system installed and tested

36.6.5 DOWNSTREAM MACHINERY

A complete extrusion line for the tire industry also consists of various downstream machinery, including the cooling line which reduces the extrudate temperature from about 100°C–115°C at the head exit to ambient plus 5% at the windup stations or at the tread-cutting station.

Principally, we differentiate between immersion and spray cooling technology, depending on the type of tire components and the desired speed of the downstream line.

36.6.5.1 Take-Off Conveyor and Shrinkage Roller Train

The first element of the downstream line is the take-off conveyor positioned right where the rubber exits the extrusion head.

This conveyor consists of the following elements:

- First part: Roller train, stainless steel rolls, diameter 50 mm, pneumatical pivoting to get acces to the extrusion head, length approximately 0.7 m
- Second part: Belt-type intralox flat top 900, first and second part driven by one AC motor

– Third part: Shrinkage roller train free aluminium rollers, roller outerdiameter: 50 mm sidewards displaceable in case of screw removal, total length approx. 3000 mm, width approx. 750 mm, height 960 mm, downstream end of the frame is adjustable in height in order to control the amount of forcing that is applied to the component shrinkage.

Including infrared temperature sensor, (1 for tread production, 2 for SW production) type TXS LTP SF, including holding device, cover window, air blast device, and programming device with software data temp TX for windows.

36.6.5.2 Marking Devices

Customarily the following equipment is applied:

– Embossing with printing letters
– Offset printer
– Color-strip marking device
– Customer made strip-marking device

36.6.5.3 Weighing Conveyor with Roller Train Scale

The running meter scale is the first control parameter for the dimensional control of the tire tread. The scale consists of evaluating electronics with a display screen. The scale is part of the quality control system and is integrated in the power line carrier (PLC) system.

36.7 CONTROL

All the hydraulic functions of the head, such as opening and closing, are controlled from a separate operator panel, located in the direct vicinity of the head itself. This panel also allows adjustments to the extrusion tool temperatures to be made, parallel to the downstream main control panel.

36.8 SUMMARY

The complete extrusion is PLC controlled and usually recipe operated.

The system can store a multitude of different product recipes (usually 200–600) depending on the number of different tire components produced on such a line.

Thus, the likelyhood of human error is largely reduced to a minimum.

Such a line with complete PLC control and automatic winding stations can be operated by 2–4 people, depending on the feeding and spoole-handling logistics.

Only a PLC-controlled extrusion line with recipe management can fulfill the stringent quality control required for high-tech radial tires.

Section VII

Eco-Friendly Technology and Recycling

37 Recent Advances in Eco-Friendly Elastomer Technology

Rabindra Mukhopadhyay, Rajatendu Sengupta, and Saikat Das Gupta

CONTENTS

37.1 INTRODUCTION

The technological and industrial growths spurred by the relentless drive of mankind to tame Nature and to make the environment conducive for the survival of humans are not without its shortcomings. With rampant industrialization, the unique ecological balance existing in Nature was severely disturbed and so the recent trend in minimizing environmental damage is of utmost concern. Now the goal of mankind is to continue technological growth without jeopardizing the unique environmental balance on our planet, Earth.

Even though the information technology industry is driving the development of other industrial sectors including agriculture, manufacturing, and service, one of the common yardsticks of measuring our technological progress is the state of the automotive industry. Currently, ~800 million automobiles are in use worldwide. Considering a global population of ~6.5 billion people, there exists the opportunity of ~88% of the world's people to receive the benefits of the automobile [1]. Taking into account the motorization of Asia, the total number of automobiles to be in use as forecasted by United Nations is expected to increase to 1.6 billion by 2030. The existing 800 million automobiles pose the following concerns as regards *environmental impact, traffic congestion*, and *traffic accidents*. With the doubling of the motor vehicles by 2030, the negative aspects are also expected to aggravate if no serious thought is given to solve the aforementioned issues. Thus, the automotive industry is currently focused on technologies aimed at mitigating these negative issues but at the same time maximizing the positive aspects such as fun, delight, excitement, and comfort appropriately summed up in Figure 37.1.

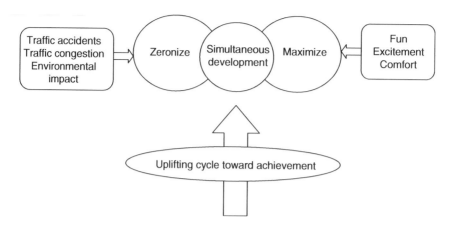

FIGURE 37.1 Zeronize and maximize (vision and philosophy). (From Haraguchi, T. *Nippon Gomu Kyokaishi*, 79, 103, 2006.)

Even though elastomer constitutes a measly 8% by component weight in a modern automobile, it is an extremely indispensable material from the engineering and design viewpoint [3]. Elastomer is the only material that possesses the unique capability of undergoing repeated deformation and restoration. This unique feature of elastomer is utilized in different automotive elastomer components like tires, bushes, insulators, and seals and the evolution of the automobiles was possible because of the significant product innovations in the field of rubber products. Surprisingly, the amount of rubber and rubbery materials used per vehicle since the 1970s has not changed significantly and this is due to the fact that no other suitable alternative material exists till date which can replace rubber. In the nontire application arena, the usage of ethylene–propylene–diene monomer (EPDM) rubber has increased due to its good heat and aging resistance and finds application as accessory drive belt and brake booster diaphragm. However, with the increase in the number of high-performance vehicles, usage of specialty polymers like hydrogenated nitrile rubber (HNBR), acrylic elastomers, and fluorocarbon elastomers are increasing with a concomitant decrease in the use of halogen-containing elastomers. The major rubber parts used in automobiles comprise tires, engine mounts, water/fuel/brake hoses, suspension bush/support/cover, wiper blade, CVJ boot, door weatherstrip, and glass run among others.

Environmental concerns of the automotive industry are centered on the following three perspectives:

- **Emission of CO_2**: The CO_2 emitted by the 800 million automobiles accounts for ~22% of all the greenhouse gas emissions. The greenhouse gas effect if left unchecked would probably cause the melting of the polar icecaps resulting in higher ocean water levels and the loss of global landmass. During driving 86% of the total CO_2 is emitted and thus any means of reducing fuel consumption would directly reduce the total CO_2 emission.
- **Disposal of automobiles**: Disposal by means of recycling is actively pursued not only for conservation of resources but also to mitigate pollution as a result of spillages, leakages, fires, etc. Recycling includes the reuse of parts, physical recycling of materials, and thermal recycling of substances to fuels.
- **Air pollution from exhaust emissions**: NO_x and particulate matter from exhaust emissions causing air pollution are ~1% at present. By adoption of better combustion control technology, better quality fuels (or even fuel cell-based motor vehicles), and more efficient catalytic converters, researchers aim to mitigate this problem.

Traffic congestion leads to vehicular pollution, loss of precious time (hampering productivity), and stress to citizens and thus has a serious negative effect on the penetration of automobiles.

To address this critical issue, town/city planners, and local government are devising ways to regulate the traffic by introduction of automated systems, periodic maintenance of roads, and development of flyovers/connectors, etc. In the developing nations, infrastructure development by creation of highways and super-highways by the build, operate, transfer (BOT) method, fast maintenance of road stretches, traffic diversion via alternate routes, etc. are being utilized to mitigate this problem.

Worldwide traffic fatalities have increased due to the motorization juggernaut sweeping the Brazil, Russia, India, and China (BRIC) economies and other countries. Traffic accidents can only be reduced by the concerted efforts of people, cars, and society as a whole. The introduction of seat belts had a positive impact on lowering passenger fatalities and the introduction of air bags among other safety developments in motor cars is expected to reduce this further.

The tires account for two-thirds by mass of the elastomer used in an automobile [1]. However, the usage of lightweight steel cord and technical improvements in elastomeric materials have reduced their weight by ~20% and this in turn has led to an overall reduction in vehicle weight, concomitant fuel savings, and better road holding property.

With the induction of new generation synthetic elastomers and novel filler materials, coupled with improvements in design areas like tire cross-sectional shape and tread pattern, the performance has improved with the added benefit of reduction in rolling resistance leading to greater fuel savings. A fully loaded truck needs only 2%–4% reduction in rolling resistance for saving 1% fuel while for passenger cars to attain a comparable fuel economy the reduction needed is 5%–7% [4]. Compounding technology and computer-aided engineering (CAE) has gone a long way in improving the braking performance of stud-less tires on icy and snowy roads.

The raw materials for tires are mostly petroleum based except natural rubber (NR), silica, steel tire cord, and bead wire. The four main petroleum materials used in tires are synthetic elastomers, carbon black (CB, used as filler), oil, and synthetic fiber (for casing). In a typical radial passenger car tire more than 70% of the raw materials used owe their genesis from petroleum either directly or indirectly [5]. The use of petroleum-based raw materials not only depletes the planet's nonrenewable natural resources but also causes environmental hazards. Globally more than three million used tires are disposed off each day. These either end up as landfill stuff or are stockpiled as tire mounds. Tires buried in landfill sites release highly toxic chemicals into the groundwater while tire mounds often catch fire releasing carcinogens and other noxious pollutants into the already fragile environment. The recent spurt in petroleum prices and an ever-increasing environmental awareness have created an impetus to search for environment-friendly nonpetroleum origin raw materials for the development of "eco-tire."

The environmental implications of rubber usage are very easily understood by considering NR, a versatile industrial material obtained from the *Hevea* trees as outlined in the tabular format below [4]:

Stage	1	2	3	4
	Production of raw material	*Transformation* into finished products	Use of products *in service*	Final product *disposal*/recycling
Benefits	CO_2 sink (NR) Timber Land conservation Employment	Employment	Reduce noise Reduce vibration Energy transfer Contain liquids	Fuel source "raw materials"
Damage	Water pollution	Odor pollution Peak power Loads	Air pollution Water pollution Traffic congestion Noise	Land loss Pollution

TABLE 37.1

Energy Consumption for Preparation of Various Elastomers

Material	Approx. Energy Consumption (GJ/t)
NR	15–16
BR	108
PP	110
SBR	130–156
EPDM	142–179
IIR	174–209
CR	120–144

Source: Jones, K.P., Rubber and the Environment, UNCTAD/IRSG Workshop, pp. 8–19, 1997.

The advantage that NR enjoys over synthetic elastomers is borne from the energy consumption pattern during preparation (Table 37.1).

These data drive home the argument for patronizing the NR-based industry since it causes minimal environmental disturbance while acting as a sink for CO_2 generated by animals (including man), the natural combustion of organic tissue, and by the burning of fossil fuels. About 40 million people are dependent directly or indirectly on NR for their livelihood and thus the global community is socially accountable for the sustenance of this industry.

People in the global elastomer industry are continuously working with non-petro-based materials to reach a significant breakthrough to reduce petroleum-based raw material to less than 30%. The trend of raw material usage in the global rubber industry is as follows:

Polymer: Synthetic rubber ⇒ NR and its derivatives
Filler: Carbon black ⇒ Natural amorphous silica, precipitated silica, nonblack nanofiller
Solvent: Organic solvent ⇒ Aqua-based solvent
Processing aid: Petroleum based ⇒ Synthetic low PCA content oil, vegetable oil
Casing: Synthetic fiber ⇒ Vegetable fiber

A simplified life cycle of elastomeric products is depicted in Figure 37.2.

37.2 TRENDS IN ECO-FRIENDLY ELASTOMER TECHNOLOGY

The trends in environmental research will be considered on the basis of polymers, fillers, oil, and other processing aids, reinforcing fabrics, other chemicals, and recycling technology. Broadly speaking, the following research trends are evident on a global scale:

– Development of various rubber components based on NR and their derivatives to replace synthetic elastomers
– Development and usage of eco-friendly filler/nanofiller
– Development of eco-friendly processing aids/multifunctional additives (MFAs)
– Usage of vegetable fiber in place of synthetic fiber as a casing material
– Regeneration of virgin material and their reuse in elastomeric products
– Enhanced use of recycled material.

37.2.1 POLYMER

Synthetic rubbers in bale or solid form may contain trace amounts of free monomer (usually <1 ppm by weight) which usually does not constitute a health hazard per se [6]. However, exposure to higher

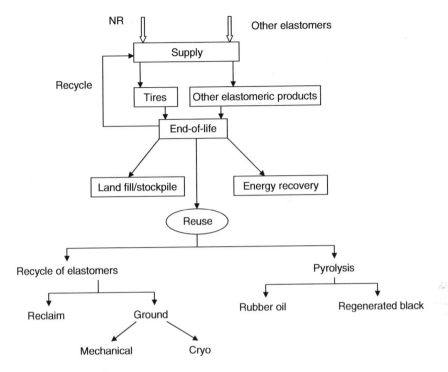

FIGURE 37.2 Flowchart showing the life cycle of elastomeric products.

levels of styrene, acrylonitrile, chloroprene, etc. is considered to be hazardous to humans. Further, the oil extended (usually aromatic oils having a high polycyclic aromatic hydrocarbon [PAH] content) and plasticizer extended elastomers may evolve toxic fumes during processing and/or vulcanization and proper safety arrangements are thus necessary during manufacturing. Although complete replacement of synthetic elastomers by NR remains a distant dream, the increasing usage of NR and thermoplastic elastomers in many applications indicate a right step in this direction. The present trends in the field of elastomers are as summarized below:

(a) Sumitomo Rubber Industries (SRI) launched the ENASAVE ES801 tire, which consists of 70% nonpetroleum materials, in the market in March 2006 [7]. This tire reduced the use of synthetic elastomers by increasing NR usage. It is well known that tires having high NR content usually exhibit poor traction during braking and cornering. The ENASAVE tire utilizes epoxidized NR (ENR) to overcome this traction problem. The molecular structure of ENR is comparable to that of the synthetic elastomers like styrene–butadiene rubber (SBR). In SBR, the benzene rings protrude like branches from the rubber backbone and absorb molecular vibrations caused by the elastomer chain touching the road surface thereby improving the traction properties. In ENR, the epoxide group performs a similar function like the benzene rings pendant from the backbone.

(b) The use of NR in tires is increasing possibly due to its use in runflat tires [8]. The use of ENR25 (25% epoxide content) and silica in car tires not only reduces fuel consumption but also improves wet grip minus the use of any specialist coupling agents. This development is definitely going to make car tires greener.

(c) The development of E-SBR and S-SBR containing nontoxic extender oils originated due to the concerns about the toxicity of highly aromatic oil [9]. The main objective was to develop replacements for the two workhorses of the tire industry, viz. SBR 1712 and 1721.

(d) Elastomers for automotive application have been changing to solve many kinds of problems and market requirements. Tables 37.2 through 37.4 summarize the trend

TABLE 37.2

Material Trends of Automotive Elastomer Parts (Hoses)

Parts	Name	Current Materials	Needs	New Materials	Trend
Engine	Fuel hose	FKM/ECO/GECO			Low cost
	Emission control hose	NBR/CR, ECO, ACM			
	Air duct	NBR + PVC	Heat resistance	TPO, ACM	
	Water hoses	EPDM	Heat resistance	EPDM (aramid)	Heat resistance
	(Radiator, etc.)	(polyester)	Heat resistance		
Body	Fuel hose	NBR/CR	Permeability	FKM/ECO	
	Filler hose	NBR + PVC		ECO	Low permeability
	AC hose	NBR/CR	Permeability	PA6/CIIR	
Chassis	A/T oil cooler hose	ACM			Heat resistance
	Power steering hose	NBR/CR	Heat resistance	ACM	Heat resistance
	Brake hose	SBR/NR/CR + EPDM	Water permeability	EPDM	
	Clutch hose	SBR/NR/CR + EPDM	Water permeability	EPDM	
	Master vacuum hose	NBR/CR	Heat resistance	ECO	

Source: Hashimoto, K., Maeda, A., Hosoya, K., and Todani, Y., *Rubber Chem. Technol.*, 71, 449, 1998.

of elastomeric materials used for various nontire automotive components. It is evident that high-performance elastomers, which have better heat resistance, long life, lower permeability, abrasion resistance, and other properties, are all being adopted.

Advances in the automobile industry are driving the research in the field of elastomers and the future research would be focused on the following aspects:

TABLE 37.3

Material Trends of Automotive Elastomer Parts (Seals)

Parts	Name	Current Materials	Needs	New Materials	Trend
Engine	Crank shaft—rear	VMQ	Long life	FKM	
	Crank shaft—front	ACM	Long life	FKM	
	Diaphragms	ECO, NBR, FVMQ			
	Valve stem seal	FKM			
	Valve cover gasket	NBR	Heat resistance	ACM	
Body	Weather strip	EPDM			Long life
	Glass run	EPDM			Low abrasion
	Filler cap O-ring	NBR + PVC			
Chassis	Transmission oil seal	NBR	Heat resistance	ACM	
	Power steering oil seal	NBR			Heat resistance
	Ball joint dust cover	CR, U			
	Constant velocity joint boot	CR	Heat, ozone	TPEE	
	Rack & pinion boot	CR	Heat, ozone	TPO, TPEE	
	Brake master cup	SBR	Heat resistance, life	EPDM	
	Caliper piston seal	SBR	Heat resistance, life	EPDM	

Source: Hashimoto, K., Maeda, A., Hosoya, K., and Todani, Y., *Rubber Chem. Technol.*, 71, 449, 1998.

TABLE 37.4

Material Trends of Automotive Elastomer Parts (Others)

Parts	Name	Current Materials	Needs	New Materials	Trend
Engine	Synchronous belt	CR	Heat resistance, life	HNBR	Long life
	Accessory drive belt	CR			Heat resistance
Body	In-tank pump cushion	NBR + PVC			
	Mufler hanger rubber	EPDM			
	Wiper blade	CR, NR			Long life
Chassis	Engine mount	NR, SBR, BR			Heat resistance

Source: Hashimoto, K., Maeda, A., Hosoya, K., and Todani, Y., *Rubber Chem. Technol.*, 71, 449, 1998.

(a) Weight reduction of automobiles is a primary requirement and the target areas are evident from Figure 37.3 [2].

(b) The futuristic automobiles would be utilizing cleaner and more efficient fuels than the ones that are being currently used. The cleanest car fuel conceivable today is based on the fuel cell technology utilizing hydrogen and oxygen with water as the product of their combustion. Gasoline and diesel are already losing ground to automobiles running on gasohol, biofuels (derived from plants), compressed natural gas (CNG), and other synthetic fuels [2]; thus, the need for elastomers that would be compatible with the alternative fuel sources. Table 37.5 gives the current assessment of future alternative energy candidates.

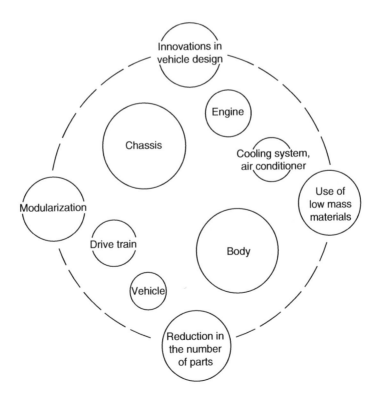

FIGURE 37.3 Weight reduction methodology in the automotive industry. (From Haraguchi, T., *Nippon Gomu Kyokaishi*, 79, 103, 2006.)

TABLE 37.5
Current Assessment of Future Alternative Energy Candidates

Target	CNG	Synthetic Fuel	Ethanol	Biodiesel	Hydrogen
CO_2 reduction			Very good	Very good	Excellent
Emission reduction	Very good	Very good		'	Excellent
Energy substitutability	Good	Good	Very good	Very good	Excellent
Marketability		Good	Good	Good	

Source: Haraguchi, T., *Nippon Gomu Kyokaishi*, 79, 103, 2006.

(c) Goodyear researchers are working on the tailored functionalization of hydrocarbon elastomers to reduce the hysteresis, decrease the rolling resistance, and thus improve fuel economy of tires [11]. The in-chain pyrrolidine functionalized polymers interact strongly with CB and/or silica leading to reduction in the Payne effect. In untreated silica compounds, a higher concentration of the functionalization monomer is needed to both wet the silica and reduce hysteresis.

(d) Bridgestone America's foray into the field of polymeric nanoparticles started way back in the 1990s [12]. Since the particles were to be used in tire formulations, styrene and butadiene monomers were chosen for synthesis for achieving good miscibility of the nanoparticle shell structure with the elastomers commonly used in tire compounds. Modification of the synthesis technique led to the production of a variety of polymeric nanoparticle structure types like spherical, string, flower, janus, composite, and hollow. In the next phase of the development, Bridgestone researchers plan to tailor the structure and properties of the polymeric nanoparticles to achieve performance improvements in different applications and proceed with their commercial use on the basis of cost/performance benefits.

37.2.2 Filler

Inhalation of any particle <10 μm in size presents a theoretical risk of damage to the lungs [13]. CBs are very fine powdered forms of elemental carbon and fall under this category. Owing to the method of production, CBs contain trace amounts of polynuclear aromatic hydrocarbons, some of which when extracted and isolated are known to be carcinogens. The International Agency for Research on Cancer (IARC), in October 1995, reclassified CB as 2B—*possibly carcinogenic to humans* [6]. In 1997, IARC reviewed the classification of silica and concluded that crystalline silica inhaled in the form of quartz or cristobalite is carcinogenic to humans whereas amorphous silica is not classifiable as to its carcinogenicity to humans. The elastomer industry mainly uses amorphous silica in the form of precipitated silica, which does not cause silicosis. The reinforcing fillers (like CB and silica) strongly influence the performance and durability of elastomeric articles and the present trends in the industry are summed up below:

(a) The success of the silica/silane reinforcing filler system in passenger car tires led to the reinitiation of CB related R&D by Degussa AG [14]. Today, the silica/silane system is regarded as the benchmark for wet grip and rolling resistance. The main focus of Degussa's research was aimed at reducing the hysteresis which would lead to lower tire rolling resistance. Even though silica/silane system is preferred in passenger car tires, when it comes to truck tread compounds, CB is preferred as it gives the best optimization in abrasion and cut-chip resistance keeping the other tire properties well balanced.

Degussa introduced the Ecorax grades of CB (nanostructure blacks) suitable for truck tread compounds. Sophistication in the process technology was essential for enhancing the surface activity of the Ecorax CBs. Degussa further introduced soft blacks with extremely low surface area to further reduce the rolling resistance of tires and for this special reactor design in combination with special process technology was adopted. These CB grades have a specific surface area of 20 m^2/g in CTAB and compared to the ASTM grades provide a huge drop in hysteresis by maintaining dynamic stiffness in tire body compounds mainly based on NR. These CB grades were tailor-made with regard to particle size, degree of aggregation, and surface energy. Replacement of the ASTM soft blacks by Ecorax grades with tailor-made morphology in body compound formulations would reduce the heat buildup of tires.

(b) Wampler et al. [15] disclosed the availability of two new CBs for the tire industry. The new tread CB (SR155) has high specific surface area and has been shown to give excellent abrasion resistance, improved traction properties, and superior tear energy while maintaining a good balance of other properties enabling it as an excellent choice for high-performance and racing tires. The non-tread CB (SR401) has very high structure, wide aggregate size distribution, and specific surface area intermediate between tread and carcass grades. Fast mixing, easy incorporation, and good processing are expected in large mixing operations. Due to its high structure it imparts a relatively high modulus and hardness while giving relatively low hysteresis due to its wide aggregate size distribution and low specific surface area.

(c) Sereda et al. [16] reported the use of rice husk ash (RHA) as a potential filler in polysiloxane compounds. Rice husk is an agricultural residue and each kilogram of husk gives 300 g of the nonbiodegradable RHA, which contributes to environmental pollution. The researchers showed that depending on the content of RHA in the formulation, comparable mechanical properties with superior processability was obtained.

(d) Novel fillers like nanoparticles are also under evaluation by the tire industry. In spite of the fact that these nanoparticles have prohibitive price levels which do not justify their bulk use in elastomers, research is currently under way all over the world to substitute the conventional fillers partly and achieve savings in weight with better overall performance.

Recently, Kumho has reported that POSS (specifically trisilanol isobutyl polyhedral oligomeric silsesquioxane) can improve the wet traction and rolling resistance of silica-filled passenger car tire tread compounds [17]. The optimum loading of POSS has been determined as between 2 and 3.5 phr. At this loading level, POSS gave physical property changes normally associated with changes in bulk filler loadings. Kumho's R&D center in Gwangju, South Korea, Avon Materials Development Center in Westbury, U.K., and the Engineering, Medicine and Elastomers Research Center at the University of West England are collaborating on this promising project.

(e) Cabot Corporation has reported the synthesis of carbon–silica dual phase fillers (CSDPFs) using a unique co-fuming technology [18]. The CSDPFs are comprised of aggregates containing both carbon and silica phases with the silica phase finely distributed in the carbon phase [19]. Cabot researchers reported that these new fillers gave significantly better overall performances in comparison with the conventional CB and silica fillers in truck tire tread compounds due to higher polymer–filler interaction and lower filler–filler interaction [20].

(f) Compounding of CB by the latex carbon black masterbatch (LCBM) technique is an environment-friendly process since it reduces the fly loss of CB while imparting better physical properties to the compound mix [21]. Figure 37.4 depicts the preparation procedure of the LCBM. The salient features of the LCBM technique are:

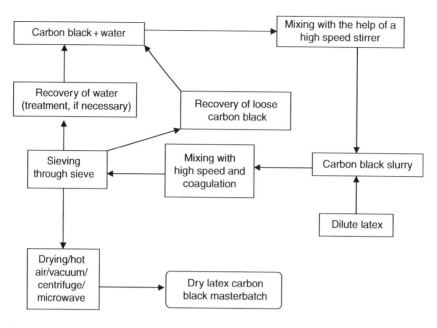

FIGURE 37.4 Flow diagram to prepare the latex carbon black (CB) masterbatch. (From unpublished results of HASETRI, India. Applied for Indian patent in 2002 [application no Del/2301/02].)

1. The filler and latex are in suspension and/or emulsion form leading to better filler distribution compared to dry blending techniques and this in turn leads to better performance properties.
2. Black incorporation time is less in case of masterbatch.
3. CB fly loss as occurs in dry blending techniques is avoided and thus this technique is environment friendly.
4. Accurate weighing of CB can be done.
5. The masterbatching procedure reduces the processing steps at the compounder's end.
6. Masterbatching would also reduce the processing steps for CB manufacturers.
      Extensive product trials using the LCBM technique in standard tire recipes have proved that the tensile strength and modulus, elongation at break, compound hardness, abrasion loss, and hot air aging properties have improved substantially compared to mixes prepared using conventional technique.

(g) Agrawal et al. [22,23] reported the utilization of marble waste in NR and SBR matrices. The marble powder originates from the marble slurry produced by gangsaw machines while cutting marble blocks in marble mines situated in the northwestern parts of India and is responsible for dust-related problems in these areas. The authors reported the use of waste marble powder as a cheap filler in place of other commercial fillers like whiting. However, neither chemical treatment nor electron beam modification of the marble powder produced any significant improvement in the compounded products. The authors concluded that waste marble powder could be used as a partial replacement of CB (up to 10 phr) in NR and SBR products, which operate under static conditions.

(h) The effect of corn powder in gum and filled compound of both NR and SBR was studied by Agrawal et al. [24]. The researchers observed that the thermal stability of NR compound was improved by the corn powder while the thermal stability remained unchanged for the SBR-based compound. The authors concluded that the corn powder acted as nonreinforcing filler in both the elastomer matrices and can partly replace CB to the extent of 5–10 phr.

(i) Ismail et al. [25] determined the possibility of using oil palm wood flour (OPWF) as a filler in NR compounds. Their investigations proved that OPWF can be a cheap, low-active filler and can be recommended to partly replace commercial CB in rubber products which do not require high strength and which are not used for dynamic applications.

The elastomer reinforcement industry is poised for technological growth in the following areas:

(a) R&D activities in Degussa AG involve the development of novel silicas, which are envisaged to reduce stress softening and enhance in-rubber structure [14]. According to the understanding of elastomer reinforcement, filler–filler network and in-rubber structure are strongly influenced by active fillers. Reducing the filler–filler network while maintaining the other performance indicators can reduce hysteresis. The stress softening at small dynamic deformations (Payne effect) shapes the energy dissipation and control of morphology of the silica particles are critical in achieving a reduced Payne effect. Research is currently under way to develop and optimize the special morphology silicas.
(b) Availability of fumed silica in granulated version from Degussa AG is expected to fuel renewed interest in suitable compound formulations [14]. The main differences of fumed silica with precipitated silica are stable primary aggregates with a narrow distribution and a reduced number of silanol groups on the surface, which would effect strong polymer–filler interaction.

37.2.3 OIL AND OTHER PROCESSING AIDS

Process and extender oils function as physical lubricants in elastomer compounds. The oils used in elastomer compounding serve three main purposes [26]:

1. Aid processing of the polymer during milling, extrusion, molding, and other fabricating operations.
2. Act as extenders in certain polymers enabling cost savings without significant reduction in physical properties.
3. Cause changes in properties of vulcanized rubber compounds such as reduced hardness or modulus.

It is customary to divide the rubber process petroleum oils into three types, viz., aromatic, naphthenic, and paraffinic. The following generalizations apply with respect to rubber/oil compatibility:

1. Aromatic oils give the best processability but cause staining, color stability problems, and give vulcanizates with poor resistance to aging.
2. Naphthenic oils give good processability and compatibility in diene elastomers and the resulting vulcanizates are relatively nonstaining.
3. Paraffinic oils are the least efficient as processing additives but the resulting vulcanizates have good aging and color retention properties and give the best low temperature performance.

Mostly, 75% of the extender oils are used in the tread, subtread, and shoulder regions of a tire. About 10%–15% are used in the sidewall, ~5% are used in the inner liner, and less than 10% are used in the remaining parts. A typical tire can contain up to 700 g of oil. All types of mineral oils should be handled and used with care, but special care is required in the handling of aromatic oils. High aromatic oils also referred to as distillate aromatic extracts (DAEs) or simply extracts have been traditionally used as extender oils for elastomeric applications [27]. Their popularity is explained by their good

TABLE 37.6

Comparison of the Properties of Various Extender Oils

Parameter	Unit	DAE	TDAE	MES	Naphthenic
Density @ 15°C	g/mL	1.015	0.95	0.91	0.93
Viscosity @ 40°C	mm²/s	1150	400	200	200
Viscosity gravity constant (VGC)		0.93–0.98	0.89	0.85	0.86
Aromatic content	%	35–50	25	10	13
Naphthenic content	%	15–50	32	35	37
Aliphatic content	%	10–25	43	55	50
DMSO extract	%	35	<3	<3	<3
Benzo[α]pyrene	ppm	>50	<1	<1	<1
Sum of PAHs	ppm	>300	<3	<3	<3

Source: Data provided by Klaus Dahleke KG (Part of Hansen and Rosenthal KG), 2005.

compatibility with the commonly used elastomers and their low price. In 1994, the Swedish National Chemicals Inspectorate, KemI published a report wherein the presence of a high content of polycyclic aromatic compounds (PCAs) in DAE was discussed. Many of these PCAs are identified or suspected carcinogens and thus the European Commission's proposal for a directive to restrict the sales and marketing of aromatic-rich tire extender oils particularly for use in tires. Since the PCAs are not chemically bound into the rubber matrix they give rise to PAHs which have toxic effects on aquatic organisms. A work group of BLIC, the association of rubber and tire manufacturers in the European Union (EU), has conducted work to identify suitable alternates to replace the high aromatic oils. The BLIC workgroup created two guideline product specifications, viz. mild extraction solvate (MES) and treated distillate aromatic extract (TDAE), and oil producers were encouraged to offer products conforming to either MES or TDAE specifications. The properties of the various types of extender oils are compared in Table 37.6 [28].

Figure 37.5 depicts the general refining techniques for obtaining DAE, MES, and TDAE. All tires imported to the EU will have to fulfill the above requirements excepting racing and aircraft tires, which can still be extended with PAH-rich extender oil.

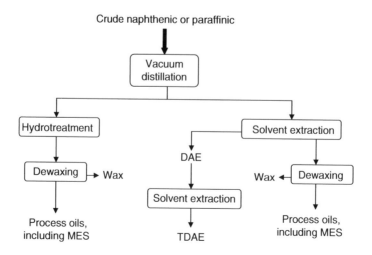

FIGURE 37.5 General refining techniques for production of distillate aromatic extract (DAE), mild extraction solvate (MES), and treated distillate aromatic extract (TDAE). (From Joona, M., High-aromatic tire extender oils implications and future, *ITEC*, OH, 2004.)

Naphthenic oils due to its chemical nature are a good option to replace the aromatic extender oils and some like the MES NYTEX 832 conforms to the MES specification ensuring low levels of PAHs [27]. The high content of naphthenic structure ensures good compatibility with the elastomers used in tire tread compounds. Hydrotreatment of the naphthenic oil further ensures consistent sulfur content and as a result no side effects during vulcanization. Further, the naphthenic character contributes to a low viscosity index than for oils of more paraffinic character. Naphthenic oil and naphthenic MES display good technical performance in conventional tire tests. Further, it was recently demonstrated that by balancing the properties gained from the naphthenic structures in the oil with a higher viscosity, it was possible to mimic the dynamic behavior of compounds extended with high aromatic oils, without compromising on the mechanical and physical performance due to the differing chemistry of the oils [29].

The effect of palm oil fatty acid additive (POFA) on curing characteristics and vulcanizate properties of silica-filled NR compounds was studied by Ismail [30]. The incorporation of POFA improved the cure rate and state of cure of the compounds. Compared to the control, the incorporation of POFA not only enhanced the vulcanizate properties but also improved the reversion resistance of the silica-filled NR compounds.

A considerable amount of work has already been successfully carried out in HASETRI with naturally occurring oils as eco-friendly process oils in conventional tire recipes [31,32]. These naturally occurring oils were found to be suitable on the basis of low PCA content. Some of the naturally occurring oils showed better processing properties, polymer–filler interaction, and dispersion properties in NR-based truck tire tread cap compound and hence better mechanical and dynamic mechanical properties. As the presently available low PCA oil in the market in the form of MES & TDAE and naphthenic oil are comparatively costly, these natural oils can act as the best alternative processing aids for the elastomer industry, especially in developing and underdeveloped countries.

Nag and Haldar reported the purification of vegetable oil extracted from the seeds of *Pongamia glabra* and its ability to substitute commercially available factices [33]. Factices are usually used for mastication of NR during production of tubing, automobile parts, and window seals. The authors reported improvement in thermal stability of the NR vulcanizate after the factice was added.

Yokohama Tire Corporation has combined citrus oil with NR to form super nanopower rubber (SNR) which is being used to reduce the use of petroleum products in the tire by 80%, [34]. Tires containing SNR are lighter than conventional equivalents, reduce rolling resistance by 18%, and are expected to hit the Japanese market later in 2007.

Noncarcinogenic mineral oil-based plasticizers and extender oils are being utilized more and more in elastomer and tire formulations and this interest is prompted by the health and environmental concerns as well as by the coming EU legislation.

37.2.4 Multifunctional Additives

In 1983, Hepburn proposed the concept of a single compounding ingredient which would possess several functions and thus the term Surfactant-Accelerator Processing Aid (SAPA) was born [26]. These SAPAs are basically cationic surfactants having the typical structure: $[R'NH_2(CH_2)_3 NH_3]^{2+} \cdot 2[C_{17}H_{35}COO^-]$. 1,2-Propanediamine dioleate, 1,2-propanediamine oleate, 1,3-propanediamine distearate, and 1,3-propanediamine stearate are some of the commonly used cationic surfactants. The term MFAs has replaced the term SAPAs to better describe the wider properties these materials possess as rubber compounding ingredients. MFAs bring about the following property changes in the field of rubber compounding:

1. Decrease compound viscosity and improve flow
2. Promote wetting and dispersion of fillers
3. Reduce temperature, power consumption, and time for filler incorporation during mixing

4. Enhance tack or autohesion
5. Help in lubrication during extrusion, calendering, and molding and thereby reduce the sticking to processing equipment
6. Produce stress-free moldings eliminating product distortion and warping after vulcanization

The MFAs on heating (during the vulcanization process) decompose into two constituents—the diamine part functions as an accelerator while the fatty acid part acts as a flow promoter-cum-lubricant.

Nandanan et al. [35] reported the utilization of linseed oil as an MFA in nitrile rubber vulcanizates. Linseed oil not only acted as a plasticizer but also as the fatty acid component of the activator in the NBR vulcanizates. Use of linseed oil gave appreciable increase in properties like tensile strength, tear resistance, etc. while the viscosity of the compound was marginally lower than that of the control compound (which used di-octyl phthalate as the plasticizer). The vulcanizates containing linseed oil also exhibited increased cure rate as well as reduced leachability compared to the control at a dosage of 2–5 phr. This loading was seen to replace ~6 phr DOP and 2 phr stearic acid in conventional NBR vulcanizates thereby reducing compound costs.

N-1,3-dimethylbutyl-N'-phenyl quinone diimine (6QDI) has been introduced as a multifunctional additive for diene rubbers and provides an advantage in mixing characteristics (functions as peptizer and improves scorch safety) as well as improved performance (better antioxidant activity than paraphenylenediamine antidegradants) of the end products [36].

The various possible roles of MFAs in the vulcanization of elastomers have been covered by Heideman et al. in a recent publication [37].

Guhathakurta et al. reported the use of waste natural gum (Bahera) as a MFA in NR and brominated isobutylene-co-paramethyl styrene (BIMS) [38]. The authors found that this renewable material not only acted as an accelerator activator but also as an antioxidant.

37.2.5 REINFORCING FABRIC

In the nonmetallic reinforcement materials, no new material was introduced in the market in 2006 [8]. Polyamide/aramid hybrids are finding use as a cap ply in high-performance tires. PEN is considered as a substitute for polyamide/aramid hybrid but is considered expensive for carcass use. SRI Group has adopted a new combination band structure using a PEN band in their LE MANS LM703 runflat tires [7]. The PEN band is ~ four times stronger than a conventional nylon band and further this design decreases cavernous resonance (air vibration inside tires) significantly. Since no new features were noticeable in the recently introduced rayon-like polyester compared to standard PET it is expected that materials based on renewable resources will gain interest.

Some of the recent developments in the field of reinforcement are as mentioned below:

1. Scientists and engineers at the University of Exeter are investigating whether natural fibers like hemp and sisal could be used to make sustainable and eco-friendly brake pads [39]. The technology of brake pads turned "green" with the replacement of asbestos by aramids (like Kevlar of DuPont) in the 1980s. Kevlar is very expensive and eco-friendly alternatives like hemp, jute, sisal, nettle, and flax are much, much cheaper. A breakthrough in this application will revolutionize brake manufacture and protect the environment.
2. Toyobo has developed high strength polyester tire cords, flame-resistant polyester (HEIM) that contribute to the improvement of car safety and helps save labor in manufacturing processes [40].

37.2.6 OTHER CHEMICALS

Several of the commonly used rubber chemicals like accelerators, retarders, antidegradants, etc. are classified as hazardous on account of values of LD_{50} (lethal dose 50%) [6]. Many of the guanidine

TABLE 37.7

Nitrosamine and Non-Nitrosamine Generating Accelerators and Sulfur Donors

S. No.	Particulars
1	**Accelerators/sulfur donors generating carcinogenic *N*-nitrosamines**
	– Dithiocarbamates: di-*n*-alkyl or -aryl dithiocarbamates (ZDMC, ZDEC, ZDBC, etc.)
	– Thiurams: tetra-*n*-alkylthiuram mono-, di-, and oligosulfides (TMTD, TMTM, TETD, etc.)
	– *N,N*-dialkylbenzothiarylsulfenamides (DIBS)
	– 2-(4-morpholinothio) benzothiazole (MBS)
	– Oxy-diethyldithiocarbamyl-oxydiethylsulfenamide (OTOS)
	– Dithiodimorpholine (DTDM)
2	**Accelerators generating noncarcinogenic *N*-nitrosamines**
	– Di-*n*-benzyldithiocarbamates (e.g., ZBEC)
	– Tetra-*n*-benzylthiuramdisulfide (e.g., TBzTD)
3	**Non-nitrosamine generating accelerators**
	– Thiophosphates (e.g., ZDBP)
	– Xanthates (e.g., ZIX)
	– Thiazoles (MBT, MBTS, ZMBT, etc.)
	– Primary amine-based benzothiazyl sulfenamides (CBS, TBBS)
	– Guanidines (DPG, DOTG, OTBG, etc.)
	– Caprolactamdisulfide (CLD)
	– Inorganic isocyanates
	– Urea (non-alkylated)
	– Bis-sulfenamides (TBSI, CBBS)

and dithiocarbamate derivatives are strongly irritant and may also damage the eyes. Some thioureas are also known to form mustard oils (alkyl isothiocyanates) which are acutely irritant to the eyes, skin, and other tissue. Accelerators form the main group of rubber chemicals from which the carcinogenic *N*-nitrosamines are produced in the presence of nitrosating agents. The retarder NDPA (*N*-nitrosodiphenylamine) has been largely withdrawn from the market since it is a powerful nitrosating agent. Among the amines, only secondary amines give rise to the potentially harmful nitrosamines. Table 37.7 lists the nitrosamine and non-nitrosamine generating accelerators and sulfur donors.

Some of the methods used for decreasing nitrosamine formation are [41]:

– Replacement of the hazard-forming chemicals in rubber vulcanizates by using amine free substances or substances that contain primary or tertiary amines in place of secondary amines.
– Prevention of nitrosamine formation at the outset and/or destruction of nitrosamines by capture reactions during vulcanization.

Nowadays, *N*-cyclohexyl-2-benzothiazylsulfenamide (CBS) or *N*-tertiary-butyl-2-benzothiazylsulfenamide (TBBS) is replacing *N*-morpholinothiobenzothiazole (MBS) from recipes. Further, the conventional dithiocarbamates are being replaced by zinc dibenzyl dithiocarbamate (ZBEC) because the nitrosamine formed from dibenzylamine is considered to be nonhazardous. In applications where thiurams are used as secondary accelerators, ZBEC or TBzTD (tetrabenzylthiuram disulfide) easily replaces them [42]. Even though phenyl-α-naphthylamine and phenyl-β-naphthylamine exhibit oral toxicity, they are considered to be safe for use in elastomeric formulations provided that the β-naphthylamine content is <1 ppm (parts per million). Further, both the *p*-phenylenediamine and phenolic antioxidants contain skin irritants and hence require precautions during handling.

The "green" technology trend is shaping the future of the chemicals used in the elastomer industry and this is easily realized from the following:

1. Goodyear introduced the BioTRED technology for environment-friendly replacement of traditional chemical compounds in tires [43]. BioTRED uses an organic compound, derived from corn starch, as a polymer filler. The filler replaces a part of the silica and CB used in tires. Goodyear's GT3 summer tire is the first tire on the market to use the BioTRED technology. The use of this technology has resulted in ~20% reduction in rolling resistance as well as a decrease in the tire weight.
2. Degussa AG introduced a new silane VP Si 363 as a successor of Si 69 for the tire industry [44]. Tire tests have proved that this new silane reduces the rolling resistance by 13% compared with mixes containing Si 69. In addition, volatile organic content (VOC) emissions were reduced by 80% compared to the use of Si 69 or Si 266.
3. The effectiveness of tea polyphenols as antioxidants in elastomeric mixes was evaluated and comparison was made with standard styrenated phenol-based antioxidant [45]. The data showed that thermal and oxidative aging resistance was comparable for both natural and synthetic antioxidants.
4. Agrawal et al. used baker's yeast as a coupling agent and evaluated the properties of corn powder–CB filler and compared the results with those obtained using the conventional silane coupling agent (Si 69) in a radial passenger tire tread compound [46]. The authors observed that the yeast increased the polymer–filler interaction by modifying the corn surface, giving rise to optimum properties for the tread compound.
5. Pysklo et al. [47] studied the effect of reducing the zinc oxide (ZnO) level in elastomeric compounds on the amount of zinc leached from elastomeric granulates and powders. The ZnO level in elastomeric compounds has become an important issue because of the harmful effect of zinc ions to aquatic organisms. The authors carried out this study using truck tire retreading formulations and commercially ground rubber obtained from end-of-life truck and car tires by ambient and cryogenic grinding. They concluded that the specific surface area of the ZnO had no significant influence on the amount of zinc leached and the amount of zinc leached increased with increase of ZnO level up to ~3 phr and then leveled off. The authors further reported that smaller amounts of zinc are leached from cryogenic rubber granulate and powder than from ambient rubber granulate and powder having the same particle size.

The health risk involved in using different materials in the elastomer industry is as summarized in Table 37.8 [48].

The increasing sophistication and detection capabilities of instruments (like GC-MS, LC-MS, etc.) used in the analysis of contaminants in a factory atmosphere are enabling the identification of chemical compounds hitherto not suspected of being present. The range of volatile species so far identified during vulcanization is shown in Figure 37.6 [49].

37.2.7 Recycling Technology

Recycling implies that the material is reprocessed from one form and made into a new product—to save resources, energy, and to control the spread of hazardous/nonbiodegradable material [5]. Recycling elastomers is not as simple as melting glass or aluminum and reusing them. Reversing the vulcanization process is often likened to unbaking a cake and then reusing the eggs. The recovery of useful, uncontaminated materials from composites like tires is very complex.

A number of reviews are already available on the above subject. Rothemeyer [50] discussed the effects of grinding and sieving on the particle size, structure, and distribution of powders obtained from waste rubber and also studied the effects of different powders on the physical properties of the

TABLE 37.8
Health Risk due to Materials in Elastomer Industry

Material	Risk Involved (Long-Term Exposure)	Remedial Measures
Synthetic rubber (SBR, CR, NBR, CSM, ACM, EPM, ECO, etc.)	Suspected carcinogenic	Free monomer concentration should be less than the threshold limit
Reinforcing agents (CB, talc, silica, and other organic and fibrous filler)	Respiratory problem, carcinogenic	Proper exhaust and ventilation system Bulk handling Granular and pellet supply form
Rubber chemicals (accelerator, antioxidant, retarder, peptizer, blowing and vulcanizing agents)	Skin and eye irritation, carcinogenic, asthma	Dust and fume extraction Improved handling change in supply form
Special chemicals (bonding agents like resorcinol–formaldehyde–latex)	Irritation to eye, skin, and respiratory system, cyanosis	Proper ventilation and exhaust system (vapor concentration less than 1 mg/m^3)
Processing aids (Oils, ester type plasticizers, etc.)	Skin irritation and respiratory problem	Proper ventilation and exhaust system Improved handling
Solvents (solvent naphtha, toluene/xylene/isopropanol)	Drowsiness and narcosis, dermatitis	Improved handling Controlled use of solvent

Source: Mukhopadhyay, R. (ed.), *Polymer Science and Rubber Technology*, IRI, Rajasthan, India, 2006.

vulcanizate. Baranwal [51] reviewed the status of characterization and specification development of recycled elastomers. Dunn [52] and Smith [53] have presented an overview on rubber recyling. Warner [54] reviewed the devulcanization techniques. Adhikari et al. [55] reviewed reclaimation and recycling of waste rubber. Mangaraj [56] presented a review on polymer recycling. Klingensmith [57] reviewed recycling, production, and reuse of reprocessed elastomers. Brown and Watson [58] presented novel concepts of environment-friendly rubber recycling. Myhre and MacKillop [59] presented a nice review on elastomer recycling.

Some of the recent developments in the field of recycling technology are summarized below:

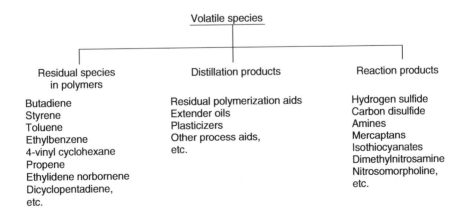

FIGURE 37.6 The volatile species liberated during vulcanization reactions and which have been identified so far. (From Lawson, G., in C.W. Evans (ed.), *Developments in Rubber and Rubber Composites-2, Applied Science*, Essex, 75–94, 1983.)

(a) Mandal et al. [60] studied the regeneration of CB from waste automobile tires and its use in carcass compound. Even though regenerated CBs had higher surface area (as a result of increase in surface roughness) compared to virgin black the structure remained unaltered. The regenerated black gave better aging properties than the virgin blacks.

(b) Bandyopadhyay et al. [61] reported the use of recycled tire material (in the form of ground crumb rubber of three different mesh sizes, viz. 40, 80, and 100) in NR-based tire tread cap compound. A marginal deterioration in tensile strength, fatigue to failure, and abrasion properties was observed. The authors found that the 100 mesh size ground crumb rubber retained the properties better than the other mesh size crumb rubbers.

(c) In a later study, Bandyopadhyay et al. [62] reported the recycling of coarse reclaim, superfine reclaim and devulcanized rubber by incorporating in NR-based bias tire tread cap compound. In all levels of usage, the superfine reclaim rubber retained the properties to a greater extent when compared to the other recycled materials.

(d) CB regenerated from waste automobile tire tread was evaluated by researchers along with virgin N330 black in a bias lug type tire tread cap compound [63]. The regenerated black exhibited marginal inferior physico-mechanical properties compared to the virgin N330. Accelerated aging properties of compounds containing regenerated black was found to be marginally better compared to the compound containing N330. However, the properties of the regenerated black did not significantly improve with the addition of Si 69.

(e) The effects of centrifuged dust stop lubricating oil collected as a liberated material from the dust stops of an internal mixer from a tire industry shop floor in an NR-based flap compound was studied by Bandyopadhyay et al. [64]. The authors found that the centrifuged dust stop oil could be used as a replacement for the aromatic oil in the flap compound. This technology holds great promise due to the cost benefits coupled with the intangible environmental benefits.

(f) Bandyopadhyay et al. [65] reported the use of ground crumb rubber of three different mesh sizes, viz. 40, 80, and 100 in an NR–BR blend-based light commercial vehicle (LCV) tire tread compound. The minimum torque and Mooney viscosity values were marginally increased with increasing dose and mesh size of crumb rubber. However, a marginal deterioration in tensile strength, fatigue to failure, and abrasion properties was observed compared to the control. The authors observed that the 100 mesh size ground crumb rubber retained the properties better than the other mesh size crumb rubbers.

(g) Bandyopadhyay et al. [66] researched the application utility of waste resorcinol–formaldehyde–latex (RFL) dip solid collected from the suction chamber of a typical tire industry dip unit. The effect of the waste material was studied in the SBR-based compound in both gum and filled state and was compared with fresh RFL dip solid obtained by drying a fresh dip solution. The authors recommended the usage of the waste RFL dipped material in the SBR-based compound at 5–10 phr loading level with minor adjustment in the cure package for deriving cost benefits and minimization of waste RFL.

(h) Cataldo compared the plasticizing properties of pyrolytic oil obtained by pyrolysis of scrap tires with DAE in rubber mixes [67]. Though the pyrolytic oil gave accelerated cure kinetics, reduced scorch safety, and lower compound viscosity in comparison to reference compounds with DAE oil, the mechanical properties were similar. However, the aging resistance of the pyrolysis oil was worse than that of the reference mix.

(i) Toyota researchers have ushered in a revolutionary technology that improves the physical properties of recycled EPDM by selectively cutting the cross-links in the three-dimensional EPDM rubber network [1,2]. This technology exhibits 30 times better productivity than conventional technologies with physical properties equivalent to those of fresh EPDM. Currently, this technology is solely used to recycle waste materials from the manufacturing processes for weathering strips and other parts into new automobile parts. In the next phase, the Toyota researchers plan to expand this technology for the recycling of other types of elastomeric materials.

37.3 CONCLUSION

The future direction of environmental research in elastomeric industry will be driven by advances in automobile industry. Automobile industry is continuously working on cutting-edge technology areas for meeting the statutory as well as regulatory requirements. Sumitomo Rubber Industries is striving for the development of tires with 97% of their weight made from nonpetroleum materials and a market launch in 2008 [7]. Michelin announced the Tweel composite tire and wheel technology in January 2005 and further breakthroughs in this research field would bring disruptive innovation to the tire industry [8]. The evolution of the automobile and elastomers are interdependent on each other and it is hoped that the aforementioned material technology areas would produce cutting-edge "green" technologies to solve the present-day engineering/technology problems and lay the foundation for the next generation futuristic technologies.

REFERENCES

1. T. Haraguchi, *Trends in Automobiles and Rubber Parts*, KGK, pp. 151–157, April 2006.
2. T. Haraguchi, Trends in rubber parts for automobiles, *Nippon Gomu Kyokaishi*, 79(3), 103, 2006.
3. R. Mukhopadhyay, Future perspective of Indian automotive rubber component industry, *Indian/International Rubber Journal*, pp. 94–104, January–February 2007.
4. K.P. Jones, Rubber and the environment, UNCTAD/IRSG Workshop, Manchester, pp. 8–19, June 13 1997.
5. R. Mukhopadhyay, Recycling of automotive systems and components and development of non-oil-based material for rubber industry, *Annals of the Indian Academy of Engineering*, III, 107–111, December 2006.
6. *Toxicity and Safe Handling of Rubber Chemicals*, 4th ed., Rapra, 1999.
7. Downloaded from the Sumitomo Rubber Industries website. (Website: http://www.srigroup.co.jp/financial/download/2005_annual_3.pdf). Site accessed on February 27, 2007.
8. A.R. Williams, Tire Technology International Annual Issue, pp. 5–6, 2006.
9. J. Bowman, M. da Via, M.E. Pattnelli, and P. Tortoreto, *KGK*, 57(1–2), 31, 2004.
10. K. Hashimoto, A. Maeda, K. Hosoya, and Y. Todani, Specialty elastomers for automotive applications, *Rubber Chemistry and Technology*, 71(3), 449, 1998.
11. B. Hsu, A. Halasa, K. Bates, Jinping Zhou, K.-C. Hua, and N. Ogata, Novel functionalized tire elastomers via new functional monomers, *Nippon Gomu Kyokaishi*, 79(3), 117, 2006.
12. G. Bohm and H. Mouri, Emerging materials—technology for new tires and other rubber products, *Tire Technology International Annual Issue*, 10, 2006.
13. R.J. McCunney, Occupational health research in the carbon black industry, in *Hazards in the European Rubber Industry, Conference Proceedings*, Rapra, 28–29 September, 1999.
14. W. Niedermeier and B. Schwaiger, Performance enhancement in rubber by modern filler system, *ITEC*, OH, 2006, Paper 2B.
15. W. Wampler, L. Nikiel, and J. Neilsen, New carbon blacks for tire applications, *ITEC*, OH, 2006, Paper 24A.
16. L. Sereda, H.L. Pereira, R.C. Reis Nunes, L.L.Y. Visconte, and C.R.G. Furtado, Rice husk ash in polysiloxane compounds, *KGK*, 9, 474, 2001.
17. G.S. Crutchley and S.J. Choi, A study of nanoparticles and their use in tires, *Tire Technology International Annual Issue*, pp. 26–29, 2006.
18. K. Mahmud, M.J. Wang, and R.A. Francis, US Patent 5830930 (Assigned to Cabot Corporation).
19. M.J. Wang, K. Mahmud, L.J. Murphy, and W.J. Patterson, *KGK*, 51, 348, 1998.
20. M.J. Wang, P. Zhang, and K. Mahmud, Carbon-silica dual phase filler, a new generation reinforcing agent for rubber Part IX. Application to truck tire tread compound, Paper 32A, Presented at a meeting of the Rubber Division, ACS, Dallas, Texas, April 4–6, 2000.
21. Unpublished results of HASETRI, India. Applied for Indian patent in 2002 (application no Del/2301/02).
22. S. Agrawal, S. Mandot, S. Bandyopadhyay, R. Mukhopadhyay, M. Dasgupta, P.P. De, and A.S. Deuri, Use of marble waste in rubber industry: Part I (in NR compound), *Progress in Rubber, Plastics and Recycling Technology*, 20(3), 229, 2004.
23. S. Agrawal, S. Mandot, S. Bandyopadhyay, R. Mukhopadhyay, M. Dasgupta, P.P. De, and A.S. Deuri, Use of marble waste in rubber industry: Part II (SBR compounds), *Progress in Rubber, Plastics and Recycling Technology*, 20(4), 267, 2004.

24. S. Agrawal, S. Mandot, N. Mandal, S. Bandyopadhyay, R. Mukhopadhyay, A.S. Deuri, R. Mallik, and A.K. Bhowmick, Effect of Corn Powder as filler in radial passenger tyre tread compound, *Journal of Material Science*, 41, 5657, 2006.

25. H. Ismail, I.N. Hasliza, and U.S. Ishiaku, The effect of oil palm wood flour as a filler in natural rubber compounds, *KGK*, 53(1–2), 24, 2000.

26. C. Hepburn, Rubber compounding ingredients—need, theory and innovation Part II—Processing, bonding, fire retardants, Report 97, Rapra, 1997.

27. M. Joona, High-aromatic tire extender oils-implications and future, *ITEC*, OH, 2004.

28. Data provided by Klaus Dahleke KG (part of Hansen and Rosenthal KG), 2005.

29. M. Joona, Non-carcinogenic tire extender oils providing good dynamic performance, *Rubber World*, 235(4), 15, 2007.

30. H. Ismail, Effect of palm oil fatty acid additive (POFA) on curing characteristics and vulcanizate properties of silica filled natural rubber compounds, *Journal of Elastomers and Plastics*, 32, 33, 2000.

31. S. Dasgupta, S. Agrawal, S. Bandyopadhyay, S. Chakraborty, R. Mukhopadhyay, R.K. Malkani, and S.C. Ameta, Characterization of eco-friendly processing aids for rubber compound, *Polymer Testing*, 26(4), 489, 2007.

32. S. Dasgupta, S. Agrawal, S. Bandyopadhyay, S. Chakraborty, R. Mukhopadhyay, R.K. Malkani, and S.C. Ameta, Characterization of eco-friendly processing aids for rubber compound: Part II, *Polymer testing* (accepted for publication).

33. A. Nag and S.K. Haldar, Studies on a newer process of purification of a vegetable oil and its utilization as factice, *KGK*, 13, 322, 2006.

34. Update section in Tire Technology International, p. 6 (March 2007).

35. V. Nandanan, R. Joseph, and D.J. Francis, Linseed oil as a multipurpose ingredient in NBR vulcanizate, *Journal of Elastomers and Plastics*, 28, 326, 1996.

36. R.N. Datta, S. Datta, and A.G. Talma, Mechanistic studies in squalene model systems, *KGK*, 57(3), 109–115, 2004.

37. G. Heideman, R.N. Datta, J.W.M. Noordermeer, and B. van Barle, Activators in accelerated sulfur vulcanization, *Rubber Chemistry and Technology*, 77(3), 512–541, 2004.

38. S. Guhathakurta, S. Anandhan, N.K. Singha, R.N. Chattopadhyay, and A.K. Bhowmick, Waste natural gum as a multifunctional additive in rubber, *Journal of Applied Polymer Science*, 102, 4897–4907, 2006.

39. Downloaded from http://www.azom.com/details.asp?newsID = 1133. Site accessed on December 6, 2006.

40. Downloaded from http://www.toyobo.co.jp/e/seihin/car/torikumi/torikumi.html. Site accessed on February 26, 2006.

41. Th. Kempermann, S. Koch, and J. Sumner (ed.), *Manual for the Rubber Industry*, Bayer AG, Germany, pp. 423–425, 1993.

42. Th. Kleiner, Prospects for accelerators based on safe amines, *KGK*, 50(1), 43, 1997.

43. Downloaded from http://www.prnewswire.co.uk/cgi/news/release?id = 69355. Site accessed on February 27, 2007.

44. O. Klockmann and A. Hasse, Application of the new rubber silane VP Si 363, *ITEC*, OH, 2006, Paper 29B.

45. Unpublished results of HASETRI, India.

46. S. Agrawal, S. Mandot, N.K. Mandal, S. Bandyopadhyay, R. Mukhopadhyay, A.S. Deuri, R. Mallik, and A.K. Bhowmick, Yeast as coupling agent for corn filler in tire tread compound, *Progress in Rubber, Plastics and Recycling Technology*, 21(3), 231, 2005.

47. L. Pysklo, P. Pawlowski, W. Parasiewicz, and M. Piaskiewicz, Influence of the zinc oxide level in rubber compounds on the amount of zinc leaching, *KGK*, 59(6), 328, 2006.

48. R. Mukhopadhyay (ed.), *Polymer Science and Rubber Technology*, IRI, Rajasthan, India, p. 243, 2006.

49. G. Lawson, Health and safety considerations in rubber processing, in C.W. Evans (ed.), *Developments in Rubber and Rubber Composites-2*, Applied Science, Essex, London, pp. 75–94, 1983.

50. F. Rothemeyer, *Kautschuk Gummi Kunststoffe*, 46, 356, 1993.

51. C.K. Baranwal, Paper presented at International Conference on Rubbers, 12–14 December 1997, Kolkata.

52. J.R. Dunn, Paper presented at a meeting of the Rubber Division, ACS, 26–29 October, 1993, Orlando, FL.

53. F.G. Smith, Paper presented at a meeting of the Rubber Division, ACS, 26–29 October 1993, Orlando, FL.

54. W.C. Warner, *Rubber Chemistry and Technology*, 67, 559, 1996.

55. B. Adhikari, D. De, and S. Maiti, *Progress in Polymer Science*, 25, 909, 2000.

56. D. Mangaraj, Paper presented at International Conference on Rubbers, 12–14 December 1997, Kolkata.
57. W. Klingensmith, *Rubber World*, p. 16, March 1991.
58. D.A. Brown and W.F. Watson, Paper presented at the International Rubber Forum 2000, 9 November 2000, Antewarp.
59. M. Myhre and D.A. MacKillop, *Rubber Chemistry and Technology*, 75, 429, 2002.
60. N. Mandal, S. Dasgupta, and R. Mukhopadhyay, Regeneration of carbon black from waste automobile tires and its use in carcass compound, *Progress in Rubber, Plastics and Recycling Technology*, 21(1), 55, 2005.
61. S. Bandyopadhyay, S. Dasgupta, N. Mandal, S.L. Agrawal, S.K. Mandot, R. Mukhopadhyay, A.S. Deuri, and S.C. Ameta, Use of recycled tire material in natural rubber based tire tread cap compound: Part I (with ground crumb rubber), *Progress in Rubber, Plastics and Recycling Technology*, 21(4), 299, 2005.
62. S. Bandyopadhyay, S.L. Agrawal, S.K. Mandot, N. Mandal, S. Dasgupta, R. Mukhopadhyay, A.S. Deuri, and S.C. Ameta, Use of recycled tire material in natural rubber based tire tread cap compound: Part I, *Progress in Rubber, Plastics and Recycling Technology*, 22(1), 45, 2006.
63. S. Bandyopadhyay, N. Mandal, S.L. Agrawal, S. Dasgupta, R. Mukhopadhyay, A.S. Deuri, and S.C. Ameta, Effect of regenerated carbon-black on a bias tire tread cap compound, *Progress in Rubber, Plastics and Recycling Technology*, 22(3), 195, 2006.
64. S. Bandyopadhyay, N. Mandal, S. Dasgupta, R. Mukhopadhyay, A.S. Deuri, and S.C. Ameta, Effect of recycled dust stop lubricating oil in rubber compounds, *Rubber World*, 233(6), 20, 2006.
65. S. Bandyopadhyay, S. Dasgupta, S.L. Agrawal, S.K. Mandot, N. Mandal, R. Mukhopadhyay, A.S. Deuri, and S.C. Ameta, Use of recycled tire material in NR/BR blend based tire tread compound: Part II (with ground crumb rubber), *Progress in Rubber, Plastics and Recycling Technology*, 22(4), 269, 2006.
66. S. Bandyopadhyay, S.L. Agrawal, P. Sajith, N. Mandal, S. Dasgupta, R. Mukhopadhyay, A.S. Deuri, and S.C. Ameta, Research on the application of recycled waste RFL (Resorcinol-Formaldehyde-Latex) dip solid in Styrene Butadiene Rubber based compounds, *Progress in Rubber, Plastics and Recycling Technology*, 23(1), 21, 2007.
67. F. Cataldo, Evaluation of pyrolytic oil from scrap tires as plasticizer of rubber compounds, *Progress in Rubber, Plastics and Recycling Technology*, 22(4), 243, 2006.

38 Waste Rubber Recycling

Anandhan Srinivasan, A.M. Shanmugharaj,
and Anil K. Bhowmick

CONTENTS

38.1 INTRODUCTION

In 1839, Charles Goodyear discovered that sulfur could cross-link polymer chains and patented the process in 1844 [1]. Since then rubber became a widely usable material. By the year 1853, natural rubber (NR) was in short supply. So attempts were made to undo what Goodyear had accomplished. Goodyear himself was involved in trying to reclaim vulcanized rubber to overcome the shortage of NR. Later, as a consequence of World War I, Germany introduced synthetic rubbers, namely the Buna rubbers, which raised the curiosity of polymer chemists all over the world. Subsequently, synthetic rubbers with tailor-made properties were born. This was followed by the discovery of new methods and chemicals for vulcanization and processing. It is obvious

TABLE 38.1

Recycling of Discarded Tire in Some Countries and Regions

Place	Year	Amount (10,000 t)	Thermal Utilizing	Making Powder	Regenerated Rubber	Renew	Export	Burying	Other
US	1992	280	23	6	4	—	3	63	1
Japan	1992	84	43	—	12	9	25	8	3
Germany	1993	55	38	14	1	18	18	2	9
British	1992	45	9	6	—	18	—	67	—
EEC	1990	197.5	30	—	20	—	—	50	—

Source: Reprinted from Fang, Y., Zhan, M., and Wang, Y., *Mater. Design*, 22, 123, 2001. With permission.

that the importance of recycling was ignored then, since there was no shortage of raw materials. At present, the annual consumption of NR is more than 15 million tons, and the output of rubber products is more than 31 million tons worldwide [2]. With the tremendous growth of the automobile industry, which is the major consumer of rubbers, a huge amount of nonbiodegradable rubber waste and scraps continues to accumulate. Worn tires are the major sources of the waste rubber vulcanizates. Presently, the amount of discarded tires reaches 10 million every year worldwide. The quantity of discarded tires in some countries and the situation for disposal of these at present are listed in Table 38.1 [3].

Recycling of scrap tire has become an environmental concern and a challenging technical problem. Scrap tire piles frequently catch fire, which is difficult to extinguish and provide ideal breeding ground for mosquitoes and rats [4,5].

Besides scrap and worn tires, other main sources of waste rubber are discarded rubber products, such as: rubber pipes, rubber belts, rubber shoes, edge scraps, automobile hoses, and various fabric-reinforced components. Scraps produced during the production of rubber goods also contribute to recycle stream [6]. Reutilization of waste rubber demanded effective techniques for its recycling.

38.2   TRADITIONAL METHODS OF RUBBER WASTE DISPOSAL

Recycling of rubber involves the breakdown of three-dimensional networks either through the cleavage of cross-links, or through carbon–carbon linkage of the chain backbone. This is much more severe process and the recycled material is entirely different to the starting precursor. Hence, the reuse is the most desirable means of recycling. Worn out rubber parts can be replaced at a fraction of manufacturing cost of the product, as is manifested by the example of tire retreading. Mainly for safety reasons, the use of retreaded tires has been limited. Search for methods of reutilization of a huge amount of waste or discarded rubber tire continues.

38.2.1   WASTE RUBBER AS LANDFILLS

Landfill is one of the early ways of disposal of discarded products. In 1977, approximately 70% of the scrap tires were discarded as landfills [7]. But, with the decreasing scope of available sites, this method is no longer feasible. Tires discarded in landfills cause mosquito breeding and leaching out of the processing additives to the nearby land and water streams. These additives are not eco-friendly and may kill advantageous bacteria and other microorganisms of the soil [8]. In this way, landfill causes serious environmental problems. Many countries have already banned the use of discarded tire for land filling [9].

38.2.2 SCRAP TIRE AS FUEL SOURCE

Tires contain more than 90% organic materials and have a calorific value of 32.6 mJ/kg compared to 18.6–27.9 mJ/kg for coal [10]. Tire derived fuel (TDF) has been used by the cement, paper and pulp plants, and power generating boilers [11]. Discarded rubber is used as a fuel to generate electricity. An environment-friendly process was developed for recycling rubber waste materials such as waste tires to generate valuable fuels or chemical feedstock in a closed oxidation process, which is free of hazardous emissions. The process involves chemical breakdown of rubber products by selective oxidative decoupling of C—C, C—S, and S—S bonds [12]. Alternatively, oil, steel, and carbon can be recovered by processing the shredded automobile tires in the presence of isocyanides, polyurethane, latex, and a vegetable oil [13].

38.2.3 CHEMICAL RECYCLING

Chemical recycling of waste rubbers means breaking down of the macromolecules into oligomers or monomers. Different chemical recycling technologies include glycolysis, methanolysis, and pyrolysis.

Pyrolysis of scrap tires was studied by several rubber, oil, and carbon black industries [14]. Pyrolysis, also known as thermal cracking is a process in which polymer molecules are heated in partial or total absence of air, until they fragment into several smaller, dissimilar, random-sized molecules of alcohols, hydrocarbons, and others. The pyrolysis temperature used is in the range of 500°C–700°C. Moreover, maintenance of partial vacuum during pyrolysis in reactors lowered the economy of the process. Several patents were issued for the pyrolysis of worn out tires to yield crude oil, monomers, and carbon black in economic ways [15–18]. The major drawback of chemical recycling is that the value of the output is normally low and the mixed oils, gases, and carbon black obtained by pyrolysis cannot compete with similar products from natural oil. Pyrolyzing plant produces toxic wastewater as a by-product of the operation [19].

38.2.4 RECLAIMING

Reclaiming of scrap rubber products, e.g., used automobile tires and tubes, hoses, conveyor belts, etc. is the conversion of a three-dimensionally interlinked, insoluble, and infusible strong thermoset polymer to a two-dimensional, soft, tacky, low modulus, processable and vulcanizable, and essentially thermoplastic product simulating many of the properties of virgin rubber. Recovery and recycling of rubber from used and scrap rubber products can, therefore, save some precious petroleum resources as well as solve scrap rubber disposal problems. Interest in rubber reclaiming processes dates back to the mid-1800s, shortly after Charles Goodyear patented the vulcanization of rubber with sulfur. These are the "thermal" process followed by the "heater/pan" and the "alkali digester" processes. There have been numerous publications [20–25] and reviews [9,26–28] on this subject. Figure 38.1 depicts the pan process along with the conditions used for reclaiming.

Reclamation technology has been relatively simple until the development of synthetic rubbers. In contrast to the softening behavior of NR under heat and pressure, hardening takes place in many synthetic rubbers. Reclaiming involves the shredding of tires, reduction of its size, followed by the application of heat and pressure, often in the presence of some chemicals, which selectively break the sulfur cross-links. The resultant mass of rubber is called the reclaim rubber (RR) or whole tire reclaim (WTR). WTR can be incorporated into tire formulation itself, but it adversely affects the mechanical properties of the resulting vulcanizate in spite of the processing advantages like decreased air entrapment, lesser nerve, reduced distortion, and lower extrudate swelling. Radial tires demand uncompromising attitude towards quality control and accordingly growth of the reclaiming industry has been slowed down. Moreover, the adverse effects of the pollutants from the reclaiming industry on the environment have caused shutdown of many plants [29].

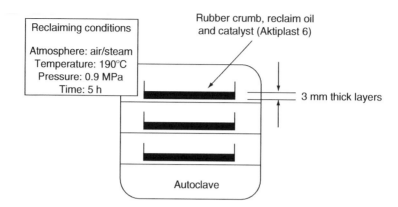

Reclaiming conditions

Atmosphere: air/steam
Temperature: 190°C
Pressure: 0.9 MPa
Time: 5 h

Rubber crumb, reclaim oil
and catalyst (Aktiplast 6)

3 mm thick layers

Autoclave

FIGURE 38.1 Schematic representation of pan process. (Reprinted from Scheirs, J., *Polymer Recycling—Science, Technology and Applications*, Wiley, New York, 1998. With permission.)

38.3 UTILIZATION OF GROUND RUBBER

This method has gained wide acceptance due to its simplicity and environmental friendliness. Research is under way to improve the economy of this method. The most important step is the preparation of ground rubber, since the size and topography of the particles vary with the different techniques of grinding [30]. The characteristics of the ground rubber play a vital role in determining the properties of the ground rubber-filled rubber composites.

Ground rubber is obtained by cryogenic grinding, ambient grinding, or wet ambient grinding. The method of grinding and sieving significantly influences the particle size, structure, and distribution of waste rubber powder [31]. Ground rubber tire (GRT) is also commonly referred to as rubber reclaim, even though the rubber has not been devulcanized. The classification of ground (powdered) rubber along with the methods of producing it and applications is given in Table 38.2.

38.3.1 Methods of Preparation of Ground Rubber

38.3.1.1 Ambient Grinding

Ambient grinding involves the comminution of scrap rubber, including tires, at ambient temperatures, typically in a toothed-wheel mill, bench grinder, shear mill, or two-roll mill, to give particles

TABLE 38.2
Classification, Manufacture, and Uses of Powdered Rubber (PR)

Classification	Particle Size (μm)	Methods of Preparation	Applications
Coarse PR	1400–500	Grinding mill, rolling mill, rotary crushing mill	Ballast mats, raw material of degrading regenerated rubber
Tiny PR	500–300	Rolling mill, rotary crushing mill	The raw material of regenerated rubber
Fine PR	300–75	Cryogenic mill	Products of molding and extrusion, rubber mats, soft pipes for irrigating, vases, modified asphalt for paving
Ultrafine PR	Below 75	Rotary colloid mill	Used in renewal of tire (under 20 μm, 30 phr)

Source: Reprinted from Scheirs, J., *Polymer Recycling—Science, Technology and Applications*, Wiley, New York, 1998, 411. With permission.

of 0.5–1.5 mm diameter (up to 40 mesh), suitable after sieving for incorporation into rubber compounds [32–36]. The main disadvantages of this method are the high cost, high energy requirement, heat, fumes, and noise production.

38.3.1.2 Cryogenic Grinding

In the cryogenic process, precooled, embrittled rubber is fractured using high strain rates (e.g. with swing or fixed-hammer impact mills) [8,37–41]. Liquid nitrogen is used to cool the rubber to very low temperatures (i.e., below the glass transition temperature of the rubber). The main advantages of this method apart from cost-effectiveness are low power requirements, improved flow characteristics, and easy segregation of metal, fabric, and powder. For the production of finer mesh-size powders, cryogenic grinding seems to be more economical than ambient grinding. Since less heat is generated in the cryogenic process, polymer degradation is also minimized. In the case of cryo-grinding technique, the particle size obtained is smaller and the size distribution is narrower than that obtained by the ambient grinding technique [26]. In cryogrinding technique, the level of oxidation that occurs on the rubber surface is low and hence fine rubber powder can be obtained.

The overall process economy depends on the rate of consumption of liquid nitrogen. It has been reported that use of 5–10 phr cryogenically ground rubber in various passenger and truck tire compounds shows some economic advantages without affecting tire performance. For extrusion compounds, the optimum level of cryoground rubber (CGR) to be added to fresh rubber is 5% and the particle size should be less than 100 mesh to avoid fracturing and rough edges of the rubber compound [32]. CGR reduces air entrapment, improves mold flow, and reduces mold shrinkage [9].

When ground rubber is incorporated into fresh rubber, flexing property deteriorates. This is the major drawback of using cryogenically ground rubber in tire industry [9].

38.3.1.3 Wet or Solution Grinding

This is a modified ambient grinding process [32] that reduces the particle size of rubber by grinding in a liquid medium. This process involves putting coarse ground rubber crumbs into a liquid medium, usually water, and grinding between two closely spaced grinding wheels [10,42,43]. Here, the particle size is controlled by the time spent in the grinding process. This process produces particle sizes of 400–500 mesh. The advantage of the fine particle wet ground rubber is that it allows good processing, thereby producing relatively smooth extrudates and calendered sheets. The disadvantage of this process is its low cost-effectiveness.

38.3.1.4 Ultrafine Ground Rubber by Extrusion

In this process, the ground polymer is produced by employing a twin screw extruder that imposes compressive shear on the polymeric material at selected temperatures [43–46]. This process has been patented under the name solid state shear extrusion (SSSE) [47–49] in which, particles of narrow size distribution with the average diameter of as low as 40–60 μm and having irregular and rough surface could be obtained by repeated passes through the extruder. Chemical analyses reveal that some of the bonds are broken during the pulverization process and partial devulcanization takes place [48].

38.3.1.5 Buffing

This method involves the removal of rubber particles from tire tread by abrasion and is confined to the preparation of tire buffings, which is obtained as a by-product of retreading. It is a normal practice in industries to reuse tire buffings to make low-technical products by the revulcanization of the powder or by using the powder-sintering process.

Abrasion involves a combination of processes including mechanical, mechano-chemical, and thermochemical. Several mechanisms of wear and abrasion have been reported. The abrasion of rubber in ordinary sliding depends upon the normal load. The abrasive wear processes are classified as cutting, flaking, wedge formation, and ploughing and the transitions between the different types of abrasive wear occurred at critical degrees of penetration, which are strongly dependent on the conditions of lubrication [50]. According to Gent [51], small particles of about 1–5 μm size are initially removed from the rubber surface, which eventually results in continued erosion when larger pieces of rubber of the order of 100 μm are torn away.

38.3.1.6 Microwave Method

Rubber recycling can be done without depolymerization, to a material capable of being recompounded and revulcanized, having physical properties essentially equivalent to the original vulcanizates, by using microwave energy with controlled dose at a specified frequency and energy level that is sufficient to break carbon–carbon bonds [52–54]. This is an economical and environment-friendly method of reusing elastomeric waste to return it to the same process and products, in which it was originally generated and produces an almost similar product with equivalent physical properties. The devulcanized rubber is not degraded when the material is being recycled [55], which normally take place in the usual commercial processes currently being practiced. For pollution-controlled microwave devulcanization process, sulfur cross-linked elastomers with polar end groups are necessary. Microwave frequency between 915 and 2450 MHz corresponding to microwave energy between 41 and 177 WH per pound is sufficient to severe all cross-links but insufficient to severe polymer chain backbone bonds. The physical properties of microwave devulcanized rubber are comparable to that of the rubber–waste rubber powder blends.

38.3.1.7 Ultrasonic Method

The first work on ultrasonic devulcanization or rubber powder was performed and patented by Pelofsky in 1973 [56]. In this process, solid rubber articles such as tires are immersed into a liquid and then put with a source of ultrasonic energy whereby the bulk rubber effectively disintegrated upon contact and dissolved into liquid. In this process, ultrasonic irradiation is in the range of about 20 kHz and at a power intensity greater than 100 W. Reclaimed rubber based on NR vulcanizates with properties almost similar to original NR can be obtained by devulcanization using ultrasonic frequency of 50 kHz for 20 min [57]. It is proposed that continuous ultrasonic treatment eventually attacks the C—S, S—S, and C—C bonds of vulcanized elastomers [58]. Through the application of certain levels of ultrasonic radiation in the presence of pressure and optionally heat, the three-dimensional network of vulcanized elastomer can be broken down. However, there may be the possibility of polymer degradation along with devulcanization [59]. The mechanical properties of the revulcanized samples obtained from ultrasonically devulcanized ground rubber tire (GRT) reach a maximum with an increasing degree of devulcanization and then drop as a result of excessive degradation. Even without optimization of cure recipes, tensile strength and elongation at break of revulcanized tire rubber of ultrasonically devulcanized rubber are 10.5 MPa and 250%, respectively [58].

38.3.2 Particulate Rubber in Asphalt and Portland Cement Concrete

It has been shown that inclusion of fine rubber particles in asphalt reduces the cracking of pavement in adverse weather conditions [60,61]. There are two methods for introducing ground waste rubber into asphalt, namely, wet and dry processes. Wet process is carried out at 170°C–220°C for 45–120 min. Rubber particles absorb components with similar value of solubility parameter (δ) from the asphalt, causing them to swell. The interaction between rubber and asphalt is mainly of physical nature. In the dry process, rubber is used as a replacement for part of the aggregate and is added to the mineral material before the latter is mixed with the asphalt binder. Addition of rubber greatly improves the elasticity of the binder and generally lowers its brittle point. Incorporation of GRT makes the handling of the asphalt mixture difficult, as it becomes sticky and increases the application time as well as expenses. GRT is being used to prepare polymer-modified bituminous materials that can be used as sealing materials for construction applications [62] and rubber concrete [63]. GRT is found to impart both flexibility and hardness to concrete. Rubber concrete has better compressive strength and Poisson's ratio than conventional cement concrete. Addition of 5 wt% of crumb rubber to a cement matrix does not imply a significant variation of the concrete mechanical features such as elastic modulus. The inclusion of rubber particulates does reduce vibration transmission in concrete. At concentrations of 33 vol% and less, mixtures might be suitable for applications such as driveways or subbases for highway pavements, sound barriers, where strength is not a high priority and where vibration reduction is desired [64].

38.3.3 Miscellaneous Applications of Ground Rubber and Its Revulcanizates

Rubber powder from scrap tires readily absorbs oil floating on water and forms a cohesive mass, which can be easily removed. The amount of oil take up and weight of oil absorbed depends on the particle size of waste tire powder, temperature, and type of oil; the weight of oil absorbed is directly proportional to the amount of rubber powder [65]. The oil uptake time decreases as the particle size decreases.

The CGR can be vulcanized with sulfur alone, but the properties are improved when CGR is vulcanized with sulfur and accelerator. Addition of stearic acid and zinc oxide causes marginal improvement in properties [66]. Rubber powder recovered from old tires might be processed into useful products [67,68]. Riedel Omni Products Inc. developed a rail crossing from scrap rubber for the British rail. The crossings are molded and the surface layer is made from tire retread compound with the bulk of the slab made from treated tire buffing. Cost-effective rubber blocks or mats can be produced by *in situ* coating of GRT with a binder [69]. Thouez [70] found that powdered waste rubber from manufacturing aerosol seals by cutting calendered rubber sheets could be recycled in fresh rubber batches of the same composition. The mechanical properties of butyl, chloroprene, and nitrile rubbers containing 6%–24% powdered waste rubber were generally comparable to those of the rubber compositions containing no powdered waste. Incorporation of the powdered waste improved the elongation to rupture of nitrile rubber.

38.3.4 Utilization of Ground Rubber in Plastics

38.3.4.1 Thermoplastics

In recent years considerable emphasis has been given to the use of ground rubber as fillers in thermoplastics, because of its potential for using large volumes of ground rubber. Furthermore, the compound can be repeatedly used and processed in conventional thermoplastics processing equipment. The thermoplastics industry and market is perhaps 100-fold larger than the nontire rubber industry. Thus, if the scrap rubber could be utilized in plastic blends and composites, even a penetration of only 10% of the materials requirement would consume a large volume of scrap rubber.

There are two approaches to the combining of scrap rubber and plastics. The initial interest was to use the crumb in minor proportions to toughen the plastics improving impact strength and reduce the overall cost. A more recent interest is to develop a type of thermoplastic elastomer (TPE) wherein the rubber is the major component bonded together by thermoplastics, which can be processed and recovered as thermoplastics.

Loading of GRT powder in polypropylene (PP–GRT blend) adversely affects the mechanical properties like tensile strength, elongation at break, and impact strength. However, dynamic vulcanization of PP–GRT blend significantly improves Izod impact strength compared to pure PP or unvulcanized PP–GRT blend [71]. Loading of ground tire powder significantly affects the melt flow properties and rheological properties of thermoplastics [71]. The effective dispersion of waste rubber powder in PP can be achieved by rheological processing in a twin-screw extruder having a process length approximately equal to an *L/D* ratio of 14, including the feeding zone [72]. Loading of GRT and compatibility between the GRT and the polymer matrix play a key role on the properties of the composite [73]. Retention of mechanical properties of the thermoplastic compounds containing GRT depends on the nature of the matrix polymer; low polar and low crystalline polymers favor better compatibility with GRT [74]. Cryogenic mechanical alloying (CMA) of GRT (70 mesh) with poly(methyl methacrylate) (PMMA) or polyethylene terephthalate (PET) produces blends with highly uniform dispersion of GRT [75].

The ductility of GRT–polyethylene blends drastically decreases at ground rubber concentration in excess of 5%. The inclusion of finely ground nitrile rubber from waste printing rollers into polyvinyl chloride (PVC) caused an increase in the impact properties of the thermoplastic matrix [76]. Addition of rubber powder that is physically modified by ultrasonic treatment leads to PP–waste ethylene–propylene–diene monomer (EPDM) powder blends with improved morphology and mechanical properties [77].

38.3.4.2 Compatibilized, Ground Rubber-Filled Plastics

From the literature it is found that incorporation of large amount of recycled rubber into virgin polymers results in materials with inferior properties. This has seriously affected the reutilization of waste rubber in polymer matrices. Loading of GRT drastically affects the properties of thermoplastics. Increase in GRT loading decreases the impact strength of compounds of thermoplastics like linear low-density polyethylene (LLDPE) compounds. This may be attributed to the poor interaction between the plastic matrix and the GRT and the dilution effect by GRT. Over the years, compatibilization including surface modification techniques has been developed to overcome this stagnancy. When the solubility parameter (δ) mismatch between the constituents of a blend is large, compatibilizers act as the interphase and bridge the different phases. GRT particles were precoated with several copolymers like ethylene–acrylic acid, styrene–ethylene–butylene–styrene (SEBS) copolymers.

Surface modification of GRT is necessary to improve the compatibility of the waste tire powders with thermoplastics. Significant work has been done in recent years by introducing active groups on the surface of GRT and thereby increasing its compatibility with thermoplastics. GRT devulcanized by ultrasonic treatment can be dispersed in PP by using maleic anhydride grafted PP as a compatibilizer in an internal mixer and a twin-screw extruder [78]. Functionalization of GRT surface by introducing active chlorine groups, using trichloroisocyanuric acid, improves the tensile strength and moduli of the PVC–GRT blends (Figure 38.2) [79]. A maximum of 40 wt% of chlorinated GRT (Cl-GRT) can be loaded in plasticized PVC without deterioration in mechanical properties of the PVC–GRT blends. PVC–Cl-GRT blends exhibit low swelling characteristics due to its enhanced dipole–dipole interaction between the PVC and Cl-GRT compared to PVC–GRT blends (Figure 38.3) [79]. Surface chlorination of GRT improves the dielectric constant of the material and its blend with plasticized PVC does not show a high dielectric loss factor. This indicates that these blends can be used as dielectrics [79].

GRT/LLDPE composites containing GRT that was irradiated with electron beam display 25% improvement in impact strength. The addition of fine rubber crumb, that has been surface treated in oxygen plasma, to both polyamide and epoxy resins results in a significant improvement in impact strength and toughness [80]. Effective functionalization of GRT can be made by grafting allylamine on to the GRT surface using ultraviolet (UV) irradiation in the presence of a photosensitizer [81]. Allylamine grafting of rubber powder significantly improves the total surface energy of the rubber powder. Surface energy of the allylamine grafted rubber (allylamine-g-GRT) powder is 27.4 mJ/m^2, whereas it is 23.6 mJ/m^2 for neat GRT. However, for achieving better dispersion of allylamine-g-GRT in PP, the system needs a compatibilizer like maleic anhydride grafted polypropylene (MA-g-PP). Loading of allylamine grafted rubber in PP (PP/allylamine-g-GRT) in the presence

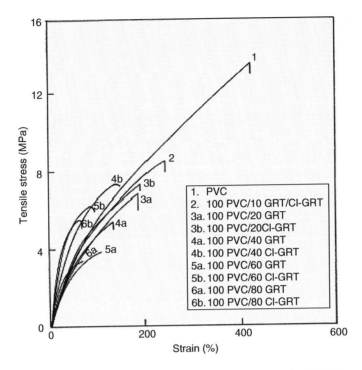

FIGURE 38.2 Stress–strain curves of polyvinyl chloride–ground rubber tire (PVC–GRT) and PVC–Cl-GRT blends. (Reprinted from Naskar, A.K., Bhowmick, A.K., and De, S.K., *J. Appl. Polym. Sci.*, 84, 622, 2002. With permission from Wiley InterScience.)

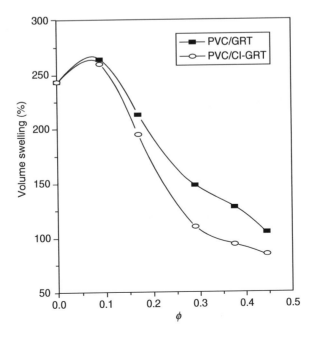

FIGURE 38.3 Volume swelling (%) of composites against volume fraction of filler. (Reprinted from Naskar, A.K., Bhowmick, A.K., and De, S.K., *J. Appl. Polym. Sci.*, 84, 622, 2002. With permission from Wiley InterScience.)

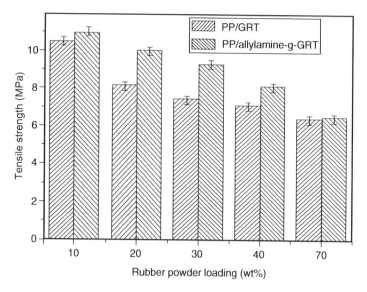

FIGURE 38.4 Tensile strength of polypropylene–maleic anhydride grafted polypropylene (PP–MA-g-PP)–rubber powder composite as a function of rubber powder content. (Reprinted from Shanmugharaj, A.M., Kim, J.K., and Ryu, S.H., *Polymer Test.*, 24, 739, 2005. With permission.)

of 50 wt% of MA-g-PP (relative to GRT) improves its mechanical properties like tensile strength and ultimate elongation compared to its unmodified counterpart (PP/GRT) (Figures 38.4 and 38.5). Improvement of mechanical properties is attributed to the chemical interaction between allylamine-g-GRT with MA-g-PP as shown in Scheme 38.1.

Alternatively, virgin NR can also be used as the best compatibilizer for PP–GRT blends [82]. Functionalization of GRT by grafting styrene onto CGR particles improves the mechanical properties without the necessity of a compatibilizer [83]. Elongational capacities of the PP–waste SBR

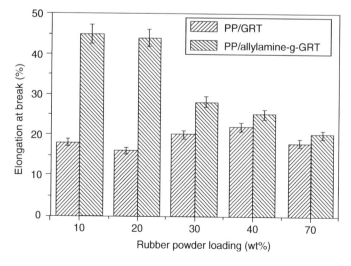

FIGURE 38.5 Elongation at break of polypropylene–maleic anhydride grafted polypropylene (PP–MA-g-PP)–rubber powder composites as a function of rubber powder content. (Reprinted from Shanmugharaj, A.M., Kim, J.K., and Ryu, S.H., *Polymer Test.*, 24, 739, 2005. Courtesy of Elsevier, U.K.)

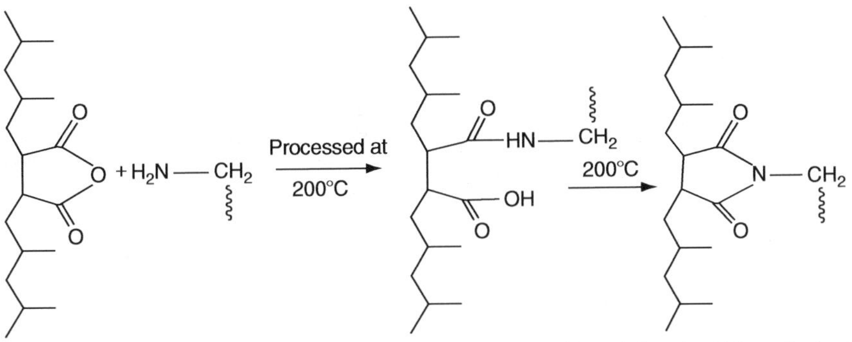

SCHEME 38.1 Reaction of allylamine-g-ground rubber tire (GRT) with maleic anhydride grafted polypropylene (MA-g-PP). (Reprinted from Shanmugharaj, A.M., Kim, J.K., and Ryu, S.H., *Polymer Test.*, 24, 739, 2005. Courtesy of Elsevier, U.K.)

reactive blends significantly improved by grafting PP to the benzene ring in the styrene–butadiene rubber (SBR). Thermoplastics containing comminuted fiber-reinforced thermoset recyclate have been shown to exhibit superior mechanical properties relative to their particulate-filled counterparts. Optimum mechanical performance of fiber-reinforced thermoset recyclate-filled thermoplastic materials is also achieved in compositions made from rubber-modified thermoplastics containing MA-g-PP in conjunction with silane coupling agent [84]. However, incorporation of GRT into a thermoplastic matrix significantly affects the mechanical properties of the resulting thermoplastic elastomeric composites, depending upon the loading and nature of GRT, polymer matrix type, and the compatibility between GRT and the polymer matrix [85–88].

38.3.4.3 Thermosets and Rubbers

Addition of scrap rubber in the form of either ground waste vulcanizates or reclaim in rubber compounds gives economic as well as processing advantages. In addition to lowering the cost of rubber compounds, the use of cross-linked rubber particles has beneficial effects such as faster extrusion rate, reduced die swell, and better molding characteristics.

Loading of ground tire powder in NR/SBR increases the rheometric property such as minimum torque (M_L), whereas, it reduces the scorch time (t_{s2}), optimum cure time (t_{90}), and maximum rheometric torque (M_H). The apparent reason for the increase in minimum torque is due to aggregation of cross-linked GRT that does not flow easily in the matrix. Hence, an increase in GRT loading may reduce the flow and consequently increase the torque. It is believed that the highly aggregated and convoluted structure of waste tire powder contains void space in which the matrix rubber is trapped, which increases the effective volume fraction of waste rubber and results in higher viscosity of the blends [89].The reduction in scorch time and optimum cure time are attributed to the curative and accelerator migration from virgin rubber matrix to ground rubber vulcanizates [90]. Devulcanization of GRT by continuous shear flow reaction treatment decreases the rheometric properties M_L and M_H, when curatives were added based on weight percentage of virgin rubber, whereas, M_L decreases while M_H increases, when curatives were added based on weight percentage of virgin and waste tire powder. In the absence of extra curatives for waste rubber powder, shortest fragments and smaller chains in the devulcanized rubber will act as plasticizers, decreasing the viscosity and torque of the compounds and increasing tack. On loading extra curatives, revulcanization and cross-link reaction occurs with the devulcanized rubber in the case of devulcanized rubber powder loaded virgin NR matrix, due to the new active sites formed during the devulcanization process and to the presence of curatives. The t_{s2} and t_{90} of the compounds did not change significantly. The scorch time decreases in all systems, with the GRT system having the lowest and the NR vulcanizates without extra curative system exhibits highest scorch time values.

TABLE 38.3

Rheometric Results of Cryoground Rubber (CGR) Filled NR Vulcanizates

Mix	Control	10% CGR	20% CGR	30% CGR	40% CGR	50% CGR	60% CGR
Initial viscosity, Nm	1.07	1.38	1.56	1.69	1.61	1.58	1.58
Thermoplasticity, Nm	0.51	0.70	0.77	0.79	0.81	0.67	0.66
Scorch time at 150°C, min	4.7	4.5	4.0	3.7	3.5	3.5	3.5
Maximum rheometric torque, Nm	6.75	6.80	6.70	6.62	6.51	6.20	5.94
Optimum cure time at 150°C, min	9.0	8.75	8.5	8.2	8.2	8.2	8.2
Cure rate, %/min	22.22	23.50	22.22	22.22	21.05	21.05	21.05

Source: Reprinted from Phadke, A.A. and De, S.K., *J. Appl. Polym. Sci.*, 32, 4063, 1986. With permission.

Compounding formulation: NR: 100; ZnO: 5.0; Stearic acid: 2.0; CBS: 1.6; Sulfur: 2.5 phr.

The shorter scorch time in the blends indicates that the cross-linking reaction started earlier. A product of a reaction between the NR matrix and the devulcanized rubber may catalyze the cross-linking reaction [91]. Also, diffusion of the accelerator from devulcanized rubber into virgin rubber would reduce scorch time [92]. Optimum cure time varies slightly between GRT and NR vulcanizates without extra curatives. The cure time is shorter in the NR vulcanizates with extra curative system, and the devulcanized rubber participates together with virgin rubber in the cross-linking reaction. Incorporation of waste rubber into the virgin NR matrix does not influence the cure time significantly.

The properties like tensile strength and modulus of NR–SBR matrices decrease with increasing loading and particle size of GRT. Flex crack resistance of the NR–SBR matrices is significantly altered on loading GRT in the virgin rubber matrix [93]. Crack growth rate decreases with increasing loading and decreasing particle size of the GRT powder.

NR vulcanizates prepared using surface-treated GRT (using nitric acid and hydrogen peroxide) retain 50% of the physicomechanical properties after aging. Similarly, loading of CGR in the virgin rubber matrix increases M_L without much increase in M_H. Also, loading of CGR reduces the scorch time, optimum cure time, and reversion time (Table 38.3). Loading of CGR shows detrimental effect on most of the vulcanizate properties such as tensile strength, flex, heat buildup, permanent set, and abrasion resistance of NR vulcanizates. However, tear strength is not adversely affected by the addition of CGR (Table 38.4) [94].

TABLE 38.4

Mechanical Properties of Cryoground Rubber (CGR) Loaded NR Vulcanizates

Mix	Control	10% CGR	20% CGR	30% CGR	40% CGR	50% CGR	60% CGR
Modulus at 300% elongation, MPa	2.49	2.10	2.39	2.79	3.02	3.18	3.15
Tensile strength, MPa	27.63	19.39	16.84	15.24	13.65	13.03	10.78
Elongation at break, %	710	640	610	590	570	550	500
Tear strength, kN/m	32.3	41.6	39.3	38.9	35.6	33.3	32.5
Volume fraction of rubber, V_r	0.219	0.222	0.219	0.221	0.222	0.222	0.222

Source: Phadke, A.A. and De, S.K., *J. Appl. Polym. Sci.*, 32, 4063, 1986. With permission.

Composite Particles, Inc. reported the use of surface-modified rubber particles in formulations of thermoset systems, such as polyurethanes, polysulfides, and epoxies [95]. The surface of the rubber was oxidized by a proprietary gas atmosphere, which leads to the formation of polar functional groups like —COOH and —OH, which in turn enhanced the dispersibility and bonding characteristics of rubber particles to other polar polymers. A composite containing 15% treated rubber particles per 85% polyurethane has physical properties similar to those of the pure polyurethane. Inclusion of surface-modified waste rubber in polyurethane matrix increases the coefficient of friction. This finds application in polyurethane tires and shoe soles. The treated rubber particles enhance the flexibility and impact resistance of polyester-based construction materials [95]. Inclusion of treated waste rubber along with carboxyl terminated nitrile rubber (CTBN) in epoxy formulations increases the fracture toughness of the epoxy resins [96].

Superior Environmental Products, Inc. introduced a product based on liquid polysulfide containing 40% of a surface-modified scrap tire rubber. The product, ER-100R, is a coating that can temporarily contain chemical, oil, and gasoline spills. Rodriguez [97] reported that an unsaturated polyester resin containing silane-treated CGR showed better mechanical properties than that containing untreated CGR.

38.3.5 Utilization of Ground Rubber in Thermoplastic Elastomers

TPEs from thermoplastics–rubber blends are materials having the characteristics of thermoplastics at processing temperature and that of elastomers at service temperature. This unique combination of properties of vulcanized rubber and the easy processability of thermoplastics bridges the gap between conventional elastomers and thermoplastics. Cross-linking of the rubber phase by "dynamic vulcanization" improves the properties of the TPE. The key factor that controls the properties of TPE is the blend morphology. It is essential that in a continuous plastic phase, the rubber phase should be dispersed uniformly, and the finer the dispersed phase the better are the properties. A number of TPEs from dynamically vulcanized rubber–plastic blends have been developed by Bhowmick and coworkers [98–102].

Utilization of scrap rubber powder in thermoplastic elastomeric blends as a partial replacement of the rubber phase provides an attractive alternative for the recycling of rubber waste. PP–latex reclaim blends were prepared and their morphology and mechanical properties were reported by Shaban and George [103] and George and Joseph [104]. Osborn [105] studied the feasibility of activated tire rubber (ATR), a patented surface-modified crumb rubber product, as a modifying ingredient in TPEs such as Santoprene. The most beneficial impact of ground tire rubber on TPE compounds may be economic, but a substantial drop in tensile strength and elongation was disappointing. Dynamic vulcanization of GRT-filled PP improves the impact properties of PP [106]. Al-Malaika and Amir studied the effect of partial replacement of NR by reclaimed rubber in NR–PP thermoplastic elastomeric blends and found that up to 50% of the NR could be replaced by RR without significantly altering the mechanical properties of the blend [107]. Naskar et al. [108] observed that in TPE formulations based on EPDM–poly(ethylene-co-acrylic acid) blend, 50% of EPDM could be substituted by GRT without significant reduction in mechanical properties and with significant reprocessing capabilities. Morphological studies show that virgin EPDM forms a coating over the GRT and the plastic phase forms the continuous matrix (Figure 38.6).

Thermoplastic elastomeric compositions from reclaimed NR and scrap LDPE with 50:50 rubber/plastic ratio shows good processability, ultimate elongation, and set properties. Polymer blends of reclaimed rubber and LDPE exhibit higher viscosity over the range of shear rate at various temperatures compared to virgin NR–LDPE blends due to the influence of filler present in the reclaimed rubber (Figure 38.7) [109].

Okamato et al. [110] developed a new rubber recycling technology to produce a polyolefin TPE-based on EPDM waste and PP. Fine rubber powder can also be obtained from the sanding process of

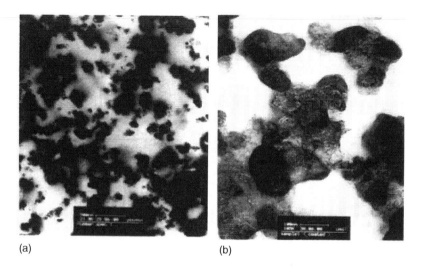

(a)   (b)

FIGURE 38.6 Morphology of (a) ethylene–propylene–diene monomer (EPDM)–poly(ethylene-*co*-acrylic acid) blend; (b) EPDM–poly(ethylene-*co*-acrylic acid)–ground rubber tire (GRT) blend. (Reprinted from Naskar, A.K., Bhowmick, A.K., and De, S.K., *Polym. Eng. Sci.*, 41, 1087, 2001. With permission from Wiley InterScience.)

polishing rubber balls and artificial eggs, to prepare PP-recycled rubber blends [111]. A similar series of blends using NR was also prepared. The results indicated that at similar rubber content, PP-recycled rubber blends have higher tensile strength and Young's modulus but lower elongation at break and thermal stability than PP–NR blends. Kumar et al. [112] used thermomechanically decomposed GRT to produce TPEs composed of LDPE, virgin rubbers like SBR, NR, and EPDM, and GRT with and without dynamic curing. Sun and Xiong [113] reported the effect of operating conditions, curing agents, blend ratio, and nature of polyethylene on the mechanical and thermal properties of RR/LDPE-based TPE.

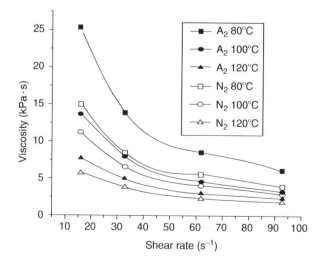

FIGURE 38.7 Viscosity versus shear rate plots of virgin natural rubber–low-density polyethylene (NR–LDPE) blends (N_2) and reclaimed rubber (RR)–LDPE blends (A_2). (Reprinted from Nevatia, P., Banerjee, T.S., Dutta, B., Jha, A., Naskar, A.K., and Bhowmick, A.K., *J. Appl. Polym. Sci.*, 83, 2035, 2002. With permission from Wiley InterScience.)

38.3.6 BIODEGRADATION OF WASTE TIRE RUBBER

NR in the latex form is quite facile to biological attack. However, loading of various ingredients including sulfur in NR minimizes the biological attack. An interesting recent approach is reported in a patent application [114] to utilize a chemolithiotrope bacterium in aqueous suspension for attacking powder elastomers on the surface only, so that after mixing with virgin rubber diffusion of soluble polymer chains is facilitated and bonding during vulcanization becomes possible again. Devulcanization of scrap rubber by holding the comminuted scrap rubber in a bacterial suspension of chemolithotropic microorganisms with a supply of air leads to the development of elemental sulfur or sulfuric acid. This is an interesting process, which obtains RR and sulfur in a simplified manner [115]. Biodegradation of the *cis*-1,4-polyisoprene chain using bacterium belonging to the genus *Nocardia*, leads to a considerable weight loss of different soft type NR vulcanizates [116–118]. Treatment of waste tire powder with various species of *Thiobacillus*, i.e., *T. ferrooxidans*, *T. thiooxidans*, and *T. thioparus* in shake flasks leads to oxidation of sulfur. However, particle size of the waste tire powder plays a major role in the biodegradation. Waste tire powder with particle size in the range of 100–200 μm having 4.7% of sulfur content can be oxidized to sulfate within 40 days [119]. Adaptation of microbial enrichment cultures with tire crumb material for several months resulted in enhanced growth of microorganisms especially for NR and SBR.

38.3.7 RECYCLING OF RUBBER FROM AUTOMOTIVE WASTE

A number of polymers have been used in making automotive components. Many parts such as tires, v-belts, seals, gaskets, oil tubes, etc. have been manufactured from rubbers and TPEs. Among all these components, tires are large volume products. Recycling of automotive tires is particularly challenging when one considers that tire is a composite of different rubber compounds reinforced with both textile and steel wire [120]. Some newer methods of recycling of rubbers used in automotive tires, seals, and weather strips are discussed in the following paragraphs.

A new recycling technology called shear flow stage reactor has high productivity and can achieve the same properties as virgin materials [121]. Automotive parts have been developed and are being produced with recycled EPDM based on this new technology. This technology can be applied to other rubber materials, such as NR, NR/SBR, and IIR with appropriate reclaim conditions. Nakahama et al. [121] developed moldings by adding thermoplastic resins to recycled pulverized vulcanized rubbers (e.g., recycled automobile weather strips), and press molding. A composition containing 120 parts 100:20 mixture of EPDM rubber and crystalline isotactic polypropylene (iPP), 50 parts carbon black, 40 parts process oil, 3.0 parts blowing agent (Neocellborn 1000S), 3.3 parts vulcanization accelerator mixture, and 1.0 part sulfur was vulcanized to give a tubular sponge showing water absorption 5%, tensile strength 3.5 MPa, and elongation at break 350%, which was cut, heated at 180°C for 3 min, mixed with 3% powdered iPP at 160°C and molded to give test pieces showing water absorption 10%, tensile strength 5.2 MPa, and elongation at break 280%. Garderes [122] patented a method for utilizing recycled rubbers in sound insulation such as in automobile, in which the material is processed by extrusion. With this objective, heavyweight insulation in the form of a thin sheet or pudding is provided, which comprises recycled or recyclable crude rubber, powdered pneumatic tire or vulcanized tire, plastics and/or recycled polyolefins, and optionally oils and/or fillers. This creates the possibility of application of waste rubber elements from used cars as an anticorrosive lining for plating equipment. Mechanical and corrosion resistance tests of waste floor lining in three selected plating baths were carried out. A good resistance of waste lining to zinc and nickel plating baths was found, while its resistance to chromium plating bath was not acceptable. It creates a possibility of the utilization of waste rubber elements for anticorrosion protection of some plating equipment. Kawakami and Udagawa [123] developed passenger car tire using recycled rubber. The tire has a cap tread and low heat-generating under tread, wherein the cap tread rubber composition contained recycled rubbers obtained by impregnation of a rubber composition with an organic solvent containing a peroxide; volume fraction of the under tread to total treads was 0.5–0.7. The tire showed enhanced gripping property at high temperature and retention of blow-out resistance. Thus, a vulcanized NR composition was soaked in toluene containing 1% benzoyl peroxide to give the recycled rubber containing polyisoprene structure, 10 parts of which was blended with 150 parts of an oil-extended 65:35 SBR emulsion, 90 parts carbon black, zinc oxide, sulfur, and vulcanization accelerators. Then, tires having cap treads made of the obtained composition and NR under tread (volume fraction 0.5) were set in a passenger car, which showed no blowout in driving.

EPDM rubber is being used in window seal compounds, which produces a lot of scrap as a factory waste. Jacob et al. [124] replaced virgin rubber by ground EPDM vulcanizate powder in a PP/EPDM-based thermoplastic vulcanizates (TPVs) and found that the mechanical properties can be improved by replacing virgin EPDM with ground EPDM vulcanizate powder up to 45% loading, beyond which processing becomes difficult. Anandhan et al. [125] reported that the effective recycling of factory waste of acrylonitrile butadiene rubber (NBR) can be done, by utilizing a model waste NBR vulcanizate powder (w-NBR) in TPVs based on dynamically vulcanized poly(styrene-co-acrylonitrile) (SAN)–NBR blends. For this purpose, they developed a model NBR compound using a standard oil seal formulation (NBR [acrylonitrile content: 34%], 100; carbon black-N770, 50; dioctyl phthalate, 5; zinc oxide, 5; stearic acid, 1; IPPD, 1; MBTS, 1.5; TMTM, 0.1; sulfur, 1; the figures are in parts per 100 g of rubber, by weight). Ambient grinding of model NBR compound shows highly aggregated and chain-like structures (Figure 38.8) with the particle size distribution of 1–135 μm (Figure 38.9), which breaks down during the mixing operation [125]. They replaced virgin NBR with w-NBR powder in SAN/NBR-based TPV and found that the mechanical properties can be improved by replacing NBR with w-NBR up to 45% [125]. Transmission electron micrograph of the blend of SAN–NBR–w-NBR shows two phase morphology. The dark phase corresponds to the rubber phase (which contains both virgin NBR and w-NBR), which has been stained by osmium tetraoxide, and the lighter phase corresponds to the SAN phase. In these blends SAN forms the matrix, while virgin NBR forms the dispersed phase encapsulating the dispersed w-NBR particles (the light-colored virgin NBR coating on the dark w-NBR particles in Figure 38.10). Thermorheological studies of the 70/30 SAN–NBR blends containing varying amounts of w-NBR reveal that viscosity of the blend increases with increase in w-NBR loading (Figure 38.11). The increase in viscosity is due to

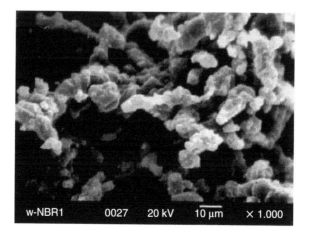

FIGURE 38.8 Ambient ground acrylonitrile butadiene rubber (NBR) waste powder. (Reprinted from Anandhan, S., De, P.P., Bhowmick, A.K., Bandyopadhyay, S., and De, S.K., *J. Appl. Polym. Sci.*, 90, 2348, 2003. With permission from Wiley InterScience.)

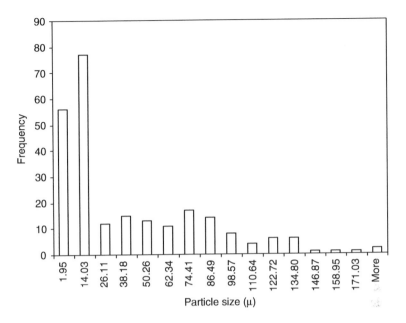

FIGURE 38.9 Particle size distribution of ambient ground acrylonitrile butadiene rubber (NBR) waste powder. (Reprinted from Anandhan, S., De, P.P., Bhowmick, A.K., Bandyopadhyay, S., and De, S.K., *J. Appl. Polym. Sci.*, 90, 2348, 2003. With permission from Wiley InterScience.)

the presence of solid particles and reinforcing effect of the filler present in w-NBR [126]. Extrudate surfaces of the SAN–NBR–w-NBR blends are relatively smoother than that of SAN–NBR blend. The filler present in w-NBR reduces the elastic turbulence of the melts and hence the surfaces of the blend containing w-NBR are smoother (Figure 38.12).

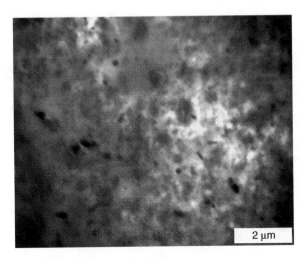

FIGURE 38.10 Transmission electron micrograph of styrene-*co*-acrylonitrile/acrylonitrile butadiene rubber/ waste NBR (SAN/NBR/w-NBR)-based thermoplastic elastomer (TPE). (Reprinted from Anandhan, S., De, P.P., Bhowmick, A.K., Bandyopadhyay, S., and De, S.K., *J. Appl. Polym. Sci.*, 90, 2348, 2003. With permission from Wiley InterScience.)

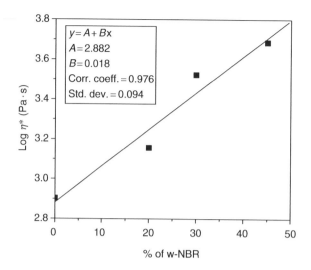

FIGURE 38.11 Log of complex viscosity (η^*) of the acrylonitrile butadiene rubber/waste NBR/styrene-*co*-acrylonitrile (NBR/w-NBR/SAN)-based thermoplastic elastomer (TPE) compositions as a function of w-NBR content at 100 rad/s. (Reprinted from Anandhan, S., De, P.P., De, S.K., Swayajith, S., and Bhowmick, A.K., *Kautsch. Gummi Kunst.*, 11, 1, 2004. With permission.)

Matsushita et al. [127] prepared compositions showing good vulcanizability containing EPDM reclaim. Weather-strip wastes comprising of sulfur-cured EPDM rubber (containing 50% carbon black) were kneaded at 300°C and 3 MPa and extruded to give a rubber (M_w 200,000; gel content 65%), 25 parts of which was mixed with unvulcanized SBR 75, carbon black 37.5, and sulfur 1.3

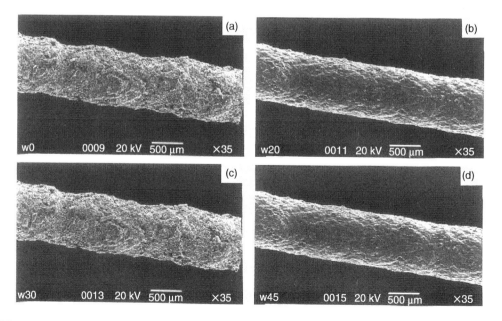

FIGURE 38.12 Extrudate roughness of acrylonitrile butadiene rubber/waste NBR/styrene-*co*-acrylonitrile (NBR/w-NBR/SAN)-based thermoplastic elastomer (TPE) compositions with (a) 0% w-NBR; (b) 20% w-NBR; (c) 30% w-NBR; and (d) 45% w-NBR. (Reprinted from Anandhan, S., De, P.P., De, S.K., Swayajith, S., and Bhowmick, A.K., *Kautsch. Gummi Kunst.*, 11, 1, 2004. Courtesy of Huthig GmbH & Co. KG.)

parts and vulcanized to give a test piece showing tensile strength 21.3 and 17.7 MPa, before and after aging at 100°C for 100 h, respectively. Hoetzeldt et al. [128] patented elastic molding and sealing compositions containing recycled rubber, which can be sprayed when cold and prepared with nonpolluting binders, containing 0.1–0.8 parts recycled rubber or flexible plastics (particle size ≤2 mm) and 0.9–0.2 parts water soluble acrylate polymer dispersion. Polysulfide sealants were reverted to a liquid sealant comprising mercaptan-terminated polysulfides by reducing them with an amount of liquid mercaptan-terminated polysulfide sufficient to provide 1%–3% —SH groups in the reclaimed sealant at 50°C–110°C. Upon addition of fresh curing agent and re-curing, the reclaimed sealant has satisfactory tensile strength and elongation properties [129]. Hecktor and coworkers [130] found that hardened polysulfide elastomers could be recycled by treating with a di- or polymercapto compound to give a mercapto-terminated prepolymer paste using plasticizer, filler, and coupling agent. Moreover, addition of sealant hardener composition gives a sealing composition with desirable mechanical properties. Sosnina et al. [131] developed fluoro rubber composition containing reclaimed fluoroelastomers for sealing applications. This composition having increased elasticity and improved heat resistance comprises 3:7 hexafluoropropylene–vinylidene fluoride copolymer rubber 10–18, reclaimed fluoroelastomers 10–20, cross-linking agent 0.9–1, barite 30–41, and asbestos. Incorporation of waste fluorocarbon rubber in virgin fluorocarbon rubber increases the Mooney viscosity, Mooney scorch time and shear viscosity, tensile strength, modulus, and hardness, whereas it decreases the tear strength of the fluoroelastomers [132]. A method for recycling of silicone rubber scraps uses heat for depolymerization in a sealed container [133]. Alternatively, recycling of silicone rubber can be done by grinding thick rubber sheets over the silicone carbide wheel of diameter 150 mm, rotating at 2900 rpm, using a bench grinder [134]. Incorporation of waste silicone rubber in virgin silicone rubber increases the Mooney viscosity, Mooney scorch time, shear viscosity, and activation energy of the viscous flow. However, tensile strength and tear strength decreases by 20%, modulus by 15%, while hardness, tension set, and hysteresis loss undergoes only marginal changes on loading waste silicone rubber in virgin rubber [134].

38.4 CONCLUSIONS

Polymer disposal in an inappropriate manner creates environmental problems such as dioxin formation, catastrophic fires, breeding of rats, and mosquitoes. Several methods have been explored to utilize plastics and rubber waste in an environment-friendly manner. Some of the recent advances in rubber recycling are reviewed in this chapter with special emphasis to waste rubber reutilization in plastics and rubbers. The utilization of waster rubber powder in polymer matrices provides an attractive strategy for polymer waste disposal.

REFERENCES

1. Goodyear, C., Improvement in India-rubber fabrics, *US patent, 3,633*, 1844.
2. The Committee of Recycling Technology. Recycling of polymer recycling, *Tokyo: Jpn. J. Publ. House*, **121**, 1982.
3. Fang, Y., Zhan, M., and Wang, Y., The status of recycling of waste rubber, *Mater. Design*, **22**, 123, 2001.
4. Bauman, B.D., High-value engineering materials from scrap rubber, *Rubber World*, **212**, 25, 1995.
5. Synder, R.H., *Scrap Tires: Disposal and Reuse*, Society of Automotive Engineers, Inc., Warrendale, PA, 1998.
6. Klingensmith, W. and Baranwal, K., in *Handbook of Elastomers—New Developments and Technology*, Bhowmick, A.K. and Stephens, H.L., Eds., 2nd ed., Marcel Dekker, New York, 2001, Chapter 34.
7. Synder, R.H., Vincent, V.R., and Querry, F.C., *Paper Presented in the National Tire Disposal Symposium*, June, Washington, DC, 1977.
8. Winkleman, H.A., The Present and Future of Reclaimed Rubber, *Ind. Eng. Chem.*, **18**, 1163, 1926.

9. Adhikari, B., De, D., and Maiti, S., Reclamation and recycling of waste rubber, *Prog. Polym. Sci.*, **25**, 909, 2000.

10. Mark, H.F., Bikales, N.M., Overberger, C.G., and Menges, G., in *Encyclopedia of Polymer Science and Engineering*, Wiley, New York, Vol. 14, p. 787, 1988.

11. Nishimura, K., Recycling of used tyres in Japan, *Kautsch. Gummi Kunst.*, **46**, 989, 1993.

12. Lee, S., Azza, F.O., and Kocher, B.S., Oxidative decoupling of scrap rubber, *US patent, 5, 516, 952*, 1996.

13. Adkins, L., Used tire process, *US patent, 5,618,852*, 1997.

14. Conessa, J.A., Fullena, A., and Font, R., Tire pyrolysis: Evolution of volatile and semivolatile compounds, *Energy Fuel*, **14**, 409, 2000.

15. Paul, J., Recycling of rubber-review, *Kirk-Othmer Encycl. Chem. Technol.*, 3rd ed., **19**, 1002, 1982.

16. Holley, C.A., Method for recovering carbon black from composites, *US patent, 5,728,361*, 1998.

17. Meador, W.R., Method and apparatus for recovering constituents from discarded tires, *US patent, 5,720, 232*, 1998.

18. Roy, C., Recovery of commercially valuable products from scrap tires, *US patent, 5,229,099*, 1993.

19. Roy, C., Chaala, A., and Darmstadt, H., The vacuum pyrolysis of used tires: End-uses for oil and carbon black products, *J. Anal. Appl. Pyrolysis*, **5**, 201, 1999.

20. Owen, E.W., Processes of reclaiming rubber and their relative merits, *Rubber Chem. Technol.*, **17**, 544, 1944.

21. Le Beau, D.S., Basic reactions occurring during rubber reclaiming. I. The influence of reclaiming media, antioxidant, and defibering agents on vulcanized natural rubber at 195 pounds per square inch gauge pressure, *Rubber Chem. Technol.*, **21**, 895, 1948.

22. Cook, W.C., Albert, H., Kilbourne, F., and Smith, G.E.P., Reclaiming agents for synthetic rubber, *Rubber Chem. Technol.*, **22**, 166, 1949.

23. Le Beau, D.S., Basic reactions occurring during rubber reclaiming. II. The influence of solvent naphtha, storage of reclaim and aging of scrap prior to reclaiming, and infrared spectra of natural rubber reclaim, *Rubber Chem. Technol.*, **22**, 560, 1949.

24. Amberlang, J.C. and Smith, G.E.P., Jr., Behavior of reclaiming agents in sulfur and nonsulfur GR-S vulcanizates, *Rubber Chem. Technol.*, **28**, 322, 1955.

25. Ball, J.M., in *Introduction to Rubber Technology*, Morton, M., Ed., van Nostrand Reinhold, New York, 1959, Chapter 7.

26. Myhre, M. and MacKillop, D.A., Rubber recycling, *Rubber Chem. Technol.*, **75**, 429, 2002.

27. Le Beau, D.S., Science and technology of reclaimed rubber, *Rubber Chem. Technol.*, **40**, 217, 1967.

28. Stafford, W.G. and Wright, R.A., in *Applied Science of Rubber*, Naunton, W.J.S., Ed., E. Arnold, London, 1961, Chapter 4.

29. Payne, E., Reclaim rubber usage and trends, *Rubber World*, **210**, 22, 1994.

30. Burford, R.P. and Pittolo, M., Characterization and performance of powdered rubber, *Rubber Chem. Technol.*, **55**, 1233, 1982.

31. Roethemeyer, F., Reprocessing of rubber wastes, *Kautsch. Gummi Kunst.*, **46**, 356, 1993.

32. Klingensmith, B., Recycling, production and use of reprocessed rubbers, *Rubber World*, **203**, 16, 1991.

33. Drozdovskii, V.F., Production of comminuted vulcanizates, *Prog. Rubber Plast. Technol.*, **14**, 116, 1998.

34. Tolonen, E.O., Method for grinding materials, *US patent, 5,115,988*, 1992.

35. Mastral, A.M., Callen, M.S., Murillo, R., and Garcia, T., Combustion of high calorific value waste material: Organic atmospheric pollution, *Environment. Sci. Technol.*, **33**, 4155, 1999.

36. Garmater, R.A., Granulating, separating and classifying rubber tire materials, *US patent, 5,299,744*, 1994.

37. Stark, J., in *The Tire Recycle Solution: Minnesota's Answer to the Scrap Tire Problem*, Tire Technology Conference, Clemson University, Oct. 28–29, 1987.

38. Murtland, W.O., Cryogrinding scrap into filler, *Elastomerics*, **110**, 26, 1978.

39. Rouse, M.W., Development and application of superfine tire powders for compounding, *Rubber World*, **206**, 25, 1992.

40. Phadke, A.A. and De, S.K., Use of cryoground reclaimed rubber in natural rubber, *Conserv. Recycling*, **9**, 271, 1986.

41. Swor, R.A., Jensen, L.W., and Budzol, M., Ultrafine recycled rubber, *Rubber Chem. Technol.*, **53**, 1215, 1980.

42. Dierkes, W., Solutions to the rubber waste problem incorporating recycled rubber, *Rubber World*, **214**, 25, 1996.

43. Kowser, M.J. and Farid, N.A., Thermochemical processing of rubber waste to liquid fuel, *Plast. Rubber Comp.* **29**, 100, 2000.

44. Enilkolopov, N.S., Some aspects of chemistry and physics of plastic flow, *Pure Appl. Chem.*, **57**, 1707, 1985.

45. Enilkolopov, N.S., Wolfson, S.A., Nepomnjaschii, A.I., and Nikolskii, V.G., Method and apparatus for pulverizing polymers, *US patent, 4,607,797*, 1986.

46. Wolfson, S.A. and Nikolskii, V.O., Powder extrusion: Fundamentals and different applications, *Polym. Eng. Sci.*, **37**, 1294, 1997.

47. Arastoopour, H., Single-screw extruder for solid state shear extrusion pulverization and method, *US patent, 5,704,555*, 1998.

48. Bilgili, E., Arastoopour, H., and Bernstein, B., Analysis of rubber particles produced by the solid state shear extrusion pulverization process, *Rubber Chem. Technol.*, **73**, 340, 2000.

49. Khait, K., Reconstituted polymeric materials derived from post-consumer waste, industrial scrap and virgin resins made by solid state pulverization, *US patent, 5,814,673*, 1998.

50. Kayaba, T., Hokkiringawa, K., and Kato, K., Analysis of the abrasive wear mechanism by successive observations of wear processes in a scanning electron microscope, *Wear*, **110**, 419, 1986.

51. Gent, A.N., A hypothetical mechanism for rubber abrasion, *Rubber Chem. Technol.*, **62**, 750, 1989.

52. Makrov, V.M. and Drozdovski, V.F., *Reprocessing of Tires and Rubber Wastes*, Ellis Horwood, New York, 1991.

53. Novotny, B.S., Marsh, R.L., Masters, F.C., and Tally, D.N., Microwave devulcanization of rubber, *US Patent 4,104,205*, 1978.

54. Fix, S.R., Microwave devulcanization of rubber, *Elastomerics*, **112**(6), 38, 1980.

55. Tantayanon, S. and Juikham, S., Enhanced toughening of polypropylene with reclaimed-tire rubber, *J. Appl. Polym. Sci.*, **91**, 510, 2003.

56. Pelofsky, A.H., Rubber reclamation using ultrasonic energy, *US Patent 3,725,314*, 1973.

57. Okuda, M. and Hatano, Y., Method of desulfurizing rubber by ultrasonic wave, *Japanese Patent, 62, 121,741*, 1987.

58. Isayev, A.I., Continuous ultrasonic devulcanization of vulcanized elastomers, *US Patent, 5,258, 413*, 1993.

59. Mangaraj, B. and Senapati, N., Multi-lever miniature fiber optic transducer, *US Patent, 4,599,711*, 1986.

60. Lewandowski, L.H., Polymer modification of paving asphalt binders, *Rubber Chem. Technol.*, **67**, 447, 1994.

61. Takkalou, H.B. and Takkalou, M.B., Recycling tires in rubber asphalt paving yields cost, disposal benefits, *Elastomerics*, **123**, 19, 1991.

62. Gebauer, M., Agsten, W.G., Mehlig, J., Salzmann, J., and Schilbach, W., Bituminous composition used e.g., in road construction, *German Patent, DE 19,716,544 A1 19,981,022*, 1998.

63. Chung, K.H. and Hong, Y., Introductory behavior of rubber concrete, *J. Appl. Polym. Sci.*, **72**, 35, 1999.

64. Raghavan, D., Study of rubber-filled cementitious composites, *J. Appl. Polym. Sci.*, **77**, 934, 2000.

65. Koutsky, J., Clark, G., and Klotz, D., The use of recycle tire rubber particles for oil spill recovery, *Conserv. Recycling*, **1**, 231, 1977.

66. Phadke, A.A. and De, S.K., Vulcanization of cryo-ground reclaimed rubber, *Kautsch. Gummi Kunst.*, **37**, 776, 1984.

67. Acetta, A. and Vergnaud, J.M., Vibration isolation properties of vulcanizates of scrap rubber powder, *Rubber Chem. Technol.*, **55**, 328, 1982.

68. Acetta, A. and Vergnaud, J.M., Upgrading of scrap rubber powder by vulcanization without new rubber, *Rubber Chem. Technol.*, **54**, 301, 1981.

69. Kim, J.K. and Lee, S.H., New technology of crumb rubber compounding for recycling of waste tires, *J. Appl. Polym. Sci.*, **78**, 1573, 2000.

70. Thouez, J.L., Use of powdered waste rubber in rubber blends for aerosol seals, *Revue Generale des Caoutchoucs & Plastiques*, **65–68**, 664, 1986.

71. Münstedt, B., Rheology of rubber-modified polymer melts, *Polym. Eng. Sci.*, **21**, 259, 1981.

72. Wagenknecht, U., Steglich, S., Wiessner, S., and Michael, H., Rubber Powder—A perspective filler of thermoplastics, *Macromol. Symp.*, **221**, 237, 2005.

73. Rajalingam, P., Sharpe, J., and Baker, W.E., Ground rubber tire/thermoplastic composites: Effect of different ground rubber tires, *Rubber Chem. Technol.*, **66**, 664, 1993.

74. Deanin, R.D. and Hashemiolya, S.M., Polyblends of reclaimed rubber with eleven thermoplastics, *Polym. Mater. Sci. Eng.*, **8**, 212, 1987.

75. Smith, A.P., Ade, H., Koch, C.C., and Spontak, R.J., Cryogenic mechanical alloying as an alternative strategy for the recycling of tires, *Polymer*, **42**, 4453, 2001.

76. Tipanna, M. and Kale, D.D., Composites of waste, ground rubber particles and poly(vinyl chloride), *Rubber Chem. Technol.*, **70**, 815, 1997.

77. Kim, J.K., Lee, S.H., and Hwang, S.H., Study on the thermoplastic vulcanizate using ultrasonically treated rubber powder, *J. Appl. Polym. Sci.*, **90**, 2503, 2003.

78. Luo, T. and Isayev, A.I., Rubber/plastic blends based on devulcanized ground tire rubber, *J. Elast. Plast.*, **30**, 133, 1998.

79. Naskar, A.K., Bhowmick, A.K., and De, S.K., Melt-processable rubber: Chlorinated waste tire rubber-filled polyvinyl chloride, *J. Appl. Polym. Sci.*, **84**, 622, 2002.

80. Xu, Z.N., Losur, S., and Gardner, S.D., Epoxy resin filled with tire rubber particles modified by plasma surface treatment, *J. Adv. Mater.*, **30**, 11, 1998.

81. Shanmugharaj, A.M., Kim, J.K., and Ryu, S.H., UV surface modification of waste tire powder: Characterization and its influence on the properties of polypropylene/waste powder composites, *Polymer Test.*, **24**, 739, 2005.

82. Phadke, A.A. and De, S.K., Effect of cryo-ground rubber on melt flow and mechanical properties of polypropylene, *Polym. Eng. Sci.*, **26**, 1079, 1986.

83. Tuchmann, D. and Rosen, S.L., The mechanical properties of plastics containing cryogenically ground tire, *J. Elast. Plast.*, **10**, 115, 1978.

84. Bream, A.E. and Hornsby, P.R., Structure development in thermoset recyclate-filled polypropylene composites, *Polym. Composites*, **21**, 417, 2000.

85. Rajalingam, P. and Baker, W.E., The role of functional polymers in ground rubber tire-polyethylene composite, *Rubber Chem. Technol.*, **65**, 908, 1992.

86. Phinyocheep, P., Axtell, F.H., and Laosee, T., Influence of compatibilizers on mechanical properties, crystallization, and morphology of polypropylene/scrap rubber dust blends, *J. Appl. Polym. Sci.*, **86**, 148, 2002.

87. Pramanik, P.K. and Baker, W.E., Toughening of ground rubber tire filled thermoplastic compounds using different compatibilizer systems, *Plastics Rubber Comp., Process. Appl.*, **24**, 229, 1995.

88. Oliphant, K. and Baker, W.E., The use of cryogenically ground rubber tires as a filler in polyolefin blends, *Polym. Eng. Sci.*, **33**, 166, 1993.

89. Li, S., Lamminmaki, J., and Hanhi, K., Improvement of mechanical properties of rubber compounds using waste rubber/virgin rubber, *Polym. Eng. Sci.*, **45**, 1239, 2005.

90. Gibala, A. and Hamed, G.R., Cure and mechanical behavior of rubber compounds containing ground vulcanizates. Part I. Cure behavior, *Rubber Chem. Technol.*, **67**, 636, 1994.

91. Ishiaku, U.S., Chong, C.S., and Ismail, H., Cure characteristics and vulcanizate properties of blends of a rubber compound and its recycled DE-VULC, *Polym. Polym. Comp.*, **6**, 399, 1998.

92. Hong, C.K. and Isayev, A.I., Blends of ultrasonically devulcanized and virgin carbon black filled NR, *J. Mater. Sci.*, **37**, 385, 2002.

93. Han, S.C. and Han, M.H., Fracture behavior of NR and SBR vulcanizates filled with ground rubber having uniform particle size, *J. Appl. Polym. Sci.*, **85**, 2491, 2002.

94. Phadke, A.A. and De, S.K., Effect of cryoground rubber on properties of NR, *J. Appl. Polym. Sci.*, **32**, 4063, 1986.

95. Bauman, A.D., High-value engineering materials from scrap rubber, *Rubber World*, **212**, 30, 1995.

96. Bagheri, R., Williams, M.A., and Pearson, R.A., Use of surface modified recycled rubber particles for toughening of epoxy polymers, *Polym. Eng. Sci.*, **37**, 245, 1997.

97. Rodriguez, A.L., The effect of cryogenically ground rubber on some mechanical properties of an unsaturated polyester resin, *Polym. Eng. Sci.*, **28**, 145, 1988.

98. Choudhury, N.R. and Bhowmick, A.K., Strength of thermoplastic elastomers from rubber-polyolefin blends, *J. Mater. Sci.*, **25**, 161, 1990.

99. Choudhury, N.R. and Bhowmick, A.K., *J. Elast. Plast.*, **28**, 161, 1996.

100. Jha, A. and Bhowmick, A.K., Thermoplastic elastomeric blends of poly(ethylene terephthalate) and acrylate rubber: I. Influence of interaction on thermal, dynamic mechanical and tensile properties, *Polymer*, **38**, 4337, 1997.

101. Jha, A. and Bhowmick, A.K., Thermal degradation and ageing behaviour of novel thermoplastic elastomeric nylon-6/acrylate rubber reactive blends, *Polym. Degrad. Stab.*, **62**, 575, 1998.

102. Chattopadhyay, S., Chaki, T.K., Bhowmick, A.K., Gao, G.J.P., and Bandyopadhyay, S., Structural characterization of electron-beam crosslinked thermoplastic elastomeric films from blends of polyethylene and ethylene-vinyl acetate copolymers, *J. Appl. Polym. Sci.*, **81**, 1936, 2001.

103. Shaban, H.I. and George, R.S., Rheological and extrudate behavior of polypropylene/latex reclaim blends, *J. Elast. Plast.*, **29**, 83, 1997.

104. George, R.S. and Joseph, R., Studies on thermoplastic elastomers from polypropylene and latex waste products, *Kautsch. Gummi Kunst.*, **47**, 816, 1994.

105. Osborn, B.D., Reclaimed tire rubber in TPE compounds, *Rubber World*, **212**, 34, 1995.

106. Michael, H., Scholz, H., and Mennig, G., Blends from recycled rubber and thermoplastics, *Kautsch. Gummi Kunst.*, **52**, 510, 1999.

107. Al-Malaika, S. and Amir, E.J., Thermoplastic elastomers: Part III—Ageing and mechanical properties of natural rubber-reclaimed rubber/polypropylene systems and their role as solid phase dispersants in polypropylene/polyethylene blends, *Polym. Degrad. Stab.*, **26**, 31, 1989.

108. Naskar, A.K., Bhowmick, A.K., and De, S.K., Thermoplastic elastomeric composition based on ground rubber tire, *Polym. Eng. Sci.*, **41**, 1087, 2001.

109. Nevatia, P., Banerjee, T.S., Dutta, B., Jha, A., Naskar, A.K., and Bhowmick, A.K., Thermoplastic elastomers from reclaimed rubber and waste plastics, *J. Appl. Polym. Sci.*, **83**, 2035, 2002.

110. Okamato, H., Fukumori, K., Matsushita, M., Sato, N., Tanaka, Y., Okita, T., Otsuka, S., and Suzuki, Y., *Proceedings of the Japan International SAMPE Symposium*, Tokyo, Japan, Nov. 13–16, 2001.

111. Ismail, H. and Suryadiansyah, S., Thermoplastic elastomers based on polypropylene/natural rubber and polypropylene/recycle rubber blends, *Polymer Test.*, **21**, 389, 2002.

112. Kumar, R.C., Fuhrmann, I., and Kocsis, J.K., LDPE-based thermoplastic elastomers containing ground tire rubber with and without dynamic curing, *Polym. Degrad. Stab.*, **76**, 137, 2002.

113. Sun, B. and Xiong, X., *Proceedings of REWAS'99 Global Symposium on Recycling, Waste Treatment and Clean Technology*, San Sebastian, Spain, Sept. 5–9, 531, 1999.

114. Straube, G., Straube, E., Neumann, W., Ruckauf, H., Forkmann, R., and Loffler, M., Method for reprocessing scrap rubber, *US patent, 5,275,948*, 1994.

115. Tsuchii, A., Suzuki, T., and Takeda, K., Microbial degradation of natural rubber vulcanizates, *Appl. Environ. Microbiol.*, **50**, 965, 1985.

116. Tsuchii, A., Takeda, K., and Tokiwa, Y., Degradation of the rubber in truck tires by a strain of Nocardia, *Biodegradation*, **7**, 405, 1997.

117. Tsuchii, A. and Takeda, K., Rubber-degrading enzyme from a bacterial culture, *Appl. Environ. Microbiol.*, **56**, 269, 1990.

118. Loffler, B., Straube, G., and Straube, E., *Biohydrometall. Technol. Proc. Int.Biohydrometall. Symp.*, **2**, 673, 1993.

119. Scheirs, J., *Polymer Recycling—Science, Technology and Applications*, Wiley, New York, p. 411, 1998.

120. Otsuka, S., Owaki, M., Suzuki, Y., Honda, H., Nakashima, K., Mouri, M., and Sato, N., Society of Automotive Engineers, Special Publication, SP-1542 (Environmental Concepts for the Automotive Industry), 1, 2000.

121. Nakahama, H., Ishii, Y., and Kawasaki, M., Moldings made of recycled vulcanized rubbers and manufacture of moldings, *Jap. Patent, 11,279,290*, 1999.

122. Garderes, P., Reusable materials made from recycled rubber, process for the preparation and use, *Euro. Patent Appl., 1,362,889 A1 20,031,119*, 2003.

123. Kawakami, S. and Udagawa, Y., Tire for passenger car, *Jap. Patent Appl., 2, 003,072,313 A2 20,030,312*, 2001.

124. Jacob, C., De, P.P., Bhowmick, A.K., and De, S.K., Recycling of EPDM waste. II. Replacement of virgin rubber by ground EPDM vulcanizate in EPDM/PP thermoplastic elastomeric composition, *J. Appl. Polym. Sci.*, **82**, 3304, 2001.

125. Anandhan, S., De, P.P., Bhowmick, A.K., Bandyopadhyay, S., and De, S.K., Thermoplastic elastomeric blend of nitrile rubber and poly(styrene-co-acrylonitrile). II. Replacement of nitrile rubber by its vulcanizate powder, *J. Appl. Polym. Sci.*, **90**, 2348, 2003.

126. Anandhan, S., De, P.P., De, S.K., Swayajith, S., and Bhowmick, A.K., Thermorheological properties of thermoplastic elastomeric blends of nitrile rubber and poly(styrene-co-acrylonitrile) containing waste nitrile rubber vulcanizate powder, *Kautsch. Gummi Kunst.*, **11**, 2004.

127. Matsushita, M., Okamoto, H., Fukumori, K., Mori, M., Sato, N., Suzuki, Y., Otsuka, S., Owaki, M., Takeuchi, K., and Nakajima, K., Rubber composition containing reclaimed EPDM rubber and production of the composition, *Jap. Patent Appl.*, *JP1,999,230,787*, 2001.

128. Hoetzeldt, K., Bochmann, G., and Passler, W., Cold sprayable, low flammability and watertight molding or sealing material used for the preparation of elastic work layers for road, play or sports surfaces, *German Patent, DE 4416570 A1 19,951,116*, 1995.

129. Hallisy, M.J., Gilmore, J.R., Hobbs, S.J., and Duncan, W.G., Process for recycling cured polysulfide sealants, *US patent, 5,349,091*, 1994.

130. Hecktor, B.J., Specht, F., Theobald, R., and Unger, G., Method of reprocessing cured polysulfide and/or mercaptan elastomers, *German Patent, DE 4,142,500 C1 19,930,519*, 1993.

131. Sosnina, A., Degtyarev, E.V., Kukushkin, V.Y., Kornev, A.E., Dontsov, A.A., Kvardashov, V.P., Kosteltsev, V.V., and Afanasenkova, Z.M., Polymeric composition for sealing articles, *Russian Patent, SU 1,081,992*, 1992.

132. Ghosh, A.K., Bhattacharya, A.K., Bhowmick, A.K., and De, S.K., Effect of ground fluororubber vulcanizate powder on the properties of fluororubber compound, *Polym. Eng. Sci.*, **43**, 267, 2003.

133. Shin-Etsu Polymer Co., Ltd., Japan, Regeneration of silicone rubber scrap, *Jap. Patent. Appl., JP 59,179,537*, 1984.

134. Ghosh, A.K., Rajeev, R.S., Bhattacharya, A.K., Bhowmick, A.K., and De, S.K., Recycling of silicone rubber waste: Effect of ground silicone rubber vulcanizate powder on the properties of silicone rubber, *Polym. Eng. Sci.*, **43**, 279, 2003.

Index

U